Methods in Cell Biology

VOLUME 84

Biophysical Tools for Biologists,
Volume One: In Vitro Techniques

Series Editors

Leslie Wilson

Department of Molecular, Cellular and Developmental Biology
University of California
Santa Barbara, California

Paul Matsudaira

Whitehead Institute for Biomedical Research
Department of Biology
Division of Biological Engineering
Massachusetts Institute of Technology
Cambridge, Massachusetts

Methods in Cell Biology

VOLUME 84

Biophysical Tools for Biologists,
Volume One: In Vitro Techniques

Edited by

Dr. John J. Correia

Department of Biochemistry
University of Mississippi Medical Center
Jackson, Mississippi

Dr. H. William Detrich, III

Department of Biology
Northeastern University
Boston, Massachusetts

ELSEVIER

AMSTERDAM • BOSTON • HEIDELBERG • LONDON
NEW YORK • OXFORD • PARIS • SAN DIEGO
SAN FRANCISCO • SINGAPORE • SYDNEY • TOKYO
Academic Press is an imprint of Elsevier

Example of the visualization of the enzyme inhibitor balanol's electrostatic potential in the binding site of protein kinase A (PDB ID 1BX6) as calculated by APBS (http://apbs.sourceforge.net/) and visualized with VMD (http://www.ks.uiuc.edu/Research/vmd/). For more information about these ligand-kinase binding calculations, see Wong CF, et al, J Med Chem 44 (10) 1530-9, 2001.

Figure prepared by Nathan Baker and David Gohara based on APBS examples provided by Chung Wong.

Academic Press is an imprint of Elsevier
84 Theobald's Road, London WC1X 8RR, UK
Radarweg 29, PO Box 211, 1000 AE Amsterdam, The Netherlands
Linacre House, Jordan Hill, Oxford OX2 8DP, UK
30 Corporate Drive, Suite 400, Burlington, MA 01803, USA
525 B Street, Suite 1900, San Diego, CA 92101-4495, USA

First edition 2008

ISBN: 978-0-12-372520-2

ISSN: 0091-679X

For information on all Academic Press publications
visit our website at books.elsevier.com

Printed and bound in USA

08 09 10 11 12 10 9 8 7 6 5 4 3 2 1

CONTENTS

SECTION 2 Computational Methods

CONTRIBUTORS

Numbers in parentheses indicate the pages on which the authors' contributions begin.

Nathan A. Baker (843), Department of Biochemistry and Molecular Biophysics, Center for Computational Biology, Washington University in St. Louis, Missouri 63110

Susan L. Bane (213), Department of Chemistry, State University of New York at Binghamton, Binghamton, New York 13902

Doug Barrick (295), T. C. Jenkins Department of Biophysics, The Johns Hopkins University, Baltimore, Maryland 21218

Dorothy Beckett (25), Department of Chemistry and Biochemistry, Center for Biological Structure and Organization, University of Maryland, College Park, Maryland 20742

Robert H. Beer (589), Department of Chemistry, Fordham University, Bronx, New York 10458

Alexandra Bonner (375), Department of Biochemistry and Molecular Biology, University College London, London, WC1E 6BT, UK

Hacène Boukari (659), Laboratory of Integrative and Medical Biophysics, National Institute of Child Health and Human Development, National Institutes of Health, Bethesda, Maryland 20892

Michael Brenowitz (589), Department of Biochemistry, Albert Einstein College of Medicine, Bronx, New York 10461

Nancy K. Burgess (181), T. C. Jenkins Department of Biophysics, Johns Hopkins University, Baltimore, Maryland 21218

Olwyn Byron (327), Division of Infection and Immunity, GBRC, 120 University Place, University of Glasgow, Glasgow, G12 8TA, United Kingdom

Anders E. Carlsson (911), Department of Physics, Washington University, St. Louis, Missouri 63130

Jonathan B. Chaires (3), James Graham Brown Cancer Center, University of Louisville, Louisville, Kentucky 40202

James L. Cole (143), Department of Molecular and Cell Biology, University of Connecticut, Storrs, Connecticut 06269; National Analytical Ultracentrifugation Facility, University of Connecticut, Storrs, Connecticut 06269

Naomi Courtemanche (295), T. C. Jenkins Department of Biophysics, The Johns Hopkins University, Baltimore, Maryland 21218

W. David Wilson (53), Department of Chemistry, Georgia State University, Atlanta, Georgia 30302

Feng Dong (843), Department of Biochemistry and Molecular Biophysics, Center for Computational Biology, Washington University in St. Louis, Missouri 63110

John F. Eccleston (445), Division of Physical Biochemistry, MRC National Institute for Medical Research, The Ridgeway, Mill Hill, London NW7 1AA, United Kingdom

Jimmy B. Feix (617), Department of Biophysics, Medical College of Wisconsin, Milwaukee, Wisconsin 53226

Anira N. Fernando (375), Department of Biochemistry and Molecular Biology, University College London, London, WC1E 6BT, UK

Karen G. Fleming (181), T. C. Jenkins Department of Biophysics, Johns Hopkins University, Baltimore, Maryland 21218

Matthew W. Freyer (79), Department of Chemistry and Biochemistry, Northern Arizona University, Flagstaff, Arizona 86011

Patricia B. Furtado (375), Department of Biochemistry and Molecular Biology, University College London, London, WC1E 6BT, UK

Nichola C. Garbett (3), James Graham Brown Cancer Center, University of Louisville, Louisville, Kentucky 40202

Bertrand E. García-Moreno (871), Department of Biophysics, The Johns Hopkins University, Baltimore, Maryland

Hannah E. Gilbert (375), Department of Biochemistry and Molecular Biology, University College London, London, WC1E 6BT, UK

Heinz Gross (425), EMEZ, Electron Microscopy Center, Swiss Federal Institute of Technology Zürich, 8093 Zürich, Switzerland

Daniel Harries (679), Department of Physical Chemistry and the Fritz Haber Center for Molecular Dynamics, The Hebrew University of Jerusalem, Jerusalem 91904, Israel

Vincent J. Hilser (871), Department of Biochemistry and Molecular Biology, University of Texas Medical Branch, Galveston, Texas 77555-1068

Andreas Hoenger (425), Department of Molecular, Cellular and Developmental Biology, University of Colorado at Boulder UCB-347, Boulder, Colorado, 80309

Michael L. Johnson (781), Department of Pharmacology and Internal Medicine, University of Virginia Health System, Charlottesville, Virginia 22908

Richard L. Karpel (517), Department of Chemistry and Biochemistry, University of Maryland Baltimore County, Baltimore, Maryland 21250

Sarah M. Keating (807), Biological and Neural Computation Group, Science and Technology Research Institute, University of Hertfordshire, College Lane, Hatfield AL10 9AB, United Kingdom

Candice S. Klug (617), Department of Biophysics, Medical College of Wisconsin, Milwaukee, Wisconsin 53226

Jeffrey W. Lary (143), National Analytical Ultracentrifugation Facility, University of Connecticut, Storrs, Connecticut 06269

Thomas M. Laue (143), Center to Advance Molecular Interaction Science, University of New Hampshire, Durham, New Hampshire 03824

Edwin A. Lewis (79), Department of Chemistry and Biochemistry, Northern Arizona University, Flagstaff, Arizona 86011

Vince J. LiCata (243), Department of Biological Sciences, Louisiana State University, Baton Rouge, Louisiana 70803

Richard M. Higashi (541), Department of Chemistry, University of Louisville, Louisville, Kentucky 40202

Stephen R. Martin (263, 445, 807), Division of Physical Biochemistry, MRC National Institute for Medical Research, The Ridgeway, Mill Hill, London NW7 1AA, United Kingdom

Somdeb Mitra (589), Department of Biochemistry, Albert Einstein College of Medicine, Bronx, New York 10461

Thomas P. Moody (143), Center to Advance Molecular Interaction Science, University of New Hampshire, Durham, New Hampshire 03824

Andrew N. Lane (541), JG Brown Cancer Center, University of Louisville, Louisville, Kentucky 40202

Binh Nguyen (53), Department of Chemistry, Georgia State University, Atlanta, Georgia 30302

Azubuike I. Okemefuna (375), Department of Biochemistry and Molecular Biology, University College London, London, WC1E 6BT, UK

Brett Olsen (843), Department of Biochemistry and Molecular Biophysics, Center for Computational Biology, Washington University in St. Louis, Missouri 63110

Stephen J. Perkins (375), Department of Biochemistry and Molecular Biology, University College London, London, WC1E 6BT, UK

Jörg Rösgen (679), Department of Biochemistry and Molecular Biology, University of Texas Medical Branch, Galveston, Texas 77555

Ioulia Rouzina (517), Department of Biochemistry, Molecular Biology and Biophysics, University of Minnesota, Minneapolis, Minnesota 55455

Dan L. Sackett (659), Laboratory of Integrative and Medical Biophysics, National Institute of Child Health and Human Development, National Institutes of Health, Bethesda, Maryland 20892

Maria J. Schilstra (263, 445, 807), Biological and Neural Computation Group, Science and Technology Research Institute, University of Hertfordshire, College Lane, Hatfield AL10 9AB, United Kingdom

David Sept (893, 911), Department of Biomedical Engineering and Center for Computational Biology, Washington University, St. Louis, Missouri 63130

Natasha Shanker (213), Department of Chemistry, State University of New York at Binghamton, Binghamton, New York 13902

Inna Shcherbakova (589), Department of Biochemistry, Albert Einstein College of Medicine, Bronx, New York 10461

Charles H. Spink (115), Department of Chemistry, State University of New York—Cortland, Cortland, New York 13045

Ann Marie Stanley (181), T. C. Jenkins Department of Biophysics, Johns Hopkins University, Baltimore, Maryland 21218

Timothy O. Street (295), T. C. Jenkins Department of Biophysics, The Johns Hopkins University, Baltimore, Maryland 21218

Farial A. Tanious (53), Department of Chemistry, Georgia State University, Atlanta, Georgia 30302

Joel Tellinghuisen (739), Department of Chemistry, Vanderbilt University, Nashville, Tennessee 37235

George J. Turner (479), Department of Chemistry and Biochemistry, Seton Hall University, South Orange, New Jersey 07079

Teresa W.-M. Fan (541), Department of Pharmacology and Toxicology, University of Louisville, Louisville, Kentucky 40202; Department of Chemistry, University of Louisville, Louisville, Kentucky 40202; JG Brown Cancer Center, University of Louisville, Louisville, Kentucky 40202

Steven T. Whitten (871), RedStorm Scientific, Inc., Galveston, Texas 77550; Department of Biochemistry and Molecular Biology, University of Texas Medical Branch, Galveston, Texas 77555-1068

Mark C. Williams (517), Center for Interdisciplinary Research on Complex Systems, 111 Dana Research Center, Northeastern University, Boston, Massachusetts 02115; Department of Physics, 111 Dana Research Center, Northeastern University, Boston, Massachusetts 02115

Andy J. Wowor (243), Department of Biological Sciences, Louisiana State University, Baton Rouge, Louisiana 70803

Xiange Zheng (893), Biomedical Engineering and Center for Computational Biology, Washington University, St. Louis, Missouri 63130

PREFACE

The era of modern Molecular Biophysics can be traced back 50 years to the first meeting of the Biophysical Society in Columbus, Ohio (March, 1957). The raison d'être of the society is to promote the application of the fundamental principles of physics, thermodynamics, and kinetics to analysis of polymeric and biological solutions, and more recently, of cells. The history of these fundamentals date to the mid-to-late 1800s, especially the works of J. Willard Gibbs, and include the concepts of chemical potential, Gibbs free energy, Gibbsian ensembles, and the Gibbs phase rule. Elegant examples of early biophysical milestones began to appear at the turn of the last century and include the demonstration of the Gibbs–Donnan equilibrium across a semipermeable membrane, A. V. Hill's analysis of the equilibrium binding of oxygen to hemoglobin (the Hill plot), and the discovery of the cooperativity of oxygen binding to hemoglobin by Adair (actually expressed as the Adair equation). Our recognition of the primary, secondary, tertiary, quaternary structures and physical properties of proteins (and other macromolecules) is based on techniques as diverse as osmotic pressure (Adair), analytical ultracentrifugation (Svedberg and Pederson), electrophoresis (Tiselius), salting out (Cohn), and hydrogen–deuterium exchange (Linderstrom-Lang). Onsager, Kirkwood, and Flory applied statistical mechanics to characterize the solution behavior and dielectric properties of polyelectrolytes, polymers, and long-chain biomolecules. Scatchard analysis of ligand binding and Wyman linkage theory have become staples in the arsenal of biophysical methods. And as legions of biochemistry students have learned, the demonstration of the semiconservative replication of DNA by Meselson and Stahl (The Most Beautiful Experiment in Biology, see book by F. L. Holmes) was based on the ability of density gradient equilibrium sedimentation (Vinograd) to resolve DNA molecules that were differentially labeled with stable isotopes of nitrogen.

Students, teachers, and researchers in biophysics in the 1960s to 1980s were introduced to the discipline's general concepts, fundamental theory, and representative applications by a number of excellent textbooks. Early classics include Biophysical Chemistry (1958) by John Edsall and Jeffries Wyman, Ultracentrifugation in Biochemistry (1959) by Howard Schachman, Introduction to Statistical Mechanics (1960) by Terrell Hill, Physical Chemistry of Macromolecules (1961) by Charles Tanford, and Mathematical Theory of Sedimentation Analysis (1962) by Hiroshi Fujita. These first generation works spawned a second wave: Interacting Macromolecules (1970) by John Cann; Physical Biochemistry by Kensal van Holde; Migration of Interacting Systems (1972) by Laurie Nichol and Donald Winzor; Data Analysis in Biochemistry and Biophysics (1972) by

Magar E. Magar; Physical Chemistry of Nucleic Acids (1974) by Victor Bloomfield, Donald Crothers, and Ignacio Tinoco; Relaxation Kinetics (1976) by Claude Bernasconi; Thermodynamics of the Polymerization of Protein (1975) by Fumio Oosawa and Sho Asakura; Physical Chemistry (1978) by Tinoco, Kenneth Sauer, and James Wang; Physical Chemistry (1979) by David Eisenberg and Crothers; Biophysical Chemistry (1980) by Charles Cantor and Paul Schimmel; Random Walks in Biology (1983) by Howard Berg; Proteins (1984) by Thomas Creighton; The Physical Chemistry of Lipids (1986) by D. M. Small; and Binding and Linkage by Wyman and Stanley Gill (1990). The ensuing decades have seen many of these books go out of print, although a few have been released as new editions [Proteins (1993) by Creighton; Random Walks in Biology (1993) by Berg; Principles of Physical Biochemistry (1998) by van Holde, Curtis Johnson, and Shing Ho; Nucleic Acid Structures (2000) by Bloomfield, Crothers, and Tinoco] and a number of third generation texts have been published, including Thermodynamic Theory of Site-Specific Binding Processes in Biological Macromolecules (1995) by Enrico Di Cera, Molecular Driving Forces, Statistical Thermodynamics in Chemistry and Biology (2003) by Ken Dill and Sarina Bromberg, Biological Physics (2004) by Philip Nelson, and very recently Van der Waals Forces (2006) by Adrian Parsegian.

To complement the evolution of available biophysical texts and to facilitate the teaching of courses in biophysics at the undergraduate and graduate levels, the Biophysics Society and other self-organizing groups have turned to new media to sustain and enhance the education of tomorrow's biophysicists. The Gibbs Society of Biological Thermodynamics was established in 1987 to promote training of biothermodynamicists (see http://mljohnson.pharm.virginia.edu/gibbs-society/ for details and a history). The Biophysical Society has continued to nurture new Special Interest Groups to expand biophysical investigation to broader areas of biology (see http://www.biophysics.org/meetings/subgroups.htm for links to these subgroups). The creation (and unfortunate demise) of an online biophysics text was an experiment that promoted web-based education; many chapters are still available on the Biophysical Society web page under educational resources (http://www.biophysics.org/education/). The American Society for Cell Biology (ASCB; founded in 1960) and the Biophysical Society have established strong interdisciplinary bonds; notable areas of cross-fertilization include muscle, motility, ion channels, and single molecule studies. To many investigators, attendance at both meetings is a prerequisite for professional development. Two articles published in the same issue of the Biophysical Journal in 1992, "Teaching molecular biophysics at the graduate level" by Norma Allwell and Victor Bloomfield, and "Graduate training in cellular biophysics" by Tom Pollard, explicitly describe the skill sets and training required for investigators who study molecular and cellular biophysics.

The past decade has witnessed the emergence of revolutionary, data-intensive disciplines, including genome biology, proteomics, bioinformatics, and systems biology, that have greatly expanded the scope of the problems we investigate. As a consequence, there has been a tremendous resurgence of interest in the application of biophysical methods to investigate macromolecular structure and function in the context of living cells. The classical methods used by physical and polymer chemists to investigate macromolecular behavior and interactions are now being applied to vastly larger data sets that encompass entire genomes worth of proteins. Goals include the prediction of protein structure from first principles, the development of gene and protein interaction maps, and the analysis of regulatory networks within cells, tissues, and organ systems. Ultimate success in these endeavors will require quantitative assessment of the mechanisms, energetics, and kinetics that underlie these global phenomena.

Volume 84 of Methods in Cell Biology, Biophysical Tools *in vitro*, is devoted to biophysical techniques and their applications to cellular biology. The volume provides, in a single resource, access to a broad range of fundamental and cutting-edge *in vitro* techniques in molecular biophysics and places a strong emphasis on problem solving. Chapters include (1) the theory and measurement of binding interactions by isothermal titration calorimetry (ITC), surface plasmon resonance (SPR), differential scanning calorimetry (DSC), spectroscopy, analytical ultracentrifugation (AUC), fluorescence anisotropy, and single molecule approaches; (2) spectroscopic methods including absorbance, fluorescence, circular dichroism (CD), spin label electron paramagnetic resonance (EPR), and fluorescence correlation spectroscopy (FCS); (3) kinetic methods including stopped flow, T-jump, P-jump, protein folding, and time-resolved RNA structural transitions; (4) structural methods including cryo-electron microscopy, X-ray, neutron scattering, and hydrodynamic shape modeling; (5) proteomic methods including metabolomic analysis by nuclear magnetic resonance (NMR) and mass spectrometry (MS); (6) thermodynamic approaches including binding theory, linkage analysis, mutational analysis, and the properties of osmolytes and their impact on macromolecules; (7) data analysis including statistics and nonlinear least squares; and (8) modeling methods including dynamics of complex systems, biomolecular electrostatics, conformational fluctuations, cytoskeletal protein interactions with drugs, and cell motility. The chapters are methods oriented, often tutorial and practical in terms of how to do it. Some also present a strong emphasis on data analysis and computational approaches, because fitting of data is typically the most difficult part of learning how to apply these methods. A few cutting-edge chapters, to paraphrase Doug Rau, rely heavily on the intelligence and the initiative of the reader.

The volume is directed toward the broad audience of cell biologists, biophysicists, pharmacologists, and molecular biologists who make use of classical and modern biophysical technologies or wish to expand their expertise to include such approaches. The volume should also interest the biomedical and biotechnology communities given the importance of biophysical characterization of drug formulations prior to Food and Drug Administration (FDA) approval. We emphasize that this volume and its contents are dedicated to biophysical applications as they are applied to problem solving. The vitality of any biological discipline depends upon its ability to reveal fundamental information, to discover new mechanisms, and to apply new ideas and concepts to the challenges of understanding life, behavior, our environment, and the treatment of disease. We trust that this volume will serve the reader as a convenient, reliable compilation of biophysical methods *in vitro* and their application to the solving of important biological problems. Our hope is that this work will stimulate increased collaboration between biophysicists, cell and molecular biologists in the years to come.

John J. Correia
H. William Detrich, III

We dedicate this volume to our mentors,
David Yphantis, Robley Williams, Jr., and Leslie Wilson,
and to our colleagues and collaborators who belong to the Gibbs
Society of Biological Thermodynamics, the Biophysical Society,
and to the American Society for Cell Biology.

SECTION 1

Solution Methods

CHAPTER 1

Binding: A Polemic and Rough Guide

Nichola C. Garbett and Jonathan B. Chaires

James Graham Brown Cancer Center
University of Louisville
Louisville, Kentucky 40202

Abstract

Binding is at the center of all of biology with cellular events being mediated by a huge array of highly orchestrated, coupled binding interactions. In order to approach a detailed understanding of the molecular forces that drive these interactions, it is essential to obtain thermodynamic information. A huge body of work already exists that describes the concepts and mathematical tools that are at the foundation of current thermodynamic techniques. The purpose of this chapter is to aid the reader in understanding how to apply the available technologies in extracting meaningful thermodynamic information for binding systems.

> The certitude that everything has been written negates us or turns us into phantoms…
> Jorge Luis Borges
> In: *The Library of Babel*

0091-679X/08 $35.00
DOI: 10.1016/S0091-679X(07)84001-5

I. Introduction

Without binding, there is no biology. Whatever that goes on in cells involves an intricate, highly orchestrated, series of binding interactions. These include protein–protein, protein–DNA, protein–RNA interactions, as well as an incredible variety of small molecule–macromolecule interactions. "Binding," in its most general sense, encompasses all such interactions. One might argue that a full understanding of cell biology will only come when the entirety of all such binding interactions is defined and quantified, allowing the "cell" to be represented as a complex collection of tightly coupled rate and equilibrium equations (see Chapter 25 by Schilstra *et al.*, this volume). The day when such a collection of equations is completed may or may not ever come. In the meantime, we can plod along by studying the binding interactions of our favorite macromolecules isolated from cells, gaining transient insight into the workings of parts of the cell, and hoping that the pieces may someday add up to complete the puzzle.

The goals of any binding study were succinctly articulated by George Scatchard (1949). These goals are to answer the questions: "How many? How tightly? Where? Why? What of it?" That is to say, binding studies should elucidate the stoichiometry and affinity of the receptor–ligand interaction, should attempt to identify the location of ligand binding on the receptor, should provide an understanding of the molecular forces that drive the binding, and, finally, should explicitly link binding to the functional response. These are ambitious goals.

The execution and analysis of binding experiments have evolved to sophisticated levels over the last century. For quantitative binding studies, analytical methods are needed that can determine the distribution of free and bound reactant concentrations within a solution. These methods include all types of spectroscopy, calorimetry, and more recently devised methods like surface plasmon resonance. Primary data that emerges from whatever technique is used must then be analyzed by postulating an appropriate and plausible reaction mechanism that describes the equilibrium under study. This statement of the equilibrium reaction mechanism then dictates the exact form of a mathematical model (binding equation) that then must be "fit" to the experimental data. Typically, this involves use of nonlinear least-squares fitting routines with appropriate use of ancillary statistical methods to evaluate the goodness and significance of any fit. A schematic representation of the general strategy for obtaining quantitative binding information is shown in Fig. 1.

Successful binding studies yield a number, the equilibrium binding constant, which, in turn, provides entry into the world of thermodynamics. None of these steps are easy and the overall process does not lend itself to simple protocols that might be mindlessly applied to every type of binding problem. Each type of reaction for which a binding constant is sought presents its own challenges and problems at every stage in the process.

Better minds than ours have described the concepts and mathematical tools that provide the underpinnings for understanding macromolecular binding interactions. There is no point in recapitulating or simplifying the original presentations

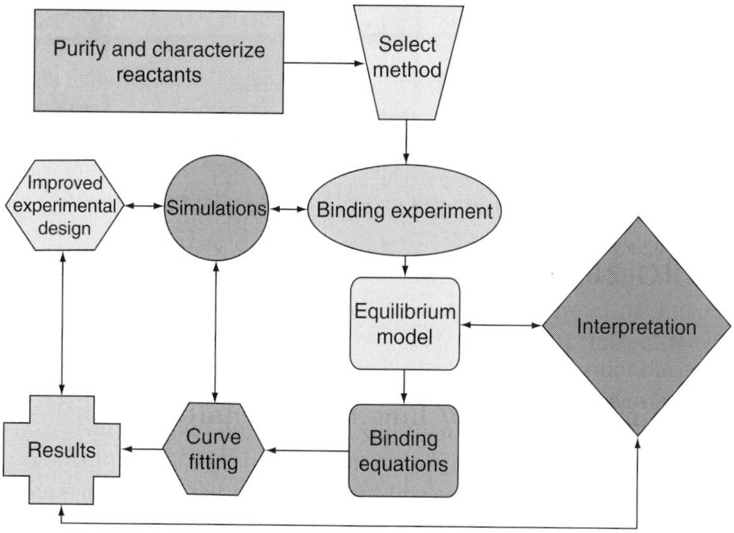

Fig. 1 General approach for determining quantitative binding information.

of such concepts. Rather than wasting further time on this chapter, the reader would be better served to delve into some of the original, uncompromising, discussions of binding. For example, there is no better starting place than Chapter 11 in Edsall and Wyman's classic *Biophysical Chemistry*, entitled "Some General Aspects of Molecular Interactions." Other primary sources include Gregorio Weber's *Protein Interactions*, Wyman and Gill's *Binding and Linkage*, Irving Klotz's *Ligand-Receptor Energetics: A Guide for the Perplexed*, and Winzor and Sawyer's *Quantitative Characterization of Ligand Binding*. Chowdhry and Harding have compiled a two-volume compendium of practical methods (with protocols) for the study of protein–ligand interactions. Bibliographic details of these publications are collected in an annotated "Suggested Reading" section at the end of this chapter, separate from specific references that are cited. Any of these will provide a better foundation in preparation for conducting a binding study than could possibly be provided here.

If, against our advice, the reader chooses not to abandon us in order to delve into the suggested reading, we will try to make a few points that will perhaps be of interest. These will be somewhat random and intended to provide only a rough guide to elements of binding studies.

II. Binding Constants Provide an Entry into Thermodynamics

Determination of a binding constant (K_{eq}) is the first step in the determination of a thermodynamic profile for a reaction of interest, as shown in the schematic in Fig. 2.

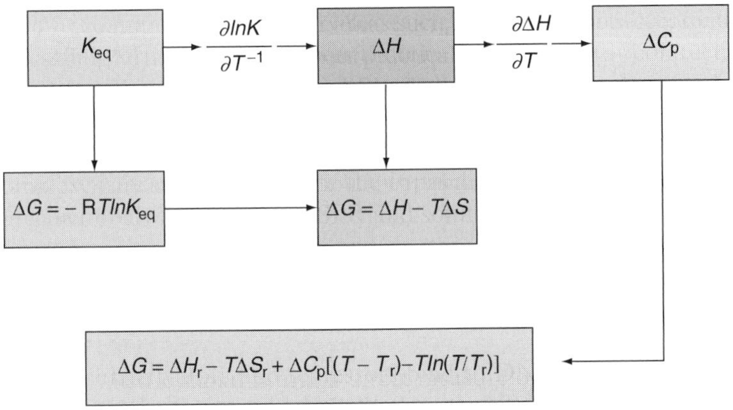

Fig. 2 Binding parameters and the interrelationships in the characterization of a complete thermodynamic profile.

The binding free energy change, $\Delta G = -RT \ln K_{eq}$, can be calculated directly once the binding constant is known. The sign and the magnitude of the free energy change specify the exact amount of energy liberated or consumed in the binding process. Free energy changes can be parsed into component enthalpy (ΔH) and entropy (ΔS) values. Enthalpy values are best determined directly by calorimetry but estimates of the van't Hoff enthalpy may be determined from the temperature dependence of binding constants:

$$\frac{\partial(\ln K_{eq})}{\partial(1/T)} = -\frac{\Delta H}{\mathrm{R}}$$

Entropy values may be calculated by difference once the free energy and enthalpy are known from the standard Gibbs equation, $\Delta G = \Delta H - T\Delta S$. If the enthalpy varies with temperature, there will be a nonzero heat capacity change, $\Delta C_p = \delta\Delta H/\delta T$. If this is the case, a more complicated expression for the free energy results, as shown in Fig. 2.

What is the utility of these thermodynamic values? First, and most fundamentally, the thermodynamic values allow one to calculate the distribution of free and bound reactant concentrations over a wide temperature range. Such information is crucial for correlating binding with function. Second, thermodynamic profiles may be interpreted in terms of the molecular forces that drive the binding process (Chaires, 1996, 1997, 2006; Haq, 2002; Haq and Ladbury, 2000; Haq *et al.*, 2000; Ren *et al.*, 2000), providing fundamental insight into the reaction mechanism. While such interpretations may amount to little more than informed speculation, they are nonetheless useful. Finally, thermodynamic studies are proving valuable in drug discovery, particularly in the optimization of lead compounds. The signs and magnitudes of enthalpy and entropy values can be used to guide medicinal chemists in the quest for the optimization of affinity and specificity of drug candidates for their binding to their receptor (Ruben *et al.*, 2006). One such example of the importance of thermodynamic information was demonstrated when attempting to understand the different binding properties exhibited by two closely related anthracycline antibiotics, doxorubicin and daunorubicin (Chaires, 2003). These compounds differ only in a single hydroxyl group at position C-13 and appear crystallographically to form isostructural ligand–DNA complexes. However, the determination of the complete thermodynamic profiles for these compounds revealed a significant difference in their thermodynamics of binding which translated into an order of magnitude tighter binding of doxorubicin through differences in solvent and ion involvement in binding.

HIV-1 protease inhibitors are attractive targets for the design of AIDS therapeutics because of the essential role of HIV-1 protease in the maturation of the virus. However, there has been a rapid appearance of viral resistance to these drugs, and there is a current push for the design of a new generation of protease inhibitors (Velazquez-Campoy *et al.*, 2001). In order to move forward with the design of more effective inhibitors, it is important to understand the molecular basis of inhibitor activity. To this end, Freire and coworkers have examined the thermodynamics of a number of inhibitors in clinical use (Velazquez-Campoy *et al.*, 2001). The binding of the inhibitors was characterized by either a slightly favorable or unfavorable enthalpy with binding being driven by a large favorable binding entropy. The entropy was found to be a consequence of the burial of a large hydrophobic surface upon binding and only a limited conformational entropy penalty because of the design of inhibitors with limited flexibility preshaped to the geometry of the binding site. In current drug design, shape complementarity and conformational rigidity are optimized to improve binding affinity and specificity. However, the appearance of viral resistance to HIV-1 protease inhibitors highlights the problem with this strategy. Viral resistance is thought to arise from amino acid mutations in the HIV-1 protease monomer that results in significantly reduced affinities toward the inhibitors but maintains sufficient affinity for the substrate. A strategy based on the design of rigid drug molecules is then problematic because of the inability of the drug molecules to accommodate changes in the target molecule. With this in mind, the Freire laboratory has developed a rapid screening approach to identify enthalpically favorable lead compounds (Velazquez-Campoy *et al.*, 2000), which can then be entropically optimized during the later stages of the drug design process (Ruben *et al.*, 2006).

III. General Properties of Binding Isotherms

If (by whatever experimental method was chosen) an investigator was able to obtain the amount of ligand bound to a receptor as a function of the free ligand concentration, a binding isotherm could be constructed. Such primary data are used to infer the correct binding model and are ultimately fit to such models to obtain binding constants. It is worthwhile to consider the general properties of binding isotherms and ways of representing the binding data. We will do so by simulations and illustrations of some simple cases. The simplest type of binding is for one ligand (L) binding to a single site on a receptor (R) to form a complex (RL):

$$\mathrm{R} + \mathrm{L} \rightleftharpoons \mathrm{RL}$$

At equilibrium, the expression for the equilibrium constant (K) for this simple reaction is:

$$K = \frac{[\mathrm{RL}]}{[\mathrm{L}]\cdot[\mathrm{R}]}$$

Equilibrium concentrations are indicated by brackets. The fractional saturation (θ) is the amount bound relative to the total receptor concentration:

$$\theta = \frac{\mathrm{RL}}{\mathrm{RL} + \mathrm{R}}$$

For this simplest of binding reactions, the binding isotherm is then described by the expression:

$$\theta = \frac{K\mathrm{L}}{1 + K\mathrm{L}} \tag{1}$$

This expression describes the shape of a rectangular hyperbola. We simulated binding data by setting $K = 10^6\ \mathrm{M}^{-1}$ and varying L from 10^{-9} to 10^{-3} M in 25 steps equally spaced on a logarithmic scale. Figure 3 shows binding data represented in three common ways.

Figure 3A shows a so-called direct plot in which the fractional saturation is shown as a function of free ligand concentration. The points define a rectangular hyperbola, as described by Eq. (1). The disadvantages of the plot are clear. Most of the points are clustered near $\mathrm{L}_{\mathrm{Free}} = 0$ and it is impossible to see the full range of the data. Simple algebraic rearrangement of Eq. (1) yields the renowned Scatchard plot (Scatchard, 1949):

$$\frac{\theta}{\mathrm{L}} = K(n - \theta)$$

where n is the number of binding sites ($n = 1$ in this case) and all other symbols have been defined above. Figure 3B shows the simulated data cast into this form. The potential advantages of this representation are that the data are linearized, yielding a plot with an intercept Kn and a slope $-K$, from which the binding constant may be readily obtained. However, it is clear that the Scatchard plot distorts the primary data in that the majority of points are clustered near the intercepts to the x- and y-axes. (The transformation also distorts the error in the

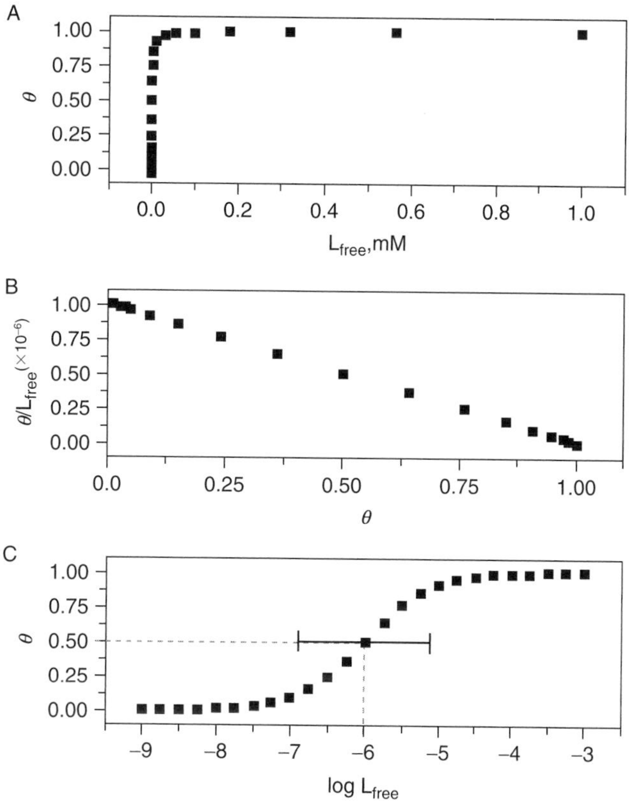

Fig. 3 Three common representations of binding data modeled for the interaction of a ligand and a receptor with an affinity of 10^6 M^{-1}: (A) a direct plot of fractional saturation as a function of free ligand concentration; (B) a Scatchard plot; and (C) a semilogarithmic plot of θ versus log L. The binding constant can be easily determined from the midpoint of the semilogarithmic plot. The bar shown in Panel C represents the range in free ligand concentration required to go from 10% to 90% saturation of a single binding site, a span of 1.8 $\log_{10}$ units. To cover this range in saturation requires at least a 100-fold variation of free ligand concentration.

primary data and also violates fundamental assumptions used for least-squares fitting.) Finally, Fig. 3C shows the data plotted in semilogarithmic form, as θ versus log L (Chaires, 2001). This representation is to be preferred. One can see the whole range of the data, running from nanomolar to millimolar free ligand concentrations. The saturation of the receptor site is clearly visible. In this representation, half saturation (the "midpoint") occurs at the free ligand concentration that is the reciprocal of the association equilibrium constant. Another notable property of this representation is the span, the range in free ligand concentration required to go from a fractional saturation of 0.1 to 0.9. The bar in Fig. 3C shows this span to be $\sim$1.8 $\log_{10}$ units. This indicates the difficulty in any binding experiment. In order to properly define a complete binding isotherm, one must conduct titrations in such a way that free ligand concentrations are varied over approximately a 100-fold range. That is a challenge.

The span provides an inviolate characterization of both the quality of a binding isotherm and the underlying complexity of the binding mechanism. For a single class of binding site with no cooperativity, the span will always be 1.8 (Klotz, 1997; Weber, 1992; Wyman and Gill, 1990). Smaller or larger values unambiguously indicate that the binding mechanism is not simple one-site binding. Values smaller than 1.8 indicate multiple binding with positive cooperativity. Values larger than 1.8 indicate multiple binding with either negative cooperativity or site heterogeneity (multiple classes of sites with different binding constants).

Figure 4 shows an example of the latter case. Data were again simulated to a simple model with two binding sites on a receptor, with binding constants that differed by a factor of 10, $K_1 = 10^6$ and $K_2 = 10^5\ \mathrm{M}^{-1}$:

$$\theta = \frac{K_1\mathrm{L}}{1 + K_1\mathrm{L}} + \frac{K_2\mathrm{L}}{1 + K_2\mathrm{L}}$$

The direct plot (Fig. 4A) remains uninformative. The Scatchard plot (Fig. 4B) now shows curvature, and the slope and intercept have lost their simple meaning. The semilogarithmic plot (Fig. 4C) provides the best indication of the complexity resulting from the additional site. The span clearly exceeds 1.8 and heterogeneity in binding is clear.

IV. Thermodynamics from Thermal Denaturation Methods

Thermal denaturation is a commonly used tool to study the stability of a complex and provides an alternate method of determining binding thermodynamics through the linkage of binding and denaturation reactions. Consider the simple case of the denaturation of a native, folded form of a receptor to its denatured, unfolded form. Binding of a ligand to the native state will stabilize this form and result in an increase in thermal stability. Conversely, if the ligand prefers to bind to and stabilize the unfolded form, this will result in a decrease in the thermal stability of the native form. Changes in thermal stability in this way reflect the operation of Le Chatelier's principle. In this example, the native-to-unfolded equilibrium is perturbed by ligand binding to one of the forms. This results in the depletion of one side of the equilibrium, which is compensated by a shift in the equilibrium toward the depleted species.

A number of detailed theories have been developed for analyzing the effects of ligands on thermal denaturation curves. Schellman reported, as early as 1958, the application of statistical mechanical theories to describe the effect of ligand binding on the helix-to-coil transition of synthetic polypeptides and the extension of these principles to the denaturation of proteins (Schellman, 1958). The utility of thermal denaturation methods was extended significantly by Brandts and Lin (1990)

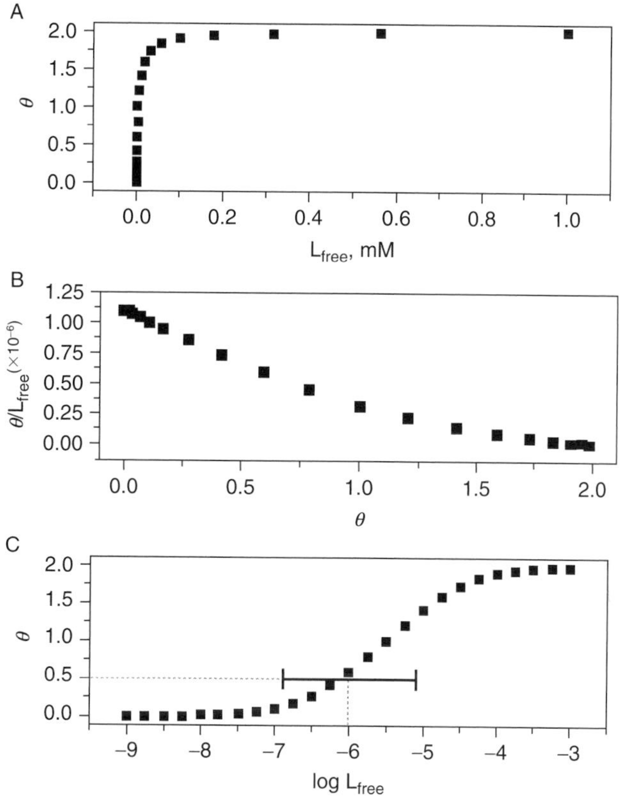

Fig. 4 Binding data modeled for two binding sites on a receptor with binding constants differing by a factor of 10, $K_1 = 10^6$ and $K_2 = 10^5\ M^{-1}$: (A) a direct plot of fractional saturation as a function of free ligand concentration; (B) a Scatchard plot; and (C) a semilogarithmic plot of θ versus log L. Only the semilogarithmic plot clearly shows the heterogeneity resulting from the second binding site with the span in free ligand concentration covering 10–90% site saturation clearly exceeding the 1.8 $\log_{10}$ units for a single binding site with no cooperativity.

through the development of models that facilitated the quantitative determination of binding constants of up to $10^{40}\ M^{-1}$ by differential scanning calorimetry (DSC). The application of DSC in this manner provided a huge advantage over optical equilibrium methods that require a fraction of the ligand (~10%) to remain in the unbound state. To achieve this fraction of unbound ligand, it is necessary to work at ligand concentrations approaching the reciprocal of the binding constant and quickly becomes impractical for ultratight binding interactions. DSC requires neither a ligand signal nor any fraction of unbound ligand. The methods discussed above were applied to the study of protein–ligand interactions but are, in fact, quite general in scope and can easily be applied to both protein and DNA systems with only subtle changes in the models to account, for example, for the lattice properties of DNA. The most useful treatments to describe the melting of ligand–DNA complexes were proposed by McGhee and Crothers (Chaires and Shi, 2006; Crothers, 1971; McGhee, 1976; Spink and Wellman, 2001).

When characterizing a new binding system, a useful first approach could be to determine a series of optical melting curves in the absence and presence of increasing amounts of a ligand of interest. Such a family of curves is shown in Fig. 5 for the binding of netropsin to poly(dA)-poly(dT). Sharp melting curves were observed for poly(dA)-poly(dT) alone and in the presence of saturating concentrations of netropsin; however, at intermediate concentrations of ligand, the curves were multiphasic and therefore difficult to analyze. By applying McGhee's equation (McGhee, 1976) to the melting curves in the absence and presence of saturating amounts of netropsin, it is possible to determine the ligand binding constant. The application of this equation requires the knowledge of melting temperatures in the absence and presence of ligand, as well as the enthalpy of DNA melting. Melting temperatures can be determined directly from the melting curves shown in Fig. 5, but the enthalpy of DNA melting must be determined independently either spectrophotometrically or calorimetrically. Ideally a model-independent calorimetric determination would be preferred (see Chapter 5 by Spink, this volume). However, even once the required parameters are known, it should be noted that McGhee's equation assumes a complete saturation of the DNA duplex with no binding to single strands and yields a binding constant at the melting temperature of the ligand–DNA complex in the case of netropsin binding

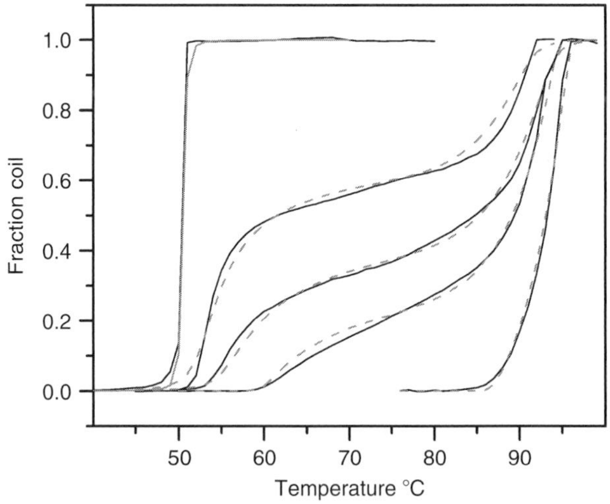

Fig. 5 Melting curves for poly(dA)-poly(dT) [45 μM(bp)] in the presence of increasing concentrations of the groove binder netropsin (from left to right, molar ratio added netropsin/bp = 0,0.08,0.14,0.18,0.28) (Chaires and Shi, 2006). The dotted lines are fits to McGhee's statistical mechanical algorithm.

to poly(dA)-poly(dT) this is at 103.6 °C. In order to equate this to some relevant temperature, it would be necessary to apply the van't Hoff equation and this requires knowledge of the ligand binding enthalpy, so again we return to the need for calorimetric studies. At this stage, the discussion has highlighted the need to apply other methods to characterize this binding system. It is important to realize that it is highly advantageous, and in most cases necessary, to apply multiple methods to achieve the complete characterization of a binding interaction.

V. Completing the Thermodynamic Profile

DSC can be applied to determine the melting enthalpies of poly(dA)-poly(dT) alone and in the presence of saturating amounts of netropsin from a direct integration of the experimental thermograms shown in Fig. 6. However, it is important to realize that the measurement of the enthalpy of melting of the saturated DNA duplex involves both the melting of the ligand–DNA complex as well as the melting of the DNA duplex to single strands, and it is necessary to apply Hess' law to determine the enthalpy of ligand binding (Chaires and Shi, 2006). Armed with enthalpies of duplex melting and ligand binding ($\Delta H_b = -12.0$ kcal/mol), it is now possible to return to the optical melting curves and determine a binding constant of 1.5×10^8 M^{-1} at the melting temperature of 103.6 °C, which can then be extrapolated to a temperature of 20 °C using the van't Hoff equation to yield a binding constant of 1.4×10^{10} M^{-1}. At this point, the

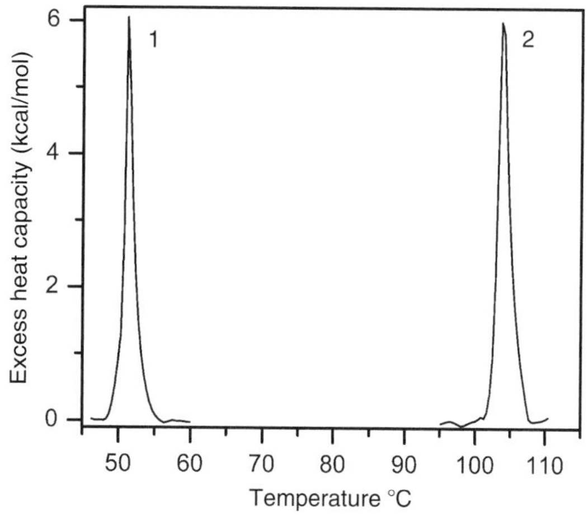

Fig. 6 DSC thermograms for poly(dA)-poly(dT) in the absence (peak 1) and presence (peak 2) of saturating amounts of the groove binder netropsin (Chaires and Shi, 2006). Experiments were conducted in BPE buffer at pH 7 at a polynucleotide concentration of 0.3 mM (bp). Integration of the thermograms directly yields the enthalpy of melting of the duplex (+10.6 kcal/mol) and the duplex–ligand complex (+13.6 kcal/mol).

complete thermodynamic profile has been determined as $\Delta G_{20} = -13.6$ kcal/mol, $\Delta H_b = -12.0$ kcal/mol, and $T\Delta S = -1.6$ kcal/mol.

The approach just described requires the determination of a binding constant from duplex melting under saturating ligand conditions and its extrapolation from the melting temperature of the saturated complex to a far removed temperature of interest. In the case of netropsin binding to poly(dA)-poly(dT), the van't Hoff extrapolation assumes a zero enthalpy change over an 83.6 °C temperature range. Such a large temperature extrapolation is unsatisfactory. A more rigorous estimate of the binding constant, as well as the binding site size, can be made with the application of McGhee's statistical mechanical algorithm (McGhee, 1976) to the complete set of melting curves shown in Fig. 5. The solid lines in Fig. 5 best fits to the McGhee algorithm yield a binding constant of 1.3×10^{10} M^{-1} and a site size of 4 bp (Chaires and Shi, 2006). Fortunately, this fitting agrees well with the binding constant calculated by application of the van't Hoff equation but this is certainly not always the case. A preferred approach must in fact be to obtain the binding constant directly at the temperature of interest by performing spectroscopic titrations or isothermal titration calorimetry (ITC) studies.

VI. Thermodynamics in the Real World: Some Useful Strategies

By this point, we hope to have convinced you of the importance of determining a complete set of thermodynamic parameters for a system of interest. Of course, this can be, in many cases, easier said than done. Figure 7 provides an overview of techniques that are commonly applied to thermodynamic measurements.

It is not our endeavor here to provide a commentary of all these techniques and how each can be applied to make specific thermodynamic measurements. The

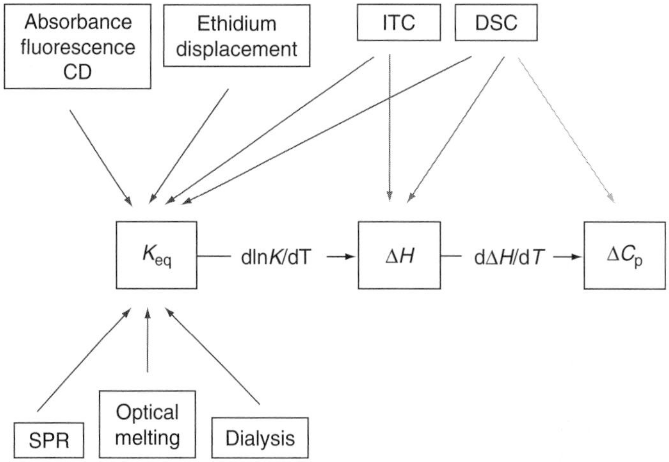

Fig. 7 Common techniques applied to the determination of binding thermodynamics.

reader has already been referred to a number of sources that will provide detailed information on these techniques and the practicalities of making thermodynamic measurements. Instead, the intent is to provide some comments concerning specific binding systems where some properties of the system have required modifications of these general techniques to obtain the desired thermodynamic measurements. For example, what strategies can be applied to study poorly soluble ligands? What approaches are appropriate for low-affinity ligands? This following commentary should provide some food for thought when attempting to apply the general approaches to the practicalities of a system of interest.

Calorimetry is often considered to be the gold standard for measuring biomolecular interactions. Since the development of nonlinear least-squares fitting methods by Wiseman *et al.* (1989), a complete thermodynamic profile can be easily determined from a single ITC experiment (followed by appropriate control and duplicate experiments). These approaches involve the application of models to describe a number of binding events, including binding to single or multiple sites, dissociation, and competitive binding. It is extremely beneficial to determine a model-independent enthalpy value from the method of excess sites (Ren *et al.*, 2000) that can then be constrained in subsequent fitting. In order to apply ITC to systems outside of the conventional 10^3–10^9 M^{-1} range of binding constants, it has been necessary to be flexible in the application of the technique. A number of useful approaches have now been developed to characterize binding of both high- and low-affinity systems (Sigurskjold, 2000; Turnbull and Daranas, 2003; Zhang and Zhang, 1998). Recently, the excess sites approach has been successfully applied to characterize the low-affinity recognition of L-argininamide by a DNA aptamer (Bishop *et al.*, 2007). Binding was found to be highly selective but energetically weak and as such presented a challenge for complete thermodynamic characterization. The ligand binding enthalpy was determined using a "reverse" model-free ITC approach. The first report of a "model-free" ITC approach by Ren *et al.* (2000) involved the titration of small aliquots of ~1-mM ligand into an ITC cell loaded with ~1 ml of a 1-mM (bp) DNA solution. To deal with the low binding affinity of the argininamide–aptamer system, the locations of the ligand and DNA were reversed such that the high concentration of ligand in the ITC cell (1 ml of 3.0-mM L-argininamide) ensured that all aptamer was bound after each injection (10-μl aliquots of 0.2-mM aptamer). In this way, a model-independent binding enthalpy was determined to be −8.7 kcal/mol at 25 °C. Bishop and coworkers exploited the significant differences in the circular dichroism (CD) spectra of the DNA aptamer and the argininamide–aptamer complex to determine the ligand binding affinity. Constructing a semilogarithmic binding plot, as discussed earlier, and following the method of Qu and Chaires (2000), a ligand binding affinity was determined to be 5998 M^{-1}, yielding a binding free energy of −5.1 kcal/mol. Determination of the binding enthalpy and binding free energy then allowed the calculation of the entropic contribution to binding as $T\Delta S = \Delta H - \Delta G = -8.7 - (-5.1) = -3.6$ kcal/mol. The weak binding free energy resulted from opposing enthalpic and entropic components. It is thought that a favorable binding enthalpy just outweighs an unfavorable entropic contribution to binding resulting from ordering of the hairpin DNA aptamer structure to form a well-ordered specific binding pocket in the presence of the argininamide ligand.

DSC, like ITC, provides a means to determine enthalpy values without recourse to binding models. In the case of DSC, enthalpy values are evaluated from the direct integration of the experimental C_p versus T scan. The enthalpy of binding is measured directly by ITC, whereas the enthalpy of melting of a ligand–DNA complex using DSC also includes the enthalpy of melting of the DNA to single strands; and, therefore, the ligand binding enthalpy must be evaluated by applying Hess' law, as described earlier. A number of interesting small molecule ligands have poor aqueous solubility, particularly at DSC concentrations, and it may be undesirable or ineffective to introduce a small percentage of DMSO to help solubilize the ligand. How can one proceed? Poor aqueous solubility was encountered for the natural product antibiotic echinomycin where the study of its DNA-binding properties was hampered by an aqueous solubility limited to just 5 μM. To facilitate the determination of a ligand-DNA binding enthalpy using DSC, Leng *et al.* (2003) developed a "solid-shake" protocol. This method involved dissolving solid echinomycin directly into a DNA solution in which the ligand had increased solubility. Undissolved ligand was then removed by low-speed centrifugation and the exact concentration of echinomycin determined spectrophotometrically. Determination of the enthalpy of melting of these preformed ligand–DNA complexes, along with the enthalpy of melting of DNA alone, then allowed calculation of the ligand binding enthalpy using the application of Hess' law as described above. The complete thermodynamic cycle was obtained by applying additional thermodynamic techniques in a manner similar to that described for the netropsin study with a series of optical melting curves obtained at less than saturating ligand conditions and analyzed using the McGhee algorithm (McGhee, 1976). The ability to work with preformed ligand–DNA complexes makes DSC a useful method for poorly soluble ligands, a property that would prohibit ITC studies or some spectroscopic titrations.

VII. Ligand–Receptor Binding in the Absence of an Optical Signal

The previous discussion has highlighted the utility of calorimetric methods in the study of receptor–ligand interactions. An additional advantage of these methods lies in the fundamental property on which these techniques are based. Calorimetry measures a universal signal, heat, which accompanies all interactions. Where a convenient optical monitor is lacking in the receptor or ligand, spectroscopic titrations are useless and calorimetry becomes a valuable technique to study the binding of these ligands. However, calorimetry can become impractical in cases when small heat changes are detected and high concentrations are required for these studies. In these cases, surface plasmon resonance (Myszka, 2000) or ethidium displacement assays could be useful methods.

Le Pecq and Paoletti (1967) first proposed in 1969 a fluorescence assay based on exploiting the change in fluorescence of ethidium upon binding to nucleic acids. When ethidium intercalates into duplex DNA, the hydrophobic environment of the binding site serves to shield ethidium from fluorescence quenching by solvent and a large fluorescence enhancement is observed. This property can be exploited in many ways as a convenient assay to study many forms of binding. In two reviews by Morgan *et al.* (1979a,b), a large number of examples were given about the application of this assay to the study of DNA, RNA, and DNA–protein interactions. Jenkins (1997) has since discussed the practical aspects of the use of the ethidium displacement assay for studying drug–DNA interactions.

The ethidium displacement assay has been employed in our laboratory in a 96-well format as a rapid initial screen of the binding selectivity of nonchromophoric ligands (unpublished results). The results can then be used to guide more rigorous and sample intensive biophysical studies such as ITC and DSC. In our hands, binding affinities of test ligands are assessed directly from changes in fluorescence resulting from their ability to displace ethidium from a preformed ethidium–DNA complex:

$$\mathrm{DNA} - \mathrm{EB} + \mathrm{L} \rightleftharpoons \mathrm{DNA} - \mathrm{L} + \mathrm{EB}$$

The first generation of the assay has focused on characterizing the binding selectivity of neomycin-based aminoglycoside ligands for duplex DNA, duplex RNA, and DNA–RNA hybrid structures. The RNA-binding properties of aminoglycosides have been extensively investigated (Walter *et al.*, 1999), but it has been shown more recently that these ligands show binding preferences for a number of A-form nucleic acids, not just RNA (Arya *et al.*, 2003). As part of an ongoing synthetic strategy to design targeted aminoglycoside ligands, the ethidium displacement assay has been employed as a rapid screen of the binding selectivity of selected ligands. Results from the assay are shown in Fig. 8.

Of the four selected aminoglycoside ligands, neomycin sulfate, paromomycin sulfate, and neomycin-free amine exhibited significant binding to poly(dG)-poly(dC), poly(dA)-poly(dT), poly(rA)-poly(rU), and poly(dA)-poly(rU), with the strongest binding observed for neomycin sulfate and neomycin free amine. Ribostamycin sulfate was revealed to be a poor ligand candidate, displaying little binding to any of the nucleic acid structures. Arya has previously demonstrated the preference of neomycin-related ligands for A-form nucleic acid structures, of which poly(dG)-poly(dC), poly(dA)-poly(dT), poly(rA)-poly(rU), and poly(dA)-poly(rU) have all been reported to exhibit a high propensity to adopt A-form conformations (Arya *et al.*, 2003). Little binding was observed to the other four nucleic acid structures, all B-form. The ethidium displacement assay was focused to include a limited number of nucleic acid structures relevant to an investigation of aminoglycoside binding but can be expanded and focused for the study of other binding systems.

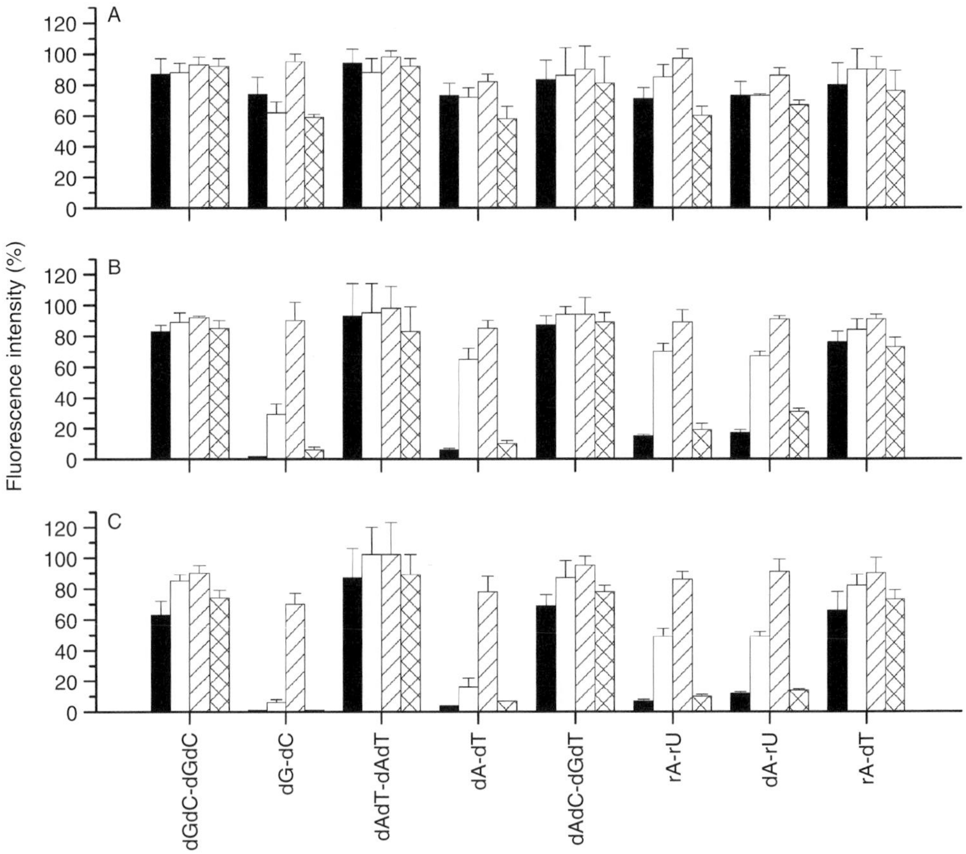

Fig. 8 Sequence selectivity off our aminoglycoside ligands as revealed by the ethidium displacement assay. Each panel shows the relative binding of neomycin sulfate (black bars), paromomycin sulfate (white bars), ribostamycin sulfate (hatched bars), and neomycin free amine (cross-hatched bars) to each of eight nucleic acid structures as a function of increasing concentration of aminoglycoside ligand: panel A, 2 μM; panel B, 4 μM; and panel C, 8 μM.

VIII. Toward High-Throughput Thermodynamics

In recent years, there has been an increasing push for high-throughput technologies. In an expanded format, the application of an ethidium displacement assay in a 96-well format offers just that. One might argue that the ethidium displacement assay is an indirect measurement of binding affinity. Research in our laboratory has attempted to answer the high-throughput demand with the application of other methods. In fact, the high-throughput methods discussed here are attempting to address Scatchard's question "Where?" By developing methods that can rapidly screen ligand binding against an array of structures and sequences, we move closer to identifying the location of ligand binding on the receptor.

We have recently described a mixture melting assay (Shi and Chaires, 2006) in which a mixture of four to five polynucleotide structures or sequences are melted together in the presence of a test ligand. Under conditions of low lattice saturation, when the binding ratio approaches zero, it is possible to observe binding selectivity to a preferred site (Chaires, 1992; Muller and Crothers, 1975). This method is in its infancy and currently only a handful of polynucleotide forms are included. However, with the expansion of the nucleic acid array and the ready availability of multicell transports, this assay offers rapid qualitative binding information that can guide more detailed thermodynamic studies.

The relatively new method of competition dialysis can provide valid binding constants for a test ligand against a vast array of potential nucleic acid targets. Competition dialysis was first employed by Muller and Crothers (1975) as a method to determine the base pair preference of a test ligand against two natural DNA samples of differing GC base pair content. This early application has now been extended in our laboratory to include 46 different nucleic acid sequences and structures (Chaires, 2002; Ragazzon *et al.*, 2007; Ren and Chaires, 1999, 2001). Chaires has shown that the method provides a rapid and rigorous determination of the distribution of free and bound ligand against an array of binding targets; these values can then be used to calculate an apparent binding constant, K_{app}, which has been shown to compare well with values of K determined by more rigorous spectrophotometric titrations (Chaires, 2005a,b). The utility of the competition dialysis assay was illustrated early on by the application of our first-generation 13-structure assay to study the binding preferences of NMM and coralyne (Ren and Chaires, 1999; Fig. 9).

NMM was shown to bind with complete selectivity to the quadruplex structure $(5'\text{-}T_2G_{20}T_2)_4$. No detectable binding was observed to any of the other nucleic acid forms (single-stranded DNA, duplex DNA, Z-DNA, duplex RNA, DNA–RNA hybrid, and triplex DNA) included in the 13 structure, first-generation competition dialysis assay. Results from the competition dialysis assay were in agreement with observations from fluorescence studies of NMM binding to quadruplex DNA but not to duplex forms (Arthanari *et al.*, 1998). Coralyne was shown to interact with almost all of the structures included in the competition dialysis assay with a preference for poly(dA)-[poly(dT)]$_2$ triplex and, interestingly, exhibited strong binding to single-stranded poly(dA). Triplex selective binding was not unexpected as this had been previously reported in the literature (Lee *et al.*, 1993); however, binding to poly(dA) was a previously unreported observation. Absorbance titration experiments were undertaken to further investigate the binding to poly(dA) and showed that coralyne binding was an order of magnitude stronger to poly(dA) $[(1.05 \pm 0.1) \times 10^5\ M^{-1}]$ than to calf thymus duplex DNA $[(1.25 \pm 0.1) \times 10^4\ M^{-1}]$, supporting the surprising competition dialysis result. In fact, subsequent reports from the Hud laboratory have showed that coralyne binds to poly(dA) with a stoichiometry of one coralyne to four adenine bases and promotes the formation of an antiparallel homoadenine duplex secondary structure (Persil *et al.*, 2004). The characterization of this novel binding mode was a very encouraging result from the first generation competition dialysis assay; and with the expansion of the

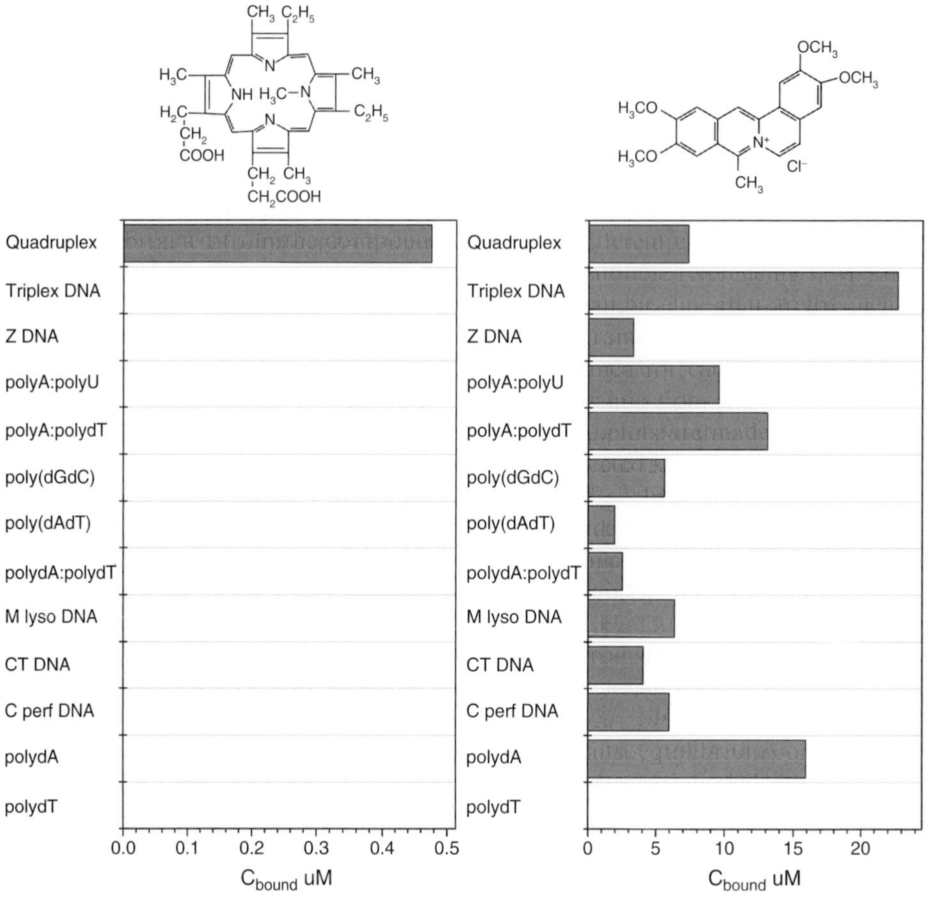

Fig. 9 Results obtained by the competition dialysis method for NMM (left) and coralyne (right) (Ren and Chaires, 1999).

assay, currently 46 structures, there exists even greater potential for the rapid identification of novel ligand binding preferences.

IX. Summary

Thermodynamic parameters are essential to properly characterize a binding interaction, provide a means toward understanding the molecular forces mediating biological processes and ultimately can be utilized to drive the rational drug design process. A vast array of techniques and approaches now exist that offer an entry into the world of thermodynamics, with an even greater number of articles and reviews that present a commentary of the collective literature. In an effort not to rehash principles and strategies that have already been adequately described elsewhere, we have attempted to provide in this chapter some insights into particular elements of conducting successful binding studies.

Acknowledgments

Supported by grant CA35635 from the National Cancer Institute (J.B.C.).

References

Arthanari, H., Basu, S., Kawano, T. L., and Bolton, P. H. (1998). Fluorescent dyes specific for quadruplex DNA. *Nucleic Acids Res.* **26,** 3724–3728.

Arya, D. P., Xue, L., and Willis, B. (2003). Aminoglycoside (Neomycin) preference is for A-form nucleic acids, not just RNA: Results from a competition dialysis study. *J. Am. Chem. Soc.* **125,** 10148–10149.

Bishop, G. R., Ren, J., Polander, B. C., Jeanfreau, B. D., Trent, J. O., and Chaires, J. B. (2007). Energetic basis of molecular recognition in a DNA aptamer. *Biophys. Chem.* **126**(1–3), 165–175.

Brandts, J. F., and Lin, L.-N. (1990). Study of strong to ultratight protein interactions using differential scanning calorimetry. *Biochemistry* **29,** 6927–6940.

Chaires, J. B. (1992). Application of equilibrium binding methods to elucidate the sequence specificity of antibiotic binding to DNA. *In* "Advances in DNA Sequence-Specific Agents" (L. H. Hurley, ed.), Vol. 1, pp. 3–23. JAI Press Inc, Greenwich, CT.

Chaires, J. B. (1996). Dissecting the free energy of drug binding to DNA. *Anti-Cancer Drug Des.* **11,** 569–580.

Chaires, J. B. (1997). Energetics of drug-DNA interactions. *Biopolymers* **44,** 201–215.

Chaires, J. B. (2001). Analysis and interpretation of ligand-DNA binding isotherms. *Methods Enzymol.* **340,** 3–22.

Chaires, J. B. (2002). Nucleic acid binding molecules: A competition dialysis assay for the study of structure-selective ligand binding to nucleic acids. *In* "Current Protocols in Nucleic Acid Chemistry" (S. L. Beaucage, D. E. Bergstrom, G. D. Glick, and R. A. Jones, eds.), Vol. 2, pp. 831–838. John Wiley & Sons, New York.

Chaires, J. B. (2003). Energetics of Anthracycline-DNA interactions. *In* "Small Molecule DNA and RNA Binders: From Small Molecules to Drugs" (M. Demeunynck, C. Bailly, and W. D. Wilson, eds.), Vol. 2, pp. 461–481. Wiley-VCH, Weinheim, Germany.

Chaires, J. B. (2005a). Competition dialysis: An assay to measure the structural selectivity of drug-nucleic acid interactions. *Curr. Med. Chem. Anticancer Agents* **5,** 339–352.

Chaires, J. B. (2005b). Structural selectivity of drug-nucleic acid interactions probed by competition dialysis. *In* "DNA Binders and Related Subjects" (M. J. Waring, and J. B. Chaires, eds.), Vol. 253, pp. 33–54. Springer-Verlag, Berlin.

Chaires, J. B. (2006). A thermodynamic signature for drug-DNA binding mode. *Arch. Biochem. Biophys.* **453,** 26–31.

Chaires, J. B., and Shi, X. (2006). Thermal denaturation of drug-DNA complexes: Tools and tricks. *In* "Sequence-specific DNA Binding Agents" (M. J. Waring, ed.), pp. 130–151. Royal Society of Chemistry, Cambridge, UK.

Crothers, D. M. (1971). Statistical thermodynamics of nucleic acid melting transitions with coupled binding equilibria. *Biopolymers* **10,** 2147–2160.

Haq, I. (2002). Thermodynamics of drug-DNA interactions. *Arch. Biochem. Biophys.* **403,** 1–15.

Haq, I., and Ladbury, J. (2000). Drug-DNA recognition: Energetics and implications for design. *J. Mol. Recognit.* **13,** 188–197.

Haq, I., Jenkins, T. C., Chowdhry, B. Z., Ren, J., and Chaires, J. B. (2000). Parsing the free energy of drug-DNA interactions. *Methods Enzymol.* **323,** 373–405.

Jenkins, T. C. (1997). Optical absorbance and fluorescence techniques for measuring drug-DNA interactions. *In* "Drug-DNA Interaction Protocols" (K. R. Fox, ed.), Vol. 90, pp. 195–218. Humana Press, Totowa, NJ.

Klotz, I. M. (1997). "Ligand Receptor Energetics: A Guide for the Perplexed." John Wiley & Sons, New York.

Lee, J. S., Latimer, L. J., and Hampel, K. J. (1993). Coralyne binds tightly to both T.A.T- and C.G.C (+)-containing DNA triplexes. *Biochemistry* **32,** 5591–5597.

Leng, F., Chaires, J. B., and Waring, M. J. (2003). Energetics of echinomycin binding to DNA. *Nucleic Acids Res.* **31,** 6191–6197.

LePecq, J. B., and Paoletti, C. (1967). A fluorescent complex between ethidium bromide and nucleic acids. Physical-chemical characterization. *J. Mol. Biol.* **27,** 87–106.

McGhee, J. D. (1976). Theoretical calculations of the helix-coil transition of DNA in the presence of large, cooperatively binding ligands. *Biopolymers* **15,** 1345–1375.

Morgan, A. R., Evans, D. H., Lee, J. S., and Pulleyblank, D. E. (1979a). Ethidium fluorescence assays. Part II. Enzymatic studies and DNA-protein interactions. *Nucleic Acids Res.* **7,** 571–594.

Morgan, A. R., Lee, J. S., Pulleyblank, D. E., Murray, N. L., and Evans, D. H. (1979b). Ethidium fluorescence assays. Part I. Physicochemical studies. *Nucleic Acids Res.* **7,** 547–568.

Muller, W., and Crothers, D. M. (1975). Interactions of heteroaromatic compounds with nucleic acids. 1. The influence of heteroatoms and polarizability on the base specificity of intercalating ligands. *Eur. J. Biochem.* **54,** 267–277.

Myszka, D. G. (2000). Kinetic, equilibrium and thermodynamic analysis of macromolecule interactions with BIACORE. *Methods Enzymol.* **323,** 325–332.

Persil, O., Santai, C. T., Jain, S. S., and Hud, N. V. (2004). Assembly of an antiparallel homo-adenine DNA duplex by small-molecule binding. *J. Am. Chem. Soc.* **126,** 8644–8645.

Qu, X., and Chaires, J. B. (2000). Analysis of drug-DNA binding data. *Methods Enzymol.* **321,** 353–369.

Ragazzon, P. A., Garbett, N. C., and Chaires, J. B. (2007). Competition dialysis: A method for the study of structural selective nucleic acid binding. *Methods* **42**(2), 173–182.

Ren, J., and Chaires, J. B. (1999). Sequence and structural selectivity of nucleic acid binding ligands. *Biochemistry* **38,** 16067–16075.

Ren, J., and Chaires, J. B. (2001). Rapid screening of structurally selective ligand binding to nucleic acids. *Methods Enzymol.* **340,** 99–108.

Ren, J., Jenkins, T. C., and Chaires, J. B. (2000). Energetics of intercalation reactions. *Biochemistry* **39,** 8439–8447.

Ruben, A. J., Kiso, Y., and Freire, E. (2006). Overcoming roadblocks in lead optimization: A thermodynamic perspective. *Chem. Biol. Drug Des.* **67,** 2–4.

Scatchard, G. (1949). The attraction of proteins for small molecules and ions. *Ann. N.Y. Acad. Sci.* **51,** 660–672.

Schellman, J. A. (1958). The factors affecting the stability of hydrogen-bonded polypeptide structures in solution. *J. Phys. Chem.* **62,** 1485–1494.

Shi, X., and Chaires, J. B. (2006). Sequence- and structural-selective nucleic acid binding revealed by the melting of mixtures. *Nucleic Acids Res.* **34,** e14.

Sigurskjold, B. W. (2000). Exact analysis of competition ligand binding by displacement isothermal titration calorimetry. *Anal. Biochem.* **277,** 260–266.

Spink, C. H., and Wellman, S. E. (2001). Thermal denaturation as a tool to study DNA-ligand interactions. *Methods Enzymol.* **340,** 193–211.

Turnbull, W. B., and Daranas, A. H. (2003). On the value of *c*: Can low affinity systems be studied by isothermal titration calorimetry? *J. Am. Chem. Soc.* **125,** 14859–14866.

Velazquez-Campoy, A., Kiso, Y., and Freire, E. (2001). The binding energetics of first- and second-generation HIV-1 protease inhibitors: Implications for drug design. *Arch. Biochem. Biophys.* **390,** 169–175.

Velazquez-Campoy, A., Todd, M. J., and Freire, E. (2000). HIV-1 protease inhibitors: Enthalpic versus entropic optimization of the binding affinity. *Biochemistry* **39,** 2201–2207.

Walter, F., Vicens, Q., and Westhof, E. (1999). Aminoglycoside-RNA interactions. *Curr. Opin. Chem. Biol.* **3,** 694–704.

Weber, G. (1992). "Protein Interactions." Chapman & Hall, New York.

Wiseman, T., Williston, S., Brandts, J. F., and Lin, L. N. (1989). Rapid measurement of binding constants and heats of binding using a new titration calorimeter. *Anal. Biochem.* **179,** 131–137.

Wyman, J., and Gill, S. J. (1990). "Binding and Linkage: Functional Chemistry of Biological Macromolecules." University Science Books, Mill Valley, CA.

Zhang, Y-L., and Zhang, Z. Y. (1998). Low-affinity binding determined by titration calorimetry using a high-affinity coupling ligand: A thermodynamic study of ligand binding to protein tyrosine phosphatase 1B. *Anal. Biochem.* **261,** 139–148.

Suggested Reading

Connors, K. A. (1987). "Binding Constants: The Measurement of Molecular Complex Stability." John Wiley & Sons, New York.

A useful survey of methods for the determination of binding isotherms and the interpretation of binding data. Written from a pharmacological perspective.

Edsall, J. T., and Wyman, J. (1958)."Biophysical Chemistry," Vol. 1. Academic Press, Inc., New York.

Chaper 11 should be read before anything else.

Klotz, I. M. (1997). "Ligand-Receptor Energetics: A Guide for the Perplexed." John Wiley & Sons, New York.

As entertaining as binding thermodynamics can be. This should be read in conjunction with the author's account of the "early days," found in: Klotz, I. M. (2004). *J. Biol. Chem.* **279,** 1–12.

Harding, S. E., and Chowdhry, B. Z. (eds.) (2001). "Protein-Ligand Interactions: Structure and Spectroscopy." Oxford University Press, Oxford.

Harding, S. E., and Chowdhry, B. Z. (eds.) (2001). "Protein-Ligand Interactions: Hydrodynamics and Calorimetry." Oxford University Press, Oxford.

These volumes provide discussions of methods used to determine primary binding data and attempt to provide protocols for the proper use of methods.

Weber, G. (1992). "Protein Interactions." Chapman and Hall, New York.

Words from one of the masters.

Winzor, D. J., and Sawyer, W. H. (1995). "Quantitative Characterization of Ligand Binding." John Wiley & Sons, New York.

A useful and insightful discussion of both the theory and practice of binding studies.

Wyman, J., and Gill, S. J. (1990). "Binding and Linkage: Functional Chemistry of Biological Macromolecule." University Science Books, Mill Valley, CA.

Words from two of the masters.

CHAPTER 2

Linked Equilibria in Regulation of Transcription Initiation

Dorothy Beckett

Department of Chemistry and Biochemistry
Center for Biological Structure and Organization
University of Maryland
College Park, Maryland 20742

Abstract

Assembly of transcriptional regulatory complexes often involves multiple binding processes and these binding processes are frequently coupled to one another. Small molecule binding can promote or inhibit DNA-binding or protein–protein interactions. DNA binding may be coupled to protein association. Finally, proteins may bind cooperatively to multiple sites in a transcriptional regulatory region. The level of transcription initiation at a promoter reflects the assembly of regulatory

0091-679X/08 $35.00
DOI: 10.1016/S0091-679X(07)84002-7

complexes in a transcription control region. Quantitative mechanistic understanding of regulatory complex assembly requires dissection of the assembly process into its constituent interactions followed by measurements of linkage between the individual binding processes. Methods and approaches to achieving this quantitative understanding of transcription regulation are outlined in this chapter.

I. Introduction

Efforts to define the macromolecular participants in cellular processes have revealed complex circuitry of interacting proteins at all levels of biology. The transcription initiation process at a single control region can reflect binding of several different proteins to DNA sequences in this region to dictate the transcription level. Moreover, the particular array of proteins that bind in one set of cellular conditions may differ significantly from those that bind under other conditions (Marr *et al.*, 2006). Such combinatorial control is exemplified by the *Escherichia coli gal* regulatory system in which the three different transcription patterns reflect the specific arrangement of transcription factors at the transcription control region (Fig. 1; Semsey *et al.*, 2006). With the development of high-throughput methods the rate of acquisition of qualitative information for these systems has increased dramatically. While availability of this information is important because it defines

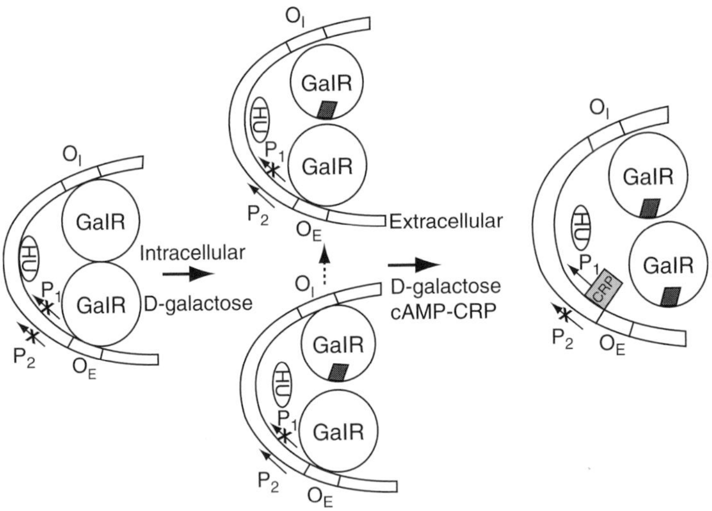

Fig. 1 Combinatorial control of transcription initiation from the two promoters, P_1 and P_2 by the proteins GalR (Galactose Repressor), HU (Prokaryotic Histone-like Protein), and CRP (Catabolite Repressor Protein). The specific arrangement of GalR, HU, and CRP can lead to either repression of initiation at both promoters, repression at only P_1, or activation of transcription at P_1 and repression at P_2 (Semsey *et al.*, 2006).

the players in a regulatory process, further elucidation of the mechanisms of regulation requires detailed biochemical and biophysical analysis of these players. In this chapter, a road map for performing such analyses is initially outlined. This outline is followed by a review of quantitative analysis of two transcriptional regulatory assemblies. The goals of the strategies described are to define the rules that govern assembly of a transcription regulatory complex and to determine the relationship of assembly to patterns and levels of transcription initiation.

II. Multiple Levels of Linkage in Transcription Regulation

Regulation of transcription initiation involves interactions between multiple proteins with regulatory sites on the genome. Moreover, small molecule binding or posttranslational modification can influence the assembly of these multiprotein–DNA complexes (Fig. 2). Therefore, quantitative mechanistic understanding of transcription regulation requires examination of these complexes at several levels. First, the complexes must be dissected into their constitutive pairwise interactions and the stoichiometry and energetics of each interaction should be characterized. Once the individual contributing interactions are understood, it is possible to investigate further how they are coupled or linked. By coupling we mean the influence, either inhibitory or activating, of one interaction on another. Classic examples of coupling or linkage in transcription regulation include the members of the galactose/lactose (GalR/LacI) family of repressor proteins (Weickert and Adhya, 1992). Sugar binding to these proteins modulates their site-specific binding to their cognate operator sequences. Depending on the protein, small molecule binding can result in relief of transcription repression by decreasing occupancy at an operator sequence or can enhance repression or activation by increasing occupancy. In eukaryotic systems, the nuclear receptors provide examples of

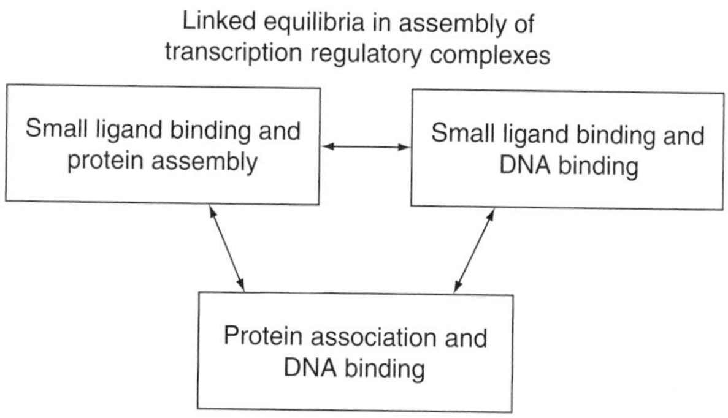

Fig. 2 Coupling of the individual pairwise interactions in assembly of a transcriptional regulatory complex.

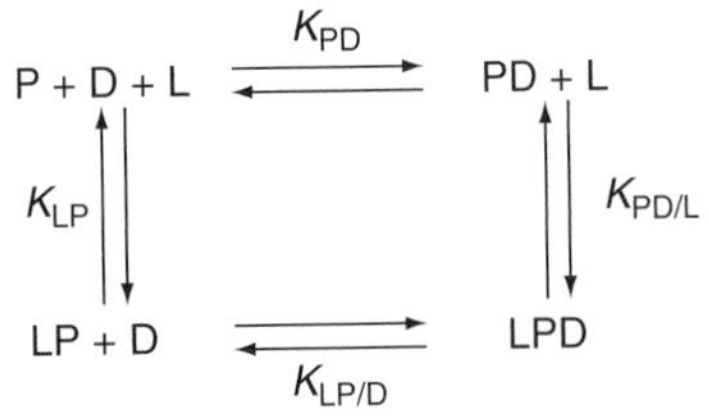

Fig. 3 A thermodynamic cycle that illustrates linkage between DNA and small ligand binding by a protein. P: protein, L: ligand, and D: DNA. Small ligand binding may precede DNA binding or vice versa. If linkage exists, the equilibrium constant for one binding reaction depends on whether the second has first occurred.

proteins to which small ligands can bind to alter DNA binding and, ultimately, transcription initiation. An illustration of linkage between small ligand and DNA binding is shown in the thermodynamic cycle in Fig. 3. The two binding processes can occur by two alternative pathways, ligand followed by DNA binding or DNA followed by ligand binding. The equilibrium expressions corresponding to these two pathways are shown in the following equations:

$$K_{\mathrm{LP}} = \frac{[\mathrm{L}][\mathrm{P}]}{[\mathrm{LP}]} \tag{1}$$

$$K_{\mathrm{PD}} = \frac{[\mathrm{P}][\mathrm{D}]}{[\mathrm{PD}]} \tag{2}$$

$$K_{\mathrm{LP/D}} = \frac{[\mathrm{LP}][\mathrm{D}]}{[\mathrm{LPD}]} \tag{3}$$

$$K_{\mathrm{PD/L}} = \frac{[\mathrm{PD}][\mathrm{L}]}{[\mathrm{LPD}]} \tag{4}$$

In this case, the expressions are presented in terms of dissociation and the constants, K, are therefore equilibrium dissociation constants. Since the two pathways start and end at the same state, the following holds true:

$$K_{\mathrm{LP}}K_{\mathrm{LP/D}} = K_{\mathrm{PD}}K_{\mathrm{PD/L}} \tag{5}$$

Furthermore, if linkage, either positive or negative, exists in the system, the following is also true:

$$K_{\mathrm{LP}} \neq K_{\mathrm{PD/L}} \tag{6}$$

$$K_{PD} \neq K_{LP/D} \tag{7}$$

In general, one is interested in determining what is referred to as the coupling free energy in a system, or the amount by which the free energy of binding of one molecule is affected by first saturating the protein with the other. The free energy of any binding interaction is calculated by using the following equation:

$$\Delta G^{\circ} = RT \ln K_{d} \tag{8}$$

where R is the gas constant, T is the temperature in Kelvin, and K_d is the equilibrium dissociation constant for the process. The coupling free energy, $\Delta G_c{}^{\circ}$ for the binding reactions shown in Fig. 2, is calculated as follows:

$$\Delta G_c{}^{\circ} = RT \ln K_{LP} - RT \ln K_{PD/L} \tag{9}$$

or

$$\Delta G_c{}^{\circ} = RT \ln K_{PD} - RT \ln K_{LP/D} \tag{10}$$

The coupling free energy may be either greater or less than zero. If it is greater, the coupling is negative or ligand binding is antagonistic to DNA binding; and if it is less than zero, the linkage is positive or ligand binding promotes DNA binding. Finally, by reciprocity, if ligand binding is antagonistic to DNA binding, then DNA binding must also be antagonistic to ligand binding. Knowledge of the coupling free energy enables prediction of the extent to which small ligand binding to a transcriptional regulator affects its affinity for DNA and thus its ability to alter transcription initiation.

In addition to the linkage between small ligand and DNA binding, several other levels of linkage can exist in assembly of transcription regulatory complex. For example, small ligand binding may be positively linked to self-association of the regulatory protein (Eisenstein and Beckett, 1999). If, as is often the case, the oligomeric species binds more tightly to a regulatory site on DNA than does the monomer, then the positive linkage between small ligand binding and protein self-association enhances DNA binding and, thus, the transcriptional readout associated with the binding event. In addition to small ligands, posttranslational modification can be positively linked to protein association and/or DNA binding (Correia *et al.*, 2001). Protein association may also be directly linked to DNA binding. In this case, the DNA binding enhances self-association or is positively coupled to self-association. Again, by reciprocity, protein self-association also promotes DNA binding in these systems. In more complex systems, formation of heterologous protein–protein interactions is linked to DNA binding. Small ligand binding or posttranslational modification may also influence these assembly processes. Control of heterologous protein–protein interactions is particularly important for combinatorial control because the ability of, for example, a single protein to participate in different protein–protein interactions can enable it to bind to distinct DNA sequences at control regions of a number of genes. Thermodynamic cycles analogous to that shown in Fig. 3 can be drawn for any of these other types of linkage, and explicit outlining of the linkage is helpful in defining the equilibria that must be measured in any system.

III. A Road Map for Quantitative Studies of Assembly of Gene Regulatory Complexes

The final readout in any transcription regulatory system is the level of transcription initiation, and this readout is directly correlated with the level of occupancy of regulatory sites on the DNA by proteins. Quantitative understanding of the complex assemblies that contribute to the regulation requires knowledge of all of the factors that influence the occupancy of the regulatory sites. Since multiple individual binding interactions feed into this occupancy, it is first critical to develop assays to measure the individual interactions. Once these assays are developed, one is positioned to measure pairwise coupling of the interactions (Fig. 4). For example, how does small ligand binding influence self-association? Finally, once the pairwise coupling is understood, one can build to higher order or multicomponent coupling (Fig. 2).

IV. Measurements of Binding Interactions in Transcription Regulation

A. Determine the Assembly State of the Protein(s) of Interest

Prior to performing any measurements of the binding interactions of a regulatory protein, it is critical to determine its assembly state in native buffer conditions (Fig. 4). This is because interpretation of any pairwise binding measurement

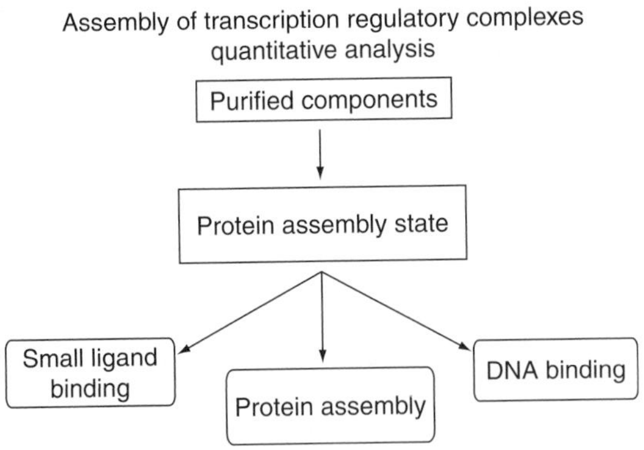

Fig. 4 Outline for characterization of the pairwise interactions that may contribute to assembly of a transcriptional regulatory complex. Details of the methods that may be used to measure each pairwise interaction are provided in the text.

requires knowledge of the protein oligomeric state. For example, binding measurements may reveal a complex mechanism involving multiple binding steps and prior knowledge of the assembly state of a protein places limits on the number of sites that can exist.

Several methods exist for determination of assembly state, the simplest of which is size exclusion chromatography (Winzor, 2003). One chromatographs the protein in a size exclusion or gel filtration resin and relates the elution volume of the protein of interest to standards of known Stokes radius. Assuming a spherical shape, the native molecular weight can be calculated from the Stokes radius. However, there are a few caveats associated with this experiment. First, although the inherent assumption in this method is that the chromatography resin is inert or noninteractive, the protein can interact with the size exclusion resin. This is usually apparent from the shape of the elution profiles in which significant tailing of the trailing edge occurs. The interaction can be decreased or eliminated by increasing the salt concentration in the elution buffer. A second potential problem is that the protein undergoes a relatively weak dynamic equilibrium between monomer and higher order species. In this case, the apparent elution volume will depend on the total protein concentration loaded onto the column with decreasing concentration indicating a lower molecular weight. Finally, the calculation of native molecular weight assumes a prior knowledge of shape, usually spherical, and this assumption may not be correct.

Other methods for determining a protein's assembly state include light scattering and analytical ultracentrifugation, which is discussed in greater detail below.

B. Small Ligand Binding

Small ligand binding can regulate the affinity of a transcriptional regulatory protein for DNA. While the prokaryotic examples such as tryptophan and purine repressors are more numerous, several eukaryotic regulatory transcription proteins are also regulated by small ligands. The majority of these are members of the nuclear hormone receptor family of transcriptional regulatory proteins (Mangelsdorf *et al.*, 1995). Measurement of small ligand binding to a regulatory protein provides information about the binding mechanism and energetics of the binding process. Mechanism refers to the number of ligands that bind and whether, for multiple ligand binding, any cooperativity exists. Knowledge of the binding energetics allows one to relate the probability of occupancy of the protein by ligand to physiological concentrations of both the regulatory protein and small ligand. There are many methods for measuring small ligand binding which range from the low-tech equilibrium dialysis method to more sophisticated methods such as isothermal titration calorimetry (ITC).

In any experiment designed to measure ligand binding, the goal is to ascertain the dependence of the fractional saturation of the binding site with ligand on the total concentration, or chemical potential, of the ligand. If one considers the simple equilibrium in which one ligand binds to a protein monomer, the relationship between the fractional saturation and free ligand concentration is derived as follows:

$$\mathrm{P} + \mathrm{L} \Leftrightarrow \mathrm{PL} \tag{11}$$

$$K_\mathrm{d} = \frac{[\mathrm{P}][\mathrm{L}]}{[\mathrm{PL}]} \tag{12}$$

$$\bar{Y} = \frac{[\mathrm{PL}]}{[\mathrm{P}]_\mathrm{TOT}} \tag{13}$$

$$[\mathrm{P}]_\mathrm{TOT} = [\mathrm{PL}] + [\mathrm{P}] \tag{14}$$

$$\bar{Y} = \frac{[\mathrm{PL}]}{[\mathrm{PL}] + [\mathrm{P}]} \tag{15}$$

$$[\mathrm{PL}] = \frac{[\mathrm{P}][\mathrm{L}]}{K_\mathrm{d}} \tag{16}$$

$$\bar{Y} = \frac{[\mathrm{P}][\mathrm{L}]/K_\mathrm{d}}{[\mathrm{P}][\mathrm{L}]/K_\mathrm{d} + [\mathrm{P}]} \tag{17}$$

$$\bar{Y} = \frac{[\mathrm{L}]}{[\mathrm{L}] + K_\mathrm{d}} \tag{18}$$

In order to be useful for measuring binding affinity, an experimental method must provide a signal that can be related to fractional saturation. Once the fractional saturation is measured as a function of added ligand concentration, the resulting data can be used to determine the equilibrium constant for the binding interaction using, in the case of simple binding, Eq. (18).

1. Equilibrium Dialysis

Equilibrium dialysis is a relatively simple and elegant method for measuring small ligand binding (Klotz, 1997). It simply requires a means of ligand detection, usually involving quantitation of a spectroscopic or radioisotopic signal. The protein is placed on one side of a dialysis compartment and the ligand on the other. The system is allowed to reach equilibrium at the point where the free ligand concentration is identical on each side of the semipermeable membrane. By quantitating the ligand concentration on each side of the membrane, one obtains information about the total ligand concentration in the presence of protein and the free ligand concentration. The bound ligand concentration is then calculated by subtraction. If the protein concentration is known, the fractional saturation of protein with ligand can be directly calculated as a function of free ligand concentration. This, of course, requires setting up several dialysis chambers, each at a different ligand concentration. As discussed below, the binding data can be analyzed to obtain the equilibrium constant(s) and mechanism of binding.

2. Fluorescence

Fluorescence techniques are also useful in measurements of small ligand binding (Eftink, 1997). The most common strategy is to monitor the change in the steady state intrinsic protein fluorescence emission spectrum as the ligand is titrated into a solution of the protein. The signal for the bound protein must, of course, differ from that of the unbound. The intensity of the signal may change, either increase or decrease, or the maximum wavelength of fluorescence may shift upon ligand binding. In either case, there must be a signal for the bound, F_B, that is distinct from that of the free protein, F_F. Monitoring of the signal as a function of ligand concentration added to a fluorescence cuvette provides fractional saturation versus ligand concentration information according to the following expression:

$$\bar{Y} = \frac{F - F_F}{F_B - F_F} \tag{19}$$

where F is the signal measured at any ligand concentration. This method is relatively simple and requires specialized but not highly specialized equipment. It does assume that the response of the fluorescence signal is linear with fractional saturation, an assumption that can be tested (Bujalowski and Lohman, 1987).

3. Isothermal Titration Calorimetry

ITC provides another means of measuring small ligand binding. The method is powerful in that it yields not only the equilibrium constant for binding but also the heat of binding and the stoichiometry (Ladbury, 2004). However, the instrumentation is both more expensive and sophisticated than that required for the first two methods. Moreover, material requirements for this measurement are significantly greater than those for either of the previous two techniques. All that is required for a protein–ligand interaction to be amenable to ITC measurement is a heat signal, either absorption or release, upon binding. Moreover, if this signal is not observed under one set of conditions of temperature, pH, or buffering agent, a change of these conditions may yield a signal. As with any other binding method, the ligand is injected into a solution of the protein and the heat evolved or taken up is monitored as a function of ligand concentration until saturation is reached. Chapter 4 by Freyer and Lewis, this volume, provides detailed information about experimental design and data analysis in ITC measurements.

C. Analysis of Binding Data

As indicated above, the goal in performing measurements of small ligand binding to a protein is to determine the binding stoichiometry and the energetics of the interaction. Regardless of the method used to perform the measurements, the data must be subjected to analysis by nonlinear least squares (NLLS) methods in order to obtain the desired information. By data we mean the fractional saturation versus ligand concentration profile. The analysis involves testing of data against different binding models to obtain the parameters that best describe, or best-fit, the binding data. By best-fit we mean the set of parameters, usually equilibrium constant and stoichiometry, that minimize the sum of the squares of the residuals. While an exhaustive description of NLLS methods is outside of the scope of this chapter, excellent introductions to the approach are contained in (Johnson and Faunt, 1992; Motulsky and Christopoulos, 2005) or readers may refer to Chapter 1 by Garbett and Chaires, and Chapter 24 by Johnson, this volume. In addition, there are many commercial software packages available for this purpose. Indeed, the Origin software package is provided with ITC instrumentation. A brief summary of the analysis procedure is as follows. One first chooses a binding model. If, for example, initial analysis of the oligomeric state of the protein indicates that the protein is monomeric, a simple model in which one ligand binds per monomer is a good place to start. The mathematical form of this model, which was presented above, is:

$$\bar{Y} = \frac{K_a[\mathrm{L}]}{1 + K_a[\mathrm{L}]} \tag{20}$$

In this case, the expression is in terms of the equilibrium association constant for the binding reaction. Using Eq. (20) and assuming a value of the equilibrium association constant for the reaction of $1 \times 10^6\ \mathrm{M}^{-1}$ the curve that one expects to observe when [L] is plotted on a logarithmic scale is shown in Fig. 5. Curves of the same shape are obtained for small ligand binding to a protein with multiple identical sites, provided that each binding event is independent of all others. This is where knowledge of the assembly state is helpful because it facilitates distinguishing multiple site noncooperative binding from single-site binding. A more complicated binding mechanism would involve cooperative binding of, for example, two ligands to two identical sites on a protein dimer. If the cooperativity is positive, then binding of the first ligand promotes binding of the second. The fractional saturation in this case is described by the following equation:

$$\bar{Y} = \frac{2k[\mathrm{L}] + k^2 k_c[\mathrm{L}]^2}{1 + 2k[\mathrm{L}] + k^2 k_c[\mathrm{L}]^2} \tag{21}$$

where the lower case k is the microscopic equilibrium association constant for binding of ligand to each site in the absence of binding to the other site and k_c is the

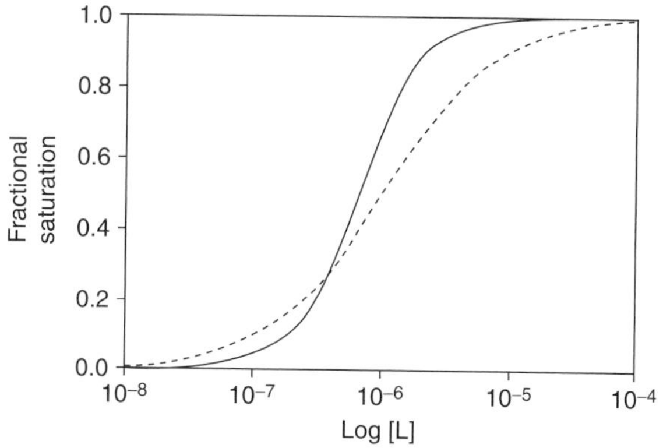

Fig. 5 Fractional saturation versus free ligand concentration for simple binding (dashed curve) or two-site cooperative binding (solid curve). The curves were generated by using Eqns. (20) and (21). The equilibrium association was assumed to be $1 \times 10^6\ M^{-1}$ for simple binding. The intrinsic equilibrium association constant for binding to each of the two sites was set at 1×10^5 M and the cooperativity constant was 100 for cooperative binding.

cooperativity constant. A curve showing the expected dependence of fractional saturation on free ligand concentration for this case is shown in Fig. 5. In this case, the cooperativity constant is assumed to be 100 and the binding constant for binding to each of the sites in the absence of binding to the other site is assumed to be $1 \times 10^5\ M^{-1}$. Inspection of the curves shown in Fig. 5 reveals that the dependence of fractional saturation on free ligand concentration is significantly steeper for cooperative binding than for noncooperative binding. In analysis of data using a cooperative versus a noncooperative model, the quality of the fit would readily indicate which model is correct. In practice, one usually starts data analysis by assuming the simplest possible model. The data are analyzed by using this model to obtain the most probable value of the equilibrium constant for binding. If, based on standard criteria of a "good fit," the data are not well described by a simple model, then more complicated models are tested. These same NLLS methods are applied to measurements of protein–protein and protein–DNA interactions described below.

D. Tools for Measuring Protein–Protein Interactions

Protein–protein interactions are a common feature of transcription control mechanisms. Protein–DNA complexes that contribute to transcription regulation are frequently formed between a homooligomeric protein and the DNA site. Additional higher order interactions of the oligomer with RNA polymerase and other proteins occur in both prokaryotes and eukaryotes. The rationale for oligomer binding to DNA is that association of the regulatory protein with multiple recognition sequences leads to tighter complexes. A second rationale is that use of oligomeric proteins also provides another point of control in assembly of the regulatory complex. In addition to homooligomeric protein–protein interactions, heterooligomers play a role in transcription regulation. This forms the basis of combinatorial control, often discussed in the context of eukaryotes but also observed in prokaryotes.

In studying the mechanism of transcription regulatory complex assembly, it is important to determine the energetics of protein–protein interactions that contribute to a transcription regulatory response. Most importantly, it provides information about the assembly state at any given total protein concentration. This is necessary for interpretation of any other binding data obtained with the protein. Again, as with small ligand binding, there are many methods for measuring protein–protein interactions. Among these are fluorescence methods including anisotropy and Förster resonance energy transfer (FRET) and hydrodynamic techniques including analytical ultracentrifugation and dynamic light scattering. One can also use techniques such as surface plasmon resonance (SPR) and ITC to measure these interactions.

1. Fluorescence Methods

Among the tools available for quantitative studies of protein–protein interactions, fluorescence is one of the most versatile. An excellent review on the topic is provided in Yan and Marriott (2003). Fluorescence approaches to measurements of protein–protein interactions include FRET and fluorescence anisotropy. In addition, fluorescence correlation spectroscopy is emerging as a powerful technique for this purpose.

FRET is a technique that takes advantage of the ability of one fluorescent molecule, the donor, to, when in the excited state, transfer its energy to another fluorophore, the acceptor. Provided that there is sufficient overlap between the emission spectrum of a FRET donor and the excitation spectrum of the acceptor and their orientation and distances are in the appropriate ranges, energy transfer will occur. In measuring a protein–protein interaction, FRET may be useful provided that the proteins can be labeled with the appropriate fluorophores and the distance between the fluorophores in the oligomer is sufficiently small to allow energy transfer. In using this technique, one must find a suitable donor–acceptor pair, and there is a large amount of literature on this topic provided by the fluorophore manufacturers. In using FRET to measure protein dimerization, one might start with a protein modified with the donor fluorophore and add the binding partner modified with acceptor. If one monitors the donor fluorescence emission versus concentration of added acceptor-labeled partner, the intensity of the signal should decrease as the partner concentration is raised through regime relevant to the protein–protein equilibrium. As the donor-modified partner becomes saturated with the binding partner, the signal intensity should level off. The expression for the fractional saturation of donor-modified protein with acceptor-modified protein is:

$$\bar{Y} = \frac{I_{\text{init}} - I}{I_{\text{init}} - I_{\text{fin}}} \tag{22}$$

where I_{init} is the initial value of the fluorescence before the binding partner is added, I_{fin} is the final value of the fluorescence intensity corresponding to the donor-bound protein when it is saturated with the acceptor-bound, and I is the intensity of donor emission at any concentration of added acceptor-bound protein. Provided that the reaction stoichiometry is simple 1:1, the data can be analyzed to obtain the equilibrium constant for binding by using Eq. (18). Alternatively, one can measure the change in the signal originating from the acceptor, which should increase as the concentration of donor-modified partner is added and level off once saturation is achieved.

Fluorescence anisotropy provides another technique for measuring protein–protein interactions. One must first label the protein of interest with a fluorophore. Starting with this labeled protein, one adds unlabeled protein (either the same protein or the distinct protein partner). As the total protein concentration is increased through the range of the protein assembly process, the labeled protein becomes part of a larger complex. The fluorescence anisotropy signal for the protein monomer should be lower than that observed for it as a member of a larger oligomer. This is because the rotational correlation time for the oligomer is longer than for the monomer. An expression analogous to Eq. (22) is used to calculate the fractional saturation of the original protein as the partner is added. In this case, the end states are A_{DIM} and A_{MON}. Thus, if one measures the anisotropy associated with the first protein as the concentration of the added protein is increased, the curve should reflect fractional oligomerization versus binding partner concentration. The success of both FRET and fluorescence anisotropy techniques depends on the success with which a protein can be labeled without perturbing its function. In addition, it is difficult to use either technique to measure interactions more complex than that corresponding to dimerization.

2. Isothermal Titration Calorimetry

If a heat signal, either exothermic or endothermic, is associated with formation of a protein–protein interaction, ITC may be used to detect the binding process (Velazquez-Campoy *et al.*, 2004). There are basically two approaches to make these measurements. The first involves titrating one binding partner into a solution containing the second and measuring the heat associated with each injection. In the second, a dilution method, the protein is diluted from a concentration at which it is primarily dimer into buffer so that at its final concentration it is primarily monomer. The method essentially measures the heat of dissociation of the dimer. Obvious limitations of the method are that it is difficult to measure protein association beyond dimerization and only for systems that are characterized by an enthalpy change. Another limitation of the method is that one must work with systems characterized by equilibrium constants in the high nanomolar to millimolar range (see Chapter 4 by Freyer and Lewis, this volume, for an expanded discussion).

3. Analytical Ultracentrifugation

Analytical ultracentrifugation provides very powerful tools for analysis of the protein assembly reactions. The reader is referred to Chapter 6 by Cole *et al.*, this volume, for a more detailed discussion of these techniques. Either equilibrium or velocity centrifugation techniques may be used. Recent advances in both the instruments used for these measurements and data analysis methods have increased the range of protein assembly processes that are amenable to these techniques.

The sedimentation equilibrium method has historically been the favored centrifugation method for measurements of protein homooligomerization (Laue, 1995). The method involves subjecting a macromolecule to sedimentation at relatively slow speeds. The macromolecule will respond to the centrifugal force by migrating toward the bottom of the cell. However, eventually, at a particular speed, the centrifugal force is balanced by the diffusive force of the macromolecule, equilibrium is reached, and no further migration of the macromolecule occurs. At this point, the concentration gradient of the protein in the cell, which is measured directly in the instrument by optical scanning along the cell length, exponentially increases with radial position. For a single protein species, this is a single exponential, the shape and position of which depends on, among other things, the molecular weight of the protein as follows:

$$C(r) = \delta + C_o e^{\sigma(r^2 - r_0^2)/2} \tag{23}$$

where $C(r)$ is the concentration at any radial position, δ is the offset of the data at the lower asymptotic limit of concentration, C_o is the concentration in absorbance units at the reference radial position, r is the radial position, and r_0 is the reference position, usually the radial position for the first data point. The parameter σ, which one obtains from NLLS analysis of the data to this model, is the reduced molecular weight which is related to the molecular weight by the following equation:

$$\sigma = \frac{M(1 - \bar{v}\rho)}{RT}\omega^2 \tag{24}$$

in which M is the molecular weight, $\bar{v}$ is the partial specific volume, ρ is the density of the solvent, R is the gas constant, T is temperature in Kelvin, and ω is the angular velocity. The molecular weight, using Eq. (24), can be calculated from the reduced molecular weight. For a protein monomer, this molecular weight should be close analytical molecular weight calculated from the sequence of the polypeptide chain. If the protein undergoes association to higher order species, the profiles in scans of the cell will reflect the presence of species in addition to monomer. In this case, the average molecular weight obtained from NLLS analysis of the data using the single species model should be higher than that expected for the monomer. Moreover, for a system in which there is an equilibrium between, for example, a monomer and dimer, the measured scans will be the sum of two exponential distributions, one for each species. The stoichiometry and energetics for self-association of the protein can be obtained from NLLS analysis of the data using models with different stoichiometries. In order to do so, one must design the sedimentation experiment to cover the range of protein concentration relevant to the entire dissociation process. This is accomplished by centrifuging samples prepared at a range of initial loading concentrations at multiple speeds (Roark, 1976). The data acquired for these multiple samples are subjected to global NLLS analysis to obtain the model that best describes the data-monomer-dimer, monomer-trimer, etc., and the equilibrium constants relevant to the process. One of the most frequently used programs for analysis of these data is Nonlin (Johnson *et al.*, 1981). Jim Cole (2004) at the University of Connecticut has made a sedimentation equilibrium analysis program, HeteroAnalysis, available as shareware. This program can be used for analysis of both hetero- and homooligomerizing proteins. SEDANAL, a program developed in the Stafford laboratory, can also be used for these systems (Stafford, 2000; Stafford and Sherwood, 2004).

Recently sedimentation velocity is being used more frequently for protein association measurements. The basic velocity experiment differs from sedimentation equilibrium in that an initially homogeneous solution is subjected to centrifugation at a relatively high rotor speed. As, for example, a single protein species migrates in the centrifugal field a boundary forms and the velocity with which the boundary moves provides information about the sedimentation coefficient. Both the molecular weight and the frictional coefficient contribute to the magnitude of the sedimentation coefficient (Van Holde, Johnson & Ho):

$$s = \frac{M(1 - \bar{v}\rho)}{Nf} \qquad (25)$$

where s is the sedimentation coefficient, M is the molecular weight, $\bar{v}$ is the partial specific volume of the protein, ρ is the density of the solvent, N is Avogadro's number, and f is the frictional coefficient. In addition to providing information about shape and molecular weight, analysis of boundary shape and position is powerful for obtaining information about sample heterogeneity.

Sedimentation velocity may also be used for measurement of protein homo- and heterooligomerization. Several methods for analyzing sedimentation velocity data to obtain this information are now available. The program SEDANAL (Stafford, 2000; Stafford and Sherwood, 2004) can be applied to analysis of both types of interacting systems. The Schuck laboratory at the National Institutes of Health has developed the program SEDPHAT for the same purposes (Dam and Schuck, 2005; Dam *et al.*, 2005; Schuck, 1998). More complete discussions of analytical ultracentrifugation methods can be found in Chapter 6 by Laue *et al.*, this volume. In addition, Dr. Fleming has a chapter devoted to the application of sedimentation equilibrium analysis to membrane proteins (Chapter 7 by Burgess *et al.*, this volume). All the analysis programs cited above are available as shareware.

E. Protein–DNA Interactions

The final pairwise interaction that must be considered in assembly of transcription regulatory complexes is between protein and DNA. These interactions can range from simple binding of a protein monomer to a single site on DNA to binding of homo- or heterooligomeric proteins to multiple sites. The goal in performing measurements of these interactions is, again, to determine the stoichiometries and energetics of the binding processes. While for the simple systems this is straightforward, for more complex systems one has to address the question of cooperativity in binding of multiple proteins to multiple sites on the DNA.

Protein–DNA interactions can be measured using any of a number of methods. In nearly all cases, the method involves determination of fractional occupancy of the site on the DNA by the protein as a function of protein concentration, as opposed to measurement of the occupancy of the protein by the DNA site. Some methods are macroscopic in approach and only allow measurement of average level of occupancy of the DNA. These methods including ITC and fluorescence anisotropy are better suited for simple, single-site binding of a protein to DNA. Other methods including footprinting and gel mobility shift assays allow the experimenter to obtain detailed thermodynamic information about binding of proteins to multiple sites on DNA.

1. Isothermal Titration Calorimetry

ITC allows one to measure the fractional occupancy of the DNA by protein by measuring the heat released or taken up as the protein is titrated into a solution of DNA (Read and Jelesarov, 2001). Alternatively, one may titrate the protein into the DNA. The current instrumentation is very sensitive and allows measurements of heat changes in the submicrocalorie range. The details of the experiment are straightforward in that one simply places the receptor (DNA or protein) into the sample cell and titrates in the appropriate binding partner. The heat associated with each injection is measured as a function of titrant concentration. Analysis of the data can yield the stoichiometry of the binding reaction, the equilibrium constant governing the process as well as the enthalpy of the reaction. Additional characterization of the binding reaction including the linkage to ion or proton binding or release can also be obtained. Thus, ITC provides an elegant and relatively simple means of obtaining a large amount of information about a binding system. The method does, however, have some limitations. First, relative to the other techniques that will be discussed below, a large amount of material is required for ITC measurements. Second, while any of these techniques requires relatively pure samples, the requirements for ITC are more stringent. This is because the heat measurement on which this technique is based is sensitive to any and all reactions that take place in the solution. This becomes particularly problematic when reactions such as protein self-association are coupled to the DNA binding process. If this is the case, one can measure the self-association reaction independently in order to deconvolute its contribution to the binding process from that deriving from DNA binding alone. Third, many DNA binding reactions are characterized by very large equilibrium association constants in the range of 10^9–10^{10} M^{-1} for simple binding, which may preclude measurements of an equilibrium constant for the binding process. Finally, as mentioned above, while for complex binding the ITC method yields information about stoichiometry and the average thermodynamics of binding, no information about individual site occupancy can be obtained.

2. Fluorescence Approaches to Study of Protein–Nucleic Acid Binding

Fluorescence anisotropy provides another macroscopic method for measuring protein–DNA interactions (Hill and Royer, 1997). In this method, one typically prepares a DNA with a fluorescent label attached by synthetic methods. If one measures steady state fluorescence emission anisotropy associated with the labeled DNA as a function of protein concentration, an increase is observed as the fractional occupancy of the DNA increases. This is due to the larger size of the complex relative to the size of the free DNA. The increased size is accompanied by an increased rotational correlation time, with an accompanying increase in the anisotropy. As with using the method for measurement of protein–protein interactions, the initial anisotropy signal is that corresponding to the free DNA; and as the protein concentration increases, the signal gradually increases to that associated with the fully bound DNA. At intermediate protein concentrations, the signal reflects the weighted contributions of the free and bound species. Advantages of this technique are severalfold. First, provided that the fluorescent probe employed is characterized by a relatively high quantum yield, one can work with very low DNA concentrations. This has the advantage that if this concentration is much lower than the K_d for the interaction, then the total protein concentration provides a reasonable measure of the free protein concentration, which simplifies data analysis. The ability to work at very low DNA concentrations also ensures that one is in the range appropriate for an equilibrium as opposed to a stoichiometric titration. In a stoichiometric titration, the DNA concentration is much greater than the equilibrium dissociation constant for the binding process, in which case all added proteins are bound by the DNA until saturation is reached. Titrations performed under these conditions yield only information about the number of protein molecules that bind to the DNA template or the stoichiometry. Another advantage of this method is its simplicity. One places the labeled double-stranded oligonucleotide in a cuvette and titrates in the protein. At each protein concentration, one measures the steady state anisotropy. Data analysis, for simple 1:1 binding system, is very straightforward. Potential disadvantages of the technique include the fact that oligonucleotides, which may not accurately mimic the large DNAs to which proteins bind in the biological context, must be used for the measurements. A second potential pitfall is that no anisotropy change may occur upon protein binding. This may be due to rapid local motion of the fluorophore, which leads to complete depolarization and loss of sensitivity to the motion of the macromolecule. This problem is exacerbated when long hydrocarbon linkers are used to attach the fluorophore to the DNA. The final shortcoming of the technique is that it is, like ITC, macroscopic in scope and, therefore, not as useful for determining individual site binding in multisite systems.

3. Footprinting and Electrophoretic Mobility Shift: Individual Site Binding Methods

Footprinting and electrophoretic mobility shift assays are very powerful techniques for measuring site-specific binding of a protein to DNA (Galas and Schmitz, 1978; Garner and Revzin, 1981). In general, footprinting simply refers to a method in which one detects the resistance of a site on DNA to chemical or enzymatic cleavage as a result of being occupied by protein. Although DNaseI is the classic cleavage reagent, other cleavage tools include restriction enzymes, hydroxyl radicals, and metal cleavage reagents. The major advantage of footprinting over other techniques is its ability to yield information about individual site occupancy. This is important for the many systems in which a protein binds to multiple sites on DNA.

Although several reviews on footprinting techniques are available, the book edited by Revzin (1993) provides comprehensive coverage of the topic. Briefly, DNA is subjected to cleavage in the absence and presence of increasing concentrations of the binding protein. Products are separated by electrophoresis on a DNA sequencing gel and the gel is imaged using phosphorimaging technology. The isotopic labeling of the DNA on one strand at one terminus with ^{32}P enables this imaging. Digitization of the image allows quantitation of the optical density in the bands on the gel. As the protein concentration is increased, the DNA region that constitutes the protein binding site becomes more protected from cleavage and, therefore, bands corresponding to the binding site become fainter. Quantitation of the density of bands generated from cleavage in the site as a function of protein concentrations yields a binding curve. This binding curve can be analyzed by using NLLS analysis to obtain the energetics of the interaction.

The application of footprinting to a multisite cooperative system can be illustrated by a simple two-site system (Brenowitz *et al.*, 1986). In this system, case two sites, 1 and 2, are characterized by different affinities for the protein that are quantitatively represented by the equilibrium association constants k_1 and k_2. Moreover, there is a favorable or cooperative interaction between two proteins as they simultaneously bind to the two sites, which is represented by the equilibrium constant k_{12}. The goal in performing DNA binding measurements is to determine the equilibrium constants for binding of the protein to each site in the

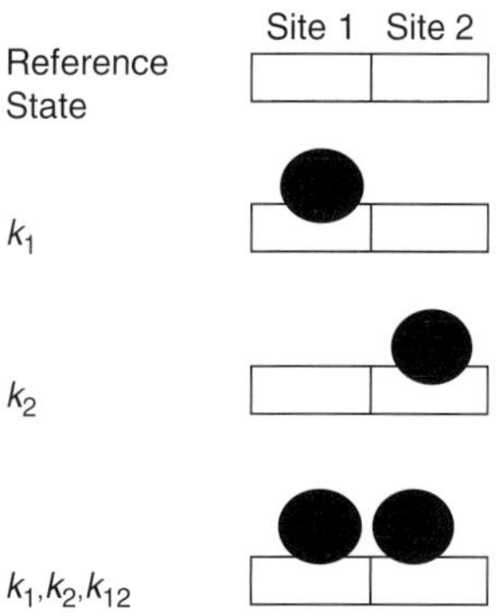

Fig. 6 Cooperative binding of a protein to two sites on DNA. Binding of the protein to each of the two sites in the absence of binding to the other is governed by k_1 and k_2 and the cooperativity constant is k_{12}.

absence of binding to the other site as well as the constant for the cooperative interaction between the two proteins (Fig. 6). This means that three equilibrium constants, k_1, k_2, and k_{12}, must be determined. One starts by measuring binding of the protein to a DNA fragment that corresponds to the wild-type sequence containing both intact binding sites. In such a measurement, one obtains information about the protein concentration required to fill each of the sites in the presence of binding to the other site. The relationship between the fractional saturation and protein concentration for this DNA molecule is:

$$\bar{Y} = \frac{(k_1 + k_2)[\mathrm{P}] + k_1 k_2 k_{12}[\mathrm{P}]^2}{1 + (k_1 + k_2)[\mathrm{P}] + k_1 k_2 k_{12}[\mathrm{P}]^2} \tag{26}$$

However, analysis of the binding curves generated from measurements with only this DNA does not allow one to obtain both intrinsic equilibrium constants and the cooperativity constant. In order to deconvolute or tease apart these three terms, independent information about the intrinsic affinities of the protein for each separate site must be obtained. This is achieved by creating mutations in the original DNA sequence that eliminate binding to one or the other site. Such mutants are referred to as reduced valency. Binding measurements performed on each mutant template provides independent information about the intrinsic equilibrium constant for binding to each site. The relationship of fractional saturation to protein concentration for these templates (Fig. 6) is:

$$\bar{Y}_1 = \frac{k_1[\mathrm{P}]}{1 + k_1[\mathrm{P}]} \tag{27}$$

$$\bar{Y}_2 = \frac{k_2[\mathrm{P}]}{1 + k_2[\mathrm{P}]} \tag{28}$$

If the data obtained from of the footprints on the wild-type template and the two reduced valencies templates are globally analyzed, all three equilibrium constants will be obtained.

Gel mobility shift or electrophoretic mobility shift assays have been used for many years in both qualitative and quantitative measurements of DNA binding. The method takes advantage of the difference in electrophoretic mobility of the DNA in its free and bound states. If samples are prepared containing a constant concentration of DNA and a range of protein concentrations, within the concentrations relevant to the equilibrium binding process, one can learn about the mechanism and energetics of the binding process.

Application of the electrophoretic mobility shift assay to single-site binding is very simple (Carey, 1991). The samples are prepared at a single DNA concentration and a range of protein concentrations and subjected to electrophoresis under native conditions for separation of the protein–DNA complexes from the free DNA. The resulting fraction bound or factional saturation can be quantitated by imaging the gel and determining the amount of signal associated with the two bands corresponding to free and bound DNA. Imaging usually relies on use of radiolabeled DNA. The data are then plotted as fraction bound versus protein concentration and can be analyzed using NLLS methods using a simple binding expression to obtain the equilibrium constant for the process.

While gel mobility shift has not been used as much as footprinting for quantitative measurements of multisite binding to DNA, it is valid for this application (Senear and Brenowitz, 1991). The method relies on the distinct electrophoretic mobilities of differentially liganded DNAs, which allows their separate quantitation. The method allows for determination of macroscopic equilibrium constants even for complicated systems. However, if one wishes to determine both intrinsic and cooperative equilibrium constants, only the simplest two-site system in which binding of the protein to each site is identical can be fully determined. As with DNaseI footprinting, complete resolution of the three terms requires the reduced valency approach that allows independent determination of the intrinsic free energy for binding of the protein to each site in the absence of binding to the second. The advantage of the gel mobility shift assay over footprinting lies in its simplicity. The gel mobility shift assay yields a binding curve in many fewer steps than does any footprinting technique.

V. Case Studies of Multiple Linked Equilibria in Transcription Regulatory Systems

The power of the quantitative analysis described above is illustrated by the results of application of these methods to assembly of two transcriptional regulatory complexes. The systems are the *E. coli* CytR-CRP and the progesterone receptor, a nuclear hormone receptor, both of which involve binding of multiple proteins to multiple sites on DNA. Small ligand binding is coupled to each assembly process and cooperativity in protein binding to the DNA also occurs in both systems. Application of the approaches described above to these two systems reveals that assembly of these regulatory complexes is governed by rules that are considerably more intricate than initially anticipated. This intricacy is undoubtedly important for modulation of biological control in these two systems.

A. CRP and CytR: Combinatorial Control of Transcription in *E. coli*

The two transcription factors, catabolite repressor protein, CRP, and the cytidine repressor, CytR, act at several promoter regions in the *E. coli* genome. CRP is a general transcription factor that acts at over 100 promoter regions in *E. coli*. CytR and CRP combine to coordinate expression at nine unlinked transcription units that code for the proteins and enzymes involved in nucleoside transport and metabolism. Here discussion of the CytR–CRP system will be limited to regulation of the *deoP2* operon (Sogaard-Andersen and Valentin-Hansen, 1993). However, while the same proteins participate in regulation at eight other regulatory regions, variability in the levels of control is observed. The structure of the control region for *deoP2*, which is shown in Fig. 7, contains two sites for binding of CRP and one for CytR binding. Three distinct levels of transcription are observed in the system; basal, activated, and repressed. The level in the absence of regulatory factors is referred to as basal. When CRP-cAMP alone is bound at the *deoP2* control region, transcription initiation is activated fourfold relative to the basal level (Shin *et al.*, 2001). When CytR is bound with two CRP dimers, transcription initiation is

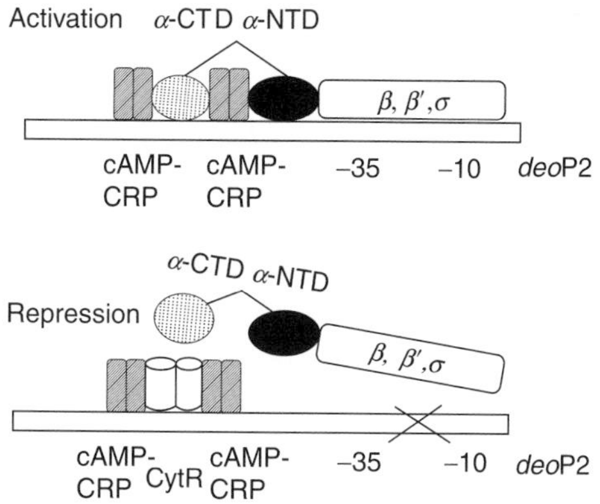

Fig. 7 Schematic diagram of the arrangement of proteins at the *deoP2* transcription regulatory region that result in activation and repression of transcription. CytR is the cytidine repressor and CRP is the catabolite repressor protein. In the activated complex, the CRP-cAMP dimers interact with the C-terminal (CTD) and N-terminal domains of the α-subunit of RNA polymerase. The other polymerase subunits are β′, β, and σ. Additional details are provided in the text (Shin *et al.*, 2001).

repressed 100-fold. This repression depends on binding of the small ligand, cytidine, to CytR.

The assembly of the transcription regulatory complex at *deoP2* can be broken down into multiple pairwise interactions. First, the small ligand, cytidine, binds to the CytR protein to regulate its function. Second, CytR binds to the CytO sequence on the DNA. CRP (CAP) binds to two different sequences, CRP1 and CRP2. In addition, the small molecule cAMP binds to CRP to regulate its binding to the CRP sites. Higher order interactions include those between CRP and CytR bound to DNA. Finally, as is often the case in cooperative DNA binding, the CRP and CytR proteins may themselves interact in the absence of DNA.

1. Binding of the Small Ligand Cytidine

Binding of cytidine to CytR has been measured by using the nitrocellulose filter binding assay (Barbier *et al.*, 1997). In this measurement, the protein is mixed with radiolabeled cytidine and the solution is filtered through a nitrocellulose membrane. While protein adheres to the membrane, the free small molecule passes through. However, if the small molecule is bound to protein, it is retained on the filter along with the protein. By subjecting the filter to scintillation counting, one can measure the amount of ligand bound to protein. Preparation of a number of samples at a single protein concentration and a range of total ligand concentrations allows measurements of the fractional saturation versus ligand concentration dependence that is required to determine the mechanism and energetics of binding. Results of these measurements indicate that while two cytidine molecules can bind to the dimeric CytR, the binding is characterized by negative cooperativity. That is, binding of the first ligand is antagonistic to binding of the second. Moreover, the effect of cytidine binding on induction of transcription is associated with binding of the first cytidine molecule.

2. Cytidine Binding and Site-Specific DNA Binding of CytR to CytO

CytR binds to the CytO sequence of the *deoP2* regulon, which lies between the two CRP binding sites. Binding of CytR to this site is well-described as a simple dimer binding to a single site. Since cytidine causes derepression at the *deoP2* regulon, it might be expected to affect the affinity of CytR for CytO. However, results of both gel mobility shift assays and DNaseI footprinting reveal that binding of the small ligand has no effect on binding of CytR to its operator sequence (Barbier *et al.*, 1997). The resolved free energy for DNA binding by CytR obtained in the presence and absence of saturating concentrations of cytidine are, within experimental error, identical. Therefore, unlike the classical allosteric regulators including the lactose, tryptophan, and purine repressors, small ligand binding to the cytidine repressor functions at a level distinct from a direct effect on DNA binding. This is despite the fact that CytR, PurR, and LacR are all members of the same protein family.

3. Protein–Protein Interactions in the CytR–CRP System

Activation of transcription initiation at *deoP2* occurs in response to CRP binding at the two sites, CRP1 and CRP2. The activator binds at each of these sites with no cooperativity. As with all DNA binding by CRP, cAMP binding influences the affinity for DNA. While the linkage between cyclic nucleotide binding and CRP affinity for DNA is outside of the scope of this work, affinity of CRP for CRP1 and CRP2 of *deoP2* is maximum at 100-μM nucleotide and decreases as nucleotide concentration is either increased or decreased.

As indicated in Fig. 7, when CytR acts to repress transcription initiation at the *deoP2* promoter, it does so in concert with two CRP dimers. In order to investigate the physical basis for the requirement of both proteins for repression in the system, measurements of DNA binding by both proteins in the presence and absence of the other have been performed. In these experiments, the CRP is bound to cAMP and CytR is free of small nucleotide. These measurements indicate that CRP and CytR interact cooperatively as they bind to DNA (Pedersen *et al.*, 1991; Perini *et al.*, 1996).

4. Mechanism of Cytidine-Mediated Induction

In vivo an increase in cytidine concentration results in derepression at genes controlled by CytR and CRP. As indicated above, cytidine binding to CytR has no effect on the affinity of CytR for the *deoP2* control region. Direct measurements of the effect of the nucleotide on DNA binding in the system reveal that cytidine functions by reducing the cooperativity associated with binding of CRP and CytR to *deoP2* (Pedersen *et al.*, 1991; Perini *et al.*, 1996).

5. DNA as a Modulator of Transcription Regulation

Cooperativity in DNA binding is usually thought to reflect protein–protein interactions. However, in the CytR–CRP system, measurement of the pairwise protein–protein interactions by sedimentation equilibrium reveals that they are very weak (Chahla *et al.*, 2003). Moreover, cytidine binding does not affect the interaction of the free proteins. Therefore, the cooperative interaction between the two proteins appears to reflect the DNA context in which it functions. The exact structural basis of this contextual effect is not known. However, it is important in that it likely serves as the basis for differential control in the system. As indicated in the introduction to this section on control at *deoP2*, the CytR–CRP combination is responsible for control at many unlinked genes. The details of the control vary among these regulons. It has been shown that binding of the different cytidine operators may exert different allosteric effects on CytR (Tretyachenko-Ladokhina *et al.*, 2006). These different allosteric effects may influence the interaction of CytR with CRP to yield a range of regulatory effects.

B. The Progesterone Receptor: A Nuclear Hormone Receptor

Nuclear hormone receptors provide another example of a system in which linkage of multiple pairwise interactions dictates levels of transcription initiation at several genes. In response to binding of small molecule hormones, these proteins can, in concert with other proteins, alter transcription initiation levels. Thus, in dissecting the nuclear hormone receptor systems into the constituent interactions significant for transcription regulation, one must consider the hormone–receptor interaction, oligomerization of the nuclear receptor, interaction of nuclear receptor with DNA, and the interaction of the nuclear receptor with transcription corepressors and coactivators. The nuclear hormone receptor systems are considerably more complex than the CytR-CRP system, and quantitative measurements of these systems are more limited in scope.

Nuclear hormone receptors are members of a very large protein family that exhibit a broad range of biochemical properties. This discussion is limited to the progesterone receptor, which is a member of the steroid hormone family of receptors (Leonhardt *et al.*, 2003; Li and O'Malley, 2003). There are two isoforms of progesterone receptor, PR-A and PR-B, which differ in the lengths of their N-terminal regions. The current understanding of these proteins is that, in the absence of steroid hormone, they are complexed with chaperones including heat shock protein (HSP90) (Fig. 8). Binding of steroid hormone induces dissociation of

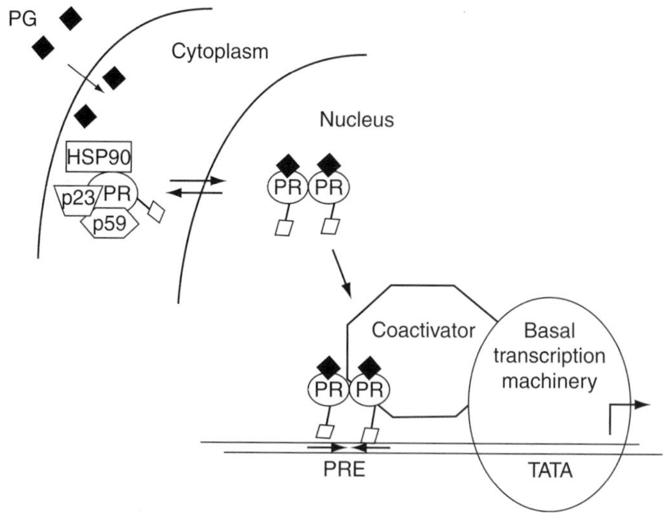

Fig. 8 Schematic diagram of progesterone (PG)-mediated activation of the progesterone receptor. In the absence of hormone, the receptor is bound to the heat shock protein, HSP90, and the lipocalins, p23 and p59. Passage of progesterone into the cytoplasm results in its binding to the receptor. This causes release of the receptor from the HSP complex, dimerization and translocation of the receptor to the nucleus where it binds to DNA. The receptor, in concert with coactivators, activates the basal transcription machinery (Leonhardt *et al.*, 2003).

the receptor–chaperone protein complex, receptor dimerization, translocation to the nucleus, and binding to the steroid responsive elements in the transcription control regions of many genes. The receptors form complexes with other protein complexes including coactivators and corepressors in these control regions and the specific array of interactions determines transcription levels. In addition to the hormone-responsive functions of progesterone receptor, activation can be initiated by cell surface receptors that result in posttranslational modification of the progesterone receptor. Thus, many layers of control exist for the progesterone receptor. This discussion is limited to the progesterone receptor B and, specifically, to self-assembly and its linkage to DNA binding. Results of recent quantitative studies of this receptor, while limited in scope, reveal deviations from the accepted paradigm for ligand-mediated activation.

1. Hormone Binding

Hormone binding by the progesterone receptor has been extensively studied. Typically, the method of choice is charcoal absorption of radiolabeled hormone in which the free ligand can be separated from the bound ligand by absorption to charcoal. The soluble bound ligand is then separated from the free, charcoal-bound ligand by centrifugation and subjected to scintillation counting for quantitation. If the total ligand concentration is known, the free ligand concentration can be calculated by subtracting the bound from the total ligand. Results of these measurements indicate that the hormone binds in the nanomolar range of concentration (Tetel *et al.*, 1997).

2. Receptor Dimerization

Detailed studies of the dimerization properties of homogeneous preparations of the intact progesterone receptor have been performed by sedimentation equilibrium (Heneghan *et al.*, 2005). Recall that the paradigm for ligand activation of nuclear receptor function is that ligand binding results in dissociation of the receptor from the chaperone, HSP90, dimerization, nuclear transport, and DNA binding. However, results of these dimerization measurements of progesterone-bound PR-B reveal that the protein dimerizes in the micromolar range of concentration, which is consistent with liganded receptor being primarily monomeric *in vivo*. Thus, these results raise questions about the model for receptor function: in particular ligand-linked dimerization and whether it is only the dimer that binds to progesterone responsive elements (PREs) on the DNA.

3. DNA Binding by PR-B

The progesterone receptor binds specifically to sequences referred to as PREs. Recently published results of DNaseI footprint titration measurements of PR-B binding to PREs indicate that the mechanism of DNA binding by this steroid hormone receptor may deviate significantly from the currently accepted mechanism (Heneghan *et al.*, 2006). These studies were carried out using DNA molecules that carried either one palindromic PRE or two. Data obtained from binding measurements performed on the single-site DNA indicate that the process is well-described both by a mechanism involving stepwise binding of two PR-B monomers and binding of the preformed dimer. Furthermore, the results also indicate that in a stepwise binding mechanism, the cooperativity between monomer binding is very low. Results of measurements obtained with the two-palindrome DNA molecule indicate that there is a large, favorable inter-PRE cooperativity in binding. Taken together, the results indicate that the DNA binding properties of PR-B deviate from simple binding of a preformed dimer to a palindromic DNA sequence.

4. Biological Implications for Progesterone Receptor Signaling

As indicated in the introduction to this section on PR-B, assembly of the receptor on transcriptional control regions involves not just the receptor but also the multiprotein coactivator or corepressor complexes as well as the proteins that form the transcription preinitiation complex. Results of sequence analysis of transcriptional control regions at which PR functions reveal that while some do contain palindromic PREs, others are characterized by a number of half-sites. The results of biophysical studies suggest that PR may have a larger repertoire of binding configurations available to it than previously thought. This larger repertoire may be important for modulating levels of transcriptional control at the large number of genes at which the receptor functions.

References

Barbier, C. S., Short, S. A., and Senear, D. F. (1997). Allosteric mechanism of induction of CytR-regulated gene expression. Cytr repressor-cytidine interaction. *J. Biol. Chem.* **272**(27), 16962–16971.

Brenowitz, M., Senear, D. F., Shea, M. A., and Ackers, G. K. (1986). Quantitative DNase footprint titration: A method for studying protein-DNA interactions. *Methods Enzymol.* **130,** 132–181.

Bujalowski, W., and Lohman, T. M. (1987). A general method of analysis of ligand-macromolecule equilibria using a spectroscopic signal from the ligand to monitor binding. Application to *Escherichia coli* single-strand binding protein-nucleic acid interactions. *Biochemistry* **26**(11), 3099–3106.

Carey, J. (1991). Gel retardation. *Methods Enzymol.* **208,** 103–117.

Chahla, M., Wooll, J., Laue, T. M., Nguyen, N., and Senear, D. F. (2003). Role of protein-protein bridging interactions on cooperative assembly of DNA-bound CRP-CytR-CRP complex and regulation of the *Escherichia coli* CytR regulon. *Biochemistry* **42**(13), 3812–3825.

Cole, J. L. (2004). Analysis of heterogeneous interactions. *Methods Enzymol.* **384,** 212–232.

Correia, J. J., Chacko, B. M., Lam, S. S., and Lin, K. (2001). Sedimentation studies reveal a direct role of phosphorylation in Smad3:Smad4 homo- and hetero-trimerization. *Biochemistry* **40**(5), 1473–1482.

Dam, J., and Schuck, P. (2005). Sedimentation velocity analysis of heterogeneous protein-protein interactions: Sedimentation coefficient distributions c(s) and asymptotic boundary profiles from Gilbert-Jenkins theory. *Biophys. J.* **89**(1), 651–666.

Dam, J., Velikovsky, C. A., Mariuzza, R. A., Urbanke, C., and Schuck, P. (2005). Sedimentation velocity analysis of heterogeneous protein-protein interactions: Lamm equation modeling and sedimentation coefficient distributions c(s). *Biophys. J.* **89**(1), 619–634.

Eftink, M. R. (1997). Fluorescence methods for studying equilibrium macromolecule-ligand interactions. *Methods Enzymol.* **278,** 221–257.

Eisenstein, E., and Beckett, D. (1999). Dimerization of the *Escherichia coli* biotin repressor: Corepressor function in protein assembly. *Biochemistry* **38**(40), 13077–13084.

Galas, D. J., and Schmitz, A. (1978). DNAse footprinting: A simple method for the detection of protein-DNA binding specificity. *Nucleic Acids Res.* **5**(9), 3157–3170.

Garner, M. M., and Revzin, A. (1981). A gel electrophoresis method for quantifying the binding of proteins to specific DNA regions: Application to components of the *Escherichia coli* lactose operon regulatory system. *Nucleic Acids Res.* **9**(13), 3047–3060.

Heneghan, A. F., Berton, N., Miura, M. T., and Bain, D. L. (2005). Self-association energetics of an intact, full-length nuclear receptor: The B-isoform of human progesterone receptor dimerizes in the micromolar range. *Biochemistry* **44**(27), 9528–9537.

Heneghan, A. F., Connaghan-Jones, K. D., Miura, M. T., and Bain, D. L. (2006). Cooperative DNA binding by the B-isoform of human progesterone receptor: Thermodynamic analysis reveals strongly favorable and unfavorable contributions to assembly. *Biochemistry* **45**(10), 3285–3296.

Hill, J. J., and Royer, C. A. (1997). Fluorescence approaches to study of protein-nucleic acid complexation. *Methods Enzymol.* **278,** 390–416.

Johnson, M. L., Correia, J. J., Yphantis, D. A., and Halvorson, H. R. (1981). Analysis of data from the analytical ultracentrifuge by nonlinear least-squares techniques. *Biophys. J.* **36**(3), 575–588.

Johnson, M. L., and Faunt, L. M. (1992). Parameter estimation by least-squares methods. *Methods Enzymol.* **210,** 1–37.

Klotz, I. (1997). "Ligand-Receptor Energetics: A Guide for the Perplexed." John Wiley & Sons, Inc., New York.

Ladbury, J. E. (2004). Application of isothermal titration calorimetry in the biological sciences: Things are heating up! *Biotechniques* **37**(6), 885–887.

Laue, T. M. (1995). Sedimentation equilibrium as thermodynamic tool. *Methods Enzymol.* **259,** 427–452.

Leonhardt, S. A., Boonyaratanakornkit, V., and Edwards, D. P. (2003). Progesterone receptor transcription and non-transcription signaling mechanisms. *Steroids* **68**(10–13), 761–770.

Li, X., and O'Malley, B. W. (2003). Unfolding the action of progesterone receptors. *J. Biol. Chem.* **278** (41), 39261–39264.

Mangelsdorf, D. J., Thummel, C., Beato, M., Herrlich, P., Schutz, G., Umesono, K., Blumberg, B., Kastner, P., Mark, M., Chambon, P., and Evans, R. M. (1995). The nuclear receptor superfamily: The second decade. *Cell* **83**(6), 835–839.

Marr, M. T., II, Isogai, Y., Wright, K. J., and Tjian, R. (2006). Coactivator cross-talk specifies transcriptional output. *Genes Dev.* **20**(11), 1458–1469.

Motulsky, H., and Christopoulos, A. (2005). "Fitting Models to Biological Data using Linear and Nonlinear Regression." GraphPad Software, Inc., San Diego, CA.

Pedersen, H., Sogaard-Andersen, L., Holst, B., and Valentin-Hansen, P. (1991). Heterologous cooperativity in *Escherichia coli.* The CytR repressor both contacts DNA and the cAMP receptor protein when binding to the deoP2 promoter. *J. Biol. Chem.* **266**(27), 17804–17808.

Perini, L. T., Doherty, E. A., Werner, E., and Senear, D. F. (1996). Multiple specific CytR binding sites at the Escherichia coli deoP2 promoter mediate both cooperative and competitive interactions between CytR and cAMP receptor protein. *J. Biol. Chem.* **271**(52), 33242–33255.

Read, C. M., and Jelesarov, I. (2001). Calorimetry of protein-DNA complexes and their components. *Methods Mol. Biol.* **148,** 511–533.

Revzin, A. (1993). "Footprinting Nucleic Acid-Protein Complexes." Academic Press, San Diego, CA.

Roark, D. E. (1976). Sedimentation equilibrium techniques: Multiple speed analyses and an overspeed procedure. *Biophys. Chem.* **5**(1–2), 185–196.

Schuck, P. (1998). Sedimentation analysis of noninteracting and self-associating solutes using numerical solutions to the Lamm equation. *Biophys. J.* **75**(3), 1503–1512.

Semsey, S., Virnik, K., and Adhya, S. (2006). Three-stage regulation of the amphibolic gal operon: From repressosome to GalR-free DNA. *J. Mol. Biol.* **358**(2), 355–363.

Senear, D. F., and Brenowitz, M. (1991). Determination of binding constants for cooperative site-specific protein-DNA interactions using the gel mobility-shift assay. *J. Biol. Chem.* **266**(21), 13661–13671.

Shin, M., Kang, S., Hyun, S. J., Fujita, N., Ishihama, A., Valentin-Hansen, P., and Choy, H. E. (2001). Repression of deoP2 in *Escherichia coli* by CytR: Conversion of a transcription activator into a repressor. *EMBO J.* **20**(19), 5392–5399.

Sogaard-Andersen, L., and Valentin-Hansen, P. (1993). Protein-protein interactions in gene regulation: The cAMP-CRP complex sets the specificity of a second DNA-binding protein, the CytR repressor. *Cell* **75**(3), 557–566.

Stafford, W. F. (2000). Analysis of reversibly interacting macromolecular systems by time derivative sedimentation velocity. *Methods Enzymol.* **323,** 302–325.

Stafford, W. F., and Sherwood, P. J. (2004). Analysis of heterologous interacting systems by sedimentation velocity: Curve fitting algorithms for estimation of sedimentation coefficients, equilibrium and kinetic constants. *Biophys. Chem.* **108**(1–3), 231–243.

Tetel, M. J., Jung, S., Carbajo, P., Ladtkow, T., Skafar, D. F., and Edwards, D. P. (1997). Hinge and amino-terminal sequences contribute to solution dimerization of human progesterone receptor. *Mol. Endocrinol.* **11**(8), 1114–1128.

Tretyachenko-Ladokhina, V., Cocco, M. J., and Senear, D. F. (2006). Flexibility and adaptability in binding of *E. coli* cytidine repressor to different operators suggests a role in differential gene regulation. *J. Mol. Biol.* **362**(2), 271–286.

Velazquez-Campoy, A., Leavitt, S. A., and Freire, E. (2004). Characterization of protein-protein interactions by isothermal titration calorimetry. *Methods Mol. Biol.* **261,** 35–54.

Weickert, M. J., and Adhya, S. (1992). A family of bacterial regulators homologous to Gal and Lac repressors. *J. Biol. Chem.* **267**(22), 15869–15874.

Winzor, D. J. (2003). Analytical exclusion chromatography. *J. Biochem. Biophys. Methods* **56**(1–3), 15–52.

Yan, Y., and Marriott, G. (2003). Analysis of protein interactions using fluorescence technologies. *Curr. Opin. Chem. Biol.* **7**(5), 635–640.

CHAPTER 3

Biosensor-Surface Plasmon Resonance Methods for Quantitative Analysis of Biomolecular Interactions

Farial A. Tanious, Binh Nguyen, and W. David Wilson

Department of Chemistry
Georgia State University
Atlanta, Georgia 30302

Abstract

The surface plasmon resonance (SPR) biosensor method has emerged as a very flexible and powerful approach for detecting a wide diversity of biomolecular interactions. SPR monitors molecular interactions in real time and provides significant advantages over optical or calorimetric methods for systems with strong binding and low spectroscopic signals or reaction heats. The SPR method simultaneously provides kinetic and equilibrium characterization of the interactions of biomolecules. Such information is essential for development of a full understanding

0091-679X/08 $35.00
DOI: 10.1016/S0091-679X(07)84003-9

of molecular recognition as well as for areas such as the design of receptor-targeted therapeutics. This article presents basic, practical procedures for conducting SPR experiments. Initial preparation of the SPR instrument, sensor chips, and samples are described. This is followed by suggestions for experimental design, data analysis, and presentation. Steady-state and kinetic studies of some small molecule–DNA complexes are used to illustrate the capability of this technique. Examples of the agreement between biosensor-SPR and solution studies are presented.

I. Introduction

Transcriptional activators and repressors bind to DNA; drugs form complexes with membrane components and in cells they target sites on nucleic acids and proteins; antibodies bind to proteins of disease organisms; and a host of other biomolecular interactions are essential for organisms and their cells to function. To understand the functional processes that drive biological systems, it is essential to have detailed information on the array of biomolecular interactions that drive and control cellular function. The binding affinity (the equilibrium constant, K, and Gibbs energy of binding, ΔG), stoichiometry (n, the number of compounds bound to the biopolymer), cooperative effects in binding, and binding kinetics (the rate constants, k, that define the dynamics of the interaction) are the basic quantitative characteristics of all biomolecular interactions. The more of these key parameters that can be determined experimentally, the better will be our understanding of the underlying interaction and how it affects cellular functions.

Because of the varied properties of biological molecules and the changes in properties that occur on complex formation, it is frequently difficult to find a method that can characterize the full array of interactions under an appropriate variety of conditions. For complexes that involve very tight binding, it is necessary to conduct experiments at very low concentrations, down to the nanomolar or lower levels, that fall below the detection limit for many systems. In such cases, radiolabels or fluorescent probes have been used for added sensitivity in detection. Biosensors with surface plasmon resonance (SPR) detection provide an alternative method, which responds to the refractive index or mass changes at the biospecific sensor surface on complex formation (Jonsson *et al.*, 1991; Karlsson *et al.*, 1994; Malmqvist and Granzow, 1994; Malmqvist and Karlsson, 1997; Myszka, 2000; Nagata and Handa, 2000). Since the SPR signal responds directly to the amount of bound compound in real time, as versus indirect signals at equilibrium for many physical measurements, it provides a very powerful method to study biomolecular interaction thermodynamics and kinetics (Karlsson and Larsson, 2004; Katsamba *et al.*, 2002a,b, 2006; Morton and Myszka, 1998; Myszka, 2000; Rich *et al.*, 2002; Svitel *et al.*, 2003; Van Regenmortel, 2003). Use of the SPR response to monitor biomolecular reactions also removes many difficulties with labeling or characterizing the diverse properties of biomolecules.

II. Rationale: Biomolecular Interactions with SPR Detection

The biosensor-SPR methods described in this chapter refer to Biacore instruments (Biacore International AB), which have been most widely used in the SPR analysis of biomolecular interactions (Rich and Myszka, 2005b). The description will be divided into three primary areas that define the SPR experiment: (i) instrument and sensor chip preparation; (ii) immobilization of one reaction component on the sensor chip surface; and (iii) data collection and analysis. For reaction of biomolecules, B1 and B2, to give a complex, the reaction is:

$$\mathrm{B1} + \mathrm{B2} \underset{k_d}{\overset{k_a}{\rightleftharpoons}} \mathrm{C} \quad K_A = \frac{k_a}{k_d} \tag{1}$$

Figure 1 shows the basic components for biosensor-SPR analysis of this bimolecular interaction with B1 in the flow solution and B2 immobilized. In this example, B2 is linked to dextran in a hydrogel matrix, a typical method for Biacore sensor chips. Reactant B1 is at a fixed concentration in the solution that flows over the biospecific surface containing B2. A number of solutions with different concentrations can be injected to cover a full binding profile. Detection of binding is through a change in refractive index that is monitored in real time by a change in the SPR resonance angle that occurs when the molecules form a complex on the surface (Fig. 1). In a typical Biacore experiment, a four channel sensor chip is used with one flow cell left blank as a control, while the remaining three cells have reaction components immobilized (such as B2 and other target biomolecules). Whether the experiment will have sufficient signal to noise for accurate data analysis depends on two primary factors that are described in the following sections: the moles of B2 immobilized and the mass of B1 bound (moles of B1 bound × MW of B1) at any time point (since the SPR signal is related to the refractive index change on binding). The molecular weight of binding molecule is thus a key consideration in SPR-biosensor experiments.

Typical Biacore sensor chips, such as the one in Fig. 1, are derivatized with a carboxymethyl-dextran (CM-dextran) hydrogel that provides many possibilities for biomolecular immobilization (BIACORE, 1994a). The Biacore web site has descriptions of a range of sensor chip surfaces and immobilization chemistries (https://www.biacore.com/lifesciences/index.html), and it is generally possible to find an appropriate surface and immobilization chemistry for any biological interaction application. It is obviously essential that the immobilization method, of whatever type, not significantly perturb the binding interactions relative to what occurs in free solution. Covalent coupling of molecules to the surface should use groups that are well away from the binding site and that do not interfere with the interaction.

As shown in Fig. 1, when a solution of a reaction component is injected into the Biacore flow system and passes over the sensor chip biospecific surface, complex

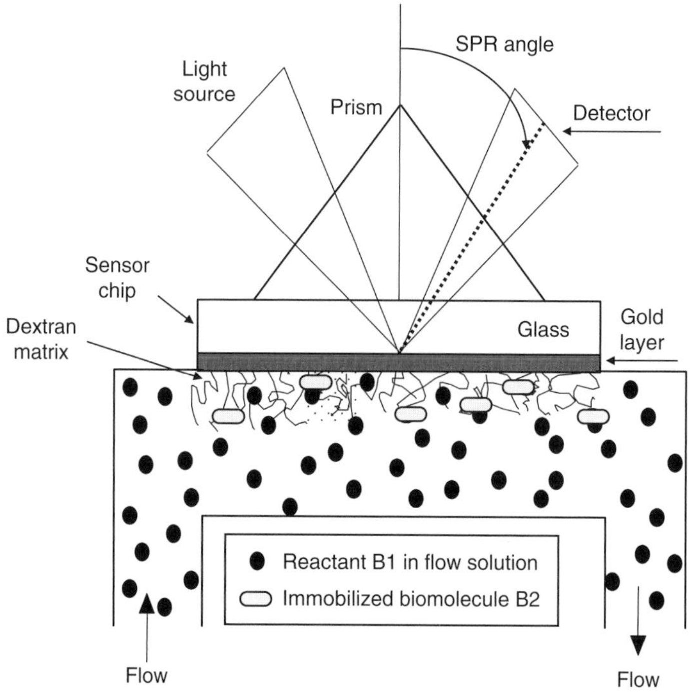

Fig. 1 The SPR signal and biosensor surfaces. The components of a biosensor-SPR experiment are illustrated: the optical unit that generates and measures the SPR angle, the sensor chip with a gold layer, the chip surface with immobilized matrix (dextran on this chip) and reaction component (B2 in this experiment), and the flow control system and solution that provide the other reaction component(s), such as B1. As more of the B1+B2 complex forms, the SPR angle changes as a function of time. Analysis of the signal change with time can provide the kinetic constants for the reaction.

formation occurs and is monitored in real time by a change in SPR angle. After a selected time, reactant flow is replaced by buffer flow and dissociation of the complex is monitored over time. The time course of the experiment shown in Fig. 1 creates a sensorgram such as the one illustrated in Fig. 2. Buffer flow establishes an initial baseline and injection of component B1 leads to the association phase. As the association reaction continues, a steady-state plateau is eventually reached such that the rate of association equals the rate of dissociation of the complex and no change of signal with time is observed. The time required for the steady state to be reached depends on the reactant concentrations and reaction kinetics. If the added molecule does not bind, the SPR angle change in the sample and reference flow cells will be the same and a zero net response, which is indicative of no binding, will be observed after subtraction. When binding does occur, the added molecule is bound at the sensor surface and the SPR angle changes more in the sample than in the reference cell to give the time course of the sensorgram (Fig. 2). Since the amount of unbound compound in the flow solution is the same

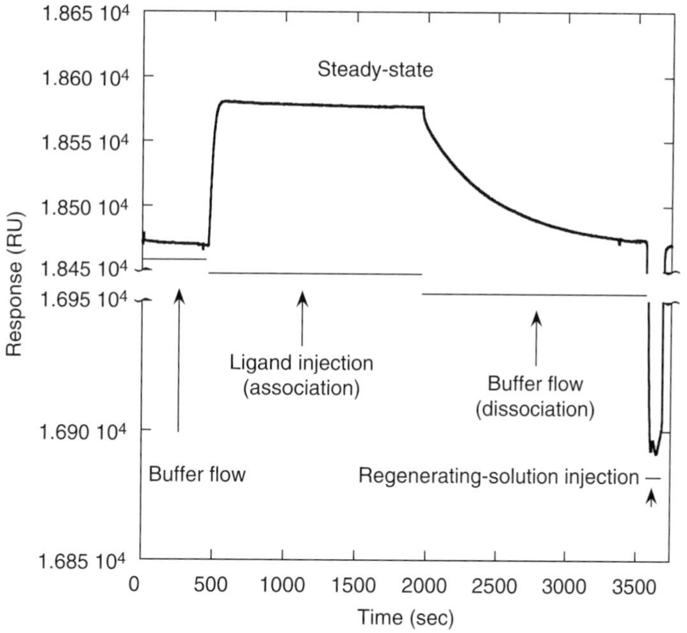

Fig. 2 An experimental sensorgram illustrating the steps in an SPR experiment. These steps include buffer flow for baseline, followed by an association phase, then another buffer flow for dissociation which is followed by injection of the regeneration solution to bring the surface back to the starting conditions. The injections are repeated at a range of concentrations to generate a set of sensorgrams.

in the sample and reference flow cells, it can be subtracted and only the bound reactant generates an SPR signal. The concentration of unbound molecule is constant and is fixed by the concentration in the flow solution.

Several sensorgrams can be obtained at different concentrations of the injected compound and they can be simultaneously fit (global fitting) to obtain the most accurate kinetic (k) and equilibrium (K) constants (Myszka, 1999a, 2000). As will be described below, equilibrium constants can be determined independently from ratios of rate constants or by fitting the steady-state response versus the concentration of the binding molecule in the flow solution over a range of concentrations. The SPR signal change is an excellent method to determine binding stoichiometry, since the refractive index change in SPR experiments generates essentially the same response for each bound molecule and depends on the molecular weight of the binding molecules. The maximum signal increase in an SPR experiment thus provides a direct determination of the stoichiometry, provided the amount of immobilized biomolecule is known. For complexes that have quite slow dissociation rates, the biosensor surface can be regenerated before complete dissociation by using a solution that causes rapid dissociation of the complex, but does not significantly degrade the surface (Fig. 2). The angle change in Biacore instruments is converted to resonance units (RU) and a 1000 RU response is equivalent to a change in surface concentration of about 1 ng/mm^2 of bound protein or DNA (the relationship between RU and ng of material bound varies with the refractive index of the bound molecule) (Davis and Wilson, 2000, 2001).

The equilibrium and kinetic constants that describe the reaction in Eq. (1) are obtained by fitting the sensorgrams or steady-state RU versus concentration plots to a 1:1 binding model. More elaborate binding models are necessary for more complex interactions, and these models are the same for all types of binding experiments and are not unique to SPR methods. It should be emphasized that to obtain accurate kinetic information about a binding reaction, it is essential that the kinetics for transfer of the binding molecule (diffusion through the hydrogel—see below for better description) to the immobilized biomolecule (mass transfer) be much faster than the binding reaction (Karlsson, 1999). When this is not true, various alternatives to deal with the mass transfer problem are available and will be described below. Annual surveys on methods, applications, and appropriate experimental approaches using SPR by Myszka and colleagues provide many additional helpful suggestions on experimental protocols to obtain high quality biosensor data (Myszka, 1999b; Rich and Myszka, 2000, 2001, 2002, 2003, 2005a,b).

III. Materials and Methods

A. Instrument Preparation

It is recommended by Biacore to run *Desorb* weekly and *Sanitize* once a month for maintaining the instrument. *Desorb* is a general method that uses a series of solutions injected through the instrument internal flow system to remove any absorbed compounds from previous experiments. *Sanitize* is a method to insure that no microbial growth is present in the liquid injection and flow system. Before beginning an experiment, it is very important to ensure that the instrument is running properly. The goal is to determine if a stable baseline can be maintained throughout a series of replicate buffer injections across a nonderivatized sensor surface. A simple method for cleaning is described below. If the baseline is not stable, for example, if the baseline drifts more than ±1.0 RU/min, additional cleanings may be needed.

1. Required Materials, Chemicals, and Solutions

- Maintenance chip with a glass flow cell surface
- CM5 chip
- Running buffer: HBS-EP buffer (10 mM HEPES pH 7.4, 150 mM NaCl, 3 mM EDTA, 0.005% (v/v) polysorbate 20)
- 0.5% SDS (BIAdesorb solution 1)

- 50 mM glycine pH 9.5 (BIAdesorb solution 2)
- 1% (v/v) acetic acid solution
- 0.2 M sodium bicarbonate solution
- 6 M guanidine HCl solution
- 10 mM HCl solution

2. Methods for Preliminary Cleaning and Checking Baseline

a. Set instrumment temperature to 25 °C

b. *Dock* a maintenance chip and *Prime* once with distilled water (*Prime* is a method for priming the liquid system by flushing pumps, integrated μ-fluidic cartridge (IFC) and autosampler with water or buffer. This procedure is used at start up, when the buffer is changed, and also to remove small air bubbles from the system).

c. Run *Desorb*.

d. After *Desorb, Prime* several times with warm water (50–60 °C).

e. *Undock* the maintenance chip and *Dock* a fresh research grade CM5 sensor chip and *Prime* once with water.

f. Switch to running buffer and *Prime* several times.

g. Prepare aliquots of 200 μl running buffer into individual vials and run Method 1 (below).

This method will collect a set of sensorgrams of replicated buffer injections across an unmodified CM5 sensor chip. These sensorgrams should overlay after double-referenced subtraction and this indicates a very stable system that is ready for experiments. Methods are written as text files, and may be created or modified with any text editor. The BIA software for instrument control converts the text file to instrument commands (BIACORE, 1994b).

Method 1:

```
MAIN
   RACK        1 thermo_c
   RACK        2 thermo_a
   FLOWCELL       1,2,3,4
   LOOP Buffer STEP
   APROG drug %sample2 %position2 %volume2 %conc2
   ENDLOOP
   APPEND Continue
END
DEFINE APROG buffer
   PARAM %sample2 %position2 %volume2 %conc2
   KEYWORD Concentration %Conc2
```

```
    CAPTION %conc2 %sample2 over (gradient surface)
    FLOW 25
    FLOWPATH 1,2,3,4
    WAIT 5:00
    KINJECT %position2 %volume2 180
    EXTRACLEAN
    EXTRACLEAN
    WAIT 5:00
  END
  DEFINE LOOP Buffer
    LPARAM %sample2 %position2 %volume2 %conc2
    TIMES 1
            Buffer  r2a1   100   0.000u
            Buffer  r2a2   100   0.000u
            Buffer  r2a3   100   0.000u
            Buffer  r2a4   100   0.000u
            Buffer  r2a5   100   0.000u
            Buffer  r2a6   100   0.000u
            Buffer  r2a7   100   0.000u
            Buffer  r2a8   100   0.000u
            Buffer  r2a9   100   0.000u
            Buffer  r2a10  100   0.000u
     END
```

3. Additional Cleaning Methods

After running the above method, if the baseline is not stable within ±1.0 RU/min (note: this specification may change as different instruments become available), the following methods, designed and provided by Biacore, may be used:

a. Super Clean (As Needed)

1. Insert a maintenance chip into the instrument and *Dock*.
2. Run *Desorb* using SDS and glycine.
3. Run the following method with warm (50–60 °C) filtered water as flowing solution.

```
main
  prime
  prime
```

```
    unclog
    rinse
    flush
    prime
    prime
    append standby
  end
```

4. Run *Desorb* using 1% (v/v) acetic acid in place of SDS and glycine.
5. *Prime* the instrument to wash out the acetic acid residuals.
6. Run *Desorb* using 0.2 M sodium bicarbonate in place of SDS and glycine.
7. *Prime* the instrument to wash out the sodium bicarbonate residuals.
8. Run *Desorb* using 6 M guanidine HCl for the SDS (solution 1) and 10 mM HCl for glycine (solution 2).
9. *Prime* the instrument a few times to thoroughly clean all residuals.

b. Super Desorb (Monthly)

1. *Dock* a maintenance chip and run *Prime* using 0.5% SDS.
2. Run *Prime* using 10 mM glycine, pH 9.5.
3. Run *Prime* at least three times using filtered water.

B. Sensor-Chip Surface Preparation

In general, there are two ways to capture biomolecules on the sensor chips: covalent and noncovalent captures. Covalent capture will be illustrated with streptavidin and this surface can then be used to immobilize biotin-labeled biomolecules. Although other immobilization techniques are available, the biotin–streptavidin coupling is popular in Biacore SPR experiments and it is particularly useful for nucleic acids immobilization (Bates *et al.*, 1995; Bischoff *et al.*, 1998; Hendrix *et al.*, 1997; Mazur *et al.*, 2000; Nair *et al.*, 2000; Nieba *et al.*, 1997; Rutigliano *et al.*, 1998; Wang *et al.*, 2000). The large affinity constant for the biotin–streptavidin complex results in a stable surface for binding studies under physiological conditions. In the example below, immobilization of streptavidin and biotin-labeled DNA will be described.

An important step in sensor chip immobilization is to decide how much biomolecule to immobilize. For kinetic experiments, it is usually best to immobilize the smallest amount of materials, while maintaining the necessary signal-to-noise ratio, in order to minimize mass transport effects. Mass transport effects of ligand to the surface will influence kinetic data when the rate of mass transport is slower than or on the same time scale as the kinetics of the interaction (BIACORE, 1994c; Karlsson, 1999; Myszka *et al.*, 1998). Since a high concentration of surface binding sites rapidly consumes the ligand at the surface, the more material immobilized the greater the contribution from mass transport. However, when the ligand is a small molecule, it becomes necessary to increase the immobilized compound surface density since the instrument response from small-molecule binding is low.

1. Immobilization of Streptavidin

For immobilizing biotin nucleic acids on a sensor chip, the sensor chip must be modified to a streptavidin surface. Biacore offers pre-made streptavidin sensor chips (SA sensor chip) that are ready for immediate use. However, it is possible and in some cases worthwhile to prepare streptavidin sensor chips (BIACORE, 1994c; Hendrix *et al.*, 1997) using standard (CM5) dextran surfaces or CM4 chips with features such as a low density carboxyl surface. The low density carboxyl surface uses dextran but has less negative charge and so may be advantageous when investigating the interactions between highly charged biomolecules. The procedure outlined below is used for immobilizing streptavidin on CM5 or CM4 sensor chips. The Biacore website has many references that describe other methods to immobilize biomolecules to different sensor chip surfaces.

a. Required Materials, Chemicals, and Solutions

- A CM5 or CM4 sensor chip that has been at room temperature for at least 30 min
- HBS-EP buffer (10 mM HEPES pH 7.4, 150 mM NaCl, 3 mM EDTA, 0.005% (v/v) polysorbate 20) (running buffer)
- 100 mM *N*-hydroxsuccinimide (NHS) freshly prepared in water
- 400 mM *N*-ethyl-*N*′-(dimethylaminopropyl)carbodiimide (EDC) freshly prepared in water
- 10 mM acetate buffer pH ~4.5 (immobilization buffer)
- 200–400 μg/ml streptavidin in immobilization buffer
- 1 M ethanolamine hydrochloride in water pH 8.5 (deactivation solution)

b. Procedures for Streptavidin Immobilization

1. *Dock* the CM4 or CM5 chip, *Prime* with running buffer. Start a sensorgram in all flow cells with a flow rate of 5 μl/min.

2. With NHS (100 mM) in one vial and EDC (400 mM) in other, use the *Dilute* command to make a 1:1 mixture of NHS/EDC.

3. *Inject* NHS/EDC for 10 min (50 μl) to activate the carboxymethyl surface to reactive esters.

4. Using *Manual Inject* with a flow rate of 5 μl/min, load the loop with ~100 μl of streptavidin in the appropriate buffer and inject streptavidin over all flow cells. Track the number of RUs immobilized which is available in real time readout and

stop the injection after the desired level is reached (typically 2500–3000 RU for CM5 chip and 1000–1500 RU for CM4 chip).

5. *Inject* ethanolamine hydrochloride for 10 min (50 μl) to deactivate any remaining reactive esters.

6. *Prime* several times to ensure surface stability.

2. Immobilization of Nucleic Acids

Derivatized nucleic acids with biotin at either the 5′ or 3′ end are ready to be immobilized on a streptavidin-coated sensor chip (SA Chip). This immobilization method provides rapid kinetics and high affinity binding of the nucleic acid to the surface. Relatively short oligonucleotide hairpins (<50 bases) do not require high salt for immobilization. A solution of ~25 nM oligonucleotide (5′-biotin nucleic acid) in HBS-EP buffer is used when immobilizing nucleic acids less than 50 bases. It may be necessary to increase the concentration when using larger nucleic acids. A concentration that is too high, however, will make control over the amount of nucleic acid immobilized very difficult. Typically, an immobilization amount of 300–450 RUs of hairpin nucleic acid (~20–30 bases in length) is immobilized for running steady-state experiments and 100–150 RUs for kinetic experiments to minimize mass transfer effects.

a. Required Chemicals, Materials, and Solutions

- Streptavidin-coated sensor chip (SA chip or prepared as outlined above)
- HBS-EP buffer (10 mM HEPES pH 7.4, 150 mM NaCl, 3 mM EDTA, 0.005% (v/v) polysorbate 20) (running buffer)
- Activation buffer (1 M NaCl, 50 mM NaOH)
- Biotin-labeled nucleic acid solutions (~25 nM of strand or hairpin dissolved in HBS-EP buffer)

b. Immobilization of Nucleic Acids on a Streptavidin Surface (or on SA Chips)

If two or more different nucleic acid hairpins are to be immobilized on different flow cells, there are two options for immobilization level: equal RU amount or equal moles. For an equal RU amount, different nucleic acid hairpins can be immobilized with the same total response units (RU) on each flow cells. For equal moles, the amount of nucleic acid to be immobilized is proportional to its molecular weight. A higher level is required for a higher molecular weight hairpin because the observed response per bound ligand (RU_{obs}) is proportional to mass bound (moles bound × MW). This option is useful to visualize and illustrate a difference in stoichiometry.

1. *Dock* a streptavidin-coated chip, *Prime* a few times with HBS-EP buffer, and start a sensorgram with a 20 μl/min flow rate.

2. *Inject* activation buffer (1 M NaCl, 50 mM NaOH) for 1 min (20 μl) five to seven times to remove any unbound streptavidin from the sensor chip.

3. Allow buffer to flow at least 5 min before immobilizing the nucleic acids.

4. Start a new sensorgram with a flow rate of 2 μl/min and select one desired flow cell on which to immobilize the nucleic acid. Take care not to immobilize nucleic acid on the flow cell chosen as the control flow cell. Generally, flow cell 1 ("fc1") is used as a control and often left blank. It is often desirable to immobilize different nucleic acids on the remaining three flow cells. A nonbinding nucleic acid may be immobilized on fc1 to provide a more similar control surface for subtraction.

5. Wait a few minutes for the baseline to stabilize. Use *Manual Inject*, load the injection loop with ~100 μl of a 25 nM nucleic acid solution and inject over the flow cell. Track the number of RUs immobilized and stop the injection after a desired level is reached.

6. At the end of the injection and after the baseline has stabilized, use the software crosshair to determine the RUs of nucleic acid immobilized and record this amount. The amount of nucleic acid immobilized is required to determine the theoretical moles of small molecule binding sites for the flow cell.

7. Repeat steps 4 to 6 for another flow cell (e.g., fc3 or fc4).

C. Sample Preparations

The solution of small molecule must be prepared in the same buffer used to establish the baseline—the running buffer. If the small molecule is not very soluble in buffer, it can be dissolved in water as a concentrated stock solution and diluted in the running buffer. If the small molecule requires the presence of a small amount of an organic solvent (e.g., <5% DMSO) to maintain solubility, the same amount of this organic solvent should be in the running buffer to minimize the refractive index difference.

Preliminary studies may be needed to obtain some information about the compounds being studied such as solubility, stability, or an estimated binding constant. Such information is useful in setting parameters for data collection. Sample concentrations should vary over a wide range (at least 100-fold). In general, the sample concentration range should vary from well below to well above $1/K_a$. If the K_a is unknown, a broad concentration range should be used in a preliminary experiment to obtain an estimate of the K_a. Ideally, the order of sample injection should be randomized. Injecting samples from low to high concentration is useful for eliminating artifacts in the data from adsorption or carry over. It is also useful to inject the same concentration twice to check for reproducibility. For binding constants of 10^6–10^9 M^{-1}, as observed with many small molecules DNA complexes, small molecule concentrations from 0.01 nM to 10 μM in the flow solution allow accurate determination of binding constants.

D. Data Collection and Processing

1. Data Collection

A sample method used to collect steady-state small molecule data on nucleic acid surfaces is shown below. This method is set for a flow rate of 10 μl/min (**FLOW 10**) over all flow cells (**FLOWPATH 1,2,3,4**). The samples are injected as written (**STEP**) from low to high concentration. Note that before any analyte is injected, buffer injections are done to enable double referencing. In addition, the volume of analyte injected is set as a variable so that the least amount of volume required to reach a steady state is used for each concentration. Much less time is required for the association reactions at high concentrations of the injected compound. In this sample method, the small molecule solution of Hoechst 33258 (or analyte) is injected over immobilized DNA (or macromolecule). Note that with the steady-state method, equilibrium, but not kinetics, constants can be obtained even when mass transfer effects dominate the observed kinetics. Much higher injection flow rates are used when collecting kinetics data.

At the end of the compound solution injection, a regeneration step may be necessary to remove any complex remaining on the surface. To subsequently remove any regeneration buffer remaining after this step, two 1-min injections of running buffer are used prior to the end of the cycle followed by a 5-min wait with running buffer flowing. After the next cycle has begun, a 5-min waiting period is set to ensure the baseline has stabilized before the next sample injection. When working with small molecules (or small responses), it is essential that the baseline does not drift significantly during the injection. To reduce carry overs (of sample and regenerating solution), a *Mix* command is added to rinse the injection tube and the injection is conducted from low to high concentrations. If needed, multiple injections of buffer at the end of the cycles are useful to check for carry over.

```
MAIN
    RACK      1 thermo_c
    RACK      2 thermo_a
    FLOWCELL 1,2,3,4
    DETECTION 2–1, 3–1, 4–1
    LOOP Hoechst33258 STEP
    APROG Flow10 %sample2 %position2 %volume2 %conc2
    ENDLOOP
    APPEND Continue
END
DEFINE APROG Flow10
    PARAM %sample2 %position2 %volume2 %conc2
    KEYWORD Conc %Conc2
    CAPTION %conc2 %sample2 over AATT_TTAA_TATA (gradient surface)
```

```
  FLOW        10
  FLOWPATH 1,2,3,4
  WAIT        5:00
  KINJECT         %position2 %volume2 300
-0:20 RPOINT       -b BASELINE
2:30 RPOINT        %sample2
  QUICKINJECT r2f3 10        ! 10 mM Glycine pH 2.5
  EXTRACLEAN
  MIX r2f7 300               ! buffer
  QUICKINJECT r2f4 10        ! buffer
  EXTRACLEAN
  QUICKINJECT r2f5 10        ! buffer
  EXTRACLEAN
  WAIT        5:00
END
DEFINE LOOP Hoechst33258
LPARAM %sample2 %position2 %volume2 %conc2
   TIMES 1
   Buffer          r2a1    200    0.0000u
   Buffer          r2a2    100    0.0000u
   Buffer          r2a3    50     0.0000u
   Buffer          r2a4    200    0.0000u
   Buffer          r2a5    100    0.0000u
   Buffer          r2a6    50     0.0000u
   Buffer          r2a7    200    0.0000u
   Buffer          r2a8    100    0.0000u
   Buffer          r2a9    50     0.0000u
   Buffer          ra10    200    0.0000u
   Buffer          r2b1    100    0.0000u
   Buffer          r2b2    50     0.0000u
   Buffer          r2b3    200    0.0000u
   Buffer          r2b4    100    0.0000u
   Buffer          r2b5    50     0.0000u

   Hoechst33258    r2c1    200    0.0001u
   Hoechst33258    r2c2    200    0.0002u
   Hoechst33258    r2c3    200    0.0004u
   Hoechst33258    r2c4    200    0.0006u
```

```
    Hoechst33258        r2c5     200     0.0008u
    Hoechst33258        r2c6     200     0.0010u
    Hoechst33258        r2c7     200     0.0020u
    Hoechst33258        r2c8     200     0.0040u
    Hoechst33258        r2c9     200     0.0060u
    Hoechst33258        r2c10    100     0.0080u
    Hoechst33258        r2d1     100     0.0100u
    Hoechst33258        r2d2     100     0.0200u
    Hoechst33258        r2d3     100     0.0400u
    Hoechst33258        r2d4     100     0.0600u
    Hoechst33258        r2d5     100     0.0800u
    Hoechst33258        r2d6     100     0.1000u
    Hoechst33258        r2d7     50      0.2000u
    Hoechst33258        r2d8     50      0.4000u
    Hoechst33258        r2d9     50      0.6000u
    Hoechst33258        r2d10    50      0.8000u
    Buffer              r2f5     200     0.0000u
    Buffer              r2f5     100     0.0000u
    Buffer              r2f5     50      0.0000u
END
```

2. Data Processing

After the data has been collected, there are several processing steps that must be performed before any quantitative information can be extracted. A number of software programs are available for processing Biacore data, including BIAevaluation (Biacore, Inc.), Scrubber2, and CLAMP (Myszka and Morton, 1998). The results can also be exported and presented in graphing software such as KaleidaGraph for either PC or Macintosh computers (Mazur *et al.*, 2000; Wang *et al.*, 2000). Zeroing on the y-axis (RU) and then x-axis (time) are the first steps in data processing. Because the flow cell surfaces are not identical to each other, the refractive index of each surface is different causing the flow cells to register at different positions on the y-axis. Zeroing the data on the y-axis is necessary to allow the responses of each flow cell to be compared. Generally the average of a stable time region of the sensorgram, prior to sample injection, should be selected and set to zero. Because the flow cells are aligned in series, sample is not injected across the flow cells simultaneously. Zeroing on the x-axis aligns the beginnings of the injections with respect to each other.

The two data-processing steps outlined below help to minimize offset artifacts and also to correct for the bulk shift that results from slight differences in injection buffer and running buffer. In the first step, the control flow cell (fc1) sensorgram is subtracted from the reaction flow cell sensorgrams (i.e., fc2-fc1, fc3-fc1, and fc4-fc1). This removes the bulk shift contribution to the change in RUs. The next step in data processing is required to remove systematic deviations that are frequently seen in the sensorgrams. In this step, the effect of buffer injection on a reaction flow cell is subtracted from the compound injections (different concentrations) on the same flow cell. These processes are referred to as "double referencing" (Myszka, 1999a), and remove the systematic drifts and shifts in baseline that are frequently observed even in control cell-subtracted sensorgrams. In the data collection method shown above, buffer injections are performed for each volume amount used for sample injection. Typically, multiple buffer injections are performed and averaged before subtraction. In double referencing, plots are made for each flow cell separately overlaying the control flow cell-corrected sensorgrams from buffer and all sample injections. The buffer sensorgram is then subtracted from the sample sensorgrams. At this point, the data should be of optimum quality and is ready for fitting to determine the thermodynamic and/or kinetic values that characterize the reaction.

IV. Results and Data Analysis

Even when it is not possible to get kinetic constants, equilibrium constants can be extracted from SPR data in a correctly performed experiment. The equilibrium constant can be obtained from fitting steady-state data, or from kinetics. The association equilibrium constant (K_A) is the ratio of the observed association (k_a) and dissociation rate (k_d) constants in Eq. (1). Comparing the K_A value obtained by different methods can help to evaluate the models used to fit the data. Kinetic constants, true k_a and k_d values, can be obtained when the reaction is not dominated by mass transfer.

Knowledge of the stoichiometry of the system is essential for obtaining correct kinetic and binding constants as well as for obtaining a complete description of the system being studied. Because the refractive index increments (R_{II}s) of small molecules can be very different from those of proteins and nucleic acids, it is essential that such a difference be accounted for during data interpretation to correctly determine stoichiometry, and subsequently kinetic and equilibrium constants. The maximum Biacore instrument response for a 1:1 binding interaction can be predicted with Eq. (2).

$$\mathrm{RU}_{\mathrm{max}} = \mathrm{RU}_{\mathrm{biopolymer}} \times \left(\frac{\mathrm{MW}_{\mathrm{compound}}}{\mathrm{MW}_{\mathrm{biopolymer}}}\right) \times R_{\mathrm{II}} \qquad (2)$$

where RU_{max} is the response for binding of one molecule to the biopolymer; $RU_{biopolymer}$ is the amount of immobilized biopolymer, in response units; MW is molecular weight of compound and biomolecule, respectively; and R_{II} is the refractive index increment ratio of compound to the immobilizing biopolymer where:

$$R_{II} = \frac{(\partial n/\partial C)_{compound}}{(\partial n/\partial C)_{biopolymer}}.$$

The R_{II} value is close to one for proteins and DNA but can deviate considerably from 1.0 for small molecules (Davis and Wilson, 2000). Reference for R_{II} values and methods for determination are given in Davis and Wilson (2000). One way to determine the R_{II} is by comparison of the predicted value from Eq. (2) to the experimental observed value RU_{max}. Small molecules may have more than a single binding site in biomolecular complexes. Nonspecific, secondary binding can occur with cationic molecules and nucleic acids for example, and the R_{II} ratio is critical for accurate determination of stoichiometry.

A. Equilibrium Analysis

After double subtraction, the average of the data in the steady-state region of each sensorgram (RU_{avg}) can be converted to r ($r = RU_{avg}/RU_{max}$) and is plotted as a function of analyte concentration. Equilibrium constants can be obtained by fitting the results with either a single site model (Eq. (3) with $K_2 = 0$) or with the two-site model in Eq. (3):

$$r = \left(\frac{K_1 \times C_{free} + 2 \times K_1 \times K_2 \times C_{free}^2}{1 + K_1 \times C_{free} + K_1 \times K_2 \times C_{free}^2}\right) \tag{3}$$

where K_1 and K_2, the macroscopic thermodynamic binding constants, are the variable parameters to fit; r is the moles of compound bound/mole DNA-hairpin $= RU_{avg}/RU_{max}$; and C_{free} is the concentration of the compound in the flow solution. Although more complex models could be used in data fitting, it is unlikely that a unique fit to the results would be obtained. In such complex cases, other experimental methods should be used to fix some of the variables before fitting the SPR results.

The monocationic Hoechst 33258 DNA minor groove binder has strong preference for A/T rich sequences (Weisblum and Haenssler, 1974). Its DNA binding affinity has been studied with different biophysical methods (Bontemps *et al.*, 1975). A crystallographic structure of the Hoechst 33258 bound to an –AATT– site is available (Pjura *et al.*, 1987; Quintana *et al.*, 1991; Teng *et al.*, 1988). Three different biotin-labeled DNA hairpin duplexes containing AATT, TTAA, TATA sites (Fig. 3) were immobilized on a streptavidin chip (as described above) and

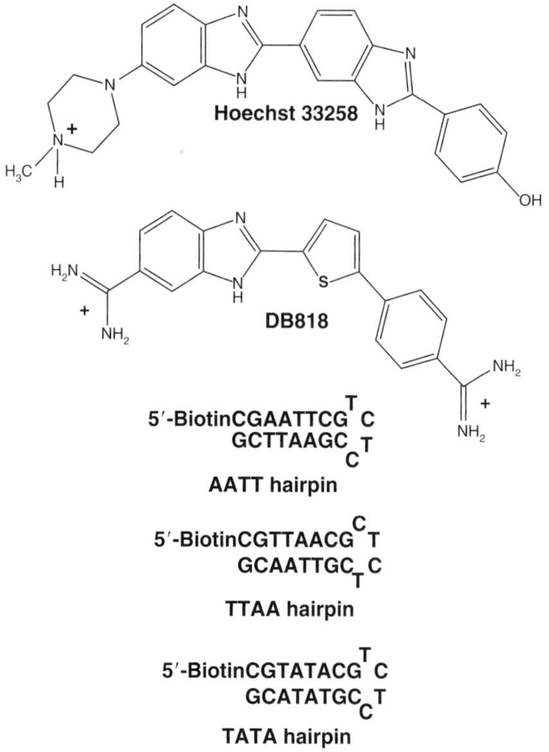

Fig. 3 Structure of Hoechst 33258, DB818, and 5′-biotin-labeled DNA hairpins.

different concentrations of Hoechst 33258 (Fig. 3) were injected onto the surface. Sensorgrams of binding of Hoechst 33258 to AATT and TTAA along with binding plots are shown in Figs. 4 and 5. The binding stoichiometry and affinity for this type of interaction are readily extracted. The binding stoichiometry can be obtained from comparing the maximum response with the predicted response per compound (Eq. (2)).

Because equal moles of DNA hairpins were immobilized, the difference in maximum responses among the sets of sensorgrams is readily seen and directly reflects the difference in binding stoichiometry (Figs. 4 and 5). Under these experimental conditions, the Hoechst ligand binds with a 1:1 ratio to the AATT site (Fig. 4) but with a 2:1 ratio to TTAA (Fig. 5) or TATA (not shown). Plotting the data in Scatchard form can visually reveal considerable information about the binding constants, stoichiometry of specific and nonspecific binding, and cooperativity (Fig. 6). In this figure, the differences in binding constants, stoichiometry and cooperativity for binding of a low molecular weight aromatic cation, Hoechst 33258, to two different DNA hairpins, AATT and TTAA are illustrated. The cooperative binding of two molecules of Hoechst 33258 to TTAA is clear.

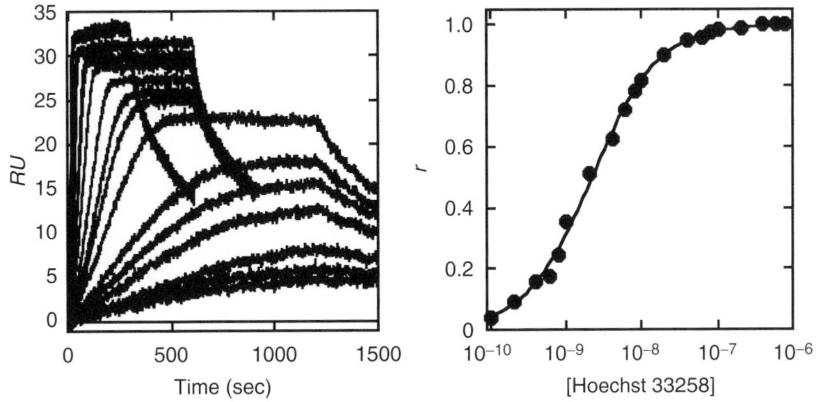

Fig. 4 Sensorgrams for the interaction of Hoechst 33258 with the 5′-biotin-labeled AATT DNA (Fig. 3). The sensorgrams (left) were collected in 0.1 M NaCl, 0.01 M MES (2-(*N*-morpholino)ethanesulfonic acid), 0.001 M EDTA, pH 6.25. The individual sensorgrams represent responses at different Hoechst concentrations; the concentrations were from 0.1 nM (lowest sensorgram) to 0.8 μM (highest sensorgram). Hoechst 33258 solutions were injected at a flow rate of 10 μl/min. The volume of Hoechst 33258 injected is set as a variable (see the method) so that the least amount of volume required to reach a steady state is used for each concentration. Much less time is required for the association reactions at high concentrations. Conversion of these sensorgrams to the binding isotherm (right) was done by dividing the averaged plateau or steady-state responses by the predicted maximum response per ligand ($RU_{pred\text{-}max} = 35$ in this case) as described in the text. The data were fitted (solid line) with a one-site model, Eq. (3), to obtain an equilibrium binding constant of $K = 4.6 \times 10^8\ M^{-1}$. This value is in excellent agreement with *K* values from solution studies (see text).

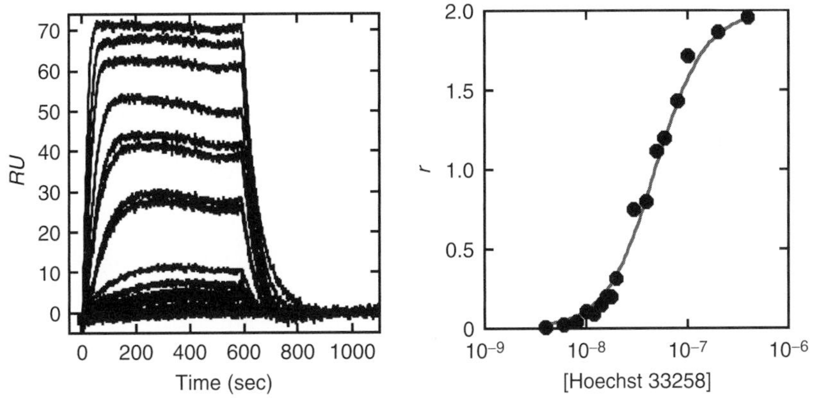

Fig. 5 Sensorgrams for the interaction of Hoechst 33258 with 5′-biotin-labeled TTAA DNA (Fig. 3). The sensorgrams (left) were collected in the same buffer as shown in Fig. 4. The concentrations were from 1.0 nM (lowest sensorgram) to 0.4 μM (highest sensorgram). Hoechst 33258 solutions were injected at a flow rate of 25 μl/min. Conversion of these sensorgrams to the binding isotherm (right) was done by dividing the averaged plateau or steady-state responses by the predicted maximum response per ligand ($RU_{pred\text{-}max} = 35$ as in this case) as described in the text. The data were fitted with a two-site model, Eq. (3), to obtain macroscopic equilibrium binding constants of $K_1 = 1.5 \times 10^6\ M^{-1}$ and $K_2 = 3.7 \times 10^8\ M^{-1}$.

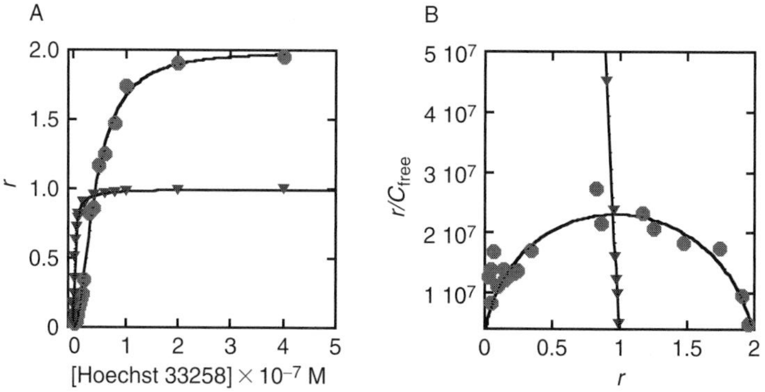

Fig. 6 Binding isotherms (A) and Scatchard plot (B) for the Hoechst 33258 complexes with the AATT and TTAA hairpins. The results are from the sensorgrams in Figs. 4 and 5. The result with the AATT hairpin is typical for AT specific minor groove agents and indicates one strong binding site. The result with the TTAA hairpin is very unusual. Two molecules bind to this oligomer with positive cooperativity. The lines in the figures were obtained by nonlinear least-squares fits of the data to one- and two-site binding equations.

The positive cooperativity in binding of Hoechst to TTAA can be easily seen from a convex shape of the Scatchard plot (Fig. 6). (See the Chapter by Garbett and Chaires for a more complete discussion of analysis of binding data.) A similar trend was observed with the TATA hairpin. The binding constants of Hoechst 33258 to the TATA hairpin are $K_1 = 6.6 \times 10^6\ \mathrm{M}^{-1}$ and $K_2 = 2.7 \times 10^7\ \mathrm{M}^{-1}$. This type of information is very difficult to obtain by other methods. Many systems involve specific binding at one or two sites followed by additional nonspecific binding at higher concentration. The SPR result of Hoechst binding to the AATT hairpin is in agreement with recent results from other methods (Breusegem *et al.*, 2002; Han *et al.*, 2005; Kiser *et al.*, 2005; Loontiens *et al.*, 1990).

B. Kinetic Analysis

Kinetic analysis was performed by global fitting of SPR data with non-mass-transport and mass transport kinetic binding models. In the non-mass-transport 1:1 binding model, Eqs. (4) and (5) are used for global fitting, while in a mass transport limitation model, Eqs. (4–7) are used for global fitting:

$$\mathrm{A} + \mathrm{B} \leftrightarrow \mathrm{AB}$$

$$[\mathrm{A}]_{t=0} = 0, [\mathrm{B}]_{t=0} = \mathrm{RU}_{\max}, [\mathrm{AB}]_{t=0} = 0$$

$$K_\mathrm{a} = \frac{[\mathrm{AB}]}{[\mathrm{A}][\mathrm{B}]} \tag{4}$$

$$\frac{d[AB]}{dt} = k_a[A][B] - k_d[AB] \tag{5}$$

$$\frac{d[A]}{dt} = k_t([A_{bulk}] - [A]) - (k_a[A][B] - k_d[AB]) \tag{6}$$

$$\frac{d[B]}{dt} = -k_a[A][B] + k_d[AB] \tag{7}$$

where [A] and [A_{bulk}] are the concentration of the compound at the sensor surface and the in the bulk solution flow, respectively; [B] is the concentration of the immobilized DNA; [AB] is the concentration of the complex; k_a is the association rate constant; k_d is the dissociation rate constant, and k_t is the mass transport coefficient, defined by Eq. (6).

The fitting can be performed with BIAevaluation software or with CLAMP (Myszka and Morton, 1998) and should be preferentially done with a global analysis method that includes fitting of association and dissociation phases of all sensorgrams (Morton and Myszka, 1998). In cases where a steady-state plateau is reached, the ratio of the rate constants (k_a/k_d) should be compared to the steady-state K_A value. An agreement between the two methods suggests that the binding constant, K_A, is correct but does not necessarily mean that the k_a and k_d values are correct due to possible mass transfer effects and possible correlation of the constants. Some considerations for kinetic fitting have been previously outlined (Nguyen *et al.*, 2007). To illustrate a kinetic fit, the interaction between a DNA minor groove binder and a DNA hairpin was studied. DB818, a DNA minor groove binding agent (Fig. 3), forms a 1:1 complex in the duplex minor groove at AATT site (Mallena *et al.*, 2004). An SPR experiment for the interaction of DB818 with a DNA hairpin containing the –AATT– site (Fig. 3) was conducted at high ionic strength (1 M NaCl) with flow rate of 50 μl/min.

From this experiment, the kinetic and steady-state analyses are obtained from the same set of sensorgrams to illustrate the agreement of the binding constants obtained from the two analysis methods. The high ionic strength in this experiment with DB818 reduced K_a and k_a to minimize the mass transfer effects. Sensorgrams for the interaction are shown in Fig. 7 and the results are analyzed by both steady-state and kinetic methods (Table I). The sensorgrams increase in response as the DB818 concentration is increased. Note that it takes longer to reach a steady-state plateau at low concentration as expected for a bimolecular reaction. The smooth lines in the figure are the best fit lines using global fitting with a single site kinetic model with a mass transport term. The steady-state RU values for DB818/DNA sensorgrams from the same experiment are converted to r and graphed directly onto a direct plot in Fig. 7 with different concentrations for fitting with Eq. (3) with $K_2 = 0$. The binding constants obtained from steady-state and kinetic analyses are in excellent agreement, and the results are summarized in Table I. The kinetic

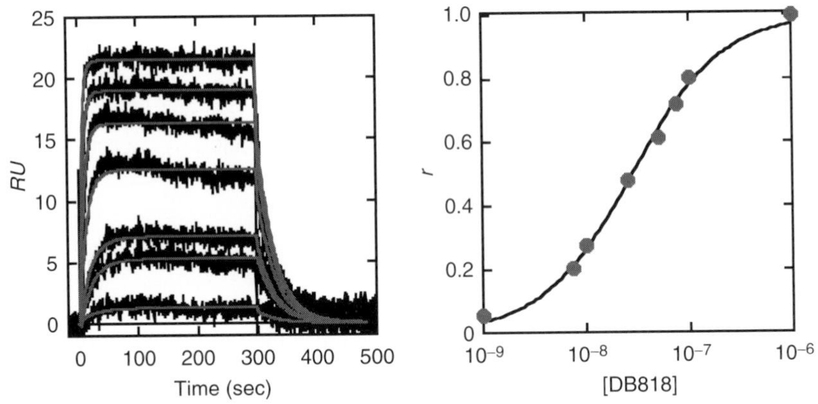

Fig. 7 Sensorgrams to evaluate the kinetics of the DB818–DNA interaction. The sensorgrams were collected with a BIACORE 2000 with flow rate of 50 μl/min at 25 °C and immobilized AATT DNA (Fig. 3) in 1.0 M NaCl, 0.01 M Tris, 0.001 M EDTA, pH 7.4. The concentrations in this experiment from the bottom to the top sensorgrams are 0 (reference line), 1, 7.5, 10, 25, 50, 75, and 100 nM. The kinetic analysis was performed by global fitting of the binding data with mass transport kinetic 1:1 binding models. Conversion of these sensorgrams to the binding isotherm (right) also was done by dividing the averaged plateau or steady-state responses by the predicted maximum response per ligand. The data were fitted with a one-site model, Eq. (3), to obtain an equilibrium binding constant of 3.6×10^7 M^{-1}.

Table I
Summary of Biacore Kinetics and Steady-State (S.S.) Results of DB818 Binding to DNA AATT Hairpin

Exp.	Flow rate (μl/min)	RU_{max} (RU)	k_a ($M^{-1} s^{-1}$)	k_d (s^{-1})	K_A (1/M) k_a/k_d	k_t [RU/(Ms)]	Chi^2 (RU)^2	$k_a \times RU_{max}/k_t$
Kinetics	50	20.6	2.9×10^6	0.065	4.5×10^7	4.3×10^7	0.287	1.4
S.S.					3.6×10^7		0.234	

The results are obtained from steady-state and kinetic analysis of sensorgrams in Fig. 7. The steady-state and kinetic analyses yield similar binding constants.

fitting results meet the criteria previously outlined ($k_a \times RU_{max}/k_t \leq 5$) (Karlsson, 1999). In addition, the half-life $t_{1/2}$ from the dissociation phase of sensorgram is close to the calculated half-life using the fitted value ($t_{1/2} = \ln 2/k_d$) suggesting the mass transport effect is minimized (Nguyen *et al.*, 2007).

V. Summary

The SPR-biosensor method is excellent for studying small molecule–macromolecule interactions and in the short time that commercial instrument have been available, it has assumed a major role in quantitative analysis of biomolecular complexes. For strong binding complexes, which are generally observed in biomolecular systems of interest, working at low concentration is required. However, many small molecules have optical properties that make low concentration measurements a clear disadvantage. The SPR method is very useful in such cases since it detects the mass change upon complex formation and can operate at very low compound concentrations. In many cases, the binding kinetics can be observed in real time and extracted from the sensorgrams. A number of studies have now shown that SPR results are comparable to those from other biophysical methods. Although this chapter has focused on small molecule–biopolymer interactions, the methods described above, with minor modifications, can be used to characterize biopolymer–biopolymer complexes.

Acknowledgments

We very much thank Professor David W. Boykin (Georgia State University, Atlanta, GA, USA) for very productive collaborations in biomolecular interaction analysis, the NIH for funding the research, and the Georgia Research Alliance for funding of Biacore instruments.

References

Bates, P. J., Dosanjh, H. S., Kumar, S., Jenkins, T. C., Laughton, C. A., and Neidle, S. (1995). Detection and kinetic studies of triplex formation by oligodeoxynucleotides using real-time biomolecular interaction analysis (BIA). *Nucleic Acids Res.* **23,** 3627–3632.

BIACORE. (1994a). "BIAapplications Handbook." Pharmacia Biosensor AB, Uppsala, Sweden.

BIACORE. (1994b). "BIACORE 2000: Instrument Handbook." Biacore AB, Uppsala, Sweden.

BIACORE. (1994c). "BIAtechnology Handbook." Pharmacia Biosensor AB, Uppsala, Sweden.

Bischoff, G., Bischoff, R., Birch-Hirschfeld, E., Gromann, U., Lindau, S., Meister, W. V., de, A. B. S., Bohley, C., and Hoffmann, S. (1998). DNA-drug interaction measurements using surface plasmon resonance. *J. Biomol. Struct. Dyn.* **16,** 187–203.

Bontemps, J., Houssier, C., and Fredericq, E. (1975). Physico-chemical study of the complexes of "33258 Hoechst" with DNA and nucleohistone. *Nucleic Acids Res.* **2,** 971–984.

Breusegem, S. Y., Clegg, R. M., and Loontiens, F. G. (2002). Base-sequence specificity of Hoechst 33258 and DAPI binding to five (A/T)4 DNA sites with kinetic evidence for more than one high-affinity Hoechst 33258-AATT complex. *J. Mol. Biol.* **315,** 1049–1061.

Davis, T. M., and Wilson, W. D. (2000). Determination of the refractive index increments of small molecules for correction of surface plasmon resonance data. *Anal. Biochem.* **284,** 348–353.

Davis, T. M., and Wilson, W. D. (2001). Surface plasmon resonance biosensor analysis of RNA-small molecule interactions. *Methods Enzymol.* **340,** 22–51.

Han, F., Taulier, N., and Chalikian, T. V. (2005). Association of the minor groove binding drug Hoechst 33258 with d(CGCGAATTCGCG)2: Volumetric, calorimetric, and spectroscopic characterizations. *Biochemistry* **44,** 9785–9794.

Hendrix, M., Priestley, E. S., Joyce, G. F., and Wong, C. H. (1997). Direct observation of aminoglycoside-RNA interactions by surface plasmon resonance. *J. Am. Chem. Soc.* **119,** 3641–3648.

Jonsson, U., Fagerstam, L., Ivarsson, B., Johnsson, B., Karlsson, R., Lundh, K., Lofas, S., Persson, B., Roos, H., Ronnberg, I., Sjolander, S., Stenberg, E., *et al.* (1991). Real-time biospecific interaction analysis using surface plasmon resonance and a sensor chip technology. *Biotechniques* **11,** 620–627.

Karlsson, R. (1999). Affinity analysis of non-steady-state data obtained under mass transport limited conditions using BIAcore technology. *J. Mol. Recognit.* **12,** 285–292.

Karlsson, R., and Larsson, A. (2004). Affinity measurement using surface plasmon resonance. *Methods Mol. Biol.* **248,** 389–415.

Karlsson, R., Roos, H., Fagerstam, L., and Persson, B. (1994). Kinetic and concentration analysis using BIA technology. *Methods* **6,** 99–110.

Katsamba, P. S., Bayramyan, M., Haworth, I. S., Myszka, D. G., and Laird-Offringa, I. A. (2002a). Complex role of the beta 2-beta 3 loop in the interaction of U1A with U1 hairpin II RNA. *J. Biol. Chem.* **277,** 33267–33274.

Katsamba, P. S., Park, S., and Laird-Offringa, I. A. (2002b). Kinetic studies of RNA-protein interactions using surface plasmon resonance. *Methods* **26,** 95–104.

Katsamba, P. S., Navratilova, I., Calderon-Cacia, M., Fan, L., Thornton, K., Zhu, M., Bos, T. V., Forte, C., Friend, D., Laird-Offringa, I., Tavares, G., Whatley, J., *et al.* (2006). Kinetic analysis of a high-affinity antibody/antigen interaction performed by multiple Biacore users. *Anal. Biochem.* **352,** 208–221.

Kiser, J. R., Monk, R. W., Smalls, R. L., and Petty, J. T. (2005). Hydration changes in the association of Hoechst 33258 with DNA. *Biochemistry* **44,** 16988–16997.

Loontiens, F. G., Regenfuss, P., Zechel, A., Dumortier, L., and Clegg, R. M. (1990). Binding characteristics of Hoechst 33258 with calf thymus DNA, poly[d(A-T)], and d(CCGGAATTCCGG): Multiple stoichiometries and determination of tight binding with a wide spectrum of site affinities. *Biochemistry* **29,** 9029–9039.

Mallena, S., Lee, M. P., Bailly, C., Neidle, S., Kumar, A., Boykin, D. W., and Wilson, W. D. (2004). Thiophene-based diamidine forms a "super" at binding minor groove agent. *J. Am. Chem. Soc.* **126,** 13659–13669.

Malmqvist, M., and Granzow, R. (1994). Biomolecular interaction analysis. *Methods* **6,** 95–98.

Malmqvist, M., and Karlsson, R. (1997). Biomolecular interaction analysis: Affinity biosensor technologies for functional analysis of proteins. *Curr. Opin. Chem. Biol.* **1,** 378–383.

Mazur, S., Tanious, F. A., Ding, D., Kumar, A., Boykin, D. W., Simpson, I. J., Neidle, S., and Wilson, W. D. (2000). A thermodynamic and structural analysis of DNA minor-groove complex formation. *J. Mol. Biol.* **300,** 321–337.

Morton, T. A., and Myszka, D. G. (1998). Kinetic analysis of macromolecular interactions using surface plasmon resonance biosensors. *Methods Enzymol.* **295,** 268–294.

Myszka, D. G. (1999a). Improving biosensor analysis. *J. Mol. Recognit.* **12,** 279–284.

Myszka, D. G. (1999b). Survey of the 1998 optical biosensor literature. *J. Mol. Recognit.* **12,** 390–408.

Myszka, D. G. (2000). Kinetic, equilibrium, and thermodynamic analysis of macromolecular interactions with BIACORE. *Methods Enzymol.* **323,** 325–340.

Myszka, D. G., and Morton, T. A. (1998). CLAMP: A biosensor kinetic data analysis program. *Trends Biochem. Sci.* **23,** 149–150.

Myszka, D. G., He, X., Dembo, M., Morton, T. A., and Goldstein, B. (1998). Extending the range of rate constants available from BIACORE: Interpreting mass transport-influenced binding data. *Biophys. J.* **75,** 583–594.

Nagata, K., and Handa, H. (eds.) (2000). "Real-Time Analysis of Biomolecular Interactions: Applications of BIACORE." Springer, New York.

Nair, T. M., Myszka, D. G., and Davis, D. R. (2000). Surface plasmon resonance kinetic studies of the HIV TAR RNA kissing hairpin complex and its stabilization by 2-thiouridine modification. *Nucleic Acids Res.* **28,** 1935–1940.

Nguyen, B., Tanious, F. A., and Wilson, W. D. (2007). Biosensor-surface plasmon resonance: Quantitative analysis of small molecule-nucleic acid interactions. *Methods* **42,** 150–161.

Nieba, L., Nieba-Axmann, S. E., Persson, A., Hamalainen, M., Edebratt, F., Hansson, A., Lidholm, J., Magnusson, K., Karlsson, A. F., and Pluckthun, A. (1997). BIACORE analysis of histidine-tagged proteins using a chelating NTA sensor chip. *Anal. Biochem.* **252,** 217–228.

Pjura, P. E., Grzeskowiak, K., and Dickerson, R. E. (1987). Binding of Hoechst 33258 to the minor groove of B-DNA. *J. Mol. Biol.* **197,** 257–271.

Quintana, J. R., Lipanov, A. A., and Dickerson, R. E. (1991). Low-temperature crystallographic analyses of the binding of Hoechst 33258 to the double-helical DNA dodecamer C-G-C-G-A-A-T-T-C-G-C-G. *Biochemistry* **30,** 10294–10306.

Rich, R. L., Hoth, L. R., Geoghegan, K. F., Brown, T. A., LeMotte, P. K., Simons, S. P., Hensley, P., and Myszka, D. G. (2002). Kinetic analysis of estrogen receptor/ligand interactions. *Proc. Natl. Acad. Sci. USA* **99,** 8562–8567.

Rich, R. L., and Myszka, D. G. (2000). Survey of the 1999 surface plasmon resonance biosensor literature. *J. Mol. Recognit.* **13,** 388–407.

Rich, R. L., and Myszka, D. G. (2001). Survey of the year 2000 commercial optical biosensor literature. *J. Mol. Recognit.* **14,** 273–294.

Rich, R. L., and Myszka, D. G. (2002). Survey of the year 2001 commercial optical biosensor literature. *J. Mol. Recognit.* **15,** 352–376.

Rich, R. L., and Myszka, D. G. (2003). A survey of the year 2002 commercial optical biosensor literature. *J. Mol. Recognit.* **16,** 351–382.

Rich, R. L., and Myszka, D. G. (2005a). Survey of the year 2003 commercial optical biosensor literature. *J. Mol. Recognit.* **18,** 1–39.

Rich, R. L., and Myszka, D. G. (2005b). Survey of the year 2004 commercial optical biosensor literature. *J. Mol. Recognit.* **18,** 431–478.

Rutigliano, C., Bianchi, N., Tomassetti, M., Pippo, L., Mischiati, C., Feriotto, G., and Gambari, R. (1998). Surface plasmon resonance for real-time monitoring of molecular interactions between a triple helix forming oligonucleotide and the Sp1 binding sites of human Ha-ras promoter: Effects of the DNA-binding drug chromomycin. *Int. J. Oncol.* **12,** 337–343.

Svitel, J., Balbo, A., Mariuzza, R. A., Gonzales, N. R., and Schuck, P. (2003). Combined affinity and rate constant distributions of ligand populations from experimental surface binding kinetics and equilibria. *Biophys. J.* **84,** 4062–4077.

Teng, M. K., Usman, N., Frederick, C. A., and Wang, A. H. (1988). The molecular structure of the complex of Hoechst 33258 and the DNA dodecamer d(CGCGAATTCGCG). *Nucleic Acids Res.* **16,** 2671–2690.

Van Regenmortel, M. H. (2003). Improving the quality of BIACORE-based affinity measurements. *Dev. Biol. (Basel)* **112,** 141–151.

Wang, L., Bailly, C., Kumar, A., Ding, D., Bajic, M., Boykin, D. W., and Wilson, W. D. (2000). Specific molecular recognition of mixed nucleic acid sequences: An aromatic dication that binds in the DNA minor groove as a dimer. *Proc. Natl. Acad. Sci. USA* **97,** 12–16.

Weisblum, B., and Haenssler, E. (1974). Fluorometric properties of the bibenzimidazole derivative Hoechst 33258, a fluorescent probe specific for AT concentration in chromosomal DNA. *Chromosoma* **46,** 255–260.

CHAPTER 4

Isothermal Titration Calorimetry: Experimental Design, Data Analysis, and Probing Macromolecule/Ligand Binding and Kinetic Interactions

Matthew W. Freyer and Edwin A. Lewis

Department of Chemistry and Biochemistry
Northern Arizona University
Flagstaff, Arizona 86011

0091-679X/08 $35.00
DOI: 10.1016/S0091-679X(07)84004-0

Abstract

Isothermal titration calorimetry (ITC) is now routinely used to directly characterize the thermodynamics of biopolymer binding interactions and the kinetics of enzyme-catalyzed reactions. This is the result of improvements in ITC instrumentation and data analysis software. Modern ITC instruments make it possible to measure heat effects as small as 0.1 μcal (0.4 μJ), allowing the determination of binding constants, *K*'s, as large as 10^8–10^9 M^{-1}. Modern ITC instruments make it possible to measure heat rates as small as 0.1 μcal/sec, allowing for the precise determination of reaction rates in the range of 10^{-12} mol/sec. Values for K_m and k_{cat}, in the ranges of 10^{-2}–10^3 μM and 0.05–500 sec^{-1}, respectively, can be determined by ITC. This chapter reviews the planning of an optimal ITC experiment for either a binding or kinetic study, guides the reader through simulated sample experiments, and reviews analysis of the data and the interpretation of the results.

I. Introduction

In biology, particularly in studies relating the structure of biopolymers to their functions, two of the most important questions are (i) how tightly does a small molecule bind to a specific interaction site and (ii) if the small molecule is a substrate and is converted to a product, how fast does the reaction take place?

Perhaps the first question we need to ask here is why calorimetry? The calorimeter, in this case an isothermal titration calorimeter (ITC), can be considered a universal detector. Almost any chemical reaction or physical change is accompanied by a change in heat or enthalpy. A measure of the heat taken up from the surroundings (for an endothermic process) or heat given up to the surroundings (for an exothermic process) is simply equal to the amount of the reaction that has occurred, n (in moles, mmoles, μmoles, nmoles, etc.) and the enthalpy change for the reaction, ΔH (typically in kcal/mol or kJ/mol). A measure of the rate at which heat is exchanged with the surroundings is simply equal to the rate of the reaction, $\partial n/\partial t$ (in moles/sec, mmoles/sec, μmoles/sec, nmoles/sec) and again the enthalpy change, ΔH. A calorimeter is therefore an ideal instrument to measure either how much of a reaction has taken place or the rate at which a reaction is occurring. In contrast to optical methods, calorimetric measurements can be done with reactants that are spectroscopically silent (a chromophore or fluorophore tag is not required), can be done on opaque, turbid, or heterogeneous solutions (e.g., cell suspensions), and can be done over a range of biologically relevant conditions (temperature, salt pH, etc.). Although not a topic covered in this chapter, calorimetric measurements have been used to follow the metabolism of cells or tissues in culture over long periods of time and under varying conditions (e.g., anaerobic or aerobic) (Bandman *et al.*, 1975; Monti *et al.*, 1986).

Titration calorimetry was first described as a method for the simultaneous determination of K_{eq} and ΔH about 40 years ago by Christensen and Izatt (Christensen *et al.*, 1966; Hansen *et al.*, 1965). The method was originally applied to a variety of weak acid–base equilibria and to metal ion complexation reactions (Christensen *et al.*, 1965, 1968; Eatough, 1970). These systems could be studied with the calorimetric instrumentation available at the time which was limited to the determination of equilibrium constant, K_{eq}, values less than about 10^4–10^5 M^{-1} (Eatough *et al.*, 1985). The determination of larger association constants requires more dilute solutions and the calorimeters of that day were simply not sensitive enough.

Beaudette and Langerman published one of the first calorimetric binding studies of a biological system using a small volume isoperibol titration calorimeter (Beaudette and Langerman, 1978). In 1979, Langerman and Biltonen published a description of microcalorimeters for biological chemistry, including a discussion of available instrumentation, applications, experimental design, and data analysis and interpretation (Biltonen and Langerman, 1979; Langerman and Biltonen, 1979). This was really the beginning of the use of titration calorimetry to study biological equilibria. It took another 10 years before the first commercially available titration calorimeter specifically designed for the study of biological systems became available from MicroCal (Wiseman *et al.*, 1989). This first commercial ITC was marketed as a device for "Determining *K* in Minutes" (Wiseman *et al.*, 1989).

ITC is now routinely used to directly characterize the thermodynamics of biopolymer binding interactions (Freire *et al.*, 1990). This is the result of improvements in ITC instrumentation and data analysis software. Modern ITC instruments make it possible to measure heat effects as small as 0.1 μcal (0.4 μJ), allowing the determination of binding constants, *K*s, as large as 10^8–10^9 M^{-1}.

Spink and Wadso (1976) published one of the first calorimetric studies of enzyme activity. Improvements in modern microcalorimeters including higher sensitivity, faster response, and the ability to make multiple additions of substrate (or inhibitors) has brought us to the point where ITC is now also routinely used to directly characterize the kinetic parameters (K_m and k_{cat}) for an enzyme (Todd and Gomez, 2001; Williams and Toone, 1993). Kinetic studies take advantage of the fact that the calorimetric signal (heat rate, e.g., μcal/sec) is a direct measure of the reaction rate and the ΔH for the reaction. Modern ITC instruments make it possible to measure heat rates as small as 0.1 μcal/sec, allowing for the precise determination of reaction rates in the range of 10^{-12} mol/sec. Values for K_m and k_{cat}, in the ranges of 10^{-2}–10^3 μM and 0.05–500 sec^{-1}, respectively, can be determined by ITC.

Ladbury has published a series of annual reviews on ITC, describing the newest applications and a year-to-year survey of the literature on ITC applications (Ababou and Ladbury, 2006; Cliff *et al.*, 2004). In order to take full advantage of the powerful ITC technique, the user must be able to design the optimum experiment, understand the data analysis process, and appreciate the uncertainties in the fitting parameters. ITC experiment design and data analysis have been the subject of numerous papers (Bundle and Sigurskjold, 1994; Chaires, 2006; Fisher and Singh, 1995; Freiere, 2004; Indyk and Fisher, 1998; Lewis and Murphy, 2005). This chapter reviews the planning of an optimal ITC experiment for either a binding or kinetic study, guides the reader through simulated sample experiments, and reviews analysis of the data and the interpretation of the results.

II. Calorimetry Theory and Operation

A. Heat Change Measurement and Theory

A calorimeter was one of the first scientific instruments reported in the early literature. Shortly after Black (1803) had measured the heat capacity and latent heat of water in the 1760s, Lavoisier designed an ice calorimeter and used this instrument to measure the metabolic heat produced by a guinea pig confined in the measurement chamber (1780s) (Lavoisier and Laplace, 1780; Fig. 1).

Thus, not only was a calorimeter the earliest scientific instrument but the first calorimetric experiment was a biologically relevant measurement.

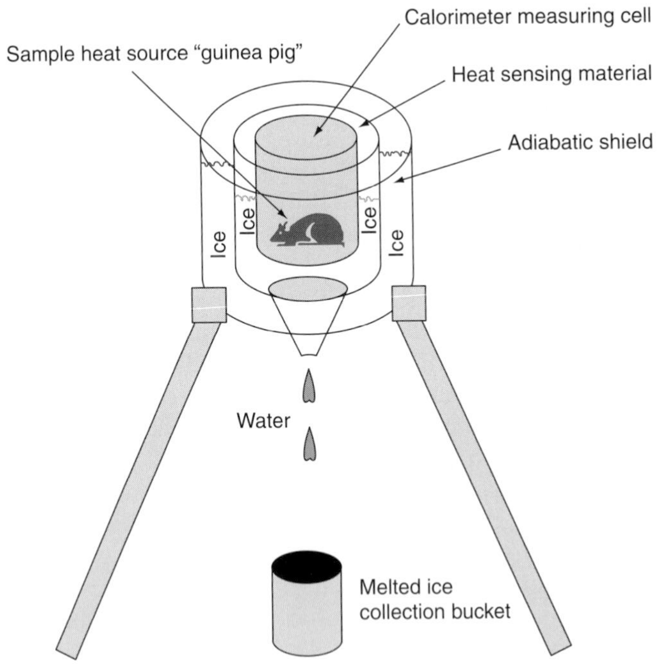

Fig. 1 Lavoisier ice calorimeter used to measure the metabolic heat produced by a guinea pig confined in a measurement chamber. An external wooden layer surrounded a layer of ice that served as an adiabatic shield. Another separate layer of ice directly surrounded the central chamber. The water produced by melting this layer was measured to calculate the metabolic heat produced by the guinea pig.

Calorimetric measurements can be made in three different ways and commercial instruments are available which employ all three techniques. The three methods of measurement are (i) temperature change (either adiabatic or isoperibol), (ii) power compensation (often called isothermal), and (iii) heat conduction· (Hansen *et al.*, 1985). It is important from the standpoint of experiment design, data collection, and data analysis to have a basic background in calorimeter design principles, especially from the standpoint of knowing what the raw signal data represent.

In a temperature change instrument, the heat produced (or consumed) by the reaction occurring in the calorimeter results in a change in temperature of the calorimeter measuring cell. The raw calorimetric signal is simply the temperature of the calorimeter cell as a function of time. With appropriate electrical or chemical calibration, the energy equivalent of the adiabatic (or isoperibol) calorimeter measuring cell can be determined. The measured temperature change is then converted to a heat change by simply multiplying the energy equivalent of the calorimeter, ε_c (in cal/°C), times the measured temperature change, ΔT in (°C).

In a power compensation instrument, the calorimeter measurement cell is controlled at a constant temperature (isothermal). This is accomplished by means of applying constant cooling to the cell and then using a temperature controller and heater to keep the cell temperature constant. As a chemical reaction takes place, any heat input from the chemical reaction is sensed and the power applied to the control heater reduced so that again the temperature remains constant. The heating power from the two sources, reaction and controlled heater, are obviously kept at a constant level so that a heat input from the reaction is compensated by a drop in the heat input from the controlled heater. The raw signal in the power compensation calorimeter is the power (μcal/sec or μJ/sec) applied to the control heater that is required to keep the calorimeter cell from changing temperature as a function of time. The heat change is then simply calculated by integrating the heater power over the time (sec) of the measurement (or more specifically the time required for the control heater power to return to a baseline value). A typical power compensation ITC is shown schematically in Fig. 2.

In a heat conduction calorimeter, the calorimeter measurement cell is passively maintained at a constant temperature by being coupled with heat flow sensors to a heat sink that is actively controlled at a constant temperature. The raw signal in the heat conduction calorimeter is typically a small voltage that is proportional to the very small ΔT that is temporarily developed across the heat flow sensors as a result of the heat produced by the chemical reaction. The Lavoisier calorimeter mentioned earlier was essentially a heat conduction calorimeter with the inner ice layer being the heat flow sensor and the melted ice being the heat signal.

B. Variations in Ligand/Macromolecule Mixing Techniques

A calorimetric experiment is begun by initiating a reaction within the calorimeter measuring cell. Historically, there have been three ways in which reagents have been brought together in the calorimeter. Other methods of initiating a reaction

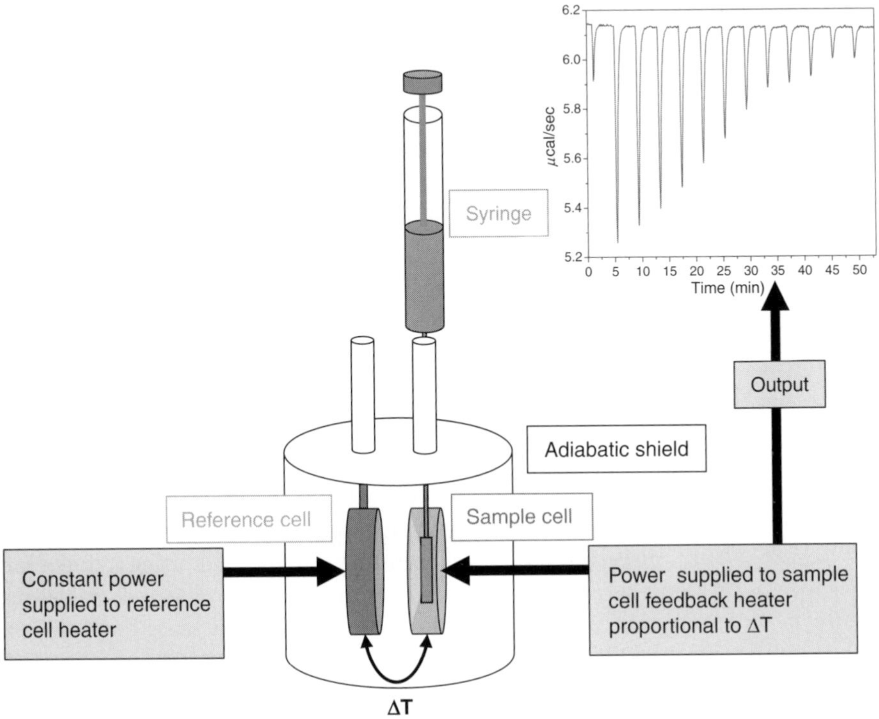

Fig. 2 Representative diagram of a typical power compensation ITC. Major features of this type of instrument such as the reference and sample cells, syringe for adding titrant, and the adiabatic shield are noted in the figure. This diagram shows an oversimplification of how the power applied by the instrument to maintain constant temperature between the reference and sample cells is measured resulting in the instrument signal.

within the calorimeter have included temperature changes (either scanning *T* or a jump in *T*) and pressure changes (either scanning *P* or a jump in *P*). The three methods for bringing reagents in contact with one another include batch, titration, and flow methods (Hansen *et al.*, 1985). Batch methods have varied from rotating the whole calorimeter to mix the contents of two separate volumes in the batch cell, breaking an ampoule again resulting in the mixing of the ampoule contents with the contents of the rest of the calorimeter cell volume, and finally injecting a volume element from outside the cell into the volume contained within the calorimeter cell. Modern ITC instruments are often employed in a batch or a direct injection mode (DIE) in which a single larger injection of a reagent solution is made to start the experiment (e.g., kinetic measurements). More commonly, modern ITC instruments are used in a titration mode in which a number of incremental injections are made at time intervals in the course of a complete titration experiment. (The slow time response of the currently available ITC instruments is such that continuous titration experiments are not possible.) Flow (and/or stopped flow) instruments having high enough sensitivity and fast enough response for most biological/biochemical studies are not currently available.

Each of the calorimeter types has its own advantages and disadvantages in terms of inherent sensitivity and time response. Adiabatic calorimeter designs are not used in the current crop of instruments designed for the study of biological systems. However, both power compensation and heat flow designs are in current use for these applications.

C. Commercial Availability

"State-of-the-art" ITC (and DSC) instruments from both MicroCal® (MA) and Calorimetry Sciences Corporation® [CSC (UT)] use the power compensation measurement method. CSC also produces isothermal calorimeters using the heat conduction measurement technology that have higher sensitivity for slow reaction (e.g., kinetic and decomposition measurements). Product information, applications, measurement specifications, data analysis procedures, and bibliographies of recent calorimetric studies can be found at http://www.calscorp.com/index.html (Calorimetry Sciences Corporation®) and http://www.microcal.com (MicroCal®).

III. Thermodynamic ITC Experiments

A. Preface and Review of Basic Thermodynamics

A typical binding interaction between a ligand and a receptor molecule is illustrated in Fig. 3.

In biological terms, the ligand could be a substrate, inhibitor, drug, cofactor, coenzyme, prosthetic group, metal ion, polypeptide, protein, oligonucleotide, nucleic acid, or any one of a number of molecules thought (or known) to non-covalently interact with a specific site of a second molecule (typically a protein or nucleic acid). As noted in the figure, there are three species in equilibrium in

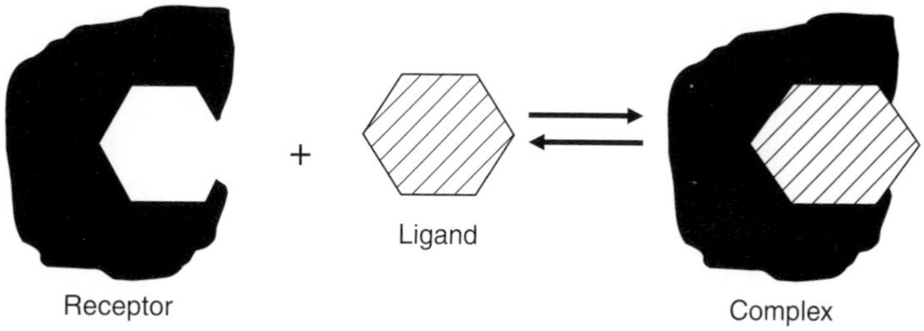

Fig. 3 A simplified model of a typical receptor/ligand-binding interaction. The ligand in this representation is shown to geometrically match the binding site on the receptor to indicate a specific binding interaction.

solution. They are the biopolymer with a vacant binding site, the free ligand, and the complex. A fundamental understanding of the pictured interaction would require at a minimum knowledge of the equilibrium constant for the binding process, K, and the binding stoichiometry, n (how many ligands are there bound to the macromolecule at saturation). A richer understanding of the ligand macromolecule interaction is established if the enthalpy (ΔH) and entropy (ΔS) change contributions to the formation of the complex are known. The following equations are provided as a brief review of the relevant thermodynamic relationships:

$$K_{\text{eq}} = \left\{ \frac{[\text{Complex}]}{[\text{Receptor}]} \times [\text{Ligand}] \right\}_{\text{equilibrium}} \tag{1}$$

$$\Delta G^{\circ} = -RT \ \ln \ K_{\text{eq}} \tag{2}$$

$$\Delta G = \Delta G^{o} + RT\ln \left\{ \frac{[\text{Complex}]}{[\text{Receptor}]} \times [\text{Ligand}] \right\}_{\text{actual}} \tag{3}$$

$$\Delta G = \Delta H - T\Delta S \tag{4}$$

where K_{eq} (K) is the equilibrium constant, [X] is the molar equilibrium (or actual) concentration of species X, ΔG° is the standard Gibbs free energy change, R is the universal gas constant, T is the temperature in Kelvin, ΔG is the actual Gibbs free energy change, ΔH is the enthalpy change, and ΔS is the entropy change for complex formation. The unique advantage of the ITC experiment is that it is possible in a single experiment, if done under optimum conditions, to obtain accurate values for K (or ΔG), ΔH, $-T\Delta S$, and n, where n is the stoichiometry of the interaction (mol ligand/mol complex).

What do we mean by optimum conditions? The ITC experiment must be done under conditions where the heat change is both measurable for each injection and where the heat change varies for subsequent injections producing a curved thermogram (a plot of heat change vs injection number, or mol ratio of ligand/macromolecule). The first condition is obvious, the instrument is a calorimeter and if there are insufficient calories produced by the reaction, then the experiment will be impossible. The second is more problematic since the curvature in the thermogram is a function of the concentration of the macromolecule, [M], and the equilibrium constant K. Figure 4 illustrates this point in that the two sets of panels with identical Brandt's "c" parameters ($c = 10$ in Fig. 4A and B, and $c = 100$ in Fig. 4C and D) (Wiseman *et al.*, 1989) exhibit the same curvature. The Brandt's "c" parameter is defined to be equal to the total macromolecule concentration

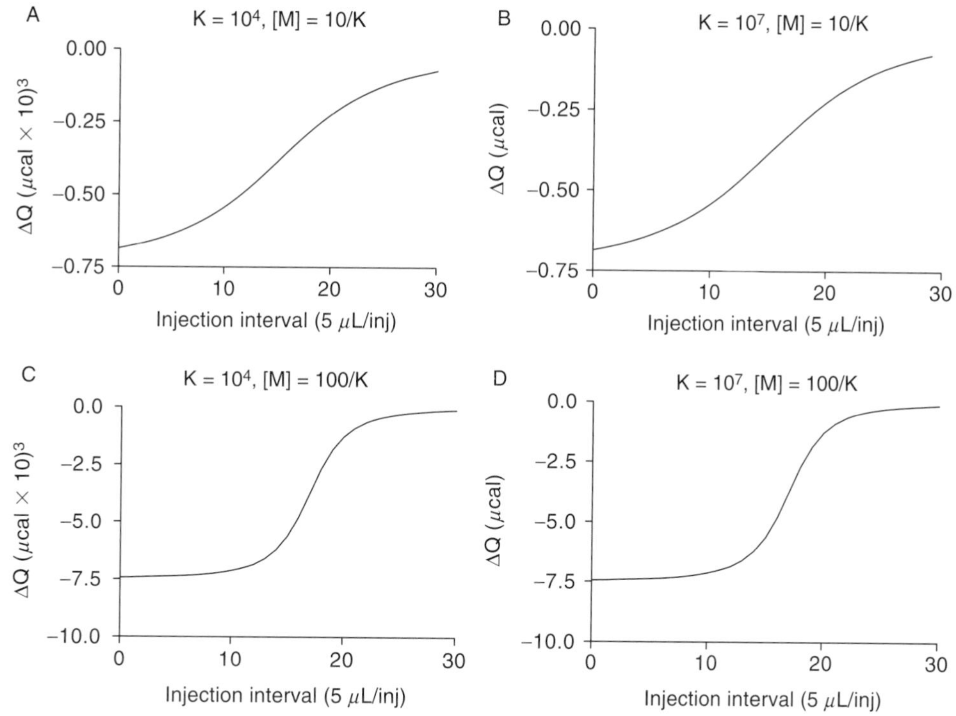

Fig. 4 Four plots demonstrating the relationship between curvature and experimental concentrations. The two top panels (A and B) are simulated with a "*c*" parameter of 10, and the two bottom panels are simulated with a "*c*" parameter of 100. Panels A and C represent a system with a fairly low *K* value of 1×10^4, while panels B and D represent a system with a more robust *K* value of 1×10^7.

multiplied by the equilibrium constant ($c = [M_{tot}] \times K$). Each of the panels in Fig. 4 exhibits a thermogram with acceptable curvature for the nonlinear regression analysis required to obtain an accurate value for *K*. However, the experiment depicted in panel B does not produce enough heat (<0.5 μcal for the largest heats), while the experiment depicted in panel C produces too much heat (>8000 μcal for the largest heats) to yield data sets that are optimal for the determination of the thermodynamic parameters *K*, ΔH, and *n*. Obviously, experiments with reactions having a very large equilibrium constant ($K > 10^8\ M^{-1}$) need to be done at low macromolecule concentrations to produce the required curvature in the thermogram but at high enough concentrations to produce measurable heats. The reverse is true for weak complexes ($K < 10^4\ M^{-1}$) in that here the problem is to achieve macromolecule concentrations where the curvature is appropriate but where the heats are not too large to be accurately measured.

There are several steps to running the ITC experiment. These are (i) planning the experiment (e.g., simulations), (ii) preparing the ligand and macromolecule solutions, (iii) collecting the raw ITC data, (iv) collecting the blank (ligand solution dilution), (v) correcting the raw ITC data, (vi) nonlinear regression of the corrected titration data to provide estimates of the thermodynamic parameter values, and (vii) interpretation of the model data. Each step will be discussed in turn below for an example ITC experiment. In our discussions of running a typical ITC experiment, we will use the binding of a hypothetical ligand, L, to a hypothetical protein, P. The approximate thermodynamic parameters for the simulated system are $K \approx 1 \times 10^5\ \mathrm{M}^{-1}$, $\Delta\mathrm{H} \approx -10$ kcal/mol, and with a stoichiometry of 1:1 at 25 °C. (The data and analysis shown in subsequent sections have been simulated for clarity.)

B. Planning the Thermodynamic ITC Experiment

The first step in running the ITC experiment is to determine the concentrations for the macromolecule and ligand solutions. If the objective of the ITC experiment is only to determine the binding enthalpy change, ΔH, then the only consideration is that the concentration of the ligand will be large enough that an accurately measurable heat effect, $\geq 10\ \mu$cal, will be observed and that the macromolecule concentration will be in excess. In the case of our hypothetical system, these conditions would be met with $[\mathrm{L}] = 2 \times 10^{-4}$ M, and $[\mathrm{P}] = 1.7 \times 10^{-4}$ M. With an injection volume of 5 μl, the heat per injection would be given by Eq. (6), and there would be no curvature in the thermogram:

$$Q_{\mathrm{inj}} = (\Delta H \times [\mathrm{L}] \times V_{\mathrm{injection}}) \tag{5}$$

$$Q_{\mathrm{inj}} = (-10\ \mathrm{kcal\ mol}^{-1}) \times (2 \times 10^{-4} M) \times (5 \times 10^{-6}) \tag{6}$$

If the concentrations of the ligand, [L], were increased to 5×10^{-3} M, the thermogram would show curvature similar to that shown in the upper panel of Fig. 4 ($c = 10$) and an endpoint would be reached after ~20 (5 μl) injections. The integrated heat values for the first injections would now be over $-500\ \mu$cal. Increasing the concentration of the protein, [P], to 1.7×10^{-3} M ($c = 100$) and the ligand concentration to 5×10^{-2} M or so would yield a thermogram showing the same curvature as that shown in the lower panels of Fig. 4. In this last case, the heat observed in the early injections would be too large, over $-5000\ \mu$cal. Figure 5 shows simulated ITC data for an experiment done under the second set of conditions ($c = 10$) where both K and ΔH would be well determined.

C. Running the Thermodynamic ITC Experiment

1. Solution Preparation and Handling

Now that we know the desired concentrations for the macromolecule and the ligand solutions, let us discuss solution preparation and handling. First, since the final results of the ITC experiment depend on exact knowledge of the titrate and titrant solution concentrations, it is imperative that the concentrations be made as

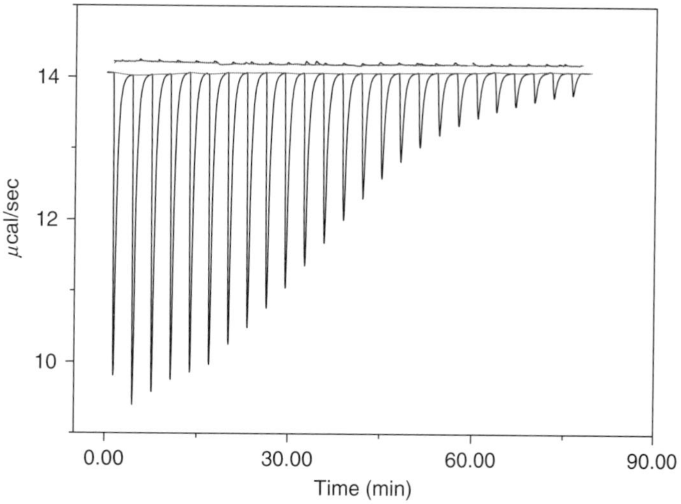

Fig. 5 Simulated ITC Raw Data showing the instrument response for a power compensation ITC instrument. The simulated data represent an exothermic reaction at concentrations producing a reasonable amount of curvature. Above the simulated "experimental" data is a smaller data set representing a typical "instrument blank."

accurately as possible. Perhaps the ITC solutions can be made by volumetric dilution of stock solutions that were made up by weight. Whenever possible the concentrations should be verified by another analytical procedure (e.g., absorbance, kinetic activity, or other analysis).

It is extremely important that the two solutions be matched with regard to composition, for example, pH, buffer, and salt concentration. If the two solutions are not perfectly matched, there may be heat of mixing (or dilution) signals that overwhelm the heat signals for the binding reaction. It is typical that the solution of the macromolecule is exhaustively dialyzed against a large volume of the buffer. The artifact heats of mixing can be minimized by using the dialysate from preparation of the macromolecule solution as the "solvent" for preparation of the ligand solution.

2. Collecting the Raw ITC Data

ITC data collection involves proper identification of optimal experimental run parameters (these should be roughly the same parameters used when simulating data to determine optimal concentrations). The number of injections required for an experiment varies. The number and volume of injections should be adjusted depending on the region of the isotherm that requires the most resolution (using the same concentrations, a titration programmed to deliver a larger number of injections of a smaller volume will result in a better nonlinear regression fit since there are more points included in the titration).

3. Correcting the Raw ITC Data

Obviously, the dialysis/dialysate approach will virtually eliminate the mixing or dilution effects for all solute species in common between the macromolecule and ligand solutions. The exception is that the heat of dilution for the ligand itself must be measured in a blank experiment. In this blank experiment, the ligand solution is titrated into buffer in the sample cell. The heat of dilution of the macromolecule should also be measured in a second blank experiment. This is done by simply injecting buffer from the syringe into the macromolecule solution in the sample cell. Usually the heat of dilution of the macromolecule measured in this way is negligible. To be completely rigorous, a third blank experiment should also be done. This buffer into buffer experiment may be thought of as an instrument blank. The equation to correct the heat data for dilution effects is

$$Q_{\mathrm{corr}} = Q_{\mathrm{meas}} - Q_{\mathrm{dil,ligand}} - Q_{\mathrm{dil,macromolecule}} - Q_{\mathrm{blank}} \tag{7}$$

The blank corrections are for the same injection volumes as used in the collection of the actual titration data. In the case of the ligand/protein titration experiment shown in Fig. 5, the only significant correction is for the dilution of the titrant (the results of the ligand dilution blank experiment are also shown in Fig. 5).

Another complicating reaction encountered in many biological binding experiments results from the release (or uptake) of protons as binding occurs. The released protons are taken up by the buffer conjugate base. The correction for this complicating reaction requires knowledge of the number of protons released (or taken up) and the heat of ionization of the buffer. The equation to correct for the ionization of the buffer is

$$Q_{\mathrm{corr}} = Q_{\mathrm{meas}} - (\Delta H_{\mathrm{ion}} \times n_p) \tag{8}$$

where ΔH_{ion} is the heat of proton ionization for the buffer and n_{p} is the number of protons released on binding 1 mol of ligand. Since n_{p} is typically unknown, this correction would be accomplished by titrations done in two buffers with different heats of ionization. In this case, the complication actually yields additional information regarding the binding reaction. This phenomenon also provides an approach to manipulating the heat signal for a reaction that is accompanied by proton release. By simply using a buffer with a large heat of ionization, the heat signal can be enhanced. Alternatively, the use of a buffer with a small heat of ionization ($\Delta H_{\mathrm{ion}} \approx 0$) could be used to minimize the "artifact signal." To determine an optimal buffer for your system, the heats of ionization (or protonation) of various buffer solutions can be found in references Christensen *et al.* (1976) and Fasman (1976).

Finally, since the generation of bubbles in the sample (or reference) solutions during an ITC experiment will generate spurious heat signals, the solutions should be degassed prior to filling the cell and injection syringe. The ITC manufacturers provide vacuum degassing accessories for this purpose. Precautions need to be taken to avoid boiling the solutions and changing the concentrations. Also ITC manufacturers supply cell loading syringes and instructions on cell filling that should be followed to avoid the problem of introducing bubbles.

4. Example ITC Experiment

The example ITC experiment described is for the binding of our hypothetical ligand, L, to our hypothetical protein, P, the same experiment that was shown in Fig. 5. In the hypothetical experiment, both the ligand and protein were purchased from commercial sources (e.g., Sigma-Aldrich, St. Louis, MO) and used without further purification. The protein solution was prepared by dissolving a weighed amount of the lyophilized powder in acetate buffer and then dialyzed for 16 h at 4 °C against 4 liter of acetate buffer using 3500 MWCO Spectrofluor dialysis tubing. The ligand solution was prepared by dissolving a weighed amount of the pure compound in the acetate buffer dialysate. The final concentrations for both the protein and ligand were determined by the appropriate analytical procedure (e.g., a spectrophotometric assay). Both the titrate and titrant solutions were degassed prior to loading the calorimeter cell and injection syringe.

The ITC experiment was run at 25 °C and was set to deliver 25 (5 μl) injections at 300-sec intervals. The raw ITC data are shown in Fig. 5. Data are shown for one titration experiment in which the ligand solution was added to the protein solution in the cell, and one titrant dilution experiment in which the ligand solution was added to buffer (dialysate) in the cell. The dilution of the titrant is slightly endothermic and contributes less than +0.6 μcal to the total heat observed for the addition of 5 μl of ligand in the protein titration. The titrant dilution represents less than 0.5% of the heat signal observed for the initial titrant additions. The dilution experiment in which buffer was added to the protein solution in the cell is not shown since the heat of dilution of the protein was even less significant than the ligand dilution under the conditions of these experiments.

The dilution-corrected and integrated heat data are shown in Fig. 6. The integrated heat data were fit with a one-site binding model using the Origin-7™ software provided with the MicroCal VP-ITC. The "best-fit" parameters resulting from the nonlinear regression fit of these data are also shown in Fig. 6 along with the fitted curve. The K and ΔH values determined in this experiment would be the appropriate values for the experiment performed under the stated conditions of temperatures, salt concentrations, buffer of choice, and at the specified pH. A more detailed discussion of the nonlinear regression fitting and data interpretation follows.

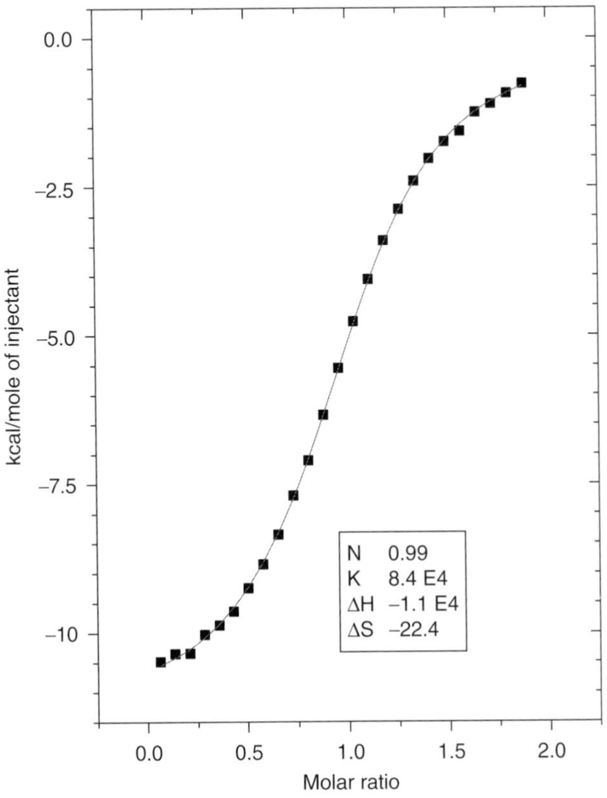

Fig. 6 Simulated data set representing the integration of the raw data shown in Fig. 5. This data has been corrected by subtraction of appropriate blank experiments and then fit with nonlinear regression. The "best-fit" parameters are given in the box in the lower right-hand corner of the plot.

D. Analyzing Thermodynamic Data

In order to analyze ITC data for the relevant thermodynamic parameters, a binding model must first be assumed [e.g., one-site (or n identical sites), two independent sites, or sequential binding]. The analysis of the thermogram is a curve fitting process in which a nonlinear regression procedure is used to fit a model to the data. The model is a mathematical description of a physical, chemical, or biological process that is taking place in the calorimeter and in which the dependent variable (e.g., heat or heat rate) is defined as a function of the independent variable (e.g., moles of titrant added) and one or more model parameters. In the case of binding experiments, the model is formed from the equilibrium constant and mass balance equations. Nonlinear regression is used along with the model equations to determine the best values of the fitting parameters (e.g., K, ΔH, and n). The goal is to model the experimental data within expected experimental error, using the simplest model, and a model that makes sense in the light of what is already known about the system (e.g., stoichiometry). The model should help to understand the actual chemistry, biology, or physics of the system being studied. It is important to note that one of the authors of this chapter has often said "All models are wrong, but some are useful."

The nonlinear regression analysis of ITC data is an iterative process. The first step in the process is to make initial estimates for each of the parameters in the model equation. Using these values, a fit or theoretical curve is generated and then compared to the actual data curve. An error function is calculated that is the sum of the squared deviations between the data and the model curve. An accepted algorithm is then used to adjust the fit parameters to move the calculated curve closer to the data points. This process is repeated over and over until the error function is minimized or insignificantly changes with subsequent iterations. If the error function cannot be minimized to an acceptable value, that is, an error square sum that is consistent with the expected experimental error, then another model must be tried. It is important to note here that "the best-fit" parameter values may depend on the starting estimates chosen in the first step or on the stopping criteria of the last step.

Figure 5 shows the raw ITC data for the hypothetical ligand protein titration, while Fig. 6 shows the integrated heat data along with a nonlinear regression fit to a one-site binding model. The line through the data points corresponds to the theoretical heat produced for 1:1 complex formation between the ligand and the protein and the best-fit values for the parameters K, ΔH, ΔS, and n are listed in the box in the lower right corner of the plot. The nonlinear regression analysis shown was performed using the simulated data in Fig. 5 and the one-site reaction model in the Origin 7 ITC software package provided by the ITC manufacturer, in this case MicroCal. Both Calorimetry Sciences and MicroCal provide for more complex binding models in their software packages, for example, single set of identical sites, two sets of independent sites, sequential and binding. They have also made it possible for the experienced user to add models within limits (i.e., the mathematics engine in Origin is unable to solve polynomials higher than third degree). In order to better understand the nonlinear regression (or curve fitting) analysis of ITC data, we will first discuss the "one-site" model in more detail.

The thermogram generated in the ITC experiment is a simple summation of all of the heat-producing reactions that occur as an aliquot of titrant is added. The initial heats are larger than the heats for subsequent additions since at the beginning of the titration there is a large excess of empty or unpopulated binding sites. Initial heats most typically are the result of complete reaction of the added ligand. As the titration proceeds, less and less of the added ligand is bound and there are three species existing in solution: free ligand, unoccupied binding sites, and the ligand/protein complex. The heat produced in the ITC experiment is linearly dependent on the ΔH or the reaction and nonlinearly dependent on the K.

The ITC thermogram for a generic binding process is modeled by Eqs. (9) and (10) (Eatough *et al.*, 1985; Freyer *et al.*, 2006, 2007a,c; Lewis and Murphy, 2005).

$$\Theta_j = \frac{[\mathrm{L}]K_j}{1 + [\mathrm{L}]K_j} \quad (9)$$

$$L_t = [\mathrm{L}] + P_t \sum_{j=1}^{k} (n_j \Theta_j) \quad (10)$$

Equations (9) and (10) describe the equilibrium and mass balance relationships for the system being studied, where Θ_j is the fraction of site *j* occupied by ligand, L_t is the total ligand concentration, [L] is the free ligand concentration, P_t is the total macromolecule concentration, K_j is the binding constant of process *j*, and n_j is the total stoichiometric ratio for process *j*. Each of the equations is defined for all potential binding sites, and solutions for any multiple site binding process can be defined. Substituting (9) into (10) and expanding the polynomial in terms of the indeterminant [L] results in a $(k + 1)$ degree polynomial. Thus, in order to determine a solution for a one-site-independent binding process, roots of a second degree polynomial must be found. Substitution of [L] into Eq. (9) allows the fraction of binding site *j* that is occupied to be calculated.

$$Q = P_t V_0 \left(\sum_{j=1}^{k} n_j \Theta_j \Delta H_j \right) \quad (11)$$

$$\Delta Q(i) = Q(i) - Q(i-1) \quad (12)$$

The total heat produced can be calculated from Eq. (11), where V_0 is the initial volume of the sample cell and ΔH_j is the molar enthalpy change for process *j*. The differential heat is defined by Eq. (12), where *i* represents the injection number. Nonlinear regression was performed on the parameters K_1, n_1, and ΔH_1 to obtain a best fit to the experimental data shown in Figs. 5 and 6. The good news is that this whole curve fitting process is transparent to the beginning or casual user of the ITC technique. It is only when more complex models are required to fit the binding data that the process becomes much more difficult.

There are many nonlinear regression algorithms available but all result in almost the same answer. In Fig. 7, we present a 3-D plot of the error square sum surface that one would expect to get from a one-site analysis. The *z*-axis represents the square of difference between experimental and theoretical points, the *x*-axis the log of the equilibrium constant *K*, and the *y*-axis the value for ΔH. It is easy to see that the minimum in the error function is at the bottom of the net. This best result is unambiguous in that single values of *K* and ΔH are found that yield the smallest error.

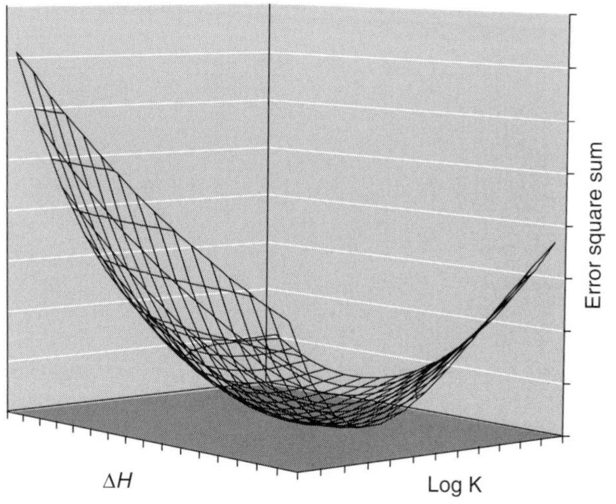

Fig. 7 Error square sum surface for a typical "one-site" nonlinear regression fit. "Best-fit" parameters for a given nonlinear regression will lie in the minimum of the error surface. Searching along the gridlines (2-D parabolas) would be useful in generating confidence intervals (see Chapter 24 by Johnson, this volume).

There are several chapters in this volume that discuss ligand-binding reactions, experimental techniques other than ITC, and data fitting in general. The information in these chapters is complimentary to the discussions of ligand binding and data fitting presented in this chapter. See for example, Chapter 1 by Garbett and Chaires, Chapter 24 by Johnson, Chapter 23 by Tellinghuisen, and Chapter 3 by Farial *et al.*, this volume.

E. Models

The equilibrium constant and mass balance expressions for the one-site (n identical sites), two-site (two independent sites), and sequential sites models are defined differently as shown below:

One-set (or n identical sites):

$$K = \frac{\Theta}{(1-\Theta)\cdot[\mathrm{L}]} \tag{13}$$

Two sets of independent sites:

$$K_1 = \frac{\Theta_1}{(1-\Theta_1)\cdot[\mathrm{L}]} \text{ and } K_2 = \frac{\Theta_2}{(1-\Theta_2)\cdot[\mathrm{L}]} \tag{14}$$

Sequential sites:

$$K_1 = \frac{[\mathrm{PL}]}{[\mathrm{P}][\mathrm{L}]}, K_2 = \frac{[\mathrm{PL}_2]}{[\mathrm{PL}][\mathrm{L}]}, \text{ and } K_3 = \frac{[\mathrm{PL}_3]}{[\mathrm{PL}_2][\mathrm{L}]} \tag{15}$$

$$L_\mathrm{t} = [\mathrm{L}] + [\mathrm{PL}] + 2[\mathrm{PL}_2] + 3[\mathrm{PL}_3] \tag{16}$$

$$P_\mathrm{t} = [\mathrm{P}] + [\mathrm{PL}] + [\mathrm{PL}_2] + [\mathrm{PL}_3] \tag{17}$$

These models are addressed in the analysis software provided by CSC® Bindworks™ and MicroCal® ITC Origin™. More complicated models, for example, three independent sites or fraction-sites would require the user to write their own analysis routines.

Theoretical ITC thermograms are shown in Fig. 8 for three different systems in which 2 mol of ligand are bound to two independent binding sites on the macromolecule.

In the upper panel, the thermogram looks as if there might be either two independent or two identical sites, each described by a similar value for K and ΔH. In reality, we can only determine the weakest value for K and a single of ΔH value for both sites (ΔH_1 and ΔH_2 must be equal) in this system. If we look at the middle panel in the figure, we can see that again the total binding stoichiometry is 2:1, but the more tightly bound ligand also has the most exothermic enthalpy change. In the lower panel, we show a simulation for two independent sites in which the weaker binding process has the more exothermic ΔH.

Let us move on to a slightly more complicated three-site simulation as shown in Fig. 9.

Although perhaps not immediately obvious, there are three different binding processes present in the data set shown. The highest affinity process has the largest exothermic ΔH, while the next two processes have overlapping K values and decreasing values for their ΔH values. The line through the simulated data points represents the nonlinear fit to the data. The best-fit values returned from the nonlinear regression analysis were within experimental error of the values used to generate the simulated data.

F. Error Analysis/Monte Carlo

The statistical significance of the best-fit parameters is always a matter of concern. It is important to know that the model parameters reported are actually meaningful. This becomes more of a concern as the models get more complicated and the number of fitting parameters increases. One way in which the certainty of the parameters can be tested is to perform a Monte Carlo analysis. In order to give the reader confidence in the fitting of ITC data to yield a large number of

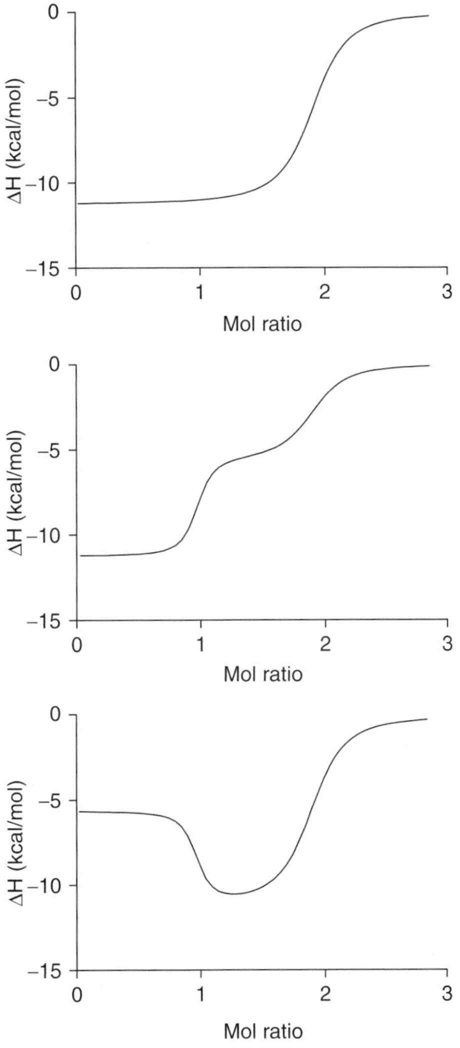

Fig. 8 Representation of three possible experimental observations for a system where two ligand molecules bind to a receptor molecule. In all three cases, the stoichiometry demonstrates that two ligand molecules are binding to the receptor molecule. The top panel represents a system where the enthalpy change and binding affinity for both ligands are close enough to be thermodynamically indistinguishable. The middle panel represents a system where the binding site with higher affinity is accompanied by a more exothermic enthalpy change. The bottom panel represents a system where the higher binding affinity site demonstrates a less exothermic enthalpy change than the lower binding affinity site.

thermodynamic parameters, we performed a Monte Carlo analysis on the simulated three-sites data shown in Fig. 9. A Monte Carlo analysis is equivalent to running a very large number of actual experiments and then comparing the

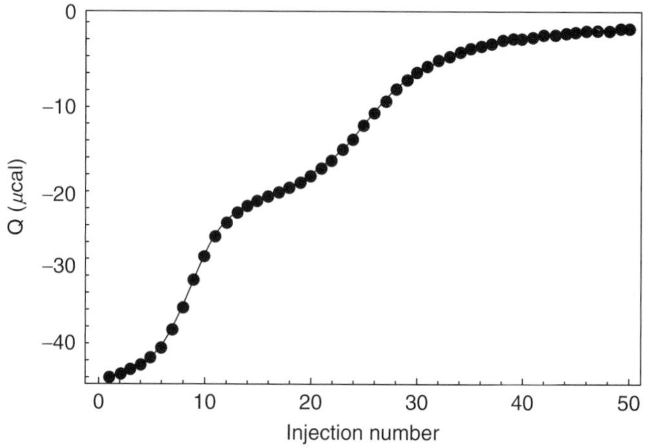

Fig. 9 A simulated plot representing a system where there are three ligand molecules binding to a receptor molecule. The highest affinity binding site has the largest exothermic enthalpy change, while the binding affinity and enthalpy changes for binding ligands to the other two binding sites are indistinguishable.

results of the fits from this large number of experiments. In our Monte Carlo analysis of the three-sites data set, we ran the equivalent of 1000 ITC virtual experiments. The Monte Carlo procedure involves several steps: (i) the generation of a perfect data set (see Fig. 9), (ii) adding random (Gaussian) noise correlated to instrumental error to the data, (iii) performing the three-sites nonlinear regression on the noisy virtual data set, and (iv) repeating the virtual experiment 1000 times. The noise level employed in this analysis was ± 0.25 μcals, which is the expected error in ITC measured heat values. This error was randomly Gaussian and added to each point in the "perfect" data set. The 1000 sets of "best-fit" parameters were placed into a statistical program and the distributions analyzed to determine the statistical error of each parameter. Table I lists the values of K_j and ΔH_j that were used to generate the perfect data set (Fig. 9) and the values that were returned from the Monte Carlo analysis described above.

Clearly, the three-sites model (and other complex models) can be applied to ITC and the nonlinear regression can yield best-fit parameters with high levels of certainty. There are a number of ways to view the parameter distributions from a Monte Carlo analysis and in Fig. 10 we show a Saroff Plot in which the listed parameter is changed over a range of values and the nonlinear regression performed on the remaining parameters.

While it is obvious that some parameters are better determined than others (e.g., K_1, K_2, ΔH_1, ΔH_2, and n_3), and some are cross correlated, most show a clearly defined minimum in the error function and thus are well determined, although the error bars may not be symmetric. The cross correlation of parameters can be visualized by plotting one parameter versus another as described by Correia and Chaires (1994).

Table I
The Results of Monte Carlo Analysis of the Simulated "Three-Sites" Data Shown in Fig. 9

Parameter	Model value	Calculated value
K_1	5.0×10^7	$5.02 \times 10^7 \pm 3.5 \times 10^6$
K_2	1.0×10^6	$1.008 \times 10^6 \pm 7.9 \times 10^4$
K_3	3.0×10^4	$3.02 \times 10^4 \pm 3.0 \times 10^3$
ΔH_1	−10 kcal/mol	−10 ± 0.04 kcal/mol
ΔH_2	−5 kcal/mol	−5.0 ± 0.05 kcal/mol
ΔH_3	−2 kcal/mol	−2.0 ± 0.12 kcal/mol

Column 2 relates the "best-fit" parameters for fitting the simulated data, and column 3 shows the result of the Monte Carlo analysis 1000 simulated experiment average ±1 standard deviation.

G. Summary

The ITC method for the simultaneous determination of K and ΔH is certainly an important technique for the characterization of biological binding interactions. The emphasis of this chapter was to guide users new to the technique through the process of performing an ITC experiment and to point out that care must be taken both in the planning of the experiment and in the interpretation of the results. The conclusions that can be drawn from the above discussions of the ITC-binding experiment, the nonlinear fitting of ITC data, and data interpretation are listed below.

- It is important in planning the ITC-binding experiment that reasonable concentrations be chosen for the macromolecule and the ligand. This is most easily done by simulating the thermogram with reasonable guesses for K and ΔH (although the guess for ΔH is less critical).
- The linear parameters ΔH and n will be better determined than the nonlinear parameter K.
- The best results will be obtained at $10/K \leq [\mathrm{M}] \leq 100/K$, and $[\mathrm{L}] \approx 20\text{–}50 \cdot [\mathrm{M}]$, subject to solubility and heat signal considerations.
- The best results will be obtained when the initial integrated heat(s) are larger than 10 μcal.
- The number of points in the titration is not critical as long as the collection of more points does not reduce the measured heat to the point where random error in the ∂q values becomes significant.
- Fitting for the fewest number of parameters is always helpful in reducing the uncertainty in the fitted parameters, for example, constrain n to fit for ΔH and K, or constrain ΔH and n to fit for K.
- Titrant and titrate concentrations must be accurately known. (Nonintegral values for n are often the result of concentration errors. Errors in titrate concentration contribute directly to a similar systematic error in n. Errors in titrant concentration or titrant delivery contribute directly to similar errors in ΔH.)

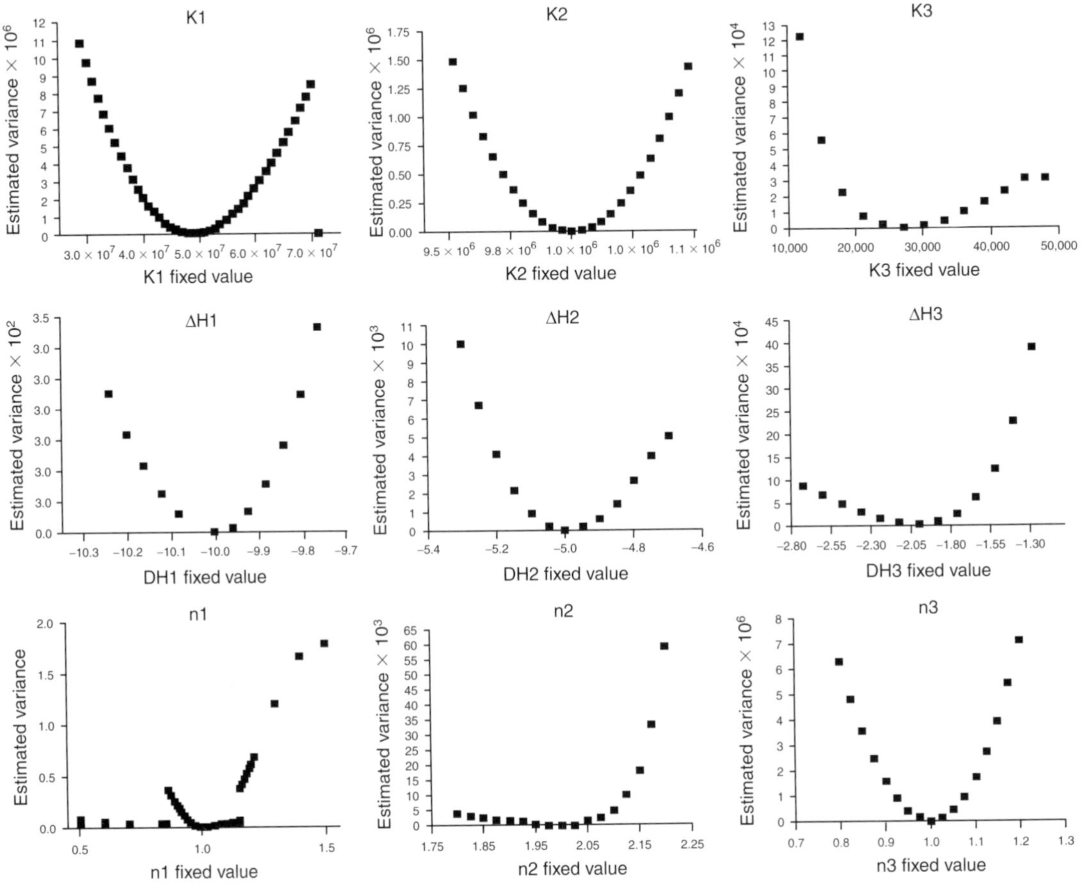

Fig. 10 Saroff distribution of error generated from the Monte Carlo analysis performed on the simulated data given in Fig. 9. These plots show the distribution of error in each of the nine fitting parameters demonstrating that some parameters can be better determined than others.

IV. Kinetic ITC Experiments

A. Reaction Rate Versus Heat Rate

A typical enzyme substrate interaction is illustrated in Fig. 11.

In biological terms, ligands of interest other than the normal substrate could be inhibitors, cofactors, coenzymes, prosthetic groups, metal ions, or other small molecules. However, since the point of these experiments is to probe the kinetics and means by which substrate is converted to product, the typical experiment involves enzyme, substrate, and possibly other reactants involved in the enzyme-catalyzed reaction. One difference, in comparison to the binding experiments,

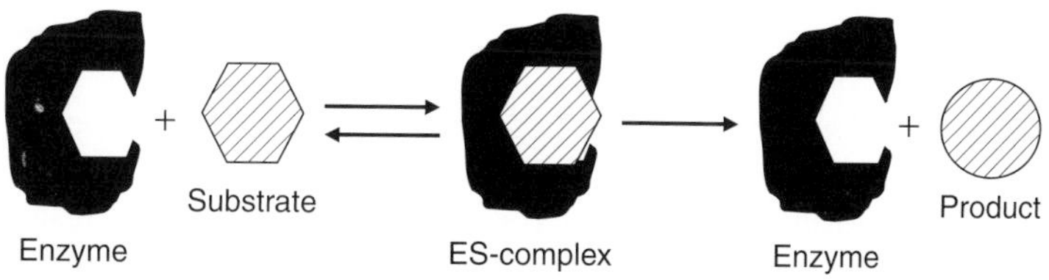

Fig. 11 A simplified model of a typical enzyme/substrate interaction. The substrate in this representation is shown to geometrically match the active site of the enzyme to indicate a specific interaction. The substrate is then converted to product leaving a regenerated enzyme.

is that the enzyme-catalyzed heats of reaction are usually much larger than the heats observed for noncovalent binding interactions. A fundamental understanding of the pictured reaction would require at a minimum knowledge of the Michaelis constant, K_m, and the turnover number, k_{cat}. Again a richer understanding of the enzyme-catalyzed reaction would be established if the substrate-binding constant (not exactly the same as $1/K_m$), the enthalpy change, ΔH, for the reaction (S→P), and any mechanistic information (e.g., hyperbolic vs sigmoidal dependence on [S], response to various types of inhibitors, effects of temperature, pH, or other solution conditions) were known. The ITC experiment can provide information in all of these areas. The following equations are provided as a review of the relevant kinetic relationships:

$$E + S \underset{k_{-1}}{\overset{k_1}{\rightleftharpoons}} ES \overset{k_2}{\rightarrow} E + P \tag{18}$$

$$K_m = \frac{(k_{-1} + k_2)}{k_1} \tag{19}$$

$$K_{eq} = \frac{[ES]}{[E][S]} \approx \frac{1}{K_m} \tag{20}$$

$$k_{cat} = \frac{v_{max}}{[E]_t} \tag{21}$$

$$v_0 = v_{max}\frac{[S]}{(K_m + [S])} = \frac{k_{cat}[E]_t[S]}{(K_m + [S])} \tag{22}$$

where k_1, k_{-1}, and k_2 are the rate constants for the forward and reverse reactions in the reaction scheme, K_m is the Michaelis constant, K is the binding constant, k_{cat} is the turnover number, v_0 is the initial velocity, v_{max} is the maximal velocity (when $[ES] = [E]_t$), and [X] is the molar concentration of species X.

The reaction rate (e.g., v_0 or v_{max}) is typically expressed in moles of product formed per unit of time (or moles of substrate consumed per unit of time). The raw calorimetric signal is expressed as a power (e.g., μcal/sec or μJ/sec). This heat rate is simply equal to the reaction rate multiplied by the enthalpy change for the reaction as shown in Eq. (23).

$$\frac{\delta Q}{\delta t} = \frac{\delta n}{\delta t} \times \Delta H \tag{23}$$

The raw calorimetric signal is thus a direct measure of the reaction rate making the calorimeter an ideal instrument for kinetic studies. The enthalpy changes for most enzyme-catalyzed reactions range from -10 to -100 kcal/mol, allowing reaction rates from 10 to100 pmol/sec to be accurately measured.

Kinetic data for an enzyme-catalyzed reaction described by the Michaelis-Menton rate equation are simulated in Fig. 12.

In these simulations, the total enzyme concentration, $[E]_t$, is 5×10^{-6} M. Substrate was added to produce the variable concentrations as listed on the x-axis. The dependence of the reaction rate on K_m or k_{cat} is illustrated by holding one variable constant (either K_m or k_{cat}) and plotting the rate curves for four different values of the other variable. In Fig. 12A, the k_{cat} value is 80 sec^{-1} for all of the rate curves shown and the values of K_m appear on the plot. In Fig. 12B, the K_m value is 0.075 M^{-1} for all of the rate curves shown and the values of k_{cat} used for the four simulations are shown on the plot. It should be obvious that calorimetric

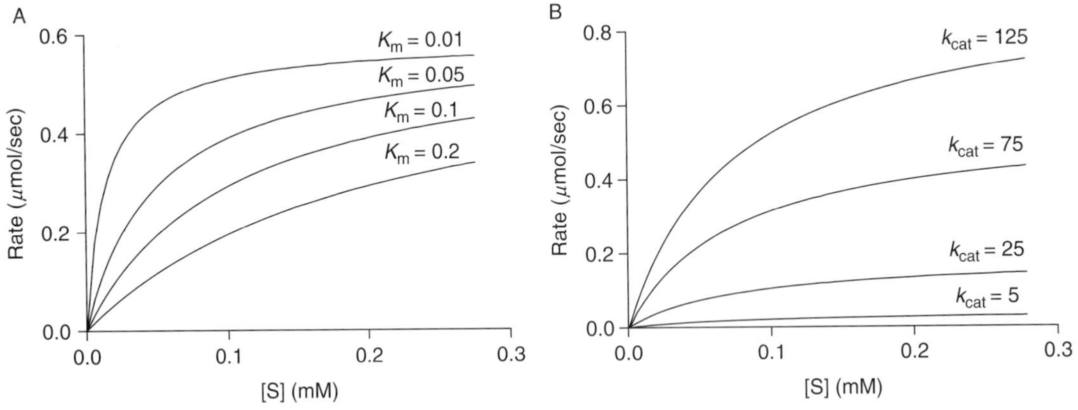

Fig. 12 Representation of the effect of K_m and k_{cat} on the hyperbolic curve shape of kinetic ITC data. The data in panel A vary with respect to K_m with a constant k_{cat} value of 80 sec^{-1}. The data in panel B vary with respect to k_{cat} with a constant K_m value of 0.075 M^{-1}. The data in both panels were simulated for an $[E]_t = 5\ \mu$M.

data (the rate curve) could easily be curve fit to yield the appropriate values for both the nonlinear parameters K_m and k_{cat} and the linear parameters $[E]_t$ and ΔH. The value given for $[E]_t$ illustrates the sensitivity of the ITC technique and is typical for these types of experiments. The ΔH value is not listed as its influence on the raw data is linear and the only concern here is that the heat rate would be detectable in the ITC.

B. Planning the Experiment

As with the ITC-binding experiments described previously, there are several steps to running an ITC kinetic experiment. These steps are (i) planning the experiment, (ii) preparing the substrate and enzyme solutions, (iii) collecting the raw ITC kinetic data, (iv) collecting the blank (substrate and protein dilutions), (v) correcting the raw ITC data, and (vi) analysis of the corrected titration data to provide estimates of the kinetic parameter values.

There are two different ITC methods for performing an enzyme kinetic experiment: [single injection (DIE) and multiple injections (or continuous)]. The first step in planning either ITC kinetic experiment is to determine optimal concentrations for the enzyme and substrate solutions. ITC kinetic experiments require a concentration of substrate large enough to produce an accurately measurable heat rate (>10 μcal/sec for single injection experiments and >2 μcal/sec for the initial injections in a multiple injection experiment). The initial heat rate produced per injection can be calculated using Eq. (23) and estimates the reaction rate and the ΔH.

When performing a single injection ITC kinetic experiment, the substrate solution is injected into the cell containing the enzyme solution producing a heat response which eventually returns to baseline after all of the substrate has reacted. A second injection of substrate can be made to collect additional information about the reaction (e.g., the presence or absence of product inhibition). Simulated data for a single injection experiment are shown in Fig. 13.

The thermogram shown is for an experiment in which the enzyme concentration in the ITC cell is 5 μM, $\Delta H = -50$ kcal/mol, $K_m = 0.075$ M^{-1}, and $k_{cat} = 80$ sec^{-1}. The substrate solution concentration was 1 mM and the injected volume was 40 μl. The substrate was completely consumed after about 1500 sec. The analysis of these data will be described in a later section. Figure 14 shows simulated data for four different experiments in which all of the parameters except ΔH are the same as those for the data shown in Fig. 13.

Satisfactory data are obtained in the three simulations with the larger ΔH values. In the case of the experiment with the lowest ΔH value (−5 kcal/mol), the enzyme concentration or the amount of substrate would need to be increased to yield an analyzable data set with an adequate initial heat rate (>10 μcal/sec).

The continuous method of ITC kinetic experimentation makes use of multiple titrations that are spaced such that subsequent titrant additions are performed when the heat rate has reached a steady state. It is important that each subsequent

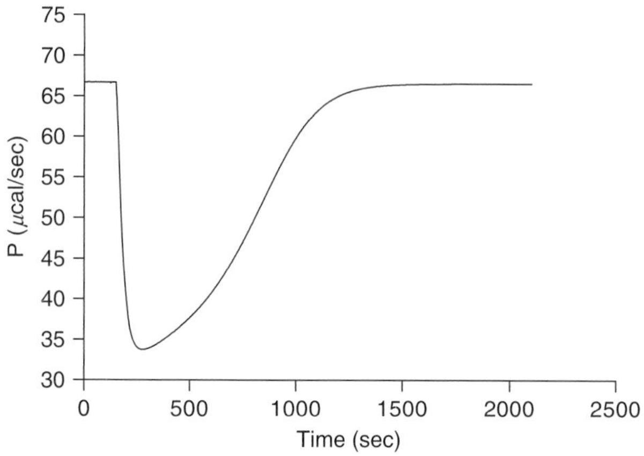

Fig. 13 Simulation of a single injection kinetic ITC experiment. This simulation was for the following parameters: $k_{cat} = 80\ sec^{-1}$, $K_m = 0.075\ M^{-1}$, $\Delta H = -50$ kcal/mol, and $[E]_t = 4.5\ \mu M$.

addition of substrate is made prior to significant reaction of the substrate. The difference in the signal plateau between each injection is used to determine the reaction rate at that step. This information is used to create a plot of the reaction rate (in units of power) versus total substrate concentration. This plot can then be fit to determine the kinetic parameters for the reaction. The analysis of the data obtained with the continuous method assumes that there is no significant substrate degradation during the time between injections. The main advantage of the

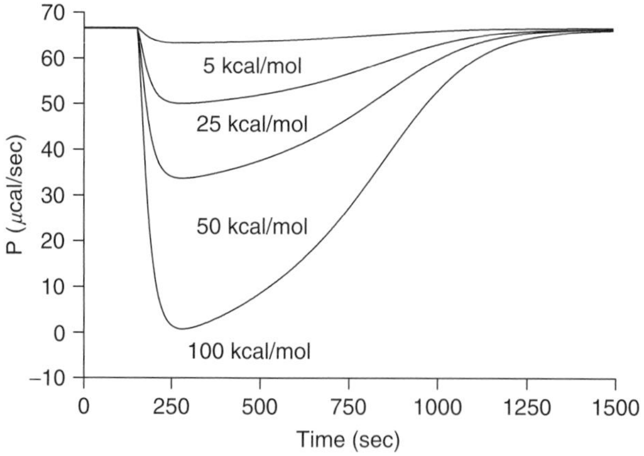

Fig. 14 Representation of the effect of ΔH on the size of single injection curves. All of these curves have the following parameters: $k_{cat} = 80\ sec^{-1}$, $K_m = 0.075\ M^{-1}$, and $[E]_t = 4.5\ \mu M$. The ΔH values vary from 5 to 100 kcal/mol.

continuous method over the single injection method is higher accuracy in determining kinetic parameters. The only disadvantage is that it is unable to determine ΔH_{app} because the titration points do not allow for the complete reaction of the substrate prior to adding more substrate. This can be overcome by performing a single injection experiment to determine the ΔH_{app}.

Simulated data for a multiple injection experiment are shown in Fig. 15.

The thermogram shown is for an experiment in which the enzyme concentration in the ITC cell is 5 μM, $\Delta H = -50$ kcal/mol, $K_m = 0.075$ M^{-1}, and $k_{cat} = 80$ sec^{-1}. The substrate solution concentration was 1 mM and there were 30 $\times$ 3 μl injections performed at 100-sec intervals. The analysis of these data will be described in a later section. Figure 16 shows simulated data for four different experiments in which all of the parameters except ΔH are the same as those for the data shown in Fig. 15.

Satisfactory data are obtained in the two simulations with the larger ΔH values. In the case of the experiments with the two lowest ΔH values (-5 and -25 kcal/mol), the enzyme concentration or the amount of substrate would need to be increased to yield an analyzable data set with an adequate initial heat rate (>2 μcal/sec).

C. Running the Kinetic ITC Experiment

1. Solution Preparation and Handling

Once the necessary concentrations have been estimated, the next step is to prepare the ligand and macromolecule solutions. The solution preparation techniques for both single injection and continuous kinetic experiments are identical. As we discussed in the section on ITC-binding experiments, it is imperative that

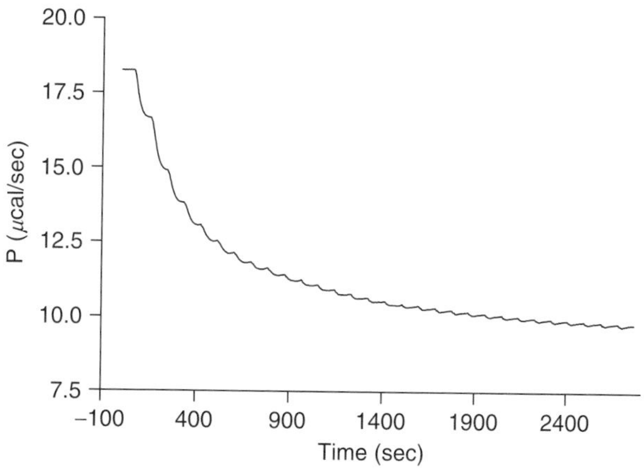

Fig. 15 Simulation of a multiple injection kinetic ITC experiment. This simulation was for the following parameters: $k_{cat} = 80$ sec^{-1}, $K_m = 0.075$ M^{-1}, $\Delta H = -50$ kcal/mol, and $[E]_t = 4.5$ μM.

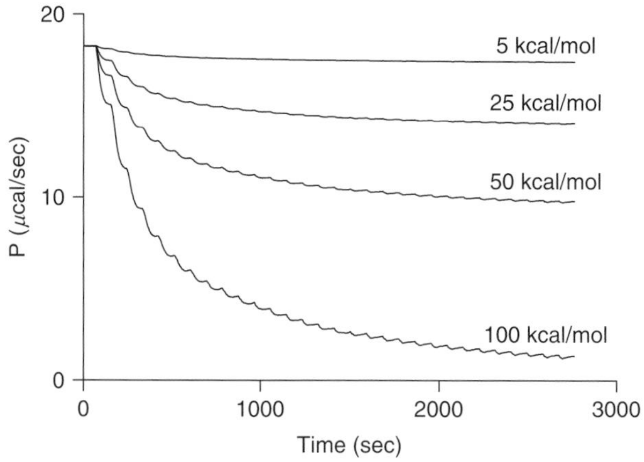

Fig. 16 Representation of the effect of ΔH on the size of multiple injection titration curves. All of these curves have the following parameters: $k_{cat} = 80\ sec^{-1}$, $K_m = 0.075\ M^{-1}$, and $[E]_t = 4.5\ \mu M$. The ΔH values vary from 5 to 100 kcal/mol.

accurate concentrations are known for both the substrate and the enzyme. Concentration errors in either [S] or $[E]_t$ will result in incorrect values for ΔH_{app}, K_m, and k_{cat}. These solutions should be prepared as accurately as possible, and whenever possible the concentrations should be verified using an analytical procedure (e.g., UV–VIS absorbance). In addition to extremely accurate knowledge of the concentrations of both the substrate and the protein, it is also essential that these two solutions are identical with respect to buffer composition (salt, pH, and so on). The best solution to this problem is to exhaustively dialyze the macromolecule (e.g., dialyzing 1 ml of concentrated protein solution in two 1-liter changes of buffer over the course of 48 h) and prepare the substrate solution in the resulting dialysate (in the dialysis example above some of the second liter of buffer would be saved to prepare substrate solutions). As discussed previously with the ITC-binding experiments, both the protein and substrate solutions should be degassed prior to filling the cell and the injection syringe. This serves to minimize the generation of bubbles and an accompanying erroneous heat signal.

2. Collecting Raw ITC Data

The collection of raw data for ITC kinetic experiments is similar to the experimental protocol for binding experiments. The main difference is the spacing between injections. If performing a single injection experiment, the instrument is programmed to perform one injection. If two injections are desired (to investigate product inhibition for example), then the spacing between injections should be quite large (e.g., 30 min) to allow ample time for the baseline to return to its starting point. If the multiple injection method of kinetic analysis is used, the instrument should be programmed to perform as many injections as needed, and the spacing between injections should be set such that the signal has achieved a steady state maximum heat rate when the next injection takes place. This will likely require the experimenter to watch the titration closely and tailor the spacing in between injections to the specific reaction and concentrations being studied (an injection interval of 100 sec was used in producing the data set shown in Fig. 15).

3. Correcting the Raw ITC Data

As discussed with ITC-binding experiments, performing exhaustive dialysis of the protein and then preparing the substrate solution with the resulting dialysate will reduce heats of dilution significantly. However, it is still important to correct for the heats of dilution of the substrate and enzyme solutions. The first blank experiment is performed by injection of the substrate solution into buffer. It is important that the volume injected here is the same as the injection volume(s) used in the kinetic experiment. To perform the blank experiment for the dilution of the enzyme, buffer is injected into the protein solution at the same concentration used in the kinetic experiment. A rigorous approach would also include a third blank experiment where buffer is injected into buffer. All of these blank heat effects would be subtracted to yield the corrected heat (heat rate) of reaction.

D. Analyzing the Kinetic ITC Data

The ITC kinetic data are analyzed using nonlinear regression analysis techniques. Commercial calorimeters come with the software (programs) that is required for the analysis of ITC kinetic raw data. The analysis is transparent to the user who only needs to determine which type of experiment to perform (single injection, multiple injection, substrate only, substrate + inhibitor, etc.) and to select the appropriate analysis routine. The linear parameters are $[\mathrm{E}]_\mathrm{t}$ and ΔH, and the nonlinear parameters are K_m and k_cat.

The analysis of single injection data (like that shown in Fig. 15) is based on first estimating the molar enthalpy change for the conversion of substrate to product. The value of the enthalpy change, ΔH, is determined by integration of the thermogram and Eq. (24).

$$\Delta H = \frac{\int_{t=0}^{\infty} \frac{\mathrm{d}Q}{\mathrm{d}t}\mathrm{d}t}{[\mathrm{S}]_{t=0} V_\mathrm{cell}} \tag{24}$$

This equation describes the molar enthalpy change for the reaction, S→P, where $\partial Q/\partial t$ is the excess power or reaction heat rate, $[\mathrm{S}]_{t=0}$ is the concentration of ligand at time zero, and V_cell is the volume of the calorimeter cell. The initial velocity (V_0) can be estimated immediately after an injection of substrate. The maximal velocity, V_{max}, can be estimated immediately after an injection of enough substrate to saturate the enzyme. K_m can be estimated form ITC-binding data under conditions where the substrate binds but is not converted to product (titrations done in the presence of a noncompetitive inhibitor or in the absence of a required reactant, cofactor, or coenzyme). The initial rate, obtained under substrate saturating conditions, can be used with the total macromolecule concentration to estimate k_{cat} and K_m for a system that follows Michaelis-Menton kinetics (see Eq. 22). Figure 17 shows a fit of the data in Fig. 13, first converted to reaction rate as a function of [S] and then fit to a Michaelis-Menton model.

In this case, the points shown represent data taken at equal intervals from the continuous curve of the thermogram from a single injection of substrate. The best-fit parameter values returned from the nonlinear regression ($K_m = 0.077\ M^{-1}$ and $k_{cat} = 89\ sec^{-1}$) are very close to the parameters that were used to generate the simulated data set ($K_m = 0.075\ M^{-1}$ and $k_{cat} = 80\ sec^{-1}$).

The analysis methods for continuous ITC kinetic experiments are slightly different than those used for single injection kinetic experiments. As mentioned previously, ΔH cannot be determined from a continuous experiment since the continuous method does not allow for the complete reaction of substrate prior to starting subsequent injections. If the ΔH value has been determined from a single injection experiment, V_0 can be calculated using the same equation used for the single injection experiment. The advantage of the continuous experiment is that it produces discrete values of V_0 as a function of total substrate concentration for well-determined values of the substrate concentration (i.e., the total substrate concentration added to the solution is used instead of a single concentration of substrate at time zero $[S]_{t=0}$). This allows iteration on the rate equation with fixed

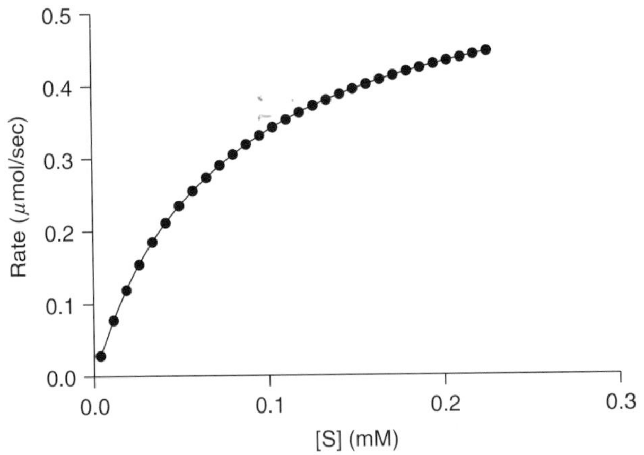

Fig. 17 Representative fit of the simulated single injection data shown in Fig. 13. The data were fit with nonlinear regression analysis resulting in "best-fit" parameters of $k_{cat} = 88.6\ sec^{-1}$ and $K_m = 0.077\ M^{-1}$.

values for the substrate concentration leading to less statistical error in the best-fit parameters for k_{cat} and K_m. Figure 18 shows a fit of the data in Fig. 15, first converted to reaction rate as a function of [S] and then fit to a Michaelis-Menton model.

In this case, the points shown are for each of the 30 substrate injections. The best-fit parameter values returned from the nonlinear regression ($K_m = 0.072\ M^{-1}$ and $k_{cat} = 85\ sec^{-1}$) are again very close to the parameters that were used to generate the simulated data set ($K_m = 0.075\ M^{-1}$ and $k_{cat} = 80\ sec^{-1}$).

E. Models

The only model that is implemented in currently available software is for systems that follow simple Michaelis–Menton kinetics (see Eq. 22). The canned programs assume, for example, that there is no product inhibition. In principle, an advanced user could write more complicated rate expressions and use a program like Mathematica 5.0™ to perform the nonlinear regression (curve fit) of the experimental thermogram to the more complex model. The calorimetric data are independent of the model and a direct measure of the reaction rate as a function of time, [S], $[E]_t$, and any competing reactions, for example, the influence of regulatory or inhibitory compounds.

F. Summary

ITC kinetic experiments take advantage of the fact that a calorimeter is a universal detector (almost all chemical reactions are accompanied by a change in heat). The recent literature has demonstrated the use of ITC in characterizing a number of enzymes by determining kinetic (as well as thermodynamic) constants. ITC experiments can be done on solutions that are either homogeneous or

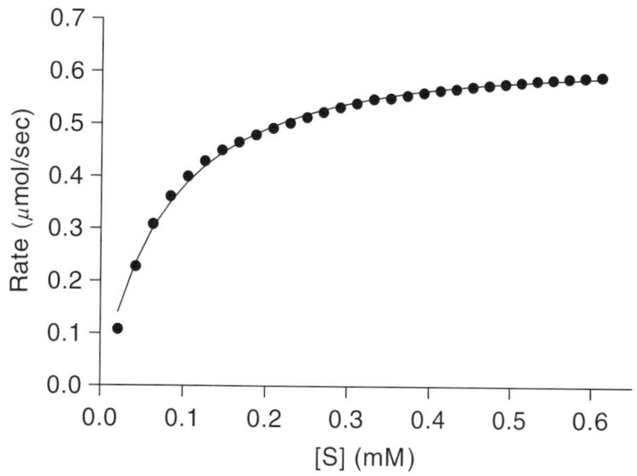

Fig. 18 Representative fit of the simulated multiple injection data shown in Fig. 15. The data were fit with nonlinear regression analysis resulting in "best-fit" parameters of $k_{cat} = 71.3\ sec^{-1}$ and $K_m = 0.085\ M^{-1}$.

heterogeneous (e.g., cell suspensions), or are turbid or opaque. ITC is also sensitive enough (noise levels >0.01 μcal/sec) that enzyme concentrations and volumes required are similar to those needed for spectrophotometric analyses. There currently exists a database of ITC enzyme studies that can be used to design experiments on as yet unstudied enzymes with minimal method development. The ITC technique generates a complete reaction rate curve in a single experiment and the values of $[E]_t$, ΔH, K_m, and k_{cat} have been shown to be in good agreement with the values for these parameters determined by other techniques.

The conclusions that can be drawn for the above discussions of the ITC kinetic experiment, the nonlinear fitting of the ITC data, and data interpretation are listed below.

- It is important in planning the ITC kinetic experiment that reasonable concentrations be chosen for the enzyme and the substrate. This may require a guess at the K_m, k_{cat}, and ΔH values and/or one or more scoping experiments in which a substrate is injected into an the enzyme solution to determine the approximate initial heat rate and total heat of reaction.
- In a single injection experiment, there must be enough enzymes in the calorimeter cell to convert all of the substrate to product in a reasonable time (e.g., 30–60 min).
- Enzyme concentrations in the calorimeter cell in the range of 1 nM to 10 μM are typical for these experiments.
- The substrate concentration (in the injection syringe) should be 10^3–10^4 more concentrated than the enzyme, larger than the K_m, and in excess of the enzyme.
- In single injection experiments, the injected volume should be in the range of 25–50 μl. In the continuous experiments, the injection volumes should be in the range of 2–5 μl for each injection of the substrate solution.
- It is possible to perform a kinetic experiment by injection of the enzyme into a limited amount of substrate in the calorimeter cell.
- The values for K_m and k_{cat} can be determined with reasonable accuracy in less than 1 h in a single experiment
- The linear parameters $[E]_t$ and ΔH will be better determined than the nonlinear parameters K_m and k_{cat}.

The only limitation of the ITC technique at present is that modeling is limited to systems obeying simple Michaelis–Menton kinetics. This is actually a limitation of the currently available software in that the ITC will generate a complete reaction rate curve for any system that produces a measurable heat.

V. Conclusions

Fundamental areas of biology, molecular biology, biochemistry, and biophysics are dedicated to determining the relationships between the structure and function of proteins and nucleic acids.

Biologists need to better understand the recognition of small molecules for specific interaction sites on larger molecules and the nature of the weak individual interactions that can result in very high affinity. Ray Salemme, chief scientific officer of 3-D Pharmaceuticals, was quoted in a *C&E News* feature article as saying, "The initial expectation of structure-based drug design, that you were going to be able to design molecules and they were going to work right out of the box, was unrealistic. We didn't understand the thermodynamics well enough" (Henry, 2001). The use of ITC methods to probe the energetics of biologically relevant binding interactions is somewhat underappreciated (Chaires, 2006; Freyer *et al.*, 2007c). ITC provides a universal approach to determining the molecular nature of noncovalent interactions involved in the binding of small molecules to biopolymers (and even biopolymers to other biopolymers). In particular, ITC-binding experiments can yield a complete set of thermodynamic parameters for complex formation in a single experiment (Christensen *et al.*, 1966; Doyle, 1997; Eatough *et al.*, 1985; Wiseman *et al.*, 1989). The parsing of the Gibbs Free Energy change into the enthalpy and entropy contributions can provide new insight into the molecular nature of the binding interaction being studied (Freyer *et al.*, 2007b; Ladbury, 1996), also see the Chapter 1 by Garbett and Chaires, this volume. The energetic information is fundamental to not only understanding naturally occurring binding interactions but is particularly useful in drug discovery studies (Freyer *et al.*, 2007a,c; Ladbury, 2001, 2004).

Biologists also need to better understand the catalytic activity of enzymes and the affinity of a substrate for a specific active site. The kinetic behavior of enzymes (an obvious biopolymer function of interest) is important to not only understanding biochemical pathways and catalytic mechanisms but is again a fruitful area for drug discovery and development. The use of ITC methods to probe the kinetics of enzymes is a rather recent development (Todd and Gomez, 2001) and actually a rediscovery of some earlier work (Morin and Freire, 1991; Spink and Wadso, 1976; Watt, 1990; Williams and Toone, 1993). Again, ITC provides a universal approach to determining the kinetic behavior of enzymes and can yield in a single experiment a complete set of kinetic parameters for an enzyme-catalyzed reaction.

Hopefully, the information provided in this chapter will encourage researchers to further explore the advantages of the ITC methods for their studies of biopolymer interactions.

Acknowledgments

We would like to thank Jonathan (Brad) Chaires, Jack Correia, Joel Tellinghuisen, Jim Thomson, and W. David Wilson for their contributions to this chapter and encouragement. Supported by an NAU Prop 301 award and ABRC grants 0014 and 0015.

References

Ababou, A., and Ladbury, J. E. (2006). Survey of the year 2004: Literature on applications of isothermal titration calorimetry. *J. Mol. Recognit.* **19,** 79–89.

Bandman, U., Monti, M., and Wadso, I. (1975). Microcalorimetric measurements of heat production in whole blood and blood cells of normal persons. *Scand. J. Clin. Lab. Invest.* **35,** 121–127.

Beaudette, N. V., and Langerman, N. (1978). An improved method for obtaining thermal titration curves using micromolar quantities of protein. *Anal. Biochem.* **90,** 693–704.

Biltonen, R. L., and Langerman, N. (1979). Microcalorimetry for biological chemistry: Experimental design, data analysis, and interpretation. *Methods Enzymol.* **61,** 287–318.

Black, J. (1803). "Lectures on the Elements of Chemistry." Mundell & Son, Edinburgh.

Bundle, D. R., and Sigurskjold, B. W. (1994). Determination of accurate thermodynamics of binding by titration microcalorimetry. *Methods Enzymol.* **247,** 288–305.

Chaires, J. B. (2006). A thermodynamic signature for drug-DNA binding mode. *Arch. Biochem. Biophys.* **453,** 26–31.

Christensen, J., Hansen, L. D., and Izatt, R. M. (1976). "Handbook of Proton Ionization Heats and Related Thermodynamic Quantities." John Wiley & Sons, Inc, New York, NY.

Christensen, J. J., Izatt, R. M., and Eatough, D. (1965). Thermodynamics of metal cyanide coordination. V. Log K, ΔH°, and ΔS° values for the Hg2 + -cn-system. *Inorg. Chem.* **4,** 1278–1280.

Christensen, J. J., Izatt, R. M., Hansen, L. D., and Partridge, J. M. (1966). Entropy titration. A calorimetric method for the determination of ΔG, ΔH and ΔS from a single thermometric titration. *J. Phys. Chem.* **70,** 2003–2010.

Christensen, J. J., Wrathall, D. P., Oscarson, J. L., and Izatt, R. M. (1968). Theoretical evaluation of entropy titration method for calorimetric determination of equilibrium constants in aqueous solution. *Anal. Chem.* **40,** 1713–1717.

Cliff, M. J., Gutierrez, A., and Ladbury, J. E. (2004). A survey of the year 2003 literature on applications of isothermal titration calorimetry. *J. Mol. Recognit.* **17,** 513–523.

Correia, J. J., and Chaires, J. B. (1994). Analysis of Drug-DNA binding Isotherms: A Monte Carlo approach. *In* "Numerical Computation Methods Part B, Methods in Enzymology" (L. Brand, and M. L. Johnson, eds.), Vol. 240, pp. 593–614. Academic Press, San Diego, CA.

Doyle, M. L. (1997). Characterization of binding interactions by isothermal titration calorimetry. *Curr. Opin. Biotechnol.* **8,** 31–35.

Eatough, D. (1970). Calorimetric determination of equilibrium constants for very stable metal-ligand complexes. *Anal. Chem.* **42,** 635–639.

Eatough, D. J., Lewis, E. A., and Hansen, L. D. (1985). Determination of ΔH and K_{eq} values. *In* "Analytical Solution Calorimetry" (K. Grime, ed.), pp. 137–161. John Wiley & Sons, New York, NY.

Fasman, G. D. (1976). "Handbook of Biochemistry: Section D Physical Chemical Data." CRC Press, Boca Raton, FL.

Fisher, H. F., and Singh, N. (1995). Calorimetric methods for interpreting proteinligand interactions. *Methods Enzymol.* **259,** 194–221.

Freiere, E. (2004). Isothermal titration calorimetry: Controlling binding forces in lead optimization. *Drug Discov. Today* **1,** 295–299.

Freire, E., Mayorga, O. L., and Straume, M. (1990). Isothermal titration calorimetry. *Anal. Chem.* **62,** 950A–959A.

Freyer, M. W., Buscaglia, R., Cashman, D., Hyslop, S., Wilson, W. D., Chaires, J. B., and Lewis, E. A. (2007a). Binding of Netropsin to several DNA constructs: Evidence for at least two different 1:1 complexes formed from an -A2T2- containing ds-DNA construct and a single minor groove binding ligand. *Biophys. Chem.* **126,** 186–196.

Freyer, M. W., Buscaglia, R., Hollingsworth, A., Ramos, J. P., Blynn, M., Pratt, R., Wilson, W. D., and Lewis, E. A. (2007b). Break in the heat capacity change at 303 K for complex binding of netropsin to AATT containing hairpin DNA constructs. *Biophys. J.* **92**(7), 2516–2522.

Freyer, M. W., Buscaglia, R., Kaplan, K., Cashman, D., Hurley, L. H., and Lewis, E. A. (2007c). Biophysical studies of the c-MYC NHE III_1 promoter: Model quadruplex interactions with a cationic porphyrin. *Biophys. J.* **92,** 2007–2015.

Freyer, M. W., Buscaglia, R., Nguyen, B., Wilson, W. D., and Lewis, E. A. (2006). Binding of netropsin and DAPI to an A2T2 DNA hairpin: A comparison of biophysical techniques. *Anal. Biochem.* **355,** 259–266.

Hansen, L. D., Christensen, J. J., and Izatt, R. M. (1965). Entropy titration. A calorimetric method for the determination of ΔG° (k), ΔH° and ΔS°. *J. Chem. Soc. Chem. Commun.* **3,** 36–38.

Hansen, L. D., Lewis, E. A., and Eatough, D. J. (1985). Instrumentation and data reduction. *In* "Analytical Solution Calorimetry" (K. Grime, ed.), pp. 57–95. John Wiley & Sons, New York, NY.

Henry, C. M. (2001). Structure-based drug design. *C&E News* **79,** 69–74.

Indyk, L., and Fisher, H. F. (1998). Theoretical aspects of isothermal titration calorimetry. *Methods Enzymol.* **295,** 350–364.

Ladbury, J. (2004). Application of isothermal titration calorimetry in the biological sciences: Things are heating up!. *Biotechniques* **37,** 885–887.

Ladbury, J. E. (1996). Just add water! The effect of water on the specificity of protein–ligand binding sites and its potential application to drug design. *Chem. Biol.* **3,** 973–980.

Ladbury, J. E. (2001). Isothermal titration calorimetry: Application to structure-based drug design. *Thermochim. Acta* **380,** 209–215.

Langerman, N., and Biltonen, R. L. (1979). Microcalorimeters for biological chemistry: Applications, instrumentation and experimental design. *Methods Enzymol.* **61,** 261–286.

Lavoisier, A. L., and Laplace, P. S. (1780). Memoire sur la Chaleur. *Mem. Acad. Roy. Sci.* 355–408.

Lewis, E. A., and Murphy, K. P. (2005). Isothermal titration calorimetry. *Methods Mol. Biol.* **305,** 1–16.

Monti, M., Brandt, L., Ikomi-Kumm, J., and Olsson, H. (1986). Microcalorimetric investigation of cell metabolism in tumour cells from patients with non-Hodgkin lymphoma (NHL). *Scand. J. Haematol.* **36,** 353–357.

Morin, P. E., and Freire, E. (1991). Direct calorimetric analysis of the enzymatic activity of yeast cytochrome c oxidase. *Biochemistry* **30,** 8494–8500.

Spink, C., and Wadso, I. (1976). Calorimetry as an analytical tool in biochemistry and biology. *Methods Biochem. Anal.* **23,** 1–159.

Todd, M. J., and Gomez, J. (2001). Enzyme kinetics determined using calorimetry: A general assay for enzyme activity? *Anal. Biochem.* **296,** 179–187.

Watt, G. D. (1990). A microcalorimetric procedure for evaluating the kinetic parameters of enzyme-catalyzed reactions: Kinetic measurements of the nitrogenase system. *Anal. Biochem.* **187,** 141–146.

Williams, B. A., and Toone, E. J. (1993). Calorimetric evaluation enzyme kinetics parameters. *J. Org. Chem.* **58,** 3507–3510.

Wiseman, T., Williston, S., Brandts, J. F., and Lin, L. N. (1989). Rapid measurement of binding constants and heats of binding using a new titration calorimeter. *Anal. Biochem.* **179,** 131–137.

CHAPTER 5

Differential Scanning Calorimetry

Charles H. Spink

Department of Chemistry
State University of New York—Cortland
Cortland, New York 13045

Abstract

Differential scanning calorimetry (DSC) has emerged as a powerful experimental technique for determining thermodynamic properties of biomacromolecules. The ability to monitor unfolding or phase transitions in proteins, polynucleotides, and lipid assemblies has not only provided data on thermodynamic stability for these important molecules, but also made it possible to examine the details of unfolding processes and to analyze the characteristics of intermediate states involved in the melting of biopolymers. The recent improvements in DSC instrumentation and software have generated new opportunities for the study of the effects of structure and changes in environment on the behavior of proteins, nucleic acids, and lipids. This review presents some of the details of application of DSC to the examination of the unfolding of biomolecules. After a brief introduction to DSC instrumentation used for the study of thermal transitions, the methods for obtaining basic thermodynamic information from the DSC curve are presented.

0091-679X/08 $35.00
DOI: 10.1016/S0091-679X(07)84005-2

Then, using DNA unfolding as an example, methods for the analysis of the melting transition are presented that allow deconvolution of the DSC curves to determine more subtle characteristics of the intermediate states involved in unfolding. Two types of transitions are presented for analysis, the first example being the unfolding of two large synthetic polynucleotides, which display high cooperativity in the melting process. The second example shows the application of DSC for the study of the unfolding of a simple hairpin oligonucleotide. Details of the data analysis are presented in a simple spreadsheet format.

I. Introduction

Differential scanning calorimetry (DSC) is an experimental technique that has contributed significantly to understanding of the stability of biomacromolecules and of macromolecular assemblies. Application of the methods of DSC to the study of protein and nucleic acid unfolding has been exploited to provide not only thermodynamic data for the denaturation transitions, but also a better understanding of the underlying complexity of the unfolding process (Chalikian *et al.*, 1999; Cooper and Johnson, 1994; Freire, 1994; Makhatadze and Privalov, 1995; Plum and Breslauer, 1995). Investigations of phase transitions in lipid membranes and model membrane systems have provided a direct way to map out the phase behavior of binary and ternary lipid mixtures (Ali *et al.*, 1989; Huang and Li, 1999; Mason, 1998). In addition to stability studies, DSC has provided insight into the interactions between biomacromolecules, such as protein–nucleic acid (Read and Jelesarov, 2001), protein–lipid (McElhaney, 1986), or nucleic acid–lipid structures (Lobo *et al.*, 2002). Some of the subtleties of drug–DNA or drug–protein binding have been described by looking into the details of the thermal behavior of the complexes (Freyer *et al.*, 2006; Haq *et al.*, 2001). Much of the success of the applications of DSC to the study of macromolecules has resulted from increase in instrumental sensitivity, and the development of automation systems that allows equilibration and scanning to proceed basically under total instrumental control. These instrumental and computer software improvements allow studies on small sample sizes and at low concentrations of protein or nucleic acid solutions or with small quantities of lipid suspension. Thus, DSC has emerged as a powerful tool to determine thermodynamic data, and to examine the solution or structural features that affect stability in macromolecular systems.

In examining the literature of DSC studies, it is clear that the majority of work in the area is on proteins, and there are a number of excellent reviews relating to the application of DSC to the problems of protein folding and stability (Freire, 1994, 1995; Lopez and Makhatadze, 2002; Makhatadze and Privalov, 1995). Investigations by DSC of lipid membrane assemblies, particularly concerning the characterization of the phase behavior of mixtures, have also been subject to relatively wide attention (Huang and Li, 1999; Mason, 1998). Perhaps, the least discussed applications in the DSC field are studies of nucleic acid thermodynamics and the problems related to DNA or RNA stability (Chalikian *et al.*, 1999; Tikhomirova *et al.*, 2004). For this reason, this chapter will consider some of the approaches and problems associated with getting information about the thermal behavior of nucleic acids using DSC. The chapter will be divided into four parts: (1) instrumentation for DSC, (2) experimental protocols and preliminary data treatment, (3) modeling DNA unfolding, and (4) summary.

II. DSC Instrumentation

Thermal scanning of solutions of macromolecules leads to transitions from the native state to intermediate, partially unfolded states, and finally to the completely denatured state of the protein or nucleic acid polymer. In many cases, particularly for protein unfolding, the thermally induced transition can approach a two-state process between native and unfolded state. Nucleic acids, on the other hand, can show high cooperativity with many domains melting in concert, or for smaller DNA fragments can melt through sequential intermediate states. In order to study these transitions, the instrument must have several important features. The calorimetric measurement requires high sensitivity to the small energy changes associated with the unfolding processes at low concentration. This high sensitivity is accomplished through use of differential power compensation between a reference and sample cell, and a carefully designed method to control temperature and scan rate during the thermal experiment. A very general diagram of the basic components of a high sensitivity DSC instrument is shown in Fig. 1, based on information

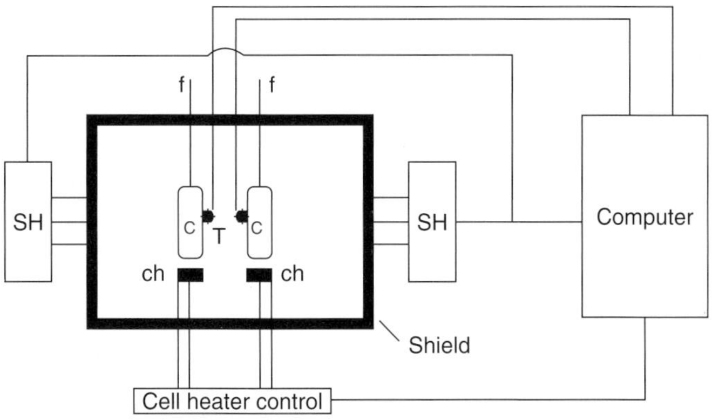

Fig. 1 General diagram for a differential scanning calorimeter. Cells (c) are located within a shield, which is in contact with shield heaters (SH). Individual cell heaters (ch) control the temperature of the sample and reference cells. Temperature sensors (T) are located on the cell surfaces, which determine if there is a temperature difference between the two cells, and through computer control apply appropriate compensating heat to the cells to keep temperature difference near zero. The compensating energy per unit time is recorded as the calorimetric signal.

from available commercial instruments. (The two prominent suppliers of calorimetry instrumentation are *MicroCal, LLC*, Northampton, MA, which makes the VP-DSC microcalorimeter and an automated capillary DSC, and *Calorimetric Sciences Corporation*, Provo, UT, which manufactures the Nano II and Nano III series of sensitive DSC calorimeters) (see Plotnikov *et al.*, 1997; Privalov *et al.*, 1995, for more detailed descriptions of the designs). The cells in these instruments are either capillary or "inverted lollipop" design and have volumes from 0.15 to 0.8 ml of solution, depending on the instrument. The cells are surrounded by a shield, which either controls the temperature of scanning (Privalov *et al.*, 1995) or is used to maintain the surrounding shield and the cells at the same temperature. This latter condition corresponds to an adiabatic scan, meaning that no heat flows from or to the cells from the surroundings during scanning the temperature of the cells (Plotnikov *et al.*, 1997). In the Nano DSC series, a computer controls the temperature program that determines the scan rate and control of scan temperature of the shield during the run. The cells are in thermal contact with the shield and thus increase in temperature as the shield increases. Temperature sensors located between the sample and reference cells determine if there is a temperature difference between the cells and if so, apply compensating power to the cells through heaters located on the cell surfaces. The power compensation signal is recorded as the calorimetric output. Because it is difficult to exactly match the thermal characteristics of the cells, the power compensation signal is adjusted through computer control to account for cell mismatch, and this adjustment will appear in the baseline of the calorimetric scan. The MicroCal VP-DSC design uses the shield to produce adiabatic control and operates by putting primary heaters on the cells to increase the temperature of the sample and reference. The shield temperature is monitored relative to the cells, and heaters in the shield are adjusted during the scan so that there is minimal temperature difference between the shield and cells, and thus the experiment is run in essentially an adiabatic mode with no heat transfer to the surroundings. In this instrument, down scanning is not adiabatic because a temperature gradient is required to cool the cells. Again, thermal sensors on the cells detect differences between the sample and reference when a thermal event occurs within the sample, and power compensation is used to keep the cells close to the same temperature during the scan. In order to account for mismatch between the cells and thermal lags between shield and cells, heater outputs are regulated by software control to maintain constant scan rate and careful temperature control.

The reported performance specifications of the above two instruments are quite similar. Both instruments have minimum thermal response times in the range of 5 sec, and baseline noise is 0.4 μcal/C for the Nano series and is 0.5 μcal/C for the VP-DSC, when scanning at 1–1.5 °C/min. Baseline repeatability is determined by doing multiple repeat scans on buffer samples, and yields standard deviations of $\pm$1.2 and $\pm$0.4 μcal/C for the VP-DSC and Nano DSC, respectively. Sample scans of a dilute protein solution of lysozyme (four scans with refill of the cells between scans, and at 0.096 mg/ml protein) gave standard deviations of 0.025 μW over the temperature range 30–90 °C, scanned at 1 deg/min for the MicroCal VP-DSC. Standard deviations for heating scans of lysozyme at 2.17 mg/ml for three scans with refilling yielded 0.56 μW deviations over a range of 25–80 °C for the Calorimetric Sciences Nano DSC instrument. Both of these instruments are capable of absolute heat capacity determinations for dilute biopolymer solutions.

MicroCal has introduced an automated version of the VP-DSC (the VP-Capillary DSC) which has an autosampler attachment that provides for unattended operation (Plotnikov *et al.*, 2002). The instrument is suitable for use with a 96-well sample plate, requiring about 0.3 ml of each sample to do complete scans at rates up to 3 °C/min. The sampler uses a robotic syringe to provide cleaning solution for the capillary cells and then to load sample from the covered well plate. The instrument can run unattended 24 h/day, and thus even at a scan rate of 1 deg/min can do 10–12 scans per day depending on the temperature range scanned. The autosampler can be programmed to change the reference solution as well as the sample. This kind of throughput now adds DSC to the techniques that can be used for screening studies, as in drug-biopolymer screening or in formulation studies, as well as in basic research in which the effects of solution conditions are important to study over a wide variety of conditions.

III. Experimental Protocols and Preliminary Data Treatment

The output data from a DSC experiment is the differential power response of the heaters (μW or μcal/sec) that keep sample and reference cells at nearly the same temperature. With concentration, cell volume, and scan rate normalization, these outputs can be converted to cal/deg-mol or J/deg-mol. In order to get the best repeatability of scanning, it is desirable to program the scans identically, including baseline scans. The thermal history of the individual scans is important in determining the baseline response, which means that the equilibration prior to the run, the scan itself, the cool-down of the calorimeter to the starting temperature, and re-equilibration should all be identical from run to run. If the sample is changed, it should be done during cool-down such as to minimally disturb the thermal protocol that will be repeated.

In order to illustrate the problems of dealing with baselines and subtraction of baseline from a scanning experiment, Fig. 2 shows the raw data for several baseline scans of buffer versus buffer, as well as repeat scans of the thermal unfolding transition for a 22-base DNA hairpin (Spink, 2005, unpublished data). These scans show that there is a natural instrumental baseline for a particular set of scan conditions that becomes quite reproducible after the first few instrumental scans. If the sample is introduced during the cool-down of the last baseline scan, it is presumed that the baseline will be the same for the sample measurement, and an average of the baseline scans can be subtracted from the DNA experimental curve. When this is done, the curves shown in Fig. 3 are generated. In principle, the unfolding data in the lower curve of Fig. 3 represent the true heat capacity

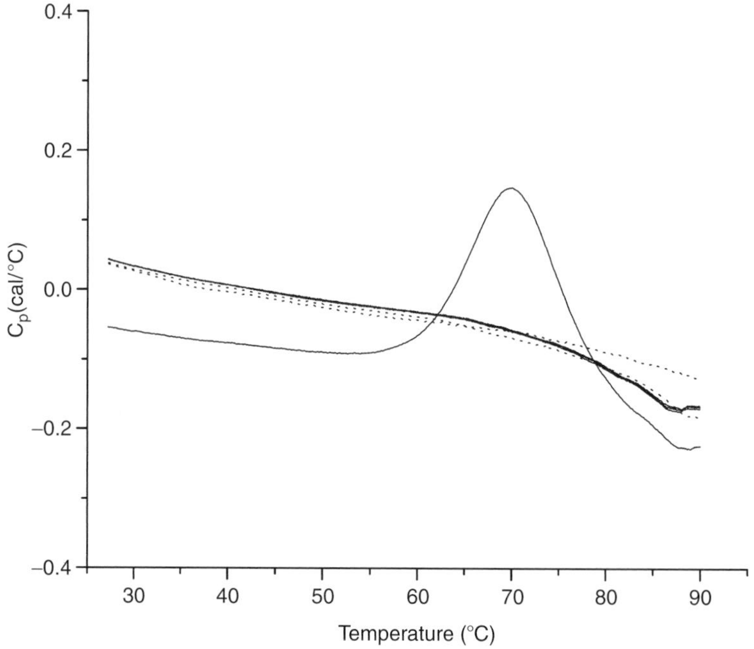

Fig. 2 Baseline scans and sample scan for a 22-base DNA hairpin, Hel2, in BPE buffer, consisting of 8 mM phosphate, 1 mM EDTA, and 16 mM Na^+ at pH 7.0. The first two baseline scans of buffer versus buffer are shown as dotted lines, and three subsequent baseline scans are superimposed as the dark black line. The sample was loaded in the cell after the last baseline scan, and consists of a 0.057 mM solution of the hairpin structure. The base sequence of Hel2, is: 5′-GGTCACGACAGCTGTCGTGACC-3′. The data were obtained using a MicroCal VP-DSC instrument. Data from Spink (2005, unpublished work).

(the heat per mole required to raise the temperature one degree) of the DNA sample with the pretransition baseline representing the change in heat capacity of the hairpin form with temperature, and the excess heat capacity during unfolding (the peak) providing a profile of the melting process to unfolded form. The posttransition baseline represents the heat capacity of the unfolded form. In the Hel2 case in Fig. 3, there is very little posttransitional data. For proteins, there is often a substantial ΔC_p for the unfolding, and the posttransition curve is quite different from the pretransition data (Freire, 1994). In the case of DNA samples changes in the pre- and posttransitional baselines are difficult to discern in a normal scan, since it has been estimated that ΔC_p is only around 100 cal/deg-mol of base pairs (Holbrook *et al.*, 1999). In order to see these transition differences, higher concentrations of the DNA samples must be used.

The next step in the analysis of the DNA unfolding curve is to determine the overall transition parameters, T_m, ΔH, and ΔS, and frequently the van't Hoff enthalpy (ΔH_{vH}) is determined for the unfolding process. If it is possible, the heat capacity change can also be evaluated at this stage. The overall parameters are determined in the region of unfolding where the excess heat capacity function is

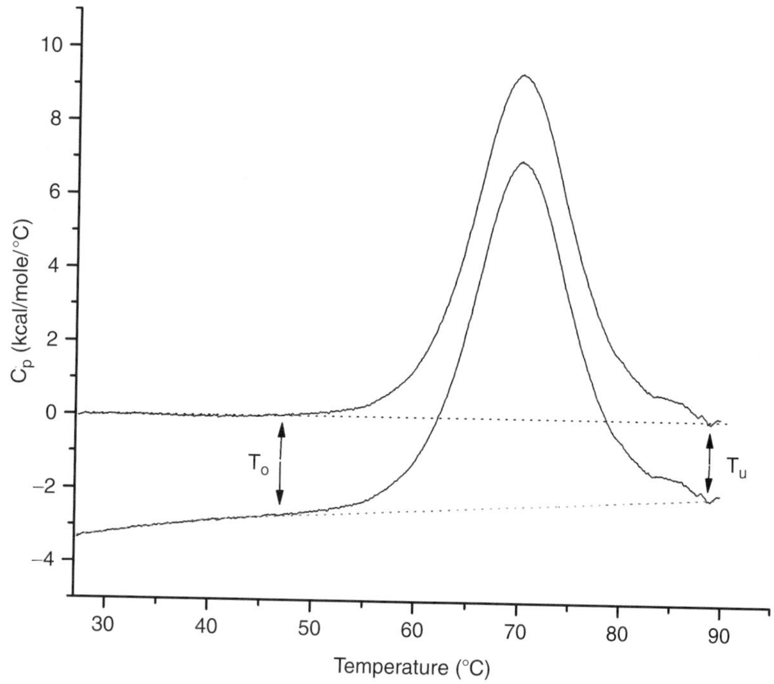

Fig. 3 DSC scan of 22-base DNA hairpin, Hel2, in BPE buffer. Lower curve is the result of subtraction of the average of the last three baselines from Fig. 2 from the sample curve in Fig. 2. The dashed line is a cubic fit baseline to the pre- and posttransitional portion of the melting curve. The upper curve is the resulting excess heat capacity after subtraction of the cubic fit baseline.

evident, and are determined from the onset of the transition, T_0, to the completion of the unfolding at T_u (see Fig. 3). The calorimetric enthalpy, the energy of unfolding of the DNA, is calculated from the area of the unfolding transition, that is, from the integral of the excess heat capacity function:

$$\Delta H = \int_{T_0}^{T_u} C_p(\text{ex})\,\mathrm{d}T \tag{1}$$

The usual procedure is to subtract a pretransition to posttransition baseline from the overall curve to leave a profile of the unfolding relative to a zero baseline, as shown in Fig. 3. The area is evaluated from the onset temperature to completion of unfolding. The baseline used is either a linear or cubic fit to the pre- and post-transitional regions. Figure 3 shows a cubic fit baseline to the original curve, and the resulting subtracted curve, which can be used to evaluate the area, T_m, and other overall parameters. It is generally good practice to try several baseline procedures to get an idea of the uncertainties that the choice of baseline makes in the results. For example, for the curve in Fig. 3, the integrated area averages

111 ± 4 kcal/mol, depending on the baseline procedure. Origin™ software, provided by MicroCal, Inc. as a part of their DSC data analysis package, offers several options for dealing with the baseline fitting.

Other transition parameters include T_m, which is the temperature of the maximum of the excess heat capacity curve. Some authors have used the temperature of one-half melting, $T_{1/2}$ to characterize the peak of the transition curve. For a two-state transition, T_m and $T_{1/2}$ are identical, however, for many DNA transitions, particularly with oligomeric forms, there are intermediate states which can cause $T_{1/2}$ to differ significantly from the maximum temperature in the curve (Freire, 1994). It is generally better to use T_m to define the temperature of the transition. The overall entropy change for unfolding is defined by Eq. (2):

$$\Delta S = \int_{T_0}^{T_u} \frac{C_p(\text{ex})}{T} \mathrm{d}T \tag{2}$$

For a well-defined two-state transition, since ΔG $(=\Delta H - T\Delta S)$ of unfolding is zero at T_m, then the entropy change is simply: $\Delta S = \Delta H/T_m$. But, in general and again for many DNA transitions, simple two-state behavior is probably not a correct interpretation of the unfolding, and Eq. (2) will give more accurate values of ΔS of melting.

A final overall parameter that can be evaluated from the unfolding transition curve is the van't Hoff enthalpy. This parameter is a general measure of the change with temperature of the equilibrium constant for unfolding:

$$\frac{\partial \ln K}{\partial T} = -\frac{\Delta H_{vH}}{RT^2} \tag{3}$$

A number of experimental techniques, which are sensitive to changes occurring upon unfolding, can monitor this parameter. For example, DNA melting is often followed from the changes in the UV absorption at 260 nm, and by appropriate modeling of the unfolding transition the van't Hoff enthalpy can be evaluated. From calorimetric data, the van't Hoff enthalpy is calculated directly from the DSC curve using the following equation:

$$\Delta H_{vH} = 4RT_m^2 \frac{C_p(m)}{\Delta H_{cal}} \tag{4}$$

Here $C_p(m)$ is the value of the excess heat capacity at the transition maximum. The van't Hoff enthalpy can be helpful in deciding the general characteristics of the unfolding transition relative to a simple two-state model. If the ratio $\Delta H_{vH}/\Delta H_{cal}$ is less than one, it is likely that there are intermediate states, while if the ratio is greater than one, cooperative melting is indicated.

Table I
Overall Thermodynamic Data for Unfolding of DNA Examples

	Hel2	p(dAdT)	p(dGdC)
T_m (°C)	70.0 ± 0.1	42.7 ± 0.07	97.7 ± 0.1
ΔH (kcal/mol)	111 ± 4	8.62 ± 0.2	11.34 ± 0.2
ΔS (cal/deg-mol)	319 ± 10	27.5 ± 0.3	30.3 ± 0.2
ΔH_{vH} (kcal/mol)	77 ± 2	708 ± 8	405 ± 2
$\Delta H_{vH}/\Delta H$	0.69	82	36

Enthalpies and entropies for Hel2 are per 22-base hairpin, while for the synthetic polynucleotides are per duplex base pair.

Table I shows the data for the overall unfolding transition for Hel2, the 22-base DNA hairpin used as an example above. A value for ΔC_p is not included since it could not be determined for this transition. The uncertainties listed are a result of the inconsistencies in zeroing the baseline for determining areas. The conclusions to be drawn from these numbers are several. First, looking at the ratio of van't Hoff to calorimetric enthalpies, it is clear that the transition is not two-state, the enthalpy ratio being significantly less than one. As has been observed in the unfolding of a number of smaller oligomeric DNA structures, it is likely that the melting occurs through intermediate states (Vallone *et al.*, 1999). The value of the calorimetric enthalpy is reasonable considering that there should be about 9–10 bp in this 22-base hairpin. Values of ~10 kcal/mol bp have been observed for other small DNA structures (Vallone and Benight, 2000). As mentioned above, it has been difficult to get values for the ΔC_p of the unfolding transitions for DNA. Some attempts have been made to measure ΔC_p for oligomeric DNA from the temperature dependence of the enthalpy of association of complementary strands by using isothermal titration calorimetry (Holbrook *et al.*, 1999; Tikhomirova *et al.*, 2004). Values determined by this approach fall in the range of 60–90 cal/deg-mol of base pairs. Thus, one can estimate for about 10 bp in the Hel2 hairpin a maximum of about 900 cal/deg-mol heat capacity change. If the enthalpy of melting is calculated not at T_m (70 °C) but at 25 °C, there would be an increase of about 40 kcal/mol, certainly not an insignificant change. $\Delta H(25) = \Delta H(70) + \Delta C_p dT = \Delta H(70) + 900 \times (-45) = \Delta H(70) - 40{,}500$ cal/mol. The problem in using DSC to measure these heat capacity effects is that the excess heat capacity change during the unfolding process is so large relative to the contribution to ΔC_p of melting. For Hel2 the estimated contribution of ΔC_p is about 60–90 cal/deg-mol in a total excess heat capacity of about 1000 cal/deg-mol bp, a quantity difficult, although not impossible, to measure.

While examining the calorimetric response to DNA melting, it is useful to consider the case of cooperative melting, which is frequently observed in the unfolding of high molecular weight DNA. Cooperativity implies that the melting unit contains many residues that melt together, and is known especially in the case of unfolding of a number of large, uniform DNAs, such as the synthetic polynucleotides (Chalikian *et al.*, 1999). In cooperative transitions the van't Hoff enthalpy, as determined from Eq. (4) or from UV-melting transitions, is generally considerably greater than the calorimetric enthalpy. For example, Fig. 4 is a scan for a mixture of synthetic poly(dAdT)·poly(dTdA) and poly(dGdC)·poly(dCdG) whose sizes are 8778 and 920 bp, respectively (Spink *et al.*, 2006). These transitions are generally sharp, the width of the transitions only a few degrees. There is an advantage to the sharpness because the evaluation of baseline and baseline extrapolations are straightforward, which can reduce error in calculating the overall thermodynamic properties of unfolding in these cases. The van't Hoff enthalpies for the two transitions calculated from Eq. (4) are 708 and 405 kcal/mol for the poly(dAdT) and the poly(dGdC) sequences, respectively, while the corresponding calorimetric enthalpies are 8.62 and 11.34 kcal/mol of base pairs in the buffer system used (Spink *et al.*, 2006). Cooperative melting units of 82 and 36 bp, respectively, result from the ratio of van't Hoff to calorimetric enthalpies. These data are also summarized in Table I, and show the marked differences in the unfolding properties of the large, synthetic DNA structures as compared with the small, oligomeric Hel2 sample also shown in the table.

Long pieces of natural DNA, such as, from calf thymus, salmon sperm, or some plasmid preparations generally show somewhat lower cooperativity, but

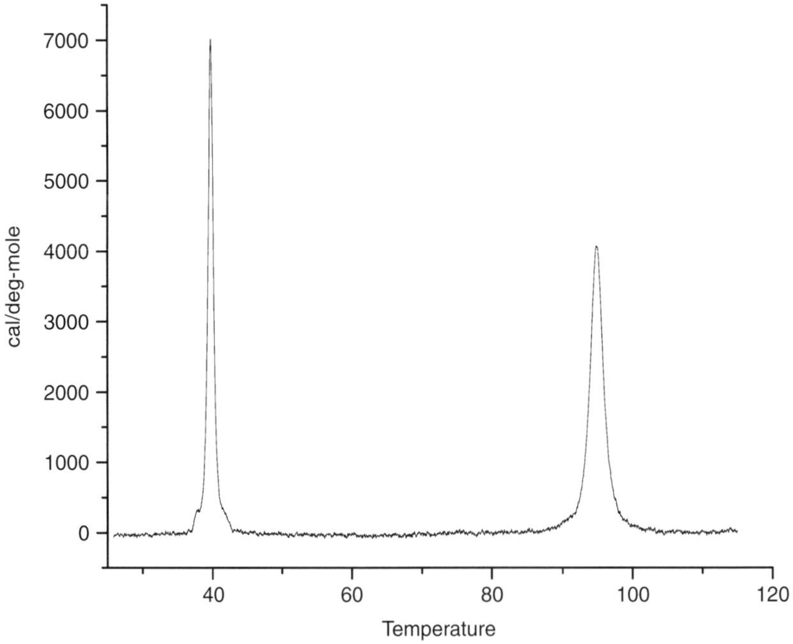

Fig. 4 DSC scan of a mixture of poly(dAdT)·poly(dTdA) (low temperature peak) and poly(dGdC)·poly(dCdG) (high temperature peak) in BPE buffer. Concentrations of the two samples were 0.28 mM bp. Data were obtained using a MicroCal VP-Capillary DSC.

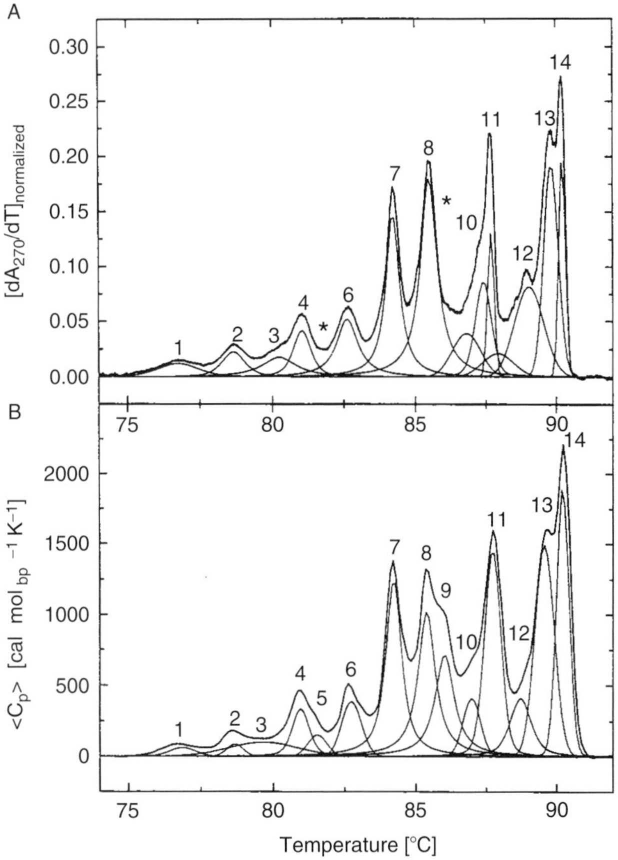

Fig. 5 (A) High resolution optical melting curve for pBR322 plasmid DNA, measured at 270 nm in 75 mM sodium ion buffer, pH 6.8. (B) Corresponding high resolution DSC melting profile for pBR322 plasmid. Peaks identified in the calorimetric curve are numbered 1–14, with the corresponding optical melting peaks labeled appropriately. There are two missing peaks (5 and 9) in the optical melting curve. (Figure reprinted from Volker *et al.*, 1999, reproduced by permission of John Wiley & Sons, Inc.)

are more complicated by the fact that many domains seem to melt independently of others, and show detailed structure in the transitions (Volker *et al.*, 1999). For example, Fig. 5 is a high resolution DSC scan of a DNA plasmid (pBR322 of 4363 bp) along with the UV-melting profile for the same sample, presented in the paper by Volker *et al.* (1999). These high resolution scans of a plasmid reveal the complexity of many natural DNA structures, the data showing many overlapping domains with their own cooperative behavior. The authors point out that the cooperativity and melting temperature of a particular sequence depend not only on the sequence, but also on the context of the bases that are undergoing cooperative unfolding. Note also that there are differences in the number of peaks in the calorimetric scan compared with the UV-melting curve. The calorimetry picked up two more transitions in the unfolding of the plasmid than the UV curve, indicating that unfolding monitored by UV absorption is not sensitive to all base pair disruptions. It is also important to recognize that the relative sizes of resolved transitions as measured by the two techniques can be quite different, since the processes being measured are based on different physical properties of the DNA.

IV. Modeling DNA Unfolding

There are few DNA samples that unfold according to a simple two-state model, that is, from a simple duplex state to single-strand random coils. High molecular weight synthetic polynucleotides generally show high cooperativity, as mentioned above. Oligomeric DNA can unfold in an almost two-state process, but pieces of DNA with even 12–40 bp indicate more complicated melting behavior, often showing evidence of intermediate states. Thus, attempts to arrive at general models of DNA unfolding have generated many papers that, while consistent in the broader sense, are often tailored to the particular type of DNA being studied. One unique aspect of the DSC calorimetric melting transition is that there is information in the data about fundamental details of the unfolding process. The important discovery by Freire and Biltonen (1978a) that the statistical thermodynamic partition function for proteins or polynucleotides can be obtained from thermal unfolding transitions has made it possible to examine in more subtle detail the process of unfolding of these important macromolecules. Deconvolution of DSC melting curves provides a direct way of getting the partition function and then properties of any intermediate states, should they exist. In the discussion below two examples of the analysis of DSC behavior will be presented, one showing the general treatment of cooperative DNA transitions, and the other will analyze the small oligomeric hairpin that we have been using as an example above. The aim is to show cases of the extremes in cooperativity, the first, an example of synthetic polynucleotides of uniform base pair sequence, and the second, a small DNA structure that appears to melt through an intermediate state when progressing from the duplex hairpin to a single strand.

A. Freire–Biltonen Deconvolution

In the deconvolution method developed by Freire and Biltonen (1978a) for the general case involving multiple intermediate states, the unfolding process is considered to be as follows:

$$I_0 \underset{K_1}{\leftrightarrow} I_1 \underset{K_2}{\leftrightarrow} I_2 \underset{\cdots}{\ldots} I_{n-1} \underset{K_n}{\leftrightarrow} I_n \tag{5}$$

The free energy of the initial state is taken as the reference state, so the individual equilibrium constants are relative to I_0:

$$K_i = e^{-\Delta G_i/RT}, \qquad \text{where } \Delta G_i = (G_i - G_0) \tag{6}$$

The partition function for the system can be written:

$$Q = 1 + \sum_{i=1}^{n-1} e^{-\Delta G_i/RT} + e^{-\Delta G_n/RT} \tag{7}$$

The fraction of a particular intermediate is then calculated from:

$$F_i = \frac{e^{-\Delta G_i/RT}}{Q} \tag{8}$$

Thus, the fraction in the initial state is $F_0 = 1/Q$, and the summation terms include all of the intermediate states. If the process is two-state, $F_n = 1 - 1/Q = K_n/(1 + K_n)$. The important observation of Freire and Biltonen is that the partition function at any temperature can be obtained directly from the calorimetric data:

$$Q(T) = \exp\left(\int_{T_o}^{T} \frac{\langle \Delta H \rangle}{RT^2} \mathrm{d}T\right) \tag{9}$$

where $\langle \Delta H \rangle$ is the average excess enthalpy, obtained from integration of the excess heat capacity function determined from scanning calorimetry:

$$\langle H \rangle = \int_{T_o}^{T} [C_p(\mathrm{ex}) - C_p(o)]\mathrm{d}T \tag{10}$$

$C_p(\mathrm{ex})$ is the excess heat capacity over the temperature range of the unfolding transition, and $C_p(o)$ is the extrapolated baseline from the initial state over the transition range. These latter two equations provide the link between the experimental excess heat capacity and the partition function, which allows fractions of intermediate states to be calculated, and thus the enthalpic contributions from the intermediates become accessible:

$$\langle H \rangle = \sum_{i=0}^{n} \Delta H_i \times F_i \tag{11}$$

ΔH_i is the enthalpy difference between the initial state and the ith intermediate state.

Two of the important state fractions, F_0 and F_n, are easily accessible from the following relationships:

$$F_0 = \frac{1}{Q} \tag{12}$$

and

$$F_n = \exp\left[-\int_T^{T_n} \frac{(\Delta H_n - \langle \Delta H \rangle)}{RT^2} \mathrm{d}T\right] \tag{13}$$

Thus, from the cumulative enthalpy, $\langle \Delta H \rangle$, and the total enthalpy of the transition, ΔH_n, the fraction in the final state can be determined by integrating from T_n across the melting curve. Then, since all the various fractions must add up to one, if there are intermediate states their total fraction can be found from: $\Sigma\, F_i = 1 - F_0 - F_n$. If there are no intermediate states, then F_0 and F_n should add up to one, otherwise there must be intermediate contributions to the unfolding process. The final problem in the analysis of the scanning calorimetry signal is to deconvolute the transition into the appropriate number of intermediate states. Freire and Biltonen (1978a) developed a set of recursion relations that provide a way to get initial estimates of the required parameters for the transitions. Keeping in mind that the free energy of a given state is:

$$\Delta G_i = \Delta H_i + \Delta C_{\mathrm{p},i}(T - T_0) - T\left[\Delta S_i + \Delta C_{\mathrm{p},i} \ln\left(\frac{T}{T_0}\right)\right] \tag{14}$$

where T_0 is the reference temperature, in this case the initial state being the reference state, it is apparent that three parameters are required in order to define each state comprising the conversion from initial to final unfolded state. The recursion relations provide access to ΔH_i for each state and to the fraction of each intermediate state as a function of temperature. Then through nonlinear least squares fitting of ΔH_i, ΔS_i, and $\Delta C_{\mathrm{p},i}$ to the excess heat capacity function, refined values of the parameters for each state are obtained. Curve fitting can be analyzed on the cumulative enthalpy or directly on the excess heat capacity:

$$\langle C_\mathrm{p} \rangle = \sum_{i=1}^{n} \Delta H_i \frac{\mathrm{d}F_i}{\mathrm{d}T} + \sum_{i=1}^{n} F_i \Delta C_{\mathrm{p},i} \tag{15}$$

The first term on the right determines the contribution from the excess enthalpy terms, and the second term from the heat capacity effects, which as mentioned above are often difficult to determine, and thus ignored. Next, we will examine two cases of DNA unfolding which make use of some of the ideas presented above.

B. Cooperative DNA Unfolding

The theories of unfolding of DNA are most often based on the breaking of base-paired residues to form internal loops which can proliferate along the polynucleotide chain, as pictured in Fig. 6. As temperature increases from that for the initial state, I_0, which is the reference state of complete base-paired duplex, loop regions form as base-pairing breaks along the chain. These intermediates are partially unfolded and partially duplex, but are not yet single-stranded random coils, the final state of the melting process. The cooperativity of the unfolding depends on the energetics and entropy effects associated with the partially unfolded intermediates. As discussed earlier, the synthetic polynucleotide duplexes, poly(dAdT) or poly(dGdC), show relatively high cooperativity and therefore the DSC transition curve is sharp, meaning that the intermediate loop structures are similar in terms of the energy and entropy associated with opening and closing of the loops.

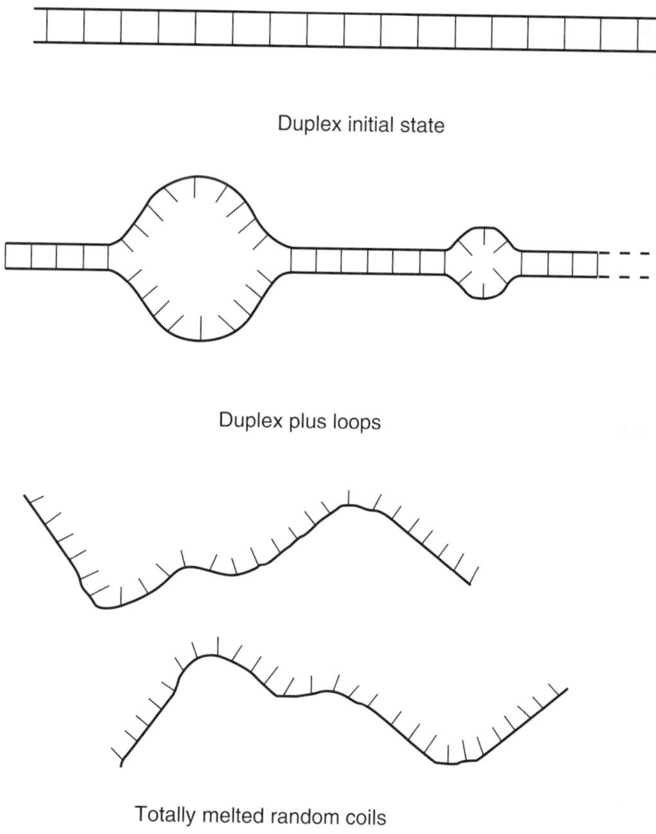

Fig. 6 Diagram depicting progress of melting from the initial duplex DNA state, to intermediate duplex—loops state, and the final state of single-strand random coils.

It is possible to analyze the experimental unfolding transition of the synthetic polynucleotides using the concepts developed by Freire and Biltonen (1978b,c). In this situation, a useful way to deconvolute the thermal transitions is to determine the fraction of the duplex form remaining at any temperature and the fraction converted to random coils. Then, since the sum of these fractions subtracted from unity would give the fraction of the intermediate loop structures, it is possible to determine the profile of melting of these combined duplex—loop configurations as temperature changes. In the discussion below, an analysis of the melting of a 900 bp poly(dGdC) and a 9000 bp poly(dAdT) will be presented to illustrate the power of DSC for the study of DNA unfolding.

The sample of duplex poly(dGdC) polynucleotide was prepared and analyzed by Amersham Biosciences Corp., Piscataway, NJ. From sedimentation and spectral data, the average molecular weight of the duplex strands is 6.0×10^5, which means that the number of base pairs is about 920 (assuming a molecular weight of 650 g/mol of base pairs). From the normalized DSC transition curve in BPE buffer, as shown in Fig. 4, the cumulative enthalpy can be calculated from the integral of the excess heat capacity curve with appropriate baseline subtractions (Eq. 10). It is convenient to do the calculations in a spreadsheet program, either manually or with script control. We use Origin™ software, which provides convenient calculation and plotting routines within the spreadsheet of data. Table II summarizes guidelines for preparing such a spreadsheet for the general case of melting with intermediate states. The normalized, baseline-corrected excess heat capacity data are entered at each temperature, and then calculations and integrations are performed within the spreadsheet. Because decisions must be made about integration limits, it is useful to plot the data for each column of information to aid in these data analysis decisions as one advances through the calculations. For the large DNA structures, it is helpful to calculate the integrals per base pair, which gives a partition function per base pair, q, that can then be related to the strand partition function, Q, by the relationship $Q = q^N$, where N is the degree of polymerization (Freire and Biltonen, 1978c). For polynucleotides, N is the number of base pairs per average strand. From the discussion above recall that the fraction of the initial state, F_0, which is the fraction of pure duplex form, is simply $1/Q$. There will be duplex portions in the loop structures, but F_0 is for the unmelted initial duplex state. Similarly, the fraction of single strands with no loop structures, that is, the completely melted state, can be obtained by integration of the heat capacity function from high to low temperature in an analogous calculation. This approach is equivalent to evaluating a partition function, z, with the reference state changed from duplex state to melted single-strand state. The fraction in the melted state, F_n, is determined from Eq. (13). Again the conversion from per base pair requires raising the partition function for the new reference state to the Nth power to get the total partition function, $Z = z^N$. Having both F_0 and F_n, it is possible to calculate the fraction of intermediate states, that is, to determine the fraction of the total strands that have loop structures: $F_i = 1 - F_0 - F_n$. Figure 7 shows the fractions calculated for the poly(dGdC) sample. Also shown is F_m, the overall

Table II
Guide for Creating Spreadsheet of Calculations from DSC Data

Column number	Description
1 and 2	Raw calorimetric data. Col 1 is the temperature (K is useful for calculations); Col 2 is the normalized excess heat capacity data (usually in cal/deg-mol).
3	Cumulative enthalpy, $\langle \Delta H \rangle$, which is the integral of the heat capacity data (Col 2), integrated from the beginning of the data to the end; should look like a sigmoid curve, leveling off to the total enthalpy of the transition, ΔH_n. Origin™ (software has an integration function which is suitable for these calculations.
4	$\langle \Delta H \rangle / RT^2$: This column is Col 3 divided by RT^2, preparing to integrate to obtain the partition function; T is from Col 1 in Kelvin degrees.
5	Integral of Col 4 = $\int(\langle \Delta H \rangle / RT^2)\mathrm{d}T$. Limits of integration are from the first discernible deviation from the zero line (T_0) and the upper limit is obtained from the point where the data deviate from a linear extrapolation from the upper portion of the curve.
6 and 7	Col 6 is the exponentiation of the data in Col 5 = exp $[\int(\langle \Delta H \rangle / RT^2)\mathrm{d}T]$, which is the partition function, Q, for the system. Col 7 is $1/Q = F_0$, the fraction of the molecules in the initial state, the duplex form. Note: if the partition function is calculated per base pair, q, then an additional calculation is required to obtain $Q = q^N$, where N is the average number of base pairs per DNA strand.
8 and 9	Col 8 is the cumulative enthalpy from high to low temperature = $(\Delta H_n - \langle \Delta H \rangle)$ and Col 9 is that enthalpy divided by RT^2.
10	This column is the integral of Col 9: $\int[(\Delta H_n - \langle \Delta H \rangle)/RT^2]\mathrm{d}T$, the integration limits determined as before from the deviation from linear extrapolations from the pre- and posttransitional baselines.
11	Because of the way the Origin™ software integrates, it is necessary to obtain the maximum value in Col 10 and create a column that is [Max Value – (Col 10)]. This effectively changes the integration limits from high to low temperature, which is necessary to obtain the melting profile in reverse.
12	This column generates F_n, the fraction melted into the final state of single strands: F_n = exp (–Col 11). Note: as above, if the data are calculated from enthalpies per base pair to get z, then an additional calculation of $Z = z^N$, is required, and $F_n = 1/Z$.
13	The fraction of intermediates, $F_i = 1 - F_0 - F_n$ or 1 – Col 7 – Col 12.

fraction of base pairs that are disrupted, calculated from: $F_{\mathrm{m}} = \langle \Delta H \rangle / \Delta H_n$ per base pair.

Several features of the fractional behavior of the various states are apparent. First, the fraction F_0 for the totally duplex state drops to zero rapidly, essentially reaching zero at about 96 °C with only about 10% of the base pairs disrupted. Second, at 50% base pairs disrupted at 97.8 °C there are no single strands formed. Single strand formation begins around 99.5 °C at which over 90% of the base pairs are disrupted. Thus, there is about a 4 °C temperature range in which all of the DNA exists as partially duplex and partially melted loop structures. It is important to recognize that these calculated fractions come directly from the calorimetric data because the partition function comes directly

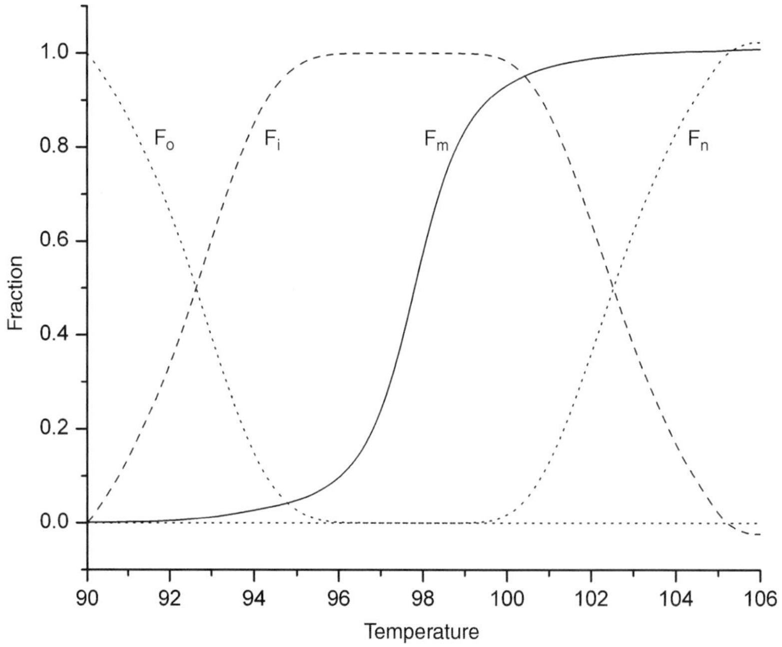

Fig. 7 Fractions of various DNA forms for poly(dGdC) transition in BPE buffer calculated from analysis of the curve as described in the text. F_0 is the fraction of totally duplex form, F_i is the fraction of intermediate with duplex and loop structures present, F_n is the fraction in the single-strand state, and F_m is the fraction of base pairs melted as determined from the cumulative enthalpy.

from the integrated calorimetric data. One outcome of this correlation of the partition function to the thermal transition data is that the ratio of the partition function of the final state to that of the initial state is the stability constant, s, of the helix as defined per base pair. That is, $s = z/q$. One can perform a van't Hoff analysis to obtain the enthalpy for helix to single-strand base pairs by plotting ln s versus $1/T$, and the result of this analysis yields the van't Hoff enthalpy, which should be equal to the integrated calorimetric enthalpy of the transition, a result of the presumption that individual base pair melting is a simple two-state process (Freire and Biltonen, 1978c). Figure 8 shows a van't Hoff plot for the poly(dGdC) transition, and from the slope of the plot $\Delta H_{vH} = 11.35$ kcal/mol bp, which is virtually identical to the integrated calorimetric enthalpy reported earlier (see Table I). This agreement confirms that the residue melting of the base pairs is a two-state process, regardless of whether the duplex state is in completely base paired structures or in already partially melted loop configurations.

For comparison, Fig. 9 shows fractional melting data for the cooperative melting transition of poly(dAdT) consisting of 8778 bp. Note that in Fig. 4, there is some asymmetry in the melting curve for this case, a result of a small

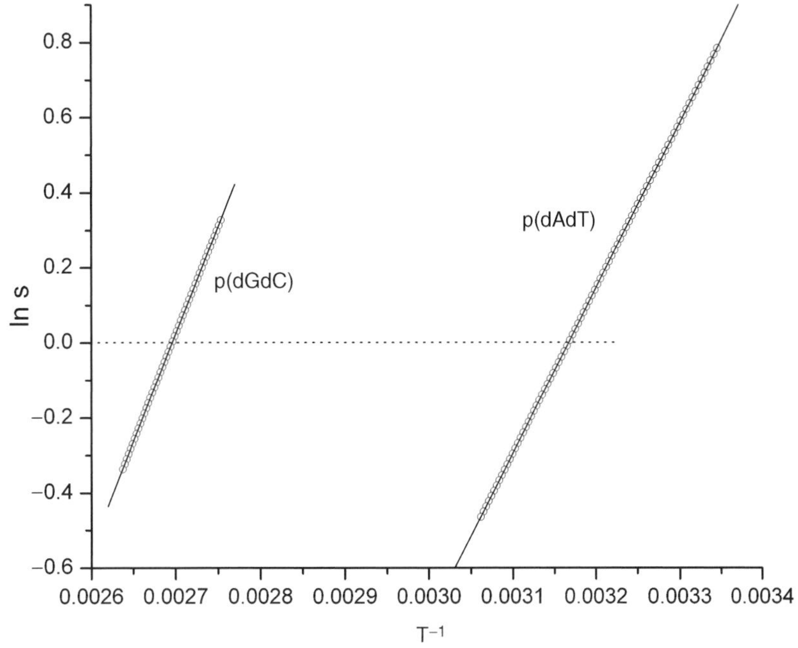

Fig. 8 van't Hoff plots for the melting transitions of poly(dGdC) (left curve) and for poly(dAdT) (right curve). The value of $s = z/q$ is the stability constant for a base pair transition from the fully base-paired condition to a disrupted base pair, the final state.

contribution on the high temperature side of the transition. The origin of this contribution for a pure synthetic polynucleotide such as poly(dAdT) is not clear, but a number of similar preparations seem to show these effects. The asymmetry in the transitions reveals itself in the fraction plots of Fig. 9. The intermediate duplex-loop structures persist to higher temperatures, and in this case almost 98% of the base pairs are disrupted before separation to single strands commences. The presence of the asymmetry means that the fractions calculated for the high temperature single-strand state may be slightly inaccurate. But, the van't Hoff enthalpy for the base pair melting in poly(dAdT) from the ln s versus $1/T$ plot in Fig. 8 is 8.81 kcal/mol, compared with the integrated calorimetric enthalpy of 8.62 kcal/mol, still remarkably close given the asymmetry in the transition curve. These analyses of cooperative transitions in DNA not only provide thermodynamic data on the unfolding process, but also show the details of melting in terms of the nature of intermediate duplex-loop structures. The total fraction of intermediate states can be calculated, but if desired, the size and number of intermediate clusters of melted and unmelted states can be calculated directly from the residue partition functions. See Freire and Biltonen (1978b,c) for details. Thus, from a single calorimetric scan, a great deal of information about the characteristics of cooperative melting can be obtained.

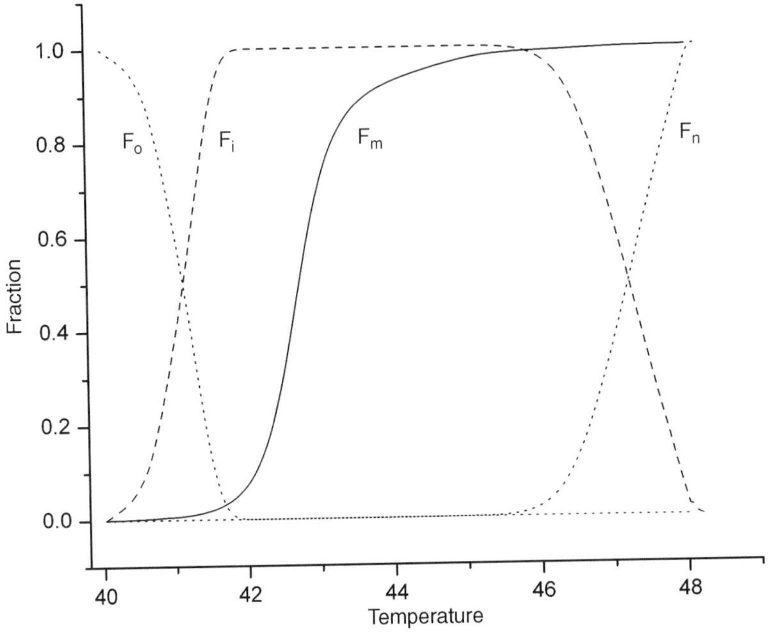

Fig. 9 Fractions of various DNA forms for poly(dAdT) transition in BPE buffer. F_0 is the fraction of totally duplex form, F_i is the fraction of intermediate with duplex and loop structures, F_n is the fraction in the single-strand state, and F_{m} is the fraction of base pairs melted as determined from the cumulative enthalpy.

The evaluation of errors in the deconvolution methods described above has been summarized by Freire and Biltonen (1978a). Basically, the quality of data obtained from the analysis of the thermal transitions depends on the uncertainties in the experimental transition curves and on the assignment of integration limits. For example, the errors in F_0 depend mostly on the starting temperature, T_0, in the integration to get q from the experimental curve. Since the melting curves are generally narrow in the case of cooperative transitions, it is not difficult to assign a baseline which can be extrapolated through the transition region. This simplifies the determination of T_0 for the low temperature limit of the integration. Even with T_0 chosen to be 20–30 °C below the peak maximum introduces insignificant error in F_0 and in the estimation of enthalpy and entropy values from the curves. Introduction of random noise into the calorimetric data also proves to cause little error in the results for simulated curves. As much as a 30% of peak height random variation yielded about 3% error in calculation of F_0 and the temperature of the maximum in the analysis (Freire and Biltonen, 1978a). Thus, for cooperative transitions with reasonable quality calorimetric data, one expects small errors as a result of the data analysis process.

C. Melting Oligomeric Hairpins

The 22-base DNA hairpin structure, Hel2, melts very differently from the cases involving cooperativity discussed above. The ratio of van't Hoff to calorimetric enthalpy is less than one, implying that rather than melting in a simple two-state transition, there are sequential intermediate states that appear between the base-paired initial state hairpin and the final state of a single, 22-base strand. As will be shown below, the thermal transition can be resolved to include a single intermediate, so that the melting sequence is a progression from hairpin, to partially melted intermediate to single strand.

$$\text{Hairpin} \rightarrow \text{Intermediate} \rightarrow \text{Singlestrand}$$

Deconvolution in this case involves first determining Q, F_0, and F_n from the integrated excess heat capacity curve using Eqs. (8–13) as described earlier. The total fraction of intermediates, $\Sigma F_i = 1 - F_0 - F_n$, is then calculated, and is shown in Fig. 10. Normally, as pointed out by Freire and Biltonen (1978a), one can

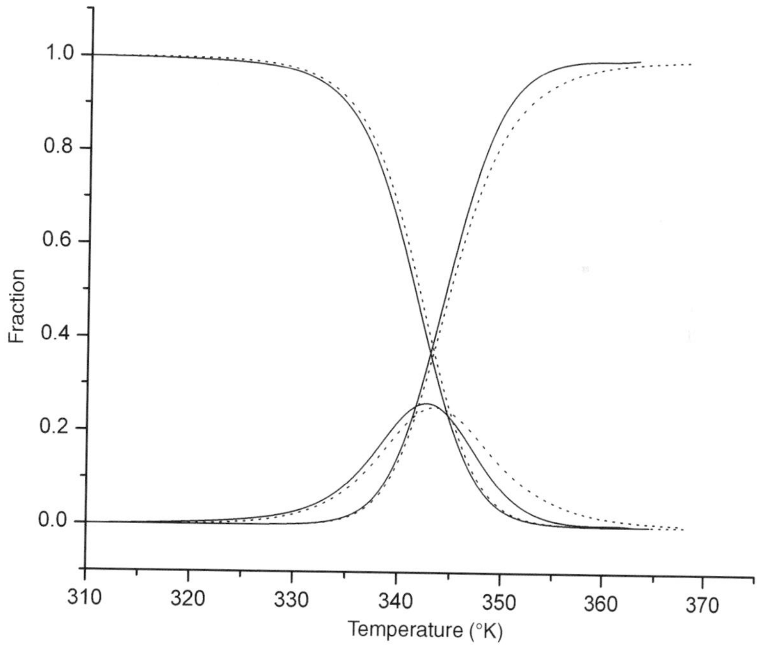

Fig. 10 Fractions of various DNA forms in the melting of the hairpin, Hel2, a 22-base hairpin structure. The upper left curve is the fraction in the initial duplex state, the upper right curve is the fraction in the final single-strand state, and the small curve in the middle is the fraction of intermediate form. Solid curve is calculated directly from the calorimetric data as described in the text, and the dashed curve is calculated from the enthalpy and entropy data determined from analysis of the melting curve.

deconvolute intermediate states through a series of recursion formulae by defining a new partition function that first factors out the contribution from the initial state. This is accomplished by noting that the quantity $\langle \Delta H \rangle/(1 - F_0)$ for a two-state transition is ΔH_n, the total enthalpy for the transition, but for a multistate transition is a curve whose minimum value is the enthalpy difference between the initial and first intermediate state, Δh_1. Thus, a new average excess enthalpy can be defined: $\langle \Delta H_1 \rangle = \langle \Delta H \rangle/(1 - F_0) - \Delta h_1$, and a new partition function, Q_1, which is the function for all remaining states save the first:

$$Q_1 = \exp\left(\int_{T_0}^{T} \frac{\langle \Delta H_1 \rangle}{RT^2} \mathrm{d}T\right) \tag{16}$$

Then, at each temperature the absolute fraction of the molecules in the first intermediate state can be calculated from:

$$F_1 = F_0 \frac{Q_0 - 1}{Q_1} \tag{17}$$

If there are additional intermediate states, this process can be recursively continued until the entire transition curve is accounted for. In addition to Δh_i, for each intermediate state the entropy change, Δs_i, can be obtained from $\Delta h_i/T_{\mathrm{m},i}$, where $T_{\mathrm{m},i}$ is obtained from the point at which $F_i = F_{(i-1)}$. It has been demonstrated that these recursion methods work very well for simulated data, allowing resolution of intermediate states that are only a few degrees separated from each other.

For the real thermal transition in the Hel2 hairpin, however, the recursive approach is problematic for a couple of reasons. First, the fraction of intermediate states never exceeds about 20% of the total, as shown in Fig. 10. This means that even if the enthalpy change for the intermediate state were fairly large, the contribution to the total enthalpy could be small and hard to detect by the recursive method. In addition, it is apparent that even if we assume a single intermediate state, the melting temperatures for the intermediate and final states are close together in magnitude. This makes it more difficult for deconvolution by the recursive approach. We have attempted to resolve the intermediate and final states for this transition and find that because the value of $\langle \Delta H \rangle/(1 - F_0)$ does not show a clear minimum value, the intermediate state enthalpy (and thus new partition function) cannot be resolved. However, if we assume there is only one intermediate state, it is possible to use nonlinear least squares fitting to the integrated average enthalpy $\langle \Delta H \rangle$ to obtain a set of parameters for the transition. This again results from the fact that the calorimetric signal contains information related directly to the partition function:

$$\langle \Delta H \rangle = \frac{\Delta H_i \times K_i + \Delta H_n \times K_i \times K_n}{1 + K_i + K_i \times K_n} = \Delta H_i \times F_i + \Delta H_n \times F_n \tag{18}$$

Here, ΔH_i is the enthalpy difference between the initial and intermediate state and ΔH_n the difference between the initial and final single-strand state, which is the total integrated area of the transition. The Ks are the equilibrium constants for the two transitions:

$$K_i = \exp\left(-\frac{\Delta g_i}{RT}\right) = \exp\left(-\frac{\Delta h_i}{RT} + \frac{\Delta s_i}{R}\right) \tag{19}$$

$$K_n = \exp\left(-\frac{\Delta g_n}{RT}\right) = \exp\left(-\frac{\Delta h_n}{RT} + \frac{\Delta s_n}{R}\right) \tag{20}$$

The Δh and Δs values are for the individual steps in the transition. Note that $\Delta h_i = \Delta H_i$, the enthalpy change for the first step equals the enthalpy difference between the initial and intermediate step, and that $\Delta h_i + \Delta h_n = \Delta H_n$, the sum of the enthalpies of the individual steps must add up to the total enthalpy for the transition. The above treatment assumes no heat capacity change in the unfolding process, so the Δh_i and Δs_i are assumed to be temperature independent. Using Eq. (18) and the experimental $\langle \Delta H \rangle$ for the fitting procedure was chosen for several reasons. One could obtain entropy and enthalpy values for each step by fitting directly on $Q = 1 + K_i + K_i \times K_n$ or on $F_0 = 1/Q$, but the experimental Q requires double integration of the heat capacity function, whereas $\langle \Delta H \rangle$ is a single integration, and thus fewer errors accumulate in the experimental data. In addition, if one were to fit to the excess heat capacity function directly, derivatives of Eq. (18) are required and can lead to complex interdependence of the fitted parameters. Thus, we chose to fit the experimental cumulative enthalpy to the calculated values using Eq. (18), rather than fitting to the excess heat capacity function. Finally, constraints can be imposed on the parameters by the requirements that $\Delta h_i = \Delta H_i$ and that $\Delta h_i + \Delta h_n = \Delta H_n$. These constraints are helpful in the convergence of the parameters on the best set of values.

Figure 11 shows the experimental and best fit curve for the cumulative enthalpy, as well as the derivative of the fitted cumulative enthalpy compared with the original excess heat capacity data. The resulting parameters describing these curves are:

$$\Delta h_i = 56.9 \pm 1.5 \text{ kcal/mol}, \qquad \Delta s_i = 165 \pm 5 \text{ cal/K-mol} \tag{21}$$

$$\Delta h_n = 53.8 \pm 1.4 \text{ kcal/mol}, \qquad \Delta s_n = 157 \pm 5 \text{ cal/K-mol} \tag{22}$$

These numbers are a result of nonlinear least squares fitting, and are the average of six trials with different starting parameters that converged to minimize the reduced chi-squared criterion using the Levenberg–Marquardt method. (In some cases simplex optimization was used prior to the final minimization with the Levenberg–Marquardt procedure.) The errors are estimated from the

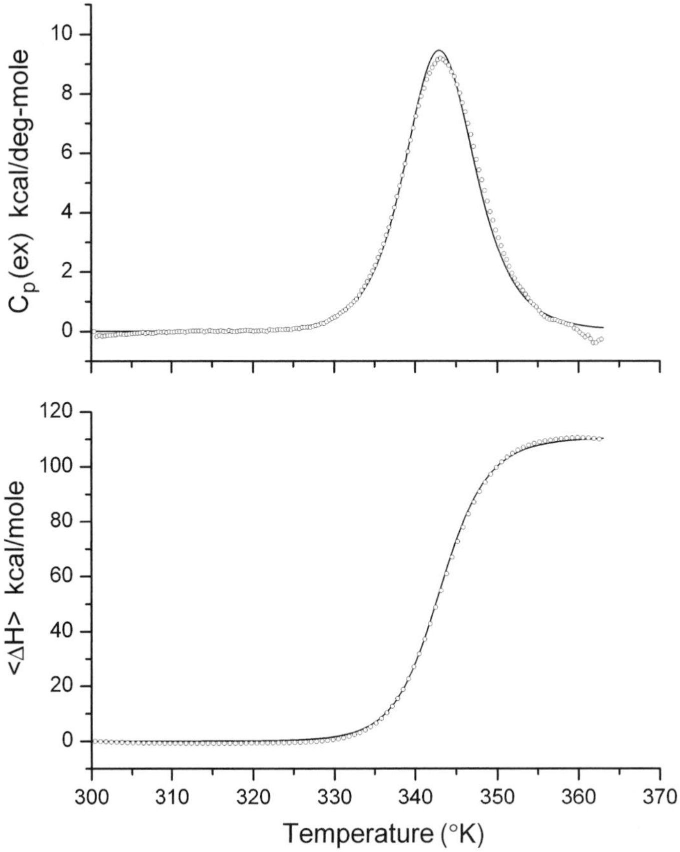

Fig. 11 Cumulative melting enthalpy for 22-base DNA hairpin as function of temperature (bottom). Circles are experimental data, and solid curve is fitted curve with the enthalpy and entropy parameters given in the text. The top curve is the experimental excess heat capacity curve (circles), and the derivative of the fitted cumulative enthalpy from the bottom figure (solid curve).

standard deviation of the six minimization trials in the curve fitting procedure that lead to the smallest chi-squared. Many other trials were made that ended up with chi-squared minimums much higher than the six trials included. These six minima gave reduced chi-squared values of 610 $\pm$ 30 ($\times 10^3$). Introduction of additional random errors of up to 5% of the total enthalpy into the experimental cumulative enthalpy increased the errors by a factor of three and the reduced chi-squared minimum was increased by 40, leading to significantly lower confidence in the parameters obtained.

An additional way to check the fitting results is to compare the calculated values of the fractions of each of the components F_0, F_i, and F_n calculated from the fitted parameters, with the experimental values obtained from the calorimetric data. This comparison helps to check on whether the assumption of a single intermediate state applies to the hairpin melt. Recall that $F_0 = 1/Q$, $F_i = K_i/Q$, $F_n = K_iK_n/Q$; and that $Q = 1 + K_i + K_i K_n$ is the partition function for the case involving a single intermediate. Using the resolved parameters from above to calculate the K values, the fractions can be compared with experimental values obtained from Eqs. (9) and (13) for F_0 and F_n, and the experimental value of $F_i = 1 - F_0 - F_n$. This comparison is shown in Fig. 10, and in general, the results from calculation are consistent with experimentally derived numbers. Because the experimental curve is a bit noisy on the high temperature side, the agreement on the higher temperature edge of the fraction curves is not as good. But, the consistency of the fraction values derived from Δh and Δs parameters with the experimental curves lends confidence in the derived thermodynamic values from the analysis using a single intermediate state.

This example of the analysis of the unfolding of the oligomeric hairpin illustrates the power of DSC for obtaining thermodynamic information on small DNA structures. While there are examples of a number of DNA oligomers that melt via a simple two-state process, it is important to check for evidence of sequential intermediates in cases for which the van't Hoff enthalpy is less than the calorimetric value. As presented here, even rather short pieces of DNA can pass through partially melted intermediate states.

V. Summary

The intent of this chapter has been to point out that DSC can provide not only thermodynamic information regarding DNA unfolding, but can also establish some of the details of the unfolding process, particularly with regard to the presence of intermediate states. Since there are a wide variety of DNA forms that can unfold differently, depending on size, base composition, and conformation, scanning calorimetry seems an ideal methodology to study thermally induced unfolding. From simple two-state to multistate transitions with many intermediates, all are accessible to analysis. The fractions of each state present as a function of temperature, and individual state thermodynamic properties are revealed by appropriate analysis. Thus, a number of problems can be addressed by application of DSC to DNA unfolding with particular interest in the distribution of intermediates in the unfolding process. The way the fraction of intermediates changes with solution conditions, base composition, and base sequence is accessible by these techniques.

Different DNA conformations respond in different unfolding patterns on thermal inducement and thus are amenable to the analyses presented earlier. Melting patterns have not been thoroughly analyzed for a number of structures relating to DNA unfolding, including triplex forms (Husler and Klump, 1995; Makube and Klump, 2000; Spink and Chaires, 1999), telemeric and quadruplex forms (Kankia and Marky, 2001; Li *et al.*, 2003; Ren *et al.*, 2002), and other oligomeric structures (Owczarcy *et al.*, 1999; Vallone *et al.*, 1999). As the sensitivity and accuracy of measurement increases, DSC offers many opportunities for research in biophysical problem solving in these and many other areas.

Acknowledgments

The author would like to thank Professor Brad Chaires from the University of Louisville Medical School for making available some of the DSC instrumentation used for several of the studies reported here.

References

Ali, S., Lin, H. N., Bittman, R., and Huang, C. H. (1989). Binary mixtures of saturated and unsaturated mixed-chain phosphatidylcholine. A differential scanning calorimetry study. *Biochemistry* **28,** 522–528.

Chalikian, T. V., Volker, J., Plum, G. E., and Breslauer, K. J. (1999). A more unified picture for the thermodynamics of nucleic acid duplex melting: A characterization by calorimetric and volumetric techniques. *Proc. Natl. Acad. Sci. USA* **96,** 7853–7858.

Cooper, A., and Johnson, C. M. (1994). Differential scanning calorimetry. *Methods Mol. Biol.* **22,** 125–136.

Freire, E. (1994). Statistical thermodynamic analysis of differential scanning calorimetry data: Structural deconvolution of heat capacity function of proteins. *Methods Enzymol.* **240,** 502–530.

Freire, E. (1995). Differential scanning calorimetry. *Methods Mol. Biol.* **40,** 191–218.

Freire, E., and Biltonen, R. L. (1978a). Statistical mechanical deconvolution of thermal transitions in macromolecules. I. Theory and application to homogeneous systems. *Biopolymers* **17,** 463–479.

Freire, E., and Biltonen, R. L. (1978b). Statistical mechanical deconvolution of thermal transitions in macromolecules. II. General treatment of cooperative phenomena. *Biopolymers* **17,** 481–496.

Freire, E., and Biltonen, R. L. (1978c). Statistical mechanical deconvolution of thermal transitions in macromolecules. III. Application to double-stranded to single-stranded transitions of nucleic acids. *Biopolymers* **17,** 497–510.

Freyer, M. W., Buscaglic, R., Nguyen, B., Wilson, W. D., and Lewis, E. A. (2006). Binding of netropsin and 4,6-diamidino-2-phenylindole to an A2T2 DNA hairpin: A comparison of biophysical techniques. *Anal. Biochem.* **355,** 259–266.

Haq, I., Chowdhry, B. Z., and Jenkins, T. C. (2001). Calorimetric techniques in the study of high-order DNA-drug interactions. *Methods Enzymol.* **340,** 109–149.

Holbrook, J. A., Capp, M., Saecker, R. M., and Record, M. T. (1999). Enthalpy and heat capacity changes for formation of an oligomeric DNA duplex: Interpretation in terms of coupled processes of formation and association of single-stranded helices. *Biochemistry* **38,** 8409–8422.

Huang, C., and Li, S. (1999). Calorimetric and molecular mechanics studies of the thermotropic phase behavior of membrane phospholipids. *Biochim. Biophys. Acta* **1422,** 273–307.

Husler, P. L., and Klump, H. H. (1995). Thermodynamic characterization of a triple helical junction containing a Hoogsteen branch point. *Arch. Biochem. Biophys.* **322,** 149–166.

Kankia, B. I., and Marky, L. A. (2001). Folding of the thrombin aptamer into G-quadruplex with Sr^{++}: stability, heat and hydration. *J. Amer. Chem. Soc.* **123,** 10799–10804.

Li, W., Miyoshi, D., Nakano, S., and Sugimoto, N. (2003). Structural competition involving G-quadruplex DNA and its components. *Biochemistry* **42,** 11736–11744.

Lobo, B. A., Rogers, S. A., Choosakoonkriang, S., Smith, J. G., Koe, G., and Middaugh, C. R. (2002). Differential scanning calorimetric studies of the thermal stability of plasmid DNA complexed with cationic lipids and polymers. *J. Pharm. Sci.* **91,** 454–466.

Lopez, M. M., and Makhatadze, G. I. (2002). Differential scanning calorimetry. *Methods Mol. Biol.* **173,** 113–119.

Makhatadze, G. I., and Privalov, P. L. (1995). Energetics of protein structure. *Adv. Protein Chem.* **47,** 307–425.

Makube, N., and Klump, H. H. (2000). A four-way junction with triple helical arms: design, characteristics, stability. *Arch. Biochem. Biophys.* **377,** 31–42.

Mason, J. T. (1998). Investigation of phase transitions in bilayer membranes. *Methods Enzymol.* **295,** 468–494.

McElhaney, R. N. (1986). Differential scanning calorimetric studies of lipid-protein interactions in model membranes. *Biochim. Biophys. Acta* **864,** 361–421.

Owczarcy, R., Vallone, P. M., Goldstein, R. F., and Benight, A. S. (1999). Studies of DNA Dumbbells VII: Evaluation of the next-nearest-neighbor sequence-dependent interactions in duplex DNA. *Biopolymers* **52,** 29–56.

Plotnikov, V. V., Brandts, J. M., Lin, L. N., and Brandts, J. F. (1997). A new ultrasensitive scanning calorimeter. *Anal. Biochem.* **250,** 237–244.

Plotnikov, V. V., Rochalski, A., Brandts, J. M., Brandts, J. F., Williston, S., Frasca, V., and Lin, L. N. (2002). An autosampling differential scanning calorimeter instrument for studying molecular interactions. *Assay Drug Dev. Technol.* **1,** 83–90.

Plum, G. E., and Breslauer, K. J. (1995). Calorimetry of proteins and nucleic acids. *Curr. Opin. Struct. Biol.* **5,** 682–690.

Privalov, G., Kavina, V., Freire, E., and Privalov, P. L. (1995). Precise scanning calorimeter for studying thermal properties of biological macromolecules in dilute solution. *Anal. Biochem.* **232,** 79–85.

Read, C. M., and Jelesarov, I. (2001). Calorimetry of protein-DNA complexes and their components. *Methods Mol. Biol.* **148,** 511–533.

Ren, J., Qu, X., Trent, J. O., and Chaires, J. B. (2002). Tiny telomere DNA. *Nucleic Acids Res.* **30,** 2307–2315.

Spink, C. H., and Chaires, J. B. (1999). Effects of hydration, ion release and excluded volume on the melting of triplex and duplex DNA. *Biochemistry* **38,** 496–508.

Spink, C. H., Garbett, N., and Chaires, J. B. (2007). Enthalpies of DNA melting in the presence of osmolytes. *Biophys. Chem.* **126,** 176–185.

Tikhomirova, A., Taulier, N., and Chalikian, T. V. (2004). Energetics of nucleic acid stability: The effect of ΔC_p. *J. Amer. Chem. Soc.* **126,** 16387–16394.

Vallone, P. M., and Benight, A. S. (2000). Thermodynamic, spectroscopic, and equilibrium binding studies of DNA sequence context effects in four 40 base-pair deoxyoligonucleotides. *Biochemistry* **39,** 7835–7846.

Vallone, P. M., Paner, T. M., Hilario, J., Lane, M. J., Faldasz, B. D., and Benight, A. S. (1999). Melting studies of short DNA hairpins: Influence of loop sequence and adjoining base pair identity on hairpin thermodynamic stability. *Biopolymers* **50,** 425–442.

Volker, J., Blake, R. D., Delcourt, S. G., and Breslauer, K. J. (1999). High-resolution calorimetric and optical melting profiles of DNA plasmids: Resolving contributions from intrinsic melting domains and specifically designed inserts. *Biopolymers* **50,** 303–318.

CHAPTER 6

Analytical Ultracentrifugation: Sedimentation Velocity and Sedimentation Equilibrium

James L. Cole,[*,†] Jeffrey W. Lary,[*] Thomas P. Moody,[‡] and Thomas M. Laue[‡]

[*]National Analytical Ultracentrifugation Facility
University of Connecticut
Storrs, Connecticut 06269

[†]Department of Molecular and Cell Biology
University of Connecticut
Storrs, Connecticut 06269

[‡]Center to Advance Molecular Interaction Science
University of New Hampshire
Durham, New Hampshire 03824

METHODS IN CELL BIOLOGY, VOL. 84

0091-679X/08 $35.00
DOI: 10.1016/S0091-679X(07)84006-4

Abstract

Analytical ultracentrifugation (AUC) is a versatile and powerful method for the quantitative analysis of macromolecules in solution. AUC has broad applications for the study of biomacromolecules in a wide range of solvents and over a wide range of solute concentrations. Three optical systems are available for the analytical ultracentrifuge (absorbance, interference, and fluorescence) that permit precise and selective observation of sedimentation in real time. In particular, the fluorescence system provides a new way to extend the scope of AUC to probe the behavior of biological molecules in complex mixtures and at high solute concentrations. In sedimentation velocity (SV), the movement of solutes in high centrifugal fields is interpreted using hydrodynamic theory to define the size, shape, and interactions of macromolecules. Sedimentation equilibrium (SE) is a thermodynamic method where equilibrium concentration gradients at lower centrifugal fields are analyzed to define molecule mass, assembly stoichiometry, association constants, and solution nonideality. Using specialized sample cells and modern analysis software, researchers can use SV to determine the homogeneity of a sample and define whether it undergoes concentration-dependent association reactions. Subsequently, more thorough model-dependent analysis of velocity and equilibrium experiments can provide a detailed picture of the nature of the species present in solution and their interactions.

I. Introduction

For over 75 years, analytical ultracentrifugation (AUC) has proven to be a powerful method for characterizing solutions of macromolecules and an indispensable tool for the quantitative analysis of macromolecular interactions (Cole and Hansen, 1999; Hansen *et al.*, 1994; Hensley, 1996; Howlett *et al.*, 2006; Scott and Schuck, 2005). Because it relies on the principle property of mass and the fundamental laws of gravitation, AUC has broad applicability and can be used to analyze the solution behavior of a variety of molecules in a wide range of solvents and over a wide range of solute concentrations. In contrast to many commonly used methods, during AUC, samples are characterized in their native state under biologically relevant solution conditions. Because the experiments are performed in free solution, there are no complications due to interactions with matrices or surfaces. Because it is nondestructive, samples may be recovered for further tests following AUC. For many questions, there is no satisfactory substitute method of analysis.

Two complementary views of solution behavior are available from AUC. Sedimentation velocity (SV) provides first-principle, hydrodynamic information about the size and shape of molecules (Howlett *et al.*, 2006; Laue and Stafford, 1999; Lebowitz *et al.*, 2002). Sedimentation equilibrium (SE) provides first-principle, thermodynamic information about the solution molar masses, stoichiometries, association constants, and solution nonideality (Howlett *et al.*, 2006; Laue, 1995). Different experimental protocols are used to conduct these two types of analyses. This chapter will cover the fundamentals of both velocity and equilibrium AUC.

A. Types of Problems That Can be Addressed

AUC provides useful information on the size and shape of macromolecules in solution with very few restrictions on the sample or the nature of the solvent. The fundamental requirements for the sample are (1) that it has an optical property that distinguishes it from other solution components, (2) that it sediments or floats at a reasonable rate at an experimentally achievable gravitational field, and (3) that it is chemically compatible with the sample cell. The fundamental solvent requirements are its chemical compatibility with the sample cell and its compatibility with the optical systems. The range of molecular weights suitable for AUC exceeds that of any other solution technique from a few hundred Daltons (e.g., peptides, dyes, oligosaccharides) to several hundred-million Daltons (e.g., viruses, organelles).

Different sorts of questions may be addressed by AUC depending on the purity of the sample. Detailed analyses are possible for highly purified samples with only a few discrete macromolecular components. Some of the thermodynamic parameters that can be measured by AUC include the molecular weight, association state, and equilibrium constants for reversibly interacting systems. AUC can also provide hydrodynamic shape information. For samples containing many components, or containing aggregates or lower molecular weight contaminants, or high concentration samples, size distributions and average quantities may be determined. While these results may be more qualitative than those from more purified samples, the dependence of the distributions on macromolecular concentration, ligand binding, pH, and solvent composition can provide unique insights into macromolecular behavior.

II. Basic Theory

Mass will redistribute in a gravitational field until the gravitational potential energy exactly balances the chemical potential energy at each radial position. If we monitor the rate at which boundaries of molecules move during this redistribution, then we are conducting a SV experiment. If we determine the concentration distribution after equilibrium is reached, then we are conducting an equilibrium sedimentation experiment.

A. Sedimentation Velocity

We can understand a SV experiment by considering the forces acting on a molecule during a SV experiment. The force on a particle due to the gravitational field is just $M_p\omega^2 r$, where M_p is the mass of the particle, ω is the rotor speed in radians per second ($\omega = 2\pi \times \text{rpm}/60$), and r is the distance from the center of the rotor. A counterforce will be exerted on the particle by the mass of solvent, M_s, displaced as the particle sediments, $M_s\omega^2 r$. The net force is $(M_p - M_s)\omega^2 r$. The mass of solvent displaced is just the M_p times partial specific volume of the particle, $\bar{v}$ (cm^3/g), times the density of the solvent, ρ (g/cm^3). So the effective or buoyant mass of the particle is $M_b = M_p(1 - \bar{v}\rho)$. The last force to consider is the frictional force developed by the motion of the particle through the solvent, which is given by fv, where f is the frictional coefficient and v is the velocity. Balancing these forces, we obtain the following relationship (see e.g., Fujita, 1975; Tanford, 1961; Williams *et al.*, 1958):

$$s \equiv \frac{v}{\omega^2 r} = \frac{M_b}{f} \tag{1}$$

which is also a definition of the sedimentation coefficient, s, as the ratio of the velocity to the centrifugal field. In terms of molecular parameters, Eq. (1) indicates that s is proportional to the buoyant molar mass, M_b, and inversely proportional to the frictional coefficient, f. Diffusion causes the sedimenting boundary to spread with time. Hence by monitoring the motion and shape of a boundary, it is possible to determine both the sedimentation coefficient and the translational diffusion coefficient, D. From the Stokes–Einstein relationship, we know that $D = RT/N_a f$, where R is the gas constant (erg/mol °K), T is the absolute temperature, and N_a is the Avogadro's number.

The time evolution of the radial concentration distribution during sedimentation is given by the Lamm equation (see e.g., Fujita, 1975; Williams *et al.*, 1958):

$$\frac{\partial c}{\partial t} = D\left[\frac{\partial^2 c}{\partial r^2} + \frac{1}{r}\frac{\partial c}{\partial r}\right] - s\omega^2\left[r\frac{\partial c}{\partial r} + 2c\right] \tag{2}$$

where c is the weight concentration of macromolecules and t is time. The optical systems on the analytical ultracentrifuge supply the radial concentration distribution at time intervals during the course of an experiment, $c(r, t)$, and the instrument provides the rotor speed, ω. The quantities sought in a velocity sedimentation experiment are s and D. There are no exact solutions to the Lamm equation: approximate (Behlke and Ristau, 1997; Philo, 1994) and numerical (Demeler and Saber, 1998; Schuck, 1998; Stafford and Sherwood, 2004) solutions form the basis of many SV analysis programs used to extract s and D from AUC data. By taking the ratio s/D, the frictional contribution to these parameters is removed and the result is proportional to the buoyant molar mass, M_b, through the Svedberg equation

$$\frac{s}{D} = \frac{M_b}{RT} \tag{3}$$

Both the Lamm and Svedberg equations, as presented above, are starting points for the equations that apply to real chemical systems. The equations for real systems are presented below, along with the assumptions and simplifications often used to extract information.

B. Sedimentation Equilibrium

When the centrifugal force is sufficiently small, an equilibrium concentration distribution of macromolecules is obtained throughout the cell where the flux due to sedimentation is exactly balanced by the flux due to diffusion. The shape of this concentration gradient can be derived using a variety of approaches (Fujita, 1975; Tanford, 1961; Williams *et al.*, 1958). For an ideal single noninteracting species, the equilibrium radial concentration gradient, $c(r)$, is given by:

$$c(r) = c_0 \exp\left[\frac{M_b\omega^2}{RT}\left(\frac{r^2 - r_0^2}{2}\right)\right] = c_0 \exp\left[\sigma\left(\frac{r^2 - r_0^2}{2}\right)\right] \tag{4}$$

where c_0 is the concentration at an arbitrary reference distance r_0. The term $M_b\omega^2/RT$ is often referred to as the reduced molecular weight, σ. SE experiments provide a very accurate way to determine M and consequently the oligomeric state of biomolecules in solution. Deviations from the simple exponential behavior described by Eq. (4) can result from the presence of either multiple noninteracting or interacting macromolecular species or thermodynamic nonideality.

III. Dilute Solution Measurements

For dilute solutions containing a single macromolecular component, detailed information is available from both SE and SV analysis (Cole and Hansen, 1999; Hansen *et al.*, 1994; Hensley, 1996; Howlett *et al.*, 2006; Laue and Stafford, 1999; Lebowitz *et al.*, 2002; Scott and Schuck, 2005). What constitutes a dilute solution depends somewhat on the nature of the macromolecule being studied and the solvent it is in. For this review, we will consider a system dilute if there is not significant hydrodynamic or thermodynamic nonideality (below), and if gradients in the solvent component concentrations are small enough to be neglected in the analysis. For globular proteins of moderate charge ($z < \sim 15$) at physiological salt concentrations, protein concentrations <2–3 mg/ml can be considered dilute. By contrast, nucleic acids or polysaccharides may form highly nonideal solutions at concentrations <0.1 mg/ml. It should be noted that only electrically neutral particles sediment. For proteins and nucleic acids in near physiological salt concentrations (ionic strengths >100 mM), there are sufficient counter ions in the immediate surroundings that the sedimentation coefficient is relatively insensitive to salt concentration. However, at lower ionic strengths (<10 mM), a greater region of solution is required to produce a neutral particle that can sediment. That is, the apparent radius of a protein or nucleic acid will increase at low ionic strength and, consequently, the sedimentation coefficient will decrease. The slowing of sedimentation at low ionic strength is called the primary charge effect (Fujita, 1975; Williams *et al.*, 1958).

Solvents that contain components at high concentrations that sediment sufficiently to form a significant gradient (e.g., 10% sucrose or 8 M urea) will affect sedimentation rates (Schuck, 2004). Even for dilute solutions, a series of experiments should be conducted at different macromolecular concentrations so that the concentration dependence of s and D may be determined. If these quantities are invariant or weakly dependent on macromolecular concentration, then the analysis below is appropriate. Under these conditions and using the analysis methods and computer programs listed below, s and D (hence M, through the Svedberg equation) may be obtained with good accuracy (s to within 2%, D to within 5%, and M to within 5%). Note that analysis using the Svedberg equation is only valid for a single, noninteracting species, or a mixture of noninteracting species.

The frictional coefficient obtained from sedimentation measurements is often interpreted in terms of the molecular size and shape through the Stokes relationship:

$$f = 6\pi\eta R_s \tag{5}$$

where η is the solution viscosity and R_S is the Stokes radius, which contains contributions from both molecular asymmetry and solvation (Williams *et al.*, 1958). In order to interpret R_S in terms of molecular asymmetry, it is necessary to have a good estimate of the solvation, usually expressed as the number of grams of solvent bound per gram of macromolecule. Although estimates of the *hydration* (i.e., bound water) of macromolecules are available (Perkins, 2001), these values neglect the amount of other solvent components that may be bound, and they do not reflect the physical meaning of R_S, which includes coupling of the macromolecular flow with flows of other solvent components. If the macromolecule is ionic, then flow-coupling with solvent ions will contribute significantly to R_S. This is particularly true at low ionic strengths where a large R_S is required to maintain electroneutrality during sedimentation. Sometimes, the frictional parameters are further interpreted using simple structural models consisting of ellipsoids of revolution or cylinders to assess molecular asymmetry (Cantor and Schimmel, 1980; Tanford, 1961). Hydrodynamic properties can also be interpreted using more complex structural models composed of assemblies of spherical beads (Byron, 2000; Garcia De La Torre *et al.*, 2000; Rai *et al.*, 2005). It is important to realize that there are pitfalls associated with interpreting s or R_S in terms of molecular dimensions determined by application of these models (see Chapter 12 by Byron, this volume). That said, *changes* in R_s (e.g., with addition of a ligand) usually reflect changes in molecular size. Using the technique of difference sedimentation, changes in R_s of only a few Angstroms may be detected (Richards and Schachman, 1957). Furthermore, the concentration dependence of the sedimentation coefficient can be useful in assessing the relative asymmetry of different molecules (Hattan *et al.*, 2001).

For a sample containing only one type of molecule, a useful quantity to report is the standard sedimentation coefficient, $s^{o}_{20,w}$. This quantity is obtained by extrapolating sedimentation coefficients determined at finite concentrations to zero concentration (i.e., $s^{o} = \lim_{c \to 0} s$), then adjusting s^{o} for the solvent density and viscosity to the density and viscosity of water at 20 °C. Values of $s^{o}_{20,w}$ are useful (e.g., 30 or 50 S ribosomal subunits) since they are a primary quantity. Thus, any differences in $s^{o}_{20,w}$, for example, due to changes in pH, reflect differences in the molecule.

One common application of AUC to dilute solutions is to determine the sedimentation coefficient distribution of macromolecules [e.g., $g(s^*)$ or $c(s)$]. Analysis methods and programs for obtaining $g(s^*)$ and $c(s)$ are described later. For solutions containing a single component, the abundance and sedimentation coefficients of irreversible aggregates or of degradation products may be determined. Often a simple relationship between s and M may be used to identify particular peaks as belonging to certain oligomers (e.g., dimer, trimer, etc.) or certain fragments of the monomer. Sedimentation coefficient distributions are used widely in the pharmaceutical industry to assess the stability of protein formulations and to characterize preparations of inherently heterogeneous samples (e.g., vaccines based on bacterial cell wall preparations).

IV. Concentrated and Complex Solutions

If a solution contains a single macromolecular component at high concentration, then one may use SE analysis to extract thermodynamic information. In particular, the concentration dependence of the apparent molecular weight, M_{app}, divided into the actual molecular weight (i.e., M/M_{app}) yields the activity coefficient, γ. The product of the activity coefficient and weight concentration yields the chemical activity (or apparent concentration). For an ideal solution, $\gamma = 1$, and the apparent concentration equals the actual concentration. For a macromolecule that undergoes self-association, γ will be <1, whereas γ will be >1 for a macromolecule that repels itself (e.g., due to excluded volume or charge–charge repulsion). While a more quantitative description of a macromolecule's behavior may be desired (e.g., what is the association stoichiometry and strength), for many questions simply knowing a macromolecule's qualitative behavior may be sufficient. More quantitative analysis at high concentration is best performed using SE (Harding *et al.*, 1992; Jiménez *et al.*, 2007; Roark and Yphantis, 1969).

The sedimentation coefficient depends on the total macromolecular concentration. In the simplest analysis, the viscosity of solutions increases with increasing concentration (above); hence the observed sedimentation coefficient decreases. However, any specific interactions between molecular species also must be considered (Fujita, 1975).

The availability of a fluorescence detector for the XLI analytical ultracentrifuge (AU-FDS, Aviv Biomedical, Lakewood, NJ) allows the rigor and power of AUC to be applied to complex, concentrated solutions such as cell lysates, serum, cerebral spinal fluid, urine, and cell culture media. As currently used, AUC is applied primarily to dilute solutions. For dilute solutions, $s^{\circ}_{20,w}$ and $D^{\circ}_{20,w}$ are considered to be properties of a molecule. In fact, however, s and D are system properties whose values depend on the concentrations of all other components in the solution. Thus, the interpretation of the data for many of the most interesting applications of the AU-FDS will require more detailed analysis than is available currently. Even now, however, phenomenological analysis of sedimentation data from complex, concentrated solutions will provide useful insights into the solution behavior of appropriately labeled molecules. For example, a mass-action association between components A^* and B, where A^* is the only labeled component, will lead to an apparent increased sedimentation coefficient of A over what would be expected simply on the basis of the viscosity (Kroe, 2005).

V. Instrumentation and Optical Systems

The analytical ultracentrifuge is similar to a high-speed preparative centrifuge in that a spinning rotor provides a gravitational field large enough to make molecular-sized particles sediment. What distinguishes the Beckman Coulter (Fullerton, CA) XLI analytical ultracentrifuge from a high-speed preparative centrifuge is the specialized rotors, sample holders and optical systems that permit the observation of samples during sedimentation. To view the sample, the analytical rotor has holes through it to hold sample containers commonly called cells. Each cell contains a centerpiece, with chambers (called channels) to hold the liquid samples. The centerpiece, in turn, is sealed between windows to permit the passage of light through the channels, thus allowing the cell contents to be viewed. Centerpieces are made out of a variety of tough, inert materials such as epoxy, anodized aluminum, or titanium. For biological materials, the epoxy-based centerpieces are used most frequently.

The epoxy contains a small amount of either charcoal or aluminum powder filler (~5 wt.%) for improved thermal conductance. With very few exceptions, either centerpiece type may be used. Depending on the type of experiment that will be performed, centerpieces are available that can hold several samples each. Rotors for the XLI are available that hold either four or eight cells, hence many samples may be analyzed at once.

The fundamental measurements in AUC are radial concentration distributions. These concentration distributions, called "scans," are acquired at intervals ranging from minutes (for velocity sedimentation) to hours (for equilibrium sedimentation). As the rotor spins, each cell passes through the optical paths of detectors capable of measuring the concentration of molecules at closely spaced radial intervals in the cell. There are three commercially available optical detectors for the XLI to measure the concentration distributions: an absorbance spectrophotometer and Rayleigh interferometer from Beckman Coulter and the fluorescence detector from Aviv Biomedical. All subsequent analysis of sedimentation data relies on the quantity and quality of data available from these detectors. A comparison of the capabilities of the three optical systems is provided in Table I. As can be seen from these data, the three optical systems are complementary. Each optical system has its strengths and weaknesses (Table II). A more detailed comparison of the absorbance and interference optical systems is available (Laue, 1996). A summary of the properties of each optical system is presented below. In addition to these real-time optical systems for SE experiments, tracer sedimentation methods have been described where the concentration gradients of labeled molecules are determined following centrifugation using a microfractionator (Howlett *et al.*, 2006; Rivas and Minton, 2003).

Table I
Capabilities of Optical Systems

	Absorbance	Interference	Fluorescence
Sensitivity[a]	0.1 OD	0.1 mg/ml	100 pM
Range[b]	2–3 logs	3–4 logs	6–8 logs
Precision[c]	Good	Excellent	Good

[a]The sensitivity is the minimum amount of signal needed to obtain good results. For the interference optical system, the signal is relatively insensitive to the type of biological material, so that a 1 mg/ml sample results in a displacement of ~3.25 fringes. Sensitivity of the fluorescence system is for fluorescein (molar extinction coefficient ~65,000 at 488 nm, quantum yield ~0.9).

[b]The range refers to the concentration range accessible by the optical system.

[c]The precision of the optical system is estimated by comparing the signal-to-noise ratio. For the absorbance and fluorescence detectors, this ratio is ~100 (e.g., the uncertainty in a 1 OD reading is about 0.01 OD, and the uncertainty in a fluorescence intensity reading is about 1% of the signal). For the current interference optical system, this ratio is closer to 1000.

Table II
Strengths and Weaknesses of Optical Systems

Characteristic	Absorbance	Interference	Fluorescence
Radial resolution[a]	20–50	10	20–50
Scan time[b]	60–300	1–10	60–90
When to use[c]	• Selectivity • Sensitivity • Nondialyzable components	• Solvent absorbs light • Solute does not absorb light • Accuracy needed • Short solution columns	• Selectivity • Sensitivity • Small sample quantities • Nondialyzable components

[a]Approximate spacing (in microns) between data points such that each measurement can be considered an independent estimate of the concentration.

[b]The minimum time (in seconds) required to complete one radial scan. The time listed for the fluorescence system is the time needed to scan all of the samples (Laue, 2006).

[c]Selectivity refers to the absorbance and fluorescence systems' ability to discriminate between components based on their spectral properties. Since the Rayleigh interference optics relies on differences in the refractive index of the sample and reference solutions, it provides no selectivity. By contrast, the interference optics do not require that samples have an appropriate chromophore, and may be used so long as the solvent does not absorb light at ~670 nm. The interference optics require that samples are at dialysis equilibrium with the reference solution; hence, they should not be used for samples containing nondialyzable components (e.g., detergents). The greater radial resolution of the interference optics allows them to be used with the eight-channel "short-column" centerpieces (Yphantis, 1960).

A. Absorbance

Absorbance is the most frequently used detector for the analytical ultracentrifuge (Laue, 1996). This optical system is the easiest to use and operates as a standard double-beam spectrophotometer. Under conditions where the Beer–Lambert law holds, the absorbance signal is directly proportional to solute concentration: $A = \varepsilon cl$, where ε is the solute's weight extinction coefficient, c is the weight concentration, and l is the sample path length (1.2 cm for standard centerpieces). The rated precision of the absorbance system is ± 0.01 OD although it is usually better than this. The noise is primarily stochastic. Hence, the noise appears as a high-frequency "fuzz" around the signal. The scans typically contain little systematic noise that is either radially independent (e.g., the entire scan is shifted up or down) or time independent (e.g., a feature, such as a scratch, that does not move from scan to scan). As described later, the other optical systems will have very different noise characteristics.

Although the absorbance optics are useable over a wavelength range from 190 to 800 nm, limited light intensity may restrict the useable range for two reasons. First, many standard biological solvent components absorb strongly at short wavelengths (e.g., disulfides, carbonyl oxygens, nitrogenous compounds, some detergents), so that solvent components should be selected with care when data collection at short wavelengths is desired. A simple rule of thumb is that the solvent absorbance at the desired operating wavelength should be less than ~0.5 OD, using water as the reference. Second, output from the Xe light source is blue-rich and "spiky," with the maximum output at 230 nm and very low red light output. If there is uncertainty about what wavelength to use, one can perform a wavelength scan using the XLI. It is best to view these data as both intensities and absorbances to ensure the data will have a good signal-to-noise ratio (Laue, 1996).

When preparing samples for the absorbance system, it is best if they have an absorbance between 0.2 and 1.0 OD. If you are interested in gathering data over a wide concentration range, you may want to scan different samples at different wavelengths. While this is permitted, the XLI wavelength selector is notoriously imprecise (±3 nm) at setting the monochromator back to the same wavelength. Consequently, if your experimental protocol involves scanning samples at different wavelengths, you should make sure the wavelengths used are in "flat" portions of the sample's absorbance spectrum, at peaks and valleys, and not in spectral regions where the absorbance is changing rapidly with wavelength. Otherwise, absorbance readings will not be reproducible from scan to scan. While some analysis programs (e.g., ULTRASCAN) have built in routines to adjust data for these variations, it is best to avoid the problem.

Of the three optical systems, the absorbance system requires the longest to complete a scan. For SE, long scan times are not a problem. However, for SV experiments, the long scan times may limit the amount of data that can be acquired over the course of an experiment. In particular, at rotor speeds above 6000 rpm, the repetition rate of the pulsed Xe lamp (100 Hz) limits the data acquisition rate. Consequently, absorbance protocols for velocity experiments typically use a fairly coarse radial step size (0.003 cm) with no data averaging. Improvements in the absorbance system are being developed to overcome the scan speed limitation, as well as the poor precision of the wavelength selection mechanism.

When used in a traditional double beam mode (each sample having a corresponding reference solution), up to three (four-hole rotor) or seven (eight-hole rotor) samples may be analyzed. It is also possible to the use intensity data for SV analysis (Kar *et al.*, 2000), thus doubling the number of samples per experiment. You should make sure the material in the sample and reference channels have approximately the same absorbance reading, and that the absorbance is not too high (<0.5 OD). Otherwise, the automatic gain control logic of the XLA may result in low intensity readings from the sample channel, or it may change the gain settings from one scan to the next, resulting in unusable data.

B. Interference

The signal from the Rayleigh interference optical system consists of equally spaced horizontal fringes whose vertical displacement, ΔY, is directly proportional to the optical path difference between light beams passing through the sample and reference solutions. Any refractive index difference, Δn, between the two solutions contributes to the optical path length so that $\Delta Y = \Delta nl/\lambda$, where l is the optical path length and λ is the wavelength of the light source (Richards and Schachman, 1959; Yphantis, 1964). For a nondialyzable solution component, the refractive index difference is proportional to the refractive index increment: $\Delta n = c(\mathrm{d}n/\mathrm{d}c)$ and the extinction coefficient ε is replaced by

$$\varepsilon \rightarrow \left(\frac{\mathrm{d}n}{\mathrm{d}c}\right)\frac{M}{\lambda} \tag{6}$$

For proteins, dn/dc is relatively independent of composition with an average value of 0.186 ml/g (Huglin, 1972). For the XLA, $\lambda = \sim$670 nm and the sample path length is 1.2 cm, so that a 1 mg/ml sample results in a fringe displacement of ~3.25 fringes (Laue, 1996).

Because the signal from interference optical system does not rely on a chromophore, colorless compounds (e.g., polysaccharides and lipids) may be characterized by AUC. Indeed, any material having a refractive index different from the reference will contribute to the signal. This is both a useful characteristic and poses possible problems if a sample contains a nondialyzable substance (e.g., detergents, lipid micelles). Thus, while the molecular weights and partial specific volumes of detergents may be characterized using the interference optics (Reynolds and McCaslin, 1985), samples containing detergents are best studied using either absorbance or fluorescence optics.

Unlike the absorbance system, the interference signal has very little stochastic noise. However, since any path length difference between the sample and reference beams contributes to the fringe displacement, even tiny optical imperfections (dust, oil, dirt, scratches on the lenses and mirrors) are visible in the signal. Consequently, there is significant time-independent systematic noise. Furthermore, the conversion of the interference image to fringe displacement measurements uses a Fourier analysis to determine the fractional fringe displacement (DeRosier *et al.*, 1972) for which the first radial position is arbitrarily assigned a zero fringe displacement. Since the fringes cannot be traced through certain image features (e.g., menisci), fringe displacement data also contain radially independent systematic noise. Both types of systematic noise must be removed prior to data analysis (Fujita, 1975; Schuck and Demeler, 1999; Stafford, 1992).

The precision and accuracy of the interference optical system places a premium on the optical components. Any variation in the window or centerpiece flatness $>0.01\ \lambda$ will cause a vertical shift in the image. A severe enough wedge ($>30\ \lambda$) will result in severe degradation or even loss of the image as the entire diffraction envelope can be displaced from the camera sensor. Stress on the optical components also may lead to refractive index changes. For this reason, sapphire windows *must* be used with the interference optical system. Also, in order to achieve the full accuracy of the interference optics, careful alignment and focusing are necessary (Richards *et al.*, 1971; Yphantis, 1964). It is not that the interference optics are particularly fussy with respect to focusing. However, they offer precision and accuracy well beyond the other optical systems, hence, require that more attention be paid to focus and alignment. Once properly aligned and focused, they remain stable. Changes that will require realignment are few (e.g., new light source mounting, new drive motor). Refocusing should be done if there is a switch from 12- to 3-mm cell path length centerpieces and high-accuracy work is desired.

C. Fluorescence

The fluorescence optical system is the most recent addition to the XLI. The AU-FDS (Aviv Biomedical) may be added to an existing XLI and is based on previously described prototypes (Laue, 2006; MacGregor *et al.*, 2004). Although the fluorescence optics are not as well characterized as the absorbance and interference systems, some features are known that impact experiment designs.

A laser light source must be used in order to achieve sufficient radial resolution (~20–50 μm). Currently, the AU-FDS laser provides excitation at 488 nm. It is likely that more excitation wavelengths will become available as solid state lasers that meet the size and power dissipation requirements become available. Because a 488-nm source is used, the fluorescence system ordinarily is used with extrinsically labeled compounds. Suitable labels include fluorescein, BODIPY, NBD, green fluorescent protein (GFP), and the many derivatives of these labels used for fluorescence microscopy. Information about specific labels and the chemistries available for attaching them to biomolecules may be found on the web (see http://probes.invitrogen.com/handbook/). In our experience, Alexa488 is an excellent choice due to its large extinction coefficient (~80,000), insensitivity to pH, resistance to photobleaching, and because of the many coupling chemistries for covalently attaching the dye to specific functional groups on proteins and nucleic acids. The many variants of GFP may be used to generate transcriptionally labeled material for the AU-FDS. Due to the extraordinary sensitivity and selectivity of fluorescence detection, it is possible to characterize the sedimentation behavior of GFP-labeled proteins in cell lysates without further purification (Kroe, 2005).

The emitted light passes through a pair of long-pass (>505 nm) dichroic filters. This choice of filters captures the maximum amount of emitted light, providing good sensitivity, but offers no opportunity to select a label by its emission characteristics. Thus, there is currently no simple way to use multiple labels in the AU-FDS (e.g., for fluorescence resonance energy transfer).

The noise characteristics of the fluorescence detector are a combination of the high-frequency stochastic noise found in the absorbance detector with the low-frequency systematic noise observed with the interference optics. The similarity of fluorescence noise to absorbance noise stems from their mutual reliance on measuring light intensities and their use of similar photo detectors. Our experience is that the stochastic noise on an intensity reading is about 1% of the value. This observation holds over a wide range of sample concentrations and detector gain settings. The systematic noise tends to be time independent and arises from two sources. First, fluorescent material may stick to the windows, particularly in places where there once was an air–liquid boundary. Hence, there can be regions where label stuck to the window while the cell was being handled (e.g., filled, put in the rotor). The severity of this problem depends strongly on the nature of the sample, with some proteins exhibiting little sticking while other proteins and other materials (especially lipids) leave an uneven coating over most of the window. While most analysis programs remove time-invariant noise, the resultant loss of materials to surfaces will affect the concentration of the labeled material (discussed later). The second source of time-invariant noise is background fluorescence from cell components (particularly epoxy centerpieces). This source of noise tends to be of lower magnitude and more uniform than that from adsorbed label and also is removed during data analysis. Sources of radially independent noise include variation in the source intensity and variation in detector sensitivity. In our experience, both of these noise sources are small.

The conversion from fluorescence intensity to concentration is not trivial. So long as the signal is directly proportional to concentration, one can determine the sedimentation coefficient, diffusion coefficient, and molecular weight without needing to convert the data. Likewise, there are many qualitative observations (e.g., the sedimentation coefficient increases or decreases in response to some stimulus) that require only relative knowledge of the concentration. For these purposes data collected using the AU-FDS may be handled in the same manner as absorbance or interference data. However, if one wishes to obtain concentration-dependent data (e.g., an association constant or nonideality coefficient), fluorescence detection poses some difficulties.

The fluorescence intensity is proportional to the concentration, $F = I_o Q\varepsilon c$, where ε is the extinction coefficient (either molar or weight, depending on the concentration units used for c), Q is the quantum yield (the fraction of photons absorbed that result in a fluorescence signal), and I_o is the incident intensity of the excitation beam. While ϵ is relatively constant, Q is sensitive to the peculiarities of the immediate surroundings of the dye (e.g., local dielectric constant, polarizability, and any dipole moments) and to the specific solution conditions (e.g., how many and how uniform are the labels attached to the molecule of interest, are quenchers present). This means that it is more difficult to relate the fluorescence intensity to concentration than it is the absorbance or fringe displacement. Comparison of fluorescence intensities to standards is one way to do this, and special calibration centerpieces are available that hold several standards (Spin Analytical, NH). Even using standards is not without problems (MacGregor *et al.*, 2004).

Collisional quenching decreases Q, hence decreases the fluorescence intensity. Removing quenchers uniformly (both sample to sample and radially) is essential for good sensitivity and good reproducibility. While most common biological solvents do not contain quenching agents, some reagents (e.g., cesium ions, acetate ions, heavy metals, iodide, acrylamide) should be avoided (see http://probes.invitrogen.com/handbook/). The most common quencher is molecular oxygen, which should be removed from samples by a nitrogen sparge or placing the samples under vacuum for a few minutes. It has been our limited experience that biological samples (e.g., serum, cell lysates) do not contain large quantities of quenchers.

One of the most common applications of AUC is the detection and characterization of molecular interactions. While it is straightforward to detect binding as changes in the sedimentation coefficient or changes in apparent molecular weight, determining an accurate association constant may be difficult. Specifically, if a label's local surroundings change on association (e.g., with respect to polarizability, dipole moments, etc.), the quantum yield may be affected, and the fluorescence intensity will not be linear with concentration. At present, only one analysis program (SEDANAL) is equipped to handle changes in the quantum yield upon molecular association. If one simply wants to get a ballpark idea of the association constant, the wide dynamic range of the AU-FDS system typically allows a complete titration curve (S_W vs c) to be obtained. The midpoint of the transition of the curve provides an estimate of the binding energy (as ln c), and it may be possible to fit the titration curve to more sophisticated models (Correia, 2000; Schuck, 2003).

While the fluorescence optics may be used over a very broad concentration range, special care must be exercised when using samples containing very low concentrations (<10 nM) or high concentrations (>5 μM) of labeled material. For low concentrations, loss of material to surfaces can be a problem. Proteins, lipids, nucleic acids, and polysaccharides can be "sticky" and form a monolayer (or thicker layer) on surfaces in contact with the solution. At low concentrations, the stuck material may be a significant fraction of the total material put in the sample cell. The degree of "stickiness" varies from substance to substance. For the AUC sample holders, there are three surfaces to consider: the walls of the centerpiece, the cell windows, and the air–liquid meniscus. The simplest way to minimize these effects is to include some nonlabeled "carrier" protein in the sample buffer. Low concentration (0.1 mg/ml) ovalbumin, serum albumin, and kappa casein have all been used as carrier proteins. It is worthwhile to try more than one type of carrier protein to make sure the carrier protein does not interact with the labeled material.

The confocal design of the AU-FDS allows the detector to provide usable data at fairly high concentrations of dye (MacGregor *et al.*, 2004). Nonetheless, absorbance of the excitation beam by dye molecules not in the observation volume will reduce I_o (inner filter effect) and lead to a nonlinear relationship between the concentration and fluorescence intensity. A similar problem will occur if the emitted light is absorbed by the fluorophore. The easiest fix for this is to reduce the concentration of the dye, either by diluting labeled material with unlabeled material or by decreasing the number of labels per molecule.

D. High Concentrations and High Concentration Gradients

It is sometimes desirable to characterize high concentration samples using AUC. The signal for both the absorbance and interference optics is dependent on the optical path length. Decreasing the sample path length is the best way to extend their concentration range to high concentrations. Special 3-mm thick centerpieces (and the adapters to use them with standard windows and cell housings) are available (Spin Analytical; Beckman Coulter) for this purpose. If accurate concentration-dependent parameters (equilibrium constants, nonideality coefficients) are sought, consideration must be given to the optical focus when using these centerpieces, particularly when high concentration gradients are present (Yphantis, 1964). Although the interference (Richards *et al.*, 1971; Yphantis, 1964) and fluorescence (http://rasmb.bbri.org/rasmb/AOS) systems may be refocused, no procedure exists to refocus the absorbance optics.

Snell's law says that light will bend from a region of lower refractive index into a region of higher refractive index. The concentration gradients developed during sedimentation also are refractive index gradients that may affect any of the optical systems. The collimated light used in the absorbance and interference optical systems will be bent toward the base of the cell (for a sedimenting boundary, but toward the meniscus for a floating boundary). Ordinarily, the imaging optics will correct this distortion and bring the deviated light back to the correct radial position. However, if the gradient is steep enough and the optics improperly focused, the correction may not be entirely accurate (Yphantis, 1964). If the gradient is steep enough, light even may be deviated entirely out of the optical path. A simple test for the absorbance system is to scan the cell at a nonabsorbing wavelength (e.g., 320 nm for a protein solution). This scan should be a flat line at 0 OD. If a too-steep gradient is present, this scan will have a "bump" in it centered at the boundary position. The height of the bump will diminish as the boundary spreads (Dhami *et al.*, 1995; Laue, 1996). The only way to obtain accurate data is to reduce the steepness of the gradient. In some cases, this may be done by sedimenting at lower rotor speeds to let diffusion spread the boundary, or just using data later in the run for analysis when the boundary has spread.

VI. Sample Requirements

Often, the first question that we face when planning an AUC experiment is "do we have enough material?" The sample requirements for AUC typically lie somewhere between crystallography/NMR and biochemical assays, but they can vary greatly depending on the nature of the experiment, the optical detection system, and the extinction coefficient. The sample volumes required for AUC analysis are quite low. SV experiments are generally performed using two-sector cells that require 420 μl/sample, but for the fluorescence detection system cells with volumes of 60 μl/sample are available (Spin Analytical). Typical SE experiments are performed in six-sector centerpieces that require 110 μl/channel; however, short-column measurements require lower volumes. In particular, the eight-channel centerpieces only use 15 μl/channel. For lower molecular weight solutes, it is often useful to perform SE measurements using longer columns (4–5 mm) in two-sector cells.

The choice of sample concentrations can be challenging and involves balancing the biological and biochemical relevance, sensitivity and linearity of the AUC detection optics, and limitations imposed by the physical chemistry of the macromolecules being investigated. One usually attempts to investigate proteins near their physiologically relevant concentrations. In many cases, however, these concentrations are not known and one simply wants to establish whether a sample is homogeneous, define the dominant association state, and possibly obtain some shape information. Here, the concentration range will be dictated by the optimal conditions for the AUC measurements. The low concentration limit for an AUC measurement is limited by the sensitivity of the detection system and the optical properties of the sample. The highest accessible concentrations are determined by the linearity of optical system, optical artifacts that occur at high concentration gradients and by thermodynamic and hydrodynamic nonideality, which become more pronounced at higher concentrations. Typical rms noise levels for the absorption system are ~0.005 OD, and for the interference system the noise is ~0.01 fringes. Thus, reasonable signal-to-noise levels require a minimum sample concentration corresponding to ~0.1 OD or 0.2 fringes. For a typical protein with a specific absorbance near 1 $(mg/ml)^{-1}$ cm^{-1}, 0.1 OD corresponds to a concentration of ~0.08 mg/ml (note that the usual centerpiece optical path is 1.2 cm). For the interference system, 0.2 fringes correspond to about 0.06 mg/ml, and the sensitivity interference optics are roughly comparable to that of the of the absorbance system operating at 280 nm. However, using the absorption optics, higher sensitivity measurements can be achieved at shorter wavelengths. In the XLI, it is useful to work at 229–230 nm where the flash lamp has a strong output, and the protein absorbance is approximately five- to sevenfold higher than at 280 nm. Reasonably good signal-to-noise can be obtained at this wavelength with protein concentrations as low as 10–15 μg/ml.

In experiments designed to measure the equilibrium constants for reversible associating systems, the concentration ranges must be chosen such that each of the species that participates in the equilibrium is present at an appreciable concentration. Thus, precise determination of K_d values for high affinity reactions requires low sample concentrations, which may lie below the detection limits discussed earlier. On the other end of the scale, weak interactions require high concentrations where nonideality and optical artifacts can become problematic. The best way to choose sample concentrations and other experimental conditions, and to determine whether the equilibrium constants are even experimentally accessible for a given system, is by simulation. Synthetic data are generated using the appropriate molecular parameters, experimental conditions and estimated equilibrium constants. Noise is added to the data to simulate the optical system being used. The data are then fit to determine whether the correct equilibrium constants can be recovered with reasonable confidence. Simulation routines are implemented in many AUC analysis software packages such as HETEROANALYSIS, SEDANAL, SEDFIT/SEDPHAT, and ULTRASCAN.

VII. Sample Preparation

The admonition from the late Efraim Racker "Don't waste clean thinking on dirty enzymes" (Schatz, 1996) applies well to AUC. Rather than trying to interpret complicated and ambiguous AUC data obtained using impure or heterogeneous samples, we find that the time is much better spent on improved purification protocols. In practice, proteins should be at least 95% pure by SDS–polyacrylamide gel electrophoresis and the mass spectrum should correspond to a single species consistent with the predicted molecular weight. Many proteins tend to form irreversible aggregates during purification or storage. Gel filtration is a good last purification step to remove such aggregates as well as low molecular weight contaminants that may not be resolved on polyacrylamide gels. Some proteins can aggregate with time or upon freeze/thaw cycles, so that it may be necessary to run a gel filtration column immediately before AUC analysis. Aggregation or proteolytic degradation can also occur during long SE experiments. These problems can be diagnosed by analysis of the sample after the AUC experiment. We have also encountered sticky samples that bind to the windows or centerpiece. This loss of soluble material can be assessed by measuring the OD at low speed (3000 rpm) after loading the sample.

Samples should be equilibrated into the experimental buffer such that the composition of the reference and sample solutions is identical. This can be accomplished by conventional gel filtration, as mentioned earlier, small volume gel filtration spin columns or by dialysis. Buffer matching is most critical when using interference optics, where any mismatch of salts or other buffer components contributes to the fringe displacement. Most of the commonly used buffer components are compatible with AUC experiments. As described earlier, the major issues to keep in mind are ionic strength, absorbance (when using absorbance detection), viscosity, and generation of density gradients (Schuck, 2004). The salt concentration should be at least 20–50 mM to shield electrostatic interactions that contribute to thermodynamic nonideality. For absorbance measurements, the OD of the buffer at the detection wavelength should be minimized. Reductants such as mercaptoethanol and dithiothreitol absorb at 280 nm upon oxidation; however, TCEP [Tris(2-carboxyethyl)phoshine] is essentially transparent at this wavelength. At shorter wavelengths, for example, 230 nm, many buffer constituents absorb and a buffer versus water spectrum should be recorded. Highly viscous buffers slow sedimentation in SV experiments and extend the time to achieve equilibrium in SE and should be avoided. Finally, density and viscosity gradients produced at high solute concentrations should be taken into account for SV experiments (Schuck, 2004).

Two critical parameters for interpretation of AUC experiments are ρ and $\bar{v}$. Typically, ρ is calculated from the composition using SEDNTERP (Laue *et al.*, 1992) or measured using a high-precision density meter (automated instruments are available from Anton-Paar). For proteins lacking prosthetic groups or posttranslational modification, $\bar{v}$ is commonly calculated from the amino acid composition (Laue *et al.*, 1992). However, these calculated values should be used with caution. Some buffer components are either excluded (e.g., glycerol) or concentrated (e.g., guanidine HCl) at the protein hydration layer, which affects $\bar{v}$ (Timasheff, 2002). The effects of glycerol (Gekko and Timasheff, 1981), salts and amino acids (Arakawa and Timasheff, 1985), and guanidine HCl (Lee and Timasheff, 1974a,b), and urea (Prakash and Timasheff, 1985) on $\bar{v}$ have been tabulated. $\bar{v}$ can also be affected by changes in the water density in the hydration layer and by changes in protein packing density, and in some cases the origin of an anomalous value of $\bar{v}$ may not be apparent from the protein structure (Philo *et al.*, 2004). Thus, in some circumstances, it may be necessary to measure partial specific volumes experimentally. Ideally, $\bar{v}$ can be obtained from the variation in solvent density with protein concentration using a high-precision density meter. In this regard, Eisenberg (2000) suggests replacing the buoyancy term used in AUC experiments $M(1 - \bar{v}\rho)$ by the more thermodynamically rigorous density increment $(\partial\rho/\partial c_2)_{p,\mu}$ where c_2 is the protein concentration and the subscript μ indicates a constant chemical potential of all other solute components. Alternatively, in the Edelstein–Schachman method, $\bar{v}$ is calculated from the linear change in the buoyant molecular weight in SE experiments performed in buffers where the density is increased by adding D_2O (Edelstein and Schachman, 1973).

VIII. Sedimentation Velocity

A. Instrument Operation and Data Collection

SV experiments are carried out in two-channel cells with sector-shaped compartments (Fig. 1) in order to prevent convection, which would occur if the cell walls were not parallel to radial lines. The usual protocol in our laboratories is to run three sample concentrations spanning at least an order of magnitude, for example, 0.1, 0.3, and 1.0 mg/ml.

For SV experiments using absorbance optics, the cells are assembled using standard double-sector centerpieces and quartz windows. The cells are filled with 430 μl of buffer in the reference sector and 420 μl of sample solution in the sample sector. The XLA (or XLI) monochromator may not reproducibly return to the same wavelength if scans are performed at multiple wavelengths. Because of this potential problem, we choose to limit SV experiments to using a single wavelength and choose concentrations of the sample that will yield ODs of 1.2, 0.4, and 0.1 at the selected wavelength. The rotor, with the cells and a correctly weighted counterbalance, is loaded into the centrifuge and the vacuum system is started. At this point the speed is set to "zero" and the run is started, though the rotor will be stationary. This procedure will turn on the diffusion pump and allow the vacuum to drop below 100 μm, at which point the temperature reading will accurately reflect the rotor temperature. Once the rotor temperature has reached the set point,

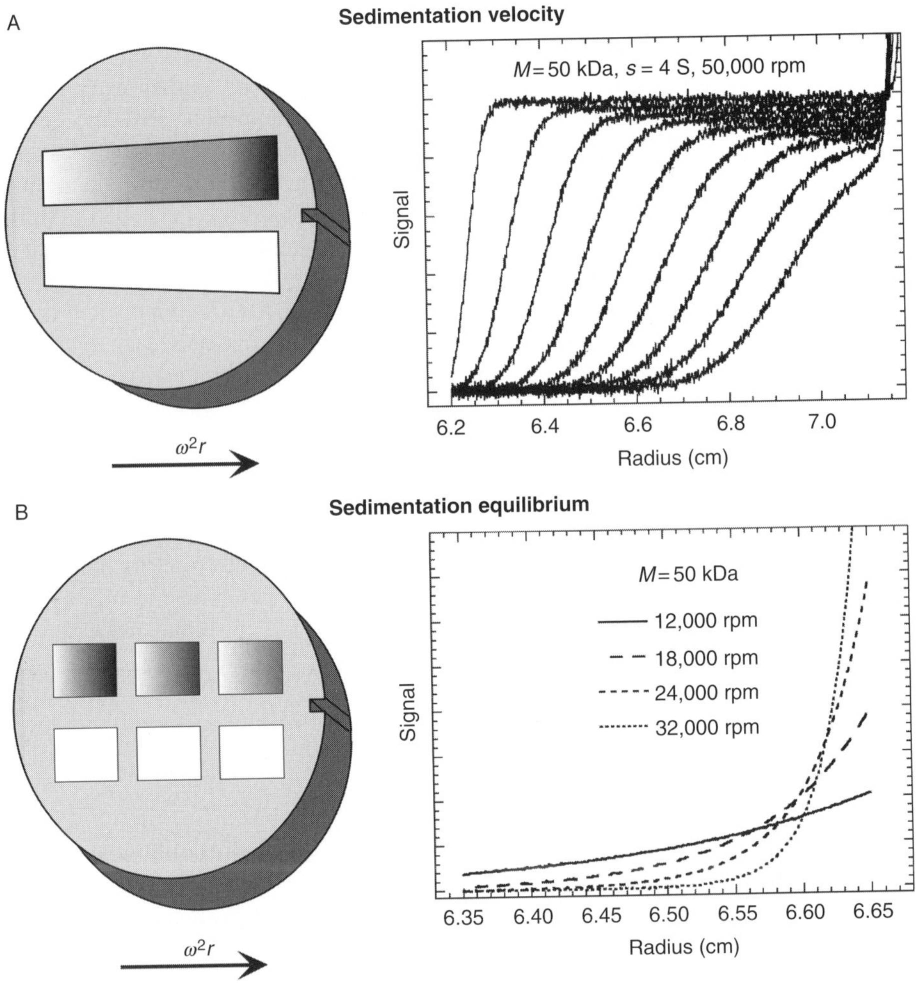

Fig. 1 Basic AUC experiments. Simulations are for a protein of 50 kDa with a sedimentation coefficient of 4 S. (A) SV experiment. Velocity sedimentation is usually performed using a two-sector cell and scans are recorded at fixed intervals during the run. The simulation is for a rotor speed of 50,000 rpm and scans are displayed at 20-min intervals. (B) SE experiment. Equilibrium measurements usually employ a six-sector cell with three loading concentrations. The equilibrium concentration gradients are simulated for four rotor speeds ranging between 12,000 and 32,000 rpm, corresponding to values of σ ranging from ~0.8 to ~6 cm^{-2}. The 32,000 rpm scan is truncated at the base.

we allow the rotor to equilibrate for an additional hour before starting the run. This, in turn, will minimize the effects of convection due to temperature gradients.

The protocol for SV experiments using interference optics is somewhat different due to the fact that interference data will reflect any refractive index differences between the sample and reference sectors including differences in the buffer gradient if the column heights are mismatched. In order to eliminate this possible problem, we use double-sector synthetic boundary, capillary-type centerpieces. The cells are assembled using sapphire windows because this optical system is focused only for this type of window, and the interference fringe pattern tends to blur at higher speeds if quartz windows are used. In addition, we perform a test run of the cells filled with water in order to preset the scan configurations for each cell and to perform a radial calibration. This test run will also allow checking of the cells for leaks, thus preventing the possible loss of sample material. It will also make it possible to start collecting data during the actual run as soon as the rotor reaches speed. Once the test run is finished the cells are removed from the rotor, the water is aspirated from the cells, and the assembled cells are dried in a vacuum chamber. One can also now replace the interference counterbalance with a fourth cell containing an additional sample dilution since the radial calibration has already been performed. For the actual run, each synthetic boundary cell is loaded with 430 μl of buffer in the reference sector and 420 μl of sample solution in the sample sector. The cells are placed in the rotor and the rotor is placed in the chamber along with the monochromator/laser assembly. The rotor is accelerated to ~12,000 rpm and the interference fringe pattern, for each cell, is checked to confirm that the excess buffer has transferred over to the sample sector from the reference side. At this point the rotor is stopped, removed from the centrifuge, and then gently inverted to thoroughly mix the contents of each cell. Now, the rotor is placed back in the centrifuge and the temperature is equilibrated as previously described. A typical concentration series for four cells would be 1.5, 0.9, 0.3, and 0.1 mg/ml. Sample dilutions may be made immediately prior to the SV run unless it is suspected that slowly reversible reactions are taking place. In that case, dilutions are made and then sufficient time allowed for complete equilibration at the experimental temperature.

The instrument operating parameters include the temperature, the rotor speed, time after speed is reached before the first scan is taken, the time interval between scans, and how many scans are to be acquired. For SV analysis, there should be no delay before data are acquired. Likewise, there is no reason to wait between scans, so there should be no interval between scans. These two parameters (scan delay and scan interval) should be set to zero in the method for either the Beckman Coulter ProteomeLab or the Aviv-AOS software to maximize the number of data sets available for analysis.

The listed operating temperature range of the XLI is 0–40 °C. However, excessive oil vapor at operating temperatures above 35 °C and difficulty maintaining temperatures below 4 °C limit the useful temperature range. Replacing the oil diffusion pump with a turbomolecular pump allows operation to 40 °C and reduces optical fouling. A kit for upgrading the XLI vacuum system is being developed (Beckman Coulter). Most experiments are conducted at 20 °C, thus simplifying correction of the sedimentation and diffusion coefficients to standard conditions. For a SV experiment, one wants to make sure the samples have stabilized at the desired temperature prior to rotor acceleration. For this reason, many people allow the system to stabilize at temperature for an hour or so before acceleration. For high accuracy work, it is desirable to calibrate the XLI temperature sensor (Liu and Stafford, 1995).

Choosing the correct rotor speed for a SV experiment depends what you want to know about your sample, what the expected component size distribution is, and which optical systems will be used. These considerations lead to competing needs. The resolution of solution components is proportional to ω^2, indicating you should use the highest rotor speed possible, especially if you are trying to determine how many components there are in a solution. Thus, for samples with $s < 10$ S (the units of s are Svedbergs (S) with 1 S $= 10^{-13}$ s), it makes sense to use the highest rotor speeds (55,000–60,000 rpm). However, with modern global analysis software, it is also beneficial to obtain a large number of scans. Thus, lower rotor speeds are required if components of interest are very large with large sedimentation coefficients. Also, the absorbance optics have long scan times and when scanning multiple samples and wavelengths it may be useful to reduce rotor speeds. Although there is no simple formula for optimizing the rotor speed, we can use the definition of the sedimentation coefficient [Eq. (2)] to determine reasonable rotor speeds. It should take a boundary at least 2 h to sediment the full length of the cell (1.5 cm maximum), to ensure sufficient scans will be acquired. Based on this criterion, the maximum recommended rotor speeds for various sedimentation coefficients are presented in Table III. In addition, when using the absorbance system, it is necessary to consider the longer scan times and adjust the rotor speed so that at least 30–40 scans are recorded during the movement of the boundary across the cell.

Table III
Maximum Rotor Speeds for Sedimentation Velocity Experiments

S[a]	M_{app}[b]	rpm[c]
10	200,000	55,000
15	400,000	50,000
30	1,000,000	30,000
90	5,000,000	20,000
270	25,000,000	10,000

[a]Maximum allowed rotor speed may be used for solutions where all components have sedimentation coefficients <10 S. However, acquiring absorbance data at multiple wavelengths will greatly increase scan times, thus decreasing the number of scans acquired at a particular wavelength over the course of an experiment. For experiments requiring multi-wavelength scanning, one may wish to spin at a lower rotor speed.

[b]These are only approximate values estimated for spherical proteins. If the molecules are asymmetric or a highly solvated, then a higher molecular weight will correspond to a given sedimentation coefficient.

[c]About 2 h of data acquisition will be available at the listed rotor speed. Be sure the maximum speed rating for the centerpiece is not exceeded. For example, the Epon-based centerpieces from Beckman Coulter have a maximum speed rating of 44,000 rpm, whereas the Spin Analytical Epon centerpieces are rated to 60,000 rpm.

B. Data Analysis

Methods for analysis of SV experiments have evolved rapidly in recent years and many alternative approaches and software packages are available. Here, we will outline an approach that we have found useful in the initial stages of analyzing an unknown system. At the early stages, it is useful to examine the data using methods that require the fewest assumptions about the nature of the system being investigated. Simply put, the goal of these "model-free" approaches is to determine how many species are present and whether they interact. Later, this information can be used to construct models and obtain starting parameters for more detailed analyses.

In the "d*c*/d*t*" method, a closely spaced group of SV scans are subtracted in pairs to approximate the time derivative of the data and thereby determine how much material is sedimenting at various rates (Stafford, 1992). This subtraction removes the systematic noise in the data, which is particularly useful for interference data. The radial variable is then transformed to an apparent sedimentation coefficient (s^*) and the data are averaged among several pairs to enhance the signal-to-noise ratio. Finally, a data transformation yields the apparent sedimentation coefficient distribution function $g(s^*)$. These algorithms have been implemented in several software packages: we find that DCDT+ is particularly easy to use and convenient. The $g(s^*)$ distributions resemble chromatographs and can be visually examined to determine whether the sample appears pure (one peak) or heterogeneous (multiple peaks or shoulders). It is important to inspect the distributions at multiple loading concentrations to check for reversible interactions. A shift in peak position to higher s^* with increasing concentration is evidence for mass-action equilibrium where the species interact the timescale of sedimentation. In this case, the peak represents a "reaction boundary" and cannot be treated as a species. Alternatively, the peak positions may remain constant or shift only slightly, but the relative area of the higher s^* feature may increase with loading concentration. This behavior is diagnostic for slowly reversible interactions and it is important to fully equilibrate such samples prior to AUC analysis. For homogeneous species or mixtures of noninteracting species, the width of each peak is related to D, and one can fit the distribution to obtain D and thus the molecular mass of each component. This fitting process can also be useful to determine whether the peak is truly homogeneous. Recent advances have improved the fitting function (Philo, 2000b) and extended the scan range (Philo, 2006) that can be used in this analysis.

The main advantage of the d*c*/d*t* method is simplicity. No models are assumed in the analysis. Also, the subtraction and averaging result in noise reduction, allowing lower sample concentrations. The chief disadvantages are the limitations in the number of scans to avoid distortion of the peak shape and the diffusional broadening of the peaks that can hide heterogeneity. Also, it is difficult to cover a large range of sedimentation coefficients using this approach, and the method does not work well with low molecular weight solutes (molar masses <10 kDa or $s < 2$ S).

Alternatively, in the $c(s)$ method implemented in the programs SEDFIT and SEDPHAT, the sedimentation coefficient distribution function is obtained from a direct fit to the data (Dam and Schuck, 2004; Schuck, 2000). Here, we describe the most basic implementation of the $c(s)$ method. First, the program creates a grid of sedimentation coefficients covering the expected range of interest. By assuming a constant shape and consequently an equal frictional ratio (f/f_0), for all species, a scaling relationship is created between s and D. The program then simulates the sedimentation boundaries for each point using a numerical solution of the Lamm equation. Finally, the data are fit to a sum of these Lamm solutions using a least-squares fitting procedure to define the concentration of each species in the grid. During this process, the systematic noise of the baseline (time-invariant noise) and the vertical displacements (jitter and integral fringe jumps) are removed by treating them as additional linear fitting parameters. The resulting $c(s)$ function is often quite "spiky," and a regularization procedure is performed to produce a smoother distribution function. Like the $g(s^*)$ distribution, the $c(s)$ distribution can be visually interpreted by looking at how many peaks are present and how they depend on loading concentration. For the $c(s)$ models, one can also check whether the model of a sum of noninteracting species provides a good fit to the experimental data. A poor fit can indicate reversible interactions. Although the $c(s)$ distribution can be converted to a distribution of molar masses [$c(M)$ distribution], the derived masses will only be accurate if there is one dominant species present or if all the species have equal frictional ratios. More complex analysis procedures that do not assume a single value of f/f_0 are also implemented in SEDFIT and SEDPHAT.

The main advantages of the $c(s)$ method are the excellent resolution and sensitivity. In contrast to the dc/dt method, there is no restriction on the number of scans that can be included in the analysis, and the diffusional broadening is deconvoluted from the $c(s)$ distribution based on the scaling relationship between s and D. The $c(s)$ method is thus very useful for characterizing homogeneity and quantitating impurities and aggregates. The main disadvantage of this approach is that it assumes a noninteracting mixture and particular care must be exercised in the analysis of self- or hetero-associating systems where the resulting distributions are developed from an incorrect model. Nonetheless, for a system undergoing *rapid* association and dissociation, the distributions are reminiscent of those expected by limiting models (Gilbert and Jenkins, 1956), and useful semi-quantitative information may be extracted (Dam and Schuck, 2005; Dam *et al.*, 2005). For interacting systems undergoing reactions on the timescale of the SV experiment, peaks in the $c(s)$ distribution may not correspond to true molecular species (Dam *et al.*, 2005). Provided that the $c(s)$ distribution is a good fit to the data, it is always feasible to extract thermodynamic parameters from the data by integration of the distribution and analyzing the dependence of weight-average sedimentation coefficients on the loading concentrations (Correia, 2000; Correia *et al.*, 2005; Schuck, 2003). The only requirement for this analysis to be accurate is that all association reactions are at equilibrium prior to the start of sedimentation. This criterion may be met by incubating the sample dilutions for a sufficient amount of time (e.g., overnight) at the sedimentation temperature prior to sedimentation.

The van Holde–Weischet approach (van Holde and Weischet, 1978) is also used for the initial, qualitative analysis of SV experiments. Because sedimentation is proportional to the first power of time whereas diffusion is proportional to the square root of time, graphic extrapolation of the boundary to infinite time yields an integral sedimentation coefficient distribution, $G(s)$ in which the diffusional contribution has been removed. This method is implemented in ULTRASCAN, SEDFIT, and the Beckman Coulter software. Recent advances have extended this method for the analysis of highly heterogeneous systems (Demeler and van Holde, 2004).

Although the information obtained from the model-free approaches may be enough to answer the relevant questions about the macromolecular system being studied, we often find it useful to analyze the system using model-dependent procedures. For analysis of mixtures, the goal is usually to obtain the concentration, sedimentation coefficient, and mass of each species. These parameters are recovered with greater precision by using a fitting model of a mixture of several discrete species rather than continuous distribution approaches. The available software uses either approximate (SVEDBERG, LAMM) or numerical (SEDANAL, SEDFIT/SEDPHAT, ULTRASCAN) solutions to the Lamm equation to fit data as a superposition of noninteracting species. We often find it useful to improve the precision of the fitted parameters by globally analyze data sets obtained at several loading concentrations using SEDANAL or SEDPHAT. The absence of systematic deviations in global fit also confirms that there are no mass-action reactions over the concentration range examined. For systems that contain a mixture of well-defined discrete species and poorly resolved aggregates or low molecular weight impurities, it can be useful to fit the data to a hybrid $c(s)$-discrete species model in SEDPHAT where the poorly resolved material is accounted for in the continuous distribution.

Analysis of reversible interactions by SV is a complex problem (Dam and Schuck, 2005; Dam *et al.*, 2005; Rivas *et al.*, 1999; Schuck, 2003; Stafford, 2000; Stafford and Sherwood, 2004). However, SV may be the only feasible approach for systems that are intrinsically unstable or that do not come to equilibrium in SE experiments. For interacting systems, the boundaries do not generally correspond to discrete species. Because the concentrations are changing throughout the cell during sedimentation, particularly where there are boundaries, the species composition is continuously varying due to the mass-action equilibria. Consequently, the apparent sedimentation coefficients and boundary shapes are complex functions of the sedimentation coefficients of the species participating in the equilibrium, their concentrations, and the equilibrium and kinetic constants governing their interactions (Cann, 1970; Dam and Schuck, 2005; Gilbert and Jenkins, 1956). As alluded to above, the traditional approach to analyzing interacting systems by SV is to measure weight-average sedimentation coefficients as a function of loading concentrations (Correia, 2000; Correia *et al.*, 2005; Schachman, 1959; Schuck, 2003). An advantage of this method is that the weight-average sedimentation coefficient is a thermodynamically valid parameter that is determined the sample composition in the plateau and is independent of the kinetics of the interactions, provided that the sample is at equilibrium prior to sedimentation. Examples of this approach can be found in studies of Cytomegalovirus protease dimerization (Cole, 1996) and the complex association reactions of tubulin (Correia, 2000; Sontag *et al.*, 2004) and HIV rev (Surendran *et al.*, 2004). Interacting systems can also be characterized by calculation of $g(s^*)$ distributions using the time-derivative method (Stafford, 2000). More recently, direct boundary fitting methods for interacting systems have been implemented in SEDANAL, SEDPHAT, and ULTRASCAN. When compared, this approach gives comparable results to those obtained using weight-average analysis (Sontag *et al.*, 2004). Some recent examples of direct boundary analysis to define the energetics of associating systems can be found in Connaghan-Jones *et al.* (2006), Correia *et al.* (2005), Dam *et al.* (2005), Gelinas *et al.* (2004), and Snyder *et al.* (2004).

IX. Sedimentation Equilibrium

The big advantage of SE is that it removes all hydrodynamic effects, so that purely thermodynamic analysis is possible. The requirements for sample purity and homogeneity are much stricter for SE measurements that for velocity experiments. In the latter case, the boundaries associated with each species separate during the sedimentation run so that it is possible to isolate contaminants from the species of interest. In contrast, different species are incompletely fractionated in an SE gradient. Furthermore, as shown below, fitting SE concentration gradients requires deconvolution of multiple exponential functions, which is a challenging mathematical operation that becomes increasingly difficult with larger number of species.

A. Instrument Operation and Data Collection

There are three commercially available centerpiece styles that are commonly used when conducting SE experiments. The choice of which style to use will be determined by the information that is being sought. The short-column centerpiece has eight channels that can hold four sample-reference pairs. Each channel requires only 15 μl of solution resulting in a column height of 700–800 μm and will typically reach equilibrium within an hour or two (Yphantis, 1960). This type of centerpiece is useful for conducting a rapid survey over a wide range of concentrations and/or conditions (Laue, 1992). The standard long-column centerpiece has six channels, which can hold three sample-reference pairs (Fig. 1B). Each of these channels requires ~120 μl of solution, which will result in a column height of ~3 mm. Long-column experiments are useful for accurately determining molecular weights, self-associations, hetero-associations, and so on, by direct fitting of data from multiple concentrations (or multiple mixing ratios) at multiple speeds using global, model-dependent, and nonlinear least squares analysis.

A version of the six-channel centerpieces is available that, along with a custom cell housing, allows the cells to be loaded and unloaded without disassembly (Ansevin *et al.*, 1970). These "external loading" cells are particularly useful with the interference optics because they facilitate blank subtraction. Prior to acquiring blanks, each cell must be "aged" in order to bring the cell, centerpiece, and windows into a mechanically stable configuration. First assemble the external loading cell according to specifications (typically sealed at between 120 and 140 inch-pounds of torque). Fill each of the reference sectors with 150 μl of water and each sample sector with 140 μl of water, and seal the filling holes with a gasket and screw. Centrifuge the cells at the maximum speed that will be used during the experiment for at least 1 h. Stop the run, remove the cells, and retorque them to specifications. Place the cells back in the rotor and centrifuge them at the same speed as before for another hour. Repeat this at least one more time for a total of three acceleration/deceleration cycles. In our experience, three or four cycles are sufficient to bring the cell into a stable state. To acquire the blank, the cells are filled with water and run at the same temperature and rotor speeds as will be used during the experiment. At each rotor speed, scans are acquired every 5 min or so until no changes in the fringe patterns are apparent. After the blanks have been acquired the water is removed, the cells dried, and the samples loaded. The "blank" scans are subtracted from the data scans to remove the systematic noise. Because they do not need disassembly, the blank correction from external loader cells (above) can result in tenfold lower noise (Ansevin *et al.*, 1970). Specialized methods for washing the external loading cells without disassembly have been described (Ansevin *et al.*, 1970). An automated cell washer recently became available (Spin Analytical). Also, Beckman Coulter produces centerpieces that facilitate cell cleaning by incorporating two holes per sector.

In order to characterize a system over a wide concentration range, different sample loading concentrations must be used. It is recommended that 1:1, 1:3, and 1:9 dilutions be used with the six-channel cells, and 1:1, 1:2, 1:4, and 1:8 dilutions be used in the eight-channel cells. Cells are loaded so that the highest concentration sample will be closest to the rotor center, and the most dilute sample will be toward the rotor's edge. This way, advantage will be taken of the gravitational field to concentrate the more dilute samples while minimizing the concentration gradients in the highest concentration sample. A layer of dense, colorless fluid should be used to create an artificial base of each sample. This layer allows data acquisition at the highest concentration region with less interference from reflections from the centerpiece base. The recommended fluid is FC-43 (3 M, Inc., MN). For two- and six-channel centerpieces, 10 μl of FC-43 is used whereas 5 μl is used for eight-channel centerpieces. It has been found that certain proteins (e.g., tubulin) will denature and aggregate at the interface between the aqueous solution and the FC-43. Thus, while it is generally inert, it is worthwhile checking to make sure that FC-43 is compatible with the solution components.

Other than the cells that are employed, there is no change in the instrumentation from SV for SE experiments. However, the operating parameters are different.

Unlike SV, it is usually not critical to allow temperature equilibration prior to starting the rotor spinning. It is important to collect data at multiple loading concentrations and rotor speeds to assess if the sample is homogeneous, if mass-action-driven self-association is occurring, or if thermodynamic nonideality is significant. The complete data set can be used subsequently in global curve fitting programs to obtain the most precise parameters from the data. Many researchers perform SE experiments using rotor speeds that are too low (e.g., the 12,000 rpm trace in Fig.1B). For a typical experiment using the standard 3 mm column heights, we recommend choosing the lowest speed such that $\sigma \sim 2\ \text{cm}^{-2}$ for the monomer (e.g., the 18,000 rpm trace in Fig. 1B). Assuming a typical protein $\bar{v} \sim 0.74\ \text{cm}^3/\text{g}$, this speed can be estimated as

$$\text{rpm} \approx 4 \times 10^6 \sqrt{\frac{1}{M_\text{p}}} \tag{7}$$

There are times when you may want to start a protocol at a rotor speed, which produces $\sigma < 2\ \text{cm}^{-2}$ (e.g., for systems exhibiting large stoichiometries or for longer solution columns). A typical experimental protocol will produce data at three or four rotor speeds using 1.2- to 1.5-fold intervals between speeds, with the highest rotor speed yielding σ as high as $10\text{–}15\ \text{cm}^{-2}$. In combination with the recommended cell loading described earlier, this protocol will produce data over a very broad concentration range that will enhance the reliability of the analysis. The experimental protocol *must* go from lowest to highest rotor speed. If a lower rotor speed is used after a higher one, the system will not reach equilibrium in a reasonable time (Roark, 1976).

B. Monitoring Approach to Equilibrium

The time to achieve equilibrium is dependent on a number of experimental factors, including the mass and shape of the particle, the solvent viscosity, and the distance between the meniscus and the base (column height). In particular, the equilibrium time is proportional to the square of column height. Although theoretical expressions are available for the simplest systems (van Holde and Baldwin, 1958), the actual time to equilibrium may be extended by slow association and dissociation rates and other factors. Thus, the approach to equilibrium is often monitored experimentally by taking the difference between successive scans and looking for the absence of systematic deviations. A better procedure is to use WINMATCH or the Match utility in HETEROANALYSIS. These programs do least-squares comparison of the scans allowing for displacements in the vertical and horizontal directions. The rms deviations decrease as a function of time until at equilibrium they reach a constant level corresponding to the noise level in the data. Although the equilibrium method in the Beckman Coulter XLI control software allows one to insert a delay prior to recording data, we recommend collecting data immediately at regular 15–30 min intervals to monitor the approach to equilibrium. When using the absorbance system, we typically record scans using a coarse point spacing of 0.003 cm with only one reading/point to monitor the approach to equilibrium. Once equilibrium is achieved, the sample is then scanned using the maximal point spacing of 0.001 cm with about 10 readings/point. Slow aggregation can cause a loss of material in successive scans and prevent achievement of equilibrium. Other potential problems in equilibrium experiments can include sample hydrolysis or denaturation. In some cases, problematic samples can be stabilized by altering the buffer composition, temperature, or changing the protein construct. However, it may be necessary to reduce the column height to achieve rapid equilibrium or use more rapid techniques such as SV.

C. Data Analysis

There are several ways to analyze SE data. The old fashioned method calculates σ as the slope of a graph of ln c versus $r^2/2$ (i.e., $\sigma = \mathrm{d}\ln c/\mathrm{d}r^2/2$). While this method is no longer widely used, it highlights a problem that must be addressed by all analysis methods, namely that one must have an accurate estimate of the concentration. It is tempting to substitute the absorbance, fringe displacement, or fluorescence intensity signal since each of these is proportional to the concentration. Before they can be used, however, it is necessary to adjust the signal to be zero at zero concentration. This adjustment is accomplished by subtracting a baseline offset. For absorbance data, one can determine the offset experimentally by increasing the rotor speed to ~40,000 rpm at the end of the run to pellet the solutes and then measuring the residual absorbance in the solution column. Alternatively, the offsets may be treated as fitting parameters in nonlinear least squares analysis software. With interference data, the offsets must be treated as fitting parameters.

Data analysis can be divided into two general methods—molecular weight moment determination and nonlinear least squares fitting. Both of these methods are useful, depending on what information is sought. Molecular weight moments can be determined directly from the data using the ratio of the different concentration moments (Harding *et al.*, 1992; Roark and Yphantis, 1969; Stafford, 1980). For example, at any point in the sample, the weight-average molecular weight (as σ_W) is the local concentration slope, $\mathrm{d}c(r)/\mathrm{d}r^2/2$, divided by the local concentration, $\sigma_W(r) = (\mathrm{d}c(r)/\mathrm{d}r^2/2c(r))$. Because the determination of $c(r)$ is subject to uncertainty due to the baseline offset, the z-average, calculated as the ratio of the curvature of the data to the slope, $[\sigma_Z(r) = (\mathrm{d}^2c(r)/\mathrm{d}r^2/\mathrm{d}c(r)/\mathrm{d}r^2)]$ is particularly useful since it does not require knowledge of the concentration. No model needs to be specified for these calculations, so they are particularly useful for the analysis of complex systems. Graphs of $\sigma_W(r)$ or $\sigma_Z(r)$ as a function of $c(r)$ or r can provide useful diagnostics about interacting systems (Roark and Yphantis, 1969). In particular, overlapping curves of $\sigma_W(r)$ versus $c(r)$ for samples at different loading concentrations and analyzed at different rotor speeds are thermodynamic proof that the system is homogeneous and undergoes reversible mass-action association. Programs are available specifically for calculating molecular weight moments (Table IV).

Nonlinear least squares analysis of sedimentation data has been performed for over 40 years (Johnson *et al.*, 1981; Yphantis, 1964). Most nonlinear least squares fitting programs directly fit the experimental data to particular models, such as a single ideal species:

$$S(r,\lambda) = \delta_\lambda + \varepsilon_\lambda c_0 \exp\left[\frac{M_b\omega^2}{RT}\left(\frac{r^2 - r_0^2}{2}\right)\right] \qquad (8)$$

where $S(r, \lambda)$ is the signal (absorbance, fringe displacement, fluorescence) at radius r and wavelength λ, δ_λ is the wavelength-dependent baseline offset, and ε_λ is the extinction coefficient. Modern SE analysis software data can incorporate data obtained at multiple rotor speeds using multiple signals for global analysis (Table IV). Notice that σ (or $M_b\omega^2/RT$) is the exponent of the fitting function. As σ gets smaller and smaller, $S(r, \lambda)$ approaches a straight line. For values of $\sigma < 2\ \text{cm}^{-2}$ or so, the correlation between c_0 and δ for individual data sets becomes so great that fitted values of σ tend to be unreliable (e.g., the trace at 12,000 rpm in Fig. 1B). Therefore, we recommend using rotors speeds such that $\sigma \geq 2\ \text{cm}^{-2}$ and, for absorbance data, fixing the baseline offsets using values of δ obtained from scans acquired after pelleting the material at the end of the run. More complex models are required for analyzing data for associating systems and for systems exhibiting thermodynamic nonideality (Johnson *et al.*, 1981). The newer analysis packages are capable of analyzing heteroassociation reactions involving two or more components. These models involve a large number of adjustable parameters, and it is often necessary to constrain the fitting process by incorporating multiple signals (Cole, 2004; Howlett *et al.*, 2006) and by invoking mass-conservation algorithms (Philo, 2000a; Vistica *et al.*, 2004).

It should be stressed that SE does not have the resolving power of SV, and reliable analysis of SE data by nonlinear least squares fitting methods requires pure samples free of aggregated material or contaminants. Depending on the size of the aggregates, it may be possible to pellet them while still analyzing the remaining sample. However, the presence of aggregates or contaminants will lead to inconsistencies in the data analysis. Thus, it is critical to characterize samples by SV prior to SE experiments. In some cases, contaminants or aggregates identified by the SV measurements can be removed by preparative gel filtration prior to SE analysis.

Table IV
AUC Analysis Programs and Utilities

Method	Application	Source[a]	References
SV			
Time derivative (dc/dt)	DCDT+	1	(Philo, 2006; Philo, 2000b; Stafford, 1992)
$c(s)$	SEDFIT	2	(Dam and Schuck, 2004; Schuck, 2000)
Van Holde–Weischet	ULTRASCAN	3	(Demeler, 2005; Demeler and van Holde, 2004; van Holde and Weischet, 1978)
	SEDFIT	2	
Discrete species: Approximate Lamm solution	SVEDBERG	1, 4	(Philo, 1997; Philo, 1994)
	LAMM	4	(Behlke and Ristau, 1997)
Discrete species: Numerical Lamm solution	SEDFIT	2	(Schuck, 1998)
	ULTRASCAN	3	(Demeler, 2005)
Global analysis of interacting systems	SEDANAL	4	(Stafford and Sherwood, 2004)
	SEDPHAT	2	(Schuck, 2003)
Hydrodynamic modeling	HYDRO, HYDROPRO	5	(Garcia De La Torre *et al.*, 2000)
	ATOB	4	(Byron, 1997)
	SOMO	6	(Rai *et al.*, 2005)
SE			
Test for equilibrium	WINMATCH	7, 4	
	HETEROANALYSIS	7, 4	(Cole, 2004)
Nonlinear least squares[b]	WINNONLIN	7, 4	(Johnson *et al.*, 1981)
	HETEROANALYSIS	7, 4	(Cole, 2004)
	ULTRASPIN	8	
Molecular weight moment analysis	SEDANAL	4	(Roark and Yphantis, 1969; Stafford and Sherwood, 2004)
	ULTRASPIN	8	
	MSTAR	4	(Harding *et al.*, 1992)
Utilities			
Data acquisition	AOS	4	
Real-time display and analysis	SEDVIEW	4	
	SEDFIT	2	
Graphics	XLGRAPH	1, 4	
Calculations	SEDNTERP	1, 4	(Laue *et al.*, 1992)
	ULTRASCAN	3	(Demeler, 2005)

[a]Websites where software can be downloaded:
1. http://www.jphilo.mailway.com
2. http://analyticalultracentrifugation.com
3. http://www.cauma.uthscsa.edu
4. http://www.rasmb.bbri.org/rasmb
5. http://leonardo.fcu.um.es/macromol
6. http://somo.uthscsa.edu
7. http://biotech.uconn.edu/auf/
8. http://ultraspin.mrc-cpe.cam.ac.uk

[b]The programs listed are specifically designed for analysis of SE data. Several general AUC analysis programs (SEDANAL, SEDPHAT, ULTRASCAN) can also be used for nonlinear least squares fitting of SE data.

X. Discussion and Summary

AUC is a versatile and rigorous technique for characterizing the molecular mass, shape, and interactions of biological molecules in solution. In particular, the size distribution analysis available with SV is more flexible, is applicable to more chemical systems, spans a much wider range of sizes, and provides higher resolution than size exclusion chromatography. The hydrodynamic information available with SV is complemented by thermodynamic analysis by SE. The availability of interference (refractive), absorbance, and fluorescence detectors makes AUC applicable to a wide variety of questions in cell biology. In particular, the fluorescence system provides a new way to extend the scope of AUC to probe the behavior of biological molecules under physiological conditions.

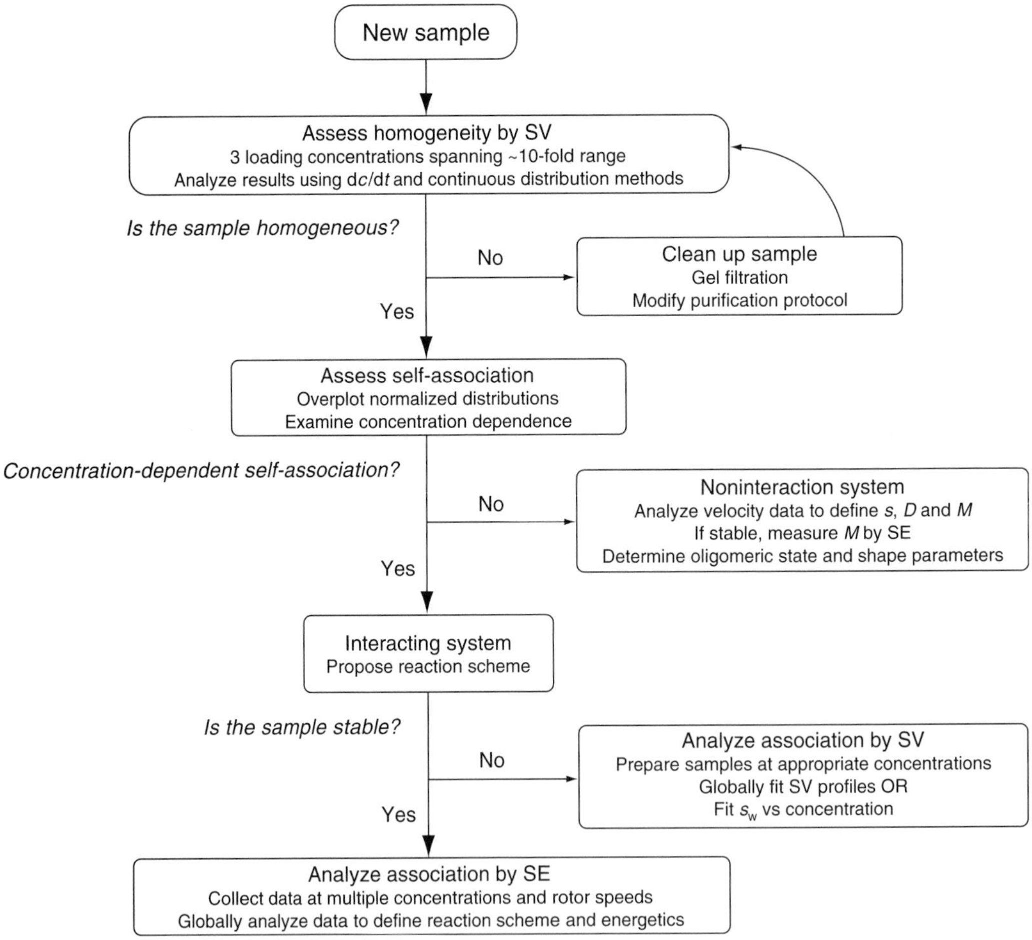

Fig. 2 Typical workflow for an AUC analysis of an unknown sample. For details see the text.

As we have described earlier, modern AUC users can choose from a broad array of experimental techniques and data analysis methods, and it can be difficult to decide how to best apply AUC methods when confronted with a new sample. Although the best strategy will depend considerably on the nature of the sample and the kinds of questions that need to be answered, Fig. 2 shows a typical workflow that we use for characterizing a new sample by AUC. It is strongly recommended that new samples are first analyzed by SV at several concentrations. These measurements are crucial for deciding whether the sample is homogeneous and suitable for more detailed analysis. The SV data should first be analyzed using model-free methods: we typically examine $g(s^*)$ and $c(s)$ distributions. If contaminants or aggregates are present that differ appreciably in size from the molecule of interest, the sample can often be purified by preparative gel filtration. In fact, we typically gel filter sample prior to AUC analysis. It should also be noted that although dynamic light scattering (DLS) lacks the resolving power of AUC, it is a fast and sensitive method to determine whether aggregates are present, and we often use DLS as a quality-control step prior to AUC.

The next step is to determine whether the sample undergoes reversible, mass-action association. A convenient method is to superimpose normalized $g(s^*)$ distributions obtained at different concentrations: reversible association will shift the distributions to higher s^* with increasing concentration, whereas hydrodynamic nonideality will shift the distributions to lower s^*. For a noninteracting system, the SV data can be analyzed to obtain s and D for the species of interest and the data can be interpreted to obtain the molar mass and shape parameters. It should be stressed that this analysis cannot be done for an interacting system: here, more sophisticated analysis is required to measure the sedimentation coefficients of the interacting species and to define the kinetics and thermodynamics of the interaction. If the system is stable, the interaction can be characterized by SE. Similarly, for a stable noninteracting system, reliable measurement of the molar mass and stoichiometry can be obtained by SE.

Throughout this review, we have described a large number of data analysis packages available for both SV and SE. Table IV lists the Web sites where this software may be obtained along with references describing the analysis algorithms and their applications. Table IV also includes a number of utility programs that perform useful calculations or graphics. It should also be mentioned that the Reversible Associations in Structural and Molecular Biology (RASMB) group also maintains an e-mail list-server that facilitates communication among researchers interested in AUC (http://www.bbri.org/RASMB/rasmb.html). The RASMB also maintains a software archive where many of the programs can be obtained.

Acknowledgment

This work was supported by grant numbers RR-18286 and AI-53615 from the NIH to J.L.C.

References

Ansevin, A. T., Roark, D. E., and Yphantis, D. A. (1970). Improved ultracentrifuge cells for high-speed sedimentation equilibrium studies with interference optics. *Anal. Biochem.* **34,** 237–261.

Arakawa, T., and Timasheff, S. N. (1985). Calculation of the partial specific volume of proteins in concentrated salt and amino acid solutions. *Methods Enzymol.* **117,** 60–65.

Behlke, J., and Ristau, O. (1997). Molecular mass determination by sedimentation velocity experiments and direct fitting of the concentration profiles. *Biophys. J.* **72,** 428–434.

Byron, O. (1997). Construction of hydrodynamic bead models from high-resolution X-ray crystallographic or nuclear magnetic resonance data. *Biophys. J.* **72,** 408–415.

Byron, O. (2000). Hydrodynamic bead modeling of biological macromolecules. *Methods Enzymol.* **321,** 278–304.

Cann, J. R. (1970). "Interacting Macromolecules." Academic Press, New York.

Cantor, C. R., and Schimmel, P. R. (1980)."Biophysical Chemistry," Part II. W.H. Freeman and Co., San Francisco.

Cole, J. L. (1996). Characterization of human cytomegalovirus protease dimerization by analytical centrifugation. *Biochemistry* **35,** 15601–15610.

Cole, J. L. (2004). Analysis of heterogeneous interactions. *Methods Enzymol.* **384,** 212–232.

Cole, J. L., and Hansen, J. C. (1999). Analytical ultracentrifugation as a contemporary biomolecular research tool. *J. Biomol. Tech.* **10,** 163–174.

Connaghan-Jones, K. D., Heneghan, A. F., Miura, M. T., and Bain, D. L. (2006). Hydrodynamic analysis of the human progesterone receptor A-isoform reveals that self-association occurs in the micromolar range. *Biochemistry* **45,** 12090–12099.

Correia, J. J. (2000). Analysis of weight average sedimentation velocity data. *Methods Enzymol.* **321,** 81–100.

Correia, J. J., Sontag, C. A., Stafford, W. F., and Sherwood, P. J. (2005). Models for direct boundary fitting of indefinite ligand-linked self-association. *In* "Analytical Ultracentrifugation: Techniques and Methods" (D. J. Scott, S. E. Harding, and A. J. Rowe, eds.), pp. 51–63. Royal Society of Chemistry, Cambridge.

Dam, J., and Schuck, P. (2004). Calculating sedimentation coefficient distributions by direct modeling of sedimentation velocity concentration profiles. *Methods Enzymol.* **384,** 185–212.

Dam, J., and Schuck, P. (2005). Sedimentation velocity analysis of heterogeneous protein-protein interactions: Sedimentation coefficient distributions *c*(*s*) and asymptotic boundary profiles from Gilbert-Jenkins theory. *Biophys. J.* **89,** 651–666.

Dam, J., Velikovsky, C. A., Mariuzza, R. A., Urbanke, C., and Schuck, P. (2005). Sedimentation velocity analysis of heterogeneous protein-protein interactions: Lamm equation modeling and sedimentation coefficient distributions *c*(*s*). *Biophys. J.* **89,** 619–634.

Demeler, B. (2005). UltraScan a comprehensive data analysis software package for analytical ultracentrifugation experiments. *In* "Modern Analytical Ultracentrifugation: Techniques and Methods" (D. J. Scott, S. E. Harding, and A. J. Rowe, eds.), pp. 210–229. Royal Society of Chemistry, Cambridge.

Demeler, B., and Saber, H. (1998). Determination of molecular parameters by fitting sedimentation data to finite-element solutions of the Lamm equation. *Biophys. J.* **74,** 444–454.

Demeler, B., and van Holde, K. E. (2004). Sedimentation velocity analysis of highly heterogeneous systems. *Anal. Biochem.* **335,** 279–288.

DeRosier, D. J., Munk, P., and Cox, D. J. (1972). Automatic measurement of interefernce photographs for the ultracentrifuge. *Anal. Biochem.* **50,** 139–153.

Dhami, R., Coelfen, H., and Harding, S. E. (1995). A comparative "Schlieren" study of the sedimentation behavior of three polysaccharides using the Beckman Optima XL-A and model E analytical ultracentrifuges. *Progr. Colloid Polym. Sci.* **99,** 187–192.

Edelstein, S. J., and Schachman, H. K. (1973). Measurement of partial specific volume by sedimentation equilibrium in H_2O-D_2O solutions. *Methods Enzymol.* **27,** 82–98.

Eisenberg, H. (2000). Analytical ultracentrifugation in a Gibbsian perspective. *Biophys. Chem.* **88,** 1–9.
Fujita, H. (1975). "Foundations of Ultracentrifugal Analysis." Wiley, New York.
Garcia De La Torre, J., Huertas, M. L., and Carrasco, B. (2000). Calculation of hydrodynamic properties of globular proteins from their atomic-level structure. *Biophys. J.* **78,** 719–730.
Gekko, K., and Timasheff, S. N. (1981). Mechanism of protein stabilization by glycerol: Preferential hydration in glycerol-water mixtures. *Biochemistry* **20,** 4667–4676.
Gelinas, A. D., Toth, J., Bethoney, K. A., Stafford, W. F., and Harrison, C. J. (2004). Mutational analysis of the energetics of the GrpE.DnaK binding interface: Equilibrium association constants by sedimentation velocity analytical ultracentrifugation. *J. Mol. Biol.* **339,** 447–458.
Gilbert, G. A., and Jenkins, R. C. (1956). Boundary problems in the sedimentation and electrophoresis of complex systems in rapid reversible equilibrium. *Nature* **177,** 853–854.
Hansen, J. C., Lebowitz, J., and Demeler, B. (1994). Analytical ultracentrifugation of complex macromolecular systems. *Biochemistry* **33,** 13155–13163.
Harding, S. E., Horton, J. C., and Morgan, P. J. (1992). MSTAR: A FORTRAN program for the model independent molecular weight analysis of macromolecules using low or high speed sedimentation equilibrium. *In* "Analytical Ultracentrifugation in Biochemistry and Polymer Science" (S. E. Harding, J. C. Horton, and A. J. Rowe, eds.), Royal Society of Chemistry, Cambridge.
Hattan, S. J., Laue, T. M., and Chasteen, N. D. (2001). Purification and characterization of a novel calcium-binding protein from the extrapallial fluid of the mollusc. *Mytilus edulis. J. Biol. Chem.* **276,** 4461–4468.
Hensley, P. (1996). Defining the structure and stability of macromolecular assemblies in solution: The re-emergence of analytical ultracentrifugation as a practical tool. *Structure* **4,** 367–373.
Howlett, G. J., Minton, A. P., and Rivas, G. (2006). Analytical ultracentrifugation for the study of protein association and assembly. *Curr. Opin. Chem. Biol.* **10,** 430–436.
Huglin, M. B. (1972). Specific refractive index increments. *In* "Light Scattering from Polymer Solutions" (M. B. Huglin, ed.), pp. 165–332. Academic Press, New York.
Jiménez, M., Rivas, G., and Minton, A. P. (2007). Quantitative characterization of weak self-association in concentrated solutions of immunoglobulin G via the measurement of sedimentation equilibrium and osmotic pressure. *Biochemistry* **46,** 8373–8378.
Johnson, M. L., Correia, J. J., Yphantis, D. A., and Halvorson, H. R. (1981). Analysis of data from the analytical ultracentrifuge by nonlinear least squares techniques. *Biophys. J.* **36,** 575–588.
Kar, S. R., Kingsbury, J. S., Lewis, M. S., Laue, T. M., and Schuck, P. (2000). Analysis of transport experiments using pseudo-absorbance data. *Anal. Biochem.* **285,** 135–142.
Kroe, R. (2005). "Application of fluorescence detected sedimentation" Ph.D. Thesis, University of New Hampshire.
Laue, T. M. (1992). Short column sedimentation equilibrium analysis for rapid characterization of macromolecules in solution Beckman Coulter Technical Report DS-835.
Laue, T. M. (1995). Sedimentation equilibrium as thermodynamic tool. *Methods Enzymol.* **259,** 427–452.
Laue, T. M. (1996). Choosing which optical system of the optima XL-I analytical centrifuge to use Beckman Coulter Technical Report A-1821-A.
Laue, T. M. (2006). A light intensity measurement system for the analytical ultracentrifuge. *In* "Progress in Colloid and Polymer Science," pp. 1–8. Springer, Berlin.
Laue, T. M., and Stafford, W. F. (1999). Modern applications of analytical ultracentrifugation. *Annu. Rev. Biophys. Biomol. Struct.* **28,** 75–100.
Laue, T. M., Shah, B. D., Ridgeway, T. M., and Pelletier, S. L. (1992). Computer-aided interpretation of analytical sedimentation data for proteins. *In* "Analytical Ultracentrifugation in Biochemistry and Polymer Science" (S. Harding, A. Rowe, and J. Horton, eds.), pp. 90–125. Royal Society of Chemistry, Cambridge.
Lebowitz, J., Lewis, M. S., and Schuck, P. (2002). Modern analytical ultracentrifugation in protein science: A tutorial review. *Protein Sci.* **11,** 2067–2079.

Lee, J. C., and Timasheff, S. N. (1974a). The calculation of partial specific volumes of proteins in guanidine hydrochloride. *Arch. Biochem. Biophys.* **165,** 268–273.

Lee, J. C., and Timasheff, S. N. (1974b). Partial specific volumes and interactions with solvent components of proteins in guanidine hydrochloride. *Biochemistry* **13,** 257–265.

Liu, S., and Stafford, W. F., III. (1995). An optical thermometer for direct measurement of cell temperature in the Beckman instruments XL-A analytical ultracentrifuge. *Anal. Biochem.* **224,** 199–202.

MacGregor, I. K., Anderson, A. L., and Laue, T. M. (2004). Fluorescence detection for the XLI analytical ultracentrifuge. *Biophys. Chem.* **108,** 165–185.

Perkins, S. J. (2001). X-ray and neutron scattering analyses of hydration shells: A molecular interpretation based on sequence predictions and modelling fits. *Biophys. Chem.* **93,** 129–139.

Philo, J. S. (1994). Measuring sedimentation, diffusion, and molecular weights of small molecules by direct fitting of sedimentation velocity concentration profiles. *In* "Modern Analytical Ultracentrifugation" (T. M. Shuster, and T. M. Laue, eds.), pp. 156–170. Birkhauser, Boston.

Philo, J. S. (1997). An improved function for fitting sedimentation velocity data for low-molecular-weight solutes. *Biophys. J.* **72,** 435–444.

Philo, J. S. (2000a). Improving sedimentation equilibrium analysis of mixed associations using numerical constraints to impose mass or signal conservation. *Methods Enzymol.* **321,** 100–120.

Philo, J. S. (2000b). A method for directly fitting the time derivative of sedimentation velocity data and an alternative algorithm for calculating sedimentation coefficient distribution functions. *Anal. Biochem.* **279,** 151–163.

Philo, J. S. (2006). Improved methods for fitting sedimentation coefficient distributions derived by time-derivative techniques. *Anal. Biochem.* **354,** 238–246.

Philo, J. S., Yang, T. H., and LaBarre, M. (2004). Re-examining the oligomerization state of macrophage migration inhibitory factor (MIF) in solution. *Biophys. Chem.* **108,** 77–87.

Prakash, V., and Timasheff, S. N. (1985). Calculation of partial specific volumes of proteins in 8 M urea solution. *Methods Enzymol.* **117,** 53–60.

Rai, N., Nollmann, M., Spotorno, B., Tassara, G., Byron, O., and Rocco, M. (2005). SOMO (SOlution MOdeler) differences between X-Ray- and NMR-derived bead models suggest a role for side chain flexibility in protein hydrodynamics. *Structure* **13,** 723–734.

Reynolds, J. A., and McCaslin, D. R. (1985). Determination of protein molecular weight in complexes with detergent without knowledge of binding. *Methods Enzymol.* **117,** 41–53.

Richards, E. G., and Schachman, H. K. (1957). A differential ultracentrifuge technique for measuring small changes in sedimentation coefficients. *J. Am. Chem. Soc.* **79,** 5324–5325.

Richards, E. G., and Schachman, H. K. (1959). Ultracentrifuge studies with Rayleigh interference optics. I. General applications. *J. Phys. Chem.* **63,** 1578–1591.

Richards, E. G., Teller, D. C., Hoagland, V. D. J., Haschemeyer, R. H., and Schachman, H. K. (1971). Alignment of Schlieren and Rayleigh optical systems in the ultracentrifuge. II. A general procedure. *Anal. Biochem.* **41,** 215–247.

Rivas, G., and Minton, A. P. (2003). Tracer sedimentation equilibrium: A powerful tool for the quantitative characterization of macromolecular self- and hetero-associations in solution. *Biochem. Soc. Trans.* **31,** 1015–1019.

Rivas, G., Stafford, W., and Minton, A. P. (1999). Characterization of heterologous protein-protein interactions using analytical ultracentrifugation. *Methods* **19,** 194–212.

Roark, D. E. (1976). Sedimentation equilibrium techniques: Multiple speed analyses and an overspeed procedure. *Biophys. Chem.* **5,** 185–196.

Roark, D. E., and Yphantis, D. A. (1969). Studies of self-associating systems by equilibrium ultracentrifugation. *Ann. N. Y. Acad. Sci.* **164,** 245–278.

Schachman, H. K. (1959). "Ultracentrifugation in Biochemistry." Academic Press, New York.

Schatz, G. (1996). Biographical Memoirs, Vol. 70, pp. 320–346. National Academy of Sciences, Washington, DC.

Schuck, P. (1998). Sedimentation analysis of noninteracting and self-associating solutes using numerical solutions to the Lamm equation. *Biophys. J.* **75,** 1503–1512.

Schuck, P. (2000). Size-distribution analysis of macromolecules by sedimentation velocity ultracentrifugation and Lamm equation modeling. *Biophys. J.* **78,** 1606–1619.

Schuck, P. (2003). On the analysis of protein self-association by sedimentation velocity analytical ultracentrifugation. *Anal. Biochem.* **320,** 104–124.

Schuck, P. (2004). A model for sedimentation in inhomogeneous media. I. Dynamic density gradients from sedimenting co-solutes. *Biophys. Chem.* **108,** 187–200.

Schuck, P., and Demeler, B. (1999). Direct sedimentation analysis of interference optical data in analytical ultracentrifugation. *Biophys. J.* **76,** 2288–2296.

Scott, D. J., and Schuck, P. (2005). A brief introduction to the analytical ultracentrifugation of proteins for beginners. *In* "Analytical Ultracentrifugation" (D. J. Scott, S. E. Harding, and A. J. Rowe, eds.), pp. 1–25. Royal Society of Chemistry, Cambridge, UK.

Snyder, D., Lary, J., Chen, Y., Gollnick, P., and Cole, J. L. (2004). Interaction of the trp RNA-binding attenuation protein (TRAP) with anti-TRAP. *J. Mol. Biol.* **338,** 669–682.

Sontag, C. A., Stafford, W. F., and Correia, J. J. (2004). A comparison of weight average and direct boundary fitting of sedimentation velocity data for indefinite polymerizing systems. *Biophys. Chem.* **108,** 215–230.

Stafford, W. F., III. (1980). Graphical analysis of nonideal monomer N-mer, isodesmic, and type II indefinite self-associating systems by equilibrium ultracentrifugation. *Biophys. J.* **29,** 149–166.

Stafford, W. F. (1992). Boundary analysis in sedimentation transport experiments: A procedure for obtaining sedimentation coefficient distributions using the time derivative of the concentration profile. *Anal. Biochem.* **203,** 295–301.

Stafford, W. F. (2000). Analysis of reversibly interacting macromolecular systems by time derivative sedimentation velocity. *Methods Enzymol.* **323,** 302–325.

Stafford, W. F., and Sherwood, P. J. (2004). Analysis of heterologous interacting systems by sedimentation velocity: Curve fitting algorithms for estimation of sedimentation coefficients, equilibrium and kinetic constants. *Biophys. Chem.* **108,** 231–243.

Surendran, R., Herman, P., Cheng, Z., Daly, T. J., and Lee, J. Ching (2004). HIV Rev self-assembly is linked to a molten-globule to compact structural transition. *Biophys. Chem.* **108,** 101–119.

Tanford, C. (1961). "Physical Chemistry of Macromolecules." John Wiley and Sons, New York.

Timasheff, S. N. (2002). Protein hydration, thermodynamic binding, and preferential hydration. *Biochemistry* **41,** 13473–13482.

van Holde, K. E., and Baldwin, R. L. (1958). Rapid attainment of sedimentation equilibrium. *J. Phys. Chem.* **62,** 734–743.

van Holde, K. E., and Weischet, W. O. (1978). Boundary analysis of sedimentation-velocity experiments with monodisperse and paucidisperse solutes. *Biopolymers* **17,** 1387–1403.

Vistica, J., Dam, J., Balbo, A., Yikilmaz, E., Mariuzza, R. A., Rouault, T. A., and Schuck, P. (2004). Sedimentation equilibrium analysis of protein interactions with global implicit mass conservation constraints and systematic noise decomposition. *Anal. Biochem.* **326,** 234–256.

Williams, J. W., van Holde, K. E., Baldwin, R. L., and Fujita, H. (1958). The theory of sedimentation analysis. *Chem. Rev.* **58,** 715–806.

Yphantis, D. A. (1960). Rapid determination of molecular weights of proteins and peptides. *Ann. N. Y. Acad. Sci.* **88,** 586–601.

Yphantis, D. A. (1964). Equilibrium ultracentrifugation of dilute solutions. *Biochemistry* **3,** 297–317.

CHAPTER 7

Determination of Membrane Protein Molecular Weights and Association Equilibrium Constants Using Sedimentation Equilibrium and Sedimentation Velocity

Nancy K. Burgess, Ann Marie Stanley, and Karen G. Fleming

T. C. Jenkins Department of Biophysics
Johns Hopkins University
Baltimore, Maryland 21218

0091-679X/08 $35.00
DOI: 10.1016/S0091-679X(07)84007-6

Abstract

Regulated molecular interactions are essential for cellular function and viability, and both homo- and hetero-interactions between all types of biomolecules play important cellular roles. This chapter focuses on interactions between membrane proteins. Knowing both the stoichiometries and stabilities of these interactions in hydrophobic environments is a prerequisite for understanding how this class of proteins regulates cellular activities in membranes. Using examples from the authors' work, this chapter highlights the application of analytical ultracentrifugation methods in the determination of these parameters for integral membrane proteins. Both theoretical and practical aspects of carrying out these experiments are discussed.

I. Introduction

As the number of high-resolution structures of membrane proteins continues to rise, it is increasingly appreciated that integral membrane proteins associate with defined stoichiometries and orientations. In some structures it is easy to rationalize why certain membrane proteins are oligomeric: the potassium channel of the KscA protein is formed only when four identical transmembrane helical subunits self-associate to create a passageway for the ion (Doyle *et al.*, 1998). In other cases, the underlying functional and physical basis for the oligomeric complex is not as easily understood from the structure. In contrast to KcsA, each monomer of the outer membrane protein F (OmpF) trimer has a pore through which ions can pass; trimerization of OmpF monomers brings them into contact with each other but does not create the physical channel. The underlying physical rationale for trimerization of OmpF is therefore not fully explained by the structural studies alone.

Thermodynamic measurements carried out in solution provide complementary information about the molecular interactions observed in structures. In particular, sedimentation equilibrium analytical ultracentrifugation (AUC) is an extremely useful and accurate method for confirming or invalidating stoichiometries observed in crystal structures (Burrows *et al.*, 1994). Moreover, if a system is reversibly associating in solution, these experiments can additionally provide access to key thermodynamic parameters for the reaction, such as the equilibrium constant; knowing this value, the oligomeric species distribution can be predicted over wide ranges of concentrations, and the biological significance of oligomeric species can be better understood.

Solution studies whose principal aim is to determine the mass or stoichiometry of a protein complex have historically been challenging to carry out on membrane proteins because they reside *in vivo* in the anisotropic, chemically heterogeneous environment of the biological lipid bilayer. Solution studies *in vitro* require manipulation of purified membrane protein samples, and solubility of integral membrane proteins requires that they be handled in the presence of a hydrophobic cosolvent.

In the vast majority of structural studies, this cosolvent is provided by detergent micelles, which introduces complexity into the analysis as this detergent binding contributes to the overall mass of a membrane protein complex and must be taken into account in order to separate the mass contributions of the protein from those of the bound detergent. Any bound lipid that may copurify with a membrane protein will also contribute to its molecular weight. In addition to the contribution of bound detergents and/or lipids to the mass of a membrane protein complex, they affect the shape of the overall complex. Therefore, any experimental method that is fundamentally dependent on transport (e.g., dynamic light scattering, gel filtration chromatography, sedimentation velocity) can be difficult to interpret because the shape contributions of the protein and the bound cosolvent can be difficult to separate from one another (see Chapter 12 by Byron, this volume). In this chapter we will discuss some of the strategies that can be used to overcome some of these technical barriers in analyzing membrane protein complexes dispersed in detergent micelle solutions to determine their molecular weights, interactions, and stoichiometries.

II. Rationale

A. Why Use AUC?

AUC, and in particular sedimentation equilibrium, is an extremely useful method for determining molecular weights of complexes in detergent micelle solutions. The principal advantage of sedimentation equilibrium is that it provides a direct measure of mass. This is in stark contrast to spectroscopic methods, such as fluorescence resonance energy transfer (FRET), which also have been used to evaluate membrane protein interactions. In FRET studies, the experimental quantity that can be obtained is the mole ratio of a particular interaction: the titration curve for a monomer–dimer reaction would appear identical to that for a dimer–tetramer reaction. Furthermore, the absence of an interaction would be experimentally observed as a lack of resonance energy transfer in a FRET experiment. Since there are many spectroscopic reasons why donor fluorophores might not efficiently transfer their energy to acceptor fluorophores, a lack of FRET is essentially a negative and noninterpretable result. In contrast, a lack of interactions (or change in molecular weight) in a sedimentation equilibrium experiment would result in a direct measurement of either the nonassociating monomeric or the nondissociating dimeric molecular weight.

B. General Considerations for Sedimentation Equilibrium Experiments of Membrane Proteins

The general considerations for both sedimentation equilibrium and sedimentation velocity experiments are described in detail in an accompanying chapter of this series (Chapter 6 by Cole *et al.*, this volume). The mechanics of carrying out experiments with membrane protein samples will be the same as described for soluble proteins in their chapter. One practical difference is that membrane protein samples in detergent solutions can be trickier to pipet into the ultracentrifugation cells as the surface tension of these solutions is lower due to the detergent content.[1] In addition to solubilizing membrane proteins, detergent solutions can also solubilize residual proteins in ultracentrifugation cells that are routinely used for soluble proteins, and it may be prudent to set up cells with detergent-only solutions overnight prior to an experiment in order to "wash" them. As is the case for soluble proteins, the purity of a membrane protein preparation is a key aspect of obtaining the best data possible, and any contaminating material can profoundly complicate the data analysis.

There are also considerations of the detection system to be used in an experiment. For the analysis of membrane protein distributions, we have exclusively used the absorbance optics of the Optima XL-A and XL-I ultracentrifuges to monitor the protein's radial distribution. We use absorbance optics because the detergents we employ do not absorb light at the wavelengths used to monitor the protein concentration, and therefore their distribution can be isolated from that of other components. While in principle the interference optics system could also be used to observe the protein distribution, the interference signal will contain contributions from both the protein and the bound detergent. Moreover, interference optics will detect any and all differences in the refractive indices between a reference and a sample chamber and may also contain a signal from free detergent. To avoid this, all buffer components should be in dialysis equilibrium with each other; however, this condition can be difficult to attain for membrane protein samples dispersed in detergent micelle solutions because detergent concentrations above the critical micelle concentration (cmc) do not always equilibrate across a dialysis membrane. In addition, for thermodynamic studies in which a membrane protein equilibrium constant is the desired information, it is essential to know the detergent concentration accurately, and measuring detergent concentrations following dialysis is not easily accomplished. Therefore, as an alternative to dialysis for our experiments, we have used a "column exchange" method to establish the solution detergent concentration (described in Section III).

In contrast to experiments with solubilized membrane proteins, interference optics is actually preferable in experiments where the detergent itself is being

[1] This is especially true for loading the two-sector cells where an air lock can cause the solution to spill over onto the outside of the cell. With both two- and six-sector cells we have found it to be better to place the pipet tip at the bottom of the cell and slowly raise it as the cell fills. In addition, when loading large volumes into two-sector cells, we have found that loading is much easier and more reliable if the entire volume can be injected into the cell at once, and we discovered that using a P1000 pipetman and "piggybacking" a long (round, not flat) gel-loading tip onto the end of the 1 ml blue tip allows the entire volume to be dispersed in one step. Alternatively, BeckmanCoulter, Inc. also manufactures long plastic loading needles that work well for fully loading cells in one step.

characterized. When scanned against a reference sector containing only buffer, detergent micelles are easily visualized by the interference optical system whereas most detergents are not easily detected using absorbance optics. As will be discussed below, when working with a new detergent or in a new buffer, we use interference optics in experiments designed to empirically confirm the buffer conditions required to match the effective density of the detergent micelle.

C. Special Considerations for Sedimentation Equilibrium Experiments in the Presence of Detergent Micelles

1. Buoyant Molecular Weight

To understand and ultimately account for the contributions from the bound detergent, it is important to recognize that the fundamental experimental quantity determined in a sedimentation equilibrium experiment is the buoyant molecular weight (Casassa and Eisenberg, 1964) defined as

$$M_{\mathrm{p}}(1-\phi'\rho) \tag{1}$$

where M_{p} is the molecular weight of only the protein portion of the sedimenting particle and excludes the molecular weight of the bound detergent, lipid, and/or water; the quantity $(1-\phi'\rho)$ is the buoyancy term; ϕ' is the effective partial specific volume (ml g^{-1}) of the protein moiety in the sedimenting particle and takes into account the contributions of the bound detergent, lipid, and water; and ρ is the solvent density. The buoyant molecular weight can be rewritten as a sum of each of the components, and the generalized form of the equation is (Reynolds and Tanford, 1976)

$$M_{\mathrm{p}}(1-\phi'\rho) = M_{\mathrm{p}}(1-\bar{v}_{\mathrm{p}}\rho) + \sum n_i M_i(1-\bar{v}_i\rho) \tag{2}$$

where M_i and $\bar{v}_i$ are the molecular weights and partial specific volumes (ml g^{-1}) of the ith component, and n_i is the number of molecules of any ith component bound to the protein. Any bound components—lipids, detergent molecules, water molecules—will contribute to the buoyant molecular weight, and each of the contributions can be explicitly stated in a specific form of Eq. (2) that can be written as follows:

$$\begin{aligned} M_{\mathrm{p}}(1-\phi'\rho) = {} & M_{\mathrm{p}}(1-\bar{v}_{\mathrm{p}}\rho) + n_{\mathrm{Lipid}}M_{\mathrm{Lipid}}(1-\bar{v}_{\mathrm{Lipid}}\rho) \\ & + n_{\mathrm{Det}}M_{\mathrm{Det}}(1-\bar{v}_{\mathrm{Det}}\rho) + n_{\mathrm{H_2O}}M_{\mathrm{H_2O}}(1-\bar{v}_{\mathrm{H_2O}}\rho) \end{aligned} \tag{3}$$

where the subscripts Lipid, Det, and H_2O indicate the contributions from the bound lipid, detergent, and water molecules, respectively. This equation is also often written in the following manner:

$$M_{\rm p}(1-\phi'\rho) = M_{\rm p}[(1-\bar{v}_{\rm p}\rho) + \delta_{\rm Lipid}(1-\bar{v}_{\rm Lipid}\rho) + \delta_{\rm Det}(1-\bar{v}_{\rm Det}\rho) + \delta_{\rm H_2O}(1-\bar{v}_{\rm H_2O}\rho)] \quad (4)$$

where δ_i represents the amount bound of the ith component in grams per gram of protein. Equations (3) and (4) can be simplified if we assume that the number of bound lipids ($n_{\rm Lipid}$) is small in a purified-membrane-protein preparation solubilized in detergent micelles at concentrations above their cmc. In this case the buoyant contribution of lipids will be much smaller than the contributions from the bound detergent and can be ignored. This assumption is further justified by the fact that many lipids have partial specific volume values that are close to unity (Durshlag, 1986), which means that the product of $\bar{v}_{\rm Lipid}\rho$ will also be unity as most buffer densities equal ~1.0 g ml^{-1}. This has the consequence of bringing the buoyancy factor $(1-\bar{v}_{\rm Lipid}\rho)$ down to a very small number that is essentially equal to zero and thus leads to a negligibly small contribution from bound lipids. This latter argument can also be made for the contribution of water to the buoyant molecular weight as long as the sedimentation equilibrium experiment is carried out in an aqueous solution lacking density additives. Eqs. (3) and (4) can thus usually be simplified in practice to Eqs. (5) and (6), respectively:

$$M_{\rm p}(1-\phi'\rho) = M_{\rm p}(1-\bar{v}_{\rm p}\rho) + n_{\rm Det}M_{\rm Det}(1-\bar{v}_{\rm Det}\rho) \quad (5)$$

and

$$M_{\rm p}(1-\phi'\rho) = M_{\rm p}[(1-\bar{v}_{\rm p}\rho) + \delta_{\rm Det}(1-\bar{v}_{\rm Det}\rho)] \quad (6)$$

The principle remaining contribution that must be accounted for is the contribution of the bound detergent, $\delta_{\rm Det}(1-\bar{v}_{\rm Det}\rho)$, to the buoyant mass of the complex.

2. Experimental Strategies to Account for Bound Cosolvent

There are three main strategies that have been used to account for the contribution of the bound detergent or lipid to the buoyant molecular weight. The choice of which strategy to use will depend on the scientific question to be addressed and on the chemical nature of the detergent micelles or lipids that must be used to solubilize the purified membrane protein. These three strategies are (i) measurement and use of the density increment; (ii) explicit accounting for the bound detergent; and (iii) density matching the detergent with the solvent.

a. Density Increment

The density-increment method is especially useful when a membrane protein is dispersed in a chemically heterogeneous detergent/lipid environment. This can occur when a protein copurifies with a significant amount of bound lipid or when the detergent environment that must be used to preserve the protein integrity cannot be density matched using the methods described below. In sedimentation equilibrium experiments, the density increment is defined as the change in solution density as a function of changing the protein concentration at constant chemical potential (Casassa and Eisenberg, 1964):

$$\left[\frac{\partial \rho}{\partial c_2}\right]_\mu = (1 - \phi' \rho) \tag{7}$$

where c_2 is the weight concentration of the protein alone and ϕ' is the effective partial specific volume of the protein in the protein–detergent complex. Note that the right side of equation (7) contains the term $(1-\phi'\rho)$, which is equal to the buoyancy factor in equation (1). Since the density increment equals the change in solution density as a function of protein concentration $(\partial\rho/\partial c_2)$, it is in principle an experimentally accessible quantity and can be measured using a high-precision density meter. The chemical potential subscript, μ, indicates that the protein must be at dialysis equilibrium with all other components; strict adherence to constant chemical potential can in practice be tricky depending on the cmc of the detergent. Nevertheless, even when constant chemical potential cannot be completely ensured, the density-increment approach appears to return molecular weight values that are sensible.

This strategy of accounting for the contribution of the bound cosolvent to membrane proteins has been used extensively by Butler and coworkers with many complex samples (Butler and Kuhlbrandt, 1988; Butler *et al.*, 2004; Konig *et al.*, 1997). In a typical experiment there are two density measurements: (i) the solvent density in the absence of protein (where the solvent contains the appropriate concentration of detergent micelles) and (ii) the density of a protein–detergent (and/or lipid) complex at a known concentration of protein. These two density points are then plotted as a function of protein concentration, fitted to a line, and the slope used as the buoyancy term in converting the experimental buoyant molecular weight into an expression for the molecular weight of the protein alone. Membrane protein complexes with density increments ranging from −0.14 to 0.734 mL g^{-1} have been analyzed using this method (Butler and Kuhlbrandt, 1988; Butler *et al.*, 2004; Konig *et al.*, 1997), and it is particularly suited for the determination of the molecular weight of membrane protein complexes with many constituents.

b. Explicitly Accounting for the Bound Detergent

A second strategy that can be used to account for the contributions of the bound detergent is to explicitly include it; the mathematical expression for this is illustrated by equation (5) above and requires knowledge of two parameters: (i) the number of detergent and/or lipid molecules that are bound and (ii) the partial specific volumes of each of these species. It is usually the case that this second parameter is much better known than the first as the partial specific volumes for a wide variety of detergents and lipids have been measured and tabulated (Durshlag, 1986). In contrast, the knowledge of the number of bound detergents for a particular membrane protein complex is generally less well known, although le Maire and colleagues have measured detergent binding for a number of proteins and describe protocols to make this measurement using radiolabeled detergent molecules (le Maire *et al.*, 1983, 2000). However, since radiolabeled detergent samples are not generally available, measuring the amount of bound detergent and using this method of disentangling the contributions of the bound detergent from that of the protein is not widely used.

c. The Density-Matching Strategy

"Density matching" is a third strategy that can be used to interpret the experimental buoyant molecular weight in terms of the protein distribution alone. We have favored this approach because the principle goal of many of our experiments has been to determine equilibrium constants for membrane protein interactions; for this we need to know the protein mass as a function of concentration. Since protein complexes of different molecular weights may also bind different amounts of detergent, there are too many parameters to be determined if we must also take into account the detergent binding of each oligomeric complex. The "density-matching" strategy avoids this complication by minimizing the effective contribution of the bound detergent on all protein oligomeric states, and the membrane protein–detergent complex can be analyzed and interpreted just as one would do for a soluble protein in an aqueous buffer.

In the "density-matching" strategy, the experimental conditions are adjusted such that the solvent density is equal to the effective density of the bound detergent molecules in the protein-detergent complex (Reynolds and Tanford, 1976). Mathematically, this is expressed as the condition where $\rho = 1/\bar{v}_{\text{Det}}$. When this is the case, the buoyancy term from the detergent contribution in equation (5) will be a very small number and essentially equal to zero:

$$(1 = \bar{v}_{\text{Det}}\rho) \cong \left[1 = \bar{v}_{\text{Det}}\left[\frac{1}{\bar{v}_{\text{Det}}}\right]\right] = (1 - 1) = 0 \tag{8}$$

When density matching is achieved, the effective contribution of the bound detergent to the experimentally observed buoyant molecular weight essentially becomes zero *no matter how many detergents are bound*, and the detergent is essentially invisible to the centrifugal field generated by the rotational force. The data can then be analyzed in the standard way and interpreted in terms of the protein mass alone.

The "density-matching" strategy works in this straightforward way only when the solvent density is adjusted by using heavy water, and both ^2H_2O and $^2H_2{}^{18}O$ have been successfully employed. The addition of other cosolvents that increase the solvent density, such as sucrose, affect the chemical potential of water and lead to preferential binding and/or exclusion of water or the additional cosolvent at the surface of the protein; in this case their contributions to the buoyant molecular weight must be taken into account by the addition of the appropriate terms in equation (2). In other words, cosolvents that increase the solvent density may in fact match the density of the bound detergent, but they simultaneously introduce a mass uncertainty from their own binding and from the preferential binding of water (Reynolds and Tanford, 1976).

This problem is largely overcome by matching the solvent density with that of heavy water; however, the need to use heavy water means that the chemical nature of detergents that can be used with this strategy is limited. The density of pure ^2H_2O is 1.1 g ml^{-1}; therefore, the partial specific volume of the detergent must be between that of water and ^2H_2O, for example, greater than 0.9 and less than unity. Unfortunately, this eliminates a simple evaluation of membrane protein complexes in the frequently employed detergents dodecylmaltoside and β-octylglucoside, since the densities of these detergents are 1.21 and 1.15 g ml^{-1}, respectively (Reynolds and McCaslin, 1985; Suarez *et al.*, 1984). Even the use of $^2H_2{}^{18}O$ cannot facilitate density matching of these detergents, although Ferguson-Miller and colleagues have shown that careful experiments coupled with a significant density extrapolation can facilitate a mass determination of membrane protein complexes in this detergent (Suarez *et al.*, 1984). It is also notable that the bile salt detergents have effective densities around ~1.3 g ml^{-1} and cannot be used at all in density-matching experiments (Reynolds and McCaslin, 1985). Nevertheless, there are several detergents that can be density matched with ^2H_2O, and these have been extremely useful in evaluating membrane protein interactions. In our early work we used the neutrally buoyant pentaoxyethylene octyl ether (C_8E_5) detergent ($\bar{v} = 0.993$ ml g^{-1}(Ludwig *et al.*, 1982)) to analyze the energetics of the dimerization of the glycophorin A (GpA) transmembrane helix (Doura and Fleming, 2004; Doura *et al.*, 2004; Fleming, 1998, 2000, 2002; Fleming and Engelman, 2001; Fleming *et al.*, 1997; Stanley and Fleming, 2005) and the human erbB transmembrane domains (Stanley and Fleming, 2005). We and others have also used the zwittergent 3-(*N,N*-dimethylmyristyl-ammonio)propanesulfonate, C14SB, and we have used this detergent in our more recent work on transmembrane β-barrels that will be discussed in this chapter (Ebie and Fleming, 2007; Fleming *et al.*, 2004; Gratkowski *et al.*, 2001; Howard *et al.*, 2002; Kobus and Fleming, 2005; Kochendoerfer *et al.*, 1999; Li *et al.*, 2004; Pinto *et al.*, 1997; Stanley *et al.*, 2006, 2007). C14SB is significantly less expensive than C_8E_5, and it preserves the native structure and function of our transmembrane β-barrel outer membrane phospholipase A (OMPLA) (Ann Marie Stanley and Karen G. Fleming, unpublished observation). In 20 mM Tris buffer with 200 mM KCl, the density of C14SB was matched by 13% ^2H_2O. In addition, DeGrado and coworkers have employed dodecylphosphocholine (DPC) detergent micelles for several studies exploring both natural and designed transmembrane helix–helix interactions (Kochendoerfer *et al.*, 1999; Li *et al.*, 2004). DPC requires 52.5% ^2H_2O to match its density in 50 mM Tris–HCl buffer with 100 mM NaCl (Kochendoerfer *et al.*, 1999).

III. Materials and Methods

A. Expression and Purification of Membrane Proteins Used in This Study

1. Expression and Purification of OMPLA

The cloning, expression, and purification of OMPLA is described in detail in Stanley *et al.* (2006). Briefly, the amino acid sequence encoding the mature OMPLA protein was cloned into a pet11A-T7 expression vector. HMS174(DE3) cells harboring the expression vector were grown to midlogarithmic phase, and expression of OMPLA was induced for three hours by the addition of 1 mM isopropyl β-D-1-thiogalactopyranoside (IPTG). Cells were harvested, lysed by French press, and inclusion bodies were isolated by centrifugation. OMPLA was refolded, as described by Dekker *et al.* (1997), except that C14SB was substituted for C12SB in all steps. Immediately prior to each sedimentation equilibrium experiment, OMPLA was exchanged into the desired detergent concentration. Instead of dialysis, we have found that the easiest way to do this was to bind the protein to a Q-sepharose column followed by washing the column with five column volumes of the desired detergent concentration in the presence of buffer and then to elute the protein in a single step with 600 mM KCl in the same detergent concentration and buffer. We typically carried out experiments in a final concentration of 200 mM KCl and could therefore dilute the high eluant in 600 mM KCl with the appropriate solution to obtain the desired final concentrations of ^2H_2O and C14SB. Since the volumes required for sedimentation equilibrium are relatively small, we carried out this final column exchange step using small (0.5 ml–1 ml) bed volumes of the ion exchange resin and manually collected the eluant samples by counting drops (typically 13–18 drops per fraction) to avoid dilution of the eluted, detergent equilibrated samples.

2. Preparation of OMPLA Q94A

A single-point mutant of OMPLA, Q94A, was generated using standard molecular biology techniques. This protein was expressed and purified using the protocol for the wild-type protein. Before any experimentation, OMPLA Q94A was applied to a Q-sepharose column and eluted in column buffer with 600 mM KCl. The extinction coefficient used for both the WT and Q94A sequence variant was 90,444 M^{-1} cm^{-1} (Dekker *et al.*, 1995).

3. Preparation of OmpF Samples

a. Expression and Purification of OmpF

The coding sequence for the mature *Escherichia coli* OmpF protein was amplified via PCR from the *E. coli* strain DH5α. To express the protein into inclusion bodies, PCR primers were designed to replace the signal sequence with a start codon (forward: 5′-G GCA GTA CAT ATG GCA GAA ATC TAT AAC AAA GAT GGC-3′; reverse: 5′-CG GGA TCC TTA CAA CTG GTA AAC GAT ACC

CAC AG-3′). The insert was cut and ligated into a pET11a expression vector (Novagen). This vector was transformed into *E. coli* BL21(DE3) cells and the nucleotide sequence was confirmed by DNA sequencing. Mature OmpF was purified from inclusion bodies using a published procedure (Dekker *et al.*, 1995). Washed inclusion body pellets were stored at −20 °C until further use.

Inclusion body pellets of OmpF were resuspended in unfolding buffer (20 mM sodium phosphate, 8 M urea, pH 7.3). According to the folding conditions described by Surrey *et al.* (1996), OmpF (13.4 μM, final concentration) was added to the folding buffer (3.7 mM dodecylmaltoside, 3.7 mM dimyristoylphosphatidylcholine (DMPC), 20 mM sodium phosphate, pH 6.5) and incubated overnight at room temperature with gentle stirring. To exchange DMPC for detergent, C14SB was added to the folding conditions (20 mM, final concentration). OmpF was then applied to a UNO Q-1 column (BioRad), washed with column buffer (5 mM C14, 20 mM Tris–HCl, pH 8.3), and eluted in column buffer with 300 mM KCl. Thin-layer chromatography confirmed that DMPC was no longer associated with the protein (data not shown). To remove unfolded protein not inserted into micelles, OmpF was incubated overnight at 37 °C in 0.02 mg/ml trypsin, followed by overnight dialysis against 300 mM KCl, 0.4 mM C14, 20 mM Tris–HCl, pH 8.3. If necessary, protein was concentrated on a Q-Sepharose column (Pharmacia) and eluted in column buffer with 1 M KCl.

The OmpF protein concentration was determined by absorbance at 280 nm using the extinction coefficient calculated by SEDNTERP (Laue *et al.*, 1992), 54,200 M^{-1} cm^{-1}. To determine the extinction coefficient of OmpF at 230 nm, protein absorbance was measured at both 230 and 280 nm and the average of the 230/280 ratio of six different samples was calculated to be 5.0. Therefore, the extinction coefficient used for OmpF at 230 nm was 271,000 M^{-1} cm^{-1}.

b. Vesicle Preparation for OmpF Folding

DMPC (Avanti Polar Lipids) dissolved in chloroform was placed under a gentle stream of N_2 gas and freeze-dried overnight to remove any residual chloroform. To hydrate the lipid, DMPC was resuspended in phosphate buffer (20 mM sodium phosphate, pH 6.5) to a final DMPC concentration of 14.75 mM and incubated at room temperature for at least 30 minutes with occasional vortexing. Unilamellar vesicles were prepared either by extrusion or sonication (Surrey *et al.*, 1996). For extrusion, hydrated DMPC was put through three freeze–thaw cycles in dry ice and 40 °C water baths, followed by extrusion through a 0.1 μM filter 11 times using the Avanti mini-extruder. Sonicated vesicles were prepared using a Branson Digital Sonifier for 50 minutes on a 50% duty cycle.

B. Determination of the Density-Matching Point for C14SB

A first step for all of these experiments was to determine the concentration of ^2H_2O, which is required to density match the C14SB detergent micelles in the background of our other buffer components. To accomplish this, we carried out sedimentation equilibrium experiments on 30 mM solutions of C14SB in 20 mM Tris–HCl, pH 8.0, and 200 mM KCl prepared in aqueous solutions containing $^{2}H_2O$ at 0%, 10%, 20%, and 30%. Reference samples contained all buffer components (including $^{2}H_2O$) but no detergent micelles. We used external loading 6-sector cells equipped with sapphire windows [(Ansevin *et al.*, 1970); also see Chapter 6 by Cole *et al.*, this volume] and collected interference buffer blanks on all cells prior to analyzing the detergent micelle samples; these scans were subsequently subtracted from the data scans and account for any radially independent, time-independent noise or slope in the interferograms. The sample volumes were 110 μl. We then removed the buffer from the sample side and loaded the detergent micelle solution. The interference optical system of a Beckman Optima XL-I was used to detect the distribution of micelles as a function of radius. We collected data at 50,000 rpm every 15 minutes and used the Windows version of Match to determine the time to equilibrium, which was generally less than 4 hours. We chose 50,000 rpm because this speed was faster than any of the speeds we anticipated using for our protein samples. Since detergents are more compressible than proteins, there is the formal possibility that any particular detergent would exhibit pressure effects; these would be reflected in a density gradient at high centrifugal forces. We reasoned that the absence of a density gradient at speeds much higher than what we would use for sedimentation equilibrium experiments would ensure its absence in our subsequent protein experiments at lower speeds. Over the course of the past 12 years of experiments, the author has never observed a pressure dependence with any protein or detergent sample; nevertheless, when initiating experiments with a new detergent it is prudent to check for an obvious presence of this in detergent samples. These detergent matching experiments were carried out at 25 °C because we anticipated using this temperature for the subsequent protein samples. We have not explored the temperature dependence of the match point for any of the detergents we have studied; however, we would recommend that an investigator determine the match point at the temperature of interest as it might be different.

Determination of the amount of $^{2}H_2O$ required to match the density of the detergent micelles requires analysis of the radial distributions of the detergent-only samples, and the shapes of the radial distributions for micelles in these density-matching experiments deserve comment. For detergents whose partial specific volumes fall between 0.9 and 1.0 ml g^{-1}, the effective molecular weights will be fairly low at the speeds attainable in the Beckman AUC instrument. The shapes of the experimentally obtained distributions are therefore extremely shallow exponentials and are in fact well described by the equation for a line. The slope of the line indicates relative densities of the micelles and the solution: if the detergent micelles are more dense than the solution, the slope will be positive; if the micelles are less dense than the solution, the slope will be negative (i.e., the micelles float); and at the isopycnic point, the slope is zero. Determination of the matching $^{2}H_2O$ concentration is therefore a simple matter of plotting the slope of the micelle distribution as a function of percent $^{2}H_2O$ and finding the point where the slope equals zero.

C. Sedimentation Equilibrium Experiments on Membrane Protein Samples Dispersed in C14SB

1. Sedimentation Equilibrium Conditions for OMPLA

Sedimentation equilibrium experiments were performed at 25 °C using a Beckman XL-A analytical ultracentrifuge. Samples were prepared in 100 mM KCl, 20 mM Tris–HCl, pH 8.3, 13% ^2H_2O (the matching amount of ^2H_2O). Initial protein concentrations corresponded to 0.9, 0.6, and 0.3 absorbance units at A_{280}, and the rotor speeds were 16,300, 20,000, and 24,500 revolutions per minute. In contrast to our earlier work on the dimeric GpA helix, we discovered in our early experiments that OMPLA has a modest propensity for self-association, and we therefore carried out the majority of our experiments in 5 mM C14SB. Increasing the concentration of detergent led to dilution of OMPLA in the detergent micelle phase and promoted dissociation of the OMPLA dimer. This detergent dependence made it impossible to determine interaction energetics as a function of detergent concentration since we could not simultaneously observe the OMPLA monomer and dimer at high detergent concentrations; under these conditions, only the monomer was populated.

2. Sedimentation Equilibrium Conditions for OmpF

Sedimentation equilibrium experiments on OmpF were performed at 25 °C using a Beckman XL-A analytical ultracentrifuge. Samples were prepared in either 100 or 200 mM KCl, 20 mM Tris–HCl, pH 8.3, 13% ^2H_2O. Because of OmpF's pronounced trimeric stability, we collected sedimentation equilibrium data on OmpF at several different detergent micelle concentrations (5, 12, and 30 mM). To increase the probability of observing OmpF monomers and/or dimers, we also collected the data by monitoring the OmpF absorbance at 230 nm, which allowed us to observe a dilute protein solution. At each detergent concentration, we collected four speeds (9,000, 11,000, 13,500, and 16,300 rpm) and set up three initial protein concentrations corresponding to A_{230} values of 0.3, 0.6, and 0.9; 12 equilibrium data sets were therefore collected in total.

Notably, the speeds used in the OmpF and OMPLA sedimentation equilibrium experiments differ. This is because the rotor speed should be chosen to optimize the exponential shape of the experimental data. As a rule of thumb, the lowest speed is chosen so that the effective molecular weight (σ) at that speed equals approximately unity, where the effective molecular weight is defined according to Yphantis (1964):

$$\sigma = \frac{M(1 - \bar{v}\rho)\omega^2}{RT} \tag{9}$$

where ω is the rotor speed in radians per second, R is the universal gas constant (8.314472 J mol^{-1} K^{-1}), and T is the temperature in degrees Kelvin. Since the mass of the OmpF trimer is much larger than that of OMPLA, sedimentation equilibrium speeds are therefore lower for this protein. Subsequent speeds were chosen such that the speed factor between any two speeds equals approximately 1.5; the speed factor is defined as

$$\text{Speed Factor} = \frac{\omega^2 \text{ for Speed2}}{\omega^2 \text{ for Speed1}} \tag{10}$$

Since the data will subsequently be globally fit to determine whether or not the macromolecules are reversibly self-associating, these speed choices ensure that the radial distributions of the protein will be significantly different from each other.

3. General Data Fitting Strategy

Typically, we collect three rotor speeds with three initial protein concentrations. This setup results in nine distinct radial distributions of the protein. Before analysis, each data set needs to be trimmed to extract the regions of the exponential distributions to be used for analysis, to remove the optical noise between the sectors in the cell, and to delete any bad points. The analyzable regions of the distributions start at a radial position just greater than that of the meniscus. Since the absorbance optical system is not linear with concentration above 1.2 on our instrument, we trim the data to end at an absorbance value no greater than this, and in cases where the concentration gradient at the bottom of the data set was high, the data were trimmed to lower maximum absorbance values. The freely available software program Winreedit is easy to use and can be employed in an iterative manner to trim data appropriately; data trimming can also be accomplished in the Sedanal preprocessor; alternatively, it can be carried out manually using almost any spreadsheet or data analysis program since the sedimentation equilibrium files are just ASCII text files. Importantly, the original files should always be backed up as the user may need to go back and reedit them at a future time.

Once data are trimmed, we globally fit each detergent concentration with either the Windows or the Mac OS9 version of NONLIN (Johnson *et al.*, 1981) by nonlinear least-squares curve fitting. The goal of the fitting process is to find the simplest model that describes the data: each set of data was initially fitted using a single ideal species model followed by fitting using equations describing increasingly complex reversible association schemes. The monomeric mass and partial specific volume of each of the proteins were calculated using the program SEDNTERP (Laue *et al.*, 1992) and the amino acid sequences as input. These values were held constant during fitting to self-association models.

D. Sedimentation Velocity Experiments on OmpF

For sedimentation velocity experiments, the absorbance optics were employed, and intensity scans were collected for protein in three different C14SB concentrations (5, 30, and 60 mM). The collection of intensity scans as opposed to default absorbance scans offered several advantages: First, it doubled the amount of data that could be collected in the same time interval since both the reference and sample sectors of the cells could be loaded with samples; for example, we collected 6 data sets in 3 two-sector cells in a single experiment. Second, this data collection method offered an increase of $\sqrt{2}$ in the signal-to-noise ratio of each collected data set because the reference intensity data were not subtracted from the sample data. We processed and analyzed the data using SedAnal (Stafford and Sherwood, 2004), which can directly read the intensity scans and can correct for any time-independent offsets in the preprocessing step. Within SedAnal, the time-derivative method was used to determine the apparent sedimentation coefficient distribution, and the weight average sedimentation coefficient (s_w) for each data set were calculated by integration over the range of this distribution. All buffers contained 300 mM KCl and 20 mM Tris–HCl, pH 8.3. Note that there is no need or advantage to density match the detergent in sedimentation velocity experiments. The protein shape, bound water, and bound detergent will all contribute to the sedimentation coefficient. Moreover, because the sedimentation coefficient contains information about mass that is coupled to the frictional coefficient of the sedimenting particle, there is no way to deconvolute these factors.

E. Viscosity Measurements

The viscosities of buffers used in AUC experiments were determined on a Brookfield HA Model DV-III viscometer. The temperature was held at 25 °C with a water cooler (VWR). Two buffer samples were made for each detergent concentration, one with and the other without 2 mM EDTA. Three separate measurements were taken at 75, 150, and 250 rpm for each buffer. The presence or absence of 2 mM EDTA had no effect on the viscosity (data not shown). Therefore, the viscosity of the buffer at each detergent concentration was determined as the average of six different measurements, for example, readings of buffer with and without 2 mM EDTA at three different speeds.

F. Density Measurements

Buffers used in the SV experiments were filtered (0.22 μm pore) and degassed before their density was measured at 25 °C using an Anton Paar DMA 5000 density meter.

IV. Results

A. Analysis of OMPLA Dimerization Energetics

OMPLA is an outer membrane phospholipase enzyme found in gram-negative bacteria; it becomes activated when these organisms experience stress, and activity requires dimerization of the enzyme as well as calcium binding. The crystal structures of OMPLA show that complete active sites and calcium binding sites are formed at the interface of the two monomers, explaining why dimerization is a prerequisite for enzyme activity. To understand the molecular basis of OMPLA activation, it is necessary to know how energetically favorable dimerization is as well as how the free energy of dimerization is affected by substrate binding and calcium binding. We therefore used sedimentation equilibrium to address this question (Stanley *et al.*, 2006, 2007), and we will use the example of OMPLA dimerization to illustrate in this chapter how a researcher would determine whether or not a system is reversibly associating in solution or alternatively is a mixture of irreversible aggregates. This distinction is important because the stoichiometry of the oligomeric species can be determined in both situations; however, the extraction of the equilibrium constant and an extrapolation of the population distribution over a wide concentration range can be carried out only when the system is reversibly associating on the timescale of the sedimentation equilibrium experiment.

We investigated the interaction energetics of OMPLA under four conditions: (1) in the absence of any cofactors; (2) in the presence of 20 mM $CaCl_2$; (3) in the presence of covalently bound substrate analog; and (4) in the presence of both covalently bound substrate analog and 20 mM $CaCl_2$ (Stanley *et al.*, 2006). The sedimentation equilibrium conditions—speeds, temperature, and detergent concentration—were identical for each of these samples; the only differences were the presence or absence of noncleavable substrate analog and calcium. Derivatization of OMPLA with the fatty acyl chain substrate analogue is described in detail in Stanley *et al.* (2006, 2007) and will not be further discussed here. For the purposes of illustrating how the analysis is carried out in this chapter, we will consider the sedimentation equilibrium profiles of OMPLA under conditions (1) and (4) from above.

1. OMPLA is Monomeric in the Absence of Cofactors

We first evaluated OMPLA in 5 mM C14SB in buffered solution with 20 mM EDTA. The extracted data sets were globally analyzed using the model for a single ideal species in which the molecular weight was a global fitting parameter that was allowed to float in the combined fit of all nine data sets. Under these buffer conditions, this mathematical model provided a good description of the OMPLA radial distributions as evidenced by a square root of the variance value on the order of the noise ($\sim$0.005 for our instrument) and residuals that were randomly distributed and centered on zero. A single data set from this global fit is shown in Fig. 1 where the residual distribution can be observed. The resultant molecular weight from this fit was within 5% of that predicted for monomeric OMPLA based on the amino acid sequence, and we therefore concluded that OMPLA was 100% monomeric at these concentrations and under these buffer conditions.

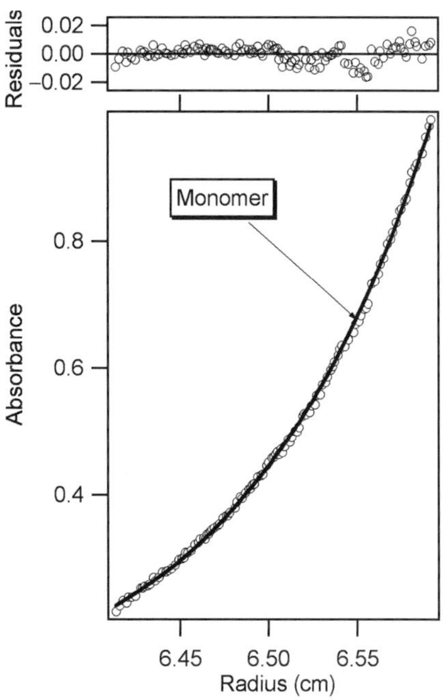

Fig. 1 Single-species fit of outer membrane phospholipase A (OMPLA) in the absence of substrate and $CaCl_2$. A representative sedimentation equilibrium scan collected at 20,000 rpm is shown. The open circles in the lower panel show the data, and the line represents the fit to a single-species equation. The open circles in the upper panel represent the residuals of the fit, which are the differences between the fit and the data at each point.

2. OMPLA Participates in a Reversible Monomer–Dimer Equilibrium Reaction When Modified with Decylsulfonylfluoride but Not When Modified with Perfluorinated Octylsulfonylfluoride

The second condition that we consider in this chapter is OMPLA modified with an acyl chain analogue and in the presence of $CaCl_2$. To explore the dependence of OMPLA dimerization on its substrate, we modified OMPLA with substrates composed of different acyl chain lengths and different chemical compositions (Stanley *et al.*, 2007). In this chapter, we will consider the data collected in the presence of two different substrates: (1) decylsulfonylfluoride (DSF), a 10-carbon acyl chain; and (2) perfluorinated octylsulfonylfluoride (pOSF), an 8-carbon acyl chain in which all hydrogen atoms have been replaced by fluorine atoms.

The setup for sedimentation equilibrium experiments of these samples was identical to that described for OMPLA in the absence of any cofactors. The data at the same three rotor speeds were collected on samples at initial protein concentrations corresponding to A_{280} values of 0.9, 0.6, and 0.3. The extracted data files were globally analyzed as before; however, the initial single ideal species fit returned values greater than that expected for a population composed entirely of monomer (data not shown). We therefore fit the data to models describing multiple species in solution (e.g., monomer–dimer, monomer–trimer, monomer–tetramer). For these fits, we fix the value of sigma to that calculated for the monomer, and we allow the equilibrium constant for a particular monomer to vary in the global fit. The monomer–dimer fits are shown in Figs. 2 and 3 for the protein modified with DSF and pOSF, respectively. While the radial profiles in the bottom panels of these two figures are not obviously different, the ability of a global monomer–dimer fit to describe the data is markedly better for the DSF sample in Fig. 2 than it is for the pOSF sample in Fig. 3. The key comparative parameter is the distribution of the residuals. Since the residuals in Fig. 3 are nonrandom, we must conclude that the global monomer–dimer fit does not describe the data; this means that the equilibrium constant returned in that "fit" is not a valid description of the reaction

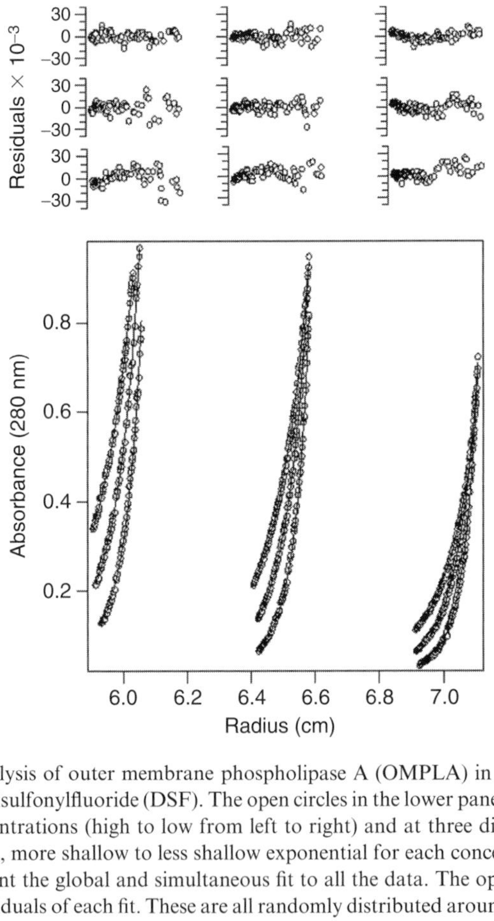

Fig. 2 Global fit analysis of outer membrane phospholipase A (OMPLA) in the presence of $CaCl_2$ and modified with decylsulfonylfluoride (DSF). The open circles in the lower panel are the data collected at three different concentrations (high to low from left to right) and at three different speeds (16,300, 20,000, and 24,500 rpm, more shallow to less shallow exponential for each concentration). The lines in the lower panel represent the global and simultaneous fit to all the data. The open circles in the upper panels represent the residuals of each fit. These are all randomly distributed around zero, suggesting that the monomer–dimer equation with a single equilibrium constant is a good description of the data.

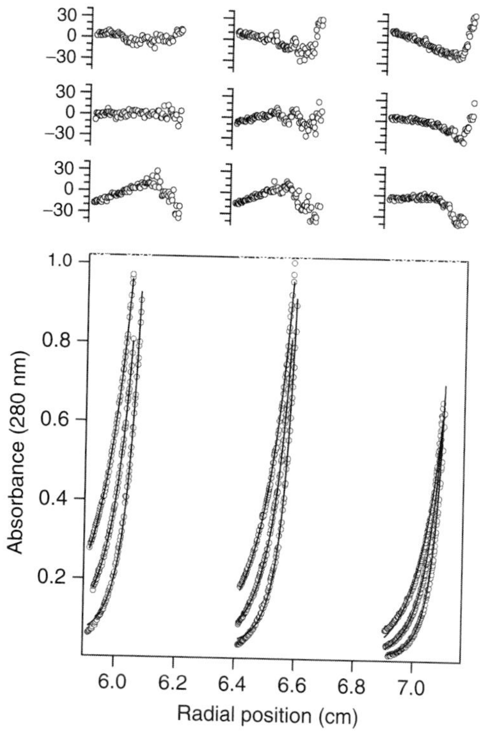

Fig. 3 Global fit analysis of outer membrane phospholipase A (OMPLA) in the presence of $CaCl_2$ and modified with perfluorinated octylsulfonylfluoride (pOSF). The open circles in the lower panel are the data collected at three different concentrations (high to low from left to right) and at three different speeds (16,300, 20,000, and 24,500 rpm, more shallow to less shallow exponential for each concentration). The lines in the lower panel represent the global and simultaneous fit to all the data. The open circles in the upper panels represent the residuals of each fit. In contrast to Fig. 2, the residuals in this figure are not randomly distributed around zero, suggesting that the monomer–dimer equation with a single equilibrium constant is not a good description of the data, which means that the monomer and dimer species are not reversibly associating on the timescale of the experiment and that no conclusions can be drawn about the thermodynamics of self-association.

and that pOSF stimulates aggregation—not reversible self-association—of OMPLA. Another possible conclusion is that the monomer–dimer model is not a correct description of the populated species when modified by pOSF. On the contrary, the residuals in Fig. 2 are randomly distributed in all data sets, suggesting that a single equilibrium constant can describe the reaction independent of the rotor speed and initial protein concentration; this means that the monomer and dimer are reversibly associating on the timescale of the experiment and that the fitted equilibrium constant is a measure of the thermodynamic activity of the solution. Notably, it is the power of the global fit that allows us to draw this conclusion. When analyzed individually, the monomer–dimer model (reversible or irreversible) is a good fit *of any single data set*, and if we had only collected sedimentation equilibrium data at a single concentration and speed, we would not have been able to globally fit the data to test whether a protein sample contained irreversible aggregates or was participating in a reversible self-association reaction.

B. Analysis of OmpF Trimer Stability in C14SB

We also carried out experiments to determine the thermodynamic stability of the OmpF trimer in C14SB detergent micelles. The experimental setup was similar to that used in the OMPLA experiments; however, the rotor speeds were slower because of the higher molecular weight as described in Section III. Since OmpF was also much more stable than OMPLA, we made two modifications to our experiments in order to try and populate monomer and/or dimers of OmpF with the goal of thermodynamically describing the assembly of the OmpF trimer: (1) We monitored the protein concentration at an absorbance wavelength of 230 nm instead of 280 nm. In general, proteins have a larger extinction coefficient at 230 nm, which allows one to experimentally access lower protein concentrations. As discussed in Chapter 6 by Cole *et al.*, this volume, it is a simple matter to change the detection wavelength on the Beckman Optima ultracentrifuges. (2) A second difference in the OmpF experiments is that we carried out sedimentation equilibrium experiments in several different detergent micelle concentrations. For each detergent concentration, we set up three initial protein concentrations and collected several speeds. Since integral membrane proteins are partitioned into the micellar phase of an aqueous solution, increasing the detergent concentration leads to dilution of the membrane protein within that phase, and we reasoned that increasing the detergent concentration would populate OmpF monomers and/or dimers if the concentration went below the dissociation constant. We previously showed using the GpA helix–helix dimer that a reversible membrane protein interaction responds to the aqueous detergent concentration in a predictable way if the solution is behaving ideally (Fleming, 2002; Fleming *et al.*, 2004), and changing the detergent concentration would further allow us to test this for OmpF.

We globally fit OmpF sedimentation equilibrium data collected for each detergent concentration (5, 12, or 30 mM) at four different speeds (9000, 11,000, 13,500, and 16,300 rpm) and three different initial protein concentrations. A representative data set from the global fit is shown in Fig. 4A and illustrates the good single ideal species fit. This fit returned a molecular weight of 105,000 ± 1,000 Da, which is 6 ± 1% lower than the calculated molecular weight for a trimer (111,000 Da) and just below the limit for the expected molecular weight accuracy of sedimentation equilibrium (±5%). There are two interpretations of these initial sedimentation equilibrium fits. First, OmpF is completely trimeric under these conditions and that the calculated molecular weight is slightly inaccurate. If so, we speculate that this inaccuracy arises from uncertainties in our calculation of the partial specific volume of the OmpF trimer, which we assume is an additive function of the partial

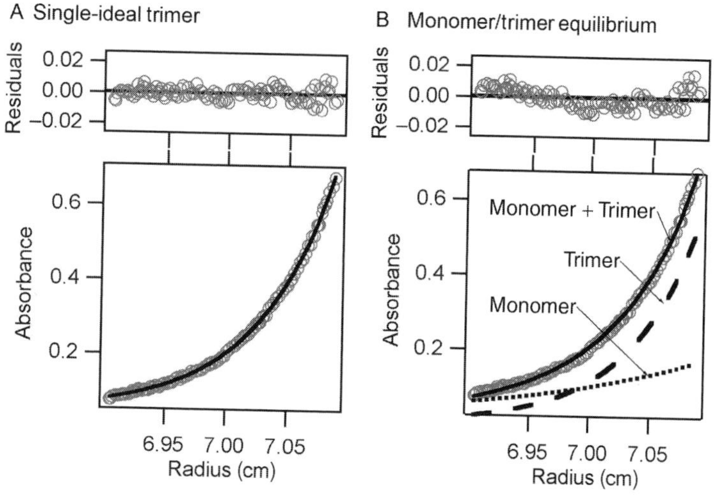

Fig. 4 Outer membrane protein F (OmpF) is a single-species trimer. A representative sedimentation equilibrium scan collected at 13,500 rpm of OmpF (11.1 μM, initial concentration) in 12 mM C14SB, 100 mM KCl, 20 mM Tris–HCl, pH 8.3. Both (A) and (B) show the same data fit with different models. The bottom panels show the observed absorbance (open circles) as a function of radius. The solid line shows the respective fits and the residuals of each fit are displayed in the top panels. (A) Data fit with a single-ideal-species model. The residuals (top panel) fall randomly and tightly about the fit, suggesting that the single-ideal-species model describes the data well. (B) Data fit with a reversibly associating monomer–trimer model. The relative contributions to the fit (solid line) by the monomer and trimer species present are shown by the dotted line (monomer) and dashed line (trimer). The residuals (top panel) of this fit show a curved trend and are greater in magnitude than the residuals of the single-ideal-species model and demonstrate that the monomer–trimer model does not describe the data better than the single-ideal-species model.

specific volumes for each of the OmpF monomers; alternatively, there could be a slight change in the density of the detergent once it is bound to OmpF. It is well known (for a $\bar{v} = 0.75\,\text{ml}\,\text{g}^{-1}$) that each 1% error in the partial specific volume propagates to a 3% error in the buoyant molecular weight; therefore, even a small deviation of this value from the calculated one can lead to an uncertainty in data interpretation. In principle, it is possible to experimentally determine the partial specific volume of a soluble protein from density measurements, but this would not work as well for a membrane protein since the density measurement would contain contributions from the bound detergent as well as the protein. A second interpretation for the slight decrease in the experimentally observed molecular weight of the OmpF trimer may be that the trimer exists in an equilibrium species of lower molecular weight. One of the advantages of sedimentation equilibrium is that it can be sensitive to components present at levels of only 5–10%. While we anticipated that the OmpF trimer would be quite stable, we were also interested to know whether we could experimentally access a reversible equilibrium between OmpF trimers and monomers or dimers.

To test this hypothesis we fit the data with models that account for multiple species. The possibility of a population of only monomers and/or dimers was eliminated because the results of a single ideal species fit demonstrated that the effective molecular weight was significantly higher than that expected for an OmpF dimer. We therefore fit the data to three other models representing association between (1) monomers and trimers, (2) monomers and tetramers, and (3) monomers, dimers, and trimers. Global fits produced with multiple-species models did not significantly decrease the square root of variance over the single-ideal-species model (data not shown). The relative quality of the fits with different models is further evident when fitting individual data sets (Fig. 4). The residuals of multiple-species models fits (Fig. 4B) either did not change from or were less random than residuals of a single-species fit (Fig. 4A). Comparing the residuals of the individual data sets and the square root of variance of the global fits reveals that the multiple-species models do not fit the data better than the single-ideal-species model. We therefore concluded that the sedimentation equilibrium data are best described by a single-species trimer.

C. Sedimentation Velocity Experiments on OmpF

To obtain independent experimental data for the single-species trimer model of OmpF, we carried out sedimentation velocity experiments. Because of the manner in which the data are analyzed, sedimentation velocity methods facilitate experiments at lower protein concentrations, which should further populate OmpF monomers and/or dimers if they are in fact in a reversible equilibrium with the OmpF trimer observed in the sedimentation equilibrium experiments. Since OmpF dimers have been observed at lower protein concentrations (Watanabe and Inoko, 2005), we postulated that sedimentation velocity experiments might allow observation of OmpF dissociation.

Furthermore, Stafford has shown that s_w is a measure of the thermodynamic activity of a macromolecule in solution (Stafford, 2000). For self-associating soluble proteins, s_w will decrease with decreasing protein concentration if the protein dissociates. A plot of s_w versus the protein concentration can be fit with an equation describing a binding isotherm to determine the association constant for the equilibrium reaction (Correia, 2000). However, it is essential to recognize that the concentration scales for soluble and membrane proteins should be fundamentally different for this type of analysis. In contrast to soluble proteins, the mass action behavior of membrane proteins will not depend on the aqueous concentration but rather will depend on the protein:detergent mole ratio (referred to as the mole fraction protein) as the concentration unit. This is extensively discussed by Fleming and has its origins in the fact that folded membrane proteins do not partition into an aqueous environment (Fleming, 2002).

Analysis of any one of the sets of sedimentation velocity scans revealed a single peak in the apparent s_w distribution. This is illustrated by the data in Fig. 5. Because of the mass and shape contribution of bound detergent molecules,

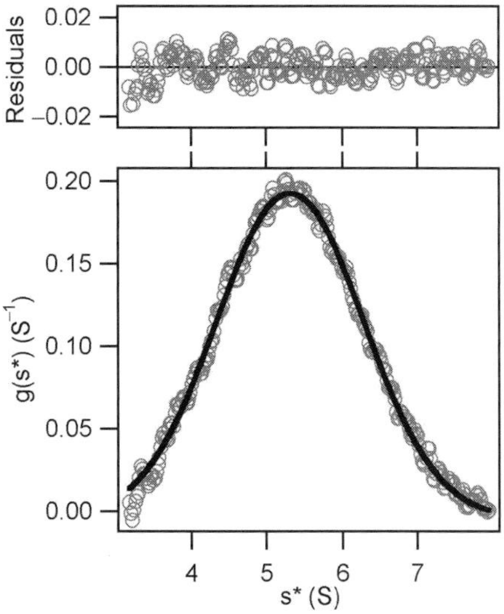

Fig. 5 Sedimentation velocity profile of outer membrane protein F (OmpF). Analysis of a representative sedimentation velocity scan of 11.1 μM OmpF in 60 mM C14SB. The bottom panel displays the apparent sedimentation coefficient distribution, g(s*) versus s* (open circles), which was calculated from the time derivative of the raw data. The error bars from this calculation were no larger than the markers used here and were therefore left out for greater graphical clarity. The single peak produced by this analysis reflects either a single species or the weight average of several species in equilibrium with each other. The mean of a Gaussian fit (black line) defines the weight average sedimentation coefficient (s_w). The top panel displays the residuals of the fit, which are small and random, demonstrating a good fit.

the theoretical sedimentation coefficient for a membrane protein cannot be calculated from either its composition or its crystal structure; this means that the single experimental peak can be interpreted as a single species or as the reaction boundary of multiple species in equilibrium. To distinguish between these single- and multiple-species models, we carried out sedimentation velocity experiments as a function of the mole fraction protein concentration. We varied both the aqueous protein and detergent concentrations to obtain several different protein mole fractions. Like the representative data in Fig. 5, each of these could be fit to a single sedimentation coefficient. The s_w values observed in our experiments ranged from 5.3 to 5.8 S (Fig. 6A). Notably, these values are consistent with previously published sedimentation coefficients of the OmpF trimer in other lipidic environments, such as in lipopolysaccharide (s=5.0 S) (Holzenburg *et al.*, 1989), in octylglucopyranoside micelles (s=6.2 S, 6.4 S, and 6.6 S) (Lustig *et al.*, 2000; Markovic-Housley and Garavito, 1986), and in sodium dodecylsulfate micelles (s=6.0 S) (Markovic-Housley and Garavito, 1986). However, a plot of the observed s_w values as a function of mole fraction protein does in fact reveal a trend that has a shape similar to a binding isotherm: s_w decreases with lower mole fraction protein (Fig. 6A). The trend may illustrate the effect of increasing buffer

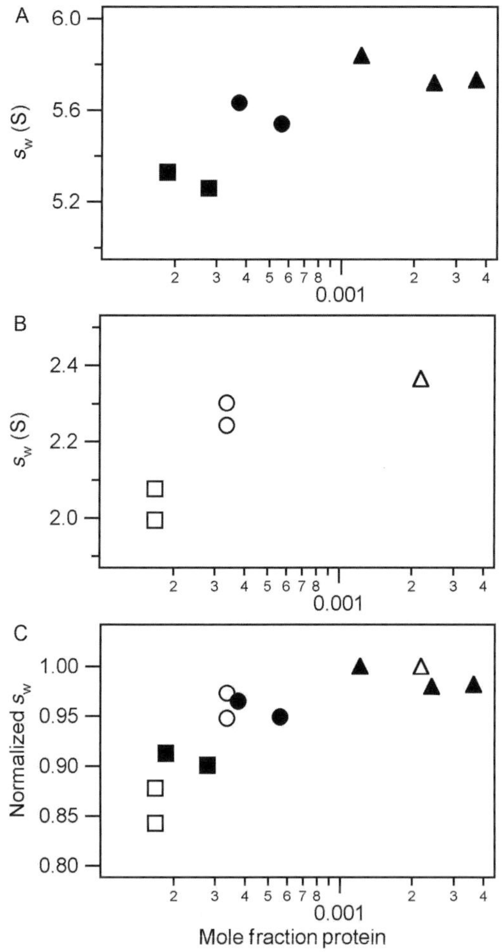

Fig. 6 Outer membrane protein F (OmpF) behaves as a single species despite the hyperbolic shape of the sedimentation coefficient concentration dependence. Results of sedimentation velocity experiments with (A) OmpF and (B) outer membrane phospholipase A (OMPLA) Q94A at various C14SB concentrations: 5 mM (triangles), 30 mM (circles), and 60 mM (squares). The apparent s_w is plotted as a function of mole fraction protein. (C) The values of s_w were normalized with respect to the largest s_w for each protein. The normalized s_w for OmpF (closed markers) and OMPLA Q94A (open markers) are plotted as a function of mole fraction protein.

density or viscosity at high detergent concentrations on the sedimentation coefficient of a single-species. Alternatively, the decrease in s_w might indicate the presence of multiple species undergoing more dissociation at greater detergent concentrations.

To determine if the variation of the s_w resulted from dissociation, we used the OMPLA sequence variant Q94A as a control for sedimentation velocity experiments. Under the same experimental conditions used for OmpF, it is known that wild-type OMPLA (Stanley *et al.*, 2006) and the sequence variant Q94A are monomeric. In the sedimentation velocity experiments with the OMPLA variant Q94A, we also observed a distribution (s_w=2.0 – 2.4 S). In Fig. 6B, s_w is plotted as a function of mole fraction protein and demonstrates that the s_w of OMPLA Q94A decreased as a function of mole fraction protein, similar to the trend observed for OmpF.

The differences in the absolute values of s_w obtained for OmpF (s_w=5.3 – 5.8 S) and OMPLA Q94A (s_w=2.0 – 2.4 S) are not surprising because of the difference in molecular weight of the OmpF trimer (111,000 Da) and the OMPLA Q94A monomer (31,000 Da). To compare the trends in the distribution of s_w, the values were normalized to the highest observed for each protein. The normalized s_w values were then plotted as a function of mole fraction protein (Fig. 6C). The trend observed for the OmpF trimer in Fig. 6C overlays that of OMPLA Q94A, a completely monomeric protein. This result indicates that the detergent dependence of s_w observed in OmpF cannot be explained by the dissociation of the OmpF trimer into monomers in dilute conditions; rather these data support the conclusions drawn from the sedimentation equilibrium data: OmpF is a single-species trimer.

The variation in s_w for a single species was unexpected, so we explored the probability that the variation resulted from changes in bulk solvent properties. Sedimentation velocity experiments quantify the hydrodynamic properties of macromolecules, so changes in the buffer viscosity or density at high detergent concentrations could contribute to the observed behavior of both OmpF and OMPLA. To determine whether bulk solvent properties affect the s_w distribution, we measured the viscosity and density of all buffers in the absence of protein. We found that the viscosity and density of the buffers do not change over the detergent concentrations used nor do they vary significantly from the calculated values of the buffer without detergent (Fig. 7). Therefore, changes in bulk solvent properties do not likely account for variation in the measured s_w.

Apart from bulk solvent properties, s_w is also sensitive to the size and shape of the macromolecule. The number of detergent molecules bound to the protein may affect both the size and the shape of a detergent–protein complex. For example, Watanabe and Inoko observed through small angle light scattering that the OmpF dimer–micelle complex and the trimer–micelle complex are of the same size. They concluded that the variation in the amount of detergent bound accounted for the difference in size between the dimer and the trimer (Watanabe and Inoko, 2005). Similarly, the number of detergents bound to OmpF may have varied between our experiments. If the number of detergent molecules varied with the detergent concentration (the only parameter that changed between sedimentation velocity experiments), then s_w would also vary with detergent concentration and account for the observed trend. However, the number of detergent molecules that solvate a given membrane protein is generally not known; neither can it be concluded whether that number varies at different detergent concentrations. Although the

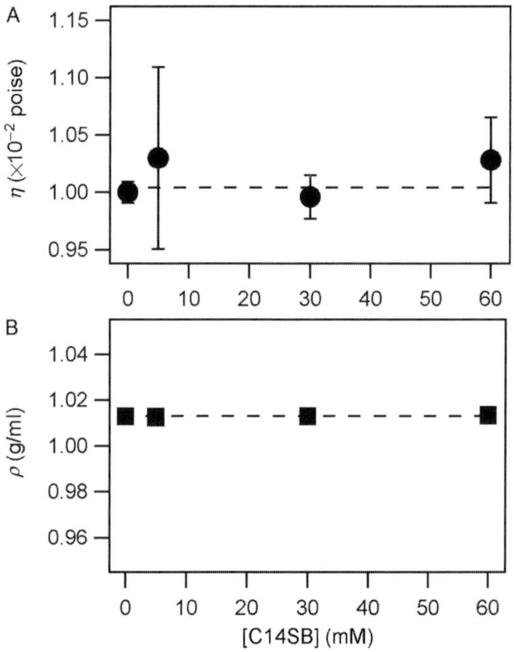

Fig. 7 Viscosity and density of buffers do not change with detergent concentration. (A) The buffer viscosity is plotted as a function of C14SB detergent concentration. The error bars represent the standard deviation of six independent measurements. (B) The buffer density of two independent measurements at each detergent concentration is plotted as a function of C14SB concentration. The viscosity and density of the buffers in the absence of detergent were determined in SEDNTERP (Laue *et al.*, 1992) and are shown in both panels as a dashed horizontal line.

structure of the protein–detergent micelle complex is not yet understood, we have shown that for β-barrel membrane proteins, the change in s_w with mole fraction protein is not necessarily a reflection of the law of mass action or an indication of the dissociation of oligomers. Sedimentation-velocity studies of detergent-solvated membrane proteins should be interpreted carefully.

V. Discussion

Knowledge of the thermodynamic stabilities of several membrane proteins makes it possible to rank a set of proteins by their interaction energies. This is illustrated by Fig. 8 in which relative oligomeric populations of several membrane proteins, OmpF, GpA (Fleming *et al.*, 2004), and OMPLA, with and without effector molecules (Stanley *et al.*, 2006) are plotted as a function of mole fraction protein. "Mole fraction protein" was chosen as the quantity for the abscissa for

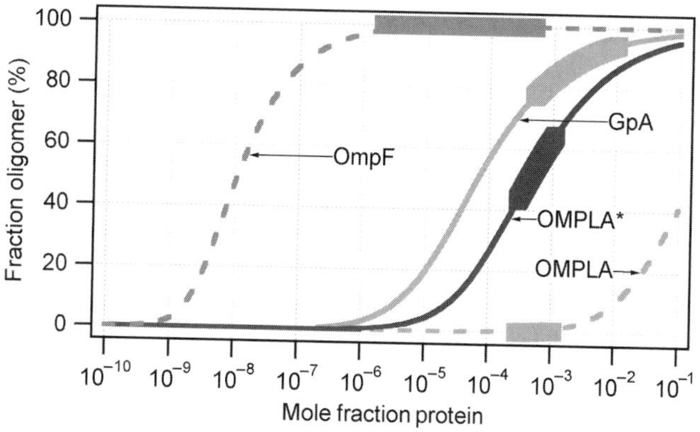

Fig. 8 A comparison of the oligomer populations for several membrane proteins. The fraction oligomers of OmpF, GpA, outer membrane phospholipase A (OMPLA) with effector molecules (*) (purple), and OMPLA without effector molecules are plotted as a function of mole fraction protein. The thick portions of the curve represent the observed oligomeric populations from which the standard-state free energies were calculated (Fleming *et al.*, 2004; Stanley *et al.*, 2006). No monomeric species for OmpF nor oligomeric species for OMPLA without effector molecules (Stanley *et al.*, 2006) were observed and therefore the dashed curves represent limits to the free energy of association. Under all observed conditions, both GpA and OMPLA self-association is much weaker than OmpF self-association.

this comparison as this concentration scale represents the distributions of the membrane proteins within the micellar phase (Fleming, 2002). Mole fraction protein is defined as the ratio of the aqueous molar concentrations of protein to micellar detergent. The mole fraction scale normalizes the observed equilibrium constant to the detergent micelle concentration used in any particular experiment and allows a direct comparison of membrane protein populations even if the data were collected in different detergent concentrations. Underlying the mole fraction representation of the data is the assumption that the membrane protein is behaving ideally within the micellar phase, and it is important to recognize that this condition has been experimentally demonstrated only for the GpA helix dimer (Fleming *et al.*, 2004); it is an assumed behavior for the other proteins on this graph. The thin lines in Fig. 8 show the predicted change in fraction oligomer as a function of mole fraction protein and were calculated from the dissociation constants of the observed oligomeric populations (thick portion of the lines) as measured by sedimentation equilibrium experiments (Fleming *et al.*, 2004; Kobus and Fleming, 2005; Stanley *et al.*, 2006). OmpF was detected as a single-species trimer over the experimentally accessible mole fraction protein, and we therefore calculated an estimate of upper limit for the dissociation constant. Assuming a cooperative trimer-to-monomer dissociation scheme, we forced the fraction oligomer to decrease from the lowest experimentally observed mole fraction protein. The theoretical curve for OmpF represents the upper limit to the mole fraction concentration that would dissociate an OmpF trimer; in reality, the OmpF trimer may be more stable than the distribution shown in Fig. 8. Similarly, there was no dimeric population observed for OMPLA without effector molecules (Stanley *et al.*, 2006); thus, the reported value for the free energy of association of OMPLA is also a limit and reflects the most favorable self-association that might exist. To date, these curves define the limits of the free energy of association for β-barrel membrane proteins.

An estimate of the increased stability of the OmpF trimer over the other membrane proteins can also be expressed numerically by calculating the free energy from the association constants that the curves in Fig. 8 represent. For OmpF, the sedimentation equilibrium data suggest that the highest limit of the standard-state free energy of association is -26 kcal mol^{-2}. The standard-state free energy of association observed for GpA is -5.7 kcal mol^{-1} (Fleming *et al.*, 2004), for OMPLA with effector molecules is -4.2 kcal mol^{-1}, and for OMPLA without effector molecules is -1.05 kcal mol^{-1} (Stanley *et al.*, 2006). These thermodynamic values illustrate the wide range of stabilities that can be encoded in membrane proteins, and it will be interesting in the future to compare these to the stabilities of additional membrane proteins.

VI. Summary

AUC is a powerful method that can be used to derive quantitative descriptions of membrane protein complexes dispersed in detergent micelle solutions. Using the density-matching method, the molecular weights of membrane proteins can be unequivocally determined, and global analysis of sedimentation equilibrium data can reveal whether membrane proteins are reversibly associating in detergent micelle solutions. When reversible association is established, the equilibrium constant can be extracted, and this quantity can be used to predict the species population over a wide range of protein concentrations. Knowing the thermodynamic stability of membrane proteins in the same detergent micelle environment allows a comparison of the population distributions and provides a basis for understanding how the sequences and the structures of membrane proteins encode their stabilities and functions.

Acknowledgments

This work was supported by a grant from the NSF (MCB0423807). Ann Marie Stanley is a Howard Hughes predoctoral fellow. We gratefully acknowledge Dr. John J. Correia at the University of Mississippi Medical Center AUC facility for the density measurements and for a critical reading of the manuscript, Dr. Richard Cone for access to a viscometer, and Elizabeth O'Hanlon for technical assistance. We also thank members of the Fleming lab for helpful discussions.

References

Ansevin, A., Roark, D., and Yphantis, D. (1970). Improved ultracentrifuge cells for high speed sedimentation equilibrium studies with interference optics. *Anal. Biochem.* **34,** 237–261.

Burrows, S., Doyle, M., Murphy, K., Franklin, S., White, J., Brooks, I., McNulty, D., Miller, S., Knutson, J., Porter, D., Young, P., and Hensley, P. (1994). Determination of the monomer-dimer equilibrium of interleukin-8 reveals it is a monomer at physiological concentrations. *Biochemistry* **33** 12741–12745.

Butler, P. J., and Kuhlbrandt, W. (1988). Determination of the aggregate size in detergent solution of the light-harvesting chlorophyll a/b-protein complex from chloroplast membranes. *Proc. Natl. Acad. Sci. USA* **85,** 3797–3801.

Butler, P. J., Ubarretxena-Belandia, I., Warne, T., and Tate, C. G. (2004). The *Escherichia coli* multidrug transporter EmrE is a dimer in the detergent-solubilised state. *J. Mol. Biol.* **340,** 797–808.

Casassa, E. F., and Eisenberg, H. (1964). Thermodynamic analysis of multicomponent systems. *Adv. Protein Chem.* **19,** 287–395.

Correia, J. (2000). Analysis of weight average sedimentation velocity data. *Methods Enzymol.* **321,** 81–100.

Dekker, N., Merck, K., Tommassen, J., and Verheij, H. M. (1995). *In vitro* folding of *Escherichia coli* outer-membrane phospholipase A. *Eur. J. Biochem.* **232,** 214–219.

Dekker, N., Tommassen, J., Lustig, A., Rosenbusch, J. P., and Verheij, H. M. (1997). Dimerization regulates the enzymatic activity of *Escherichia coli* outer membrane phospholipase A. *J. Biol. Chem.* **272,** 3179–3184.

Doura, A. K., and Fleming, K. G. (2004). Complex interactions at the helix-helix interface stabilize the glycophorin A transmembrane dimer. *J. Mol. Biol.* **343,** 1487–1497.

Doura, A. K., Kobus, F. J., Dubrovsky, L., Hibbard, E., and Fleming, K. G. (2004). Sequence context modulates the stability of a GxxxG mediated transmembrane helix-helix dimer. *J. Mol. Biol.* **341,** 991–998.

Doyle, D. A., Cabral, J. M., Pfuetzner, R. A., Kuo, A., Gulbis, J. M., Cohen, S. L., Chait, B. T., and MacKinnon, R. (1998). The structure of the potassium channel: Molecular basis of K^+ conduction and selectivity. *Science* **280,** 69–77.

Durshlag, H. (1986). Specific volumes of biological macromolecules and some other molecules of biological interest. *In* "Thermodynamic Data for Biochemistry and Biotechnology" (H. J. Hinz, ed.), pp. 45–128. Springer-Verlag, Berlin.

Ebie, A. Z., and Fleming, K. G. (2007). Dimerization of the erythropoietin receptor transmembrane domain in micelles. *J. Mol. Biol.* **366,** 517–524.

Fleming, K. G. (1998). Measuring transmembrane α-helix energies using analytical ultracentrifugation. *In* "Chemtracts: Biological Applications of the Analytical Ultracentrifuge" (J. C. Hanson, ed.), pp. 985–990. Springer-Verlag, New York.

Fleming, K. G. (2000). Probing the stability of helical membrane proteins. *Methods Enzymol.* **323,** 63–77.

Fleming, K. G. (2002). Standardizing the free energy change of transmembrane helix-helix interactions. *J. Mol. Biol.* **323,** 563–571.

Fleming, K. G., and Engelman, D. M. (2001). Specificity in transmembrane helix-helix interactions defines a hierarchy of stability for sequence variants. *Proc. Natl. Acad. Sci. USA* **98,** 14340–14344.

Fleming, K. G., Ackerman, A. L., and Engelman, D. M. (1997). The effect of point mutations on the free energy of transmembrane α-helix dimerization. *J. Mol. Biol.* **272,** 266–275.

Fleming, K. G., Ren, C. C., Doura, A. K., Kobus, F. J., Eisley, M. E., and Stanley, A. M. (2004). Thermodynamics of glycophorin A transmembrane helix-helix association in C14 betaine micelles. *Biophys. Chem.* **108,** 43–49.

Gratkowski, H., Lear, J. D., and DeGrado, W. F. (2001). Polar side chains drive the association of model transmembrane peptides. *Proc. Natl. Acad. Sci. USA* **98,** 880–885.

Holzenburg, A., Engel, A., Kessler, R., Manz, H. J., Lustig, A., and Aebi, U. (1989). Rapid isolation of OmpF porin-LPS complexes suitable for structure-function studies. *Biochemistry* **28,** 4187–4193.

Howard, K. P., Lear, J. D., and DeGrado, W. F. (2002). Sequence determinants of the energetics of folding of a transmembrane four-helix-bundle protein. *Proc. Natl. Acad. Sci. USA* **99,** 8568–8572.

Johnson, M. L., Correia, J. J., Yphantis, D. A., and Halvorson, H. R. (1981). Analysis of data from the analytical ultracentrifuge by nonlinear least-squares techniques. *Biophys. J.* **36,** 575–588.

Kobus, F. J., and Fleming, K. G. (2005). The GxxxG-containing transmembrane domain of the CCK4 oncogene does not encode preferential self-interactions. *Biochemistry* **44,** 1464–1470.

Kochendoerfer, G. G., Salom, D., Lear, J. D., Wilk-Orescan, R., Kent, S. B., and DeGrado, W. F. (1999). Total chemical synthesis of the integral membrane protein influenza A virus M2: Role of its C-terminal domain in tetramer assembly. *Biochemistry* **38,** 11905–11913.

Konig, N., Zampighi, G. A., and Butler, P. J. (1997). Characterisation of the major intrinsic protein (MIP) from bovine lens fibre membranes by electron microscopy and hydrodynamics. *J. Mol. Biol.* **265,** 590–602.

Laue, T. M., Shah, B., Ridgeway, T. M., and Pelletier, S. L. (1992). Computer-aided interpretation of analytical sedimentation data for proteins. *In* "Analytical Ultracentrifugation in Biochemistry and Polymer Science" (S. E. Harding, A. J. Rowe, and J. C. Horton, eds.), pp. 90–125. Royal Society of Chemistry, Cambridge, UK.

le Maire, M., Champeil, P., and Møller, J. V. (2000). Interaction of membrane proteins and lipids with solubilizing detergents. *Biochim. Biophys. Acta* **1508,** 86–111.

le Maire, M., Kwee, K., Anderson, J. P., and Møller, J. V. (1983). Mode of interaction of polyoxyethyleneglycol detergents with membrane proteins. *Eur. J. Biochem.* **129,** 525–532.

Li, R., Gorelik, R., Nanda, V., Law, P. B., Lear, J. D., DeGrado, W. F., and Bennett, J. S. (2004). Dimerization of the transmembrane domain of Integrin alphaIIb subunit in cell membranes. *J. Biol. Chem.* **279,** 26666–26673.

Ludwig, B., Grabo, M., Gregor, I., Lustig, A., Regenass, M., and Rosenbusch, J. P. (1982). Solubilized cytochrome c oxidase from *Paracoccus denitrificans* is a monomer. *J. Biol. Chem.* **257,** 5576–5578.

Lustig, A., Engel, A., Tsiotis, G., Landau, E. M., and Baschong, W. (2000). Molecular weight determination of membrane proteins by sedimentation equilibrium at the sucrose or nycodenz-adjusted density of the hydrated detergent micelle. *Biochim. Biophys. Acta* **1464,** 199–206.

Markovic-Housley, Z., and Garavito, R. M. (1986). Effect of temperature and low pH on structure and stability of matrix porin in micellar detergent solutions. *Biochim. Biophys. Acta* **869,** 158–170.

Pinto, L. H., Dieckmann, G. R., Gandhi, C. S., Papworth, C. G., Braman, J., Shaughnessy, M. A., Lear, J. D., Lamb, R. A., and DeGrado, W. F. (1997). A functionally defined model for the M2 proton channel of influenza A virus suggests a mechanism for its ion selectivity. *Proc. Natl. Acad. Sci. USA* **94,** 11301–11306.

Reynolds, J. A., and McCaslin, D. R. (1985). Determination of protein molecular weight in complexes with detergent without knowledge of binding. *Methods Enzymol.* **117,** 41–53.

Reynolds, J. A., and Tanford, C. (1976). Determination of molecular weight of the protein moiety in protein-detergent complexes without direct knowledge of detergent binding. *Proc. Natl. Acad. Sci. USA* **73,** 4467–4470.

Stafford, W. F. (2000). Analysis of reversibly interacting macromolecular systems by time derivative sedimentation velocity. *Methods Enzymol.* **323,** 302–325.

Stafford, W. F., and Sherwood, P. J. (2004). Analysis of heterologous interacting systems by sedimentation velocity: Curve fitting algorithms for estimation of sedimentation coefficients, equilibrium and kinetic constants. *Biophys. Chem.* **108,** 231–243.

Stanley, A. M., and Fleming, K. G. (2005). The transmembrane domains of the ErbB receptors do not dimerize strongly in micelles. *J. Mol. Biol.* **347,** 759–772.

Stanley, A. M., Chauwang, P., Hendrickson, T. L., and Fleming, K. G. (2006). Energetics of outer membrane phospholipase A (OMPLA) dimerization. *J. Mol. Biol.* **358,** 120–131.

Stanley, A. M., Treubodt, A. M., Chauwang, P., Hendrickson, T. L., and Fleming, K. G. (2007). Lipid chain selectivity by outer membrane phospholipase A. *J. Mol. Biol.* **366,** 461–468.

Suarez, M. D., Revzin, A., Narlock, R., Kempner, E. S., Thompson, D. A., and Ferguson-Miller, S. (1984). The functional and physical form of mammalian cytochrome c oxidase determined by gel

filtration, radiation inactivation, and sedimentation equilibrium analysis. *J. Biol. Chem.* **259,** 13791–13799.
Surrey, T., Schmid, A., and Jahnig, F. (1996). Folding and membrane insertion of the trimeric beta-barrel protein OmpF. *Biochemistry* **35,** 2283–2288.
Watanabe, Y., and Inoko, Y. (2005). Physicochemical characterization of the reassembled dimer of an integral membrane protein OmpF porin. *Protein J.* **24,** 167–174.
Yphantis, D. A. (1964). Equilibrium ultracentrifugation of dilute solutions. *Biochemistry* **3,** 297–317.

CHAPTER 8

Basic Aspects of Absorption and Fluorescence Spectroscopy and Resonance Energy Transfer Methods

Natasha Shanker and Susan L. Bane

Department of Chemistry
State University of New York at Binghamton
Binghamton, New York 13902

0091-679X/08 $35.00
DOI: 10.1016/S0091-679X(07)84008-8

Abstract

Absorption and fluorescence spectroscopy are the heart of life science laboratory techniques. These two spectroscopic tools are sensitive, are easy to use, and can provide a wide range of information. The goal of this chapter is to familiarize the reader with some basic principles and applications of absorption and steady-state fluorescence spectroscopy. Selected applications are illustrated with examples.

I. Introduction

This chapter is written for the novice and is intended to introduce some basic concepts of absorption and steady-state fluorescence spectroscopy. Both types of spectroscopy are widely used in biological research; just a few of the multitude of applications are discussed here. For additional information, one of the many excellent books and monographs should be consulted. A few leading sources are listed under Suggested Reading.

II. Absorption Spectroscopy

Absorption spectroscopy, also referred to as UV-Visible (UV-Vis) spectroscopy, is one of the simplest techniques of optical spectroscopy. The technique is most frequently employed in life science laboratories as a method to determine concentration or to monitor changes in states (such as DNA melting) or chromophore environment (such as receptor binding).

Absorption spectroscopy can be used to evaluate molecules that undergo electronic transitions excited by ultraviolet (UV) and visible light (~190–800 nm). Biologically relevant chromophores in this region include the peptide bond (~210 nm), nucleic acid bases (~250–260 nm), aromatic amino acid side chains (~260–280 nm), heme (~400 and 600 nm), and flavin (~450 nm).

The basic principle behind absorption spectroscopy is that light is absorbed by a molecule when the energy of the light matches the energy difference between two electronic states. In a typical absorption spectrophotometer, a solution of the chromophore of interest is placed between the light source and a detector. The sample is illuminated, and the intensity of the light at each wavelength is measured before (I_0) and after (I) it passes through the solution. The spectrophotometer measures the light absorbed by the solution, which is expressed as percent transmittance (I/I_0) or more frequently as absorbance (the negative log of transmittance). The term "optical density" (O.D.) is an older expression that is synonymous to absorbance. The amount of light absorbed by a sample is proportional to the concentration of the substance based on the Beer–Lambert law:

$$\log\left(\frac{I_0}{I}\right)_\lambda = A_\lambda = \varepsilon_\lambda \times c \times l$$

where I_0 is the intensity of light of wavelength λ incident on the sample, I is the intensity of light of wavelength λ transmitted through the sample, A_λ is the absorbance at wavelength λ, ε_λ is the absorption coefficient at wavelength λ, c is the concentration of the chromophore, and l is the path length.

The absorbance of a molecule at a particular wavelength (A_λ) is therefore a function of the concentration of the molecule (c), the path length of the cell containing the solution (l, units in cm), and the absorption coefficient (ε), which is also known by the older term "extinction coefficient." The absorption coefficient can be loosely described as the efficiency by which the electron moves from the ground state to the excited state. It is a property of the structure of the molecule and can be influenced by the environment of the molecule in solution. Highly efficient transitions will have large absorption coefficients. When the concentration of a substance is measured in moles/liter, the term molar absorptivity is used (units in $M^{-1}cm^{-1}$). Absorption coefficients of proteins are frequently expressed in terms of mass rather than moles [e.g., $(mg/ml)^{-1}cm^{-1}$].

The detection limits of absorption spectroscopy are generally a function of the absorption coefficient, since a fixed path, usually 1.0 cm, is defined by the cuvette used to measure the spectrum in a spectrophotometer (although cuvettes with smaller paths are also available). Thus, molecules with high absorption coefficients will be detectable at lower concentrations.

A. Absorption Spectrophotometers

A UV-Vis spectrophotometer is a device that measures the amount of light absorbed by the chromophore as a function of the wavelength of the electromagnetic radiation. Conventional spectrophotometers consist of the following: a light source (deuterium lamp for the UV region: 190–350 nm and tungsten lamp for the wavelength region above 350 nm); a collimator, for focusing the beam of light; a monochromator, for wavelength selection; a sample compartment; and a detector. Conventional spectrophotometers have been largely replaced by diode array spectrophotometers, which have an advantage of speed over the former. A diode array spectrophotometer uses reverse optics, that is, the polychromatic light passes through the sample first and is then dispersed onto the diode array. The array comprises a series of photodiode detectors juxtaposed on a silicon chip, with each diode designed to measure a finite but narrow band of the spectrum. Thus, the entire spectrum can be collected in a matter of seconds. The short exposure to the incident light can also minimize photodecomposition of samples.

B. Measuring Absorption Spectrum

It is very simple to measure an absorption spectrum with just a few precautions to be considered. Each spectrophotometer has a sensitivity range also known as the photometric range. The photometric range is the linear range of the instrument. Absorbance values must be kept within this range for the data to be valid. It is good to keep in mind that the absorption value provided by the instrument is a logarithmic ratio: an absorbance of 1.0 unit means that 10% of the incident light has been absorbed by the sample, while a sample that has an absorbance of 2.0 units absorbs 99% of the incident light. The photometric precision and accuracy should also be noted.

Spectra in the UV region of the electromagnetic spectrum (190–350 nm) are obtained in quartz cuvettes, which are available in many different sizes. Glass or plastic cuvettes can be used for measurements in the visible region. When more than one cuvette is used in an absorption measurement, the cuvettes should be "matched (i.e., have the same optical properties)." Cuvettes should not be handled by touching the polished walls. The tips of Pasteur pipettes are sharp enough to scratch cuvettes; these pipettes should be briefly fire polished before using them with the cuvettes.

A "blank" is recorded first, which is a spectrum of all of the solution components except the substance(s) of interest. Buffer components may contain impurities or may absorb in the same region as the substance of interest. Therefore, it is wise to take a spectrum of the buffer using a blank of distilled water to make sure that no strongly absorbing components interfere with the spectral region of interest.

A spectrophotometer fails to distinguish between absorption and scattering. The scattering of light by a turbid solution will also be read as absorbance by the instrument. It is possible to obtain true absorption spectra of chromophores within turbid samples using instruments that possess specialized accessories for the task (such as a diffuse reflectance accessory). Manipulation of the data from an unmodified spectrophotometer can also be attempted. Taking the second derivative of the spectrum may eliminate the contribution of scatter to the spectrum, particularly for samples with narrow absorption bands (Castanho *et al.*, 1997). If turbidity is not an intrinsic property of the sample, then the best approach is to clarify the solution prior to the spectral measurement. If the sample is intrinsically turbid, micelles for example, use of a spectrophotometer that places the cuvette very close to the photomultiplier tube is advised to reduce scattering effects.

It is essential that the beam passes entirely through the solution and not through cuvette walls or air above the liquid. Self-masking cuvettes have black walls to prevent the beam from passing through the sides of the cuvette. They are more expensive than plain cuvettes, but the investment is worthwhile if highly accurate measurements are required. The volume of liquid in a cuvette must be enough to cover the entire aperture of the cell holder. A dental or mechanic's mirror is a convenient tool for checking the level of the sample once the cuvette is on the instrument.

When cold samples are measured, the cuvette should be checked for condensation. Condensation on the windows of a cuvette will also result in inaccurate absorption spectral data, and it should be removed by a gentle wipe with a tissue. If the experiment has to be performed at low temperature, then a continuous flow of dry nitrogen or argon gas can prevent condensation and minimize the error in measurement.

C. Common Applications

1. Concentration Determination

This is a direct application of the Beer–Lambert law. If the absorption coefficient for a molecule is known, then the concentration of the molecule in solution can be obtained by taking an absorption spectrum of the solution and solving the equation:

$$c = \varepsilon_{\lambda} \times \frac{l}{A_{\lambda}}$$

If a literature value for the absorption coefficient is used, it is important to know the buffer concentration, concentration of any additives, and the pH of the solution. Differences in solution variables can affect an absorption coefficient. An absorption coefficient is always reported with the wavelength and the solvent/buffer solution. For instance, the two commonly used absorption coefficients for measurement of concentration of anticancer drug paclitaxel for laboratory purposes are $\varepsilon_{228\ \mathrm{nm,\ ethanol}} = 2.79 \times 10^4\ \mathrm{M}^{-1}\ \mathrm{cm}^{-1}$, $\varepsilon_{273\ \mathrm{nm,\ DMSO}} = 1.70 \times 10^3\ \mathrm{M}^{-1}\ \mathrm{cm}^{-1}$ (Li *et al.*, 2000; Wani *et al.*, 1971). Absorption coefficients for many proteins, nucleotides, and other biomolecules are compiled by Fasman (1989).

In the absence of a literature value, an absorption coefficient for a biological molecule can be experimentally determined or estimated by calculation. A detailed procedure for determining absorption coefficients for proteins, experimentally and computationally, can be found in Pace *et al.* (1995). Comparison of calculation methods for absorption coefficients for DNA can be found in Kallansrud and Ward (1996). It should be noted that absorption coefficients for proteins and nucleic acid polymers may be reported using different terminology. Absorption coefficients for proteins are frequently encountered as percent solution absorption coefficients ($\varepsilon_{\mathrm{percent}}$), which has units of $(\mathrm{g}/100\ \mathrm{ml})^{-1}\mathrm{cm}^{-1}$. Absorption coefficients for DNA and RNA are frequently expressed as $(\mathrm{g/L})^{-1}\mathrm{cm}^{-1}$. Absorption coefficient data for nucleic acid polymers are also expressed in terms of O.D. units at 260 nm. (1 O.D. unit is equivalent to an absorbance of 1.0.) For example, an O.D. unit of 1 at 260 nm for dsDNA corresponds to a concentration of 50 μg/ml (Fasman, 1989).

Absorption coefficients for small molecules can be determined experimentally, and it is convenient to know an absorption coefficient for a frequently used small molecule. The procedure is based on measurement of the absorbance and determining the absorption coefficient from the Beer–Lambert law. A solution of the ligand at a known concentration is subjected to serial dilutions until the absorption spectrum of the molecule falls in the linear range of the instrument. The concentration of this solution is calculated based on the dilution factor, and the absorption coefficient for the substance is calculated from the Beer–Lambert relationship: $\varepsilon_\lambda = c \times l/A_\lambda$.

The most critical aspect of this procedure is that the concentration of the original stock solution must be very accurate and the subsequent dilutions must be carefully performed. For this purpose, highly pure, dry solid is carefully weighed and dissolved completely in a known volume of buffer or solvent. It is a good practice to use an analytical balance to determine the mass of solvent added. For pure solvents, the volume of the solvent can be determined precisely by dividing the mass of the solvent by the density of the solvent. Measuring the dilution factors by mass as well as volume is also prudent, since an error in pipetting or in the pipette calibration can be detected and corrected from the mass measurements.

2. Ligand–Receptor Interactions

Absorption spectra can be sensitive to the environment of the chromophore, and this sensitivity may be exploited to evaluate ligand–receptor interactions, particularly when the ligand absorbs light in a region of the spectrum that is different from that of the receptor. Environmental conditions that affect the ionization state of a chromophore can result in large changes in the absorption maximum and molar absorptivity. For example, the absorption spectrum of tyrosine at pH above and below the pK_a of the phenol is shown in Fig. 1. Ionization of the phenol results in 18 nm shift in the lowest energy absorption maximum and an increase in the molar absorptivity.

The shifts in absorption bands are described by various terms. A shift in an absorption band to longer wavelength is a shift to lower energy, and is also called a red shift or a bathochromic shift. A shift in an absorption band to shorter wavelength is a shift to higher energy, and is also called a blue shift or a hypsochromic shift. Wavelength shifts may or may not be accompanied by changes in intensity. If the intensity of a band increases, such as in DNA melting, the increase is called a hyperchromic shift. If the intensity of an absorption band decreases, it is a hypochromic shift.

Small changes in absorption spectra are most easily observed by difference spectroscopy. An absorption spectrum of the unperturbed chromophore, such as a ligand in the absence of the receptor, is recorded. The absorption spectrum of an identical concentration of the ligand in the presence of the receptor is then recorded, and the first (unperturbed) spectrum is subtracted from the second (perturbed) spectrum to yield the difference spectrum. Such measurements can be repeated with varying concentrations of one of the components until saturation in

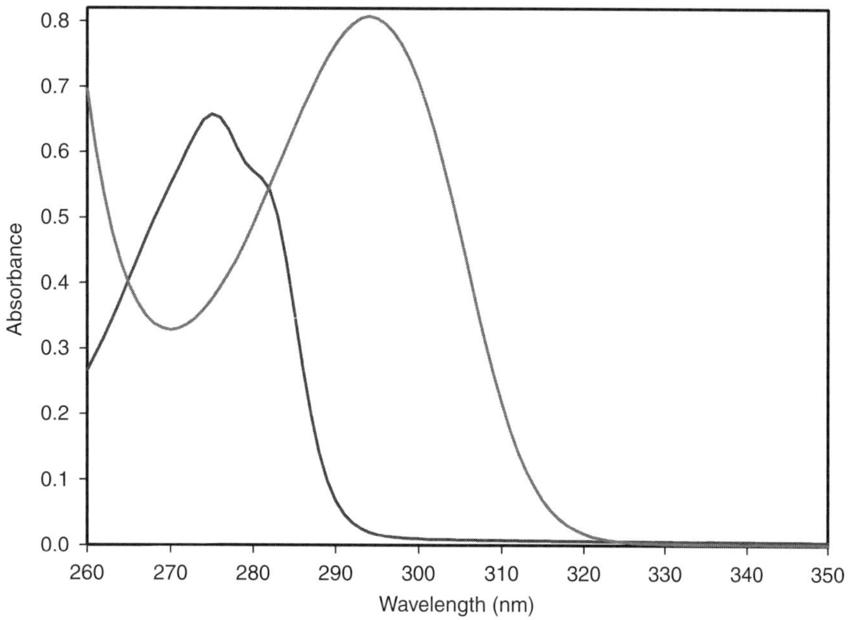

Fig. 1 Absorption spectra of L-tyrosine below and above the pK_a. The absorption maximum shifts from 275 nm in acidic to 294 nm in basic medium with a concomitant increase in the absorption intensity.

the signal change is reached. These data can then be used to construct a binding curve (Connors, 1987).

Difference spectra can be generated computationally by subtracting digitized spectra from one another. Alternatively, difference spectra may be directly recorded using tandem absorption cuvettes. These cuvettes contain a partition that divides the sample compartment precisely in half but is shorter than the exterior walls of the cuvette. Identical volumes of ligand and receptor are placed in each compartment of the cuvette and the instrument is blanked with this cell. The contents of the cuvette are then mixed by stoppering and inverting the cuvette. The resulting solution has half the initial concentration of each component distributed over twice the path length. If there is no interaction between the components, the spectrum of the mixed components is identical to that of the unmixed sample. Since the instrument is blanked with the unmixed sample, the spectrum will be identical to the baseline.

If an interaction between the components affects the absorption spectrum of one or more components, a difference spectrum will be observed. Figure 2 shows a difference spectrum for binding of thiocolchicine to tubulin. Thiocolchicine possesses an absorption band with a maximum of ~380 nm. Tubulin binding yields a difference spectrum that resembles a sine curve in the lower energy region of the spectrum. The positive lobe at longer wavelength coupled with a negative lobe at

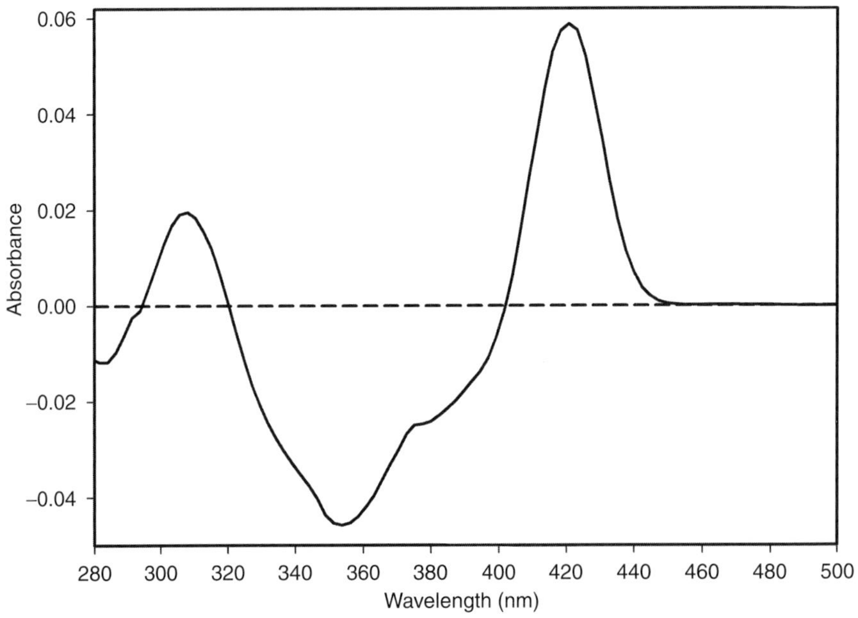

Fig. 2 Difference spectrum for binding of thiocolchicine to tubulin in PME buffer (0.1 M PIPES, 1 mM $MgSO_4$, 2 mM EGTA, pH 6.90). The binding results in the red shift of the thiocolchicine band.

shorter wavelength indicates that tubulin binding induces a bathochromic shift in the thiocolchicine absorption maximum. A negative lobe at the higher wavelength side of the curve and a positive lobe at the lower wavelength side of the curve would indicate that a hypsochromic shift in the absorption maximum occurred.

The tandem cell method is a particularly convenient way to measure the kinetics of a reaction or a ligand–receptor association. An example of a chemical reaction between an aldehyde and a hydrazine to form a hydrazone is shown in Fig. 3. The absorption maximum of the hydrazone is at longer wavelength than the absorption maximum of the hydrazine, which is reflected in the shape of the difference spectrum.

The arrows in Fig. 3 identify isosbestic points. An isosbestic point is the wavelength at which two components have identical absorptivities. The existence of an isosbestic point in a mixture is normally an indication of a two-state transition, such as one reactant to a single product, as shown in Fig. 3, or one type of receptor binding site for a ligand. It is possible that other species are present in solution that may have no absorption at the wavelength of the isosbestic point. The lack of an isosbestic point, however, indicates that a more complicated situation exists. For example, Fig. 4 shows the same hydrazone formation reaction performed at higher concentrations. Under these conditions, product formation is followed by

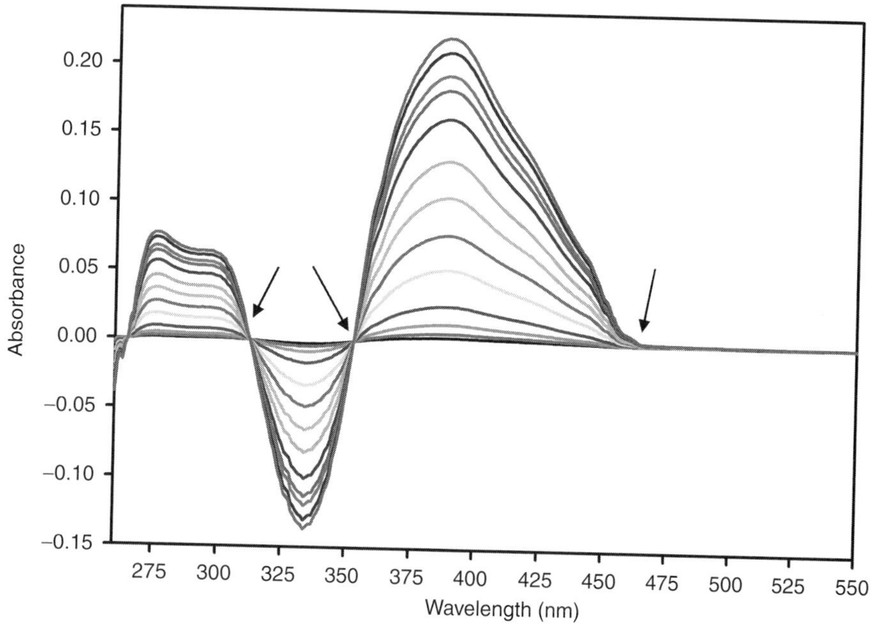

Fig. 3 Difference spectrum for hydrazone formation reaction as a function of time. Coumarin hydrazine (65 μM) and salicyladehyde (650 μM) in 0.1 M phosphate buffer at pH 7.0 were placed in separate compartments of a tandem cell. A baseline was collected. The reaction was initiated by stoppering the cuvette and mixing the contents by inversion. The absorption spectrum of the mixed solutions was collected at 60-sec time intervals. The arrows on the graph identify the isosbestic points, indicating the two-state transition in the formation of the hydrazone from the aldehyde and hydrazine.

precipitation, which scatters the visible light and increases the apparent baseline. The sharp isosbestic points observed in Fig. 3 are indiscernible in Fig. 4.

3. Turbidity Measurements

Figure 4 illustrates another function of an absorption spectrophotometer, which is to indirectly detect light scattering. Particles in a solution will scatter light in a manner dependent on the size of the particle and the wavelength of the light. The wavelength dependence of the turbidity of a solution of biological macromolecules can be used to estimate the size and shape of the macromolecule (Camerini-Otero and Day, 1978). Since turbidity is directly proportional to absorbance, measurements performed on the absorption spectrophotometer can provide information about molecular weight and dimensions and the concentrations of particles or macromolecules in the solution. For example, bacteria scatter light as if they were small particles, and suspensions of bacteria at sufficient concentration will appear turbid (Koch, 1968). Measurement of turbidity (apparent absorption) is common laboratory practice to monitor bacterial growth and estimate bacteria concentration (Murray *et al.*, 1979). Reactions that proceed with protein aggregation or polymer

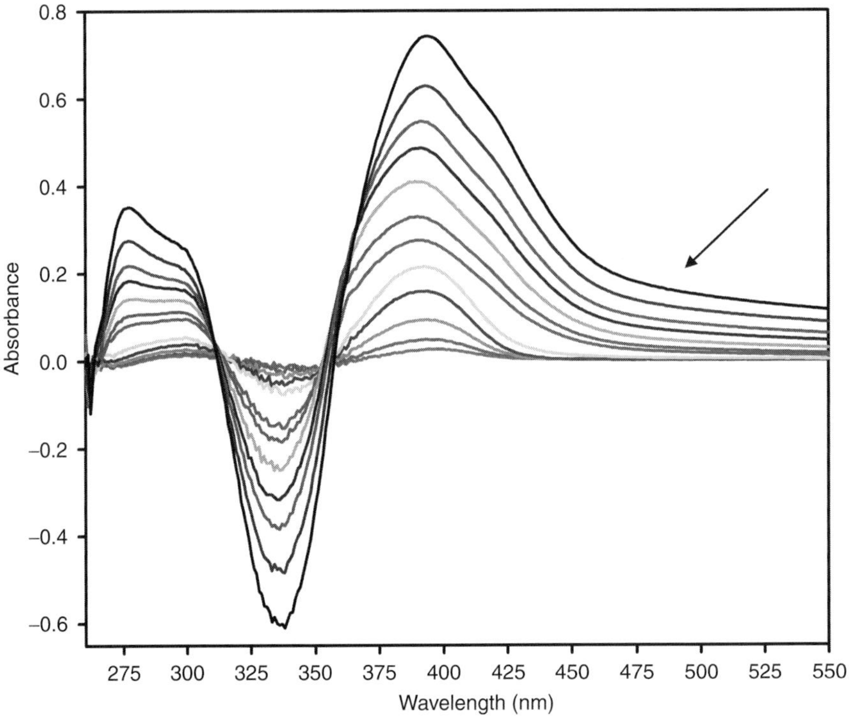

Fig. 4 Difference spectrum for hydrazone formation reaction as a function of time. Solutions of coumarin hydrazine (200 μM) and salicyladehyde (2 mM) in 0.1 M phosphate buffer at pH 7.0 were treated as described in Fig. 3. Comparing the spectra to those in Fig. 3, note the lack of isosbestic points and the increase in scatter observed at longer wavelength. The hydrazone precipitates out of solution during the reaction.

formation may be monitored by measuring turbidity changes (Findeis, 2000; Jobling *et al.*, 2001; Kanzaki *et al.*, 2006). For example, the cytoskeletal protein tubulin polymerizes to form tubular structures known as microtubules, resulting in a milky white appearance of a previously clear protein solution. The extent of light scattering, measured as apparent absorption in the region of 350–400 nm, is directly proportional to the mass of polymerized protein. This property has been used to determine the extent and rate of polymerization of the protein under various conditions (Berne, 1974; Gaskin *et al.*, 1974; Hall and Minton, 2005).

D. Microplate Reader Spectrophotometers

The use of plate readers for measuring optical properties of multiple samples is increasingly common in life science. Microtiter plates are commonly available in 6-, 12-, 24-, 48-, 96-, 384-, and 1536-well formats, and a majority of the commercial instruments will read 96- and 384-well plates. The instrumental specifications vary with manufacturer and model, so it is necessary to consult the manufacturer's specifications to determine the capabilities of a particular instrument. Many instruments use filters rather than monochromators, which limit applications (e.g., reading at fixed wavelengths rather than collection of the entire spectrum). The low wavelength limit of many instruments is ~320 nm, so protein and nucleic acid signals cannot be monitored. A significant difference between absorption data collected in plates compared to those collected in a standard spectrophotometer is that the path length in the plate is a function of the sample volume, while the path length when measured in spectrophotometer is physically defined by the dimensions of the cuvette regardless of the sample volume. Since the absorbance of a solution is a function of the path length, it is necessary to use identical sample volumes in each well in a plate and/or know how the plate reader responds to samples of varying volumes. There are commercially available plate readers that can correct the output data for variations in sample volume.

The most common application using microtiter plates are assays involving multiple samples of similar composition and identical volumes, such as colorimetric assays for cytotoxicity or enzyme-linked immunosorbent assays (ELISAs) using alkaline phosphatase-conjugated secondary antibodies. Many protocols that require absorption measurements on multiple samples are much easier to perform using plate readers than conventional spectrophotometers. However, miniaturization of absorption assays to 1536-well plates or low volume 384-well plates is not necessarily straightforward. The short optical path and variations in that path due to the meniscus become important factors in such assays (Zuck *et al.*, 2005).

III. Fluorescence Spectroscopy

Fluorescence spectroscopy is one of the most widely used optical techniques in biochemistry and cell biology. Highly sensitive and tremendously versatile, fluorescence spectroscopy can be found in virtually all areas of life science research. The common instruments encountered in life science laboratories are steady-state spectrofluorimeter and fluorescence plate reader. This chapter will be limited to the types of information available from the basic versions of these instruments. Even with these limitations, the coverage here is far from comprehensive. The reader is referred to the Suggested Reading and References section for leading sources to more specific and detailed information.

A. Introduction to Fluorescence

Absorption of a photon of appropriate energy causes an electronic transition from ground state to an excited state. Once a molecule is in an electronically excited state, the energy must eventually be dissipated for the molecule to return

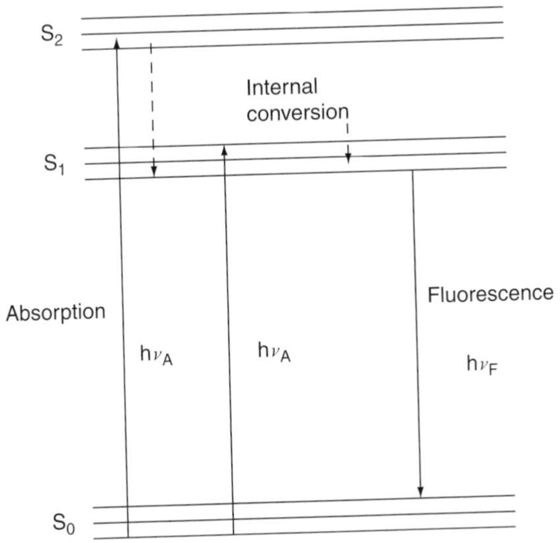

Fig. 5 Simplified Jablonski diagram depicting the excitation of an electron by absorption of a photon ($h\nu_A$) to higher electronic states S_1 or S_2. The electron returns to the ground electronic state, S_0, from the lowest vibrational state of S_1. Fluorescence is observed with the release of photon ($h\nu_F$). The dotted and solid arrows in the diagram represent nonradiative and radiative electronic transitions, respectively.

to its ground state. These processes are frequently illustrated by a Jablonski diagram (Fig. 5).

The absorption of a photon occurs very quickly ($\sim10^{-15}$ sec) from the lowest vibrational level of the ground state (S_0) to higher electronic and vibrational states (e.g., S_1, S_2). The electron relaxes quickly from higher energy states ($\sim10^{-10}$ to 10^{-12} sec) to the lowest vibrational level of the first excited state (S_1). This is known as internal conversion. In the absence of photochemical reactions, there are two general paths for loss of excited state energy: radiative and nonradiative. Loss of energy in a radiative pathway involves release of a photon (Fig. 5). When the photon comes from the first excited singlet state to the ground state, the light released is fluorescence. An electron can also undergo intersystem crossing, that is, move to an excited triplet state from the excited singlet state (not shown in Fig. 5). The return of the electron from this triplet state to the ground state may be accompanied by release of a photon. This emission is referred to as phosphorescence and will not be discussed in this chapter.

Excited state energy may also be dissipated by nonradiative paths (without emission of a photon) through mechanisms such as release of heat, interactions with solvent molecules, collisions with other molecules, or resonance energy transfer to another chromophore.

B. Fluorophores

A fluorophore may be endogenous to the biological system, such as an aromatic amino acid residue in a protein. Endogenous or intrinsic fluorophores frequently absorb and emit in the UV region. Visible region fluorophores are generally exogenous, and may be introduced to a biological system through the use of fluorescently labeled antibodies, chemical labeling, or expression of green fluorescent protein or one of its variants. Exogenous or extrinsic fluorophores can be designed to absorb and emit virtually any frequency of light, including near infrared radiation. Detailed information on the use of the exogenous fluorophores in life science laboratories can be found in Mason (1993). Both endogenous and exogenous fluorophores may be characterized by the following experimental observables.

1. Excitation and Emission Spectra

An emission spectrum is a plot of number of photons emitted at a particular wavelength when the molecule is irradiated at a single wavelength in an absorption band. In general, the shape of an emission spectrum is a mirror image of the absorption spectrum. Absorption of a photon primarily occurs from the lowest vibrational level of the ground state to multiple vibrational levels within the excited state. Vibrational relaxation of the excited state results in the emitted photon originating from the lowest vibrational level of the excited state. Emission of a photon returns the fluorophore to various vibrational levels within the ground state. The vibrational energy levels of the ground and excited state are about equally spaced, therefore, the absorption and emission spectra appear to be reflected through a mirror plane.

An excitation spectrum is also a plot of the number of photons emitted at a particular wavelength; however, in this case, the emission wavelength is held constant and the excitation wavelength is varied. The shape of an excitation spectrum of a molecule is generally the same as its absorption spectrum.

Normally, the shape of an excitation or emission spectrum is constant regardless of excitation wavelength, although the overall intensity will vary depending on the absorption coefficient at the excitation wavelength.

2. Stokes' Shift

The emission maximum of a fluorophore is observed at a lower energy (longer wavelength) than the absorption maximum (Fig. 6).

This is a consequence of the loss of vibrational energy in the excited state before emission of the photon (Fig. 5). The energy difference between the absorption maximum (ν_A) and the emission maximum (ν_F) is the Stokes' shift. The Stokes' shift is expressed in wavenumbers (cm^{-1}). The magnitude of the Stokes' shift of a fluorophore is a function of its molecular structure and, for many fluorophores,

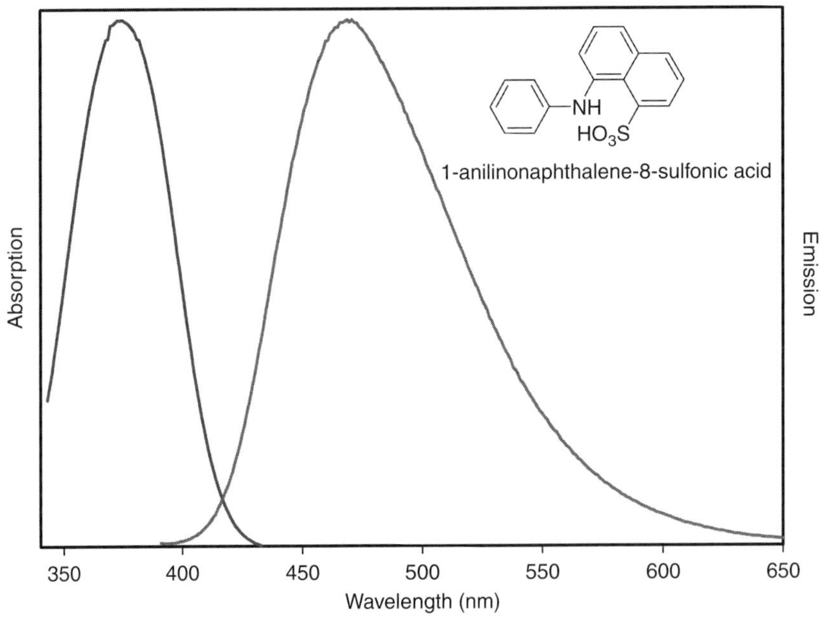

Fig. 6 Absorption and emission spectrum of 8-anilino-1-naphthalenesulfonic acid (ANS) in dimethyl sulfoxide.

the nature of its surroundings. Environmental effects on the Stokes' shift of a fluorophore can be useful to probe some aspects of biological systems (see Section III.H).

3. Fluorescence Lifetime

The lifetime of a fluorophore is the average time the molecule spends in the excited state. Loss of excited state energy is due to radiative and nonradiative processes. Therefore, lifetime (τ) is the inverse of the sum of the rate constant for radiative emission (fluorescence, k_r) and the rate constants for all nonradiative dissipation of excited state energy (k_{nr}):

$$\tau = \frac{1}{k_r + k_{nr}}$$

The lifetime of a single electronic transition is characterized by a single exponential decay of fluorescence and is the time required for the fraction of the population of molecules in the excited state to decrease by a factor of $1/e$, or $\sim 37\%$:

$$F(t) = F_0 e^{-t/\tau}$$

In the above equation, t is the time, τ is the fluorescence lifetime, F_0 is the initial fluorescence at $t = 0$. Fluorescence lifetimes are measured by either pulse fluorometry or phase-modulated fluorometry, both of which are beyond the scope of this chapter. It may be noted that the fluorescence lifetime is not directly affected by the energy or intensity of the light emitted. Thus, fluorophores that have similar spectral characteristics can be distinguished if their lifetimes are different. Discrimination between fluorescence lifetimes is the basis for the technique known as fluorescence lifetime imaging microscopy (FLIM; see *Quantifying Protein Activity Using FRET and FLIM Microscopy* by Pralle and Kalab in Volume 2 of this series, in press).

4. Quantum Yield

The quantum yield (ϕ) of a fluorophore is the ratio of the number of photons emitted as fluorescence to the number of photons absorbed by the molecule:

$$\phi = \frac{\text{photons emitted}}{\text{photons absorbed}}$$

The quantum yield can also be expressed in terms of the fluorescence lifetime:

$$\phi = \frac{k_r}{k_r + k_{nr}} = k_r \tau$$

A fluorophore with a quantum yield of 1.0 emits all absorbed photons as fluorescence. Relative "brightness" of fluorophores can be assessed by multiplying the absorption coefficient at the excitation wavelength by the fluorescence quantum yield (Waggoner, 1995). A comparison of a molecule's fluorescence intensity with the emission intensity of a fluorophore whose quantum yield is available in the literature is another common way to express relative quantum yield. Lists of standard fluorophores with their quantum yields can be found in a number of sources (Crosby and Demas, 1971; Lakowicz, 2006; Valeur, 2002).

5. Fluorescence Anisotropy

Fluorophores absorb photons that have their electric dipoles aligned parallel to the transition moment of the fluorophore. When freely diffusing fluorophores are excited with polarized light, the emitted light will normally be depolarized as a result of rotational diffusion during the lifetime of the excited state. The extent of depolarization is assessed by determining the intensity of light emitted parallel ($I_{\parallel}$) and perpendicular ($I_{\perp}$) to the excitation light. The fluorescence anisotropy (r) is calculated from these measurements by the following equation:

$$r = \frac{I_{\parallel} - I_{\perp}}{I_{\parallel} + 2I_{\perp}}$$

Applications of anisotropy measurements include determination of equilibrium binding constants for ligand–receptor interactions, particularly if the fluorescence of the ligand is monitored. This topic is further elaborated in Chapter 9 by LiCata, this volume.

C. Fluorescence Instrumentation

Two types of instruments frequently available in life science laboratories are fluorescence spectrophotometers and fluorescence plate readers. Fluorescence spectrophotometers have some resemblance to absorption spectrophotometers: both have light sources and accessories to control the energy of light incident on the sample, which is typically in a quartz cuvette. In an absorption spectrophotometer, the detector is in a straight path to the light source, but in a fluorescence spectrophotometer, the detector is at a right angle to the light source. Fluorescence spectrophotometers usually have two monochromators, hence the wavelength of light that hits the sample and the detector can be modulated independently. These instruments can collect two types of spectra: excitation spectra and emission spectra. Emission spectra are the more common. The sample is irradiated with light of a particular energy, selected with the excitation monochromator. The light emitted from the sample passes through the emission monochromator, which is usually scanned over a range sufficient to collect data from the entire emission band.

Some fluorescence instruments and many fluorescence plate readers use filters rather than monochromators for wavelength selection. Such instruments are suitable for measurements in which relative fluorescence intensities are measured (such as fluorescence-based ELISAs), but are not useful when changes in the energy (wavelength) of emission are to be observed.

Fluorescence plate readers vary from simple steady-state units to sophisticated systems capable of measuring fluorescence polarization, fluorescence lifetime, and time-resolved fluorescence. For steady-state measurements, important information includes: light source, wavelength range, whether monochromators or filters are used, availability of temperature control, and types of plates accepted. As in absorption spectrophotometers, it is important to know the optical specifications for the instrumentation. The linear (or dynamic) range of the instrument is particularly important to note, as measurements outside the linear range of the instrument are a frequently encountered error.

D. Absorption Versus Fluorescence

The absorption intensity of a molecule in solution will be the same regardless of the absorption spectrophotometer used. The same cannot be said for fluorescence intensity of the same solution. Absorption intensity is defined as the ratio of photons transmitted per photons absorbed, whereas fluorescence intensity is proportional to photons emitted. The number of photons emitted will depend on the number of photons incident on the sample, which will depend on instrumental parameters such as lamp intensity and slit width. With the exception of quantum yield measurements, fluorescence spectra are not ratioed, so the absolute value of fluorescence intensity from the same sample is not necessarily constant from day to day even on the same instrument under the same experimental conditions. Fluorescence spectra are therefore shown with arbitrary units in the legend on the *y*-axis, although other identifying legends may be employed (number of photons, fluorescence intensity, etc.).

Unlike absorption spectrum, the intensity of fluorescence is largely influenced by the temperature. Fluorescence quantum yields are sensitive to temperature because the processes by which excited state energy is dissipated (vibrations, collisions) are affected by temperature to a greater extent than the process by which the excited state is formed. Therefore, the solution in the cell compartment should be controlled at a constant temperature, even when the spectra are obtained at "room temperature."

E. Measuring Emission and Excitation Spectra

It is always a good idea to take an absorption spectrum of the sample that will be examined by fluorescence spectroscopy prior to recording fluorescence spectra.

1. Collecting Emission Spectra

The most common steady-state fluorescence spectrum is the emission spectrum. To collect the spectrum, a cuvette containing the fluorophore is placed in the instrument and the solution is equilibrated to the desired temperature. The excitation monochromator is adjusted to the wavelength chosen for excitation. This is frequently the absorption maximum for the fluorophore, but another wavelength within the absorption band may also be chosen. The emission monochromator is set to collect a range of wavelengths. The lower limit should be a value greater than the excitation wavelength to avoid collecting stray excitation light. The upper limit should be at a wavelength beyond the end of the emission band. This latter value is determined empirically.

An initial scan of the emission spectrum is performed, and the wavelength of the maximum emission intensity is noted. The intensity should be compared to the acceptable range provided with the instrument documentation. The intensity of the emission can be adjusted by methods described in the instrument documentation, which frequently consists of increasing or decreasing the slits on the excitation or emission monochromator until a suitable value is achieved. The sample is scanned again using the adjusted slits and wavelength limits.

A blank spectrum of the solution components without the fluorophore must also be collected using the same parameters employed for the emission spectrum. The blank emission spectrum is subtracted from the sample emission spectrum to yield the emission spectrum of the fluorophore.

Emission spectra may be corrected for fluctuations in the emission intensity due to features of emission monochromator and emission photomultiplier tube. Many modern instruments have correction factors available in the software. If all comparisons are going to be performed on the same instrument, it is normally not necessary to correct the emission spectrum.

2. Collecting Excitation Spectra

The procedure for collecting an excitation spectrum is similar to the procedure for collecting emission spectra, except that the emission wavelength is held constant and the excitation wavelength region is scanned. A major difference is that excitation spectra *must* be corrected to be meaningful because the intensity of light from the excitation source varies with wavelength. The standard method for correcting excitation spectra is to use a quantum counter in a reference channel. The specific procedure for a particular instrument should be found in the instrument manual.

F. Common Experimental Problems and Their Solutions

Most steady-state fluorescence experiments are straightforward to perform. There are, however, some trivial sources of error that are frequently encountered by amateur researchers. "Trivial" is used in the sense of simple, not unimportant.

1. Inner Filter Effect

Emission spectra are collected assuming that the same intensity of light hits the front and the back of the sample (Fig. 7A). When the sample absorbs the excitation light strongly, the intensity of light diminishes as it passes through the cell. The emission intensity emanating from the back of the cell is less than that emanating from the front of the cell (Fig. 7B). Therefore, the overall emission intensity will be lower in a sample with higher absorption at the excitation wavelength. If the absorption of the sample is 0.05 units or less over the effective path of the sample cell, then no inner filter effect is observed. If the absorption of the sample is >0.05, then a linear relationship between concentration of fluorophore and emission intensity cannot be assumed.

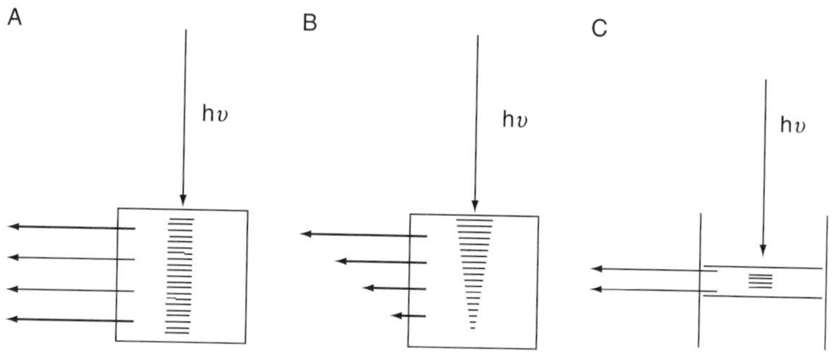

Fig. 7 Fluorescence of a sample is observed at right angle to the excitation. (A) The excitation at the front and back of the cuvette is same, so no inner filter effect is observed. (B) For a concentrated solution, the excitation of the sample in the cuvette is not uniform and hence lesser amount of emission is observed. Inner filter correction needs to be performed for this sample. (C) The cuvette has a smaller excitation path (2 mm) and the emission path is 10 mm. The use of this dual path length cuvette decreases the inner filter effect.

Two ways of managing the inner filter effect are avoiding it or correcting for it. If possible, avoiding an inner filter effect is the better choice. A simple way to decrease the light absorbed by the sample is to move the excitation wavelength to a region of the band with lower molar absorptivity. If this procedure is not feasible, the effective path can be decreased by using smaller dimension cells. Dual path length cells (e.g., 2 mm $\times$ 10 mm) can be oriented such that the excitation light passes through the shorter dimension (Fig. 7C). The inner filter effect in a 2-mm $\times$ 10-mm-cell will be at least fivefold less than that of the same solution in the standard 10-mm $\times$ 10-mm-cell. These cells have an added advantage that less sample is required to collect the data.

Correction for an inner filter effect can be done using the equation:

$$F_{\text{corr}} \cong F_{\text{obs}}\text{antilog}\left(\frac{A_{\text{ex}} + A_{\text{em}}}{2}\right)$$

where A_{ex} and A_{em} are the absorbance of the sample at excitation and emission wavelengths, respectively. F_{obs} is the observed fluorescence intensity and F_{corr} is the fluorescence intensity after correcting for the inner filter effect (Fig. 8). The equation should be applied with care. The effective path length depends on the placement of the sample compartment in the instrument, and it is not always half the length of the cuvette. The effective path length of the cuvette can be determined empirically. Samples containing a standard fluorophore are prepared and their emission intensities are recorded. The data are manipulated using the inner filter effect equation mentioned above, except that the denominator is varied until a straight line is obtained. The path that provides the best fit to the data is then used in correction for the inner filter effect in the experimental data of interest.

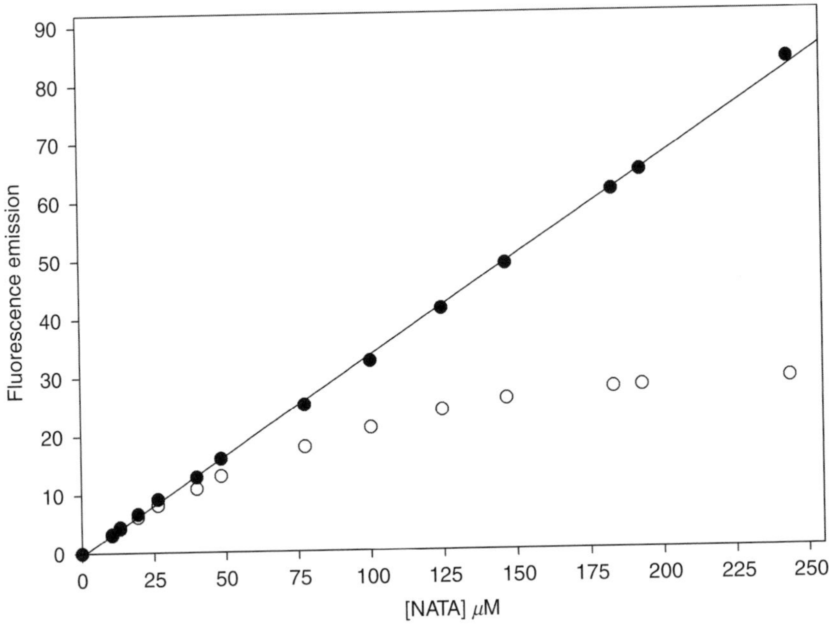

Fig. 8 The fluorescence intensity of aqueous solution of *N*-acetyl-L-tryptophanamide (NATA) deviates from linearity with increase in concentration (o). After correction for the inner filter effect, the emission intensity displays a linear trend (•). The samples were excited at 290 nm and fluorescence intensity was recorded at 360 nm.

2. Secondary Absorption Effect

A secondary absorption effect occurs when a component in the sample absorbs the emission light. This is less frequently encountered, and typically occurs when more than one chromophore is in the sample. The secondary absorption effect will depend on the concentration of the species, absorption coefficient of the acceptor (A), and quantum yield of the donor (D). The absorption spectrum of the sample should be examined to ensure that there is little to no absorptivity in the spectral region in which the emission data are collected. There is no satisfactory way to correct emission spectra for secondary absorption effects, so they should be avoided. Secondary absorption effects can often be eliminated by sample dilution.

3. Photobleaching

Molecules in the excited state can also undergo photochemical reactions. If these reactions lead to products with different emissive properties, the emission intensity in the sample will appear to decrease over time. For most fluorophores, this process is irreversible. Photobleaching has been used in fluorescence microscopy as fluorescence recovery after photobleaching (FRAP). In *in vitro* assays, however, photobleaching is normally undesirable. Photobleaching is decreased by limiting the amount of time a fluorophore is exposed to excitation energy. Many instruments have some mechanism to block the light from the excitation source when spectra are not being actively collected. Even kinetic data may be collected with minimal bleaching if the instrument is capable of closing the excitation shutter between data points. If supplies and instrumentation permit, the effect of bleaching on a sample can be minimized by employing relatively large volumes in the cuvette and stirring the cuvette during data acquisition. (Many instruments have magnetic cell stirrers included or as optional accessories.)

4. Light Scattering

The presence of particles and bubbles may result in the scattering of light; these should be removed from a sample whenever possible. A Raman band from the solvent may be observed in an emission scan. In biological systems, most commonly employed solvent is water and the Raman signal is observed 3600 cm^{-1} lower in energy than the excitation wavelength. Therefore, for an excitation wavelength of 280 nm, the Raman peak will be observed at 311 nm. When the sample fluorescence is intense, the contribution of the Raman band is negligible.

Identifying a peak as a scatter rather than a fluorescence peak can be accomplished by changing the excitation wavelength and rescanning the solution. If the peak is fluorescence, the emission intensity of the peak should change, but the wavelength of the emission maximum will not. If it is a scatter peak, the emission maximum will change as the excitation wavelength changes, and the intensity of the peak will remain approximately the same. For example, the water Raman peak will move from 311 to 362 nm if the excitation wavelength is changed from 280 to 320 nm.

G. Fluorescence Quenching

When the fluorescence intensity of a fluorophore is decreased by its interaction with its environment, the fluorescence is said to be "quenched." Collisional encounters with other molecules in solution that result in deactivation of the excited state result in collisional quenching. Contact between the fluorophore and the quencher is required for collisional quenching to occur, therefore, measurements of collisional quenching can be useful for determining the accessibility of a fluorophore on a biological macromolecule (France and Grossman, 2000). There are two common types of collisional quenching: dynamic quenching and static quenching. In the former, the quencher collides with the fluorophore in its excited state, dissipating its energy without release of a photon. Dynamic quenching of fluorescence is described by the Stern–Volmer equation:

$$\frac{F_0}{F} = 1 + k_q\tau_0[Q] = 1 + K_D[Q]$$

where F_0 and F are the fluorescence intensities in the absence and presence of quencher, respectively. k_q is the bimolecular quenching constant, τ_0 is the lifetime of fluorophore in the absence of quencher, [Q] is the concentration of the quencher, and K_D is the Stern–Volmer constant for dynamic quenching.

A plot of F_0/F against [Q] gives K_D as the slope. A linear Stern–Volmer plot is usually indicative of a single class of fluorophores, all equally accessible to the quencher.

In static quenching, a nonfluorescent complex is formed between the fluorophore and the quencher. The quenching equation then becomes:

$$\frac{F_0}{F} = 1 + K_S[Q]$$

where K_s is the association constant for the formation of the complex in static quenching.

Note that the form of the two equations is the same. In order to determine whether a process is due to static or dynamic quenching, additional experiments must be performed. One simple method is to repeat the quenching experiment at higher temperature. The slope of the plot F/F_0 versus [Q] should increase if the process is dynamic quenching and decrease if the loss of fluorescence is due to a static quenching mechanism.

H. Environmental Effects on Fluorescence

The distribution of electron density in each electronic state is defined by the molecular structure. If the distribution of electron density in the ground and excited states is very similar, the transition dipole, which describes the difference in electron density distribution, will be small. Absorption and emission spectra of fluorophores with this characteristic will be little affected by environment; that is, the absorption and emission maxima will be similar whether the fluorophore is in an aqueous environment or in an apolar pocket of a protein. These are called environmentally insensitive fluorophores. Environmentally insensitive probes are particularly useful in imaging. By contrast, when the difference in electron density distribution is pronounced, the absorption and emission spectra can be severely affected by the molecule's milieu. These are referred to as environmentally sensitive fluorophores. Fluorophores that are environmentally sensitive can also be used for imaging but are frequently used as sensors or to monitor ligand–receptor interactions.

Many of the environmentally sensitive fluorophores used as biological probes have an excited state that is more polar than the ground state. The effect of changing the polarity of the environment for such a probe is illustrated in Fig. 9. An apolar environment will stabilize the ground state but destabilize the excited state. The energy difference between the two states will increase as the polarity of the solvent decreases. Thus, the absorption and emission maxima will shift to shorter wavelength (higher energy). A more polar environment will stabilize

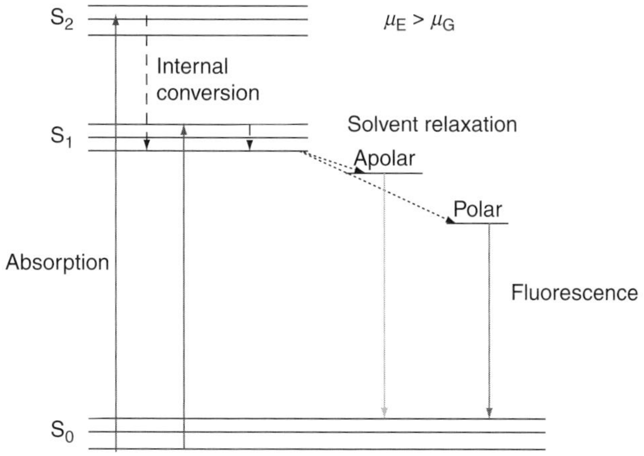

Fig. 9 Jablonski diagram depicting the influence of solvent on fluorescence of the fluorophore. In this representation, the dipole moment of the fluorophore in excited state is greater than the ground state ($\mu_E > \mu_G$), and hence a polar environment stabilizes the excited state better.

the excited state and destabilize the ground state, and thus the absorption and emission maxima will shift to longer wavelength (lower energy).

The nature of the environment of a fluorophore in a biological system can therefore be assessed by examining the Stokes' shift of the fluorophore as a function of solvent and comparing those data to the Stokes' shift of the fluorophore in the biological system. Details on this type of experiment can be found in Lakowicz (2006).

The quantum yield of environmentally sensitive fluorophores also may be affected by the environment. The relationship between environment and quantum yield is less defined than the relationship between fluorophore environment and Stokes' shift. Many of the environmentally sensitive fluorophores routinely used in biological systems undergo an increase in quantum yield when the polarity of the environment is decreased. Since receptor sites are typically less polar than the medium on the exterior of the binding site, the increase in fluorescence intensity can be used to quantitatively assess ligand–receptor interactions (Li *et al.*, 2000). A few examples of environmentally sensitive fluorophores are shown in Fig. 10.

Some environmentally sensitive probes have structural features that will be affected in a specific way by the environment, and therefore such molecules can be used as sensors. For example, molecules that possess ionizable groups that strongly affect the fluorescence have many biological applications as pH sensors. Fluorophores can also be designed to detect specific entities such as biologically

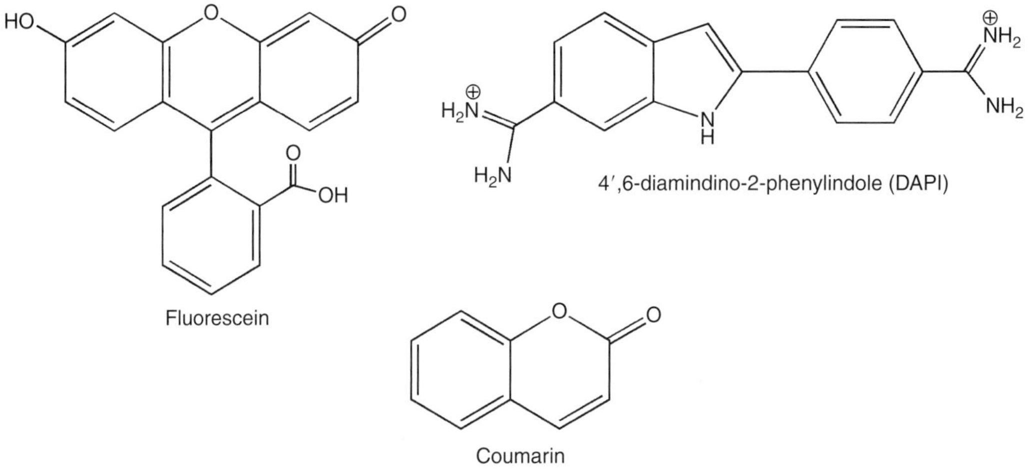

Fig. 10 Some examples of environment sensitive fluorophores. Fluorescein emission is pH dependent, 4′,6-diamidino-2-phenylindole (DAPI) emission intensity increases upon binding to DNA, Coumarin emission intensity increases in apolar environment with a concomitant shift in the emission maximum to lower energy.

important cations (Ca^{2+}, Mg^{2+}, and Zn^{2+}), reactive oxygen species, and inorganic anions and biological events such as changes in membrane potential (Altschuh *et al.*, 2006; Katerinopoulos and Foukaraki, 2002).

I. Fluorescence Resonance Energy Transfer

Fluorescence resonance energy transfer (FRET or simply referred to as RET) occurs when a molecule in its excited state transfers energy to another molecule through dipole–dipole interactions, without the appearance of a photon (Fig. 11). The transfer is highly dependent on the distance between the D and A species. The efficiency of energy transfer (E) is described by the following equation:

$$E = \left[1 + \left(\frac{R}{R_0}\right)^6\right]^{-1}$$

where R is the distance between the D and A and R_0 is the Förster distance, which is the distance at which energy transfer is 50% efficient. The sharp dependence of RET efficiency on distance means that energy transfer between D and A will be observed only when the pair is within a limited range of distances. Figure 12 illustrates the relationship between R_0 and RET efficiency. A D–A pair with a Förster distance of 20 Å will undergo transfer with 15% efficiency at 27 Å and

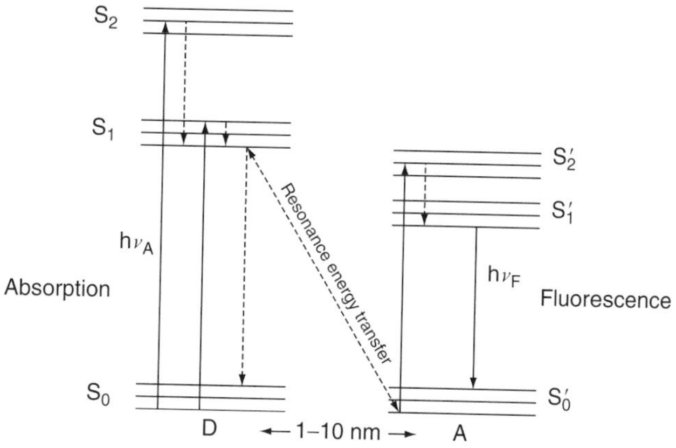

Fig. 11 Fluorescence resonance energy transfer (FRET or RET) between a donor (D) and acceptor (A). The dipole–dipole interaction of the electron in the D excited state with the A, results in A excitation.

85% efficiency at 15 Å; thus, in order for energy transfer to be readily observed, the D and A need to be within about 12 Å of one another. The dynamic range is larger when the Förster distance is larger: a D–A pair with a Förster distance of 60 Å will undergo transfer with 15% efficiency at 80 Å and 85% efficiency at 45 Å. A good rule of thumb is that the D–A distance (R) should be within a factor of 2 of R_0 for reliable distance measurements (Lakowicz, 2006).

The Förster distance is characteristic of the particular D–A pair. Some examples of organic molecule D–A pairs are listed in Table I. It should be noted that the A need not be a fluorescent molecule. The trinitrophenyl (TNP) group frequently appended to nucleotides is one example of a nonemissive A. A number of variants of green fluorescent protein are also available for RET experiments (Pollok and Heim, 1999; Shaner *et al.*, 2005). In principle, the range of distances detectable between two species can be tuned by changing the nature of the D or A.

The Förster distance is calculated from the spectral properties of the D and the A. One form of the equation, used when wavelength is expressed in nanometers, is:

$$R_0 = 0.211[\kappa^2 \eta^{-4} Q_D J(\lambda)]^{1/6}$$

where Q_D is the quantum yield of the D. $J(\lambda)$ is the overlap integral, which measures the degree of spectral overlap between the D emission spectrum and the A absorption spectrum (Lakowicz, 2006). Good overlap between D and A spectra and a high quantum yield for the D produce larger R_0 values. The other two terms are the refractive index of the medium (η) and a factor describing the orientation of the D and A dipoles (κ^2). Since neither of these is normally known, particularly in a complex biological system, standard values of 1.4 and 2/3, respectively, are used.

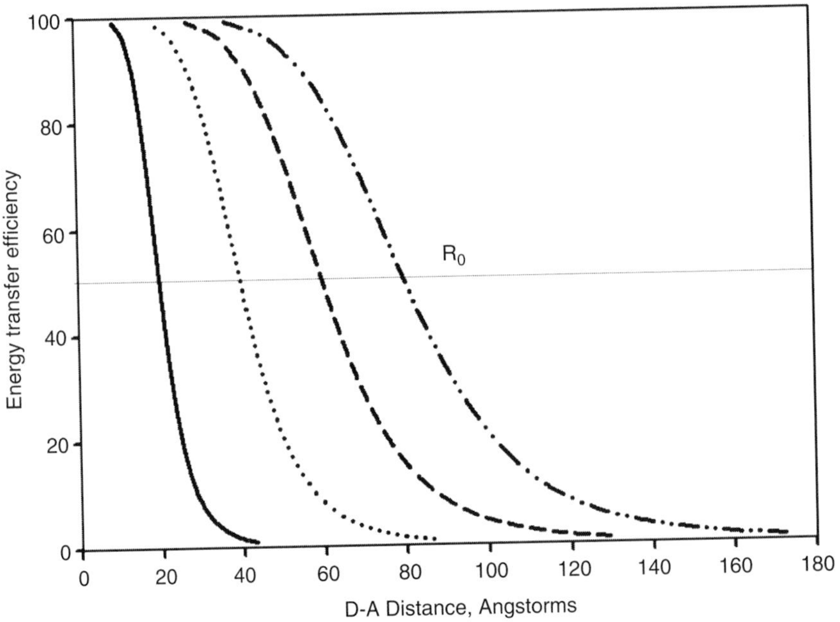

Fig. 12 The efficiency of resonance energy transfer is determined by the distance between the D–A pair. The figure represents D–A pair Förster distance, R_0, of: 20 Å (—), 40 Å(· · ·), 60 Å(– –), and 80 Å (– ·· –).

Table I
Calculated R_0 Values for RET for Some Organic Donor-Acceptor Pairs

	Donor			Acceptor			
	Excitation (nm)	Emission (nm)	R_0 (Å)		Excitation (nm)	Emission (nm)	References
Naphthalene	280	340	22	DANSYL	340	520	Waggoner, 1995
BODIPY	505	512	57	BODIPY	505	512	Karolin *et al.*, 1994
CF	495	517	51	Texas red	595	615	Wu and Brand, 1994
EDANS	336	470	33	Dabcyl[a]	472	–	Matayoshi *et al.*, 1990
FITC	494	518	54	EITC	524	548	Wu and Brand, 1994
FITC	494	518	49–54	TMR	555	580	Wu and Brand, 1994
CY3	548	562	40	CY5	650	670–700	Massey *et al.*, 2006
Fluorescein	495	517	56	CY3	548	560	Norman *et al.*, 2000
NBD-DMPE	340	520	59–61	Rh-DMPE	570	610	Loura *et al.*, 2001

[a]Dabcyl is a nonfluorescent acceptor and the transfer is also referred to as "Dark transfer." It has the advantage of eliminating the background fluorescence resulting from direct acceptor excitation.

CF, carboxyfluorescein, succinimidyl ester; Dabcyl, 4-((4-(dimethylamino)phenyl)azo)benzoic acid; DANSYL, 5-dimethylamino naphthalene-1-(*N*-(5-aminopentyl)) sulfonamide; EDANS, *N*-(aminoethyl)-5-naphthylamine-1-sulfonic acid; EITC, eosin-5-isothiocyanate; FITC, fluorescein-5-isothiocyanate; NBD, 7-nitro-benz-2-oxa-1,3-diazol-4-yl; NBD-DMPE, *N*-(7-nitrobenz-2-oxa-1,3-diazol-4-yl)-dimyristoylphosphatidylethanolamine; Rh-DMPE, *N*-(lissamine™-rhodamine B)-dipalmitoylphosphatidylethanolamine.

The dependence of the energy transfer efficiency on the negative sixth power of distance provides a sensitive method to measure molecular and atomic distance relations at nanometer level. This is the reason that the technique is at times referred to as a spectroscopic ruler. RET therefore can be used in structural studies, such as to determine distances between two fluorophores on a biological macromolecule. RET is frequently used as a tool rather than a method for measuring distances. The appearance or disappearance of RET can signal a biochemical event. Virtually any assay that involves association or dissociation of molecular species may be adaptable to monitor RET. For example, protease activity can be measured by monitoring the disappearance of RET that accompanies cleavage of a peptide substrate labeled with a D at one terminus and an A at the other end (Giepmans *et al.*, 2006; Stockholm *et al.*, 2005). Other examples of RET-based molecular assays include nucleic acid hybridization reactions, detection of single nucleotide polymorphism, studies of membrane microdomains, immunoassays, and ligand–receptor associations. RET technology has been expanded into chromatographic assays, electrophoresis, microscopy, and flow cytometry to understand intermolecular and intercellular interactions (Altschuh *et al.*, 2006; Asseline *et al.*, 2006; Croney *et al.*, 2003; Giepmans *et al.*, 2006; Khanna and Ullman, 1980; Loura *et al.*, 2001; Matayoshi *et al.*, 1990; Pollok and Heim, 1999; Selvin, 2000; Silvius and Nabi, 2006; Wu and Brand, 1994).

IV. Summary

Absorption and fluorescence spectroscopy have a huge number of applications in biological research. Many of the techniques are simple to perform and require only basic instrumentation. In this chapter, the fundamental processes that occur in absorption and fluorescence spectroscopy have been touched upon. A very small fraction of the life science applications of steady-state absorption and fluorescence techniques have been noted. Descriptions of some of the most basic experimental techniques (and common ways that these techniques can go awry) will hopefully provide the reader with a good start toward collecting meaningful data with steady-state spectrophotometers and plate readers.

Acknowledgments

The work has been supported by NIH grant CA-69571. We thank Abhijit Banerjee and Shubhada Sharma for providing us with illustrations (Figs. 3 and 4, and Fig. 8, respectively).

References

Altschuh, D., Oncul, S., and Demchenko, A. P. (2006). Fluorescence sensing of intermolecular interactions and development of direct molecular biosensors. *J. Mol. Recog.* **19,** 459–477.

Asseline, U., Chassignol, M., Aubert, Y., and Roig, V. (2006). Detection of terminal mismatches on DNA duplexes with fluorescent oligonucleotides. *Org. Biomol. Chem.* **4,** 1949–1957.

Berne, B. J. (1974). Interpretation of the light scattering from long rods. *J. Mol. Biol.* **89,** 755–758.

Camerini-Otero, R. D., and Day, L. A. (1978). The wavelength dependence of the turbidity of solutions of macromolecules. *Biopolymers* **17,** 2241–2249.

Castanho, M., Santos, N. C., and Loura, L. M. S. (1997). Separating the turbidity spectra of vesicles from the absorption spectra of membrane probes and other chromophores. *Eur. Biophys. J.* **26,** 253–259.

Connors, K. A. (1987). "Binding Constants. The Measurement of Molecular Complex Stability." John Wiley & Sons, Inc., New York.

Croney, J. C., Cunningham, K. M., Collier, R. J., and Jameson, D. M. (2003). Fluorescence resonance energy transfer studies on anthrax lethal toxin. *FEBS Lett.* **550,** 175–178.

Crosby, G. A., and Demas, J. N. (1971). Measurement of photoluminescence quantum yields. *J. Phys. Chem.* **75,** 991–1024.

Fasman, G. D. (1989). "Practical Handbook of Biochemistry and Molecular Biology." CRC Press, Inc., Boca Raton, Florida.

Findeis, M. A. (2000). Approaches to discovery and characterization of inhibitors of amyloid beta-peptide polymerization. *Biochim. Biophys. Acta* **1502,** 76–84.

France, R. M., and Grossman, S. H. (2000). Acrylamide quenching of apo- and holo-alpha-lactalbumin in guanidine hydrochloride. *Biochem. Biophys. Res. Commun.* **269,** 709–712.

Gaskin, F., Cantor, C. R., and Shelanski, M. L. (1974). Turbidimetric studies of the *in vitro* assembly and disassembly of porcine neurotubules. *J. Mol. Biol.* **89,** 737–755.

Giepmans, B. N., Adams, S. R., Ellisman, M. H., and Tsien, R. Y. (2006). The fluorescent toolbox for assessing protein location and function. *Science* **312,** 217–224.

Hall, D., and Minton, A. P. (2005). Turbidity as a probe of tubulin polymerization kinetics: A theoretical and experimental re-examination. *Anal. Biochem.* **345,** 198–213.

Jobling, M. F., Huang, X., Stewart, L. R., Barnham, K. J., Curtain, C., Volitakis, I., Perugini, M., White, A. R., Cherny, R. A., and Masters, C. L. (2001). Copper and zinc binding modulates the aggregation and neurotoxic properties of the prion peptide PrP106–126. *Biochemistry* **40,** 8073–8084.

Kallansrud, G., and Ward, B. (1996). A comparison of measured and calculated single- and double-stranded oligodeoxynucleotide extinction coefficients. *Anal. Biochem.* **236,** 134–138.

Kanzaki, N., Uyeda, T. Q., and Onuma, K. (2006). Intermolecular interaction of actin revealed by a dynamic light scattering technique. *J. Phys. Chem. B. Condens. Matter Mater. Surf. Interfaces Biophys.* **110,** 2881–2887.

Karolin, J., Johansson, L. B. A., Strandberg, L., and Ny, T. (1994). Fluorescence and absorption spectroscopic properties of dipyrrometheneboron difluoride (Bodipy) derivatives in liquids, lipid-membranes, and proteins. *J. Am. Chem. Soc.* **116,** 7801–7806.

Katerinopoulos, H. E., and Foukaraki, E. (2002). Polycarboxylate fluorescent indicators as ion concentration probes in biological systems. *Curr. Med. Chem.* **9,** 275–306.

Khanna, P. L., and Ullman, E. F. (1980). 4′,5′-Dimethoxy-6-carboxyfluorescein: A novel dipole-dipole coupled fluorescence energy transfer acceptor useful for fluorescence immunoassays. *Anal. Biochem.* **108,** 156–161.

Koch, A. L. (1968). Theory of the angular dependence of light scattered by bacteria and similar-sized biological objects. *J. Theor. Biol.* **18,** 133–156.

Lakowicz, J. R. (2006). "Principles of Fluroescence Spectroscopy," 3rd edn. Springer, New York, Berlin.

Li, Y., Edsall, R., Jr., Jagtap, P. G., Kingston, D. G., and Bane, S. (2000). Equilibrium studies of a fluorescent paclitaxel derivative binding to microtubules. *Biochemistry* **39,** 616–623.

Loura, L. M., Fedorov, A., and Prieto, M. (2001). Fluid-fluid membrane microheterogeneity: A fluorescence resonance energy transfer study. *Biophys. J.* **80,** 776–788.

Mason, W. T. (1993). "Fluorescent and Luminescent Probes for Biological Activity." Academic Press, London, San Diego.

Massey, M., Algar, W. R., and Krull, U. J. (2006). Fluorescence resonance energy transfer (FRET) for DNA biosensors: FRET pairs and Forster distances for various dye-DNA conjugates. *Anal. Chim. Acta* **568,** 181–189.

Matayoshi, E. D., Wang, G. T., Krafft, G. A., and Erickson, J. (1990). Novel fluorogenic substrates for assaying retroviral proteases by resonance energy transfer. *Science* **247,** 954–958.

Murray, J., Hukins, D. W., and Evans, P. (1979). Application of Mie theory and cubic splines to the representation of light scattering patterns from bacteria in the logarithmic growth phase. *Phys. Med. Biol.* **24,** 408–415.

Norman, D. G., Grainger, R. J., Uhrin, D., and Lilley, D. M. (2000). Location of cyanine-3 on double-stranded DNA: Importance for fluorescence resonance energy transfer studies. *Biochemistry* **39,** 6317–6324.

Pace, C. N., Vajdos, F., Fee, L., Grimsley, G., and Gray, T. (1995). How to measure and predict the molar absorption coefficient of a protein. *Protein Sci.* **4,** 2411–2423.

Pollok, B. A., and Heim, R. (1999). Using GFP in FRET-based applications. *Trends Cell Biol.* **9,** 57–60.

Selvin, P. R. (2000). The renaissance of fluorescence resonance energy transfer. *Nat. Struct. Biol.* **7,** 730–734.

Shaner, N. C., Steinbach, P. A., and Tsien, R. Y. (2005). A guide to choosing fluorescent proteins. *Nat. Methods* **2,** 905–909.

Silvius, J. R., and Nabi, I. R. (2006). Fluorescence-quenching and resonance energy transfer studies of lipid microdomains in model and biological membranes. *Mol. Membr. Biol.* **23,** 5–16.

Stockholm, D., Bartoli, M., Sillon, G., Bourg, N., Davoust, J., and Richard, I. (2005). Imaging calpain protease activity by multiphoton FRET in living mice. *J. Mol. Biol.* **346,** 215–222.

Valeur, B. (2002). "Molecular Fluorescence: Principles and Applications." Wiley-VCH, Weinheim, New York.

Waggoner, A. (1995). Covalent labeling of proteins and nucleic acids with fluorophores. *Methods Enzymol.* **246,** 362–373.

Wani, M. C., Taylor, H. L., Wall, M. E., Coggon, P., and McPhail, A. T. (1971). Plant antitumor agents. VI. The isolation and structure of taxol, a novel antileukemic and antitumor agent from Taxus brevifolia. *J. Am. Chem. Soc.* **93,** 2325–2327.

Wu, P., and Brand, L. (1994). Resonance energy transfer: Methods and applications. *Anal. Biochem.* **218,** 1–13.

Zuck, P., O'Donnell, G. T., Cassaday, J., Chase, P., Hodder, P., Strulovici, B., and Ferrer, M. (2005). Miniaturization of absorbance assays using the fluorescent properties of white microplates. *Anal. Biochem.* **342,** 254–259.

Suggested Reading

Some of the sources consulted in the preparation of this chapter are listed below.

Andrews, D. L., and Demidov, A. A. (1999). "Resonance Energy Transfer." Wiley, New York.

Bohren, C. F., and Huffman, D. R. (1983). "Absorption and Scattering of Light by Small Particles." Wiley, New York.

Brown, G. H. (1971). "Photochromism." Wiley-Interscience, New York.

Burgess, C., and Knowles, A. (1981). "Standards in Absorption Spectrometry." Chapman and Hall, London, New York.

Dyer, J. R. (1965). "Applications of Absorption Spectroscopy of Organic Compounds." Prentice-Hall, Englewood Cliffs, New Jersey.

Fasman, G. D. (1989). "Practical Handbook of Biochemistry and Molecular Biology." CRC Press, Boca Raton, Florida.

Gore, M. G. (2000). "Spectrophotometry & Spectrofluorimetry: A Practical Approach." Oxford University Press, Oxford, New York.

Hammes, G. G. (2005). "Spectroscopy for the Biological Sciences." Wiley-Interscience, Hoboken, New Jersey.
Haugland, R. P. (2002). "Handbook of Fluorescent Probes and Research Chemicals," 9th edn. Molecular Probes, Oregon.
Lakowicz, J. R. (2006). "Principles of Fluroescence Spectroscopy," 3rd edn. Springer, New York, Berlin.
Mason, W. T. (1993). "Fluorescent and Luminescent Probes for Biological Activity: A Practical Guide to Technology for Quantitative Real-Time Analysis." Academic Press, London, San Diego.
Sandorfy, C., and Theophanides, T. (1983). "Spectroscopy of Biological Molecules: Theory and Applications—Chemistry, Physics, Biology and Medicine." D. Reidel Publishing Company, Dordrecht, Boston.
Sauer, K. (1995). Biochemical Spectroscopy. *In* "Methods in Enzymology," Vol. 246. Academic Press, San Diego.
Smith, B. C. (2002). "Quantitative Spectroscopy: Theory and Practice." Academic Press, Amsterdam, Boston.
Tkachenko, N. V. (2006). "Optical Spectroscopy, Methods and Instrumentations," 1st edn. Elsevier, Amsterdam, Boston.
Turro, N. J. (1965). "Molecular Photochemistry." W. A. Benjamin, New York.
Valeur, B. (2002). "Molecular Fluorescence: Principles and Applications." Wiley-VCH, Weinheim, New York.

CHAPTER 9

Applications of Fluorescence Anisotropy to the Study of Protein–DNA Interactions

Vince J. LiCata and Andy J. Wowor

Department of Biological Sciences
Louisiana State University
Baton Rouge, Louisiana 70803

Abstract

The use of fluorescence anisotropy to monitor protein–DNA interactions has been on the rise since its introduction by Heyduk and Lee in 1990. As a solution-based, true-equilibrium, real-time method, it has several advantages (and a few

0091-679X/08 $35.00
DOI: 10.1016/S0091-679X(07)84009-X

disadvantages) relative to the more classical methods of filter binding and the electrophoretic mobility shift assay (gel shift). This chapter discusses the basis for monitoring protein–DNA interactions using fluorescence anisotropy, as well as the advantages and disadvantages of the method, but the bulk of the chapter is devoted to experimental tips and guidance meant to augment existing reviews of the method. The focus is on the current primary use of the method: direct measurement of binding isotherms for protein–DNA interactions *in vitro*. A short summary of emerging applications of the method is also included.

I. Introduction and General Background

Since its introduction in 1990, the use of fluorescence anisotropy as a method for monitoring protein–DNA interactions (Heyduk and Lee, 1990) has been steadily on the rise. As a real-time, solution-based assay, it has several advantages over its closest "competitors": filter binding (Riggs *et al.*, 1970) and the electrophoretic mobility shift assay (Fried and Crothers, 1981; Garner and Revzin, 1981). The methodology has been reviewed several times, both in the context of general overviews of fluorescence-based methods (Brown and Royer, 1997; Eftink, 1997; Hill and Royer, 1997) and as the sole focus of particular reviews (Chin *et al.*, 2004; Heyduk *et al.*, 1996; Jameson and Sawyer, 1995; Lundblad *et al.*, 1996). The latter four, more specific reviews contain a wealth of practical information on experimental design, and the reader interested in utilizing this technique is advised to consult these four reviews. The method is also the subject of three US patents (Royer, 1995, 1998, 2001). The present review largely seeks to augment and expand on the practical information in the four method-specific reviews noted above, adding new insights, advice, and guidance obtained from our laboratory's use of the technique over a number of years. Overlap with previous reviews, except where necessary to avoid large gaps in coherence, is avoided as much as possible.

A. Fluorescence Anisotropy in a Nutshell

The chapter by Shanker and Bane in this volume provides an excellent overview of the fundamentals of fluorophore excitation and emission. For fluorescence anisotropy, the two key fundamental properties to keep in mind are (1) that fluorescent molecules have both an excitation and an emission dipole and (2) that there is a short time delay between absorbance of the exciting photon and release of the fluorescent photon (the fluorescent lifetime).

The excitation dipole of the molecule dictates that, for any solution of fluorophores, polarized light will excite only those molecules in the solution that are in the proper orientation, that is, those fluorophores that just happen to be oriented so that their excitation dipole aligns with the polarized incident light. Since illumination is constant in fluorescence anisotropy, fluorophores in the solution will continuously be tumbling into and out of alignment with the polarized incident light.

Similar to the excitation dipole, the emission dipole of a fluorophore determines the polarity of the light released by that fluorophore. Imagine a solution of randomly diffusing and tumbling fluorophores that is suddenly immobilized. Now shine plane polarized light through this immobilized collection of fluorophores. As noted above, only those fluorophores that have their excitation dipoles aligned with the polarized incident light will absorb that light, and this will be a small subset of the total population. Since these fluorophores are immobilized, all of the light emitted by the subpopulation of excited fluorophores will also be coherently polarized. The angular relationship between the emission and excitation dipoles of the molecule will also determine the angular relationship between the polarization of the emitted light and the polarization of the incident light.

Now, instead of shining light on an immobilized collection of fluorophores, consider the more typical experimental situation: shining polarized light on a normal liquid solution of fluorophores. Here is where the time delay between excitation and emission becomes important. The exact time delay between excitation and emission follows an exponential decay law, and the average time delay is denoted by the fluorescence lifetime for that fluorophore (see also Shanker and Bane, this volume). Fluorescent lifetimes for common biochemical fluorophores are typically in the 1–25 nsec time range.

If an excited fluorophore molecule tumbles (rotationally diffuses) within its fluorescent lifetime, then the polarization of its emitted light will be determined by its new position. If a whole population of fluorophores randomizes within the fluorescent lifetime, the emitted light will become completely depolarized, because the positions of all the emission dipoles will have effectively been randomized.

The foundation for using fluorescence anisotropy to monitor molecular interactions is based on the fact that larger molecules tumble more slowly (and thus retain more emission polarization), while smaller molecules tumble more quickly (and thus depolarize the emission more effectively). A small stretch of fluorescently labeled DNA will tumble (and spin on its axis) faster when alone in solution than when bound to a protein. Figure 1 summarizes this effect in cartoon form. Thus, the increase in anisotropy due to slower rotational diffusion of the protein–DNA complex relative to the free DNA is the dependent signal that translates directly into the fraction of DNA bound in this technique.

B. Anisotropy and Polarization

In biochemistry, the methods of "fluorescence anisotropy" and "fluorescence polarization" are often considered equivalent, and the terms are frequently used interchangeably. Spectroscopically, they do differ slightly, although they are mathematically easily interconvertable. Just the words anisotropy and polarization have different meanings. Anisotropy is a more general term for directional dependence of a physical property. For example, a piece of plywood is anisotropic, since its strength is different along the grain versus against the grain. Polarization (when not being used in a political/ideological context) is most commonly used

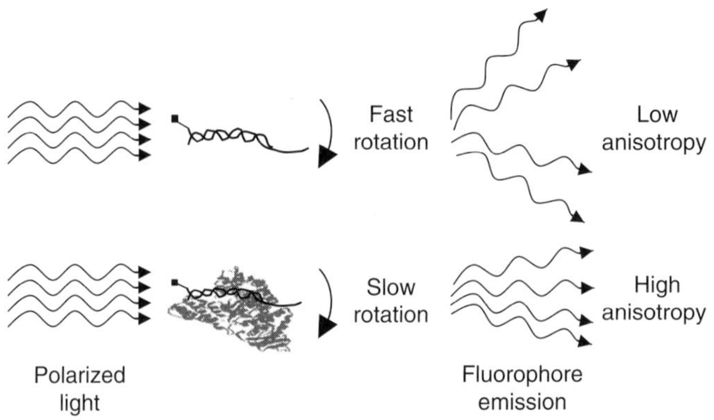

Fig. 1 Schematic illustration of the effect of rotational diffusion rate (tumbling and spinning on axis) on the anisotropy of emitted light from fluorescently labeled DNA. Both the free DNA and the complex-bound DNA are illuminated by polarized light. Since the DNA in the complex tumbles more slowly, a larger proportion of its emitted light remains polarized or anisotropic.

to describe a particular property of light waves or other electromagnetic waves. Polarization is the degree of uniformity of direction of the perpendicular transverse oscillation of a collection of light waves relative to their direction of travel. However, since a plane polarized light wave is, by definition, also anisotropic, the terms have long been used interchangeably.

Quantitatively, polarization, denoted P, is measured as

$$P = \frac{(I_{\parallel} - I_{\perp})}{(I_{\parallel} + I_{\perp})} \tag{1}$$

while anisotropy, denoted A, is measured as

$$A = \frac{(I_{\parallel} - I_{\perp})}{(I_{\parallel} + 2I_{\perp})} \tag{2}$$

the only difference being the doubling of the perpendicular intensity in the denominator for anisotropy. Polarization and anisotropy can be interconverted using

$$\left[\left(\frac{1}{P}\right) - \left(\frac{1}{3}\right)\right]^{-1} = \frac{3A}{2} \tag{3}$$

Most commercial fluorometers available today provide anisotropy (A) directly, whereas in the past it was common to find manufacturer-specific use of either A or P. For further discussion of polarization versus anisotropy, see Cantor and Schimmel (1980).

II. Advantages and Disadvantages of Anisotropy in Monitoring DNA Binding

The inherent advantages and disadvantages of monitoring protein–DNA interactions by fluorescence anisotropy have been extensively discussed in different reviews and several original research papers (see especially Heyduk *et al.*, 1996, and Lundblad *et al.*, 1996). The main advantage of the technique is the fact that it is a solution-based equilibrium technique. This allows one to make measurements without fear that the detection technique is altering the reaction equilibrium. Furthermore, it allows one to alter solution conditions for measurements quite easily without fear of altering the direct relationship between the signal provided by the detection method and the progress of the reaction. Separation methods, such as filter binding (Riggs *et al.*, 1970) or electrophoretic mobility shift assays (Fried and Crothers, 1981; Garner and Revzin, 1981), can easily perturb the reaction equilibrium. The separation process itself pulls reactants (DNA and protein) away from products (complex). This creates concentration gradients for each component, and so each subenvironment (i.e., region of specific concentrations of all reactants and products) during the separation will be thermodynamically pushed toward its own equilibrium, unless such rearrangement can be fully quenched during the separation process. The result is that the fractions of each component seen separated on the final gel, or retained on the filter, may not be the same as the fractions of each component in the original equilibrium mixture. In fact, some researchers prefer to refer to these assays as "nonequilibrium" methods in general (Hey *et al.*, 2001).

A further problem with separation-based assays arises if one changes solution conditions in the sample mixture (salt, temperature, pH, osmolytes, etc.). One must then perform controls to ensure that such changes have not perturbed the separation method itself. For example, varying the amount of salt or adding an osmolyte to a protein-DNA reaction can directly alter the efficiency with which the protein–DNA complex sticks to a nitrocellulose filter or enters a gel. If one changes the reaction conditions in fluorescence anisotropy, one may alter the value for the absolute anisotropy, but one will not alter the fact that the normalized change in anisotropy (ΔA) will still scale directly with fractional saturation ($\bar{Y}$).

Other advantages of fluorescence anisotropy include the following: (1) The fact that it is a real-time assay. One does not wait for a gel to run or radioactivity on filters to be counted to obtain the result. (2) The data produced are almost always of much higher precision (much lower random data scatter) than those typically obtained via filter binding or electrophoretic mobility shift assays. This allows discrimination among reactions of very similar affinity. Figures 2 and 3 show the clean, clear resolution among binding reactions that differ from one another by less than 0.5 kcal/mole. The technique can easily and reproducibly resolve between binding reactions that differ from one another by <10% (e.g., a binding curve with a 9 nM K_d is distinct from one with a 10 nM K_d; see Datta *et al.*, 2006, for examples). Much current biophysical research on protein–DNA interactions is

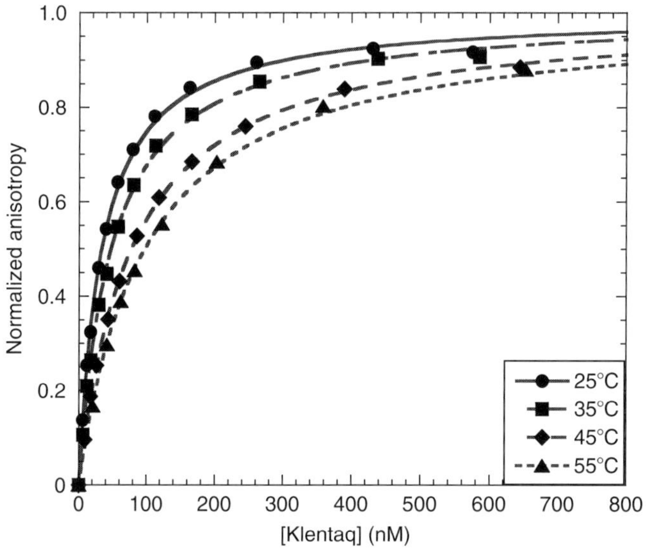

Fig. 2 The temperature dependence of ROX-labeled single-stranded DNA (63-mer) binding to Klentaq DNA polymerase, illustrating the ability to resolve binding reactions with very similar affinities. Equilibrium titrations are shown at 25 °C (●), 35 °C (■), 45 °C (◆), and 55 °C (▲). All titrations were performed in 10 mM Tris, 5 mM $MgCl_2$, 5 mM KCl, at pH 7.9. Increasing temperature decreases the binding affinity of Klentaq polymerase to single-stranded DNA. At 25 °C, the K_d is 33.6 nM ($\Delta G = -10.2$ kcal/mole). At 35 °C, the K_d is 47.6 nM ($\Delta G = -10.3$ kcal/mole). At 45 °C, the K_d is 77.6 nM ($\Delta G = -10.3$ kcal/mole). At 55 °C, the K_d is 97.0 nM ($\Delta G = -10.5$ kcal/mole). (Wowor and LiCata, unpublished data).

focused on how solvent components (ions, protons, osmolytes, water activity, etc.) act to regulate these interactions. Because of its precision and its reliability across a wide range of solution conditions, fluorescence anisotropy is an extremely well-suited method for studying such aspects of protein–DNA interactions. (3) The titration process can readily be automated.

One of the main disadvantages of the technique stems from the instrumental constraints. Fluorescence anisotropy is infrequently used to measure binding constants that are tighter than 1 nM, simply because the anisotropy signal from <1 nM concentration of most fluorophores dips below the detection limit for most commercial fluorometers. This problem is discussed further below in the "Equipment" section.

III. Equipment

The experimental setup for measurement of fluorescence anisotropy involves significant loss of light intensity at numerous points along the optical path. Incident light first passes through a monochromator to select the incident

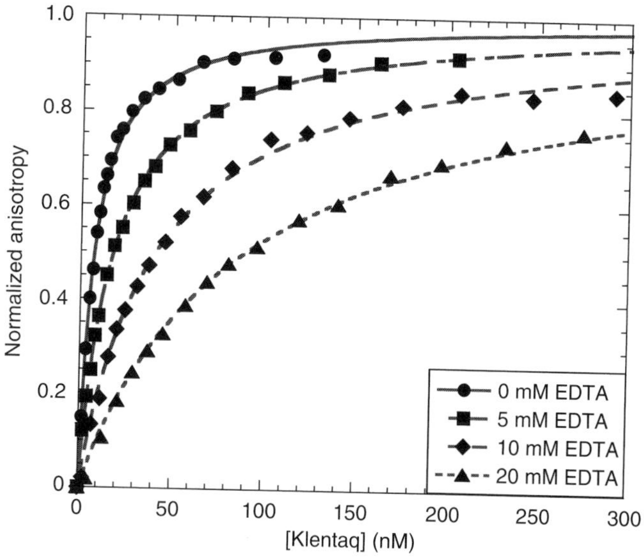

Fig. 3 The effects of EDTA on the binding of Klentaq DNA polymerase to primed-template DNA (13/20-mer DNA). As with Fig. 2, the data illustrate the high precision possible when monitoring binding with fluorescence anisotropy. Equilibrium titrations are shown in the absence of EDTA (●) and in the presence of 5 mM EDTA (■), 10 mM EDTA (◆), and 20 mM EDTA (▲). EDTA, a metal chelator, decreases the affinity of Klentaq polymerase to DNA. In the absence of EDTA, the K_d is 7.5 nM ($\Delta G = -11.1$ kcal/mole). In 5 mM EDTA, the K_d is 18.0 nM ($\Delta G = -10.6$ kcal/mole). In 10 mM EDTA, the K_d is 41.3 nM ($\Delta G = -10.1$ kcal/mole). In 20 mM EDTA, the K_d is 89.9 nM ($\Delta G = -9.6$ kcal/mole). All titrations were performed at 25 °C in 10 mM Tris, 50 mM KCl, at pH 7.9. (Wowor and LiCata, unpublished data).

wavelength(s) (even at the selected wavelength, 30% losses in intensity are not uncommon in a standard monochromator). The incident light then passes through the vertical polarizer. Figure 4 shows a potential arrangement of polarizers in a fluorometer measuring anisotropy. Just as with polarized sunglasses, the loss in intensity through a polarizer is tremendous as all incident light except that in the vertical plane is filtered out by the polarizer. As the light passes through the sample, only those fluorophores with properly aligned absorbance dipoles will absorb the vertically polarized light. This is a very small fraction of the total fluorophore population at any "steady state" instant. Only those fluorophores that absorb can fluoresce, so the total outgoing fluorescence intensity will be significantly lower than if all the fluorophores in the solution had absorbed. Finally, as the light leaves the sample it must pass through another polarizer and usually another monochromator. This loss of intensity at so many steps can mean that, in some cases, a fluorophore that might have a steady state emission intensity of several million photons per second in a particular machine can display a steady state anisotropic emission intensity in either the vertical or horizontal detection

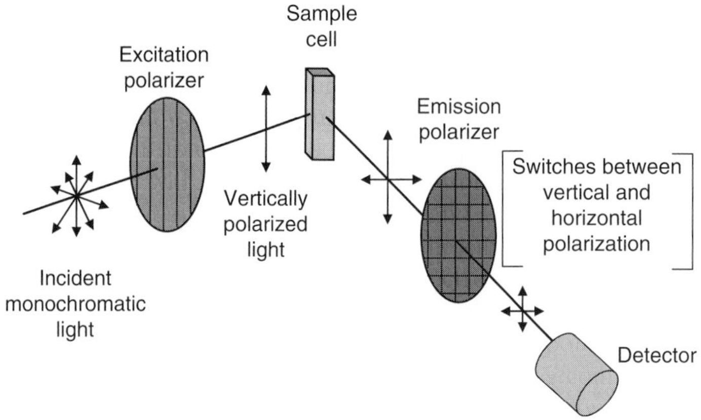

Fig. 4 Schematic of the sample compartment and polarizers in a fluorometer measuring anisotropy. The excitation polarizer remains at vertical at all times, while the emission polarizer switches between vertical and horizontal in order to measure $I_{||}$ and $I_{|}$, respectively. Anisotropy is calculated (automatically in most instruments) using Eq. (2) in the text.

mode [$I_{||}$ and $I_{\perp}$ in Eq. (2), respectively] of only a few thousand photons per second. Thus, common biological fluorophores such as fluorescein- or rhodamine-based dyes, which might give strong standard fluorescence signals at 1 nM concentration, may give no discernable signal at all once the polarizers are placed in the light path.

For this reason, photon-counting fluorometers are the instrument of choice for fluorescence anisotropy relative to analogue fluorometers, due to their enhanced sensitivity to low intensity emission. A machine called the Beacon 2000, designed specifically for this application (fluorescence anisotropic measurements), is also available from Invitrogen. The major modifications in the Beacon 2000, relative to most other commercially available instruments, are as follows: (1) the use of thin-film polarizers and (2) the use of filters instead of monochromators. These changes significantly improve light throughput and allow many common fluorophores to be used at concentrations approaching 10 pM. Standard fluorometers can also be straightforwardly similarly modified but are typically sold with monochromators and more precise Glan-Thompson polarizers. Some machines simultaneously detect vertical and horizontal emission by placing fixed emission polarizers on two sides of the sample. Since these polarizers never move, lower precision polarizers (usually with higher light throughput) can be used and still retain reproducibility. In machines where horizontal and vertical emission are measured by rotating a single polarizer lens, high optical precision/performance (and typically much larger light loss) is usually necessary for high measurement reproducibility. Some researchers have also replaced standard light sources with lasers, laser diodes, or LEDs for higher intensity and which, in the case of laser sources, emit polarized light, thus eliminating the need for an excitation polarizer (Lakowicz *et al.*, 1999; Royer, 2001).

IV. Experimental Design and Performance

Since the DNA is much more easily fluorescently labeled, the usual titration mode for a fluorescence anisotropy experiment is to titrate protein into a solution of labeled DNA. Also, generally, the DNA fragment is smaller than the protein, and thus one will see a larger relative change in rotational diffusion (and hence anisotropy) for the DNA when it forms a complex. The DNA concentration in many published studies is near 1 nM (often the lowest usable concentration, as discussed above), and titrations are performed by adding protein to a partially filled cuvette containing the DNA solution. After each incremental addition of protein, the solution is allowed to stir for several minutes, and then the fluorescence anisotropy is measured. A good plot of anisotropy (A) versus protein concentration has a clear plateau at high protein concentration and contains several points below the K_d value for the reaction. The following sections contain a variety of advice on the individual steps in the experimental procedure.

A. Reagents

The DNA should be as clean as possible. Commercially obtained DNA oligomers should be HPLC or PAGE purified. Generally, it is good to try to label the DNA at a point farthest from the anticipated protein-binding site, to avoid protein interactions with the fluorophore. Most commercial DNA oligomer synthesis companies will attach a fluorophore onto either end of an oligomer, and many fluorophores can be attached in the middle of an oligomer. The reviews by Heyduk *et al.* (1996) and by Waggoner (1995) contain instructions and further references for in-house attachment of different fluorophores to DNA, if commercial preparation is undesirable or unavailable.

A number of different fluorophores have been successfully used. Fluorescein-based dyes remain the popular favorite. Rhodamine-X (ROX), introduced for use in this application by Beechem and associates (Perez-Howard, 1995), has the advantage of displaying a lower tendency to interact directly with the protein. Figure 5 shows the structure of rhodamine-X and its excitation and emission spectra. The complication of protein–fluorophore interactions has been observed for several systems with fluorescein-labeled DNA (see Section IV.G below). Rusinova *et al.* (2002) describe the use of newer Alexa and Oregon Green fluorophores as fluorescent labels in fluorescent anisotropy. Their publication also contains an excellent discussion of the requirements for determining the suitability of a fluorophore for use in the method.

Filtering the reagents is generally a good idea, as light scattering can be a significant problem in fluorescent anisotropy. The baseline fluorescence before and after filtration should be checked, however, as we have found that filtration of some fluorophores actually causes problems in some fluorometric assays. A few control experiments usually quickly help determine an acceptable filter and

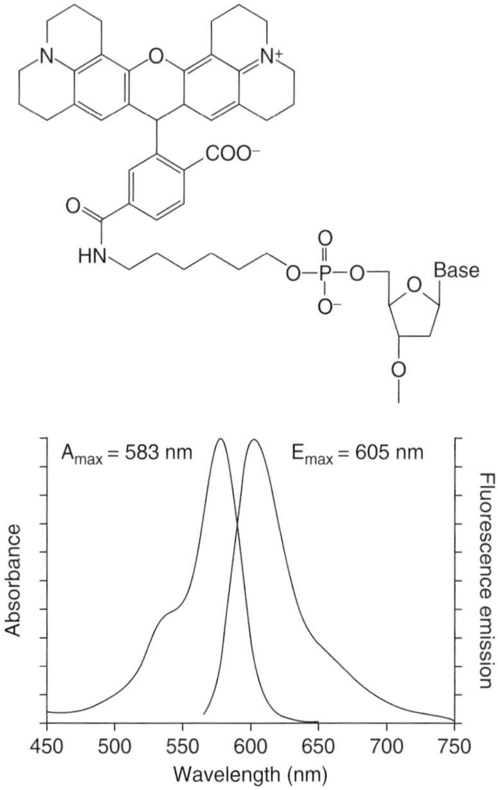

Fig. 5 The structure of ROX shown attached to the α phosphate at the 5′ end of a DNA oligomer (top panel). The bottom panel shows the excitation and emission spectra of the fluorophore.

filtration protocol. Also, as with any fluorophore, fluorescently labeled DNA should be kept away from bright light as much as possible, both during storage and during titration.

It is important that the DNA not be too large, otherwise binding of the protein will not alter its rotational diffusion enough to produce an observable change in the anisotropy signal. Several reviews quote 40 bp as an upper limit on the size of the DNA, but our laboratory has used oligomers in the 70 bp range without problems. Heyduk *et al.* (1996) predict that oligomers up to $\sim 10^5$ Da (about 140 bp) should work based on estimated rotational diffusion rates.

Since most fluorescence anisotropy titrations used to examine protein–DNA interactions involve titrating protein into DNA, it is best if the DNA concentration is far below the K_d value (however, sometimes the detection limit of the fluorometer precludes this). If the DNA concentration is kept far below the K_d value, then even significant errors in the DNA concentration will not propagate into the final data. Any error in the protein concentration, however, will be directly reflected in the K_d value, since the protein concentration comprises the x-axis of the titration. For this reason we typically determine our protein concentrations by two different methods [in our laboratory we use absorbance at 280 nm and the Bradford assay (Bradford, 1976)].

Many enzyme solutions, on long-term storage, especially frozen, will accumulate insoluble particulates, often invisible to the naked eye, but which can significantly interfere with both standard fluorescence measurements and fluorescence anisotropy measurements. A 5-min, full-speed spin in a standard microfuge will usually clear such particulates. The enzyme concentration must be redetermined after such treatment, but often, the enzyme concentration before and after such particulate clearance does not even measurably change.

B. Polarizer Calibration and G-Factor

If the fluorometer being used requires calibration of the polarizers, one should determine the frequency of calibration that works best for one's own machine. The fluorometer manual often provides no guidance on this. For our fluorometer, weekly calibration seems to work best. For many years, nondairy coffee creamer has been used by researchers in the fluorescence and light scattering community as a "perfect scatterer." We prefer an in-house, freshly made glycogen solution, as it seems to provide higher reproducibility of results. A 0.1 g/ml stock solution is added dropwise to a buffer-filled cuvette, with both polarizers set at vertical until we obtain a strong scattering signal from the fluorometer (1.5–2 million photons per second). The polarizers can then be aligned using the procedures specific to each machine.

In some fluorometers a G-factor must be measured and used. For an isotropic or "perfect" scatterer such as glycogen or coffee creamer, the emitted light should be equivalent regardless of whether the emission polarizer is in the vertical or the horizontal position. If this is true for a fluorometer, then the G-factor is 1.0, and need not be included in any calculations of anisotropy. For many fluorometers manufactured within the past decade, the G-factor is 1.0. If, however, these values are not equivalent for a particular machine, the G-factor simply corrects the vertical and horizontal emission values such that they are equivalent for isotropic scatter. It is usually measured with horizontally polarized excitation ($\perp$) and calculated as the ratio of $I_{\parallel}/I_{\perp}$ (Lakowicz, 1999; Lakowicz *et al.*, 1999). When required it enters into the calculation of A as follows:

$$A = \frac{(I_{\parallel} - GI_{\perp})}{(I_{\parallel} + 2GI_{\perp})} \tag{4}$$

C. Sample Compartment Control

Temperature control is extremely crucial during fluorescence anisotropy experiments. Even at "room temperature," it is difficult to obtain high quality data without a circulating water bath or Peltier control of the cuvette temperature. We have even noted decay of data quality when the circulation rate of the water bath decreases because of neglected maintenance.

The sample should be constantly stirred during the experiment (both during and between measurements) although, for most of our in-lab systems, we have noted that turning off the stirrer during the actual measurement (but after the sample has fully mixed for each titrant addition) does not alter the measured anisotropy.

D. Excitation and Emission Parameters

Because so much of both the incident and emitted light is lost in an anisotropy measurement, the setup of the excitation and emission parameters requires some caution and iterative empirical testing. If monochromators are used, the band pass should be opened as widely as possible, to capture as much of the excitation and emission peaks as possible, without risking overlap of the two. Since most steady state fluorescence signals are so strong, the use of band pass widths <1 nm are typical in normal fluorescence measurements, but for anisotropy this is not the case. For example, with rhodamine-X, the excitation peak is at 583 and the emission peak is at 605 nm (see Fig. 5), and we use an 8 nm band pass width around each peak maximum. Similar considerations should be used if choosing band pass filters: letting through as much light as possible without cross contaminating the excitation and emission signals.

Again, because the signal is so low, integration times should be maximized. Instead of the typical 0.1–1 sec integration times used with steady state fluorescence, we use 10 sec integration times with a minimum of five averaged measurements to obtain maximal precision under low signal conditions.

An odd, but experimentally necessary element in fluorescence anisotropy measurements in protein–DNA interactions is the need for an exceedingly long "wait time" after each addition of protein for the anisotropy signal to stabilize. For protein–DNA interactions with nanomolar K_d's the actual time till equilibrium will generally be less than a second. It is typical, however, to wait 4–10 min after each addition of protein before the next anisotropy measurement is taken to achieve maximal precision and stability of the measurement (in our laboratory, 8 min is used). This might be either a mixing effect or a temperature effect. Even in a well-stirred cuvette there is only a slow approach to absolute homogeneity of mixing. Evidence for such an effect can be seen if one adds titrant to the top of the solution in the cuvette versus inserting the pipette farther into the cuvette and adding titrant near the bottom. Additionally, since precise temperature control is so tightly linked to signal precision in anisotropy, adding even small amounts of titrant to the cuvette may necessitate a slow return to the set temperature. Similar "solution settling" effects are seen in dynamic light scattering measurements, and since any stray light scattering will interfere with anisotropy measurements, the exact mechanism for this effect in fluorescence anisotropy may be similar. One can empirically determine the best "wait" time for one's own system, but virtually all published studies simply use a consistent wait time of at least 4 min (Boyer *et al.*, 2000; Heyduk *et al.*, 1996).

E. Data Collection

For any ligand-binding titration, one wants to achieve maximal consistency of spacing of data along the *y*-axis. It is naturally much easier to achieve equal spacing along the independent axis, since one knows how much protein one is adding at each step. However, uniform spacing along the dependent axis is more crucial for successful data analysis. A typical total anisotropy change (ΔA_T) for a protein–DNA interaction might be in the range of 0.1–0.15. One typically wants 15–20 points spanning that range for a single-site binding isotherm. More points may be necessary for more complex multisite binding situations, where subtle curve shape changes need to be accurately quantitated. Because *y*-axis spacing is more important to data analysis, the amount of protein added to the cuvette will typically increase as the titration continues. One must plan for this, sometimes by using two or more "stock" titrant concentrations.

On the *x*-axis, it is important that some points must be below the K_d value, otherwise one is determining a K_d value using data that does not even overlap the K_d. Although this is certainly possible, it significantly reduces the reliability of the K_d determination. Since one generally does not know the actual K_d value when starting a titration, this frequently means that "one" titration will actually involve collecting an iterative set of titrations until a data set is obtained that includes about 3–5 data points below the K_d value and about 4 points on the plateau.

In our laboratory the titration is performed in a 4-ml cuvette, starting with 3 ml of labeled DNA in buffer, and then adding protein until the reaction plateaus or until the capacity of the cuvette is reached. One has to correct the protein concentration at each point by the dilution factor. Generally, however, one does not have to worry about dilution effects on the fluorophore. An alternate strategy is also commonly used, where one removes volume from the cuvette as the titration proceeds, in order to avoid an overfilled cuvette. With this procedure, however, it is easier to dilute the fluorophore to the point where it might become problematic. One diagnostic for such a problem is obtaining a sloped plateau region. A simple background titration (with buffer but no protein) will confirm whether a sloped plateau is a fluorophore dilution problem. If fluorophore dilution seems to be the source of the problem, it can be corrected by simultaneously adding fluorophore with each protein addition so that the fluorophore concentration remains constant. A positively sloped plateau may, however, be indicative of higher order oligomerization or aggregation, while a negatively sloped plateau may be indicative of a contaminating nuclease activity.

F. Data Analysis

We typically normalize all data prior to analysis. This makes it more straightforward to simultaneously graph and compare isotherms from different conditions, where the ΔA_{T} might change slightly. Large changes in ΔA_{T}, however, should not be ignored, as they can be diagnostic of linked processes such as oligomerization of the DNA or protein or large conformational changes. So, for example, one might examine a protein–DNA interaction over an 800 mM salt range and observe a range of ΔA_{T} values between 0.1 and 0.15. If one observes large changes in ΔA_{T} as one changes solution conditions, one should suspect a linked reaction.

Most published studies fit the resultant isotherm (ΔA versus [protein]) to the full quadratic expansion of the binding polynomial derived for total concentrations of reactants:

$$\Delta A = \frac{\Delta A_{\mathrm{T}}}{2D_{\mathrm{T}}}\left\{(E_{\mathrm{T}} + D_{\mathrm{T}} + K_{\mathrm{d}}) - \left[(E_{\mathrm{T}} + D_{\mathrm{T}} + K_{\mathrm{d}})^2 - 4E_{\mathrm{T}}D_{\mathrm{T}}\right]^{1/2}\right\} \tag{5}$$

where ΔA is the change in anisotropy, ΔA_{T} is the total anisotropy change, E_{T} is the total polymerase concentration at each point in the titration, D_{T} is the total DNA concentration, and K_{d} is the dissociation constant (this is a slight rearrangement of the equation as used by Heyduk and Lee).

In many protein-DNA binding titrations, however, the concentration of the DNA is far below the K_{d} value for the reaction. If this is the case, it is easier to use the binding polynomial derived for free reactant concentrations, and assume that $E_{\mathrm{free}} = E_{\mathrm{total}}$

$$\Delta A = \left\{\frac{\Delta A_{\mathrm{T}}(E/K_{\mathrm{d}})}{(1 + E/K_{\mathrm{d}})}\right\} \tag{6}$$

where ΔA is the change in fluorescence anisotropy, ΔA_{T} is the total change in anisotropy, E is the total polymerase concentration at each point in the titration, and K_{d} is the dissociation constant for polymerase-DNA binding. If the [DNA] is $10\times$ lower than the K_{d} value, the error incurred by using this equation and making this approximation is 10%. If the [DNA] is $100\times$ below the K_{d} value, the incurred error is 1%, and so on. One advantage of using Eq. (6) is that it is easily modified to include a Hill coefficient to test for cooperative/multisite binding:

$$\Delta A = \left\{\frac{\Delta A_{\mathrm{T}}(E^{nH}/{K_{\mathrm{d}}}^{nH})}{(1 + E^{nH}/{K_{\mathrm{d}}}^{nH})}\right\} \tag{7}$$

where nH is the fitted Hill coefficient. It is also easily modified for competitive binding.

A particular hazard of using Eq. (5) is its too frequent use to obtain K_d values for a binding reaction under stoichiometric binding conditions. If the concentrations of both reagents (protein and DNA) are far above the K_d value for their association, the binding is stoichiometric. When Eq. (5) is used to analyze stoichiometric binding curves, both D_T and K_d are allowed to vary during the nonlinear regression, and the binding stoichiometry is determined as the ratio of the fitted D_T and the known D_T (see Fig. 6). While determining the stoichiometry of the reaction, as in Fig. 6, is an important control, the K_d values obtained from fits to stoichiometric data are frequently unreliable. This is because even small errors in the concentrations of the reactants (D_T and E_T) are propagated into large errors on K_d. For example, if the concentrations of reactants are in the 10 μm range, and the true K_d is in the nanomolar range, and the error on determining the protein concentration is a standard $\pm 10\%$, then the error on the fitted K_d is not $\pm 10\%$ of the K_d, it is $\pm 10\%$ of the protein concentration. So one might obtain a fitted K_d of 10 nM, but the true error on that value is $\pm$ several micromolar. Error in the DNA concentration propagates into the K_d in the same manner. Variants of Eq. (5) are frequently used to obtain K_d values under stoichiometric titration conditions in titration calorimetry. Such K_d values must be eyed with extreme caution in that technique as well. Under conditions where the reactant concentrations are near the K_d value, or more typically where the concentration of one reactant is below the K_d value and the other is titrated through the K_d value, Eq. (5) is perfectly applicable. This

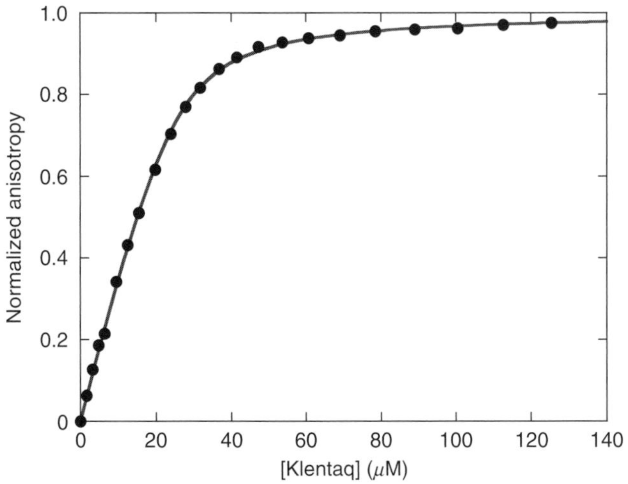

Fig. 6 Determination of binding stoichiometry of Klentaq polymerase to double-stranded DNA (63/63-mer). The titration was performed at 25 °C in 10 mM Tris, 5 mM $MgCl_2$, and 75 mM KCl at pH 7.9. The DNA concentration used in the titration was 20 μM ([DNA] $\gg K_d$). The binding constant for Klentaq polymerase binding to 63/63-mer under "equilibrium titration" conditions is 29.2 ± 1.8 nM. The data were fit to Eq. (5) in the text. The ratio of bound Klentaq polymerase to the 63/63-mer double stranded DNA is 1.2 (Wowor and LiCata, unpublished data).

section has discussed only the most straightforward approaches for analyzing single-site or simple multisite isotherms. For a detailed discussion of more complex analyses, the reader is directed to reviews such as those by Lohman and Bujalowski (1991) and by Bujalowski (2006).

G. Other Controls

Ideally, there should be no change in steady state fluorescence when the labeled DNA binds protein. This is easy to test and affords assurance that one is monitoring DNA-protein binding and not protein–dye interactions. Changes in steady state fluorescence could be either due to protein–dye interactions or due to propagated conformational changes in the DNA on binding. One does not want to be in the situation of studying binding to the fluorophore instead of binding to the DNA. One way to troubleshoot this possibility is to label the DNA with different dyes. Often a fluorophore can be found that shows an anisotropy change on complex formation but does not show a steady state fluorescence change. Alternately, if the same DNA labeled with several chemically different fluorophores yield the same results (i.e., the same K_d), it is highly likely that one is observing protein–DNA interactions and not protein–dye interactions.

We have found that the storage life of fluorescently labeled DNA oligomers can vary from several months to several years. The most reliable diagnostic of a problem with a stock of labeled DNA is a change in the initial fluorescence anisotropy (before addition of any protein). If the initial anisotropy changes by more than about 20% from when the labeled DNA was first used, it probably should be discarded.

Another useful control, but one often not mentioned in published studies, is a test of the ability of unlabeled DNA to compete effectively with the fluorescently labeled DNA. This test can be performed as a stoichiometric titration or an equilibrium titration. In the stoichiometric competition, labeled DNA is supplemented by exactly the same concentration of unlabeled DNA, and then this mixture is titrated with protein. If the protein binds both DNAs, the apparent stoichiometric breakpoint will exactly double. See Datta and LiCata (2003) for an example of this control. This unlabeled DNA competition control can also be performed as an equilibrium titration. Figure 7 shows an example where unlabeled DNA is titrated into an equilibrium of labeled DNA + protein that is at its K_d value. The fitted K_I for the unlabeled DNA should be similar to the previously determined K_d for the labeled DNA. There are numerous published studies where these simple unlabeled DNA competition controls are either not performed or not mentioned. One risks studying the binding of protein to the fluorophore instead of the DNA in such cases. For example, in the case of the DNA polymerases studied in our laboratory, fluorescein-labeled DNA is not equivalently displaced by unlabeled DNA, whereas ROX-labeled DNA is (Datta and LiCata, 2003).

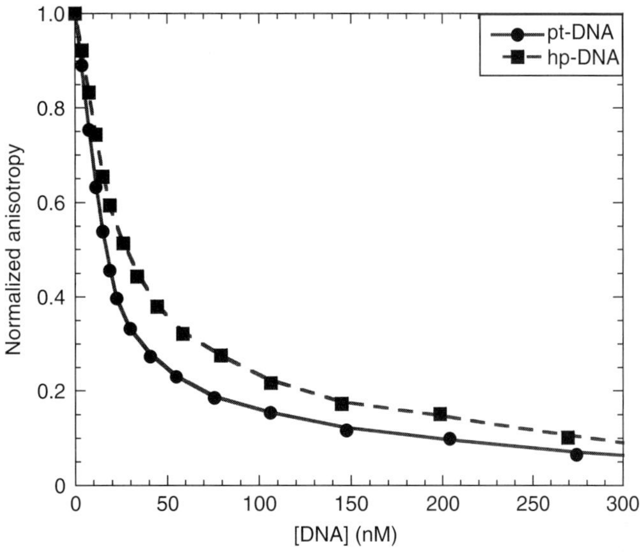

Fig. 7 Fluorescently labeled primed-template DNA (a 13/20-mer labeled with ROX) being displaced from Klentaq DNA polymerase by identical, but unlabeled, primer-template DNA (circles) and an unlabeled hairpin DNA structure (squares). The competition titrations were performed at 25 °C in 10 mM Tris and 5 mM $MgCl_2$ at pH 7.9. The DNA sequences of the primer-template DNA (pt-DNA) and the hairpin DNA (hp-DNA) are very similar. The cuvette initially contained 1-nM-labeled primer-template DNA (13/20-mer DNA) plus Klentaq polymerase at the K_d. The unlabeled competitor DNA was then titrated into the cuvette. Additional Klentaq polymerase and 1-nM-labeled primer-template DNA are included in each addition so that their concentrations remain constant throughout the titration. K_I values obtained from fits of the data are 17.5 nM ($\Delta G = -10.6$ kcal/mole) for the unlabeled primed-template and 29.1 nM ($\Delta G = -10.25$ kcal/mole) for the hairpin DNA (Wowor and LiCata, unpublished data).

H. Competition Experiments

A natural extension of the unlabeled DNA competition controls described above is the ability to use the fluorescence anisotropy assay in competitive mode to measure the K_d's of a series of unlabeled DNA oligomers for a protein. In this application, the cuvette initially contains 1 nM labeled DNA, plus protein either at its K_d value (50% saturation) or just at saturation ($\bar{Y} \approx 0.95$). The unlabeled competitor DNA is then titrated into the cuvette. Additional protein and 1 nM labeled pt-DNA are included in each titrant addition so that their concentrations remain constant at all times. The anisotropy will decrease as the unlabeled DNA competes with labeled DNA to bind the protein. Figure 7 also shows an example of such competitive binding experiments, where labeled DNA is displaced by an unlabeled oligomer with a different sequence or structure. The method of analysis depends on the exact procedure used (i.e., whether protein was at the K_d value or near saturation with the labeled DNA, whether labeled DNA and protein concentrations maintained constant, and so on.). Discussions of experimental design and analysis of competitive binding experiments can be found in Zetner (1973) and Jezewska and Bujalowski (1996).

V. Other Applications of Fluorescence Anisotropy to the Study of Protein–DNA Interactions

Several groups have reported the use of fluorescence anisotropy in a high throughput mode for use in drug screening by adapting the method for use in fluorescent plate readers (Burke *et al.*, 2003; Parker *et al.*, 2000; Rishi *et al.*, 2005; Wang *et al.*, 2004). DNA as well as other small, fluorescently labeled ligands have been used as the anisotropic probe in such assays.

Time-resolved anisotropy has also been used, mostly by Millar and associates (Bailey *et al.*, 2001; Guest *et al.*, 1991; Millar, 2000) to study protein–DNA interactions. In this application, decay of the anisotropy is monitored versus time after a single pulse of polarized light, similar to the way one might perform a fluorescence lifetime experiment. Deviations from a single exponential decay can be indicative of multiple binding modes. Resolution of the number of different anisotropic decays observed, and their relative fractions of the total decay, can provide information on the relative populations of the different binding modes or mixed conformer subpopulations. This potentially promising application is only beginning to see widespread use.

A relatively new application of fluorescence anisotropy to ligand binding is its adaptation into solid phase assays (McCauley *et al.*, 2003). Since DNA or RNA that is immobilized at one end can still rotate and pivot, one can still obtain an increase in anisotropy signal if a protein binds to the immobilized nucleic acid. This new adaptation is already seeing widespread application.

References

Bailey, M. F., Thompson, E. H., and Millar, D. P. (2001). Probing DNA polymerase fidelity mechanisms using time-resolved fluorescence anisotropy. *Methods* **25,** 62–77.

Boyer, M., Poujol, N., Margeat, E., and Royer, C. A. (2000). Quantitative characterization of the interaction between purified human estrogen receptor alpha and DNA using fluorescence anisotropy. *Nucleic Acids Res.* **28,** 2492–2502.

Bradford, M. M. (1976). A rapid and sensitive method for the quantitation of microgram quantities of protein utilizing the principle of protein-dye binding. *Anal. Biochem.* **72,** 248–254.

Brown, M. P., and Royer, C. (1997). Fluorescence spectroscopy as a tool to investigate protein interactions. *Curr. Opin. Biotechnol.* **8,** 45–49.

Bujalowski, W. (2006). Thermodynamic and kinetic methods of analyses of protein-nucleic acid interactions. From simpler to more complex systems. *Chem. Rev.* **106,** 556–606.

Burke, T. J., Loniello, K. R., Beebe, J. A., and Ervin, K. M. (2003). Development and application of fluorescence polarization assays in drug discovery. *Comb. Chem. High Throughput Screening* **6,** 183–194.

Cantor, C. R., and Schimmel, P. R. (1980). "Biophysical Chemistry," pp. 454–459. W.H. Freeman and Company, San Francisco, CA.

Chin, J., Langst, G., Becker, P. B., and Widon, J. (2004). Fluorescence anisotropy assays for analysis of ISWI-DNA and ISWI-nucleosome interactions. *Methods Enzymol.* **376,** 3–16.

Datta, K., and LiCata, V. J. (2003). Salt dependence of DNA binding by *Thermus aquaticus* and *Escherichia coli* DNA polymerases. *J. Biol. Chem.* **278,** 5694–5701.

Datta, K., Wowor, A. J., Richard, A. J., and LiCata, V. J. (2006). Temperature dependence and thermodynamics of Klenow polymerase binding to primed-template DNA. *Biophys. J.* **90,** 1739–1751.

Eftink, M. R. (1997). Fluorescence methods for studying equilibrium macromolecule-ligand interactions. *Methods Enzymol.* **278,** 221–257.

Fried, M., and Crothers, D. M. (1981). Equilibria and kinetics of lac repressor-operator interactions by polyacrylamide gel electrophoresis. *Nucleic Acids Res.* **9,** 6505–6525.

Garner, M. M., and Revzin, A. (1981). A gel electrophoresis method for quantifying the binding of proteins to specific DNA regions: Application to components of the *Escherichia coli* lactose operon regulatory system. *Nucleic Acids Res.* **9,** 3047–3060.

Guest, C. R., Hochstrasser, R. A., Dupuy, C. G., Allen, D. J., Benkovic, S. J., and Millar, D. P. (1991). Interaction of DNA with the Klenow fragment of DNA-polymerase-I studied by time-resolved fluorescence spectroscopy. *Biochemistry* **30,** 8759–8770.

Hey, T., Lipps, G., and Krauss, G. (2001). Binding of XPA and RPA to damaged DNA investigated by fluorescence anisotropy. *Biochemistry* **40,** 2901–2910.

Heyduk, T., and Lee, J. C. (1990). Application of fluorescence energy transfer and polarization to monitor *Escherichia coli* cAMP receptor protein and lac promoter interaction. *Proc. Natl. Acad. Sci. USA* **87,** 1744–1748.

Heyduk, T., Ma, Y. X., Tang, H., and Ebright, R. H. (1996). Fluorescence anisotropy: Rapid, quantitative assay for protein-DNA and protein-protein interaction. *Methods Enzymol.* **274,** 492–503.

Hill, J. J., and Royer, C. A. (1997). Fluorescence approaches to study of protein-nucleic acid complexation. *Methods Enzymol.* **278,** 390–416.

Jameson, D. M., and Sawyer, W. H. (1995). Fluorescence anisotropy applied to biomolecular interactions. *Methods Enzymol.* **246,** 283–300.

Jezewska, M. J., and Bujalowski, W. (1996). A general method of analysis of ligand binding to competing macromolecules using the spectroscopic signal originating from a reference macromolecule. Application to *Escherichia coli* replicative helicase DnaB protein-nucleic acid interactions. *Biochemistry* **35,** 2117–2128.

Lakowicz, J. R. (1999). "Principles of Fluorescence Spectroscopy," 2nd edn. Kluwer Academic/Plenum Publishers, New York.

Lakowicz, J. R., Gryczynski, I., Gryczynski, Z., and Dattelbaum, J. D. (1999). Anisotropy-based sensing with reference fluorophores. *Anal. Biochem.* **267,** 397–405.

Lohman, T. M., and Bujalowski, W. (1991). Thermodynamic methods for model-independent determination of equilibrium binding isotherms for protein-DNA interactions: Spectroscopic approaches to monitor binding. *Methods Enzymol.* **208,** 258–290.

Lundblad, J. R., Laurance, M., and Goodman, R. H. (1996). Fluorescence polarization analysis of protein-DNA and protein-protein interactions. *Mol. Endocrinol.* **10,** 607–612.

McCauley, T. G., Hamaguchi, N., and Stanton, M. (2003). Aptamer-based biosensor arrays for detection and quantification of biological macromolecules. *Anal. Biochem.* **319,** 244–250.

Millar, D. P. (2000). Time-resolved fluorescence methods for analysis of DNA-protein interactions. *Methods Enzymol.* **323,** 442–459.

Parker, G. J., Law, T. L., Lenoch, F. J., and Bolger, R. E. (2000). Development of high throughput screening assays using fluorescence polarization: Nuclear receptor-ligand-binding and kinase/phosphatase assays. *J. Biomol. Screen.* **5,** 77–88.

Perez-Howard, G. M., Weil, P. A., and Beehem, J. M. (1995). Yeast TATA binding protein interaction with DNA: Fluorescence determination of oligomeric state, equilibrium binding, on-rate, and dissociation kinetics. *Biochemistry* **34,** 8005–8017.

Riggs, A., Suzuki, H., and Bourgeois, S. (1970). Lac repressor-operator interaction. I. Equilibrium studies. *J. Mol. Biol.* **48,** 67–83.

Rishi, V., Potter, T., Lauderman, J., Reinhart, R., Silvers, T., Selby, M., Stevenson, T., Krosky, P., Stephen, A. G., Acharya, A., Moll, J., Oh, W. J., *et al.* (2005). A high-throughput fluorescence-anisotropy screen that identifies small molecule inhibitors of the DNA binding of B-ZIP transcription factors. *Anal. Biochem.* **340,** 259–271.

Royer, C. A. (1995). Quantitative detection of macromolecules with fluorescent nucleotides. US Patent # 5,445,935.

Royer, C. A. (1998). Quantitative detection of macromolecules with fluorescent oligonucleotides. US Patent # 5,756,292.

Royer, C. A. (2001). Quantitative detection of macromolecules with fluorescent oligonucleotides. US Patent # 6,326,142.

Rusinova, E., Tretyachenko-Ladokhina, V., Vele, O. E., Senear, D. F., and Ross, J. B. A (2002). Alexa and Oregon Green dyes as fluorescence anisotrophy probes for measuring protein-protein and protein-nucleic acid interactions. *Anal. Biochem.* **308,** 18–25.

Waggoner, A. (1995). Covalent labeling of proteins and nucleic acids with fluorophores. *Methods Enzymol.* **246,** 362–373.

Wang, S. Y., Ahn, B. S., Harris, R., Nordeen, S. K., and Shapiro, D. J. (2004). Fluorescence anisotropy microplate assay for analysis of steroid receptor-DNA interactions. *Biotechniques* **37,** 807–817.

Zetner, A. (1973). Principles of competitive binding assays (saturation analyses). I. Equilibrium techniques. *Clin. Chem.* **19,** 699–705.

CHAPTER 10

Circular Dichroism and Its Application to the Study of Biomolecules

Stephen R. Martin* and Maria J. Schilstra†

*Division of Physical Biochemistry
MRC National Institute for Medical Research
The Ridgeway, Mill Hill, London NW7 1AA
United Kingdom

†Biological and Neural Computation Group
Science and Technology Research Institute
University of Hertfordshire, College Lane, Hatfield AL10 9AB
United Kingdom

0091-679X/08 $35.00
DOI: 10.1016/S0091-679X(07)84010-6

Abstract

Circular dichroism (CD) is an excellent method for the study of the conformations adopted by proteins and nucleic acids in solution. Although not able to provide the beautifully detailed residue-specific information available from nuclear magnetic resonance (NMR) and X-ray crystallography, CD measurements have two major advantages: they can be made on small amounts of material in physiological buffers and they provide one of the best methods for monitoring any structural alterations that might result from changes in environmental conditions, such as pH, temperature, and ionic strength. This chapter describes the important basic steps involved in obtaining reliable CD spectra: careful instrument and sample preparation, the selection of appropriate parameters for data collection, and methods for subsequent data processing. The principal features of protein and nucleic acid CD spectra are then described, and the main applications of CD are discussed. These include: methods for analyzing CD data to estimate the secondary structure composition of proteins, methods for following the unfolding of proteins as a function of temperature or added chemical denaturants, the study of the effects of mutations on protein structure and stability, and methods for studying macromolecule–ligand and macromolecule–macromolecule interactions.

I. Introduction

Circular dichroism (CD), the differential absorption of left- and right-handed circularly polarized light, is a spectroscopic property uniquely sensitive to the conformation of molecules, and so has been very widely used in the study of biomolecules. CD often provides important information about the function and conformation of biomolecules that is not directly available from more conventional spectroscopic techniques, such as fluorescence and absorbance. The experimentally measured parameter in CD is the *difference* in absorbance for left- and right-handed circularly polarized light, ΔA ($=A_L - A_R$). Because CD is an absorption phenomenon, the chromophores that contribute to the CD spectrum are exactly the same as those contributing to a conventional absorption spectrum. In order to show a CD signal, a chromophore must be either inherently chiral (asymmetric) or must be located in an asymmetric environment. Chromophores that contribute to protein CD (see below) are generally achiral because they have a plane of symmetry. It is the interaction between the chromophores in the chiral field of the protein that introduces the perturbations leading to optical activity (Sreerama and Woody, 2004a).

The near-UV CD bands of proteins (310–255 nm) derive from Trp, Tyr, Phe, and cystine (Note: cysteine does not absorb in this region) and reflect the tertiary, and occasionally quaternary, structure of the protein. Although several amino acid side chains (notably Tyr, Trp, Phe, His, and Met) absorb light strongly in the far-UV region of the spectrum (below 250 nm), the most important contributor here is the peptide bond (amide chromophore), with $n \rightarrow \pi^*$ and $\pi \rightarrow \pi^*$ transitions at ~220 and ~190 nm, respectively. The far-UV CD bands of proteins reflect the secondary structure of the protein (α-helix, β-sheet, β-turn, and unordered content). In the case of nucleic acids and oligonucleotides, the aromatic bases are the principal chromophores, with absorption beginning at around 300 nm and extending far into the vacuum UV region. The electronic transitions of the ether and hydroxyl groups of the sugars begin at 200 nm, but their intensity is much weaker than that of the bases, and the electronic transitions of the phosphate groups begin further still into the vacuum UV (Johnson, 1996a).

Although CD spectroscopy generally provides only low-resolution structural information, it does have two major advantages. First, it is extremely sensitive to *changes* in conformation, whatever their origin, and second, an extremely wide range of solvent conditions is accessible to study with relatively small amounts of material. The principal applications of CD spectroscopy in the study of biomolecules are

a. The estimation of protein secondary structure *content* from far-UV CD spectra

b. The detection of conformational changes in proteins and nucleic acids brought about by changes in pH, salt concentration, and added cosolvents (simple alcohols, tri-fluoroethanol, and so on), and the structural analysis of recombinant native proteins and their mutants

c. Monitoring protein or nucleic acid unfolding brought about by changes in temperature or by the addition of chemical denaturants (such as urea and guanidine hydrochloride)

d. Monitoring protein–ligand, protein–nucleic acid, and protein–protein interactions

e. Studying (in favorable cases) the kinetics of macromolecule–macromolecule, macromolecule–ligand interactions (particularly slow dissociation processes), and the kinetics of protein folding reactions. The general principles of the most common kinetic methods are discussed in Chapter 15 by Eccleston *et al.*, this volume and will not be considered in detail here

There are numerous excellent reviews that describe the basic principles of CD spectroscopy and its applications in the study of different biomolecules (Bishop and Chaires, 2002; Gray, 1996; Greenfield, 1996, 2004; Johnson, 1985, 1988, 1990, 1996a; Kelly *et al.*, 2005; Strickland, 1974; Venyaminov and Yang, 1996; Woody, 1985, 1995, 1996; Woody and Dunker, 1996; Yang *et al.*, 1986).

II. Instrumentation and Sample Preparation

A. Instrumentation

CD instruments are commercially available from several sources: Jasco Inc. (http://www.jascoinc.com/), Aviv Biomedical Inc. (http://www.avivbiomedical.com/), OLIS Inc. (http://www.olisweb.com/), and Applied Photophysics (http://www.photophysics.com/). A Peltier system for temperature control and thermal ramping is an invaluable accessory, particularly for studies of the thermal unfolding of proteins and nucleic acids. The only other significant requirement is for a set of high-quality quartz cuvettes with good far-UV transmission (either rectangular or cylindrical) with path lengths ranging from 0.1 to 10 mm. Self-masking (black-walled) micro- or semimicro cuvettes with 10 mm path length are particularly useful for near-UV CD measurements with small volumes (~0.25 ml). Cuvettes are obtainable from several suppliers (e.g., Hellma; http://www.hellma-worldwide.de/). Cuvettes with path lengths of less than 1 mm should always be calibrated. This is easily done by using the cuvette to record a conventional absorption spectrum of any solution with accurately known absorbance, or as described by Johnson (1996b). Cuvettes may have some intrinsic strain that can give significant CD artifacts, and although moderate strain can be tolerated, it is probably sensible to eliminate any possible strain effects by always orienting the cuvette the same way in the instrument. Cuvettes should always be cleaned immediately after use in order to avoid the buildup of hard-to-remove protein deposits. This can be done using a preparation such as Hellmanex II cuvette cleaning solution. After cleaning, the cuvettes should be rinsed extensively with distilled water, then ethanol, and dried using an air pump or by evaporation. Cuvettes should be stored in the cases generally provided by the manufactures. All standard reagents used should be of the highest purity available. It is particularly important that any organic solvents used should be of spectroscopic purity and should be checked for the absence of absorbing impurities.

B. Instrument Care and Calibration

The CD instrument should always be purged with high-purity, oxygen-free, nitrogen (generally run at ~3–5 l/min) for at least 20 min before starting the light source and throughout the measurements. If oxygen is present, it may be converted to ozone by the far-UV light from the high-intensity arc, and ozone will damage the expensive optical surfaces. Higher nitrogen flow rates will generally be necessary for measurements made at very short wavelengths.

The calibration of the instrument should be checked periodically. Although several CD standards are available, the one used most frequently is d10 camphor sulfonic acid (d10-CSA). The exact concentration (C) of a solution of d10-CSA in water (at ~2.5 mM) should be determined from an absorption spectrum (using $\varepsilon_{285} = 34.5\ \mathrm{M}^{-1}\ \mathrm{cm}^{-1}$) and not by weight because the solid is hygroscopic.

The differential absorption (ΔA) recorded at 290.5 nm in a 10 mm path length cuvette should be $C\Delta\varepsilon_{M,290.5}$ (or 32,980$C\Delta\varepsilon_{M,290.5}$ millidegrees—see Section IV.A.), where $\Delta_{M,290.5} = 2.36\ M^{-1}\ cm^{-1}$ (Johnson, 1990). If the intensity is not within 1% of the expected value, the user should refer to the manufacturer's handbook for details of the appropriate adjustment procedure. It is also advisable to check the wavelength calibration of the instrument and its general transmission performance in the far-UV from time to time. Because d10-CSA has a second CD band at 192.5 nm ($\Delta\varepsilon_{M,192.5} = -4.72\ M^{-1}\ cm^{-1}$), one can check the far-UV performance by recording the spectrum of the standard d10-CSA solution using a 1 mm path length cuvette. If the (absolute) value of the ratio of the intensities of the two d10-CSA peaks is significantly less than 1.95, then the machine is no longer performing correctly; this is probably due to the age of the lamp and/or degradation of the mirrors, most probably the first mirror after the light source. This measurement also provides a useful check on the wavelength calibration of the instrument, although this can also simply be done by scanning with a holmium oxide filter in the light path and monitoring the voltage on the instrument's photomultiplier.

C. Sample Preparation

All samples should, of course, be of the highest possible purity. Misleading results can be obtained even with relatively low levels of impurities if these have strong CD signals. For example, the weak near-UV signals of proteins can be swamped by the strong signals from relatively small levels of contaminating nucleic acids. One major problem in CD measurements is that the signals become seriously distorted if too little light reaches the photomultiplier and, in practical terms, this means that one cannot make reliable measurements on samples with an absorbance (sample *plus* solvent) much greater than about 1. The absorption spectrum of the sample should always be checked to see if and where this absorbance limit is going to be exceeded. In far-UV measurements the absorbance of the sample itself is generally rather small, and the major problems arise from absorption by buffer components, almost all of which will limit far-UV penetration to some extent. The majority of simple buffer components will generally permit CD measurements to below 200 nm (see Johnson, 1996b). However, high concentrations of chloride and (especially) nitrate (use perchlorate or fluoride salts where possible), certain solvents (dioxane, DMSO), high concentrations (>25 mM) of some biological buffers (Hepes, Pipes, Mes), high concentrations (>0.25 mM) of common chelators (EGTA/EDTA), and high concentrations (>1 mM) of reducing agents (dithiothreitol and 2-mercaptoethanol) should be avoided whenever possible. It is also worth noting that distilled water stored in a polyethylene bottle will generally develop poor far-UV transparency owing to the presence of eluted polymer additives. The CD spectra of membrane proteins are often recorded in detergent solubilized form in order to avoid artifacts arising from differential light scattering and absorption flattening (Fasman, 1996).

The far-UV CD spectra of proteins (260–178 nm) are intense, and relatively small amounts of material are required to record them. Because all peptide bonds contribute to the observed spectrum, the amount of material required (measured in mg/ml) is effectively the same for any protein. Measurements are almost invariably made in short path length cuvettes in order to reduce absorption by buffer components (see above). Typical quantities are 200 μl of a 0.1–0.15 mg/ml solution when using 1-mm path length cuvette or 30 μl of a 1.0–1.5 mg/ml solution when using a 0.1-mm (demountable) cuvette. The latter is preferable for good far-UV penetration, but the material is not generally recoverable.

The near-UV CD spectra of proteins are generally more than an order of magnitude weaker than the far-UV CD spectra. Recording them therefore requires more concentrated material and/or longer optical path lengths. Spectra are usually recorded under conditions similar to those used for measuring a conventional absorption spectrum, for example, use a 10-mm path length cuvette and aim for a peak absorbance in the range 0.7–1.0 (The optimal absorbance for best signal to noise ratio (S/N) in a CD measurement is, in fact, 0.869—see Johnson, 1996b). Less concentrated solutions may be used if the CD signals are intense and more concentrated samples may, of course, be examined using short path length cuvettes when necessary. The near-UV CD spectra of nucleic acids, which are significantly stronger than those of proteins, should also be recorded with a peak absorption in the range 0.7–1.0.

D. Determination of Sample Concentration

Accurate sample concentrations are absolutely essential for the analysis of far-UV CD spectra for secondary structure content and whenever one wishes to make meaningful comparisons between different protein or nucleic acid samples. Lowry's or Bradford's analyses for proteins are not sufficiently accurate for use with CD measurements unless they have been very carefully calibrated for the protein under investigation using concentrations determined with a more direct method, such as quantitative amino acid analysis. We routinely determine protein and nucleic acid concentrations using absorption spectroscopy. When the extinction coefficient is known, the concentration can be calculated with considerable accuracy. The absorption spectrum should ideally be recorded with temperature control and careful attention should be given to correct baseline subtraction, especially when buffers containing reducing agents are being used. Highly scattering samples should always be clarified by low-speed centrifugation or filtration prior to concentration determination. If the spectrum still shows significant light scattering, that is, significant background absorption above ~315 nm, a correction must be applied. In most cases, it is reasonable to assume that the scattering is Rayleigh in nature and that the absorbance due to scatter is then proportional to λ^n (where the exponent n is generally close to 4). The light scattering contribution to be subtracted at 280 nm, for example, would then be $(A_{350\text{ nm}})(350/280)^4 = 2.442 \times A_{350\text{ nm}}$. A more elaborate method, easily implemented in a spreadsheet program, is to plot $\ln(A_\lambda)$ against $\ln(\lambda)$ for $\lambda > 315$ nm and perform a least-squares fit to the straight line. This method has the advantage that significant deviations from linearity may indicate the presence of contaminants rather than light scattering, and the actual value of the wavelength exponent (n) can be calculated. The scattering contribution for the full wavelength range can then be constructed and subtracted from the measured spectrum.

Although it is possible to calculate the extinction coefficient of a protein with reasonable accuracy from its aromatic amino acid content (Gill and von Hippel, 1989; Pace *et al.*, 1995), it is, of course, much more reliable to measure it. This is best done using the Edelhoch method (see Pace *et al.*, 1995). Make identical dilutions of the protein stock in the experimental buffer and in the same buffer containing 6 M guanidine hydrochloride and record absorption spectra with appropriate buffer subtraction. Correct for light scattering if necessary and measure the absorbance at the chosen wavelength. Then, for example, the extinction coefficient at 280 nm is calculated from the amino acid composition as (Pace *et al.*, 1995):

$$\varepsilon_{280,\mathrm{buffer}} = \frac{(A_{280,\mathrm{buffer}})(\varepsilon_{280,\mathrm{GuHCl}})}{(A_{280,\mathrm{GuHCl}})} \tag{1}$$

$$\text{where } \varepsilon_{280,\mathrm{GuHcl}}(\mathrm{M}^{-1}\ \mathrm{cm}^{-1}) = (\#\mathrm{Trp})(5685) + (\#\mathrm{Tyr})(1285) + (\#\mathrm{cystine})(125) \tag{2}$$

Whatever method of concentration determination is used, it is probably wise in any subsequent analyses to assume that it might be in error by up to 5%.

III. Data Collection

Having chosen the appropriate sample concentration and cuvette path length for the measurement (see Section II.C), the user will need to select suitable instrument settings. In addition to choosing the appropriate wavelength range for the measurement, it will generally be necessary to select scanning speed, instrumental time constant (or response time), spectral band width, and number of scans to be averaged. Consideration should also be given to the selection of an appropriate temperature for the measurement.

A. Wavelength Range

Far-UV spectra of proteins should generally be scanned from 260 nm to the lowest attainable wavelength. This low-wavelength limit will depend largely on the composition of the buffer being used (see Section II.C). Near-UV spectra are routinely scanned over the range 340–255 nm for proteins and from 340 nm to the lowest attainable wavelength for nucleic acids.

B. Scanning Speed and Time Constant (or Response Time)

In the case of analogue instruments, the product of the scanning speed (nm min^{-1}) and the time constant (sec) should be less than 20 nm min^{-1} sec. If the instrument uses a response time (equal to three time constants), then the product of scanning speed and response time should be less than 60 nm min^{-1} sec (Johnson, 1996b). If significantly higher values are used, there will be potentially serious errors in both the positions and intensities of the observed CD bands. Several good general discussions of the potential sources of errors in CD measurements have been published (Hennessey and Johnson, 1982; Johnson, 1996b; Kelly *et al.*, 2005; Manning, 1989). The S/N of a CD measurement is proportional to the square root of the number of scans and to the square root of the time constant. Good S/N ratios can therefore be achieved either by averaging multiple fast scans recorded with a short time constant or by recording a small number of slow scans with a long time constant. The choice is largely one of personal preference.

C. Spectral Bandwidth

The spectral bandwidth is generally set to 1 nm, but it may occasionally be necessary to use lower values in order to resolve fine structure in near-UV spectra of proteins. Increasing the spectral bandwidth will reduce the noise by increasing light throughput, but it should always be 2 nm or less in order to avoid distorting the spectrum.

D. Temperature Control

It is good practice to always record CD spectra with temperature control. This is particularly important for the far-UV CD spectra of proteins, which often show quite pronounced temperature dependence, even outside the range of any thermally induced unfolding of the protein (see Section V.C.2). These small changes in the signal from the folded protein with temperature reflect a true change in conformation and are not simply due to changes in the optical properties of a helix or strand. The changes, which are often linear with temperature, are probably due to fraying of the ends of a helix or to changes in helix–helix interactions (Greenfield, 2004).

Having set the parameters described above, it is advisable to run a single scan in order to check that the selection is appropriate. CD spectra will be seriously distorted if the photomultiplier voltage rises above a certain limit, generally of the order of 600 V. The low-wavelength limit selected for far-UV spectra should be reset to a higher value if this photomultiplier voltage limit is exceeded. If the voltage is too high in the near-UV region, either the sample concentration or the cuvette path length should be reduced. The near-UV CD spectra of proteins generally have no significant intensity in the 315–340-nm region. If there is a significant signal in this region (often becoming increasingly negative toward lower wavelengths), this may indicate that there is a disulfide contribution to the spectrum (see Section IV.B.1). The upper wavelength limit for the scan should then be extended (400 nm is generally sufficient in most cases) in order to allow for correct baseline alignment (see Section IV.A).

The full measurement should then be made with enough repeat scans to increase the S/N ratio to acceptable levels. If necessary, for example in performing a titration, make the required additions to the cuvette and repeat the measurement. Making additions (especially to short path length cuvettes) poses several problems. The small volume sample has to be mixed thoroughly, either by inversion or using a long thin pipette tip. This should be done with great care in order to minimize the almost inevitable small loss of solution that will occur during these manipulations. Another frequently encountered problem, especially with dilute protein samples, is that material is lost through absorption onto cuvette walls and pipette tips. In practice, there is little to be done about this, but it is advisable to check for the problem by performing a "dummy" titration in which one simply adds buffer to the sample to check for intensity changes that cannot be accounted for solely by dilution. Finally, since additions may increase the total absorption, it is always worth estimating (or better measuring) what the final absorbance will be before beginning an experiment in which additions are to be made.

The final step is to record the baseline using the same cuvette, buffer, and instrument settings. Strictly speaking, the baseline should be recorded using a buffer containing something that has the same normal absorption as the sample but no CD signal (Johnson, 1996b). However, this is seldom done and is unlikely to be a major problem except with very weak signals. One should not be tempted to reduce the number of scans used to record the baseline, since any noise in the baseline scan will simply be added to the sample scan in subsequent numerical processing.

IV. Data Processing and Spectral Characteristics

A. Data Processing

The first step is to subtract the baseline scan from the sample scan. All spectra should have been collected with a starting wavelength (260 nm for far-UV spectra; 340 nm for near-UV spectra—see Section III) that gives *at least* 15–20 nm at the start of the scan where there should be no signal. After baseline subtraction this region should be, and generally is, flat, but the signal may not be zero. The slow vertical drift in signal that is common to CD spectrometers is usually the cause of this problem. The solution is to average the apparent signal over the first 15–20 nm of the spectrum and subtract this average value from the whole of the curve.

The spectrum should be converted to the desired units. In the case of proteins, the observed CD signal, S in millidegrees (Note: 1 millidegree = 32,980 × ΔA), is generally converted to either the molar CD extinction coefficient ($\Delta\varepsilon_M$) or to the mean residue CD extinction coefficient ($\Delta\varepsilon_{MRW}$) using:

$$\begin{aligned} \Delta\varepsilon_M &= \frac{S}{32,980 \times C_M \times L} \text{ or} \\ \Delta\varepsilon_{MRW} &= \frac{S \times \text{MRW}}{32,980 \times C_{mg/ml} \times L} (\text{units} : \text{M}^{-1}\text{cm}^{-1}) \end{aligned} \quad (3)$$

where L is the path length (in cm), C_M is the molar concentration, $C_{mg/ml}$ is the concentration in mg/ml, and MRW is the mean residue weight (molecular weight divided by the number of residues). Although large globular proteins generally have a mean residue weight of approximately 111, the actual value must always be calculated in order to avoid potentially large errors in the calculated intensities. Calculating far-UV intensities is almost invariably done on a per residue basis in order to facilitate comparison between proteins and peptides with different molecular weights. Near-UV CD intensities should generally be reported on a molar rather than a per-residue basis because only four of the amino acid side chains contribute to the CD signals in this region (see Section IV.B.1). In the case of nucleic acids, the CD intensities can be calculated using the base, base pair, or molar concentrations.

CD intensities are also sometimes reported as molar ellipticity ($[\theta]_M$) or mean residue ellipticity ($[\theta]_{mrw}$), which may be directly calculated as:

$$\begin{aligned} [\theta]_M &= \frac{S}{10 \times C_M \times \text{L}} \text{ or} \\ [\theta]_{mrw} &= \frac{S \times \text{MRW}}{10 \times C_{mg/ml} \times L} (\text{units} : \text{degrees} \times \text{cm}^2\text{dmol}^{-1}) \end{aligned} \quad (4)$$

$[\theta]$ and $\Delta\varepsilon$ values may be interconverted using the relationship $[\theta] = 3298\Delta\varepsilon$.

B. Spectral Characteristics

1. Near-UV Spectra of Proteins

Near-UV CD bands from individual residues in a protein may be either positive or negative and may vary dramatically in intensity, with residues that are immobilized and/or have coupled-oscillator interactions with neighboring aromatic residues producing the strongest signals. There is therefore little correlation between the number of aromatic residues and the aromatic CD intensity, and the near-UV CD spectrum of a protein does not generally allow one to say anything explicit about its tertiary structure. Knowledge of the position and intensity of CD bands expected for a particular chromophore is helpful in understanding the observed near-UV CD spectrum of a protein and the principal characteristics of the four chromophores are therefore summarized below (Strickland, 1974; Woody and Dunker, 1996):

- Phenylalanine has sharp fine structure in the range 255–270 nm with peaks generally observed close to 262 and 268 nm ($\Delta\varepsilon_M \pm 0.3$ M^{-1} cm^{-1}).
- Tyrosine generally has a maximum in the range 275–282 ($\Delta\varepsilon_M \pm 2$ M^{-1} cm^{-1}), possibly with a shoulder some 6 nm to the red.[1]
- Tryptophan often shows fine structure above 280 nm in the form of two ^1L_b bands [one at 288 to 293 and one some 7 nm to the blue, with the same sign ($\Delta\varepsilon_M \pm 5$ M^{-1} cm^{-1})] and a ^1L_a band (around 265 nm) with little fine structure ($\Delta\varepsilon_M \pm 2.5$ M^{-1} cm^{-1}).
- Cystine CD begins at long wavelength (>320 nm) and shows one or two broad peaks above 240 nm ($\Delta\varepsilon_M \pm 1$ M^{-1} cm^{-1}); the long-wavelength peak is frequently negative.

Some of these features are illustrated in Fig. 1, which shows near-UV CD spectra of *Drosophila* calmodulin and its complex with a 25-residue peptide (IKKN-FAKSKWKQAFNATAVVRHMRK) corresponding to the target sequence from CaM-dependent kinase I (Clapperton *et al.*, 2002). This calmodulin contains nine phenylalanines (giving the sharp bands at 262 and 268 nm in panel A), a single tyrosine in the C-terminal domain (giving the broad band around 275 nm in panel A that changes sign on calcium binding), and no tryptophans. Complex formation with the target peptide introduces a single tryptophan (and two further phenylalanines). Panel B shows that the free peptide, which is mobile and unstructured, has only a very weak CD signal but that immobilization in the complex generates the intense peaks characteristic of tryptophan.

2. Far-UV Spectra of Proteins

Far-UV CD spectra of proteins depend on secondary structure content and simple inspection of a spectrum will generally reveal information about the structural class of the protein. The characteristic features of the spectra of different protein classes may be summarized as follows (Venyaminov and Yang, 1996):

- All-α proteins show an intense negative band with two peaks (at 208 and 222 nm) and a strong positive band (at 191–193 nm). The intensities of these bands reflect α-helical content. $\Delta\varepsilon_{mrw}$ values for a *totally* helical protein would be of the order of -11 M^{-1} cm^{-1} (at 208 and 222 nm) and $+21$ M^{-1} cm^{-1} (at 191–193 nm).
- The spectra of regular all-β proteins are significantly weaker than those of all-α proteins. These spectra usually have a negative band (at 210–225 nm, $\Delta\varepsilon_{mrw}$: -1 to -3.5 M^{-1} cm^{-1}) and a stronger positive band (at 190–200 nm, $\Delta\varepsilon_{mrw}$: 2–6 M^{-1} cm^{-1}).

[1] Shifted to the red means shifted to longer wavelengths; shifted to the blue means shifted to shorter wavelengths.

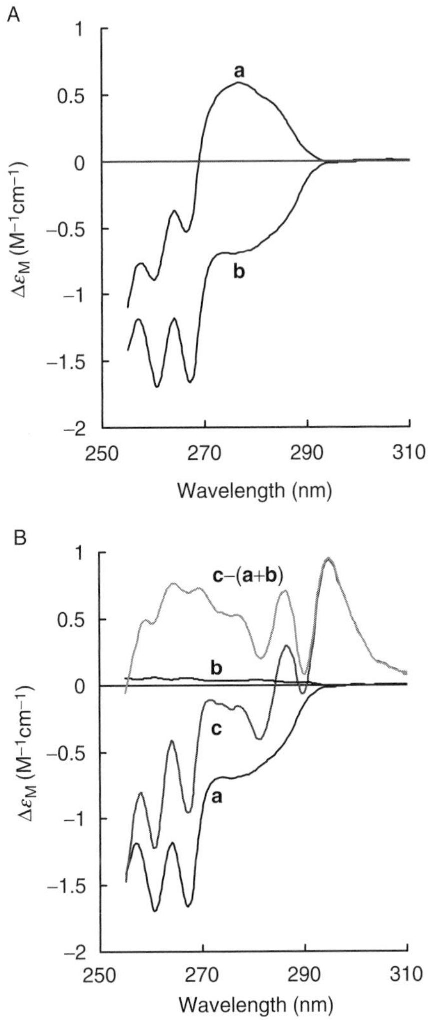

Fig. 1 Typical near-UV CD spectra. (A) Near-UV CD spectra of calcium-free (**a**) and calcium-saturated (**b**) *Drosophila* calmodulin. The nine phenylalanines in this protein produce the sharp bands at 262 and 268 nm; the single tyrosine in the C-terminal domain produces the broad band around 275 nm. Note that calcium binding reverses the sign of the tyrosine band. (B) Near-UV CD spectra of calcium-saturated *Drosophila* calmodulin (**a**) and the target peptide from CaM-dependent kinase I (**b**). The spectrum of the 1:1 complex (**c**) is clearly not equal to the sum of the component spectra and the difference spectrum [**c** − (**a** + **b**)] shows the characteristic features of tryptophan CD (see text and Clapperton *et al.*, 2002).

- Unordered peptides and denatured proteins have a strong negative band (at 195–200 nm, $\Delta\varepsilon_{mrw}$: −4 to −8 M^{-1} cm^{-1}) and a much weaker band (which can be either positive or negative) between 215 and 230 nm ($\Delta\varepsilon_{mrw}$: +0.5 to −2.5 M^{-1} cm^{-1}).

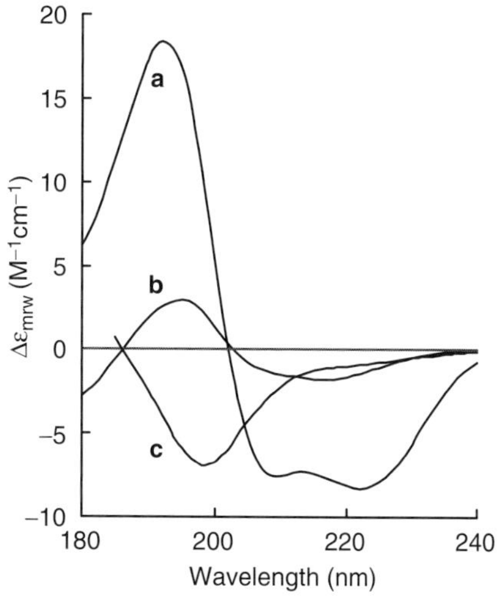

Fig. 2 Typical far-UV CD spectra far-UV CD spectra of (**a**) myoglobin (All-α: 4mbn.pdb); (**b**) prealbumin (All-β: 2pab.pdb), and (**c**) acid denatured staphylococcal nuclease at pH 6.2 and 6 °C (unordered). These spectra are available on the Internet (at http://lamar.colostate.edu/~sreeram/CDPro/).

- $\alpha + \beta$ and α/β proteins almost always have spectra dominated by the α-helical component and therefore often show bands at 222, 208, and 190–195 nm. In some cases, there may be a single broad minimum between 210 and 220 nm because of overlapping α-helical and β-sheet contributions.

Some of these features are illustrated in Fig. 2, which shows the far-UV CD spectra of myoglobin (4mbn.pdb), prealbumin (2pab.pdb), and acid denatured staphylococcal nuclease at pH 6.2 (6 °C). These spectra and those of many other proteins are available on the Internet (at http://lamar.colostate.edu/~sreeram/CDPro/).

3. Nucleic Acids

Because the aromatic bases themselves are planar, they do not possess any intrinsic CD signals; it is the presence of the sugars that creates the asymmetry which leads to the small CD signals of the monomeric nucleotides. Likewise, the stacking of the bases in the different polymeric forms results in the close contact and electronic interactions that produce the intense CD signals of the nucleic acids and oligonucleotides. CD is sensitive to secondary structure because the precise nature of these interactions determines the shape of the spectrum. The principal conformational forms of the nucleic acids are the A- and B-forms. In neutral aqueous buffers at moderate salt concentrations, DNA is usually in the B-form, while RNA adopts the A-form. These conformations have characteristic CD spectra that depend on base composition and somewhat on sugar type (Johnson, 1996a). Variation in spectral shape with base composition is, of course, significantly more important with short oligonucleotides. For this reason, CD is usually used empirically in the study of nucleic acid conformation by comparing an experimentally measured spectrum to that of a nucleic acid of known structure (Bishop and Chaires, 2002; Johnson, 1996a).

V. Applications

A. Secondary Structure Content of Proteins

The estimation of protein secondary structure *content* from far-UV CD spectra is one of the most widely used applications of CD. If the chiroptical properties of the different secondary structure elements are assumed to be additive then the measured far-UV CD spectrum of a protein at any wavelength (λ) can be expressed as:

$$S(\lambda) = \sum_{i=1}^{n} f_i S_i(\lambda) \qquad (5)$$

where $S(\lambda)$ is the observed mean residue CD, f_i is the fraction of the ith conformation, and $S_i(\lambda)$ is the reference CD signal for the ith conformation. If the reference spectra are known, then the f_is can be solved from a series of simultaneous equations (one for each λ) by a least-squares method. The original approach to this problem used synthetic polypeptides as model compounds to provide reference spectra for the α-helix, β-form, and unordered form (Greenfield and Fasman, 1969). The approach was later extended by Brahms and Brahms (1980) to include a reference for β-turns. An alternative approach is to compute the appropriate reference spectra from the spectra of proteins of known secondary structure using Eq. (5). Saxena and Wetlaufer (1971) used spectra of just three proteins to derive reference spectra for α-helix, β-form, and unordered form. The same reference spectra were subsequently derived from 5 proteins (Chen *et al.*, 1972) and then 8 proteins (Chen *et al.*, 1974). Chang *et al.* (1978) used 15 proteins and added a general term for β-turn. In addition to the assumption of additivity [see Eq. (5)], these methods depend on four other major assumptions (Sreerama *et al.*, 2000): (1) that the effect of tertiary structure on CD is negligible, (2) that the ensemble-averaged solution structure is equivalent to the X-ray structure, (3) that the contributions from nonpeptide chromophores do not influence the analysis, and (4) that the geometric variability of secondary structures need not be explicitly considered. Although assumptions (1) and (2) are probably valid, assumptions (3) and (4) are probably invalid in many cases (Sreerama *et al.*, 2000). For example, several recent experiments, particularly those using site-directed mutagenesis, have shown that aromatic residues can make surprisingly large contributions to the far-UV CD spectra of some proteins and this obviously has serious implications for secondary structure estimation (see Woody and Dunker (1996) and references therein). Assumption (4) is even more likely to be incorrect because it is highly unlikely that the secondary structure of a protein will be ideal. Thus, for example, a single reference spectrum for the α-helix cannot account for variable contributions from regular helices, 3_{10}-helices, and distorted helices, and the question of chain-length dependence of the α-helix intensity (see Chen *et al.*, 1974) is generally not explicitly considered. Likewise, a single reference spectrum for the β-form cannot account for variable contributions from parallel and antiparallel forms, sheet width, and chain-length dependence of the CD signal, and the fact that this form is often twisted and nonplanar, with a degree of twisting that varies from protein to protein.

Two approaches designed to circumvent the problem of defining suitable reference spectra were introduced in 1981. Provencher and Glöckner (1981) developed a program (CONTIN), which uses ridge regression analysis to fit the spectrum of the unknown directly by a linear combination of the spectra of proteins with known structure. In this program, the contribution of each reference spectrum is kept small unless it contributes to a good fit to the raw data. In the second approach, introduced by Hennessey and Johnson (1981), singular value decomposition (SVD) was applied to extract orthogonal basis curves from a set of spectra of proteins with known structure. The deconvoluted basis curves, each of which has a unique shape, can be associated with a known mixture of secondary structures and can therefore be used to analyze the conformation of an unknown sample. Convex constraint analysis (CCA) has also been used to extract basis spectra (Perczel *et al.*, 1991).

The reference or basis curves obtained from linear regression, SVD or CCA, will depend strongly on the proteins included in the data set, and an inadequate representation of the spectrum of the unknown will result in a failed analysis (Sreerama and Woody, 2004a). For this reason, most of the methods currently employed achieve greater flexibility by using some form of selection process to create several different subsets of reference spectra in an attempt to include proteins that are important for the analysis and exclude those that have an adverse effect. This approach, known as variable selection, is implemented in different ways by different programs.

The basic method (implemented in VARSLC; Manavalan and Johnson, 1987) combines variable selection with SVD in the following way. Starting from a large database of reference proteins, the program *sequentially* eliminates one or more spectra to create a large series of new databases. SVD is used on these reduced data sets to evaluate the unknown conformation; all the results are then examined and those fulfilling certain selection criteria (see Sreerama and Woody, 2004a) for a good fit are averaged. The two other approaches for introducing variable selection with SVD are known as the minimal basis and locally linearized approaches. In the minimal basis approach (CDSSTR; Johnson, 1999), *small* subsets of reference spectra are *randomly* selected to create the minimal basis set. This creates a large number of different combinations and the process is stopped when an acceptable number of valid solutions has been obtained. In the locally linearized approach, spectra are *systematically* removed from the reference set on the basis of their similarity to the analyzed spectrum, with the least similar reference spectra being removed first. Local linearization is employed in SELCON (Sreerama and Woody, 1993), which uses a modification of the variable selection method called the self-consistent method. The program includes the spectrum of the unknown in the database and makes an initial guess of the structure. The database is decomposed using SVD and the unknown is analyzed. The solution replaces the initial guess and the process is repeated until the results are self-consistent. Variable selection has also been explicitly introduced into CONTIN by Sreerama and Woody (2000b) in the locally linearized implementation known as CONTIN/LL. Finally, neural networks are programs that are used to detect patterns in data, and two such programs have been described for the analysis of far-UV CD data of proteins (K2D; Andrade *et al.*, 1993 and CDNN; Böhm *et al.*, 1992), but these do not include variable selection.

The various methods available have been extensively discussed in the literature (Greenfield, 1996, 2004; Johnson, 1990; Sreerama and Woody, 1993, 1994, 2000a,b, 2004a; van Stokkum *et al.*, 1990; Venyaminov and Yang, 1996) and Greenfield (2004) has made useful recommendations for selecting the program that is most appropriate for a particular purpose. Two important points are worth remembering (Manavalan and Johnson, 1987; Venyaminov and Yang, 1996): these are that CD analysis methods cannot recognize a failed analysis and that the solution selected by a particular program may not necessarily be the best of those examined.

Three very popular programs (SELCON, CDSSTR, and CONTIN/LL) provided with several reference sets with different wavelength ranges are available on the Internet (at http://lamar.colostate.edu/~sreeram/CDPro/) with the CDPro software package (a more extensive list of Web sites where CD analysis programs are available is given in Greenfield, 2004). The question of whether it is appropriate to use the standard soluble protein reference data sets for analyzing the CD spectra of membrane proteins has been addressed by Wallace *et al.* (2003) and Sreerama and Woody (2004b). The latter authors have shown that the data sets available with the CDPro package do perform reasonably well and that this performance is improved still further by including membrane proteins in the reference set. Finally, methods available for obtaining additional structural information by calculating the number of secondary structure segments in a protein and identifying its tertiary structure class have been described (see Sreerama and Woody, 2004a and references therein).

B. Detecting Altered Conformation

1. Solution Conditions

The stability of any particular secondary structure element in a protein or nucleic acid will depend on several different factors. For example, the stability of an α-helix is determined by the number and strength of interhelical hydrogen bonds, by interactions (potentially either stabilizing or destabilizing) between ionized side chains, and by interactions with other secondary structural elements in the protein. These factors, and therefore the protein's structure, are often influenced by changes in solution conditions. For example, changes in the pH and ionic strength of the solution will generally affect interactions between charged side chains in proteins, and this can lead to significant changes in secondary and/or tertiary structure. Such changes are readily studied using CD. For example, CD has been extensively used in the study of conformational transitions in peptides (see Fig. 3) and in studies of the α-helix to β-sheet conversion in recombinant prion proteins induced by the addition of low concentrations of denaturants and/or salt at low pH (Morillas *et al.*, 2001; Swietnicki *et al.*, 2000).

Alcohols with low-dielectric constants, such as trifluoroethanol (TFE) and hexafluoroisopropanol, are often used to promote the formation of stable conformations in peptides that are unstructured in aqueous solution. There may also be a significant effect of even relatively simple alcohols on the structure of a protein (Griffin *et al.*, 1986). The low-dielectric constant of these structure-inducing cosolvents is thought to resemble that of the interior of a protein. Although this low-dielectric environment presumably strengthens favorable electrostatic interactions,

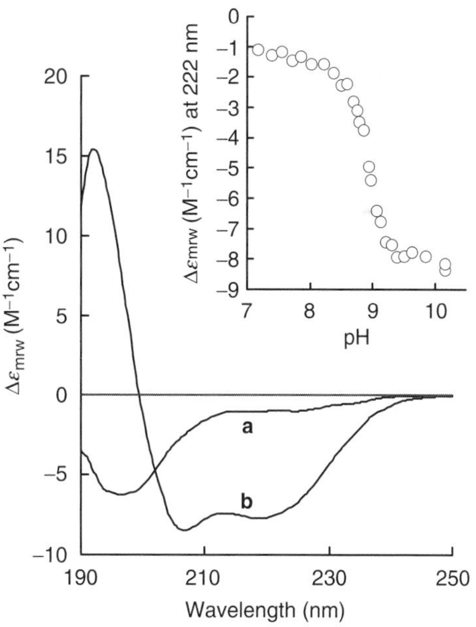

Fig. 3 pH-induced conformational transitions. The short peptide LKLKKLLKLLKKLLKLG was designed to form a perfect amphipathic helix. However, at low pH it is unordered (curve **a**), probably because of charge repulsion between the lysines. Neutralization of the lysines at high pH permits formation of the α-helix (curve **b**) and the unordered to α-helical transition is readily monitored using CD (inset) (Schilstra, M. J., unpublished observations).

this particular effect may be partially offset by increased counterion binding. The major effect is generally attributed to weaker hydrogen bonding of the amide protons to the solvent with concomitant strengthening of intramolecular hydrogen bonds and stabilization of the helix.

The effects of high temperature will be discussed further in Section V.C.2. CD spectra can, when necessary, be recorded at temperatures below zero by using suitable water–alcohol or water–glycerol mixtures. It is essential in such studies to check for any direct effect of the solvent itself on the conformation of the protein. This is done by measuring the CD spectrum in the solvent at room temperature and comparing it with the spectrum measured in an aqueous buffer at this temperature.

2. Changes Accompanying Complex Formation

Protein–protein and protein–nucleic acid interactions are often accompanied by changes in the intrinsic CD of one or both of the components owing to changes in secondary structure (Greenfield and Fowler, 2002) and/or the environment of aromatic groups (Clapperton *et al.*, 2002). Such changes may also be caused by the binding of small ligands, such as drugs and metal ions (Martin and Bayley, 1986), to macromolecules, and in certain cases the ligand itself may change its optical activity or become optically active. Because CD is a quantitative technique, these changes can be used to determine equilibrium dissociation constants for complex formation (see Section V.D). The source of the change in CD is immaterial for this application. Even when it is not possible to use these changes to determine dissociation constants they do provide unequivocal evidence that two species interact, and frequently provide useful information about the nature of the interaction. For example, stoichiometric titrations of calmodulin with metal ions, in which the signal from tyrosine-138 in the C-terminal domain (see Fig. 1A) is monitored, can be used to establish the preference of different metals for binding to the individual calmodulin domains (see Fig. 4).

3. Structural Analysis of Recombinant Native Proteins and Their Mutants

When a wild-type protein is produced using recombinant DNA technology, it is important to demonstrate that it has the same overall structure as a protein isolated by traditional methods. This is especially important if inclusion body formation necessitated the addition of denaturation and refolding steps to the isolation procedure. Comparison of the near- and far-UV CD spectra is ideal for this purpose.

When working with mutant proteins, it is, of course, a good practice to test for any effect of the mutation on the general conformation of the protein and, here again, CD provides a convenient means of doing this with the limited amounts of material that are sometimes available. A significant difference in *shape* between the far-UV CD spectra of the wild-type and mutant proteins can be an indication that the mutation has produced some change in the secondary structure. However, this may not be true if the mutation involves the introduction or replacement of an

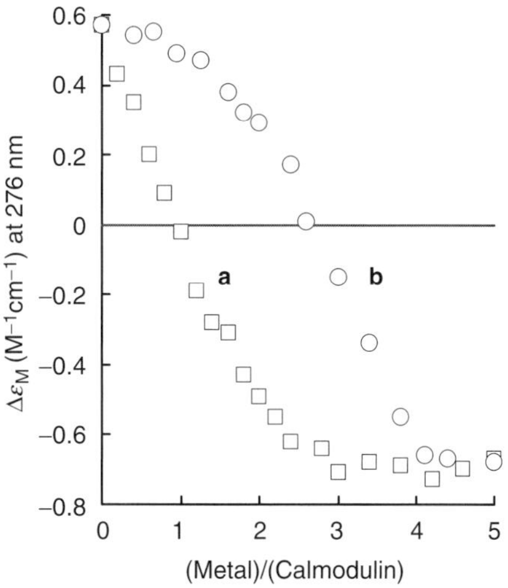

Fig. 4 Binding of metal ions to calmodulin. Binding of calcium (**a**) and terbium (**b**) to *Drosophila* calmodulin monitored using tyrosine CD (see Fig. 1A). Calmodulin is a two-domain protein with a total of four metal ion-binding sites, two in each domain. Because this particular calmodulin contains a single tyrosine located in the C-terminal domain these stoichiometric titrations show that calcium binds preferentially to the C-terminal tyrosine-containing domain (>75% saturation at [Ca]/[CaM] = 2), while terbium (a calcium analogue) binds preferentially to the N-terminal domain (<25% saturation at [Tb]/[CaM] = 2) (Martin, S. R., unpublished observations).

aromatic residue (Tyr or Trp) because these residues may, in certain cases, contribute significantly to the far-UV spectrum (see Section V.A). A change in shape may also occur if the mutation destabilizes the protein to such an extent that it exists as a mixture of folded (with wild-type conformation) and unfolded forms under the conditions of the measurement. This is a distinct possibility with small proteins, which frequently have relatively low free energies for unfolding and are therefore only just stable at room temperature (see Section V.C.2). When a difference in shape is observed, especially if the mutant looks to be less well folded, it is useful to repeat the measurements at a lower temperature, where the protein would be expected to be more stable (see Section V.C.2). If the difference between the wild-type and mutant spectra is smaller at the lower temperature, then it is highly likely that the mutation has had a significant destabilizing effect and the stability properties should be fully characterized using one of the methods described in Section V.C.

Changes in the near-UV region of the spectrum may indicate that the mutation has caused a change in tertiary structure but could again result from destabilization of the protein. In addition, it is important to remember that significant changes may often derive from rather subtle changes in the environment of one or more of the aromatic residues that are not necessarily associated with any major structural change. Mutations that involve the introduction or removal of tyrosine or tryptophan would, of course, be expected to change the spectrum and serial replacement of aromatic residues by site-directed mutagenesis can be used to investigate the contributions of individual aromatic residues to the near-UV CD spectra of proteins (Craig *et al.*, 1989).

It is generally wise to be suspicious if the wild-type and mutant spectra differ in *intensity* but not in shape. If the spectra can be superimposed by applying a simple multiplication factor, then there has most probably been an error in concentration determination (see Section II.D), particularly if the same factor serves to superimpose both the near- and far-UV spectra.

C. CD in the Study of Protein Stability

Unfolding of macromolecules is generally studied by using an optical method (absorbance, fluorescence, or CD) or by using differential scanning calorimetry (DSC). CD is very widely employed in the study of protein stability because unfolding is almost invariably accompanied by major changes in both the near- and far-UV CD spectra. The unfolding mechanism for many small globular proteins is essentially a two-state process, where only the native or folded (N) and denatured or unfolded (D) states are significantly populated at equilibrium (see Chapter 11 by Street *et al.*, this volume). Conformational stability is defined in terms of the free energy of unfolding:

$$\Delta G^\circ = -RT\ln K \tag{6}$$

where K is the unfolding constant (= [D]/[N]),[2] R is the universal gas constant, and T is the absolute temperature.

The free energy of unfolding is determined by monitoring an appropriate CD signal as a function of the concentration of a chemical denaturant (urea or guanidine hydrochloride) or as a function of temperature. Because the object is to measure thermodynamic parameters, the unfolding must be reversible and this must be confirmed by demonstrating that the signal of the native protein is recovered on lowering the temperature or on removing the denaturant. It is also essential that equilibrium is reached at each point in the unfolding curve. In general, the nearer the midpoint of the transition the longer will be the time taken to reach equilibrium.

1. Chemical Denaturation

Chemical denaturation experiments are generally performed in one of two ways: (1) by measuring the CD of a large number of individually prepared samples, each with a different concentration of denaturant but the same concentration of protein;

[2] The unfolding constant can equally well be defined as K = [N]/[D] (see Chapter 11 by Street *et al.*, this volume). With the definition adopted here positive ΔG° values correspond to lower free energies for N than for D.

or (2) by titrating a solution of protein with concentrated denaturant containing protein at the same concentration. The second approach can, of course, only be used when the unfolding is reversible. When the unfolding is reversible, the titration can also be performed in the reverse direction; that is, adding protein in buffer to protein in concentrated denaturant. In most cases, one only requires values of CD intensity at a single wavelength for the analysis. To allow proper baseline alignment at higher wavelength, these values should be taken from full wavelength scans whenever the signal is weak. When the signal is strong, the baseline alignment problems should be minimal and observation at a single wavelength may be adequate. For far-UV titrations, it may be helpful to use a solution at the normal concentration for a far-UV measurement (i.e., 0.1–0.15 mg/ml) but use a longer path length cuvette. This restricts the accessible lower wavelength range, but normally permits measurements in the region of interest (generally 220 nm).

With both of the approaches described above, one obtains the variation of the CD signal with denaturant concentration. It is quite common to observe that the CD signal of the native and denatured forms shows some dependence on denaturant concentration, $[x]$, *outside* the region of the unfolding transition. The variation of the CD signal, Y, with $[x]$ must therefore be described by:

$$Y = (Y_{\mathrm{N,H_2O}} + \beta_{\mathrm{N}}[x])f_{\mathrm{N}} + (Y_{\mathrm{D,H_2O}} + \beta_{\mathrm{D}}[x])f_{\mathrm{D}} \tag{7}$$

where $Y_{\mathrm{N,H_2O}}$ and β_{N} are the intercept and slope of the pretransition baseline, $Y_{\mathrm{D,H_2O}}$ and β_{D} are the intercept and slope of the posttransition baseline, and f_{N} and f_{D} are the fractions of protein present in the native and denatured forms, respectively. The dependence of the free energy of unfolding on $[x]$ is most often described using the linear extrapolation model,

$$\Delta G^{\circ}(x) = \Delta G^{\circ}{}_{\mathrm{H_2O}} + m[x] \tag{8}$$

where $\Delta G^{\circ}{}_{\mathrm{H_2O}}$ is the free energy of unfolding in the absence of denaturant and m describes the dependence on the denaturant concentration. Other extrapolation models may be used (see Chapter 11 by Street *et al.*, this volume) but they will not be described here. For simple two-state unfolding $K = [\mathrm{D}]/[\mathrm{N}] = f_{\mathrm{D}}/f_{\mathrm{N}}$. Substituting $f_{\mathrm{N}} = 1/(1 + K)$ and $f_{\mathrm{D}} = K/(1 + K)$ in Eq. (7) with K expressed as $\exp(-\Delta G^{\circ}(x)/RT)$ and $\Delta G^{\circ}(x)$ from Eq. (8) gives:

$$\frac{Y = (Y_{\mathrm{N,H_2O}} + \beta_{\mathrm{N}}[x]) + (Y_{\mathrm{D},H_2O} + \beta_{\mathrm{D}}[x])\mathrm{e}^{-(\Delta G^{\circ}\mathrm{H_2O}+m[x])/RT}}{1 + \mathrm{e}^{-(\Delta G^{\circ}_{\mathrm{H_2O}}+m[x])/RT}} \tag{9}$$

Fitting Eq. (9) using standard nonlinear least-squares methods (see Chapter 24 by Johnson, this volume, for details of these methods) should yield $\Delta G^{\circ}{}_{\mathrm{H_2O}}$ and the m value.[3] Guanidine hydrochloride is a more powerful denaturant than urea because it has a larger m value for any particular protein. [Myers *et al.* (1995) have provided a useful compilation of m values for a large number of proteins.]

2. Thermal Unfolding

In thermal unfolding experiments, the CD signal at a single wavelength is recorded while the temperature is slowly increased (preferably at no more than 1 °/min). This is generally done using a Peltier unit or a programmable water bath. In either case it is advisable to check the actual temperature profile of the device by measuring the temperature in the cuvette using an immersible electronic probe. Buffers with high thermal coefficients (e.g., Tris) should be avoided if possible in order to avoid large changes in pH as the temperature is increased.

The CD signal of the folded and unfolded forms will generally show some dependence on temperature so that there will be pre and posttransition slopes, as with chemical denaturation. The variation of the CD signal, S, with temperature must therefore be described by:

$$Y = (Y_{\mathrm{N}} + \beta_{\mathrm{N}}T)f_{\mathrm{N}} + (Y_{\mathrm{D}} + \beta_{\mathrm{D}}T)f_{\mathrm{D}} \tag{10}$$

where Y_{N} (Y_{D}) and $\beta_{\mathrm{N}}(\beta_{\mathrm{D}})$ denote the slopes and intercepts of the pre- and posttransition slopes. The variation of ΔG° with temperature is described by the modified Gibbs–Helmholtz equation (Becktel and Schellman, 1987):

$$\Delta G^{\circ}(T) = \Delta H_{T_{\mathrm{m}}}\left(1 - \frac{T}{T_{\mathrm{m}}}\right) + \Delta C_{\mathrm{p}}\left\{(T - T_{\mathrm{m}}) - T\ln\left(\frac{T}{T_{\mathrm{m}}}\right)\right\} \tag{11}$$

where T_{m} and ΔH_{T_m} are the midpoint melting temperature and the enthalpy at T_{m} respectively, and ΔC_{p} is the difference in heat capacity between the folded and unfolded protein (see Robertson and Murphy, 1997).

Substituting $f_{\mathrm{N}} = 1/(1 + K)$ and $f_{\mathrm{D}} = K/(1 + K)$ in Eq. (10) with K expressed as $\exp(-\Delta G^{\circ}(T)/RT)$ and $\Delta G^{\circ}(T)$ from Eq. (11) gives:

$$Y = \frac{(Y_{\mathrm{N}} + \beta_{\mathrm{N}}T) + (Y_{\mathrm{D}} + \beta_{\mathrm{D}}T)\mathrm{e}^{-(\Delta H_{T\mathrm{m}}(1-(T/T_{\mathrm{m}}))+\Delta C_{\mathrm{p}}\{(T-T_{\mathrm{m}})-T\ln(T/T_{\mathrm{m}})\})/RT}}{1 + \mathrm{e}^{-(\Delta H_{T\mathrm{m}}(1-(T/T_{\mathrm{m}}))+\Delta C_{\mathrm{p}}\{(T-T_{\mathrm{m}})-T\ln(T/T_{\mathrm{m}})\})/RT}} \tag{12}$$

Although it is theoretically possible to fit the CD data to this equation using nonlinear least-squares methods, there is generally not enough information to

[3] An alternative approach that should avoid possible errors in writing the code for Eq. (9) is to fit Eq. (7) directly with f_{N} and f_{D} calculated in separate steps, i.e., $f_{\mathrm{N}} = 1/(1 + K)$ and $f_{\mathrm{D}} = K/(1 + K)$, with K expressed as $\exp(-\Delta G^{\circ}(x)/RT)$ and $\Delta G^{\circ}(x)$ from Eq. (8).

determine ΔC_p accurately (because it is the second derivative of $\Delta G°$ with respect to temperature). Greenfield (2004) has noted that trying to include the ΔC_p term can, in fact, actually worsen the agreement between T_m values determined by CD and DSC. It is therefore common practice to set ΔC_p equal to zero in the analysis. If ΔC_p can be determined using another method, such as DSC, then ΔH_{Tm} and T_m can be used to calculate ΔG at any temperature using the modified Gibbs–Helmholtz equation [Eq. (11)]. The variation of the free energy for unfolding of proteins with temperature has several interesting features (Fig. 5). $\Delta G°$ will not vary linearly with temperature because the ΔC_p term is large and positive for protein unfolding (Robertson and Murphy, 1997). In addition to the well-characterized high temperature unfolding (midpoint = T_m, where $\Delta G° = 0$), there will also be a temperature of maximum stability, below which the protein will again become less stable (curves **b** and **c**). In certain cases, the free energy for unfolding will become small enough at low temperatures for one to observe the phenomenon of cold-induced denaturation (curve **c**), which is easily monitored by using CD (Chen and Schellman, 1989). Another feature illustrated in Fig. 5 is that proteins with a high T_m for high-temperature unfolding do not necessarily have high overall stability at room temperature (compare curves **a** and **b**). Hence, the relative stability of two *different* proteins at ambient temperatures cannot be estimated solely from T_m measurements; a full analysis yielding ΔH_{T_m} and ΔC_p values is required for this [using Eq. (11)].

As noted above, any full thermodynamic analysis is only valid if the unfolding is reversible. Nevertheless, even when the unfolding is irreversible (usually because

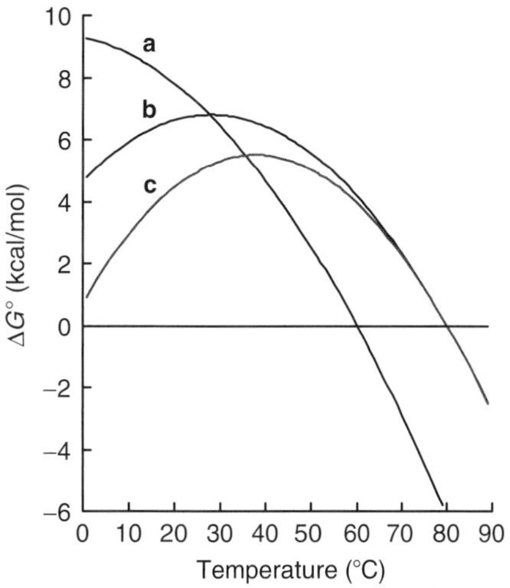

Fig. 5 Protein stability curves. Calculated variation of $\Delta G°$ as a function of temperature for three proteins with the same ΔH°_{Tm} (90 kcal/mol) but different values of T_m and ΔC_p. (**a**) $T_m = 60$ °C, $\Delta C_p = 1.2$ kcal/mol K; (**b**) $T_m = 80$ °C, $\Delta C_p = 1.6$ kcal/mol K; (**c**) $T_m = 80$ °C, $\Delta C_p = 2.0$ kcal/mol K (see text for further details).

the protein aggregates at elevated temperatures), the methods described in this section can still give useful information about relative stabilities. However, this is true only if the same unfolding behavior is obtained with different rates of heating, showing that the results are not being perturbed by the kinetics of aggregation.

3. Unfolding Intermediates

The two-state nature of the unfolding process can often be confirmed by monitoring the unfolding using different probes. If the unfolding is truly two-state, then the transition should be identical whether monitored by near-UV CD (reporting on tertiary structure) or far-UV CD (reporting on secondary structure). Under some conditions (generally mildly denaturing conditions such as low pH) some proteins, for example α-lactalbumin (Kuwajima, 1996), form an intermediate molten globule state that is neither fully folded nor fully unfolded (see Arai and Kuwajima (2000) for a review of the role of the molten globule state in protein folding). This intermediate state retains most of the native secondary structure and generally has a native-like far-UV CD spectrum, but has a highly fluid tertiary structure and therefore has a near-UV CD spectrum characteristic of the fully unfolded form.

Some proteins, particularly those containing more than a single domain, may not unfold in a single transition. In such cases, the data can be fit to a model in which it is assumed that the unfolding is represented by the sum of two or more independent transitions. The equation used to fit the data is then the appropriate sum of those for the individual transitions given above (see, for example, Masino *et al.*, 2000).

D. Determination of Equilibrium Dissociation Constants

1. Direct Methods

As noted in Section V.B.2, the interaction of a macromolecule with another macromolecule or small molecule is often associated with a change in the CD signal of one or both of the components. Although not generally the method of choice owing to its relatively poor S/N ratio, these changes can sometimes be used to determine equilibrium dissociation constants. While a detailed discussion of the methods available for characterizing such interactions is beyond the scope of this chapter [see Greenfield, 1975, 2004 for general reviews], it is appropriate to make some remarks here. We will do this with reference to the simplest possible binding model, formation of a 1:1 complex according to the following scheme.

$$\mathrm{A} + \mathrm{B} \Leftrightarrow \mathrm{AB} \text{ with } K_\mathrm{d} = \frac{[\mathrm{A}][\mathrm{B}]}{[\mathrm{AB}]} \qquad (13)$$

The primary requirement for a direct titration, as with any spectroscopic method, is that the sum of the spectra of the components (A and B) should be different from that of the complex (AB) at some wavelength (i.e., $\Delta\varepsilon_A + \Delta\varepsilon_B \neq \Delta\varepsilon_{AB}$). If this requirement is fulfilled, then a titration of A with B (or B with A; see below) can, in suitable cases, allow determination of the equilibrium dissociation constant, K_d. The observed CD signal for *any* mixture of A and B is given by:

$$\text{Signal} = [A]\Delta\varepsilon_A + [B]\Delta\varepsilon_B[B] + [AB]\Delta\varepsilon_{AB} \tag{14}$$

Eliminating [A] and [B] using the mass balance relationships ($[A] = A_T - [AB]$ and $[B] = B_T - [AB]$) and substituting the following expression for [AB]

$$[AB] = \frac{([A_T] + [B_T] + K_d) - \left(([A_T] + [L_T] + K_d)^2 - 4[A_T][B_T]\right)^{0.5}}{2} \tag{15}$$

gives

$$\text{Signal} = \Delta\varepsilon_A A_T + \Delta\varepsilon_B B_T + (\Delta\varepsilon_{AB} - \varDelta\varepsilon_A - \Delta\varepsilon_B)\left(\frac{(A_T + B_T + K_d) - \left((A_T + B_T + K_d)^2 - 4A_T B_T\right)^{0.5}}{2}\right) \tag{16}$$

where A_T and B_T are the total concentrations of A and B. Because $\Delta\varepsilon(A)$ and $\Delta\varepsilon(B)$ can be measured separately, there are only two unknowns in this equation, the K_d and $\Delta\varepsilon(AB)$. Several important factors must be considered in designing the experiment.

1. *Titration mode*: The titration can, of course, be performed by adding A to B or by adding B to A. The choice may, of course, be dictated by availability and/or solubility of one of the components. If both components have a CD signal, then the component being added should have the weaker signal of the two as this permits more accurate determination of the end point. For example, in studies of protein–nucleic acid interactions, the signal from the protein will generally be much weaker than that of the nucleic acid, and the titration is best performed by adding the protein to the nucleic acid (Gray, 1996). The other factor to be considered here is that the total absorbance at the observation wavelength (see below) must be kept below 1.2.

2. *Choice of wavelength*: In the simplest cases, the wavelength selected is where the signal change is largest. However, two factors may dictate otherwise. First, particularly in far-UV measurements, it may be advisable to sacrifice amplitude for improved S/N ratio at another (longer) wavelength. Second, in experiments with strongly absorbing samples moving away from the absorption maximum will permit higher concentrations to be used.

3. *Concentration range*. It is clear from inspection of Eq. (16) that a successful analysis requires that the variation of the CD signal with concentration accurately defines both K_d and $\varDelta\varepsilon(AB)$. In addition, it must be possible to demonstrate that it

is indeed a 1:1 complex that is being formed. Consider for example, a titration of A with B performed at a constant concentration of A ($=A_T$) by "spiking" the stock solution of B with A (at A_T). One needs to select an appropriate value for A_T and estimate what final concentration of B ($=B_{T(Final)}$) will be required. The following discussion is based on the assumption that a preliminary examination has allowed an approximate value of the dissociation constant to be determined. The estimate of $B_{T(Final)}$ is easy to deal with since $\Delta\varepsilon(AB)$ can clearly only be determined accurately if a large fraction of A (ideally >0.8) is saturated with B at the end of the titration. Fractional saturation of A ($= f_A$) is related to the concentration of unbound B by the relationship $f_A = [B]/(K_d + [B])$; therefore [B] at the end of the titration should be $>5 \times K_d$. The minimum total concentration of B required is then calculated as $B_{T(Final)} = (0.8 \times A_T + 5 \times K_d)$. The experiment should be designed with this minimum concentration in mind, but in practice, it is, of course, best to reach the highest final concentration possible. This is especially true if one is not using "spiking", as dilution during the titration can lead to the appearance of a false plateau and the titration can be ended prematurely.

The choice of an appropriate value of A_T can be illustrated by inspection of Fig. 6. This figure shows f_A as a function of the B_T/A_T ratio for a system with $K_d = 1\ \mu M$ and different values of A_T. In the ideal case (Curve a, $A_T = 2 \times K_d$), the K_d and the $\Delta\varepsilon(AB)$ values will be well defined and the shape of the binding curve

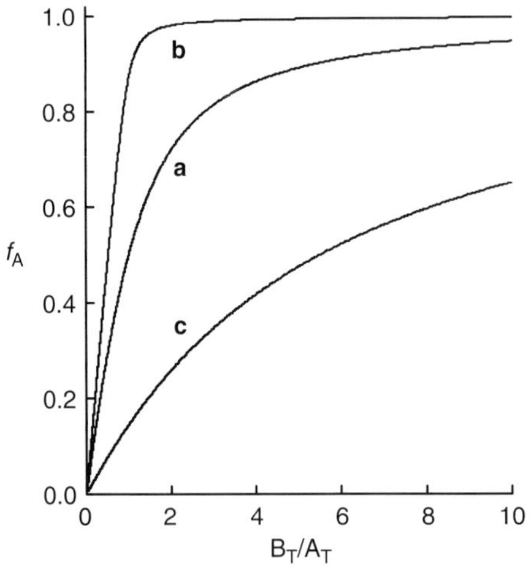

Fig. 6 Titration saturation curves. Observed fractional saturation for a titration of A with B in which a simple 1:1 complex (AB) is formed (see text). The fractional saturation of A (defined as $f_A = [AB]/A_T$) is plotted as a function of the B_T/A_T ratio for a system with $K_d = 1\ \mu M$ and different values of A_T: (**a**) $A_T = 2\ \mu M$ ($2 \times K_d$), (**b**) $A_T = 25\ \mu M$ ($25 \times K_d$), and (**c**) $A_T = 0.2\ \mu M$ ($0.2 \times K_d$). *Note*: A_T and B_T indicate total concentrations of A and B (see text for further details).

will clearly be consistent with 1:1 stoichiometry. If A_T is too high, one observes what is essentially a stoichiometric titration with a very sharp transition at the 1:1 point (Curve b, $A_T = 25 \times K_d$) and the K_d cannot be accurately determined because there will be no point during the titration where A, B, and AB coexist at significant levels. In practical terms, this means that CD measurements require that $A_T < 5\text{–}10 \times K_d$, depending on the S/N. It should be noted, however, that titrations performed at high concentrations are useful for demonstrating 1:1 stoichiometry. If A_T is too low, then the binding curve will be very flat (Curve c, $A_T = 0.2 \times K_d$) and even if $\Delta\varepsilon(AB)$ can be determined it will not be possible to verify that the stoichiometry is 1:1. In practical terms, this means that CD measurements should ideally be performed with $A_T > 0.5 \times K_d$. Although this range of useful A_T values is relatively narrow, the availability of cuvettes with different path lengths means that a large range of K_d values can, at least in principle, be determined using CD methods.

2. Indirect Methods

This section contains a brief survey of the methods that can be used when there is no suitable signal change accompanying the interaction.

When a ligand binds to a protein only (or preferentially) when it is in the native-folded form, then the protein will be more stable in the presence of the ligand. This is because the native and denatured forms are in equilibrium; complex formation removes some of the native form and some of the denatured protein must refold to maintain the equilibrium. The increase in stability in the presence of the ligand can be used to determine the equilibrium dissociation constant. In the case of thermal unfolding experiments, the effect of ligand on the midpoint for unfolding can be used. If the effect of ligand binding on the enthalpy for unfolding is small, then (Schellman, 1975):

$$\Delta T_m = \frac{T_m T_m^* R}{\Delta H_{Tm^*}} \ln\left(1 + \frac{[L]}{K_d}\right) \tag{17}$$

where ΔT_m is the change in the midpoint of the unfolding curve, T_m^* and T_m are the midpoint values in the presence and absence of ligand, $\Delta H_{T_m^*}$ is the enthalpy of unfolding in the presence of the ligand at T_m^*, and $[L]$ is the free ligand concentration. A disadvantage of this approach is that one gets an estimate of the K_d at an elevated temperature. A more detailed evaluation of this approach has been described recently by Mayhood and Windsor (2005).

The change in stability of a protein to chemical denaturation can be used in the same way by measuring the free energies of unfolding in the presence and absence of the ligand (Pace and McGrath, 1980). The relevant equation here is:

$$\Delta G_L - \Delta G = RT \ln\left(1 + \frac{[L]}{K_d}\right) \tag{18}$$

where ΔG_L and ΔG are the free energies of unfolding in the presence and absence of the ligand, respectively. When the interaction involves charged species, these experiments are better performed with urea rather than guanidine hydrochloride because of the high ionic strength of the latter. If the interaction does not have simple 1:1 stoichiometry, then the expressions are more complicated (see, for example, Masino *et al.*, 2000). It should be emphasized that this type of experiment is useful even if precise constants cannot be determined because a change in stability is an unequivocal demonstration that an interaction does exist.

Another method that can sometimes be used to determine a dissociation constant for a ligand that binds with no observable signal change is to use a displacement (or competition) assay in which the optically "silent" ligand is used to displace a ligand that does give a signal. One major advantage of this approach is that it depends on the *difference* in affinity between the two ligands, and not on the absolute values of the affinities. It is therefore possible to use this approach to study very high-affinity interactions (Martin and Bayley, 2002).

VI. Summary

In summary, we hope to have shown that CD remains an excellent technique for studying the structures of proteins and nucleic acids in solution. CD measurements allow one to quickly determine if a protein is folded and, if so, to characterize its secondary structure content. Such measurements therefore enable one to compare the structures of different mutants of a protein, and of different samples of a protein obtained from different species or using different expression systems. The great sensitivity of CD to changes in conformation also makes it an excellent technique for studying the effects of changes in environmental conditions (such as pH, temperature, and ionic strength), for determining protein stability using either chemical or thermal denaturation studies, and (in appropriate cases) for determining equilibrium dissociation constants. One final remark is worth making. A recent review (Kelly *et al.*, 2005) emphasized the unfortunate fact that CD measurements are often severely compromised by inappropriate experimental design or by a lack of attention to important key aspects of instrument calibration and sample characterization. We hope that this chapter shows how reliable CD data can be obtained, analyzed, and understood.

Acknowledgments

S.R.M. is grateful to Dr. Peter Bayley (N.I.M.R.) for introducing him to CD and for his encouragement and support. The preparation of this review was supported in part by Wellcome Trust Grant 072930/Z/03/Z to M.J.S.

References

Andrade, M. A., Chacón, P., Merelo, J. J., and Morán, F. (1993). Evaluation of secondary structure of proteins from UV circular dichroism spectra using an unsupervised learning neural network. *Protein Eng.* **6,** 383–390.

Arai, M., and Kuwajima, K. (2000). Role of the molten globule state in protein folding. *Adv. Protein Chem.* **53,** 209–282.

Becktel, W. J., and Schellman, J. A. (1987). Protein stability curves. *Biopolymers* **26,** 1859–1877.

Bishop, G. R., and Chaires, J. B. (2002). Characterization of DNA structures by circular dichroism. *In* "Current Protocols in Nucleic Acid" (R. A. Jones, ed.), Vol. 2, pp. 7.11.1–7.11.8. Wiley & Sons, Hoboken, New Jersey.

Böhm, G., Muhr, R., and Jaenicke, R. (1992). Quantitative analysis of protein far UV circular dichroism spectra by neural networks. *Protein Eng.* **5,** 191–195.

Brahms, S., and Brahms, J. (1980). Determination of protein secondary structure in solution by vacuum ultraviolet circular dichroism. *J. Mol. Biol.* **138,** 149–178.

Chang, C. T., Wu, C.-S. C., and Yang, J. T. (1978). Circular dichroic analysis of protein conformation: Inclusion of the β-turns. *Anal. Biochem.* **91,** 13–31.

Chen, B.-L., and Schellman, J. A. (1989). Low-temperature unfolding of a mutant of phage T4 lysozyme. 1. Equilibrium studies. *Biochemistry* **28,** 685–691.

Chen, Y.-H., Yang, J. T., and Chau, K. H. (1974). Determination of the helix and β form of proteins in aqueous solution by circular dichroism. *Biochemistry* **13,** 3350–3359.

Chen, Y.-H., Yang, J. T., and Martinez, H. M. (1972). Determination of the secondary structures of proteins by circular dichroism and optical rotatory dispersion. *Biochemistry* **11,** 4120–4131.

Clapperton, J. A., Martin, S. R., Smerdon, S. J., Gamblin, S. J., and Bayley, P. M. (2002). Structure of the complex of calmodulin with the target sequence of calmodulin-dependent protein kinase I: Studies of the kinase activation mechanism. *Biochemistry* **41,** 14669–14679.

Craig, S., Pain, R. H., Schmeissner, U., Virden, R., and Wingfield, P. T. (1989). Determination of the contributions of individual aromatic residues to the CD spectrum of IL-1 beta using site directed mutagenesis. *Int. J. Pept. Protein Res.* **33,** 256–262.

Fasman, G. D. (1996). Differentiation between transmembrane helices and peripheral helices by deconvolution of circular dichroism spectra of membrane proteins. *In* "Circular Dichroism and the Conformational Analysis of Biomolecules" (G. D. Fasman, ed.), pp. 381–412. Plenum Press, New York.

Gill, S. C., and von Hippel, P. H. (1989). Calculation of protein extinction coefficients from amino acid sequence data. *Anal. Biochem.* **182,** 319–326.

Gray, D. M. (1996). Circular dichroism of protein-nucleic acid interactions. *In* "Circular Dichroism and the Conformational Analysis of Biomolecules" (G. D. Fasman, ed.), pp. 469–500. Plenum Press, New York.

Greenfield, N. J. (1975). Enzyme ligand complexes: Spectroscopic studies. *CRC Crit. Rev. Biochem.* **3,** 71–110.

Greenfield, N. J. (1996). Methods to estimate the conformation of proteins and polypeptides from circular dichroism data. *Anal. Biochem.* **235,** 1–10.

Greenfield, N. J. (2004). Analysis of circular dichroism data. *Meth. Enzymol.* **383,** 282–317.

Greenfield, N. J., and Fasman, G. D. (1969). Computed circular dichroism spectra for the evaluation of protein conformation. *Biochemistry* **8,** 4108–4116.

Greenfield, N. J., and Fowler, V. M. (2002). Tropomyosin requires an intact N-terminal coiled coil to interact with tropomodulin. *Biophys. J.* **82,** 2580–2591.

Griffin, M. C. A., Price, J. C., and Martin, S. R. (1986). The effect of alcohols on the structure of caseins. Circular dichroism studies of kappa-casein A. *Int. J. Biol. Macromol.* **8,** 367–371.

Hennessey, J. P., Jr., and Johnson, W. C., Jr. (1981). Information content in the circular dichroism spectra of proteins. *Biochemistry* **20,** 1085–1094.

Hennessey, J. P., Jr., and Johnson, W. C., Jr. (1982). Experimental errors and their effect on analyzing circular dichroism spectra of proteins. *Anal. Biochem.* **125,** 177–188.

Johnson, W. C., Jr. (1985). Circular dichroism and its empirical application to biopolymers. *Methods Biochem. Anal.* **31,** 61–163.

Johnson, W. C., Jr. (1988). Secondary structure of proteins through circular dichroism spectroscopy. *Ann. Rev. Biophys. Biochem.* **17,** 145–166.

Johnson, W. C., Jr. (1990). Protein secondary structure and circular dichroism: A practical guide. *Proteins: Struct. Funct. Genet.* **7,** 205–214.

Johnson, W. C., Jr. (1996a). Determination of the conformation of nucleic acids by electronic CD. *In* "Circular Dichroism and the Conformational Analysis of Biomolecules" (G. D. Fasman, ed.), pp. 433–468. Plenum Press, New York.

Johnson, W. C., Jr. (1996b). Circular dichroism instrumentation. *In* "Circular Dichroism and the Conformational Analysis of Biomolecules" (G. D. Fasman, ed.), pp. 635–652. Plenum Press, New York.

Johnson, W. C., Jr. (1999). Analysing protein circular dichroism spectra for accurate secondary structure. *Proteins: Struct. Funct. Genet.* **35,** 307–312.

Kelly, S. M., Jess, T. J., and Price, N. C. (2005). How to study proteins by circular dichroism. *Biochim. Biophys. Acta* **1751,** 119–139.

Kuwajima, K. (1996). The molten globule state of alpha-lactalbumin. *FASEB J.* **10,** 102–109.

Manavalan, P., and Johnson, W. C., Jr. (1987). Variable selection method improves the prediction of protein secondary structure from circular dichroism spectra. *Anal. Biochem.* **167,** 76–85.

Manning, M. C. (1989). Underlying assumptions in the estimation of secondary structure content in proteins by circular dichroism spectroscopy—a critical review. *J. Pharm. Biomed. Anal.* **7,** 1103–1119.

Martin, S. R., and Bayley, P. M. (1986). The effects of Ca^{2+} and Cd^{2+} on the secondary and tertiary structure of bovine testis calmodulin. *Biochem. J.* **238,** 485–490.

Martin, S. R., and Bayley, P. M. (2002). Regulatory implications of a novel mode of interaction of calmodulin with a double IQ-motif target sequence from *murine* dilute myosin V. *Protein Sci.* **11,** 2909–2923.

Masino, L., Martin, S. R., and Bayley, P. M. (2000). Ligand binding and thermodynamic stability of a multidomain protein, calmodulin. *Protein Sci.* **9,** 1519–1529.

Mayhood, T. W., and Windsor, W. T. (2005). Ligand binding affinity determined by temperature-dependent circular dichroism: Cyclin-dependent kinase 2 inhibitors. *Anal. Biochem.* **345,** 187–197.

Morillas, M., Vanik, D. L., and Surewicz, W. K. (2001). On the mechanism of alpha-helix to beta-sheet transition in the recombinant prion protein. *Biochemistry* **40,** 6982–6987.

Myers, J. K., Pace, C. N., and Scholtz, J. M. (1995). Denaturant *m* values and heat capacity changes: Relation to changes in accessible surface areas of protein unfolding. *Protein Sci.* **4,** 2138–2148.

Pace, C. N., and McGrath, T. (1980). Substrate stabilization of lysozyme to thermal and guanidine hydrochloride denaturation. *J. Biol. Chem.* **255,** 3862–3865.

Pace, C. N., Vajdos, F., Fee, L., Grimsley, G., and Gray, T. (1995). How to measure and predict the molar absorption coefficient of a protein. *Protein Sci.* **4,** 2411–2423.

Perczel, A., Hollósi, M., Tusnády, G., and Fasman, G. D. (1991). Convex constraint analysis: A natural deconvolution of circular dichroism curves of proteins. *Protein Eng.* **4,** 669–679.

Provencher, S. W., and Glöckner, J. (1981). Estimation of globular protein secondary structure from circular dichroism. *Biochemistry* **20,** 33–37.

Robertson, A. D., and Murphy, K. P. (1997). Protein structure and the energetics of protein stability. *Chem. Rev.* **97,** 1251–1267.

Saxena, V. P., and Wetlaufer, D. B. (1971). A new basis for interpreting the circular dichroism spectra of proteins. *Proc. Natl. Acad. Sci. USA* **68,** 969–972.

Schellman, J. A. (1975). Macromolecular binding. *Biopolymers* **14,** 999–1018.

Sreerama, N., and Woody, R. W. (1993). A self-consistent method for the analysis of protein secondary structure from circular dichroism. *Anal. Biochem.* **209,** 32–44.

Sreerama, N., and Woody, R. W. (1994). Protein secondary structure from circular dichroism spectroscopy. Combining variable selection principle and cluster analysis with neural network, ridge regression and self-consistent methods. *J. Mol. Biol.* **242,** 497–507.

Sreerama, N., and Woody, R. W. (2000a). Estimation of protein secondary structure from circular dichroism spectra: Inclusion of denatured proteins with native proteins in the analysis. *Anal. Biochem.* **287,** 243–251.

Sreerama, N., and Woody, R. W. (2000b). Estimation of protein secondary structure from circular dichroism spectra: Comparison of CONTIN, SELCON, and CDSSTR methods with an expanded reference set. *Anal. Biochem.* **287,** 252–260.

Sreerama, N., and Woody, R. W. (2004a). Computation and analysis of protein circular dichroism spectra. *Meth. Enzymol.* **383,** 318–351.

Sreerama, N., and Woody, R. W. (2004b). On the analysis of membrane protein circular dichroism spectra. *Protein Sci.* **13,** 100–112.

Sreerama, N., Venyaminov, S. Y., and Woody, R. W. (2000). Estimation of protein secondary structure from circular dichroism spectra: Inclusion of denatured proteins with native proteins in the analysis. *Anal. Biochem.* **287,** 243–251.

Strickland, E. H. (1974). Aromatic contributions to circular dichroism spectra of proteins. *CRC Crit. Rev. Biochem.* **2,** 113–175.

Swietnicki, W., Morillas, M., Chen, S. G., Gambetti, P., and Surewicz, W. K. (2000). Aggregation and fibrillization of the recombinant human prion protein $huPrP_{90-231}$. *Biochemistry* **39,** 424–431.

van Stokkum, I. H. M., Spoelder, H. J. W., Bloemendal, M., van Grondelle, R., and Groen, F. C. A. (1990). Estimation of protein secondary structure and error analysis from circular dichroism spectra. *Anal. Biochem.* **191,** 110–118.

Venyaminov, S. Y., and Yang, J. T. (1996). Determination of protein secondary structure. *In* "Circular Dichroism and the Conformational Analysis of Biomolecules" (G. D. Fasman, ed.), pp. 69–107. Plenum Press, New York.

Wallace, B. A., Lees, J. G., Orry, A. J. W., Lobley, A., and Janes, R. W. (2003). Analyses of circular dichroism spectra of membrane proteins. *Protein Sci.* **12,** 875–884.

Woody, R. W. (1985). Circular dichroism of peptides. *In* "The Peptides" (V. J. Hruby, ed.), Vol. 7, pp. 15–114. Academic Press, New York.

Woody, R. W. (1995). Circular dichroism. *Methods Enzymol.* **246,** 34–71.

Woody, R. W. (1996). Theory of circular dichroism of proteins. *In* "Circular Dichroism and the Conformational Analysis of Biomolecules" (G. D. Fasman, ed.), pp. 25–67. Plenum Press, New York.

Woody, R. W., and Dunker, A. K. (1996). Aromatic and cystine side-chain circular dichroism in proteins. *In* "Circular Dichroism and the Conformational Analysis of Biomolecules" (G. D. Fasman, ed.), pp. 109–157. Plenum Press, New York.

Yang, J. T., Wu, C.-S. C., and Martinez, H. M. (1986). Calculation of protein conformation from circular dichroism. *Meth. Enzymol.* **130,** 208–269.

CHAPTER 11

Protein Folding and Stability Using Denaturants

Timothy O. Street, Naomi Courtemanche, and Doug Barrick

T. C. Jenkins Department of Biophysics
The Johns Hopkins University
Baltimore, Maryland 21218

METHODS IN CELL BIOLOGY, VOL. 84

0091-679X/08 $35.00
DOI: 10.1016/S0091-679X(07)84011-8

Abstract

Measurements of protein folding and thermodynamic stability provide insight into the forces and energetics that determine structure, and can inform on protein domain organization, interdomain interactions, and effects of mutations on structure. This chapter describes methods, theory, and data analysis for the most accessible means to determine the thermodynamics of protein folding: chemical denaturation. Topics include overall features of the folding reaction, advances in instrumentation, optimization of reagent purity, mechanistic models for analysis, and statistical and structural interpretation of fitted thermodynamic parameters. Examples in which stability measurements have provided insight into structure and function will be taken from studies in the author's laboratory on the Notch signaling pathway. It is hoped that this chapter will enable molecular, cell, and structural biologists to make precise measurements of protein stability, and will also provide a strong foundation for biophysics students who wish to undertake experimental studies of protein folding.

I. Introduction

A major focus of modern molecular and cell biology is understanding the functional and structural properties of proteins that shape the cell, relay signals between the cell and its surroundings, and control dynamics of cell physiology and behavior. The majority of the proteins fulfilling these functions in eukaryotes can be divided, at least conceptually, into multiple folded domains, often separated by natively disordered regions (Romero *et al.*, 2004). In many cases, these domains act as independent units, both in a structural and a functional sense. In other cases, these domains interact with one another structurally and functionally. Dividing large, multidomain proteins into smaller, structurally autonomous units can provide functional insight, and is often an important step in obtaining high levels of pure protein needed for structural analysis.

Unfortunately, it is not always obvious from analysis of primary sequence where structural domains begin and end, and whether a protein segment will fold into a stable tertiary structure or will be natively disordered under physiological conditions. Sequence similarity among homologous proteins is often decreased at domain boundaries; conversely, sequences outside of typical "structural" domains often contribute to structural integrity and improve behavior in solution. Although high-resolution structure determination unambiguously defines domain boundaries, such analysis is often very time- and effort-intensive, and may require the screening of a large number of constructs that differ in the position of their termini. Moreover, high-resolution structure determination is neither feasible nor routine for the majority of cell biologists interested in addressing functional questions.

Measurement of protein stability is a much easier way to obtain information on domain boundaries, and provides a simple means to test whether a segment excised from a larger polypeptide can fold into a rigid, well-defined structure. Stability studies can identify elements outside of regions of high sequence similarity that are critical for obtaining large quantities of well-behaved material for high-resolution structural determination. Even when the stabilities, structures, and functional contributions of "individual domains" of a large multidomain protein cannot be separated from each other, studying the properties (and aberrances) of the parts can provide key insights into the often complex relationship between structure and function.

Stability studies also provide a direct means to analyze the effects of sequence substitutions on stability. Such information helps to define the physicochemical interactions that stabilize a folded structure, and often provides insight into evolutionary relationships among protein sequences. In addition, stability studies provide insight into the molecular basis underlying functional deficiencies resulting from sequence substitutions.

II. Rationale

This chapter is meant to make methods for studying protein stability and folding accessible to cell biologists as well as biophysicists. There are a number of different methods used for such studies, including thermal scanning calorimetry, high-pressure denaturation studies, and hydrogen-exchange methods monitored by high-resolution NMR spectroscopy (Grimsley *et al.*, 2003; Pace and Scholtz, 1997; Privalov, 1979; Royer, 2002). Although all of these methods provide insight into protein folding and stability, they require specialized instrumentation, and in some cases, a high level of expertise and effort. Here we focus on "chemical denaturation" (the isothermal disruption of protein structure using agents like urea and guanidine), because sophisticated instrumentation is not required, because data analysis is straightforward, and because high-precision numerical estimates of protein stability can easily be obtained. In addition, chemical denaturation is the most common method for determining folding energies. Finally, equilibrium chemical denaturation studies are key to interpreting most studies of protein folding kinetics, which rely on denaturants to produce rapid, synchronous population shifts.

It is hoped that this chapter will serve two purposes. The first is to guide molecular and cell biologists in determining free energies of protein folding using chemical denaturation. Examples will show how measurements of protein folding energetics can inform on the effects of sequence substitutions on protein structure and stability, and on domain–boundaries and domain–domain interactions in large, complex proteins that are often of interest to cell and structural biologists alike.

The second purpose of this chapter is to review chemical denaturation methods for biophysicists. A similar methods paper written 20 years ago by Pace (1986) was one of the most influential, enjoyable, and useful papers that one of us (DB) read as a beginning graduate student many years ago. Although this paper remains a classic, much has changed in the areas of instrumentation, data acquisition, and data analysis. In addition, a large increase in the number of proteins for which folding energies are now available has led to advances in understanding the mechanisms of protein folding and denaturation. This chapter will incorporate these advances, and will attempt to identify some of the potential pitfalls associated with modern methods of acquisition and analysis of chemical denaturation data. This chapter will also investigate the efficacies of various protocols intended to purify denaturant solutions for spectroscopic studies of unfolding, including some adaptations we have found to be effective in our own laboratory. It is our hope that this chapter will provide a useful starting point for both students and established biophysicists wishing to investigate issues connected to protein folding energetics. Readers in this group should also see two recent chapters detailing methods to determine thermodynamic stabilities of proteins by Scholtz, Pace, and coworkers (Grimsley *et al.*, 2003; Pace and Scholtz, 1997).

III. Methods

A. General Features of Chemical Denaturation

The chemical denaturation experiment is simple both in its design and in the appearance of the data. A structure-sensitive probe [usually spectroscopic, such as fluorescence or circular dichroism (CD; see Chapter 10 by Martin and Schilstra, this volume for a more detailed discussion), although probes of hydrodynamic properties are also satisfactory] is selected that can distinguish the native (folded) and denatured (less folded) conformational ensembles. Then a series of samples are prepared that contain varying amounts of denaturant (typically urea or guanidine hydrochloride), but have the same protein concentration, and the probe is used to determine the amount of structure in each sample at a fixed temperature.

By plotting the degree of structure as a function of molar denaturant concentration, a sigmoidal plot is obtained for most single-domain proteins (Fig. 1A). Such plots can be conceptually broken into three regions. At low denaturant concentration, where the native structure persists, there is little (or sometimes no) change in the structural variable being monitored. This region will be referred to as a native baseline (labeled "N" in Fig. 1A). As denaturant concentration increases, the flat or linearly varying native baseline gives way to a steep "transition" region, where the native and denatured forms coexist in measurable proportions (labeled "N + D" in Fig. 1A). Thermodynamic parameters relating to protein stability depend critically on this region—the baselines do not directly contain information on the position of equilibrium (although they are essential for interpreting the transition, see below). Determination of thermodynamic quantities outside of this region necessarily involves extrapolation, which can magnify the uncertainties in these quantities. As denaturant concentration increases further, the transition region

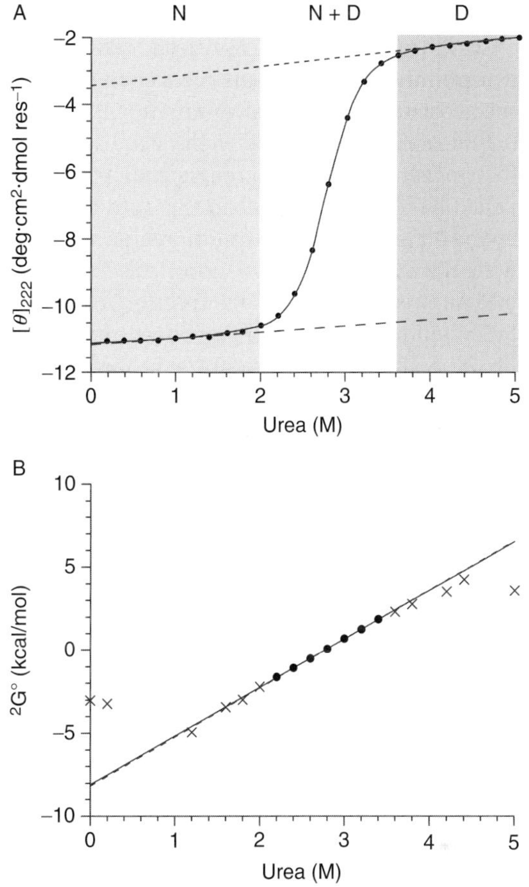

Fig. 1 Chemical denaturation of a single-domain monomeric protein. (A) Urea-induced unfolding of the ankyrin domain of the Drosophila Notch receptor, monitored by CD at 222 nm. A transition region (denoted N + D) separates the native and denatured baseline regions (shaded areas; boundaries between these regions are somewhat arbitrary). Dashed and dotted lines show linear fits to native and denatured baselines, respectively. The solid line shows the results of nonlinear least-squares fitting of the linear free energy relation [Eq. (9)] to the CD data. Measurements were made on a 5 μM protein sample containing 25 mM Tris, 150 mM NaCl, pH 8.0, 14 °C. (B) Folding free energy estimates using fitted baseline values [Eq. (4)], inside the transition region (filled circles) and in the baseline regions (*X*s). The solid line results from a linear fit to free energy estimates from the transition region. In this case, the line closely approximates that derived from a nonlinear least-squares fit to the entire transition [dashed line, corresponding to the fitted curve in (A)]. Note that estimates of ΔG° from the baseline regions are highly scattered; moreover, equilibrium constants calculated using Eq. (4) for roughly half of the baseline points had negative values, confirming that such estimates are meaningless in the baseline regions.

gives way to a second baseline region, in which there is a vast excess of denatured protein, with an undetectable amount of native protein (labeled "D" in Fig. 1A). Like the native baseline region, there is often a modest, linear change in the structural variable in the denatured baseline region.

Although equilibration of native and denatured proteins is only directly observed in the transition region,[1] all three regions are critically important for determining thermodynamic parameters for protein stability. This is because the spectroscopic signal used to monitor the position of equilibrium in the transition region must be compared to the signals expected for native and denatured protein to convert the spectroscopic signal to equilibrium populations. These limiting values, which are defined in the native and denatured baseline regions, must be extrapolated into the transition region to account for the variation in these signals with denaturant. Thus, a good rule of thumb is to make a similar number of measurements in each of the three regions. Because the transition region is typically limited to a narrow range of denaturant concentration, and because the modest baseline slopes often require a broad range of denaturant concentration to be accurately determined, a different density of points (in terms of denaturant concentration step sizes) in these three regions is sometimes required, with a higher density in the transition region than in the baseline regions.

B. Equilibration and Reversibility

Of central importance in determining thermodynamic quantities for protein folding is that chemical equilibrium be achieved and maintained through the course of measurement. To achieve equilibrium, two conditions must be met. First, enough time must pass after mixing the protein (either denatured or native) with denaturant for the reaction to reach its endpoint, that is, the rates (but not the rate constants!) for folding and unfolding balance. Second, unfolding must be reversible. The most common cause of irreversible unfolding is aggregation in the denatured state, a problem that is often more severe at high temperatures and protein concentrations (increased aggregation at higher temperatures often results in lower reversibility in thermal denaturation compared to chemical denaturation).

Reversibility can be checked by a series of experiments in which denaturant concentration is changed rapidly (refolding by dilution of denaturant with buffer, unfolding by addition of denaturant), and the progress of the reaction is monitored spectroscopically as a function of time. If the reaction is reversible, progress curves should come to the same final value regardless of the starting conditions, that is, regardless of whether reaction started from native or denatured material (Chen *et al.*, 1989). Although a qualitative comparison of stabilities may be made for proteins that fail to come to equilibrium, results will always have a time dependence that is not a characteristic of system at true chemical equilibrium: reversible thermodynamic parameters can only be quantified for a system in reversible equilibrium.

[1] Amide hydrogen-deuterium exchange provides an exception to this, being sensitive to the very low concentrations of denatured material populated within the native baseline region (Englander *et al.*, 2002; Li and Woodward, 1999). Thus, this method can be used to obtain thermodynamic properties outside the transition region, although considerably more material, instrumentation, expertise, and effort must be applied than for the bulk methods described here.

Equilibration times can also be checked by monitoring reaction progress as described above. As a loose guideline, samples should be equilibrated for five or more halftimes. For reversible first-order folding reactions, rates are slowest near the denaturant concentration midpoint (referred to as C_m) for the unfolding transition. Thus, unfolding and refolding to the midpoint provides the most conservative estimate of equilibration time. Equilibration times often show a large temperature dependence, being longest at low temperatures. Equilibration times also vary from protein to protein, ranging from a second or less, to many minutes (see Junker *et al.*, 2006 for an example of an extraordinarily slowly equilibrating folding reaction). The halftime can be estimated by fitting a first-order rate constant to a kinetic progress curve using the following three-parameter equation:

$$Y(t) = Y_{\infty} + \Delta Y \times e^{-k_{app}t} \tag{1}$$

where Y_{∞} is the (equilibrium) signal at long times, ΔY is the difference between the starting and equilibrium signal, k_{app} is the apparent rate constant for the observed folding step, and t is time. The halftime is obtained through the formula $t_{1/2} = 0.693/k_{app}$.

C. Automated Titrations

One of the most important advances in measuring protein stability by chemical denaturation in the last decade has been the availability of automatic, real-time titrators that can interface with CD and fluorescence spectrometers. These devices are sold as accessories from some spectrometer manufacturers. One particularly reliable package that includes instrumentation and computer-controlled software is bundled with Aviv spectrometers (Lakewood, NJ). The Aviv titration system has provided highly reliable titration curves in our laboratory with a minimum of sample waste. All titrations in this chapter were collected with an Aviv automated titrator. These devices use a pair of motor-driven or pneumatically controlled syringes to change the denaturant concentration of a protein sample in the spectroscopy cell, allowing an entire denaturation experiment to be collected without manual intervention and from only two protein samples. Typically, a buffered protein solution is placed in the cell, and is fitted with two narrow-diameter pieces of tubing, each connected to a pump syringe. Following signal measurement, a small amount of the starting material is removed from the cuvette with one syringe, and is replaced with an equal volume of denaturant (typically containing protein at the same concentration as is in the cuvette[2]) from the other syringe. After an

[2] Although protein can be omitted from the titrating solution and the data can be scaled to account for the dilution that results, there are two major disadvantages to this approach. First, dilution of protein degrades signal-to-noise in the denatured baseline, resulting in an increased error in the fitted thermodynamic parameters (in fact, the assumption of uniform error required for least-squares fitting no longer holds). Second, on scaling, nonzero spectroscopic baseline signals (resulting from photomultiplier offsets, buffer impurities, or strain in the cell wall) will be stretched nonlinearly through the course of titration, will distort the titration curve, and will produce systematic errors in the fitted parameters.

appropriate equilibration time, in which the sample is magnetically stirred to ensure mixing, the spectroscopic signal is recorded, and the process is repeated. The entire titration schedule is programmed in advance to give the desired density of measurements in the transition and baseline regions.

Automatic titrators result in greater pipetting accuracy and greater mechanical stability than are accessible from manual titration, resulting in higher signal-to-noise. In addition, automatic titrations require less material than do manual titrations. Further, automatic titrators greatly decrease the time expended in preparing and changing samples, resulting in higher throughput (for rapidly equilibrating samples). However, for samples with long equilibration times, automatic titration becomes impractical because equilibrium must be reestablished after each injection, resulting in very lengthy titrations.

D. Modeling the Unfolding Reaction

The thermodynamic quantities obtained from chemical denaturation experiments describe energies of reaction. Thus, estimating such energies requires a model for the reaction mechanism. Once a mechanism is specified, thus defining relevant equilibrium constant(s), the extent of reaction in the transition region can be evaluated from the progress of the observed signal from the native baseline toward the denatured baseline. The simplest model is one in which monomeric native and denatured forms of the protein (N and D, respectively) convert directly, without highly populated intermediates:

$$\mathrm{D} \rightleftharpoons \mathrm{N} \tag{2a}$$

$$K_{\mathrm{eq}} = \frac{[\mathrm{N}]}{[\mathrm{D}]} \tag{2b}$$

This simple scheme, which is referred to as "equilibrium two-state," does not imply that there are only two specific conformations. Rather, N and D can be regarded as separate ensembles of structures, one largely folded, the other largely unfolded. The two-state scheme simply requires that conversion between these two ensembles is substantially slower than conversion within each ensemble. For an excellent description of ensembles and how they influence protein folding, see Hilser *et al.* (2006). Note also that the equilibrium two-state scheme does not preclude transient kinetic intermediates, as long as such intermediates are high in energy compared to either N or D (Sanchez and Kiefhaber, 2003a,b).

Although Eq. (2a) does not apply to all proteins, particularly those with multiple structural domains, the scheme does a surprisingly good job describing the folding transitions of many single-domain globular proteins. The free energy of protein folding at each denaturant concentration through the transition can then be described using the standard relationship:

$$\Delta G^\circ = -RT \ln K_{eq} \tag{3}$$

where R is the universal gas constant (8.314 J mol^{-1} K^{-1}) and T is temperature on the Kelvin scale. Here ΔG° is the free energy of change when one mole of D folds to N, at a constant concentration of one molar each of N and D, assuming an ideally dilute solution (i.e., activity coefficients for N and D of one). Although at these high concentrations the activity coefficients of N and D would likely be very different from one,[3] the unimolecular nature of Scheme (2) and the associated equilibrium constant [Eq. (2b)] results in equivalent reaction free energies if lower standard state concentrations (such as micromolar) of N and D are chosen.

The "direction" of reaction defined in Scheme (2) differs from that in many published studies and from the historical development of protein folding, which treated the reaction as a disruption of folded structure isolated from biological materials (i.e., from N to D). While the direction of the reaction is arbitrary for reactions in reversible equilibrium, it determines the sign of the reaction free energy. With the direction of the reaction adopted here (Scheme 2), negative ΔG° values correspond to lower free energies for N than for D. A negative reaction energy in going from D to N is consistent with energy landscape descriptions of protein folding that have emerged in the last decade.

For some proteins, unfolding transitions are more complicated than as in Fig. 1A. In such cases, models must be expanded to include additional species. In rare instances, these species are clearly resolved between separate transitions (as for integer repeat insertions into the Notch ankyrin domain; Tripp and Barrick, 2006). More often, however, the population of additional intermediate species has a more subtle effect, increasing the breadth of the transition without resolving into two separate transitions.

E. Calculating Folding Energies Within Transition Regions

As described in Section III.A , Methods, with knowledge of baseline spectroscopic signals, relative concentrations of N and D, and thus the equilibrium constant, can be determined within the transition region. This can be achieved with the formula:

$$K_{eq} = \frac{[N]}{[D]} = \frac{Y_{obs} - Y_D}{Y_N - Y_{obs}} \tag{4}$$

where Y_{obs} is the observed spectroscopic signal, and Y_N and Y_D are signals of pure N and D, determined from extrapolation of the linear baselines into the transition region. The numerator on the right-hand side is simply a measure of the extent that the observed signal deviates from the denatured baseline, and is thus proportional to the amount of protein that is not in the denatured state (N, according to the two-state model). The denominator references this to the corresponding deviation from

[3] Moreover, obtaining concentrations of large macromolecules of 1 M would be impossible.

the native baseline, and represents the amount of material that is not native (i.e., D). In Fig. 1B, Eq. (4) has been applied to the unfolding data in the transition region of Fig. 1A to estimate free energies of folding (circles).

Consideration of Eq. (4) makes clear the reason that K_{eq} (and ΔG°) can only be evaluated in the transition: only in this region is Y_{obs} significantly different from both Y_N and Y_D. Outside the transition region, either the numerator (at high denaturant concentrations) or the denominator (at low denaturant concentrations) is dominated by the noise associated with the spectroscopic measurement, and have no information about the concentration of N and D, respectively. Figure 1B shows the (inappropriate) use of Eq. (4) to estimate equilibrium constants outside the transition (*X*s), demonstrating the inability of the baseline regions to report on energetics with any precision.

Although free energies in the transition region estimated from Eq. (4) are valid from a thermodynamic perspective, they have several practical shortcomings. First, they are free energies of folding in high concentrations of denaturant, which limits their physiological relevance. Second, regardless of protein stability under defined conditions, folding energies from the transition region will all be within the same range of ± 1 kcal mol^{-1}. Thus, such transition free energies do not directly allow comparison of the stabilities of different proteins. While it is possible to get some sense of the stability from the concentration midpoint of the transition, midpoint comparisons between proteins can be complicated by differing slopes in the transition region. Instead, what is needed is an analytical expression that can be used to evaluate protein stabilities at a common denaturant concentration. The most popular (and physiologically relevant) denaturant concentration is zero molar. The free energy of folding in the absence of denaturant (referred to as $\Delta G^{\circ}_{H_2O}$) is simply the *y*-intercept in Fig. 1B.

F. Modeling the Effects of Denaturant on Folding Energies

Several relations have been proposed to describe the variation in folding free energies with denaturant concentration (Pace, 1986; Schellman, 2006). The most commonly used of these is one in which the folding free energy varies linearly with denaturant concentration (Greene and Pace, 1974):

$$\Delta G^{\circ}(x) = \Delta G^{\circ}_{H_2O} + m[x] \tag{5}$$

Here $[x]$ is the molar denaturant concentration, $\Delta G^{\circ}_{H_2O}$ is the *y*-intercept mentioned above, and m is a proportionality constant related to the steepness of the transition. Physical interpretations of the *m*-value will be discussed below (Section B, Discussion).

Based on the free energies determined from the transition region in Fig. 1B, the linear dependence in Eq. (5) seems quite reasonable. However, determination of $\Delta G^{\circ}_{H_2O}$ as a *y*-intercept from ΔG° values in the transition region involves a long extrapolation, and small errors in quantifying the slope of the transition can be magnified through this extrapolation. A more serious error in $\Delta G^{\circ}_{H_2O}$ may arise if ΔG° varies nonlinearly with denaturant. Although values within the transition appear to be linear, any well-behaved function is approximately linear over a narrow range, and the transition is indeed narrow.

One detraction from the linear expression [Eq. (5)] is that it is not easily justified based on simple chemical principles; rather, it is postulated as a phenomenological expression. Although the implications of linearity on underlying thermodynamics and mechanism have been extensively discussed (e.g., see Schellman, 1978), no simple mechanistic explanation for linearity has been provided. A linear relationship would be more likely between the *chemical potential* of denaturant, which is approximately logarithmic in denaturant molarity, and the free energy of folding. Indeed, a logarithmic relationship is often observed between the concentrations of small molecules and free energies of macromolecular conformational transitions (Wyman, 1964).

One such relationship has been arrived at by treating the denaturant–protein interaction using mass action. In this treatment, it is assumed that denaturant molecules can bind to specific sites on the native and denatured proteins. The effect of denaturants on the conformational transition then results from a difference in the number of binding sites (Δn) in N and D, a difference in affinities, or both. In this treatment, the free energy of binding varies with the log of the ratio of the binding polynomials. Under the rather extreme assumption that all sites have identical affinities, the expression for the denaturant dependence of the folding free energy simplifies to:

$$\Delta G^{\circ}(x) = \Delta G^{\circ}_{H_2O} - \Delta nRT\ln(1 + k[x]) \tag{6}$$

where k is a microscopic (i.e., single-site) binding constant. This relationship has been in the literature (Aune and Tanford, 1969; Schellman, 1955) longer than the linear relation [Eq. (5)]. Although this logarithmic relationship might seem incompatible with chemical denaturation data owing to the apparent linear (rather than logarithmic) variation in transition free energies with denaturant (Fig. 1B), Eq. (6) is compatible with apparent linearity as long as denaturant interactions are weak but numerous (i.e., small k, large Δn). One disadvantage of the binding model [Eq. (6)], particularly in the weak-binding (i.e., linear) limit, is that the denaturant dependence is controlled by two parameters (Δn and k), rather than one [m, Eq. (5)], and these parameters are highly correlated. Another disadvantage of the binding model is the assumption that the denaturant effect can be described through mass action (rather than by altering the structure of water or its activity). The difficulties in using mass-action concepts in solutions in which the denaturant concentration approaches that of the solvent have been discussed extensively by Schellman (Schellman, 1990). An additional shortcoming of the binding model is the unlikely simplification that there is only one type of (thermodynamic) site (see Schellman, 1994 for a detailed discussion of the implications of this approximation). Although modeling binding site diversity is straightforward, it necessarily introduces additional parameters.

More rigorous thermodynamic approaches have come from measurements of preferential interaction coefficients and preferential hydration studies (Courtenay *et al.*, 2000; Parsegian *et al.*, 1995; Schellman, 2002; Timasheff, 1998). While such studies are of great interest to protein physical chemists, the measurements associated with this type of analysis often require high-precision measurements of refractive index and vapor pressure, limiting their use to nonspecialists. Other noteworthy additions to theories of chemical denaturation include treatment of exchange reactions of denaturant with water through mass action (Schellman, 1990), and accounting for excluded volume effects between the denaturant and protein (which is roughly linear in denaturant concentration, but usually favors the native state; Schellman, 2003).

In between the entirely phenomenological linear relation and the highly approximate binding model are solvent transfer models [Tanford, 1970; see Pace, 1986 for a discussion]. These models are based on measurement of free energies of transfer of the 20 amino acids and backbone mimics from water into denaturant, and are approximately linear in denaturant concentration. Solvent transfer models have recently been applied with considerable success to analyze the effects of stabilizing osmolytes such as trimethylamine-*N*-oxide (TMAO) on protein stability (Wang and Bolen, 1997).

Several studies have tested which of the different analytical equations above best describes the denaturant dependence of ΔG° outside of the folding transition. One test, described in Pace's review (Pace, 1986), compares values of $\Delta G^\circ_{H_2O}$ obtained using different denaturants. This parameter should be independent of the denaturant used to bring about a transition. Values of $\Delta G^\circ_{H_2O}$ estimated from different denaturants using the linear relation [Eq. (5)] agree within error, whereas discrepancies have been noted with the binding model [Eq. (6)].

Other tests supporting the linear relationship involve combining chemical denaturation with other independent means of modulating the free energy of unfolding. Santoro and Bolen (1988) found $\Delta G^\circ_{H_2O}$ values obtained from the linear relationship to complete a closed thermodynamic cycle when combined with changes in pH (which can be treated in a thermodynamically rigorous way using potentiometric methods), demonstrating the path independence expected of a thermodynamic state function (Bolen and Santoro, 1988). Other studies have combined chemical denaturation with temperature variation (which can be modeled using thermodynamic relationships that take into account heat capacity changes between N and D) and have provided support for extrapolation using the linear equation (Agashe and Udgaonkar, 1995; Santoro and Bolen, 1992; Swint and Robertson, 1993; Zweifel and Barrick, 2002). Insight into denaturant mechanisms has also been provided by calorimetric measurement of heats of denaturant interactions (Makhatadze and Privalov, 1992; Zou *et al.*, 1998).

Thus, although the linear relationship is based neither on mechanistic nor solution thermodynamic relationships, it has received more experimental support than other relationships (also see Gupta and Ahmad, 1999). Combined with its simplicity and common use, it will likely be the method of choice for most studies of protein stability, and will be treated exclusively in the remainder of this chapter.

G. Analyzing Denaturation Curves to Extract Thermodynamic Parameters

Eqs. (3–5) provide a simple multistep procedure to estimate the two thermodynamic parameters ($\Delta G^{\circ}_{H_2O}$ and m-value) associated with a denaturation transition. First, baseline regions are selected, fitted with straight lines, and are extrapolated into the transition. These extrapolated baseline signals are then combined with signals in the transition to yield equilibrium constants using Eq. (4), which are converted to free energies using Eq. (3). Finally, these transition free energies are fitted by Eq. (5) using a second round of linear regression (solid line, Fig. 1B) to determine $\Delta G^{\circ}_{H_2O}$ (the y-intercept) and the m-value (the slope). Although conceptually simple, there are a number of drawbacks to this method. First, selection of which measurements are part of the baseline and which are part of the transition is rather subjective. There is a tradeoff between taking points far from the midpoint that have large errors (Xs, Fig. 1B) and omitting points near the ends of the transition that could help define the m-value (and thus the intercept, $\Delta G^{\circ}_{H_2O}$). Finally, both the transformations from Eqs. (3) and (4) distort errors, greatly compromising the ability of least-squares methods to accurately determine fitted parameters.

These drawbacks can all be avoided by using nonlinear least-squares fitting directly to the data (Johnson, 1992). Details of nonlinear least-squares fitting can be found in this volume (see Chapter 24 by Johnson, this volume). Numerous programs implement nonlinear least-squares fitting routines, including KaleidaGraph (Synergy Software), ProFit (Quantum Soft), Matlab (Mathworks, Inc.), and also the free statistics package *R* (http://www.r-project.org/). Santoro and Bolen (1988) presented one of the first examples of fitting chemical denaturation data with Eq. (9) by nonlinear least-squares.

Direct fitting requires an analytical expression relating the observed signal to the population-weighted averages of the signals of all species present. For a two-state mechanism:

$$Y_{obs} = f_N \times Y_N + f_D \times Y_D \qquad (7)$$

where f_N and f_D are the fraction of native and denatured protein, and sum to one.[4]

[4] Summation of f_N and f_D to one is another expression of the two-state model. To account for additional species, Eq. (7) is modified to include analogous terms for each distinct species.

Expressing these fractions explicitly in terms of N and D allows Eq. (7) to be expressed in terms of the folding free energy:

$$\begin{aligned} Y_{\text{obs}} &= \frac{[\text{N}]}{[\text{N}]+[\text{D}]} \times Y_{\text{N}} + \frac{[\text{D}]}{[\text{N}]+[\text{D}]} \times Y_{\text{D}} \\ &= \frac{K_{\text{eq}}}{1+K_{\text{eq}}} \times Y_{\text{N}} + \frac{1}{1+K_{\text{eq}}} \times Y_{\text{D}} \\ &= \frac{Y_{\text{N}}\text{e}^{-\Delta G^\circ/RT}}{1+\text{e}^{-\Delta G^\circ/RT}} + \frac{Y_{\text{D}}}{1+\text{e}^{-\Delta G^\circ/RT}} \\ &= \frac{Y_{\text{D}} + Y_{\text{N}}\text{e}^{-\Delta G^\circ/RT}}{1+\text{e}^{-\Delta G^\circ/RT}} \end{aligned} \tag{8}$$

Denaturant dependence of Eq. (8) is introduced via the folding free energy using the linear model [Eq. (8); although, any other analytical relation can be used] and via the baselines, as simple linear dependences:

$$Y_{\text{obs}} = \frac{Y_{\text{D,H}_2\text{O}} + \beta_{\text{D}}[x] + (Y_{\text{N,H}_2\text{O}} + \beta_{\text{N}}[x])\text{e}^{-(\Delta G^\circ_{\text{H}_2\text{O}}+m[x])/RT}}{1+\text{e}^{-(\Delta G^\circ_{\text{H}_2\text{O}}+m[x])/RT}} \tag{9}$$

where β_{D} and β_{N} are slopes of the denatured and native baseline signals with denaturant, and $Y_{\text{D,H}_2\text{O}}$ and $Y_{\text{N,H}_2\text{O}}$ are denatured and native baseline values in the absence of denaturant. In total, Eq. (9) has six adjustable parameters that must be fitted. Though this may seem a large number of parameters to fit from a single unfolding transition, as long as the two baselines and the transition region are adequately sampled, each of these parameters can be fitted with high confidence. Although the multistep fitting procedure outlined above appears to only involve two-parameter fits, three separate fits are required to obtain free energies, bringing the total to six parameters.

Equation (9) can be used to demonstrate the effects of the two thermodynamic parameters for folding ($\Delta G^\circ_{\text{H}_2\text{O}}$ and the m-value) on unfolding transitions. As expected from Eq. (5), variation in $\Delta G^\circ_{\text{H}_2\text{O}}$ shifts the midpoint of the transition, but has no effect on the shape (Fig. 2A). Changes in shape result from changes in the m-value, with larger m-values corresponding to steeper slopes [greater sensitivity to denaturant; Eq. (5)]. However, if $\Delta G^\circ_{\text{H}_2\text{O}}$ remains constant, changes in the m-value result in variation in the midpoint of the transition (Fig. 2B).[5] To hold the transition midpoint constant, $\Delta G^\circ_{\text{H}_2\text{O}}$ must increase in proportion to the m-value (Fig. 2C).

Since one of the major goals of analyzing chemical denaturation experiments is to determine the thermodynamic parameters for folding ($\Delta G^\circ_{\text{H}_2\text{O}}$ and the m-value), it is important to determine the uncertainties associated with these parameters. Many of the software packages listed above report "errors," and offer various levels of explanation of what these quantities represent. Such values are typically determined

[5] This is because at the transition midpoint, $\Delta G^\circ = 0$, and thus from Eq. (5), $\Delta G^\circ_{\text{H}_2\text{O}} = m[C_{\text{m}}]$, where C_{m} is the denaturant concentration at the transition midpoint.

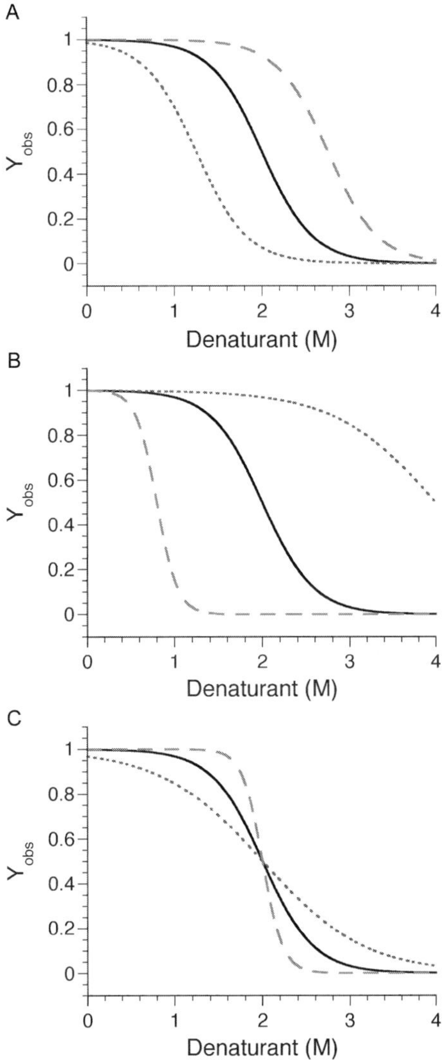

Fig. 2 Sensitivity of denaturation curves to underlying thermodynamic parameters. Chemical denaturation curves were generated assuming a linear dependence of free energy on denaturant concentration [Eq. (9)]. (A) Decreasing the value of $\Delta G^{\circ}_{H_2O}$ (−2.5, −4, and −5.5 kcal mol^{-1} in dotted, solid, and dashed lines, respectively) with constant *m*-value (2 kcal mol^{-1} M^{-1}) shifts the transition midpoint to higher values, but leaves the shape unchanged. (B) Increasing the *m*-value (1, 2, and 5 kcal mol^{-1} M^{-1} in dotted, solid, and dashed lines, respectively) with a constant value of $\Delta G^{\circ}_{H_2O}$ (set to −4 kcal mol^{-1}) increases the slope of the transition, but also decreases the transition midpoint. (C) Increasing the *m*-value as in (B) with proportionate decreases in $\Delta G^{\circ}_{H_2O}$ (−2, −4, and −10 kcal mol^{-1}) fixes the transition midpoints, emphasizing the changes in slope.

by asking how much the parameters can be perturbed from their optimized values while maintaining the goodness of fit above some threshold. Although such "fitting errors" provide a feel for how well a particular data set constrains the adjustable parameters, they are blind to experiment-to-experiment variations that may result from preparation of solutions, protein samples, pipetting, and slow instrument drift. In our opinion, the best way to determine the accuracy of the fitted parameters is to perform multiple separate chemical denaturation experiments, fit parameters for each data set separately, and compare the results.

H. Spectroscopic Tests of the Two-State Model

The use of Eqs. (2–9) assumes that only two thermodynamically distinct states are populated at all denaturant concentrations. As noted above, equilibrium intermediates can be significantly populated in the transition region without producing two clearly resolved unfolding steps. One way to test the two-state model using simple chemical denaturation methods is to compare unfolding transitions monitored by independent spectroscopic signals. If the transition is two-state, the same thermodynamic parameters should be obtained regardless of the signal. This test is most stringent when at least one of the spectroscopic measurements is specific to a local region of the macromolecule such as fluorescence of a single tryptophan. If the spectroscopic measurements are both global in nature (e.g., far-UV CD and peptide infrared spectroscopy), both measurements will likely produce the same transition regardless of the number of states involved. One complementary pair of measurements is far- and near-UV CD, which probe secondary structure and tertiary structure surrounding aromatic side chains. This pair of probes has identified molten globule intermediates in the unfolding of several proteins (Kuwajima, 1989). It is important to note that although deviation of independent spectroscopic probes (and fitted thermodynamic parameters) is sufficient to invalidate the two-state mechanism, coincidence of independent probes does not prove a two-state mechanism. Rather, such coincidence is supportive of such a mechanism, and can be strengthened by other tests (e.g., thermal and kinetic).

IV. Materials

An advantage of chemical denaturation as a means to determine protein stability is that it is both conceptually and experimentally simple, requiring little more than a sensitive optical detection system (preferably thermostatted), some inexpensive denaturant, and some accurate pipettes (a collection of gas-tight Hamilton syringes are preferable, although as described above, an automated titrator interfaced with the optical system greatly decreases the effort involved, and increases the precision of the data). However, some simple precautions should be taken to ensure that the data and resulting stability estimates are of high quality.

A. Preparation of Denaturant Solutions

There are two issues related to the use of denaturants that must be considered. First, the most common denaturants, urea and guanidine hydrochloride, have carbonyl moieties. As a result, they have high absorbance in the far-UV. Although this is not an issue when monitoring unfolding transitions by tryptophan fluorescence (e.g., the absorbance of 6 M urea in a 1 cm cuvette is negligible above 250 nm; Fig. 3A), it presents a challenge when monitoring transitions through secondary structure loss by CD. At 220 nm, where α-helical structure is detected, a 6 M urea solution has an absorbance between 0.5 and 1.0 in a 1 cm cuvette (Fig. 3A), depending on how the solution is prepared (see below). This high absorbance tends to degrade signal-to-noise at high denaturant concentrations, and limits the concentration of protein that can be used in chemical denaturation experiments.

A second issue related to the use of denaturants is that commercial preparations of denaturants often contain significant amounts of impurities. Urea often contains trace metal ions and cyanates (Pace, 1986). Such contaminants increase the absorbance of urea solutions, further complicating spectroscopic measurements in the far-UV. Cyanates are particularly problematic because they covalently modify lysine side chains, and are formed (along with ammonia) from urea via a decomposition reaction.

Fortunately, these impurities can be removed from urea solutions with little effort. Published procedures for purification of urea solutions include treatment with deionizing resin [e.g., AG 501-X8 (D); Biorad], treatment with activated charcoal, and recrystallization (Prakash *et al.*, 1981). Of these, the first two are the least time-consuming, and both decrease significantly the background absorbance of urea solutions in the far-UV (by as much as 40% at 220 nm; Fig. 3A). Because deionizing resin produces a larger decrease in absorbance than charcoal, and treatment with charcoal provides no added benefit to resin-treated urea, we simply treat urea solutions with resin for 1 h. Following treatment with deionizing resin, we carry out all titrations within 24 h to avoid cyanate buildup.[6] Alternatively, deionized urea stock solutions can be stored frozen at −80 °C without significant degradation. When deionization is used to prepare urea for protein denaturation studies, it is important to add buffer and salt components after deionization, to avoid depletion by the resin.

Owing to its high pK_a, guanidine dissociates to a salt in aqueous solution. As a result, treatment with deionizing resin is not an appropriate means of purification. Although in principle, activated charcoal could be used to absorb nonionic organic impurities, we find such treatment to increase absorbance in the far-UV. Although crystallization would also seem a fairly straightforward means to remove impurities, we have not found published protocols (Nozaki, 1972) to lead to significant decreases in absorbance in the far-UV region (Fig. 3B).

[6] Although cyanate concentration will eventually be reformed through equilibrium with urea, the reaction is quite slow, taking several days to reestablish, and lies far in the direction of urea (Hagel *et al.*, 1971).

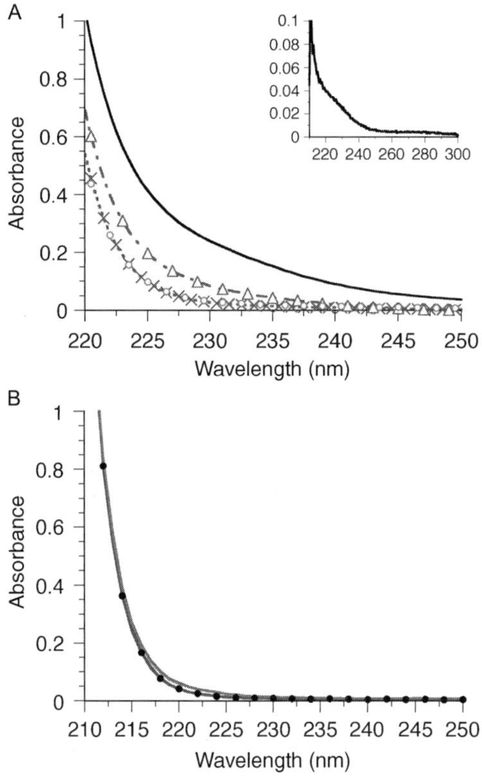

Fig. 3 Effects of deionizing resin on far-UV absorbance of urea. (A) By exposing 6.0 M urea solutions to deionizing resin (dotted curve with *X*s), absorbance can be significantly lowered at wavelengths used to detect protein secondary structure by CD spectroscopy (230 nm and below; compare to untreated curve, solid line). Reduced absorbance in this region can also be achieved by treating with activated charcoal (dot-dashed curve with triangles), although this decrease is less pronounced than that produced by treatment with deionizing resin, and provides no added reduction in absorbance in combination with deionizing resin (circles). The inset shows the absorbance difference between untreated and deionized urea solutions at 1.0 M concentration. For these spectra, ultrapure urea was obtained from Amresco, deionizing resin [5%, w/v AG 501-X8 (D)] was obtained from Biorad, and activated charcoal was obtained from Sigma. Solutions were prepared using distilled, deionized water (ddH_2O), and were stirred with 5% (w/v) added resin or activated charcoal for 1 h, and were filtered through 0.22 μm filters. (B) By comparison, recrystallization of guanidine HCl (Nozaki, 1972) does not significantly decrease absorbance in this wavelength range (compare starting solution, black with solid circles, to the three curves, each the result of a separate recrystallization trial). For crystallization, ultrapure guanidine HCl from Invitrogen was dissolved in 50% (w/v) ddH_2O, was heated to 40 °C, and water was removed by rotary evaporation until the precipitation point was reached. Trace precipitate was removed by adding ddH_2O dropwise, and crystals were allowed to form at room temperature and at 4 °C for a few days. Mother liquor was removed by vacuum filtration. Crystals were then dissolved to 0.5 M (using refractometry to determine concentration) for spectral acquisition.

As can be seen from comparing the absorbance spectra in Fig. 3, guanidine HCl and (deionized) urea solutions have similar absorbances at 220 nm, contributing about 0.1 AU M^{-1} cm^{-1}. Thus, better signal-to-noise ratios might be expected from guanidine titrations in the far-UV since guanidine is a stronger denaturant on a molar basis than urea. However, titration with guanidine salts have the added complication that the ionic strength of the solution becomes variable. This must be kept in mind in studies of electrostatic interactions in proteins. For proteins with high unfolding midpoints, long-range coulombic interactions should be effectively screened through the entire transition, which simplifies the transition but may result in errors in extrapolation to water, if the background ionic strength is low (Santoro and Bolen, 1992).

B. Determination of Concentrations of Denaturant Solutions

Because denaturants are often hygroscopic, determination of denaturant concentrations by weight can be inaccurate. Moreover, treatment with deionizing resin may alter the concentration of urea solutions. Thus, concentrations of urea and guanidine solutions are best determined using refractometry. Following dissolution (and deionization for urea), the refractive index of concentrated denaturant solutions can be measured with high precision using a bench-top refractometer. Refractive index readings are then converted to molar denaturant concentrations using third-order polynomial approximations (Gordon, 1966; Nozaki, 1972).[7] For denaturant solutions to which salts and buffer components have been added, the refractive index increment between the solutions with and without denaturant should be used in the polynomial equations.

The ability to accurately measure denaturant concentrations in solution refractometrically allows the concentration of denaturant to be confirmed at the end of a titration. This is particularly important during automated titration experiments, where small pipetting errors are difficult to detect and can be compounded through the course of the experiment. To avoid such errors, we routinely compare the refractive index of the material in the cuvette at the end of automated titration to the target value.

V. Discussion

The methods and technical details given above are intended to provide an adequate background for researchers at all levels to generate high-quality chemical denaturation curves, quantitatively evaluate the thermodynamic parameters associated with protein folding, and avoid common pitfalls and systematic errors. In the following section, we will discuss structural interpretations of folding free energies and m-values, application of such methods to learning about domain

[7] For urea, concentration in moles per liter is given as: $[\text{urea}] = 117.66\Delta n + 29.753\Delta n^2 + 185.56\Delta n^3$, where Δn is the refractive index increment over buffer. For guanidine hydrochloride, concentration in moles per liter is given as $[\text{guanidine·HCl}] = 57.147\Delta n + 38.68\Delta n^2 - 91.60\Delta n^3$.

boundaries, interdomain interaction, and the effects of point substitutions on structural integrity. Examples of these applications will be selected from our own research on protein domains involved in the Notch signaling pathway. However, before turning to these applications, we will discuss some important but subtle issues that can affect the precision with which thermodynamic parameters can be estimated.

A. Baselines, Parameter Correlation, and Influence on Stability Estimates

As mentioned in Methods, although the thermodynamic parameters describing protein folding from chemical denaturation ($\Delta G^{\circ}_{H_2O}$ and m-value) are directly determined by the transition region, the native and denatured baselines also play a critical role in determining these parameters. Denatured baselines can typically be adequately determined by exposure to high concentrations of denaturant (guanidine hydrochloride, a strong denaturant, is sometimes preferable for proteins of high stability). For extremely stable proteins (those from hyperthermophiles or resulting from protein design), the denatured baseline can be better defined by increasing the temperature of the titration.

In contrast, the native baseline can be more difficult to resolve, depending on the stability of the protein. To illustrate how the length of the native baseline influences fitted $\Delta G^{\circ}_{H_2O}$ and m-values, we have simulated chemical denaturation curves and varied the position of the transition by incrementing $\Delta G^{\circ}_{H_2O}$, while holding the m-value constant at a value typical for globular proteins. Points were sampled at a uniform interval from curves generated using Eq. (9). Errors in the spectroscopic signal (Y_{obs}) were randomly selected from a Gaussian distribution with a standard deviation of 1% of the total unfolding amplitude ($Y_{N,H_2O} - Y_{D,H_2O} = 1$). Examples at moderate and low stability are shown in Fig. 4A. $\Delta G^{\circ}_{H_2O}$ and m-values were fitted along with the four baseline parameters using Eq. (9), and the process was repeated 100 times with new random errors at each value of $\Delta G^{\circ}_{H_2O}$. The standard deviation of $\Delta G^{\circ}_{H_2O}$ is roughly constant for proteins of moderate stability ($\Delta G^{\circ}_{H_2O} < -2.0$ kcal mol^{-1}, $C_m > 0.75$ M), but for marginally stable proteins ($\Delta G^{\circ}_{H_2O} > -1.5$ kcal mol^{-1}, $C_m > 0.6$ M) the standard deviation of $\Delta G^{\circ}_{H_2O}$ rises sharply (Fig. 4B). Fitted m-values also show a larger spread at lower denaturant concentration, although the spread develops less abruptly, beginning at higher denaturant concentrations (Fig. 4C). This increased uncertainty is related to the short, poorly defined native baselines for marginally stable proteins.

Additional insight into the origin and magnitude of errors in $\Delta G^{\circ}_{H_2O}$ and m-values can be seen by examining the correlations between parameters. For stable proteins,[8] there is a strong, roughly linear negative correlation between individual fitted values of $\Delta G^{\circ}_{H_2O}$ and m-value, as expected for a fitted slope and intercept (Fig. 5A). Because the native baseline is well determined for stable proteins, there is essentially no correlation between fitted values of $\Delta G^{\circ}_{H_2O}$ and Y_{N,H_2O}, with the latter being determined with high relative precision (Fig. 5B). In contrast, for

[8] Note that in Fig. 4, more negative fitted $\Delta G^{\circ}_{H_2O}$ values are plotted to the right, to reflect a greater C_m (and stability).

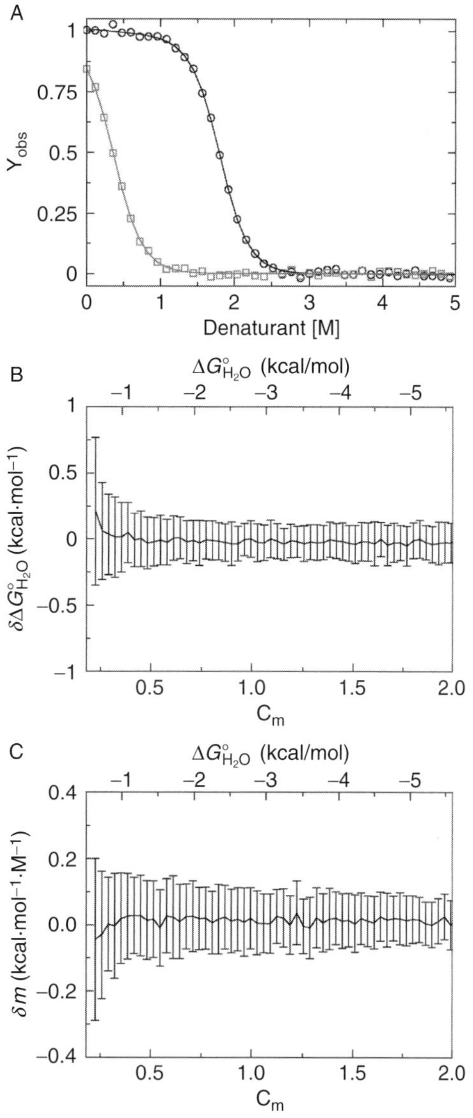

Fig. 4 Relationship between the transition midpoint and uncertainties in fitted thermodynamic parameters. Denaturation transitions generated using the linear model [Eq. (9)] were subjected to random errors (see text) at different values of $\Delta G^{\circ}_{H_2O}$. The other five parameters in Eq. (9) were kept constant at the following values: $m = 2.8$ kcal mol^{-1} M^{-1}, $Y_{N,H_2O} = 1.0$, $\beta_N = 0.0$, $Y_{D,H_2O} = 0.0$, $\beta_D = 0.0$. (A) Two transitions with random errors, with $\Delta G^{\circ}_{H_2O}$ values of -5.0 (circles) and -1.0 kcal mol^{-1} (squares). (B and C) Distribution of fitted $\Delta G^{\circ}_{H_2O}$ and m-values as a function of transition midpoint (C_m, and also $\Delta G^{\circ}_{H_2O}$, upper scale). To better compare the spread in the fitted parameters at different midpoints, differences between fitted and actual values are shown ($\delta X = X_{actual} - X_{fitted}$). The solid line shows mean differences. Error bars show root-mean-square deviations of fitted parameters from the actual value.

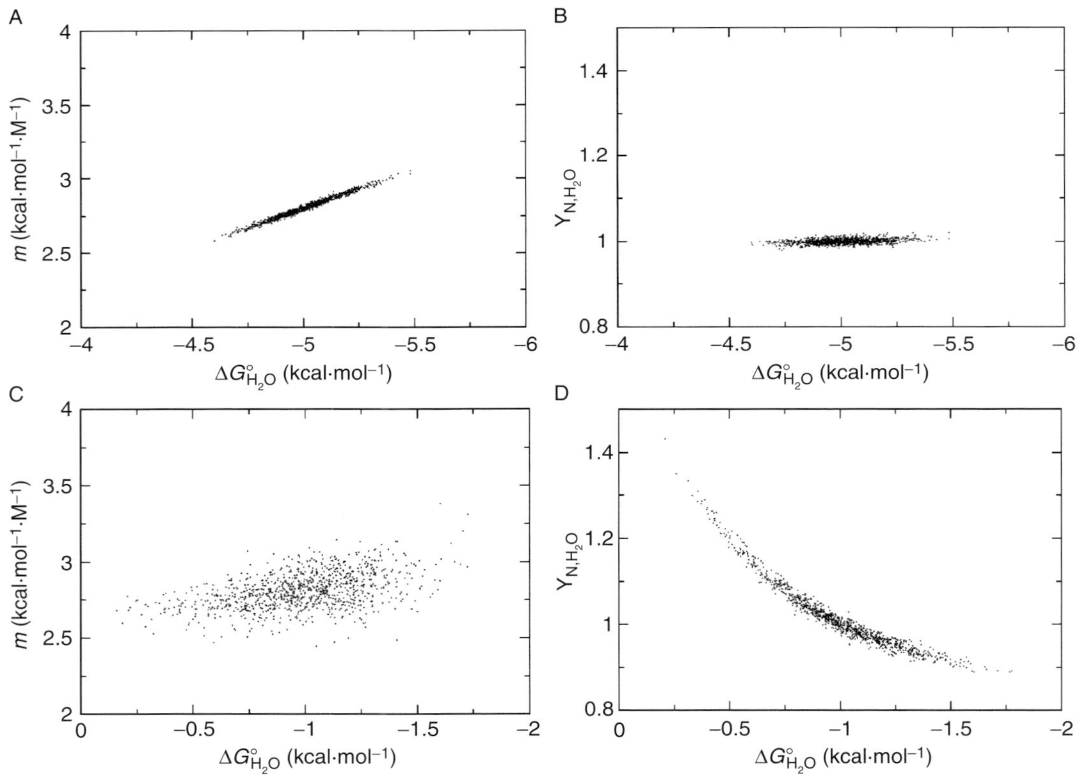

Fig. 5 Correlations between fitted parameters from chemical denaturation transitions of stable and marginally stable proteins. Synthetic transitions generated using the linear model [Eq. (9)] were perturbed with random error (see Fig. 4 legend and discussion section). Individual points represent a fitted parameter pair from a single transition. (A and B) Fitted parameter pairs for a stable protein ($\Delta G^{\circ}_{H_2O} = -5$ kcal mol^{-1}). (C and D) Fitted parameter pairs for a marginally stable protein ($\Delta G^{\circ}_{H_2O} = -1$ kcal mol^{-1}). More negative $\Delta G^{\circ}_{H_2O}$ values are plotted to the right, to reflect high C_m values (and stability). Coefficients of linear correlation are (A) 0.990, (B) 0.340, and (C) 0.464.

marginally stable proteins, although there is still a modest correlation between $\Delta G^{\circ}_{H_2O}$ and *m*-value (Fig. 5C), it is overwhelmed by a strong, positive nonlinear correlation between $\Delta G^{\circ}_{H_2O}$ and Y_{N,H_2O} (Fig. 5D), which results from a lack of information on the native baseline. For data sets where Y_{N,H_2O} is overestimated, the unfolding transition appears be advanced past the true equilibrium position, increasing the fitted value of $\Delta G^{\circ}_{H_2O}$ (to a more positive value, i.e., lower stability). For data sets where Y_{N,H_2O} is underestimated, the unfolding transition appears be lag behind the true equilibrium position, decreasing the fitted value of $\Delta G^{\circ}_{H_2O}$.

These simulations demonstrate two points. First, for proteins of high stability, fitted $\Delta G^{\circ}_{H_2O}$ and *m*-values from individual transitions are highly correlated (Fig. 5A). At high stability, this correlation accounts for nearly all of the variation in these two parameters (hence the high correlation coefficient), demonstrating that a much more precise measure of $\Delta G^{\circ}_{H_2O}$ and *m*-values can be obtained from averaging parameters from independent transitions. Second, resolving the native baseline is very important for determining $\Delta G^{\circ}_{H_2O}$. The importance of the native baseline for quantifying thermal unfolding transitions has been similarly emphasized by Pielak and coworkers (Allen and Pielak, 1998).

For proteins of low stability, where $\Delta G^{\circ}_{H_2O}$ is highly correlated with Y_{N,H_2O} (Fig. 5D), a large uncertainty in Y_{N,H_2O} produces a large uncertainty in $\Delta G^{\circ}_{H_2O}$. Because of this tight correlation, constraining the value of Y_{N,H_2O} using independent information would greatly decrease the uncertainty in $\Delta G^{\circ}_{H_2O}$. In Fig. 5D for example, although values of $\Delta G^{\circ}_{H_2O}$ are spread from -0.5 to -1.5 kcal mol^{-1}, prior knowledge of a Y_{N,H_2O} value of 1.0 would restrict $\Delta G^{\circ}_{H_2O}$ to a much narrower range of values (-0.85 to -1.15 kcal mol^{-1}). One means by which independent information on native baseline signals has been obtained is by adding stabilizing agents. One of the more popular of these stabilizing "osmolytes" is TMAO, which has been shown to promote folding through poor solvation of the exposed backbone of the denatured state (Wang and Bolen, 1997). Bolen and coworkers have combined urea denaturation from partly structured proteins with TMAO titration to drive folding (Baskakov and Bolen, 1998). These authors have shown that a combined analysis using these two separate partial transitions yields much more accurate estimates of $\Delta G^{\circ}_{H_2O}$ than does analysis of the urea transitions alone (Wu and Bolen, 2006). In our studies of fragments of the Notch ankyrin domain, we have simultaneously analyzed the effects of urea and TMAO in a mixed cosolvent system to quantify the stability of marginally stable constructs (Mello and Barrick, 2003). We find that to good approximation, the effects of these two solvents can be treated as independent and additive, using a version of Eq. (5) with separate linear terms for urea and TMAO (with *m*-values of opposite sign).

B. The Physical Interpretation of *m*-Values

Aside from providing a means to compare free energies of folding in the absence of denaturant, the *m*-value has a clear connection to the steepness of the unfolding transition (Fig. 2B). In this regard, the *m*-value is loosely analogous to the Hill coefficient and ligand binding cooperativity. Indeed, *m*-values are often regarded as a measure of the cooperativity of equilibrium folding transitions. However, there is an important difference between the *m*-value and the Hill coefficient. While different values for Hill coefficients can be interpreted in terms of differences in strength of coupling between sites on a macromolecule, the two-state mechanism [Eq. (2)] demands that all parts of the protein be coupled in folding. Once the two-state mechanism is invoked, the issue of cooperativity is settled: the transition is fully cooperative, regardless of the slope. Thus, the size of the *m*-value is not a measure of the degree to which different parts of the protein couple. Instead the *m*-value measures the interaction strength, in terms of free energy, of the denaturant with the native versus denatured state of the protein.

For a particular denaturant, different proteins can have significantly different *m*-values. Thus, the strength of denaturant interaction must vary from protein to protein. Although the binding model [Eq. (6)] can in principle provide mechanistic insight into this variation (number of binding sites, binding constant, or both), the shortcomings of this model (in particular, the strong correlation of these two parameters) limit such an approach. A significant advance in understanding this variation was made by Myers et al. in a correlative study of *m*-values for a large number of proteins that appear to undergo two-state transitions. These authors found a strong positive linear correlation between *m*-values for urea- and guanidine-induced denaturation and the amount of solvent exposure in the unfolding reaction (Myers *et al.*, 1995). This observation is consistent with a nonspecific interaction of denaturant with newly exposed sites (both nonpolar and polar, see Nozaki and Tanford, 1963, 1970, Robinson and Jencks, 1965, Schellman, 1955, and Scholtz *et al.*, 1995) that result from unfolding. The relationship between *m*-values and the amount of surface area exposed on unfolding has been used to interpret the sensitivity of the former to point substitution, pH variation, and salt sensitivity (Pace *et al.*, 1992; Pradeep and Udgaonkar, 2004; Shortle and Meeker, 1986).

The correlation between *m*-value and exposed surface area has important practical implications. Since estimates of surface area exposure on unfolding are very strongly correlated with protein size, the surface area correlation to *m*-value results in a similar correlation between polypeptide length and *m*-value. Although estimating surface area exposure on unfolding requires knowledge of the three-dimensional structure, chain length does not. Thus, the *m*-value of a protein can be estimated from primary sequence (or more precisely, from chain length).

This calculated *m*-value can then be compared to the experimental value as a loose check of the folding mechanism. If partly folded intermediates are populated during unfolding, the measured *m*-value will be lower than the calculated value. If instead the native protein is oligomerized in the native state, the measured *m*-value will be higher than the calculated value. In this regard, this test is similar to a calorimetric test of the two-state mechanism in which the van't Hoff enthalpy of unfolding, determined from the slope of a temperature melt, is compared with a calorimetric measurement of the enthalpy obtained as the heat of reaction (Privalov, 1979).

Owing to the imperfect correlation between *m*-values and chain length seen by Myers *et al.*, the above check is not sensitive to low populations of intermediates or weak association that would change m-values by 10–20%. Rather, close agreement between the measured and calculated m-values can be regarded as supportive of a two-state mechanism, whereas large ($\geq$50%) deviations can be regarded as indicative of a breakdown in the two-state mechanism [Eq. (2)]. As with other tests of the two-state mechanism, results from the m-value comparison described here are strengthened when they are combined with the results of other tests like the spectroscopic test described above.

C. Identification of Domain Boundaries

As described in the Introduction, functional domains in proteins are often embedded within larger polypeptide chains, and are frequently flanked by linker sequences that are of high flexibility. In such cases, the boundaries of embedded structural domains cannot always be determined from primary sequence. Protein stability measurements can provide a simple means to assess domain boundaries.

In one example, urea denaturation of overlapping constructs of different lengths was used to define the boundaries of an ankyrin domain from the Notch transmembrane receptor (Zweifel and Barrick, 2001b). Analysis of primary sequence identified six tandem ankyrin repeats [33-residue imperfect sequence repeats (Bork, 1993) that form a linear array of antiparallel helix bundles] in a region of the receptor known to be important for a number of genetic and biochemical interactions. Comparison of the stability of a construct containing these six repeats with a construct with an additional 33 C-terminal residues showed the additional C-terminal sequence to greatly increase stability (decreasing folding free energy by 4 kcal/mol), suggesting this extra segment to be an important part of the ankyrin domain (Fig. 6B). Moreover, the m-value was around 35%larger with the C-terminal segment, indicating that this segment couples a substantially larger region of the chain to the folding transition. In contrast, there was little effect from deleting the sixth repeat, and in particular, the m-value was unchanged (Fig. 6B). This observation suggests that in the absence of the C-terminal stabilizing segment, the sixth ankyrin repeat is unfolded, a model supported by spectroscopic and hydrodynamic comparisons (Zweifel and Barrick, 2001a) and mutational studies (TOS, C. Bradley, and DB, unpublished). Once the stabilizing effects of the additional C-terminal segment were recognized, diffraction-quality crystals of constructs including this segment were readily obtained, and revealed that this C-terminal region adopts an ankyrin repeat structure (Fig. 6A), despite significant sequence differences from the consensus.

D. Assessing the Effects of Missense Mutations on Structure and Stability

Molecular genetic studies often connect phenotypic variation with specific missense mutations in coding sequences. One particularly informative class of missense mutations involves structurally conservative amino acid substitutions. For such substitutions, phenotypic variation often results from perturbation of

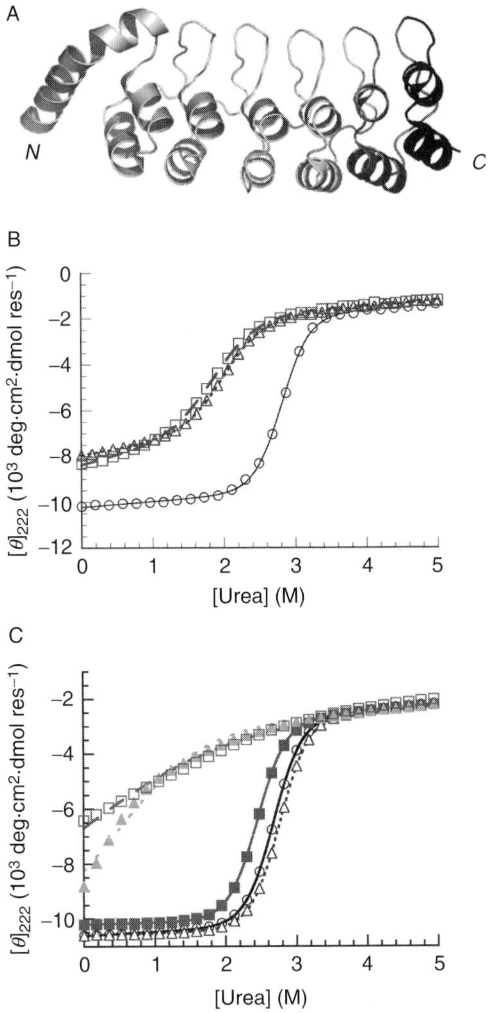

Fig. 6 Mapping the domain boundaries and effects of point substitutions on the Notch ankyrin domain with urea denaturation. (A) Ribbon diagram of the Notch ankyrin domain (chain A of 1o8t. pdb; Zweifel *et al.*, 2003), with repeats six and seven colored in grey and black, respectively. (B) Inclusion of a 33-residue segment C-terminal to the sixth ankyrin repeat (circles, solid line) greatly stabilizes the domain and increases the steepness compared to the six-repeat construct (Zweifel and Barrick, 2001b; open triangles, dotted line). In contrast, deletion of the sixth repeat in the absence of this C-terminal segment does not significantly alter the transition (squares, dashed line). (C) Point substitutions within the Notch ankyrin domain that perturb signaling separate into two classes. Surface substitutions (closed squares, open triangles) have stability similar to wild type (open circles), whereas core substitutions (open squares, closed triangles) are strongly destabilized (Zweifel *et al.*, 2003). Panel B is adapted from (Zweifel and Barrick, 2001b), panel C is adapted from (Zweifel *et al.*, 2003), with permission from the publishers.

specific interactions with individual binding partners. Such substitutions provide a rough map of surface binding sites. Alternatively, point substitutions may prevent folding across the entire domain, indirectly disrupting interactions with all binding partners. These two classes of point substitutions can be distinguished by measuring protein stability, as structurally conservative substitutions often have little or no effect on folding transitions, whereas disruptive substitutions have greatly destabilized transitions.

Point substitutions in the Notch ankyrin domain have been identified from clinical studies of human diseases, and from targeted substitutions designed to perturb Notch signaling. We have examined the effects of four such substitutions on the structural integrity of the Notch ankyrin domain using urea denaturation (Fig. 6C). We find two of the four substitutions to be structurally conservative, producing little change in the overall unfolding transition. Thus, the phenotypes produced by these substitutions are likely a result of the loss of specific interactions between the Notch ankyrin domain and partner proteins. In contrast, two sets of substitutions designed to modify the ankyrin consensus sequence greatly destabilize the entire domain, resulting in what appears to be the second half of an unfolding transition (Fig. 6C). Although the thermodynamic unfolding parameters of these proteins cannot be quantified with high precision because native baselines are not well defined, it seems likely that the functional consequences of these substitutions arise from the inability of these variants to fold to the native state.

E. Identification of Interdomain Interactions

For proteins in which multiple sequence domains are encoded in a single polypeptide chain, adjacent domains may be independent of one another, in which case the covalent linker between the domains has no effect on stability. Alternatively, adjacent domains may stabilize each other. Protein stability measurements provide a convenient means to probe interdomain interaction. Such interactions, which are likely to have a specific structural origin, may have important functional consequences wherein activities are a property of the tandem domain pair (or potentially of a higher order array) rather than of isolated domains.

One example where protein stability measurements uncovered interaction between adjacent sequence modules is in the Notch-binding domain of the Deltex protein (Zweifel *et al.*, 2005). This domain contains two adjacent WWE modules. WWE modules are ~80-residue sequence motifs that are often found in single copy, and when two modules are found, they are sometimes separated by long polypeptide chains. Thus, primary sequence comparisons do not provide a compelling argument for the independence (vs autonomy) of the two WWE modules of Deltex. Urea denaturation of the Deltex WWE tandem shows a single transition, rather than two independent transitions (Fig. 7A; Zweifel *et al.*, 2005). Although this could result from two independent domains that have similar midpoints, the steepness of the transition argues for one large transition. The fitted *m*-value,

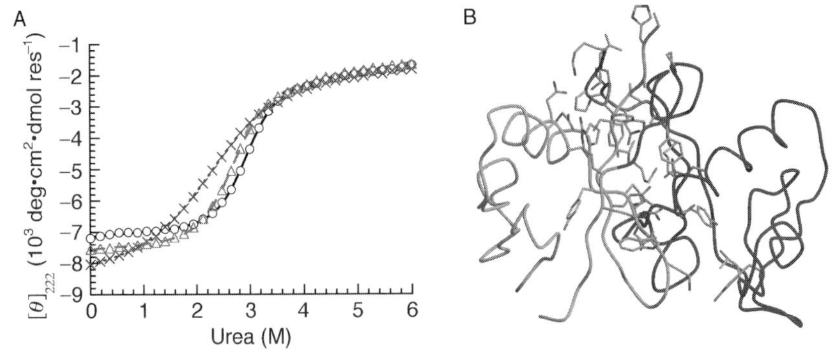

Fig. 7 Probing intermodule interactions in the Deltex tandem WWE domain using urea denaturation. (A) Urea denaturation of the tandem WWE domain of Drosophila Deltex shows a single, steep transition (m-value = 2.04 kcal mol^{-1} M^{-1}; open circles), suggesting the two modules to be thermodynamically coupled. Deletion of a C-terminal segment results in a shallow transition (m-value = 0.99 kcal mol^{-1} M^{-1}; *X*s), suggesting the two modules have become uncoupled. By replacing a 12-residue C-terminal segment that continues a β-sheet with the N-terminal module, the transition is sharpened (open triangles), suggesting this conserved segment to be important for coupling the modules (Zweifel *et al.*, 2005). (B) The two WWE modules of the *Drosophila* Deltex protein (in light and dark shading; Zweifel *et al.*, 2005). Side chains involved in contacts between modules are shown as sticks. Adapted from (Zweifel *et al.*, 2005), with permission from the publisher. (See Plate no. 1 in the Color Plate Section.)

which is identical when determined by CD and tryptophan fluorescence, is *twice* that predicted for a single WWE module, suggesting that the WWE pair comprises the cooperative unit. This high denaturant sensitivity is consistent with the X-ray structure of the Deltex WWE pair, which shows a very large and intricately packed interface between the two domains (Fig. 7B; Zweifel *et al.*, 2005). Interestingly, coupling of the two WWE modules can be lost in large part by deleting a short C-terminal segment that appears to act as a clasp between the two modules (Zweifel *et al.*, 2005; Fig. 7).

VI. Summary

In addition to providing quantitative data for understanding the forces and energetics that determine protein structure and dynamics, experimental studies of protein folding can provide valuable information regarding the organization of domains and the interactions between domains in large, multidomain proteins. In addition, studies of protein stability can provide information on the underlying structural and mechanistic effects of point substitutions. The experimental methods, basic theory, and analytical approaches described above should be sufficient to enable molecular, cell, and structural biologists to quantify the thermodynamics of protein folding to gain insight into the structure of proteins of interest.

Acknowledgments

We thank Mr. Kristos Moshos (Johns Hopkins University) for assistance in recrystallization of guanidine HCl. We thank past and present members of our laboratory for shaping our thinking about various experimental aspects of chemical denaturation of proteins.

References

Agashe, V. R., and Udgaonkar, J. B. (1995). Thermodynamics of denaturation of barstar: Evidence for cold denaturation and evaluation of the interaction with guanidine hydrochloride. *Biochemistry* **34,** 3286–3299.

Allen, D. L., and Pielak, G. J. (1998). Baseline length and automated fitting of denaturation data. *Prot. Sci.* **7,** 1262–1263.

Aune, K. C., and Tanford, C. (1969). Thermodynamics of the denaturation of lysozyme by guanidine hydrochloride. II. Dependence on denaturant concentration at 25 degrees. *Biochemistry* **8,** 4586–4590.

Baskakov, I., and Bolen, D. W. (1998). Forcing thermodynamically unfolded proteins to fold. *J. Biol. Chem.* **273,** 4831–4834.

Bolen, D. W., and Santoro, M. M. (1988). Unfolding free energy changes determined by the linear extrapolation method. 2. Incorporation of $\Delta G^{\circ}_{N\text{-}U}$ values in a thermodynamic cycle. *Biochemistry* **27,** 8069–8074.

Bork, P. (1993). Hundreds of ankyrin-like repeats in functionally diverse proteins: Mobile modules that cross phyla horizontally? *Proteins* **17,** 363–374.

Chen, B. L., Baase, W. A., and Schellman, J. A. (1989). Low-temperature unfolding of a mutant of phage T4 lysozyme. 2. Kinetic investigations. *Biochemistry* **28,** 691–699.

Courtenay, E. S., Capp, M. W., Saecker, R. M., and Record, M. T., Jr. (2000). Thermodynamic analysis of interactions between denaturants and protein surface exposed on unfolding: Interpretation of urea and guanidinium chloride *m*-values and their correlation with changes in accessible surface area (ASA) using preferential interaction coefficients and the local-bulk domain model. *Proteins* **41,** 72–85.

Englander, S. W., Mayne, L., and Rumbley, J. N. (2002). Submolecular cooperativity produces multi-state protein unfolding and refolding. *Biophys. J.* **101–102,** 57–65.

Gordon, J. A. (1966). On the refractive indices of aqueous solutions of urea. *J. Phys. Chem.* **70,** 297.

Greene, R. F., Jr., and Pace, C. N. (1974). Urea and guanidine hydrochloride denaturation of ribonuclease, lysozyme, alpha-chymotrypsin, and beta-lactoglobulin. *J. Biol. Chem.* **249,** 5388–5393.

Grimsley, G. R., Huyghues-Despointes, B. M. P., Pace, C. N., and Scholtz, J. M. (2003). Measuring the conformational stability of a protein. *In* "Purifying Proteins for Proteomics: A Laboratory Manual" (R. J. Simpson, ed.), pp. 535–566. Cold Spring Harbor Press, Cold Spring Harbor.

Gupta, R., and Ahmad, F. (1999). Protein stability: Functional dependence of denaturational Gibbs energy on urea concentration. *Biochemistry* **38,** 2471–2479.

Hagel, P., Gerding, J. J., Fieggen, W., and Bloemendal, H. (1971). Cyanate formation in solutions of urea. I. Calculation of cyanate concentrations at different pH and temperature. *Biochim. Biophys. Acta.* **243,** 366–373.

Hilser, V. J., Garcia-Moreno, E. B., Oas, T. G., Kapp, G., and Whitten, S. T. (2006). A statistical thermodynamic model of the protein ensemble. *Chem. Rev.* **106,** 1545–1558.

Johnson, M. L. (1992). Why, when, and how biochemists should use least squares. *Anal. Biochem.* **206,** 215–225.

Junker, M., Schuster, C. C., McDonnell, A. V., Sorg, K. A., Finn, M. C., Berger, B., and Clark, P. L. (2006). Pertactin beta-helix folding mechanism suggests common themes for the secretion and folding of autotransporter proteins. *Proc. Natl. Acad. Sci. USA* **103,** 4918–4923.

Kuwajima, K. (1989). The molten globule state as a clue for understanding the folding and cooperativity of globular-protein structure. *Proteins* **6,** 87–103.

Li, R., and Woodward, C. (1999). The hydrogen exchange core and protein folding. *Protein Sci.* **8,** 1571–1590.

Makhatadze, G. I., and Privalov, P. L. (1992). Protein interactions with urea and guanidinium chloride. A calorimetric study. *J. Mol. Biol.* **226,** 491–505.

Mello, C. C., and Barrick, D. (2003). Measuring the stability of partly folded proteins using TMAO. *Protein Sci.* **12,** 1522–1529.

Myers, J. K., Pace, C. N., and Scholtz, J. M. (1995). Denaturant *m* values and heat capacity changes: Relation to changes in accessible surface areas of protein unfolding. *Protein Sci.* **4,** 2138–2148.

Nozaki, Y. (1972). The preparation of guanidine hydrochloride. *Meth. Enzymol.* **26**(PtC), 43–50.

Nozaki, Y., and Tanford, C. (1963). The solubility of amino acids and related compounds in aqueous urea solutions. *J. Biol. Chem.* **238,** 4074–4081.

Nozaki, Y., and Tanford, C. (1970). The solubility of amino acids, diglycine, and triglycine in aqueous guanidine hydrochloride solutions. *J. Biol. Chem.* **245,** 1648–1652.

Pace, C. N. (1986). Determination and analysis of urea and guanidine hydrochloride denaturation curves. *Meth. Enzymol.* **131,** 266–280.

Pace, C. N., Laurents, D. V., and Erickson, R. E. (1992). Urea denaturation of barnase: pH dependence and characterization of the unfolded state. *Biochemistry* **31,** 2728–2734.

Pace, C. N., and Scholtz, J. M. (1997). Measuring the conformational stability of a protein. *In* "Protein Structure: A Practical Approach" (T. E. Creighton, ed.), pp. 299–321. IRL Press, Oxford.

Parsegian, V. A., Rand, R. P., and Rau, D. C. (1995). Macromolecules and water: Probing with osmotic stress. *Meth. Enzymol.* **259,** 43–94.

Pradeep, L., and Udgaonkar, J. B. (2004). Effect of salt on the urea-unfolded form of barstar probed by m value measurements. *Biochemistry* **43,** 11393–11402.

Prakash, V., Loucheux, C., Scheufele, S., Gorbunoff, M. J., and Timasheff, S. N. (1981). Interactions of proteins with solvent components in 8 M urea. *Arch. Biochem. Biophys.* **210,** 455–464.

Privalov, P. L. (1979). Stability of proteins: Small globular proteins. *Adv. Prot. Chem.* **33,** 167–241.

Robinson, D. R., and Jencks, W. P. (1965). The effect of compounds of the urea-guanidinium class on the activity coefficient of acetyltetraglycine ethyl ester and related compounds. *J. Am. Chem. Soc.* **87,** 2462–2470.

Romero, P., Obradovic, Z., and Dunker, A. K. (2004). Natively disordered proteins: Functions and predictions. *Appl. Bioinformat.* **3,** 105–113.

Royer, C. A. (2002). Revisiting volume changes in pressure-induced protein unfolding. *Biochim. Biophys. Acta* **1595,** 201–209.

Sanchez, I. E., and Kiefhaber, T. (2003a). Evidence for sequential barriers and obligatory intermediates in apparent two-state protein folding. *J. Mol. Biol.* **325,** 367–376.

Sanchez, I. E., and Kiefhaber, T. (2003b). Non-linear rate-equilibrium free energy relationships and Hammond behavior in protein folding. *Biophys. Chem.* **100,** 397–407.

Santoro, M. M., and Bolen, D. W. (1988). Unfolding free energy changes determined by the linear extrapolation method. 1. Unfolding of phenylmethanesulfonyl alpha-chymotrypsin using different denaturants. *Biochemistry* **27,** 8063–8068.

Santoro, M. M., and Bolen, D. W. (1992). A test of the linear extrapolation of unfolding free energy changes over an extended denaturant concentration range. *Biochemistry* **31,** 4901–4907.

Schellman, J. A. (1955). The stability of hydrogen-bonded peptide structures in aqueous solution. *C. R. Lab. Carlsberg, Ser. Chim.* **29,** 230–259.

Schellman, J. A. (1978). Solvent denaturation. *Biopolymers* **17,** 1305–1322.

Schellman, J. A. (1990). A simple model for solvation in mixed solvents. Applications to the stabilization and destabilization of macromolecular structures. *Biophys. Chem.* **37,** 121–140.

Schellman, J. A. (1994). The thermodynamics of solvent exchange. *Biopolymers* **34,** 1015–1026.

Schellman, J. A. (2002). Fifty years of solvent denaturation. *Biophys. Chem.* **96,** 91–101.

Schellman, J. A. (2003). Protein stability in mixed solvents: A balance of contact interaction and excluded volume. *Biophys. J.* **85,** 108–125.

Schellman, J. A. (2006). Destabilization and stabilization of proteins. *Q. Rev. Biophys.* **38,** 1–11.

Scholtz, J. M., Barrick, D., York, E. J., Stewart, J. M., and Baldwin, R. L. (1995). Urea unfolding of peptide helices as a model for interpreting protein unfolding. *Proc. Natl. Acad. Sci. USA* **92,** 185–189.

Shortle, D., and Meeker, A. K. (1986). Mutant forms of staphylococcal nuclease with altered patterns of guanidine hydrochloride and urea denaturation. *Proteins* **1,** 81–89.

Swint, L., and Robertson, A. D. (1993). Thermodynamics of unfolding for turkey ovomucoid third domain: Thermal and chemical denaturation. *Prot. Sci.* **2,** 2037–2049.

Tanford, C. (1970). Protein denaturation. C. Theoretical models for the mechanism of denaturation. *Adv. Prot. Chem.* **24,** 1–95.

Timasheff, S. N. (1998). Control of protein stability and reactions by weakly interacting cosolvents: The simplicity of the complicated. *Adv. Prot. Chem.* **51,** 355–432.

Tripp, K. W., and Barrick, D. (2006). Enhancing the stability and folding rate of a repeat protein through the addition of consensus repeats. *J. Mol. Biol.* **365,** 1187–1200.

Wang, A., and Bolen, D. W. (1997). A naturally occurring protective system in urea-rich cells: Mechanism of osmolyte protection of proteins against urea denaturation. *Biochemistry* **36,** 9101–9108.

Wu, P., and Bolen, D. W. (2006). Osmolyte-induced protein folding free energy changes. *Proteins* **63,** 290–296.

Wyman, J., Jr. (1964). Linked functions and reciprocal effects in hemoglobin: A second look. *Adv. Prot. Chem.* **19,** 223–286.

Zou, Q., Habermann-Rottinghaus, S. M., and Murphy, K. P. (1998). Urea effects on protein stability: Hydrogen bonding and the hydrophobic effect. *Proteins* **31,** 107–115.

Zweifel, M. E., and Barrick, D. (2001a). Studies of the ankyrin repeats of the *Drosophila melanogaster* Notch receptor. 1. Solution conformational and hydrodynamic properties. *Biochemistry* **40,** 14344–14356.

Zweifel, M. E., and Barrick, D. (2001b). Studies of the ankyrin repeats of the *Drosophila melanogaster* Notch receptor. 2. Solution stability and cooperativity of unfolding. *Biochemistry* **40,** 14357–14367.

Zweifel, M. E., and Barrick, D. (2002). Relationships between the temperature dependence of solvent denaturation and the denaturant dependence of protein stability curves. *Biophys. Chem.* **101–102,** 221–237.

Zweifel, M. E., Leahy, D. J., and Barrick, D. (2005). Structure and Notch receptor binding of the tandem WWE domain of Deltex. *Structure (Camb)* **13,** 1599–1611.

Zweifel, M. E., Leahy, D. J., Hughson, F. M., and Barrick, D. (2003). Structure and stability of the ankyrin domain of the *Drosophila* Notch receptor. *Prot. Sci.* **12,** 2622–2632.

CHAPTER 12

Hydrodynamic Modeling: The Solution Conformation of Macromolecules and Their Complexes

Olwyn Byron

Division of Infection and Immunity
GBRC, 120 University Place
University of Glasgow, Glasgow, G12 8TA
United Kingdom

0091-679X/08 $35.00
DOI: 10.1016/S0091-679X(07)84012-X

Abstract

Hydrodynamic bead modeling (HBM) is the representation of a macromolecule by an assembly of spheres (or beads) for which measurable hydrodynamic (and related) parameters are then computed in order to understand better the macromolecular solution conformation. An example-based account is given of the main stages in HBM of rigid macromolecules, namely: model construction, model visualization, accounting for hydration, and hydrodynamic calculations. Different types of models are appropriate for different macromolecules, according to their composition, to what is known about the molecule or according to the types of experimental data that the model should reproduce. Accordingly, the construction of models based on atomic coordinates as well as much lower resolution data (e.g., electron microscopy images) is described. Similarly, several programs for hydrodynamic calculations are summarized, some generating the most basic set of solution parameters (e.g., sedimentation and translational diffusion coefficients, intrinsic viscosity, radius of gyration, and Stokes radius) while others extend to data determined by nuclear magnetic resonance, fluorescence anisotropy, and electric birefringence methods. An insight into the topic of hydrodynamic hydration is given, together with some practical suggestions for its satisfactory treatment in the modeling context. All programs reviewed are freely available.

I. Introduction

Structural biology aims to obtain the high-resolution coordinates for all biological macromolecules and their complexes, as a basis for understanding their function. However, many entire classes of molecules (e.g., multidomain and modular proteins) continue to evade study at this resolution. In addition, the ternary conformations (Nakasako *et al.*, 2001; Trewhella *et al.*, 1988; Vestergaard *et al.*, 2005) and quaternary structures (Svergun *et al.*, 1997, 2000) adopted by biomacromolecules in the crystalline state can be quite different from those observed in solution. In such cases hydrodynamic modeling offers a way of obtaining structural data for systems in solution, albeit at a lower resolution.

This chapter will primarily focus on hydrodynamic *bead* modeling (hereafter HBM—the method most commonly used to compute hydrodynamic parameters. Reference will be made, however, to related methods including ellipsoidal and finite element modeling. HBM is the representation of a macromolecule by an assembly of spheres (or beads) for which hydrodynamic (and related) parameters are then computed. The model that successfully reproduces experimentally determined parameters [such as the sedimentation (s), and diffusion (D_t) coefficients, the primary hydrodynamic parameters against which HBMs are often assessed] is not a unique structure for the macromolecule; instead, it represents one plausible solution conformation: there will be other HBMs, perhaps with quite different arrangements of constituent beads, that can give rise to exactly the same hydrodynamic profile. In fact, HBM is most effective when undertaken in conjunction with other methods, the most common of which is the modeling of data from small-angle X-ray or neutron scattering.

HBM is of particular utility when a molecule does not crystallize and has a mass in excess of the upper limit for detailed structural study via nuclear magnetic resonance (NMR) spectroscopy (about 50 kDa). The following types of questions can be answered with HBM: Is the solution conformation of a macromolecule or its noncovalent complex consistent with measured hydrodynamic data? Is a low-resolution structure (emanating perhaps from an electron micrograph) consistent with a measured s and D_t? How well does a high-resolution homology model reproduce hydrodynamic parameters? What is the shape of a particular large macromolecular complex?

The structure of this chapter is as follows:

Background to HBM—setting the scene—is this the right type of modeling for your system?

Model construction—how to build an HBM–freely available programs are described.

Model visualization—once you have constructed your model, you will want to see it.

Hydration—how do you deal with the hydrodynamic hydration?

Hydrodynamic calculations—now you have a hydrated HBM, you can compute its hydrodynamic profile.

Advanced hydrodynamic calculations—for more serious applications.

Throughout this chapter, lysozyme will be used as an example because it is small, extremely well characterized and is used as an example by some of the groups providing the freely available HBM software reviewed herein. The pdb file from which the models are constructed is 6lyz.pdb, downloadable from, for example, http://leonardo.fcu.um.es/macromol/programs/hydropro/hydropro.htm.

II. Background to HBM

The sedimentation coefficient of a macromolecule describes its frictional behavior and, as a consequence, its shape in solution. From the following well-known relationship

$$s = \frac{M(1 - \bar{v}\rho)}{N_A f} \tag{1}$$

where M is the molecular mass (g/mol), $\bar{v}$ the partial specific volume (ml/g), ρ the solvent density (g/ml), N_A Avogadro's number (6.02214×10^{23} mol^{-1}), the frictional coefficient f (g/sec) can be derived. The frictional coefficient of a sphere with volume equal to the macromolecule is f_0. The difference between the two, or rather their ratio, reveals a measure of molecular elongation (and/or hydration) and was used, before the advent of HBM, as an indicator of solution shape, albeit at ultralow resolution. f/f_0 is still a useful parameter since it is not always possible to propose a bead model for a macromolecule, yet one seeks to characterize its shape in some way (see e.g., Garnier *et al.*, 2002; Kar *et al.*, 1997, 2001; Stafford *et al.*, 1995; Toedt *et al.*, 1999; Waxman *et al.*, 1993). Apart from describing some form of simple elongation, f/f_0 can also be interpreted in terms of prolate or oblate ellipsoids, since a given axial ratio will result in a corresponding f/f_0 (see e.g., Zarutskie *et al.*, 1999).

In HBM the molecule is represented as a collection of hard spheres. For the calculation of the sedimentation coefficient, the spheres are frictional elements for which hydrodynamic interaction tensors may be calculated on the basis of their Cartesian coordinates and radii, given the mass- and partial-specific volume of the macromolecule and the temperature, density, and viscosity of the solvent in which it is suspended.

HBM is possible because of decades of theoretical and computational development by Kirkwood, Riseman, Bloomfield, García de la Torre, collaborators, and coworkers. The body of relevant literature is extensive, and its review is outwith the remit of this chapter. Instead, the reader is directed toward two excellent reviews of this topic (Carrasco and García de la Torre, 1999; García de la Torre and Bloomfield, 1981). In brief though, given the Cartesian coordinates and radii of beads in an HBM, it is possible to compute the frictional force exerted on the solvent by each bead, based on Stokes's law and on the hydrodynamic interaction tensors ($\mathbf{T}_{ij}$) between beads i and j in the HBM. This is the frictional force experienced by bead i due to its movement through a solvent, perturbed by the other beads (j) in the HBM. The resulting set of N linear equations is used to build a $3N \times 3N$ supermatrix (where N is the number of beads in the HBM) that is then inverted to obtain the translational, rotational, and roto-translational coupling frictional tensors. Via a series of subsequent partitionings, inversions, and other calculations, the observable translational and rotational properties [and the intrinsic viscosity (η)] of the HBM are then computed. The required computer time for the hydrodynamic calculations is proportional to N^3. For pairs of beads with equal radius, $\mathbf{T}_{ij}$ is computed using the Rotne–Prager–Yamakawa (RPY) (Rotne and Prager, 1969; Yamakawa, 1970) modification to the Oseen tensor (Oseen, 1927) that accounts for the non-point-like nature of the beads and allows for bead overlapping. When beads i and j have different radii, $\mathbf{T}_{ij}$ is given by a further modified form of the RPY tensor (García de la Torre and Bloomfield, 1981); but there is no exact tensor to describe the hydrodynamic interaction between overlapping beads of differing radii. In this case, an *ad hoc* expression [derived by Zipper and Durchschlag (1997, 1998) and implemented in programs from the García de la Torre group] may be used that provides a satisfactory correction. However, this "patch" should only be used occasionally and overlapping beads of differing size should be avoided whenever possible (García de la Torre *et al.*, 2000a).

In some papers on HBM theory and applications mention is made of a "volume correction." This correction was devised by García de la Torre and coworkers for the rotational friction tensor and the intrinsic viscosity. Accordingly, it is not strictly relevant to HBM of the sedimentation and diffusion coefficients, but an awareness of its existence is important.

The most familiar HBMs are those in which the molecular volume is filled with beads. But there are other types of HBMs (and indeed hydrodynamic models not involving beads). These are very well reviewed by García de la Torre *et al.* (2000a). Of particular mention are shell models, first proposed in the late 1960s by Bloomfield *et al.* (e.g., 1967) wherein the molecular surface is reproduced by a shell of very small beads. Atoms (either in the macromolecule or in the solvent trapped within it) not exposed to bulk solvent do not contribute to the frictional properties of a particle. Therefore, the hydrodynamic properties of a macromolecule can be computed from a shell model alone. This approach was taken further by Teller *et al.* (1979) who used shell modeling to calculate f from the atomic coordinates of eight proteins. More recently, García de la Torre *et al.* (2000a) developed this method so that in the modeling process the shell bead size is extrapolated to zero and the hydrodynamic (and other) parameters calculated approach those for a particle with a smooth surface. Shell modelling accuracy depends on the ratio of shell bead size to the average particle size. Large molecules can thus be safely modelled with shells of larger beads. Shell modeling is not appropriate for the computation of parameters that depend on the distribution of mass within the particle, for example the small-angle X-ray or neutron-scattering curve (and the associated radius of gyration, R_g), though some shell-modeling programs [e.g., HYDROPRO (García de la Torre *et al.*, 2000a)] compute these properties *en route* from a filling model to the shell model.

Beads are not the only elements with which a macromolecule may be constructed. Its surface can instead be decomposed into flat plates whose hydrodynamic (and electrostatic) properties can be computed with finite boundary element (BE) modeling techniques. For example, Allison (1999) and, more recently, Aragon (2004) have represented macromolecules with arrays of flat triangular plates. In common with shell modeling, hydrodynamic parameters are calculated for several resolutions of the model and then extrapolated to infinity. Calculations are performed under two sets of microhydrodynamic conditions: "stick" (the normal boundary condition applied in HBM wherein at the particle–fluid interface the velocity of fluid is the same as that of the particle) and "slip" [only the normal components of the fluid and particle velocity match at the interface—slip conditions describe better the rotational diffusion of some small (e.g., simple aromatic) molecules for which hydrogen bonding with the (e.g., organic) solvent is minimal]. BE modeling successfully reproduces with high accuracy the transport properties of ellipsoids, cylinders, and toroids (Allison, 1999) and has been applied to describe the hydrodynamics (and electrophoresis) of small and large pieces of DNA and capped cylinders (Allison, 2001). While the FORTRAN code for Dr Allison's BE programs is not freely available, he collaborates with those wishing to undertake the considerable computations required.

A by-no-means-comprehensive list of the parameters generated for HBMs of rigid macromolecules by seven different programs is given in Table I so that the reader can choose which software is more suited to the problem in hand.

The following sections describe the main stages of modeling: model construction, model visualization, and hydrodynamic calculations. Each section includes summaries of some of the freely available software. The coverage is not exhaustive—there are other programs available, but this chapter has focused on those most familiar to the author, those most prevalent in the literature, and those most relevant to the readership of this text. The summaries are precisely that and users are strongly advised to read the papers and manuals that pertain to the software before and during their use. Note that certain programs, particularly those written in FORTRAN, are very sensitive to the formatting of the input files. The files should generally be in text only format (probably compatible with MS-DOS), and the user should pay attention to the spacings between data in the files, since FORTRAN programs sometimes expect data to follow particular number width and decimal place formats.

III. Model Construction

While it is in principle possible to compute hydrodynamic parameters from atomic coordinates directly, either by parameterizing the atoms or by long molecular dynamics simulations (Smith and van Gunsteren, 1994; Venable and Pastor, 1988) the computation time to do so is huge. Instead, it is more efficient to condense these coordinates to a lower resolution HBM for which hydrodynamic (and other) parameters can be more rapidly computed. This is particularly important if some type of rigid body modeling is being undertaken. There are a number of freely available HBM construction programs. Some of these are described and evaluated in Section III.

A. AtoB

AtoB is FORTRAN source code that, once compiled, generates a program to construct HBMs either from atomic resolution coordinates or *de novo*, via a series of user-coded instructions in the input file (Byron, 1997) (Table II). A very detailed

Table I
Solution Parameters Generated by 7 Hydrodynamic Computation Programs

Parameter	Definition	Programs(s) that compute parameter
D_t	Translational diffusion coefficient	*a, b, c, d, e, f, g*
R_S^t	Translational Stokes radius	*a, b, d, e, g*
R_g	Radius of gyration	*a, b, c, d, e, g*
V	Anhydrous volume	*a, b, d, e*
D_r	Rotational diffusion coefficient	*a, b, c, d, e, f, g*
τ_{1-5}	Relaxation times	*a, b, c, d, e, f, g*
τ_h	Harmonic mean relaxation time	*a, b, c, d, e, f, g*
$[\eta]$	Intrinsic viscosity (with various corrections)	*a, b, c, e, d, g*
s	Sedimentation coefficient	*a, b, d, e, g*
D_{max}	Maximum dimension of the molecule	*a, b, c, d, e*
Distance distribution	Analogous to the p(r) function generated by indirect Fourier transform of small-angle X-ray or neutron scattering data	*a, b, d, e*
Scattering form factor	Angular dependence of small-angle scattering	*a, b, d, e*
$D^r_{II}/D^r_{\perp}$	Anisotropy of rotational diffusion	*c, f*
D_{1-3}	Eigenvalues of rotational diffusion	*a, c, d, e*
U	Covolume	*a, c, d*
u_{red}	Reduced covolume	*c*
ν	Viscosity increment	*c*
P	Translational friction ratio	*c*
τ_{1-5}/τ_0	Ratios of relaxation times 1–5 with that of equivalent sphere	*c*
τ_h/τ_0	Relaxation harmonic ratio	*c*
τ_{ini}/τ_0	Initial relaxation time ratio (fluorescence, NMR, birefringence)	*c*
τ_{mean}/τ_0	Mean relaxation time ratio (fluorescence, NMR, birefringence)	*c*
G	Reduced radius of gyration function	*c*
Φ	Flory viscosity function	*c*
P_0	Flory P_0 function	*c*
β	Scheraga–Mandelkern function	*c*
R	Wales–van Holde function	*c*
Ψ	Psi function	*c*
Π	Harding pi function (reduced covolume + viscosity increment)	*c*
$K_{\tau r1-5}$	$K_{\tau r}$ functions 1–5	*c*
$K_{\tau rh}$	$K_{\tau r}$ harmonic function	*c*
$K_{\tau rini}$	$K_{\tau r}$ initial function (fluorescence, NMR, birefringence)	*c*
$K_{\tau rmean}$	$K_{\tau r}$ mean function (fluorescence, NMR, birefringence)	*c*
Ψ_{1-5}	Psi translation + rotation functions 1–5	*c*
Ψ_h	Psi harmonic function	*c*

(*continues*)

Table I (*continued*)

Parameter	Definition	Programs(s) that compute parameter
Ψ_{ini}	Psi initial function (fluorescence, NMR, and birefringence)	*c*
Ψ_{mean}	Psi mean function (fluorescence, NMR, and birefringence)	*c*
Λ_{1-5}	Lambda rotation + viscosity functions 1–5	*c*
Λ_{h}	Lambda harmonic function	*c*
Λ_{ini}	Lambda initial function (fluorescence, NMR, and birefringence)	*c*
Λ_{mean}	Lambda mean function (fluorescence, NMR, and birefringence)	*c*
H	H ratio ($D_{\mathrm{max}}/R_{\mathrm{g}}$)	*c*
T_1, T_2	NMR relaxation times	*f*
T_1/T_2	Ratio of NMR relaxation times	*f*
∇	Relative deviation of T_1/T_2 with respect to the average over the number of residues in a protein	*f*
SA	Surface area	*g*
f_{t}	Translational frictional coefficient	*g*
f_{r}	Global rotational frictional coefficient	*g*
$f_{\mathrm{r}}^{\mathrm{k}}$	Rotational frictional coefficient ($k = x, y$, and z components)	*g*
$D_{\mathrm{r}}^{\mathrm{k}}$	Rotational diffusion coefficient ($k = x, y$, and z components)	*g*
$R_{\mathrm{S}}^{\mathrm{r,k}}$	Rotational Stokes radius ($k = x, y$, and z components)	*g*
CR	Center of resistance	*g*
CM	Center of mass	*g*
CD	Center of diffusion	*g*
CV	Center of viscosity	*g*
R_{E}	Einstein's radius (with various corrections)	*g*
a, b, c	Maximum extensions along x, y, and z	*g*
a/c, a/b, b/c,	Axial ratios	*g*

[a]HYDRO.
[b]HYDROPRO.
[c]SOLPRO.
[d]HYDROSUB.
[e]HYDROMIC.
[f]HYDRONMR.
[g]SUPCW/SUPCWIN.

account of its utility is given by Byron (2000). When supplied with atomic coordinates, AtoB determines the spatial extremes of the molecule and encloses it in a cuboid of those dimensions. The cuboid is subdivided into cubes of user-specified dimension (the nominal resolution of the resultant HBM), and a sphere of radius proportional to the cube root of the mass contained within that cube is positioned at its center of gravity. AtoB also allows the user to generate *de novo* HMBs based on the shape of a macromolecule observed, for example on an electron micrograph.

When an HBM is constructed by filling the molecular volume with beads, the beads have to overlap in order to reproduce the molecular volume by the sum of their respective volumes. The validity of hydrodynamic computation for bead models with overlapping beads of equal and unequal sizes was discussed above. To reiterate, exact hydrodynamic parameters can be computed for an HBM comprising overlapping beads only if they are of equal radius. This means that if the mixed bead radii option is selected in program AtoB, the hydrodynamic calculations for that HBM will be incorrect; the fixed bead radius option should instead be used. Part of AtoB (generation of HBMs from atomic coordinates and overlap removal) is also implemented within the program SOMO (see below).

Example 1: Generation of lysozyme bead model from 6lyz.pdb with AtoB (Byron, 1997).

1. Download AtoB and example files from http://www.rasmb.bbri.org/rasmb/ms_dos/atob-byron/.

2. Compile the program (e.g., with F77 compiler) on preferred platform (e.g., PC, Mac, Linux, and UNIX).

3. Download pdb file (e.g., 6lyz.pdb). Cut and paste coordinate lines into example input file (e.g., ex.atfix, downloaded from the AtoB site) and edit header lines accordingly.

4. Here, a low-resolution bead model is generated (with a resolution of 10 Å) in order to demonstrate the utility of AtoB for reducing large coordinate files to more compact bead models [remember that this function can be accomplished within SOMO (Rai *et al.*, 2005) (Section III.D)].

```
6lyz
'xtal'       !beads (bead) or crystallographic (xtal)?
'leave'      !manipulate final model (manip) or not (leave)?
1001         !number of lines of coordinates to be read in (ibdn)
0.727        !vbar (bead model)
129, 10.0     !number of residues, resolution (atomic)
1            !fixed (1) or mixed (2) bead radii? (atomic)
'atoms'      !atom (atoms) or resdiue (resid) coordinates? (atomic)
ATOM      1  N   LYS     1       3.287  10.092  10.329  5.89  1.50      6LYZ 247
ATOM      2  CA  LYS     1       2.445  10.457   9.182  8.16  1.50      6LYZ 248
ATOM      3  C   LYS     1       2.500  11.978   9.038  8.04  1.50      6LYZ 249
(section of coordinates omitted for brevity here)
ATOM    998  CG  LEU   129     -13.595  21.390   7.436  3.74  1.50      6LYZ1244
ATOM    999  CD1 LEU   129     -12.954  22.586   8.097  3.02  1.50      6LYZ1245
ATOM   1000  CD2 LEU   129     -12.974  20.117   7.959  3.02  1.50      6LYZ1246
ATOM   1001  OXT LEU   129     -17.840  19.891   8.551  4.69  1.50      6LYZ1247
```

Example 1a.

5. Run executable via a run script [such as that available from the AtoB site ("run")] such that names of input file and output file are requested and the program is executed.

```
> run atob
Name of file where input is stored: 6lyz.atfix
Name of file where output is stored: 6lyz.out
Wed Apr  4 10:52:21 BST 2007

number of lines   1001
there are   129 residues and resolution is   10.0
fixed (1) or mixed (2) radii? 1
reading in   129 residues
vbar calculated from atomic data is       0.720 ml/g
hydn calculated from atomic data is       0.387 g/g
mass calculated from atomic data is   14297.297 g/mole
X-ray scattering length is            21494.381 fm
neutron scattering length (2H2O) is    6131.020 fm

contains      31 beads
  n         x          y          z          r
    1    -9.5632    16.8392     6.3187     4.8528
    2    -0.2276    13.5154     7.9666     4.8528
    3     5.0570    15.5684    10.9475     4.8528
(section of coordinates omitted for brevity here)
   29    -4.7859    24.5985    31.5447     4.8528
   30     7.6662    20.6277    32.6438     4.8528
   31    -4.6384    28.8875    31.3687     4.8528
Wed Apr  4 10:52:29 BST 2007
[1]   + Terminated                  run atob
```

Example 1b.

Table II
Summary Description of AtoB (Byron, 1997, 2000)

Web site	http://rasmb.bbri.org/rasmb/ms_dos/atob-byron/
Summary of method	2 modes: (i) generation of HBM from atomic coordinates (ii) *de novo* HBM generation.
Good for	Reducing the number of beads in a model; modeling DNA/RNA as well as (glyco) proteins; creating low-resolution HBMs *de novo* (i.e., without atomic coordinates).
Limitations	Hydration via uniform expansion; models do not reproduce intrinsic viscosity and rotational diffusion as well as higher resolution models. Requires all non-H atoms for high-resolution model construction.
Input	For high resolution: atomic coordinates (of all non-H atoms or just α-carbons). For *de novo* low-resolution: $\bar{v}$, bead coordinates and radii, instructions for manipulations of primary bead model.
Output	For high resolution: $\bar{v}$, δ (hydration), mass, X-ray and neutron scattering lengths, bead coordinates and radii. For *de novo* low-resolution: coordinates and radii of final bead model.
Operating system(s)	Any: program is downloaded as FORTRAN code that must be compiled to generate the executable.
Language	FORTRAN
Example model	Fig. 1

6. This will generate the final bead coordinate file (called 6lyz.out, in this example).

7. The model can be visualized in a number of ways. Figure 1 was generated using VisualBeads (http://leonardo.fcu.um.es/macromol/programs/visualbeads0b/visualbeads.htm) and a vrml viewer.

8. VisualBeads generates .vrml files from two main input files (colors.txt and coorconn.txt), easily generated in a text editor to incorporate the bead coordinates and radii (from AtoB) and colors (chosen by the user).

9. The resultant .vrml file can then be read by freely available viewers such as FreeWRL (http://freewrl.sourceforge.net/) or Cortona (http://www.parallelgraphics.com/products/cortona/).

Example 2: Generation of *de novo* bead model with AtoB (Byron, 1997).

A model used previously (Byron, 2000) as an example of *de novo* model building with AtoB is reprised here.

1. Download AtoB and example files from http://www.rasmb.bbri.org/rasmb/ms_dos/atob-byron/.

2. Compile the program (e.g., with F77 compiler) on preferred platform (e.g., PC, Mac, Linux, and UNIX).

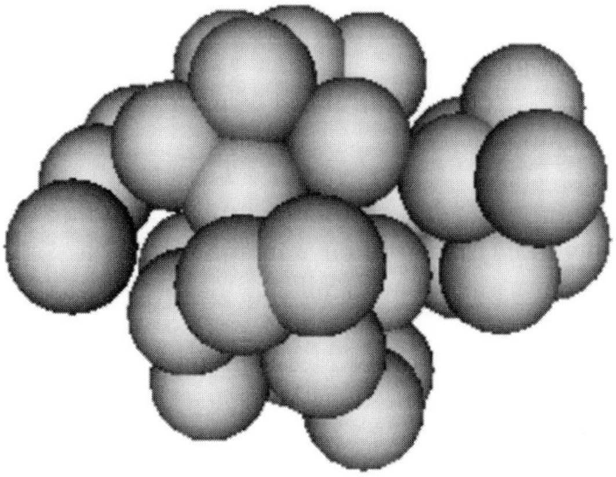

Fig. 1 Bead model of lysozyme generated with AtoB (Byron, 1997) as described in Example 1. The coordinates of the model were converted to .vrml format with the program VisualBeads (Section IV.B); the final model was visualized using the program FreeWRL (Section IV.B).

3. Generate input file, such as the following:

```
Demonstration
'bead'
'manip'
1
0.73
1, 5.0
2
'atoms'
   1     -10.0     10.0     0.0     10.0
a
3
1,  10.0,  10.0, 0.0, 10.0
2, -10.0, -10.0, 0.0, 10.0
3,  10.0, -10.0, 0.0, 10.0
y
c
160.0
     0.000,     0.000,     0.000
20
y
d
3,3
y
e
1,22
0.3
y
r
z
90.0
1,2
y
s
10.0,10.0,10.0
5
1,24
n
```

Example 2a.

4. Run executable via a run script [such as that available from the AtoB site ("run")] such that names of input file and output file are requested and the program is executed.

```
> run atob
Name of file where input is stored: test.i
Name of file where output is stored: test.o
Wed May  2 17:30:20 BST 2007
                  Demonstration
number of lines      1
vbar is  0.730 ml/g
manipulating coordinates
selected mode is  a
adding      3 beads to model
selected mode is  c
circle has diameter    160.000
coordinates of circle centre are (x,y,z)      0.000     0.000     0.000
there are    20 beads in the circle
selected mode is  d
deleting beads     3 to     3 inclusive
selected mode is  e
hydrating beads     1 to     22 by 0.300 g solvent per g solute
selected mode is  r
rotating around z axis
rotation angle is  90.00
rotating beads     1 to     2 inclusive
selected mode is  s
translational vectors     10.000    10.000    10.000
performing     5 translations
moving beads     1 to    24 inclusive
move number     1
move number     2
move number     3
move number     4
move number     5
                 Demonstration
contains   145 beads
  n       x          y          z         r
    1  -10.0000   10.0000    0.0000   11.2160
    2   12.4320   10.0000    0.0000   11.2160
    3   12.4320  -12.4320    0.0000   11.2160
```

(section of coordinates omitted for brevity here)

```
  143  123.8075   -3.9569   50.0000   14.0366
  144  126.0845   25.2786   50.0000   12.5148
  145   40.0000   40.0000   50.0000   11.2160
```

Example 2b.

5. Visualize the model (Fig. 2), as for Example 1.

B. PDB2AT, PDB2AM and MAP2GRID

Atomic coordinate structures can also be condensed to HMBs using the running mean, cubic grid, or hexagonal grid approaches espoused by Zipper and Durchschlag (1997, 2000). In fact the cubic grid approach is equivalent to that used by AtoB (Byron, 1997) above, whereas the hexagonal grid, while based on the same philosophy, is more appropriate for molecules with a natural hexagonal symmetry (Zipper and Durchschlag, 2000). The running mean method generates a

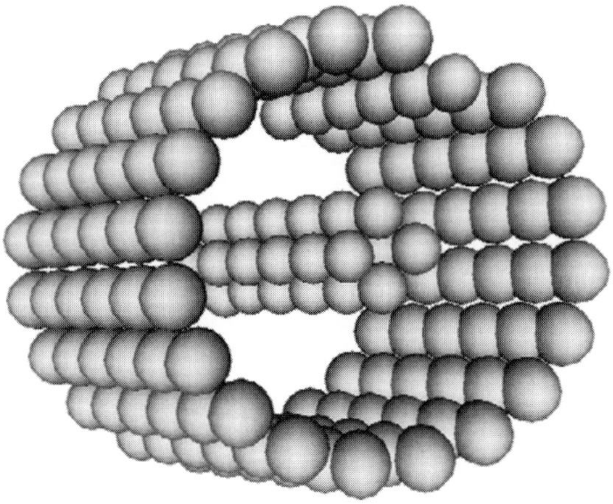

Fig. 2 Bead model generated *de novo* with AtoB (Byron, 1997) as described in Example 2. The coordinates of the model were converted to .vrml format with the program VisualBeads (Section IV.B); the final model was visualized using the program FreeWRL (Section IV.B).

reduced model by proceeding along the peptide chain, determining the center of gravity for every N residues, and placing a sphere of appropriate radius (in proportion with the cube root of the mass of the N residues) at that position. As with the cubic and hexagonal grid methods (and AtoB) the user can opt to assign all the spheres identical radii. By comparison of experimental data with hydrodynamic values computed for a series of proteins via the three approaches, the authors have shown that these methods yield HBMs for which satisfactory hydrodynamic calculations can be made. The authors have noted however that the use of too large a reduction factor with the running mean method distorts the original protein structure more than the cubic or hexagonal grid methods. They also recommend replacement of unequally sized spheres with those of an equal size in order to avoid inaccuracies in subsequent hydrodynamic computations (see discussion in Section II). This, however, results in distortions of models computed with the grid methods! So the user has to exercise judgment in each case. Several versions of Zipper and Durchschlag's "in-house" programs for generating HMBs via the running mean, cubic grid, and hexagonal grid methods are available on request from the authors. For example, the programs PDB2AT and PDB2AM generate HBMs from atoms/atomic groups or amino acid residues, respectively, while program MAP2GRID encodes the cubic grid and hexagonal grid algorithms (Table III, Fig. 3).

C. MAKEPIXB (and HYDROPIX)

MAKEPIXB (García de la Torre, 2001) generates a file describing a solid, geometrical shape that is then converted to a shell-bead model by the related program HYDROPIX [(García de la Torre, 2001), an analogue of HYDRO

Table III
Summary Description of PDB2AT, PDB2AM and MAP2GRID (Zipper and Durchschlag, 1997, 2000)

Web site	None
Summary of method	Reduction of atomic coordinate structures via running mean, cubic grid or hexagonal grid method.
Good for	Smaller proteins for which the running mean method may be used with a small-to-moderate reduction factor, without distortion of the structure.
Limitations	When a large reduction factor is used, cubic or hexagonal grid results in less distortion to the original structure (compared with running mean). Models of a quality similar to those generated by the running mean approach can be obtained provided beads of unequal size are located at centers of gravity; location of beads on lattice points yields less satisfactory results, irrespective of whether the beads are of equal or unequal size. For proteins with pronounced cubic or hexagonal symmetry, the cubic grid or hexagonal grid algorithms are preferable.
Input	Atomic coordinates, reduction factor
Output	Bead model
Operating system(s)	Unknown
Language	Unknown
Example model	Fig. 3

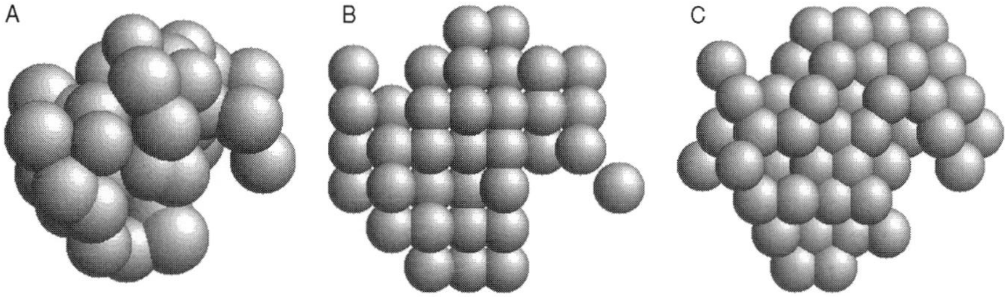

Fig. 3 Bead models generated from 6lyz.pdb by the (A) running mean (program PDB2AM), (B) cubic grid (program MAP2GRID), and (C) hexagonal grid (program MAP2GRID) algorithms (Zipper and Durchschlag, 1997, 2000). These models were kindly provided by the program authors.

(García de la Torre *et al.*, 1994)] that also computes hydrodynamic and related properties for the model. In order to describe the geometrical shape (e.g., an ellipse, a toroid an icosahedron), the user must be able to represent its spatial limits in a few lines of FORTRAN code within one of the two required input files. The user also provides the program with the *x*, *y*, and *z* dimensions of the cuboid that most efficiently contains the shape in question and the dimension of the volume element of which the cuboid is comprised. MAKEPIXB then scans the volume elements (called pixels by the author), assigning them a binary digit 1 if they are within the shape, 0 otherwise. The output file (a .pxb file) is in binary format and can be read by HYDROPIX. The program also generates an ASCII file of the Cartesian coordinates of the shape that can be read by molecular graphics programs such as RasMol (Sayle and Milner-White, 1995) (http://www.openrasmol.org/). HYDROPIX translates the pixbits to a three-dimensional array of (overlapping) spheres that serves as the primary hydrodynamic model from which a shell model and its corresponding hydrodynamic (and related) properties are calculated, as with HYDROPRO (Section VI.B) (García de la Torre *et al.*, 2000a) (Table IV; Fig. 4).

D. SOMO and ASAB1

SOMO (Rai *et al.*, 2005) generates medium-resolution HBMs from the atomic coordinates of proteins by placing a bead of volume equal to the sum of the constituent atom volumes at the center of mass (CM) of the main-chain segment of each residue and a second bead at a defined position for the side-chain segment depending on its chemical characteristic (Table V). The volumes of the beads are

Table IV
Summary Description of MAKEPIXB (& HYDROPIX) (García de la Torre, 2001)

Web site	http://leonardo.fcu.um.es/macromol/programs/hydropix/hydropix.htm
Summary of method	Macromolecule is represented by a regular geometrical shape (or an assembly of shapes). MAKEPIXB determines volume elements that comprise the shape and encodes them in a compact format readable by related program HYDROPIX.
Good for	Macromolecules that can be or have to be satisfactorily represented by geometrical shapes (e.g., ellipsoids and toroids) or assemblies of shapes.
Limitations	The user has to determine the equations that define whether a point in space belongs within or outwith the desired shape.
Input	For MAKEPIXB: Maximum values for x, y, and z that define one extreme corner of the cuboid in which the molecule can just be contained (the other extreme corner being (0,0,0)), the spacing (i.e., the dimension of the volume elements comprising the shape), equations that define whether a volume element lies within or outwith the shape. For HYDROPIX: MAKEPIXB output pixbit file (.pxb) plus user-defined parameters controlling shell modeling process (e.g., radius of beads filling the primary hydrodynamic model), solution physical parameters (e.g., viscosity, density and so on).
Output	From MAKEPIXB: pixbit binary file encoding the geometrical shape and ASCII file with corresponding Cartesian coordinates. From HYDROPIX: hydrodynamic and (optionally) scattering distance distribution and form factor.
Operating system(s)	MAKEPIXB: Any (FORTRAN code supplied—must be compiled); HYDROPIX: Linux, MS-DOS, and Windows
Language	FORTRAN
Example model	Fig. 4

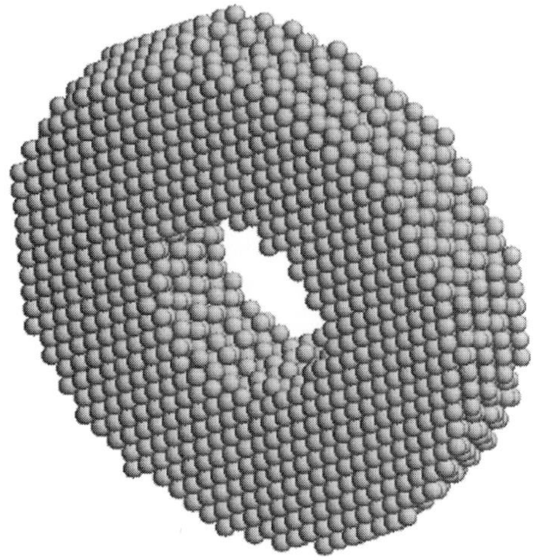

Fig. 4 Hollow disk generated by MAKEPIXB, translated to spheres by HYDROPIX (García de la Torre, 2001a) and visualized using RasMol (Sayle and Milner-White, 1995).

increased in order to include water of hydration (Section V). Overlaps between beads are removed in a hierarchical process that maintains the topography of the original outer surface. Certain carbohydrate moieties can be included in models as can the detergent β-octyl glucoside. Although the number of beads in SOMO models is about one quarter the number of atoms in the original structure file, SOMO also includes an implementation of part of the AtoB algorithm to permit the conversion of high- or medium-resolution models to lower resolution—useful for very large complexes in order to economize on CPU time in subsequent hydrodynamics computations. Finally, because SOMO identifies beads as being solvent exposed or not, buried beads can be excluded from core hydrodynamic calculations, providing a further saving of computer time. Included in the output from SOMO is a file that, in conjunction with an ancillary program (RAZ), generates a RasMol (Sayle and Milner-White, 1995) model of the HBM with categories of side-chains colored according to their chemistry. This means that it is possible to make informed decisions about creating oligomers from SOMO HBMs, rather than having to continually revert to the original atomic structure for this purpose.

SOMO generates coordinate files that can be input directly either to HYDRO (García de la Torre *et al.*, 1994) (Section VI.A) or to SUPCW/SUPCWIN (Section VI.F) (Spotorno *et al.*, 1997) for calculation of hydrodynamic parameters. SOMO models for a range of well-studied proteins were able to reproduce at worst to within 5% all hydrodynamic parameters computed by HYDRO or SUPCW/SUPCWIN. Prior to the generation of the HBM by SOMO, the pdb file is "prepared" by the ancillary program ASAB1 whose primary job is to identify exposed and buried atoms within the original structure. A full outline of the operation of SOMO and ASAB1 is given in Supplemental Appendix II of Rai *et al.* (2005).

Table V
Summary Description of SOMO and ASAB1 (Rai *et al.*, 2005)

Web site	http://somo.uthscsa.edu
Summary of method	Represents each residue with 2 beads (one for main-chain segment, one for side chain).
Good for	(Glyco)proteins; large complexes; realistic hydration; retaining correspondence between beads and residues.
Limitations	Models are medium-resolution; only for glyco(proteins) and β-octyl glucoside.
Input	pdb file edited to just the lines containing coordinates of the protein (and/or sugar, detergent). User is prompted for additional input during computation.
Output	ASAB1 produces 3 files [provaly [atomic coordinates plus name of second file generated (i.e., provaly1)], provaly1 (radius, mass and accessible surface area information for each residue) and provaly2 (atom and residue list)]. SOMO reads the provaly files and generates the HBM coordinates and radii in formats suitable for subsequent hydrodynamic computation by either HYDRO (García de la Torre *et al.*, 1994) (note, the dimension and the number of beads have to be added in the first two lines of this output file prior to submission to HYDRO) or SUPC (Spotorno *et al.*, 1997) (with same format as provaly). In addition, SOMO generates an .rmc (radius, mass and color) file and a similar, rmcoresp file (including the correspondence between original residues and beads). It also generates a script file (.spt) and the coordinates and radii (.bms) for visualization by RasMol (Sayle and Milner-White, 1995).
Operating system(s)	Linux
Language	C++, binaries supplied.
Example model	Fig. 5

Example 3: Generation of lysozyme bead model from 6lyz.pdb with ASAB1 and SOMO (Rai *et al.*, 2005).

1. Download SOMO and ancillary files from http://somo.uthscsa.edu.
2. Install programs in a Linux partition according to instructions on the SOMO web site.
3. Download RasMol for Linux (from e.g., http://openrasmol.org) and install it either in the same directory as SOMO (e.g., somo/bin) or edit your .bashrc file so that RasMol can be activated from any directory.

4. Check (and repair if necessary), for example with WHATIF (http://swift.cmbi.kun.nl/WIWWWI/), that there are no missing atoms within residues in the pdb file (otherwise SOMO will crash). Be aware that missing residues in pdb structures, especially at the surface, will affect the computed hydrodynamic parameters.

5. Edit your pdb file so that it comprises coordinate lines only (remove TER fields).

6. Generate a bead model in a terminal (command line) window as follows. At the command prompt, type asab1 to prepare the pdb file for bead model construction with SOMO. (Note, the text has been condensed here, to save space).

```
########################################################
#    National Institute for Cancer Research (IST)      #
#          Advanced Biotechnologies Center (CBA)       #
#                    Genova, ITALY                     #
########################################################
#    ASAB1 - Preparing PDB files for bead modelling    #
#          - Re-checking bead models for ASA           #
#                                                      #
#                 Version 3.0, March 2003              #
########################################################

 To process a PBD file, enter 2
 To re-check a bead model, enter 3
 --> 2

PROBE RADIUS [>=0.0]___1.4

- Insert the ASA threshold level in Angstroms [usually 10] : 10

Input tabella: /home/olwyn/somo/bin/tabella1.cor
        Insert the PDB filename : 6lyz.pdb
        Output files: Enter '0' for default (provaly, provaly1, provaly2, asaris )
                            '1' for new filenames
                      -->1

        ** Insert the root output filename :___ lyz

Molecule's extension along the z-axis = 48.636999 [angstrom]
Insert the integration step in angstroms : 1
Number of resulting iterations: 49
Confirm ? [yes=1;no=0] 1

Iteration number  = 49  Number of atoms in this iteration = 2
```

Example 3a.

7. Next, use SOMO to generate the bead model.

```
----------------------------------------------------------------------------------------
Welcome to SOMO (SOlution MOdeller) by N.Rai, M. Nollmann, M. Rocco and O. Byron
Last Modification: 09.02.05 (by M. Nollmann, any trouble mailto::marcnol@chem.gl a.ac.uk)
----------------------------------------------------------------------------------------

Do you want to use default values in somo.cfg (0) or specify them yourself (1)?1

READING CONFIGURATION FILE
--------------------------
Using /home/olwyn/somo/bin
Do you want to use somo.par in /home/olwyn/somo/bin (0:yes, 1:no)? 0
somo.par file read: </home/olwyn/somo/bin/somo.par>

Choose Algorithm: for AtoB type 1, for trans type 0: 0
For surface side chain beads select 1 of the following options:
0    = remove overlaps asynchronously
1    = pop beads but don't remove overlaps
2    = pop beads and remove overlaps asynchronously
5    = neither
>> 2

If you would like outward translation of surface beads,
then type '1', otherwise type '0' >>1
select percentage OVERLAP threshold %: 70

For surface side chain + peptide bond beads select 1 of the following options:
0    = remove overlaps asynchronously
1    = pop beads but don't remove overlaps
2    = pop beads and remove overlaps asynchronously
5    = neither
>> 2
select percentage OVERLAP threshold %: 70

For buried beads joined to surface beads select 1 of the following options:
0    = remove overlaps asynchronously
1    = pop beads but don't remove overlaps
2    = pop beads and remove overlaps asynchronously
5    = neither
>> 0

if you would like to rename output file type 1;
otherwise type 0 for default output: 1
type in the new name - lyz70
if you would like to output a HYDRO formatted file type 2, 0 for raw BEAMS, 1 for reordered
BEAMS formatted output:
bear in mind though that, depending on your computer, option 1 may make it crash for models
with more than about 4000 beads!
1
>> Do you want to calculate the sedimentation coefficient with BEAMS afterwards (0:yes, 1:NO):
1
the output filename is lyz70

create_bondsNchains> # of peptide beads= 54
create_bondsNchains> #  of buried beads = 106
create_bondsNchains> # of surface side chain beads = 85
create_bondsNchains> tot # of beads = 245
creates_bondsNchains> Elapsed time(0): 0 secs
```

(continues)

```
TransBMol> processing stage= 0
--------------------------------
choice = 2

pop_beads1> Total beads popped: 1
ProcessBeads> Number of overlapping beads = 112, Maximum overlap = 4.39915
No beads completed [|115|102|78|62|40|22|14|6
TransMol> Elapsed time(1): 0 secs

TransBMol> processing stage= 1
--------------------------------
choice = 2

pop_beads1> Total beads popped: 1
ProcessBeads> Number of overlapping beads = 173, Maximum overlap = 3.76281
No beads completed [|148|126|107|87|67|53|42|28|23|16|12|7|3
TransMol> Elapsed time (2): 1 secs

TransBMol> processing stage= 2 choice = 0
ProcessBeads> Number of overlapping beads = 632, Maximum overlap = 3.2564
No beads completed
[|604|528|468|414|365|313|278|240|201|178|132|106|82|69|52|44|39|33|25|22|20|15|10|3

PRODUCING OUTPUT FILES with BEAD MODELS
---------------------------------------
bondsNchains::print_reorderedprov> Total output Beads: 245
print_reorderedprov> Total Number of beads indexed: 245
print_reorderedprov> 245 atoms written to PDB file
print_reorderedprov> Number of GLY modified = 12
print_reorderedprov> Printing lyz70.rmc
print_reorderedprov> Printing lyz70.rmc
print_reorderedprov> Printing lyz70.rmcoresp

TransMol> Elapsed time(3): 10 secs
TransBMol> finished processing all beads


RUNNING RAZ
-----------

and now what?
type 1 to run raz
1
l=file lyz70
molt = 0.524429
RasMol Molecular Renderer
Roger Sayle, August 1995
Copyright (C) Roger Sayle 1992-1999
Version 2.7.2.1.1 January 2004
Copyright (C) Herbert J. Bernstein 1998-2004
*** See "help notice" for further notices ***
[32-bit version]

Unable to find RasMol help file!
RasMol>
```

Example 3b.

At this point RasMol should have started up and the bead model should be visible (Fig. 5).

8. Submit the output coordinates to ASAB1 and choose option "3" (recheck) to identify and color-code correctly beads originally labeled as buried that may have become exposed during model generation and overlap removal.

9. Because of an as yet unfixed small bug in SOMO, to compute the sedimentation coefficient using SUPCW/SUPCWIN the first line of the final coordinates file must be edited to change a flag value from "0.000" to "– 2.000," and the value of the partial specific volume should be added at the end of the line.

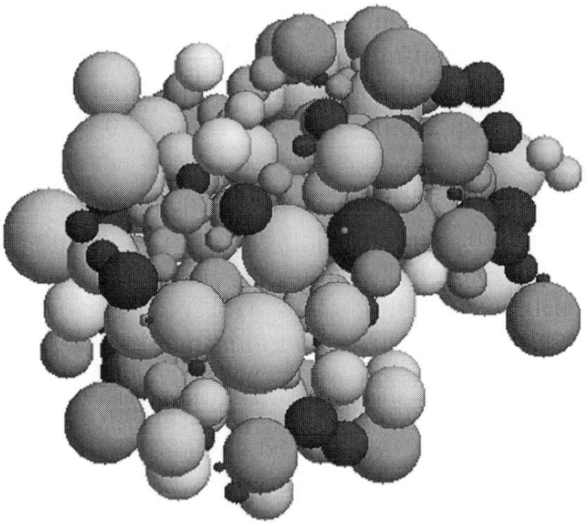

Fig. 5 Bead model of lysozyme generated from 6lyz.pdb with ASAB1 and SOMO (Rai *et al.*, 2005). For an explanation of bead colors, consult the SOMO web site or the original reference. (See Plate no. 2 in the Color Plate Section.)

1. Problems you may experience with SOMO and ASAB1 and how you may solve them
    a. The program does not run.
    Check that you have the correct paths specified in, for example, your .bashrc file.
    b. ASAB1 keeps asking for the name of the pdb file.
    Check that your pdb file is in text format and remove lines other than those containing coordinate data. Check also that you are entering the correct and complete filename (i.e., including the .pdb suffix).
    c. RasMol does not open at the end.
    Check that either you have the RasMol executable in the same directory as SOMO or you have edited the .bashrc file so that RasMol can be activated from any directory.

IV. Model Visualization

Invariably, the next step in the HBM process is to view the model generated with the methods above. Strangely, this is the step of the process for which there are the fewest options! In fact, only two will be covered here.

A. RasMol and RAZ

RasMol (Sayle and Milner-White, 1995) is widely used molecular graphics software. It is easy to use and is compatible with most operating systems (although not with newer ones, such as Macintosh OS X). Provided the coordinates of an HBM are written in a protein data bank-like format, RasMol will display them

(Table VI). However, pdb-style files do not include radii values, and therefore RasMol will not be able to directly use them. To overcome this problem, many of the HYDRO (Section VI) programs output RasMol compatible coordinate files (see below) in addition to their primary output. The program RAZ, an ancillary of SOMO (http://somo.uthscsa.edu) (Rai *et al.*, 2005), generates two files that can be used in conjunction with RasMol to view the HBMs produced by SOMO. The .bms file contains the bead coordinates and radii, and the .spt file refers to this file and is run as a script by RasMol to produce the final, interrogable image. Be aware that RasMol has an upper radius limit for drawing, so the models may be rescaled by an arbitrary factor (for instance, measured distances between beads may be incorrect).

B. VisualBeads

VisualBeads, in a fashion akin to RAZ, takes files that comprise the coordinates and radii of beads (and optional cylindrical connectors) and generates a .vrml file that can be read and visualized by freely available programs or internet browser plug-in tools such as Cortona (http://www.parallelgraphics.com/products/cortona/) and FreeWRL (http://freewrl.sourceforge.net/) (Table VII).

V. Hydration

Hydrodynamic hydration is difficult to quantify, and yet it is bound up in key parameters such as the sedimentation coefficient: the higher the hydration for a molecule of given shape and mass, the more its sedimentation coefficient will be

Table VI
Summary Description of RasMol (Sayle and Milner-White, 1995) and RAZ (Rai *et al.*, 2005)

Web sites	http://www.umass.edu/microbio/rasmol/index2.htm; http://somo.uthscsa.edu
Summary of method	RAZ generates files encoding the coordinates, radii and colors (usually reflecting the chemical nature) of the beads within the HBM generated by SOMO. The files are read by Ras Mol to produce the image.
Good for	SOMO-generated models.
Limitations	Only SOMO-generated models can be visualized in this way.
Input	SOMO/BEAMS format files: one with and a header containing the number of beads (N), a flag, a second file filename and (optionally) the partial specific volume, followed by N lines of coordinates; a second file with N lines containing the radii, masses and colors of the beads.
Output	Movable, interrogable molecular graphics
Operating system(s)	RasMol: Linux, MS-DOS, Windows, Mac PPC, and others RAZ: Linux
Language	precompiled executables
Example model	Fig. 5

Table VII
Summary Description of VisualBeads

Web site	http://leonardo.fcu.um.es/macromol/programs/visualbeads0b/visualbeads.htm
Summary of method	User inputs bead coordinates, radii and colors. Program generates .vrml file to then be read by freely downloadable vrml viewer (e.g., Cortona or FreeWRL).
Good for	All HYDRO-type bead models.
Limitations	Does not automatically represent chemical nature of beads in HBMs generated from atomic coordinates.
Input	Coordinates and radii (coorconn.txt) file and bead colors file (colors.txt).
Output	.vrml file.
Operating system(s)	Linux, MS-DOS, and Windows
Language	Executable
Example model	Figs. 1 and 2

depressed. In order to assess whether a hydrodynamic bead model does correctly represent the molecule of interest, some level of hydration has to be assigned and modeled. It is different from the water visualized in crystal, NMR, small-angle X-ray scattering (SAXS) and small-angle neutron scattering (SANS) structures and is, in the case of sedimentation, the (probably nonuniform) shell of water that moves with the macromolecule. A thorough account of protein hydration is given by Kuntz and Kauzmann (1974) and is revisited with particular reference to hydrodynamics in a seminal paper by Squire and Himmel (1979). However, the static picture of hydration espoused by these authors is now inconsistent with NMR data and molecular dynamics simulations that both reveal a highly dynamic interface at which the rotation and exchange of most water molecules are several orders of magnitude faster than biomolecular diffusion. A more recent perspective on hydrodynamic hydration (Halle and Davidovic, 2003) proposes a dynamic hydration model that explicitly links protein hydrodynamics to hydration dynamics. In this model, the first layer of hydration has different viscous properties from that of the bulk solvent, justifying an otherwise unsound static treatment of hydration. Accordingly, it is not unreasonable to compute protein hydration, based on amino acid composition (see below), and use this value in subsequent hydrodynamics calculations. Typical proteins are thought to include about 0.3–0.4 g water/g protein. Glycosylation has a slight effect on this value.

In order to reproduce the experimental sedimentation coefficient, an HBM should incorporate hydration. The total volume of the beads should be equal to the hydrated volume (V_h):

$$V_\mathrm{h} = \frac{M}{N_\mathrm{A}}(\bar{v} + \delta v^0) \tag{2}$$

where M is the molecular mass (g/mol), N_A Avogadro's number (6.02214×10^{23} mol^{-1}), $\bar{v}$ the partial specific volume (ml/g), δ the hydration (g water per g macromolecule), and v^0 the specific volume of solvent (ml/g). All terms on the right-hand side of Eq. (2) will be known or can be calculated for a macromolecule under scrutiny, except for δ.

Given the mass and sedimentation coefficient of a macromolecule, it is possible at least to determine the maximum possible hydration $\delta_{\max}$ by assigning all hydrodynamic deviation from anhydrous sphericity to hydration alone (i.e., making the particle in question a sphere). Then

$$\delta_{\max} = \frac{1}{\nu^0}\left[\frac{4\pi N_A}{3M}\left(\frac{M(1-\bar{v}\rho)}{6\pi\eta N_A s}\right)^3 - \bar{v}\right] \tag{3}$$

An estimate for δ can be made based on the amino acid composition of a protein [using data such as tabulated by Durchschlag (1986) and included in the program SEDNTERP (Laue *et al.*, 1992) (http://rasmb.bbri.org/rasmb/ms_dos/sednterp-philo/)], but this estimate is based on empirical data and does not explicitly account for the surface composition and distribution of amino acids, a key determinant in hydrodynamic hydration.

A measure of hydration may be obtained experimentally if the second virial coefficient [B (ml mol/g^2)] of the macromolecule can be determined. B can be decomposed into two terms:

$$B = B_{ex} + B_z \tag{4}$$

where B_{ex} results from the excluded volume of the particle and B_z from its net charge. B can be measured via analytical ultracentrifugation or static light scattering. B_{ex} and B_z are as follows:

$$B_{ex} = \frac{uN_A}{2M^2}; \quad B_z = \frac{1000Z^2}{4M^2 I}\left(\frac{1+2\kappa r_s}{(1+\kappa r_s)^2} + \ldots\right) \tag{5}$$

where u is the excluded volume (or covolume) of the macromolecule (ml), Z is its net charge or valence, I is the ionic strength (mol/liter), κ is the inverse screening length of the particle (cm^{-1}) and r_s its solvated radius. According to Harding *et al.* (1999), the Stokes radius [R_s, which can be derived from the experimentally determined sedimentation coefficient: $R_s = (M(1-\bar{v}\rho))/(6\pi\eta N_A s)$] provides an acceptable estimate of r_s in this context. The authors also advise that the magnitude of κ can be evaluated as follows: $\kappa = 3.27 \times 10^7\sqrt{I}$ at 20 °C. Therefore, an estimate of B_z can be made so that B_{ex} can be extracted from the experimental value of B. The excluded volume of a particle includes hydrodynamic hydration δ, in the form of V_h, the hydrated volume (Eq. 2), that is

$$u = V_h u_{red} \tag{6}$$

where u_{red} is the reduced covolume—a normalized parameter solely dependent on molecular shape (Rallison and Harding, 1985). u_{red} is calculated by the program COVOL (Harding *et al.*, 1999) based on the values of the three semi-axes describing the particle (or on their ratios). The program also calculates B_z and, therefore, with a few assumptions pertaining to surface charge and the description of the particle in terms of an ellipsoid, a value for δ can be deduced. This approach is described fully by Harding *et al.* (1999). The covolume is also computed by SOLPRO (Section VI.C). An alternative approach is what has become termed "crystallohydrodynamics" in which domains within a macromolecule are represented by ellipsoids that are then shell modeled and for which hydration-independent parameters are computed (Carrasco *et al.*, 2001; Longman *et al.*, 2003).

So, how can hydration actually be built into the HBM process? The answer to this question depends on the type of HBM. If the HMB is ultralow resolution, *s* may be calculated using HYDRO (García de la Torre *et al.*, 1994) (Section VI.A). The HBM could be hydrated prior to its submission to HYDRO (García de la Torre *et al.*, 1994) by a process of uniform expansion [an option (e), described in Table IV of Byron (2000) encoded in the bead model construction program AtoB (Byron, 1997)] such that the volume of its constituent beads equals that of the hydrated particle and their coordinates are modified to represent the necessarily larger particle. This approach works reasonably well for representing the hydration of globular particles but does not perform well for elongated particles (wherein the process results in the overlengthening of the particle). Equivalently, the anhydrous sedimentation coefficient calculated by HYDRO (García de la Torre *et al.*, 1994) for the unmodified HBM can be "hydrated" via the following equation

$$s_\delta = s_0\left(\frac{\bar{v}}{\bar{v}+\delta v^0}\right)^{1/3} \tag{7}$$

where s_δ is the hydrated sedimentation coefficient (that one aims to have match the experimentally determined value) and s_0 is the value calculated [by e.g., HYDRO (García de la Torre *et al.*, 1994)] for the anhydrous HBM. The bracketed term in Eq. (7) is the inverse of the radial expansion factor used in the uniform expansion process by AtoB (Byron, 1997) (Section III.A).

Hydration of an HBM based on atomic coordinates is more certain. The value of *s* calculated using HYDROPRO (García de la Torre *et al.*, 2000a) (Section VI.B) is automatically that for a particle hydrated to some level by virtue of the use of the atomic element radius (AER). Extrapolation to radius equal to zero for the shell beads yields the final values for the hydrodynamic parameters. García de la Torre and colleagues posit that the difference between the van der Waals radius and the AER could be regarded as the thickness of the consequent "hydration" layer. Accordingly, this "hydration" can be calculated from the difference between the anhydrous volume ($M\bar{v}/N_A$) and the volume of the HBM [given in the output from HYDROPRO (García de la Torre *et al.*, 2000a)].

Of course, the *shape* of the frictional surface affects the sedimentation coefficient. This is the shape of the macromolecule and the layer of hydrodynamic hydration. Although the shape of the macromolecule may be known at some level (either at high resolution, in the case of an HBM based on atomic coordinates or at low resolution when the HBM is drawn from, say, EM images or a molecular envelope restored from small-angle scattering data), the distribution of hydration and hence the true hydrodynamic surface is unlikely to be well known.

The programs HYDCRYST and HYDMODEL (Durchschlag and Zipper, 2002) account for the distribution of hydration on the macromolecular surface by hydrating HBMs originating from atomic coordinates according to detailed surface topography and a variable number of bound water molecules per particular amino acid. In a similar but more simplified manner the TRANS subroutine of program SOMO (Rai *et al.*, 2005) (Section III.D) expands the radii of the beads used to represent protein residues (two per residue—one for each of the main-chain and side-chain segments) to include the water of hydration for particular residues according to the data of Kuntz and Kauzmann (1974). This means that the HBM has the correct starting hydrated volume *and* its surface topography approximates that of a realistically hydrated particle, rather than a merely uniformly expanded form. It should be pointed out that, owing to the peculiar packing characteristics of nonoverlapping spheres, the final total volume of the beads will be substantially less than the starting hydrated volume. However, SOMO-generated models can be considered as in between shell [e.g., generated by HYDROPRO (García de la Torre *et al.*, 2000a)] and full [e.g., generated by AtoB (Byron, 2000)] bead models, and it is the correct representation of the hydrated macromolecular *surface* that allows for the accurate computation of the hydrodynamic parameters.

VI. Hydrodynamic Calculations

Once a satisfactory approach to incorporation of hydration has been adopted, the final stage in the modeling process is the calculation of hydrodynamic and other solution properties. In this section, several general and more specialized programs for this purpose are described. The more advanced software is only briefly mentioned, while focus is placed on programs more widely used and likely to be of more interest to the reader.

A. HYDRO

HYDRO (García de la Torre *et al.*, 1994) is the most fundamental of the programs for computation of hydrodynamic (and related) parameters for rigid HBMs. It is the core on which a number of successive programs from the group of José García de la Torre are based (Sections VI.B, VI.C, VI.D, VI.E, and VI.G). The starting point for modeling with HYDRO is generation of an HBM, as outlined in Section III.

The resultant Cartesian coordinates and radii then form one of the input files for HYDRO, the other comprising additional, largely user-defined data concerning solvent and solute properties. The calculation of hydrodynamic parameters entails the inversion of a $3N \times 3N$ supermatrix (N being the number of beads in the HBM) by HYDRO to compute the hydrodynamic interactions between all the beads. This has been discussed in more detail in Section II. Here it is important to make the point that the length of time taken to perform these hydrodynamic calculations by HYDRO and the other related programs (see below) increases in proportion with N^3 and that the supermatrix requires $9N^2$ memory positions for storage (García de la Torre and Bloomfield, 1981) (Table VIII).

Example 4: Computation by HYDRO (García de la Torre *et al.*, 1994) of hydrodynamic parameters for the HBM of 6lyz.pdb generated with AtoB (Example 1, Fig. 1).

1. Download the HYDRO executable from http://leonardo.fcu.um.es/macromol/programs/hydro/hydro.htm together with the files hydro.dat and hydroigg3.dat.

Table VIII
Summary Description of HYDRO (García de la Torre *et al.*, 1994)

Web site	http://leonardo.fcu.um.es/macromol/programs/hydro/hydro.htm
Summary of method	Computes hydrodynamic and related parameters for HBMs from their Cartesian coordinates and radii and additional solvent and solute data.
Good for	Straightforward, rapid calculation of hydrodynamic and related properties for HBMs.
Limitations	Not designed for atomic coordinate models (i.e., pdb files) for which HYDROPRO (García de la Torre *et al.*, 2000a) is better suited or which could be first converted to HBMs using AtoB (Byron, 1997) (Section III.A) or SOMO (Rai *et al.*, 2005) (Section III.D). Inaccurate for HBMs incorporating overlapping beads of unequal size. Hydration has to be accounted for somehow (Section V). There is an upper limit to the number of beads handled by HYDRO (currently 2000).
Input	Main input file (including user defined modeling parameters, temperature, solvent viscosity, molecular weight, partial specific volume, solution density, number of values of scattering angle, termed h (although known more commonly as s or Q) and maximum value of scattering angle (h_{max}), number of intervals for computation of the distance distribution function and R_{max}, again more commonly known as D_{max}, the maximum dimension in the particle, number of trials in the Monte Carlo computation of the covolume, an integer to indicate whether detailed diffusion tensors are to be calculated). Cartesian coordinates and radii file.
Output	D_t, R_s, R_g, V, D_r, τ_{1-5}, τ_h, $[\eta]$, s, D_{max}, distance distribution, scattering form factor, u, diffusion tensors, and input file for SOLPRO (García de la Torre *et al.*, 1997) (below). See Table I for full definitions. Coordinates of primary bead model in two formats and a summary file.
Operating system(s)	Linux, MS-DOS, and Windows
Language	FORTRAN executable supplied
Example model	Not applicable, although the model in Fig. 1 was used as input in Example 4.

2. Edit the .dat files so that they comprise input data for the lysozyme model, that is hydro.dat becomes:

```
lysozyme atob model                 Title
lyzatob                            filename for output files
lyz.dat                            Structural filename
293.                                Temperature, Kelvin
0.010                               Solvent viscosity
14297.                             Molecular weigth
0.720                               Specific volume of macromolecule
1.0                                 Solution density
26,                Number of values of H
3.e+6,             HMAX
30,                Number of intervals for the distance distribution
-3                 RMAX
10000,             (ONLY IF ISCA IS NOT ZERO) NTRIALS
1                  IDIF=1 (yes) for full diffusion tensors
*                  End of file
```

Example 4a.

Note that setting R_{max} to any negative number induces HYDRO to use the value of R_{max} it computes for the HBM. The coordinates and radii are in the file defined on line 3 of hydro.dat (i.e., lyz.dat):

```
1.E-07,     !Unit of length for coordinates and radii, cm (10 A)
31,         !Number of beads
 -9.5632   16.8392     6.3187     4.8528
 -0.2276   13.5154     7.9666     4.8528
  5.0570   15.5684    10.9475     4.8528
```

(section of coordinates omitted for brevity here)

```
 -4.7859   24.5985   31.5447   4.8528
  7.6662   20.6277   32.6438   4.8528
 -4.6384   28.8875   31.3687   4.8528
```

Example 4b.

3. Save the .dat files in text only (MS-DOS) format.

4. Place the executable and the .dat files in the same directory. Make sure that you have assigned sufficient virtual memory on your PC.

5. Double-clicking on the .exe icon (in Windows) will start the program and an MS-DOS-style command line window will (briefly, depending on the number of beads in the HBM) appear, listing the progress of the computation.

6. Upon completion of the computation several new files will appear in the HYDRO folder: lyzatob.res (the results of the hydrodynamic computations), lyzatob.bea [coordinates of the HBM that can be visualized using RasMol (Section IV.A)], lyzatob.sol [a file that can serve as input for the program SOLPRO (García de la Torre *et al.*, 1997) (Section VI.C) to facilitate computation of additional parameters for the HBM], summary.txt (a summary file), and lyzatob. vrml [the coordinates of the HBM in a format that can be read by a vrml reader (e.g., Cortona (http://www.parallelgraphics.com/products/cortona/) or FreeWRL (http://freewrl.sourceforge.net/) (see above) (Fig. 1)].

7. The crucial data are in lyzatob.res (some blank lines have been skipped, for brevity):

```
------------------------------------------------------------
                  HYDRO Version 7.C
J. Garcia de la Torre, S. Navarro, M.C. Lopez Martinez, F.G.
Diaz, J. Lopez Cascales. "HYDRO. A computer software for the
prediction of hydrodynamic properties of macromolecules".
Biophys. J. 67, 530-531 (1994).
------------------------------------------------------------
                     SUMMARY OF DATA AND RESULTS

                          This file:  lyzatob.res
                               Case:  lysozyme atob model
                    Structural file:  lyz.dat

                        Temperature:  293.0 K
                  Solvent viscosity:  0.01000 poise
                   Molecular weight:  1.430E+04 Da
         Specific volume of solute:  0.720 cm3/g
                   Solution Density:  1.000 g/cm3

Translational diffusion coefficient:  1.306E-07 cm2/s
      Stokes (translational) radius:  1.644E-06 cm
                 Radius of gyration:  1.404E-06 cm
                             Volume:  1.484E-17 cm3
   Rotational diffusion coefficient:  1.965E+04 s-1
                 Relaxation time (1):  9.113E-06 s
                 Relaxation time (2):  8.844E-06 s
                 Relaxation time (3):  8.727E-06 s
                 Relaxation time (4):  7.938E-06 s
                 Relaxation time (5):  7.935E-06 s
Harm. mean relax.(correlation) time:  8.483E-06 s
                 Intrinsic viscosity:  3.478E+03 cm3/g
         Sedimentation coefficient:  2.146E-01 svedberg

                  Longest distance :  4.744E-06 cm

           Calculation of scattering form factor, P vs h
                          h             P(h)
                     0.00E+00         1.00E+00
                     1.20E+05         9.91E-01
                     2.40E+05         9.63E-01
```

(section of form factors omitted for brevity here)

```
                     2.76E+06         1.04E-02
                     2.88E+06         1.12E-02
                     3.00E+06         1.23E-02

       Distribution of distances
     30 intervals between RMIN and RMAX centered at R;
       Values of p(R)

    RMIN(cm)     RMAX(cm)     R(cm)     p(R)(cm^(-1))
    0.000E+00   1.581E-07   7.906E-08   6.295E+03
    1.581E-07   3.163E-07   2.372E-07   4.267E+04
    3.163E-07   4.744E-07   3.953E-07   9.549E+04
```

(section of intervals omitted for brevity here)

```
    4.269E-06   4.428E-06   4.349E-06   1.113E+03
    4.428E-06   4.586E-06   4.507E-06   2.134E+02
    4.586E-06   4.744E-06   4.665E-06   7.909E+00

       Check for normalization:
       (RMAX-RMIN)*SUM(p(R))=   1.000E+00   SHOULD BE= 1
```

(continues)

```
          Monte Carlo simulation of covolume of bead models
          Number of trials=     10000 divided into 10 subsets
               Covolume=  1.884E-16 +-  5.345E-18 cm3

               Center of diffusion (x):  1.560E-08 cm
               Center of diffusion (y):  2.149E-06 cm
               Center of diffusion (z):  1.998E-06 cm

           Generalized (6x6) diffusion matrix:  (Dtt  Dtr)

                                                (Drt  Drr)

 1.316E-07 -2.685E-09  5.521E-09    -1.879E-04  4.503E-04  2.037E-04
-2.685E-09  1.269E-07 -1.366E-09     4.503E-04  7.667E-05 -1.286E-04
 5.521E-09 -1.366E-09  1.331E-07     2.037E-04 -1.286E-04  6.641E-05

-1.879E-04  4.503E-04  2.037E-04     2.005E+04 -8.630E+02  1.836E+03
 4.503E-04  7.667E-05 -1.286E-04    -8.630E+02  1.847E+04 -5.874E+02
 2.037E-04 -1.286E-04  6.641E-05     1.836E+03 -5.874E+02  2.041E+04
```

Example 4c.

B. HYDROPRO

HYDROPRO (García de la Torre *et al.*, 2000a) generates an HBM from an atomic coordinate file (i.e., a pdb file) and computes hydrodynamic parameters for it by constructing a primary hydrodynamic particle (PHP) from the input atomic coordinates, representing all nonhydrogen atoms with a bead (radius = AER, a user-defined input variable greater than the atomic radius). The surface of the PHP is then reproduced with much smaller beads to generate a shell model for which hydrodynamic parameters are calculated [using HYDRO (García de la Torre *et al.*, 1994) (Section VI.A) incorporated within HYDROPRO]. Included in the output from HYDROPRO is a file containing the coordinates of the PHP (Table IX; Fig. 6). The coordinates of the shell models are not output to the user. Hydration of HYDROPRO models is discussed in Section V.

Example 5: Computation of hydrodynamic parameters for (the shell-model limit of) 6lyz.pdb with HYDROPRO.

1. Download the HYDROPRO executable from http://leonardo.fcu.um.es/macromol/programs/hydropro/hydropro.htm together with the files hydropro.dat and 6lyz.pdb.

Table IX
Summary Description of HYDROPRO (García de la Torre *et al.*, 2000a)

Web site	http://leonardo.fcu.um.es/macromol/programs/hydropro/hydropro.htm
Summary of method	Replaces each non-H atom with a bead of radius = AER. Resulting surface then coated with small beads to make shell model. Hydrodynamics calculated for shell model.
Good for	Small-to-moderate-sized atomic coordinate protein (or nucleic acid) structures.
Limitations	Computationally quite intensive for large molecules. Hydration modeled as uniform layer.
Input	Main .dat file, including the usual inputs and the name of the pdb file containing the atomic coordinates on which the HBM is based.
Output	D_t, R_s, R_g, V, D_r, τ_{1-5}, τ_h, $[\eta]$, s, D_{max}, distance distribution, scattering form factor, input file for SOLPRO (García de la Torre *et al.*, 1997) (below). See Table I for full definitions. Coordinates of primary bead model in two formats and a summary file.
Operating system(s)	Linux, MS-DOS, and Windows
Language	Precompiled executable
Example model	Fig. 6

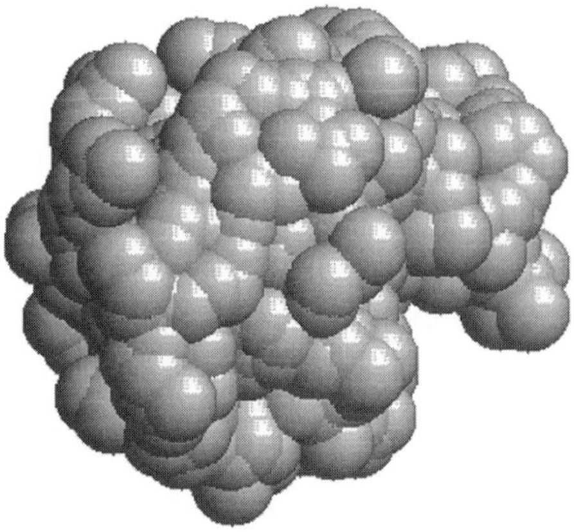

Fig. 6 Primary hydrodynamic model of lysozyme generated from 6lyz.pdb by HYDROPRO (García de la Torre *et al.*, 2000a), visualized using RasMol (Sayle and Milner-White, 1995).

2. Place the executable, the .dat and .pdb files in the same directory. Make sure that you have assigned sufficient virtual memory on your PC. Here is the content of hydropro.dat:

```
3.1-lysozyme                  !Name of molecule
lysozyme33                    !Name for output file
6lyz.pdb                      !Strucutural (PBD) file
3.3,                          !AER, radius of the atomic elements
6,              !NSIG
1.0,            !Minimum radius of beads in the shell (SIGMIN)
2.0,            !Maximum radius of beads in the shell (SIGMAX)
293.,           !T (temperature, K)
0.01,           !ETA (Viscosity of the solvent in poises)
14298.,         !RM (Molecular weigth)
0.720,      !Partial specific volume, cm3/g
1.0,            !Solvent density, g/cm3
21              !Number of values of H
2.e+7,          !HMAX
30,             !Number of intervals for the distance distribution
-1.,            !RMAX
1000,           !Number of trials for MC calculation of covolume
1               !IDIF=1 (yes) for full diffusion tensors
*                             !End of file
```

Example 5a.

3. hydropro.dat tells HYDROPRO which pdb file contains the coordinates from which the HBM is to be constructed (6lyz.pdb, in this case). Explanations of the other parameters are given in the accompanying hydropro.pdf, which is the original HYDROPRO paper (García de la Torre *et al.*, 2000a).

4. Provided sufficient virtual memory is available, double-clicking on the .exe icon (in Windows) will start the program and, provided the .pdb file contains sufficient atoms, an MS-DOS-style command line window will appear, listing the progress of the computation.

5. Be patient: the computation is CPU intensive. Upon completion of the computation, the command line window will vanish and several new files will appear in the HYDRPRO folder: lysozyme31.res (the results of the hydrodynamic computations for the shell model), lysozyme31-pri.bea [coordinates of the primary bead model that can be visualized using RasMol (Section IV.A) (Fig. 6)], lysozyme31.sol [a file that can serve as input for the program SOLPRO (García de la Torre *et al.*, 1997) (Section VI.C) to facilitate computation of additional parameters for the HBM], summary.txt (a summary file), lysozyme31-pri.vrml [the coordinates of the primary bead model in a format that can be read by a vrml reader (e.g., Cortona (http://www.parallelgraphics.com/products/cortona/) or FreeWRL (http://freewrl.sourceforge.net/) (Section IV.B)].

6. The crucial data are in lysozyme31.res (some blank lines have been skipped for brevity):

```
-----------------------------------------------------------------
              HYDROPRO Version 7.C
     J. Garcia de la Torre, M.L. Huertas and B. Carrasco,
     "Calculation of hydrodynamic properties of globular proteins
     from their atomic-level structure". Biophys. J. 78, 719-730
     (2000).
     -----------------------------------------------------------------
                           SUMMARY OF DATA AND RESULTS
                              This file:  lysozyme31.res
                                   Case:  3.1-lysozyme
                        Structural file:  6lyz.pdb

                            Temperature:  293.0 K
                      Solvent viscosity:  0.01000 poise
                       Molecular weight:  1.432E+04 Da
             Specific volume of solute:  0.702 cm3/g
                       Solution Density:  1.000 g/cm3

             Radius of atomic elements:  3.1 Angs
    Translational diffusion coefficient:  1.079E-06 cm2/s
           Stokes (translational) radius:  1.988E-07 cm
                      Radius of gyration:  1.527E-07 cm
                                  Volume:  2.416E-20 cm3
       Rotational diffusion coefficient:  1.951E+07 s-1
                     Relaxation time (1):  9.642E-09 s
                     Relaxation time (2):  9.112E-09 s
                     Relaxation time (3):  8.984E-09 s
                     Relaxation time (4):  7.639E-09 s
                     Relaxation time (5):  7.637E-09 s
    Harm. mean relax.(correlation) time:  8.523E-09 s
                    Intrinsic viscosity:  3.488E+00 cm3/g
             Sedimentation coefficient:  1.891E+00 svedberg

                      Longest distance :  5.356E-07 cm

          Distribution of distances
        30 intervals between RMIN and RMAX centered at R;
          Values of p(R)
       RMIN(cm)      RMAX(cm)      R(cm)      p(R)(cm^(-1))
       0.000E+00    1.785E-08    8.927E-09    2.256E+05
       1.785E-08    3.571E-08    2.678E-08    3.701E+05
       3.571E-08    5.356E-08    4.464E-08    1.097E+06
```
(section of intervals omitted for brevity here)
```
       4.821E-07    4.999E-07    4.910E-07    0.000E+00
       4.999E-07    5.178E-07    5.089E-07    0.000E+00
       5.178E-07    5.356E-07    5.267E-07    0.000E+00

          Check for normalization:
          (RMAX-RMIN)*SUM(p(R))=   1.001E+00   SHOULD BE= 1
```

(continues)

```
                    Calculation of scattering form factor, P vs h
                              h              P(h)
                          0.00E+00         1.00E+00
                          1.00E+06         9.94E-01
                          2.00E+06         9.74E-01
```

(section of form factors omitted for brevity here)

```
                          1.80E+07         1.09E-01
                          1.90E+07         8.37E-02
                          2.00E+07         6.32E-02

              Monte Carlo simulation of covolume of bead models
              Number of trials=       1000 divided into 10 subsets
                   Covolume=  2.954E-19 +-. 2.533E-20 cm3

                   Center of diffusion (x):  2.085E-07 cm
                   Center of diffusion (y):  1.762E-07 cm
                   Center of diffusion (z):  2.396E-07 cm

               Generalized (6x6) diffusion matrix:  (Dtt  Dtr)
                                                    (Drt  Drr)

 1.078E-06 -1.040E-08  4.643E-08     1.833E-02  3.131E-02  1.629E-02
-1.036E-08  1.062E-06 -1.022E-08     3.122E-02 -3.087E-04 -1.193E-02
 4.639E-08 -1.019E-08  1.098E-06     1.630E-02 -1.183E-02 -1.387E-02

 1.833E-02  3.122E-02  1.630E-02     1.986E+07 -9.130E+05  3.432E+06
 3.131E-02 -3.087E-04 -1.183E-02    -9.130E+05  1.770E+07 -9.384E+05
 1.629E-02 -1.193E-02 -1.387E-02     3.432E+06 -9.384E+05  2.097E+07
```

Example 5b.

C. SOLPRO

SOLPRO (García de la Torre *et al.*, 1997, 1999) is an extension to the HYDRO family of programs that enables computation of time-dependent properties of HBMs, such as fluorescence anisotropy, electric birefringence, and electric dichroism decay and decay of the $P_2(t)$ correlation function used, for example, in NMR for determining relaxations (Table X). SOLPRO also generates the shape-dependent, size-independent quantities that were the focus of a series of papers by Steve Harding and Arthur Rowe in the 1980s, for example Harding (1987), Harding and Rowe (1982), Harding and Rowe (1983). The importance of this is that it offers the chance to eliminate hydration from the modeling process (Section V).

D. HYDROSUB

For macromolecules or assemblies for which there are few high-resolution data available, HYDROSUB (García de la Torre and Carrasco, 2002) may prove useful. The program, another from the HYDRO family, generates bead-shell models of the ellipsoids and/or cylinders from which the molecular model is

Table X
Summary Description of SOLPRO (García de la Torre *et al.*, 1997, 1999)

Web site	http://leonardo.fcu.um.es/macromol/programs/solpro/solpro.htm
Summary of method	Calculation of time-dependent properties and "universal" functions (dependent on molecular shape alone, i.e., hydration independent).
Good for	Extending parameters by which molecular shape is modeled.
Limitations	More complex to utilize than, e.g., HYDROPRO (García de la Torre *et al.*, 2000a).
Input	Main input file (.dat) plus .sol file output by HYDRO (García de la Torre *et al.*, 1994), HYDROPRO (García de la Torre *et al.*, 2000a), HYDROSUB (García de la Torre and Carrasco, 2002), and some of the other HYDRO programs.
Output	D_t, R_g, D_r, τ_{1-5}, τ_h, $[\eta]$, D_{max}, $D^r_{II}/D^r_{\perp}$, D_{1-3}, u, v, P_0, τ_{1-5}/τ_0, τ_h/τ_0, τ_{ini}/τ_0, τ_{mean}/τ_0, G, Φ, P, β, R, u_{red}, Ψ, Π, $K_{\tau r1-5}$, $K_{\tau rh}$, $K_{\tau ini}$, $K_{\tau mean}$, Ψ_{1-5}, Ψ_h, Ψ_{ini}, Ψ_{mean}, Λ_{1-5}, Λ_h, Λ_{ini}, Λ_{mean}, plus a lot more that is difficult to summarize. See Table I for full definitions.
Operating system(s)	Linux, MS-DOS, and Windows
Language	FORTRAN, executable supplied
Example model	Fig. 6

composed (together with optional additional spherical beads, not shell modeled) and then computes [via HYDRO (García de la Torre *et al.*, 1994) (Section VI.A)] solution parameters for said model. The models are constructed on the basis of user input parameters, such as the subunit dimensions, the coordinates of subunit CM, and the two polar angles that define the orientation of the major symmetry axis (Table XI; Fig. 7).

E. HYDROMIC

HYDROMIC (García de la Torre *et al.*, 2001) reads 3-D reconstruction (e.g., spider) files generated from cryo electron microscopy data and, according to a selected threshold, assigns voxels to the particle and calculates their Cartesian coordinates. The voxels are then converted into spherical beads, yielding a primary hydrodynamic model. Then, in a fashion analogous to the procedure followed by HYDROPRO (García de la Torre and Carrasco, 2002) (Section VI.B), a shell model is constructed for which solution parameters are computed (Table XII; Fig. 8).

F. SUPCW/SUPCWIN

Part of a larger suite of programs for the generation, visualization, and computation of hydrodynamic and conformational properties of bead models [BEAMS, (Spotorno *et al.*, 1997)], SUPCW (operating under Linux), and SUPCWIN (for the Windows operating system) are programs that compute hydrodynamic and related

Table XI
Summary Description of HYDROSUB (García de la Torre and Carrasco, 2002)

Web site	http://leonardo.fcu.um.es/macromol/programs/hydrosub/hydrosub.htm
Summary of method	Bead-shell models constructed for multi-subunit particles. Solution properties computed using HYDRO (García de la Torre *et al.*, 1994).
Good for	Particles for which high-resolution data are not available.
Limitations	Represents subunits as cylinders, ellipsoids, or spheres.
Input	Main .dat file, including the usual inputs and the name of the file defining the structure which in turn is very simple, compared with the lengthy coordinate files of standard bead models.
Output	D_t, R_s, R_g, V, D_r, τ_{1-5}, τ_h, $[\eta]$, s, D_{max}, u, distance distribution, scattering form factor, diffusion tensors, input file for SOLPRO (García de la Torre *et al.*, 1997) (below). See Table I for full definitions. Coordinates of bead model in two formats and a summary file.
Operating system(s)	Linux, MS-DOS, and Windows
Language	FORTRAN, executable supplied
Example model	Fig. 7

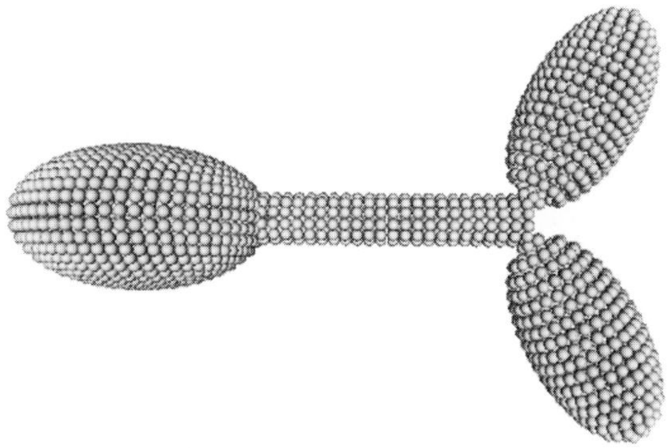

Fig. 7 Cylinder-and-ellipsoid shell model of IgG3 constructed with HYDROSUB (García de la Torre and Carrasco, 2002). The coordinates (igg100.vrml) were taken from the HYDROSUB web site and visualized using the program FreeWRL (Section IV.B).

parameters for rigid bead models under standard conditions (20 °C, water). Collectively known as SUPC (Table XIII) hereafter, for convenience, the program differs from HYDRO (García de la Torre *et al.*, 1994) (Section VI.A) in several useful regards: the user is given the choice of (i) computation of hydrodynamic parameters at the Cartesian origin or at the diffusion center; (ii) stick or slip boundary conditions (Section II); (iii) the first and last bead to be included in order to perform the computation for part of the HBM; (iv) excluding buried beads from the computation (apart from the calculation of R_g)—this substantially reduces the computational time and allows for bigger models to be analyzed, useful in checking for the effect of buried beads on the values generated for

Table XII
Summary Description of HYDROMIC (García de la Torre *et al.*, 2001)

Web site	http://leonardo.fcu.um.es/macromol/programs/hydromic/hydromic.htm
Summary of method	HYDROMIC constructs a primary hydrodynamic model based on the supplied 3-D reconstruction file, then follows the procedure used by HYDROPRO (García de la Torre and Carrasco, 2002) (Section VI.B) to compute solution properties for a shell-bead model.
Good for	(Macro)molecules for which medium- and high-resolution 3-D reconstructions from (cryo) electron microscopy data are available.
Limitations	Only suitable for 3-D reconstruction data-based modeling.
Input	Main .dat file, including the usual inputs and the name of the file in which the microscopy output data on which the HBM is to be based are found.
Output	D_t, R_s, R_g, V, D_r, τ_{1-5}, τ_h, $[\eta]$, s, D_{max}, scattering form factor, distance distribution, diffusion tensors. See Table I for full definitions. Coordinates of bead model in RasMol format and a summary file.
Operating system(s)	MS-DOS and Windows
Language	FORTRAN
Example model	Fig. 8

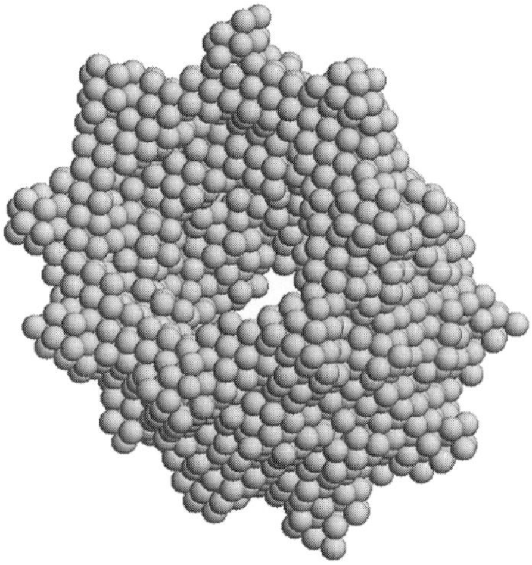

Fig. 8 Bead model of cytosolic chaperonin generated by HYDROMIC (García de la Torre *et al.*, 2001), based on a 3-D reconstruction from cryo electron microscopy data. The coordinates (apo_cct-37-spi-she.bea) were taken from the HYDROMIC Web site and visualized using RasMol (Sayle and Milner-White, 1995).

rotational diffusion parameters and $[\eta]$; (v) multiple models (up to 100) can be analyzed sequentially, and their parameters then averaged (useful, for instance, for NMR-derived models). In common with HYDRO (García de la Torre *et al.*, 1994), SUPC performs the inversion of a $3N \times 3N$ supermatrix, but the computation of certain specific parameters thereafter differs. This is apparent in the results presented in Table XV.

Table XIII
Summary Description of SUPCW/SUPCWIN (Spotorno *et al.*, 1997)

Web site	http://somo.uthscsa.edu
Summary of method	Computes hydrodynamic and related parameters for HBMs (under standard conditions, water at 20 °C) from their Cartesian coordinates, radii, masses, and partial specific volumes.
Good for	Can select stick or slip boundary conditions; can include or exclude coded (e.g., buried) beads from the global computations, and optionally also from the volume correction for either D_r and/or [η]. Can sequentially analyze up to 100 models and return also the means of the parameters.
Limitations	Must first construct HBM using e.g. AtoB (Byron, 1997) (Section III.A) or SOMO (Rai *et al.*, 2005) (Section III.D). Will not compute parameters for HBMs incorporating overlapping beads.
Input	User-defined computational parameters, scale factor to convert any dimension units (e.g., μm, Å) into the program's units (nm). Cartesian coordinates (with optional partial specific volume) and separate radii, masses and colors files.
Output	File containing results of hydrodynamic computations.
Operating system(s)	Linux (SUPCW) and Windows (SUPCWIN)
Language	C++, executable supplied
Example model	Not applicable, although the model in Fig. 5 was used as input in Example 6.

Abbreviations: μm: micrometer; Å: angstrom; nm: nanometer.

G. HYDRONMR (and Fast-HYDRONMR)

Although the earlier programs of the García de la Torre group calculated the rotational diffusion tensor ($\mathbf{D}_{rr}$) and the coordinates of the center of diffusion (CD), the full utility of these parameters remained unrealized until the release of HYDRONMR (García de la Torre *et al.*, 2000b) (Table XIV) in which they have been combined with the atomic coordinates of the molecule in question in order to predict the NMR relaxation times (T_1 and T_2) for each residue. T_1 and T_2 are measurable and can thus be used to confirm or refute the proposed molecular structure. From the ratio of T_1 and T_2 the rotational correlation time (τ_c) can be derived. HYDRONMR represents the molecule of interest as a rigid body and assumes that relaxation stems only from the modulation of dipolar couplings and chemical shift anisotropy by global tumbling. Full calculation of NMR relaxation times for one structure takes about 5 min with a desktop PC of 2005 vintage (Ortega and García de la Torre, 2005), though reversion to a double-sum approximation method (an optional mode in HYDRONMR) reduces the calculation time to about 1 sec.

HYDRONMR functions like HYDROPRO (García de la Torre *et al.*, 2000a) in that calculations are based on a shell model derived from a primary hydrodynamic model. As such, the AER is an adjustable parameter, though interestingly García de la Torre and colleagues have noticed that for a series of globular proteins a value of 3.3 Å is optimal, as for more conventional hydrodynamic calculations. In a subsequent paper (Bernadó *et al.*, 2002b), an optimization protocol for the objective determination of a value for the AER that maximizes the agreement between calculated and experimental T_1/T_2 ratios was described. The resultant AER value is then diagnostic of "problems" with the system: larger values indicating oligomerization or aggregation; smaller values stemming from models that do not adequately describe molecular conformation in solution (underrepresenting large-scale, low-frequency motions from extended conformers).

A very nice example of the use of HYDRONMR was given by Bernadó and colleagues (2002a) who presented a new method for determining the association constants, and in favorable cases the stoichiometries, of higher order complexes normally only observable transiently and at high concentrations. In fact, this is a consequence of the sensitivity of the AER to oligomerization processes. The authors found that experimentally determined relaxation rate ratios (R_2/R_1) could not be fitted for the system in question (a protein tyrosine phosphatase) with an AER of 3.3 Å at any of the concentrations used when crystal structures of the monomer and dimer alone were used. Instead, they introduced R_1 and R_2 for nuclei in the tetramer (for which there is no crystal structure) as adjustable parameters in the fitting process and, in this way, determined that the relaxation data are much better fitted with the introduction of a weakly associating tetrameric component. So here we see relaxation rates measured by NMR compared with relaxation rates computed with HBM to yield association constants, stoichiometries, and prediction of residues involved in the dimer–dimer interface (not reviewed here).

H. How Do Differently Constructed Models and Different Computations Compare?

Three types of model and three hydrodynamic computation programs are compared in Table XV. Models of lysozyme generated from 6lyz.pdb (i) at 4 "resolutions" by AtoB (Byron, 1997), (ii) by SOMO (Rai *et al.*, 2005) and (iii) (in the shell limit) by HYDROPRO (García de la Torre *et al.*, 2000a) were evaluated by HYDRO (García de la Torre *et al.*, 1994), SUPCWIN (Spotorno *et al.*, 1997), and HYDROPRO (García de la Torre *et al.*, 2000a), as appropriate to the type of model. Recall that SUPCWIN cannot handle models with overlapping beads—hence the AtoB models have not been evaluated by SUPCWIN. Note also that the AtoB models are not hydrated, which is why parameters such as the sedimentation coefficient are higher than for the other models. The purpose of Table XV is to illustrate that all three programs compute parameters differently and that different types of model perform better for some parameters than others.

VII. Advanced Hydrodynamic Calculations

In Section VI.H , the computation of solution properties for individual HBMs was considered. But it is important to bear in mind that often the purpose of hydrodynamic modeling is to come up with a plausible model describing the

Table XIV
Summary Description of HYDRONMR (García de la Torre *et al.*, 2000b)

Web site	http://leonardo.fcu.um.es/macromol/programs/hydronmr/hydronmr.htm
Summary of method	Solution properties of proteins (or nucleic acids) are computed from their atomic coordinates. Dipole orientations are also extracted and combined with the hydrodynamic parameters [computed by HYDRO (García de la Torre *et al.*, 1994)] to predict, for each residue in the protein, NMR relaxation times.
Good for	Proteins and nucleic acids for which NMR relaxation rates are known.
Limitations	NMR data required.
Input	Main .dat file, including the usual inputs and the name of the pdb file containing the atomic coordinates on which the HBM is based. Note: the .dat file also includes some specific user-defined NMR experimental parameters.
Output	General results, including D_t, D_r, $D^r_{II}/D^r_{\perp}$, τ_{1-5}, τ_h, diffusion tensors, and NMR-specific results for each user-defined applied magnetic field: the NMR relaxation times T_1 and T_2, their ratio together with their relative deviations with respect to the average over the number of residues in the protein (∇), the NOEs, and a table of spectral densities. See Table I for full definitions. And an additional file containing only the ratios of the NMR relaxation times T_1 and T_2 together with their relative deviations with respect to the average over the number of residues in the protein, for the user-defined applied magnetic fields. Coordinates of primary bead model in two formats and a summary file.
Operating system(s)	Linux, MS-DOS, and Windows
Language	FORTRAN, executable supplied
Example model	As for HYDROPRO

solution conformation of the rigid macromolecule in question. How can one be satisfied that an individual HBM does *uniquely* reproduce the measured parameters? The answer is that one cannot. However, by moving toward conformational search procedures, such as those included in Section VII.A, a more realistic solution is surely more likely to be obtained.

A. MULTIHYDRO, MONTEHYDRO and Rayuela

MULTIHYDRO (García de la Torre *et al.*, 2005) (http://leonardo.fcu.um.es/macromol/programs/multihydro/multihydro1c.htm) allows the evaluation of many possible conformations of a single rigid bead model in one run. In order to use MULTIHYDRO, the user has to insert two pieces of FORTRAN code defining solution physical data, simulation parameters (e.g., the number of conformations), and construction of the Cartesian coordinates and radii. Hydrodynamic computations are undertaken by HYDRO automatically. Therefore, to use MULTIHYDRO, one needs a certain facility with the FORTRAN language. Models can be simple (e.g., *de novo*) or can be based on atomic coordinates. The MULTIHYDRO

Table XV
Selected Hydrodynamic Parameters Calculated by Three Programs for Different Types of Lysozyme HBM, Generated from the File 6lyz.pdb with a Mass of 14298 and a Partial Specific Volume of 0.720 ml/g

Program		HYDRO[a]					SUPCWIN[b]	HYDROPRO[c]		
Parameter	Experimental value	AtoB (12 Å)[d]	AtoB (10 Å)[d]	AtoB (5 Å)[d]	AtoB (3 Å)[d]	SOMO model[e]	SOMO model[e]	3.0 Å[f]	3.3 Å[f]	3.6 Å[f]
D_t ($\times 10^{-6}$ cm^2 sec^{-1})		1.357	1.306	1.268	1.263	1.196	1.19	1.082	1.068	1.061
R_s ($\times 10^{-7}$ cm)	1.88 ± 0.06[g]	1.581	1.644	1.693	1.699	1.794	1.79	1.983	2.009	2.022
R_g ($\times 10^{-7}$ cm)		1.422	1.404	1.396	1.400	1.513	1.44	1.521	1.544	1.559
D_r ($\times 10^{7}$ sec^{-1})	1.95 ± 0.28[g]	2.142	1.965	1.777	1.753	2.040	2.119	1.965	1.889	1.859
τ_1 ($\times 10^{-9}$ sec)		8.33	9.11	10.16	10.34	9.13	8.83	9.61	9.99	10.14
τ_2 ($\times 10^{-9}$ sec)		8.11	8.84	9.79	9.94	8.68	8.26	9.05	9.43	9.56
τ_3 ($\times 10^{-9}$ sec)		7.98	8.73	9.72	9.87	8.56	8.38	8.94	9.29	9.46
τ_4 ($\times 10^{-9}$ sec)		7.30	7.94	8.71	8.80	7.40	7.09	7.54	7.88	8.03
τ_5 ($\times 10^{-9}$ sec)		7.30	7.94	8.71	8.80	7.40	7.09	7.53	7.88	8.03
τ_h ($\times 10^{-9}$ sec)	7.76 ± 0.47[g]	7.78	8.48	9.38	9.51	8.17	7.87	8.45	8.81	8.96
$[\eta]$ (cm^3 g^{-1})	2.99 ± 0.01[g]	3.197	3.478	3.832	3.886	3.288	2.50	3.463	3.613	3.692
s (S)	1.86 ± 0.06[g]	2.231	2.146	2.084	2.076	1.966	2.00	1.779	1.755	1.744

[a](García de la Torre *et al.*, 1994).
[b](Spotorno *et al.*, 1997).
[c](García de la Torre *et al.*, 2000a).
[d]Resolution of model generated by AtoB (Byron, 1997).
[e]Model generated in Example 1 and depicted in Fig. 1.
[f]Value of AER used to generate primary hydrodynamic model (see Section VI.B).
[g]For original source of data, refer to Rai *et al.* (2005).

approach can be extended to Monte Carlo simulations and also importance sampling Monte Carlo methods, suitable for the description of flexible particles. A similar approach is described by Nöllmann *et al.* (2004, 2005) who used their program *rayuela* to combine rigid body modeling with Monte Carlo/simulated annealing methods to search a large conformational space for models that best fit a wide range of solution data (from small-angle scattering, fluorescence resonance energy transfer and analytical ultracentrifugation).

B. BROWNRIG and BROWNFLEX

While parameters such as the diffusion coefficient of a bead model calculated by programs such as HYDRO and HYDROPRO accurately reflect the hydrodynamic properties measured under the influence of a relatively weak external field (e.g., in the analytical ultracentrifuge), recent Brownian dynamics (BD) simulations have revealed that this is not the case for stronger fields in which a particle becomes preferentially oriented. In the program BROWNRIG (Fernandes and García de la Torre, 2002) (http://leonardo.fcu.um.es/macromol/programs/brownrig1c/brownrig.htm), the Brownian trajectory of a particle, described as a bead model or an ellipsoid of revolution, for instance, is computed (for typically a total of 0.1 msec). The coordinates of CD and the generalized diffusion tensor of the bead model are calculated, using for example HYDRO, and BROWNRIG performs the computations needed to transform from the particle-fixed system to the laboratory-fixed system at each, for example, 10^{-13} sec time-step of the BD simulation. Hydrodynamics plays a key role in BD simulations, as it predicts the effect of collisions by solvent molecules with the macromolecule and the friction exerted by the solvent on the macromolecule as it moves. Fernandes and García de la Torre describe the utility of this approach in three environments: (i) free diffusion [wherein the computed diffusion coefficient is almost identical with that computed by HYDRO (García de la Torre *et al.*, 1994), as one would expect for free diffusion]; (ii) limited diffusion—representative of the quasi-two-dimensional diffusion of a macromolecule in a lipid bilayer; and (iii) diffusion in a weak and strong external electric field.

The flexibility of the protein Pin1, comprising two globular domains connected by a flexible linker, was interpreted recently using BD (Bernadó *et al.*, 2004) because it was otherwise not possible to model experimental relaxation rates satisfactorily. In order to reduce the computational expense [and hence extend the BD simulation to a longer (microsecond) timescale], the protein was represented as a simplified model that reproduced the hydrodynamic properties of Pin1. On this timescale, the global tumbling and slower, large-scale interdomain motions that occur on the nanosecond timescale can be satisfactorily simulated. The modeling was undertaken by Bernadó and colleagues at different, advancing levels of complexity, some of which exceeds the scope of this chapter. Nonetheless, for anyone interested in hydrodynamic modeling, a reprise of the more basic findings is useful here. Using HYDRONMR (García de la Torre *et al.*, 2000b) (Section VI.G)

Pin1 bound to its target peptide could be modeled with an AER = 3.3 Å, indicating that it adopts a rigid structure in solution, unlike the protein in its free, unliganded state that could only be modeled with AER = 2.5 Å, indicative of significant large-scale motions in solution. Furthermore, different AER values were required to describe the two domains, indicative of interdomain motion. Interestingly, the authors modeled the BD trajectory with and without inclusion of hydrodynamic interactions (using an in-house procedure BROWNFLEX). Because adjacent beads shield each other from collisions with solvent molecules, hydrodynamic effects result in more compact structures as random solvent collisions tend to push the two domains together. Additionally, interdomain constraints were modeled by the inclusion of elastic "strings" (of variable spring constant, K) between certain of the domain constituent beads. The resultant model reproduced the characteristic solution behavior of a flexible, two-domain protein but the authors note that, in common with all hydrodynamic modeling, the model is not unique.

VIII. Concluding Comments

In their 1979 paper, Teller *et al.* remarked:

> It has become a custom to assume a specific hydration of proteins of 0.25 g H_2O per gram of protein and report resulting axial ratios of equivalent rotational ellipsoids. Frequently the axial ratios resulting from such a treatment are absurd in the light of the present knowledge of protein structure. Unfortunately, the situation is often no better when viscosity data are considered at the same time. The reason why these absurd axial ratios are accepted into the modern literature is that they are rarely interpreted further, i.e., they are deadend numbers. It appears, however, that if a more rigorous treatment of the problem would be available, friction coefficients could indeed provide insights into protein structure in solution.

We are well and truly in the era foreseen by Teller *et al.*

Acknowledgments

I thank Jack Correia for his patience in awaiting this chapter and for his encouragement during its completion; Mattia Rocco for his careful reading of the manuscript, his input to the chapter and his helpful comments; Helmut Durchschlag and Peter Zipper for contribution of their models; José García de la Torre for continuing advice and extensive comments on this chapter and the wider topic; Neil Paterson, Mathis Riehle, and Scott Arkison for assistance with computing matters.

References

Allison, S. A. (1999). Low Reynolds number transport properties of axisymmetric particles employing stick and slip boundary conditions. *Macromol.* **32,** 5304–5312.

Allison, S. A. (2001). Boundary element modeling of biomolecular transport. *Biophys. Chem.* **93,** 197–213.

Aragon, S. (2004). A precise boundary element method for macromolecular transport properties. *J. Comp. Chem.* **25,** 1191–1205.

Bernadó, P., Åkerud, T., García de la Torre, J., Akke, M., and Pons, M. (2002a). Combined use of NMR relaxation measurements and hydrodynamic calculations to study protein association.

Evidence for tetramers of low molecular weight protein tyrosine phosphatase in solution. *J. Am. Chem. Soc.* **125,** 916–923.

Bernadó, P., Fernandes, M. X., Jacobs, D. M., Fiebig, K., García de la Torre, J., and Pons, M. (2004). Interpretation of NMR relaxation properties of Pin1, a two-domain protein, based on Brownian dynamic simulations. *J. Biomol. NMR* **29,** 21–35.

Bernadó, P., García de la Torre, J., and Pons, M. (2002b). Interpretation of 15N NMR relaxation data of globular proteins using hydrodynamic calculations with HYDRONMR. *J. Biomol. NMR* **23,** 139–150.

Bloomfield, V., Dalton, W. O., and van Holde, K. E. (1967). Frictional coefficients of multi-subunit structures. I. Theory. *Biopolymers* **5,** 135–148.

Byron, O. (1997). Construction of hydrodynamic bead models from high-resolution X-ray crystallographic or nuclear magnetic resonance data. *Biophys. J.* **72,** 408–415.

Byron, O. (2000). Hydrodynamic bead modelling. *Methods Enzymol.* **321,** 278–304.

Carrasco, B., and García de la Torre, J. (1999). Hydrodynamic properties of rigid particles: Comparison of different modeling and computational procedures. *Biophys. J.* **75,** 3044–3057.

Carrasco, B., García de la Torre, J., Davis, K. G., Jones, S., Athwal, D., Walters, C., Burton, D. R., and Harding, S. E. (2001). Crystallohydrodynamics for solving the hydration problem for multi-domain proteins: Open physiological conformations for human IgG. *Biophys. Chem.* **93,** 181–196.

Durchschlag, H. (1986). Specific volumes of biological macromolecules and some other molecules of biological interest. *In* "Thermodynamic Data for Biochemisty and Biotechnology" (H.-J. Hinz, ed.), pp. 45–128. Springer-Verlag, Berlin, Heidelberg, New York, Tokyo.

Durchschlag, H., and Zipper, P. (2002). Modelling of protein hydration. *J. Phys. Condens. Matter* **14,** 2439–2452.

Fernandes, M. X., and García de la Torre, J. (2002). Brownian dynamics simulation of rigid particles of arbitrary shape in external fields. *Biophys. J.* **83,** 3039–3048.

García de la Torre, J. (2001). Building hydrodynamic bead–shell models for rigid bioparticles of arbitrary shape. *Biophys. Chem.* **94,** 265–274.

García de la Torre, J., and Bloomfield, V. A. (1981). Hydrodynamic properties of complex rigid biological macromolecules: Theory and applications. *Q. Rev. Biophys.* **14,** 81–139.

García de la Torre, J., and Carrasco, B. (2002). Hydrodynamic properties of rigid macromolecules composed of ellipsoidal and cylindrical subunits. *Biopolymers* **63,** 163–167.

García de la Torre, J., Carrasco, B., and Harding, S. E. (1997). SOLPRO: Theory and computer program for the prediction of SOLution PROperties of rigid macromolecules and bioparticles. *Eur. Biophys. J.* **25,** 361–372.

García de la Torre, J., Harding, S. E., and Carrasco, B. (1999). Calculation of NMR relaxation, covolume, and scattering-related properties of bead models using the SOLPRO computer program. *Eur. Biophys. J.* **28,** 119–132.

García de la Torre, J., Huertas, M. L., and Carrasco, B. (2000a). Calculation of hydrodynamic properties of globular proteins from their atomic-level structure. *Biophys. J.* **78,** 719–730.

García de la Torre, J., Huertas, M. L., and Carrasco, B. (2000b). HYDRONMR: Prediction of NMR relaxation of globular proteins from atomic-level structures and hydrodynamic calculations. *J. Magn. Reson.* **147,** 138–146.

García de la Torre, J., Llorca, O., Carrascosa, J. L., and Valpuesta, J. M. (2001). HYDROMIC: Prediction of hydrodynamic properties of rigid macromolecular structures obtained from electron microscopy images. *Eur. Biophys. J.* **30,** 457–462.

García de la Torre, J., Navarro, S., Lopez Martinez, M. C., Díaz, F. G., and Lopez Cascales, J. J. (1994). HYDRO: A computer program for the prediction of hydrodynamic properties of macromolecules. *Biophys. J.* **67,** 530–531.

García de la Torre, J., Ortega, A., Pérez Sánchez, H. E., and Hernández Cifre, J. G. (2005). MULTIHYDRO and MONTEHYDRO: Conformational search and Monte Carlo calculation of solution properties of rigid or flexible bead models. *Biophys. Chem.* **116,** 121–128.

Garnier, C., Barbier, P., Devred, F., Rivas, G., and Peyrot, V. (2002). Hydrodynamic properties and quaternary structure of the 90 kDa heat-shock protein: Effects of divalent cations. *Biochemistry* **41,** 11770–11778.

Halle, B., and Davidovic, M. (2003). Biomolecular hydration: From water dynamics to hydrodynamics. *Proc. Nat. Acad. Sci. USA* **100,** 12135–12140.

Harding, S. E. (1987). A general method for modelling macromolecular shape in solution: A graphical (Π-G) intersection procedure for triaxial ellipsoids. *Biophys. J.* **51,** 673–680.

Harding, S. E., and Rowe, A. J. (1982). Modelling biological macromolecules in solution: 3. The Λ-R intersection method for triaxial ellipsoids. *Int. J. Biol. Macromol.* **4,** 357–361.

Harding, S. E., and Rowe, A. J. (1983). Modeling biological macromolecules in solution. II. The general triaxial ellipsoid. *Biopolymers* **22,** 1813–1829.

Harding, S. E., Horton, J. C., Jones, S., Thornton, J. M., and Winzor, D. J. (1999). COVOL: An interactive program for evaluating second virial coefficients from the triaxial shape or dimensions of rigid macromolecules. *Biophys. J.* **76,** 2432–2438.

Kar, S. R., Adams, A. C., Lebowitz, J., Taylor, K. B., and Hall, L. M. (1997). The cyanobacterial repressor smtB is predominantly a dimer and binds two Zn^{2+} ions per subunit. *Biochemistry* **36,** 15343–15348.

Kar, S. R., Lebowitz, J., Blume, S., Taylor, K. B., and Hall, L. M. (2001). SmtB-DNA and protein-protein interactions in the formation of the cyanobacterial metallothionein repression complex: Zn^{2+} does not dissociate the protein-DNA complex *in vitro*. *Biochemistry* **40,** 13378–13389.

Kuntz, I. D., and Kauzmann, W. (1974). Hydration of proteins and polypeptides. *In* "Advances in Protein Chemistry" (C. B. Anfinsen, J. T. Edsall, and F. M. Richards, eds.), Vol. 28, pp. 239–345. Academic Press, New York.

Laue, T. M., Shah, D. D., Ridgeway, T. M., and Pelletier, S. L. (1992). Computer-aided interpretation of analytical sedimentation data for proteins. *In* "Analytical Ultracentrifugation in Biochemistry and Polymer Science" (S. E. Harding, A. J. Rowe, and J. C. Horton, eds.), pp. 90–125. Royal Society of Chemistry, Cambridge, UK.

Longman, E., Kreusel, K., Tendler, S. B., Fiebrig, I., King, K., Adair, J., O'Shea, P., Ortega, A., García de la Torre, J., and Harding, S. E. (2003). Estimating domain orientation of two human antibody IgG4 chimeras by crystallohydrodynamics. *Eur. Biophys. J.* **32,** 503–510.

Nakasako, M., Fujisawa, T., Adachi, S.-i., Kudo, T., and Higuchi, S. (2001). Large-scale domain movements and hydration structure changes in the active-site cleft of unligated glutamate dehydrogenase from *Thermococcus profundus* studied by cryogenic x-ray crystal structure analysis and small-angle x-ray scattering. *Biochemistry* **40,** 3069–3079.

Nöllmann, M., Stark, W. M., and Byron, O. (2004). Low-resolution reconstruction of a synthetic DNA Holliday junction. *Biophys. J.* **86,** 3060–3069.

Nöllmann, M., Stark, W. M., and Byron, O. (2005). A global multi-technique approach to study low-resolution solution structures. *J. App. Cryst.* **38,** 874–887.

Ortega, A., and García de la Torre, J. (2005). Efficient, accurate calculation of rotational diffusion and NMR relaxation of globular proteins from atomic-level structures and approximate hydrodynamic calculations. *J. Am. Chem. Soc.* **127,** 12764–12765.

Oseen, C. W. (1927). Neure Methoden und Ergebnisse in der Hydrodynamik. *In* "Mathematik und ihre Anwendungen in Monographien und Lehrbuchern" (E. Hilb, ed.), p. 337. Academisches Verlagsgellschaft, Leipzig.

Rai, N., Nöllmann, M., Spotorno, B., Tassara, G., Byron, O., and Rocco, M. (2005). *SOMO* (*SO*lution *MO*deler): Differences between x-ray and NMR-derived bead models suggest a role for side chain flexibility in protein hydrodynamics. *Structure* **13,** 723–734.

Rallison, J. M., and Harding, S. E. (1985). Excluded volume for pairs of triaxial ellipsoids at dominant Brownian motion. *J. Colloid Interface Sci.* **103,** 284–289.

Rotne, J., and Prager, S. (1969). Variational treatment of hydrodynamic interaction in polymers. *J. Chem. Phys.* **50,** 4831–4837.

Sayle, R. A., and Milner-White, E. J. (1995). Rasmol- Biomolecular graphics for all. *Trends Biochem. Sci.* **20,** 374–376.

Smith, P. E., and van Gunsteren, W. F. (1994). Translational and rotational diffusion of proteins. *J. Mol. Biol.* **236,** 629–636.

Spotorno, B., Piccinini, L., Tassara, G., Ruggiero, C., Nardini, M., Molina, F., and Rocco, M. (1997). BEAMS (BEAds Modelling System): A set of computer programs for the generation, the visualisation and the computation of the hydrodynamic and conformational properties of bead models of proteins. *Eur. Biophys. J.* **25,** 373–384.

Squire, P. G., and Himmel, M. E. (1979). Hydrodynamics and protein hydration. *Arch. Biochem. Biphys.* **196,** 165–177.

Stafford, W. F., III, Mabuchi, K., Takahashi, K., and Tao, T. (1995). Physical characterisation of calponin. *J. Biol. Chem.* **270,** 10576–10579.

Svergun, D. I., Barberato, C., Koch, M. H. J., Fetler, L., and Vachette, P. (1997). Large differences are observed between the crystal and solution quaternary structures of allosteric aspartate transcarbamylase in the R state. *Protein Struct. Funct. Genet.* **27,** 110–117.

Svergun, D. I., Petoukhov, M. V., Koch, M. H. J., and König, S. (2000). Crystal versus solution structures of thiamine diphosphate-dependent enzymes. *J. Biol. Chem.* **275,** 297–302.

Teller, D. C., Swanson, E., and de Haën, C. (1979). The translational frictional coefficients of proteins. *Meth. Enzymol.* **61,** 103–124.

Toedt, J. M., Braswell, E. H., Schuster, T. M., Yphantis, D. A., Taraporewala, Z. F., and Culver, J. N. (1999). Biophysical characterisation of a designed TMV coat protein mutant, R46G, that elicits a moderate hypersensitivity response in *Nicotiana sylvestris*. *Protein Sci.* **8,** 261–270.

Trewhella, J., Carlson, V. A. P., Curtis, E. H., and Heidorn, D. B. (1988). Differences in the solution structures of oxidised and reduced cytochrome c measured by small-angle x-ray scattering. *Biochemistry* **27,** 1121–1125.

Venable, R. M., and Pastor, R. W. (1988). Frictional models for stochastic simulations of proteins. *Biopolymers* **27,** 1001–1014.

Vestergaard, B., Sanyal, S., Roessle, M., Mora, L., Buckingham, R. H., Kastrup, J. S., Gajhede, M., Svergun, D. I., and Ehrenberg, M. (2005). The SAXS solution structure of RF1 differs from its crystal structure and is similar to its ribosome bound cryo-EM structure. *Mol. Cell* **20,** 929–938.

Waxman, E., Laws, W. R., Laue, T. M., Nemerson, Y., and Ross, J. B. A. (1993). Human factor VIIa and its complex with soluble tissue factor: Evaluation of asymmetry and conformational dynamics by ultracentrifugation and fluorescence anisotropy decay methods. *Biochemistry* **32,** 3005–3012.

Yamakawa, H. (1970). Transport properties of polymer chains in dilute solution: Hydrodynamic interaction. *J. Chem. Phys.* **53,** 435–443.

Zarutskie, J. A., Sato, A. K., Rushe, M. M., Chan, I. C., Lomakin, A., Benedek, G. B., and Stern, L. J. (1999). A conformational change in the human major histocompatibility complex protein HLA-DR1 induced by peptide binding. *Biochemistry* **38,** 5878–5887.

Zipper, P., and Durchschlag, H. (1997). Calculation of hydrodynamic parameters of proteins from crystallographic data using multibody approaches. *Prog. Colloid Polym. Sci.* **107,** 58–71.

Zipper, P., and Durchschlag, H. (1998). Recent advances in the calculation of hydrodynamic parameters from crystallographic data by multibody approaches. *Biochem. Soc. Trans.* **26,** 726–731.

Zipper, P., and Durchschlag, H. (2000). Prediction of hydrodynamic and small-angle scattering parameters from crystal and electron microscopic structures. *J. App. Cryst.* **33,** 788–792.

CHAPTER 13

X-Ray and Neutron Scattering Data and Their Constrained Molecular Modeling

Stephen J. Perkins, Azubuike I. Okemefuna, Anira N. Fernando, Alexandra Bonner, Hannah E. Gilbert, and Patricia B. Furtado

Department of Biochemistry and Molecular Biology
University College London
London, WC1E 6BT, UK

0091-679X/08 $35.00
DOI: 10.1016/S0091-679X(07)84013-1

Abstract

X-ray and neutron solution scattering methods provide multiparameter structural and compositional information on proteins that complements high-resolution protein crystallography and NMR studies. We describe the procedures required to (1) obtain validated X-ray and neutron scattering data, (2) perform Guinier analyses of the scattering data to extract the radius of gyration R_G and intensity parameters, and (3) calculate the distance distribution function $P(r)$. Constrained modeling is important because this confirms the experimental data analysis and produces families of best-fit molecular models for comparison with crystallography and NMR structures. The modeling procedures are described in terms of (4) generating appropriate starting models, (5) randomizing these for trial-and-error scattering fits, (6) identifying the final best-fit models, and (7) applying analytical ultracentrifugation (AUC) data to validate the scattering modeling. These procedures and pitfalls in them will be illustrated using work performed in the authors' laboratory on antibodies and the complement proteins of the human immune defense system. Four different types of modeling procedures are distinguished, depending on the number and type of domains in the protein. Examples when comparisons with crystallography and NMR structures are important are described. For multidomain proteins, it is often found that scattering provides essential evidence to validate or disprove a crystal structure. If a large protein cannot be crystallized, scattering provides the only means to obtain a structure.

I. Introduction

Solution scattering is a diffraction technique that studies the overall structure of biological macromolecules in random orientations in solution (Glatter and Kratky, 1982; Perkins, 1988a,b). A sample is irradiated with a collimated, monochromatic beam of X-rays or neutrons. The physical principles are same in both types of scattering experiments. As a result of the constructive interference process in the scattered beam (Fig. 1A), an intense circularly symmetric diffraction pattern is observed on a two-dimensional (2D) area detector at low scattering vectors Q (Fig. 1B) that diminishes as Q increases. The radially averaged intensities $I(Q)$ are

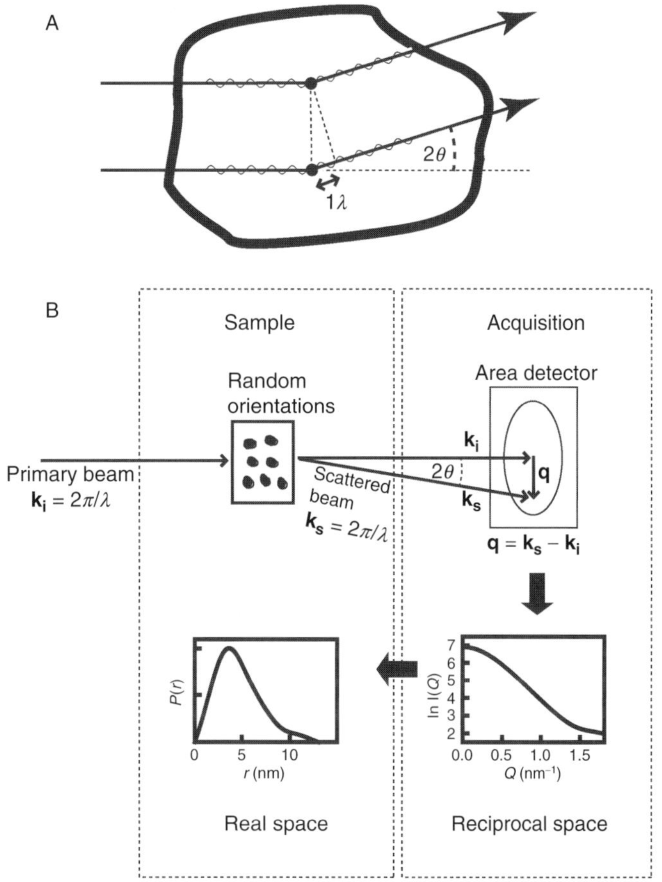

Fig. 1 Schematic representations of the scattering experiment. (A) Scattering from two point scatterers (•) within a globular macromolecule. The diffracted rays are in phase with each other but out of step by λ at the scattering angle 2θ shown. This causes constructive interference. The accumulation of these events at low 2θ values gives rise to the scattering pattern of the macromolecule. (B) The diffraction from high-scattering density macromolecules in a low-scattering density solution gives rise to a scattering pattern on an area detector. **q** is the scattering vector $\mathbf{k}_s - \mathbf{k}_i$, whose magnitude $Q = |\mathbf{q}| = 4\pi \sin \theta/\lambda$. The radial average of the scattering pattern about the position of the direct main beam (masked by a beam stop) gives rise to the scattering curve $I(Q)$ in reciprocal space. Fourier transformation of this to the $P(r)$ curve gives real-space information.

measured as a function of Q, where $Q = 4\pi \sin \theta/\lambda$ (2θ = scattering angle; λ = wavelength). Classical solution scattering views structures at a low structural resolution of about 2–4 nm from data obtained in reciprocal space in a Q range of about 0.05–2 nm^{-1}. Analyses of $I(Q)$ by Guinier plots lead to the overall molecular weight and the radius of gyration R_G (and in certain cases those of the cross section and the thickness). The Fourier transform of $I(Q)$ gives the distance distribution function $P(r)$, from which the maximum dimension of the macromolecule and its shape in real space can be deduced (Fig. 1B). Modeling curve fit analyses provide interpretations of the data.

Solution scattering can be applied to many biological macromolecules. Proteins of molecular weight 10 kDa upward can be studied, and this includes glycoproteins in which Asn or Ser/Thr residues provide *N*- or *O*-glycosylation sites. Other examples include DNA and RNA on their own, which are more difficult because of their extended polyelectrolyte structures, and their scattering densities are less optimal to work with than those of proteins. Scattering is also very useful to study protein–protein, protein–DNA, and protein–lipid complexes, especially if the unbound form can be studied as well.

Solution scattering results from the density contrast between the solute and solvent (Fig. 1B). The scattering density is the total of "scattering lengths" in the macromolecule divided by its molecular volume. X-ray scattering is based on the scattering event that occurs at electrons, and the "scattering length" of atoms increases with atomic number (Table I). Neutrons differ from X-rays in that scattering occurs at nuclei, and their "scattering lengths" are positive and similar to each other with the important exception of that for ^{1}H nuclei that are negative (Table I). X-rays are suitable for studying protein–protein complexes in which the scattering densities are similar throughout. The effect of the contrast is mostly vividly shown by neutron contrast variation experiments that study the internal structure of the macromolecule if this is strongly inhomogeneous in scattering density (e.g., protein–DNA or protein–lipid complexes). Since the proportion of nonexchangeable ^{1}H atoms varies for different macromolecular classes, lipids, proteins, carbohydrates, and nucleic acids, each possesses distinct neutron scattering densities (Table II). They fall between the very different neutron scattering densities of H_2O and $^{2}H_2O$. The appropriate choice of the ratio of H_2O and $^{2}H_2O$ in neutron scattering is used to select the contrast(s) that will reveal the internal structure of these components within the macromolecule. A special case of neutron scattering involves the specific deuteration of a given protein to change its scattering density. This means that the internal structure of its complex with a protonated protein can be revealed by contrast variation. In terms of contrast

Table I
Scattering Lengths *f* (X-Rays) and *b* (Neutrons) of Biologically Important Atoms

		Atomic number	$f(2\theta = 0°)$ (fm)	b (fm)
Hydrogen	^{1}H	1	2.81	−3.742
	^{2}H	1	2.81	6.671
Carbon	^{12}C	6	16.9	6.651
Nitrogen	^{14}N	7	19.7	9.40
Oxygen	^{16}O	8	22.5	5.804
Phosphorus	^{31}P	15	42.3	5.1

Table II
Scattering Densities of Solvents and Biological Macromolecules[a]

	X-rays (e.nm^{-3})	Neutrons (% ^2H_2O)
H_2O	334	0
^2H_2O	334	100
50% (w/w) sucrose in H_2O	402	13
Lipids	310–340	10–14
Detergents	300–430	6–23
Proteins	410–440	40–45
Hydrophobic residues	410	38
Hydrophilic residues	440	54
Carbohydrates	490	47
DNA	590	65
RNA	600	72

[a]Adapted from Perkins (1988b), in which a detailed compilation is given for all classes of biological macromolecules.

variation, X-ray data are closely equivalent to neutron data in 0% ^2H_2O, although it should be noted that the X-ray data correspond to hydrated macromolecules, while neutron data corresponds to unhydrated ones (Perkins, 2001).

High-flux sources make possible many new applications of solution scattering in biology through being able to access low sample concentrations. For X-ray scattering, a high flux means that X-ray scattering cameras avoid instrumental scattering curve distortions through the use of a focused monochromatized beam and idealized pin-hole optics. As signal-noise ratios are dramatically improved, these make possible the use of (1) the study within one experimental beam time session (1 day) of many samples; (2) the need for nonphysiological high sample concentrations that is minimized and macromolecules of low solubility can be studied; (3) time-resolved scattering can follow the rate of conformational changes or oligomerization/dissociation processes (Narayanan, 2007). In neutron scattering, the high beam fluxes also make possible: (4) the use of full contrast variation experiments using mixtures of H_2O and ^2H_2O to reveal information on internal structures; and (5) neutron kinetic experiments can be performed in ^2H_2O buffers. To handle the abundance of high-quality scattering data from these newer cameras, a powerful method of automated constrained solution scattering modeling has been developed by our group in recent years (Perkins *et al.*, 1998). By this, curve fits based on randomizing arrangements of the subunits in known crystal structures extract structural information to a precision of 0.5–1.0 nm. These correspond to medium-resolution structure determinations that can then be deposited in the Protein Data Bank (PDB).

This chapter outlines scattering theory, the experiment, and data modeling. The X-ray and neutron scattering instrumentation that were previously described (Perkins, 1994, 2000) are updated here to include newer cameras. The automated constrained modeling procedures (Perkins *et al.*, 1998) are updated in terms of the molecular interpretation of hydration shells (Perkins, 1986, 2001) and four strategies that have emerged for the modeling of scattering and sedimentation data for multidomain proteins (Perkins *et al.*, 2005).

II. Rationale

A. Complementary Structural Approaches

The two main strengths of solution scattering are the provision of a multiparameter description of a protein structure in near-physiological conditions and the ability to produce medium-resolution models of the scattering data from known atomic structures. Solution scattering is complementary to other methods of structure determination:

1. X-ray crystallography gives electron density maps at atomic resolution (0.1–0.3 nm). Fitting the maps to atomic models gives the macromolecular structure. The results can be affected by artifacts caused by intermolecular contacts in the crystal lattice packing or the high-salt concentration used to obtain crystals. Scattering is able to identify any large-scale conformational changes between the crystal and solution states and the oligomerization state in solution. If the protein cannot be crystallized and atomic models are available for all the subunits in the protein, scattering will provide a medium-resolution structure by constrained modeling.
2. 2D-NMR spectroscopy gives atomic resolution structures in solution, but is limited to smaller proteins of size less than about 30 kDa. Scattering leads to solution structures for macromolecules of size 15 kDa upward.
3. Electron microscopy visualizes macromolecular structures directly, but the measurement conditions *in vacuo* using stains can be harsh. The macromolecular dimensions determined by scattering are useful to confirm those from electron microscopy. If electron microscopy has been used to derive structures, these can be tested by scattering in the same way as for crystal structures.
4. AUC and light scattering provide sedimentation or diffusion coefficients as single-parameter measures of macromolecular shape. As they are equivalent to the radius of gyration R_G from scattering, they are useful to confirm the R_G values. Unlike scattering, AUC is able to resolve sedimentation coefficients of the individual components in mixtures (see below).

B. Properties of X-Ray Scattering

The main practical features of X-ray scattering are as follows:

1. Most biological macromolecules are studied in high positive solute–solvent contrasts (Table II). Only lipids have an electron density less than that of water. As this positive contrast usually corresponds to a much higher macromolecular

scattering density compared to that of the solvent, this large difference has the effect of minimizing systematic errors in the scattering modeling fits because internal density fluctuations can be neglected. So X-ray scattering often corresponds to a single contrast measurement.

2. Good counting statistics are usually obtained in this contrast if the background level of extraneous instrumental scattering is minimal. X-ray background scattering levels can sometimes be high but only at low Q values. Compared to neutron data for samples in H_2O buffers, which are affected by a high uniform incoherent scattering background, X-rays have the advantage that there is no such high background (unless working in high salt: see below).

3. Instrumental errors caused by wavelength polychromicity and beam divergence are minimal in synchrotron X-ray scattering. Thus, the Guinier and wide-angle X-ray analyses are not affected by systematic instrumental errors, and curve corrections do not have to be applied.

4. Radiation damage effects are common in X-ray scattering. As samples may aggregate rapidly after short irradiation times with synchrotron X-rays, exposure times have to be optimized in test runs, and postirradiated samples are usually discarded.

5. X-ray scattering reveals the hydrated dimensions of the macromolecule. Hydration means that a monolayer of water molecules is hydrogen bonded to the protein surface, and the electron density of this bound water is higher than that of bulk water (Perkins, 2001). The macromolecule appears larger by the thickness of this water monolayer, and this is detectable by X-rays.

6. Provided that the sample concentrations are known from spectrophotometer optical density measurements and a suitable X-ray intensity standard is available, X-ray scattering leads to relative molecular weights. This is useful to check for oligomer formation or to confirm the composition of the macromolecule being studied.

C. Properties of Neutron Scattering

The corresponding features of neutron scattering are summarized:

1. Contrast variation experiments using mixtures of H_2O and ^2H_2O buffers permit the structural analysis of hydrophobic and hydrophilic regions within proteins and glycoproteins, the location of detergents or lipids when complexed with solubilized membrane proteins, and the location of DNA or RNA when complexed with protein. Deuteration of a component in a multicomponent complex means that this component can be located within the complex.

2. The neutron buffer background is very low in ^2H_2O solvents, even in the presence of high-salt concentrations. This is useful when studying macromolecules at very low concentrations (0.5 mg/ml) in a range of salt concentrations, as X-rays are absorbed by high-salt buffers.

3. Guinier analyses at low scattering angles are not affected by significant beam divergence or wavelength polychromicity effects. However, neutron intensities at large Q are noticeably affected by both of these. Neutron curve fit modeling requires these to be considered. In addition, neutron modeling for proteins in ^2H_2O buffers also requires a flat background correction for a small residual incoherent scattering background.

4. No radiation damage effects occur in neutron scattering. Generally, neutron samples can be recovered for other structural studies. However, protein or glycoprotein samples can aggregate in a nonspecific manner in heavy water buffers if protein solvation is important for solubility. This problem is overcome by the use of gel filtration and low protein concentrations.

5. The hydration shell is not visible in neutron scattering for reason of the exchange of H and ^2H atoms with bulk solvent. Hence, the unhydrated dimensions of the macromolecule are studied. Because this corresponds directly to the water-free protein coordinates obtained from crystallography or NMR, this simplifies the neutron curve modeling from crystallography or NMR structures.

6. Provided that the sample concentrations are known, the absolute molecular weight can be calculated to within $\pm 5\%$ from the neutron data using H_2O buffers, although only instruments D11 and D22 at the Institut Laue Langevin (ILL: see below) offer sufficient flux to do this. The reason why H_2O buffers are used is that the neutron molecular weight calculation becomes insensitive to the partial specific volume (Jacrot and Zaccai, 1981).

III. X-Ray and Neutron Facilities

A. High-Flux Sources

The probability of a diffraction event when an X-ray photon (or a neutron) approaches an electron or a nucleus is similar and very low at 10^{-25} (or 10^{-23}), respectively. Compared to scattering experiments in other disciplines such as chemistry or metallurgy, biological experiments usually involve dilute samples and the available signal-noise ratios can be poor. The use of high-flux sources, such as European Synchrotron Radiation Facility (ESRF) or the ILL, overcome this significant limitation.

There are about 70 X-ray synchrotron sources in the world (April, 2007), mostly in Europe (25 facilities), North America (16 facilities), and Japan (15 facilities). The most reliable lists of facilities are on the internet and can be found on the clickable maps provided at the Synchrotron Radiation Source website at http://www.srs.ac.uk/srs/SRworldwide/index.htm at Daresbury, UK and the "light source facilities" menu listing available at http://www.lightsources.org. While the links often become outdated, searches will reveal up-to-date lists. A history of synchrotrons is available at http://xdb.lbl.gov/Section2/Sec_2-2.html. The earliest machines known as the "first-generation" sources took X-ray beams parasitically from particle physics experiments. The "second-generation" machines became dedicated to the production of synchrotron radiation, and sources such as that of the 2 GeV Synchrotron Radiation Source facility at Daresbury (the first one of this type) provides sufficient flux for scattering data collection. The most powerful synchrotrons are fully dedicated "third-generation" machines such as the first one in 1994 at the 6 GeV ESRF in France, followed by the 8 GeV Super Photon Ring (SPring8) near Kyoto, Japan, and the 7 GeV Advanced Photon Source (Illinois). They specialize in either short-wavelength (high energy or hard) X-rays or vacuum-ultraviolet and long-wavelength (low energy or soft) X-rays. "Third-generation" machines have multiplied in recent years, and examples of these include the Advanced Light Source at Berkeley, California (1.9 GeV), the Synchrotrone Trieste (2.0 GeV) in Italy, the Synchrotron Radiation Research Center (1.3 GeV) in Hsinchu, Taiwan, and the Pohang Light Source (2.0 GeV) in Pohang, Korea. The most recent machines starting in 2007 include the 3 GeV Diamond facility in Oxfordshire, UK and the 2.75 GeV Soleil facility at Paris, France.

There are about 40 neutron sources in the world (April, 2007), mostly reactor sources, but including spallation sources. They are located in Europe (16 facilities), North America (13 facilities), and Japan (6 facilities). Facility lists are available at links such as http://www.isis.rl.ac.uk/neutronSites/index.htm and http://www.ill.fr/index_sc.html. The most powerful reactor source is the 58 MW High Flux Reactor at the ILL in Grenoble, France (thermal neutron flux of 1×10^{15} neutrons/cm^2/sec), in operation since 1967. The most successful spallation (pulsed) source is that of the ISIS facility adjacent to Diamond, Oxfordshire, UK, operational since 1977 and currently being upgraded by the addition of its Second Target Station (TS2) due to be completed in 2007. TS2 will significantly increase the available neutron flux for scattering. The Spallation Neutron Source is another accelerator-based neutron source at Oak Ridge, Tennessee. At full power in late 2007, the Spallation Neutron Source will provide the most intense pulsed neutron beams in the world.

State-of-the art multiuser X-ray and neutron facilities are expensive, but the cost of an individual experiment is much reduced if there is a large international base of users. The full economic cost of beam time on a scattering camera is of the order £7500 (X-rays) to £15,000 (neutrons) per day. It is very much in a user's interests to list the total annual cost of his/her beam time awards in any summary of grants that is requested by a University or a funding body. Usually 2–4 days of beam time taken in two sessions is sufficient for a given project. On a per user basis, beam time represents good value for money when compared to the cost of a new analytical ultracentrifuge costing £0.25 million or an NMR spectrometer costing over £1 million, for which there are hidden extra costs for salaries, instrument consumables and maintenance, and building infrastructure. X-ray or neutron facilities usually provide at least one scattering camera for multiple users. Scattering cameras are shared between workers in metallurgy, polymers, and other fields as well as biology, and are generally oversubscribed and operated by ever-patient, overworked instrument scientists. This means that beam time schedules are tight, and there is a high user turnover. Users are not only responsible for providing their samples but also for providing minor apparatus and complying with local procedures, all which are ultimately reduced to the application of common sense and much foresight. The User Liaison Office at each facility and the instrument scientist responsible for the scattering camera are useful first contact points. The web pages for each facility are usually very informative.

Scattering curves intensities $I(Q)$ are measured as a function of the scattering vector Q (Fig. 1B). An alternative name for Q is the momentum transfer, and other symbols for this include h and q. The scattering vector is sometimes redefined as s, the reciprocal of the Bragg spacing d, where $s = Q/2\pi$. Dimensions are commonly reported in nanometer to reflect the low and medium structural resolutions of the method, but Ångström units (Å) (where 1 nm = 10 Å, and 1 Å is the approximate diameter of an H atom) are also used as their usage is widespread in crystallography.

B. X-Ray Instrumentation

Typical synchrotron X-ray cameras include those at Instrument ID02 at the ESRF (Narayanan *et al.*, 2001; Panine *et al.*, 2006). Electrons circulating at relativistic speeds around the storage ring of the synchrotron emit a white beam composed of all wavelengths including X-rays tangentially from the ring. The electron lifetime is of the order of 12–24 h, so storage ring refills are performed several times a day. Instruments such as ID02 receive this white beam from the ring, which is then focused and monochromated before it reaches the sample. The ID02 beamline optics consists of a cryogenic (liquid nitrogen) cooled Si-111 channel-cut monochromator and a focusing toroidal mirror (Fig. 2). The Si-111 crystal is cooled in order to withstand the high heat load resulting from the absorbed power of the white beam. Primary, secondary, and guard slits collimate the beam (dimensions 0.2 mm × 0.4 mm) to reduce parasitic background scattering. The whole camera is inside a radiation-shielded hutch, protected by safety interlocks to avoid accidental lethal X-ray exposure to the users. ID02 is, in fact, composed of two experimental cameras (Fig. 2). Conventional small-angle scattering with pin-hole optics is achieved with a detector in a vacuum within a 10-m-long detector tank, together with a second detector mounted outside the tank at the side of the sample to record the wide-angle scattering curve. Alternatively, a Bonse-Hart camera is used to perform ultra-low-angle scattering. The X-ray flux at scattering cameras at SRS Daresbury is of the order 4×10^{10} to 5×10^{11} photons/sec (Perkins, 2000). At ID02, the maximum photon flux at the sample position is of the order of 3×10^{13} photons/sec/100 mA with $\Delta\lambda/\lambda = 0.015\%$ at 12.4 keV. All these are adequate for biological X-ray scattering work, although suitable beam exposure times are about 10 min at SRS Daresbury but about 10 sec at ESRF. Beam time applications do need to justify the importance of the flux, such as the need to measure low concentrations or to perform stopped-flow measurements.

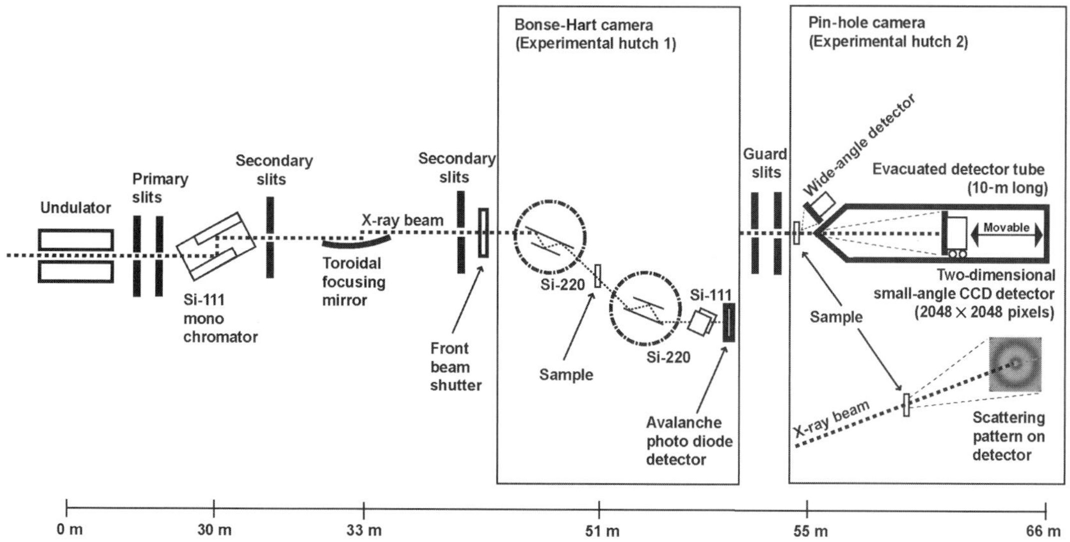

Fig. 2 Layout of the solution scattering camera ID02 at the ESRF, Grenoble. The undulator provides a high photon flux with a low divergence from the storage ring. The camera optics uses a cryogenic Si-111 monochromator and toroidal mirror optical system to produce a focused X-ray beam with wavelength 0.1 nm. The beam cross section is 0.2 mm × 0.4 mm. The optics hutch is in vacuum. There are two separate experimental stations, the first one being for ultra-small-angle scattering using a Bonse-Hart camera with a 2-m-detector tube (not shown), and the other is a pinhole camera for small-angle and wide-angle work. The two cameras are switched around in a few minutes when required. The incident and transmitted main beam intensities are monitored.

The sample holder is usually aligned in the beam using the detector electronics to monitor the main beam intensity as the sample is moved into position. The setup of the camera (detector positioning, beam alignment and focusing, sample and detector alignment) is usually completed in 2–4 h. The optimum sample thickness is a trade-off between thinner samples that absorb less and thicker samples that scatter more. This optimum is about 1 mm for water and dilute protein solutions. Samples can be static ones of 1 mm path length that are mounted in Perspex water-cooled cells with 10- to 20-μm-thick mica windows within a brass cell holder connected to a water bath for temperature control. Alternatively, the samples can be translated across the beam through a 1-mm-wide quartz capillary controlled by a mechanically operated syringe. This means that fresh sample can be continuously exposed during data collection in order to reduce radiation damage effects. A double-syringe cells with much larger sample volumes permit sample mixing to enable time-resolved experiments to be performed.

The X-ray scattering to be measured depends on the sample-detector distance in use (0.5–8 m) in order to yield the desired Q range. The detector needs to cover a large dynamic range in intensity of over 10^6 and in Q of about 100. Note, however, that the scattered intensity will decrease according to the inverse of the sample-detector distance squared. 2D area detectors have mostly replaced the original linear one-dimensional detectors: (1) The detector design at ID02 is based on an image-intensified charged coupled device (CCD) that is capable of handling the high-scattered flux. Scattered X-ray photons fall upon a phosphor surface, and they are converted into visible photons by means of an image intensifier device. The latter are then detected by the CCD detector of resolution up to 2048 × 2048 pixels. Both the dark image (with no X-ray beam exposure) and the raw-scattered image need to be measured, then the dark image is subtracted from the raw image. Exposure times can be as short as 1 msec (Fig. 2). (2) Another design is that of the quadrant detector, which corresponds to a multiwire linear detector that is ideal for isotropic (circularly symmetric) scattering patterns. The scattered intensities are recorded using 512 channels in a 2D angular sector of a circle (70°), and the main beam (blocked by the beam stop) is notionally located at the center of this circle. This design leads to much improved counting statistics at large scattering angles, where the hardware itself performs a radial integration of the intensities. The intensities of the incident and transmitted main beam are also important to measure. On ID02, these are obtained using a diode device placed on the beam stop used to protect the detector from the main beam at zero scattering angle. The detector is interfaced with a computer for data accumulation, which is in turn networked to a workstation for data storage and processing. During the X-ray experiment, it is essential to monitor the data acquisition. Thus, online data reduction using automated software will convert the 2D data set to a 1D curve for display, buffer background subtractions, and visual inspection. This is particularly important to avoid radiation damage or to monitor time-resolved scattering.

Similar scattering cameras with pin-hole optics are generally available at synchrotron facilities. Thus, at the new Diamond facility, the low-beam divergence optics on the scattering camera at beamline I22 utilizes five sets of slits, a Si-111 double crystal monochromator, and a Kirkpatrick-Baez mirror pair to deliver the beam to the sample. A refined ADC per input detector (RAPID) gas wire detector system with high-time resolution and low noise will be used to record small-angle and wide-angle curves. The X-ray flux will be of the order of 10^{14} photons/sec.

C. Neutron Reactor Instrumentation

A neutron camera at a high-flux reactor such as D22 and D11 at the ILL Grenoble (Ghosh *et al.*, 2006; Lindner *et al.*, 1992) maximizes the incident flux on the sample by the use of physically large designs and large samples (Fig. 3), unlike the X-ray camera.

The main scattering instrument at the ILL is D22. Neutrons enter a beam guide from the reactor after moderation by a cold source (liquid $^2H_2$ at 25 K). The cold source maximizes the number of neutrons in a wavelength range of 0.1–1 nm (Fig. 3). Beam monochromatization is achieved by the use of a velocity selector based on a rotating drum with a helical slit in it. The speed at which this rotates allows only the neutrons of the desired mean wavelength (velocity) to pass through, but at the price of a modest wavelength spread $\Delta\lambda/\lambda$ of 10%. Wavelengths

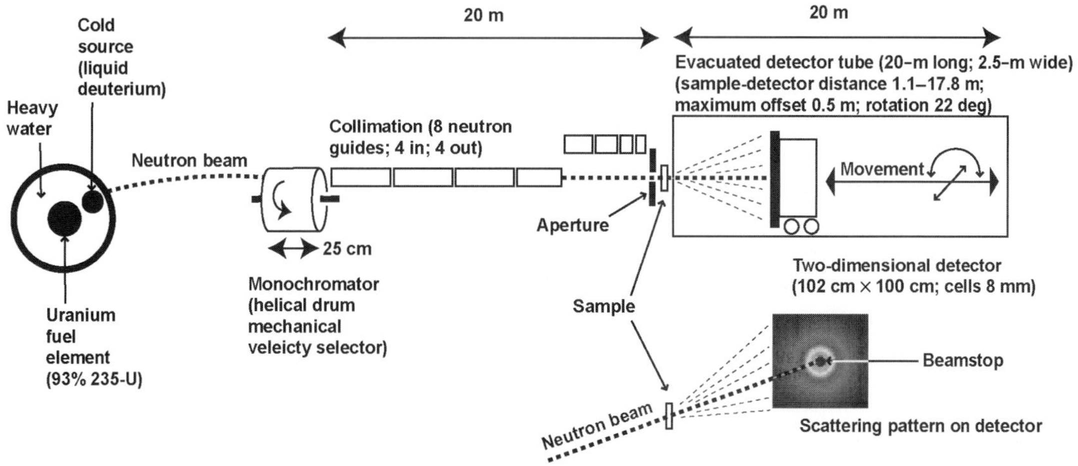

Fig. 3 Layout of the solution scattering camera D22 at the ILL, Grenoble. Neutrons from the horizontal cold source at the reactor are monochromated using a velocity selector and the source-sample detector is selected by a choice of eight collimation sections. Samples are mounted in a multiple holder rack with a choice of environments. The detector is movable along the length of a large detector tube, or can be offset sideways or rotated.

range from 0.45 to 4 nm. Beam collimation employs a series of eight movable straight beam guides of individual lengths between 0.6 and 3.2 m. The beam size is defined by an aperture. The maximum neutron flux is 1.2×10^8 neutrons/cm^2/sec. A large ^{3}He area detector of size 102 cm × 100 cm with cells of side 0.8 cm (totaling 128 × 128 or 16k cells) is movable in a 20-m evacuated tube and can be offset sideways to access a larger Q range. The available sample-detector distances are 1.35–18 m, which gives a maximum possible Q range of 0.004–4.4 nm^{-1} depending on the sample-detector distance. Typically at least two sample-detector distances are used in a biology experiment. A typical biology experiment can involve two sample-detector/collimation distances of 5.6/5.6 m and 1.4/8.0 m, a beam aperture of 7 mm × 10 mm, and a wavelength of 1.0 nm. Note that sample transmissions are wavelength dependent. If the same wavelength is used at these different detector positions, and spectra are normalized using H_2O runs at the same wavelength at both positions, the merger of data at both configurations is straightforward. This D22 configuration avoids the saturation of the detector and the need for dead-time corrections that arises with the very high flux obtained with short collimation distances, and gives a useful Q range of 0.07–2.5 nm^{-1}. Alignment of the beam stop is performed by the use of a 0.5-mm-thick Teflon strip as the sample. This is a strong isotropic scatterer that reveals the shadow of the beam stop on the detector. Since the neutron main beam intensity is monitored using a small sensor in the incident beam, data normalization requires a separate set of transmission measurements. No radiation hutches are required; however, users are prevented from approaching the sample area when the beam is on. Data acquisition, storage, and instrument control are performed by computer, and online data analyses can be automated to monitor the progress of the experiment.

D11 is the archetypal neutron scattering camera at the ILL, constructed in 1972 and upgraded many times since then (Lindner *et al.*, 1992). The maximum flux at the sample position is 3.2×10^7 neutrons/cm^2/s. Neutrons from the cold source are collimated by a series of moveable glass guides. Wavelengths range from 0.45 to 2.0 nm ($\Delta\lambda/\lambda = 9\%$). A 2D ^{3}He CERCA area detector is constructed from 10 mm × 10 mm cells (total of 64 × 64 = 4096 cells) and gives a maximum accessible Q range of 0.0045–4 nm^{-1}. A new fast D11 detector with improved resolution and large counting area of 100 cm × 100 cm is being constructed. With the detector mounted in an evacuated 40-m detector tube, sample-detector distances between 1.1 m and 36 m are available. For biology work on D11, a typical configuration is based on 2.5/2.5 m and 10/10 m sample-detector/collimation distances and a wavelength of 1.0 nm (although the flux is much reduced at 10 m distances). This configuration gives a useful Q range of 0.04–1.6 nm^{-1} with good spectral overlap. Up to 22 samples (1- or 2-mm path lengths; surface area 7 mm × 10 mm; total volume 150 or 300 μl) in rectangular Hellma quartz cells can be loaded onto an automatic sample changer under temperature control (Fig. 3).

D. Spallation Neutron Instrumentation

A neutron camera at a spallation source such as LOQ at ISIS, Rutherford-Appleton Laboratory (Heenan *et al.*, 1997) is designed to accommodate a pulsed main beam, not a continuous beam as with a reactor source (Fig. 4). Neutrons are generated at 50 pulses/sec that are emitted from a uranium or tantalum target after bombardment with an intense high-energy proton beam generated using a synchrotron, cooling the target using heavy water. The neutrons are slowed by a liquid H_2 moderator positioned 11.1 m from the sample rack in order that each neutron pulse contains wavelengths in the range 0.2–1.0 nm. In normal usage, a disk chopper spinning at 25 Hz removes every other pulse to prevent frame overlap between consecutive pulses on LOQ. Two apertures define the pin-hole optics of LOQ. Heavy shielding is used to protect LOQ from external background neutron levels. Up to 20 samples in rectangular quartz Hellma cells are loaded onto an automatic sample changer, which is under temperature control. Unlike the other X-ray and neutron cameras described above, the collimation on LOQ is fixed in length at 4.5 m from the source (moderator) to the sample, and the sample-detector distance is fixed at 4.1 m. The area detector is a ^{3}He-CF_4 ORDELA type with an active surface of size 64 cm × 64 cm and individual cells of side 0.5 cm each. Monochromatization is achieved by time-of-flight techniques based on the total distance of 15.5 m that each pulse travels from the uranium target to the ORDELA detector. The neutrons in each pulse will reach the detector at different times depending on their wavelength (velocity), and this enables a wavelength to be assigned to each neutron. The standard LOQ configuration employs 102 time frames for the assignment of wavelengths in each pulse. This is an efficient

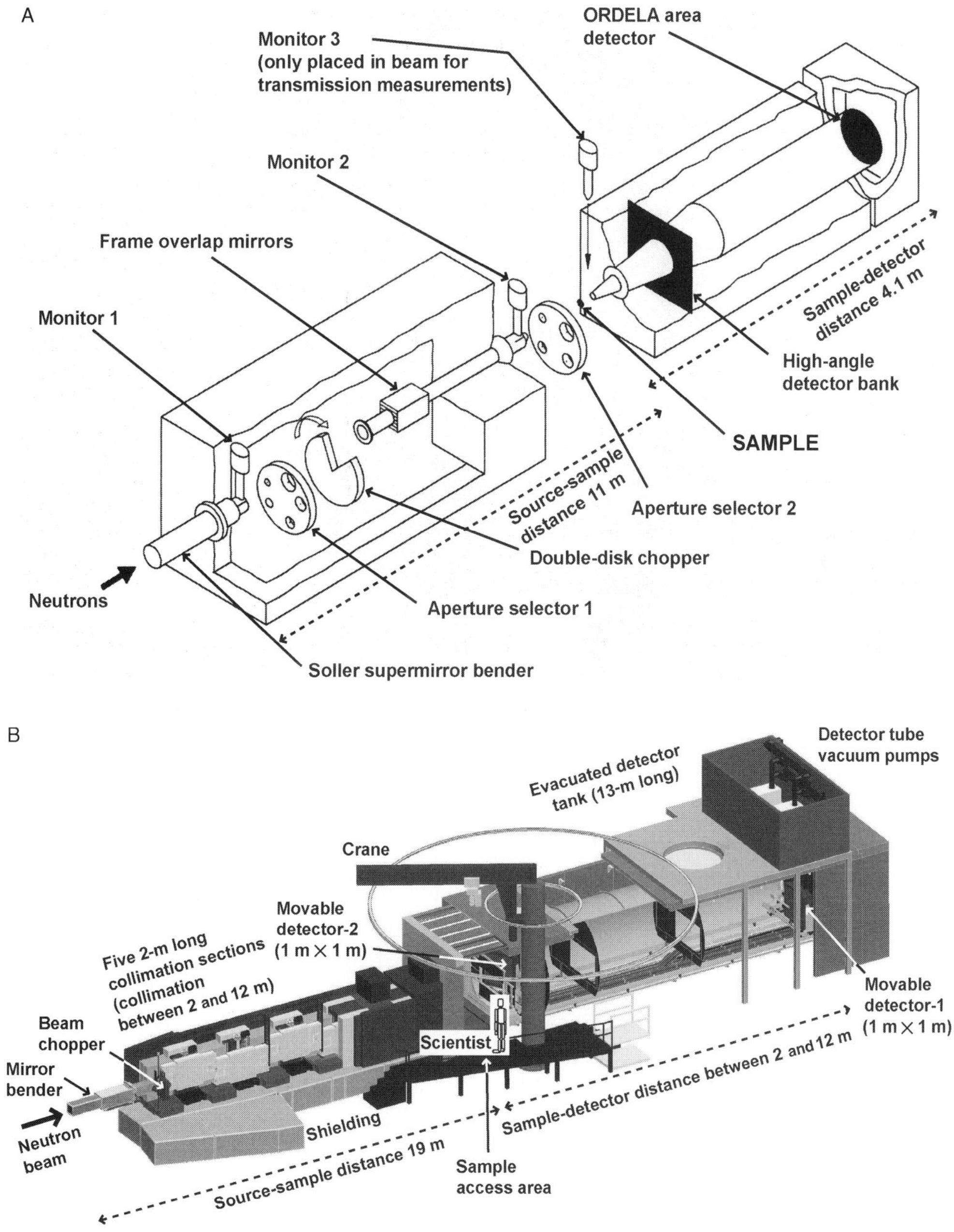

Fig. 4 Schematic cutaway views of the neutron solution scattering cameras at ISIS. (A) In the LOQ camera, the chopper removes every other pulse of neutrons. After passing through the sample position, the diffracted neutrons are detected on the ORDELA area detector and on a high-angle detector bank. The detectors are interfaced with a computer for time frame data collection and storage. (B) An outline view of the new SANS2D camera with two area detectors is shown. (Figures kindly provided courtesy of Dr. R. K. Heenan. Reproduced with permission. Copyright 2007 Science & Technology Facilities Council.

arrangement as this means that all the neutrons in each pulse are used for data acquisition. Wavelength and instrumental effects are similar in magnitude to those seen with D22 and D11 (Ashton *et al.*, 1997). A typical time-averaged flux on LOQ is 2×10^5 neutrons/cm^2/s. This flux is adequate for working with 5–10 mg/ml proteins in ^2H_2O buffers (acquisition times of 2 h) or with strong scatterers such as lipoproteins, which have a high lipid content (acquisition times of 10 min). Transmissions at each wavelength are measured either during scattering data acquisition using a semitransparent beam stop or separately from the scattering data using an attenuated beam. The major advantage of LOQ is that the entire scattering curve in the Q range of 0.06–2.2 nm^{-1} can be measured simultaneously, and this is ideal for both $P(r)$ calculations and kinetic experiments. A side benefit of this camera is that neutrons of shorter wavelengths are the most abundant in every pulse, and these contribute the most to the scattered curve at large Q, which is where the scattered intensities are lower and signal-noise ratios are poorer. To acquire wide-angle data in a second Q range of 1.5–14 nm^{-1}, an annular high-angle area detector is located at 0.5 m from the sample. Computer script files control the data collection and sample changer, and the time-of-flight data is reduced to Q space within the COLETTE software package.

The ISIS second target station TS2 will operate at 10 Hz by taking 1 in 5 proton pulses from the ISIS synchrotron to a target optimized for production of low-energy neutrons. At TS2, the new SANS2D scattering camera (Fig. 4) will view a coupled, grooved, methane moderator with a water premoderator as the neutron source, and will have 20–30 times the neutron flux of LOQ. Two 1 m^2 multiwire gas detectors can be positioned between 2 and 12 m from the sample within a 13-m-long vacuum tank. The flux and Q range on SANS2D will be adjusted in individual experiments by the ability to vary the collimation length and move the two detectors. In a single configuration, a Q range of around 0.02–20 nm^{-1} could, for example, be obtained with one detector at 12 m and the other at 2 m but offset to one side. The gains in performance compared to LOQ will make SANS2D comparable in performance with D11 and D22 and will lead to world-class biology experiments.

IV. Experimental Methods

A. Applications for X-Ray and Neutron Beam Time

Long forward planning is implicit in applying for beam time. Web pages and the User Liaison Office at the synchrotron or neutron facility should be contacted 6–9 months in advance to download electronic application forms and make contact with the instrument scientist. Outside this formal procedure, Director's discretionary beam time may sometimes be available at short notice for exceptional cases, or it may be possible to negotiate trials with an established user or the facility if feasible (e.g., in relation to safety aspects). Beam time applications are competitive with a success rate of perhaps 50%. Hence, a well-justified proposal that clearly explains the aims, background, any preliminary work, and details of the new experiment is essential. A relevant strong track record is helpful.

B. Sample Preparation for Scattering

Many samples need to be ready in time for a short but intensive one- or two-day experiment, and this can overload purification equipment in the home laboratory. Pure, monodisperse samples at a high enough concentration are required to measure an observable scattering curve in the required solute–solvent contrast. Large-scale preparations with 20 mg at concentrations of 5 mg/ml should ensure that most projects are completed in 2 or 3 beam time sessions, although it is possible to work with less material. For a one-day session, 0.5 ml of material at 5 mg/ml is ideal for X-ray work, and 1.5 ml at 5 mg/ml for neutron work in three contrasts. If stopped-flow methods are to be used, these amounts become considerably larger.

Biochemical standards of purity as assessed by clean bands in reducing and nonreducing SDS-PAGE are adequate for scattering, but physical monodispersity is more important. Because the intensities at low Q are proportional to the square of the molecular weight, scattering curves are sensitive to nonspecific aggregation. Guinier R_G analyses of ln $I(Q)$ versus Q^2 normally give a linear plot of intensities at low Q values (Fig. 5C). If 1–2% aggregates are present, the R_G plots are curved upward at the lowest Q values and are often unusable. Smaller amounts of aggregates can also compromise Guinier R_G fits, even when curvature is not obvious. It is thus essential to remove all traces of aggregates prior to measurement by size-exclusion gel filtration and reconcentration. Microfiltration is not sufficient. It is good practice to subject all scattering samples to gel filtration immediately prior to scattering data collection. Other approaches include the use of laser light scattering to test for aggregates before beam time is taken.

Sample dialysis in the buffer used for the background runs is essential, as this background is used to subtract the X-ray or neutron scattering from the solvent, the cell windows, and any constant camera background. Usually samples are dialyzed with stirring and several buffer changes in the cold room, and the final dialysate buffer is used as the background. If gel filtration has been used, the postsample column eluate can be used as the buffer background for X-ray scattering. For X-rays, slight variations in the electron density of the buffer caused by changes in the salt concentration can compromise an accurate buffer subtraction. The closer the buffer is to pure water, the higher the sample X-ray transmission becomes, and the better the counting statistics. Phosphate buffer saline (12-mM-phosphate; 140-mM-NaCl; pH 7.4) is commonly used, with an X-ray transmission close to about 40% for a path length of 1 mm. The use of 1-M-NaCl reduces the X-ray transmission to about 10%. For neutrons, slight differences in the exchangeable proton content of the buffer can likewise invalidate the buffer subtraction. The strong incoherent neutron scattering of ^{1}H reduces

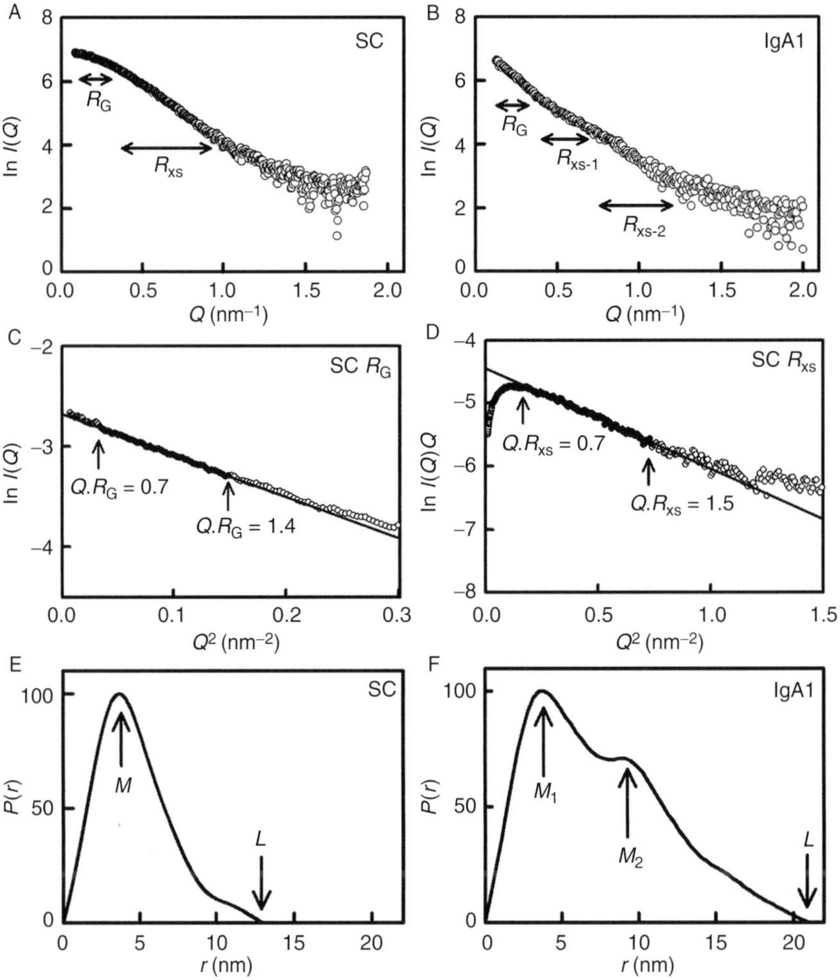

Fig. 5 Representative scattering curve analyses for human secretory component (SC) and human immunoglobulin A (IgA1). (A,B) Scattering curves for SC and IgA1 are compared to show the more rapid decrease in $I(Q)$ with Q as the molecular weight increases from SC (80 kDa) to IgA1 (164 kDa). The Q ranges used for the Guinier R_G and R_{XS} analyses are indicated by arrowed ranges. (C) The Guinier R_G fit of the SC scattering curve at low Q is shown. (D) The Guinier R_{XS} fit of the SC scattering curve at low Q is shown. (E) The distance distribution function $P(r)$ for SC. M is the peak maximum at 3.8 nm, and L is the maximum macromolecular length (13 nm). (F) The $P(r)$ curve for IgA1. M_1 and M_2 are the peak maxima at 3.7 and 8.9 nm, and L is the maximum macromolecular length at 21 nm.

the sample transmission. Thus, H_2O buffers have a neutron transmission of about 45% for a path length of 1 mm and a wavelength λ of 1.0 nm, while ^2H_2O buffers have a neutron transmission of about 88% for a path length of 2 mm at λ of 1.0 nm. Hence, neutron samples and buffers must be sealed against atmospheric exchange with moisture during dialysis and data collection. Parafilm is effective. Neutron curve fits using ^2H_2O buffers require correction for a flat background from this effect (see below).

Immediately prior to X-ray and neutron measurements, concentrations (c) must be determined to permit molecular weight and neutron match point calculations. The neutron match point is that percentage ^2H_2O in which the macromolecular scattering density is the same as that of the solvent (Table II). For proteins, concentrations are measured from the optical density at 280 nm in a UV spectrophotometer, converting this to concentrations using the absorption coefficient calculated from the sequence composition (Perkins, 1986). Protein concentrations are usually measured in the same quartz cell used for neutron data collection. Lowry or Bradford assays can be performed if optical densities cannot be measured, and commercial kits are available. For DNA, concentrations are measured from the absorbance at 260 nm, converting this using an absorption coefficient of 1 for a path length of 1 cm to correspond to 50 μg of DNA as in fully paired duplex DNA (Sambrook *et al.*, 1989). For lipids, concentrations are measured in individual color reagent assays for each lipid present (phospholipid, triglyceride, free and esterified cholesterol), for which commercial kits are available (Converse and Skinner, 1992). Biochemical sample assays for functional activity will validate the scattering data, this often being important when publishing the results.

C. Data Collection Strategies

Data collection is best designed on the assumption that there will be an equipment breakdown during the experiment. Hence, basic data essential for subsequent analysis should be collected before the sample and buffer runs, and samples should be measured in order of priority with the easiest ones first. Beam time may not be available again until many months later. The beam can be lost due to the weather, major accident, power surge, vacuum leak, electronic failure, or operator error. The camera itself can fail due to hardware or software issues. The instrument scientist should be aware that biological samples are not easy to prepare and are not stable indefinitely. In practice, improvements in reliability have reduced the scope for breakdown in recent years. One-day X-ray or two-day neutron beam time sessions are usually viable.

Modern scattering cameras such as ID02 or LOQ routinely provide simultaneous access to relatively large Q ranges (Fig. 5A and B). Others such as D22 and D11 are more restricted in Q range, in which case the camera should be initially set up for a low Q range for the Guinier analyses. Preplanning is necessary. If the Q range is not appropriate, the detector will need to be moved, meaning that the camera will need to be recalibrated and fresh runs are needed, which is costly in time. For D11 and D22, when short sample-detector distances are needed for a large enough Q range, it takes only minutes to move the detector inside its vacuum tube (Fig. 3).

Guinier analyses provide a good starting point in an experiment as they provide both the radius of gyration R_G and the molecular weight M_r from $I(0)$ and c (Fig. 5C and D). The smallest Q required for an R_G determination is given by the relationship $Q_{min}L_{max} \leq \pi$, in which L_{max} is the maximum length or dimension of the protein (Glatter and Kratky, 1982). The Q range is usually determined by setting the sample-detector distance, sometimes by adjusting the wavelength. If the protein is known to be compact in structure, the R_G value can be estimated from the following relationship with M_r (Perkins, 1988b):

$$\log R_G = 0.365 \log M_r - 1.342$$

If the protein is known to be elongated, the R_G value will increase. An approximate R_G value can be estimated from the overall length L from the approximation $R_G = L/(\sqrt{12}$, provided that the two cross-sectional axes (leading to the R_{XS} value: see below) are negligibly low. L can be estimated from electron microscopy data or hydrodynamic calculations of sedimentation coefficients assuming a rigid rod model. Large R_G values require larger sample-detector distances in order to access a low enough Q value, however, at the cost of decreased counting rates, which are proportional to $(\text{distance})^2$. If in doubt, it is preferable to err on the side of lower Q. The distance distribution function $P(r)$ is a key output of scattering (Fig. 5E and F). For this, $I(Q)$ needs to be measured over at least two orders of magnitude of Q, including the Guinier R_G region.

X-ray data collection includes: (1) determining the accessible detector Q range, (2) measuring the detector response, (3) evaluating radiation damage and exposure times, and (4) performing the scattering measurements on the samples and buffers. The procedure for X-ray data at Daresbury was previously discussed (Perkins, 2000). The more brilliant X-ray beams at ESRF have resulted in the following adaptations:

1. The Q range is defined using the sharp peaks in the powder diffraction pattern of silver behenate ($C_{21}H_{43}COOAg$), a salt with a d-spacing of 5.84 nm (Huang *et al.*, 1993). This is a reproducible standard that avoids the need to prepare fresh rat tail tendons containing collagen as a standard with a d-spacing of 67 nm (Perkins, 2000).

2. X-ray detectors have evolved from multiwire devices toward CCD sensors during the past decade, and the nonuniform efficiency of individual detector channels is generally not an issue. Instead, the exposure of a Lupolen polymer standard is used to provide a calibration of the measured detector intensities in order to provide absolute measurements of the $I(Q)$ values.

3. Radiation damage is significant. This is all too often neglected by experimentalists, yet its explicit consideration is important for securing the acceptance of a publication. At ESRF, to minimize damage, one option is to reduce the beam intensity by the use of single-bunch or 16-bunch modes at beam currents of 16 or 90 mA, respectively (i.e., the latter corresponds to 16 highly populated and equally

spaced bunches in the storage ring), as alternatives to the uniform fill mode at 200 mA (992 bunches equally distributed around the storage ring). Alternatively, a motorized syringe that pushes fresh solution continuously through a quartz capillary through the beam permits four acquisitions from 100 μl of sample or buffer. Trials at the start of each new sample with exposures ranging from 0.1 to 2 sec and monitoring the scattering curve online establishes the optimal exposure time. Data are then recorded in 10 equal time frames, and the first time frame is compared with all 10 time frames merged. Should small time-dependent changes still be seen, only the first time frame is used for data analyses. A detailed written log record must be taken of all runs.

4. Sample concentration series are essential to confirm that concentration-dependent effects are absent. Sample and buffer data collection needs to allow for alterations in beam intensities (e.g., from a slow accumulation of protein precipitate on capillary walls) or the decay of the storage ring beam intensity. This effect is minimized by measuring samples in duplicate with buffer runs between them, all for equal periods of time (usually 1–10 sec), starting with the most dilute ones. While dependent on the molecular weight, the lowest measurable sample concentration at ESRF is around 0.1 mg/ml, compared to tenfold higher values at Station 2.1 at SRS Daresbury. The final X-ray scattering curve is calculated by subtracting the buffer curve from the sample curve.

In neutron data collection, the Q range is defined by the camera dimensions; hence, this does not need calibration. A neutron experiment thus involves (1) background runs, (2) transmissions, and (3) the sample and buffer runs.

1. For D11 or D22, the basic background runs (Ghosh *et al.*, 2006) are cadmium (to monitor the ambient neutron and electronic noise background in the detector), Teflon (for defining the beam center on the detector and the detector area masked by the beam stop), a 1-mm-thick H_2O standard sample (for determination of the detector response and the absolute scale), an empty cell (the H_2O background), and an empty cell holder (a check for stray reflections or scattering). The H_2O standard is often replaced by the buffer in 0% ^2H_2O if this is a dilute salt solution such as phosphate buffer saline. The H_2O standard should be remeasured several times during the experiment to check detector reproducibility. If large sample-detector distances are used, the H_2O standard will give a very low count rate. This can be compensated by measuring this at a short sample-detector distance where the count rate is much higher, and adapting this for the data reduction involving the large distance. For LOQ (spallation source), the background runs are only those for a standard deuterated polymer used to determine absolute intensities and the empty position as the background for this polymer run.

2. Neutron transmission measurements (using an attenuated beam) are required for all samples and buffers in order to perform molecular weight and match point calculations and to confirm that the dialysis in ^2H_2O buffers is complete. They are

required for merging the LOQ timeframes. Transmissions vary from about 0.5 for 1-mm-thick H_2O samples to about 0.9 for 2-mm-thick ^2H_2O samples, depending on the wavelength and temperature. Consequently, samples are usually measured in 1-mm-thick cells for 0–30% ^2H_2O buffers and 2-mm-thick cells for 40–100% ^2H_2O buffers.

3. Sample counting times depend on the amount of ^2H_2O in the buffer. On D11 and D22, proteins and buffers in H_2O each require about 1–2 h of acquisition, while those involving ^2H_2O require 5–10 min each. DNA samples are more difficult to measure as the match point is high, and about 1 h is needed for both H_2O and ^2H_2O buffers. On LOQ, proteins and buffers in ^2H_2O buffer each require about 1–3 h. In all cases, buffer background subtractions are usually straightforward for reason of instrumental stability. The lack of radiation damage and instrumental stability mean that neutron runs can be added to improve signal-noise ratios. For the D11 or D22 data, the raw scattering curves are corrected by subtracting the cadmium background. The buffer curve is subtracted from the sample curve, and then the final reduced curve is calculated by dividing this by the water background (corrected for the transmission of water) minus the empty cell background (Ghosh *et al.*, 2006). For LOQ data, (sample – buffer) subtractions are also performed, but the intensities are now normalized using data from a standard polymer corrected using the empty position run (King, 2006).

D. Guinier Analyses

Guinier fits of the scattering curve ln $I(Q)$ as a function of Q^2 at low Q values (Fig. 5C) give the radius of gyration R_G and the forward scattered intensity $I(0)$ from the slope and intercept of linear plots (Glatter and Kratky, 1982):

$$\ln I(Q) = \ln I(0) - \frac{R_G^2 Q^2}{3}$$

A satisfactory Guinier R_G analysis requires that data is measured in order to include Q values below $Q.R_G$ of 1 for the approximation to be valid. For both X-ray and neutron cameras, R_G analyses at low Q are not affected by instrumental corrections for wavelength spread or beam divergence as the cameras generally employ an ideal pin-hole optical configuration. For elongated macromolecules, the mean cross-sectional radius of gyration R_{XS} and the cross-sectional intensity at zero angle $[I(Q)Q]_{Q\to 0}$ (Hjelm, 1985) are obtained using fits in a larger Q range that does not overlap with that used for the R_G determinations (Fig. 5D):

$$\ln [I(Q)Q] = \ln [I(Q)Q]_{Q\to 0} - \frac{R_{XS}^2 Q^2}{2}$$

Note that the intensities in these R_{XS} plots are typically lower at low Q before the fit region.

Plots of the $I(0)/c$ values as a function of c (c = sample concentration in mg/ml) will validate the Guinier R_G analyses. There should be no or a small concentration dependence, thereby confirming the absence of sample association or dissociation. Molecular weights can be deduced from $I(0)/c$ values either as relative values from the X-ray data or as absolute values from neutron data, if $I(0)$ has been referenced to a standard (Jacrot and Zaccai, 1981; Kratky, 1963). If LOQ samples are measured in ^2H_2O buffers, molecular weights M_r can be calculated using a calibration graph (Fig. 6) in which $M_r = I(0)/c \times 9 \times 10^5$ (Boehm *et al.*, 1999), and absolute molecular weights are calculable if the partial specific volume is accurately known (Perkins, 1986). Neutron contrast variation match points are determined by plotting $\sqrt{(I(0)/ctT_s)}$ values (t: sample thickness; T_s: sample transmission) as a function of the volume percentage ^2H_2O in the buffer, remembering that $\sqrt{I(0)}$ is negative above the match point. The intercept at zero $\sqrt{I(0)}$ gives the match point. This should be within 1% ^2H_2O of the match point calculated from the composition for an unhydrated macromolecule (Perkins, 1986) (i.e., the partial specific volume will be higher than that conventionally measured by densitometry).

The R_G values measure the degree of particle elongation. Once the R_G data have been validated from the $I(0)$ values, these are summarized as the mean ± standard deviation. A potential shortcoming is the presence of trace aggregates that imperceptibly increase the R_G value. This can only be detected at the final stage of constrained modeling. If the R_G values increase with dilution, interparticle effects (when the protein molecules are too close together) may be important. If the R_G values decrease with dilution, a monomer–oligomer equilibrium may be present.

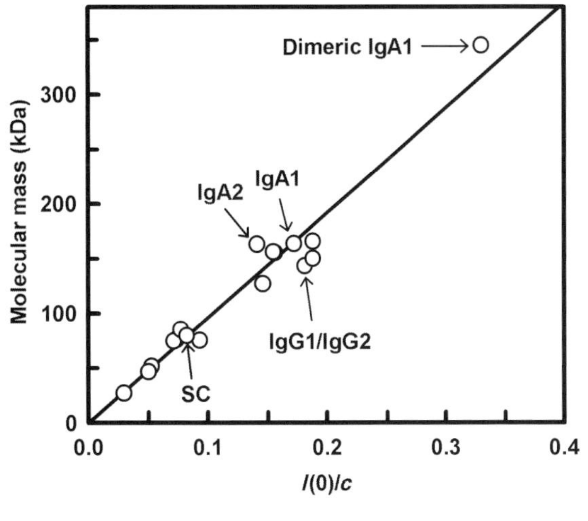

Fig. 6 Linear relationship between the molecular weight and the neutron $I(0)/c$ values for 17 glycoproteins in 100% ^2H_2O buffer measured on LOQ. Earlier versions of this graph were published in Boehm *et al.* (1999) and Perkins (2000). Data points for SC, IgA1, IgA2, bovine IgG1/2, and dimeric IgA1 are arrowed.

X-ray R_G values correspond to a hydrated structure in a high positive solute–solvent contrast. If the X-ray R_G values are assumed to be independent of the internal structure, the interpretation of the X-ray R_G values often uses the ratio of the observed R_G value to the R_G calculated for a sphere of the same hydrated or unhydrated volume. This R_G/R_O anisotropy ratio indicates the degree of macromolecular elongation. Globular proteins have a R_G/R_O ratio close to 1.28 (Perkins, 1988b). If both are available, the combination of the R_G and R_{XS} analyses will give the longest macromolecular dimension L.

In neutron contrast variation R_G analyses, neutron R_G values correspond to an unhydrated structure and depend on the solute–solvent contrast. In a full contrast variation study involving at least three contrasts, this R_G^2 dependence is described by the Stuhrmann equation (Ibel and Stuhrmann, 1975):

$$R_G^2 = R_C^2 + \alpha\Delta\rho^{-1} - \beta\Delta\rho^{-2}$$

where R_C is the R_G at infinite contrast (when $\Delta\rho^{-1}$ is zero), α is the radial distribution of scattering density fluctuations within the macromolecule, and β measures the effect on the R_G if the center of gravity of the scattering density varies with the contrast. A similar analysis can be made using the R_{XS}^2 values. For two-component centrosymmetric complexes with two very different scattering densities (e.g., protein and DNA), linear Stuhrmann R_G^2 and R_{XS}^2 plots as a function of $\Delta\rho^{-1}$ will result since β is negligible. In Fig. 7, the symmetric arrangement within the RuvA protein–DNA Holliday junction complex resulted in such a linear dependence of R_G^2 (Chamberlain *et al.*, 1998). The value of the Stuhrmann α can be compared with other known values (Perkins, 1988b), based on the proportionality of α to R_C^2. If two components of similar size in a complex are asymmetrically distributed with respect to the center of mass, and have very different scattering densities, the Stuhrmann plot becomes parabolic. Its curvature β provides the distance between the centers of mass of the two components (e.g., protein and lipid) (Glatter and Kratky, 1982; Perkins, 1988b).

E. Distance Distribution Function Analyses

If the scattering curve $I(Q)$ in reciprocal space is measured over two orders of magnitude in Q, $I(Q)$ can be converted by a Fourier transform into the distance distribution function $P(r)$, which represents the structure in real space with units of nm (Fig. 1B). The solution structure is now more directly visualized because $P(r)$ corresponds to the distribution of all the distances r between all the volume elements within the macromolecule (Fig. 5E and F). The transformation step is ordinarily unstable, since data points are missing at low Q value because of the beam stop, and are truncated at high Q at the edge of the detector (Figs. 2 and 3). The classic Indirect Transformation Procedure program of Glatter (Glatter and Kratky, 1982) dealt with this by fitting the curve as 10–20 B-spline

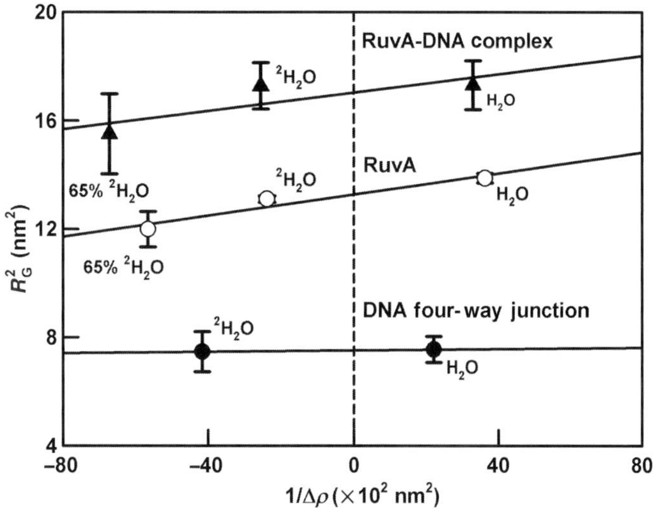

Fig. 7 Stuhrmann contrast variation R_G analysis for RuvA, DNA four-way junction and their complex in 0%, 65%, and 100% ^2H_2O buffers. The R_G^2 values from D22 at the ILL are shown as a function of the reciprocal solute–solvent contrast difference $1/\Delta\rho$. Note that the intercepts are well separated and that the slopes of the graphs are positive and small for both the RuvA and the complex. This indicated that the DNA component is buried within the center of two RuvA tetramers which have moved apart to accommodate the DNA. Adapted from Chamberlain *et al.* (1998).

mathematical functions, after which the B-splines are transformed to give $P(r)$. This program has been superseded by the GNOM program of Semenyuk and Svergun (1991), in which an automated regularization procedure stabilizes the $P(r)$ calculation.

By GNOM, $P(r)$ curve calculations test a range of maximum assumed dimensions D_{max} for the macromolecule. For proteins, the final D_{max} is based on the knowledge that the $P(r)$ curve should exhibit positive intensities and should be stable as this dimension D_{max} is increased beyond the macromolecular length L. The $P(r)$ calculation gives an alternative calculation of the R_G and $I(0)$ values from the full Q range of the scattering curve, which should agree with those from the Guinier analysis. The point at which $P(r)$ becomes zero at large r gives the length L of the protein (Fig. 5E and F). However, as the intensity of the $P(r)$ curve is lowest at large Q, the determination of L can be imprecise. Any peaks M in the $P(r)$ curve give the r value of the most commonly occurring distances within the macromolecule. The interpretation of M values can be highly informative, and is facilitated by modeling calculations. For example, if the r value at M is half that of L, the macromolecule is clearly spherical. If a multidomain protein gives rise to two or more peaks M_1 and M_2 at different r values, it is possible that the protein has a relatively inflexible structure or that its domain structure is on average bent (see below).

V. Constrained Scattering Modeling

Modeling extends the scattering analyses by determining the three-dimensional (3D) structural model that accounts for the observed scattering curve. Even though unique structure determinations are not possible for the reason of the random molecular orientations in solution and the spherical averaging of scattering data, modeling is able to rule out structures that are incompatible with the scattering curves. The use of known atomic structures for subunits or even the entire structure from NMR or crystallography constitutes a strong constraint of the scattering curve modeling. Using this constraint in fit searches means that a large number of conformationally randomized but stereochemically correct structures are generated, but relatively few of these will fit the scattering curve. These best-fit models can provide biologically useful information. The constrained modeling approach is also known as "rigid body" modeling (Svergun and Koch 2003). This strategy was first applied to model manually pentameric immunoglobulin M in terms of its Fab and Fc structures (Perkins *et al.*, 1991). The strategy was first automated to model human immunoglobulin A1 (Boehm *et al.*, 1999; Perkins *et al.*, 1998). Note that the full stereochemically complete macromolecular structure is computed prior to the curve fitting, not afterward.

A. Modeling Using Atomic Structures

While the 43,000 coordinate files (April, 2007) in the PDB at http://www.rcsb.org/pdb/home/home.do can be directly used to calculate scattering curves using the Debye equation, thousands of atoms in these slow the calculation unnecessarily. The use of several hundred small spheres to replace atoms provides a sufficiently good representation of structural details (Fig. 8), and remains the most flexible and powerful approach to model scattering curves (Glatter and Kratky, 1982). The scattering curve $I(Q)$ is calculated using Debye's Law adapted to spheres, essentially by computing all the distances r from each sphere to the remaining spheres and summing the results. For a single-density macromolecule, the X-ray and neutron scattering curve $I(Q)$ can be calculated assuming a uniform scattering density for the spheres using the Debye equation adapted to spheres (Perkins and Weiss, 1983):

$$\frac{I(Q)}{I(0)} = g(Q)\left(n^{-1} + 2n^{-2}\sum_{j=1}^{m} A_j \frac{\sin Qr_j}{Qr_j}\right)$$

$$g(Q) = \frac{3(\sin QR - QR\cos QR)^2}{Q^6 R^6}$$

where $g(Q)$ is the squared form factor for the sphere of radius r, n is the number of spheres filling the body, A_j is the number of distances r_j for that value of j, r_j is the distance between the spheres, and m is the number of different distances r_j. If it is

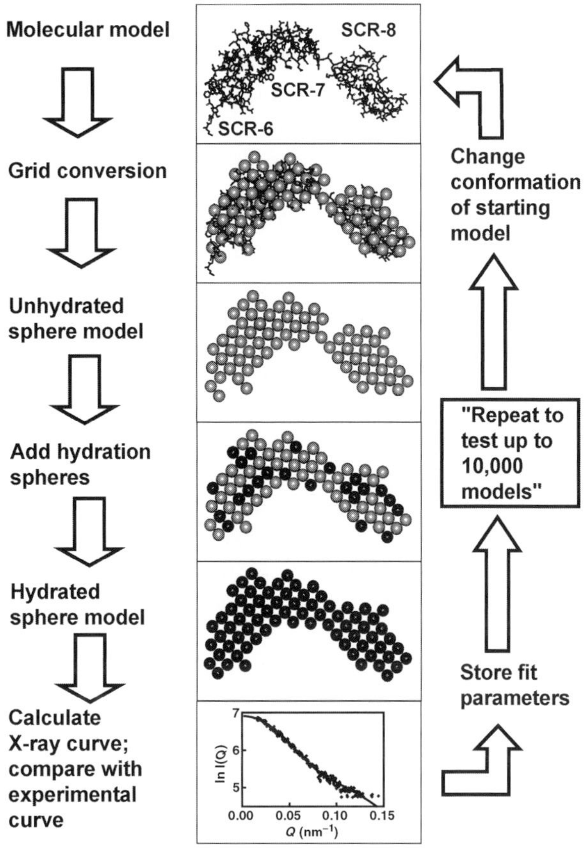

Fig. 8 Flow diagram of the constrained modeling algorithm used for multidomain proteins. The atomic coordinates in a small protein (FH SCR-6/8) are converted into spheres using a 3D grid of cubes. A Debye sphere is assigned to each cube containing sufficient atoms, and this resulted in 160 spheres for the example shown. An additional 56 spheres are added for the hydration shell. The neutron scattering curve is calculated using the unhydrated sphere model. The X-ray scattering curve and the sedimentation coefficient are calculated from the hydrated sphere model.

necessary to incorporate two different scattering densities 1 and 2 in the model for neutron modeling, the above expression is modified as follows (Perkins and Weiss, 1983):

$$\frac{I(Q)}{I(0)} = g(Q)\left[n_1\rho_1^2 + n_2\rho_2^2 + 2\rho_1^2\sum A_j^{11}\left(\frac{\sin Qr_j}{Qr_j}\right)\right.$$
$$\left.+2\rho_2^2\sum A_j^{22}\left(\frac{\sin Qr_j}{Qr_j}\right) + 2\rho_1\rho_2\sum A_j^{12}\left(\frac{\sin Qr_j}{Qr_j}\right)\right](n_1\rho_1 + n_2\rho_2)^{-2}$$

The model is constructed from n_1 and n_2 spheres of two different densities ρ_1 and ρ_2. A_j^{11}, A_j^{22}, and A_j^{12} are the number of distances r_j for that increment of j between the spheres 1 and 1, 2 and 2, and 1 and 2 in that order. The summations $\sum$ are performed for $j = 1$ to m, where m is the number of different distances r_j.

No instrumental corrections are required at low Q values. At large Q on synchrotron X-ray cameras, no further corrections are needed (Narayanan *et al.*, 2001; Nave *et al.*, 1985; Towns-Andrews *et al.*, 1989). At large Q on neutron cameras, instrumental corrections are now required for reason of the physically large dimensions of the camera (Ghosh *et al.*, 2006; Heenan *et al.*, 1997; Lindner *et al.*, 1992). The neutron corrections lead to wavelength spread and beam divergence effects. A Gaussian function is used to convolute the calculated modeled curve for both the D11/D22 and the LOQ fits. In addition, it is noticeable that, at large Q, the neutron fits deviate upward for reason of a flat background from the incoherent scatter from the protons within the macromolecule. This is readily corrected using a flat baseline correction which has a magnitude of 1.7–2.7% of the $I(0)$ value (Bonner *et al.*, 2007; Furtado *et al.*, 2004).

The conversion of atomic coordinates to two types of sphere models, unhydrated and hydrated, is straightforward. The diameter of the spheres has to be significantly less than the resolution of the scattering experiment (Fig. 8), so that the form factor term $g(Q)$ becomes almost invariant as Q increases. Hence, the coordinates are converted by placing the atoms within a 3D grid of cubes of side about 0.55 nm using BRKTOS (panel 2 of Fig. 8). A cube is allocated as a sphere if it contains sufficient atoms above a specified cutoff (usually about four atoms) such that the total volume of all the converted cubes equals to that of the unhydrated protein calculated from the full sequence (including carbohydrate) (Perkins, 1986). If residues are missing in the crystal structure, or a nonidentical but related structure is used, the resulting volume discrepancy is compensated by adjustment of the cube side and atom cutoff value in order to reach the correct total volume. This unhydrated sphere model is used for neutron curve modeling. For X-ray curve modeling, a hydrated sphere model is needed. A hydration shell is well represented by 0.3 g of H_2O/g (glyco)protein and an electrostricted volume of 0.0245 nm^3 per bound water molecule (Perkins, 2001). This corresponds to a monolayer of water surrounding the protein surface (Perkins, 1986). In comparison, the volume of a free water molecule is 0.0299 nm^3, and is therefore less electron dense. Computationally, hydration is achieved by adding extra spheres around every sphere in the model, then reapplying the atom-to-sphere conversion to remove duplicated spheres such that the total volume of the spheres now matches the hydrated volume (panel 4 of Fig. 8). This procedure uses HYPRO to add these spheres to the unhydrated model, then BRKTOS to remove the unwanted duplicates (Ashton *et al.*, 1997). The outcome can be visually verified by overlaying displays of the coordinate and sphere models using INSIGHT II (Accelrys, San Diego).

For Types 2, 3, and N proteins, where the domains are covalently joined by two, three, or more peptide linkers, respectively, the scattering modeling of a multidomain protein is constrained by the use of (1) known molecular structures for each domain, (2) known steric connections between the domains, and (3) the known total volume from its sequence-derived composition (Perkins *et al.*, 1998). An automated search takes several days to run on a SGI IRIX O2 or OCTANE platform with 1 Gb of memory (Fig. 8). First, libraries of conformations are generated using molecular dynamics simulations for each linker using DISCOVER in INSIGHT II (Accelrys, San Diego). Next, one linker per library is selected at random to be combined with the domain models (including oligosaccharide where present) to generate about 2000–5000 randomized domain arrangements for the intact (glyco)protein. These are saved in PDB format. The optimal cube side and atom cutoff for the sphere conversion is determined using one of the most extended models. The models are then used to calculate the X-ray and neutron $I(Q)$ scattering curves using SCT and automated script files. The Guinier fits for the R_G and R_{XS} values of each model and the number of spheres in each model are computed. The experimental curve to be fitted is selected from that used for the final GNOM $P(r)$ analysis. The goodness-of-fit R-factor ($R\text{-factor} = 100 \times \sum|I(Q)_{exp} - I(Q)_{cal}|/\Sigma|I(Q)_{exp}|$) of the calculated curve with the experimental X-ray and neutron curves are computed using SCTPL. Distances that define the appearance of each model are computed (e.g., the N-terminal and C-terminal separation). This output for all the models is imported into a large EXCEL spreadsheet.

Type 1 modeling only involves the association of oligomeric proteins with no covalent linkers. For Type 1, constraint (2) above is replaced by the consideration of macromolecular symmetry. Only about 200–900 models now require analysis.

The modeled curves are sorted within EXCEL to identify the best fit ones. Three filters are used to reject poor models, rank the models in order of fit, and identify the best-fit structure. (1) The creation of conformationally randomized models can result in steric overlap between the domains in the models. If overlap occurs, the number of spheres in the unhydrated model will be reduced. Hence, models are retained only if they possess at least 95% of the expected number of spheres. (2) Models are retained if their Guinier-fitted R_G and R_{XS} values are within 5% or $\pm$ 0.3 nm from the experimental values. (3) Models are ranked according to their R-factors. If the $I(0)$ value of the calculated curve is set to be 1000, typical satisfactory R-factors are less than 10% and in good cases are less than 5%. Note that the R-factors increase with experimental noise; hence, good signal-noise ratios facilitate smaller R-factors. Once the best models have been filtered, SIGMAPLOT graphs of the R_G, R_{XS}, and R-factor values plotted against selected distances in the models are then used to confirm that a sufficient number of randomized structures have been screened. If the curve fitting has progressed well, a V-shaped distribution of R-factors versus R_G values is obtained. The R_G value at the minimum should be close to the experimental R_G value. If the experimental R_G is higher than expected, slight aggregation may have affected the Guinier fits.

The final outcome is confirmed by visual inspection of the experimental and modeled $I(Q)$ and $P(r)$ curves overlaid upon each other. The outcome also includes visual inspection of the best-fit models using INSIGHT II to see that they are stereochemically reasonable (e.g., no steric overlap; oligosaccharide chains are located on the surface), and the preparation of artwork showing how the best-fit models relate to biological function (e.g., comparisons with related macromolecules; calculation of electrostatic maps). The α-carbon coordinates of the best models are deposited in the PDB at http://www.rcsb.org/pdb/, initially following the procedure for crystal structures.

B. Scattering Modeling by Other Approaches

Despite the advantage of constrained scattering modeling in terms of PDB structures, "ab initio" scattering approaches remain popular. These alternatives model the scattering curve using 3D mathematical functions that reproduce the external shape, surface, or envelope of the scattering macromolecule. The "ab initio" methods assume that the protein can be represented by a compact structure of uniform scattering density. When automated, the fits are relatively quick compared to constrained modeling. The major difference is that no PDB models are involved in the fits, unlike the constrained approaches, so there are no structural constraints. Afterward the use of appropriate graphics software enables the surface envelope to be superimposed with related PDB crystal structures for visual comparisons. The website at EMBL, Hamburg provides useful articles and software at http://www.embl-hamburg.de/ExternalInfo/Research/Sax/software.html (Svergun and Koch, 2003).

The first of the "ab initio" methods was that of spherical harmonics devised by Stuhrmann (1970) and implemented in the SASHA program (Svergun and Stuhrmann, 1991; Svergun *et al.*, 1996). The macromolecular surface is defined by a molecular envelope function $F(\theta,\varphi)$ that has a scattering density of 1 inside it and 0 outside it. $F(\theta,\varphi)$ can be expanded in a series of spherical harmonics $Y_{lm}(\theta,\varphi)$:

$$F(\theta,\varphi) = R_{\text{o}} \sum_{l=0}^{H} \sum_{m=-1}^{1} f_{lm} Y_{lm}(\theta,\varphi)$$

where R_{O} is a scale factor and f_{lm} are complex multipole coefficients that are related to coefficients of the power series describing the experimental scattering curve. The f_{lm} coefficients can be determined by the automated minimization of an R-factor between the calculated and the observed data points in the scattering curve (Svergun and Stuhrmann, 1991). The number of harmonics terms H that can be used in the above summation is determined by the number of degrees of freedom (or the minimum number of independent parameters) that describes the experimental scattering curve. This is given by Shannon's sampling theorem as $D_{\text{max}}Q_{\text{max}}/\pi$ (Glatter and Kratky, 1982). This is 20 for a protein of length $L = 25$ nm for which a scattering curve was measured out to $Q = 2.5$ nm^{-1}. If H is 4, which is typical for the determination of a molecular envelope for a protein, there are $(H + 1)^2$ coefficients to be determined, of which 6 are arbitrary rotations and translations that do not affect the scattering curve, leaving 19 coefficients. If molecular symmetry is present, the number of coefficients to be determined is two- or threefold less in the cases of a dimer or trimer, respectively, and higher values of H are permitted. By this approach, the scattering curve can be represented by relatively simple shapes (in particular without holes inside the shape). It is possible to detect conformational changes on the binding of a small ligand by inspection of the molecular surfaces before and after ligation.

A second "ab initio" method is that based on small sphere models and the Debye equation for spheres (Glatter and Kratky, 1982). This was used to define macromolecular shapes for complement proteins such as C3, C4, C1q, C4bBP, and others (Perkins and Furtado, 2005). Simple triaxial objects were constructed from assemblies of small spheres, sometimes with guidance from electron micrograph images. The dimensions and arrangement of these assemblies were adjusted manually until they accounted for the scattering curve. First, the R_G value of the sphere model was adjusted to fit the scattering curve at low Q (i.e., the Guinier region). Next, the fits were progressively extended to large Q by trial and error. This approach was subsequently automated by rotating and translating groups of spheres to determine the best curve fits for IgE-Fc and bovine IgG (Beavil *et al.*, 1995; Mayans *et al.*, 1995). Note that the resulting sphere models were compared with relevant crystal structures after curve fitting, not before.

This second "ab initio" method was automated by specifying mathematical rules for the assembly of small spheres, which was then refined to fit the scattering curve. The first of these was the DALAI-GA genetic algorithm (Chacón *et al.*, 1998, 2000) (http://sbg.cib.csic.es/Software/Dalai_GA/index.html). Genetic algorithms start from a 3D array of thousands of spheres usually contained within a spherical volume. Inside this, the spheres are assigned as weights of 1 (macromolecule) or 0 (solvent). The arrangement of 1- and 0-assigned spheres is refined by a random Monte Carlo approach to obtain a fit to the observed scattering curve. Constraints including the physical connectivity of the macromolecular spheres are applied to force the algorithm to converge to a solution (Chacón *et al.*, 1998, 2000). A second program DAMMIN uses a simulated annealing algorithm and an explicit penalty term to ensure the compactness and connectivity of the resulting shape (Svergun, 1999). A third program SAXS3D (http://www.cmpharm.ucsf.edu/~walther/saxs/) defines the starting sphere model in terms of a hexagonal lattice, and each step of the refinement adds, remove, or relocates a sphere to improve the agreement with the scattering curve (Walther *et al.*, 2000). The best-fit arrangements of spheres from all three automated approaches compare well with known structures where these are available.

A third "ab initio" method is based on representing a protein of known sequence as an assembly of dummy residues. The starting point is a randomly distributed arrangement of residues in a spherical search volume of diameter D_{max}. The number of amino acid residues is known from the DNA sequence, and it is also known that two adjacent residues are separated by about 0.38 nm. Hence, in the GASBOR program (Svergun *et al.*, 2001), a simulated annealing algorithm determines the coordinates of the residues that best fit the experimental scattering curve in order to build a compact protein-like structure. This is constrained by the requirement that the model has a "chain-compatible" spatial arrangement of dummy residues.

Different random runs of the three "ab initio" methods give different spatial results that give similar good curve fits. It can be informative to rerun the "ab initio" software to appreciate the sphere arrangements that will give good curve fits. This is not necessary in automated constrained fit modeling, as the structural assessment of a best-fit family already provides 10, 50, and 100 best-fit structures derived from the original randomized models, and these can be used to assess how unique the modeling is.

C. Comparison with Sedimentation Coefficients

AUC provides sedimentation coefficients (Cole *et al.*, 2007). These are analogous to R_G values in being single-parameter values that measure macromolecular elongation. Unlike X-ray scattering, sedimentation coefficients do not lead to structural models as this is a single-parameter measurement. There are three main ways to model sedimentation coefficients:

1. Simple shape models (e.g., several ellipsoids or large spheres of different sizes) can be used. These simple models are not favored, as their shapes do not follow the detailed surface of the protein to be modeled, and therefore are unconvincing.

2. The atomic coordinate model can be converted into small spheres (also known as "bead" models). The model must include the hydration shell, typically at 0.3 g H_2O/g (glyco)protein. The small sphere approach utilizes software such as BRKTOS or SOMO to generate the sphere model. BRKTOS generates the unhydrated sphere model using a cubic grid, then this is accurately hydrated without macromolecular distortions (Fig. 8). SOMO replaces each residue by two different-sized spheres whose size is increased to allow for hydration (Byron, 2007).

3. The atomic coordinate model is inputted directly into the software. HYDROPRO represents the protein as a hollow shell of equally sized spheres at the protein surface that represents the macromolecular surface. Hydration is considered by empirically assigning the surface sphere radius as an effective value of 0.31 nm (Garcia de la Torre *et al.*, 2000).

GENDIA and HYDRO calculate the frictional coefficient from the sphere model GENDIA does not permit sphere overlap in the model but is computationally fast (Garcia de la Torre and Bloomfield, 1977a,b), while HYDRO permits sphere overlap but takes longer to compute (Garcia de la Torre *et al.*, 1994). GENDIA and HYDRO require substantial computing power compared to the calculation of scattering curves and R_G values. Constrained scattering modeling searches do not therefore use GENDIA or HYDRO for all the trial models, only for those short-listed as best-fit models after filtering. HYDROPRO outputs the sedimentation coefficient directly. These approaches are discussed in more detail elsewhere (Byron, 2007).

The use of identical sphere models for both the scattering and the ultracentrifugation modeling provides strong support for the outcome of the constrained modeling. This is only limited by computer memory requirements, as the hydrodynamic programs will only accept up to about 2000 spheres, unlike the scattering ones. The agreement between experiment and model should be within 0.3 S. If discrepancies are larger than this, several options are available: (1) reinvestigation of the experimental sedimentation coefficient using at least two different approaches (e.g., DCDT+ and SEDFIT: Cole *et al.*, 2007); (2) use at least two different modeling programs such as HYDRO and HYDROPRO; and (3) visual inspections of the sphere model directly overlaid on the starting atomic coordinates (e.g., using INSIGHT II, Accelrys, San Diego) and the number of spheres in the model. If the discrepancy cannot be resolved, a possible explanation is that the macromolecule possesses an unusually high surface hydration. The comparison of HYDRO and HYDROPRO for seven proteins whose solution structures were fitted by constrained scattering modeling showed that HYDROPRO gave more accurate numbers that HYDRO in six cases and was more able to handle larger proteins including antibodies (Almogren *et al.*, 2006).

VI. Examples

Recent constrained scattering analyses are summarized in Table III. As noted above, it is conceptually useful to classify the modeling into one of four types (Perkins *et al.*, 2005). Type 1 proteins involve the simple association of oligomeric proteins with no covalent linkers between the subunits. Types 2, 3, and *N* proteins involved the analyses of multidomain proteins with two, three, or more covalently linked domains or subunits, respectively. These four types each involve distinct procedures and lead to distinct outcomes. Their basic features are described in the examples below.

A. Type 1: Self-Association of Complement C3d and Other Proteins

The importance of dilution series in scattering is well illustrated by the unexpected oligomerization of the complement C3 fragment C3d. C3d is a 35 kDa cleavage fragment of the central innate immunity complement protein C3 and contains the thiolester active site region of C3. C3d is a ligand for complement receptor type 2 (CR2), acting to stimulate an immune response. X-rays showed a concentration dependence in the R_G values from 1.8 nm at 0.3 mg/ml to 3.5 nm at 11.2 mg/ml with a dissociation constant K_D that was refined to 23 ± 3 μM using AUC data (Perkins *et al.*, 2005; Fig. 9A). As C3 occurs in plasma at 1.3 mg/ml, this implies that C3d is normally monomeric. The $P(r)$ curve for C3d showed a single peak M_1 at $r = 2.8$ nm at low concentration which shifted to $r = 3.3$ nm at high

Table III
Protein Structures Determined by Constrained Scattering Modeling

Procedure	Protein	Summary of domains	PDB code[a]
Type 1	C3d	Monomer–dimer	
	MFE-23	Monomer–dimer	
	AmiC	Monomer–trimer	
	RuvA	Tetramer–octamer	
	SAP	Pentamer–decamer	
	Human IgA1 dimer	2 IgA1 monomers	
Type 2	CR2 SCR-1/2	2 SCR	1w2r
	CR2 SCR-1/2 + C3d	2 SCR + C3d	1w2s
	Cellulases	1 Cel6B, 1 Cel6A	
	Human IgA1-HSA	1 IgA1, 1 HSA	1esg
Type 3	Factor H SCR-6/8	3 SCR	2ic4
	Human IgA1	2 Fab/1 Fc	1iga
	Human IgA2	2 Fab/2 Fc	1r70
	Human IgD	2 Fab/1 Fc	1zvo
	CR2-Ig	2 CR2 SCR-1/2/1 Fc	2aty
	Murine Crry-Ig	2 Crry/1 Fc	1ntl
Type *N*	PIWF1	1 Cys, 1 WAP, 1 FnIII	1zlg
	Crry	5 SCR	1ntj
	Secretory component	5 Ig	2ocw
	CEA	7 Ig	1e07
	PIWF4	1 Cys, 1 WAP, 4 FnIII	1zlg
	Properdin dimer	2 × 7 TSR	1w0r
	Factor H SCR-1/20	20 SCR	1haq
	CR2 SCR-1/15	15 SCR	2gsx
	Properdin trimer	3 × 7 TSR	1w0s

[a]Structures are listed here if they were determined using the constrained modeling procedure of Fig. 8. The PDB code for the α-carbon coordinates of the scattering model are shown when available (http://www.rcsb.org).

concentration, at which a second peak M_2 appeared at $r = 7.0$ nm (Fig. 9B). The length of C3d increased from 7.0 to 10.5 nm. The first peak M_1 is attributed to all the distances within a single C3d molecule, and the second peak M_2 to all the distance vectors between the two monomers.

The constrained modeling of C3d is limited at low concentration by the poor signal-noise ratios of the scattering curve, nonetheless an excellent curve fit was obtained for the C3d monomer from its crystal structure (Fig. 9C). C3d dimers were arbitrarily created by performing symmetry operations about each of the 307 α-carbon atoms in the C3d crystal structure to create three putative dimers by 180° inversions of the monomer along the x-, y- and z-axes. The 921 modeled scattering curves were ranked using their R-factors. The best-fit dimer successfully reproduced an inflexion in the scattering curve at Q of 1 nm^{-1} (Fig. 9C) and gave a predicted $P(r)$ curve with a double peak that resembled M_1 and M_2. The fit implied that the C3d dimer is formed from relatively few contacts between the two monomers.

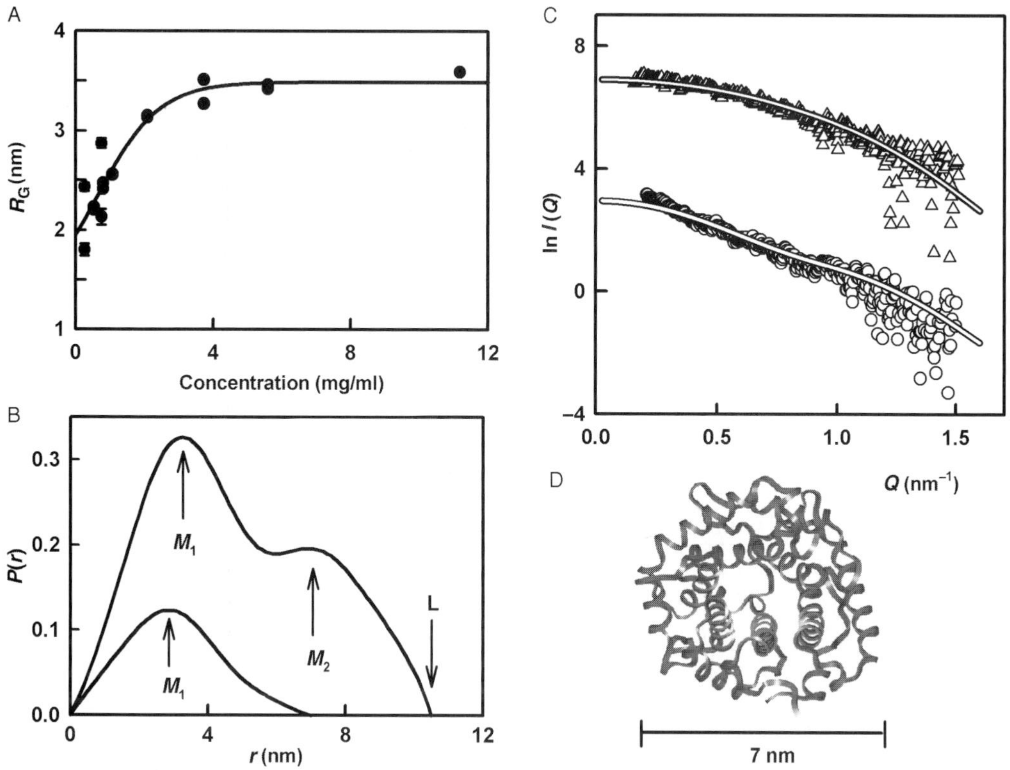

Fig. 9 Type 1 analyses of protein oligomerization: the C3d monomer–dimer equilibrium. (A) Concentration dependence of the C3d X-ray R_G values. (B) $P(r)$ curves for the monomer with one peak M_1 and the dimer with two peaks M_1 and M_2. The length of the dimer is denoted by L. (C) X-ray curve fits to the monomer (upper curve: Δ) and dimer (lower curve: ○). (D) Crystal structure of C3d, whose maximum dimension L of 7 nm corresponds to that shown in (B).

Other proteins show scattering curves that are explained in terms of oligomers; however, the constrained modeling will generally differ in each case. AmiC is the amide sensor protein and negative regulator of the five-gene amidase operon in an important opportunistic pathogen *Pseudomonas aeruginosa*. Even though it is a structural analogue of the monomeric periplasmic binding proteins that mediate small molecule transport in bacteria, dilution series in the presence of ligands showed pronounced effects, with interparticle interference effects at the highest concentrations and dissociation at the lowest ones (Chamberlain *et al.*, 1997). AmiC was found to exist as a monomer–trimer equilibrium in solution. Modeling using a monomeric crystal structure showed good curve fits, resulted when monomers were used in conjunction with a compact symmetric back-to-back trimer. The trimer is necessarily symmetric as otherwise AmiC would aggregate indefinitely in solution.

Human serum amyloid P component (SAP) is a pentameric plasma glycoprotein related with abnormal tissue deposits in amyloidosis. The pentamer is formed from five identical glycosylated subunits assembled in a flat disk (Ashton *et al.*, 1997). In the presence of Ca^{2+} and a monosaccharide, SAP forms pentamers. In the presence of EDTA, SAP forms tight decamers in which the two pentamers make face-to-face contact. Pentamers and decamers were readily identified by the appearance of the scattering curve (Ashton *et al.*, 1997). Gel filtration was not effective for this, as their shapes were too similar to be distinguished by this qualitative method. In addition, it was possible to identify the orientation of the two flat disks within the decamer. The $I(0)/c$ value for the decamer gave an abnormally low molecular weight. As the crystal structure showed that 20 Trp residues were close to one face of the disk, the low $I(0)/c$ could be explained by the proximity of 40 Trp residues in the decamer, which would alter the 280 nm absorption coefficient of the decamer relative to the pentamer. Constrained modeling gave good curve fits for the decamer. This used the central axis of symmetry of the pentamer to translate one pentamer through the other to generate 120 trial decamer models (Ashton *et al.*, 1997).

B. Type 1: Dimeric IgA

A different type of constrained dimer modeling occurs in the case of immunoglobulin A (IgA). IgA is the most abundant antibody in humans, and offers a major mucosal defense against external microorganisms. IgA exists in two subclasses IgA1 and IgA2, each possessing the classic two Fab and one Fc fragment structure with 12 domains in each monomer. Predominantly T-shaped structures were determined for monomeric IgA1 and IgA2 by constrained X-ray and neutron modeling (Boehm *et al.*, 1999; Furtado *et al.*, 2004). These opened the way for the scattering modeling of the IgA1 dimer, which is formed from the disulphide-linked covalent assembly of two monomers and a J chain domain positioned between them. Here, a Type 1 modeling strategy was the most appropriate to generate trial models for the dimer from two monomers. Unlike the examples described in the previous section, it was not necessary to assume symmetry in the dimer for reason of the J chain and the disulphide bridges linking the monomers. However, the position of the covalent links is not known, accordingly the modeling proceeded without this constraint. A search positioned the two monomers end-to-end, and then one was reorientated in 34,295 *x*- *y*- and *z*-axis rotations about the other, which was held fixed. This generated all possible orientations between the two IgA1 monomers. The outcome was successful, and resulted in a small single family of near-planar structures that gave good X-ray and neutron fits (A. Bonner, P. B. Furtado, A. Almogren, M. A. Kerr, and S. J. Perkins, unpublished data).

C. Type 2: SCR-1 and SCR-2 in Complement Receptor Type 2

CR2 is a membrane glycoprotein expressed primarily on mature B lymphocytes and follicular dendritic cells. It belongs to the short complement regulator (SCR) superfamily of small 61-residue domains, where each SCR domain is $\sim$2 nm $\times$ 2 nm $\times$ 4 nm in dimensions, and CR2 is formed from 15 or 16 SCR domains. The

SCR-1/2 domains contain the C3d binding site, and their crystal structures (Fig. 10C and F) proved to be controversial (Prota *et al.*, 2002; Szakonyi *et al.*, 2001). Both when free and in complex with C3d, SCR-1 and SCR-2 formed close side-by-side contacts with each other. Functional studies and AUC data, however, suggested that such a domain arrangement was not able to explain its solution properties. Accordingly constrained X-ray scattering was applied to resolve the discrepancy (Gilbert *et al.*, 2005). For SCR-1/2, X-rays revealed a R_G value of 2.12 $\pm$ 0.05 nm from Guinier analyses and a maximum length of 10 nm from $P(r)$ analyses. For its complex with C3d, the R_G value of 2.44 $\pm$ 0.1 nm and its length of 9 nm showed that its structure was not much more elongated than that of C3d.

Type 2 modeling is based on the structural randomization of the eight-residue linker peptide joining SCR-1 and SCR-2 to give 9950 trial models. In effect the linker becomes the only conformational variable to be fitted, and this simplicity results in a relatively well-defined structural outcome. The calculated scattering parameters from these SCR-1/2 models were filtered and sorted to identify a single best-fit family of structures. The most favored arrangements for the two SCR domains corresponded to an open V-shaped structure at an angle of 69° with no contacts between the SCR domains (Fig. 10A and B). Calculations with 9950 models of CR2 SCR-1/2 bound to C3d through SCR-2 showed that SCR-1 formed an open V-shaped structure at an angle of 39° with SCR-2 and was capable of interacting with the surface of C3d (Fig. 10D and E). Thus, the open V-shaped structures formed by SCR-1/2, both when free and when bound to C3d, appeared optimal for the formation of a tight two-domain interaction with its ligand C3d (Gilbert *et al.*, 2005). This result was supported by accompanying mutagenesis experiments (Hannan *et al.*, 2005). This analysis is a good example of when crystal structures have had to be significantly revised in the light of scattering modeling, as the interdomain linker flexibility had influenced the way in which the crystals form.

D. Type 3: Antibody Structural Modeling of IgD

Monomeric antibodies are composed of two Fab and one Fc fragments joined by a hinge region. The hinge is key to the antibody solution structure. Only a handful of crystal structures for intact IgG antibodies in PEG or high salt (e.g., 0.8- or 1.5-M ammonium sulphate) are known to date. These display single snapshots of symmetric or asymmetric IgG hinge structures frozen by the intermolecular contacts in the crystallographic unit cell. The hinge is structurally diverse in the five human antibody classes IgG, IgA, IgM, IgD, and IgE. The hinges in the four IgG subclasses vary in length, where IgG1 contains 15 residues, IgG2 and IgG4 contain 12 residues, and IgG3 contains 62 residues. The IgD hinge is the longest one, being 64 residues in length (Sun *et al.*, 2005). The function of the hinges in IgM and IgE is replaced or substituted by an extra pair of Ig domains. What is not clear is the extent to which hinge flexibility is important for antibody interactions with its ligands. Constrained scattering modeling is well suited to

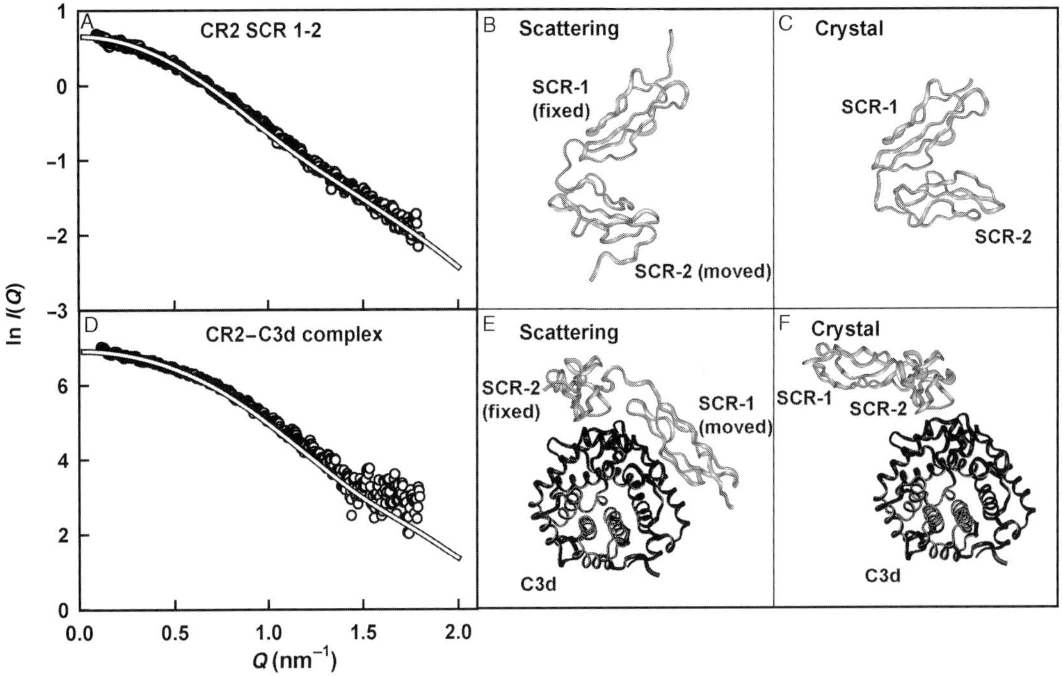

Fig. 10 Type 2 modeling of the CR2 SCR-1/2 domain structure when free and bound to C3d. (A) X-ray scattering curve of CR2 SCR-1/2 (○) in comparison with the best-fit model curve (white line). (B) One of the best-fit CR2 models showing the open V-shaped structure of SCR-1 and SCR-2 (PDB code 1w2r). (C) The crystal structure of CR2 SCR-1/2 showed a closed V-shaped structure in which SCR-1 and SCR-2 are in contact with each other (PDB code 1ly2). SCR-1 is viewed in the same orientation in both (B) and (C). (D) X-ray scattering curve of the CR2–C3d complex (○) in comparison with the best-fit model curve (white line). (E) One of the best-fit models for the complex showing that both SCR-1 and SCR-2 are in contact with the surface of C3d (PDB code 1w2s). (F) The crystal structure of the CR2–C3d complex showing that only SCR-2 is in contact with the surface of C3d (PDB code 1ghq). C3d and SCR-2 are viewed in the same orientation in both (E) and (F).

elucidate the Fab and Fc structures mediated by these different hinges. Type 3 constrained modeling benefits from the twofold symmetry of the antibody structure. This means that the same hinge conformation can be used for the two hinges that link the two Fab and Fc fragments, thus simplifying the outcome of the modeling.

Human IgD occurs most abundantly as a membrane-bound antibody on the surface of mature B cells. Soluble myeloma IgD gave a X-ray R_G of 6.9 ± 0.1 nm showing that this is more extended in solution than IgA1 (R_G of 6.1–6.2 nm). Its distance distribution function $P(r)$ showed a single peak at 4.7 nm and a maximum dimension of 23 nm. Molecular dynamics was used to generate randomized IgD hinge structures, to which homology models for the Fab and Fc fragments were connected. The dependence of the R-factor values on the R_G values showed that the lowest R-factor corresponded to the R_G value closest

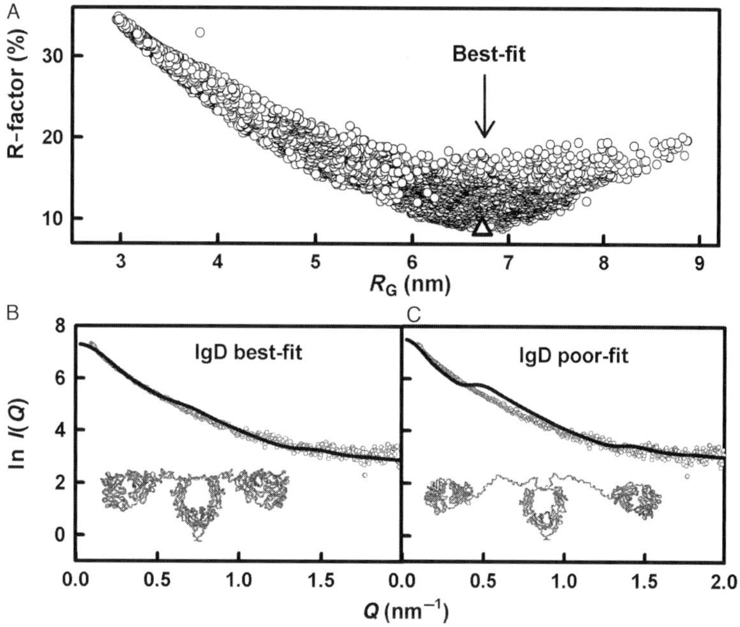

Fig. 11 Type 3 modeling of the IgD antibody structure. (A) The dependence of the goodness-of-fit *R*-factor on the R_G value shows that the *R*-factor is at a minimum close to the experimental R_G at 6.7 nm (arrowed). The best-fit model of panel (B) is denoted by Δ. (B) The best-fit model of IgD is shown together with the experimental (○) and modeled (line) X-ray curves. (C) A poor-fit IgD structure is shown in which the hinge region is extended compared to that in (B).

to the experimental value (Fig. 11A). Good curve fits were obtained with 18 semi-extended IgD models starting from 8500 trial models (Fig. 11B). The best-fit models showed that the hinge does not correspond to an extended polypeptide (which gives a poor curve fit: see Fig. 11C), but instead to a principally T-shaped arrangement of the Fab and Fc fragments (Sun *et al.*, 2005). Comparison with the constrained scattering modeling of IgA1 (Boehm *et al.*, 1999) suggested that the hinges of IgA1 and IgD are more similar than that might have been expected. Both possess flexible T-shaped solution structures, probably reflecting the presence of restraining O-linked sugars.

E. Type 3: Factor H SCR-6/8

Factor H (FH) is a major complement regulatory protein in serum and contains 20 SCR domains. The seventh of these (SCR-7) is associated with age-related macular degeneration, the most common cause of blindness in the Western world, through a Tyr402His polymorphism. The three-domain recombinant SCR-6/8 fragment of FH containing either His402 or Tyr402 and their complexes with a heparin decasaccharide were studied by constrained X-ray scattering modeling (Fernando *et al.*, 2007). First, AUC was used to identify a weak self-association, which was slightly greater for the His402 allotype than the Tyr402 allotype. As the monomer–dimer equilibrium has a dissociation constant of 40 μM for the His402 form, the scattering data corresponded to 25–31% dimers in the concentrations studied. The Guinier R_G of 3.1–3.3 nm and the R_G/R_O ratio of 2.0–2.1 showed that SCR-6/8 is relatively extended in solution. Interestingly, no concentration dependence of the R_G or the $P(r)$ curve was observed, showing that dimer formation did not affect these parameters. The maximum dimension of 10 nm is less than the length expected for a linear arrangement of SCR domains, hence the SCR-6/8 fragment has a bent conformation in solution. Heparin caused the formation of a more linear structure, possibly by binding to linker residues.

The SCR-6/8 fragment is too big for molecular structure determination by NMR. The constrained scattering modeling of FH SCR-6/8 is a Type 3 analysis because of the three domains (Fernando *et al.*, 2007). Unlike the Type 3 antibody modeling, there is no twofold symmetry. The filtering of 2000 randomized SCR-6/8 models identified best-fit bent models for the monomer (Fig. 12A). Joining two models face-to-face at SCR-7 was able to account for dimer formation. A moderate movement of the SCR-6 and SCR-8 domains toward each other gave a $P(r)$ curve with two peaks M_1 and M_2 at r values of 1.8 and 4.5 nm as observed experimentally, unlike linear SCR models that gave only one peak M_1 at 1.7 nm. The M_2 peak is assigned to the increased numbers of vectors between SCR-6 and SCR-8 in relatively static bent models. The relative intensities of the M_1 and M_2 peaks reproduced that seen experimentally when the presence of both monomer and dimer was considered (Fig. 12B). Hence, Type 3 modeling is simple enough to be informative about the bending in three-domain structures, although of course it is not possible to specify the degree of twist between the SCR domains.

F. Type *N*: Secretory Component of IgA and Other Multidomain Proteins

Many multidomain proteins contain more than three domains, and the constrained scattering modeling of these Type *N* structures are useful, even though there is less available information when compared to Type 2 and 3 analyses. Scattering will show whether these possess extended domain structures or are folded back upon themselves.

An interesting example is secretory component (SC) that associates with dimeric IgA to form secretory IgA (SIgA), the major antibody active at mucosal surfaces. SC exists in the free form, and has five heavily glycosylated variable (V)-type Ig domains D1–D5. The SC structure was determined by X-ray and neutron scattering to give a R_G of 3.53–3.63 nm and a $P(r)$ length of 12.5 nm. This was unexpectedly compact as the length L should have been closer to 20 nm. The neutron data verified the absence of X-ray radiation damage and any possible density variations because of the 19% carbohydrate content in SC. The SC structure was accordingly probed in greater detail using its fortuitous cleavage during storage into two D1-D3 and D4-D5 fragments, each of which were subjected to X-ray scattering.

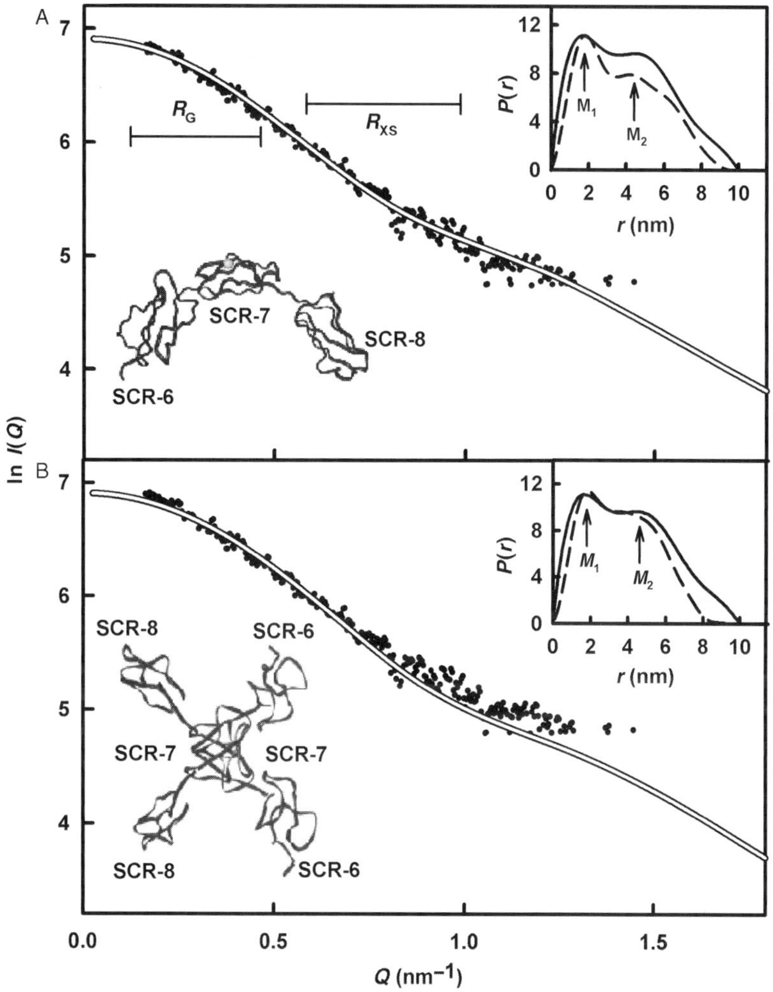

Fig. 12 Type 3 modeling of the FH SCR-6/8 structure. (A) Modeling of the X-ray scattering curve for the His402 allotype in terms of the bent three-domain monomer structure shown as inset. The R_G and R_{XS} regions of the curve are denoted by horizontal lines. The experimental (continuous line) and modeled (dashed line) $P(r)$ curves with two peaks M_1 and M_2 are shown as inset. Note the discrepancy in the peak M_2 intensity. (B) Modeling in terms of 75% monomer and 25% dimer, where the dimer model is shown as inset. Note the better agreement for peak M_2 in the $P(r)$ curve.

Both D1-D3 and D4-D5 showed overall lengths similar to that of SC, suggesting that SC has a J-shaped structure with D1-D3 folded back on D4-D5 (Bonner *et al.*, 2007).

Constrained scattering modeling was first applied to each of D1-D3 and D4-D5, in which D4-D5 was readily modeled as a Type 2 structure with a small bend in it, while the modeling of D1-D3 was complicated by the presence of additional degradation products and was completed only by postulating the presence of single domains as well as D1-D3. AUC technology was important as size distribution analyses provided the necessary evidence to support this postulate. Finally, constrained modeling was applied to the intact SC structure to show that J-shaped structures fitted the X-ray and neutron data well (Fig. 13B). Searches in which all five domains or only the linker between fixed best-fit structures for the D1-D3 and D4-D5 fragments was randomized were required to ensure that a full range of conformations had been assessed (Fig. 13A).

Multidomain proteins do not generally show folded-back structures. Scattering studies often benefit from the study of fragments as well as the intact protein. That for anosmin-1 with six domains was determined to be extended from the study of three- and six-domain forms (Hu *et al.*, 2005). Carcinoembryonic antigen (CEA) with seven Ig domains is also extended with slight bends between the domains (Boehm and Perkins, 2000). Large SCR proteins such as FH with 20 SCR domains and CR2 with 15 SCR domains were most successfully modeled as partially folded-back structures that probably show a degree of inter-SCR flexibility (Aslam and Perkins, 2001; Gilbert *et al.*, 2006). An understanding of the FH structure is central for an understanding of complement regulation, yet this is too big to be crystallized. Thus, the FH SCR-1/5 and SCR-16/20 fragments were also studied by scattering (A. I. Okemefuna, H. E. Gilbert, K. Griggs, R. Ormsby, D. L. Gordon, and S. J. Perkins, unpublished data). This showed interestingly that both SCR fragments exhibit similar partial folding back as that seen in the starting FH structure.

VII. Discussion

A. Summary and Future Considerations

Solution scattering is a low- to medium-resolution structural discipline applicable to a broad range of proteins and protein–ligand complexes. Modern solution scattering benefits from the availability of powerful cameras with large Q ranges and good count rates at high-flux X-ray and neutron sources. Data processing using UNIX script files or EXCEL macros to handle many experimental data files considerably eases the Guinier and $P(r)$ analyses. Key controls of the data must include explicit checks for radiation damage and molecular weight calculations to check for oligomers. Data interpretation is much improved by the ability to perform molecular modeling using tight constraints based on available atomic structures in the PDB. The key step is to ensure that a full range of stereochemically correct conformations have been generated for fitting. However, it should be remembered that unique structures cannot be determined by scattering for reason of the randomized orientations of the macromolecule in solution. Scattering modeling is at its most effective in rejecting poor models (Fig. 11). Constrained modeling uses this concept to reject about 99% of the conformationally possible

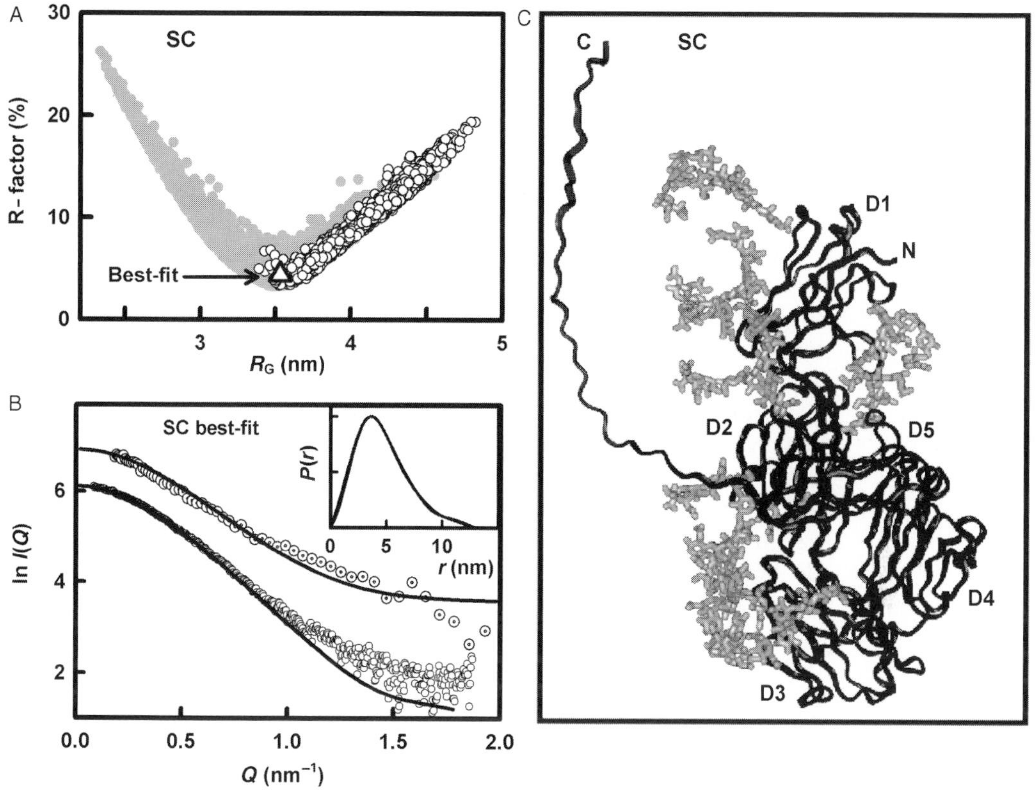

Fig. 13 Type N constrained modeling of SC by X-ray and neutron scattering. (A) The R-factors are compared with their R_G values. The 5000 SC models represented by gray circles correspond to the SC models in which all four linker peptides were varied. The final 5000 SC models represented by unfilled circles correspond to SC models in which the best-fit D1-D3 model was randomized relative to the best-fit D4-D5 model. (B) The neutron (upper) and X-ray (lower) experimental curves are compared with the best-fit SC model (line). The inset shows the comparison between the experimental and modeled X-ray $P(r)$ curves; the two curves are indistinguishable. (C) The best-fit model is shown with the D1-D3 domains running from top to bottom, and the D4-D5 domains are bent back against these to the right to form an overall J-shaped structure. The seven oligosaccharide chains are shown in gray.

models. The advantage of constrained modeling over the more commonly used "ab initio" methods is that medium-resolution best-fit structures can be determined and deposited in the PDB. It is conceivable that constrained modeling may grow in importance and impact. As evidenced by scattering publications in mainstream biochemical journals, this technology provides significant new insights on structure–function relationships. This is especially so when no crystal structures are available or are apparently not achievable, and even after a crystal structure determination has already provided much information. Once the data collection is completed and interpreted, the timescale for project completion is a matter of weeks.

Each scattering study is different. The concept of Types 1, 2, 3, or N modeling clarifies how the constrained modeling should be performed. Examples of oligomer formation were shown for C3d in Fig. 9, where dimers were obvious from the scattering data, and for FH SCR-6/8 in Fig. 12, where dimer only became apparent during the modeling. The need to confirm crystal structures by solution structures is shown for both CR2 SCR-1/2 and its complex with C3d in Fig. 10. Examples where scattering provided structures for which there are no crystal structures include the IgA1 dimer, IgD, FH SCR-6/8, and SC (Figs. 11–13).

A side benefit of constrained modeling is the ability to interpret the $P(r)$ curve. Prior to constrained modeling, simple model calculations showed, for example, that M occurs at half the value of L for the scattering from a sphere, and that M moves to smaller r values when the scatterer becomes elongated. The position of M in relation to the macromolecular length L, especially when two peaks are seen and not one, is of great interest. In the case of IgA1, the second peak M_2 (Fig. 5F) suggested that the hinges between the Fab and Fc are sufficiently rigid to hold the two Fab fragments apart from the Fc fragment, while limited flexibility is present to enable the antibody to dock onto its antigenic targets. In the case of FH SCR-6/8, the second peak M_2 (Fig. 12) indicated that the three-domain structure is on average bent in solution.

B. Comparison of Scattering with Crystallography and NMR

A combination of methods is best applied to a given protein, whether these are immunologic (this chapter) or from another discipline such as the cell cycle proteins, chaperones, or otherwise. The major difference between methods lies in the structural resolution, ranging from atomic level provided by NMR spectroscopy and X-ray crystallography to overall shape information obtained directly by electron microscopy or indirectly by solution scattering. None of the methods can generally be applied in totally physiological conditions. For example, they usually require proteins to be highly purified, and in many instances only portions of larger molecules can be studied. Nonetheless, used in combination, these methods have greatly increased our understanding of protein function.

The strength of crystallography is in the level of detail provided. The weaknesses are that it is only possible to determine the X-ray structure of proteins that crystallize and the structure of the protein packed in the crystal unit cell may not accurately represent the native protein especially where several domains are present in an extended conformation (see the CR2 example discussed above). However, many structures determined by multiple methodologies are identical in domain structure and orientation, suggesting that crystal structures often do approximate native structures. Sometimes crystallography can be very rapid with proteins going from initial trials to structures in a few weeks. However, crystallization remains unpredictable and many key structures have taken years to complete.

While NMR-derived structures are of slightly lower resolution than X-ray crystal structures, NMR can offer several advantages over crystallography. The NMR structure is obtained in solution although the protein concentration can be quite high. Atomic details are forthcoming of the effect of pH, ionic strength, and small ligand-binding sites, and any flexible or mobile regions within the protein can be localized. A major limitation of NMR is that it is only applicable to smaller proteins of up to 30 kDa in size.

Scattering offers the possibility of studying intact, large molecules in near-physiological conditions including low concentrations. They are generally applicable to proteins of molecular weight above 10 kDa, only requiring high protein purity. The first overall views of many protein structures have been obtained by X-ray and neutron scattering (or by AUC as a related but less informative technology), followed by constrained modeling to identify best-fit structures at medium structural resolutions.

C. Biological Relevance of Scattering Structures

The structural resolution of constrained scattering modeling is sufficient to identify the most probable domain arrangement in multidomain proteins. The structural precision of this approach can be sufficient to identify functional roles. For example, the use of homology modeling for domains and knowledge of the putative domain arrangement revealed the first analyses of structural effects of mutations in FH SCR-19 and SCR-20 that lead to atypical haemolytic uraemic syndrome (kidney failure after an immunologic insult, mostly in young children). See our website at http://www.fh-hus.org. While the scattering modeling opens up the first structural view of FH and its mutations, the follow through by crystallography and NMR approaches is nonetheless necessary. The reanalysis of the SCR-19 and SCR-20 domains using their subsequent NMR and crystal structures revealed that the interpretations of the mutations were correctly made (Saunders *et al.*, 2007).

The medium resolution of the scattering modeling means that further experiments will be needed to confirm any important results. The case of the CR2 SCR-1/2 complex with C3d is an example of this, where mutagenesis work with this system was required to confirm that both SCR-1 and SCR-2 made contact with the surface of C3d in solution. In the absence of the scattering modeling, it would have been difficult to argue that the crystal structure required correction (Gilbert *et al.*, 2005; Hannan *et al.*, 2005). In other cases, the study of fragments becomes necessary to place any scattering modeling conclusions on a more firm basis. This is well illustrated by the study of SC and its D1-D3 and D4-D5 fragments (Fig. 13) both by X-ray scattering and by AUC (Bonner *et al.*, 2007).

Acknowledgments

The work in the authors' laboratory has been generously supported by funding provided by the Wellcome Trust, the Biotechnology and Biological Sciences Research Council, the Fight for Sight and Henry Smith Charities, and Graduate Research Scholarships from University College London. We are particularly grateful to the Instrument Scientists at the facilities that we have recently used to acquire data sets, in particular Dr. Theyencheri Narayanan, Dr. Stephanie Finet, and Dr. Pierre Panine at the ESRF, and Dr. Richard K. Heenan and Dr. Stephen M. King at ISIS.

References

Almogren, A., Furtado, P. B., Sun, Z., Perkins, S. J., and Kerr, M. A. (2006). Biochemical and structural properties of the complex between human immunoglobulin A1 and human serum albumin by X-ray and neutron scattering and analytical ultracentrifugation. *J. Mol. Biol.* **356,** 413–431.

Ashton, A. W., Boehm, M. K., Gallimore, J. R., Pepys, M. B., and Perkins, S. J. (1997). Pentameric and decameric structures in solution of the serum amyloid P component by X-ray and neutron scattering and molecular modelling analyses. *J. Mol. Biol.* **272,** 408–422.

Aslam, M., and Perkins, S. J. (2001). Folded-back solution structure of monomeric Factor H of human complement by synchrotron X-ray and neutron scattering, analytical ultracentrifugation and constrained molecular modelling. *J. Mol. Biol.* **309,** 1117–1138.

Beavil, A. J., Young, R. J., Sutton, B. J., and Perkins, S. J. (1995). Bent domain structure of recombinant human IgE-Fc in solution by X-ray and neutron scattering in conjunction with an automated curve fitting procedure. *Biochemistry* **34,** 14449–14461.

Boehm, M. K., and Perkins, S. J. (2000). Structural models for carcinoembryonic antigen and its complex with the single-chain Fv antibody molecule MFE23. *FEBS Lett.* **475,** 11–16.

Boehm, M. K., Woof, J. M., Kerr, M. A., and Perkins, S. J. (1999). The Fab and Fc fragments of IgA1 exhibit a different arrangement from that in IgG: A study by X-ray and neutron solution scattering and homology modelling. *J. Mol. Biol.* **286,** 1421–1447.

Bonner, A., Perrier, C., Corthésy, B., and Perkins, S. J. (2007). Solution structure of human secretory component and implications for biological function. *J. Biol. Chem.* **282,** 16969–16980.

Byron, O. (2007). Hydrodynamic modelling: The solution conformation of macromolecules and their complexes. *In* "Biophysical Tools for Biologists" (J. J. Correia, and H. W. Dietrich, III, eds.), pp. 327–373. Academic Press, San Diego.

Chacón, P., Díaz, J. F., Morán, F., and Andreu, J. M. (2000). Reconstruction of protein form with X-ray solution scattering and a genetic algorithm. *J. Mol. Biol.* **299,** 1289–1302.

Chacón, P., Moran, F., Díaz, J. F., Pantos, E., and Andreu, J. M. (1998). Low-resolution structures of proteins in solution retrieved from X-ray scattering with a genetic algorithm. *Biophys. J.* **74,** 2760–2775.

Chamberlain, D., Keeley, A., Aslam, M., Arenas-Licea, J., Brown, T., Tsaneva, I. R., and Perkins, S. J. (1998). A synthetic Holliday junction in sandwiched between two tetrameric *Mycobacterium leprae* RuvA structures in solution: new insights from neutron scattering contrast variation and modelling. *J. Mol. Biol.* **285,** 385–400.

Chamberlain, D., O'Hara, B. P., Wilson, S. A., Pearl, L. H., and Perkins, S. J. (1997). Oligomerization of the amide sensor protein AmiC by X-ray and neutron scattering and molecular modelling. *Biochemistry* **36,** 8020–8029.

Cole, J. L., Lary, J. W., Moody, T., and Laue, T. M. (2007). Analytical ultracentrifugation: Sedimentation velocity and sedimentation equilibrium. *In* "Biophysical Tools for Biologists" (J. J. Correia, and H. W. Dietrich, III, eds.), pp. 143–179. Academic Press, San Diego.

Converse, C. A., and Skinner, E. R. (1992). "Lipoprotein Analysis: A Practical Approach." IRL Press, Oxford, UK.

Fernando, A. N., Furtado, P. B., Clark, S. J., Gilbert, H. E., Day, A. J., Sim, R. B., and Perkins, S. J. (2007). Associative and structural properties of the region of complement Factor H encompassing the Tyr402His disease-related polymorphism and its interactions with heparin. *J. Mol. Biol.* **368,** 564–581.

Furtado, P. B., Whitty, P. W., Robertson, A., Eaton, J. T., Almogren, A., Kerr, M. A., Woof, J. M., and Perkins, S. J. (2004). Solution structure determination of human IgA2 by X-ray and neutron scattering and analytical ultracentrifugation and constrained modelling: A comparison with human IgA1. *J. Mol. Biol.* **338,** 921–941.

Garcia de la Torre, J., and Bloomfield, V. A. (1977a). Hydrodynamic properties of macromolecular complexes. I. Translation. *Biopolymers* **16,** 1747–1761.

Garcia de la Torre, J., and Bloomfield, V. A. (1977b). Hydrodynamics of macromolecular complexes. III. Bacterial viruses. *Biopolymers* **16,** 1779–1793.

Garcia de la Torre, J., Huertas, M. L., and Carrasco, B. (2000). Calculation of hydrodynamic properties of globular proteins from their atomic-level structure. *Biophys. J.* **78,** 719–730.

Garcia de la Torre, J., Navarro, S., Martinez, M. C. L., Diaz, F. G., and Cascales, J. L. (1994). HYDRO: A computer program for the prediction of hydrodynamic properties of macromolecules. *Biophys. J.* **67,** 530–531.

Gilbert, H. E., Asokan, R., Holers, V. M., and Perkins, S. J. (2006). The flexible 15 SCR extracellular domains of human complement receptor type 2 can mediate multiple ligand and antigen interactions. *J. Mol. Biol.* **362,** 1132–1147.

Gilbert, H. E., Eaton, J. T., Hannan, J. P., Holers, V. M., and Perkins, S. J. (2005). Solution structure of the complex between CR2 SCR 1–2 and C3d of human complement: An X-ray scattering and sedimentation modelling study. *J. Mol. Biol.* **346,** 859–873.

Ghosh, R. E., Egelhaaf, S. U., and Rennie, A. R. (2006). "A Computing Guide for Small-Angle Scattering Experiments." 5th edn., Institut Laue Langevin Publication ILL06GH05T, Grenoble, France.

Glatter, O., and Kratky, O. (1982). "Small Angle X-ray Scattering." Academic Press, New York.

Hannan, J. P., Young, K. A., Guthridge, J. M., Asokan, R., Szakonyi, G., Chen, X. S., and Holers, V. M. (2005). Mutational analysis of the complement receptor Type 2 (CR2/CD21)–C3d interaction reveals a putative charged SCR1 binding site for C3d. *J. Mol. Biol.* **346,** 845–858.

Heenan, R. K., Penfold, J., and King, S. M. (1997). SANS at pulsed neutron sources: Present and future prospects. *J. Appl. Cryst.* **30,** 1140–1147.

Hjelm, R. P. (1985). The small-angle approximation of X-ray and neutron scatter from rigid rods of non-uniform cross section of finite length. *J. Appl. Cryst.* **18,** 452–460.

Hu, Y., Sun, Z., Eaton, J. T., Bouloux, P. M. G., and Perkins, S. J. (2005). Extended domain solution structure of the extracellular matrix protein anosmin-1 by X-ray scattering, analytical ultracentrifugation and constrained modeling. *J. Mol. Biol.* **350,** 553–570.

Huang, T. C., Toraya, H., Blanton, T. N., and Wu, Y. (1993). X-ray powder diffraction analysis of silver behanate, a possible low-angle diffraction standard. *J. Appl. Cryst.* **26,** 180–184.

Ibel, K., and Stuhrmann, H. B. (1975). Comparison of neutron and X-ray scattering of dilute myoglobin solutions. *J. Mol. Biol.* **93,** 255–265.

Jacrot, B., and Zaccai, G. (1981). Determination of molecular weight by neutron scattering. *Biopolymers* **20,** 2413–2426.

King, S. M. (2006). Using COLETTE: A step-by-step guide. Rutherford-Appleton Laboratory Report RAL-95–005 .

Kratky, O. (1963). X-ray small angle scattering with substances of biological interest in diluted solutions. *Prog. Biophys. Chem.* **13,** 105–173.

Lindner, P., May, R. P., and Timmins, P. A. (1992). Upgrading of the SANS instrument D11 at the ILL. *Physica B* **180,** 967–972.

Mayans, M. O., Coadwell, W. J., Beale, D., Symons, D. B. A., and Perkins, S. J. (1995). Demonstration by pulsed neutron scattering that the arrangement of the Fab and Fc fragments in the overall structures of bovine IgG1 and IgG2 in solution is similar. *Biochem. J.* **311,** 283–291.

Narayanan, T. (2007). Synchrotron small-angle X-ray scattering. *In* "Soft Matter: Scattering, Imaging and Manipulation" (R. Borsali, and R. Pecora, eds.), Springer, in press.

Narayanan, T., Diat, O., and Bosecke, P. (2001). SAXS and USAXS on the high brilliance beamline at the ESRF. *Nucl. Instrum. Methods Phys. Res. A* **467,** 1005–1009.

Nave, C., Helliwell, J. R., Moore, P. R., Thompson, A. W., Worgan, J. S., Greenall, R. J., Miller, A., Burley, S. K., Bradshaw, J., Pigram, W. J., Fuller, W., Siddons, D. P., *et al.* (1985). Facilities for solution scattering and fibre diffraction at the Daresbury SRS. *J. Appl. Cryst.* **18,** 396–403.

Panine, P., Finet, S., Weiss, T. M., and Narayanan, T. (2006). Probing fast kinetics in complex fluids by combined rapid mixing and small-angle X-ray scattering. *Adv. Colloid Interf Sci.* **127,** 9–18.

Perkins, S. J. (1986). Protein volumes and hydration effects: The calculation of partial specific volumes, neutron scattering matchpoints and 280 nm absorption coefficients for proteins and glycoproteins from amino acid sequences. *Eur. J. Biochem.* **157,** 169–180.

Perkins, S. J. (1988a). Structural studies of proteins by high-flux X-ray and neutron solution scattering. *Biochem. J.* **254,** 313–327.

Perkins, S. J. (1988b). X-ray and neutron solution scattering. *New Comprehen. Biochem.* **18B**(Pt. II), 143–264.

Perkins, S. J. (1994). High-flux X-ray and neutron solution scattering. *In* "Physical Methods of Analysis" in "Methods in Molecular Biology" (C. Jones, B. Mulloy, and A. H. Thomas, eds.), Vol. 22, pp. 39–60. Humana Press, Inc., New Jersey.

Perkins, S. J. (2000). High-flux X-ray and neutron scattering studies. *In* "Protein–Ligand Interactions: A Practical Approach." (B. Chowdhry, and S. E. Harding, eds.), Chapter 9, Vol. 1, pp. 223–262. Oxford University Press, Oxford, UK.

Perkins, S. J. (2001). X-ray and neutron scattering analyses of hydration shells: A molecular interpretation based on sequence predictions and modelling fits. *Biophys. Chem.* **93,** 129–139.

Perkins, S. J., Ashton, A. W., Boehm, M. K., and Chamberlain, D. (1998). Molecular structures from low angle X-ray and neutron scattering studies. *Int. J. Biolog. Macromol.* **22,** 1–16.

Perkins, S. J., and Furtado, P. B. (2005). Complement and immunoglobulin protein structures by X-ray and neutron solution scattering and analytical ultracentrifugation. *In* "Structural Biology of the Complement System" (D. Morikis, and J. D. Lambris, eds.), Chapter 13, pp. 293–315. Taylor and Francis, Boca Raton, Florida, ISBN: 0824725409.

Perkins, S. J., Gilbert, H. E., Lee, Y. C., Sun, Z., and Furtado, P. B. (2005). Relating small angle scattering and analytical ultracentrifugation in multidomain proteins. *In* "Modern Analytical Ultracentrifugation: Techniques and Methods" (D. J. Scott, S. E. Harding, and A. J. Rowe, eds.), Chapter 15, pp. 291–319. Royal Society of Chemistry, London, UK, ISBN: 0854045473.

Perkins, S. J., Nealis, A. S., Sutton, B. J., and Feinstein, A. (1991). The solution structure of human and mouse immunoglobulin IgM by synchrotron X-ray scattering and molecular graphics modelling: A possible mechanism for complement activation. *J. Mol. Biol.* **221,** 1345–1366.

Perkins, S. J., and Weiss, H. (1983). Low resolution structural studies of mitochondrial ubiquinol-cytochrome c reductase in detergent solutions by neutron scattering. *J. Mol. Biol.* **168,** 847–866.

Prota, A. E., Sage, D. R., Stehle, T., and Fingeroth, J. D. (2002). The crystal structure of human CD21: Implications for Epstein-Barr virus and C3d binding. *Proc. Natl. Acad. Sci. USA* **99,** 10641–10646.

Sambrook, E. F., Fritsch, E. F., and Maniatis, T. (1989). "Molecular cloning: A laboratory manual." 2nd edn., Cold Spring Harbor Laboratory Press, New York.

Saunders, R. E., Abarrategui-Garrido, C., Frémeaux-Bacchi, V., Goicoechea de Jorge, E., Goodship, T. H. J., López Trascasa, M., Noris, M., Ponce Castro, I. M., Remuzzi, G., Rodríguez de Córdoba, S., Sánchez-Corral, P., Skerka, C., *et al.* (2007). The interactive Factor H—atypical haemolytic uraemic syndrome mutation database and website: Update and integration of membrane cofactor protein and factor I mutations with structural models. *Hum. Mutat.* **28,** 222–234.

Semenyuk, A. V., and Svergun, D. I. (1991). *GNOM*—a program package for small-angle scattering data processing. *J. Appl. Cryst.* **24,** 537–540.

Stuhrmann, H. B. (1970). Interpretation of small-angle scattering functions of dilute solutions and gases. A representation of the structures related to a one-particle scattering function. *Acta. Crystallogr. A* **26,** 297–306.

Sun, Z., Almogren, A., Furtado, P. B., Chowdhury, B., Kerr, M. A., and Perkins, S. J. (2005). Semi-extended solution structure of human myeloma immunoglobulin D determined by constrained X-ray scattering. *J. Mol. Biol.* **353,** 155–173.

Svergun, D. I. (1999). Restoring low resolution structure of biological macromolecules from solution scattering using simulated annealing. *Biophys. J.* **76,** 2879–2886.

Svergun, D. I., and Koch, M. H. J. (2003). Small-angle scattering studies of biological macromolecules in solution. *Rep. Prog. Phys.* **66,** 1735–1782.

Svergun, D. I., Petoukhov, M. V., and Koch, M. H. J. (2001). Determination of domain structure of proteins from X-ray solution scattering. *Biophys. J.* **80,** 2946–2953.

Svergun, D. I., and Stuhrmann, H. B. (1991). New developments in direct shape determination from small-angle scattering. 1. Theory and model calculations. *Acta Crystallogr. A* **47,** 736–744.

Svergun, D. I., Volkov, V. V., Kozin, M. B., and Stuhrmann, H. B. (1996). New developments in direct shape determination from small-angle scattering. 2. Uniqueness. *Acta Crystallogr. A* **52,** 419–426.

Szakonyi, G., Guthridge, J. M., Li, D., Young, K., Holers, V. M., and Chen, X. S. (2001). Structure of complement receptor 2 in complex with its C3d ligand. *Science* **292,** 1725–1728.

Towns-Andrews, E., Berry, A., Bordas, J., Mant, G. R., Murray, P. K., Roberts, K., Sumner, I., Worgan, J. S., Lewis, R., and Gabriel, A. (1989). Time-resolved X-ray diffraction station: X-ray optics, detectors, and data acquisition. *Rev. Sci. Instrum.* **60,** 2346–2349.

Walther, D., Cohen, F. E., and Doniach, S. (2000). Reconstruction of low-resolution three-dimensional density maps from one-dimensional small-angle X-ray solution scattering data for biomolecules. *J. Appl. Cryst.* **33,** 350–363.

CHAPTER 14

Structural Investigations into Microtubule-MAP Complexes

Andreas Hoenger* and Heinz Gross†

*Department of Molecular, Cellular and Developmental Biology
University of Colorado at Boulder
UCB-347, Boulder, Colorado, 80309

†EMEZ, Electron Microscopy Center
Swiss Federal Institute of Technology Zürich
8093 Zürich, Switzerland

Abstract

Microtubules interact with a large variety of factors commonly referred to as either molecular motors (kinesins, dyneins) or structural microtubule-associated proteins (MAPs). MAPs do not exhibit motor activity, but regulate microtubule dynamics and their interactions with molecular motors, and organelles such as kinetochores or centrosomes. Structural investigations into microtubule-kinesin motor complexes are quite advanced today and by helical three-dimensional (3-D) analysis reveal a resolution of the motor-tubulin interface at <1.0 nm. However, due to their flexible structure MAPs like tau or MAP2C cannot be visualized in the

0091-679X/08 $35.00
DOI: 10.1016/S0091-679X(07)84014-3

same straightforward manner. Helical averaging usually reveals only the location of strong binding sites while the overall structure of the MAP remains unsolved. Other MAPs such as EB1 bind very selectively only to some parts of the microtubule lattice such as the lattice seam. Thus, they do not reveal a stoichiometric tubulin:MAP-binding ratio that would allow for a quantitative helical 3-D analysis. Therefore, to get a better view on the structure of microtubule-MAP complexes we often used a strategy that combined cryo-electron microscopy and helical or tomographic 3-D analysis with freeze-drying and high-resolution unidirectional surface shadowing. 3-D analysis of ice-embedded specimens reveals their full 3-D volume. This relies either on a repetitive structure following a helical symmetry that can be used for averaging or suffers from the limited resolution that is currently achievable with cryotomography. Surface metal shadowing exclusively images surface-exposed features at very high contrast, adding highly valuable information to 2-D or 3-D data of vitrified structures.

I. Introduction

Structural biology and the capabilities of studying protein complexes in three dimensions evolved into an integral part of modern biology and biomedical research. Many researchers would like to know at great detail how an organism functions and to this end we must ultimately understand the molecular basis of supramolecular and cellular structures. While X-ray crystallography and high-field NMR spectroscopy clearly dominate the atomic resolution end of structural biology, the major thrust of cryo-electron microscopy (cryo-EM, Dubochet *et al.*, 1988) and cryo-electron tomography (cryo-ET, reviewed in Lucic *et al.*, 2005; McIntosh *et al.*, 2005; Steven and Aebi, 2003) provides the direct link to cell structure and function while trading some resolution with physiological relevance. Cryo-EM technology omits fixation and staining that opened the range of molecular EM to near atomic resolution, at least in theory. On the practical side, however, the real power of EM lies in reconstructing large macromolecular and cellular assemblies at a resolution range between 1 and 3 nm that can be subsequently docked with atomic resolution data from X-ray or electron crystallography and NMR (Volkmann and Hanein, 1999; Wriggers *et al.*, 1999; for examples on microtubule-motor structures see Nogales *et al.*, 1999; Hoenger *et al.*, 2000; Kikkawa *et al.*, 2001; Hirose *et al.*, 2006).

Microtubules are highly dynamic tubular structures composed of α,β-tubulin dimers that polymerize head-to-tail into so-called protofilaments under the control of guanosine triphosphate (GTP). The α,β-tubulin structure has been solved to near-atomic resolution by electron crystallography (Löwe *et al.*, 2001; Nogales *et al.*, 1998). So far, the best resolved structure of an intact microtubule has been reported by Li *et al.* (2002; see also Krebs *et al.*, 2004). *In vivo*, protofilaments associate laterally and form a hollow tube. Most commonly, microtubules found *in vivo* are composed of 13 protofilaments (see Fig. 1). These microtubules do not exhibit helical symmetry because they contain so-called seams that interrupt the lattice at one or more lateral contacts between protofilaments [visualized by Kikkawa *et al.* (1994) by freeze-fracturing (Heuser, 1989)] and [Sosa and Milligan (1996) by back-projection reconstruction (Bluemke *et al.*, 1988)]. At the seam, protofilaments interact laterally in a so-called A-lattice conformation while the rest of the tube forms a B-lattice (Amos and Klug 1974; Song and Mandelkow, 1993, 1995). *In vitro* it is possible to form tubulin polymers that contain between 9 and ~16 protofilaments (Chrétien *et al.*, 1996). This variability in protofilament number can be controlled to some extent by modifying the tubulin polymerization

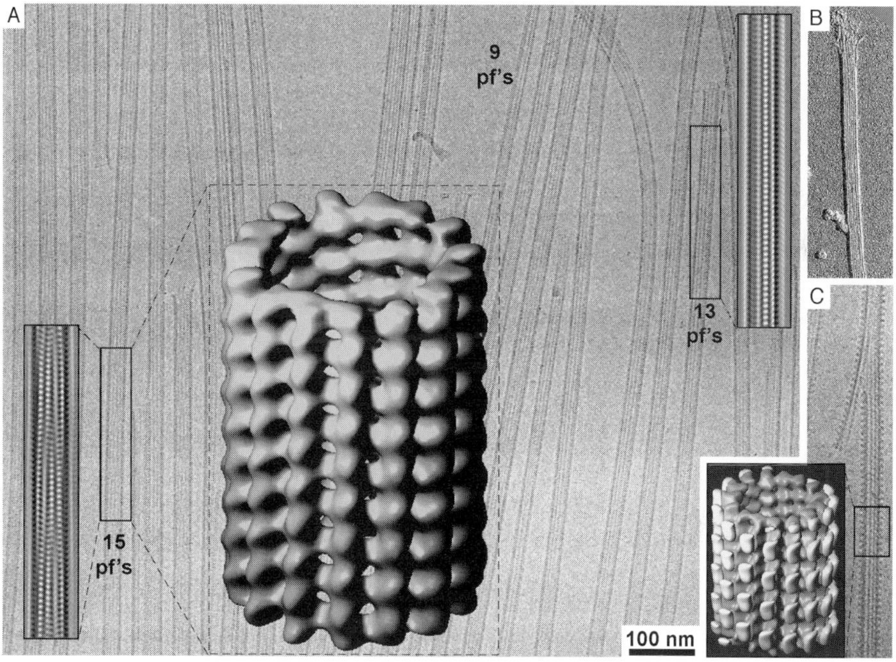

Fig. 1 Summary of cryo-EM techniques applied to microtubules and microtubule-kinesin motor head complexes: We are frequently using vitrification and either direct imaging in ice (frozen-hydrated specimens: A and C) or freeze-drying and subsequent metal shadowing (B). (A) Cryomicrograph of plain microtubules reveals the structural variability of *in vitro* polymerized microtubules. They may range in protofilament number from 9 to about 16 protofilaments. Insets show Fourier-filtered segments of a 13- and a 15-protofilament microtubule. 13 protofilaments is the most common version of microtubules *in vivo*, and the protofilaments run perpendicular to the tubular axis. However, the lattice of these microtubules contains a so-called seam that interrupts the helical symmetry in these tubes (see also Fig. 6). 15 protofilament microtubules exhibit a truly helical lattice and are used as templates for helical 3-D reconstructions of microtubules (blue volume in A) and microtubules complexed with motors (panel C) and other MAPs (see also Fig. 4D). (B) As an alternative cryomethod freeze-drying and metal shadowing is used to strongly enhance surface related features. Panel (B) shows a freeze-dried and shadowed plain microtubule as a direct comparison to vitrified and ice-embedded microtubules shown in panel (A). (See Plate no. 3 in the Color Plate Section.)

buffers. Arnal *et al.* (1996) were the first to generate 15-protofilament microtubules with a right-handed protofilament supertwist (see Fig. 1). These microtubules are truly helical structures and therefore suitable for three-dimensional (3-D) reconstruction based on helical averaging (see Fig. 2; DeRosier and Moore, 1970). Since their introduction, they have been widely used for many structural studies of motors and microtubule-associated proteins (MAPs) as they interact with microtubules (kinesins: e.g., see Hirose *et al.*, 2006; Hoenger *et al.*, 1998, 2000a; Kikkawa *et al.*, 2001; Sosa *et al.*, 1997; Wendt *et al.*, 2002; tau: Al-Bassam *et al.*, 2002; Kar *et al.*, 2003; Santarella *et al.*, 2004; Ncd-tail: Wendt *et al.*, 2003; dynein microtubule binding domain: Mizuno *et al.*, 2004).

In our labs we often complement our cryo-EM 3-D volume data with surface metal shadowing. Unlike frozen-hydrated preparations, surface metal shadowing does not reveal full 3-D data, but it does constitute a fast and reliable approach for direct investigations into the surface and other properties of complexes and large macromolecular assemblies such as (1) structural determination of inner and outer surfaces (e.g., see Hoenger *et al.*, 2000; Steinmetz *et al.*, 1998), (2) a preliminary image of the 3-D arrangement of an entire complex, and (3) a direct determination of symmetry related features, crucial for further 3-D reconstructions, such as the hand of helical assemblies (see Fig. 2). Surface shadowing has also been used very successfully when applied to freeze-fractured specimens, elucidating the surface properties of cellular organelles etc. after creation of a surface replica (Heuser, 1989). However, in this chapter we would like to concentrate on high-resolution approaches for surface shadowing and surface relief reconstruction which is achieved most successfully on regular assemblies such as two-dimensional (2-D) crystalline arrays or helical tubular structures (Fuchs *et al.*, 1995; Guckenberger, 1985; Smith and Kistler, 1977). Newer developments even reached subnanometer resolution (Walz *et al.*, 1996). Surface data may also be further combined with 3-D volume data to complete missing 3-D information in either of the reconstructions (Dimmeler *et al.*, 2001).

II. Rationale

Cryo-EM-derived 3-D data constitute the best possible link between atomic resolution structures and cellular organelles. In addition, cryo-EM and more recently cryo-ET allow reconstructing large marcromolecular and cellular complexes to a resolution of ca. 1–3 nm. Cryo-EM generally deals with samples that are embedded in a thin layer of vitrified ice. Vitrified ice is an ice form that can be generated by ultrarapid freezing that prevents the formation of ice crystals and hence, the ice remains amorphous. Vitrification warrants the preservation of molecular detail in large complexes to near-atomic dimensions and allows for structural and functional interpretations at atomic scale by docking individual components, solved by X-ray and electron crystallography or high-field NMR spectroscopy, into EM-derived 3-D envelopes. The true advantage of cryo-EM

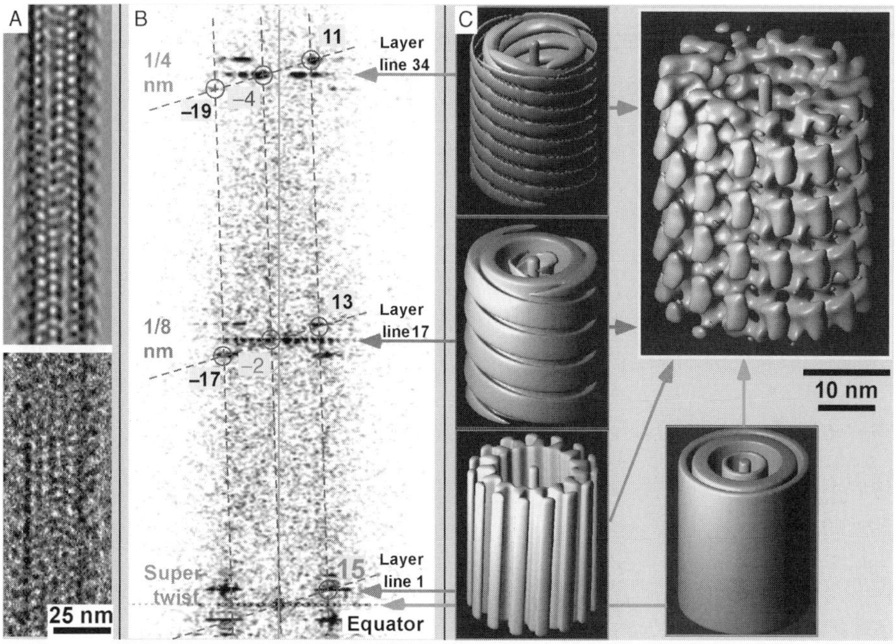

Fig. 2 Dissecting the helical lattice of a 15-protofilament microtubule decorated with kinesin motor domains: (A) Raw data and Fourier-filtered image of a 15-protofilament microtubule decorated with kinesin motor domains. (B) Diffraction pattern of a motor decorated 15-protofilament microtubule shows three distinct clusters of layer lines: Closest to the center lies the equator and layer line 1. The equator (green lines and green framed volume at the bottom in C) corresponds to the radial density distribution of the tube but does not form any helical pattern. Layer line 1 (orange lines and volume in C) marks the longest possible helical path in the lattice, and here this is the protofilament supertwist. Since there are 15 protofilaments following a right-handed helical path that layer line has a Bessel order of +15. The second cluster of strong layer lines forms around layer line 17 (red lines and frame in the middle of C) that marks the left-handed helical path of the motor domains roughly perpendicular to the supertwist. Each path is 8 nm apart (the length of the $\alpha\beta$-tubulin dimer) but there are two $\alpha\beta$-independent paths twisting around each other and therefore this pattern forms to a left-handed, two-start helix with the corresponding Bessel order −2. Finally, the layer line cluster at 1/4 nm is directly related to the 1/8 nm cluster (cyan lines and frame at the top in C) and corresponds to the α-β-α-β tubulin monomer repeat that is exactly half of the dimer repeat. Accordingly, the handedness is the same (left) and the number of starts is doubled to four (= >Bessel order −4). (C). Back-Fourier transform of the individual layer line data according to its helical parameters forms the volumes with continuous helical paths. Convolution of all the layer line data reveals the full structure in 3-D (yellow frame). (See Plate no. 4 in the Color Plate Section.)

and cryo-ET is that they may target large and flexible complexes that are beyond high-resolution methods, either because averaging is not applicable (which eliminates crystallographic, helical, and single-particle reconstructions) or particles are simply too large (often the case for NMR investigations). Helical 3-D reconstructions or other averaging-based approaches revealed (sub-) nanometer resolution on plain microtubules (Li *et al.*, 2002) or on microtubules complexed with motor domains (Hirose *et al.*, 2006; Kikkawa *et al.*, 2001; Krebs *et al.*, 2004) resulting in accurate molecular models with near-atomic detail.

The situation for helical averaging of MAP-microtubule complexes is quite a bit different with MAPs that show high structural flexibility and/or do not decorate the microtubule surface in a stoichiometric or otherwise regular fashion. In these cases quantitatively accurate 3-D maps are not achievable and the resulting maps have to be interpreted accordingly. In these cases simply collecting 3-D data by helical averaging (Fig. 2) is not sufficient. Tomographic reconstructions (e.g., see Nicastro *et al.*, 2006) and surface relief data often constitute valuable and complementary information for an accurate structural and functional interpretation. In this chapter, we demonstrate the power and limitations of helical averaging on motors (Fig. 2) and MAP-microtubule complexes on selected examples from our own work [tau: Fig. 4, hepatoma upregulated protein (HURP): Fig. 5, Mal3p: Fig. 6]. In all of these cases, only the combination of helical 3-D reconstruction of frozen-hydrated specimens and high-resolution surface metal shadowing (Fig. 3) revealed a clear picture on how these MAPs interact with microtubules.

III. Methods

A. Helical Reconstructions of MAP-Microtubule Complexes

The preparation and helical 3-D reconstruction of motor-microtubule complexes have been extensively described in Beuron and Hoenger (2001) and in Hoenger and Nicastro (2007). Therefore, I will only briefly repeat the basics. Helical arrays are either naturally occurring or they are sometimes artificially produced by repetitive arrangements of macromolecules or complexes of macromolecules: They are often of great use to structural biologists. Helical reconstruction approaches have yielded near-atomic resolution on either naturally occurring helices, such as bacterial flagella (Yonekura *et al.*, 2003), or artificially reconstituted ones, such as tubes formed by the acetylcholine receptor (Miyazawa *et al.*, 2003). Under certain conditions, microtubules may also form true helices, but often their lattices show one or more discontinuities by so-called seams (see Fig. 6). To avoid interference of helical symmetry, we use microtubules that are composed of 15 protofilaments. They can be easily generated by adding 5–10% of dimethyl sulfoxide (DMSO) to the polymerization buffer (Beuron and Hoenger, 2001).

Helical reconstruction methods combine features of tilt-series reconstruction from 2-D crystalline arrays with methods used for icosahedral and single-particle reconstruction (DeRosier and Moore, 1970). The regularity of the helical packing generates distinct layer-line patterns in Fourier space that can be analyzed by methods resembling those used for crystallography. These methods take advantage of the discrete layer line spacing that reflect the axial repeats of helical turns, which are analogous to the patterns of distinct spots generated by 2-D crystals. The rotation along a helical path projects the repeating building blocks at all the angles that are required for an isotropic 3-D reconstruction. In the real part of Fourier space (the visible diffraction pattern where amplitudes are transformed into intensities, Fig. 2B), this information is reflected in the appearance of continuous layer lines, running perpendicular to the helix axis, rather than rows of distinct spots, as in 2-D crystal diffraction patterns. A 2-D projection of a helix is a 1-D crystal that features repeating elements along the helical axis, but perpendicular to the axis the image features are continuous, so in Fourier space the information that corresponds to these features is not sampled in discrete spots. As demonstrated in Fig. 2, each layer line corresponds to a particular helical path that is reflected in the Bessel order. Left- and right-handedness are expressed in the prefix – and +, respectively. The number of simultaneous starts is reflected in the Bessel order number (e.g., −2 describes a left-handed 2-start helix). Reaching isotropic resolution (i.e., same resolution in all directions) requires projecting an image element at small intervals between 0° and 180°. Hence, the projected image of a helical array that includes at least one-half a helical turn is theoretically sufficient for two independent isotropically resolved (equivalent resolution in all directions of the helical repeat) helical 3-D reconstructions. In practice, however, the low contrast in cryo-EM micrographs requires that several helical repeats must be averaged together to increase the signal-to-noise ratio to a useful level. A significant signal-to-noise ratio improvement is typically achieved with around 100 repeats. In an ideal world the signal-to noise ratio increases by the square root of units averaged (i.e., with 100 repeats = >tenfold increase of signal-to-noise ratio). However, one helical repeat may include many asymmetric units which here is one tubulin dimer plus a bound motor or MAP. A reconstruction of a microtubule-motor complex that reaches 2 nm resolution typically includes about 40,000–50,000 such asymmetric units.

B. High-Resolution Surface Shadowing

Surface imaging and reconstruction of heavy metal-shadowed specimens have a long tradition in transmission electron microscopy (reviewed in Gross, 1987). To this end a fine layer of metal is applied, either rotationally or unidirectionally, to the specimens, modulating the surface topography. Conventional techniques employ various coating materials such as platinum/carbon (Pt/C) or tantalum/tungsten (Ta/W) and make use of an additional carbon layer applied subsequently to the shadowing process. This layer of carbon prevents collapsing of a freeze-dried structure when warmed up at room temperature, and avoids oxidation of air and humidity. This layer is required to protect the sample during the transfer through the atmosphere into the microscope. The imaging is performed at room temperature and under high-dose conditions. These procedures had led to surface reproductions in a range of 3 nm lateral resolution. Gross *et al.* (1990) at the ETH-Zürich (Switzerland) constructed the Midilab, a high-vacuum/cryo-transfer system by which samples can be transferred from the preparation chamber into the microscope without changing the ambient pressure (high-vacuum) and temperature conditions. Consequently, the protecting carbon layer can be avoided and it could be shown that an increase in resolution is gained. The Midilab is directly hooked up to a CM-12 electron microscope (Fig. 3A). Furthermore, by imaging freeze-dried, unstained, and Ta/W-stained samples at different electron dosages they could show that the preservation of the material that supports the metal coat attains high relevance for an improved reproduction of the metal layer during imaging. Therefore, optimal prerequisites for the imaging of shadowed biological molecules are an omission of the carbon layer and image recording under low-dose and cryo conditions. With these techniques, in combination with image averaging (averaging out the granularity of the ~0.5-nm-thin Ta/W films =>"optimal granularity approach"), surface reconstructions of up to 0.8 nm lateral and 0.4 nm axial resolution can be obtained for 2-D crystalline objects (Walz *et al.*, 1996). Comparisons of freeze-dried, stained and freeze-dried, unstained samples have shown identical resolution limits, confirming that the resolution of metal-shadowed specimens is mainly affected by the object preservation and less by the quality of the metal film.

The Midilab and its cryo-transfer system allowed for Ta/W shadowing without the need for carbon backing in a unidirectional mode at an elevation angle of 45°. This approach is creating the illusion of having shadows in the images and giving them a pseudo-3-D) appearance (see Figs. 3–6), which can be very helpful for a preliminary assessment of a complex structure.

Figures 3–6 show several examples of specimens created with the Midilab preparation chamber, which illustrate the power of high contrast and signal-to-noise ratio, achieved by unidirectional Ta/W metal shadowing. We call this kind of approach used here an "optimal visibility approach," as opposed to an "optimal granularity approach" used for high-resolution surface topography reconstruction where subnanometer resolution was achieved (e.g., see Walz *et al.*, 1996). The preparations shown here typically diffract to about 2 nm or even beyond that, but the main issue here was the details visible without using further image enhancement methods such as averaging over image elements or imposing helical or crystallographic symmetry.

Technical procedures: Samples were adsorbed for 30–60 sec onto carbon-coated copper grids (400 mesh), briefly washed, and subsequently quick-frozen in liquid nitrogen or liquid ethane. After transferring the grids into the evaporation (preparation) chamber, the samples were freeze-drying at −80 °C, pressure $\leq 10^{-5}$ Pa for ~90 min in close proximity to a shutter kept at −150 °C and acting as a cryopump. After freeze-drying the specimens were kept at −80° for shadowing with Ta/W at an elevation angle of 45° to a thickness between 0.5 (optimal granularity) and 1.0 nm (optimal visibility). After shadowing, grids were then transferred under high-vacuum and cryo conditions into the microscope onto a modified Gatan cryoholder and imaged at −170 °C.

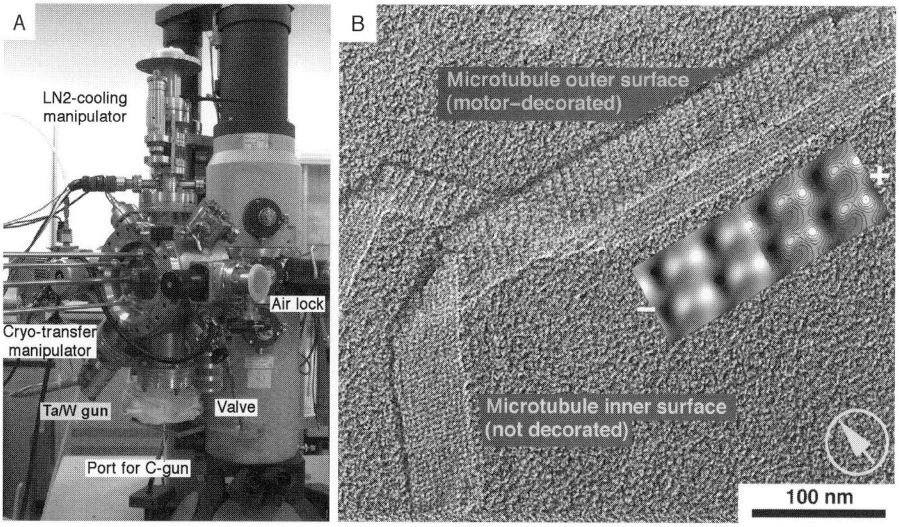

Fig. 3 High-resolution metal surface shadowing and cryo-vacuum transfer unit at the ETH-Zürich, Switzerland (Gross *et al.*, 1990). A second, modernized unit of this kind is in planning for the Boulder 3-D lab. (A) All the metal shadowing data shown in this chapter have been prepared by the "Midilab" shadowing unit that is directly mounted to a vacuum port of an FEI-CM12 electron microscope. This setup allows for unidirectional shadowing of specimens with Ta/W. The specimens are cooled with liquid nitrogen and typically kept at −80 °C during the freeze-drying process. The unit features a metal gun port at an elevation angle of 45°, and an optional carbon-gun port at a 90° angle to the specimen. The specimens are quick-frozen in liquid nitrogen or ethane, mounted on a transfer table, and inserted via the airlock into the shadowing chamber in an upside-down position. After drying and shadowing the specimens are directly transferred into the microscope by the manipulator. (B) Unidirectionally shadowed tubulin sheet decorated with kinesin motor domains. The shadowing technique reveals the two differently structured surfaces of the inner and outer microtubule wall. The outer surface shows a roughly perpendicular 8-nm striation caused by the stoichiometric binding of one motor head per tubulin dimer. Inner surfaces do not bind motors and therefore reveal the 4-nm α-β-α-β tubulin pattern.

C. 3-D Analysis of Microtubules Complexed with Tau

Tau proteins, a family of six isoforms generated by alternative splicing in the human central nervous system, are prominent in neurons and promote the outgrowth of axons. Tau may also aggregate into pathologically relevant "paired helical filaments" (PHFs) that are hallmarks of Alzheimer's disease and other dementias (Lee *et al.*, 2001). Tau tightly regulates microtubule-based traffic in axons (Ebneth *et al.*, 1998; Seitz *et al.*, 2002). Tau is a "natively unfolded" protein that does not have a well-defined shape, as judged by structural, spectroscopic, and biochemical evidence (Schweers *et al.*, 1994). It can be subdivided into two major domains, the N-terminal "projection domain" and the C-terminal "assembly domain" which binds and stabilizes microtubules. The assembly domain contains three or four pseudorepeats of ~31 residues important for microtubule binding, flanked by proline-rich regions which enhance microtubile binding. The absence of

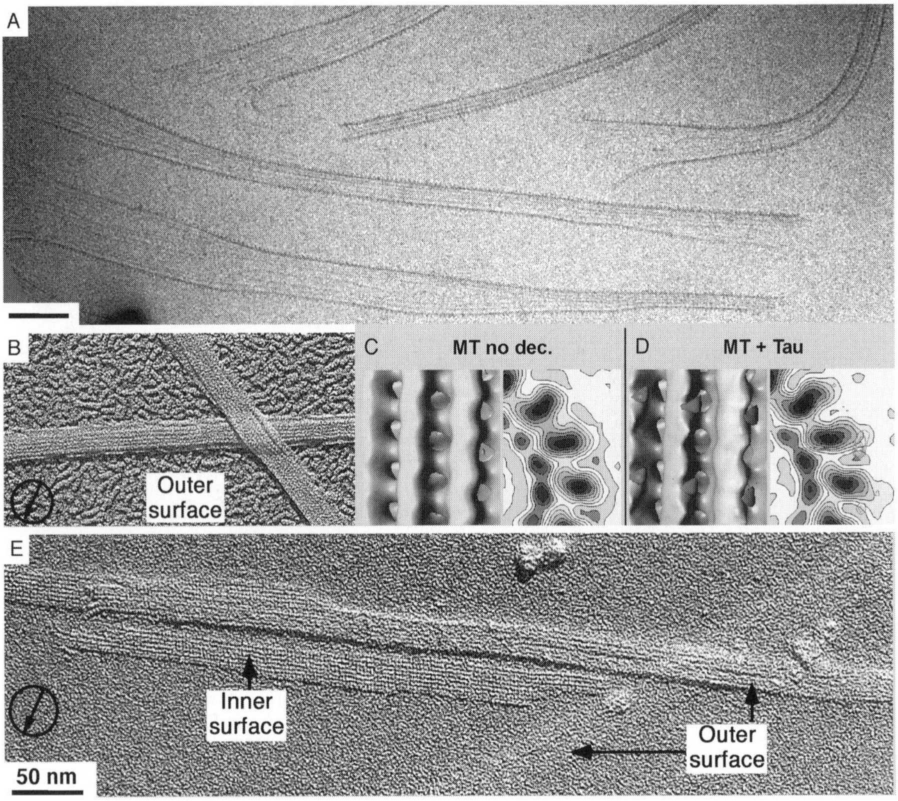

Fig. 4 Cryo-EM on microtubule-tau complexes: tau is a very unstructured molecule, and 3-D image reconstruction reveals more qualitative results such as identifying the microtubule-binding site, than quantitative maps of the entire molecule. (A) Codecoration of microtubules with tau and kinesin motor domains under depolymerizing conditions reveals two complementary effects. Kinesin stabilizes the axial protofilament contacts (Krebs *et al.*, 2004) while tau links the protofilaments laterally. Hence, long filamentous assemblies form that resemble microtubules with loosened lateral protofilament contacts that are still held together by the tau molecule (Santarella *et al.*, 2004). (B and E) Plain microtubules (B) and microtubule-tau complexes imaged after freeze-drying and unidirectional metal shadowing. Plain microtubules show an intrinsic instability that is reflected in the background of tubulin oligomers next to intact tubes. In the presence of tau these oligomers disappear, indicating a stronger stabilization effect of tau on microtubules (tau also stabilizes oligomers vs dimer). The tau molecule itself is only visible as a diffuse mass randomly overlaying the protofilament structure. Tau does not bind to the microtubule inner surface and therefore these surfaces are unobstructed and reveal the tubulin pattern seen from the inner side of a microtubule. (C and D) Difference mapping between helical 3-D reconstructions of ice-embedded plain microtubules (C) and tau-decorated microtubules reveal the strong binding site of tau to microtubules (red volumes in D). According to our studies in Santarella *et al.* (2004) these binding sites are on α-tubulin and face the opposite site of a kinesin motor binding site. Hence, kinesin and tau may bind simultaneously to tubulin protofilaments. (Adapted from Santarella *et al.*, 2004). (See Plate no. 5 in the Color Plate Section.)

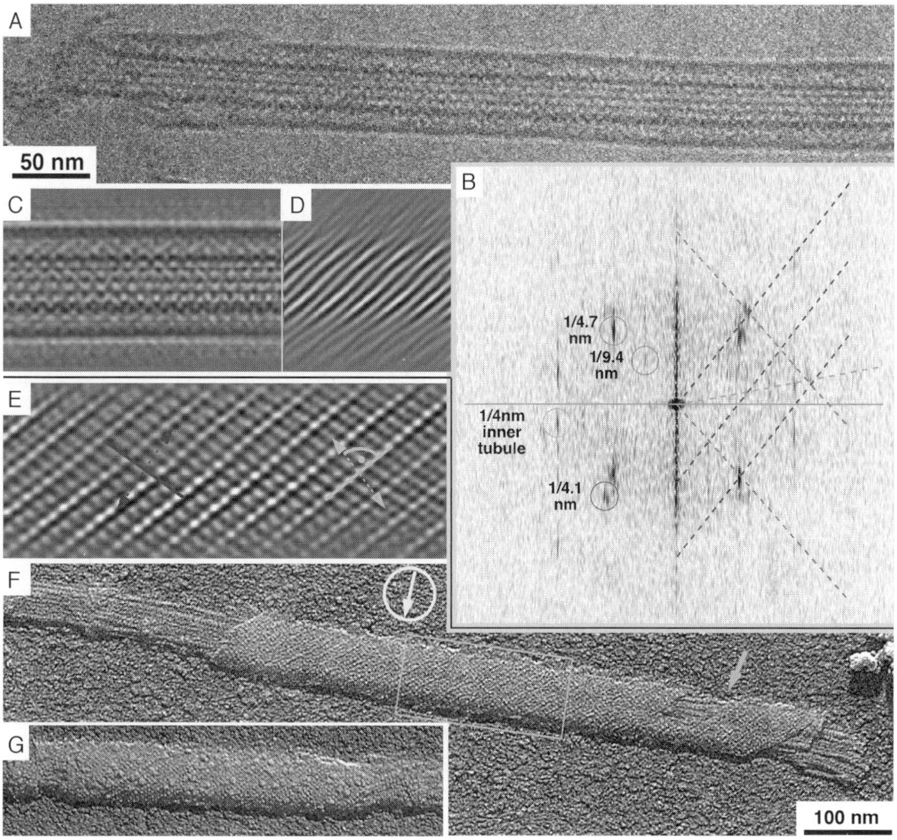

Fig. 5 Induction of a novel tubulin sheet conformation by HURP (HURP: Koffa *et al.*, 2006): (A) HURP induces a tubulin sheet with antiparallel P2-symmetry protofilament arrangement that wraps itself around an intact microtubule (Santarella *et al.*, 2007). Cryo-EM reveals the diameter of the wrapping sheet to be about 50 nm in diameter. (B) The calculated optical diffraction pattern of these complexes reveal both layer lines, the regular tubulin layer line at 1/4 nm (orange) and the reflections from the wrapping sheets at a ~47° angle at 1/9.4 nm (red) perpendicular to the antiparallel protofilaments. Along the wrapping protofilaments the α-β-α-β tubulin repeat is visible as a reflection at 1/4.1 nm (blue). (C) Fourier filtering including all visible reflections. (D) Fourier filtering including only the reflections marked in red, emphasizing the antiparallel nature of the protofilaments in the wrapping sheet. (E) Fourier filtering including all visible reflections originating from the wrapping sheet (red and blue) adding the tubulin repeats to the image obtained in (D and F): Unidirectional surface metal shadowing revealed the handedness of the wrapping sheet, indicated the orientation of the protofilaments in the sheets. This is particularly well visible on a broken piece of the sheet (arrow) revealing the underlying structure of the wrapped microtubule. (G) Kinesin motor heads were used to determine the arrangement of the antiparallel protofilaments with regard to exposing inner or outer microtubule faces. The binding efficiency of kinesin motor domains to the wrapping sheet was surprising low, but revealed unambiguously that all protofilament expose their outer (kinesin-binding) surface. (Adapted from Santarella *et al.*, 2007). (See Plate no. 6 in the Color Plate Section.)

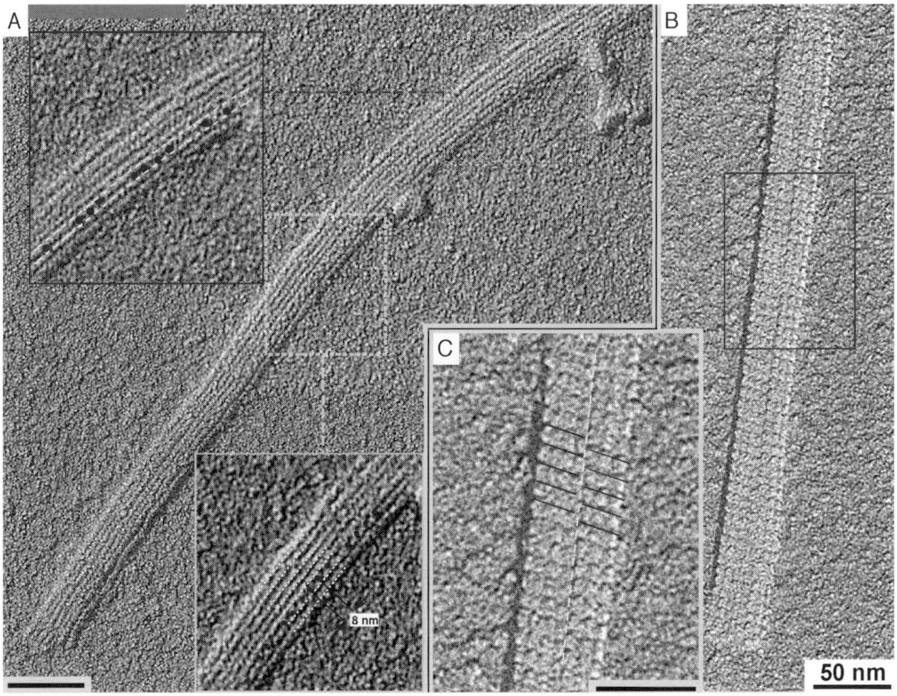

Fig. 6 Binding properties of the *Schizosaccharomyces pombe* EB1 homologue Mal3p to microtubules (Sandblad *et al.*, 2006). Due to the small size of the globular domains of Mal3p the findings shown here were only visible with high-resolution metal shadowing and could not be reproduced with helical or tomographic 3-D analysis on ice-embedded specimens. (A) The Mal3p particles bind along the microtubule axis just to one single groove but leave the rest of the surface free. In Sandblad *et al.* (2006) we could demonstrate that this groove represents the microtubule lattice seam, a typical feature of 13- and 14-protofilament microtubules. Mal3p clearly stabilizes the seams, and after codecoration with kinesin motor domains seams were clearly visible on the surface of microtubule-Mal3p-kinesin complexes. Panel (C) emphasizes the axial 8 nm repeat of the motor units, and the 4 nm stagger at one position on the lattice (red lines). The seam itself is marked with a dotted yellow line. (Adapted from Sandblad *et al.*, 2006). (See Plate no. 7 in the Color Plate Section.)

compact folding has precluded a detailed structural analysis so far. Three independent groups (including our own) investigated frozen-hydrated microtubule-tau complexes by helical analysis with mostly consistent results (Al-Bassam *et al.*, 2002; Kar *et al.*, 2003; Santarella *et al.*, 2004). However, due to the flexible nature of tau none of the 3-D structures published was able to outline the full extent of the entire molecule, but revealed more qualitative results such as the interaction sites between tau and tubulin (see Fig. 4C and D). Thus, averaging-based 3-D analysis misses large portions of the molecule as they arrange irregularly in space. Cryotomography produces 3-D without the need for averaging but nevertheless, the technique is equally unsuited as the currently achievable resolution of this method prevents visualizing flexible random coil structures.

However, averaging and 3-D structure determination is not the only way to gain insight into the structure and function of a protein such as tau. Decoration of tau onto preformed microtubule-kinesin motor domain complexes at stoichiometric binding conditions revealed two structurally important aspects: (1) tau and kinesin can coexist on the microtubule lattice, and (2) tau binding may extend laterally over several protofilaments. This experiment took advantage over the axial protofilament stabilization of kinesin while lateral contacts are less affected (e.g., see Hoenger *et al.*, 2000). Figure 4A nicely demonstrates that under depolymerizing conditions microtubules fall apart but not into little oligomers as they would do in the absence of any stabilizing factors (Fig. 2B: tubulin oligomers cover the entire background, visualized with surface shadowing). In the presence of kinesin alone protofilaments release laterally but extend much longer than the oligomers shown in Fig. 4B (see Hoenger *et al.*, 2000; Krebs *et al.*, 2004). Tau and kinesin together now stabilize the protofilaments axially and in addition loosely link the protofilaments laterally (see Fig. 4A). This action is clearly provided by tau and demonstrates the protofilament cross-linking potential of tau.

High-resolution unidirectional surface shadowing with Ta/W (Fig. 4B and E) confirmed the results obtained on frozen-hydrated specimens (Fig. 4A). While in the absence of tau the microtubule surface shows clearly the outer structure of protofilaments that mostly form an axially oriented continuous rim (Fig. 4B). In the presence of tau the surface appears obstructed by overlaying mass somewhat resembling a landscape after snowfall (Fig. 4E, outer surface). Hence, individual molecules are not discernible, but the cross-linking nature observed in Fig. 4A can be confirmed. Part of the upper left microtubule broke off during preparation and now reveals the inner surface of a microtubule wall (see also Fig. 3B). This surface is different from the outer surface and shows an unobstructed view onto the 4 nm tubulin repeat that is much better visible on inner rather than outer surfaces (Fig. 4B). Despite the presence of ample amounts of tau in this preparation, the inner surface shows no binding of tau.

D. 3-D Analysis of a Novel Tubulin Conformation Induced by HURP

The so-called "hepatoma upregulated protein" (HURP) (Koffa *et al.*, 2006; Tsou *et al.*, 2003) is a newly discovered MAP required for correct spindle formation both *in vitro* and *in vivo*. HURP is part of a multicomponent complex required for the transition of RanGTP-induced microtubule aster-like structures into polar, spindle-like structures (Koffa *et al.*, 2006). HURP has also been shown to play an important role in stabilizing microtubules *in vivo*, specifically the kinetochore microtubules within the mitotic spindle apparatus (Koffa *et al.*, 2006; Sillje *et al.*, 2006).

The isolated HURP molecule shows a similarly unstructured and random folding pattern as tau (see above). There are essentially no globular elements and very few secondary structure elements. Hence, the full extent of the protein is again very difficult to be visualized directly by cryo-EM on frozen-hydrated or metal-shadowed specimens (Santarella *et al.*, 2007) or by any other structure determination methods. However, this MAP shows a very peculiar specialty: it induces the formation of a tubulin sheet with antiparallel protofilaments, exhibiting P2 symmetry. This sheet then, either after it has been preformed or during formation, wraps around an intact microtubule in a very reproducible way (Fig. 5). Hence, what is visible in Fig. 5 is not the HURP itself, but the wrapping tubulin sheet. According to gold-labeling experiments (see Santarella *et al.*, 2007), the HURP molecule is located in the space between inner microtubule and the wrapping sheet.

The combined data from frozen-hydrated (Fig. 5A) and freeze-dried/metal-shadowed specimens (Fig. 5F) allowed creating a complete picture of this assembly. Frozen-hydrated data would allow separating the inner from the outer tube by selectively collecting the corresponding layer line data (Fig. 5B) as long as there are no superimposed layer lines. The Fourier-filtered image in Fig. 5D was created by a back transform on the layer line data marked with a red circle in Fig. 5B. Unfortunately, the outer sheet was too flexible for a meaningful helical 3-D reconstruction.

However, even without actually calculating a 3-D structure of the wrapping sheet, we got a complete picture of its configuration. The metal-shadowing data revealed the geometry and the symmetry of these sheets. Metal shadowing unambiguously revealed the helical hand of the wrapping sheet as left-handed (Fig. 5F), a property that is invisible in frozen-hydrated preparations (Fig. 5A and C). If only frozen-hydrated data were available, Fig. 5D could have been the near or the far side (toward the observer and away from the observer, respectively). With the knowledge about the handedness we now know this is the near side analogous to the shadowed surface shown in Fig. 5E. The initial assumption about the direction of protofilaments is supported by shadowed preparations where the wrapping sheet partially broke off (arrow in Fig. 5F), and at the ends of the sheets.

The actual packing of the protofilaments was tested by adding molecular motors and analyzing their binding patterns on the surface of the sheets (Fig. 5G). Both freeze-dried/metal-shadowed and frozen-hydrated preparations revealed the antiparallel nature of the protofilaments (Fig. 5D and E). However, neither of the preparations could clearly determine if the wrapping sheet adopted a Zn-sheet configuration (Wolf *et al.*, 1993) with P21 symmetry (alternating inner and outer surfaces exposed) or if it would exhibit P2 symmetry (either inner or outer surfaces exposed). To this end we used kinesin motor domains as a marker of the outer protofilament surface. Surprisingly, the affinity of motors to the wrapping sheet was much lower than to intact microtubules, but nevertheless, the presence of motors on every protofilament, and not only on every other one, made clear that all protofilaments expose their outer motor-binding surface. Interestingly, in this case the protofilaments bend in the opposite direction as the oligomeric products from depolymerizing microtubules. The bending of the protofilaments in HURP-induced sheets is also opposite to the artificial tubes produced by Wang and Nogales (2005).

E. 3-D Analysis of a MAP That Locates to a Highly Specific Structure on the Microtubule Surface

This section and Fig. 6 describe and illustrate a situation in which helical 3-D reconstruction of frozen-hydrated MAP-microtubule complexes failed completely and only the metal shadowing approach visualized the binding properties of this MAP. In this example we investigated Mal3p (Beinhauer *et al.*, 1997), a fission yeast homologue of the human EB1 (Su *et al.*, 1995). Like its human counterpart, Mal3p is known to accumulate at growing microtubule plus-ends and also to interact with the microtubule lattice (Busch and Brunner, 2004). The protein promotes the initiation and maintenance of microtubule growth (Busch and Brunner, 2004), recruits the kinesin Tea2p to the microtubules, and promotes Tea2p motor activity (Browning *et al.*, 2003). Mal3p consists of a globular calponin-homology (CH) domain (Hayashi and Ikura, 2003) and a-helical tail capable of coiled-coil formation (Honnappa *et al.*, 2005). Hence, the molecule functions in a dimeric form, which is also the proposed functional unit for the mammalian EB1 homologues.

Our first attempts to create helical 3-D maps of 15-protofilament microtubules decorated with Mal3p failed badly, and at a later stage in our work it became very clear why they failed. The full-length, dimeric Mal3p complex is capable of selectively binding to the microtubule lattice seam (Fig. 6) while otherwise leaving the surface of microtubules free for other factors to bind (Sandblad *et al.*, 2006). The lattice seam is a lateral contact between microtubules that forms according to a so-called A-lattice interaction while the reminder of protofilaments in a microtubule interact through B-lattice interactions (Amos and Klug, 1974). Mal3p is capable of codecorating the microtubule surface with kinesin motor domains that generate a clear view of the seam (Fig. 6B and C). According to our findings Mal3p strongly stabilizes the seam that otherwise represents the weakest lateral contacts in the microtubule lattice, which is the main reason why there are rarely multiple seams and why the A-lattice does not occur. Hence, there are two reasons why a helical averaging approach did not work: First, the density of Mal3p on the microtubule surface is way below a stoichiometric ratio to tubulin dimers, and second, the presence of a seam breaks the helical symmetry. Vice versa, the 15-protofilament microtubules that we have to use for helical averaging because of symmetry requirements do not have a seam and therefore show a very low binding affinity for Mal3p.

IV. Discussion

One of the main purposes of this chapter is to illustrate the power of a combined structural approach to the 3-D structure of microtubules complexed with MAPs and motor domains. While in the case of kinesin motor domains helical 3-D reconstruction alone may reveal qualitative and quantitative volume maps where the tubulin portion and the attached molecular motors may both be visualized at their actual size and shape. These 3-D maps are often good enough to reveal nucleotide-specific conformational states (e.g., see Al-Bassam *et al.*, 2002; Hirose *et al.*, 2006; Hoenger *et al.*, 1998, 2000; Kikkawa *et al.*, 2001; Krzysiak *et al.*, 2006; Skiniotis *et al.*, 2003; Sosa *et al.*, 1997; Wendt *et al.*, 2002). Accordingly, these volume data are most suitable for molecular docking attempts using atomic resolution data of the individual components such as crystallographic data on motor heads (e.g., see Kozielski *et al.*, 1997; Kull *et al.*, 1996; Sablin *et al.*, 1996, 1998; Turner *et al.*, 2001) and tubulin dimer (Nogales *et al.*, 1998). However, that is typically applicable only to relatively small globular domains like kinesin head domains and may not be feasible for flexible MAPs and MAPs with very distinct binding patterns. In these cases helical reconstruction may fail to accurately reproduce the molecular shape and binding patterns. Even microtubules complexed with dimeric motor constructs may suffer from a lack of symmetry (e.g., compare Hoenger *et al.*, 2000 and Krzysiak *et al.*, 2006 with Wendt *et al.*, 2002). Furthermore, MAPs like tau protein or MAP2C have been analyzed by helical methods (Al-Bassam *et al.*, 2002; Kar *et al.*, 2003; Santarella *et al.*, 2004; see also Fig. 4) and produced very valuable qualitative data about binding patterns, but the entire molecules could not visualize in their full extent, as they did not fully adopt the underlying helical symmetry of a 15-protofilament microtubule (see Fig. 2). This chapter provides guidance on how these problems could be addressed.

Rarely is one imaging method self-sufficient and therefore we would like to place a strong argument for a comprehensive methodic approach that combines 3-D volume reconstructions of macromolecular assemblies with high-resolution unidirectional metal surface shadowing for two reasons: (1) by reducing image data to the surface of a macromolecular complex all underlying structures are eliminated and do not obstruct the surface pattern and reveal a high-contrast pseudo-3-D image resembling a late-afternoon aerial photograph of a landscape on a sunny day. In contrast, images of frozen-hydrated samples are a 2-D reduction of a full 3-D volume, somewhat comparable to 2-D projections of a glassy, semitransparent object where near- and far-side density information merges together and could only be deconvoluted again by a series of different projections (tomography or averaging-based 3-D reconstruction methods). (2) As demonstrated here with three examples on microtubule-MAP complexes, surface metal shadowing generates very valuable additional 3-D structural data that would otherwise remain invisible. For example the handedness of a helical structure is directly visible (see Figs. 3B and 5F). 3-D surface and volume data can be combined for further structural interpretations (Dimmeler *et al.*, 2001). Although metal shadowing has been around for many years, particularly also on freeze-fractured samples (Heuser, 1989), little progress has been made over the last two decades and the value of surface shadowing for 3-D analysis of macromolecular assemblies has been underused. The construction of the Midilab shadowing unit (Fig. 3A) constitutes an important exception and added a new dimension to metal shadowing and produced unique surface data at resolutions in the subnanometer range (Walz *et al.*, 1996). This was achieved on 2-D crystalline arrays that were suitable for averaging and surface relief reconstruction (Fuchs *et al.*, 1995; Guckenberger, 1985;

Smith and Kistler, 1977). However, as demonstrated here, even without averaging, the Midilab with its high-resolution Ta/W gun and high-vacuum/cryo-transfer system produced high-contrast surface data with remarkable detail, directly interpretable without further computational image enhancement approaches. Most of the details shown here are currently beyond the resolving power of cryo-ET (Lucic *et al.*, 2005; McIntosh *et al.*, 2005; Steven and Aebi, 2003) that may not reach beyond 2 nm resolution for some more time due to technical and physical limitations. Therefore, we are convinced that surface metal shadowing will remain an important component in 3-D structural investigations of macromolecular and cellular specimens and should be considered for flexible and filamentous structures that are difficult to assemble in a regular array such as a helical complex or a 2-D crystalline array. Hence, while the capabilities of averaging based 3-D reconstruction methods of macromolecular complexes are widely recognized and enjoy a high amount of visibility in molecular and cellular biology, metal shadowing is currently somewhat underrated. However, this could change in the near future as the combination of high-resolution metal layers such as Ta/W and high-vacuum cryo-transfer system clearly have the potential to revive shadowing for structural biology and add a valuable tool to the existing approaches for 3-D investigations of biological specimens. We are now in the process of installing a cryo-vacuum transfer system for shadowed specimens in the NCRR facility at the University of Colorado at Boulder. While the old unit in Zürich only allows for elevation angles of 45° and unidirectional shadowing, we would like to add more flexibility to the elevation angle and add the option for rotational shadowing.

Acknowledgments

We thank all the key authors involved in producing the data shown here [tau: Rachel Santarella (EMBL-Heidelberg, Germany) and Eva and Eckhard Mandelkow (Max Planck Unit for Struct. Biology, DESY-Hamburg, Germany; HURP: Rachel Santarella (EMBL-Heidelberg, Germany) and Maria Koffa (Democritus University of Thrace, Alexandroupolis, Greece); Mal3p: Linda Sandblad and Damian Brunner (both EMBL-Heidelberg, Germany); microtubule images: Kenneth N. Goldie]. We very grateful to Peter Tittmann (ETH-Zürich, Switzerland) for operating the Midilab unit.

References

Al-Bassam, J., Ozer, R. S., Safer, D., Halpain, S., and Milligan, R. A. (2002). MAP2 and tau bind longitudinally along the outer ridges of microtubule protofilaments. *J. Cell Biol.* **157,** 1187–1196.

Amos, L. A., and Klug, A. (1974). Arrangement of subunits in flagellar microtubules. *J. Cell Sci.* **14,** 523–549.

Arnal, I., Metoz, F., DeBonis, S., and Wade, R. H. (1996). Three-dimensional structure of functional motor proteins on microtubules. *Curr. Biol.* **6,** 1265–1270.

Beinhauer, J. D., Hagan, I. M., Hegemann, J. H., and Fleig, U. (1997). Mal3, the fission yeast homologue of the human APC-interacting protein EB-1 is required for microtubule integrity and the maintenance of cell form. *J. Cell Biol.* **139,** 717–728.

Beuron, F., and Hoenger, A. (2001)."Structural Analysis of the Microtubule-Kinesin Complex by Cryo-Electron Microscopy," Vol. 164. Humana Press Inc., Totowa, NJ.

Bluemke, D. A., Carragher, B., and Josephs, R. (1988). The reconstruction of helical particles with variable pitch. *Ultramicroscopy* **26,** 255–270.

Browning, H., Hackney, D. D., and Nurse, P. (2003). Targeted movement of cell end factors in fission yeast. *Nat. Cell Biol.* **5,** 812–818.

Busch, K. E., and Brunner, D. (2004). The microtubule plus end-tracking proteins mal3p and tip1p cooperate for cell-end targeting of interphase microtubules. *Curr. Biol.* **14,** 548–559.

Chrétien, D., Kenney, J. M., Fuller, S. D., and Wade, R. H. (1996). Determination of microtubule polarity by cryo-electron microscopy. *Structure* **4,** 1031–1040.

DeRosier, D., and Moore, P. B. (1970). Reconstruction of three-dimensional images from electron micrographs of structures with helical symmetry. *J. Mol. Biol.* **52,** 355–369.

Dimmeler, E., Marabini, R., Tittmann, P., and Gross, H. (2001). Correlation of topographic surface and volume data from three-dimensional microscopy. *J. Struct. Biol.* **136,** 20–29.

Dubochet, J., Adrian, M., Chang, J. J., Homo, J. C., Lepault, J., McDowall, A. W., and Schultz, P. (1988). Cryo-electron microscopy of vitrified specimens. *Q. Rev. Biophys.* **21,** 129–228.

Ebneth, A., Godemann, R., Stamer, K., Illenberger, S., Trinczek, B., and Mandelkow, E. (1998). Overexpression of tau protein inhibits kinesin-dependent trafficking of vesicles, mitochondria, and endoplasmic reticulum: Implications for Alzheimer's disease. *J. Cell Biol.* **143,** 777–794.

Fuchs, K. H., Tittmann, P., Krusche, K., and Gross, H. (1995). Reconstruction and representation of surface data from two-dimensional crystalline, biological macromolecules. *Bioimaging* **3,** 15–24.

Gross, H. (1987). High resolution metal replication of freeze-dried specimens. *In* "Cryotechniques in Biological Electron Microscopy" (R. A. Steinbrecht and K. Zierold, eds.), pp. 205–215. Springer-Verlag Berlin, Heidelberg, New York, London, Paris, Tokyo.

Gross, H., Krusche, K., and Tittmann, P. (1990). Recent progress in high resolution shadowing for biological samples. *In* "Proceedings of the XIIth International Congress for Electron Microscopy" (L. D Peachey, and D. B Williams, eds.) San Francisco Press, Inc., San Francisco.

Guckenberger, R. (1985). Surface reliefs derived from heavy-metal-shadowed specimens—Fourier space techniques applied to periodic objects. *Ultramicroscopy* **16,** 287–304.

Hayashi, I., and Ikura, M. (2003). Crystal structure of the amino-terminal microtubule-binding domain of end-binding protein 1 (EB1). *J. Biol. Chem.* **278,** 36430–36434.

Heuser, J. E. (1989). Protocol for 3-D visualization of molecules on mica via the quick-freeze, deep-etch technique. *J. Electron Microsc. Tech.* **13**(3), 244–263.

Hirose, K., Akimaru, E., Akiba, T., Endow, S. A., and Amos, L. A. (2006). Large conformational changes in a kinesin motor catalyzed by interaction with microtubules. *Mol. Cell* **23**(6), 913–923.

Hoenger, A., and Nicastro, D. (2007). Electron microscopy of microtubule-based cytoskeletal machinery. *J. Methods Cell Biol.* **79,** 437–462.

Hoenger, A., Sack, S., Thormählen, M., Marx, A., Muller, J., Gross, H., and Mandelkow, E. (1998). Image reconstructions of microtubules decorated with monomeric and dimeric kinesins: Comparison with X-ray structure and implications for motility. *J. Cell Biol.* **141,** 419–430.

Hoenger, A., Thormählen, M., Diaz-Avalos, R., Doerhoefer, M., Goldie, K. N., Muller, J., and Mandelkow, E. (2000). A new look at the microtubule binding patterns of dimeric kinesins. *J. Mol. Biol.* **297,** 1087–1103.

Honnappa, S., John, C. M., Kostrewa, D., Winkler, F. K., and Steinmetz, M. O. (2005). Structural insights into the EB1-APC interaction. *EMBO J.* **24,** 261–269.

Kar, S., Fan, J., Smith, M. J., Goedert, M., and Amos, L. A. (2003). Repeat motifs of tau bind to the insides of microtubules in the absence of taxol. *EMBO J.* **22,** 70–77.

Kikkawa, M., Ishikawa, T., Nakata, T., Wakabayashi, T., and Hirokawa, N. (1994). Direct visualization of the microtubule lattice seam both *in vitro* and *in vivo*. *J. Cell Biol.* **127,** 1965–1971.

Kikkawa, M., Sablin, E. P., Okada, Y., Yajima, H., Fletterick, R. J., and Hirokawa, N. (2001). Switch-based mechanism of kinesin motors. *Nature* **411,** 439–445.

Koffa, M. D., Casabova, C. M., Santarella, R., Kocher, T., Wilm, M., and Mattaj, I. W. (2006). HURP is part of a Ran-dependent complex involved in spindle formation. *Curr. Biol.* **16,** 743–754.

Kozielski, F., Sack, S., Marx, A., Thormählen, M., Schönbrunn, E., Biou, V., Thompson, A., Mandelkow, E.-M, and Mandelkow, E. (1997). The crystal structure of dimeric kinesin and implications for microtubule-dependent motility. *Cell* **91,** 985–994.

Krebs, A., Goldie, K. N., and Hoenger, A. (2004). Complex formation with kinesin motor domains affects the structure of microtubules. *J. Mol. Biol.* **335,** 139–153.

Krzysiak, T. C., Wendt, T., Sproul, L. R., Tittmann, P., Gross, H., Gilbert, S. P., and Hoenger, A. (2006). A structural model for monastrol inhibition of dimeric kinesin Eg5. *EMBO J.* **25**(10), 2263–2273.

Kull, F. J., Sablin, E., Lau, P., Fletterick, R., and Vale, R. (1996). Crystal structure of the kinesin motor domain reveals a structural similarity to myosin. *Nature* **380,** 550–554.

Lee, V. M., Goedert, M., and Trojanowski, J. Q. (2001). Neurodegenerative tauopathies. *Annu. Rev. Neurosci.* **24,** 1121–1159.

Li, H., DeRosier, J., Nicholson, W. V., Nogales, E., and Downing, K. H. (2002). Microtubule structure at 8 Å resolution. *Structure* **10,** 1317–1328.

Löwe, J., Li, H., Downing, K. H., and Nogales, E. (2001). Refined structure of alpha beta-tubulin at 3.5 A° resolution. *J. Mol. Biol.* **313,** 1045–1057.

Lucic, V., Forster, F., and Baumeister, W. (2005). Structural studies by electron tomography: From cells to molecules. *Annu. Rev. Biochem.* **74,** 833–865.

McIntosh, R., Nicastro, D., and Mastronarde, D. (2005). New views of cells in 3D: An introduction to electron tomography. *Trends Cell Biol.* **15,** 43–51.

Miyazawa, A., Fujiyoshi, Y., and Unwin, N. (2003). Structure and gating mechanism of the acetylcholine receptor pore. *Nature* **423,** 949–955.

Mizuno, N., Toba, S., Edamatsu, M., Watai-Nishii, J., Hirokawa, N., Toyoshima, Y. Y., and Kikkawa, M. (2004). Dynein and kinesin share an overlapping microtubule-binding site. *EMBO J.* **23,** 2459–2467.

Nicastro, D., Schwartz, C., Pierson, J., Gaudette, R., Porter, M. E., and McIntosh, J. R. (2006). The molecular architecture of axonemes revealed by cryoelectron tomography. *Science* **313,** 944–948.

Nogales, E., Whittaker, M., Milligan, R. A., and Downing, K. H. (1999). High-resolution model of the microtubule. *Cell* **96,** 79–88.

Nogales, E., Wolf, S. G., and Downing, K. H. (1998). Structure of the alpha beta tubulin dimer by electron crystallography. *Nature* **391,** 199–203.

Sablin, E. P., Case, R. B., Dai, S. C., Hart, C. L., Ruby, A., Vale, R. D., and Fletterick, R. J. (1998). Direction determination in the minus-end-directed kinesin motor Ncd. *Nature* **395,** 813–816.

Sablin, E. P., Kull, F. J., Cooke, R., Vale, R. D., and Fletterick, R. J. (1996). Crystal structure of the motor domain of the kinesin-related motor ncd. *Nature* **380,** 555–559.

Sandblad, L., Busch, K. E., Tittmann, P., Gross, H., Brunner, D., and Hoenger, A. (2006). The *Schizosaccharomyces pombe* EB1 homolog Mal3p localizes preferentially to the microtubule lattice seam. *Cell* **127**(7), 1415–1424.

Santarella, R. A., Koffa, M. D., Tittmann, P., Gross, H., and Hoenger, A. (2007). HURP wraps microtubule-ends with an additional tubulin sheet that has a novel conformation of tubulin. *J. Mol. Biol.* **365**(5), 1587–1595.

Santarella, R. A., Skiniotis, G., Goldie, K. N., Tittmann, P., Gross, H., Mandelkow, E. M., Mandelkow, E., and Hoenger, A. (2004). Surface-decoration of microtubules by human tau. *J. Mol. Biol.* **339,** 539–553.

Schweers, O., Schonbrunn-Hanebeck, E., Marx, A., and Mandelkow, E. (1994). Structural studies of tau protein and Alzheimer paired helical filaments show no evidence for beta-structure. *J. Biol. Chem.* **269,** 24290–24297.

Seitz, A., Kojima, H., Oiwa, K., Mandelkow, E.-M., Song, Y.-H., and Mandelkow, E. (2002). Single-molecule investigation of the interference between kinesin, tau and MAP2c. *EMBO J.* **21,** 4896–4905.

Sillje, H. H., Nagel, S., Korner, R., and Nigg, E. A. (2006). HURP is a Ran-importin beta-regulated protein that stabilizes kinetochore microtubules in the vicinity of chromosomes. *Curr. Biol.* **16,** 731–742.

Skiniotis, G., Surrey, T., Altmann, S., Gross, H., Song, Y.-H., Mandelkow, E., and Hoenger, A. (2003). Nucleotide-induced conformations in the neck region of dimeric kinesin. *EMBO J.* **22,** 1518–1528.

Smith, P. R., and Kistler, J. (1977). Surface reliefs computed from micrographs of heavy metal-shadowed specimens. *J. Ultrastruct. Res.* **61,** 124–133.

Song, Y.-H., and Mandelkow, E. (1993). Recombinant kinesin motor domain binds to ab-tubulin and decorates microtubules with a B surface lattice. *Proc. Natl. Acad. Sci. USA* **90,** 1671–1675.

Song, Y.-H., and Mandelkow, E. (1995). The anatomy of flagellar microtubules: Polarity, seam, junctions and lattice. *J. Cell Biol.* **128,** 81–94.

Sosa, H., Dias, D. P., Hoenger, A., Whittaker, M., Wilson-Kubalek, E., Sablin, E., Fletterick, R. J., Vale, R. D., and Milligan, R. A. (1997). A model for the microtubule-Ncd motor protein complex obtained by cryo- electron microscopy and image analysis. *Cell* **90,** 217–224.

Sosa, H., and Milligan, R. A. (1996). Three-dimensional structure of ncd-decorated microtubules obtained by a back-projection method. *J. Mol. Biol.* **260,** 743–755.

Steinmetz, M. O., Hoenger, A., Tittmann, P., Fuchs, K., Gross, H., and Aebi, U. (1998). An atomic model of actin tubes: Combining electron microscopy and X-ray crystallography. *J. Mol. Biol.* **278,** 703–711.

Steven, A. C., and Aebi, U. (2003). The next ice age: Cryo-electron tomography of intact cells. *Trends Cell Biol.* **13,** 107–110.

Su, L. K., Burrell, M., Hill, D. E., Gyuris, J., Brent, R., Wiltshire, R., Trent, J., Vogelstein, B., and Kinzler, K. W. (1995). APC binds to the novel protein EB1. *Cancer Res.* **55,** 2972–2977.

Tsou, A. P., Yang, C. W., Huang, C. Y., Yu, R. C., Lee, Y. C., Chang, C. W., Chen, B. R., Chung, Y. F., Fann, M. J., Chi, C. W., Chiu, J. H., and Chou, C. K. (2003). Identification of a novel cell cycle regulated gene, HURP, overexpressed in human hepatocellular carcinoma. *Oncogene* **22,** 298–307.

Turner, J., Anderson, R., Guo, J., Beraud, C., Fletterick, R., and Sakowicz, R. (2001). Crystal structure of the mitotic spindle kinesin Eg5 reveals a novel conformation of the neck-linker. *J. Biol. Chem.* **276,** 25496–25502.

Volkmann, N., and Hanein, D. (1999). Quantitative fitting of atomic models into observed densities derived by electron microscopy. *J. Struct. Biol.* **125,** 176–184.

Walz, T., Tittmann, P., Fuchs, K. H., Muller, D. J., Smith, B. L., Agre, P., Gross, H., and Engel, A. (1996). Surface topographies at subnanometer-resolution reveal asymmetry and sidedness of Aquaporin-1. *J. Mol. Biol.* **264,** 907–918.

Wang, H. W., and Nogales, E. (2005). Nucleotide-dependent bending flexibility of tubulin regulates microtubule assembly. *Nature* **435,** 911–915.

Wendt, T., Karabay, A., Krebs, A., Gross, H., Walker, R. A., and Hoenger, A. (2003). A structural analysis of the interaction between ncd tail and tubulin protofilaments. *J. Mol. Biol.* **333,** 541–552.

Wendt, T. G., Volkmann, N., Skiniotis, G., Goldie, K. N., Müller, J, Mandelkow, E., and Hoenger, A. (2002). Microscopic evidence for a minus-end directed power stroke in the kinesin motor ncd. *EMBO J.* **21,** 5969–5978.

Wolf, S. G., Mosser, G., and Downing, K. H. (1993). Tubulin conformation in zinc-induced sheets and macrotubes. *J. Struct. Biol.* **111,** 190–199.

Wriggers, W., Milligan, R. A., and McCammon, J. A. (1999). Situs: A package for docking crystal structures into low-resolution maps from electron microscopy. *J. Struct. Biol.* **125,** 185–195.

Yonekura, K., Maki-Yonekura, S., and Namba, K. (2003). Complete atomic model of the bacterial flagellar filament by electron cryomicroscopy. *Nature* **424,** 643–650.

CHAPTER 15

Rapid Kinetic Techniques

John F. Eccleston,* Stephen R. Martin,* and Maria J. Schilstra†

*Division of Physical Biochemistry
MRC National Institute for Medical Research
The Ridgeway
Mill Hill, London NW7 1AA,
United Kingdom

†Biological and Neural Computation Group
Science and Technology Research Institute
University of Hertfordshire
College Lane
Hatfield AL10 9AB
United Kingdom

METHODS IN CELL BIOLOGY, VOL. 84

0091-679X/08 $35.00
DOI: 10.1016/S0091-679X(07)84015-5

Abstract

The elementary steps in complex biochemical reaction schemes (isomerization, dissociation, and association reactions) ultimately determine how fast any system can react in responding to incoming signals and in adapting to new conditions. Many of these steps have associated rate constants that result in subsecond responses to incoming signals or externally applied changes. This chapter is concerned with the techniques that have been developed to study such rapidly reacting systems *in vitro* and to determine the values of the rate constants for the individual steps. We focus principally on two classes of techniques: (1) flow techniques, in which two solutions are mixed within a few milliseconds and the ensuing reaction monitored over milliseconds to seconds, and (2) relaxation techniques, in which a small perturbation to an existing equilibrium is applied within a few microseconds and the response of the system is followed over microseconds to hundreds of milliseconds. These reactions are most conveniently monitored by recording the change in some optical signal, such as absorbance or fluorescence. We discuss the instrumentation that is (commercially) available to study fast reactions and describe a number of optical probes (chromophores) that can be used to monitor the changes. We discuss the experimental design appropriate for the different experimental techniques and reaction mechanisms, as well as the fundamental theoretical concepts behind the analysis of the data obtained.

I. Introduction

A large number of important biological processes must obviously occur on timescales of very much less than a second. By way of illustration, consider the position of a cricket batsman (or baseball batter); for very fast deliveries as little as 400 msec may elapse between the ball leaving the bowler's (pitcher's) hand and it reaching the batter's bat. During this short time, a whole series of processes occur in the batter's eyes, brain, nerves, and muscles, which all involve the interaction of a wide variety of different molecules. The first steps toward understanding what happens in these processes include identification of the participants and sketching out reaction pathways and interaction networks. The next step consists of a thorough characterization of the dynamics of the interactions. This information is essential if one wishes to create a quantitative picture of such processes, evaluate the "dynamic range" of their constituents, assess the conditions under which systems will function optimally, and, most important, predict under which conditions they will fail.

The majority of biochemical reactions are reversible and, when studied in isolation in the "test tube," will reach an equilibrium state, with reactant and product concentrations determined by the equilibrium constant, *K*. Once this equilibrium state has been reached, the concentrations of the individual species involved will remain constant, unless the system is perturbed in some way. Reaction systems may be perturbed not only by the influx or efflux of reactants but also by changes in some external parameter such as temperature. A change in temperature induces a perturbation because equilibrium constants are generally dependent on temperature ($K = e^{-\Delta G/RT}$, and $\Delta G^\circ = \Delta H^\circ - T\Delta S^\circ$, where ΔG°, ΔH°, and ΔS° are the changes in free energy, enthalpy, and entropy, respectively, R is the universal gas constant, and T is the absolute temperature).

The speed at which any individual reaction or reaction network can respond following such a perturbation depends on the rates of the different elementary steps. Generally speaking, the more dynamic a system, that is, the faster its forward and reverse reactions, the more quickly it will reach the new equilibrium position. Although the cell is, of course, a nonequilibrium system, it is important to remember that any reaction network that can reach equilibrium rapidly will also be able to rapidly reach other "target" states. Thus, it is the rates of the elementary steps that ultimately determine how fast a system can react, respond to incoming signals, and adapt to new conditions. In any biochemical reaction, these elementary steps are either unimolecular (an isomerization of a single entity, or the dissociation of a complex) or bimolecular processes (association of two molecules or complexes) that occur almost exclusively on subsecond, often submillisecond, timescales. To study these reactions, instrumentation is required that allows the investigator (1) to rapidly mix the reactants or perturb the system in a controlled way and (2) to monitor the concentrations of one or more reaction partners with sufficient time resolution as the reaction proceeds.

There are various ways of monitoring the concentration of participants during a reaction. For example, samples may be taken from the reaction volume, mixed with a chemical quenching agent to stop the reaction, and their contents assessed by chromatography, electrophoresis, or mass spectrometric techniques. Such methods can directly determine the concentrations of different reaction participants in a relatively straightforward way, but they are discontinuous and have a limited time resolution that depends on the sampling rate that can be achieved. A more common approach is to take advantage of changes in optical signals, such as absorbance, circular dichroism, or fluorescence, which often accompany a reaction. Although spectroscopic methods do not generally permit the direct determination of concentrations of individual species, they are continuous and can achieve high precision and time resolution. Unfortunately, most conventional spectrophotometers cannot be used to study reactions that are complete within less than about 10 sec, as it takes at least that amount of time to mix the reagents, close the sample compartment, and activate the instrument, or to increase the temperature of the reaction mixture. This chapter is therefore concerned with the application of techniques that have been developed to overcome these problems. We will focus on the two principal classes of methods:

a. *Flow* (or *rapid mixing*) techniques, which are essentially an extension of the classical "mix and observe" approach, and

b. *Relaxation* techniques, in which a system at equilibrium is perturbed by applying a rapid change in an external parameter such as temperature or pressure

Methods such as X-ray crystallography, NMR, and equilibrium binding studies, which yield the necessary (static) structural and thermodynamic information, are complemented by flow and relaxation methods, which yield the kinetic information that is indispensable for understanding the dynamics of biochemical processes.

II. Basic Theory

In the following sections, we shall use the symbols P and L to indicate the participants in simple first- and second-order processes (reaction steps whose kinetics are determined by the concentration of one or two reactants, respectively). P and L stand for "protein" and "ligand"—as many intracellular interactions are between proteins and smaller molecules—but may represent any two reactants, proteins, DNA, lipids, biomolecular assemblies, and so on. PL denotes a complex between P and L, and P*, L*, and PL* indicate different conformational states of P, L, and PL, respectively.

A. First-Order Reactions

The simplest reversible reaction is one where the forward and reverse steps are both unimolecular processes with first-order rate constants k_{+1} and k_{-1} (units s^{-1}) for the forward and reverse steps.

Scheme 1 $$\mathrm{P} \underset{k_{-1}}{\overset{k_{+1}}{\rightleftharpoons}} \mathrm{P^*}$$

The equilibrium constant, K, for this reaction is defined as $K = k_{+1}/k_{-1} = [\mathrm{P}^*]/[\mathrm{P}]$, which is dimensionless. There is no net influx or efflux of material, so that $[\mathrm{P}] + [\mathrm{P}^*] = P_{\mathrm{tot}}$, where square brackets indicate instantaneous (or current) concentrations, and P_{tot} is the total concentration of protein. The rate $\mathrm{d}[\mathrm{P}^*]/\mathrm{d}t$ at which P* is formed is given by

$$\frac{d[\mathrm{P}^*]}{dt} = k_{+1}[\mathrm{P}] - k_{-1}[\mathrm{P}^*] = k_{+1}P_{\mathrm{tot}} - (k_{+1} + k_{-1})[\mathrm{P}^*] \tag{1}$$

Note that this rate may be positive (P* is being formed) as well as negative (P* is disappearing). Equation (1) has an analytical solution that expresses the concentration of P* as a function of time:

$$[\mathrm{P}^*](t) = P^*_{\mathrm{eq}} + (P^*_0 - P^*_{\mathrm{eq}})\mathrm{e}^{-k_{\mathrm{OBS}}t} \tag{2}$$

Here k_{OBS}, which is equal to $k_{-1} + k_{+1}$, is called the *observed rate* of the reaction. P^*_0 and P^*_{eq} are the initial and equilibrium (or final) concentrations of P* (see below). Figure 1 shows the change in [P*] with time for several different combinations of P^*_0 and k_{OBS}. These curves are known as transients and illustrate the following concepts.

a. **Equilibrium.** If P^*_0 is not equal to P^*_{eq} (e.g., if the system is subjected to a sudden change in pH, which alters the equilibrium constant), the concentrations of P and P* will change until the altered equilibrium position is reached and the concentrations of P and P* are constant. P^*_{eq}, the equilibrium concentration of P*, is equal to $(P_{\mathrm{tot}}K)/(1 + K)$, which is equal to $k_{+1}P_{\mathrm{tot}}/k_{\mathrm{OBS}}$.

b. **Amplitude.** The difference between the concentration of P* at the beginning and the end of the reaction (at equilibrium), $P^*_{\mathrm{eq}} - P^*_0$, is called the reaction *amplitude*, Δ[P*].

c. **Observed rate.** Equation (2) is a single exponential function whose shape and amplitude are fully defined by k_{OBS}, P^*_{eq}, and P^*_0. The larger k_{OBS}, the more rapidly the system will reach equilibrium. Because k_{OBS} is equal to the sum of the individual rate constants, two reversible reactions with the same k_{+1} but different k_{-1} values will reach equilibrium at different rates, and the one with the *largest* k_{-1} will equilibrate fastest (compare curves D and E in Fig. 1C).

d. **Rate constants from observed rates.** Analysis of the change of [P*] with time will yield k_{OBS}, not the individual rate constants. However, when the equilibrium constant K is known, the individual rate constants can be calculated as follows:

$$k_{-1} = \frac{k_{\mathrm{OBS}}}{1+K}, \quad k_{+1} = \frac{K \cdot k_{\mathrm{OBS}}}{1+K} \tag{3}$$

e. **Relaxation time and half-life.** The reciprocal of k_{OBS} is called the *relaxation time*, or *time constant*, τ, of the system. At $t = \tau$, the difference between [P*] and $[P^*_{\mathrm{eq}}]$ will have decreased to $1/e$ of the total amplitude Δ[P*], *independent* of the value of P^*_0. The *half-life*, $t_{1/2}$, of the reaction is defined as the time taken for the difference between [P*] and $[P^*_{\mathrm{eq}}]$ to decrease to half of the total amplitude and relates to the relaxation time and observed rate as $t_{1/2} = 0.693\tau = 0.693/k_{\mathrm{OBS}}$ [ln(0.5) = −0.693]. It should be emphasized that although k_{OBS} and τ^{-1} are

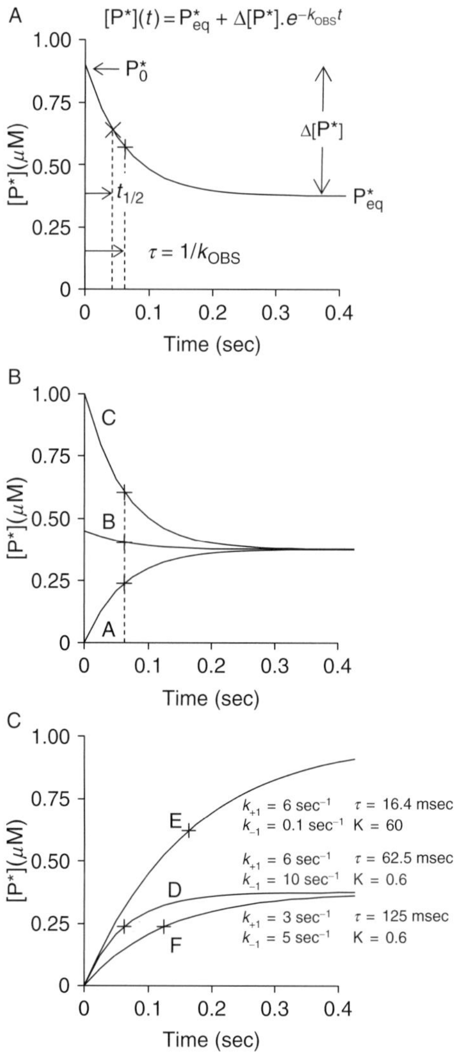

Fig. 1 Transient for the reaction P ⇔ P* (Scheme 1). Values of k_{+1} and k_{-1} were 6 sec^{-1} and 10 sec^{-1}, respectively, unless otherwise indicated; the total concentration of P, P_{tot}, was 1 μM. (A) Definition of P^*_0 and P^*_{eq} (initial and equilibrium concentrations of P*), Δ[P*] (reaction amplitude), $t_{1/2}$ (half-life), and τ (relaxation time), which is equal to the reciprocal of k_{OBS} (observed rate). (B) The relaxation time is equal to $1/k_{OBS} = 1/(k_{+1} + k_{-1})$ (62.5 msec for the rate constants used here), independent of the initial concentration of P* (0, 0.45, and 1 μM in curves A, B, and C, respectively). (C) If two systems have equal forward rate constants, the system with the largest reverse rate constant will reach equilibrium first [curves D ($k_{+1} = 6$; $k_{OBS} = 16$) and E ($k_{+1} = 6$; $k_{OBS} = 6.1$)], whereas two systems with identical equilibrium constants may have different observed rates [curves D ($K = 0.6$, $k_{OBS} = 16$) and F ($K = 0.6$, $k_{OBS} = 8$)].

identical, the former is generally used to describe transients observed in flow experiments, whereas the latter is generally used to describe relaxation (or small perturbation) experiments.

B. Second-Order Reactions

Reversible binding reactions, such as those in which a ligand associates with a protein, include a second-order association process and a first-order dissociation and are described by

Scheme 2
$$\mathrm{P} + \mathrm{L} \underset{k_{-1}}{\overset{k_{+1}}{\rightleftharpoons}} \mathrm{PL}$$

Here k_{+1} is the second-order association rate constant (units: $M^{-1}s^{-1}$) and k_{-1} is the first-order dissociation rate constant (units: s^{-1}). The *equilibrium dissociation constant* for this reaction, K_d, is equal to k_{-1}/k_{+1} (units: M), whereas the *equilibrium association constant,* K_a, is its reciprocal (k_{+1}/k_{-1}, units: M^{-1}). There are now two conservation equations that must be obeyed: $[P] + [PL] = P_{tot}$ and $[L] + [PL] = L_{tot}$, where P_{tot} and L_{tot} are the total concentrations of P and L, respectively. The rate d[PL]/dt at which PL is formed is given by

$$\frac{d[PL]}{dt} = k_{+1}[P][L] - k_{-1}[PL] = k_{+1}(P_{tot} - [PL])(L_{tot} - [PL]) - k_{-1}[PL] \quad (4)$$

This equation does not have a simple analytical solution [a general bimolecular equivalent of Eq. (2)] that describes the change in [PL] as a function of time for given initial and total concentrations of P, L, and PL. However, under certain special conditions, the solution to Eq. (4) can be shown to be very similar to Eq. (2).

1. Pseudo-First-Order Conditions

If one of the reactants is in large excess over the other ($L_{tot} \gg P_{tot}$ or $L_{tot} \gg P_{tot}$), the concentration of the component in excess remains effectively constant during the whole time course, because $X_{tot} - [PL] \approx X_{tot}$, where X_{tot} is the concentration of the component present in excess. Eq. (4) simplifies to one with the same form as Eq. (1), with $k_{+1}X_{tot}$ instead of k_{+1}. As a result, the formation of PL is said to follow pseudo-first-order kinetics, with an observed rate k_{OBS} given by

$$k_{OBS} = \tau^{-1} = k_{+1}X_{tot} + k_{-1} \quad (5)$$

Such conditions are generally employed in flow experiments.

2. Near Equilibrium Conditions

It can be shown that the change in concentration of PL also approaches first-order kinetics when the concentrations of P and L are very close to their equilibrium values (Caldin, 1964). This is because the changes in the concentrations of all species are then very small compared to the values of the concentrations themselves and therefore may be regarded as being constant. Such conditions can be introduced by a small perturbation of the system, for example, by increasing the temperature or pressure of the reaction volume by a small amount (see Section III.B), thereby slightly changing the equilibrium constant for the system. Before the change, P and L are at their equilibrium concentrations under the current conditions. If the change is applied virtually instantaneously, the concentrations will still be at their old values immediately after the jump. At that point, the system is out of equilibrium but will immediately begin to relax toward the new equilibrium, where [P] and [L] are slightly different. In this case, the expression for the reciprocal relaxation time is (Bernasconi, 1976):

$$\tau^{-1} = k_{+1}([\mathrm{P}] + [\mathrm{L}]) + k_{-1} \tag{6}$$

III. Techniques

In this section, we introduce the two major groups of techniques used in the study of rapid reactions: flow and relaxation techniques. These techniques differ in their timescales of applicability: flow methods are generally used to study reactions that occur on timescales varying from a few milliseconds to tens of seconds, whereas relaxation techniques are applicable to reactions that happen within microseconds to hundreds of milliseconds.

A. Flow Techniques

All flow techniques use special mixing chambers that are designed to mix two solutions containing the appropriate reactants. The solutions are generally driven at high velocity into the mixing chamber in order to achieve mixing that is both rapid and complete. Reactions cannot, in general, be monitored inside the mixing chamber itself, so that the mixed solution can be observed only at some point "downstream" from the mixing chamber. Because the mixing and subsequent flow to the point of observation take a finite amount of time, the mixed solution already has a certain average "age"[1] before it can be observed. The time during which the reaction cannot be monitored is called the *dead time* of the instrument.

[1] Because mixing is not instantaneous, there will be a certain spread in the age of the mixed solution. An effective mixing chamber produces a mixed solution in which the spread is relatively small.

1. Stopped Flow

Of all the rapid reaction techniques, stopped flow comes closest to being a standard laboratory technique. As with all flow techniques, the two reactant solutions are rapidly mixed by driving them from "drive" syringes into an appropriately designed mixing chamber. Following mixing, the solution flows into an observation chamber, and the flow is stopped using a "stopping" syringe (Eccleston *et al.*, 2001). The reaction is then followed by monitoring the change in some suitable optical signal, generally absorption or fluorescence, as a function of time. The effective dead time of the instrument is determined by the time it takes to start and stop the flow and is typically 1–2 msec. Reactions with half times shorter than this dead time cannot be studied. Reactions with very long half times (>10 sec) can, at least in principle, be studied using the stopped-flow technique. In practice, the study of very slow reactions may be complicated by lamp instabilities and, in the case of fluorescence, by photobleaching of the chromophore. The latter effect can generally be reduced by the use of an automatic shutter so that the reaction mixture is not constantly illuminated by light.

Commercial equipment is available from several sources (see Section IV). These devices range from stand-alone models with a variety of detection modes to small, hand-driven devices that can be used in conjunction with regular spectrophotometers. The latter are relatively inexpensive and permit the study of reactions with half times as short as 10 msec (although this also depends on the response time of the spectrometer) and are useful for studying many reactions. Most stopped-flow instruments are designed to mix equal volumes of the two reactants, but some will allow different volumes to be used. This particular technique is most widely used in studies of protein folding using chemical denaturants, where large and rapid changes in the concentration of denaturant are required (see Chapter 11 by Street *et al.*, this volume). Devices that permit double mixing experiments are also available: two reactants are mixed and this mixed solution is then rapidly mixed with a third solution after a variable preselected time. This approach allows the study of the reactions of short-lived intermediates and has been used to provide important information in the study of several enzyme mechanisms.

2. Quenched Flow

In the quenched-flow method the two reactant solutions are rapidly mixed as in the stopped-flow method and then flow down an "aging tube" at constant velocity before being mixed with a "quenching agent," which stops the reaction. The age of the quenched sample is determined by the flow rate and the volume of the flow tube, so by doing a series of experiments with different flow rates and flow tube volumes a series of time points can be built up. The most frequently used quenching agent is acid, but alternative quenchers, such as metal chelators (EDTA or EGTA), can be used in the case of some metal-dependent reactions. The quenched reaction mixture is collected and analyzed using an appropriate method, such as

HPLC, to separate reactant and product. Time points between ~5 and ~150 msec can usually be obtained in this way. In principle, points at longer times could be obtained by using longer flow tubes and/or slower flow rates. However, in the former case, excessively large amounts of material would be used, and in the latter case, the flow would not be fast enough to obtain the turbulent flow necessary for rapid and efficient mixing. An alternative approach is to operate the instrument in the "pulsed flow" mode where the mixed solution passes down the flow tube, but before reaching the quenching solution, flow is stopped for an electronically controlled time before being resumed. In this case, the age of the quenched sample is a combination of flow rate, flow tube volume, and the delay time, although the first two become negligible at times of longer than a few seconds. Although quenched-flow methods are much more labor intensive than stopped-flow methods, they do have the advantage that they can be used when no optical signal is available (Barman *et al.*, 2006). In addition, it is one of the few experimental techniques that can be used to give unequivocal information about steps involving the formation or cleavage of covalent bonds (see Chapter 19 by Shcherbakova *et al.*, this volume).

3. Continuous Flow

In this method, the two reactants are pumped at high velocity into a small volume mixing chamber. The optical signal is monitored at different points downstream from the mixer in the direction of the flow and translated into the time-dependent signal change on the basis of the known flow rate. Historically, stopped flow was first developed from the earlier continuous flow method and has almost completely been preferred to continuous flow because of its better sample economy and the ability to measure the kinetics out to long times. However, continuous flow can measure reactions on a much faster timescale than stopped flow, and Shastry *et al.* (1998) have recently shown that the efficiency of the method can be greatly improved by using a charge-coupled device (CCD) camera to image the flow tube and the distance down the flow tube is converted to time from the linear flow rate. By combining this improved detection method with a very efficient capillary mixer, they were able to achieve mixing times of ~15 μsec and dead times as short as 45 μsec.

B. Relaxation Techniques

Reactions that are too fast to be studied using rapid mixing techniques can be studied using techniques in which a system at equilibrium is perturbed by applying a rapid change in some external parameter. In each of the techniques discussed below, the response of the system to the new conditions is then monitored by recording a suitable optical signal.

1. Temperature Jump

The best known relaxation technique is the temperature-jump method, which has been available for several decades. In this technique, the perturbation consists of a sudden increase in the temperature of the solution. In the most common setup, the temperature jump is brought about by Joule heating, the temperature increase in a conducting solution resulting from the resistance to an electric current flowing through it. Discharge of a high-voltage capacitor through the solution is used to produce temperature jumps of up to 5 °C in as little as 1–2 μsec. Therefore, reactions with half times shorter than this temperature rise time cannot be studied using Joule heating devices. However, laser heating devices can be used to produce large increases in temperature in relatively small irradiated volumes in much shorter times, as little as 10 nsec (see, e.g., Turner *et al.*, 1972; Williams *et al.*, 1996).

In Joule heating devices, the solution will begin to cool immediately after the jump, owing to equilibration with the cell body, optical windows, electrodes, and so on. The temperature of the heated volume decays to ambient temperature, with a half-life of the order of 50 sec depending on the precise geometry of the device. Therefore, reactions with half times longer than several hundred milliseconds are difficult to study. However, in some cases applying the cooling corrections described by Rabl (1979) can extend this upper limit. In laser heating devices, the small heated volume decays to ambient temperatures with a much shorter half-life.

2. Pressure Jump

In the pressure-jump method, which has also been available for several decades, the equilibrium is perturbed by applying a rapid change in pressure rather than temperature. Recent technical advances (see, e.g., Pearson *et al.*, 2002) have led to the development of devices that use piezoelectric crystals to generate large pressure increases (of up to 200 atm) in small samples ($\sim$50 μl) in as little as 50 μsec. Although pressure-induced perturbations are very much smaller than those produced by temperature changes, high repetition rates can be used (because there is no equivalent of a cooling phase in a pressure-jump experiment), and this allows the collection of data with very good signal-to-noise ratio. In addition, the pressure-jump technique has several distinct advantages over the temperature-jump method: (1) reactions can be followed on long timescales when required because the pressure remains constant after the jump, (2) samples do not require the high ionic strength needed for efficient Joule heating in temperature-jump devices, and (3) jumps can be recorded in both directions. In addition, intrinsic fluorescence signals (such as those from tryptophan in proteins) show little pressure sensitivity over the range used for perturbation kinetics, so there is no transient associated with the jump itself, in contrast to the effect of a temperature change.

C. Flash Photolysis

Flash photolysis, in which a reaction is triggered by a pulse of light, was previously limited to the study of intrinsically light-dependent reaction systems, such as those involved in visual processes. In recent years, the applicability of the technique has been significantly extended by the development of methods in which a ligand for a particular reaction is converted to an inactive form by the addition of a caging group that can be converted by light into the natural, active form. Caged compounds release the active ligand species, generally on a millisecond or faster timescale, on flash photolysis with near-UV light. They are used principally in studies of rapid biological processes to enable the application of a particular ligand at or near its site of action. Many caged compounds have now been developed, ranging from caged nucleotides such as ATP to caged forms of neuroexcitatory amino acids such as L-glutamate. It is also possible to produce a pH jump by flashing an aqueous solution containing a suitable photolabile caged compound, such as *O*-nitrobenzaldehyde, with a nanosecond UV laser (see, for instance, Abbruzzetti *et al.*, 2000; Gutman and Nachliel, 1990).

A somewhat different philosophy was used in the development of the widely used caged calcium compounds. The three commercially available caged calcium compounds (DM-nitrophen, nitrophenyl-EGTA, and nitr-5) bind calcium with very high affinity (K_d: 5–150 nM) but can be rapidly photolyzed into photoproducts with very much lower affinity for calcium (K_d: 0.01–1 mM). Locally applied flashes of light release calcium from the cage and thus produce rapid and large increases in calcium concentration.

Further detailed discussion of these particular techniques is beyond the scope of this chapter. However, the analysis of the results obtained using such approaches is based on the same general principles as those outlined in subsequent sections.

IV. Instrumentation

Instruments for the study of fast reactions are commercially available from a number of different suppliers: TgK Scientific Ltd. (Supplier of HiTech instruments: http://www.tgkscientific.com/), KinTek Corporation (http://www.kintek-corp.com/), OLIS (http://olisweb.com/), Applied Photophysics (http://www.photophysics.com/), and Biologic Science Instruments (http://www.bio-logic.info/). The principal optical detection modes employed are fluorescence and absorbance. Fluorescence detection is now very widely employed because it is intrinsically much more sensitive than absorption and therefore permits measurements to be made at very low concentrations in many cases. Although circular dichroism (CD) detection is widely employed in studies of protein unfolding (see Chapter 10 by Martin and Schilstra, this volume), the inherently poor signal-to-noise ratios of CD signals limits its use in the study of protein–ligand interactions. Fluorescence measurements will therefore be the main focus of this section.

A. Instrument Characteristics

As with any scientific instrument, it is important that the user understand the characteristics and limitations of the equipment being used.

In the case of stopped flow, for example, it is important (1) to demonstrate that mixing is efficient and (2) to determine the dead time of the instrument. Mixing efficiency can be tested by rapidly mixing a solution of a pH indicator above its pK_a with a buffer at a pH below its pK_a. For example, a 1 μM solution of 4-methylumbelliferone in 0.1 M sodium pyrophosphate, pH 8.7, is initially mixed with 0.1 M sodium pyrophosphate, pH 8.7, alone and the signal observed. The solution in the syringe containing only buffer is then changed to 0.1 M pyrophosphate at pH 6.2, and this is then mixed with the 4-methylumbelliferone solution. Because proton transfer reactions are effectively instantaneous on the timescale of a stopped-flow experiment, the *absence* of any detectable reaction is taken to indicate that mixing is efficient. The instrumental dead time can be measured using any well-characterized second-order reaction that has a large change in an appropriate optical signal. The reaction of *N*-acetyl tryptophanamide (NATA) with *N*-bromosuccinimide in 0.1 M sodium phosphate, pH 7.5 can be used for fluorescence measurements. The true starting signal for the reaction is determined by mixing 10 μM NATA in 0.1 M sodium phosphate, pH 7.5, with buffer alone. With excitation at 280 nm, emission is observed through a 320 nm cutoff filter and a suitable signal obtained. Then the NATA solution is mixed with 600 μM *N*-bromosuccinimide. The reaction is therefore performed under pseudo-first-order conditions with concentrations chosen to give a known half time of the order of 20 msec (see Section VI). Extrapolation of the observed signal back to the true starting signal using the known half time gives the dead time. Detailed protocols for performing dead time measurements and for the investigation of mixing efficiency are given by Eccleston *et al.* (2001).

In order to do a temporal calibration of a quenched-flow instrument, the alkaline hydrolysis of 2,4-dinitrophenyl acetate (DNPA) can be used. A 50 mM solution of DNPA in ethanol is prepared and diluted 50-fold with 2 mM HCl. The other reactant is 1 M NaOH and the quencher is 2 M HCl. Initially, a sample of time equal to zero is obtained manually by mixing 1 ml of the DNPA solution with 5 ml of 2 M HCl before adding 1 ml of 1 M NaOH and making up to 10 ml with 2 M HCl. Then a time infinity sample is prepared by mixing 1 ml of DNPA with 1 ml of 1 M NaOH, allowing a few seconds for the hydrolysis to occur and then making up to 10 ml with 2 M HCl. The absorbance spectra of both samples are then measured between 240 and 450 nm. It will be seen that a new shoulder occurs in the time infinity sample at about 294 nm and an isosbestic point (a wavelength at which two chemical species have the same molar extinction coefficient) is at 260 nm. The absorbances at both wavelengths are measured and the ratio $A_{294\ \mathrm{nm}}/A_{260\ \mathrm{nm}}$ is calculated. Having made these measurements on the zero and infinity time points of the reaction, equal volumes of the DNPA and 1 M NaOH are mixed together in a quenched-flow instrument and quenched with 2 M HCl with points taken over the time range of 0–150 msec. By measuring the $A_{294\ \mathrm{nm}}/A_{260\ \mathrm{nm}}$ ratio of each quenched sample, the extent of the reaction is calculated. The data are then fitted to a single exponential. The fitted line should pass through zero time and an end point of 100% reaction. In addition, it should follow the same time course as when the DNPA solution is mixed with 1 M NaOH in a stopped-flow instrument with the absorbance being monitored at 420 nm.

In the case of temperature jump, the most important parameters to be determined are (1) the rise time of the heating pulse, (2) the magnitude of the temperature increase, and (3) the cooling characteristics of the cell (see Section III.B.1). These can easily be determined by performing temperature jumps on a solution of a pH indicator, such as phenolphthalein, at a pH close to its pK_a.

B. Instrument Settings

Several important factors must be considered in choosing the appropriate instrument settings for optical methods, and these are discussed briefly here, with particular emphasis on fluorescence methods.

a. **Lamp selection.** Fluorescence measurements generally require high intensity light sources such as mercury, xenon, or xenon/mercury arc lamps. Xenon arc lamps have a relatively smooth emission spectrum, whereas mercury or xenon/mercury lamps have a series of intense emission bands, which can sometimes be used to advantage. The emission from deuterium or quartz halide lamps is significantly less intense but is also less noisy. These lamps are frequently used in absorbance measurements and can sometimes be used for fluorescence excitation in the visible region if the fluorophore has a high quantum yield and extinction coefficient (note: fluorescence intensity is proportional to the product of the quantum yield and the extinction coefficient).

b. **Slit widths** Selection of slit widths is a balance between light intensity and spectral purity. If the fluorophore has a large Stokes shift (the wavelength difference between the excitation and emission maxima), a large slit width can be used to increase the light intensity. If the Stokes shift is small then the slit width may have to be reduced in order to exclude scattered light from the photomultiplier. Alternatively, the wavelength of the exciting light may be set to a shorter wavelength than the excitation maximum. These choices need to be made in conjunction with the choice of detection conditions. In some cases, slit widths may also need to be reduced to minimize photobleaching of the fluorophore. Photobleaching is not usually a problem on timescales of <1 sec and can be quantified by mixing the fluorophore with buffer and recording the decrease in fluorescence intensity caused by the photobleaching.

c. **Detection of emission.** The emitted light is generally detected by a photomultiplier after it has passed through a suitable optical filter that should be selected to pass fluorescence emission light and exclude any exciting light that might be scattered by the solution. Scattered light arises from three sources: Rayleigh scattering of the exciting light (observed at the excitation wavelength λ_{Ex}),

Rayleigh scattering of the first harmonic of the exciting light (observed at $2\lambda_{Ex}$), and Raman scattering from water. The wavelength (in nanometers) of the Raman scattering peak (λ_R) depends on the excitation wavelength according to $1/\lambda_R = 1/\lambda_{Ex} - 0.00034$. Filters should be chosen to maximize the fluorescence signal and minimize these other signals. This can be done using the appropriate cutoff and/or band pass filters. The other equally important issue is that one should maximize the signal change relative to the total background intensity in a rapid kinetic experiment. For this reason, it is essential that a steady-state investigation of the optical signal changes is used to determine the appropriate choice of emission filters.

d. **Selection of time constant.** The signal-to-noise ratio in a rapid kinetic measurement is proportional to the square root of the instrumental time constant. In general, the time constant should be selected to be <10% of the half time of the fastest process being observed. This gives the biggest reduction in noise without affecting the rate of the process being observed. In some instruments, the noise can also be reduced by collecting data at the fastest possible rate and averaging appropriate blocks of data to give the individual time points.

e. **Recording and analyzing kinetic transients** In the simplest cases, data are usually collected using linear timescales. In more complex systems, the observable processes may occur on very different timescales, and it is then generally more appropriate to collect data with a logarithmic time base that allows data to be collected at progressively longer time intervals as the reaction proceeds (see Eccleston *et al.*, 2001). Although the time constant will need to be set to be less than the fastest process the data can often be collected in the oversampling mode to improve the signal-to-noise ratio for long time points. Although most stopped-flow fluorescence studies are performed at a single emission wavelength (or range of wavelengths selected by interference or band pass filters), it is now also possible to use rapid scanning monochromators or intensified diode array detectors to collect complete fluorescence spectra as a function of time.

Analysis of the data obtained from rapid kinetic experiments by fitting one or more exponential terms to the curves obtained is usually straightforward in simple situations (see Section VI). In this case, the software supplied with commercially available equipment is often adequate. However, fitting to exponentials is not always the appropriate approach, for example, if second or higher order processes occur—see Section VII.

V. Probes

In most cases, an essential prerequisite for a successful rapid kinetic study is that there should be a suitable change in some optical signal, generally either absorbance or fluorescence, accompanying the reaction. The ideal case is, of course, where the optical signal is intrinsic to the system. There are, however, several approaches that may be used when there is no suitable change in an intrinsic signal, and these are the major topic of this section.

A. Intrinsic Probes

Tryptophan is the major contributor to both the absorption and fluorescence properties of proteins, and large changes, particularly in fluorescence, may accompany protein–ligand interactions. Although tryptophan does have unique advantages as an intrinsic probe, it can be difficult to use for studying protein–nucleic acid interactions because the absorption spectra of the nucleic acids completely overlap that of tryptophan. It may also be difficult to use in the study of protein–protein interactions because many proteins contain more than one tryptophan, and the overall signal change accompanying the interaction can then be small. In certain cases, it may be possible to use site-directed mutagenesis to replace some of the tryptophan residues and thereby increase the size of the signal change (Málnási-Csizmadia *et al.*, 2001; Wakelin *et al.*, 2003). However, this procedure may, of course, alter the structure and/or function of the protein. Intrinsic optical signals associated with nucleic acids are generally much less useful in rapid kinetic studies, although major changes in the circular dichroism signals of the nucleic acids may result from interactions with proteins.

There are also numerous naturally occurring chromophoric cofactors and coenzymes such as NADH and pyridoxal phosphate that provide useful optical signals for rapid kinetic studies.

B. Extrinsic Probes

An extrinsic probe is most often introduced by covalent attachment of a suitable chromophoric probe to one of the reaction partners.

There is a very wide range of commercially available fluorescent probes that react with either thiol or amine groups in proteins and oligonucleotides. A comprehensive guide to fluorescent probes and suitable labeling procedures is available from Molecular Probes (2006). Thiol or amine groups can be incorporated in a chemically synthesized oligonucleotide. These groups can then be conjugated to a thiol-reactive or amine-reactive fluorophore. Labeling proteins can be more problematic. If the protein contains more than a single site for the label, it may be difficult to obtain a reproducible product. Even when only a single site is available for labeling, this may be far from the binding site for the reaction partner, and not therefore report on the interaction. In the latter case, it may still be possible to study the interaction using anisotropy measurements (see Section V.D). An alternative approach is to use genetic engineering to create a protein containing a single cysteine residue that can then be specifically labeled (Brune *et al.*, 1998, 2001; see also Chapter 20 by Klug and Feix, this volume). In this case, knowledge of the secondary structure of the protein is invaluable because it permits the cysteine to be introduced at a position where the label would be expected to be perturbed by ligand binding. For all of these approaches, it is essential that the modified protein is fully characterized, ideally using a combination of mass spectrometry and limited proteolysis, and that the ratio of probe to protein should be determined. It should also be demonstrated that the modification does not affect any biological activity of the protein. Finally, equilibrium binding measurements should be performed to determine the affinity of the modified protein for the ligand, and this should be compared with that of the native protein. This can be done using suitable competition or displacement experiments (Martin and Bayley, 2002).

Another possible approach that can be used with proteins is to replace particular tryptophan residues with analogs that have absorption spectra whose absorption maxima are shifted toward the red (i.e., long wavelength) end of the spectrum, for example, 5-hydroxytryptophan and 7-azatryptophan (Ross *et al.*, 1997). The major experimental advantage of this approach is that the red-shifted absorption spectrum allows selective excitation of the fluorescence of the analog-containing protein in the presence of nucleic acids or other tryptophan-containing proteins. There are also highly fluorescent analogs of nucleotide bases, such as 2-amino-purine, that may provide correspondingly sensitive probes for studying the dynamics of protein–nucleic acid interactions (Hill and Royer, 1997). These analogs can be incorporated site-specifically into oligonucleotides using standard automated synthetic methods. A major advantage of using such analogs is their similarity in constitution and chemical properties to the natural compounds. Unlike large chromophores (see below), incorporation of these analogs into proteins or nucleic acids can normally be accomplished without introducing significant structural or chemical changes that might alter the measurement. It should not, however, be assumed that this is always the case.

Modification of the smaller reaction partner, such as a cofactor or other ligand for a protein, is attractive for two reasons. First because it will generally be much easier to produce a well-characterized product and second because the fluorophore will necessarily be closer to the site of action. For example, many ribose-modified derivatives of ATP and GTP have been synthesized to study the kinetic mechanisms of ATPases and GTPases (Cremo, 2003; Jameson and Eccleston, 1997). One potential disadvantage is that the properties of the labeled compound may differ significantly from those of the unlabeled one. For example, the 2′(3′)-*O*-(*N*-methylanthraniloyl)-derivative of ADP binds to myosin subfragment 1 ten times more tightly than does ADP (Woodward *et al.*, 1991). However, their fundamental mode of action is likely to remain the same as the parent nucleotide, and they also remain useful for studying the unlabeled nucleotide by the use of competition and displacement experiments (see below).

C. Indicators, Linked Assays, and Biosensors

There are numerous applications in which a reaction may be monitored by linking it to a second process that provides the appropriate spectroscopic signal. The essential requirement is that the linked process is very much faster than the one being studied, and does not affect the rate of the process being investigated. The simplest example is in the use of pH indicators to monitor reactions in which protons are either taken up or released. There are also several indicators available for Ca^{2+} and other metal ions (see Molecular Probes, 2006). Classical linked-enzyme assays are generally not fast enough for monitoring fast reactions but can be used to monitor those reactions occurring with sufficiently long half times. For example, the release of ADP can be monitored by linking it to the pyruvate kinase and lactate dehydrogenase system in which the conversion of NADH to NAD provides the optical signal (either fluorescence or absorbance). The last decade has seen the development of a large number of biosensors for studying rapid reactions. For example, Webb and colleagues have produced sensors for phosphate and purine nucleoside diphosphates by fluorescently labeling a phosphate-binding protein (Brune *et al.*, 1998) and a nucleoside diphosphate kinase (Brune *et al.*, 2001), respectively.

Resonance energy transfer can also be used if a suitable donor/emission pair is available with a combination of intrinsic and/or extrinsic fluorophores. The emission spectrum of tryptophan overlaps the excitation spectrum of 2′(3′)-*O*-(*N*-methylanthraniloyl)-adenine nucleotides and this has been taken advantage of in stopped-flow studies of the myosin subfragment 1 ATPase mechanism (Woodward *et al.*, 1991). By exciting the tryptophan at 280 nm and observing the methylanthraniloyl emission, the bound fluorophore is preferentially excited over free fluorophore. This allows higher concentrations of the excess fluorophore to be used compared to exciting the methylanthraniloyl fluorophore directly.

D. Fluorescence Anisotropy

Fluorescence anisotropy measurements may be used to monitor reactions in which there is no change in fluorescence intensity. In these measurements, the fluorophore is excited with vertically polarized light, and the intensity of the emitted light polarized parallel ($I_{\parallel}$) and perpendicular ($I_{\perp}$) to the plane of the exciting light is recorded. The total fluorescence intensity is given by $I_{\parallel} + 2\,I_{\perp}$, and the anisotropy is calculated as $r = (I_{\parallel} - I_{\perp})/(I_{\parallel} + 2\,I_{\perp})$. The anisotropy is related to fluorophore's rotational correlation time (τ_c) by the equation $r = r_o/(1 + \tau/\tau_c)$, where r_o is the limiting anisotropy of the fluorophore and τ is its excited state lifetime. Anisotropy measurements are particularly appropriate in the study of the binding of small fluorescent ligands to large macromolecules because τ_c is related to size and such reactions will therefore be accompanied by large increases in anisotropy. However, because anisotropy can be measured with high precision, it is also possible to use this approach in the study of protein–protein interactions.

Measurements of anisotropy require an instrument that is equipped with a polarizer filter in the excitation path that can be rotated to give light polarized either vertical or horizontal to the laboratory axis. It is generally best to make the measurements in what is known as the "*T*" format, with two detection photomultipliers equipped with polarizers positioned at right angles to the incident light direction being used to measure $I_{\parallel}$ and $I_{\perp}$. Because the two photomultipliers will respond differently to the parallel and perpendicular light, they must be normalized. This is done by exciting the fluorophore with horizontally polarized light. Because the amount of light depolarized to either the parallel or perpendicular plane should be equal, the high voltage on each photomultiplier must be adjusted to give the same output signal. Methods for the analysis of anisotropy data are discussed in Section VI.D (see also Chapter 9 by LiCata and Wowor, this volume).

VI. Experimental Design and Data Analysis

In this section, we shall concentrate mainly on the determination of rate constants in reversible second-order processes, as introduced in Section II.B. Many protein–ligand and protein–protein interactions are of this type. It is invariably true that knowledge of the equilibrium dissociation constant K_d (k_-/k_+) is helpful in designing transient kinetic experiments and in interpreting the data obtained from them. Methods for determining these constants are described elsewhere in this volume and in several published reviews (e.g., Eftink, 1997).

A. General Considerations

1. Flow Methods

Stopped-flow kinetic studies of reactions of the type shown in Scheme 2 are generally performed under pseudo-first-order conditions, that is, with one of the reagents (X) in large excess over the other (see Section II.B.1). As $k_{OBS} = k_+ X_{tot} + k_-$ [Eq. (5)], k_{OBS} must be determined from experimental transients recorded at several different values of X_{tot}. A plot of k_{OBS} versus X_{tot} should then be linear with slope k_{+1} and intercept k_{-1}, provided the binding is indeed a simple second-order process.

As noted above, it is important in transient kinetic experiments that one should maximize the ratio of signal change to total background signal. Thus, for example, if the ligand (L) and the complex (PL) are fluorescent, but the protein is not, then the protein should be the component used in excess. In the ideal case, the concentration of the reagent in excess should be at least tenfold higher than the other. If lower ratios are used, the measured value of k_{OBS} will be significantly different from that predicted by Eq. (5).

It is clearly important that the largest possible concentration range should be covered in experiments of this type. This permits accurate determination of the kinetic constants, and the demonstration that k_{OBS} varies linearly over an extended concentration range is necessary to confirm that Scheme 2 is an adequate description of the process. Consider, for example, the case where the protein is the reactant used in excess. In a typical situation with an association rate constant of the order of $10^7\ M^{-1}\ sec^{-1}$, a dissociation rate constant of 10 sec^{-1}, and an instrument dead time of 2 msec, the largest usable value of P_{tot}, the total concentration of P, would be about 50 μM ($k_{OBS} = 510\ sec^{-1}$). The lower concentration limit is determined by the lowest concentration of ligand that can be used while still maintaining the relationship $P_{tot} \gg L_{tot}$ (L_{tot} is the total concentration of L). As the ligand concentration is decreased, the photomultiplier gain will have to be increased, and at some point this will result in an unacceptably high noise level. In the case of low affinity interactions, where the dissociation rate constant (k_{-1}) is likely to be large, it may be possible to cover only a limited concentration range before k_{OBS} becomes too fast to measure. In such cases, it is generally possible to extend the available range by performing the experiment at a lower temperature. In cases where the association rate constant cannot be accurately determined, it can be estimated from the measured dissociation rate constant and a known dissociation constant using $k_{+1} = k_{-1}/K_d$. This is, of course, true only if the reaction conforms to Scheme 2. Figure 2 illustrates under which conditions k_{+1} and k_{-1} can be determined accurately from this type of experiment.

The measured values of the rate constants should be used to calculate a value for K_d for the interaction using the relationship $K_d = k_{-1}/k_{+1}$, and this should agree with the K_d determined from equilibrium titrations. A significant difference between the values may indicate that Scheme 2 is not an adequate description of the process. The observed reaction amplitudes should also be shown to be consistent with the K_d measured independently. The concentration of protein–ligand complex formed following stopped-flow mixing is, of course, readily calculated from the total concentrations of protein and ligand present after mixing and the known K_d using

$$[PL] = \frac{(P_{tot} + L_{tot} + K_d) - \sqrt{(P_{tot} + L_{tot} + K_d)^2 - 4P_{tot}L_{tot}}}{2} \tag{7}$$

2. Relaxation Experiments

a. Relaxation Times

In the case of a relaxation measurement, such as temperature jump, analysis of the observed exponential time course gives the relaxation time τ, which depends on the concentrations of P and L according to Eq. (6) ($\tau^{-1} = k_{+1}([P] + [L]) + k_{-1}$). Note [P] and [L] are the concentrations of free protein and free ligand present at equilibrium *prior* to application of the temperature jump, whereas the rate constants are those for the temperature reached *after* the temperature jump. A plot of the reciprocal relaxation time versus ([P] + [L]) should be linear with slope k_{+1} and intercept k_{-1}. The values of ([P] + [L]) can be calculated from the known total concentrations (P_{tot} and L_{tot}) and an experimentally determined overall dissociation constant K_d ($=k_{-1}/k_{+1}$) using Eq. (7). There may, however, be instances where it is not possible to determine an accurate value of the dissociation constant and [P] and [L] cannot then be calculated. There are three ways around this problem:

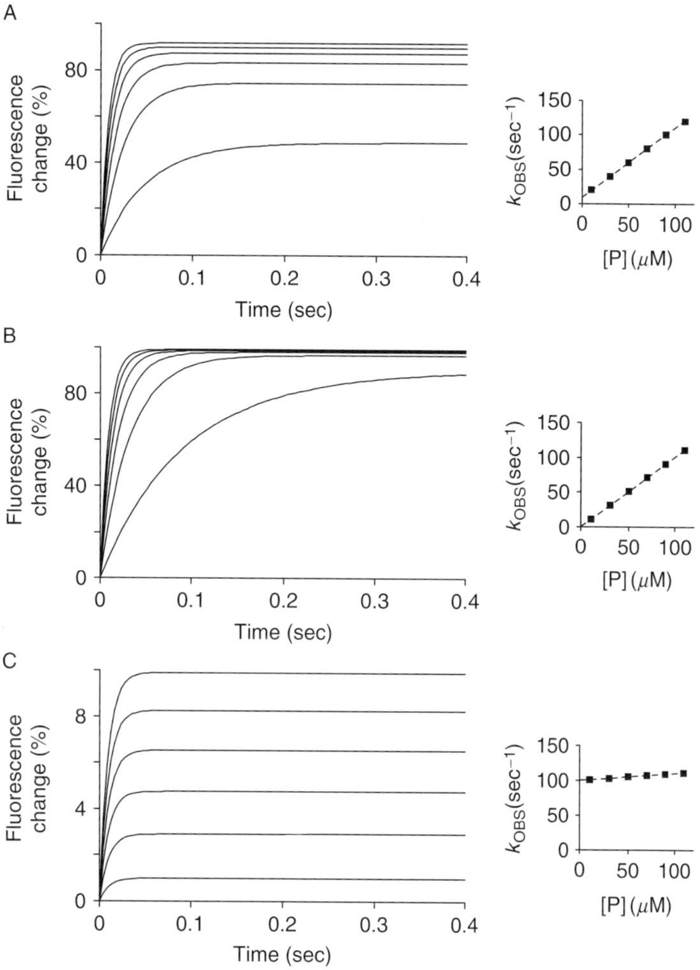

Fig. 2 Left panels: Typical transients (simulated) in a set of stopped-flow experiments in which a fluorescent ligand L (at 1 μM) is mixed with a large excess (10–110 μM) of nonfluorescent protein P. The specific fluorescence of the complex PL is higher than that of L, so that the overall fluorescence of the mixture increases as the system approaches equilibrium. The vertical axis indicates the change in fluorescence observed as a percentage of the maximal change (the difference in fluorescence between 1 μM L and 1 μM PL). Right panels: Plots of the observed rates obtained from the transients versus the concentration of the component in excess (P). As $k_{OBS} = k_{+1}[P] + k_{-1}$, the slope and intercept of these plots are k_{+1} and k_{-1}, respectively. (A) Both slope and intercept (1 μM^{-1} sec^{-1} and 10 sec^{-1}, so that $K_d = 10$ μM) are well defined (have small associated errors) under the experimental conditions. (B) k_{+1} is well defined (1 μM^{-1} sec^{-1}), but k_{-1} is too small (because $K_d = 1$ μM) to be obtained with any accuracy. (C) k_{-1} is well defined (100 sec^{-1}), but k_{+1} is very small (as $K_d = 1$ mM in this system). Note that the affinity of P for L is so low that at 100 μM P, only about 10% of the total L has formed PL.

1. Use a large excess of one reagent (tenfold or greater) so that the sum of [P] and [L] is effectively equal to the total concentration of the reagent in excess. In order to maximize the ratio of signal change to background signal the reagent used in excess should be the one with the smallest contribution to the total optical signal. As with stopped-flow measurements, a potential disadvantage of this method is that it may require the use of a low concentration of the component with the optical signal and this may result in unacceptably high noise levels. A further significant disadvantage is that the greatest perturbation of the equilibrium is achieved when the concentrations of protein and ligand are equal (see below).

2. Use the relationship

$$\tau^{-2} = 2k_{+1}k_{-1}([P_{\mathrm{tot}}] + [L_{\mathrm{tot}}]) + k_{-1}^{2} \tag{8}$$

under conditions where the total concentrations of protein and ligand are the same ($P_{\mathrm{tot}} = L_{\mathrm{tot}}$). This method has the advantage that the largest perturbations of the equilibrium are obtained when $P_{\mathrm{tot}} = L_{\mathrm{tot}}$ (see below). It does, however, have two potential disadvantages. If there is an error in the estimation of either of the concentrations, then $P_{\mathrm{tot}} \neq L_{\mathrm{tot}}$ and the plot will show upward curvature whose magnitude will depend on the magnitude of the difference. Even relatively small curvature can, under certain conditions, result in large errors in the calculated dissociation rate constant. Given the difficulties of estimating protein concentrations, this is quite a likely source of error. The other problem is that errors in the reciprocal relaxation time translate into larger errors in τ^{-2}.

3. Equations. (6) and (7) can be combined to give a single equation describing the dependence of the reciprocal relaxation time on the *total* concentrations of the two reactants:

$$\tau^{-1} = k_{+1}\sqrt{\left(P_{\mathrm{tot}} + L_{\mathrm{tot}} + \frac{k_{-1}}{k_{+1}}\right)^{2} - 4P_{\mathrm{tot}}L_{\mathrm{tot}}} \tag{9}$$

b. Amplitudes

The magnitude of the concentration perturbation for the simple equilibrium shown in Scheme 2 is Δ[L], where Δ[L] = Δ[P] = −Δ[PL]. For temperature-jump experiments, it has been shown (Malcolm, 1972) that

$$\Delta[\mathrm{L}] = \alpha\cdot\delta \ln K_{\mathrm{d}} = \alpha\left(\frac{-\Delta H}{RT^{2}}\right)\delta T \tag{10}$$

The amplitude of the observed relaxation process therefore depends on the enthalpy (ΔH), on the size of the temperature jump (δT), and on the magnitude of α, which is given by

$$\alpha = -0.5K_\mathrm{d} + 0.5K_\mathrm{d}/\sqrt{1 - 4\left(\frac{S}{K_\mathrm{d}+S}\right)^2\left(\frac{\beta}{(1+\beta)^2}\right)} \tag{11}$$

where $S = P_\mathrm{tot} + L_\mathrm{tot}$ and $\beta = P_\mathrm{tot}/\,L_\mathrm{tot}$. When the term after the minus sign in the square root term is small, this expression may be simplified (Malcolm, 1972) to

$$\alpha = K_\mathrm{d}\left(\frac{S}{K_\mathrm{d}+S}\right)^2\left(\frac{\beta}{(1+\beta)^2}\right) \tag{12}$$

The term $\beta/(1+\beta)^2$ has a maximum when $\beta = 1$. Therefore, the magnitude of β, and hence the total observed amplitude, increases as S increases and as β approaches 1. This is illustrated in Fig. 3. In cases where the affinity is reduced at lower pH values, it is possible to produce larger perturbations by using a buffer such as Tris, which has a high temperature coefficient.

B. Displacement Experiments

Small dissociation rate constants may not be well determined using the above approaches because the intercept will be very close to the origin (see Fig. 2). Stopped-flow methods can sometimes be used to make an independent measurement of the dissociation rate constant if a displacement experiment can be performed. For example, an excess of a nonfluorescent ligand (N) can be used to dissociate a fluorescent ligand (L) from its complex with the protein. The relevant reactions are shown in Scheme 3:

Scheme 3 $\quad \mathrm{P + L} \underset{k_{-1}}{\overset{k_{+1}}{\rightleftharpoons}} \mathrm{PL} \qquad \mathrm{P + N} \underset{k_{-2}}{\overset{k_{+2}}{\rightleftharpoons}} \mathrm{PN}$

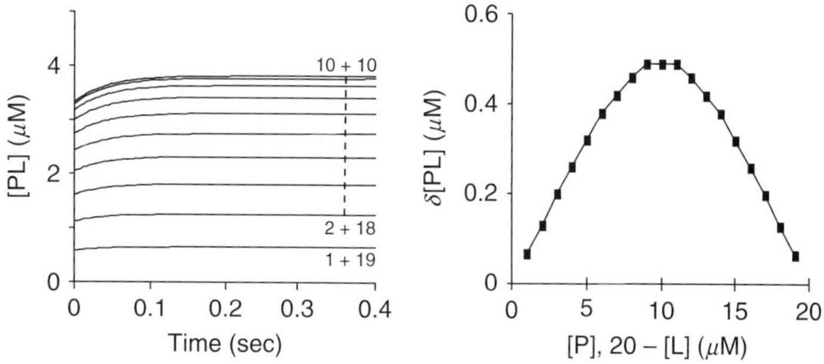

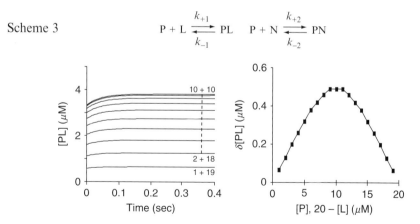

Fig. 3 A set of transients (simulated) that might be observed in a temperature-jump experiment. A rapid temperature jump (from 20 to 25 °C) is applied to mixtures of P and L in which the sum of [P] and [L] is 20 μM, but their ratio varies. The right panel, in which the reaction amplitude Δ[PL] is plotted against [P], shows that the amplitude is maximal when [P]/[L] = 1.

The experiment involves rapidly mixing one solution containing P *and* L (at concentrations chosen to give a reasonable saturation of P) with a second solution containing an excess of N. High concentrations of N are required so that when L dissociates from the protein it cannot reassociate before N binds. The rate constant of the observed process should then be equal to k_{-1}. However, depending on the relative values of the other rate constants this may not always be the case, and the observed rate may be either higher or lower than the true dissociation rate constant (see Wu *et al.*, 1992). To confirm that the true dissociation rate constant is being measured, it must be demonstrated that the observed rate is independent of the concentration of the displacing ligand, N. If the observed rate does vary with the concentration of N then the observed rate will plateau at the true value of k_{-1} at sufficiently high [N]. Wu *et al.* (1992) have also described ways in which the truly dissociative mechanism represented by Scheme 3 (L must dissociate before N binds) can be distinguished from an associative mechanism in which a ternary complex is formed between the incoming ligand (N) and the complex containing the leaving ligand (PL) prior to dissociation of L. In the associative mechanism the observed rate will also generally plateau at high [N] but, in most cases, this value will be higher than k_{-1}.

In some cases, it is also possible to use displacement experiments to determine dissociation rate constants for an optically silent ligand by using a fluorescent (or absorbing) ligand to induce the displacement. This may, however, be technically difficult because the strong fluorescent background will result in poor S/N ratios.

Dissociation of a protein–ligand complex can, in certain cases, be induced by rapid mixing with an excess of a compound that reacts with the ligand rather than the protein. For example, the dissociation of calcium from calcium-binding proteins such as calmodulin and calbindin D_{9k} can be studied by rapid mixing with an excess of a fluorescent calcium chelator such as Quin 2 (Martin *et al.*, 1990), which forms a strongly fluorescing, high affinity 1:1 complex with calcium.

C. Competition Experiments

The interaction of a nonfluorescent ligand with a protein can, in certain cases, be studied using competition with a second, fluorescent, ligand. The reactions involved are again those shown in Scheme 3. The experiment involves rapidly mixing the protein with different premixed solutions of N *and* L. If the dissociation rate constants are very small and both L and N are in large excess over P then the observed first-order rate constant will be given by

$$k_{\mathrm{OBS}} = k_{+1}L_{\mathrm{tot}} + k_{+2}N_{\mathrm{tot}} \tag{13}$$

If the experiment is performed using different concentrations of N at a fixed concentration of L then a plot of k_{OBS} versus $[\mathrm{N}_{\mathrm{tot}}]$ should give a straight line with slope k_{+2} and intercept $k_{+1}L_{\mathrm{tot}}$. More complex behavior will be observed if the dissociation rate constants are not small enough to be ignored. Competition experiments can be used even if one of the reactions consists of two steps [e.g., as in Scheme 4 (Engelborghs and Fitzgerald, 1987)].

Competition experiments can also be performed using relaxation measurements. If conditions are chosen such that the reaction with the fluorescent ligand is much faster than the reaction with the nonfluorescent ligand then two relaxation processes will be observed. The reciprocal relaxation times for the fast (τ_F) and slow (τ_S) processes will be given by Guillain and Thusius (1970)

$$\tau_F^{-1} = k_{+1}([P] + [L] + k_{-1}) \tag{14}$$

$$\tau_S^{-1} = k_{+2}\left([P] + \frac{[N]([P] + K_{dl})}{[P] + [L] + K_{dl}}\right) + k_{-2} \tag{15}$$

One of the potential advantages of relaxation methods over flow methods is that such expressions for relaxation times are generally much easier to derive than the rate expressions for stopped-flow measurements (Bernasconi, 1976), and the experiments do not have to be performed under fixed (e.g., pseudo-first-order) conditions. The major disadvantage is that accurate equilibrium measurements must be done to permit calculation of the equilibrium concentrations. For this particular example, it should be possible to determine K_{d1} by direct equilibrium titration and K_{d2} by a displacement titration in which N is used to displace L from a preformed PL complex (see Martin and Bayley, 2002).

D. Analysis of Stopped-Flow Anisotropy Data

If there is no change in intensity accompanying complex formation, then the kinetic transients can be analyzed as simple exponentials and the data plotted using Eq. (5). However, if there is a change in fluorescence intensity then the time-dependent change in anisotropy must be described by Eccleston *et al.* (2001)

$$r(t) = f_L(t)r_L + f_{PL}(t)r_{PL} \tag{16}$$

where r_L and r_{PL} are the anisotropies of L and P, and $f_L(t)$ and $f_{PL}(t)$ are the fractional intensities of light coming from L and PL at time t:

$$f_L(t) = \frac{[L](t)}{[L](t) + D[PL](t)} \quad \text{and} \quad f_{PL}(t) = 1 - f_L(t)$$

Here D is the fluorescence intensity of PL divided by that of L. Substituting in Eq. (16) gives

$$r(t) = \frac{[\mathrm{L}](t)(r_{\mathrm{L}} - r_{\mathrm{PL}})}{[\mathrm{L}](t) + D[\mathrm{PL}](t)} + r_{\mathrm{PL}}$$

Substituting the expressions for the disappearance of L and appearance of PL, $[\mathrm{L}](t) = [\mathrm{L}_0]\mathrm{e}^{-k_{\mathrm{OBS}}t}$ and $[\mathrm{PL}](t) = [\mathrm{L}_0] - [\mathrm{L}](t)$ gives the expression that must be used to analyze the time dependence of the anisotropy change in order to obtain k_{OBS}:

$$r(t) = \frac{r_{\mathrm{L}} - r_{\mathrm{PL}}}{1 - D + D\mathrm{e}^{k_{\mathrm{OBS}}t}} \tag{17}$$

VII. Complex Reactions

A. Two-Step Mechanisms

Although many protein–ligand interactions conform to Scheme 2, this is by no means always the case. The most obvious indication of additional complexity in a rapid kinetic investigation is the observation of more than a single kinetic phase. When only a single kinetic phase is observed, complexity will most likely be indicated by the observation that the variation of the observed rate with concentration is not linear. One of the most commonly encountered complexities is the presence of an additional step that involves an isomerization. This can be a first-order isomerization of a protein–ligand complex following an initial second-order binding event:

Scheme 4 $$\mathrm{P} + \mathrm{L} \underset{k_{-1}}{\overset{k_{+1}}{\rightleftharpoons}} \mathrm{PL} \underset{k_{-2}}{\overset{k_{+2}}{\rightleftharpoons}} \mathrm{PL}^*$$

or a first-order isomerization of the protein (or the ligand) followed by a second-order binding event.

Scheme 5 $$\mathrm{P} \underset{k_{-1}}{\overset{k_{+1}}{\rightleftharpoons}} \mathrm{P}^* + \mathrm{L} \underset{k_{-2}}{\overset{k_{+2}}{\rightleftharpoons}} \mathrm{PL}^*$$

It is important to realize that neither of these schemes would be identified by simple equilibrium binding measurements. Analysis of binding curves for these schemes will always appear to conform to Scheme 2, with the experimentally measured dissociation constant given by

$$K_{\mathrm{d}} = \frac{K_{\mathrm{d1}}K_{\mathrm{d2}}}{1 + K_{\mathrm{d2}}} \quad \text{for Scheme 4} \tag{18}$$

and

$$K_{\mathrm{d}} = K_{\mathrm{d2}}(1 + K_{\mathrm{d1}}) \quad \text{for Scheme 5} \tag{19}$$

with the individual equilibrium dissociation constants defined as $K_{\mathrm{d1}} = k_{-1}/k_{+1}$ and $K_{\mathrm{d2}} = k_{-2}/k_{+2}$ for both schemes.

1. Flow Methods

The analytical solutions to the rate equations for these simple two-step mechanisms can be derived for flow experiments only when one of the concentrations is in large excess, so that it remains effectively invariant over the time course of the reaction. If the second-order binding step in Scheme 4 is very much faster than the isomerization step, and L is in large excess over P, then a stopped-flow record will have two kinetic phases, with the fast process varying linearly with L_{tot} according to Halford (1971)

$$k_{OBS}(F) = k_{+1}L_{tot} + k_{-1} \tag{20}$$

and the slow process varying hyperbolically with L_{tot} according to

$$k_{OBS}(S) = \frac{k_{+2}L_{tot}}{K_{d1} + L_{tot}} + k_{-2} \tag{21}$$

What will actually be observed experimentally will clearly depend on the relative contributions of the different species to the optical signal being monitored, as well as on the magnitudes of the individual rate constants (see Fig. 4). Only in the most favorable cases, where two easily resolvable kinetic events are observed over a wide range of L_{tot} values (with the bimolecular step *always* remaining very much faster than the isomerization for all L_{tot}), will it be possible to extract all four rate constants by analyzing Eqs. (20) and (21). Thus, although inspection of Eq. (21) shows that $k_{OBS}(S)$ should *increase* from k_{-2} when $L_{tot} \ll K_{d1}$ to $(k_{-2} + k_{+2})$ when $L_{tot} \gg K_{d1}$, this will not always be observable. Consider, for example, a typical situation where the bimolecular rate constant (k_{+1}) is 10^7 M^{-1} sec^{-1} and the

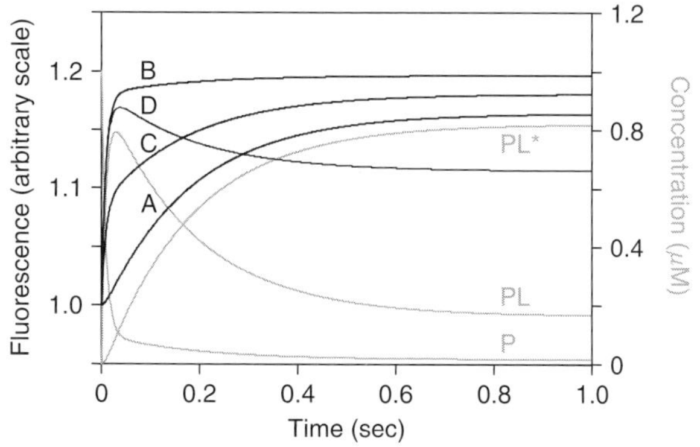

Fig. 4 Relationship between the simulated time-dependent changes in concentrations of P, PL, and PL* (grey lines, right axis) for the reaction scheme P + L ⇔ PL ⇔ PL* (Scheme 4, with $P_{tot} = 1$ μM, $L_{tot} = 10$ μM, $k_{+1} = 10$ $\mu M^{-1} sec^{-1}$, and k_{-1}, k_{+2}, k_{-1} = 10, 5, and 1 sec^{-1}, respectively) and the observed fluorescence of the mixture (black lines, left axis). The specific fluorescence of PL and PL* relative to that of P are, respectively, 1.0 and 1.2 (A), 1.2 and 1.2 (B), 1.1 and 1.2 (C), and 1.2 and 1.1 (D).

overall equilibrium dissociation constant as defined in Eq. (18) is 0.1 μM. Two extreme situations can then be considered:

- If, on the one hand, K_{d1} is low, then, given the nature of the typical stopped-flow experiment, it is unlikely that it will be possible to work under conditions where $L_{tot} \ll K_{d1}$ and $k_{OBS}(S)$ will vary significantly with L_{tot} only when $k_{-2} < k_{+2}$ and k_{-2} will generally be difficult to determine. In extreme cases, K_{d1} may be so low that $k_{OBS}(S)$ may well be completely independent of L_{tot} under all attainable experimental conditions and only the sum of the rate constants for the isomerization step will be measurable. In such cases, it may be possible to perform displacement experiments (see Section VI.B) in order to determine one or both of the dissociation rate constants.
- If, on the other hand, the bimolecular step is the fast diffusion controlled formation of an encounter complex with very low overall affinity (high $K_{d1} \sim 1$ mM), then the value of K_{d2} would need to be 100 μM to give the overall equilibrium dissociation constant of 0.1 μM [see Eq. (18)]. A typical stopped-flow experiment could then show only a single transient process because occupancy of the intermediate PL would always be very low. The single observed rate might then vary linearly with L_{tot} [since deviations from linearity would be observed only when L_{tot} approached K_{d1}, see Eq. (21)] with an apparent second-order rate constant of k_{+2}/K_{d1} and dissociation rate constant of k_{-2}. When L_{tot} does approach K_{d1}, some curvature may, of course, be observed and the *initial* slope can then be taken as equal to k_{+2}/K_{d1} (see De La Cruz *et al.*, 2001, for a good discussion of these effects).

The association rate constants measured for protein–ligand and protein–protein interactions are, in fact, often significantly lower than the values predicted using theoretical calculations based on diffusion coefficients, shape, and viscosity. Scheme 4, with a high value of K_{d1} and a low value of k_{+2}, is frequently invoked as an explanation for the observation of these unexpectedly low values.

In the case of Scheme 5, with the bimolecular step very much faster than the isomerization step, the rate expressions for experiments performed under the condition that $L_{tot} > (P_{tot} + P^*_{tot})$ are (Halford, 1971)

$$k_{OBS}(F) = k_{+2}L_{tot} + k_{-2} \tag{22}$$

$$k_{OBS}(S) = \frac{k_{-1}K_{d2}}{K_{d2} + L_{tot}} + k_{+1} \tag{23}$$

Thus, Scheme 5 can, at least in principle, be distinguished from Scheme 4 by the fact that the observed rate for the slow process should *decrease* from $(k_{-1} + k_{+1})$ when $L_{tot} \ll K_{d2}$ to k_{+1} when $L_{tot} \gg K_{d2}$. However, as for Scheme 4, only in the most favorable cases will it be possible to extract all four rate constants for the reaction.

2. Relaxation Methods

The relaxation times in a multistep process are the eigen values of a system of coupled differential equations. There will be as many relaxations as there are *independent* concentration variables, though not all of these relaxation times will necessarily be observable. The number of independent concentration variables (two for Schemes 4 and 5) will be equal to or smaller than the number of steps, depending on the concentration conservation conditions for the scheme. For Scheme 4, Eigen (1968) has shown that the sums and products of the reciprocal relaxation times are given by

$$\tau_F^{-1} + \tau_S^{-1} = k_{+1}([P] + [L]) + k_{-1} + k_{+2} + k_{-2} \tag{24}$$

$$\tau_F^{-1}\tau_S^{-1} = k_{+1}(k_{+2} + k_{-2})([P] + [L]) + k_{-1} + k_{-1}k_{-2} \tag{25}$$

and in favorable cases, these two combinations can be plotted versus ([P] + [L]) to yield all four rate constants (Eigen, 1968). If the relaxation times are sufficiently well separated, these equations can be uncoupled to relate the individual relaxation times to particular steps. For example, if the first (bimolecular) step in Scheme 4 is very much faster than the isomerization step then it can be treated as an isolated process (namely Scheme 2), but an additional concentration factor has to be included with the isomerization term to account for coupling with the faster step.

Thus, because $k_{+1}([P] + [L]) + k_{-1} \gg k_{+2} + k_{-2}$, one can write (Halford, 1972):

$$\tau_F^{-1} = k_{+1}[(P] + [L]) + k_{-1} \tag{26}$$

and

$$\tau_S^{-1} = \frac{\tau_F^{-1}\tau_S^{-1}}{k_{+1}([P] + [L]) + k_{-1}} = \frac{k_{+2}([P] + [L])}{[P] + [L] + K_{d1}} + k_{-2} \tag{27}$$

The reciprocal relaxation time will then vary from k_{-2} when ([P] + [L]) $\ll K_{d1}$ to a limiting plateau value of $k_{+2} + k_{-2}$ when ([P] + [L]) $\gg K_{d1}$. As with flow experiments it may not always be possible to work under conditions that permit determination of all four rate constants.

For Scheme 5 with a fast bimolecular step, uncoupling leads to the following expressions for the relaxation times:

$$\frac{1}{\tau_F} = k_{+2}([P^*] + [L]) + k_{-2} \tag{28}$$

$$\frac{1}{\tau_s} = \frac{k_{-1}([P^*] + K_{d2})}{[P^*] + [L] + K_{d2}} + k_{+1} \tag{29}$$

In the case of Scheme 4, solving Eq. (7) using the measured overall dissociation constant gives ([PL] + [PL*]), so that [P] and [L] for use in Eqs. (26) and (27) can be calculated. This is not the case for Scheme 5; solving Eq. (7) gives [PL*], but the equilibrium between P and P* is not known and [P*] cannot be calculated. Two solutions are possible: work under conditions where L is in large excess or use an iterative procedure in the fitting.

B. Multistep Mechanisms

Multistep mechanisms will almost invariably consist of a series of first-and second-order reactions and in some simple cases, particularly for relaxation experiments, it is possible to derive the analytical solutions necessary for a kinetic analysis (see, e.g., Halford, 1972). However, a more common approach is to try to study the individual steps in isolation (e.g., De La Cruz *et al.*, 1999, 2001; Eccleston *et al.*, 2006).

VIII. Data Analysis in Practice

As noted above, most commercial equipment will come with software for analyzing kinetic transients as sums of exponential terms, and this aspect of data analysis will not be discussed further here. When analyzing rate expressions, such as that given in Eq. (23), it is, in general, not good practice to transform them into linear functions, because the associated errors transform accordingly (see Chapter 24 by Johnson, this volume, for further discussion). There are now numerous mathematical procedures available for χ^2-minimization of nonlinear functions such as these; for example, the widely used Marquardt procedure is both efficient and relatively robust (Press *et al.*, 1990). Whenever possible, it is best to determine the variance in k_{OBS} values for each value of the independent concentration variable. The resulting sample variances may then be used to weight each k_{OBS} value by the inverse of its estimated variance. In some cases, it may not be possible to obtain variances for individual samples, and it is then reasonable to assume that the *relative* error in k_{OBS} is constant. The fitting should then be done to the logarithm of k_{OBS} since the error in log(k_{OBS}) will be constant. This is particularly important in cases where k_{OBS} values vary by more than an order of magnitude. Fitting using the logarithms of rate and equilibrium constants has the additional advantage that it forces them to be physically meaningful (positive) values.

As noted elsewhere, most stopped-flow fluorescence studies are performed at a single emission wavelength (or range of wavelengths), but it is also possible to collect complete fluorescence spectra as a function of time. This permits a more rigorous analysis of kinetic data and allows the measurement of the spectra of intermediates in the reaction pathway. Furthermore, the application of principal component analysis may permit the direct determination of the number of fluorescing species in the reaction scheme on a completely model-independent basis, and therefore may aid in model selection.

The traditional approaches to analyzing rapid kinetic data described in the previous sections generally involve fitting the time dependence of an observed optical signal to one or more exponential terms. The requirement for an explicit analytical solution to the rate equations for the reaction generally necessitates simplifying assumptions and places what are often severe constraints on the experimental conditions that can be used. In many cases, it will not be possible to work within these constraints. For example, if it is not possible to work under pseudo-first-order conditions, it will be necessary to analyze progress curves for two-step reactions using an iterative method based on numerical integration of the appropriate differential rate equations.

Global analysis allows the fitting of multiple kinetic data sets obtained under different concentration conditions. The simultaneous analysis of the different data sets has the potential to achieve better definition of the rate constants common to all the sets. In favorable cases, it may allow the determination of kinetic constants not attainable by traditional methods and can be used to distinguish between different kinetic models. Another strong point of global analysis is that the different data sets can be obtained using different methods, for example, fluorescence intensity and anisotropy data, in which the kinetic constants are nevertheless the same. In such cases, it is important to weight the different data sets correctly. This can be done by determining the standard deviation for a signal that has reached equilibrium. One potential drawback is that if the kinetic constants to be determined are not adequately constrained by the data there will be a large range of constants giving equally good fits to the data. The extent to which a particular rate constant is defined by the data can be tested by simulating the mechanism and varying each rate constant in turn (see Chapter 25 by Schilstra *et al.*, this volume). Rate constants that are poorly constrained may need to be held constant, either at an estimated value or at a value determined in an independent approach. We recommend that a conventional kinetic analysis should always be attempted before embarking on global analysis. No mathematical treatment, however sophisticated, can make up for less than adequate data collection.

Having extracted rate constants, it is generally instructive to simulate the reaction by computer methods (see Chapter 25 by Schilstra *et al.*, this volume) in order to see how well the data fits the assumed mechanism. This is most often done at the level of simulating how the observed rate of a particular process depends on the concentrations of the reagents, but it is also instructive to simulate individual reaction traces. Computer simulation is also invaluable as a teaching tool and a useful aid in the design of experiments. In our experience, intuitive arguments can frequently be wrong, even in apparently simple situations.

References

Abbruzzetti, S., Crema, E., Masino, L., Vecli, A., Viappiani, C., Small, J. R., Libertini, L. J., and Small, E. W. (2000). Fast events in protein folding: Structural volume changes accompanying the early events in the n→l transition of apomyoglobin induced by ultrafast pH jump. *Biophys. J.* **78,** 405–415.

Barman, T. E., Bellamy, S. R., Gutfreund, H., Halford, S. E., and Lionne, C. (2006). The identification of chemical intermediates in enzyme catalysis by the rapid quench-flow technique. *Cell. Mol. Life Sci.* **63,** 2571–2583.

Bernasconi, C. F. (1976). "Relaxation Kinetics." Academic Press, New York.

Brune, M., Corrie, J. E. T., and Webb, M. R. (2001). A fluorescent sensor of the phosphorylation state of nucleoside diphosphate kinase and its use to monitor nucleoside diphosphate concentrations in real time. *Biochemistry* **40,** 5087–5094.

Brune, M., Hunter, J. L., Howell, S. A., Martin, S. R., Hazlett, T. L., Corrie, J. E. T., and Webb, M. R. (1998). Mechanism of inorganic phosphate interaction with phosphate binding protein from *Escherichia coli*. *Biochemistry* **37,** 10370–10380.

Caldin, E. F. (1964). "Fast Reactions in Solution." Blackwell Scientific Publications, Oxford.

Cremo, R. C. (2003). Fluorescent nucleotides: Synthesis and characterization. *Meth. Enzymol.* **360,** 128–177.

De La Cruz, E. M., Ostap, E. M., and Sweeney, H. L. (2001). Kinetic mechanism and regulation of myosin VI. *J. Biol. Chem.* **276,** 32373–32381.

De La Cruz, E. M., Wells, A. L., Rosenfeld, S. S., Ostap, E. M., and Sweeney, H. L. (1999). The kinetic mechanism of myosin V. *Proc. Natl. Acad. Sci. USA* **96,** 13726–13731.

Eccleston, J. F., Hutchinson, J. P., and White, H. D. (2001). "Protein-Ligand Interactions: Structure and Spectroscopy" (S. E. Harding, and B. Z. Chowdhry, eds.), pp. 201–237. Oxford University Press, Oxford.

Eccleston, J. F., Petrovic, A., Davis, C. T., Rangachari, K., and Wilson, R. J. M. (2006). The kinetic mechanism of the SufC ATPase: The cleavage step is accelerated by SufB. *J. Biol. Chem.* **281,** 8371–8378.

Eftink, M. R. (1997). Fluorescence methods for studying equilibrium macromolecule-ligand interactions. *Meth. Enzymol.* **278,** 221–257.

Eigen, M. (1968). New looks and outlooks on physical enzymology. *Q. Rev. Biophys.* **1,** 3–33.

Engelborghs, Y., and Fitzgerald, T. J. (1987). A fluorescence stopped flow study of the competition and displacement kinetics of podophyllotoxin and the colchicine analog 2-methoxy-5-(2′,3′,4′-trimethoxyphenyl) tropone on tubulin. *J. Biol. Chem.* **262,** 5204–5209.

Guillain, F., and Thusius, D. (1970). The use of proflavin as an indicator in temperature-jump studies of the binding of a competitive inhibitor to trypsin. *J. Am. Chem. Soc.* **92,** 5534–5536.

Gutman, M., and Nachliel, E. (1990). The dynamic aspects of proton transfer processes. *Biochim. Biophys. Acta* **1015,** 391–414.

Halford, S. E. (1971). *Escherichia coli* alkaline phosphatase. An analysis of transient kinetics. *Biochem. J.* **125,** 319–327.

Halford, S. E. (1972). *Escherichia coli* alkaline phosphatase. Relaxation spectra of ligand binding. *Biochem. J.* **126,** 727–738.

Hill, J. J., and Royer, C. A. (1997). Fluorescence approaches to the study of protein-nucleic acid complexation. *Meth. Enzymol.* **278,** 390–416.

Jameson, D. M., and Eccleston, J. F. (1997). Fluorescent nucleotide analogs: Synthesis and applications. *Meth. Enzymol.* **278,** 363–390.

Malcolm, A. D. (1972). Coenzyme binding to glutamate dehydrogenase. A study by relaxation kinetics. *Eur. J. Biochem.* **27,** 453–461.

Málnási-Csizmadia, A., Pearson, D. S., Kovács, M., Woolley, R. J., Geeves, M. A., and Bagshaw, C. R. (2001). Kinetic resolution of a conformational change and the ATP hydrolysis step using relaxation methods with a Dictyostelium myosin II mutant containing a single tryptophan residue. *Biochemistry* **40,** 12727–12737.

Martin, S. R., and Bayley, P. M. (2002). Regulatory implications of a novel mode of interaction of calmodulin with a double IQ-motif target sequence from murine dilute myosin V. *Protein Sci.* **11,** 2909–2923.

Martin, S. R., Linse, S., Johansson, C., Bayley, P. M., and Forsén, S. (1990). Protein surface charges and Ca^{2+} binding to individual sites in calbindin D_{9k}: Stopped flow studies. *Biochemistry* **29,** 4188–4193.

Molecular Probes (2006). "The Handbook—A Guide to Fluorescent Probes and Labeling Technologies." Invitrogen, http://probes.invitrogen.com/handbook/.

Pearson, D. S., Holtermann, H., Ellison, P., Cremo, C., and Geeves, M. A. (2002). A novel pressure-jump apparatus for the microvolume analysis of protein-ligand and protein-protein interactions: Its application to nucleotide binding to skeletal-muscle and smooth-muscle myosin subfragment-1. *Biochem. J.* **366,** 643–651.

Press, W. H., Flannery, B. P., Teukolsky, B. P., and Vetterling, W. T. (1990). "Numerical Recipes. The Art of Scientific Computing. Fortran Version." Cambridge University Press, Cambridge, UK.

Rabl, C. R. (1979). High-resolution temperature-jump measurements with cooling correction. *In* "Techniques and Applications of Fast Reactions in Solution" (W. J. Gettins, and E. Wyn-Jones, eds.), pp. 77–82. Reidel, Dordrecht.

Ross, J. B. A., Szabo, A. G., and Hogue, C. W. V. (1997). Enhancement of protein spectra with tryptophan analogs: Fluorescence spectroscopy of protein-protein and protein-nucleic acid interactions. *Meth. Enzymol.* **278,** 151–190.

Shastry, M. C. R., Luck, S. D., and Roder, H. (1998). A continuous-flow capillary mixer to monitor reactions on the microsecond time scale. *Biophys. J.* **74,** 2714–2721.

Turner, D. H., Flynn, G. W., Sutin, N., and Beitz, J. V. (1972). Laser Raman temperature-jump study of the kinetics of the triiodide equilibrium. Relaxation times in the 10^{-8}–10^{-7} second range. *J. Am. Chem. Soc. USA* **94,** 1554–1559.

Wakelin, S., Conibear, P. B., Woolley, R. J., Floyd, D. N., Bagshaw, C. R., Kovács, M., and Málnási-Csizmadia, A. (2003). Engineering Dictyostelium discoideum myosin II for the introduction of site-specific fluorescence probes. *J. Musc. Res. Cell Motil.* **23,** 673–683.

Williams, S., Causgrove, T. P., Gilmanshin, R., Fang, K. S., Callender, R. H., Woodruff, W. H., and Dyer, R. B. (1996). Fast events in protein folding: Helix melting and formation in a small peptide. *Biochemistry* **35,** 691–697.

Woodward, S. K. A., Eccleston, J. F., and Geeves, M. A. (1991). Kinetics of the interaction of 2′(3′)-O-(N-methylanthraniloyl)-ATP with myosin subfragment 1 and actomyosin subfragment 1: Characterization of two acto.S1.ADP complexes. *Biochemistry* **30,** 422–430.

Wu, X., Gutfreund, H., and Chock, P. B. (1992). Kinetic method for differentiating mechanisms for ligand exchange reactions: Application to test for substrate channeling in glycolysis. *Biochemistry* **31,** 2123–2128.

CHAPTER 16

Mutagenic Analysis of Membrane Protein Functional Mechanisms: Bacteriorhodopsin as a Model Example

George J. Turner

Department of Chemistry and Biochemistry
Seton Hall University
South Orange, New Jersey 07079

METHODS IN CELL BIOLOGY, VOL. 84

0091-679X/08 $35.00
DOI: 10.1016/S0091-679X(07)84016-7

I. Introduction

Membrane proteins are ubiquitous components in pathways that regulate cell physiology. Despite their critical importance, knowledge of membrane protein high-resolution structures and mechanisms of action has lagged far behind our understanding of these properties for soluble proteins (White, 2004). Much of that has to do with difficulties in obtaining membrane protein quantities required for physical studies of structure–function relationships. Genetic engineering can be used as an alternative probe of mechanisms of function. As a uniquely tractable system for investigating membrane protein structure–function relationships bacteriorhodopsin (bR) will be used to illustrate how random, site-directed, and scanning mutagenesis can be used, in combination with physical analyses, to demonstrate how individual amino acids, as well as functional domains, are energetically coupled to membrane protein function. While the technical aspects of mutagenesis have been greatly simplified, rationalization of structural and functional consequences requires careful interpretation and extensive verification. Indeed, mechanistic insights beyond the purview of the highest resolution structural analyses can be obtained by mutagenic strategies.

All cells sense their environment and transport solutes across the lipid bilayer via integral membrane proteins. Understanding the molecular mechanism of these activities requires not only knowledge of the receptor, the transporter, the channel structure, and conformational changes but also the physical forces that regulate structural transitions coupled to membrane protein function. Even for soluble proteins, this type of information is difficult to obtain due to the cooperative nature of intermediate-state conformational transitions. Model systems that have provided comprehensive descriptions of structures, molecular interactions, and energetics coupled to protein function include myoglobin (Frauenfelder *et al.*, 2003; Garcia-Moreno, 1995; Matthew *et al.*, 1985), hemoglobin (Ackers and Holt, 2006), thymidylate synthetase (Finer-Moore *et al.*, 2003), staph nuclease (Shortle, 1995), lysozyme (Matthews, 1996; Merlini and Bellotti, 2005), arc repressor (Sauer *et al.*, 1996), and barnase (Daggett and Fersht, 2003; Fersht, 1993)—all soluble proteins. While membrane proteins make up ~30% of the coding region of the human genome (Liu and Rost, 2001), no membrane protein has yet been sufficiently characterized to allow description of its mechanism of action at a comparable level of detail.

Of particular interest are the seven-transmembrane alpha-helical proteins that comprise the largest gene family in higher eukaryotes, the G protein-coupled receptors (GPCRs, Baldwin *et al.*, 1997). GPCRs regulate biological activities as diverse as cell growth, cardiac activity, smooth muscle activity, nociception, cognition, cell development, taste, and sexual function. Aberrant GPCR activity has been linked to numerous pathologies and it is estimated that 50% of drugs on the market today target a GPCR (Wise *et al.*, 2002). The molecular events leading to GPCR function are largely unknown due to our inability to characterize the structural and energetic correlates of ligand binding specificity, receptor activation, and coupling to trimeric G proteins.

GPCR activation likely involves ligand-dependent conformational changes. Classically, the pharmacological activity of GPCRs has been modeled as a transition between liganded and unliganded states. However, the diversity of ligand and effector interactions and cellular responses, even for a single GPCR, indicates that two-state models of receptor activation are an oversimplification (Kenakin, 1997). Recent studies have provided clues that Adrenergic and Muscarinic GPCR activation proceeds through multiple conformational states, but defining their numbers and structures remains challenging (Han *et al.*, 2005; Kobilka, 2004; Liapakis *et al.*, 2004). Without knowledge of all conformational states and concomitant functional linkages, the full therapeutic potential of GPCRs will be difficult to maximize.

Due to the lack of high-resolution structural analyses of GPCR (the only X-ray structure is the inactive form of bovine rhodopsin, Palczewski *et al.*, 2000) some of the most insightful observations regarding receptor structure–function relationships come from mutational studies (Schoneberg *et al.*, 2004; Tao, 2006). This chapter illustrates how mutational studies have contributed to determining the functional mechanism of a membrane protein model system, bR. The combination of mutagenic and physical analyses can identify the amino acid interactions that participate in the membrane protein function. It will be demonstrated that single amino acid mutations of can be used to isolate otherwise inaccessible individual (or subsets of) intermediate-state conformations and to identify molecular domains that are energetically coupled to membrane protein functional properties.

II. Rationale

A. bR as a Model System

bR is a small, polytopic membrane protein that serves as a light-activated proton pump in the halophilic Archaeon, *Halobacterium salinarum*. bR consists of seven alpha-helical transmembrane segments and is thus a structural homologue of the eukaryotic GPCR (Fig. 1, Kimura *et al.*, 1997; Luecke *et al.*, 1999b; Pebay-Peyroula *et al.*, 1997). There are numerous advantages for using bR as a model system for membrane protein structure–function analyses. First is the facility with which bR can be obtained. *H. salinarum* strains are readily available and can be maintained with microbiological techniques commonly used for culturing *Escherichia coli* (DasSarma and Fleischmann, 1995). In wild-type and overproducing strains (S1 and ET1001, respectively), bR can occupy from 50 to 70% of the membrane surface area, producing 15 mg of protein per liter of culture, an extremely high density of receptors. This naturally high level of accumulation results in the formation of specialized patches of membrane called the

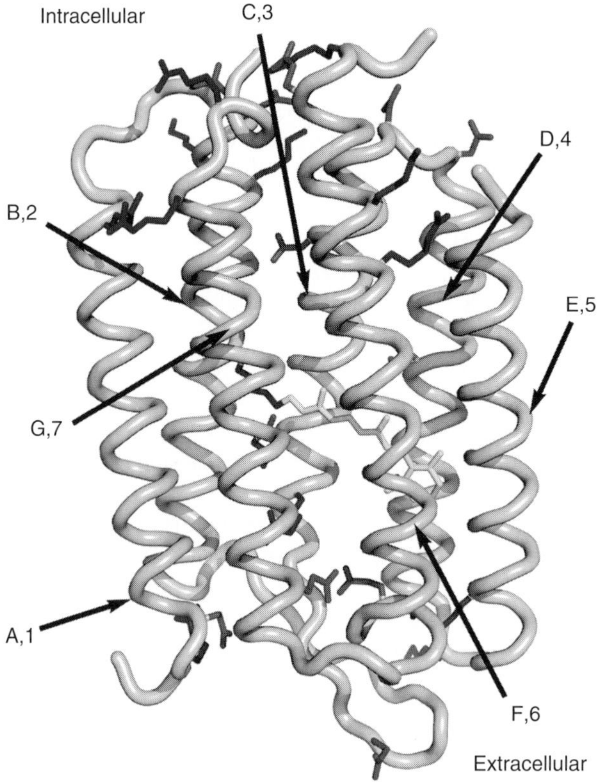

Fig. 1 Three-dimensional model for the ground-state structure of bR (1C3W, Luecke *et al.*, 1999b). The backbone tracing of the seven-transmembrane spanning alpha helices and extramembranous linking regions is shown in pink. Each helix is identified by standard letter and number designations. The retinal chromophore is shown in yellow. All carboxylate (red) and amine (blue) containing amino acid side chains are indicated. (See Plate no. 8 in the Color Plate Section.)

purple membrane (PM) in which trimers of bR assemble in two-dimensional crystalline lattices. The high protein-to-lipid ratio causes the PM to have a higher buoyant density than the rest of the cell membrane (Belrhali *et al.*, 1999; Oesterhelt and Stoeckenius, 1974). This high-density PM is purified by sucrose density ultracentrifugation of the membrane fraction of cell lysates. bR is the only protein in the PM. bR is a one-to-one complex of the bacterio-opsin apoprotein and a Schiff base-coupled chromophore, retinal. Amino acid–chromophore interactions lead to convenient absorption properties in the visible region of the electromagnetic spectrum. *H. salinarum* produces so much bR that wild-type cell cultures become intensely purple. bR's chromophoric properties are a key component of the structure–function analytical strategy to be discussed.

A second advantage is the wealth of physical analyses that have been brought to bear to define the number and structural properties of intermediate-states contributing to bR function (Subramaniam *et al.*, 1999; Varo and Lanyi, 1991). This dramatically contrasts the lack of knowledge regarding GPCR intermediate-state conformations and conformational changes. Changes in protein–chromophore interactions during proton pumping alter bR visible absorbance properties, such that the accumulation of intermediate-states can be monitored spectroscopically. The bR proton-pumping mechanism consists of sequential transition through seven unique spectral components [bR, J, K, L, M, N, and O (Lozier *et al.*, 1992; Mathies *et al.*, 1988; Varo and Lanyi, 1991) Fig. 2]. Atomic resolution structural models exist for a number of those states (Lanyi and Schobert, 2002, 2003; Luecke *et al.*, 1999a,b; Pebay-Peyroula *et al.*, 1997; Rouhani *et al.*, 2001; Royant *et al.*, 2001; Vonck, 1996). Structural models of the ground-state and intermediate-state bR conformations comprise a significant advantage in the design and analysis of mutational studies. Indeed, bR has served as a useful model of the arrangement of GPCR transmembrane domains and ligand-binding sites (Marjamaki *et al.*, 1999; Nikiforovich *et al.*, 2001; Trumpp-Kallmeyer *et al.*, 1992; Zhao *et al.*, 1998). Based on analogy with the bR and rhodopsin photomechanisms, the essence of GPCR activation likely lies within transitions between multiple ligand-dependent substates.

A further advantage of the bR model system is the facility with which genetic engineering can be accomplished. The *bacterio-opsin* gene was isolated and sequenced in 1981 (Dunn *et al.*, 1981) and every conceivable mutagenesis strategy has subsequently been applied in effort to unravel bR's molecular mechanism of function. Taking advantage of bR's chromophore and photosynthetic function,

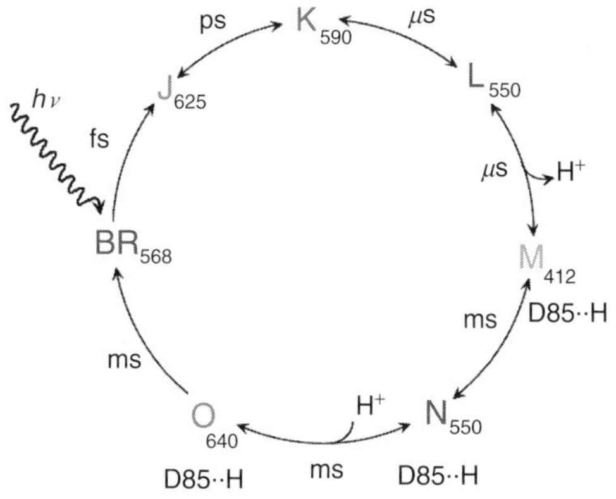

Fig. 2 Wild-type bacteriorhodopsin reaction scheme. The color of the letter denotes that of the intermediate and the subscript denotes its wavelength at maximal absorbance (λ_{max}). The kinetic scheme is essentially that described by Lozier *et al.* (1975) with the following exceptions: The J intermediate was described by Mathies *et al.* (1988); the inclusion of back reactions in the kinetic scheme was described by Lozier *et al.* (1992). The transitions in which the proton is released to the extracellular compartment and taken up from the cytoplasmic compartment are indicated. The intermediates in which aspartate 85 is protonated are indicated by D85··H. The time domains of the kinetic transitions are indicated.

the first mutagenic studies exploited naturally occurring mutants, random mutagenesis and *in vivo* screens and selections to identify determinants of bR expression and amino acids essential for light-activated proton pumping (DasSarma *et al.*, 1983; Oesterhelt and Krippahl, 1983). The complete coding region was chemically synthesized (Nassal *et al.*, 1987) and heterologous expression systems developed, allowing subsequent site-directed mutagenic analyses (Hildebrandt *et al.*, 1989; Shand *et al.*, 1991). Ultimately, a convenient homologous expression system was developed affording large quantities of bR mutants in native lipid environments (Turner *et al.*, 1993). As a result, an extensive library of bR mutants has been generated and characterized (for review see Brown, 2000), including all aspartate (Mogi *et al.*, 1988), threonine, serine (Marti *et al.*, 1991), proline (Mogi *et al.*, 1989), arginine (Lin *et al.*, 1991; Stern and Khorana, 1989), tryptophan (Wu *et al.*, 1992), and tyrosine residues (Mogi *et al.*, 1987). Insights derived from this body of work are likely not unique to bR and should serve as a guide in experimental design and analysis of less well-understood membrane proteins. To illustrate the degree of mechanistic insight that can be obtained by mutagenic strategies, this chapter will focus on the role of two critical aspartates in the bR proton-pumping mechanism. When rationalized with the combination of careful physical studies surprising mechanistic insights are revealed.

B. Mutagenic Rationale

Amino acid perturbations can be engineered into membrane protein coding sequences by random or directed strategies. Either approach can be biased by functional or structural intuition.

Random mutagenesis has been referred to as the "American Wild West Perturbation Strategy" (shoot first and ask questions later). The method relies on the introduction of numerous random (or semi-random) amino acid substitutions over extensive regions of gene coding sequences. The perturbation techniques can be directed at functional domains and structural elements or be totally unbiased (saturating). This strategy is limited only by the ability to evaluate the large numbers of mutants that can be generated (the "shooting first"). For example, saturation mutagenesis of a tripeptide can yield 20^3 unique molecules. The ultimate success of random mutagenesis depends on the availability of efficient phenotypic screens or selections to identify mutations of interest (the "questions asked later"). The examples discussed will demonstrate how random and structure-directed semirandom mutagenesis defined roles for aspartic acid 96 (D96) and aspartic acid 85 (D85) in the bR proton-pumping mechanism and the linkage of the protonated form of these amino acids to the conformational stability of a subset of bR intermediate-states.

Site-directed mutagenesis is useful for identifying functional roles of individual amino acids and requires structural (at least primary amino acid sequence) and/or chemical intuition for its application. bR contains 19 transbilayer and/or membrane interfacial aspartic (Asp, D) and glutamic (Glu, E) amino acids. bR contains no histidines (His, H). Intuitively, the Asp and Glu are candidate proton-binding and-release sites in the bR photosynthetic mechanism and each has been subjected to site-directed mutagenesis. A subset has been confirmed to have an essential role in the proton pumping. Analysis of mutations introduced at amino acid positions D96 and D85 will be used to illustrate how site-directed perturbation can result in mechanistic information regarding membrane protein structure–function relationships in both intuitive and surprising ways.

C. Analytical Rationale

Due to the combined success of genome sequencing projects and the advent of "kit" molecular biology, genetic engineering of a favorite membrane protein may be the most straightforward part of any mutagenic structure–function project. Frequently, it is much more challenging to rationalize the functional and structural consequences of amino acid perturbation. Sound interpretation relies on careful comparison with the wild-type structure and function via as many biochemical and biophysical approaches as can be applied. Functional response to amino acid mutation can take three general forms: local, long-range coupled, or global. Each type of perturbation response will be illustrated by consideration of asparagine (Asn, N) substitutions of the aspartic acids at amino acid positions 96 and 85 of bR (D96N and D85N, respectively).

Interpretation of the vast majority of site-directed mutagenesis studies of membrane proteins has relied on evaluation of the local, noncovalent interactions in the immediate environment of the substituted amino acid. Rationalization is based on an assessment of changes in local stereochemistry and its influence on ionic and H-bonds as well as hydrophobic and Van der Waals interactions. More detailed analysis including calculated electrostatic interactions and estimated solvent exposure effects have an absolute requirement for structural data to model these interactions. These analytical strategies assume that the amino acid substitution produces only a local effect. The lack of membrane protein crystallographic structures severely limits interpretation of the consequences of amino acid perturbation by mutagenesis.

In addition to local effects, a single amino acid substitution may produce changes that are propagated throughout the membrane protein, or specific allosteric effects that are transmitted to sites remote from the mutation. The transmitted perturbation may be nonspecific or global in nature or disrupt highly coupled molecular interactions essential to functional mechanisms. This added complexity of cooperative and compensating interactions can confound the analysis of site-directed mutagenesis.

It is feasible to discriminate between local and global perturbations by an analysis of the patterns of functional responses that arise from mutations distributed throughout the protein's structure (Martinez *et al.*, 2002; Turner *et al.*, 1992). This analysis requires three-dimensional structural models for its application. In this approach, the chemistry of the individual substitutions is not considered. Rather, the mutations that perturb the function in specific ways are analyzed as a function of their position in the 3D structure. Thus, if a subset of mutations that perturb the function in similar ways cluster in a discrete structural domain, that domain is considered to be a part of the functional pathway. These amino acids may either participate directly in the function (e.g., active site) or may be involved in structural roles (e.g., orienting active site residues). Using a large numbers of mutations combined with structural data, it is feasible to determine the pathways through which local changes in a protein's structure are energetically coupled to its function (Ackers and Smith, 1986).

III. Materials and Methods

A. Prokaryotic Strains

The *E. coli* strain used was DH5α [F^-, *recA1*, *endA1*, *gyrA96*, *thi-1*, *hsdR17* (r^-_k, m^+_k), *supE44*, 1^-)]. The *H. salinarum* strains used were L33 (Vac^-, Rub^-, bR^-; 33, Wagner *et al.*, 1983) and S1 (bR^+).

B. Media and Growth Conditions

All salts and chemicals were reagent grade. Lennox Broth (LB) and Bacto-Agar were from Difco Laboratories (Detroit, MI). Complex haloarchaeal media was basal salts (DasSarma and Fleischmann, 1995) plus peptone (Oxoid, Unipath Ltd., Hampton, England). *H. salinarum* cells were transformed as described (DasSarma and Fleischmann, 1995). Selective media included 10- to 25-μM mevinolin (a gift from Merck and Co., Rahway, NJ). Growth of *H. salinarum* cultures was monitored spectrophotometrically at λ_{660}. Complex *E. coli* medium was LB. *E. coli* cells were transformed by Hanahan's high efficiency calcium shock method (Hanahan *et al.*, 1991). Selective media included 50 μg/ml ampicillin (Sigma, St. Louis, MO). *E. coli* and *H. salinarum* cells were grown at 37 °C in an Infors Multitron II Refrigerated Shaking Incubator (Laurel, MD) at 225 rpm.

C. Reagents

All restriction enzymes and DNA-modifying enzymes were from New England BioLabs, Inc., (Beverly, MA). Custom oligo-deoxynucleotides were synthesized by GIBCO Life Technologies (Rockville, MD). *Taq* Polymerase and deoxynucleotide triphosphates were from Perkin-Elmer (Norwalk, CT) and *Pfu* Polymerase was from Stratagene (La Jolla, CA). Wizard DNA miniprep kit was from Promega Corporation (Madison, WI), Qiaquick PCR purification kit was from Qiagen (Valencia, CA), and Freeze 'n Squeeze DNA gel extraction spin columns were from Bio-Rad (Hercules, CA). Electrophoresis grade agarose was from FMC Corporation (Rockland, ME).

D. Site-Directed Mutagenesis

The wild-type *bacterio-opsin* gene was isolated from the *H. salinarum* strain RI by the polymerase chain reaction (PCR) using oligos 1 and 2 as forward and reverse primers, respectively (Table I). The PCR oligos added upstream and downstream cloning sites recognized by the DNA restriction enzymes PstI and *Bam*HI, respectively. The resultant 1.2-kb PCR fragment was cloned into pUC19 and resultant plasmid (pENDS bop) formed the basis for all subsequent mutagenesis. A silent *Xho*I restriction site was introduced into the pENDS bop vector (Turner *et al.*, 1999) downstream of the G helix coding region at amino acid position 225 (oligo 3, Table I). The point mutations D85N and D96N were constructed by oligo-directed mutagenesis (Shand *et al.*, 1991).

The D85N and D85N:D96N mutants were constructed by a combination of restriction fragment subcloning and PCR as described (Martinez and Turner, 2002). The D96N mutant (Shand *et al.*, 1991) was cloned on a KpnI/NotI fragment into pENDS bop I (Turner *et al.*, 1999). Using pENDS bop I:D96N as a template, oligos 4 and 5 (Table I) were used to amplify a DNA fragment containing both the D85N and D96N mutations. Along with the D96N mutation, the DNA template introduces the silent restriction site SpeI within the codons for amino acids 99 and 100. Oligo 5 introduces the D85N mutation and the silent restriction site, *Bss*HII, within the codons for amino acids 81 and 82. The product was digested with

Table I
Oligodeoxynucleotides Used in Mutagensis

Oligo No.	Sequence (5′–3′)
1	TAATCTGCAGGATGGGTGCAACCGTGA
2	AAAAGGATCCGAGTACAAGACCGAGTG
3	CTCATCCTCCTGCGCTCGAGGGCGATCTTCGGCGAA
4	AAAAGGATCCGAGTACAAGACCGAGTG
5	TTGCGCGCTACGCTAACTGGCTG
6	CCTCGAAGGCCGAAAGCATGCGCCCCGAGGTC
7	TACTTTGAACGTGGATGCGACCTCGGGGCGCATGC
8	GCATCCACGTTCAAAGTACTGCGTAACGTTACCGTT
9	ATACGCGGACCACAACACAACGGTAACGTTACGCAG
10	GTGTTGTGGTCCGCGTATCCCGTCGTGTGGCTGACT
11	TCCCGCACCTTCGCTGCCGATCAGCCACACGACGGG
12	GGCAGCGAAGGTGCGGGAATCGTGCCGCTGAACATC
13	CATGAACAGCAGCGTCTCGATGTTCAGCGGCACGAT
14	GAGACGCTGCTGTTCATGGTGCTTGACGTGAGCGCG
15	GAGCCCGAAGCCGACCTTCGCGCTCACGTCAAGCAC
16	AAGGTCGGCTTCGGGCTCATCCTCCTGCGCTCGAGG
17	CTTCGCCGAAGATCGCCCTCGAGCGCAGGAGGAT
18	TTTCTGCAGCCTCGAAGGCCGAAAGCATG
19	TGGATCCCTTCGCCGAAGATCGCCCTC

DNA cloning sites described in the text are underlined.

*Bam*HI and *Bss*HII, purified and ligated into the pENDS:D85N vector. Desired clones were identified by restriction analysis and confirmed by DNA sequencing.

E. Scanning Mutagenesis

Scanning mutagenesis of the *bop* gene coding region was performed using *in vitro* gene assembly (Martinez *et al.*, 2002; Stemmer, 1994). Twelve overlapping oligonucleotides were designed to reconstruct the 200-bp region of the F and G helices and intervening linking region (oligos 6–17, Table I). Two additional oligonucleotides were designed (one for each strand) for each cysteine mutation, and replaced the wild-type codon with the sequence TGC. Fifty-nine cysteine mutants were individually synthesized by primerless PCR (Fig. 3, Martinez *et al.*, 2002).

Each of the pENDS D85N/cysteine double mutants was digested with *Pst*I/*Bam*HI and subcloned into the *H. salinarum* expression vector, pHex (Turner *et al.*, 1999). 1.5 μg of DNA for each of the pHex vectors was dried in a Savant Speed-Vac (Holbrook, NY) and resuspended in 15 μl of Spheroplasting Solution (0.5 M EDTA, 2 M NaCl, 27 mM KCl, 50 mM Tris–HCl, 15% sucrose, DasSarma and Fleischmann, 1995) for transformation into *H. salinarum*. The *bop* gene was isolated from transgenic *H. salinarum* cells and sequenced to confirm the anticipated mutations.

F. Purification of bR Mutants

H. salinarum cultures expressing the bR mutants were grown from single colonies in 5 ml cultures with 10 μM mevinolin. Following three subcultures, a 50 ml culture (10 μM mevinolin) was inoculated and grown to midlog phase (OD_{660} 0.5–0.7). This culture was used to inoculate a 1.5 l culture to an OD_{660} of 0.01. Cells were harvested, lysed, and bR purified as described previously (Turner *et al.*, 1993). Purified membranes were stored at −70 °C, in 25% sucrose.

G. Cellular Screen

To expedite the analysis of the numbers of bR mutants that can be generated by random and semirandom mutagenesis a cellular screen was developed. *H. salinarum* cultures expressing the bR constructions were grown to late stationary phase in 5 ml Rich Halo Media containing 10 μM mevinolin. Cultures were pelleted and resuspended in 1 ml of basal salts. Four 200 μl aliquots of cell suspension were transferred to a nylon membrane (Hybond N, Amersham) using a vacuum manifold, designed in house for this purpose. To prevent crystal formation upon drying, salt was removed from the membrane by equilibrating with three changes of deionized water. While this procedure likely lysed the *H. salinarum* cells, the resultant whole cell pastes (e.g., total membrane and cytoplasmic fractions) were retained on the membrane. Equilibration was accomplished by immersing a blotting pad (VWR, So. Plainfeild, NJ) in water (or buffer) such that the blotting pad was thoroughly wet, but not submerged. Nylon membranes were placed, sample side up on the blotting pad for 15 min. Cell samples were subsequently equilibrated in titration buffer (100 mM NaCl, 20 mM CAPS, 20 mM BTP, and 40 mM NaH_2PO_4) at pH 7.5, 9.0, and 10.5. The pH of the nylon membranes was checked with pH paper. Samples were allowed to dry, at room temperature, for 1 hour before spectra were

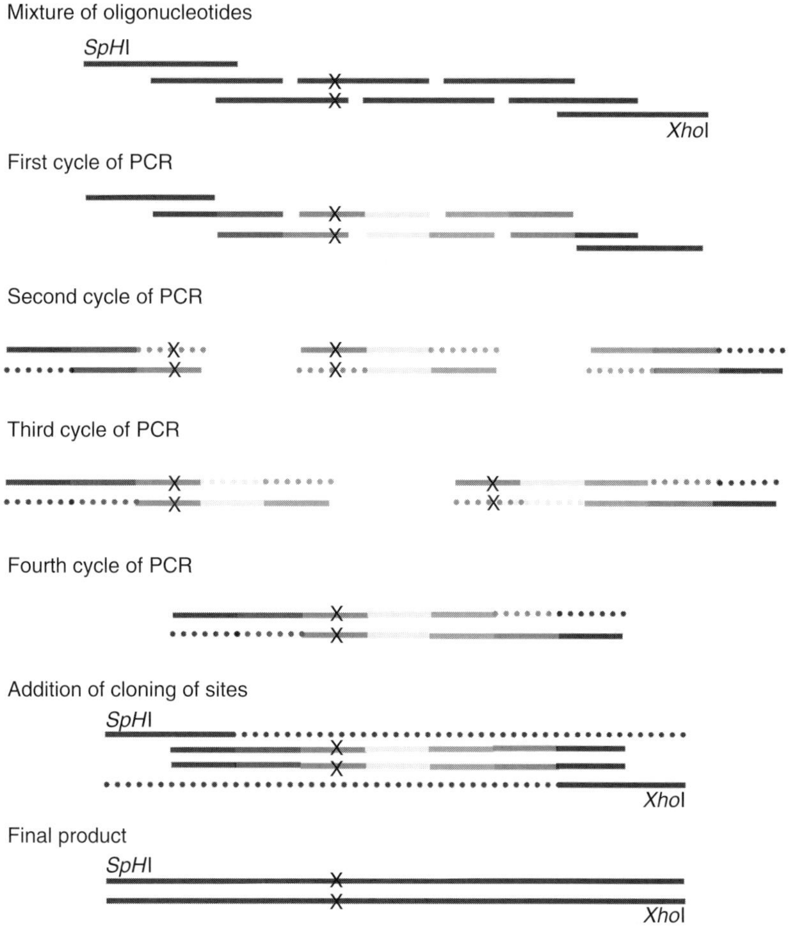

Fig. 3 PCR-based *in vitro* gene assembly. The scanning mutagenesis was performed using *in vitro* gene assembly. Twelve overlapping oligonucleotides were designed to reconstruct the 200 bp region of the wild-type *bacterio-opsin* gene that codes for the F and G helices, and the intervening linking region (only six overlapping oligonucleotides and their self-annealing-extension reactions are shown for clarity). Two additional oligonucleotides were designed (one for each strand) for each cysteine mutation (X) that replaced each wild-type codon with the codon TGC. One hundred twenty-eight unique oligonucleotides were constructed to create the 59 cysteine scanning mutants. A final PCR step used outside primers containing *Sph*I (upstream primer) and *Xho*I (downstream primer) restriction sites, allowing cloning into the same sites previously engineered into the *bop* gene. (See Plate no. 9 in the Color Plate Section.)

collected. Spectra were recorded at the reflectance port of an RSA 150 integration sphere (Labsphere, North Sutton, NH) coupled to a PE λ18 UV/V is spectrophotometer (Perkin-Elmer, Norwalk, CT). Four spectra (350–800 nm) were recorded of each sample. Spectra were normalized to 0.000 OD at 750 nm and smoothed using a 10-point moving average. λ_{max} values were determined using the Perkin-Elmer UV WinLab software's peak finder function. The λ_{max} at each pH value was reported as an average of the λ_{max} values determined for the four spectra. The N/O peak heights were determined by plotting one normalized spectrum from each pH value on the same graph and measuring from the λ_{max} to the baseline. Percent changes were reported relative to the peak height measured at pH 7.5.

To further assist in the assignment of whole-cell spectral phenotypes a qualitative assessment of the spectra was also used. This assessment was based on peak height, broadness, and the presence of shoulders or scattering in the spectra that would influence the interpretation of the spectral transitions.

H. Assignment of Phenotypes and Mapping

The single-site mutant D85N was used to isolate the linear transitions between three spectral states (M, N, and O, Turner *et al.*, 1993). The N- and O-states have unique absorbance maxima (560 and 600 nm, respectively) but exhibit significant spectral overlap. The M spectrum is well resolved from N and O. The effect of mutation on accumulation of the M-, N-, and O-states was evaluated by their pH-dependent spectral transitions. In all cases, the D85N/Cys mutant spectra could be assigned to be M-, N-, or O-like. The pH-dependence of the spectral changes was used to evaluate the apparent pK_a of the accumulation of M, N, and O. The D85N/Cys mutants were grouped into at least one of six categories based on the quantitative and qualitative analysis of their pH-dependent absorbance spectra (Martinez *et al.*, 2002). Mutants which affected the accumulation of the M-like state were mapped to the 3D structure of the ground-state protein (1C3W, Luecke *et al.*, 1999b) using the WebLab-Viewer Lite software (Molecular Simulations, Inc.).

I. Spectra of Purified D85N:Cysteine Mutants

Purified high-density membranes were pelleted and washed three times in titration buffer (100 mM NaCl, 20 mM CAPS, 20 mM BTP, and 40 mM NaH_2PO_4) at pH 7.2. Membranes were dark-adapted for at least 16 h at 4°C. All subsequent manipulations were performed under dim red light. Absorbance spectra were recorded at the diffuse transmittance port of the integration sphere. Samples (0.5–0.8 OD/ml) were placed in a 3 ml quartz cuvette (1 cm pathlength). The cuvette holder was thermostated at 20 °C using a Lauda RC20 circulator (Brinkmann Instruments, Westbury, NY). Samples were titrated by delivering (via a Microlab 500B dual syringe pump, Hamilton, Reno, NV) 1–5 μl of 5 N HCl or 5 N NaOH (in titration buffer) to the sample cuvette. The cuvette was then capped and inverted 15 times. Following the addition of titrant, the temperature was measured using a temperature probe (Radiometer America, Westlake, OH). When the temperature had equilibrated (steady reading for 20 s at 20 ± 0.3 °C) the pH was recorded on a Radiometer PHM240 pH meter using a Radiometer calomel electrode (pHC4000-8). At each pH value, the volume of titrant was tabulated and absorbance spectra were collected from 800 to 250 nm at 1 nm intervals (2 nm slit width).

IV. Results

bR is a useful model system for demonstrating how membrane protein structural elements contribute to their function. Mutagenesis has been an indispensable tool for obtaining insight on how individual amino acids contribute to the bR-pumping mechanism. An analysis of bR containing mutations at amino acid positions 96 and 85 will be used to illustrate the complex roles amino acids can play in functional mechanisms of membrane proteins. Fig. 4 summarizes the protonation state of D96 and D85 in the bR-pumping mechanism. A brief summary of the mechanism follows.

bR consists of seven alpha-helical transmembrane segments with short inter-helical loops (Fig. 1). Retinal is covalently bound to bR via a Schiff base linkage to Lysine 216 near the middle of the seventh (G) helix. Interactions between retinal and the chromophore binding-pocket amino acids lead to convenient and characteristic absorbance properties. In the ground-state, light-adapted bR contains an

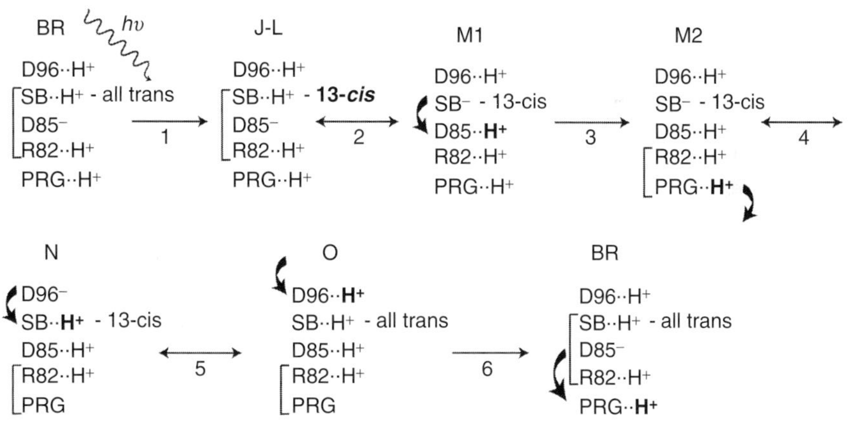

Fig. 4 Proton transfers in wild-type bacteriorhodopsin reaction scheme. Intermediate-state designations are listed above grouped amino acid interactions. Arrows indicate the direction of proton transfer. Brackets indicate the residues interacting with the side chain of arginine 82 (R82). Water molecules participating in these transfers are omitted for clarity. Briefly, the order of events are (1) light-dependent retinal isomerization; (2) Schiff base proton transfer to D85; (3) proton release group deprotonates to the extracellular medium; (4) Schiff base reprontonation by D96; (5) D96 reprotonation from cystosol; and (6) D85 deprotonation and reprotonation of the proton release group. The Schiff base complex counterion is restored to the ground-state condition.

all-trans retinal isomer and absorbs maximally at 568 nm. Light absorption isomerizes retinal to *13-cis* and generates an excited-state conformation. Changes in the interactions between amino acid side chains and retinal alter bR visible absorbance properties, such that the accumulation of intermediate-states can be monitored spectroscopically. Sequential transition through seven unique spectral components ultimately regenerates the ground-state (Fig. 2). The time constants linking the intermediate-states span twelve orders of magnitude (Varo and Lanyi, 1991). A number of transitions are near equilibrium (e.g., $M \leftrightarrow N \leftrightarrow O$, Lozier *et al.*, 1992; Turner *et al.*, 1993). Transitions between the intermediate-state conformations are coupled to protonation–deprotonation reactions. For each photon absorbed a proton is released to the extracellular side (the deprotonation reaction) and a proton is taken up from the intracellular side of the membrane (the reprotonation reaction). Understanding the bR mechanism requires knowledge of how the proton-binding and -release site pK_as are coupled to conformational transitions between intermediate-state components.

Owing to the extreme hydrophobic character of the transmembrane elements, it is feasible that the dominant physical and chemical forces that drive functional transitions may not be the same as those for water-solvated proteins (Haltia and Freire, 1994). Hydrophobic interactions likely play a significant role whereas ionic and hydration interactions less so in the mechanisms of membrane bound proteins. However, ionization reactions do contribute and it is critical to probe conformational energetics and determine how the proton binding and release is partitioned in the functional mechanism of bR. While the pK_as of transbilayer amino acid side chains are likely to be uniquely constrained versus those for soluble proteins (Harris and Turner, 2002), mutagenesis can help to identify key proton-binding sites in bR and to gain insight on regulation of their affinities.

A. *In Vivo* Random Mutagenesis of bR

Random mutagenesis is a totally unbiased amino acid perturbation strategy. As such it may be employed in the absence of structural models, the reality for the vast majority of membrane proteins. A critical constraint in the application of random mutagenesis is that there is a phenotype (a functional filter) that can be used to evaluate the consequence of the mutagenesis. bR's photosynthetic activity allows its use as a functional filter. Under phototrophic growth conditions (intense light and low oxygen) *H. salinarum* cells without a functional bR die (Oesterhelt and Krippahl, 1983). Soppa and Oesterhelt exploited this property to design a positive selection to isolate nonfunctional bR mutants. Cells were randomly mutagenized with X-ray or UV irradiation and phototrophic negative cells were isolated based on their ability to survive 5-bromo-2′-deoxyuracil incorporation under periods of phototrophic growth conditions (Soppa and Oesterhelt, 1989). Six different amino acid substitutions at five different codons were recovered, providing the first insights into functionally critical amino acids. The low diversity of mutants recovered indicates that only these sites are critical for bR-dependent photosynthesis or that the filter is not optimal. The selection used would not allow recovery of mutations that subtly affect the proton-pumping mechanism, so regulatory contributions are likely missed. However, mutants at amino acid positions 96 and 85, aspartic acids in transmembrane helix C (3) in the wild-type protein, were common isolates and are instructive to evaluate.

1. Aspartic Acid 96

bR mutants with aspartic acid 96 replaced by glycine (D96G) and separately by asparagine (D96N) were recovered from the phototrophic selection. The ground-state absorbance properties of both proteins were indistinguishable from the wild type (Soppa *et al.*, 1989). The visible absorbance properties demonstrate that D96G and D96N are able to bind retinal and are therefore not misfolded. It also appears that the size of the side chain of amino acid 96 is not critical for protein–chromophore interactions in the ground-state. The two substitutions suggest that an uncharged side chain satisfies the electrostatic environment in the vicinity of amino acid 96, in the ground-state. Therefore, it is logical to assume that D96 in the wild-type bR ground state is protonated. Since cells containing bR with the D96G,N mutations are photosynthetically defective, D96 is likely to deprotonate at a critical point in the wild-type pumping mechanism.

2. Aspartic Acid 85

Aspartic acid 85 replaced by asparagine (D85N) was also recovered from the phototrophic selection. The D85N ground-state absorbance was very perturbed (e.g., red shifted by greater than 40 nm at neutral pH, Fig. 5). The visible absorbance properties demonstrate that D85N is able to bind retinal and is therefore not misfolded. The chromophoric perturbation observed must be electrostatic in origin since a protonated aspartic acid and asparagine are isostructural. The carboxlyate form of D85 and/or its counterion(s) must have a critical interaction with the chromophore in the ground state of wild-type bR. The extreme ground-state absorbance red shift indicates there are significant secondary effects due to the D85N mutation. D85N does not pump protons (Soppa *et al.*, 1989).

The mechanistic interpretations offered so far are based only on local chemical intuition and the preliminary photochemical characterizations performed concomitant with the *in vivo* studies. More sophisticated insights are discussed in the context of the physical analysis of the D96 and D85 site-directed mutagenesis studies (below).

B. Site-Directed Mutagenesis of bR

Site-directed mutagenesis requires a structural or functional bias for implementation. Mutagenesis can be targeted to specific amino acids based on knowledge of amino acid sequence (the minimal requirement), chemical reaction at an active site,

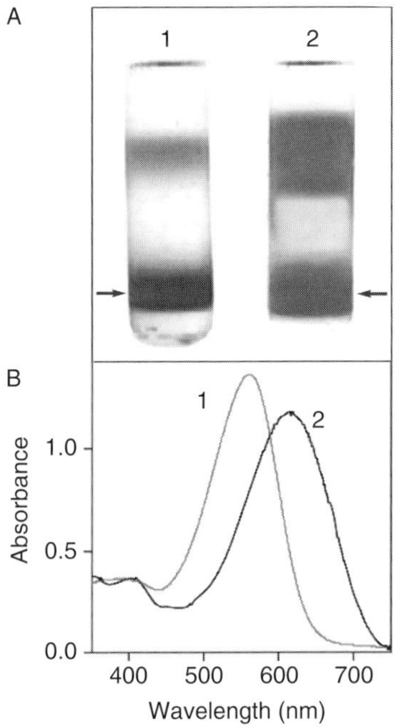

Fig. 5 Ground-state spectral properties of the wild-type bR and the single-site mutant D85N (aspartate at position 85 changed to asparagine), at pH 7.0. (A) Sucrose density gradients (linear gradients: 25–45%, 50% cushion) used to isolate the high-density membrane (colored fraction at arrow) containing pure bR proteins. 1 is wild-type bR and 2 is bR containing the mutation D85N. On visual inspection, the wild-type PM appears purple and the D85N PM appears blue (at neutral pH). (B) Ground-state absorbance spectra of the purified high-density membranes of wild-type bR (1) and D85N (2). (See Plate no. 10 in the Color Plate Section.)

structure of ion or solute being transported, or the chemical nature of a known ligand. Hydropathy analyses of the bR amino acid sequence directed the earliest site-directed mutagenesis. Numerous ionizable amino acids were predicted to be transbilayer, subsequently confirmed by EM and X-ray structural analyses. Carboxylate containing side chains were considered candidate proton-binding and release sites and a subset of the native aspartic and glutamic acids have since been shown to be essential for proton pumping. Site-directed mutagenesis was instrumental for defining key proton-binding sites and establishing their linkage to the pumping mechanism.

Since the *in vivo* studies indicated that aspartic acids 96 and 85 were critical for bR's photosynthetic function, a combination of site-directed mutagenesis and spectroscopic analyses was used to define their specific roles in the proton-pumping mechanism. The amino acid substitutions asparagine (Asn, N) and glutamic acid (Glu, E) have been particularly instructive. bR mutants were initially expressed in *E. coli*, purified in SDS and organic solvents, and reconstituted in mixed detergent/lipid micellar systems (Nassal *et al.*, 1987; Shand *et al.*, 1991). Subsequently, homologous expression allowed functional analysis in the native lipid environment (Needleman *et al.*, 1991; Turner *et al.*, 1993).

Without a structural model, the interpretation of the consequence of any mutagenesis must rely on chemical intuition of putative local interactions. Fortunately, of the handful of atomic resolution structures of all classes of membrane proteins solved to date, one is the bR ground state (1.55 Å, Luecke *et al.*, 1999b) and three others are the conformations accessed during the reprotonation phase of the pumping mechanism: M (2.0 Å, Luecke *et al.*, 1999a), N (3.5 Å by 10.5 Å, Vonck, 1996), and O (2.4 Å, Rouhani *et al.*, 2001). In addition, owing to its convenient chromophoric properties bR has been probed in nearly every region of the electromagnetic spectrum. Significant contributions have derived from UV/Abs, FTIR, Raman, NMR, and EPR spectroscopy. Extensive photochemical analyses have been performed on wild-type bR and the D96 and D85 mutations allowing evaluation of the local and global consequences of amino acid perturbation at these key proton-binding sites. Our mechanistic interpretation of the critical roles played by D96 and D85 is only possible via synthesis of the work from many laboratories over a period spanning 30 years.

1. Aspartic Acid 96

The mutant D96N was engineered separately by at least four laboratories, and the *in vitro* analyses support the *in vivo* studies (Cao *et al.*, 1991; Gerwert *et al.*, 1989; Mogi *et al.*, 1988; Shand *et al.*, 1991). Flash spectroscopy demonstrated that the mechanistic defect due to the D96 mutations is kinetic; all mutants pump protons, but very slowly. D96N possesses a pH-dependent photocycle; at high proton activities the mechanism is similar to wild type (Stern *et al.*, 1989). At low proton (near neutral pH) activity, the proton uptake phase of the mechanism is extremely slow. These observations demonstrate that the proton transfer reactions perturbed by D96N are accessible to bulk solvent (Cao *et al.*, 1991; Soppa *et al.*, 1989; Stern *et al.*, 1989). Proton transfer, in D96N, has a component similar to that for an aqueous channel. This may or may not hold for the wild-type mechanism, a caveat of all mutagenesis strategies. In D96N, the rate of decay of the M-state and the coupled proton uptake reactions are very slow. D96N is able to compensate (albeit slowly) by a direct transition between the M- and O-states. Subsequent work demonstrated that for wild-type bR D96 is the internal proton donor to the Schiff base nitrogen and that this occurs during the M to N transition (Fig. 2). D96 functions as a catalyst in the reprotonation of the Schiff base. Removing this key proton-binding site did not stop the pump, indicating that the bR-pumping mechanism exhibits "plasticity," with alternative pathways available to ensure proton transport.

Since D96N can pump protons it was not surprising that D96E was also found to be a functional proton pump. However, the photocycle was again significantly slower than that for wild-type bR (Stern *et al.*, 1989). The deprotonation reactions are similar to that for the wild-type bR, up to formation of the M-state. The rates and amplitudes of the reprotonation reactions are perturbed (e.g., the transitions involving the M-, N-, and O-states). Extending the carboxylate by one methyl group perturbs the apparent pK_a of the proton transfer reactions. Therefore, critical steric and/or electrostatic constraints exist in the vicinity of the side chain at 96 and its counterion(s).

2. Aspartic Acid 85

Site-directed mutagenesis demonstrated that glutamic acid is a nonconservative substitution at amino acid position 85. D85E possesses a slow photocycle with very reduced M accumulation (Mogi *et al.*, 1988; Otto *et al.*, 1990; Subramaniam *et al.*, 1990). The stereochemistry environment in the vicinity of amino acid 85 is essential in the regulation of amino acid–retinal interactions. These interactions are perturbed by extending the carboxylate by one methyl group.

Asparagine and a protonated aspartic acid are isostructural. It was initially a surprise when the mutant D85N was observed to be nonfunctional. At neutral pH, D85N has a redshifted, ground-state absorbance maximum (Fig. 5). The M-state does not accumulate following photoexcitation (Otto *et al.*, 1990; Thorgeirsson *et al.*, 1991). At elevated pH an extremely small amplitude M-state was observed (opposite the trend for the D96 mutants). It has subsequently been shown that D85 is the primary counterion to Schiff base in the bR ground-state (Needleman *et al.*, 1991; Richter *et al.*, 1996) and that the pK_a of D85 is highly perturbed, ~8.0. The global consequence of the D85N mutation rendered mechanistic interpretation of the role of D85 in proton pumping problematic.

Why is the ground-state of the D85N mutant so dramatically redshifted (Fig. 5)? The answer came from a pH-dependent analysis of the D85N ground state (Fig. 6). pH titrations of the D85N absorbance spectrum revealed that, in the absence of photoexcitation, three chromophoric species are present in equilibrium and that their relative abundance depends on solvent proton activities: D85 is solvent accessible. The absorbance spectra of these three species are indistinguishable from those of the M (λ_{max} = 410 nm), N (λ_{max} = 550), and O (λ_{max} = 610 nm) intermediates of the wild-type photocycle (Fig. 2, Turner *et al.*, 1993). The pH dependence of the relative stabilities of the M-, N-, and O-states is shown in Fig. 6B and C. Neutralization of the carboxylate at position 85 (by protonation or the D85N mutation) is energetically coupled to the transitions between the M-, N-, and O-intermediate-states. Instead of being a "dead protein," D85N is useful for isolating a subset of linear transitions between bR substates and allows for a thermodynamic analysis of a classical kinetically control reaction (Fig. 7).

Analysis of the D85N equilibrium leads to numerous insights into the bR-pumping mechanism:

1. The transitions between the D85N substates are light *independent*. Light is required for the first step in the reaction cycle only. Photon absorption generates an excited state and the subsequent substate transitions are thermal decay reactions.

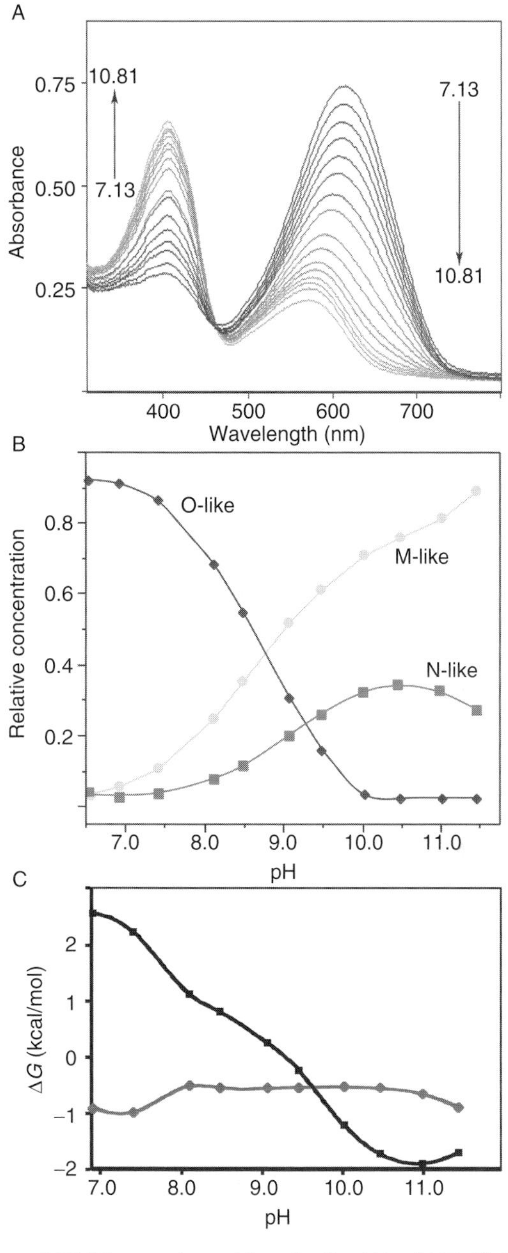

Fig. 6 pH-dependence of D85N spectral transitions (buffer: 100-mM NaCl, 20-mM CAPS, 20-mM BTP, and 40-mM NaH_2PO_4). (A) Absorbance spectra of dark-adapted (DA) high-density membranes containing D85N. The spectra were obtained at pH 7.13 (greatest amplitude at 615 nm and lowest amplitude at 410 nm), 7.51, 7.77, 7.95, 8.15, 8.34, 8.72, 9.06, 9.50, 9.94, 10.20, 10.42, and 10.81 (lowest amplitude at 615 nm and greatest amplitude at 410 nm). The spectral changes are completely

This insight is consistent with the Arrhenius analysis of the bR photocycle performed by Lanyi's group (Varo and Lanyi, 1991).

2. The transfer of a proton from the Schiff base to D85 is a critical switch point between the deprotonation and reprotonation phases of the pumping mechanism. Protonation of D85 (as modeled by D85N) is sufficient to isolate the reprotonation phase substates.

3. Some of the substate transitions are near equilibrium. A few "irreversible" steps (e.g., negligible back reactions) drive the reaction in the forward direction. This insight is consistent with the requirement for the inclusion of back reactions in the global analysis of photokinetic analyses (Lozier *et al.*, 1992; Varo *et al.*, 1990).

4. Transitions between the M↔N↔O conformations are "paid" for proton-binding and release reactions at sites other than D85.

5. bR is a retinal isomerase; the transitions between M-, N-, and O-are coupled to the regeneration of the all-trans retinal chromophore. The D85N ground-state transitions demonstrate that retinal *cis*-to-*trans* isomerization is light independent (Turner *et al.*, 1993).

6. The molecular interactions that stabilize M-, N-, and O-substates are different. As seen in Fig. 6C, the O↔N equilibrium is pH dependent, the N↔M is pH independent. This leads to the prediction that the structures of at least the O- and M-states are different, and that unique noncovalent binding interactions contribute to their relative stabilities. This prediction is consistent with spectroscopic and crystallographic studies of the structures of the M-, N-, and O-substates (Rouhani *et al.*, 2001; Sass *et al.*, 2000; Vonck, 1996).

As illustrated by the D96 and D85 examples, the combination of careful physical analysis and site-directed mutagenesis can define the roles specific amino acids contribute to membrane protein functional mechanisms. It may be straightforward to interpret the consequence of site-directed mutagenesis if the influence of the

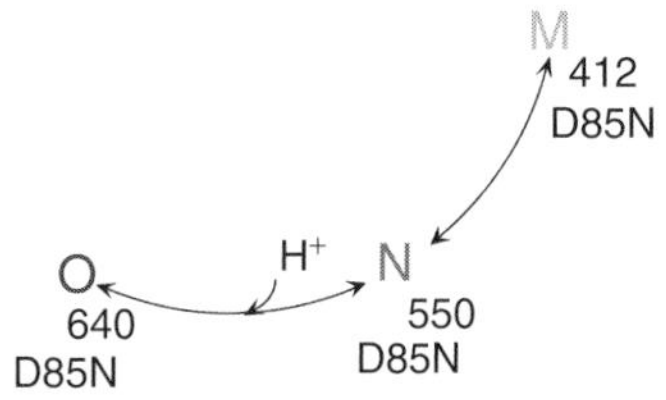

Fig. 7 The D85N pH-dependent species reaction scheme. The color of the letter denotes that of the intermediate and the subscript denotes its wavelength at maximal absorbance (λ_{max}).

reversible. (B) Singular value decomposition and nonlinear least squares fitting of the titrated absorbance spectra were used to calculate the relative concentrations of the M-, N-, and O-like intermediates as a function of pH. (C) The species concentrations in (B) were used to determine the ΔG values for the O↔N and N↔M transitions.

perturbed amino acid is narrow in focus (local). Alternatively, mutagenesis of single amino acids may uncover unanticipated linkages in functional mechanisms. Site-directed mutagenesis of single amino acids cannot, by itself, supply subtle information about regulation of ligand-binding affinities or identify pathways of coupled interactions participating in functional mechanisms. This requires combinatorial mutagenesis strategies, discussed below.

C. Scanning Mutagenesis of bR

Proton pumping, like many enzymatic mechanisms, relies on coordinated perturbation of pK_a values of critical catalytic groups (Harris and Turner, 2002). Unlike a water-solvated protein where ionization and hydration energies may dominate protein folding and conformational changes, hydrophobic interactions are likely to play a more significant role in membrane proteins. The low dielectric within the lipid bilayer may also increase the energetic significance of the charge–charge interactions within the protein (Harris and Turner, 2002). At interfacial regions of the bilayer, where the dielectric varies steeply according to the degree of hydration, hydrophobic and electrostatic interactions may both be important (White and Wimley, 1999). Conformational changes that alter the microenvironment and consequently the dielectric around interacting groups may regulate which of these forces dominate the energetics. Due to the existence of diverse microenvironments, membrane proteins present unique challenges in the study of their mechanisms of action.

The bR proton-pumping mechanism relies on vectoral proton transfer between binding and release sites. In the previous sections, D96 and D85 were identified as two such critical sites. Proton occupancy at D96 and D85 is determined by the intrinsic pK_a of the carboxylate and the microenvironment created by each substate conformation. The intrinsic pK_a of the D96 and D85 carboxylates will be driven to their functional or apparent value by interactions with other fully or partially charged groups as well as by the polarity or dielectric of the medium that surrounds it. The pK_a and pK_a changes of D96 and D85 have been experimentally determined. After photoexcitation and retinal isomerization, the pK_a of Asp-85 increases from 2.2 to 6.9 (Richter *et al.*, 1996) resulting in the transfer of a proton from the Schiff base to Asp-85. The Schiff base is reprotonated by Asp-96 when the pK_a of Asp-96 drops from >12 to 7.1 (Zscherp *et al.*, 1999). After reprotonation, retinal isomerizes back to *all-trans*, and the Asp-96 pK_a increases back to its initially very high value. Asp-96 is reprotonated from the intracellular side of the membrane. The extreme perturbations observed for the bR proton-binding site pK_as are accomplished by both charge–charge interactions and desolvation effects.

The essence of bR-pumping mechanism is the coordinated regulation of at least these pK_as. The D96 and D85 site-directed mutants, by themselves, yield little insight on the mechanism of regulation of these pK_as. Combinatorial mutagenesis can be helpful in this regard. It will be demonstrated that scanning mutagenesis of D85N can be used to define the role of individual amino acids in the conformational transitions of the reprotonation mechanism. D85N will be used as a surrogate "wild type" and the M↔N↔O equilibrium exploited to dissect global linkages that regulate the pK_a of the transitions between substates in this reprotonation phase model.

bR has 246 amino acids that can influence the D85N substate transitions. Understanding how the rest of the protein contributes to substate transitions could, in principle, be accomplished by additional site-directed mutagenesis. Engineering 246 mutations into the D85N background (with 19 possible substitutions at each amino acid position) is a daunting proposition. An alternative strategy was pursued. The fortuitous availability of bR structures served to guide a mutagenic investigation of important amino acid interactions. The D96N mutation was essential in this regard. Recall that the structural transitions isolated by D85N include the M-, N-, and O-states. Spectroscopic and structural analyses indicate that the O-state closely resembles bR ground state. There is a 1.55 Å resolution X-ray structure for the bR ground state (Luecke *et al.*, 1999b). As discussed earlier, the D96N mutant has a very slow M decay. Photoexcitation of D96N (at cryo-temperatures) was used to trap the M-state, allowing EM and X-ray structural analyses (Luecke *et al.*, 1999a; Nilsson *et al.*, 1995; Sass *et al.*, 2000). Comparison of the ground- and M-state structures suggests that formation of M-state is coupled to a major reorientation of the cytoplasmic end of helices 6(F) and 7(G) (Fig. 8). More limited structural analyses for mutagenic models of the N- and O-states support a key role for F(6) and G(7) helical movements in substate formation (Rouhani *et al.*, 2001; Vonck, 1996). It is likley that changes in amino acid interactions involving the F and G helices are energetically coupled to the structural rearrangements of the reprotonation mechanism. Unfortunately, the crystallographic analysis cannot identify specific amino acid interactions that have changed since the cytoplamsic ends of the 6 and 7 transmembrane helices are disordered in the M-state model. Therefore, site-directed mutagenesis targeting specfic amino acid interactions could not be pursued. Instead, a scanning mutagenesis on the entire 6 and 7 helices and the extracellular loop region joining them was pursued.

Having decided on a structural target the next question was: what amino acid to scan with? Alanine is a logical choice in such a scanning strategy. The alanine side chain is relatively small and is well tolerated in both hydrophobic and hydrophilic proteinaceous environments (Faham *et al.*, 2004; Hristova and White, 2005; Kyte and Doolittle, 1982). However, cysteine was chosen for two reasons. First, cysteine is a fairly conservative substitution for most amino acids and has been shown to be useful in membrane protein structure–function analyses (Akabas *et al.*, 1992; Dahl and Pfahnl, 2001). Second, bR contains no native cysteines. Therefore, the mutants generated are a resource for future site-directed spectroscopy, following labeling with sulfhydryl reactive fluorescence or electron paramagnetic probes (Farrens, 1999, see also Chapter 20 by Klug and Feix, this volume, on site-directed spin labeling).

Primerless PCR gene synthesis (Stemmer, 1994) was used to replace all 59 amino acid positions, in the F(6) and G(7) helices of D85N, with a cysteine (Martinez *et al.*, 2002). The success of this ambitious undertaking relied heavily on a fortuitous

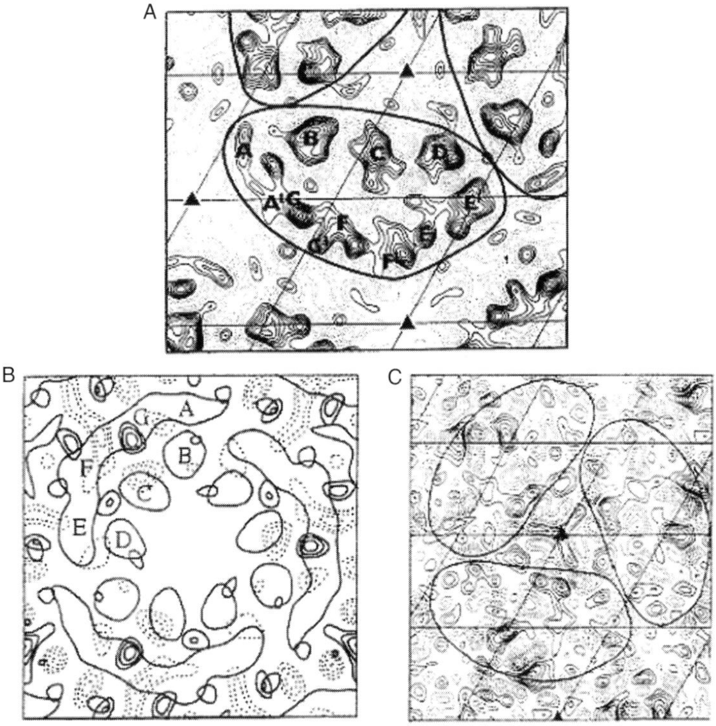

Fig. 8 Electron diffraction projection maps indicating the major bR conformational transition. (A) Projection map wild-type bR with the seven helices identified (Bullough and Henderson, 1999). bR monomers within the crystallographic trimer are indicated by outlined areas. Transmembrane helices B, C, and D are orthogonal to the plane of the membrane while helices A, G, F, and E are tilted as indicated by the overlapping densities. Difference projection maps of bR structural changes (B and C). Solid contour lines indicate density increases while the dashed lines indicate density loss. (B) Density of D85N at pH 9.2 minus density of D85N at pH 6.0 (Brown *et al.*, 1997). (C) Density of a model of the M intermediate-state minus the density of the wild-type bR (Subramaniam *et al.*, 1993). In both (B) and (C), the major conformational change appears to be an increase in order of the G helix and an outward tilting of the F helix.

functional filter, the bR chromophore. Therefore, the Lysine at amino acid position 216 was not mutated since it is the attachment point of retinal. Mutational phenotypes were screened by comparison of the pH-dependent absorbance properties of the D85N second-site cysteine mutants with those of D85N. The *H. salinarum* cells expressed so much of the mutant proteins that this screen can be accomplished by visual inspection (Fig. 9). The D85N cells appear blue at neutral pH (the O-state predominates) and yellow-green at elevated pH (the result of significant M-state accumulation). Second-site cysteine mutations that perturb the "wild-type" (e.g., D85N) intermediate-state populations will have perturbed color (substate) distributions.

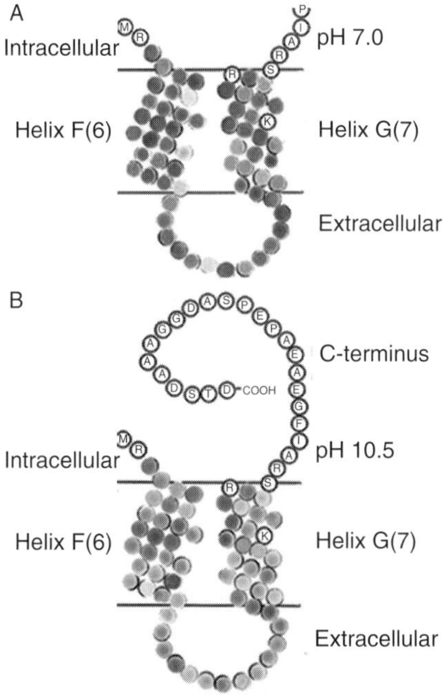

Fig. 9 Whole cell samples of *Halobacterium salinarum* expressing the D85N:Cysteine mutants. One milliliter of whole cell samples were transferred to nylon membranes and equilibrated with buffer at pH 7 (A) and pH 10.5 (B). The nylon membranes mounted cell samples were pasted onto the position of the corresponding amino acid mutation in the two-dimensional representation of the bR F and G helices. The colors observed are due exclusively to the mutation introduced into the bR mutant D85N and reflect a weighted average of equilibrium populations of the blue (O-like), purple (N-like), and yellow (M-like) species. The solid black lines denote the borders of the lipid bilayer. Transbilayer helices F(6) and G(7) are indicated. The cytoplasmic and extracellular sides of the membrane are labeled. The extramembranous C-terminus is shown in (B) for clarity. No cysteine mutation was made at position lysine 216 (in the G helix) as this amino acid forms the Schiff base linkage to the retinal chromophore. (See Plate no. 11 in the Color Plate Section.)

While the fortuitous formation of the PM in *H. salinarum* has facilitated the purification and structural analysis of bR and bR mutants, PM purification of the entire scan would be prohibitively time consuming. Therefore, a spectroscopic method was developed to screen the large numbers of mutants generated, whole-cell reflectance spectroscopy (Martinez and Turner, 2002). The screening procedure was calibrated against well-established spectroscopic protocols used for the characterization of purified bR (described below).

bR absorbance properties were quantified using a Perkin-Elmer 118 spectrophotometer and LabSphere RSA 150-ml light scattering attachment. In the absence of the integration sphere light scattering by the *H. salinarum* cells rendered the spectrum featureless over the wavelength range characteristic for bR absorbance (Fig. 10A, spectrum a). Spectra b and c (Fig. 10A) demonstrate the signal improvement

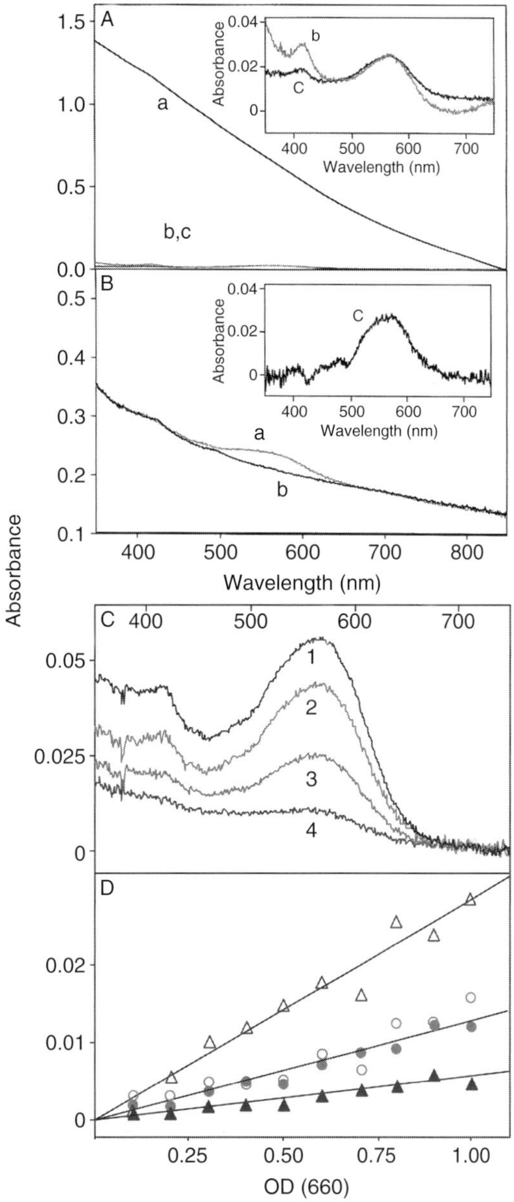

Fig. 10 Reflectance spectroscopy of whole cells. (A) Fifty-milliliter cultures of L33 and L33 transgenically expressing bR (wild type) were grown to stationary phase. (B) Cell pellets were collected and washed with basal salts and resuspended to 50 ml. (C) Reflectance spectra of L33 cells expressing bR for the following volume dilutions with basal salts: (1) undiluted, (2) 80% cell suspension:20% basal salts, (3) 40% cell suspension:60% basal salts, and (4) 10% cell suspension:90% basal salts. The absorbance maximum at 568 nm is due to bR; the absorbance maximum at 410 nm is due to a membrane-associated cytochrome. (D) Reflectance measurements of dilutions of L33 (solid symbols) and L33 expressing bR

obtained with the scattering attachment for a whole cell sample (b) and purified membrane fraction (c). Shown in Fig. 10B are the whole cell difference spectra used for calculating bR concentration [bR expressing strain S1 (spectrum a) minus strain L33 (bR^-, spectrum b)]. Samples for spectral measurements were prepared by dilution (in basal salts) of washed whole cells to equivalent absorbance at 410 nm (Fig. 10C). The absorbance at 410 nm is due to a membrane-associated cytochrome and is linearly proportional to the amount of total cells in the sample whether bR was expressed (Fig. 10D, solid triangles) or not (Fig. 10D, open triangles). The bR and cytochrome spectral properties obeyed standard Beer–Lambert relationships.

While the quality of the absorbance spectra was much improved by use of the integration sphere, the cytochrome absorbance and the scattering component remaining below 500 nm were found to complicate the analysis of the M-state spectral properties for whole cell preparations containing the D85N mutants. These components were removed by difference spectroscopy (Fig. 11). A reflectance spectral library of dilutions of L33 cells (bR^-) was generated. The L33 cells

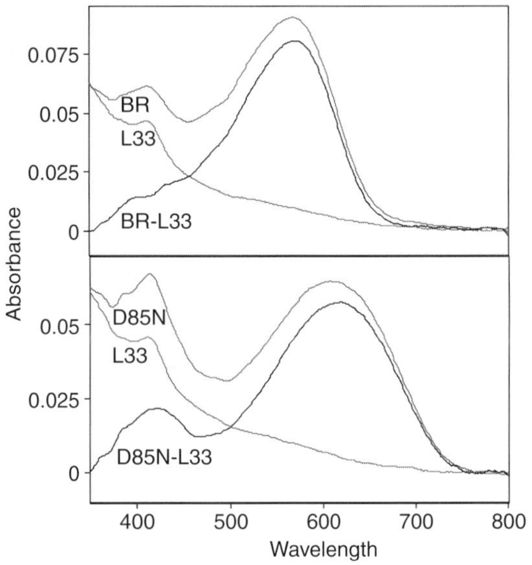

Fig. 11 Reflectance spectroscopic screen used to quantify λ_{max} and M peak areas for the D85N: second-site mutants. bR and D85N were expressed in *Halobacterium salinarum* strain L33 (bR^-). Reflectance spectra of whole cells expressing bR and the second-site mutants were collected from 350 to 800 nm. Reflectance spectra of identical concentrations of L33 cells were subtracted from the bR containing samples. The difference spectra (e.g., bRL33 and D85N-L33) used to compare the M, N, and O species populations, as described in the text.

(open symbols). Triangles are measurements taken at 568 nm (bR absorbance maximum) and circles are measurements taken at 410 nm (cytochrome absorbance maximum). The cytochrome absorbance exhibits the same linear dependence with cell number in the presence or absence of bR accumulation.

contain only the residual scattering and cytochrome absorbance signals. Spectra of L33 cells were matched to within 0.005 absorbance units at 750 nm to spectra of cells containing bacteriorhodopsin (or D85N and the D85N cysteine mutants). Difference in spectra were calculated and used to obtain absorbance maxima (λ_{max}) and peak areas.

The ability of reflectance spectroscopy to reproduce absorbance properties characteristic of purified bR was then established. PM containing wild-type bR was applied to nylon filters and equilibrated at neutral pH, as described for the whole cells. The λ_{max}, determined by reflectance spectroscopy, for dark-adapted bR was 568 nm while the λ_{max} for light-adapted bR was 558 nm (Fig. 12B). The same analysis was performed for whole cells expressing wild-type bR and the λ_{max} determined were 566 ± 1 and 558 ± 3 nm, for the light-adapted and dark-adapted, respectively (Fig. 12A). This exercise demonstrated the precision and accuracy of the whole cell spectroscopic method.

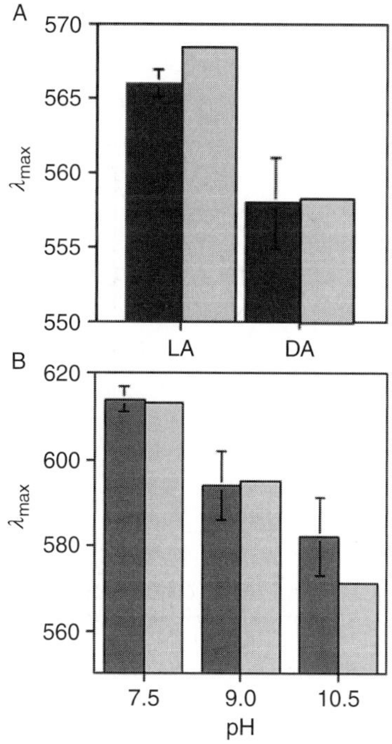

Fig. 12 Comparison of λ_{max} values determined by reflectance spectroscopy. (A) Spectral changes during light-adaptation for wild-type bR. Whole cells, or purified high-density membranes, were coated on to nylon filter paper by vacuum. Samples were dark-adapted (DA) at room temperature for 5 days and subsequently light-adapted (LA) by exposure to light greater than 500 nm for 20 min. Wild-type whole cell samples are indicated by the dark gray bars; wild-type high-density membranes containing purified bR are indicated by the light gray bars. (B) λ_{max} values for the N/O peak changes during pH titration for D85N. Whole cell samples are indicated by the dark gray bars; high-density membranes containing purified bR are indicated by the light gray bars.

To evaluate the utility of reflectance spectroscopy in characterizing the D85N cysteine scanning mutants, the spectral transitions of D85N and the D85N/D96N double mutant were evaluated. As shown in Fig. 13A and B, the λ_{max} for the absorbance peak between 550 and 650 nm ($^{550\text{-}650}\lambda_{max}$) for D85N becomes blue-shifted as pH increased. The blueshift originates from increasing concentrations of the N-state as pH is raised. The pH-dependence of the blueshift is evident in both the whole cell pastes and the purified protein samples. The 400-nm absorbance also increased with increasing pH, indicating M-substate accumulation. The pH-dependent spectra were then compared for the D85N/D96N double mutant. X-ray diffraction of D85N/D96N revealed that, at neutral pH, the double mutant has a conformation like that of D85N at alkaline pH; an M-state homologue is present in both mutants (Kataoka *et al.*, 1994). In D85N/D96N, the N-intermediate does

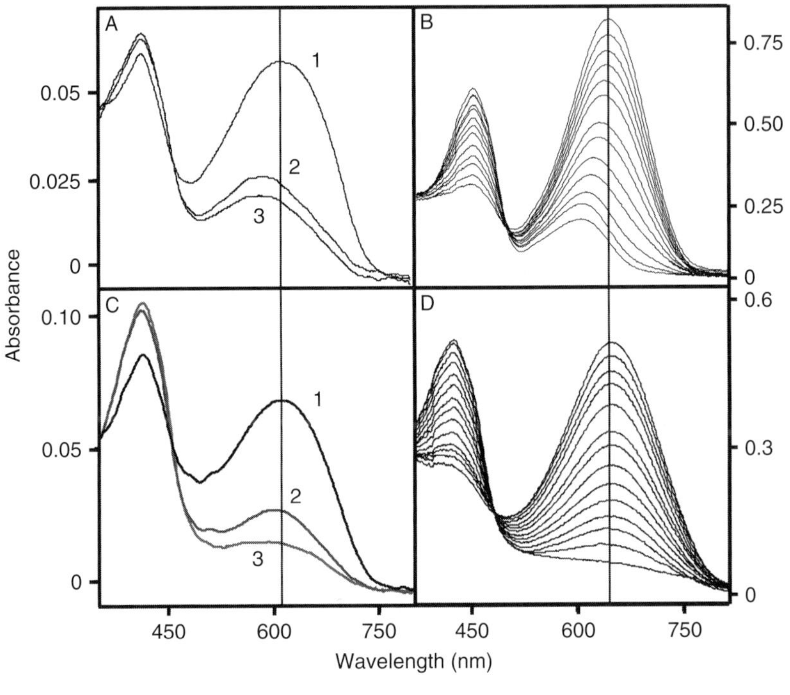

Fig. 13 pH titrations of bR mutants. pH-dependence of D85N spectra changes (100 mM NaCl, 20 mM CAPS, 20 mM BTP, 20 mM, and 20 mM NaH2PO4). (A) Reflectance spectra of dark-adapted whole cells expressing D85N: (1) pH 7.0, (2) pH 9.0, and (3) pH 10.5. (B) Absorbance spectra of dark-adapted high-density membranes containing D85N. The spectra were obtained at pH 7.13 (greatest amplitude at 615 nm and lowest amplitude at 410 nm), 7.51, 7.77, 7.95, 8.15, 8.34, 8.72, 9.06, 9.50, 9.94, 10.20, 10.42, and 10.81 (lowest amplitude at 580 nm and greatest amplitude at 410 nm). The spectral changes are completely reversible (data not shown). (C) Reflectance spectra of dark-adapted whole cells expressing D85N:D96N: (1) pH 7.0, (2) pH 9.0, and (3) pH 10.5. (D) Absorbance spectra of dark-adapted high-density membranes containing D85N:D96N. The spectra were obtained at pH 6.75, 7.26, 7.45, 7.75, 8.05, 8.22, 8.44, 8.86, 9.08, 9.22, 9.53, 9.75, and 10.97.

not form as the pH is raised (Brown *et al.*, 1998; Tittor *et al.*, 1994). Therefore, no shift in $^{550\text{-}650}\lambda_{max}$ is expected as pH was increased for D85N/D96N whole cell samples. In addition, due to the decrease in the pK_a of the Schiff base, M was predicted to accumulate at lower pH values. The cell paste reflectance spectra demonstrated that $^{550\text{-}650}\lambda_{max}$ of D85N/D96N only shifted by 10 nm as the pH was increased from 7.5 to 10.5, compared to a 32 nm shift in D85N (Fig. 13C and D). The change in N/O peak height is ~10% greater for D85N/D96N than for D85N, consistent with increased M accumulation in the double mutant. The whole-cell reflectance analysis is consistent with both predictions and support a critical role for D96 deprotonation in the accumulation of the N-state. In sum, there is a clear phenotypic consensus on comparing spectral phenotypes of PM containing purified bR and the reflectance spectroscopy of the whole cell pastes.

The reflectance screen was then applied to the cysteine scanning mutations and revealed that amino acid changes can perturb the D85N conformational equilibrium (Martinez *et al.*, 2002). Representative whole cell spectra for 12 consecutive amino acid positions at the cytoplasmic end of helix F(6) are shown in Fig. 14. The mutants contain three spectral species with absorbance maxima similar to those of D85N. Cells containing the mutant bR's appeared blue (O) near neutral pH and, as the pH

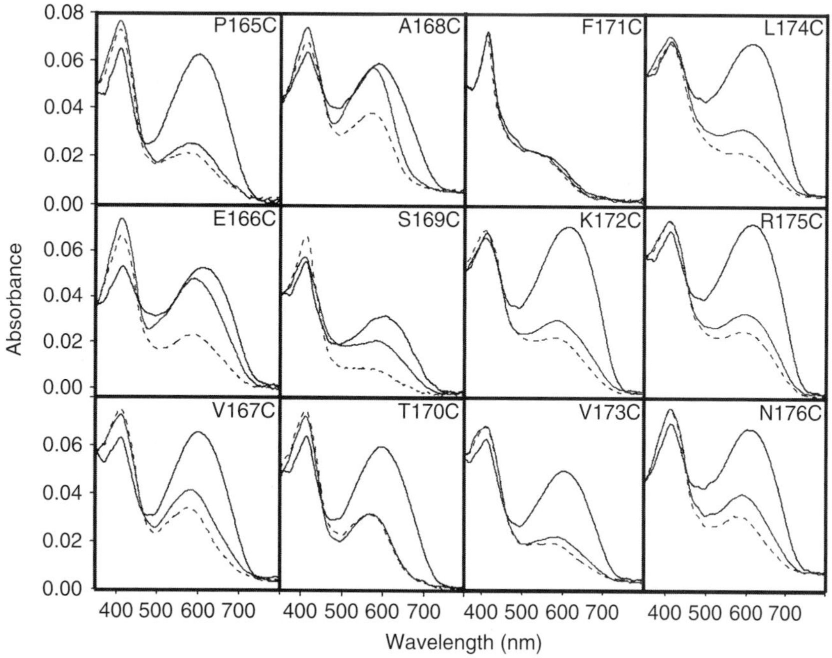

Fig. 14 Whole-cell reflectance spectra of F helix D85N:Cysteine mutants at pH 7.5 (solid line with highest amplitude λ_{max} near 600 nm), 9.0 (solid line with intermediate amplitude λ_{max} near 600 nm), and 10.5 (dashed line).

is raised, change to a yellow-green color, due to the mixture of the yellow (M) and purple (N) species. The pH-dependence of the spectral equilibrium was sensitive to amino acid perturbation indicating that amino acid interactions involving the F(6) and G(7) helices contribute to the relative stabilities of M-, N-, and O-substates.

We analyzed the scan by evaluating patterns of similar transition perturbations instead of attempting to evaluate the local stereochemical perturbation introduced by each of the 59 mutants. If cysteine mutant altered the pH-dependent spectral transition, it is likely that the site of the perturbation participates in the coupling of substate transitions and proton-binding reactions. The collection of mutants that lower the pK_a of M-substate accumulation indicate the degree of coupling. Since a significant number of the cysteine mutants possessed this phenotype it is unlikely that a single local chemical interaction critically influences the stability of the M-state. When mapped to the bR ground-state structure, this collection of mutants formed discernable clusters interpreted to be domains of coupled interactions (Fig. 15). A major conclusion is that these domains (not critically restricted local interactions) are energetically linked to substate stabilities. The locations of these domains do not

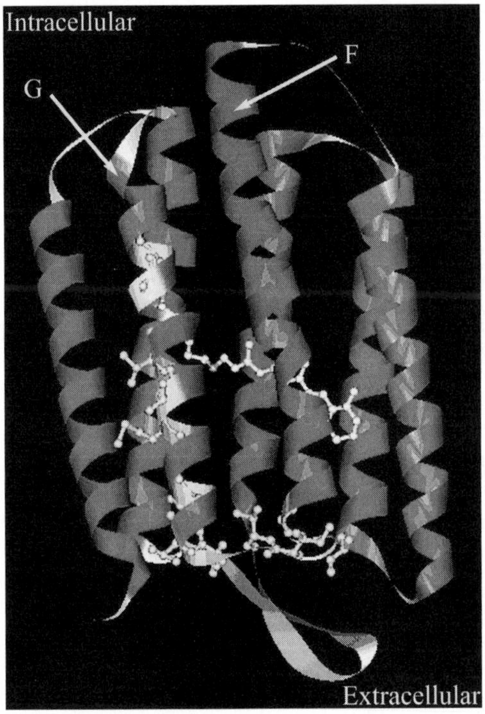

Fig. 15 Structure–function map of the D85N:Cysteine mutants. The D85N:Cysteine mutants were classified according to their effect on the pH-dependent changes in the absorbance spectrum of D85N mutants. The positions of mutations that lower the apparent pK_a of M formation are shown mapped to the 3D structure of the bR ground-state structure of (1C3W, Luecke, 1999b). The side chains that produce the phenotypes are shown in yellow. The retinal chromophore is shown in yellow. (See Plate no. 12 in the Color Plate Section.)

correspond to where our crystallographic intuition might have anticipated; the amino acid interactions energetically coupled to the M-, N-, and O-substate transitions are not located at the cytoplasmic ends of the (F)6 and (G)7 helices.

Numerous mechanistic insights were obtained from the scanning mutagenesis:

1. Mutagenesis by cysteines is not inherently perturbing to membrane protein function. As seen in Figs. 9, 14, and 15, a large number of the cysteine mutants possess absorbance properties that are indistinguishable from D85N (at all pHs evaluated). The experiment contains its own negative controls. In addition, the single-site cysteines introduced are a valuable tool for future site-directed labeling studies.
2. There are numerous mutations that perturbed the substate populations. There is not a single (or even a few) critical local amino acid interaction(s) that uniquely regulate the transitions between functional substates. It is not a matter of disrupting a critical local, all or nothing, chemical interaction that leads to a switch between conformational states.
3. The distribution of mutations that produce similar phenotypes is not random, they appear to cluster in unique molecular locations (domains, Fig. 15). Domains of highly coupled (allosteric) intermolecular interactions are the critical features regulating the transitions between bR substates.
4. The critical domains are remote from the crystallograhically identified structural changes. Perturbation of amino acid interactions at the extracellular surface of bR is coupled to conformational perturbation at the cytoplasmic end of the F(6) and G(7) helices. Long-range and highly coupled intramolecular interactions exist that are capable of "transducing" structural perturbations (e.g., signals) across the cellular membrane. The implications for the mechanisms of action of cellular receptors are obvious.

The molecular mechanisms of membrane protein mediated signal transduction and solute transport are only beginning to be probed. bR is frequently used as a model system for conceptualizing the structure and ligand-binding sites of other receptors predicted to contain seven transmembrane helical bundles. Given the complex nature of the bR photomechanism, the extensive structural rearrangements linked to proton pumping, and our demonstration of allosteric domains, make it difficult to consider mechanistic roles at the level of single amino acids. However, using a combination of systematic mutagenesis and careful physical analyses it is possible to gain significant insight into the mechanisms of membrane protein function.

V. Conclusions

Understanding the mechanisms of action of membrane proteins requires not only knowledge of the structures (conformations) accessed function but also knowledge of the physical forces that regulate the transitions between these structures. Since, there are very few crystal structures of membrane proteins and even less structural, and energetic, information about the intermediate-states in their reaction pathways our mechanistic intuitions are fairly naive. Mutagenesis remains a tractable approach for probing structural linkages to membrane protein function. The facility with which genetic engineering can be accomplished is responsible for its popularity in probing protein structure–function relationships in all aspects of modern molecular sciences. However, the complex nature of the intra- and intermolecular interactions involved in the mechanisms of membrane protein function and the high degree of coupling between these interactions may cause the analysis of mutational studies to be misleading.

Site-directed mutational analysis of membrane protein structure/function relationships can been useful for identifying critical amino acids as well as functional domains such as interfaces, channels, and binding sites. In the absence of detailed structural information, analysis of mutational perturbations can only be performed in terms of putative local, noncovalent interactions in the immediate environment of the substituted amino acid. However, it must be considered that in addition to local effects, a single-site substitution may also produce global changes in conformation that may be propagated throughout the molecule. This added complexity of cooperative and compensating interactions can confound the analysis of single-site mutants. The extent of coupling can be appreciated by an analysis that focuses on the patterns of responses that arise from mutations distributed throughout the protein's structure. This mutagenic strategy requires an extensive commitment to both engineering and functional characterization phases of the project. It of course also requires significant structural knowledge.

The bR examples used demonstrate that the combination of mutagenic and physical analyses (here spectroscopy and crystallography) can identify key amino acid interactions that participate in mechanisms of action. The most useful insights relied on mutations that produced functional perturbations and knowledge of their position in the 3D structure of bR. Granted, a similar synthesis is not possible for the vast majority of membrane proteins since atomic resolution structures are not routinely available. An objective of this chapter was to indicate the analyses required to obtain deep insight into membrane protein functional mechanisms. bR's experimental advantages render it a useful model system for probing the molecular mechanism of membrane proteins.

Extracellular signals interacting with receptor proteins are likely transduced through highly coupled sets of amino acid interactions that span extra- and intramembranous domains. It is appealing to consider that complex interactions regulating bR substate transitions mimic the allosteric interactions that contribute to GPCR activation, where ligand binding (on the extracellular side of the GPCR) is coupled through conformational changes leading to G-protein activation on the intracellular side of the lipid bilayer. Systematic mutagenesis and integration with the best available structural models (bR and/or rhodopsin) can unravel the molecular mechanisms of transbilayer signal transduction and solute transport.

I have attempted to illustrate how three different mutagenic strategies have been used to identify intramolecular interactions essential for a subset of conformational changes involved in membrane protein function. The mechanistic insights obtained rely heavily on the extensive, careful physical characterizations performed by many laboratories and are subject to refinement as more detailed structural insights are obtained. The approach of combining mutagenic, spectroscopic, and thermodynamic analyses will produce new knowledge about the physical forces and chemical participants that make up the bR reprotonation reaction (e.g., M↔N↔O). The energetic dissection of bR conformational changes will serve as a blueprint for understanding the dominant physical forces and chemical participants that couple transbilayer structural elements and ligand-binding site affinities in this important class of proteins.

References

Ackers, G. K., and Holt, J. M. (2006). Asymmetric cooperativity in a symmetric tetramer: Human hemoglobin. *J. Biol. Chem.* **281,** 11441–11443.

Ackers, G. K., and Smith, F. R. (1986). Resolving pathways of functional coupling within protein assemblies by site-specific structural perturbation. *Biophys. J.* **49,** 155–165.

Akabas, M. H., Stauffer, D. A., Xu, M., and Karlin, A. (1992). Acetylcholine receptor channel structure probed in cysteine-substitution mutants. *Science* **258,** 307–310.

Baldwin, J. M., Schertler, G. F., and Unger, V. M. (1997). An alpha-carbon template for the transmembrane helices in the rhodopsin family of G-protein-coupled receptors. *J. Mol. Biol.* **272,** 144–164.

Belrhali, H., Nollert, P., Royant, A., Menzel, C., Rosenbusch, J. P., Landau, E. M., and Pebay-Peyroula, E. (1999). Protein, lipid and water organization in bacteriorhodopsin crystals: A molecular view of the purple membrane at 1.9 Å resolution. *Struct. Fold Des.* **7,** 909–917.

Brown, L. S. (2000). Reconciling crystallography and mutagenesis: A synthetic approach to the creation of a comprehensive model for proton pumping by bacteriorhodopsin. *Biochim. Biophys. Acta* **1460,** 49–59.

Brown, L. S., Dioumaev, A. K., Needleman, R., and Lanyi, J. K. (1998). Local-access model for proton transfer in bacteriorhodopsin. *Biochemistry* **37,** 3982–3993.

Brown, L. S., Kamikubo, H., Zimányi, L., Kataoka, M., Tokunaga, F., Verdegem, P., Lugtenburg, J., and Lanyi, J. K. (1997). A local electrostatic change is the cause of the large-scale protein conformation shift in bacteriorhodopsin. *Proc. Natl. Acad. Sci. USA* **94,** 5040–5044.

Bullough, P. A., and Henderson, R. (1999). The projection structure of the low temperature K intermediate of the bacteriorhodopsin photocycle determined by electron diffraction. *J. Mol. Biol.* **286,** 1663–1671.

Cao, Y., Varo, G., Chang, M., Ni, B. F., Needleman, R., and Lanyi, J. K. (1991). Water is required for proton transfer from aspartate-96 to the bacteriorhodopsin Schiff base. *Biochemistry* **30,** 10972–10979.

Daggett, V., and Fersht, A. (2003). The present view of the mechanism of protein folding. *Nat. Rev. Mol. Cell Biol.* **4,** 497–502.

Dahl, G., and Pfahnl, A. (2001). Mutagenesis to study channel structure. *Methods Mol. Biol.* **154,** 251–268.

DasSarma, S., and Fleischmann, E. M. (1995). "Halophiles." Cold Spring Harbor Laboratory Press, Cold Spring Harbor.

DasSarma, S., RajBhandary, U. L., and Khorana, H. G. (1983). High-frequency spontaneous mutation in the bacterio-opsin gene in Halobacterium halobium is mediated by transposable elements. *Proc. Natl. Acad. Sci. USA* **80,** 2201–2205.

Dunn, R., McCoy, J., Simsek, M., Majumdar, A. S. C., RajBhandary, U., and Khorana, H. G. (1981). The bacteriorhodopsin gene. *Proc. Natl. Acad. Sci. USA* **78,** 6744–6748.

Faham, S., Yang, D., Bare, E., Yohannan, S., Whitelegge, J. P., and Bowie, J. U. (2004). Side-chain contributions to membrane protein structure and stability. *J. Mol. Biol.* **335,** 297–305.

Farrens, D. L. (1999). Site-directed spin labeling (SDSL) studies of the G protein-coupled receptor rhodopsin. *In* "Structure-Function Analysis of G Protein-Coupled Receptors." (J. Wess, ed.), Wiley-Liss, New York, pp. 289–314.

Fersht, A. R. (1993). The sixth Datta lecture. Protein folding and stability: The pathway of folding of barnase. *FEBS Lett.* **325,** 5–16.

Finer-Moore, J. S., Santi, D. V., and Stroud, R. M. (2003). Lessons and conclusions from dissecting the mechanism of a bisubstrate enzyme: Thymidylate synthase mutagenesis, function, and structure. *Biochemistry* **42,** 248–256.

Frauenfelder, H., McMahon, B. H., and Fenimore, P. W. (2003). Myoglobin: The hydrogen atom of biology and a paradigm of complexity. *Proc. Natl. Acad. Sci. USA* **100,** 8615–8617.

Garcia-Moreno, B. (1995). Probing structural and physical basis of protein energetics linked to protons and salt. *Meth. Enzymol.* **259,** 512–538.

Gerwert, K., Hess, B., Soppa, J., and Oesterhelt, D. (1989). Role of aspartate-96 in proton translocation by bacteriorhodopsin. *Proc. Natl. Acad. Sci. USA* **86,** 4943–4947.

Haltia, T., and Freire, E. (1994). Forces and factors that contribute to the structural stability of membrane proteins. *Biochim. Biophys. Acta* **1228,** 1–27.

Han, S. J., Hamdan, F. F., Kim, S. K., Jacobson, K. A., Bloodworth, L. M., Li, B., and Wess, J. (2005). Identification of an agonist-induced conformational change occurring adjacent to the ligand-binding pocket of the M(3) muscarinic acetylcholine receptor. *J. Biol. Chem.* **280,** 34849–34858.

Hanahan, D., Jessee, J., and Bloom, F. R. (1991). Plasmid transformation of *Escherichia coli* and other bacteria. *Meth. Enzymol.* **204,** 63–113.

Harris, T. K., and Turner, G. J. (2002). Structural basis of perturbed pKa values of catalytic groups in enzyme active sites. *IUBMB Life* **53,** 85–98.

Hildebrandt, V., Ramezani-Rad, M., Swida, U., Wrede, P., Grzesiek, S., Primke, M., and Buldt, G. (1989). Genetic transfer of the pigment bacteriorhodopsin into the eukaryote *Schizosaccharomyces pombe*. *FEBS Lett.* **243,** 137–140.

Hristova, K., and White, S. H. (2005). An experiment-based algorithm for predicting the partitioning of unfolded peptides into phosphatidylcholine bilayer interfaces. *Biochemistry* **44,** 12614–12619.

Kataoka, M., Kamikubo, H., Tokunaga, F., Brown, L. S., Yamazaki, Y., Maeda, A., Sheves, M., Needleman, R., and Lanyi, J. K. (1994). Energy coupling in an ion pump. The reprotonation switch of bacteriorhodopsin. *J. Mol. Biol.* **243,** 621–638.

Kenakin, T. (1997). Agonist-specific receptor conformations. *Trends Pharmacol. Sci.* **18,** 416–417.

Kimura, Y., Vassylyev, D. G., Miyazawa, A., Kidera, A., Matsushima, M., Mitsuoka, K., Murata, K., Hirai, T., and Fujiyoshi, Y. (1997). Surface of bacteriorhodopsin revealed by high-resolution electron crystallography. *Nature* **389,** 206–211.

Kobilka, B. (2004). Agonist binding: A multistep process. *Mol. Pharmacol.* **65,** 1060–1062.

Kyte, J., and Doolittle, R. F. (1982). A simple method for displaying the hydropathic character of a protein. *J. Mol. Biol.* **157,** 105–132.

Lanyi, J. K., and Schobert, B. (2002). Crystallographic structure of the retinal and the protein after deprotonation of the Schiff base: The switch in the bacteriorhodopsin photocycle. *J. Mol. Biol.* **321,** 727–737.

Lanyi, J. K., and Schobert, B. (2003). Mechanism of proton transport in bacteriorhodopsin from crystallographic structures of the K, L, M1, M2, and M2' intermediates of the photocycle. *J. Mol. Biol.* **328,** 439–450.

Liapakis, G., Chan, W. C., Papadokostaki, M., and Javitch, J. A. (2004). Synergistic contributions of the functional groups of epinephrine to its affinity and efficacy at the beta2 adrenergic receptor. *Mol. Pharmacol.* **65,** 1181–1190.

Lin, G. C., el-Sayed, M. A., Marti, T., Stern, L. J., Mogi, T., and Khorana, H. G. (1991). Effects of individual genetic substitutions of arginine residues on the deprotonation and reprotonation kinetics of the Schiff base during the bacteriorhodopsin photocycle. *Biophys. J.* **60,** 172–178.

Liu, J., and Rost, B. (2001). Comparing function and structure between entire proteomes. *Protein Sci.* **10,** 1970–1979.

Lozier, R. H., Bogomolni, R. A., and Stoeckenius, W. (1975). Bacteriorhodopsin: A light-driven proton pump in *Halobacterium halobium. Biophys. J.* **15,** 955–962.

Lozier, R. H., Xie, A., Hofrichter, J., and Clore, G. M. (1992). Reversible steps in the bacteriorhodopsin photocycle. *Proc. Natl. Acad. Sci. USA* **89,** 3610–3614.

Luecke, H., Schobert, B., Richter, H. T., Cartailler, J. P., and Lanyi, J. K. (1999a). Structural changes in bacteriorhodopsin during ion transport at 2 angstrom resolution. *Science* **286,** 255–261.

Luecke, H., Schobert, B., Richter, H. T., Cartailler, J. P., and Lanyi, J. K. (1999b). Structure of bacteriorhodopsin at 1.55 Å resolution. *J. Mol. Biol.* **291,** 899–911.

Marjamaki, A., Frang, H., Pihlavisto, M., Hoffren, A. M., Salminen, T., Johnson, M. S., Kallio, J., Javitch, J. A., and Scheinin, M. (1999). Chloroethylclonidine and 2-aminoethyl methanethiosulfonate recognize two different conformations of the human alpha(2A)-adrenergic receptor. *J. Biol. Chem.* **274,** 21867–21872.

Marti, T., Otto, H., Mogi, T., Rosselet, S., Heyn, M., and Khorana, H. G. (1991). Bacteriorhodopsin mutants containing single substitutions of serine or threonine residues are all active in proton translocation. *J. Biol. Chem.* **266**(11), 6919–6927.

Martinez, L. C., Thurmond, R. L., Jones, P. G., and Turner, G. J. (2002). Subdomains in the F and G helices of bacteriorhodopsin regulate the conformational transitions of the reprotonation mechanism. *Proteins* **48,** 269–282.

Martinez, L. C., and Turner, G. J. (2002). High-throughput screening of bacteriorhodopsin mutants in whole cell pastes. *Biochim. Biophys. Acta* **1564,** 91–98.

Mathies, R, Brito-Cru, C, Pollar, W, and Shan, C (1988). Direct observation of the femptosecond excited state *cis-trans* isomerization in bacteriorhodopsin. *Science* **240,** 777–779.

Matthews, B. W. (1996). Structural and genetic analysis of the folding and function of T4 lysozyme. *FASEB J.* **10,** 35–41.

Matthew, J. B., Gurd, F. R., Garcia-Moreno, B., Flanagan, M. A., March, K. L., and Shire, S. J. (1985). pH-dependent processes in proteins. *CRC Crit. Rev. Biochem.* **18,** 91–197.

Merlini, G., and Bellotti, V. (2005). Lysozyme: A paradigmatic molecule for the investigation of protein structure, function and misfolding. *Clin. Chim. Acta* **357,** 168–172.

Mogi, T., Stern, L. J., Chao, B. H., and Khorana, H. G. (1989). Structure-function studies on bacteriorhodopsin. VIII. Substitutions of the membrane-embedded prolines 50, 91, and 186: The effects are determined by the substituting amino acids. *J. Biol. Chem.* **264,** 14192–14196.

Mogi, T., Stern, L. J., Hackett, N. R., and Khorana, H. G. (1987). Bacteriorhodopsin mutants containing single tyrosine to phenylalanine substitutions are all active in proton translocation. *Proc. Natl. Acad. Sci. USA* **84,** 5595–5599.

Mogi, T., Stern, L. J., Marti, T., Chao, B. H., and Khorana, H. G. (1988). Aspartic acid substitutions affect proton translocation by bacteriorhodopsin. *Proc. Natl. Acad. Sci. USA* **85,** 4148–4152.

Nassal, M., Mogi, T., Karnik, S. S., and Khorana, H. G. (1987). Structure-function studies on bacteriorhodopsin III. Total synthesis of a gene for bacterio-opsin and its expression in *Escherichia coli. J. Biol. Chem.* **262,** 9264–9270.

Needleman, R., Chang, M., Ni, B., Varo, G., Fornes, J., White, S., and Lanyi, J. K. (1991). Properties of Asp 212—Asn bacteriorhodopsin suggest that Asp 212 and Asp 85 both participiate in a counterion and proton acceptor complex near the Schiff base. *J. Biol. Chem.* **266,** 11478–11484.

Nikiforovich, G. V., Galaktionov, S., Balodis, J., and Marshall, G. R. (2001). Novel approach to computer modeling of seven-helical transmembrane proteins: Current progress in the test case of bacteriorhodopsin. *Acta Biochim. Pol.* **48,** 53–64.

Nilsson, A., Rath, P., Olejnik, J., Coleman, M., and Rothschild, K. J. (1995). Protein conformational changes during the bacteriorhodopsin photocycle. A Fourier transform infrared/resonance Raman study of the alkaline form of the mutant Asp-85->Asn. *J. Biol. Chem.* **270,** 29746–29751.

Oesterhelt, D., and Krippahl, G. (1983). Phototrophic growth of halobacteria and its use for isolation of photosynthetically-deficient mutants. *Ann. Microbiol. (Paris)* **134B,** 137–150.

Oesterhelt, D., and Stoeckenius, W. (1974). Isolation of the cell membrane of *Halobacterium halobium* and its fractionation into red and purple membrane. *Meth. Enzymol.* **31,** 667–678.

Otto, H., Marti, T., Holz, M., Mogi, T., Stern, L. J., Engel, F., Khorana, H. G., and Heyn, M. P. (1990). Substitution of amino acids Asp-85, Asp-212, and Arg-82 in bacteriorhodopsin affects the proton release phase of the pump and the pK of the Schiff base. *Proc. Natl. Acad. Sci. USA* **87,** 1018–1022.

Palczewski, K., Kumasaka, T., Hori, T., Behnke, C. A., Motoshima, H., Fox, B. A., Le Trong, I., Teller, D. C., Okada, T., Stenkamp, R. E., Yamamoto, M., and Miyano, M. (2000). Crystal structure of rhodopsin: A G protein-coupled receptor [see comments]. *Science* **289,** 739–745.

Pebay-Peyroula, E., Rummel, G., Rosenbusch, J. P., and Landau, E. M. (1997). X-ray structure of bacteriorhodopsin at 2.5 angstroms from microcrystals grown in lipidic cubic phases [see comments]. *Science* **277,** 1676–1681.

Richter, H. T., Brown, L. S., Needleman, R., and Lanyi, J. K. (1996). A linkage of the pKa's of asp-85 and glu-204 forms part of the reprotonation switch of bacteriorhodopsin. *Biochemistry* **35,** 4054–4062.

Rouhani, S., Cartailler, J. P., Facciotti, M. T., Walian, P., Needleman, R., Lanyi, J. K., Glaeser, R. M., and Luecke, H. (2001). Crystal structure of the D85S mutant of bacteriorhodopsin: Model of an O-like photocycle intermediate. *J. Mol. Biol.* **313,** 615–628.

Royant, A., Edman, K., Ursby, T., Pebay-Peyroula, E., Landau, E. M., and Neutze, R. (2001). Spectroscopic characterization of bacteriorhodopsin's L-intermediate in 3D crystals cooled to 170 K. *Photochem. Photobiol.* **74,** 794–804.

Sass, H. J., Buldt, G., Gessenich, R., Hehn, D., Neff, D., Schlesinger, R., Berendzen, J., and Ormos, P. (2000). Structural alterations for proton translocation in the M state of wild-type bacteriorhodopsin [see comments] [comment]. *Nature* **406,** 649–653.

Sauer, R. T., Milla, M. E., Waldburger, C. D., Brown, B. M., and Schildbach, J. F. (1996). Sequence determinants of folding and stabilityfor the P22 Arc repressor dimer. *FASEB J.* **10,** 42–48.

Schoneberg, T., Schulz, A., Biebermann, H., Hermsdorf, T., Rompler, H., and Sangkuhl, K. (2004). Mutant G-protein-coupled receptors as a cause of human diseases. *Pharmacol. Ther.* **104,** 173–206.

Shand, R. F., Miercke, L. J. W., Mitra, A. K., Fong, S., Stroud, R. M., and Betlach, M. C. (1991). Wild-type and mutant bacterio-opsins D85N, D96N and R82Q: High level expression in *Escherichia coli. Biochemistry* **30,** 3082–3088.

Shortle, D. (1995). Staphylococcal nuclease: A showcase of m-value effects. *Adv. Protein Chem.* **46,** 217–247.

Soppa, J., and Oesterhelt, D. (1989). Bacteriorhodopsin mutants of *Halobacterium* sp. GRB. I. The 5-bromo-2′deoxyuridine selection as a method to isolate point mutants in Halobacteria. *J. Biol. Chem.* **264,** 13043–13048.

Soppa, J., Otomo, J., Straub, J., Tittor, J., Meesen, S., and Oesterhelt, D. (1989). Bacteriorhodopsin mutants of *Halobacterium* sp. GRB II. Characterization of mutants. *J. Biol. Chem.* **264,** 13049–13056.

Stemmer, W. P. C. (1994). Rapid evolution of a protein *in vitro* by DNA shuffling. *Nature* **370,** 389–391.

Stern, L. J., Ahl, P., Marti, T., Mogi, T., Duñach, M., Berkowitz, S., Rothschild, K., and Khorana, H. G. (1989). Substitution of membrane-embedded aspartic acids in bacteriorhodopsin causes specific changes in different steps of the photochemical cycle. *Biochemistry* **28,,** 10035–10042.

Stern, L. J., and Khorana, H. G. (1989). Structure-function studies on bacteriorhodopsin. X. Individual substitutions of arginine residues by glutamine affect chromophore formation, photocycle, and proton translocation. *J. Biol. Chem.* **264,** 14202–14208.

Subramaniam, S., Gerstein, M., Oesterhelt, D., and Henderson, R. (1993). Electron diffraction analysis of structural changes in the photocycle of bacteriorhodopsin. *EMBO J.* **12,** 1–8.

Subramaniam, S., Lindahl, M., Bullough, P., Faruqi, A. R., Tittor, J., Oesterhelt, D., Brown, L., Lanyi, J., and Henderson, R. (1999). Protein conformational changes in the bacteriorhodopsin photocycle. *J. Mol. Biol.* **287,** 145–161.

Subramaniam, S., Marti, T., and Khorana, H. G. (1990). Protonation state of Asp (Glu)-85 regulates the purple-to-blue transition in bacteriorhodopsin mutants Arg-82 $\rightarrow$ Ala and Asp-85 $\rightarrow$ Glu: The blue form is inactive in proton translocation. *Proc. Natl. Acad. Sci. USA* **87,** 1013–1017.

Tao, Y. X. (2006). Inactivating mutations of G protein-coupled receptors and diseases: Structure-function insights and therapeutic implications. *Pharmacol. Ther.* **111**(3), 949–973.

Thorgeirsson, T. E., Milder, S. J., Miercke, L. J., Betlach, M. C., Shand, R. F., Stroud, R. M., and Kliger, D. S. (1991). Effects of Asp-96 → Asn, Asp-85 → Asn, and Arg-82 → Gln single-site substitutions on the photocycle of bacteriorhodopsin. *Biochemistry* **30,** 9133–9142.

Tittor, J., Schweiger, U., Oesterhelt, D., and Bamberg, E. (1994). Inversion of proton translocation in bacteriorhodopsin mutants D85N, D85T, and D85,96N. *Biophys. J.* **67,** 1682–1690.

Trumpp-Kallmeyer, S., Hoflack, J., Bruinvels, A., and Hibert, M. (1992). Modeling of G-protein-coupled receptors: Application to dopamine, adrenaline, serotonin, acetylcholine, and mammalian opsin receptors. *J. Med. Chem.* **35,** 3448–3462.

Turner, G. J., Galacteros, F., Doyle, M. L., Hedlund, B., Pettigrew, D. W., Turner, B. W., Smith, F. R., Moo-Penn, W., Rucknagel, D. L., and Ackers, G. K. (1992). Mutagenic dissection of hemoglobin cooperativity: Effects of amino acid alteration on subunit assembly of oxy and deoxy tetramers. *Proteins* **14,** 333–350.

Turner, G. J., Miercke, L., Thorgeirsson, T., Kliger, D., Betlach, M. C., and Stroud, R. M. (1993). Bacteriorhodopsin D85N: Three spectroscopic species in equilibrium. *Biochemistry* **32,** 1332–1337.

Turner, G. J., Reusch, R., Winter-Vann, A. M., Martinez, L., and Betlach , M. C. (1999). Heterologous gene expression in a membrane-protein-specific system. *Prot. Expr. Purif.* **17,** 312–323.

Varo, G., Duschl, A., and Lanyi, J. K. (1990). Interconversions of the M, N, and O intermediates in the bacteriorhodopsin photocycle. *Biochemistry* **29,** 3798–3804.

Varo, G., and Lanyi, J. K. (1991). Thermodynamics and energy coupling in the bacteriorhodopsin photocycle. *Biochemistry* **30,** 5016–5022.

Vonck, J. (1996). A three-dimensional difference map of the N intermediate in the bacteriorhodopsin photocycle: Part of the F helix tilts in the M to N transition. *Biochemistry* **35,** 5870–5878.

Wagner, G., Oesterhelt, D., Krippahl, G., and Lanyi, J. (1983). Bioenergetic role of halorhodopsin in *Halobacterium halobium* cells. *FEBS Lett.* **131,** 341–345.

White, S. H. (2004). The progress of membrane protein structure determination. *Prot. Sci.* **13,** 1948–1949.

White, S. H., and Wimley, W. C. (1999). Membrane protein folding and stability: Physical principles. *Annu. Rev. Biophys. Biomol. Struct.* **28,** 319–365.

Wise, A., Gearing, K., and Rees, S. (2002). Target validation of G-protein coupled receptors. *Drug Discov. Today* **7,** 235–246.

Wu, S., Chang, Y., el-Sayed, M. A., Marti, T., Mogi, T., and Khorana, H. G. (1992). Effects of tryptophan mutation on the deprotonation and reprotonation kinetics of the Schiff base during the photocycle of bacteriorhodopsin. *Biophys. J.* **61,** 1281–1288.

Zhao, M. M., Gaivin, R. J., and Perez, D. M. (1998). The third extracellular loop of the beta2-adrenergic receptor can modulate receptor/G protein affinity. *Mol. Pharmacol.* **53,** 524–529.

Zscherp, C., Schlesinger, R., Tittor, J., Oesterhelt, D., and Heberle, J. (1999). In situ determination of transient pKa changes of internal amino acids of bacteriorhodopsin by using time-resolved attenuated total reflection Fourier-transform infrared spectroscopy. *Proc. Natl. Acad. Sci. USA* **96,** 5498–5503.

CHAPTER 17

Quantifying DNA–Protein Interactions by Single Molecule Stretching

Mark C. Williams,*,† Ioulia Rouzina,‡ and Richard L. Karpel§

*Department of Physics
111 Dana Research Center
Northeastern University
Boston, Massachusetts 02115

†Center for Interdisciplinary Research on Complex Systems
111 Dana Research Center
Northeastern University
Boston, Massachusetts 02115

‡Department of Biochemistry
Molecular Biology and Biophysics
University of Minnesota
Minneapolis, Minnesota 55455

§Department of Chemistry and Biochemistry
University of Maryland Baltimore County
Baltimore, Maryland 21250

METHODS IN CELL BIOLOGY, VOL. 84

0091-679X/08 $35.00
DOI: 10.1016/S0091-679X(07)84017-9

Abstract

In this chapter, we discuss a new method for quantifying DNA–protein interactions. A single double-stranded DNA (dsDNA) molecule is stretched beyond its contour length, causing the base pairs to break while increasing the length from that of dsDNA to that of ssDNA. When applied in a solution containing DNA binding ligands, this method of force-induced DNA melting can be used to quantify the free energy of ligand binding, including the free energy of protein binding. The dependence of melting force on protein concentration is used to obtain the equilibrium binding constant of the ligand to DNA. We have applied this method to a well-studied DNA-binding protein, bacteriophage T4 gene 32 protein (gp32), and have obtained binding constants for the protein to single-stranded DNA (ssDNA) under a wide range of solution conditions. Our analysis of measurements conducted at several salt concentrations near physiological conditions indicates that a salt-dependent conformational change regulates DNA binding by gp32.

I. Introduction

Over the past decade, several sophisticated techniques have been developed for examining the properties of single macromolecules. In one method, DNA molecules can be stretched. The force required to extend a single DNA molecule to a set distance is determined, or, alternatively, the distance a molecule moves on application of a specific constant force is measured (Allemand *et al.*, 2003; Bustamante *et al.*, 2000, 2003; Strick *et al.*, 2000; Williams and Rouzina, 2002; Williams *et al.*, 2002b). In this article we will focus on measurements in which long, single double-stranded DNA (dsDNA) molecules are stretched using optical tweezers. The resulting force–extension data yield the amount of mechanical work needed to extend the DNA molecule. When this mechanical work is used to force structural changes in the molecule, such as a conversion of dsDNA into single-stranded DNA (ssDNA), the energy calculated from the mechanical work required to alter the DNA structure can be used to determine the equilibrium free energy required to form the structure initially. Thus, when we use force to break DNA base pairs, the energy required to do this is equal to the energy holding the base pairs together. As shown below, the mechanical work required to break the DNA base pairs is equal to the equilibrium melting free energy (difference in free energy between the double- and single-stranded states) when the mechanical process is reversible. Even when this does not hold, it is often possible to obtain the equilibrium melting free energy from nonequilibrium measurements (Collin *et al.*, 2005; Liphardt *et al.*, 2002).

In this chapter, we discuss a method for characterizing DNA stretching experiments utilizing optical tweezers. We then show that this force-induced melting technique, when applied in a solution containing DNA-binding ligands, can be used to quantify the free energy of ligand binding, including the free energy of protein binding. As an example, we outline the results we have obtained for DNA-stretching experiments in the presence of the well-studied DNA-binding protein, T4 gene 32 protein (gp32). We show that we can use this method to obtain association constants for gp32—ssDNA binding over a wide range of solution conditions. Our analysis of measurements conducted at several salt concentrations near physiological conditions indicates that a salt-dependent conformational change regulates DNA binding by gp32.

II. Stretching Single DNA Molecules with Optical Tweezers

There are several ways by which force can be applied to single DNA molecules, including atomic force microscopy, magnetic tweezers, and optical tweezers. All of these methods can be used to obtain the measurements outlined in this chapter. However, each technique has a different set of usable force ranges and accuracies, and optical tweezers appear to be best suited to force-induced DNA-melting experiments (Williams and Rouzina, 2002). In the experiments discussed here, we use a dual beam optical tweezers instrument to stretch single λ-DNA molecules.

A DNA stretching experiment is shown schematically in Fig. 1. The DNA molecule, of bacteriophage λ, contains 48,500 base pairs and is labeled with several

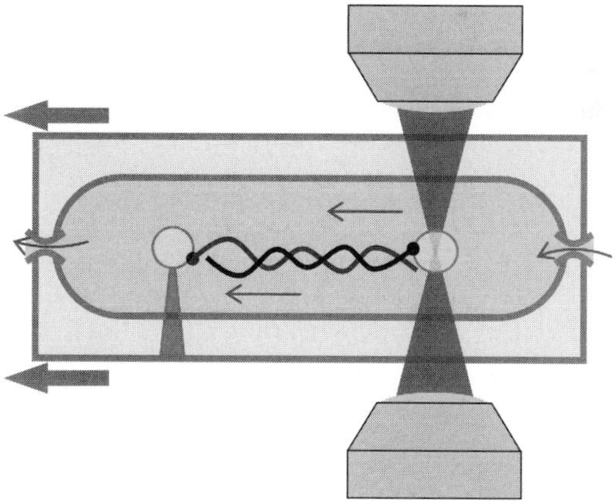

Fig. 1 In a dual beam optical tweezers instrument, two laser beams are focused to a small spot, creating an optical trap that attracts polystyrene beads. Single DNA molecules are attached at one end to a bead in the trap, while the other end is attached to another bead held by a glass micropipette. As the DNA molecule is stretched by moving the micropipette, the resulting force on the bead in the trap is measured.

biotin molecules on the 3′ end of each strand. Two 830 nm wavelength diode lasers with approximately 200 mW of vertically polarized continuous wave laser power are convergently directed using high efficiency polarizing beam splitting cubes (Melles Griot). Before entering the first microscope objective, the beam enters a quarter wave plate, which converts the linearly polarized light into circularly polarized light. The beams are focused to a small spot size less than one micron in diameter using 60×, 1.0 numerical aperture water immersion objectives (Nikon). The two counter propagating beams are focused to the same spot in a flow cell, which forms the optical trap. When the beams exit the cell and are collimated by the second microscope objective, the light passes through another quarter wave plate, which is oriented in such a way that the polarization becomes horizontal. The horizontally polarized light is directed by another beam splitting cube into a lateral effect photodiode detector. These diodes determine any deflection of each beam to within a few microns, and output a voltage that is directly proportional to the force being exerted on the bead in the optical trap. White light sources and CCD cameras provide images of the tip and the beams to guide the instrument user (Smith *et al.*, 2003).

To capture beads in the optical trap, micron-sized streptavidin-coated polystyrene beads are made to flow through the cell, and are subsequently trapped by the laser beam. The glass micropipette, which is attached to the flow cell, is also used to capture a single polystyrene bead. When single beads are captured at both the end of the pipette and in the trap, the other beads are rinsed out of the cell. A very dilute solution of labeled DNA molecules, typically in 10 mM HEPES, pH 7.5, and varying NaCl concentrations, is then introduced into the cell until one molecule is captured between the two polystyrene beads. The flow cell, and thus the glass micropipette, is then moved a fixed distance, thereby exerting a force that begins to pull the other bead out of the center of the optical trap. Because the distance that the bead moves inside the optical trap is proportional to the force exerted by the trapping laser beams, we obtain a direct measurement of the force required to extend the DNA molecule a fixed distance (Vladescu *et al.*, 2005).

The results of a typical DNA force–extension measurement are shown as data points in Fig. 2. The solid lines represent theoretical force–extension curves for dsDNA (leftmost line) and ssDNA (rightmost line). The stretched DNA initially follows the theoretical dsDNA curve, but at a force of about 65 picoNewton (pN), there is a transition from the dsDNA curve to near the ssDNA curve. This transition occurs at almost constant force, and we refer to it as the DNA force-induced melting transition, which will be discussed in Section III. The dsDNA theoretical curve follows a characteristic shape determined by the elasticity of the DNA molecule. At very low forces (a few pN) and extensions less than 0.34 nm/bp (nanometers per base pair), the stretching force merely causes the randomly bent, flexible molecule to straighten. As the extension of the DNA nears 0.34 nm, the DNA double helix is being stretched, and much more force is required to extend the DNA molecule a small distance.

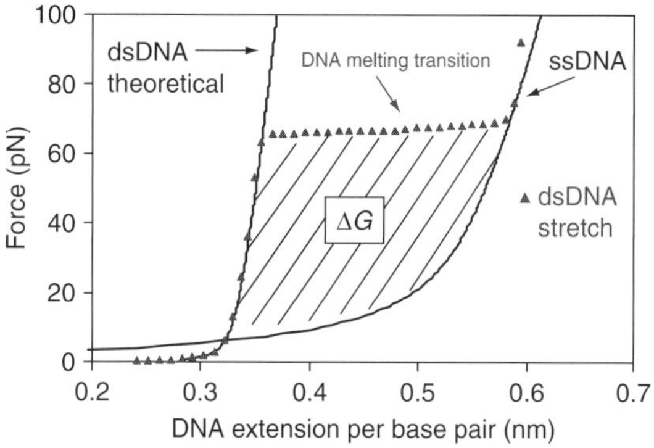

Fig. 2 Typical force–extension curve for stretching dsDNA at pH 7.5 and 500 mM ionic strength (triangles). Theoretical force–extension curves for dsDNA and ssDNA are shown as thin solid lines. When the transition is reversible, the melting free energy is obtained from the area between the dsDNA stretching curve and the ssDNA curve.

III. Force-Induced Melting of Single DNA Molecules

When the DNA force-induced melting transition of ~65 pN was originally discovered, it was referred to as DNA overstretching. Rouzina and Bloomfield proposed that this behavior is a transition from dsDNA to ssDNA, or a force-induced melting transition (Rouzina and Bloomfield, 2001a,b). The model was used to quantitatively predict the dependence of the overstretching force on changes in solution conditions, such as temperature, ionic strength, pH, and the presence of DNA-binding ligands (Williams *et al.*, 2002b). These predictions were quantitatively verified in subsequent experimental studies. For example, the pH dependence of the transition followed that of thermal melting studies (Williams *et al.*, 2001b). In addition, measurements of the temperature dependence of the transition were used to obtain the free energy of DNA melting as a function of temperature, and the resulting measurements of the changes in heat capacity and entropy of DNA on melting agreed with calorimetry measurements (Williams *et al.*, 2001c). Similarly, the observed dependence of the transition on salt concentration was consistent with that expected from polyelectrolyte theory for DNA melting (Wenner *et al.*, 2002). Finally, recent studies have demonstrated that in the presence of DNA-binding drugs, the results of DNA overstretching experiments parallel those obtained in thermal melting experiments, consistent with the force-induced melting model (Mihailović *et al.*, 2006; Vladescu *et al.*, 2005). In addition to the experimental evidence in support of force-induced melting, several recent theoretical studies are also consistent with this model (Harris *et al.*, 2005; Heng *et al.*, 2006; Piana, 2005).

These results suggest that single-molecule DNA stretching is a powerful new technique for measuring thermodynamic parameters that describe the stability of double-stranded nucleic acids, in many ways analogous to the information derived from conventional DNA thermal melting studies. However, DNA stretching has several advantages over thermal melting. First, stretching experiments can be obtained under a wide range of solution conditions, including physiological and room temperatures. This capability is particularly useful for studying the effect of protein binding on DNA melting, since at physiological ionic strength, the thermal denaturation of the DNA often occurs at a temperature above the point where the protein denatures. Thus, DNA stretching is a valuable tool for studying a protein's effect on DNA stability under physiologically relevant conditions (Pant *et al.*, 2003; Williams *et al.*, 2001a, 2002a). Examples of such proteins include nucleic acid chaperone proteins such as HIV-1 Gag and nucleocapsid (NC) protein (Cruceanu *et al.*, 2006; Rein *et al.*, 1998; Williams *et al.*, 2001a, 2002a), or ssDNA-binding proteins, such as T4 gp32 (Pant *et al.*, 2003, 2004, 2005; Rouzina *et al.*, 2005), T7 gene 2.5 protein (He *et al.*, 2003), or *E. coli* SSB (Lohman and Ferrari, 1994).

By stretching a long polymeric DNA molecule, such as λ-DNA, one obtains the sequence-averaged thermodynamic characteristics of DNA melting. Thus, the area between the experimental dsDNA stretching curve and the ssDNA stretching curve, and below the DNA melting plateau, can be used to obtain the DNA melting free energy. When this melting free energy is obtained in the presence of DNA-binding ligands, the melting force (the force measured at the midpoint of the melting transition) can change dramatically, depending on the DNA binding characteristics of the ligand. In general, if the ligand binds more strongly to dsDNA than ssDNA, the DNA melting force should increase, as was recently observed for DNA stretching in the presence of the intercalator ethidium as well as several ruthenium compounds (Mihailović *et al.*, 2006; Vladescu *et al.*, 2005). In contrast, ligands that bind more strongly to ssDNA should induce a decrease in the DNA melting force, as observed in the presence of gp32 and several truncated forms of this protein.

In Section IV we will discuss our recent studies of the binding thermodynamics of T4 gp32, a classic ssDNA-binding protein that has been well studied by bulk biochemical methods for over 30 years. The very high affinities of gp32 for single-stranded nucleic acids necessitated that most of the thermodynamic data was obtained at very high salt concentrations, so that very little binding information at physiological conditions is available from bulk experiments. This was not the case for our DNA force-induced melting experiments in the presence of gp32, where we have interpreted the results in terms a salt-dependent conformational change in the protein, required for DNA binding.

IV. T4 gp32 Interactions with DNA

A. Background

We investigated full length gp32 as well as two truncated forms, denoted *I and *III, as shown in Fig. 3 (Pant *et al.*, 2003, 2004, 2005; Rouzina *et al.*, 2005). *I lacks the C-terminal domain (CTD) (residues 254–301) and *III lacks both the CTD and

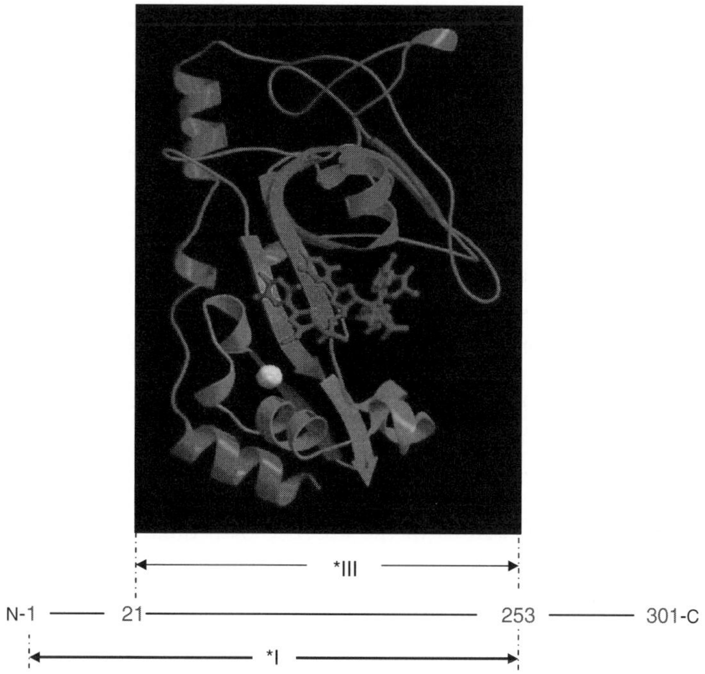

Fig. 3 Proteolytic fragments of gp32. *I was originally obtained by trypsin cleavage of full length gp32 at residue 253, while *III results from cleavage at residues 21 and 253. A MOLSCRIPT (Kraulis, 1991) representation of a *III-oligonucleotide complex is shown at its location within the protein sequence. The protein is pictured in ribbon mode, with the major lobe green, the minor (Zn-containing) lobe blue, and the residue 198–239 flap red. The bound oligonucleotide, in sticks mode, is red, and the coordinated Zn^{2+}, in space-filling mode, is yellow. The position of the oligodeoxynucleotide, pTTAT, is approximate; it was modeled by Shamoo *et al.* to maximally overlap excess electron density in the trough (Shamoo *et al.*, 1995). The Protein Data Bank entry for core domain (without the oligonucleotide) is 1gpc.pdb. (See Plate no. 13 in the Color Plate Section.)

the N-terminal domain (NTD) (residues 1–21). Previous studies have shown that *III binds ssDNA noncooperatively and displays helix-destabilizing activity at low salt levels (Waidner *et al.*, 2001). Due to experimental limitations, binding measurements of *I and gp32 to ssDNA have generally been obtained at high ionic strengths (greater than 300 mM Na^+). Both proteins displayed strongly salt-dependent binding to ssDNA. The $\log(K_{ss})/\log([Na^+])$ plots, where K_{ss} is the noncooperative or intrinsic equilibrium binding constant of the protein to ssDNA, yielded slopes of approximately −7 in each case (Lonberg *et al.*, 1981). Based on extrapolation of this linear dependence, the low salt (~10 mM Na^+) binding affinities of both proteins would be expected to be in the order of 10^9 M^{-1}. These studies also showed that the *I truncate has a similar site size and cooperativity of ssDNA binding. At the high-salt conditions used, *I was shown to bind two to three times stronger to ssDNA relative to gp32.

Based on the strong preferential binding of gp32 to ssDNA and extrapolation of the binding data to low [salt], it was predicted that gp32 and *I should lower the thermal melting temperature (T_m) of natural dsDNA under these conditions by ~50 °C. However, gp32 had no effect on the DNA melting temperature (Jensen *et al.*, 1976). In contrast, *I was subsequently shown to lower the T_m significantly, consistent with predictions based on the extrapolation of high-salt equilibrium binding data (Lonberg *et al.*, 1981; Waidner *et al.*, 2001). The single molecule studies outlined here have largely resolved this apparent contradictory behavior of the two forms of the protein.

B. T4 gp32 Destabilizes Single DNA Molecules

The results of a typical force-induced melting experiment in the presence of gp32 or *I, as well as in the absence of protein are shown in Fig. 4. As the dsDNA molecule is stretched beyond its B-form contour length, the force required to stretch the DNA increases rapidly until a plateau is reached, which represents the force-induced DNA melting transition. In the absence of protein, the stretching at any point in this transition region can be reversed and the force–extension curve follows a similar curve during relaxation, except for a small portion showing some hysteresis near the B-form contour length of about 0.34 nm/bp. The observed reversibility of the stretching suggests that the DNA melting force represents an equilibrium melting force. Thus, in the absence of protein, the area between

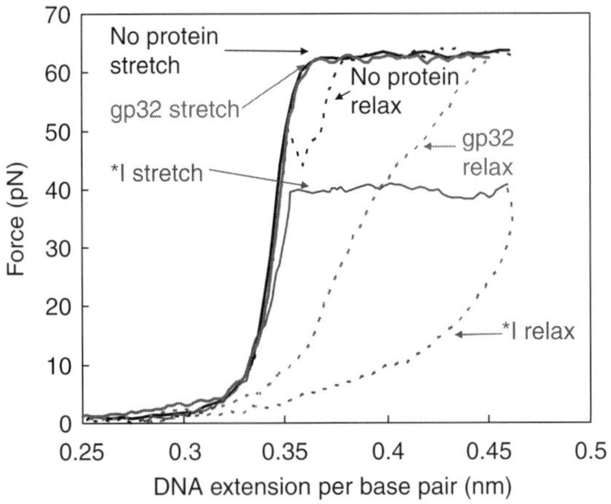

Fig. 4 DNA force-induced melting experiments in the absence of protein and in the presence of 200 nM gp32 and 200 nM *I. Stretching curves are shown as solid lines, while relaxation curves are shown as dashed lines.

the dsDNA and ssDNA stretching curves gives the DNA melting free energy (see Fig. 2).

In the presence of 200 nM gp32, the force–extension curve shown in Fig. 4 resembles that obtained in the absence of protein. However, at the stretching rates employed (about 2 min per experiment), the relaxation curve does not follow the stretching curve at any point in the cycle. Thus, the melting force observed in this experiment does not represent an equilibrium melting value. To distinguish the two types of forces, we denote the equilibrium melting force F_m, and the nonequilibrium melting force F_k, which represents the kinetically determined melting force, observed in the presence of protein. The same strong hysteresis is also observed in the presence of *I, but in contrast to the full-length protein, the nonequilibrium force is significantly smaller than the melting force observed in the absence of protein. Since the DNA stretching and relaxation curves are not superimposable for either protein, the area under the observed stretching curve cannot be interpreted as an equilibrium melting free energy in the presence of the protein.

It is nevertheless possible to obtain a measurement of the equilibrium melting force in the presence of protein. (Pant *et al.*, 2003, 2005) To do this, we stretch dsDNA to 0.42 nm/bp, an extension that is clearly in the force-induced melting plateau, as illustrated in Fig. 5A. At this point, the DNA extension is held constant, while the force is measured as a function of time, as shown in Fig. 5B. The resulting force decays exponentially over time, and approaches a constant value, although there are still significant fluctuations around this force. We fit this time-dependent force data to the following relation:

$$F(t)_{\text{stretch}} = F_{\text{eq}}^{\text{p}} + (F_{\text{k}} - F_{\text{eq}}^{\text{p}})\ \exp\left(-\frac{t}{\tau_{\text{melt}}}\right) \tag{1}$$

After the initial melting force, F_k, the force obtained at times much greater than the decay time constant is the equilibrium melting force in the presence of protein, F_{eq}^{p}. The equilibrium nature of this force is confirmed by the fact that the force measured when starting from the relaxation curve also approaches the same value at long times (Pant *et al.*, 2003). The constant, τ_{melt}, represents the characteristic time for additional DNA melting as the system approaches equilibrium binding and an equilibrium fraction of melted base pairs in the presence of protein. The initial force, F_k, depends strongly on pulling rate and can be used to determine the kinetics of protein binding, but a detailed discussion of this effect is beyond the scope of this chapter (Pant *et al.*, 2004; Sokolov *et al.*, 2005a,b).

C. Quantifying Equilibrium Protein Interactions with DNA

We can directly determine the reduction in equilibrium melting force in the presence of protein, F_{eq}^{p}, relative to that observed in the absence of protein, as a function of protein concentration (Fig. 5). This measurement is analogous to the dependence of the T_m of DNA on protein concentration

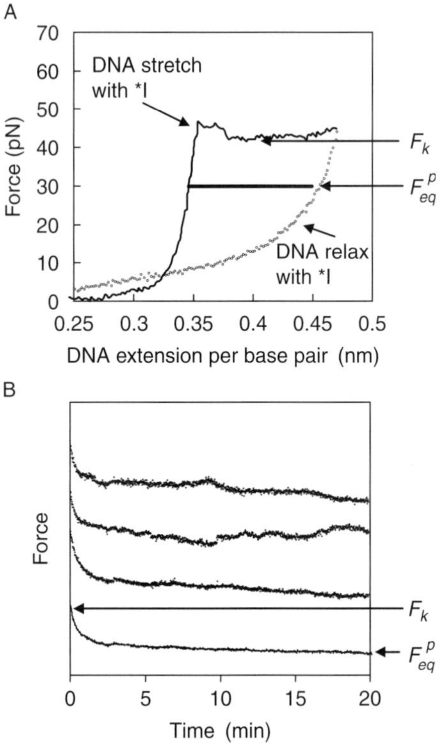

Fig. 5 (A) Stretching and relaxation curves for λ-DNA in 10 mM HEPES pH 7.5, 100 mM $[Na^+]$ in the presence of 200 nM *I. The kinetic melting force F_k is identified on the figure. To measure DNA denaturation, the DNA is melted by force at an extension of 0.42 nm/bp and the force is measured as a function of time. The force decreases initially in this measurement. (B) Representative curves for time dependence of DNA stretching force at constant position in the presence of *I. The kinetic and equilibrium melting forces corresponding to the forces on the force–extension diagram on part (A) are identified on the figure. Three separate time dependence measurements are shown as the top three lines. The bottom dataset represents an average of the three measurements, which is used to determine F_{eq}^p. The curves are shifted arbitrarily along the force axis to allow for ease of comparison.

(Frank-Kamenetskii *et al.*, 1987; Jensen and von Hippel, 1976; Jensen *et al.*, 1976; Kelly and von Hippel, 1976), which has previously been shown to depend on the DNA-binding constant according to the relation (McGhee, 1976):

$$\frac{1}{T_m^0} - \frac{1}{T_m^p} = \frac{k_B}{\Delta H} \ln \left\{ \frac{(1 + K_{ds}C)^{1/n_{ds}}}{(1 + K_{ss}\omega C)^{2/n_{ss}}} \right\} \quad (2)$$

where T_m^0 and T_m^p are the melting temperature in the absence and presence of protein, K_{ds} is the affinity of the protein for an isolated binding site on dsDNA of occluded length n_{ds}, K_{ss}, and n_{ss} are the corresponding quantities for protein binding ssDNA, ω is the cooperativity parameter for binding ssDNA, which represents the probability of protein binding adjacent to an already bound protein relative to its probability of binding anywhere, and C is the (free) protein concentration. Here we also assume that binding to dsDNA is noncooperative, which is the case for gp32 (Jensen *et al.*, 1976).

The analogy between thermal- and force-induced melting leads to the following relationship between the protein-effected change in melting force and the corresponding change in the melting temperature (Rouzina and Bloomfield, 2001b):

$$F_m^p - F_m^0 = (T_m^p - T_m^0)\frac{\Delta S}{\Delta x} \tag{3}$$

where Δx is the difference in residue length per base pair between the protein-bound ssDNA and dsDNA. Thus, after combining Eqs. (2) and (3), and substituting $T_m^0 = \Delta H/\Delta S$, the shift in melting force is given by:

$$F_m^p = F_m^0 + \frac{k_B T}{\Delta x}\ \ln\left\{\frac{(1 + K_{ds}C)^{1/n_{ds}}}{(1 + K_{ss}\omega C)^{2/n_{ss}}}\right\} \tag{4}$$

In general, Eq. (4) could be used to obtain both K_{ds} and $K_{ss}\omega$ by measuring F_m as a function of protein concentration C, where double-stranded binding tends to increase the melting force and single-stranded binding tends to decrease the force. However, because we have two unknown variables, it would be difficult to obtain these numbers without additional information. Fortunately, for ssDNA-binding proteins, such as gp32, the binding to ssDNA, $K_{ss}\omega$, is generally much greater than the binding constant to dsDNA, K_{ds}, so we can neglect the term involving K_{ds}, obtaining:

$$F_m \approx F_m^0 - \frac{2k_B T}{n_{ss}\Delta x}\ \ln\ (1 + K_{ss}\omega C) \tag{5}$$

Thus, while Eq. (4) is more general, Eq. (5) is likely all that is needed to describe ssDNA-binding proteins, such as gp32, as we discuss here.

The fundamental result is that measuring F_m as a function of C using optical tweezers in the presence of ssDNA-binding proteins and fitting the data to Eq. (5) allows for a quantitative determination of $K_{ss}\omega$ for each protein under a variety of solution conditions. Several measurements of F_m as a function of C for both gp32 and *I and at varying salt concentration are shown in Fig. 6. Under all conditions tested, the data fit well to Eq. (5). The results clearly show that DNA force-induced melting can be used to determine $K_{ss}\omega$ for ssDNA-binding proteins under a wide variety of solution conditions.

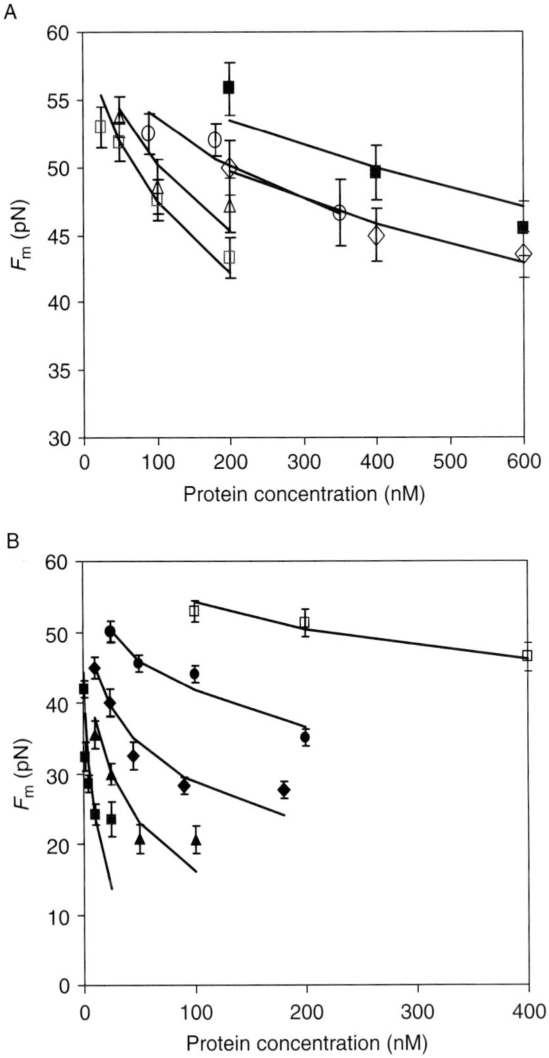

Fig. 6 (A). The measured (symbol) dependence of equilibrium DNA melting force F_m in the presence of protein as function of protein concentration (C) for gp32. Measurements are shown in 0.05 M Na^+ (open square), 0.075 M Na^+ (open triangle), 0.1 M Na^+ (open diamond), 0.15 M Na^+ (open circle), and 0.2 M Na^+ (solid square). Lines are fit to data using Eq. (4). (B) The measured (symbol) dependence of equilibrium DNA melting force F_m in the presence of protein as function of protein concentration (C) for *I. Measurements are shown in 0.05 M Na^+ (closed square), 0.075 M Na^+ (closed triangle), 0.1 M Na^+ (closed diamond), 0.15 M Na^+ (closed circle), and 0.2 M Na^+ (open square). Lines are fit to the data using Eq. (5).

D. Salt Dependence of T4 Gene 32 Binding to ssDNA

Despite the fact that gp32 has been studied extensively for over 30 years, there remain a number of important questions concerning the protein's behavior. For example, gp32 was assumed to bind much more strongly to ssDNA than dsDNA over a wide range of salt conditions. At low salt, the affinities for ssDNA are too high to be directly determined. At high salt ($\geq$300 mM NaCl), log K_{ss} varies linearly with log[Na^+], so that affinities at lower salt can be calculated by extrapolation of the log–log plots, assuming uninterrupted linearity. The binding to dsDNA is much weaker, and the limited amount of available dsDNA binding data is restricted to low-salt conditions (Jensen *et al.*, 1976). Since most proteins denature at temperatures where dsDNA denatures at physiological conditions, thermal DNA melting experiments are necessarily conducted at low salt. The predicted differential magnitudes of the protein's affinities for ssDNA versus dsDNA should lead to a reduction of the T_m of dsDNA to near room temperature. However, no effect on melting was observed (Jensen *et al.*, 1976). The inability of bulk experiments to directly measure the protein affinities for DNA at physiological salt, and the absence of the predicted T_m depression, led us to examine the salt dependence of binding via the equilibrium force-induced melting method.

The results of our measurements of $K_{ss}\omega$, the overall binding constant of gp32 and *I to ssDNA, as a function of salt concentration are shown in Fig. 7. These results from our DNA stretching experiments extend from 50 mM Na^+ to 200 mM Na^+, while the majority of earlier bulk measurements of $K_{ss}\omega$ were obtained above 200 mM Na^+. For comparison, we show both the data obtained in single molecule measurements and the bulk data from earlier work on the same graph. There is excellent agreement between the salt dependence of gp32 binding to single-stranded nucleic acids in bulk and single molecule experiments, providing strong evidence in support of our method for using force-induced DNA melting to quantitatively characterize ssDNA binding. For comparison of data obtained below 200 mM Na^+, the only available results from bulk experiments are of gp32 binding the polyribonucleotide, poly(A) (Villemain and Giedroc, 1996). The overall affinity ($K_{ss}\omega$) of gp32 to poly(A) is expected to be slightly lower than its affinity to natural ssDNA, so the shift of that data to slightly lower values is expected. In addition, the salt dependence of the poly(A) binding data very strongly resembles that obtained in our equilibrium force-induced melting experiments for binding of gp32 to λ-DNA.

While it is important that the single molecule data agrees with bulk data, what is even more significant is that we are able to obtain important new information about gp32 binding to DNA that was not available from bulk experiments. First, we see from Fig. 7 that gp32 binding to ssDNA has a $d\log(K_{ss}\omega)/d\log([Na^+])$ slope of -7 above 200 mM Na^+, but the slope approaches zero below this salt concentration. While *I binding to ssDNA in high salt also has a slope of -7, at low salt the slope is reduced to -3, significantly greater in magnitude than the near-zero slope for gp32 binding under these conditions. The divergence in the behavior of gp32 and *I at low

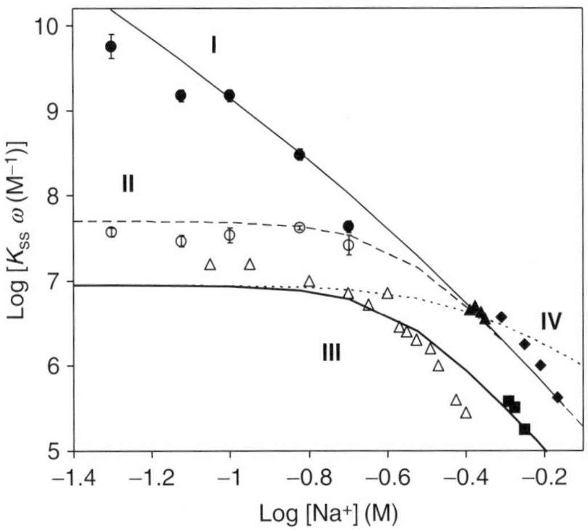

Fig. 7 Binding constants and their fits to our binding model for *I (solid circles and squares, and curves I and III) and gp32 (the rest of the symbols, and curves II–IV) as function of NaCl (curves I–III) or NaF salt (curve IV). The closed circles associated with curve I and open circles with curve II were obtained for ss λ-DNA in low salt in a single-molecule study (Pant *et al.*, 2005; Rouzina *et al.*, 2005), and the diamonds for ϕx174 ssDNA are from Newport *et al.* (Newport *et al.*, 1981) in high salt. The open triangles associated with curve III are from Villemain and Giedroc (Villemain and Giedroc, 1996) for gp32 binding to poly(A) and the solid squares are from Lonberg *et al.* (Lonberg *et al.*, 1981) for *I binding to poly(rA). The closed triangles associated with curve IV are from Villemain and Giedroc (Villemain and Giedroc, 1996) measured for gp32 binding to poly(A) in NaF. These two sets of curves were fit using Eqs. (7), (11), (12), and (16), with a common set of parameters: $n_{\mathrm{Na}} = 2.95 \pm 0.30$, $n_{\mathrm{Cl}} = 3.70 \pm 0.30$, $K_{\mathrm{Cl}}=2 \pm 0.5\ \mathrm{M}^{-1}$, and $\Delta G_{\mathrm{0CTD}} = 3.2 \pm 0.2\ k_{\mathrm{B}}T$. The only difference between the binding curves for ss λ-DNA and poly(A) is the vertical shift due to the difference in $(K_{*\mathrm{I},0}\omega)$, which was $\log(K_{*\mathrm{I},0}\omega) = 6.5 \pm 0.1$ for ss λ-DNA and $\log(K_{*\mathrm{I},0}\omega) = 5.75 \pm 0.1$ for poly(A).

salt was not previously observed, and this important result provides strong clues concerning the nature of gp32 and *I binding to ssDNA.

In general, the salt dependence of ligand binding to nucleic acids can be described by counterion condensation theory, first derived by Oosawa (1968, 1971) and Manning (1978) for the idealized case of an infinite line charge. They showed that counterions in solution are effectively bound to a DNA molecule when the charge density parameter of a macromolecule (such as DNA) is greater than $k_{\mathrm{B}}T$. The charge density parameter is given by $\xi = q^2/\epsilon b$, where ϵ is the dielectric constant and b is the distance between charges of magnitude q. Counterion binding will occur when the distance between charges is small, as it is for both dsDNA and ssDNA. When this is the case, these counterions must be removed in order for a charged ligand to bind the DNA. The subsequent release of counterions into solution results in an entropy change of $\Delta S = n_{\mathrm{q}} k_{\mathrm{B}} \ln\ (N_{\mathrm{s}}/I)$, where n_{q} is the number of counterions removed, N_{s} is the surface charge density of counterions on the DNA, and I is the solution ionic strength. The application of this concept to the interpretation of the salt dependence of nucleic acid binding of cationic ligands was developed by Record and coworkers (Record *et al.*, 1976, 1998). It was later shown that similar counterion condensation can exist on a macroion of arbitrary size and shape, within a certain range of solution ionic strength (Frank-Kamenetskii *et al.*, 1987; Gueron and Weisbuch, 1981; Rouzina and Bloomfield, 1997; Safran *et al.*, 1991). The only requirement for counterion condensation is that there must be a local high surface charge density on the macroion, which creates a nearby region with counterion energy larger than $k_B T$ (Rouzina and Bloomfield, 1996, 1997). This situation can be represented phenomenologically by assuming that the protein or nucleic acid has a binding site for the neutralizing counterions with a strong, salt-dependent binding constant. In most cases, this results in a linear $d \log(K_{ss}\omega)/d \log([Na^+])$ dependence, where the slope is equal to the number of counterions removed from the DNA on ligand binding. This in turn reveals information about the cationic nature of the ligand-binding site (Bloomfield *et al.*, 2000). If, as in the case of gp32, this salt dependence is not always linear, one must consider possible cationic and anionic binding sites that may be perturbed on protein–DNA binding. In Section V we describe a model that explains the effects observed in bulk and single molecule experiments, and provides a quantitative understanding of the salt dependence of gp32 and *I binding to ssDNA.

V. Model for Salt-Dependent Regulation of T4 Gene 32 Binding to DNA

To construct our model, we first note that *I differs from gp32 only by the absence of the 48 C-terminal residues. These residues are not present in the crystal structure of the core domain of gp32 (*III), which is missing both C- and N-terminal residues. The CTD of gp32 is highly acidic, and it is possible that it can function as an effective DNA mimic, binding to the core of gp32 in low salt, and thus preventing binding of DNA. In such a case, decreasing the $[Na^+]$ simply increases the affinity of the CTD for the DNA-binding site, preventing the expected further increase in DNA binding affinity as the salt concentration is lowered. Since *I lacks the CTD, its binding to DNA continues to increase as the salt concentration is lowered, even in low salt. A schematic diagram of this situation is depicted in Fig. 8.

In the case of *I, the salt dependence is relatively easily described since there are no effects due to the CTD. Nevertheless, the change in the slope $d \log(K_{ss}\omega)/d \log([Na^+])$ as the salt concentration is lowered requires an explanation. We have suggested that this effect is due to the presence of Cl^- ions in solution. Villemain and Giedroc showed that replacement of NaCl with NaF significantly altered the binding of gp32 to single-stranded nucleic acids (Villemain and Giedroc, 1996). Since Cl^- ions generally bind much more strongly than F^- ions to cationic binding sites on proteins, the Villemain and Giedroc data likely indicate that Cl^- ions can bind to amino acid residues of the DNA binding surface on gp32. Occupancy of these sites increases with chloride concentration,

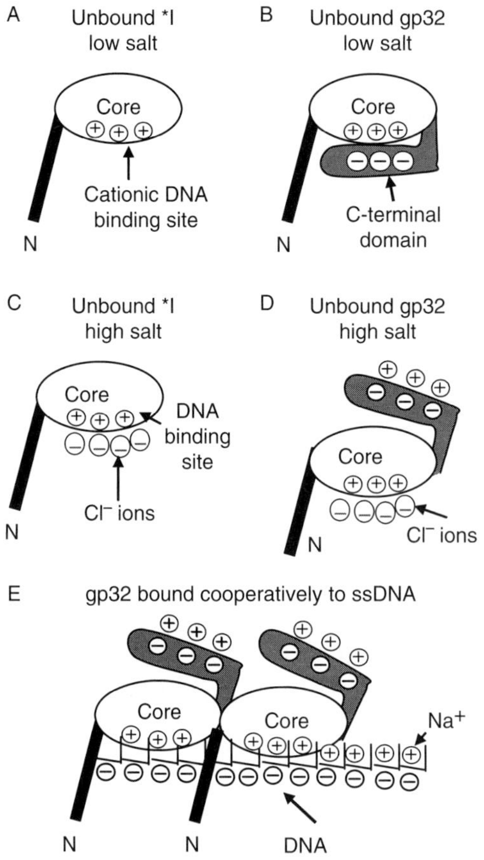

Fig. 8 Schematic depiction of model for electrostatic regulation of DNA binding. (A) *I lacks the CTD, so in low salt its DNA-binding site is always available for binding to DNA. (B) In low salt, the gp32 CTD spends a significant amount of time bound to the DNA-binding site, thus preventing binding to DNA. (C) In high salt, four Cl^- ions are condensed onto the binding site of *I. (D) In high salt, the CTD of gp32 is released from the core, so gp32 resembles *I, with four Cl^- ions bound to the DNA-binding site on the core. (E) When full length gp32 is bound to DNA, the gp32 CTD is exposed to solution. Three sodium ions are condensed onto the gp32 CTD (as in panel D).

so that in high salt, Cl^- must be removed in order for *I to bind. The probability of having all n_{Cl} Cl^- ions released simultaneously to allow *I to bind DNA is $1/(1 + K_{Cl}[Cl^-])^{n_{Cl}}$. Thus, if we assume that there are no anions bound to the cationic DNA-binding site of *I in the presence of NaF, the binding constant of *I to DNA in the presence of NaCl is given by:

$$K_{*I,Cl} = K_{*I,F}\frac{1}{(1 + K_{Cl}[Cl^-])^{n_{Cl}}} \tag{6}$$

where K_{Cl} is the binding constant of Cl^- ions to *I. Therefore, according to Eq. (6), the binding constant of *I to DNA in NaCl should be weaker than in NaF in high salt, but in sufficiently low salt the two binding constants should be the same.

In a solution containing Cl^- ions that can bind to *I, binding of the protein to DNA requires the removal of all ions bound to the interactive sites on protein or DNA, so the binding constant should have the following salt dependence:

$$\log K_{*I,Cl} = \log K_{*I,0,F} - n_{Na} \log([Na^+]) - n_{Cl} \log(1 + K_{Cl}[Na^+]) \tag{7}$$

where $K_{*I,0,F}$ is the binding constant of *I to DNA at 1 M NaF, and we have assumed $[Cl^-] = [Na^+]$. Thus, in the high salt limit, the last term in Eq. (7) is given by:

$$\log(1 + K_{Cl}[Na^+])\Big|_{\substack{\text{high}\\ \text{salt}}} \simeq \log(K_{Cl}[Na^+]) = \log(K_{Cl}) + \log([Na^+]) \tag{8}$$

Therefore, in the high salt limit, Eq. (7) becomes:

$$\log K_{*I,Cl}\Big|_{\substack{\text{high}\\ \text{salt}}} = \log K_{*I,0,Cl} - (n_{Na} + n_{Cl}) \log([Na^+]) \tag{9}$$

where $\log K_{*I,0,Cl} = \log K_{*I,0} - n_{Cl} \cdot \log(K_{Cl})$, and the apparent slope becomes $d\log(K_{ss}\omega)/d\log(Na^+) = -(n_{Na} + n_{Cl})$. According to the data in Fig. 7, $n_{Na} + n_{Cl} \simeq 7$ in high salt. In low salt:

$$\log K_{*I,Cl}\Big|_{\substack{\text{low}\\ \text{salt}}} = \log K_{*I,0,F} - n_{Na} \log([Na^+]) \tag{10}$$

so the apparent slope becomes $d\log(K_{ss}\omega)/d\log(Na^+) = -n_{Na}$. The slope of the low-salt data in Fig. 7 yields $n_{Na} \simeq 3$. Combining this with the high salt results yields $n_{Cl} \simeq 4$. Thus, for *I, the slope in high salt is due to the removal of three sodium ions and four chloride ions, and in low salt the slope is due solely to the removal of three sodium ions.

Taking into account the above understanding of the salt dependence of *I binding, we can now explain the unusual results obtained for gp32. As shown in Fig. 8A and C, *I lacks the CTD, so the cationic DNA-binding site is exposed in low salt, and it has four Cl^- ions bound to it in high salt. Fig. 8B and D show the situation for gp32. In low salt, the CTD may be bound to the cationic DNA-binding site, thus preventing gp32 from binding DNA. In high salt, the interaction between the CTD and the cationic DNA-binding site is weaker, so the CTD spends most of its time open, and the protein resembles *I. Thus, relative to *I, the binding affinity of gp32 is reduced by the probability that the flap is open, P_{op}:

$$K_{gp32} = K_{*I} P_{op} \tag{11}$$

where

$$P_{\mathrm{op}} = \frac{\mathrm{e}^{\Delta G_{\mathrm{CTD}}/k_{\mathrm{B}}T}}{\mathrm{e}^{\Delta G_{\mathrm{CTD}}/k_{\mathrm{B}}T} + 1} \tag{12}$$

Here ΔG_{CTD} is the free energy of CTD binding to its protein binding site.

As shown in Fig. 7, in low salt the slope $d\log(K_{\mathrm{ss}}\omega)/d\log(\mathrm{Na}^+) \sim 0$, so under these conditions the net number of ions released in solution must be zero. The corresponding results for *I indicate that, on binding DNA, the protein must release three sodium ions. In this reaction for gp32, the CTD must also be exposed, so we proposed that the loss of three sodium ions from DNA is balanced by the binding of the same number of Na^+ to the CTD, that is, there is counterion condensation on the CTD. In high salt, the CTD is open and already has sodium ions condensed on it, so this does not affect the salt dependence of the reaction. This is depicted in Fig. 8D, where the flap is open and four Cl^- ions are bound to gp32. If the CTD and DNA-binding site are electrostatically equivalent, such that they both take up the same number of ions on binding, then the electrostatic components of their binding constants must be proportional to each other. Thus, at any salt concentration:

$$\frac{K_{\mathrm{CTD}}}{K_{0\mathrm{CTD}}} = \frac{K_{*\mathrm{I}}}{K_{*\mathrm{I},0}} \tag{13}$$

Here K_{CTD} is the binding constant of the CTD to the gp32 DNA-binding site and $K_{0\mathrm{CTD}}$ and $K_{*\mathrm{I},0}$ are the binding constants at 1 M Na^+. This can be expressed as:

$$\ln(K_{\mathrm{CTD}}) = \ln(K_{*\mathrm{I}}) + \ln(K_{0\mathrm{CTD}}) - \ln(K_{*\mathrm{I},0}) \tag{14}$$

or

$$\Delta G_{\mathrm{CTD}} = \Delta G_{0\mathrm{CTD}} + k_{\mathrm{B}}T[\ln(K_{*\mathrm{I}}) - \ln(K_{*\mathrm{I},0})] \tag{15}$$

Combining Eqs. (7) and (15) gives

$$\Delta G_{\mathrm{CTD}} = \Delta G_{0\mathrm{CTD}} + n_{\mathrm{Na}}k_{\mathrm{B}}T\ \ln([\mathrm{Na}^+]) + n_{\mathrm{Cl}}k_{\mathrm{B}}T\ \ln(1 + K_{\mathrm{Cl}}[\mathrm{Na}^+]) \tag{16}$$

Together, Eqs. (11), (12), and (16) allow us to calculate K_{gp32} with the same parameters used to calculate $K_{*\mathrm{I}}$, and with an additional parameter $\Delta G_{0\mathrm{CTD}}$. Thus, using five fitting parameters, we can simultaneously fit all available data for the binding of gp32 and *I to DNA over both high- and low-salt conditions in the presence of NaCl and NaF. The results of our fitting are shown as lines in Fig. 7. The data points in the figure represent measurements from both bulk and single molecule experiments.

The free energy of CTD binding to the core domain of gp32 can also be calculated directly from our data, by solving Eqs. (11) and (12) to obtain:

$$\Delta G_{\mathrm{CTD}} = -k_{\mathrm{B}} T \ln \left(\frac{K_{*\mathrm{I}}}{K_{\mathrm{gp32}}} - 1 \right) \tag{17}$$

For comparison, we can obtain ΔG_{CTD} from Eq. (16) using the parameters obtained from our fits in Fig. 7. The result of this theoretical calculation is given as a solid line in Fig. 9A, along with the measured data points obtained by using Eq. (17). The calculated free energy change associated with CTD binding to the core of gp32 in high salt is positive, so the nonelectrostatic portion of this interaction (evaluated at 1 M NaCl) is repulsive. Thus, electrostatic interactions regulate this protein conformational change, which must occur prior to the binding of gp32 to DNA. This also explains why the effect of the CTD is only seen in low salt. In high salt, the CTD spends most of its time unbound because of the positive ΔG_{CTD}. This is illustrated in Fig. 9B, which shows the probability of finding the CTD open, P_{op}, as a function of salt concentration. At about 200 mM salt, the CTD flap is open half the time, and above this salt level, it is mainly open. The fact that this interaction becomes important near physiological salt concentration suggests that it plays a critical role in regulating the activity of gp32 *in vivo*. The five parameters resulting from the fitted data should also describe the salt dependence of the affinities of gp32, *I, and *III for various DNA molecules and polynucleotides as well as of the corresponding binding kinetic measurements. A recent analysis of data from several bulk studies using the parameters from this study supports our model for salt-dependent gp32, *I, and *III interactions with DNA (Rouzina *et al.*, 2005).

We began our single molecule study of DNA force-induced melting as a test of a new method for studying DNA–protein interactions. gp32 was chosen because it is a very well-studied, classic ssDNA-binding protein. Nevertheless, a number of the *in vitro* experimental results with this protein were not well understood. In particular, binding studies of gp32 to ssDNA appeared to predict that the protein should be capable of lowering the thermal melting temperature of dsDNA by 50 °C, but there was essentially no effect on the T_{m} of natural DNA. In contrast, *I, which lacked the CTD of gp32, but exhibited very similar binding to ssDNA, lowered the T_{m} by the predicted degree (Lonberg *et al.*, 1981; Waidner *et al.*, 2001).

The results reviewed here help to resolve this seemingly contradictory behavior. We used our single molecule technique to show that the bulk ssDNA-binding experiments that were performed in high salt cannot be extrapolated to low salt by assuming a constant $d \log (K_{\mathrm{ss}}\omega)/d \log ([\mathrm{Na}^+])$. In fact, gp32 binding saturates at about 200 mM Na^+, while *I binding continues to increase as the salt concentration is decreased, although at a slightly lower slope of 3 rather than 7. Therefore, the prediction (based on linear extrapolation of high salt measurements) that gp32 should lower the thermal melting temperature by the same amount as *I is not

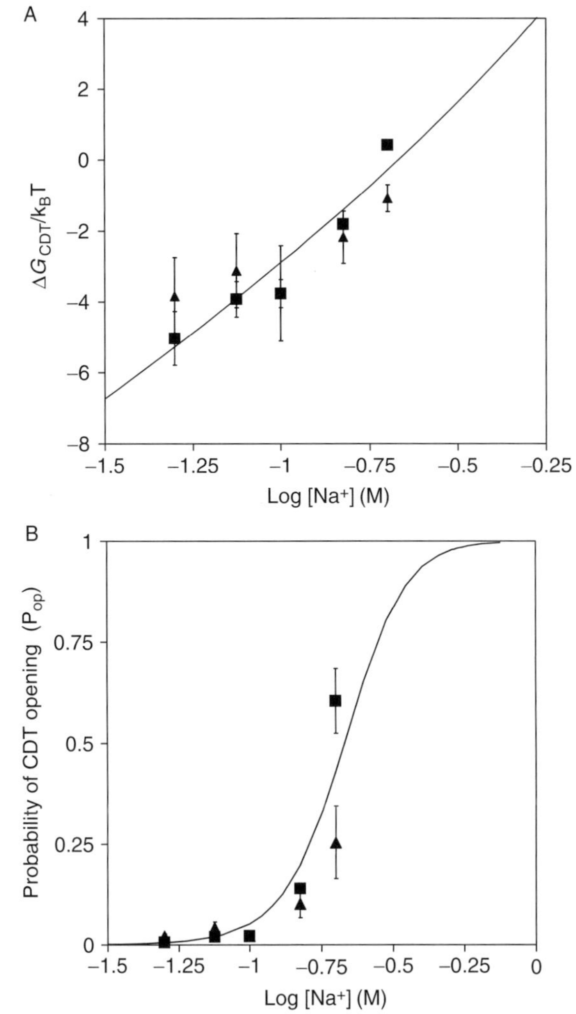

Fig. 9 (A) Free energy of CTD binding to a cationic surface on gp32 as a function of solution ionic strength in NaCl. Data for gp32 and *I binding to ss (square) and ds (triangle) λ-DNA, derived from our measurements according to Eq. (17). The solid line is calculated from Eq. (16) with the parameters taken from the optimal global fit of the data on Fig. 7: $n_{Na} = 2.95 \pm 0.30$, $n_{Cl} = 3.70 \pm 0.30$, $K_{Cl} = 2 \pm 0.5\ M^{-1}$, and $\Delta G_{0CTD} = 3.2 \pm 0.2\ k_BT$. (B) The measured (symbol) probability of CTD release from the gp32 binding groove (P_{op}) as a function of salt [NaCl]. Measurements are shown from our experimental $\Delta G_{CTD}/k_BT$ to ss (square) and ds (triangle) for λ-DNA, according to Eq. (17). Line is fit to Eq. (16).

valid in low salt. In fact, the predicted binding constant of gp32 to ssDNA at low salt is off by several orders of magnitude. Our results show gp32 should be much less effective at lowering T_m than *I in low salt, as was observed in thermal melting experiments. There is also an additional kinetic factor that influences the effect of gp32 on T_m, which is discussed in detail in our earlier work (Pant *et al.*, 2004). Briefly, by using our DNA stretching measurements to determine protein binding kinetics, we find that the rate at which gp32 and *I destabilize dsDNA is determined by one-dimensional diffusion of the protein on dsDNA. Because *I exhibits stronger binding to dsDNA, the rate at which *I destabilizes dsDNA is strongly enhanced, and this effect likely contributes to the inability of gp32 to destabilize dsDNA in thermal melting experiments (Pant *et al.*, 2004; Sokolov *et al.*, 2005a,b; Williams *et al.*, 2006).

VI. Conclusions

We have reviewed the development of a new biochemical and biophysical method for measuring DNA–protein interactions using single molecule force spectroscopy. In this method, single dsDNA molecules are stretched beyond their B-form contour length until the DNA is melted by force, causing the base pairs to break and increasing the length of the DNA from that of dsDNA to that of ssDNA. When the force-induced melting transition is reversible, the area between the dsDNA and ssDNA stretching curves yields the DNA melting free energy. If this experiment is then performed in the presence of DNA-binding ligands, the change in melting force can be used to quantify the DNA–ligand binding free energy. When the force-induced melting process is not reversible, an equilibrium free energy change may be obtained by measuring the melting force at constant position until an equilibrium force is obtained.

To demonstrate the capabilities of this new method, we have examined the binding of T4 gp32 to single DNA molecules using equilibrium force-induced melting. We measured the binding constants of gp32 and its truncate *I to ssDNA as a function of salt concentration from 50 mM Na^+ to 200 mM Na^+. While this range was easily accessible to our single molecule experiment, there was very little previously published data on gp32 binding to ssDNA below 200 mM Na^+, and no previously published data on *I binding to ssDNA in this range. However, where bulk binding data was available, our single molecule data has agreed very well. Thus, this new single molecule method has extended the range of solution conditions over which binding data could be obtained to include physiological salt concentration.

The results of our salt-dependent binding measurements have allowed us to construct a model for the electrostatic regulation of gp32 binding by its acidic CTD. We showed that the salt dependence of binding was governed by counterion condensation of sodium ions on the gp32 CTD. *I, which lacks the CTD, does not exhibit the unusual pattern of salt-dependent binding that is a consequence of this counterion condensation. By combining bulk and single molecule data, fits of our model have allowed us to calculate the number of sodium and chloride ions released on gp32 or *I binding to ssDNA, the binding constant of chloride ions to the gp32 or *I cationic DNA-binding site, and the nonelectrostatic free energy of CTD binding to gp32. The remaining parameter is the nonelectrostatic binding constant for protein binding to ssDNA, which changes with the type of DNA molecule. These parameters were subsequently shown to be consistent with a wide variety of additional equilibrium and kinetic measurements on gp32, *I, and other truncations and mutants (Rouzina *et al.*, 2005). The experiments we have described demonstrate that DNA force-induced melting is a powerful new technique for quantitative examination of DNA–protein interactions under a wide range of solution conditions.

Acknowledgments

The authors wish to thank Kiran Pant for her work on the development of the method of DNA force-induced melting in the presence of gp32 and Micah McCauley for the original drawing of Fig. 1. Funding for this work was provided by NIH (GM 52049, RLK and GM 72462, MCW) and NSF (MCB-0238190, MCW).

References

Allemand, J. F., Bensimon, D., and Croquette, V. (2003). Stretching DNA and RNA to probe their interactions with proteins. *Curr. Opin. Struct. Biol.* **13,** 266–274.

Bloomfield, V. A., Crothers, D. M., and Ignacio Tinoco, J. (2000). "Nucleic Acids: Structures, Properties, and Functions." University Science Books, Sausalito, California.

Bustamante, C., Bryant, Z., and Smith, S. B. (2003). Ten years of tension: Single-molecule DNA mechanics. *Nature* **421,** 423–427.

Bustamante, C., Smith, S. B., Liphardt, J., and Smith, D. (2000). Single-molecule studies of DNA mechanics. *Curr. Opin. Struct. Biol.* **10,** 279–285.

Collin, D., Ritort, F., Jarzynski, C., Smith, S. B., Tinoco, I., Jr., and Bustamante, C. (2005). Verification of the Crooks fluctuation theorem and recovery of RNA folding free energies. *Nature* **437,** 231–234.

Cruceanu, M., Urbaneja, M. A., Hixson, C. V., Johnson, D. G., Datta, S. A., Fivash, M. J., Stephen, A. G., Fisher, R. J., Gorelick, R. J., Casas-Finet, J. R., Rein, A., Rouzina, I., *et al.* (2006). Nucleic acid binding and chaperone properties of HIV-1 Gag and nucleocapsid proteins. *Nucleic Acids Res.* **34,** 593–605.

Frank-Kamenetskii, M. D., Anshelevich, V. V., and Lukashin, A. V. (1987). Polyelectrolyte model of DNA. *<Translation> Soviet Physics—Uspekhi* **151,** 595–618.

Gueron, M., and Weisbuch, G. (1981). Polyelectrolyte theory of charged-ligand binding to nucleic acids. *Biochimie* **63,** 821–825.

Harris, S. A., Sands, Z. A., and Laughton, C. A. (2005). Molecular dynamics Simulations of duplex stretching reveal the importance of entropy in determining the biomechanical properties of DNA. *Biophys. J.* **88,** 1684–1691.

He, Z. G., Rezende, L. F., Willcox, S., Griffith, J. D., and Richardson, C. C. (2003). The carboxyl-terminal domain of bacteriophage T7 single-stranded DNA-binding protein modulates DNA binding and interaction with T7 DNA polymerase. *J. Biol. Chem.* **278,** 29538–29545.

Heng, J. B., Aksimentiev, A., Ho, C., Marks, P., Grinkova, Y. V., Sligar, S., Schulten, K., and Timp, G. (2006). The electromechanics of DNA in a synthetic nanopore. *Biophys. J.* **90,** 1098–1106.

Jensen, D. E., Kelly, R. C., and von Hippel, P. H. (1976). DNA "melting" proteins. II. Effects of bacteriophage T4 gene 32-protein binding on the conformation and stability of nucleic acid structures. *J. Biol. Chem.* **251,** 7215–7228.

Jensen, D. E., and von Hippel, P. H. (1976). DNA "melting" proteins. I. Effects of bovine pancreatic ribonuclease binding on the conformation and stability of DNA. *J. Biol. Chem.* **251,** 7198–7214.

Kelly, R. C., and von Hippel, P. H. (1976). DNA "melting" proteins. III. Fluorescence "mapping" of the nucleic acid binding site of bacteriophage T4 gene 32-protein. *J. Biol. Chem.* **251,** 7229–7239.

Kraulis, P. J. (1991). MOLSCRIPT: A program to produce both detailed and schematic plots of protein structures. *J. Appl. Cryst.* **24,** 946–950.

Liphardt, J., Dumont, S., Smith, S. B., Tinoco, I., Jr., and Bustamante, C. (2002). Equilibrium information from nonequilibrium measurements in an experimental test of Jarzynski's equality. *Science* **296,** 1832–1835.

Lohman, T. M., and Ferrari, M. E. (1994). Escherichia coli single-stranded DNA-binding protein: Multiple DNA-binding modes and cooperativities. *Ann. Rev. Biochem.* **63,** 527–570.

Lonberg, N., Kowalczykowski, S. C., Paul, L. S., and von Hippel, P. H. (1981). Interactions of bacteriophage T4-coded gene 32 protein with nucleic acids. III. Binding properties of two specific proteolytic digestion products of the protein (G32P*I and G32P*III). *J. Mol. Biol.* **145,** 123–138.

Manning, G. S. (1978). The molecular theory of polyelectrolyte solutions with applications to the electrostatic properties of polynucleotides. *Q. Rev. Biophys.* **11,** 179–246.

McGhee, J. D. (1976). Theoretical calculations of the helix-coil transition of DNA in the presence of large, cooperatively binding ligands. *Biopolymers* **15,** 1345–1375.

Mihailović, A., Vladescu, I., McCauley, M., Ly, E., Williams, M. C., Spain, E. M., and Nuñez, M. E. (2006). Exploring the interaction of ruthenium(II) polypyridyl complexes with DNA using single-molecule techniques. *Langmuir* **22,** 4699–4709.

Newport, J. W., Lonberg, N., Kowalczykowski, S. C., and von Hippel, P. H. (1981). Interactions of bacteriophage T4-coded gene 32 protein with nucleic acids. II. Specificity of binding to DNA and RNA. *J. Mol. Biol.* **145,** 105–121.

Oosawa, F. (1968). Interaction between parallel rodlike macroions. *Biopolymers* **6,** 1633–1647.

Oosawa, F. (1971). "Polyelectrolytes." Marcel Dekker, New York.

Pant, K., Karpel, R. L., Rouzina, I., and Williams, M. C. (2004). Mechanical measurement of single-molecule binding rates: Kinetics of DNA helix-destablization by T4 gene 32 protein. *J. Mol. Biol.* **336,** 851–870.

Pant, K., Karpel, R. L., Rouzina, I., and Williams, M. C. (2005). Salt dependent binding of T4 gene 32 protein to single- and double-stranded DNA: Single molecule force spectroscopy measurements. *J. Mol. Biol.* **349,** 317–330.

Pant, K., Karpel, R. L., and Williams, M. C. (2003). Kinetic regulation of single DNA molecule denaturation by T4 gene 32 protein structural domains. *J. Mol. Biol.* **327,** 571–578.

Piana, S. (2005). Structure and energy of a DNA dodecamer under tensile load. *Nucleic Acids Res.* **33,** 7029–7038.

Record, M. T., Lohman, T. M., and deHaseth, P. L. (1976). Ion effects on ligand-nuclei acid interactions. *J. Mol. Biol.* **107,** 145–158.

Record, M. T. J., Zhang, W., and Anderson, C. F. (1998). Analysis of effects of salts and uncharged solutes on protein and nucleic acid equilibria and processes: A practical guide to recognizing and interpreting polyelectrolyte effects, Hofmeister effects, and osmotic effects of salts. *Adv. Protein Chem.* **51,** 281–353.

Rein, A., Henderson, L. E., and Levin, J. G. (1998). Nucleic-acid-chaperone activity of retroviral nucleocapsid proteins: Significance for viral replication. *Trends Biochem. Sci.* **23,** 297–301.

Rouzina, I., and Bloomfield, V. A. (1996). Competitive electrostatic binding of charged ligands to polyelectrolytes: Planar and cylindrical geometries. *J. Phys. Chem.* **100,** 4292–4304.

Rouzina, I., and Bloomfield, V. A. (1997). Competitive electrostatic binding of charged ligands to polyelectrolytes—Practical approach using the non-linear Poisson-Boltzmann equation. *Biophys. Chem.* **64,** 139–155.

Rouzina, I., and Bloomfield, V. A. (2001a). Force-induced melting of the DNA double helix. 1. Thermodynamic analysis. *Biophys. J.* **80,** 882–893.

Rouzina, I., and Bloomfield, V. A. (2001b). Force-induced melting of the DNA double helix. 2. Effect of solution conditions. *Biophys. J.* **80,** 894–900.

Rouzina, I., Pant, K., Karpel, R. L., and Williams, M. C. (2005). Theory of electrostatically regulated binding of T4 gene 32 protein to single- and double-stranded DNA. *Biophys. J.* **89,** 1941–1956.

Safran, S. A., Pincus, P. A., Andelman, D., and MacKintosh, F. C. (1991). Stability and phase behavior of mixed surfactant vesicles. *Phys. Rev. A* **43,** 1071–1078.

Shamoo, Y., Friedman, A. M., Parsons, M. R., Konigsberg, W. H., and Steitz, T. A. (1995). Crystal structure of a replication fork single-stranded DNA binding protein (T4 gp32) complexed to DNA. *Nature* **376,** 362–366.

Smith, S. B., Cui, Y., and Bustamante, C. (2003). Optical-trap force transducer that operates by direct measurement of light momentum. *Methods. Enzymol.* **361,** 134–162.

Sokolov, I. M., Metzler, R., Pant, K., and Williams, M. C. (2005a). First passage time of N excluded-volume particles on a line. *Phys. Rev. E.* **72,** 041102.

Sokolov, I. M., Metzler, R., Pant, K., and Williams, M. C. (2005b). Target search of N sliding proteins on a DNA. *Biophys. J.* **89,** 895–902.

Strick, T., Allemand, J., Croquette, V., and Bensimon, D. (2000). Twisting and stretching single DNA molecules. *Prog. Biophys. Mol. Biol.* **74,** 115–140.

Villemain, J., and Giedroc, D. (1996). Characterization of a cooperativity domain mutant $Lys^3 \rightarrow$ Ala (K3A) T4 gene 32 protein. *J. Biol. Chem.* **271,** 27623–27629.

Vladescu, I. D., McCauley, M. J., Rouzina, I., and Williams, M. C. (2005). Mapping the phase diagram of single DNA molecule force-induced melting in the presence of ethidium. *Phys. Rev. Lett.* **95,** 158102.

Waidner, L., Flynn, E., Wu, M., Li, X., and Karpel, R. L. (2001). Domain effects on the DNA-interactive properties of bacteriophage T4 gene 32 protein. *J. Biol. Chem.* **276,** 2509–2516.

Wenner, J. R., Williams, M. C., Rouzina, I., and Bloomfield, V. A. (2002). Salt dependence of the elasticity and overstretching transition of single DNA molecules. *Biophys. J.* **82,** 3160–3169.

Williams, M. C., Gorelick, R. J., and Musier-Forsyth, K. (2002a). Specific zinc finger architecture required for HIV-1 nucleocapsid protein's nucleic acid chaperone function. *Proc. Natl. Acad. Sci. USA* **99,** 8614–8619.

Williams, M. C., and Rouzina, I. (2002). Force spectroscopy of single DNA and RNA molecules. *Curr. Opin. Struct. Biol.* **12,** 330–336.

Williams, M. C., Rouzina, I., and Bloomfield, V. A. (2002b). Thermodynamics of DNA interactions from single molecule stretching experiments. *Acc. Chem. Res.* **35,** 159–166.

Williams, M. C., Rouzina, L., and Karpel, R. L. (2006). Thermodynamics and kinetics of DNA-protein interactions from single molecule force spectroscopy measurements. *Curr. Org. Chem.* **10,** 419–432.

Williams, M. C., Rouzina, I., Wenner, J. R., Gorelick, R. J., Musier-Forsyth, K., and Bloomfield, V. A. (2001a). Mechanism for nucleic acid chaperone activity of HIV-1 nucleocapsid protein revealed by single molecule stretching. *Proc. Natl. Acad. Sci. USA* **98,** 6121–6126.

Williams, M. C., Wenner, J. R., Rouzina, I., and Bloomfield, V. A. (2001b). The effect of pH on the overstretching transition of dsDNA: Evidence of force-induced DNA melting. *Biophys. J.* **80,** 874–881.

Williams, M. C., Wenner, J. R., Rouzina, I., and Bloomfield, V. A. (2001c). Entropy and heat capacity of DNA melting from temperature dependence of single molecule stretching. *Biophys. J.* **80,** 1932–1939.

CHAPTER 18

Isotopomer-Based Metabolomic Analysis by NMR and Mass Spectrometry

Andrew N. Lane,* Teresa W.-M. Fan,*,†,‡ and Richard M. Higashi†

*JG Brown Cancer Center
University of Louisville
Louisville, Kentucky 40202

†Department of Chemistry
University of Louisville
Louisville, Kentucky 40208

‡Department of Pharmacology and Toxicology
University of Louisville
Louisville, Kentucky 40202

Abstract

Nuclear magnetic resonance (NMR) and mass spectrometry (MS) together are synergistic in their ability to profile comprehensively the metabolome of cells and tissues. In addition to identification and quantification of metabolites, changes in

0091-679X/08 $35.00
DOI: 10.1016/S0091-679X(07)84018-0

metabolic pathways and fluxes in response to external perturbations can be reliably determined by using stable isotope tracer methodologies. NMR and MS together are able to define both positional isotopomer distribution in product metabolites that derive from a given stable isotope-labeled precursor molecule and the degree of enrichment at each site with good precision. Together with modeling tools, this information provides a rich functional biochemical readout of cellular activity and how it responds to external influences.

In this chapter, we describe NMR- and MS-based methodologies for isotopomer analysis in metabolomics and its applications for different biological systems.

I. Introduction

Since the inception of metabolomics, there have been numerous definitions posted (e.g., http://en.wikipedia.org/wiki/Metabolomics; www.grants.nih.gov/grants/guide/rfa-files/RFA-ES-04-008.html; www.ansci.de/fileadmin/img/pb3/pdf/info01_0604.pdf; www.highchem.com/showpage.php?name=The_Objective; www.geocities.com/bioinformaticsweb/definition.html). For the purposes of this chapter, we define metabolomics as "a systematic analysis of metabolite structures, concentrations, pathways and fluxes, and molecular interactions within and among cells, organs, and organisms as a function of their environment." The information generated from a metabolomics approach can be connected to gene and protein expression networks determined from transcriptomics and proteomics approaches, respectively (Fan *et al.*, 2005; Fan *et al.*, 2004a; Heijne *et al.*, 2005). This third "omics" is essential insofar as it represents the functional biochemistry of organisms.

In the last decade, extensive metabolomics analyses have become feasible due to the rapid development in analytical chemical tools, most notably nuclear magnetic resonance (NMR) spectroscopy and hyphenated mass spectrometry (MS) (Fan *et al.*, 2006; Hirai *et al.*, 2004; Lindon *et al.*, 1999; Lutz, 2005; Ruzsanyi *et al.*, 2005). This development is revolutionizing our conceptual framework and approaches to metabolic studies (Fan *et al.*, 2004a) by enabling global inquiries into regulation occurring at the enzyme/protein levels (e.g., allosteric control, product inhibition, ligand–protein interactions), thereby completing the regulatory gap between genes and the phenotype.

NMR and MS methods have been extensively applied to metabolic analyses of cells and tissues both *in vivo* and *in vitro* (Anousis *et al.*, 2004; Fan, 2005; Fan *et al.*, 1986, 1992, 1993, 1997, 2003, 2004a; Gradwell *et al.*, 1998; Lloyd *et al.*, 2004). Considerable information can also be obtained by analyzing the metabolite profiles of biofluids such as urine, blood, or saliva (Bollard *et al.*, 2005; Foxall *et al.*, 1993; Lindon *et al.*, 2004), especially in the detection of disease states and drug metabolism, which has been extensively reviewed (Lindon *et al.*, 2004).

In this chapter, we outline methods for determining metabolite structure and concentration as well as isotopomer distributions in extracts taken from cells grown in culture or from tissues. There are numerous reviews and textbooks on the general subject of structure identification that can be consulted for specific experimental details (Cavanagh *et al.*, 1996; Claridge, 1999; Ernst *et al.*, 1990; Fan *et al.*, 1986; Sanders and Hunter, 1987; Siuzdak, 2003; Watson, 1985). A brief glossary of terms commonly used in this chapter is given at the end of the chapter. Similarly, details of the biological experimental design are beyond the scope of the present chapter.

II. Rationale

The metabolome and how it responds to perturbations provides a detailed biochemical and functional readout of a cellular status or tissue activity. A detailed knowledge of metabolic pathways and relative fluxes provides essential information about cellular demands in response to varied external conditions and can be systematically related to regulation at the protein and/or gene levels, that is, mechanism(s) of molecular regulation.

The metabolome is very complex and is not independent of protein and gene expression. In order to make biological sense of a cellular function, any modeling process is ultimately dependent on both experimentally derived knowledge and the extent and quality of the data pertaining to the problem under study. In order to define the utilization of major metabolic pathways, their fluxes, and how they are regulated, it is necessary to acquire as much quantitative data about the system as possible.

NMR and MS are two complementary techniques that provide high-quality information about molecular structure, concentration, and flux both in pure states and in complex mixtures. These methods, and others, have been extensively used for metabolite profiling, which is a common approach of metabonomics (Bollard *et al.*, 2005; Craig *et al.*, 2006; Fan, 1996a,b; Fernie *et al.*, 2004; Harrigan *et al.*, 2005; Lindon *et al.*, 2003; Nicholson, 2005; Wang *et al.*, 2006; Whitfield *et al.*, 2004; Wolfender *et al.*, 2003). In classical biochemistry, radioisotope tracers were used in conjunction with extensive chemical analysis to identify metabolic pathways, such as in the pioneering work that led to the discovery of the Emden-Meyerhoff pathways and the Krebs cycle (Koshland, 1998; Kresge *et al.*, 2005). A less common but more powerful atom-based approach is stable isotopomer-based metabolomics where metabolic pathways and fluxes are deduced from the transformations of stable isotope tracers through metabolic pathways. These transformations are determined by analyzing the isotopic enrichment at different atomic positions of relevant metabolite products (isotopomer analysis). Isotopomers are versions of the same compound in which they differ by the distribution of isotopes of each element. In a three-carbon compounds such as lactate, there are eight possible stable isotopomers (^{12}C and ^{13}C) corresponding to ^{12}C-3, ^{13}C-3, and three each of ^{12}C-1^{13}C-2 and ^{13}C-1^{12}C-2. In NMR, these correspond to eight distinguishable compounds.

InMS, there are only four mass isotopomers—$m0$, $m + 1$, $m + 2$, and $m + 3$—because the three isotopomers of $m + 1$ and $m + 2$ have identical mass.

This is where NMR and MS together hold a major advantage over other biophysical techniques. By measuring isotopomer distributions, it is feasible to estimate the relative contributions of various pathways that lead to the production of metabolites of interest, and metabolic fluxes through those pathways can be determined from time courses. By using different labeled precursors, numerous aspects of metabolism can be probed. Table I summarizes some commonly available labeled metabolic precursors that are used in isotopomer-based metabolomics.

III. Materials

A. Extractions

Although it is possible to monitor metabolites directly in intact tissues or whole organisms by NMR (also known as magnetic resonance spectroscopy or MRS in the biomedical field), in general the resolution and metabolic coverage are substantially lower than in the corresponding extracts. For analysis of extracts by MS and NMR, it is essential to remove macromolecules and any materials that form

Table I
Some Common Readily Available Stable Isotope-Labeled Precursors

Source	Label/position	Application	References
Glucose	[U-^{13}C]	General metabolism; surveys	(Fan *et al.*, 2004a; Fan, 2005)
	^{13}C-1	Glycolysis; glycogen synthesis; TCA	(Mason *et al.*, 2002, 2003)
	^{13}C-1,2	Pentose phosphate + glycolysis	(Lee *et al.*, 1998; Vizan *et al.*, 2005)
	^{13}C-1,6	Pentose phosphate + glycolysis	(Henry *et al.*, 2003; Yang *et al.*, 2005)
Glutamine	[U-^{13}C]	Glutaminolysis; amino acid metabolism; TCA; pyrimidine biosynthesis	(Haberg *et al.*, 1998)
	^{15}N2	Nitrogen metabolism	(Street *et al.*, 1993)
Serine	[U-^{13}C]	Serinolysis; amino acid metabolism	
Glycine	^{13}C/^{15}N	1-carbon metabolism	(Brennan *et al.*, 2003)
Palmitic acid	[U-^{13}C]	Fatty acid oxidation	(Ventura *et al.*, 1999)
Acetate	^{13}C-1/^{13}C-2	Fatty acid biosynthesis; TCA	(Fan *et al.*, 2003)
	[U-^{13}C]	Fatty acid biosynthesis; TCA	(Diraison *et al.*, 2003; Koeberl *et al.*, 2003)

micelles or colloids. However, there are several concerns to be addressed with extract analysis, including recovery and stability of the metabolites. Given the wide range of classes of molecules present in cells, usually several extraction methods are required for extensive coverage. In addition, labile metabolites may require special procedures to stabilize them. Extraction procedures also need to be fast at least in the early stages where enzymes may still be active.

In addition to extracts from the cells, valuable information can be obtained from analyzing the cell culture medium or extracellular fluid of tissues. For example, cancer cells typically have a very high glycolytic rate even under aerobic conditions, resulting in secretion in large amounts of lactate; this is the so-called Warburg effect (Dang *et al.*, 2005; Garber, 2004; Robey *et al.*, 2005; Warburg, 1956). A number of other metabolites may also be secreted that are unique to transformed or malignant cells, including those involved in cell–cell communication and migration. Thus, monitoring the medium as a function of time not only can provide kinetic information about cellular metabolism but can also reveal means via which cells interact with or influence their environment.

B. Sample Preparation

To prepare cells or tissues for metabolite analysis, it is necessary to wash them in compatible inorganic buffers, such as ice-cold phosphate buffered saline, to remove the influence of treatment or growth media. Then, the samples are immediately flash-frozen in liquid N_2 to minimize further changes in the metabolic status. The frozen samples can be stored at $-80\,^{\circ}C$ or lower temperatures before further processing is performed, including lyophilization or pulverization.

1. Extraction of Stable Metabolites from Cells

To maximize extraction efficiency, it is imperative that cells or tissues be pulverized into fine powders (e.g., $\leq$5-μm particles) under liquid N_2 temperature. This can be achieved by grinding with liquid N_2 in a mortar and pestle or in a mechanical device such as a freezer mill (SpexCertiPrep, Inc., www.spexcsp.com) or micro ball mill (Retsch, Inc., Newtown, PA). Alternatively, samples can be lyophilized before pulverization.

Sample powders can be extracted for polar metabolites with ice-cold 10% trichloroacetic acid (TCA) (Gradwell *et al.*, 1998) or ice-cold 5% perchloric acid (PCA) (Fan *et al.*, 1986). Both acid extractants precipitate proteins and nucleic acids in addition to inactivating enzymes. Extraction should be performed twice on a given sample, each with excess extractants (e.g., w:v of 1:40) to be quantitative. It is also necessary to maintain ice-cold conditions (e.g., $4\,^{\circ}C$) throughout the extraction procedure (including centrifugation for removing insoluble materials). TCA or PCA should then be removed immediately following extraction to minimize metabolite degradation. TCA is readily removed with lyophilization while PCA is precipitated by titration with K_2CO_3 to pH 3.5–4 or with KOH to pH 7 and lyophilized for subsequent NMR or MS analysis. In case of excessive interference from paramagnetic ions (commonly encountered for tissues), these ions can be removed by passing the extracts through a cation exchanger such as Chelex 100 (BioRad, Inc., Hercules, CA). Alternatively, a metal ion chelator such as EDTA can be added to the extract. However, the presence of excess chelator may interfere with the analysis of certain metabolites.

For both NMR and MS analysis, it is important to normalize the final pH of the extract. This is because NMR parameters, such as chemical shifts, are pH dependent and an appropriate pH is important for metabolite derivatization or ionization for MS analysis. pH adjustment is essential for the PCA extraction whereas lyophilized TCA extracts result in an acidic pH that is directly suitable for NMR analysis or for derivatization such as silylation before gas chromatography-mass spectrometry (GC-MS) analysis.

We have been successful in applying the above extraction strategies for preparing widely different cell and tissue samples for NMR and MS analyses (Fan, 1996; Fan *et al.*, 1986, 1993, 1998, 2006; Gradwell *et al.*, 1998).

Low-polarity metabolites such as triglycerides and phospholipids can be extracted with excess chloroform/methanol (v/v) 2:1 for total lipids (Fan *et al.*, 1994) and methanol for phospholipids, sphingolipids, and their metabolites (Rujoi *et al.*, 2003; Yappert *et al.*, 2003). The simple methanolic extraction avoids losses of lipid metabolites into the aqueous phase and is more selective for extracting phospholipids and shingolipids. Unsaturated acyl chains are prone to oxidation, so a suitable antioxidant should be added during extraction. Butylated hydroxytoluene (BHT) at 1–5 mM is one such additive commonly employed for minimizing the oxidation of polyunsaturated lipids in solvent extracts.

2. Extraction of Labile Metabolites

Not all metabolites survive the acidity of TCA or PCA. To circumvent this problem, acid-labile metabolites (e.g., fructose-2,6-bisphosphate, some deoxynucleotides, AcCoA, NADPH, and NADH) can be extracted with ice-cold aqueous acetonitrile (e.g., 50–80%) in 0.1 M ammonium bicarbonate (pH 7.8) to precipitate the bulk of the macromolecules, followed by lyophilization and ultrafiltration through 3-kDa molecular weight cutoff (MWCO) filter (e.g., Vivaspin centrifugal concentrator, Vivascience) to remove small proteins. For labile metabolites with active hydrogen functional groups (e.g., prostaglandin A1), the MTBSTFA (*N*-methyl-*N*-[*tert*-butyl-dimethylsilyl]trifluoroacetamide):acetonitrile (1:1, v/v) mixture may be used to derivatize and extract the metabolites directly from lyophilized sample powders using sonication for GC-MS analysis (see below). The same procedure can also be applied to labile thiols and selenols (e.g., selenocysteine) suitable for GC-MS analysis (Fan *et al.*, 1998). For thiol- and selenol-containing peptides (e.g., GSSeH) or metabolites not suitable for GC-MS analysis, fluorogenic mBrB (Sigma Chemicals Co., St. Louis, MO), can be employed to derivatize and extract the metabolites in aqueous acetonitrile (e.g., 50%) mixture as described previously (Fan *et al.*, 2004b). This mBrB derivatization procedure can also be performed on extracts pretreated with 5 mM DTT to obtain the total pool of reduced + oxidized thiols or selenols. The oxidized pool is then calculated by subtracting the total from the reduced pool.

Lipid hydroperoxides (e.g., phosphatidylcholine hydroperoxide) are stabilized by derivatization with luminal, followed by LC-APCI-MS or LC-chemiluminescence detection (Zhang *et al.*, 1995). Aldehydic lipid oxidation products (e.g., 4-HNE) may be profiled by derivatizing cell or tissue samples with dinitrophenylhydrazine (DNPH) (Kolliker *et al.*, 1998), followed by extraction with acetonitrile and analyses by LC-APCI-MS and/or UV detection. For a more sensitive detection, aldehydic products in lyophilized sample powders are reacted with NBD methylhydrazine [*N*-methyl-4-hydrazino-7-nitrobenzofurazan (Zurek *et al.*, 2000) or BODIPY® FL hydrazide (Katayama *et al.*, 1998)] and centrifuged to recover the derivatives. The derivatives are then analyzed by LC-APCI-MS for structure and quantified by liquid chromatography (LC) with fluorescence detection for example. Aldehydic sugars or oligosaccharides can be derivatized with 2-aminoacridone for matrix-assisted laser desorption ionization (MALDI)-MS analysis (North *et al.*, 1997). For MS identification and quantification, these derivatization approaches have the added advantages of minimizing salt interference by desalting via C-18 cartridges, predictable tandem MS fragmentation, and improved sensitivity by increasing metabolite mass. Ultra trace levels of the fluorescent or luminescent derivatives can also be quantified by fluorescence or chemiluminescence after chromatographic separation, if MS quantitation proves to be problematic.

IV. Methods

A. Application of NMR to Isotopomer Analysis

1. Metabolite Structure Elucidation/Confirmation by NMR

NMR is well suited for metabolite structure elucidation since the majority of the biologically active nuclei (e.g., ^{1}H, ^{13}C, ^{14}N, ^{15}N, ^{23}Na, ^{31}P, and ^{77}Se) are NMR observable. The basic NMR strategy for structure identification is well established and in general makes use of a number of one-dimensional (1-D) and two-dimensional (2-D) experiments that reveal covalent bonding pattern and spatial configuration of atoms in a given compound. These include direct detection 1-D ^{1}H NMR, ^{31}P, ^{14}N, ^{13}C, and ^{77}Se, 2-D homonuclear (e.g., ^{1}H TOCSY, DQF-COSY, NOESY, ROESY) and 2-D heteronuclear experiments both as direct and as indirect detection (e.g., HSQC, HMBC, Hetero TOCSY, HSQC-TOCSY, HCCH-TOCSY). All of these experiments are standard and have been described in numerous textbooks (Cavanagh *et al.*, 1996; Claridge, 1999; Ernst *et al.*, 1990; Sanders and Hunter, 1987). As described in detail below, this suite of experiments each alone or in combination typically provides unequivocal structure identification of metabolites directly in crude extracts by comparison with in-house multinuclear NMR database, public databases, or proprietary databases as appropriate.

In cell or tissue extracts of metabolites, each molecule behaves essentially independently of all the others, that is, the system is an ideal mixture (though under some circumstances there may be exceptions to this general rule). Thus the NMR spectra are a simple weighted superposition of the individual molecular spectra, where the weights are simply the relative concentrations.

To optimize the database matching, it is crucial to standardize the conditions for NMR acquisition so that the NMR parameters obtained are consistent for comparison across different samples. To this end, we have adopted the procedure of lyophilizing extracts in 10% TCA, which protonates all functional groups containing active hydrogen atoms. This minimizes the variation of chemical shifts as a function of pH and other exchangeable cations and is amenable for high-throughput operations. In addition, 1-D spectra are typically acquired using a 90° pulse angle in D_2O at 20°C with a recycle time of 5 sec; while for observing exchangeable protons, spectra are obtained similarly except in 90% H_2O plus 10% D_2O at 10°C with Watergate solvent suppression (Piotto *et al.*, 1992). These conditions minimize saturation effects from insufficient T_1 relaxation while optimizing signal intensity; both are also important for quantification purpose. To facilitate database matching and searching, it is essential to record spectra of standards under the same conditions. This is the approach taken by commercial vendors of data analysis software such as Chenomx (Weljie *et al.*, 2006). In addition to commercial databases, there are public databases, such as BioMedResBank (BMRB, http://www.bmrb.wisc.edu/metabolomics/), that post 1-D and 2-D NMR spectra of metabolite standards on the Web site for NMR assignment. Where unknowns are detected, which is rather common in metabolomics research, standards-free methods for structure identification and elucidation are needed, especially for peaks that respond significantly to altered conditions. Such peaks might be revealed by automated decomposition approaches or principal components analysis, for example (Ladroue *et al.*, 2003; Lopez-Diez *et al.*, 2005), though appropriate bioinformatics tools are still being developed.

The spectral resolution afforded by the high magnetic field strength, the use of different nuclei, and multiple dimensions is often sufficient such that no physical fractionation is required for many of the metabolites (Fan, 1996a,b, 1986, 1993, 1997, 1998, 2001, 2003; Gradwell *et al.*, 1998). This greatly improves the analysis throughput over those approaches that require sample fractionation. In addition, all of the experiments can be performed unattended under automation mode, which reduces the labor involved. However, overall sample throughput is limited by raw sensitivity, which is determined by the amount and quality of the sample (see above), and the intrinsic sensitivity of the NMR spectrometer. With modern cold probe technology, a threefold gain in sensitivity over equivalent conventional probes can be obtained in aqueous solutions (Kelly *et al.*, 2002; Lane and Arumugam, 2005; Styles *et al.*, 1989). This in turn allows up to a ninefold reduction in sample requirement or in acquisition time. For example, a good quality TOCSY spectrum can now be acquired in <2 h while natural abundance ^{13}C-^{1}H HSQC and heternonuclear multiple bond correlation (HMBC) spectra can be recorded in 2–3 h from <5 mg of biomass.

2. Structure Elucidation of Unknowns

The same set of NMR techniques described above can be employed to elucidate structures of unknown metabolites or metabolites with no standards available, directly from crude extracts (Fan *et al.*, 1997, 2001). The ^{1}H TOCSY, ^{1}H DQF-COSY, and ^{1}H-^{13}C HSQC patterns reveal the proton and carbon covalent linkages uninterrupted by heteroatoms while the ^{1}H-^{13}C HMBC pattern provides those linkages across the interruptions such as N and carbonyl groups. ^{31}P-^{1}H HSQC is used to determine whether phosphorus to proton linkages that are three or four bonds apart (typically found in phosphorylated metabolites) are present in the unknown metabolite. If the metabolite is labeled with ^{13}C, ^{1}H-^{13}C HSQC-TOCSY or HCCH-TOCSY can be performed to assist in assigning carbon and proton linkages, while for ^{15}N-labeled targets, ^{1}H-^{15}N HSQC-TOCSY (Fig. 3), HNCO, and HNCA experiments (Kupce *et al.*, 2003) provide proton, nitrogen, and carbon linkages, and these are particularly useful for peptidic structure assignment (Cavanagh *et al.*, 1996).

If the unknown structure is still elusive after the suite of NMR characterization, fractionation of crude extracts using LC may be required. The fractions can then be analyzed by the total correlation spectroscopy (TOCSY) or other 2-D NMR experiments in automation mode [e.g., using a robotic sample changer or by HPLC-NMR (Lindon, 2003)]. These experiments provide spectral "fingerprints" of molecular fragments of unknowns in the extract, which cannot be easily obtained from 1-D NMR spectra. This provides precise molecular information about parts of the molecules of interest. Such fragments can be matched with libraries that are arranged to extract fragment data. The elemental composition or molecular formula of the fraction that contains the unknown can be determined by MS methods described below.

3. Concentration Determination

NMR is not an absolute quantitation technique. Although the detector response is strictly proportional to the number of nuclei present in the sample, so that within a sample, relative concentrations are directly proportional to peak areas, normalized to the number of protons that make up each resonance. However, the detected emf is subject to a large number of factors. Therefore, an internal reference compound must be used to determine the absolute concentration of each analyte. This can either be an added standard which has a convenient resonance well resolved from all the molecules of interest, or of a metabolite whose concentration has been independently determined, such as by MS (see below). It should be noted that quantitation is more complicated for X nuclei (e.g., ^{13}C, ^{15}N) than ^{1}H when there is a nuclear overhauser enhancement (NOE); it is often necessary to suppress the NOE effect.

The optimal sensitivity in 1-D NMR with signal averaging is achieved when the repetition time of the experiment is 1.27 times the T_1 value (Ernst *et al.*, 1990). Under these conditions, the resonances will be partly saturated, according to the following relation:

$$M(\text{obs}) = M^0[1 - \exp(-D/T_1)] \tag{1}$$

where M(obs) is the observed magnetization, M^0 is the true (equilibrium) magnetization, D is the recycle time, and T_1 is the longitudinal spin-relaxation time. For a complex mixture, there will be a range of T_1 values (cf. Table II for measured T_1 of assigned peaks in a TCA extract). Hence, there cannot be an optimal repetition rate for all compounds present in the extract, and some resonances will show a higher degree of saturation than others. The greater the ratio D/T_1, lesser the error. Clearly for accurate concentration determination, saturation factors must be accounted for.

4. Reference Compounds

Reference compounds are usually added for chemical shift determination for which there are several choices according to the nucleus being examined. 2,2′-Dimethylsilapentane-5-sulfonate (DSS), trimethylsilylpropionic acid (TSP), and tetramethylsilane (TMS) are all ^{1}H and ^{13}C chemical shift references that rely on the electropositive Si atom. When the methyl protons and carbons attached

Table II
Proton $1/T_1$ Values of Selected Resonances of a TCA Extract of Rhabdomyosarcoma Cells

Species	$1/T_1(=R_1)$ (sec^{-1})
Lac $^{12}CH_3$	0.61 ± 0.01
Lac $^{13}CH_3$	0.7 ± 0.03
Thr $^{12}CH_3$	1.1 ± 0.03
Ala $^{12}CH_3$	0.65 ± 0.03
ATP ^{12}CH1′	0.79 ± 0.06
ATP ^{13}CH1′	1.96 ± 0.07
UTP ^{12}CH5	0.82 ± 0.07
UTP ^{13}CH5	1.98 ± 0.06

Note: T_1 values were determined by nonlinear regression to inversion recovery data recorded at 14.1 T, 20 °C.

to the Si are assigned a chemical shift of zero, almost all biological metabolites appear at positive chemical shifts.

The main differences between these compounds are solubility, pH dependence, and interference. DSS and TSP are both water-soluble compounds. DSS has no significant pH dependence but does have proton resonances near 3 ppm that can overlap with resonances of important metabolites. TSP is pH sensitive near pH 4. TMS is soluble only in nonpolar solvents, such as chloroform, and can be used as an internal reference only for lipid extracts. $CDCl_3$ and CD_3OD are reasonable chemical shift references but are not suitable concentration standards. For ^{31}P NMR, the choices are more limited because most compounds are pH dependent. We commonly use an external reference, methylene diphosphonate under defined conditions (100 mM Tris, pH 8.8), though 85% phosphoric acid is also commonly used. For absolute concentration determination by ^{31}P NMR, it is usually necessary to spike a sample with a known concentration of an appropriate stable phosphate compound, such as AMP, or alternatively use a compound whose concentration has been independently determined, such as by LC.

5. Detection Sensitivity

Of the nonradioactive NMR-active nuclei, ^{1}H is by far the most sensitive by virtue of its large gyromagnetic ratio γ. As sensitivity in terms of signal detection varies with cube of γ (and see below), proton detection is often preferred (Cavanagh *et al.*, 1996; Ernst *et al.*, 1990). Most biological molecules contain hydrogen atoms directly bonded to carbon and nitrogen, which can then be detected indirectly via the attached proton. In general, a proton-detected experiment, especially in two dimensions, is much more sensitive than direct X-nucleus detection. For ^{13}C, proton detection is up to 64-fold more sensitive than the direct detection and for ^{15}N it is almost 1000-fold more sensitive. These values are very much upper limits of signal enhancement by proton detection, as sensitivity of the X detection can be considerably enhanced using either the NOE or the INEPT transfer (up to 1 γ equivalent). Different relaxation times for X and H may also affect the sensitivity differences under actual experimental conditions. The sensitivity gain for proton detection of ^{31}P is much lower, as γ_H/γ_X is only about 2.5 and most biological molecules have a small three-bond scalar coupling to the ^{31}P. In this case, there is not so much to be gained compared with ^{31}P detection, especially as one generally wishes to determine the concentration of ATP, ADP, and inorganic phosphate, which do not have convenient protons for indirect detection. However, as ^{31}P is essentially 100% natural abundance and no other NMR-observable isotopes exist, the issue of isotopomer analysis does not arise. However, there are good reasons to detect an X nucleus (e.g., ^{13}C, ^{15}N, or ^{31}P) directly, especially if it does not have a directly attached proton such as carbonyl and quaternary carbons, quaternary nitrogen atoms, and many biological phosphates.

1-D NMR experiments have the advantage of speed; a single transient can be recorded in 1–2 sec, and even with signal averaging, only a few minutes of spectrometer time are needed for acquisition. In automation mode, however, additional time considerations come into play, including sample changing, equilibrating and for quantitative work, the time taken to lock, shim, and determine the 90° pulse width. Even in a flow probe (see below), this can take several minutes, implying a throughput in automation mode of 30–60 samples per hour, where the samples are sufficiently strong, or one is unconcerned about the less abundant species.

B. 2-D NMR and Isotope Editing

The resolution in 1-D is often too poor for structure identification and isotope quantitation in crude cell or tissue extracts. The largest increase in resolution is achieved by increasing the dimensionality of the experiment. Thus spreading signals over a plane results in an enormous increase in resolution, with the additional advantage that whole molecular fragments can be visualized according to the type of experiment used. The cost of the increase in accessible information content is time, as a 2-D experiment is effectively a series of 1-D experiments. Therefore, the 2-D experimental time is determined by the sensitivity of detection (see above), which include the consideration for metabolite concentrations or the amount of biological material used for extraction (and see below for technical developments in this area).

The 2-D methods that rely on scalar coupling, that is, through the bonding network, therefore correlate atoms that are bonded to one another in the same molecule. The proton correlation spectroscopy (COSY) and TOCSY experiments make correlations primarily between protons that are two to four bonds apart. The COSY experiment detects only direct scalar couplings, such as between the CH_3 and βCH, and between the β and α CH of threonine. In contrast, the TOCSY experiment will not only produce these two three-bond couplings but also show the interaction between the Hα and the methyl protons. The specific patterns of cross peaks and characteristic chemical shifts make it possible to assign a large number of metabolites directly in crude extracts with high reliability. The construction of a searchable library of standards containing information on scalar connectivity and chemical shift of the coupled peaks can be the basis for automated structure assignment.

Although many common metabolites can be reliably identified with a single TOCSY experiment, there are others that remain ambiguous, at least in new tissue types. For these, additional experiments may be needed to find correlations with other atoms not necessarily sampled in the simple proton TOCSY experiment. These experiments include HSQC, HMBC, NOESY, and ROESY, which correlate protons with ^{13}C (HSQC, HMBC) via scalar coupling or protons via dipolar (through space) interactions. The application of these experiments to structure identification is well established for pure compounds (Claridge, 1999; Sanders and Hunter, 1987) and is not fundamentally different for mixtures of compounds such as in an extract. For metabolomics analyses, however, more reliance is made on spectral libraries, which can be used to match whole molecule or molecular fragments with library compounds.

1. Isotope Editing

Simplification of complex NMR spectra can be achieved by filtering the proton spectrum via a rare heteronucleus, such as ^{31}P or ^{77}Se. A ^{31}P-^{1}H correlation experiment, such as HSQC, selects only those molecules that contain a phosphorus atom scalar coupled to a proton and therefore will filter out the nonphosphorylated metabolites in a crude extract (Gradwell *et al.*, 1998). Spectral editing techniques may also be used to select for isotopically enriched molecules such as ^{13}C or ^{15}N, and depending on the isotope precursor used, for newly synthesized molecules that have adjacent labels within a molecular fragment.

There is a large variety of possible isotope-editing techniques (Burgess *et al.*, 2001; Cavanagh *et al.*, 1996; Cordier *et al.*, 1999; Riek *et al.*, 2001; Zhang and Gmeiner, 1996). The choice of the experiment to use is in large measure determined by the information that is desired. In our laboratories, we make use of a comparatively small number of such editing experiments in one or two dimensions, including HSQC, HMBC, HSQC-TOCSY, and HCCH-TOCSY (Fan, 2005). These are all proton-detected experiments that have the advantage of very high sensitivity compared with the direct detection methods. This is of increasing importance the lower gyromagnetic ratio of the heteroatom to be observed, for example, in the order $^{15}N > {}^{13}C > {}^{31}P$. Additional correlation experiments can be used for more specific questions, such as HCCs experiments (Cavanagh *et al.*, 1996) to detect labeled carbonyl groups for example. In the following, we briefly describe the use of some of the more common editing experiments for metabolomics analyses.

a. ^{1}H-X HSQC and HMQC

HSQC or HMQC correlate an X-nucleus frequency with that of a proton via the scalar coupling between the two nuclei. The sensitivity compared with the directly detected correlation experiment (Hetcor) is proportional to $(\gamma H/\gamma X)^3$ ignoring relaxation effects and variable transfer efficiencies. The latter depends on the value of the one-bond coupling constant, which is significantly variable for C–H bonds. For example, the $^{1}J_{CH}$ of methyl group in alanine and lactate is about 127 Hz, whereas for the methine CH it is ~145 Hz. Anomeric sugar CH have $^{1}J_{CH}$ of around 160–170 Hz, and aromatic CH typically have $^{1}J_{CH}$ in the range 180 Hz and higher (cf. Table III). Thus, in a single experiment, the transfer efficiency cannot be optimized for all groups simultaneously, which calls for two separate experiments each with a $^{1}J_{CH}$ setting optimized for aliphatic or aromatic/anomeric protons. Recording two spectra increases the total experimental time needed, and absolute quantitation is only possible if the coupling constants are known. However, with a setting of J = 140 Hz (or a delay in the INEPT period of around 3.6 msec), the variation in the INEPT transfer for coupling constant in the range 120–160 Hz is only 2.5%. A greater distortion of intensity (peak volumes) may

Table III
$^{1}J_{CH}$ Values of Selected Metabolites

Molecule	C–H	$^{1}J_{CH}$ (Hz)
Lactate	CH_3	127.5
	CH	146
Alanine	CH_3	128
	CH	146
Glutamate	CγH	130
	CβH	131
	CαH	147
rATP	C1′H	168.3
	C2′H	150
rUTP	C1′H	170.8
	C2′H	151.6
Glucose	C1Hα	169.5
	C1Hβ	159.5
UTP	C5H	179
	C6H	183

Note: Values were measured at 20 °C from 1-D or 2-D spectra.

arise from the relaxation delay. Such spectra are usually recorded with relatively short acquisition time in t_2 owing to the active X-decoupling. An acquisition time of <0.15 sec is commonly used which limits the resolution to 6.7 Hz/point (or 3.3 Hz/point after 1 zero filling) in F2. To improve overall sensitivity, the spectra are often acquired with quite short relaxation delays, less than 1.5 sec. Under these conditions the ^{1}H, whose T_1 values determine the net magnetization transfer, are significantly saturated (cf. Table II), which can distort intensities if not accounted for in any relative quantification.

For mixtures of unlabeled compounds, moderate resolution in the indirect dimension is adequate because all peaks are singlets (the probability of two adjacent carbons in the same molecule at natural abundance is 1.1% of 1.1%, i.e., very low). However, adjacent carbons in an enriched metabolite may be simultaneously labeled in which case there will be a scalar coupling between the two (or more) carbons with a $^{1}J_{CC}$ (and possible longer range coupling) of 40–60 Hz depending on the functional group fragment (Fig. 1). This gives rise to peak broadening at low resolution, and at sufficiently high resolution in the indirect dimension, the C–C splitting becomes observable (Fig. 1B). There are two approaches to dealing with this. One is to record constant time heteronuclear single (multiple) quantum coherence [HS(M)QC] spectra, which effectively decouples the carbons from one another, giving rise to singlets whose widths are determined by the constant time period (Cavanagh *et al.*, 1996; Homans, 1992; Legault *et al.*, 1995). Alternatively, the spectra can be recorded with sufficient resolution to resolve the couplings, which then gives a direct readout of the number of scalar-coupled adjacent carbons in the same molecule.

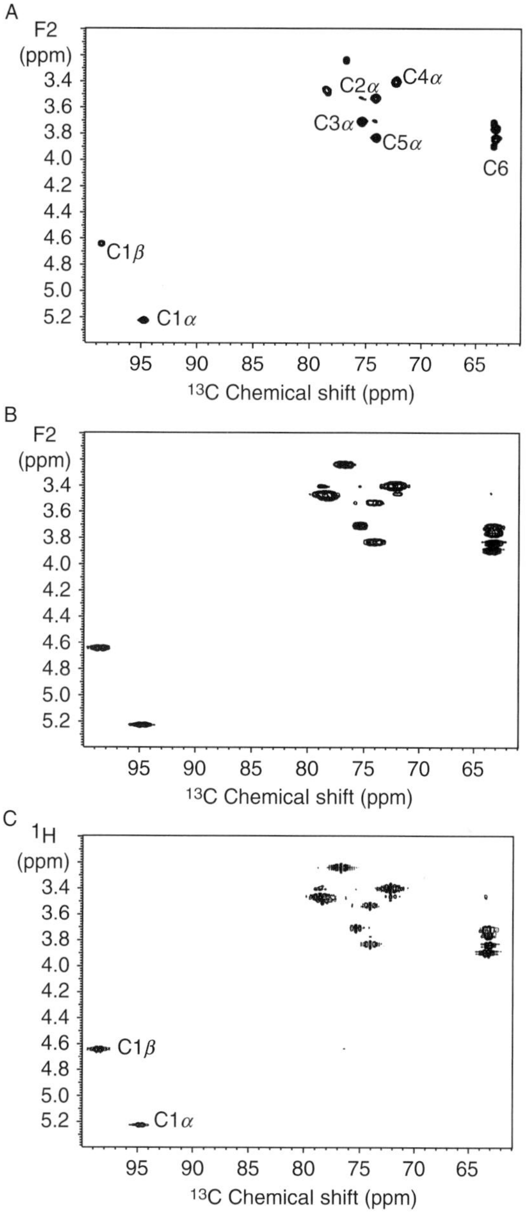

Fig. 1 HSQC spectra of glucose: effects of CC coupling and resolution. The spectra were at 14.1 T with 256 complex pairs in t_1 and 4 scans/increment (total time = 58 min) with spectral widths of 6 kHz in F2 and 15 kHz in F1. The acquisition times were 0.14 sec in t_2 and 0.0085 sec in t_1. The free induction decays were linear predicted to 512 points in t_1, and zero filled to 2048 points. The glucose was dissolved in 100 mM KCl in D_2O. Panel (A) shows the HSQC spectrum of D-glucose at natural isotopic

For example, to acquire an HSQC spectrum with a spectral width of 100 ppm, or 15 kHz at 14.1 T, then 1000 points (resolution = 15 Hz/point) are needed to resolve the C–C couplings into doublets and triplets, etc. This is shown in Fig. 1. However, with linear prediction plus zero filling, the resolution is readily achieved with 256 increments in the indirect dimension. Assuming that the protons attached to ^{12}C have been properly suppressed (and this is very efficient in modern spectrometers using TANGO and gradient purge sequences), then a C–C doublet will be easy to detect in the presence of the singly labeled species at 50 times the concentration. This is because the singly labeled species is at only 1% of the concentration, whereas the doubly labeled species is at 2%.

With a strong sample (or one that is highly enriched), a spectrum can be acquired in less than an hour, though for limited tissue samples, 8–12 h may be more typical. Such an experiment can make full use of the proton sensitivity inherent in an inverse detection cold probe.

b. ^{1}H-X HSQC-TOCSY

The 2-D version of this 3-D experiment is an X-edited proton TOCSY and selects for molecules containing scalar-coupled protons, at least one of which is ^{13}C. This experiment is often recorded as a ^{1}H-^{13}C correlation (equivalent to a projection of the 3-D experiment), so that each ^{13}C site shows interactions with its directly bonded proton (HSQC) and to other protons that are scalar coupled to that proton, as in the molecular fragment Ha-^{13}C-^{12}C-Hb. Magnetization is transferred between Ha and Hb via isotropic mixing (TOCSY), filtered through the ^{13}C. At natural abundance, this experiment shows all correlations, and therefore does not show extensive editing in ^{13}C, though it is useful for structure identification as the ^{13}C chemical shifts are recorded in addition to the ^{1}H shifts. For isotopomer analysis, it is very effective where a relatively small number of ^{13}C sites are substantially enriched, as those fragments are very much more intense than the background natural abundance, as shown in Fig. 2. Even an enrichment of a few percent gives rise to a very large increase in signal intensity over background. For example, in the human myocyte extract the protonated carbons of lactate, alanine, glutamate, glutamine, and the nucleotide riboses show prominent sets of correlations due to ^{13}C enrichment in their carbons. Since a complete set of scalar correlations from ^{13}C to ^{1}H is observed in Fig. 2, the identity of these metabolites is readily determined. HSQC-TOCSY is also valuable for discerning metabolites whose resonances fall in a crowded spectral region. For example, phosphorylated sugars are typically difficult to resolve but with ^{31}P editing, only the phosphorylated compounds are detected, and with the TOCSY relays, it is practical to identify these metabolites more reliably (Gradwell *et al.*, 1998).

abundance. All peaks are singlets. Panel (B) shows the [U-^{13}C]-glucose spectrum—the anomeric peaks are doublets owing to the large C1-C2 coupling, whereas the C2, C3, C4, and C5 peaks are triplets owing to the coupling to the two adjacent carbon atoms. Panel (C) shows a mixture of the natural abundance spectrum (97%) and a 3% admixture of the ^{13}C glucose. The anomeric peaks appear as doublets.

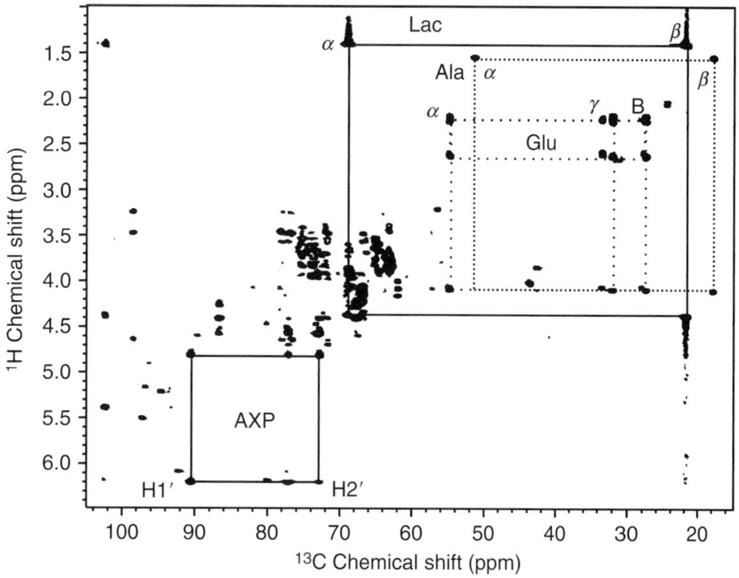

Fig. 2 HSQC-TOCSY of normal myocytes. The HSQC-TOCSY spectrum was recorded at 14.1 T, 20 °C on TCA extracts of normal human myocytes grown in the presence of [U-^{13}C]-glucose (T. W. Fan, A. N. Lane, and M. Z. Ratacjzak, unpublished data), with an isotropic mixing time of 50 msec and a B_1 field strength of 8 kHz.

HSQC-TOCSY experiments are also useful for detecting different types of isotopic enrichment. If tissues or cells are provided with two types of isotope sources, such as ^{15}N and ^{13}C, the ^{15}N-^{1}H HSQC-TOCSY experiment edits the spectrum by selecting for ^{15}N to ^{1}H scalar correlations. If the ^{13}C is also present in the same molecule, a pair of satellite peaks will be observed for the protons attached to that carbon, making it possible to identify those metabolites that contain both ^{15}N and ^{13}C as well as the label position(s) in the metabolites (Fig. 3A). For example, when rice coleoptiles were treated with ^{15}N-nitrate and ^{13}C-acetate, the ^{15}N-^{1}H HSQC-TOCSY spectrum of the PCA extract was dominated by ^{15}N-labeled alanine, ammonium, and γ aminobutyrate (GAB). This spectrum was substantially simplified due to ^{15}N editing, which suppressed a large number of signals from metabolites that were not enriched in ^{15}N. In addition, the ^{15}N-labeled GAB showed a pair of split ^{13}C satellite peaks at the β proton position (peak 3) due to the presence of the β ^{13}C derived from ^{13}C-acetate. Thus, GAB was synthesized from ^{15}N-nitrate and ^{13}C-acetate via nitrate assimilation (Fig. 3B), citric acid cycle (not shown), Glu dehydrogenase (GDH), and Glu decarboxylase (GDC) reactions (Fig. 3B) [adapted from Fan *et al.* (1997) with permission]. The ^{13}C satellites can be integrated to determine the degree of labeling, which was about 50% (and see below). This experiment demonstrates the simplification of a very complex spectrum of a crude extract by isotope editing while

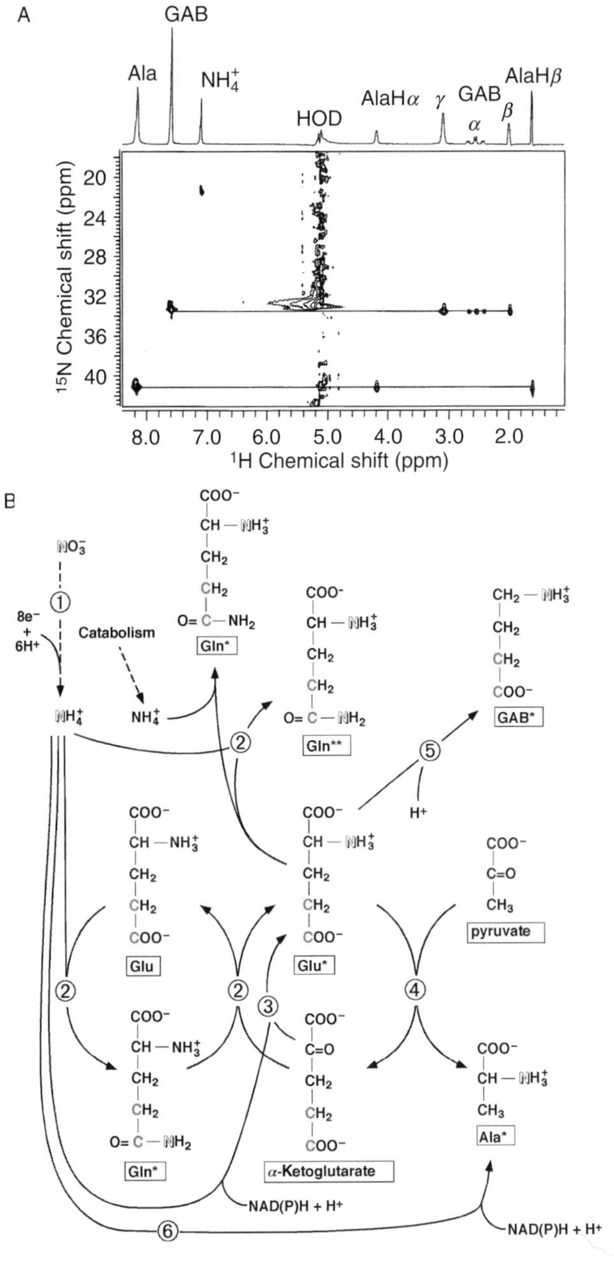

Fig. 3 ^{15}N-^{1}H HSQC-TOCSY rice coleoptiles. (A) 2-D ^{1}H-^{15}N HSQC-TOCSY spectrum of a PCA extract of rice coleoptiles grown in the presence of ^{15}N-nitrate and ^{13}C-acetate. Traces covalent linkages from ^{15}N to the proton network that is undisrupted by other heteroatoms, carbonyls, or 4° carbons. This is achieved by ^{15}N-editing coupled with magnetization transfer from protons attached to ^{15}N to the rest of the network. Highlighted here are two sets of connectivities: one from ^{15}N to the γ-, α-, and β-CH_2

providing information about metabolic transformation and biosynthetic pathways of specific labeled products from precursors.

c. HCCH-TOCSY

This is a 2-D proton correlation experiment in which protons attached to an NMR active heteronucleus, for example ^{13}C, are correlated via magnetization transfer between scalar-coupled ^{13}C nuclei such as in the fragment H-^{13}Ca-^{13}Cb-H. Magnetization is transferred between Ca and Cb using isotropic mixing (TOCSY). This editing experiment therefore detects only molecular fragments of this kind where both carbons in the same molecule are labeled, whereas molecules in which either Ca or Cb only are labeled are not detected. For example, for rhabdomyosarcoma cells (Rh30) grown in uniformly ^{13}C-labeled glucose ([U-^{13}C]-glucose), the HCCH-TOCSY spectrum (Fig. 4) was dominated by scalar correlations

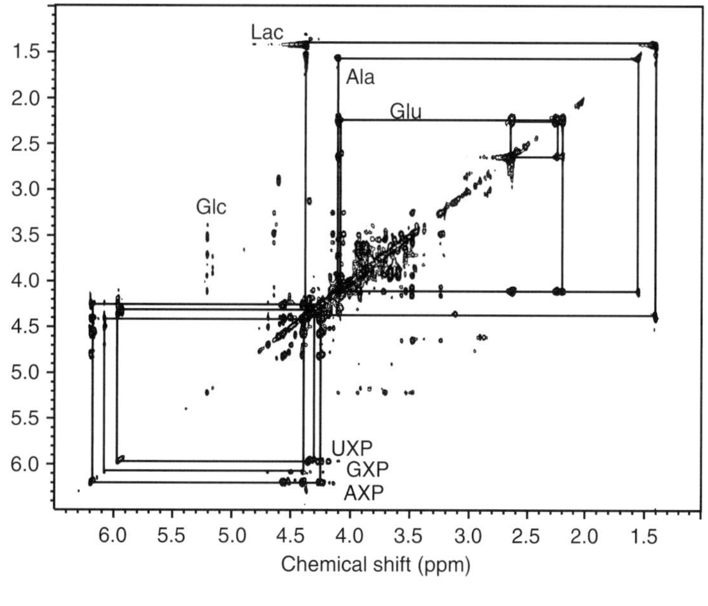

Fig. 4 HCCH-TOCSY spectrum of rhabdomyosarcoma cell extract. The HCCH-TOCSY spectrum was recorded at 14.1 T, 20 °C on TCA extracts of rhabdomyosarcoma cells grown in the presence of [U-^{13}C]-glucose (T. W. Fan, A. N. Lane, and M. Z. Ratacjzak, unpublished data) with a 12 msec mixing time, with a spin-lock field strength of 8 kHz. Boxes denote connectivities on adjacent carbons of selected metabolites.

---

of γ-aminobutyrate (GAB) and the other from ^{15}N to the α-CH and β-CH_3 of Ala. Also readily discerned is the dual enrichment of ^{13}C and ^{15}N in GAB as ^{13}C labeling at the α position is evident from the ^{13}C satellite cross-peak pattern. (B) Metabolic pathways giving rise to the observed isotopomer distribution in GAB. (See Plate no. 14 in the Color Plate Section.)

from Lac, Ala, Glu, and riboses of adenine, guanine, and uracil nucleotides (AXP, GXP, and UXP). Also evident were the scalar correlations from the original labeled glucose and some other unidentified sugars derived from glucose. The substantial ^{13}C editing in this spectrum made it practical to identify metabolites that were otherwise hidden under other, intense resonances. The presence of HCCH correlations indicates that these metabolites contained molecular fragments with at least two consecutive ^{13}C-labeled carbons, which in turn suggest that these fragments remain intact through metabolic transformation from [U-^{13}C]-glucose. This experiment therefore complements the HSQC-TOCSY experiment in discerning multiple, consecutive isotope-labeled metabolites.

2. Determination of Isotopic Distribution

The editing experiments described above provide unambiguous information on the positional enrichment of particular isotopes. For flux or pathways analyses, the enrichment at each position also needs to be determined.

The natural abundance level of ^{13}C is 1.1%, which means that the attached proton appears as a central peak of 99% intensity and two satellite peaks at 0.5% intensity each. If the labeling doubles the ^{13}C level to 2%, the effect on the proton spectrum is small (doubles the satellite intensities) but is much greater in the HSQC experiment, as now the detected species has twice the concentration. Furthermore, in high-resolution HSQC experiment the scalar coupling between adjacent ^{13}C sites (if present) can be readily determined (see above).

2-D ^{1}H TOCSY and HSQC are particularly useful for isotopomer analysis. For TOCSY analysis of ^{13}C-labeled metabolites, characteristic satellite cross-peak patterns can be used for diagnosing the ^{13}C labeling pattern in a given metabolite. Figure 5A shows the ^{13}C isotopomer-dependent patterns of central and satellite cross peaks in a TOCSY spectrum of a TCA extract of rhabdomyosarcoma cells grown in the presence of [U-^{13}C]-glucose. Red lines trace the cross peaks of several metabolites corresponding to the ^{12}C isotopomer. The green boxes show the satellite peaks surrounding each cross peak. The patterns of lactate and glutamate for example are quite different. The lactate methyl-a cross peak shows a simple square pattern, which shows that both of the carbon atoms are labeled in the same molecule. If only the Cα group were labeled (e.g., using ^{13}C-2 glucose as the source), then the cross-peak pattern would be a vertical pair (cf. Fig. 5B). The observed pattern implies the existence of two and only two forms of lactate, with no metabolic mixing (scrambling).The cross peaks of the glutamate α to γ and β CH_2 are substantially more complex (Fig. 5A). The central cross peak arises from the (fully unlabeled) isotopomer of ^{12}Cγ-^{12}Cβ. The observed satellite patterns are actually a superposition of several individual patterns as shown in Fig. 5B. Thus the α–γ cross peak shows the square pattern that indicates the presence of the doubly labeled ^{13}Cα-^{13}Cγ isotopomer, and the two singly labeled species ^{12}Cα-^{13}Cγ and ^{13}Cα-^{12}Cγ. The superposition then gives rise to the square + greek cross-peak pattern at the bottom of Fig. 5B. Similar remarks apply to the α–β cross peak (Fig. 5A) indicating

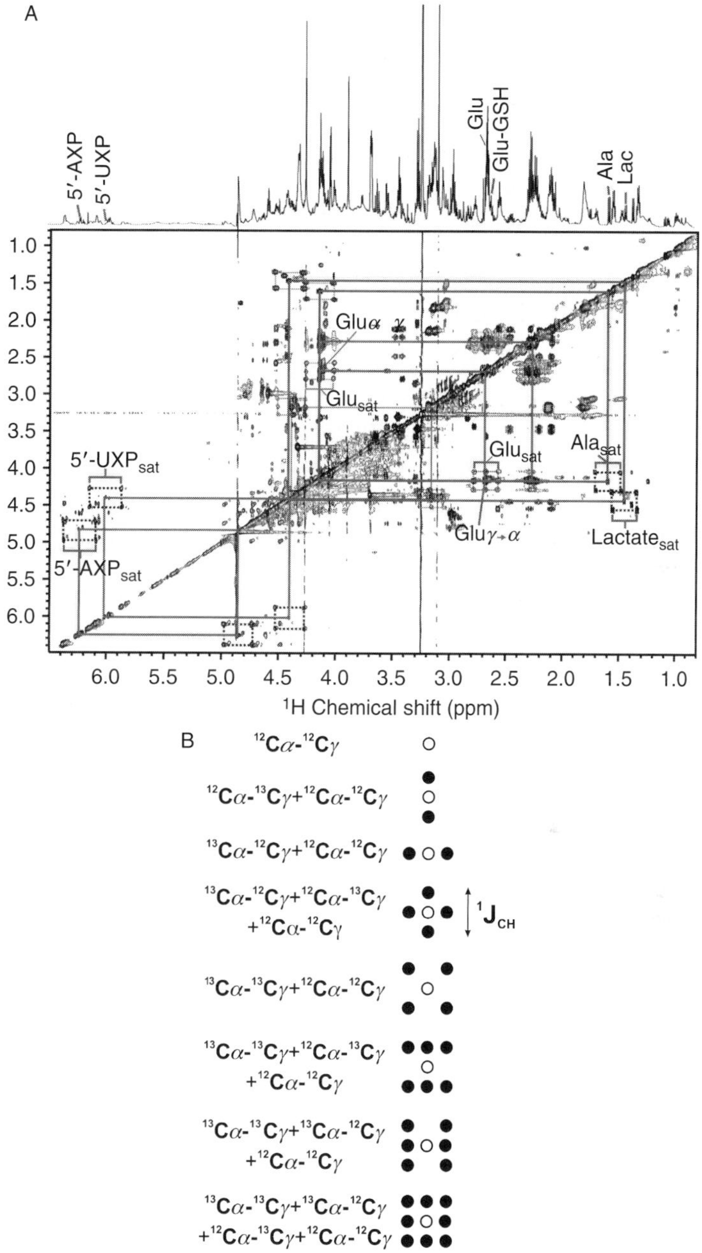

Fig. 5 TOCSY spectrum of rhabdomyosarcoma cell extract (A) The TOCSY spectrum was recorded at 14.1 T, 20 °C on a TCA extract of rhabdomyosarcoma cells grown in the presence of [U-^{13}C]-glucose (T. W. Fan, A. N. Lane, and M. Z. Ratacjzak, unpublished data) with a 50 msec mixing time, with a spin-lock field strength of 8 kHz. (B) Expected cross-peak patterns for glutamate labeling. All possible isotopomers are shown and the patterns arising from superpositions of individual isotopomer patterns. Open circles denote protons attached to ^{12}C and filled circles denote protons attached to ^{13}C.

that all possible isotopomers of Glu are present, which provides atom-specific information about the pathways of synthesis of glutamate starting from labeled glucose. We have recently analyzed the positional enrichment in several metabolites in extracts from A549 cells (Fan, 2005), including lactate, alanine, eight different isotopomers of Glu, Gln, and GSH, Asp, the ribose moieties of free pyrimidine and purine nucleotides, and four isotopomers of the uracil base in UTP after providing [U-^{13}C]-glucose to the cell growth medium. The medium was also analyzed for glucose depletion and lactate and alanine secretion. The changes in enrichments due to treatment of the cells with selenium compounds provided essential information for understanding the biochemical responses of the cells to the cytotoxic agents and helped understand the gene array data obtained under the same conditions.

3. Isotopomer Quantification

The X-nucleus editing experiments described above provide unambiguous information on the structure and labeled position(s) of metabolites. In this section, we discuss how isotopic enrichment can be quantified.

There is a large literature on isotopomer quantification and analysis in metabolism, both *in vivo* and with extracts (Anousis *et al.*, 2004; Bederman *et al.*, 2004; Boren *et al.*, 2001; Burgess *et al.*, 2001; Carvalho *et al.*, 2001; Cline *et al.*, 2004; Des Rosiers *et al.*, 2004; Fan, 2005; Goddard *et al.*, 2004; Henry *et al.*, 2003; Lloyd *et al.*, 2004; London *et al.*, 1999; Lu *et al.*, 2002; Marin *et al.*, 2004; Mason and Rothman, 2004; Mason *et al.*, 2002; Mollney *et al.*, 1999; Zwingmann *et al.*, 2003). Much of this work has been carried out with direct detection, which has the advantage that all atoms are directly observed. However, where material is in short supply, the sensitivity can be limiting, and for quantitative analyses, long relaxation delays may be needed. An alternative is to use indirect proton detection, which is intrinsically much more sensitive (see above). This approach relies on the detection of satellite peaks; thus, ^{12}C is magnetically inactive, so does not cause any splitting of the attached proton, whereas ^{13}C has a spin 1/2, so splits the attached protons into a doublet, symmetrically displaced either side of the central (^{12}C) resonances.

For well-resolved spectra, it is sometimes possible to determine the extent of labeling at one or more position from a simple 1-D experiment. Figure 6 shows a 1-D ^{1}H NMR spectrum of a TCA extract of rhabdomyosarcoma cells grown in the presence of [U-^{13}C]-glucose. The methyl region (1.2–1.7 ppm) is relatively well resolved and is dominated by resonances from threonine, lactate, and alanine. The lactate and alanine resonances have satellite peaks separated by a splitting of 127 Hz that correspond to the $^1J_{CH}$ of the ^{13}C- labeled molecule. The complexity of the splitting pattern is due to the addition scalar coupling between the methyl protons with the α carbon and the carbonyl carbon as well as the α proton (Lloyd *et al.*, 2004). As these coupling constants are similar, a characteristic six-line pattern rather than the expected eight-peak pattern is observed. This actually shows that there are two isotopomers of lactate present in the mixture, namely unlabeled lactate, which arise from reduction of unlabeled pyruvate, and fully

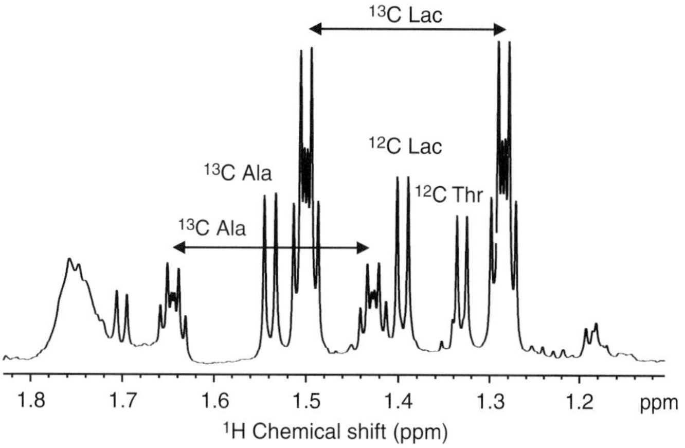

Fig. 6 1-D NMR spectrum of rhabdomyosarcoma cell extract showing satellite peaks. Rhabdomyosarcoma cells Rh30 were grown in the presence of [U-^{13}C]-glucose for 24 h. The cells were extracted with cold 10% TCA, lyophilized, and redissolved in D_2O. The spectrum of the TCA extract was recorded at 14.1 T, 20 °C using a recycle time of 5 sec. 256 transients were co-added (22 min acquisition). The methyl region shows Lac, Ala, and Thr and the satellite peaks of Lac and Ala.

labeled lactate (i.e., in which all three carbon atoms are labeled) deriving from fully labeled pyruvate. The latter must have originated exclusively from the labeled glucose supplied to the medium, as opposed to alternative sources such as glutamine via glutaminolysis (Mazurek *et al.*, 1999; Mazurek and Eigenbrodt, 2003).

The fractional amount of label, F, is given by

$$F = \frac{A(\text{satellites})}{[A(\text{satellites}) + A(\text{central resonance})]} \tag{2}$$

where A is the peak area.

As a proton attached to ^{13}C relaxes faster than one attached to ^{12}C, the fraction label will be overestimated under conditions of partial saturation. This problem can be resolved by collecting spectra with sufficiently long relaxation delays or by correcting for saturation using the measured T_1. We routinely record 1-D spectra with a 5-sec recycle delay. Under these conditions, methyl protons attached to ^{12}C have a T_1 value in the range of 1–2 sec (see Table II), so that the protons are at least 90% relaxed. Figure 7 shows how the apparent fraction varies with the ratio of the recycle time to the effective T_1, at different fractions of label. As can be seen, the error is small once the ratio exceeds 2.5–3. As expected [cf. Eq. (2)], the apparent fraction of ^{13}C label is overestimated at short recycle times. The weak dependence of F for the lactate methyl peaks reflects the similarity of the T_1 for the proton attached to ^{12}C or ^{13}C (cf. Table II). In contrast, the value of F varies strongly with relaxation delay for the ATP H1′ where there is a large difference in the T_1 between the unlabeled and labeled molecules.

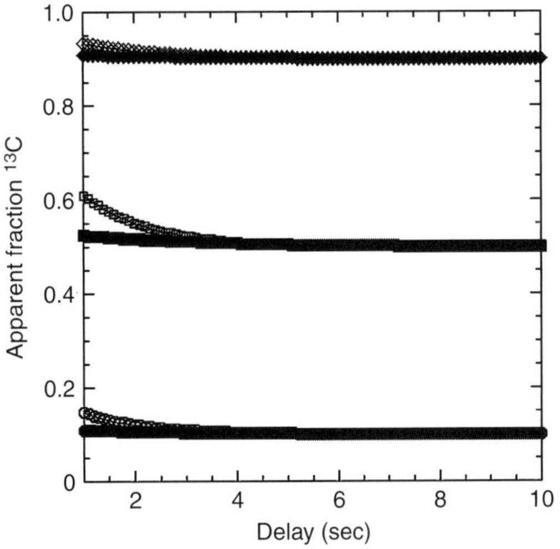

Fig. 7 Dependence of fractional enrichment on recycle time in 1-D experiments. The fraction ^{13}C was calculated using Eq. (4) using measured values of the spin-lattice relaxation rate constants of lactate methyl (0.61 sec^{-1} for ^{12}C and 0.7 sec^{-1} for ^{13}C) and ATP H1′ (0.79 sec^{-1} for ^{12}C and 1.96 sec^{-1} for ^{13}C) in a TCA extract of rhabdomyoscarcoma cells grown in the presence of [U-^{13}C]-glucose, as a function of the recycle time in a simple 1-pulse experiment. Fraction was calculated at 10%, 50%, and 90% ^{13}C. Filled symbols, lactate methyl; open symbols, ATP H1′.

Determining isotope enrichment from 1-D NMR is not always possible because of spectral overlap. For example, depending on pH, the threonine and lactate methyl resonances may be coincidental. To improve the resolution, and also identify more rigorously the isotopomer distributions, TOCSY is a sensitive 2-D correlation technique we have found to be especially useful. Figure 8 shows a TOCSY spectrum of a lactate standard in D_2O solution. The cross-peak pattern reveals the interactions of the β methyl protons attached to ^{13}C and the α proton attached to ^{13}C (each as a pair of satellite peaks to the central resonance, cf. Fig. 5B). The satellite peaks are symmetrically displaced from the central peak by one half of the one-bond coupling constant, which is 127 Hz for the methyl carbon and 146 Hz for the methine carbon.

The fraction of ^{13}C label is calculated in an analogous fashion to the 1-D analysis using peak volumes, V, as

$$F(\text{2-D}) = \frac{V(^{13}\text{C satellites})}{[V(^{13}\text{C satellites}) + V(^{12}\text{C peak})]} \tag{3}$$

In TOCSY spectra, although the peaks are absorptive, there can be a small dispersive component remaining despite careful use of purge pulses and/or z-filters. The determination of the peak volume requires careful adjustment of the base-plane

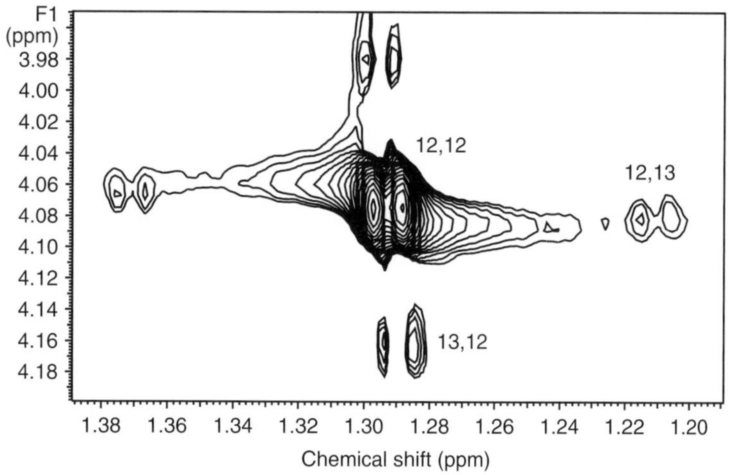

Fig. 8 TOCSY spectrum of a lactate standard. The solution contained 2.3 mM sodium lactate in 100 mM KCl, 10 mM sodium phosphate, pH 7. The spectrum was recorded at 18.8 T, 20 °C using a spin-lock strength of 8 kHz and an isotropic mixing time of 50 msec. The acquisition times were 0.34 sec in t_2 and 0.05 sec in t_1 with a recycle time of 2 sec. The total experimental time was 8 h. The β-Me cross peak is shown, with the satellites for the natural abundance ^{13}C species. The double label would account for 1% of the satellite peaks and does not appear at this level. From peak integration, the satellites indicated 1.3 ± 0.3% ^{13}C for both the methyl and the methine groups (no correction was made for differential relaxation). This can be compared with the expected 1.1% at natural abundance. The peaks labeled as 12,12; 12,13; and 13,12 denote the carbon isotopes in the C2 and C3 positions of lactate, respectively.

and phase, and the peaks should be adequately digitized. As small metabolites give rise to sharp peaks, this may require relatively long acquisition times in t_2 (at least 0.4 sec) with zero filling to achieve a digital resolution of ca. 1 Hz/point, and sufficient increments in t_1 to ensure that several points are sampled across the peak in the F1 dimension. Linear prediction works well in these cases as the T_2 values are typically long.

With adequate digitization, phase and base-plane correction, quite accurate volume integration of the cross peaks can be obtained using the simple approach of Simpson's rule (which is how most spectrometer software works) rather than attempting to use least squares to fit the complex peak shapes. The major limitation on precision then is the signal-to-noise ratio, which is determined by the concentration of the molecules and the degree of enrichment. In favorable cases, such as where a metabolite is present at high concentration and the peaks are well resolved, even natural abundance ^{13}C (ca. 1.1%) can be detected. In some tissue extracts, we have measured ^{13}C satellites at the 2% level, with a probable error of 0.5%, and we can detect the natural abundance with a similar probable error (T.W.-M.F. and A.N.L., unpublished data). This was verified with a pure solution of 2.3 mM lactate in D_2O (cf. Fig. 5B), for which the estimate ^{13}C was 1.3 ± 0.3% (actual = 1.1%), measured in 4 h.

Hence the accuracy is potentially quite high, and the precision is adequate for most modeling purposes. 1% ^{13}C at 2.3 mM in 0.35 ml is only 8 nmol ^{13}C.

A remaining factor that determines the accuracy of the isotopomer quantitation from TOCSY spectra is differential relaxation. The protons bonded to ^{13}C relax faster in both the longitudinal and rotating frames than protons attached to ^{12}C. For a simple AX spin system, and assuming a strong spin-lock field (negligible tilt), the magnetization transfer, which determines the cross-peak volume, is approximately given by

$$M_{\text{obs}} = M^0 a_{ij} \exp(-R_{1\rho} t_m)[1 - \exp(-R_1 D)] \tag{4}$$

$R_{1\rho}$ is the spin-lattice relaxation rate constant in the rotating frame, t_{m} is the spin-lock mixing time, R_1 is the longitudinal spin-lattice relaxation rate constant, and D is the relaxation delay time.

The last term is the saturation factor due to incomplete longitudinal relaxation [cf. Eq. (1)]. a_{ij} is the mixing coefficient for spins i and j which is [approximately equal to $\sin^2(\pi J t_m)$ for a two spin system under ideal isotropic mixing conditions] and J is the spin-spin coupling constant. The ^{1}H-^{1}H coupling constant is essentially the same for both ^{13}C and ^{12}C so that in forming the volume ratios in Eq. (3) the a_{ij} terms cancel, and the value of F depends on the longitudinal saturation factors modulated by the exponential relaxation along the spin-lock axis. As $R_{1\rho}$ for the ^{13}C proton is larger than for ^{12}C, this causes the ^{13}C satellite peaks to decay faster than the protons attached to ^{12}C, leading to an underestimate of the enrichment. In contrast, as described above, the longitudinal saturation factor works in the opposite direction, partially compensating for the spin-lock relaxation. We have found by experiment that for commonly used mixing times (ca. 50 msec) and typical metabolites, the error in the estimate of the enrichment without correction is small when using a recycle time of 2 sec.

C. Application of MS to Isotopomer Analysis

Because of the complexity of the metabolome, NMR alone is by no means a comprehensive profiling method. Despite its power for structural analysis, there are several limitations in metabolite profiling by NMR, most notably a relative lack of sensitivity (>1–2 nmol metabolite for ^{1}H NMR detection), lack of resolution of some classes of compounds, and the inability to detect NMR-inactive, unsuitable (e.g., O, S), or paramagnetically influenced nuclei. For example, many metabolites that can be detected by NMR can be difficult to quantify due to insufficient spectral resolution even using the highest performance instruments. This is especially so for sugars and the acyl chains of lipids which are crowded together in a narrow regions of the spectrum. MS is a highly complementary sensitive technique that can be used to fill these gaps, providing both confirmation and quantification of the NMR-identified metabolite changes.

Even the least expensive GC-MS has extraordinarily high resolving power, and the sensitivity far exceeds that of NMR. Thus many more low abundance metabolites such as organic acids can be readily detected by GC-MS that are not usually detected by NMR. Furthermore, the presence of mass isotopomers is readily apparent from the departure of the mass spectrum from that obtained from natural isotopic abundance samples. The departures can be quantified, giving the enrichment of mass isotopomers containing 1,2 or more additional labeled atoms (Fan *et al.*, 1997, 2005; Lee *et al.*, 1998; Vizan *et al.*, 2005).

For decades, GC-MS has reigned as probably the most stable and robust structure-confirming technology for quantification of metabolites, but is strongly dependent on correspondingly robust derivatization technology. Along these lines, silylation reagents, such as MTBSTFA, provide effective derivatization across many classes of metabolites, imparting on them excellent characteristics for GC separation while simultaneously adding useful properties for identification and quantification in the MS (Fan *et al.*, 1993). Therefore, the long-stated adage remains true: if a GC-MS method suits the analytes of interest, it is the method of choice over other types of MS.

More recently, the coupling of high-performance liquid chromatography (HPLC) to MS is a complementary—and in the long run more powerful—tool for metabolomics studies. The impetus for using LC-MS ranges from avoiding limitations of derivatization chemistry that tethers GC-MS, or the need to analyze metabolites that are not amenable, to GC even after derivatization. LC-MS can resolve and quantify multiple components in crude biological extracts using very low analyte consumption down to picomole levels for hundreds of compounds, even for "untargeted" analyses with the goal to obtain "global" metabolite profiles (Hirai *et al.*, 2004; Saghatelian *et al.*, 2004). For the element-selective GC- and LC-inductively coupled plasma (ICP)-MS, analysis down to low femtomole levels is feasible, for example for Se compounds (Larsen *et al.*, 2001) and other metals and metalloids.

The last two decades have seen an explosion of new types of MS technologies with a variety of innovative interfaces to LC, as well as the coupling of MS to each other to multiply their analytical power; the latter is generally termed "tandem" MS or "MS*n*" (Siuzdak, 2003). Virtually all of these MS technologies are of great value in metabolic profiling. However, it should be noted that despite recent great strides in improvement ion source techniques continue to limit all MS techniques. For example, the highly popular electrospray (ES) and MALDI techniques exhibit variable ion yields depending on the composition of the sample and other conditions, thereby impeding precise metabolite quantification. This is particularly problematic in attempts at "global" metabolite analyses that are necessarily untargeted, such as those cited above (Hirai *et al.*, 2004; Saghatelian *et al.*, 2004). Conversely, by destroying chemical structure, the ICP source can produce highly quantitative results independent of structure and under a wide variety of sample compositions and conditions (Larsen *et al.*, 2001). Thus, the ion source is a major determinant of the type of MS, which in turn determines the analytical emphasis toward (1) structure and identification, (2) accurate quantification, (3) low detection limits, (4) wide variety of metabolites, or (5) rapid analysis. Two and occasionally three of these are achieved at once by a given MS method, but usually at the sacrifice of the others.

1. Structure Identification

As a tool that is complementary to NMR, the structure elucidation and confirmation efforts by MS can focus on establishing linkage of molecular ion targets to NMR data, functional group classification, generation of substructure molecular formula, and MS chemoinformatics.

First, it is necessary to establish which ions in the MS data, usually molecular ions, correspond to the particular metabolites being studied by NMR or other techniques. In cases where several structures can be hypothesized, analyses using versatile, "soft" ion sources, such as ES and MALDI, coupled to various types of MS can be performed to detect the expected molecular ions. If there are no matches and/or no structure hypotheses available, then the nondestructive LC-NMR can yield fractions matched with NMR characteristics of target analytes. Alternatively, taking advantage of the sensitivity of MS, >90% of an LC-MS analysis can be diverted to fraction collection. These partially purified fraction(s) can then be analyzed by NMR and/or by either MALDI or nanospray-ES (nanoES) MS. There are at least two commercially available instruments that automate such fraction-collection coupled to MALDI- or nanoES-MS. NanoES is preferred for metabolites because MALDI imparts a very large background from the requisite chemical matrix. Such interfaces between LC and nanoES can simultaneously achieve four of the five analytical emphases of MS, at the expense of analytical speed. The use of ultra-high-resolution MS, for example FT-ICR-MS (Fourier transform ion cyclotron mass spectrometry), with fraction collection nanoES, yields possible molecular formulae on many compounds per sample. Often, the candidate molecular formulae narrow to just a few or even one (Marshall *et al.*, 1998). Once a particular molecular ion is confirmed to be the target analyte, classification can proceed by linking molecular ions with functional groups identified by NMR. A simple example of this is the well-known detection of phosphate groups, consisting of neutral loss of 98 Da. The data mining for phosphate and numerous other biologically prevalent substructures is either already built-in or readily programmed into modern MS software. In the case of derivatized metabolites, this capability can be used to verify the existence and number of derivative tags on the metabolite. This is a traditional technique in GC-MS where examination of the appropriate isotopic ion pattern can yield the number of functional groups present; for example, the deconvolution of the Si isotopic pattern reveals the number of active-hydrogen functional groups when MTBSTFA derivatization is used (Anderson *et al.*, 1986).

A far more complex analysis that can be performed entirely within "trap" MS instruments, such as ion trap MS or ICR-MS, is the generation of "ion trees."

These can yield substructure molecular formula and functional group information important for isotopomer analysis. Under continuous introduction of sample (typically nanoES), an ion trap MS*n* automatically obtains MS3 spectra from each of the ions in the MS2 spectrum. The process is repeated for each of the ions in the MS3 and higher-order spectra until the ion yields fall below detection (Fig. 9). A complete ion tree can be obtained in this fashion with <5 min of sample introduction. The resulting data set can be readily data-mined for a host of computer-recognizable features such as methylene chains, aromatics having various substitution patterns, steroidal or other multiring structures, or functional groups such as the phosphate loss mentioned earlier. Here again, the basic MS*n* operation and data-mining features are supported by software packages, although application to complex metabolomic extracts typically needs in-house custom development of both MS operation and data mining. Several software packages are under development to address these needs.

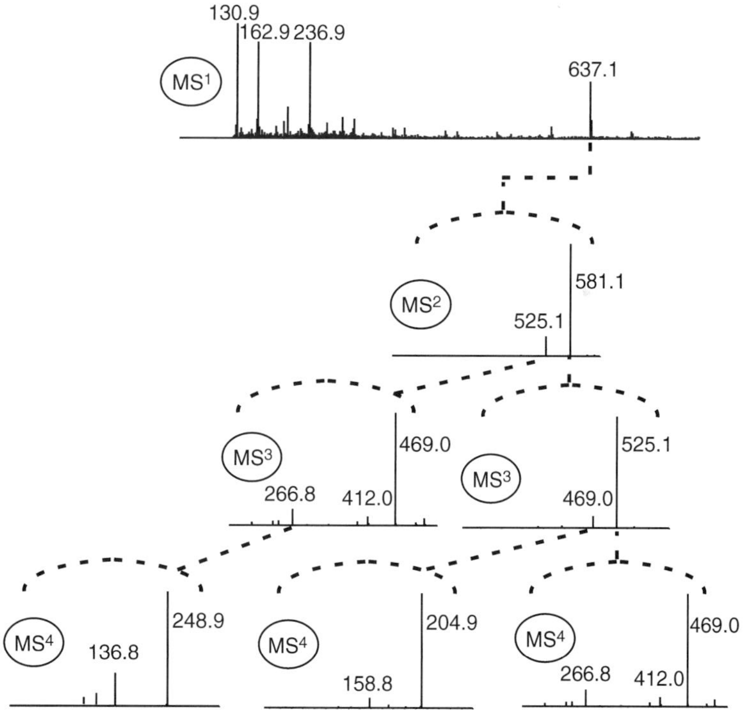

Fig. 9 MS*n* ion tree of an unknown plant metabolite. The spectrum was obtained directly from a crude extract. Under continuous introduction of sample (e.g., AP/MALDI), an ion trap MS*n* automatically obtains MS3 spectra from each of the ions in the MS2 spectrum. The process is repeated for each of the ions in the MS3 and higher-order spectra until the ion yields fall below detection. This simple ion tree was obtained in <30 sec.

In fact, powerful new chemoinformatics tools have been recently released, and others are constantly being developed both by instrument manufacturers as well as third-party vendors, to help with the types of data mining needed for metabolomics. For example, HighChem's Mass Frontier uses known reaction mechanisms in the MS and hypothesized structures to generate *in silico* ion trees to be compared with the actual ion trees. A simple capability of this software is illustrated in Fig. 9. Mass Frontier also has extensive built-in capabilities for linear multivariate (e.g., principal components analysis) and nonlinear (Kohonen, or self-organizing-map neural net) association of mass spectra chemical structures. In essence, spectral databases, such as the 108,000-entry National Institute for Standards and Technology (NIST) Mass Spectral Library or in-house spectral libraries, are initially pattern-related by Mass Frontier to the known chemical structure. An unknown spectrum can then be classified according to its "similarity" along principal component axes, which can suggest the chemical classification, for example, a steroidal structure. The use of Kohonen neural nets is claimed by HighChem to be far more powerful than such simple statistical relations and is also being developed for use in metabolomics structure elucidation. Regardless of the particular claims, there is no doubt as to the great value of annotating actual MS ion trees with chemical structures from *in silico*-generated ion trees. Mass Frontier is just one example of several packages evolving to become powerful biochemoinformatic tools.

Moreover, there are important links between structure elucidation, stable isotope tags, and isotopomer analysis. The presence of stable isotope derivative tags or enrichment by stable isotopes can be vital evidence of the biological validity of the target metabolite. For example, the ^{15}N NMR structure elucidation of Gln initially showed differential labeling at the amino and amido positions, even the most basic GC-MS can verify and quantify this isotopomer pattern (Fan *et al.*, 1997). Sophisticated MS*n* techniques and ion-tree approaches, together with isotopomer-enabled software, such as Mass Frontier, can verify the isotope positional information in far more complex molecules, and use it to assist in structure elucidation of fragments.

2. Concentration Determination

As described above, derivatization is a powerful tool for quantitative metabolomics, especially in cases where authentic standards are not available, or the metabolite is unstable, or to impart physicochemical properties amenable to the MS techniques. For example, single-step reaction of MTBSTFA with exchangeable hydrogen functional groups (e.g., –COOH, –OH, –NHR, –PO_4, –SH) produces stable gas chromatographable *tert*-butyldimethylsilyl esters, which yields characteristic electron ionization (EI) mass spectra allowing deduction of the molecular ion for unknown, unexpected, or otherwise difficult metabolites. In fact, we have combined this GC-MS analysis with both *in vitro* and *in vivo* NMR for metabolite profiling for nearly two decades (Fan *et al.*, 1986). Modern "soft" ionization methods in GC- and LC-tandem MS further improve yield of the molecular weight information that is desirable for tandem MS experiments (see above) for further structural elucidation.

For metabolite quantification, there are numerous derivatization methods available for preparing metabolite derivatives for GC- and LC-tandem MS analysis, as in the case of MTBSTFA discussed above. For instance, the formation of the silyl esters mentioned above also stabilizes otherwise labile structures, significantly improving quantification. Other examples include formation of methyl esters of carboxylates, formation of methoximes of carbonyls for differentiation of tautomeric pairs, alkylsilylation, and a variety of chromogenic and fluorogenic derivatives, just to name a few. As a general guide, commercial catalogs provide an important resource to help choose the appropriate derivatives and often provide extensive literature references. For elucidating metabolic pathways, the focus should be on those derivatization methods that are suitable for stable isotopomer analysis. For LC-based separations, diode-array spectrophotometric or fluorescence quantification can be used with various ion sources for MS analysis. Introduction of isotopomer calibration standards (surrogate) of given metabolites helps to eliminate ionization variability or other artifacts associated with ion sources and other MS performance issues.

Derivatization of both extracts and *in situ* by stable-isotope-labeled reagents is an example of the well-known "mass-tagging" technique (Watson, 1985). In its simplest use, control and treated sample extracts are derivatized with either an unlabeled or a multiply ^{2}H-labeled reagent, respectively. The two extracts are then mixed together and analyzed at once by the appropriate MS technique. Metabolites originating from the treated sample are manifest in the presence of the deuteriated mass tags. Because a given metabolite in both samples is analyzed simultaneously, under identical instrument conditions, the relative quantification between the two samples is excellent. A bonus is that the analytical throughput is doubled.

3. Analysis of Polar Metabolites

As discussed above, GC-MS can be the method of choice for a wide variety of polar metabolites. For stable metabolites with active hydrogen groups (e.g., glyceraldehyde-3-phosphate, amino acids, organic acids, carbohydrates, alcohols, amines, and amides), the silylating agent MTBSTFA in acetonitrile (1:1, v/v) is used to derivatize the metabolites in lyophilized extracts by sonication and the derivatives injected directly into the GC-MS using EI, as routinely performed in our laboratories (Fan *et al.*, 1998). Likewise, many labile metabolites (e.g., PGA1 and selenols) are stabilized by MTBSTFA derivatization and analyzed by GC-MS. Other silyl derivatizations, such as MSTFA, are widely used to achieve somewhat different goals (Fiehn *et al.*, 2000; Nikiforova *et al.*, 2005).

Many compounds such as polyamines, GSSeG, are not compatible with GC-MS analysis even after derivatization due to insufficient volatility and/or thermal instability. Direct infusion (no LC) and LC-MS is often used without derivatization, unless specific detection such as fluorescence is desired. For polar metabolites and LC separations, the choice of ion source is usually ES or nanoES and less frequently atmospheric pressure chemical ionization (APCI). Usually direct infusion (no LC) is not performed with these ion sources due to lack of quantification from complex mixtures, generally due to the phenomenon of "ion suppression." In order to avoid this, the common "cure" is to turn to LC, which strongly limits the number of compounds in the ES source at any given moment. LC separation methods are far too numerous to mention here, but the trend is clearly toward increasingly larger number and classes of metabolites in a single run. In all cases of GC- and LC-MS, authentic standards are included in the crude mixture to calibrate the MS response.

4. Analysis of Phospholipids and Lipid Metabolites

The PL extracts are particularly amenable to analysis by ^{31}P NMR spectroscopy and MALDI with *para*-nitroaniline (PNA) and CsCl as the matrix or ES-MS. ^{31}P NMR analyses are carried out without or with the addition of internal standards, for example, dimyristoyl phosphatidic acid. A similar sequence is performed for the MS analysis without or with the internal standards of short-chain PLs not present in the natural samples, for example SM(6:0) and PC(14:0/14:0). The metabolites of glycero- and sphingophospholipids are readily measured by MALDI-MS. Because the ionization efficiency of diacylglycerols, ceramides, and sphingosines is greater in 2,5-dihydroxybenzoic acid (DHB) than in PNA, the lipid extracts are also analyzed with DHB as matrix. As for the absolute quantification of lipid metabolites, short-chain diacylglycerols, ceramides, and sphingosines not present in these extracts can be added as internal standards. To better resolve lipid positional isomers, the same lipid extracts can be analyzed by direct flow-injection with ES-or nanoES-MS*n* analysis coupled with ion-tree analysis in both [+] and [−] ion modes (Fridriksson *et al.*, 1999). The nanoES is the ion source of choice for detailed studies such as generating ion trees because of its ability to sustain sample introduction for many tens of minutes while keeping sample consumption small (e.g., $<1\ \mu$l) and still achieving high sensitivity.

5. Isotopomer Analysis

Many MS applications have been reported using stable isotopes, including various ^{13}C isotopomers of glucose, deuterium, and N-15 (Birkemeyer *et al.*, 2005; Boros, 2005; Des Rosiers *et al.*, 2004; Fan *et al.*, 2003; Lee *et al.*, 1998; Marin *et al.*, 2004; Mashego *et al.*, 2004; Turner *et al.*, 2003; Vogt *et al.*, 2005). The basic principle is that enrichment within a molecule increases the mass by $n \times dm$ where n is the number of atoms substituted and dm is the mass difference between the natural abundance state and the enriched molecule. This causes a change in the isotope distribution of the analyte so that the number of different isotopomers produced can be determined (Hellerstein and Neese, 1999; Turner and Hellerstein, 2005; Turner *et al.*, 2005; van Winden *et al.*, 2002; Wahl *et al.*, 2004).

Figure 10 shows isotopomers in rhabdomyosarcoma cells grown on [U-^{13}C]-glucose, and the derivatized cell extract separated and detected by GC-MS. The figure shows how the enrichment of ^{13}C in particular molecules affects the isotope distribution. For example, the lactate shows an enhancement of the $m + 3$ peak compared with the natural abundance case, but no increase of the $m + 2$ and $m + 1$ isotopomers, implying that there are only two pools of lactate, namely natural

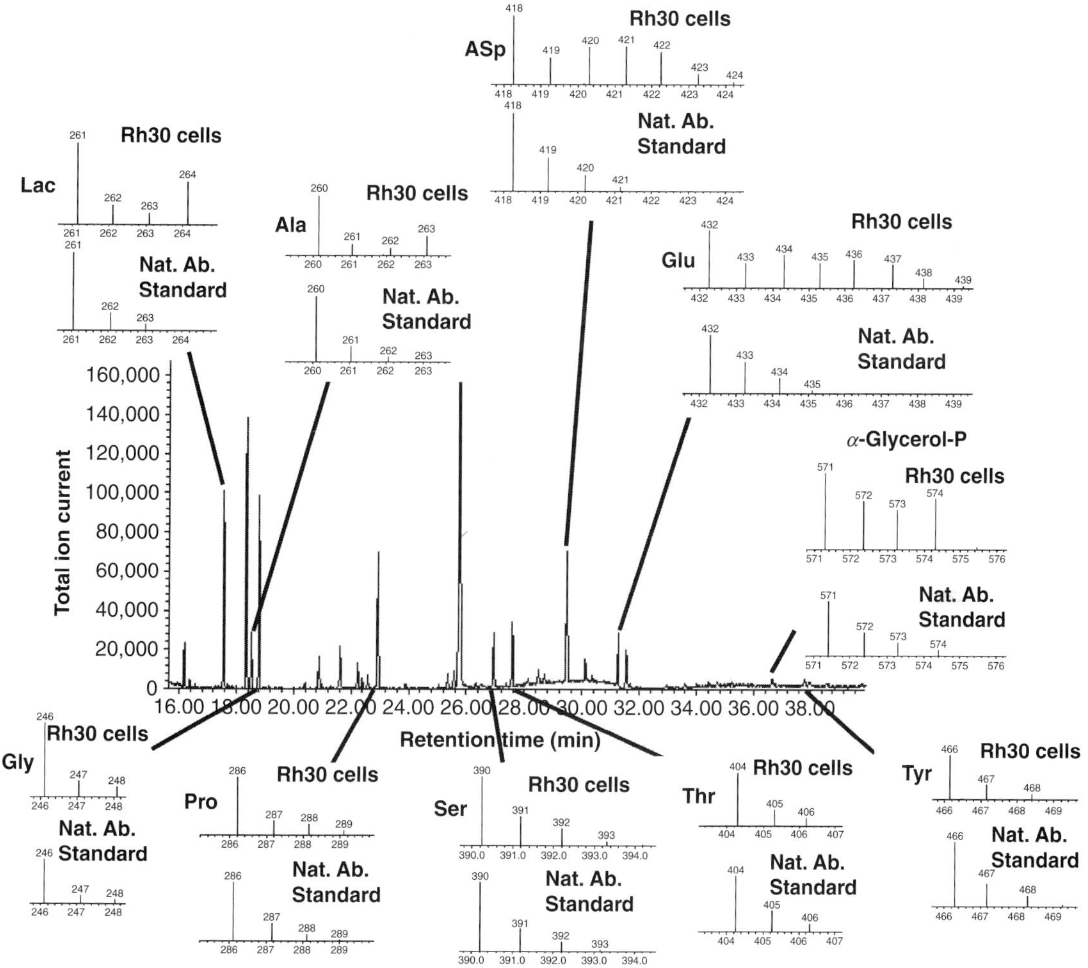

Fig. 10 MS and isotopomer patterns in metabolites from rhabdomyosarcoma cells. Extracts of cells as in Figs. 4 and 5 were derivatized with MTBSTFA as described by Fan *et al.* (1986). Individual MS spectra of resolved peaks in the chromatogram were identified, and the mass isotopomer distributions were compared with those of unlabeled authentic standards.

abundance (not metabolically enriched) and lactate in which all three carbons are labeled in the same molecule. This is in agreement with the NMR analyses (cf. Fig. 5). Alanine shows a similar pattern to lactate, as both derive from pyruvate. Other amino acids, however, show more complex patterns, including glutamate, which displays a large number of isotopomers, including $m + 2$ to $m + 5$. The complementarity of using MS and NMR is illustrated by amino acids such as Ser, which is generally difficult to quantify by NMR. In Fig. 10, Thr as well as Ser showed a pattern indistinguishable from the natural abundance pattern implying no metabolic labeling from glucose. As Thr is an essential amino acid for mammalian cells, this result is fully expected, and therefore acts as an internal control for the isotopomer analysis.

To speciate the positional isotopomers (i.e., determine which atoms are actually labeled) requires additional analysis by ion-tree MS*n*. Because of the time needed to perform MS*n*—many seconds to a few minutes—this is not compatible with GC or LC. As discussed above, the ion source choice with ion-tree analysis is therefore nanoES.

The various MS tools are generally considered adjunctive, given the power of NMR in isotopomer analysis. However, even the simplest GC-MS has long supported fruitful isotopomer work (Anderson *et al.*, 1986; Christensen and Nielsen, 1999; Dauner and Sauer, 2000; Fan *et al.*, 2003; Lee *et al.*, 1998), although we previously always coupled such analysis with 2-D NMR (Fan *et al.*, 1997, 2001). The main advantage of performing isotopomer analysis by MS is high sensitivity, especially for directly quantifying "non-NMR-observable" isotopes such as ^{12}C and ^{14}N, which are important for obtaining the extent of enrichment. Structure elucidation of fragments in the MS is critical to determining the position of the labels. In Fig. 11, it is apparent that some fragments contain only the tertiary (ring) N while others have only the secondary amino N, which enables positional isotopomer analysis. For full control over the production of isotopomer fragments, it is best to use ion-tree MS*n* in continuous-introduction modes such as nanoES, combined with interpretation from *in silico* ion trees can elucidate the label positions of isotopomers.

Elegant GC-MS analysis of glucose metabolism has been carried out by Boros and coworkers (Lee *et al.*, 1998; Vizan *et al.*, 2005). Using the isotopomer [1,2-$^{13}C_2$]-glucose, they were able to determine the relative fluxes through the oxidative and nonoxidative branches of the pentose phosphate pathway, and further used this method to probe the effect of different K-ras mutants on the pentose phosphate pathway and glycolytic flux in NIH3T3 cells.

6. Elemental Analysis

Elemental analysis is important not only for structure determination but also for metabolomics. Metal ions, for example, are important components of enzymes and the cellular milieu, as are several other essential elements (other than C, H, and O) in biochemical form, such as P, S, and Se. The ICP ion source for MS heats liquid,

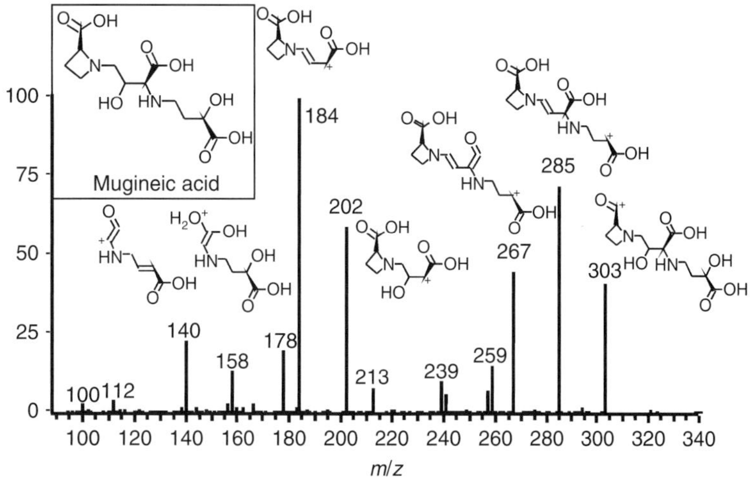

Fig. 11 MS/MS spectrum of mugineic acid from a crude extract. The root exudate was collected and treated as described by Fan *et al.* (1997). The mass spectra are annotated with Mass Frontier chemoinformatics interpretation. Fragment ion structures were assigned to every one of the major (>5% intensity) fragments, using reaction mechanism-based pathways.

aerosol, or gas stream to >6500 K, sufficient to dissociate chemical structures completely into elemental ions. These ions are detected by a quadrupole or magnetic sector MS. Analysis is not always straightforward, as formation of "unusual" polyatomic interferences, such as Ar_2 can interfere with the elemental ions. This problem is dealt with by collision/reaction cells to remove the polyatomic species or by high-resolution MS to avoid the interferences.

The ability to quantitatively broad screen elements in complex samples can provide key information about the changes on metal content of critical proteins related to redox stress or changes in seleno protein activity, for example (Fan *et al.*, 2002; Larsen *et al.*, 1997, 2001, 2003; Sloth *et al.*, 2003).

An interesting development using an ICP-MS as the elemental detector system is in the area of proteomics in which a 2-D gel can be scanned with a UV laser, and the ablated aerosol is introduced into the ICP-MS (Fan *et al.*, 2002). This LA-ICP-MS method makes it possible to determine the elemental composition of individual protein spots, such as to define the selenoproteome (Fan *et al.*, 2002), detect phosphorylation, or the presence of other elements.

7. Isotope Ratio MS for Global Isotope Balance

Organic MS has been shown to give adequate quantitative analysis of isotope incorporation into selected compounds (Lee *et al.*, 1998). However, under conditions of high isotopic dilution, such as isotopes distributing throughout the body fluids, the ability to determine quantitatively very small degrees of enrichment becomes paramount. Neither NMR nor most MS methods have sufficient analytical precision to determine small variations of isotope enrichment at low total isotope levels. However, the isotope ratio MS (irMS) is designed specifically for this purpose and can readily discriminate between very small degrees of enrichment to better than 0.01 per mil (0.001%). These spectrometers have been long used in geology, forensic, agriculture, food and beverage, quality control, doping, pharmaceuticals or medical, and diagnostics industries.

The precision of irMS arises from converting organic compounds to CO_2 or N_2O, which the irMS is designed to measure. Unlike "organic" MS, it is not intrinsically capable of chemical structure determination, thus chemical speciation is entirely dependent on prior procedures such as GC or LC. Important applications of this technique include CO_2 respiration from cells or tissue. The $^{13}CO_2$ that is released from oxidative metabolism accounts for a significant fraction of ingested carbon, and the $^{13}C/^{12}C$ ratio is a sensitive indicator of pathways when different ^{13}C-labeled precursors are used (Yang *et al.*, 2004, 2005). Furthermore, as part of isotope inventories (see below), macromolecule sinks such as proteins and nucleic acids are difficult to analyze quantitatively by NMR or MS, yet are likely to account for a significant amount of ingested stable isotope if the cells are supplied with [U-^{13}C]-glucose, for example. For such analyses, irMS is the means to obtain the total ^{13}C incorporation into the major metabolic fractions following suitable extraction procedures.

8. Isotope Inventory

In an ideal world, it would be desirable to obtain a complete inventory of the supplied isotopes, that is, how much of the precursors is utilized, and what fraction of the total is converted into each molecules. This is very difficult, and there have been relatively few attempts at a global analysis of this nature. For example, using ^{14}C isotope tracing to determine the uptake and conversion of labeled precursors by MCF-7 breast cancer cells in culture, it was possible to relate these observations to rate of ATP synthesis and oxygen consumption (Guppy *et al.*, 2002). They concluded that as much as 65% of the ATP production arose from oxidation of sources other than glucose, glutamine, and fatty acids. However, a common feature of cancer cells is that they convert a high fraction of glucose to lactate, which is secreted into the medium. It is straightforward to determine the amount of labeled glucose consumed from the medium, and how much label is secreted into the medium in the form of lactate and alanine, for example, by either NMR or MS. It is commonly observed that cancer cells in culture, and also tumors *in vivo*, have an enhanced rate of glycolysis, resulting in secretion of a large amount of lactate and protons (Helmlinger *et al.*, 2002; Petch and Butler, 1994; Portais *et al.*, 1993; Sanfeliu *et al.*, 1997; Vriezen and van Dijken, 1998). Measuring the time course is rather valuable, as the consumption/production rates give valuable information about the metabolic demands of the cells under defined conditions (Artemov *et al.*, 1998).

In cell culture, the amount of biomass is small compared with the extracellular pool of metabolites. Thus, the fraction, *f*, of ^{13}C glucose that is converted to lactate (or other secreted metabolites) can be easily determined as

$$f = \frac{\text{No. moles } ^{13}\text{C lactate produced}}{2 \times \text{no. moles } ^{13}\text{C Glucose consumed}} \tag{5}$$

The factor 2 arises because, in principle, two molecules of lactate can be produced from one molecule of glucose. The complement, $1 - f$, reflects the glucose consumed by the cells or tissue that is converted into all other components, which may be biomass as well as free metabolites. A *significant* sink of glucose carbon, at least in cancer cells, appears to be protein (which represents ~20% of the wet weight of cells) followed by lipids and complex carbohydrates (including glycoproteins). The smallest sink is likely to be the DNA fraction as it accounts for a small amount of the cell weight (0.2%). As described above, irMS is probably the best general technique for determining the incorporation of label into major biochemical classes and pools (Hellerstein and Neese, 1999). Although NMR can be used for protein and lipids, it is a less reliable technique for total label determination, though may be valuable for measuring positional isotopomer enrichment in individual amino acids in protein digests.

V. Discussion

The combination of NMR, various MS techniques with isotope tracing methods, promises to give very detailed metabolic information of cells. Acquiring the data is in itself not the major bottleneck to progress rather the functional analysis of the data sets, and the experimental design. Once metabolic fluxes have been measured experimentally, the hard part is in the modeling to reflect the biological reality. This is often considered part of systems biology. There are several main approaches, ranging from the global modeling of an entire organismal network (Weitzke and Ortoleva, 2003) in which the differential equations for the enzyme catalyzed reaction are explicitly solved numerically (Maher *et al.*, 2003; Marin *et al.*, 2004; Mendes and Kell, 2001; Orosz *et al.*, 2003). This approach is very demanding and requires very careful experimental design, including a chemostat and multiple labels (Mollney *et al.*, 1999; Vriezen and van Dijken, 1998). Such ambitious technologies have been applied mainly to microorganisms, which can be manipulated readily (Arita, 2003).

For other cases, an empirical, *ad hoc* approach is generally used in which specific areas of metabolism, usually assumed to be known, are simulated, and the simulations compared with the data sets (Boros *et al.*, 2002; Lu *et al.*, 2002; Maher *et al.*, 2003; Marin *et al.*, 2004; Mason and Rothman, 2004). At present, this appears to be the only hope for mammalian systems owing to the enormous complexity not only at the cellular level but also because of the reciprocal interactions with neighboring cells and the transport of nutrients to cells. Even with single cells in suspension culture, the assumption that clusters of reactions within the cell are functionally uncoupled from other clusters may not be valid (Arita, 2004b). A further major challenge is the integration of metabolomics information with that obtained from genomics and proteomics, as it is clear too that these levels cannot be functionally independent (Fan, 2005; Fan *et al.*, 2004a; Hirai *et al.*, 2004). The ability to project metabolomics data onto pathway maps is an important part of understanding the results (and also in the design of new experiments) (Jenkins *et al.*, 2004; Lange and Ghassemian, 2005). Such tools continue to be developed (Karp *et al.*, 1999; Krieger *et al.*, 2004; Romero and Karp, 2003, 2005; Zhang *et al.*, 2005) and include the ability to work at the atomic level of representation (Arita, 2003, 2004a,b; Hirai *et al.*, 2004), which is essential for isotopomer-based metabolomics.

In addition to the large effort being put into streamlining data collection and analysis, the computational aspects are a major growth area, and ultimately we can expect to see convergence of the "omics" to bring deeper understanding of biological regulation, and how errors in regulation lead to disease states.

Short Glossary of Terms

Chemical Shift: Relative resonant frequency, defined (in ppm) as $10^6 (\nu_{obs} - \nu_{ref})/\nu_{Fref}$. Depends on chemistry, such as type of atom, nearest neighbors, bonding, and therefore is characteristic of local functional groups. It is also very sensitive to inductive effects from nearby ionizable groups such as carboxyls and amino functions.

COSY: **CO**rrelation **S**pectroscop**Y:** Usually refers to correlation of resonances of scalar-coupled nuclei, used for determining through-bond interactions within a covalent network. Defines molecular fragments via proton–proton connectivity.

Coupling Constant: Interaction strength between two scalar- (through-bond) coupled magnetic nuclei. The scalar coupling constant (denoted J) is a measure of nuclear interactions transmitted through the bonding network. Three-bond J values provide information on the torsion angle.

Dipolar Coupling: Interaction between magnetic nuclei via dipole–dipole interactions. This depends on the inverse sixth power of the internuclear separation and the intrinsic magnetic strength of the two nuclei.

FID (Free Induction Decay): Rotating magnetization induces a voltage in the receiver coil (magnetic induction), which decays exponentially by *relaxation processes*. The time-dependent voltage is digitized and converted into a frequency response by *Fourier transformation* (qv). The two signals in quadrature are

$$S(\text{real}) = a_0 \exp(-t/R_2)\cos(\omega t)$$
$$S(\text{imaginary}) = a_0 \exp(-t/R_2)\sin(\omega t)$$

Gas Chromatography-MS (GC-MS): Complex mixtures of volatile compounds are separated by gas-liquid chromatogahy and analyzed by MS, usually at unit mass resolution to obtain a pseudomolecular ion. Nonvolatile analytes can be made volatile by a variety of derivatization agents. With electron impact (EI) ionization, the technique is quantitative and reproducible.

Fourier Transform (FT): Mathematical trick for converting one periodic function into another function. In NMR, the raw data comprise an exponentially decaying sum of cosinusoidal functions. The FT converts time into frequency. The decaying part of the signal becomes manifest as a broadening of the resonance. The FT of a single resonance is

$$\mathrm{Re}\left\{a_0 \int \mathrm{d}t\, \exp(-i\Omega t)\, \exp\left(i\omega - \frac{1}{T}\right)t\right\} = \left(\frac{2a_0}{T}\right) \Big/ \left[(\Omega - \omega)^2 + \frac{1}{T_2}\right]$$

width at half-height $L = 1/\pi T$. T is the decay time constant (also called "T_2").

Gyromagnetic ratio, γ is a property of nuclei that determines the strength of the nuclear magnetic moment and is characteristic for each isotope. The fundamental frequency is proportional to the product of γ and the applied magnetic field strength.

HMQC: (**H**eteronuclear **M**ultiple **Q**uantum **C**oherence): Formation of multiple-quantum coherence via scalar coupling between nuclei, for example, C–H or N–H occurs in any multipulse experiment. The coherence common evolution of the coupled spins is not directly observable, but its effects are detectable on reconversion to SQC. Such experiments are commonly used in inverse detection of low γ nuclei. MQC also has different relaxation properties (involves terms at higher frequencies) than SQC.

HSQC: **H**eteronuclear **S**ingle **Q**uantum **C**oherence. Homologue of HMQC, but only single quantum coherences are selected. This experiment requires more pulses than HMQC but has more favorable relaxation properties and is the experiment of choice for ^{15}N-H systems.

HSQC-TOCSY and HMQC-TOCSY: Combination of two experiments that can be used in 3-D mode or as 2-D. The former correlates a heteronucleus (e.g., ^{13}C with the directly attached proton, and other protons that are in the same molecular fragment). This is useful for assignments of compounds in complex mixtures when some atoms are enriched or otherwise rare (e.g., ^{15}N in amino or amido compounds or ^{31}P in phosphorylated compounds). HCCH-TOCSY specifically selects for two or more directly bonded heteroatoms such as ^{13}C in molecular fragments of the type HC–CH, and the protons are correlated with one another. This experiment edits complex spectra and shows where multiply labeled compounds retain their bonds or of new bonds are formed between labeled compounds.

Inverse detection: Detection of a "weak" X nucleus (low γ) via a strong nucleus (usually ^{1}H). Usually applies to proton detection of ^{13}C, ^{15}N. Requires scalar coupling between the X nucleus and the ^{1}H atom, generally (but not exclusively) by one bond. The sensitivity enhancement (SNR) possible is given by the ratio $(\gamma_H/\gamma_X)^{5/2}$. This is >300-fold for ^{15}N.

Ion Trap and Ion Cyclotron Resonance (ICR)-MS: Ions are trapped in a magnetic field and can be stored at precise numbers (e.g, ion trap) or made to circulate (ICR). The frequency in the ICR can be measured with very high precision, giving rise to ultrahigh mass resolution (ppm). Ultrahigh mass resolution can often be used to identify metabolites by mass alone. ICR also uses the Fourier Transform to convert from frequency to mass.
Isotopomers are versions of the same compound in which differ by the distribution of isotopes of each element. In a 3-carbon compounds, such as lactate, there are eight possible stable isotopomers (^{12}C and ^{13}C) corresponding to ^{12}C-3, ^{13}C-3, and three each of ^{12}C-1, ^{13}C-2, and ^{13}C-1^{12}C-2. In NMR, these correspond to eight distinguishable compounds. In MS, there are only four mass isotopomers $m0$, $m + 1$, $m + 2$, and $m + 3$ because the three isotopomers of $m + 1$ and $m + 2$ have identical mass.
Liquid Chromatography-MS (LC-MS): Usually HPLC interfaced to a mass detector. Resolves nonvolatile analyses in mixtures for MS analysis.
MALDI (Matrix Assisted Laser Desorption Ionization): Ionization method that involves transfer of excitation energy from an absorbing species to the analytes of interest embedded in the organic matrix. Less sensitive to salt than electrospray but less suited to low molecular weight analytes where the molecules comprising the matrix interfere.
NOE (Nuclear Overhauser Enhancement): Increase in magnetization of a spin when the magnetization of a neighboring spin is perturbed from equilibrium. The NOE depends on r^{-6} and is therefore the primary source of distance information within a molecule.
NOESY: (**N**uclear **O**verhauser **E**nhancement **S**pectroscop**Y**): 2-D experiment to measure NOE effects.
Relaxation Time: Characteristic time for the return of bulk magnetization to the equilibrium value. Comes in two main flavors, T_1, spin-lattice relaxation and T_2, spin-spin relaxation. T_1 is the relaxation time that described return of z-magnetization and is a process that is associated with changes in enthalpy. T_2 is a loss of phase coherence among spins in the x-y plane and is an entropic process.
Rotating Frame: Larmor frequencies are in the hundreds of megahertz and vary by a few tens of kilohertz for similar spins. It is convenient to remove this high-frequency rotation from consideration by working in a frame rotating at the Larmor frequency. In this frame, the spins appear stationary.
ROESY: (**R**otating frame **O**verhauser **E**nhancement **S**pectroscop**Y**): ROEs are positive for all molecular correlation times. Exchange has opposite sign, and so this experiment is useful for distinguishing exchange reactions from dipolar interactions.
Scalar Interactions: Through-bond interactions. Nuclei interact via bonding electrons, which transmit the information about nuclear spin state through the covalent network. This gives rise to the coupling constant (qv) which is a measure of the interaction strength.

Spin: Property of particles that has a classical analogue (a spinning top has spin angular momentum etc.). In the original Schrödinger wave mechanics, there were only three quantum numbers, and no spin. Spin arises naturally as a consequence of a relativistic treatment of wave mechanics (cf. P.A.M. Dirac). Spin is also quantized—only discrete values are possible. For the common case, the spin quantum number is 1/2, that is, there are but two spin states (up and down, α andβ). Magnetic nuclei are also referred to as spins.

TOCSY: (**T**otal **C**orrelation **S**pectroscopy): Also known as HOHAHA. Provides (scalar) correlations within an entire spin system. Especially useful for identification of compounds in mixtures, and also for quantitative analysis of isotopomer distributions.

Acknowledgments

This work was supported in part by grants from the Susan G. Komen Foundation Grant BCTR0503648, Kentucky Science and Engineering Foundation, KSEF-296-RDE-3, NSF EPSCoR EPS-0447479, NCI 1RO1 CA101199, and the Brown Foundation.

References

Anderson, L. W., Zaharevitz, D. W., and Strong, J. M. (1987). Glutamine and glutamate—Automated quantification and isotopic enrichments by gas-chromatography mass-spectrometry. *Anal. Biochem.* **163,** 358–368.

Anousis, N., Carvalho, R. A., Zhao, P. Y., Malloy, C. R., and Sherry, A. D. (2004). Compartmentation of glycolysis and glycogenolysis in the perfused rat heart. *NMR Biomed.* **17,** 51–59.

Arita, M. (2003). In silico atomic tracing by substrate-product relationships in *Escherichia coli* intermediary metabolism. *Genome Res.* **13,** 2455–2466.

Arita, M. (2004a). Computational resources for metabolomics. *Brief. Funct. Genomic. Proteomic.* **3,** 84–93.

Arita, M. (2004b). The metabolic world of *Escherichia coli* is not small. *Proc. Natl. Acad. Sci. USA* **101,** 1543–1547.

Artemov, D., Bhujwalla, Z. M., Pilatus, U., and Glickson, J. D. (1998). Two-compartment model for determination of glycolytic rates of solid tumors by *in vivo* C-13 NMR spectroscopy. *NMR Biomed.* **11,** 395–404.

Bederman, I. R., Reszko, A. E., Kasumov, T., David, F., Wasserman, D. H., Kelleher, J. K., and Brunengraber, H. (2004). Zonation of labeling of lipogenic acetyl-CoA across the liver—Implications for studies of lipogenesis by mass isotopomer analysis. *J. Biol. Chem.* **279,** 43207–43216.

Birkemeyer, C., Luedemann, A., Wagner, C., Erban, A., and Kopka, J. (2005). Metabolome analysis: The potential of *in vivo* labeling with stable isotopes for metabolite profiling. *Trends Biotechnol.* **23,** 28–33.

Bollard, M. E., Stanley, E. G., Lindon, J. C., Nicholson, J. K., and Holmes, E. (2005). NMR-based metabonomic approaches for evaluating physiological influences on biofluid composition. *NMR Biomed.* **18,** 143–162.

Boren, J., Cascante, M., and Marin, S., (2001). Gleevec (ST1571) influences metabolic enzyme activities and glucose carbon flow toward nucleic acid and fatty acid synthesis in myeloid tumor cells. *J. Biol. Chem.* **276,** 37747–37753.

Boros, L. G. (2005). Metabolic targeted therapy of cancer: Current tracer technologies and future drug design strategies in the old metabolic network. *Metabolomics* **1,** 11–15.

Boros, L. G., Cascante, M., and Lee, W. N. P. (2002). Metabolic profiling of cell growth and death in cancer: Applications in drug discovery. *Drug Discov. Today* **7,** 364–372.

Brennan, L., Corless, M., Hewage, C., Malthouse, J. P. G., McClenaghan, N. H., Flatt, P. R., and Newsholme, P. (2003). C-13 NMR analysis reveals a link between L-glutamine metabolism, D-glucose metabolism and gamma-glutamyl cycle activity in a clonal pancreatic beta-cell line. *Diabetologia* **46,** 1512–1521.

Burgess, S. C., Carvalho, R. A., Merritt, M. E., Jones, J. G., Malloy, C. R., and Sherry, A. D. (2001). C-13 isotopomer analysis of glutamate by J-resolved heteronuclear single quantum coherence spectroscopy. *Anal. Biochem.* **289,** 187–195.

Carvalho, R. A., Zhao, P., Wiegers, C. B., Jeffrey, F. M. H., Malloy, C. R., and Sherry, A. D. (2001). TCA cycle kinetics in the rat heart by analysis of C-13 isotopomers using indirect H-1 C-13 detection. *Am. J. Physiol. Heart Circ. Physiol.* **281,** H1413–H1421.

Cavanagh, J., Fairbrother, W. J., Palmer, A. G., and Skelton, A. G. N. J. (1996). "Protein NMR Spectroscopy Principles and Practice." Academic Press, San Diego.

Christensen, B., and Nielsen, J. (1999). Isotopomer analysis using GC-MS. *Metab. Eng.* **1,** 282–290.

Claridge, T. D. W. (1999). "High-Resolution NMR Techniques in Organic Chemistry." Elsevier, San Diego.

Cline, G. W., LePine, R. L., Papas, K. K., Kibbey, R. G., and Shulman, G. I. (2004). C-13 NMR isotopomer analysis of anaplerotic pathways in INS-1 cells. *J. Biol. Chem.* **279,** 44370–44375.

Cordier, F., Dingley, A. J., and Grzesiek, S. (1999). A doublet-separated sensitivity-enhanced HSQC for the determination of scalar and dipolar one-bond J-couplings. *J. Biomol. NMR* **13,** 175–180.

Craig, A., Cloareo, O., Holmes, E., Nicholson, J. K., and Lindon, J. C. (2006). Scaling and normalization effects in NMR spectroscopic metabonomic data sets. *Anal. Chem.* **78,** 2262–2267.

Dang, D. T., Knock, S. A., Chen, S. W., Chen, F., and Dang, L. H. (2005). HIF-1 alpha mediates aerobic glycolyis (the Warburg effect) and normoxic growth in colorectal cancer cells. *Gastroenterology* **128,** A479–A479.

Dauner, M., and Sauer, U. (2000). GC-MS analysis of amino acids rapidly provides rich information for isotopomer balancing. *Biotechnol. Prog.* **16,** 642–649.

Des Rosiers, C., Lloyd, S., Comte, B., and Chatham, J. C. (2004). A critical perspective of the use of C-13-isotopomer analysis by GCMS and NMR as applied to cardiac metabolism. *Metab. Eng.* **6,** 44–58.

Diraison, F., Yankah, V., Letexier, D., Dusserre, E., Jones, P., and Beylot, M. (2003). Differences in the regulation of adipose tissue and liver lipogenesis by carbohydrates in humans. *J. Lipid Res.* **44,** 846–853.

Ernst, R. R., Bodenhausen, G., and Wokaun, A. (1990). "Principles of Nuclear Magnetic Resonance in One and Two Dimensions." Clarendon Press, Oxford.

Fan, T. W.-M. (1996a). Metabolite profiling by one- and two-dimensional NMR analysis of complex mixtures. *Prog. Nucl. Magn. Reson. Spectrosc.* **28,** 161–219.

Fan, T. W.-M. (1996b). "Recent Advancement in Profiling Plant Metabolites by Multi-Nuclear and Multi-Dimensional NMR." American Society of Plant Physiologists, Rockville, MD.

Fan, T. W.-M., Bandura, L. L., Lane, A. N., and Higashi, R. M. (2005b). Metabolomics-edited transcriptomics analysis of Se anticancer action in human lung cancer cells. *Metabolomics* **1,** 325–339.

Fan, T. W.-M., Clifford, A. J., and Higashi, R. M. (1994). *In vivo* 13C NMR analysis of acyl chain composition and organization of perirenal triacylglycerides in rats fed vegetable and fish oils. *J. Lipid Res.* **35,** 678–689.

Fan, T. W.-M., Colmer, T. D., Lane, A. N., and Higashi, R. M. (1993). Determination of metabolites by H-1-Nmr and Gc—analysis for organic osmolytes in crude tissue-extracts. *Anal. Biochem.* **214,** 260–271.

Fan, T. W.-M., Higashi, R. M., Frenkiel, T. A., and Lane, A. N. (1997). Anaerobic nitrate and ammonium metabolism in flood-tolerant rice coleoptiles. *J. Exp. Botany* **48,** 1655–1666.

Fan, T. W.-M., Higashi, R. M., and Lane, A. N. (1998). Biotransformations of selenium oxyanion by filamentous cyanophyte-dominated mat cultured from agricultural drainage waters. *Environ. Sci. Technol.* **32,** 3185–3193.

Fan, T. W.-M., Higashi, R. M., Lane, A. N., and Jardetzky, O. (1986). Combined use of H-1-NMR and GC-Ms for metabolite monitoring and *in vivo* H-1-NMR assignments. *Biochim. Biophys. Acta* **882,** 154–167.

Fan, T. W.-M., Lane, A. N., and Higashi, R. M. (2003). *In vivo* and *in vitro* metabolomic analysis of anaerobic rice coleoptiles revealed unexpected pathways. *Russ. J. Plant Physiol.* **50,** 787–793.

Fan, T. W.-M., Lane, A. N., and Higashi, R. M. (1992). Hypoxia does not affect rate of Atp synthesis and energy-metabolism in rice shoot tips as measured by P-31 NMR *in vivo*. *Arch. Biochem. Biophys.* **294,** 314–318.

Fan, T. W.-M., Lane, A. N., and Higashi, R. M. (2004a). The promise of metabolomics in cancer molecular therapeutics. *Curr. Opin. Mol. Ther.* **6,** 584–592.

Fan, T. W.-M., Lane, A. N., and Higashi, R. M. (2004b). An electrophoretic profiling method for thiol-rich phytochelatins and metallothioneins. *Phytochem. Anal.* **15,** 175–183.

Fan, T. W.-M., Lane, A. N., Pedler, J., Crowley, D., and Higashi, R. M. (1997). Comprehensive analysis of organic ligands in whole root exudates using nuclear magnetic resonance and gas chromatography-mass spectrometry. *Anal. Biochem.* **251,** 57–68.

Fan, T. W.-M., Lane, A. N., Shenker, M., Bartley, J. P., Crowley, D., and Higashi, R. M. (2001). Comprehensive chemical profiling of gramineous plant root exudates using high-resolution NMR and MS. *Phytochemistry (Oxford)* **57,** 209–221.

Fan, T. W.-M., Pruszkowski, E., and Shuttleworth, S. (2002). Speciation of selenoproteins in Se-contaminated wildlife by gel electrophoresis and laser ablation-ICP-MS. *J. Anal. At. Spectrom.* **17,** 1621–1623.

Fernie, A. R., Trethewey, R. N., Krotzky, A. J., and Willmitzer, L. (2004). Metabolite profiling: From diagnostics to systems biology. Nature reviews. *Mol. Cell Biol.* **5,** 1–7.

Fiehn, O., Kopka, J., Dormann, P., Altmann, T., Trethewey, R. N., and Willmitzer, L. (2000). Metabolite profiling for plant functional genomics. *Nat. Biotechnol.* **18,** 1157–1161.

Foxall, P. J. D., Parkinson, J. A., Sadler, I. H., Lindon, J. C., and Nicholson, J. K. (1993). Analysis of biological-fluids using 600 Mhz proton NMR-spectroscopy—application of homonuclear 2-dimensional J-resolved spectroscopy to urine and blood-plasma for spectral simplification and assignment. *J. Pharm. Biomed. Anal.* **11,** 21–31.

Fridriksson, E. K., Shipkova, P. A., Sheets, E. D., Holowka, D., Baird, B., and McLafferty, F. W. (1999). Quantitative analysis of phospholipids in functionally important membrane domains from RBL-2H3 mast cells using tandem high-resolution mass spectrometry. *Biochemistry* **38,** 8056–8063.

Garber, K. (2004). Energy boost: The Warburg effect returns in a new theory of cancer. *J. Natl. Cancer Inst.* **96,** 1805–1806.

Goddard, A. W., Mason, G. F., Appel, M., Rothman, D. L., Gueorguieva, R., Behar, K. L., and Krystal, J. H. (2004). Impaired GABA neuronal response to acute benzodiazepine administration in panic disorder. *Am. J. Psychiatry* **161,** 2186–2193.

Gradwell, M. J., Fan, T. W. M., and Lane, A. N. (1998). Analysis of phosphorylated metabolites in crayfish extracts by two-dimensional 1H-31P NMR heteronuclear total correlation spectroscopy (hetero TOCSY). *Anal. Biochem.* **263,** 139–149.

Guppy, M., Leedman, P., Zu, X., and Russell, V. (2002). Contribution by different fuels and metabolic pathways to the total ATP turnover of proliferating MCF-7 breast cancer cells. *Biochem. J.* **364,** 309–315.

Haberg, A., Qu, H., Bakken, I. J., Sande, L. M., White, L. R., Haraldseth, O., Unsgard, G., Aasly, J., and Sonnewald, U. (1998). *In vitro* and *ex vivo* C-13-NMR spectroscopy studies of pyruvate recycling in brain. *Dev. Neurosci.* **20,** 389–398.

Harrigan, G. G., Brackett, D. J., and Boros, L. G. (2005). Medicinal chemistry, metabolic profiling and drug target discovery: A role for metabolic profiling in reverse pharmacology and chemical genetics. *Mini-Rev. Med. Chem.* **5,** 13–20.

Heijne, W. H. M., Lamers, R., van Bladeren, P. J., Groten, J. P., van Nesselrooij, J. H., and van Ommen, B. (2005). Profiles of metabolites and gene expression in rats with chemically induced hepatic necrosis. *Toxicol. Pathol.* **33,** 425–433.

Hellerstein, M. K., and Neese, R. A. (1999). Mass isotopomer distribution analysis at eight years: Theoretical, analytic, and experimental considerations. *Am. J. Physiol. Endocrinol. Metabol.* **276,** E1146–E1170.

Helmlinger, G., Schell, A., Dellian, M., Forbes, N. S., and Jain, R. K. (2002). Acid production in glycolysis-impaired tumors provides new insights into tumor metabolism. *Clin. Cancer Res.* **8,** 1284–1291.

Henry, P. G., Oz, G., Provencher, S., and Gruetter, R. (2003). Toward dynamic isotopomer analysis in the rat brain *in vivo*: Automatic quantitation of C-13 NMR spectra using LCModel. *NMR Biomed.* **16,** 400–412.

Hirai, M. Y., Yano, M., Goodenowe, D. B., Kanaya, S., Kimura, T., Awazuhara, M., Arita, M., Fujiwara, T., and Saito, K. (2004). Integration of transcriptomics and metabolomics for understanding of global responses to nutritional stresses in Arabidopsis thaliana. *Proc. Natl. Acad. Sci. USA* **101,** 10205–10210.

Homans, S. W. (1992). "A Dictionary of Concepts in NMR." Clarendon Press, Oxford.

Jenkins, H., Hardy, N., Beckmann, M., Draper, J., Smith, A. R., Taylor, J., Fiehn, O., Goodacre, R., Bino, R. J., Hall, R., Kopka, J., Lane, G. A.,, *et al.* (2004). A proposed framework for the description of plant metabolomics experiments and their results. *Nat. Biotechnol.* **22,** 1601–1606.

Karp, P. D., Riley, M., Paley, S. M., Pellegrini-Toole, A., and Krummenacker, M. (1999). Eco Cyc: Encyclopedia of *Escherichia coli* genes and metabolism. *Nucleic Acids Res.* **27,** 55–58.

Katayama, M., Nakane, R., Matsuda, Y., Kaneko, S., Hara, I., and Sato, H. (1998). Determination of progesterone and 17-hydroxyprogesterone by high performance liquid chromatography after pre-column derivatization with 4,4-difluoro-5,7-dimethyl-4-bora-3a,4a-diaza-s-indacene-3-propionohydrazide. *Analyst* **123,** 2339–2342.

Kelly, A. E., Ou, H. D., Withers, R., and Dotsch, V. (2002). Low-conductivity buffers for high-sensitivity NMR measurements. *J. Am. Chem. Soc.* **124,** 12013–12019.

Koeberl, D. D., Young, S. P., Gregersen, N., Vockley, J., Smith, W. E., Benjamin, D. K., An, Y., Weavil, S. D., Chaing, S. H., Bali, D., McDonald, M. T., Kishnani, P. S., *et al.* (2003). Rare disorders of metabolism with elevated butyryl- and isobutyryl-carnitine detected by tandem mass spectrometry newborn screening. *Pediatric Res.* **54,** 219–223.

Kolliker, S., Oehme, M., and Dye, C. (1998). Structure elucidation of 2,4-dinitrophenylhydrazone derivatives of carbonyl compounds in ambient air by HPLC/MS and multiple MS/MS using atmospheric chemical ionization in the negative ion mode. *Anal. Chem.* **70,** 1979–1985.

Koshland, D. E. (1998). The era of pathway quantification. *Science* **280,** 852–853.

Kresge, N., Simoni, R. D., and Hill, R. L. (2005). JBC Centennial—1905–2005—100 years of biochemistry and molecular biology—Otto Fritz Meyerhof and the elucidation of the glycolytic pathway. *J. Biol. Chem.* **280**(4), e3.

Krieger, C. J., Zhang, P. F., Mueller, L. A., Wang, A., Paley, S., Arnaud, M., Pick, J., Rhee, S. Y., and Karp, P. D. (2004). MetaCyc: A multiorganism database of metabolic pathways and enzymes. *Nucleic Acids Res.* **32,** D438–D442.

Kupce, E., Muhandiram, D. R., and Kay, L. E. (2003). A combined HNCA/HNCO experiment for N-15 labeled proteins with C-13 at natural abundance. *J. Biomol. NMR* **27,** 175–179.

Ladroue, C., Howe, F. A., Griffiths, J. R., and Tate, A. R. (2003). Independent component analysis for automated decomposition of in vivo magnetic resonance spectra. *Magn. Reson. Med.* **50,** 697–703.

Lane, A. N., and Arumugam, S. (2005). Improving NMR sensitivity in room temperature and cooled probes with dipolar ions. *J. Magn. Reson.* **173,** 339–343.

Lange, B. M., and Ghassemian, M. (2005). Comprehensive post-genomic data analysis approaches integrating biochemical pathway maps. *Phytochemistry* **66,** 413–451.

Larsen, E. H., Hansen, M., Fan, T., and Vahl, M. (2001). Speciation of selenoamino acids, selenonium ions and inorganic selenium by ion exchange HPLC with mass spectrometric detection and its application to yeast and algae. *J. Anal. At. Spectrom.* **16,** 1403–1408.

Larsen, E. H., Quetel, C. R., Munoz, R., Fiala-Medioni, A., and Donard, O. F. X. (1997). Arsenic speciation in shrimp and mussel from the Mid-Atlantic hydrothermal vents. *Marine Chem.* **57,** 341–346.

Larsen, E. H., Sloth, J., Hansen, M., and Moesgaard, S. (2003). Selenium speciation and isotope composition in Se-77-enriched yeast using gradient elution HPLC separation and ICP-dynamic reaction cell-MS. *J. Anal. At. Spectrom.* **18,** 310–316.

Lee, W.-N. P., Boros, L. G., Puigjaner, J., Bassilian, S., Lim, S., and Cascante, M. (1998). Mass isotopomer study of the nonoxidative pathways of the pentose cycle with [1,2-13C2]glucose. *Am. J. Physiol. Endocrinol. Metab.* **274,** E843–E851.

Legault, P., Jucker, F. M., and Pardi, A. (1995). Improved measurement of C-13, P-31 J-coupling-constants in isotopically labeled RNA. *Febs Lett.* **362,** 156–160.

Lindon, J. C. (2003). HPLC-NMR-MS: Past, present and future. *Drug Discov. Today* **8,** 1021–1022.

Lindon, J. C., Holmes, E., and Nicholson, J. K. (2003). So whats the deal with metabonomics? Metabonomics measures the fingerprint of biochemical perturbations caused by disease, drugs, and toxins. *Anal. Chem.* **75,** 384A–391A.

Lindon, J. C., Holmes, E., and Nicholson, J. K. (2004). Metabonomics: Systems biology in pharmaceutical research and development. *Curr. Opin. Mol. Ther.* **6,** 265–272.

Lindon, J. C., Nicholson, J. K., and Everett, J. R. (1999). NMR spectroscopy of biofluids. "Annual Reports on NMR Spectroscopy," Vol. 38, pp. 1–88. Academic Press Inc, San Diego.

Lloyd, S. G., Zeng, H. D., Wang, P. P., and Chatham, J. C. (2004). Lactate isotopomer analysis by H-1 NMR spectroscopy: Consideration of long-range nuclear spin-spin interactions. *Magn. Reson. Med.* **51,** 1279–1282.

London, R. E., Allen, D. L., Gabel, S. A., and DeRose, E. F. (1999). Carbon-13 nuclear magnetic resonance study of metabolism of propionate by *Escherichia coli*. *J. Bacteriol.* **181,** 3562–3570.

Lopez-Diez, E. C., Winder, C. L., Ashton, L., Currie, F., and Goodacre, R. (2005). Monitoring the mode of action of antibiotics using Raman spectroscopy: Investigating subinhibitory effects of amikacin on Pseudomonas aeruginosa. *Anal. Chem.* **77,** 2901–2906.

Lu, D. H., Mulder, H., Zhao, P. Y., Burgess, S. C., Jensen, M. V., Kamzolova, S., Newgard, C. B., and Sherry, A. D. (2002). C-13 NMR isotopomer analysis reveals a connection between pyruvate cycling and glucose-stimulated insulin secretion (GSIS). *Proc. Natl. Acad. Sci. USA* **99,** 2708–2713.

Lutz, N. W. (2005). From metabolic to metabolomic NMR spectroscopy of apoptotic cells. *Metabolomics* **1,** 251–268.

Maher, A. D., Kuchel, P. W., Ortega, F., de Atauri, P., Centelles, J., and Cascante, M. (2003). Mathematical modelling of the urea cycle—A numerical investigation into substrate channelling. *Eur. J. Biochem.* **270,** 3953–3961.

Marin, S., Lee, W. N. P., Bassilian, S., Lim, S., Boros, L. G., Centelles, J. J., Maria Fernández-Novell, J., Guinovart, J. J., and Cascante, M. (2004). Dynamic profiling of the glucose metabolic network in fasted rat hepatocytes using 1,2-C-13(2) glucose. *Biochem. J.* **381,** 287–294.

Marshall, A. G., Hendrickson, C. L., and Jackson, G. S. (1998). Fourier transform ion cyclotron resonance mass spectrometry: A primer. *Mass Spectrom. Rev.* **17,** 1–35.

Mashego, M. R., Wu, L., Van Dam, J. C., Ras, C., Vinke, J. L., van Winden, W. A., van Gulik, W. M., and Heijnen, J. J. (2004). MIRACLE: Mass isotopomer ratio analysis of U-C-13-labeled extracts. A new method for accurate quantification of changes in concentrations of intracellular metabolites. *Biotechnol. Bioeng.* **85,** 620–628.

Mason, G. F., Petersen, K. F., de Graaf, R. A., Kanamatsu, T., Otsuki, T., and Rothman, D. L. (2002). A comparison of C-13 NMR measurements of the rates of glutamine synthesis and the tricarboxylic acid cycle during oral and intravenous administration of 1-C-13 glucose. *Brain Res. Protoc.* **10,** 181–190.

Mason, G. F., Petersen, K. F., de Graaf, R. A., Kanamatsu, T., Otsuki, T., and Rothman, D. L. (2003). A comparison of C-13 NMR measurements of the rates of glutamine synthesis and the tricarboxylic acid cycle during oral and intravenous administration of 1-C-13 glucose (vol 10, pg 181, 2003). *Brain Res. Protoc.* **10,** 181–190.

Mason, G. F., and Rothman, D. L. (2004). Basic principles of metabolic modeling of NMR C-13 isotopic turnover to determine rates of brain metabolism *in vivo*. *Metab. Eng.* **6,** 75–84.

Mazurek, S., and Eigenbrodt, E. (2003). The tumor metabolome. *Anticancer Res.* **23,** 1149–1154.

Mazurek, S., Eigenbrodt, E., Failing, K., and Steinberg, P. (1999). Alterations in the glycolytic and glutaminolytic pathways after malignant transformation of rat liver oval cells. *J. Cell. Physiol.* **181,** 136–146.

Mendes, P., and Kell, D. B. (2001). MEG (Model Extender for Gepasi): A program for the modelling of complex, heterogeneous, cellular systems. *Bioinformatics* **17,** 288–289.

Mollney, M., Wiechert, W., Kownatzki, D., and de Graaf, A. A. (1999). Bidirectional reaction steps in metabolic networks: IV. Optimal design of isotopomer labeling experiments. *Biotechnol. Bioeng.* **66,** 86–103.

Nicholson, J. (2005). Metabonomics and global systems biology approaches to molecular diagnostics. *Drug Metab. Rev.* **37,** 10–10.

Nikiforova, V. J., Kopka, J., Tolstikov, V., Fiehn, O., Hopkins, L., Hawkesford, M. J., Hesse, H., and Hoefgen, R. (2005). Systems rebalancing of metabolism in response to sulfur deprivation, as revealed by metabolome analysis of arabidopsis plants. *Plant Physiol.* **138,** 304–318.

North, S., Okafo, G., Birrell, H., Haskins, N., and Camilleri, P. (1997). Minimizing cationization effects in the analysis of complex mixtures of oligosaccharides. *Rapid Commun. Mass Spectrom.* **11,** 1635–1642.

Orosz, F., Wagner, G., Ortega, F., Cascante, M., and Ovadi, J. (2003). Glucose conversion by multiple pathways in brain extract: Theoretical and experimental analysis. *Biochem. Biophys. Res. Commun.* **309,** 792–797.

Petch, D., and Butler, M. (1994). Profile of energy-metabolism in a murine hybridoma—glucose and glutamine utilization. *J. Cell. Physiol.* **161,** 71–76.

Piotto, M., Saudek, V., and Sklenar, V. (1992). Gradient-tailored excitation for single-quantum NMR spectroscopy of aqueous solutions. *J. Biomol. NMR* **2,** 661–665.

Portais, J.-C., Schuster, R., Merle, M., and Canioni, P. (1993). Metabolic flux determination in C6 glioma cells using carbon-13 distribution upon (1-13C)glucose incubation. *Eur. J. Biochem.* **217,** 457–468.

Riek, R., Pervushin, K., Fernandez, C., Kainosho, M., and Wuthrich, K. (2001). [C-13,C-13]- and [C-13,H-1]-TROSY in a triple resonance experiment for ribose-base and intrabase correlations in nucleic acids. *J. Am. Chem. Soc.* **123,** 658–664.

Robey, I. F., Lien, A. D., Welsh, S. J., Baggett, B. K., and Gillies, R. J. (2005). Hypoxia-inducible factor-1 alpha and the glycolytic phenotype in tumors. *Neoplasia* **7,** 324–330.

Romero, P., and Karp, P. (2003). PseudoCyc, a pathway-genome database for Pseudomonas aeruginosa. *J. Mol. Microbiol. Biotechnol.* **5,** 230–239.

Romero, P., Wagg, J., Green, M. L., Kaiser, D., Krummenacker, M., and Karp, P. D. (2005). Computational prediction of human metabolic pathways from the complete human genome. *Genome Biol.* **6**(1), R2.

Rujoi, M., Jin, J., Borchman, D., Tang, D., and Yappert, M. C. (2003). Isolation and lipid characterization of cholesterol-enriched fractions in cortical and nuclear human lens fibers. *Invest. Ophthalmol. Visual Sci.* **44,** 1634–1642.

Ruzsanyi, V., Baumbach, J. I., Sielemann, S., Litterst, P., Westhoff, M., Freitag, L., and Lehotay, J. (2005). Detection of human metabolites using multi-capillary columns coupled to ion mobility spectrometers. *J. Chromatogr. A* **1084,** 145–151.

Saghatelian, A., Trauger, S. A., Want, E. J., Hawkins, E. G., Siuzdak, G., and Cravatt, B. F. (2004). Assignment of endogenous substrates to enzymes by global metabolite profiling. *Biochemistry* **43,** 14332–14339.

Sanders, J. K. M., and Hunter, B. K. (1987). "Modern NMR Spectroscopy. A Guide for Chemists." Oxford University Press, Oxford, UK.

Sanfeliu, A., Paredes, C., Cairo, J. J., and Godia, F. (1997). Identification of key patterns in the metabolism of hybridoma cells in culture. *Enzyme and Microbial Technol.* **21,** 421–428.

Siuzdak, G. (2003). "The Expanding Role of Mass Spectrometry in Biotechnology." MCC Press, San Diego.

Sloth, J. J., Larsen, E. H., Bugel, S. H., and Moesgaard, S. (2003). Determination of total selenium and Se-77 in isotopically enriched human samples by ICP-dynamic reaction cell-MS. *J. Anal. At. Spectrom.* **18,** 317–322.

Street, J. C., Delort, A. M., Braddock, P. S., and Brindle, K. M. (1993). A 1H/15N n.m.r. study of nitrogen metabolism in cultured mammalian cells. *Biochem. J.* **291,** 485–492.

Styles, P., Soffe, N. F., and Scott, C. A. (1989). An improved cryogenically cooled probe for high-resolution NMR. *J. Magn. Reson.* **84,** 376–378.

Turner, S. M., and Hellerstein, M. K. (2005). Emerging applications of kinetic biomarkers in preclinical and clinical drug development. *Curr. Opin. Drug Discov. Dev.* **8,** 115–126.

Turner, S. M., Linfoot, P. A., Neese, R. A., and Hellerstein, M. K. (2005). Sources of plasma glucose and liver glycogen in fasted ob/ob mice. *Acta Diabetol.* **42,** 187–193.

Turner, S. M., Murphy, E. J., Neese, R. A., Antelo, F., Thomas, T., Agarwal, A., Go, C., and Hellerstein, M. K. (2003). Measurement of TG synthesis and turnover *in vivo* by (2HO)-O-2 incorporation into the glycerol moiety and application of MIDA. *Am. J. Physiol. Endocrinol. Metab.* **285,** E790–E803.

van Winden, W. A., Wittmann, C., Heinzle, E., and Heijnen, J. J. (2002). Correcting mass isotopomer distributions for naturally occurring isotopes. *Biotechnol. Bioeng.* **80,** 477–479.

Ventura, F. V., Costa, C. G., Struys, E. A., Ruiter, J., Allers, P., Ijlst, L., de Almeida, T., Duran, M., Jakobs, C., and Wanders, R. J. A. (1999). Quantitative acylcarnitine profiling in fibroblasts using [U-C-13] palmitic acid: An improved tool for the diagnosis of fatty acid oxidation defects. *Clin. Chim. Acta* **281,** 1–17.

Vizan, P., Boros, L. G., Figueras, A., Capella, G., Mangues, R., Bassilian, S., Lim, S., Lee, W.-N. P., and Cascante, M. (2005). K-ras codon-specific mutations produce distinctive metabolic phenotypes in human fibroblasts. *Cancer Res.* **65,** 5512–5515.

Vogt, J. A., Hunzinger, C., Schroer, K., Holzer, K., Bauer, A., Schrattenholz, A., Cahill, M. A., Schillo, S., Schwall, G., Stegmann, W., and Albuszies, G. (2005). Determination of fractional synthesis rates of mouse hepatic proteins via metabolic C-13-labeling, MALDI-TOF MS and analysis of relative isotopologue abundances using average masses. *Anal. Chem.* **77,** 2034–2042.

Vriezen, N., and van Dijken, J. P. (1998). Fluxes and enzyme activities in central metabolism of myeloma cells grown in chemostat culture. *Biotechnol. Bioeng.* **59,** 28–39.

Wahl, S. A., Dauner, M., and Wiechert, W. (2004). New tools for mass isotopomer data evaluation in C-13 flux analysis: Mass isotope correction, data consistency checking, and precursor relationships. *Biotechnol. Bioeng.* **85,** 259–268.

Wang, Y. L., Utzinger, J., Xiao, S. H., Xue, J., Nicholson, J. K., Tanner, M., Singer, B. H., and Holmes, E. (2006). System level metabolic effects of a Schistosoma japonicum infection in the Syrian hamster. *Mol. Biochem. Parasitol.* **146,** 1–9.

Warburg, O. (1956). On the origin of cancer cells. *Science* **123,** 309–314.

Watson, J. T. (1985). "Introduction to Mass Spectrometry." Raven Press, New York.

Weitzke, E. L., and Ortoleva, P. J. (2003). Simulating cellular dynamics through a coupled transcription, translation, metabolic model. *Comput. Biol. Chem.* **27,** 469–480.

Weljie, A. M., Newton, J., Mercierr, P., Carlson, E., and Slupsky, C. M. (2006). Targeted profiling: Quantitative analysis of ^{1}H NMR metabolomics data. *Anal. Chem.* **78**(13), 4430–4442.

Whitfield, P. D., German, A. J., and Noble, P. J. M. (2004). Metabolomics: An emerging post-genomic tool for nutrition. *Br. J. Nutr.* **92,** 549–555.

Wolfender, J.-L., Ndjoko, K., and Hostettmann, K. (2003). Liquid chromatography with ultraviolet absorbance-mass spectrometric detection and with nuclear magnetic resonance spectroscopy: A powerful combination for the on-line structural investigation of plant metabolites. *J. Chromatogr.* **1000,** 437–455.

Yang, T. H., Heinzle, E., and Wittmann, C. (2005). Theoretical aspects of C-13 metabolic flux analysis with sole quantification of carbon dioxide labeling. *Comput. Biol. Chem.* **29,** 121–133.

Yang, T. H., Wittmann, C., and Heinzle, E. (2004). Membrane inlet mass spectrometry for the on-line measurement of metabolic fluxes in the case of lysine-production by Corynebacterium glutamicum. *Eng. Life Sci.* **4,** 252–257.

Yappert, M. C., Rujoi, M., Borchman, D., Vorobyov, I., and Estrada, R. (2003). Glycero- versus sphingo-phospholipids: Correlations with human and non-human mammalian lens growth. *Exp. Eye Res.* **76,** 725–734.

Zhang, J. R., Cazers, A. R., Lutzke, B. S., and Hall, E. D. (1995). Hplc chemiluminescence and thermospray lc/ms study of hydroperoxides generated from phosphatidylcholine. *Free Radic. Biol. Med.* **18,** 1–10.

Zhang, P. F., Foerster, H., Tissier, C. P., Mueller, L., Paley, S., Karp, P. D., and Rhee, S. Y. (2005). MetaCyc and AraCyc. Metabolic pathway databases for plant research. *Plant Physiol.* **138,** 27–37.

Zhang, W. X., and Gmeiner, W. H. (1996). Improved 3D gd-HCACO and gd-(H)CACO-TOCSY experiments for isotopically enriched proteins dissolved in H_2O. *J. Biomol. NMR* **7,** 247–250.

Zurek, G., Buldt, A., and Karst, U. (2000). Determination of acetaldehyde in tobacco smoke using N-methyl-4-hydrazino-7-nitrobenzofurazan are liquid chromatography/mass spectrometry. *Fresenius J. Anal. Chem.* **366,** 396–399.

Zwingmann, C., Chatauret, N., Leibfritz, D., and Butterworth, R. F. (2003). Selective increase of brain lactate synthesis in experimental acute liver failure: Results of a H-1-C-13 nuclear magnetic resonance study. *Hepatology* **37,** 420–428.

CHAPTER 19

Following Molecular Transitions with Single Residue Spatial and Millisecond Time Resolution

Inna Shcherbakova,* Somdeb Mitra,* Robert H. Beer,† and Michael Brenowitz*

*Department of Biochemistry
Albert Einstein College of Medicine
Bronx, New York 10461

†Department of Chemistry
Fordham University
Bronx, New York 10458

0091-679X/08 $35.00
DOI: 10.1016/S0091-679X(07)84019-2

Abstract

"Footprinting" describes assays in which ligand binding or structure formation protects polymers such as nucleic acids and proteins from either cleavage or modification; footprinting allows the accessibility of individual residues to be mapped in solution. Equilibrium and time-dependent footprinting links site-specific structural information with thermodynamic and kinetic transitions, respectively. The hydroxyl radical (•OH) is a uniquely insightful footprinting probe by virtue of it being among the most reactive chemical oxidants; it reports the solvent accessibility of reactive sites on macromolecules with as fine as a single residue resolution. A novel method of millisecond time-resolved •OH footprinting is presented based on the Fenton reaction, $Fe(II) + H_2O_2 \rightarrow Fe(III) + \bullet OH + OH^-$. It is implemented using a standard three-syringe quench-flow mixer. The utility of this method is demonstrated by its application to the studies on RNA folding. Its applicability to a broad range of biological questions involving the function of DNA, RNA, and proteins is discussed.

I. Introduction

Every biological process can be described as a chemical reaction at the molecular level. Unlike simple chemical reagents, the protein, and nucleic acid components of the most biochemical reactions are long polymers consisting of hundreds if not thousands of monomers. In order to understand the chemical reactions that involve or process these macromolecules during cellular metabolism, it is often essential to know which parts of the macromolecules are involved in the interaction or reaction of interest. Significant insight into biological processes can be gained when *local* changes in macromolecular structure are quantitatively followed (Ackers *et al.*, 1983). Quantitative "footprinting" allows macromolecular structural transitions to be followed simultaneously at many discrete sites along a polymer as a function of an effector or ligand concentration or time (Ackers *et al.*, 1982; Brenowitz *et al.*, 1986; Hsieh and Brenowitz, 1996).

A. What Is Footprinting?

Footprinting refers to assays that examine ligand binding and/or conformational changes by determining the accessibility of the backbone or residues of macromolecules through their sensitivity to chemical or enzymatic modification or cleavage (Fig. 1) (Galas and Schmitz, 1978). The key characteristics of a

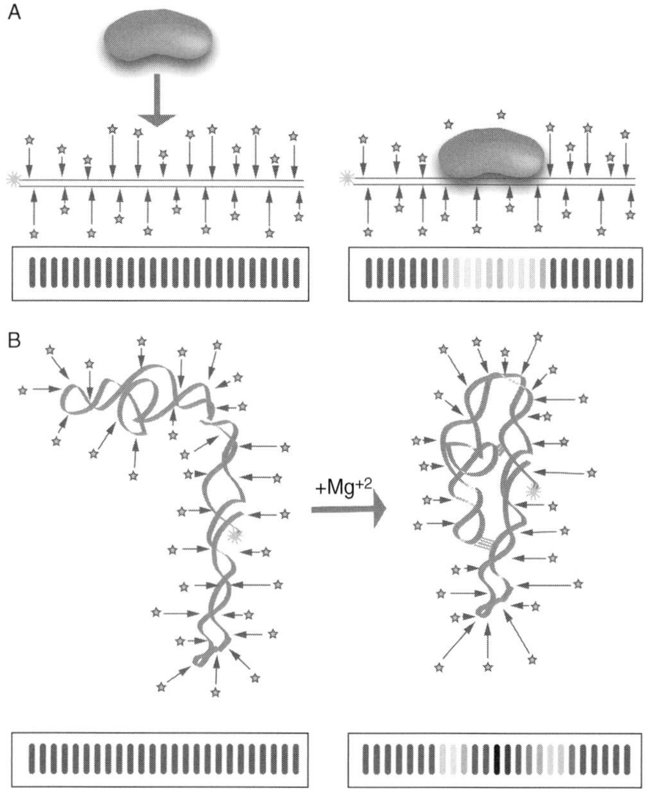

Fig. 1 (A) Schematic representation of a footprinting experiment where a protein binds to a specific site on DNA. The DNA backbone is protected from the "footprinting" probe at this site resulting in a decrease in the reaction products terminating at that site. Detection of a unique set of cleavage products is accomplished by labeling one strand of the DNA duplex at one end. (Below) A simulated gel electrophoretogram depicts the decrease in cutting ("footprint") of DNA associated with protein binding. (B) Schematic representation of a footprinting experiment monitoring Mg^{2+}-induced folding of RNA into its three-dimensional structure that results into formation of regions less (or more) accessible to the footprinting probe. The asterisks and arrows indicate a solvent accessible backbone position potentially cut by •OH.

footprinting assay are that (1) the reaction of the footprinting probe with the polymer is limited, such that each position along the chain is sampled with comparable low probability, and (2) the cleavage or modification products are uniquely identified. A variety of footprinting probes and approaches have been successfully used to analyze DNA, RNA, and proteins, including chemical and enzymatic nucleases and modification reagents (Brenowitz *et al.*, 2002; Feng *et al.*, 1999; Guan and Chance, 2005; Heyduk and Heyduk, 1994; Petri and Brenowitz, 1997; Silverman and Harbury, 2002; Takamoto and Chance, 2006; Tullius and Greenbaum, 2005; Wilkinson *et al.*, 2005; Zarrinkar and Williamson, 1994).

The separation of the products of DNA or RNA backbone cleavage in footprinting is routinely achieved by denaturing gel or capillary electrophoresis. The products of peptide backbone cleavage can be analyzed by gel electrophoresis while side chain modification is analyzed by mass spectrometry. The details of these techniques are discussed in Sections I.E and I.F.

B. Quantitative Footprinting

Footprinting assays can provide solution structural information with single residue resolution coupled to thermodynamic and kinetic transitions. The individual-site isotherms (Ackers *et al.*, 1982, 1983) and kinetic progress curves (Hsieh and Brenowitz, 1996, 1997) determined from thermodynamic and kinetic footprinting studies, respectively, provide an ensemble of local measures of macromolecular transitions from which detailed energetic and mechanistic portraits can be painted. In principle and now in practice (Das *et al.*, 2005; Shadle *et al.*, 1997; Takamoto *et al.*, 2002, 2004), the change in the solvent accessibility of each nucleotide of nucleic acids hundreds of residue in length can be separately quantitated as a function of either a thermodynamic variable or time. Quantitative protocols have been used to determine thermodynamic and kinetic constants describing protein–DNA interactions and RNA assembly reactions. Representative studies from our group include Brenowitz *et al.* (2002), Laederach *et al.* (2006), Mollah and Brenowitz (2000), Nguyenle *et al.* (2006), Petri and Brenowitz (1997), Sclavi *et al.* (2005), and Sprouse *et al.* (2006). Of particular interest to this chapter is the development of quantitative footprinting protocols for the hydroxyl radical (•OH).

C. The Hydroxyl Radical (•OH) as a Footprinting Probe

The hydroxyl radical (•OH) has proven itself as a particularly insightful footprinting reagent by virtue of it being among the most reactive chemical oxidants (Buxton *et al.*, 1988). Its small radius reports the solvent accessible surface of macromolecules. Multiple methods have been used to •OH footprint DNA, RNA, and proteins, including Fe-EDTA (Tullius and Dombroski, 1985, 1986), peroxonitrite (Chaulk and MacMillan, 2000; King *et al.*, 1992, 1993), photolysis (Aye *et al.*, 2005; Hambly and Gross, 2005; Sharp *et al.*, 2004), radiolysis (Hayes *et al.*, 1990; Maleknia *et al.*, 1999; Ottinger and Tullius, 2000), and synchrotron X-ray radiolysis (Sclavi *et al.*, 1997). Fe-EDTA footprinting uses the Fenton–Haber–Weiss reaction (Fenton, 1894; Haber and Weiss, 1934),

$$\mathrm{Fe(II)} + \mathrm{H_2O_2} \rightarrow \mathrm{Fe(III)} + \bullet\mathrm{OH} + \mathrm{OH^-}, \qquad (1)$$

to generate •OH in solution from peroxide solutions by iron oxidation (Pogozelski *et al.*, 1995). Chelation of Fe(II) by EDTA prevents the transition metal ion from binding to the macromolecules being studied and prevents hydrolysis of the iron at biological values of pH (Pogozelski *et al.*, 1995). This method is widely applied and inexpensive to perform. A convenient implementation of this chemistry for footprinting is to reductively cycle Fe(III) back to Fe(II) with ascorbate (Tullius *et al.*, 1987). This catalysis allows reagent concentrations that are micromolar in Fe-EDTA and millimolar in H_2O_2 and ascorbate to be used with reaction times of several to tens of minutes (Dixon *et al.*, 1991; Tullius and Dombroski, 1986; Tullius and Greenbaum, 2005). The oxidation of Fe(II) to Fe(III) by H_2O_2 in Eq. (1) produces a burst of a high concentration of •OH that is the basis of the time-resolved footprinting method (Section I.F).

D. Nucleic Acid •OH Footprinting

Hydroxyl radicals cleave the polynucleotide backbone by abstracting the solvent accessible sugar hydrogens (Balasubramanian *et al.*, 1998). Cleavage of RNA and DNA by •OH is relatively insensitive to base sequence and whether a nucleic acid is single or double stranded (Celander and Cech, 1990). That •OH cleavage of nucleic acids is quantitatively correlated with the solvent accessibility of the phosphodiester backbone has been demonstrated through comparisons of •OH footprints with solvent accessibility calculations from crystal structures. Protein–DNA complexes so analyzed include the λ cI-repressor (Dixon *et al.*, 1991) and the TATA binding protein (Pastor *et al.*, 2000). For RNA tertiary structure, •OH cleavage pattern correlates with the solvent accessibility calculated from structure of the P4-P6 domain of the *Tetrahymena* ribozyme (Cate *et al.*, 1996; Sclavi *et al.*, 1997). This correspondence also holds true for the full-length *Tetrahymena* ribozyme (Celander and Cech, 1991; Golden *et al.*, 1998; Lehnert *et al.*, 1996). Backbone cleavage of DNA by •OH correlates with the accessible surface of the hydrogen atoms of the nucleotide sugar (Balasubramanian *et al.*, 1998). Thus, •OH footprinting yields robust and readily interpretable measures of the structure and interactions of nucleic acids with as fine as single nucleotide spatial resolution. As discussed above and below, protocols for thermodynamic and kinetic implementations of nucleic acid •OH footprinting have been published.

E. Protein •OH Footprinting

Hydroxyl radical footprinting was first extended to proteins by monitoring cleavage of the peptide backbone by gel electrophoresis (Heyduk and Heyduk, 1994; Zhong *et al.*, 1995). However, peptide bond cleavage is inefficient (King *et al.*, 1992) and further development of protein •OH footprinting has focused on the oxidation of amino acid side chains [reviewed in (Guan and Chance, 2005) and (Aye *et al.*, 2005; Hambly and Gross, 2005)]. Mass spectrometric analysis of proteolytic fragments is used to quantitate the oxidation rate of individual or groups of amino acid side chains. The differential reactivity of the amino acid side chains to oxidation is addressed in thermodynamic and kinetic analyses by quantitating the *relative* change in residue reactivity (Guan and Chance, 2005; Maleknia *et al.*, 1999). A comparable relationship between •OH reactivity and solvent accessibility is beginning to emerge for proteins [reviewed in (Guan and Chance, 2005)]. Protocols for thermodynamic, but not kinetic, protein •OH footprinting have been published (Kiselar *et al.*, 2003).

F. Generation of Hydroxyl Radicals for Millisecond Resolution Time-Resolved Studies

Concentrations of •OH sufficient for an RNA folding kinetics analysis were achieved with reaction times as short as several seconds using H_2O_2, ascorbate, and an Fe-EDTA concentration of 500 μM to mediate the Fenton reaction (Hampel and Burke, 2001). Time-resolved •OH footprinting with seconds time resolution has been accomplished with peroxonitrite (Chaulk and MacMillan, 1998; King *et al.*, 1992, 1993; Swisher *et al.*, 2002). Until recently, synchrotron X-ray radiolysis was unique among available millisecond •OH footprinting methods (Sclavi *et al.*, 1997). However, it was recently shown that UV laser photolysis of H_2O_2 (Aye *et al.*, 2005; Hambly and Gross, 2005) and a novel implementation of the Fenton–Haber–Weiss reaction [(Shcherbakova *et al.*, 2006); the subject of this chapter] can produce •OH sufficient for footprinting proteins and nucleic acids on millisecond or shorter timescales. The strengths and weaknesses of the millisecond methods of •OH production are briefly discussed below.

1. Synchrotron Radiolysis

Radiolysis of water by a high flux "white" synchrotron X-ray beam produces sufficient •OH for fast time-resolved footprinting (Brenowitz *et al.*, 2002; Ralston *et al.*, 2000; Sclavi *et al.*, 1998a). Beamline X-28C at the National Synchrotron Light Source was established for conducting footprinting experiments. Synchrotron footprinting has proved itself as an advanced experimental technique that not only reveals the sites and extent of these interactions but also does so in a time-resolved manner, allowing one to study intermediate states of complex biochemical reactions (Brenowitz *et al.*, 2002; Dhavan *et al.*, 2003; Maleknia *et al.*, 1999; Ralston *et al.*, 2000; Sclavi *et al.*, 1998b; Xu *et al.*, 2003, 2005). Exposure to unfocused white beam of $\sim$10 msec for nucleic acids and $\sim$50 msec for proteins (Brenowitz *et al.*, 2002; Guan and Chance, 2005) have been reduced to $\leq$1 msec by the installation of focusing mirror (unpublished results). The main advantage of the method is that no chemical reagent is added to interfere with the macromolecular reaction being studied. Disadvantages include the need to travel to a facility to expose samples to the X-ray beam, rigorous safety restrictions on the use of dispersible radioactivity, and potential sample heating by the high flux beam.

2. UV Laser Photolysis of H_2O_2

It was recently demonstrated that a microsecond flash of UV light effectively decomposes H_2O_2 to produce sufficient •OH for kinetic footprinting studies (Aye *et al.*, 2005; Hambly and Gross, 2005). Unlike the synchrotron and fast Fenton methods discussed in this section, UV laser photolysis has not yet been used in time-resolved studies. Thus, its general applicability is untested. Strengths of UV photolysis of H_2O_2 would seem to be the potentially accessible timescales and the absence of potentially disruptive reagents. Weaknesses include the efficiency with which UV radiation damages nucleic acids in ways (e.g., covalent adducts such as thymine dimers) that are likely to be incompatible with electrophoretic separation of nucleic acid reaction products and the high cost of purchasing and installing a UV laser.

3. Fast Fenton Footprinting

We have developed a novel time-resolved •OH footprinting approach based on Eq. (1) that we call fast Fenton footprinting (Shcherbakova *et al.*, 2006). Fast Fenton footprinting is a laboratory-based method of time-resolved •OH footprinting able to reveal the sites and extent of macromolecular interactions with a detection limit of 1–2 msec when implemented with quench-flow mixing technology. This time resolution reveals transient, often short-lived, intermediate states of complex biochemical processes. The method utilizes inexpensive chemical reagents [H_2O_2, $Fe(NH_4)_2(SO_4)_2$, EDTA, thiourea, or ethanol] and widely available quench-flow mixers (e.g., KinTek® three-syringe mixers). It is applicable to DNA, RNA, and protein. Less than 2 msec exposure to 0.75 mM Fe(II)-EDTA is necessary and sufficient to footprint DNA and RNA with the single-hit kinetics required for quantitative analysis under ideal reaction conditions. The production of the •OH is easily scalable to achieve single-hit kinetics by adjusting concentrations of the reagents involved into •OH production (see Section II.C). This feature of the method is especially valuable when the necessary experimental conditions include high concentrations of radical scavengers. The following section presents a general experimental protocol for the analysis of DNA or RNA that can be readily tailored to particular applications.

II. Acquisition of •OH Footprinting Time-Progress Curves

A. Reagents and Solutions

A stock solution of 100 mM $Fe(NH_4)_2(SO_4)_2$ (Sigma-Aldrich) was prepared and stored in small aliquots at −70 °C. Stock solutions of 500 mM Na_2-EDTA (Ambion) and 30% H_2O_2 (Fluka) were kept at room temperature and +4 °C, respectively. 10× "assay buffer" (200 mM sodium cacodylate and 2 M NaCl at pH 7.4) and 2× thiourea quench solution (50 mM thiourea, 40 mM EDTA, and

200 mM NaCl) were prepared and stored at +4 °C. An "alternative quench solution" is absolute ethanol that facilitates subsequent precipitation of nucleic acids. The studies described below were performed with the L-21 ScaI ribozyme from *Tetrahymena thermophila* 5′-end radiolabeled with [γ-^{32}P]ATP (Zaug *et al.*, 1988) or the HindIII/NdeI DNA restriction fragment of 282 bp from plasmid ppUMLP 3′-end labeled with ^{32}P at the HindIII site (Patikoglou *et al.*, 1999; Petri *et al.*, 1998). The radiolabeled nucleic acids were stored in 10 mM sodium cacodylate, pH 7.3 at −70 °C. The key consideration for the production and storage of ^{32}P-labeled footprinting substrates is to minimize nicks in the polynucleotide backbone so as to provide a clean background over which the •OH-mediated cleavage products can be visualized.

B. Minimizing RNase Contamination When Working With RNA

Contamination by RNase is the bane of the RNA research community. Below is described the protocol we use to clean the KinTek® RQF-3 mixer used in our RNA studies. This protocol can be adapted by consideration of the particular plumbing and control program being used:

1. Be sure that the water in the circulating bath used for temperature control is free of bacterial growth. Use of a bacterial growth inhibitor is recommended.
2. Prepare all solutions with ultrapure RNase-free water.
3. Load the drive and quench syringes with a solution of RNaseZap® (Ambion) and flush them three times by moving the liquid between the 5-ml fill and drive syringes.
4. Repeat step 3 with a fresh RNaseZap® solution.
5. Load the drive and quench syringes with a fresh RNaseZap® solution. Turn all the valves to the "fire" position. In the control software menu, choose the "adjust position" option and "down" as a direction of the platform movement. Start the motor and let the entire "fire" path to be washed by the detergent.
6. Fill two new 5-ml syringes with the RNaseZap® solution and insert them into the sample ports. Move the sample valves to the "load" position. Fill the sample and reaction loops with the detergent solution and allow them to soak for 5–10 min.
7. Repeat steps 3–5 using ultrapure RNase-free water. Multiple repetitions may be necessary to remove all trace of the RNaseZap®.
8. Connect a vacuum to the mixer's exit line and flush the detergent out of loading and reaction loops by drawing ultrapure RNase-free water through sample inlets following turning the sample valves to the "load" position. Turn the sample valves to "flush" position and wash all the reaction loops by drawing clean water through the flush lines. Repeat the last step using ethanol. Dry the tubing and valves by continuing to apply a vacuum since alcohols are good •OH scavengers.

C. Determine the Correct Amount of •OH Production

Single-hit •OH cleavage kinetics is essential to successful time-resolved footprinting studies. The presence of •OH scavengers (e.g., Tris and HEPES buffers, DTT and glycerol) (Tullius *et al.*, 1987) requires either higher concentrations of Fe(II)-EDTA and H_2O_2 and/or longer reaction times. Therefore, dose-response determinations should be conducted as described below using ^{32}P-labeled DNA or RNA to calibrate •OH production for particular experimental conditions. Extensive analyses of •OH production are conveniently conducted by following the quenching of the fluorescence emission of fluorescein on oxidation of the dye (Chen *et al.*, 2002; Ou *et al.*, 2002) as described elsewhere (Shcherbakova *et al.*, 2006). The protocol described below was implemented using the KinTek® RQF-3 three-syringe mixer following the standard "quench-flow run" control option (Hsieh and Brenowitz, 1996; Ralston *et al.*, 2000; Sclavi *et al.*, 1998b):

1. Load the drive syringes with the chosen assay buffer and the quench syringe with the quench solution (Fig. 2).
2. Prepare the following solutions for the sample syringes:
   2.1. A ^{32}P-DNA or ^{32}P-RNA (typically ~100,000 dpm/data point) solution in assay buffer containing H_2O_2 at twice the desired concentration. The total solution volume, that is prepared, depends on the number of data points to be

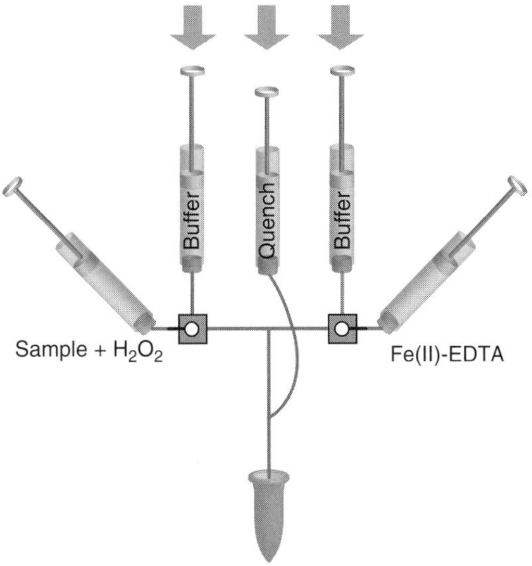

Fig. 2 Schematic representation of an experimental setup to conduct a dose-response experiment in the quench-flow mixer to find appropriate Fenton reaction time. The three vertical syringes are the drive syringes; the arrows denote the movement of the drive plate. The angled syringes denote the two sample loading syringes. The boxes indicate the sample loading valves.

collected at 15–20 μl per point. A small excess of the solution allows for mistakes and acquisition of additional data points. H_2O_2-containing solutions should be protected from light to minimize •OH production by photolysis thus diminishing its concentration (Sharp *et al.*, 2004).

2.2. The same volume of a solution containing $Fe(NH_4)_2(SO_4)_2$ at twice the desired concentration and a 1.1 M excess of EDTA in assay buffer.

2.3. A volume of assay buffer without Fe-EDTA sufficient for acquisition of several data points.

3. Load solutions 2.1 and 2.3 in the sample syringes. Collect several zero-time point samples by running the "load-fire-flush" cycle using the KinTek® "quench-flow run" option of the control program the desired number of times. It is advisable to acquire replicate zero-time point samples.

4. Replace solution 2.3 with solution 2.2 and run a series of "load-fire-flush" cycles to mix the ^{32}P-DNA- or ^{32}P-RNA-containing samples with the Fe(II)-EDTA solution. The delay time is incremented to acquire the desired range of Fenton reaction times, typically 2–20 msec. Each expelled solution is separately collected and stored on ice until the completion of the time course.

5. The collection of samples are then prepared for analysis by gel electrophoresis as described elsewhere (Ralston *et al.*, 2000; Sclavi *et al.*, 1997).

6. The products of •OH cleavage are separated by denaturing gel electrophoresis and visualized by exposure of the dried gels to a storage phosphor screen (Ralston *et al.*, 2000; Sclavi *et al.*, 1997) (Fig. 3A). Electrophoresis is typically carried out so that the smallest fragments migrate to the bottom of the gel.

7. The fraction of nucleic acid cleaved is calculated from the ratio of the density of the fragments to the total density of cleaved and uncut samples (Fig. 3A, the intense band at the top of the gel represents the intact molecules). Plotting the fraction cut versus time yields a dose-response graph (Fig. 3B). Single-hit kinetics is typically achieved when 85–90% of the input nucleic acid remains intact (Brenowitz *et al.*, 1986).

While steps 1–7 can be conducted at a series of reagent concentrations, a convenient alternative is to sample ^{32}P-nucleic acid cleavage as a function of Fe(II)-EDTA concentration at constant reaction time and H_2O_2 concentration (Fig. 3C). The advantage of this approach is that the •OH production can be tuned for a particular study in a single experiment.

D. A General Protocol for Fast Fenton Footprinting

In footprinting experiments, samples A and B are mixed to initiate the reaction being studied. After a defined aging time, the distribution of reactants and products is sampled by generating a pulse of •OH by the addition of Fe(II)-EDTA

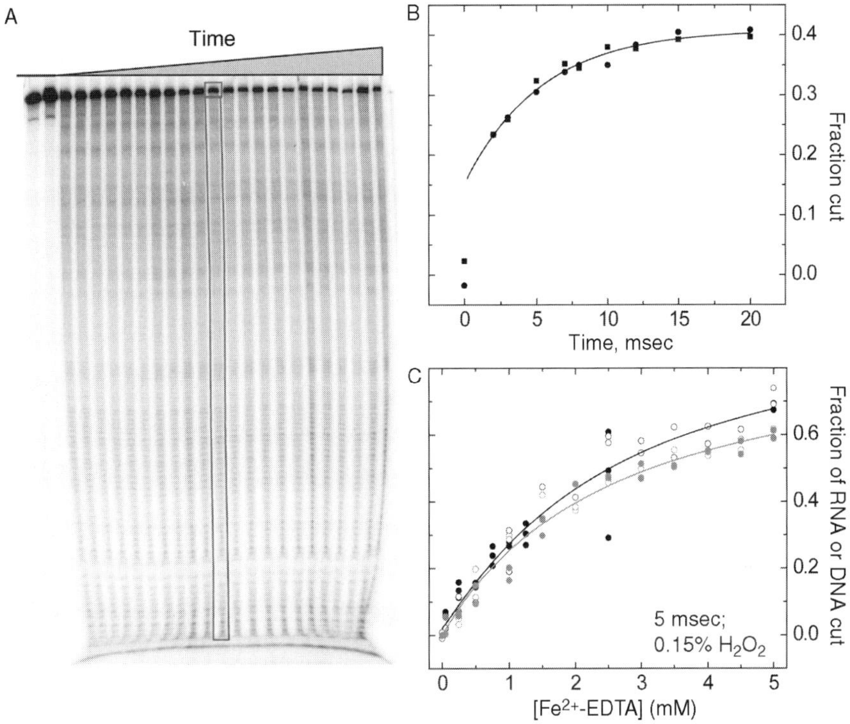

Fig. 3 (A) Autoradiogram of the cleavage products of RNA derived from the mixing of Fe(II)-EDTA and H_2O_2 to final concentrations of 1 and 44 mM, respectively, to produce •OH for the times spanning 0–20 msec. The black boxes indicate the boundaries drawn to quantitate the ratio of the uncut RNA (top band) and the •OH cleavage products; (B) The extent of •OH-mediated cleavage of RNA determined for 1mM Fe(II)-EDTA and 44 mM H_2O_2 as a function of the reaction time; (C) The extent of •OH-mediated cleavage of DNA (black) and RNA (gray) determined for 5 msec and 44 mM H_2O_2 as a function of [Fe(II)-EDTA]. Open and filled symbols indicate replicate experiments.

from the "quench" syringe (Fig. 4). The •OH footprinting reaction takes place within the exit tube of the KinTek® mixer until the expulsion of the sample into the collection tube that contains an excess volume of radical scavenger. Selecting the "constant quench volume" option from the "alter quench parameters" menu of the KinTek® control program keeps the delivered volume constant during acquisition of a dataset.

The duration of the •OH footprinting reaction is proportional to the flow rate of, and distance traveled by, the solution containing both Fe(II)-EDTA and H_2O_2. Either flow rate, distance, or both parameters can be varied to control the reaction time. Both approaches have been used in our laboratory. The exit tube of our KinTek® mixer was trimmed to ~5 cm and the minimal motor speed increased

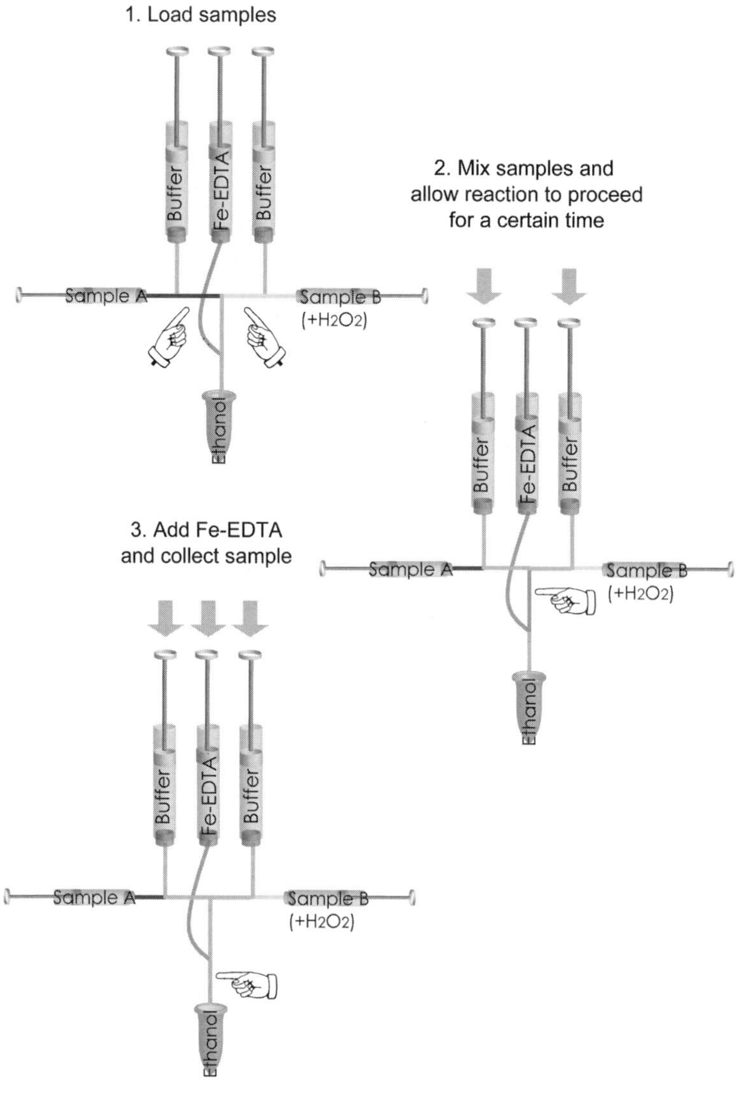

Fig. 4 Schematic representation of a time-resolved fast Fenton footprinting experiment conducted in the quench-flow mixer. The fickle fingers point to the position of the samples in the mixer's flow train: (1) The quench syringe is loaded with appropriate concentration of Fe(II)-EDTA. H_2O_2 is added to the nucleic acid solution. Samples A and B are loaded and the mixer is prepared for firing; (2) The samples are mixed by the motor push to allow the reaction being studied to age for a certain time; (3) Following aging for the desired time, the programmed volume of the Fe(II)-EDTA solution mixes with the sample so as to •OH footprint it. The "A/B/Fe(II)-EDTA" solution continues through the flow path and is vigorously ejected into quench solution present in the collection tube.

resulting in a footprinting reaction time of <4 msec.[1] A protocol suitable for conducting time-resolved •OH footprinting experiments to follow folding or ligand binding by nucleic acids is described below (see Fig. 4):

1. Prepare the following solutions:
   1.1. A solution of ^{32}P-DNA or ^{32}P-RNA at 600,000 dpm/data point in assay buffer. This solution will be "Sample A" in Fig. 4. The solution volume to be prepared depends on the number of data points to be collected (15–20 μl per point, depending on the calibrated values of sample loops). It is prudent to prepare a little excess solution to allow for mistakes and addition of extra data points.
   1.2. A comparable volume of "Sample B" (Fig. 4) at *twice* its desired final concentration and H_2O_2 at *triple* its desired final concentration in assay buffer.
   1.3. A volume of H_2O_2 solution at triple the desired final concentration in assay buffer sufficient for the acquisition of several zero-time point samples.
   1.4. Approximately 5 ml of Fe(II)-EDTA solution at triple the desired final concentration in assay buffer should be freshly prepared before each experiment. Sample B can be included at its final desired concentration to prevent its dilution on mixing the aged reactants with this solution.
   1.5. A volume of thiourea quench solution or absolute ethanol enough to quench the Fenton reaction for all the time points in the set.
2. Aliquot a volume of 2× thiourea quench solution, equal to the volume of the expelled sample, or three times that volume of ethanol into each of the 1.5 ml microfuge tubes in which the expelled sample will be collected (see Fig. 4). Place each spent collection tube on ice and affix a fresh tube to the exit line for each data point.
3. Load the two drive syringes with assay buffer. Load the quench syringe with solution 1.4, prepared as described above. Load solution 1.1 as Sample A.
4. Load assay buffer in the Sample B port and run the "load-fire-flush" cycle to mix the nucleic acid sample with assay buffer and collect the effluent into a tube containing the quench solution. These samples document the integrity of the ^{32}P-labeled DNA or RNA.
5. Flush the Sample B port and load it with solution 1.3, prepared as described above. Run the "load-fire-flush" cycle to obtain a zero point for the transition being studied. Two or three zero point replicates should be collected.

[1] The desired •OH footprinting reaction time is achieved with the KinTek®RQF-3 by increasing the default motor speed Shcherbakova, I., Mitra, S., Beer, R. H., and Brenowitz, M. (2006). Fast Fenton footprinting: A laboratory-based method for the time-resolved analysis of DNA, RNA, and proteins. *Nucleic Acids Res.* **34,** e48. Modified control software that allows the user to set the default motor speed is available from either the manufacturer or our laboratory.

6. Flush the Sample B port and load it with solution 1.2, prepared as described above. Run the "load-fire-flush" cycle to obtain a time point for the transition being studied by incrementing the delay time until the desired range of aging times is acquired. Collect each expelled solution separately into a tube containing the quench solution and store on ice until the time course is complete.

7. When all the desired data points are collected, the samples are processed for denaturing gel electrophoresis. One or more gels are run in order to clearly separate the regions of interest (Fig. 5). The dried gels are visualized by a storage phosphor screen and analyzed as discussed below (Ralston *et al.*, 2000; Sclavi *et al.*, 1997).

8. The upper limit of the kinetic transition being measured is set by the •OH probing of a sample equilibrated at the final solution condition. This sample is prepared by mixing equal volumes of samples A and B to 45–50 μl and incubating the solution at the reaction condition. After the whole time course of the transition is acquired, the Sample A port is flushed and loaded with the equilibrated sample. Run the "load-fire-flush" cycle to obtain final points for the transition using the smallest delay time achievable by the fast mixer.

III. Data Reduction and Production of Time-Progress Curves

A. Quantitation of the •OH Footprinting Reaction Products

The insight afforded by •OH footprinting into macromolecular function is its ability to *separately* follow the change in the solvent accessibility of each residue. For this level of insight to be achieved it is necessary that each band on an autoradiogram be individually quantitated. Quantitation of most of the discernible bands on gels such as that shown in Fig. 5A can be efficiently achieved using contemporary peak fitting software such as Semi-Automated Footprinting Analysis (SAFA) software (Das *et al.*, 2005). SAFA features include correction of geometric gel distortion and annotation of the base sequence when appropriate nucleotide-specific marker lanes are present. The output of SAFA is a spreadsheet whose columns represent the lanes on the gel and whose rows represent the density of the individual bands corresponding to the RNA/DNA fragments. SAFA also implements the automatic "normalization" of footprint transitions that corrects for variation in the ^{32}P-DNA or ^{32}P-RNA loading among the lanes of an autoradiogram (Takamoto *et al.*, 2004). Groups of bands can also be quantitated by "box analysis" in which the group is enclosed with a single contour (Brenowitz *et al.*, 1986, 2002). Details of these procedures can be found in the referenced papers. A caveat to box analysis or collecting individual bands into a group is that the included bands should change equivalently as a function of time.

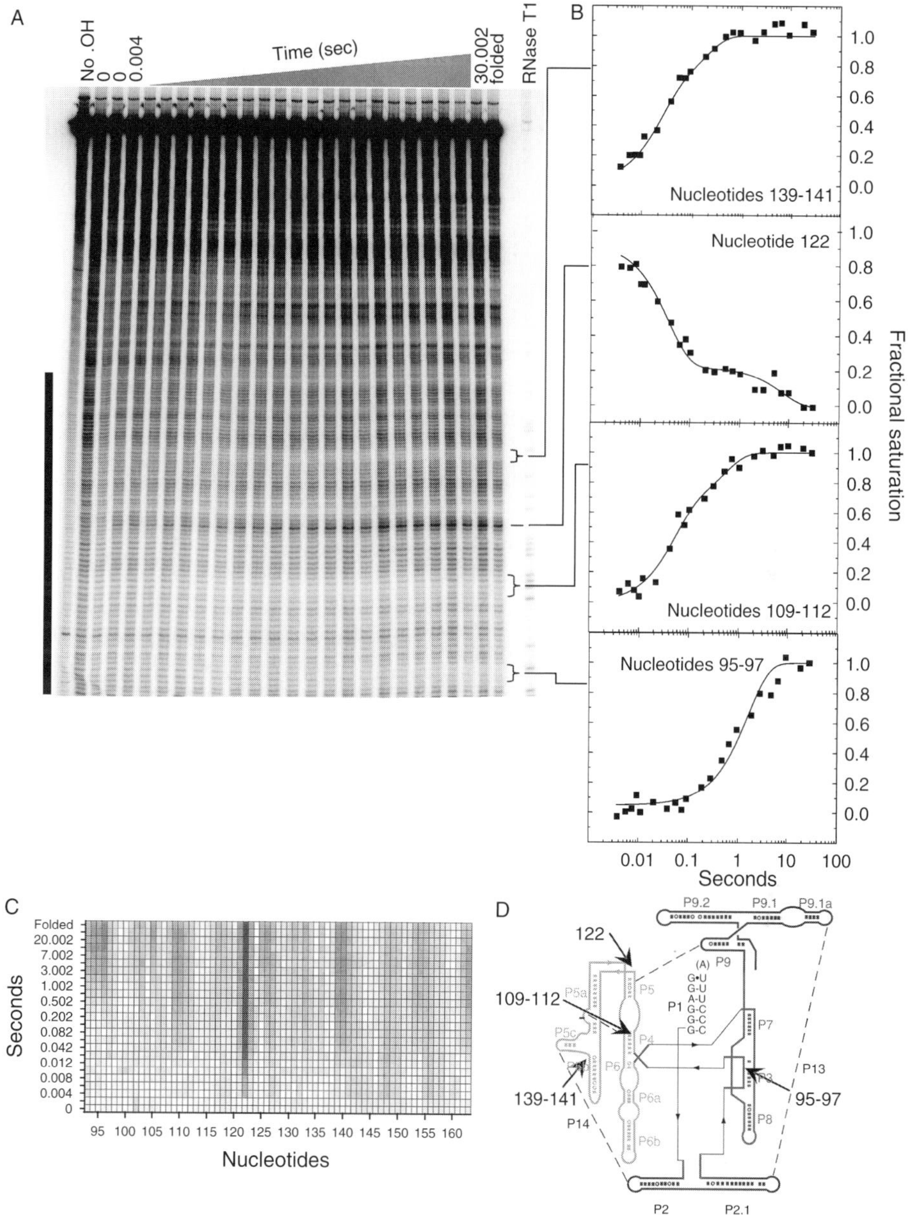

Fig. 5 (A) Autoradiogram of the cleavage products of the *Tetrahymena* ribozyme on its folding by Mg^{2+} in the presence of Na^{+}. Region of the gel included into single-band analysis is shown by the black bar; (B) Changes in solvent accessibility of protected regions and the regions of enhanced reactivity were quantitated as described in the methods description as a function of folding time. The changes in band density have been transformed to fractional saturation. Best-fit rate constants (k_i) and amplitudes (A_i) are: for nucleotides 95–97 $k_1 = 0.71 \pm 0.10$ sec^{-1}, $A_1 = 0.94 \pm 0.03$; for nucleotides 109–112 $k_1 = 19 \pm 10$ sec^{-1},

B. Assembling Nucleotides into Sites of •OH Protection or Hypersensitivity

Inspection of autoradiograms such as that shown in Fig. 5A is often sufficient for the assembly of a family of •OH reactivity changes that track local changes in solvent accessibility with time. The goal is to assemble a collection of time-progress curves that can minimally be related to the polymer sequence and ultimately to initial and/or final three-dimensional structures (Fig. 5D) (e.g., Dhavan *et al.*, 2002; Sclavi *et al.*, 1998a; Uchida *et al.*, 2003). Indeed, choice of the sites to be analyzed is often guided by knowledge of the structure of either the initial or final state of the system. The term protection denotes a decrease in the observed •OH reactivity that reflects decreased solvent accessibility. The term hypersensitivity denotes an increase in •OH reactivity as a consequence of local surface exposure. For example, nucleotide 122 is more reactive in the Mg^{2+}-folded RNA (Fig. 5A). While the cause of hypersensitivity to •OH in footprinting studies is often not clearly defined, our experience is that it also reflects changes in solvent accessibility. In our example, nucleotide 122 is less accessible to solvent in the initial compared with the final state.

Analysis of autoradiograms with single nucleotide resolution allows the investigator to objectively visualize the time evolution of *all* of the nucleotides. A false-color plot of the quantitated band intensities (Takamoto *et al.*, 2002) can be an effective tool for the identification of bands whose •OH reactivity changes with time. Such plots can be generated by SAFA. Figure 5C shows such a representation for the portion of the Fig. 5A autoradiogram, denoted with the black bar. It is immediately evident from this plot that groups of nucleotides become less reactive to •OH with time (blue coloring) and that the reactivity of nucleotide 122 increases (red coloring). It is also evident that the time dependence of these reactivity changes is different. The investigator selects individual or groups of bands from either autoradiograms or false color density plots from which individual-site time progress curves will be generated as described below.

$A_1 = 0.62 \pm 0.16$, $k_2 = 2.0 \pm 1.3$ sec^{-1}, $A_2 = 0.38 \pm 0.17$; for nucleotide 122 $k_1 = 26 \pm 5$ sec^{-1}, $A_1 = 0.71 \pm 0.03$, $k_2 = 0.14 \pm 0.05$ sec^{-1}, $A_2 = 0.23 \pm 0.03$; for nucleotides 139–141 $k_1 = 43 \pm 13$ sec^{-1}, $A_1 = 0.57 \pm 0.11$, $k_2 = 5.8 \pm 2.0$ sec^{-1}, $A_2 = 0.41 \pm 0.11$; (C) A false-color diagram that depicts changes in •OH reactivity for every nucleotide shown by black bar on the Panel A as a function of time for Mg^{2+}-induced folding of the *Tetrahymena* ribozyme. The •OH reactivity is normalized to the initial state of the ribozyme. Decreased •OH reactivity ("protection") is colored blue; increased reactivity ("enhancement") is colored red; (D) Locations of the reporters of Mg^{2+}-induced folding of the ribozyme shown on Panel (B) are indicated by black arrows on the schematic secondary structure of the ribozyme. Different colors indicate different structural domains of the ribozyme: green (light gray in the monochrome reproduction) is for the P4-P6 domains, blue (gray in the monochrome reproduction) is for peripheral helices, and red (black in the monochrome reproduction) is for the catalytic core. (See Plate no. 15 in the Color Plate Section.)

C. Generation of Individual-Site Time-Progress Curves

Single bands or groups of bands can be used to define transition curves of band density versus time. These transitions are individually scaled to fractional saturation, $\bar{Y}$, by

$$p = p_{\text{lower}} + (p_{\text{upper}} - p_{\text{lower}}) \times \bar{Y} \tag{2}$$

where p denotes the integrated density of the bands, and p_{lower} and p_{upper} represent the lower and upper limits to the transition, respectively. The lower limit p_{lower} usually is the average of the zero-time point values. The upper limit, p_{upper}, can be set from the average value of the equilibrated samples. Values of $\bar{Y}$ of 0 and 1 define the initial and equilibrated final states of the system, respectively. The fractional values of $\bar{Y}$ indicate the extent of progress along the reaction coordinate for each of the protections analyzed as a function of either a thermodynamic variable or time. While SAFA can implement Eq. (2) automatically, it is important for the investigator to carefully review the assignment of transition endpoints for accuracy and consistency.

The datasets individually scaled to fractional saturation by Eq. (2) are next fit to a single exponential or if necessary multiple exponentials

$$\bar{Y} = 1 - \sum_{i=1} \alpha_i \exp(-k_i \times t) \tag{3}$$

where α_i and k_i are the amplitude and rate constant, respectively, of the ith kinetic phase. The reaction time used in the analysis is the sum of the delay time plus half of the Fenton reaction time since the macromolecular reaction is ongoing during the Fenton footprinting. Figure 5B shows the time-progress curves generated for three sites of •OH protection and one hypersensitive band from quantitation of the autoradiogram shown in Fig. 5A. The location of these local reporters of the Mg^{2+}-induced folding of the *Tetrahymena* ribozyme is mapped on the secondary/tertiary structure diagram shown in Fig. 5D.

IV. Interpretation of Individual Nucleotide Time-Progress Curves

Time-resolved footprinting is most effectively applied to systems in which the initial, final, or both structures are known. In such cases, the collection of local solvent accessibility measures can "connect the dots" between the initial and final states and thus define the pathway(s) of the transition and provide structural constraints for the intermediate species that are populated along the way. Below is described a graduated approach to the analysis and interpretation of time-resolved •OH footprinting data that yields increasing insight into the mechanism underlying the macromolecular process being studied.

A. Analysis by Inspection

Much can be learned about a transition by "simply" comparing collections of time-progress curves. Folding or binding hierarchies and kinetic mechanisms have been deduced from sets of time-resolved •OH footprinting studies (Dhavan *et al.*, 2002; Sclavi *et al.*, 1997, 1998a; Uchida *et al.*, 2003). The information readily gleaned by inspecting a set of time-progress curves is illustrated for the four curves shown in Fig. 5B (of the 25–30 curves that can be obtained for this RNA) describing the Mg^{2+}-mediated folding of the *Tetrahymena* ribozyme.

That folding of the RNA is heterogeneous is immediately apparent (Fig. 5B). Formation of the tertiary contacts located within the P4-P6 domain (top three panels; Fig. 5D) is significantly faster than nucleotides 95–97 (bottom panel), which reports the docking of the P2 helix against the catalytic core (Fig. 5D). Thus, folding under this condition is hierarchical. Additional heterogeneity is observed within the time-progress curves; the three curves shown for P4-P6 are best described by the sum of two exponentials. This kinetic behavior may reflect either multiple kinetic processes or multiple populations of molecules. Finally, it should be noted that while nucleotide 122 increases in its reactivity with time, its time course is comparable to the two P4-P6 protections. Together these progress curves reflect the overall folding of the P4-P6 domain.

B. Time-Progress Curve Clustering

While the subset of local measures shown in Fig. 5B outlines the sequence of folding events for part of the ribozyme (Fig. 5D), it represents only a small part of the 25–30 individual measures that can be acquired for this macromolecule. Inspecting collections of such curves is tedious at best. Depending on the complexity of the system, the time-resolved •OH footprinting can routinely provide tens of individual-site time-progress curves reporting on the changes in the solvent accessibility of the biopolymer backbone caused by reaction-associated structure changes. Collection of similar curves by clustering distills a general picture from overwhelming amounts of information.

Many statistical analysis software packages are commercially available that are suitable for clustering the results of •OH footprinting studies. Much insight can be gained by fitting time-progress curves to one or more exponentials and clustering the resolved rate constants. Alternatively, the curves themselves can be clustered. We will briefly summarize the approach used in our laboratory in order to illustrate how we approach this analysis (Laederach *et al.*, 2006). Matlab® scripts and data examples for the cluster analysis and modeling of individual-site time-progress curves can be downloaded from http://simtk.org/home/KinFold.

The initial question to ask is what should be the reasonable number of clusters to adequately describe the system under study. While more clusters provide smaller intra-cluster dispersion and thus a more accurate description of the data, the precision of the data limits the fineness of the unique grouping that is possible.

The Gap-statistic (Tibshirani *et al.*, 2001) is an algorithm that validates the number of clusters that are able to explain the precision gain of the data description caused by nonuniformity in the data distribution. It is used herein to estimate the minimum number of clusters sufficient to describe the data. Once the number of clusters is established, the *k*-means clustering algorithm with a Manhattan distance metric determines the affiliation of each time-progress curve to a cluster by comparison of inter- with intra-cluster dispersion by calculating distance between corresponding points on time-progress curves as the sum of horizontal and vertical segments. Figure 6 illustrates the application of cluster to a simple model system.

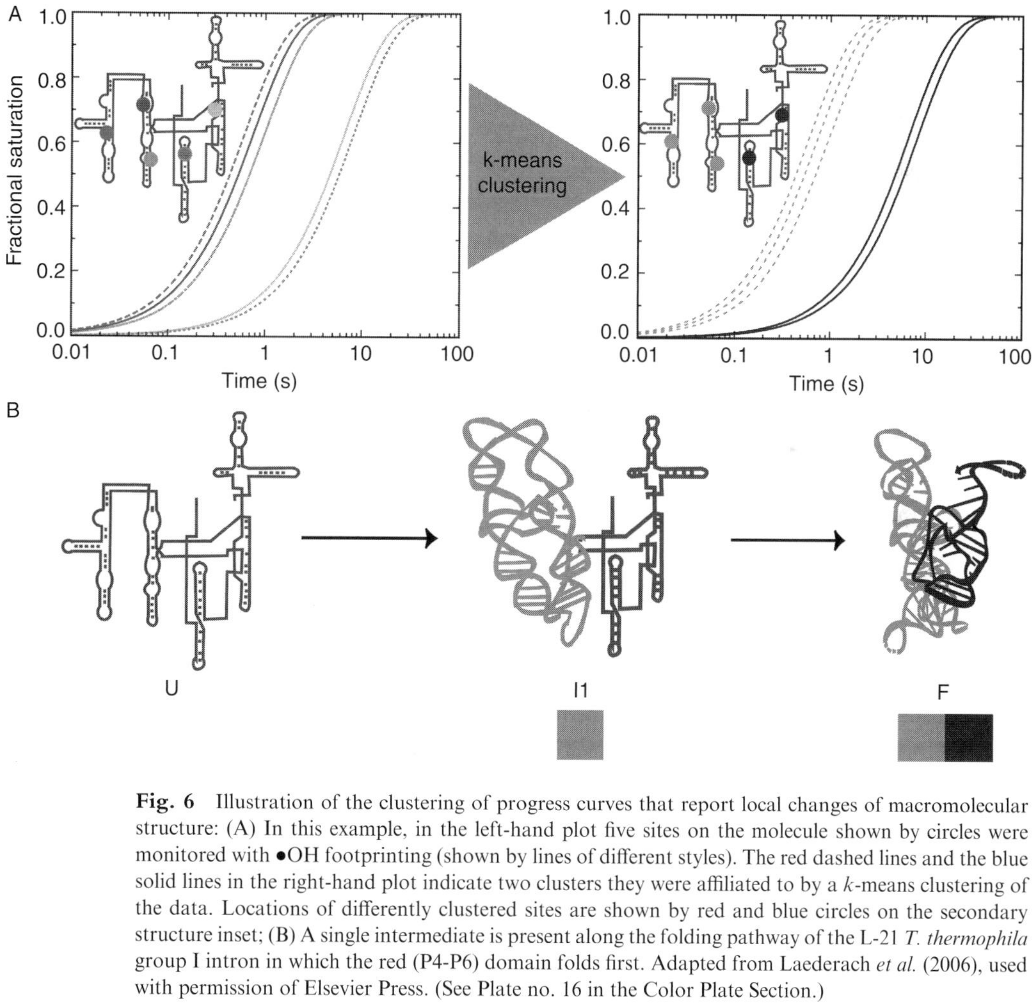

Fig. 6 Illustration of the clustering of progress curves that report local changes of macromolecular structure: (A) In this example, in the left-hand plot five sites on the molecule shown by circles were monitored with •OH footprinting (shown by lines of different styles). The red dashed lines and the blue solid lines in the right-hand plot indicate two clusters they were affiliated to by a *k*-means clustering of the data. Locations of differently clustered sites are shown by red and blue circles on the secondary structure inset; (B) A single intermediate is present along the folding pathway of the L-21 *T. thermophila* group I intron in which the red (P4-P6) domain folds first. Adapted from Laederach *et al.* (2006), used with permission of Elsevier Press. (See Plate no. 16 in the Color Plate Section.)

The left panel of Fig. 6A shows simulated kinetic progress curves for five local reporters of the *Tetrahymena* ribozyme folding. The location of these reporters on the schematic of the ribozyme's structure is indicated by the circles (Fig. 6A, inset). The Gap-statistic for these data proposes two clusters; the result of the *k*-mean clustering is shown on the right panel of Fig. 6A by the grouping of the curves into red and blue clusters. Structural insight into the folding mechanism is gleaned by connecting the location of reporters to the initial or final structure. In this way, the regions of the macromolecule characterized by similar kinetic behavior are visualized. Inspecting the location of the red cluster's reporters on the ribozyme secondary structure, we can conclude that formation of the P4-P6 domain for the presented case precedes formation of the catalytic core. The structure of the kinetic intermediate can be predicted as a molecule with formed P4-P6 domain and unfolded catalytic core resulting in a predicted folding mechanism such as shown in Fig. 6B.

The value of clustering and its importance to the kinetic modeling described in the next section is illustrated with a published study (Laederach *et al.*, 2006) conducted under experimental conditions different than those used in the experiments shown in Fig. 5. The 18 time-progress curves for this Mg^{2+}-induced folding reaction of the *Tetrahymena* ribozyme clustered into four clusters (Fig. 7; see Plate 17 in the color insert section for the full color figure); the color coding of the cluster centroids (Panel A) is reflected on the maps of the ribozyme secondary (Panel B) and tertiary structure (Panel C). The relationship between structure and folding behavior is clearly evident. The cyan cluster reflected formation of the three-helix junction in P5abc domain and A-bulge, a defined structural element (Zheng *et al.*, 2001), which slightly precedes formation of the P4-P6 domain (green cluster). The peripheral helices (red cluster) that wrap around the outside of the folded molecule form at intermediate rates. The catalytic core folds the slowest under this experimental condition. This analysis shows that the folding of the ribozyme is quasi-hierarchical: formation of the periphery (red cluster) starts before folding of the P4-P6 domain (green cluster) is complete, a similar hierarchy is observed for the periphery and the catalytic core (red and blue clusters); only folding of P4-P6 domain and the catalytic core (cyan + green and blue clusters) are completely separated. While these results can be rationalized by a sequential model (Sclavi *et al.*, 1998a), such an interpretation oversimplifies the folding process by failing to account for a well-documented property of RNA folding, the prevalence of parallel folding pathways. The modeling approach described below allows kinetic models to be proposed and critically tested that in turn allow insight into the fundamental nature of complex kinetic processes.

C. Kinetic Modeling

While much can be learned by inspecting collections of footprinting curves and their clusters, the full potential of the data can only be unlocked through the development of quantitative kinetic models (Laederach *et al.*, 2006; Sclavi *et al.*, 2005).

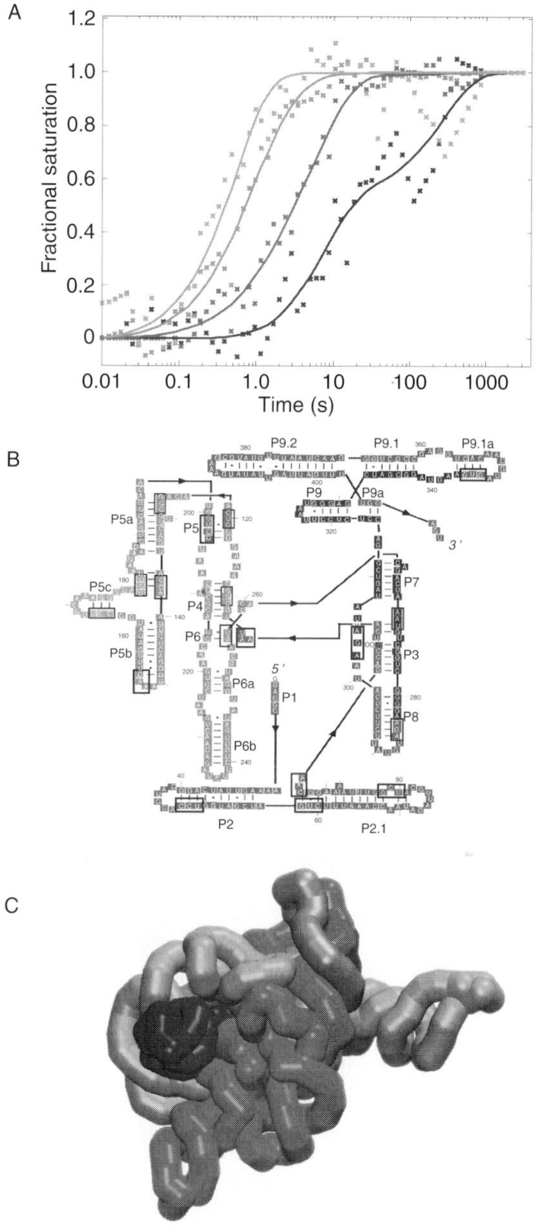

Fig. 7 Clustering of the time-progress curves. (A) Cluster centroids calculated by Kinfold algorithm from time-progress curves are plotted with symbols. The best-fitting kinetic model predictions to time-progress curves for the Mg^{2+}-mediated folding are shown by lines. Colors correspond to the different regions of the molecule shown in Panel B. (B) Colored secondary structure diagram representation of the L-21 *T. thermophila* group I intron. Colors represent regions of molecule exhibiting similar time-progress curves during Mg^{2+}-mediated folding, as determined by *k*-means clustering of site-specific progress curves. Boxes indicate sites of protection that were monitored on the molecule. From Laederach *et al.* (2006), used with permission of Elsevier Press. (C) Coarse-grained model of the folded structure of the *Tetrahymena* ribozyme with color coding as for Panel B. (See Plate no. 17 in the Color Plate Section.)

Our approach to kinetic modeling of time-resolved footprinting data starts with the cluster determination as described in the preceding section (Laederach *et al.*, 2006). It was empirically concluded that to describe the various kinetic behaviors of n clusters of curves, at least $n - 1$ intermediates are required. Thus, three intermediates are required to accommodate four clusters in our example (Fig. 8A). The combinatorial explosion of possible kinetic models as the number

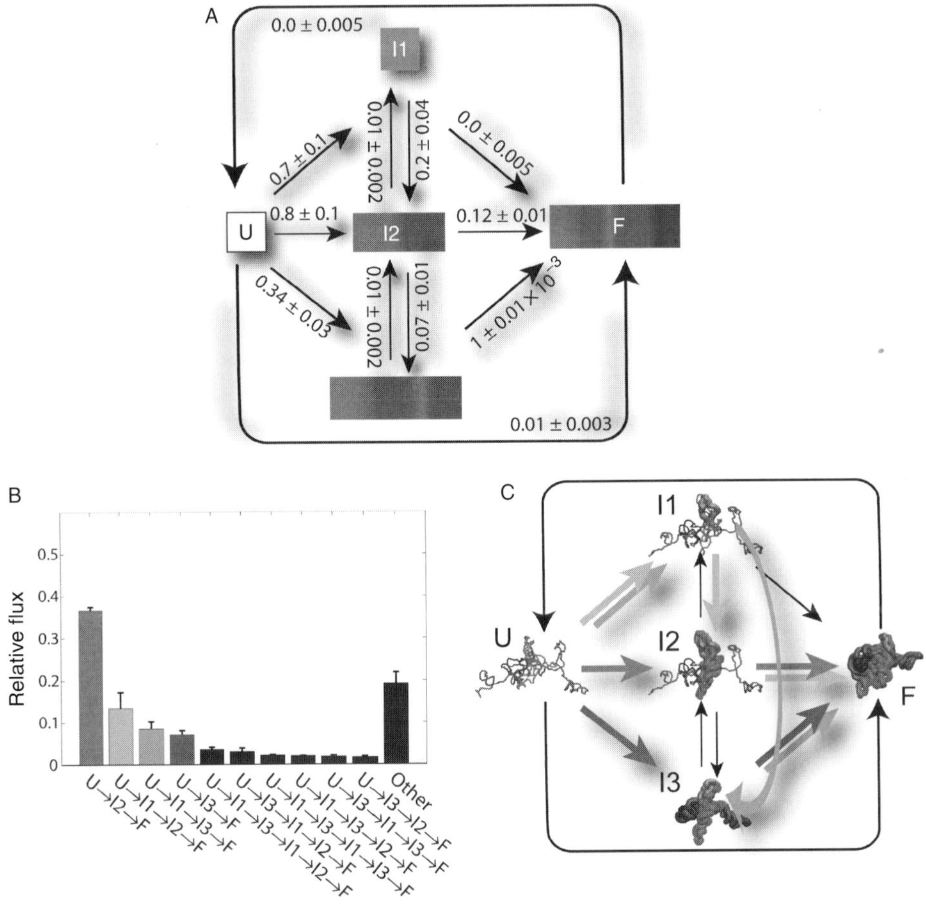

Fig. 8 Kinetic modeling of the *Tetrahymena* ribozyme folding. (A) Best-fitting kinetic model diagrams for Mg^{2+}-mediated folding of the *Tetrahymena* ribozyme at low salt condition (rate constants are in sec^{-1}). Reverse rates were constrained to zero as an equivalently good fit to the data was obtained with and without this constraint. The I1 → I3 rate is 0.3 ± 0.02 sec^{-1} but not shown for clarity. All other rates not shown are ≤ 0.01 sec^{-1}; (B) The relative flux through the different major folding pathways of the *T. thermophila* group I intron sorted from the highest to the lowest for the same folding conditions; (C) Structural cartoon models of possible conformations of the folding intermediates are shown, thicker bonds represent regions of the molecule that adopt a native-like structure. The dominant folding pathways are illustrated by arrows of corresponding color and style. From Laederach *et al.* (2006), used with permission of Elsevier Press. (See Plate no. 18 in the Color Plate Section.)

of independent local measures increases is attenuated by analyzing the clustering of time-progress curves rather than the complete collection of separate curves. We utilize parallel supercomputing to systematically test the possible models and determine the best-fit kinetic topology and corresponding rate constants.

The "winning" kinetic model topology for the dataset shown on Fig. 7A is presented on Fig. 8A. The structure of the intermediates is shown by colors (see Plate 18 in the color insert section for the full color figure), corresponding to the local reporters united as a cluster of the corresponding color to be folded for the intermediate. Definition of the structures of reaction intermediates along with the rates of their interconversion is unique to site-specific structural dynamic methods (Fig. 8C). For example, the I1 intermediate has only the P5abc three-helix junction and A-bulge formed, whereas I2 is characterized by folded P4-P6 domain, I3 has periphery formed in addition to the P4-P6 domain. The local •OH protections provide constraints for course-grained structural models such as shown on Fig. 8C.

The reaction flux through each intermediate can be calculated from the best-fit rate constants (Fig. 8B). In our example of Mg^{2+}-induced folding of the *Tetrahymena* ribozyme, the reaction flux is dispersed among the four most populated pathways (Fig. 8B). The flux analysis provides a reasonable estimation of the molecular fraction pursuing one or another pathway from unfolded to folded molecular ensemble quantitatively demonstrating the importance of parallel folding pathways. This approach to kinetic modeling is applicable to any set of time-resolved local measures, including footprinting, fluorescence, and NMR to name but a few possibilities. Systems amenable to such study include protein and RNA folding, protein–protein and protein–nucleic acid interactions, and assembly of complexes and enzymological processes. The invaluable insight gained by this approach is the distillation of overwhelming amount of local changes of solvent accessibility into a kinetic model with structurally defined intermediates.

Acknowledgments

This work was supported by National Institutes of Health grants PO1–GM066275 and RO1-GM39929 from the Institute of General Medical Sciences. R.H.B. acknowledges the financial support of Fordham University and the Research Corporation for a Cottrell College Science Grant (CC5650). We thank Jörg Schlatterer for critically reading the manuscript.

References

Ackers, G. K., Johnson, A. D., and Shea, M. A. (1982). Quantitative model for gene regulation by lambda phage repressor. *Proc. Natl. Acad. Sci. USA* **79,** 1129–1133.

Ackers, G. K., Shea, M. A., and Smith, F. R. (1983). Free energy coupling within macromolecules. The chemical work of ligand binding at the individual sites in co-operative systems. *J. Mol. Biol.* **170,** 223–242.

Aye, T. T., Low, T. Y., and Sze, S. K. (2005). Nanosecond laser-induced photochemical oxidation method for protein surface mapping with mass spectrometry. *Anal. Chem.* **77,** 5814–5822.

Balasubramanian, B., Pogozelski, W. K., and Tullius, T. D. (1998). DNA strand breaking by the hydroxyl radical is governed by the accessible surface areas of the hydrogen atoms of the DNA backbone. *Proc. Natl. Acad. Sci. USA* **95,** 9738–9743.

Brenowitz, M., Chance, M. R., Dhavan, G., and Takamoto, K. (2002). Probing the structural dynamics of nucleic acids by quantitative time-resolved and equilibrium hydroxyl radical "footprinting." *Curr. Opin. Struct. Biol.* **12,** 648–653.

Brenowitz, M., Senear, D. F., Shea, M. A., and Ackers, G. K. (1986). Quantitative DNase footprint titration: A method for studying protein–DNA interactions. *Meth. Enzymol.* **130,** 132–181.

Buxton, G. V., Greenstock, C. L., Helman, W. P., and Ross, A. B. (1988). Critical review of rate constants for reactions of hydrated electrons, hydrogen atoms and hydroxyl radicals in aqueous solution. *J. Phys. Chem. Ref. Data* **17,** 513–886.

Cate, J. H., Gooding, A. R., Podell, E., Zhou, K., Golden, B. L., Kundrot, C. E., Cech, T. R., and Doudna, J. A. (1996). Crystal structure of a group I ribozyme domain: Principles of RNA packing [see comments]. *Science* **273,** 1678–1685.

Celander, D. W., and Cech, T. R. (1990). Iron(II)-ethylenediaminetetraacetic acid catalyzed cleavage of RNA and DNA oligonucleotides: Similar reactivity toward single- and double-stranded forms. *Biochemistry* **29,** 1355–1361.

Celander, D. W., and Cech, T. R. (1991). Visualizing the higher order folding of a catalytic RNA molecule. *Science* **251,** 401–407.

Chaulk, S. G., and MacMillan, A. M. (1998). Caged RNA: Photo-control of a ribozyme reaction. *Nucleic Acids Res.* **26,** 3173–3178.

Chaulk, S. G., and MacMillan, A. M. (2000). Characterization of the Tetrahymena ribozyme folding pathway using the kinetic footprinting reagent peroxynitrous acid. *Biochemistry* **39,** 2–8.

Chen, F., Ma, W., He, J., and Zhao, J. (2002). Fenton degradation of malachite green catalyzed by aromatic additives. *J. Phys. Chem.* **106,** 9485–9490.

Das, R., Laederach, A., Perlman, S. M., Herschlag, D., and Altman, R. B. (2005). SAFA: Semi-Automated Footprinting Analysis software for high-throughput quantification of nucleic acid footprinting experiments. *RNA* **11,** 344–354.

Dhavan, G. M., Chance, M. R., and Brenowitz, M. (2003). Kinetics analysis of DNA–protein interactions by time resolved synchrotron X-ray footprinting. *In* "Analysis of Macromolecules: A Practical Approach" (K. A. Johnson, ed.)., pp. 75–86. IRL Press at Oxford University Press, Oxford.

Dhavan, G. M., Crothers, D. M., Chance, M. R., and Brenowitz, M. (2002). Concerted binding and bending of DNA by Escherichia coli integration host factor. *J. Mol. Biol.* **315,** 1027–1037.

Dixon, W. J., Hayes, J. J., Levin, J. R., Weidner, M. F., Dombroski, B. A., and Tullius, T. D. (1991). Hydroxyl radical footprinting. *Meth. Enzymol.* **208,** 380–413.

Feng, Z., Ha, J. H., and Loh, S. N. (1999). Identifying the site of initial tertiary structure disruption during apomyoglobin unfolding. *Biochemistry* **38,** 14433–14439.

Fenton, H. J. H. (1894). What species is responsible for strands scission in the reaction of $[FeIIEDTA]^{2-}$ with H_2O_2 with DNA? *J. Chem. Soc.* **6,** 899.

Galas, D. J., and Schmitz, A. (1978). DNAse footprinting: A simple method for the detection of protein-DNA binding specificity. *Nucleic Acids Res.* **5,** 3157–3170.

Golden, B. L., Gooding, A. R., Podell, E. R., and Cech, T. R. (1998). A preorganized active site in the crystal structure of the Tetrahymena ribozyme [see comments]. *Science* **282,** 259–264.

Guan, J. Q., and Chance, M. R. (2005). Structural proteomics of macromolecular assemblies using oxidative footprinting and mass spectrometry. *Trends Biochem. Sci.* **30,** 583–592.

Haber, F., and Weiss, J. (1934). The catalytic decomposition of hydrogen peroxide by iron salts. *Proc. R. Soc. Lond. A* **147,** 332–351.

Hambly, D. M., and Gross, M. L. (2005). Laser flash photolysis of hydrogen peroxide to oxidize protein solvent-accessible residues on the microsecond timescale. *J. Am. Soc. Mass. Spectrom.* **16,** 2057–2063.

Hampel, K. J., and Burke, J. M. (2001). Time-resolved hydroxyl-radical footprinting of RNA using Fe(II)-EDTA. *Methods* **23,** 233–239.

Hayes, J. J., Kam, L., and Tullius, T. D. (1990). Footprinting protein-DNA complexes with gamma-rays. *Meth. Enzymol.* **186,** 545–549.

Heyduk, E., and Heyduk, T. (1994). Mapping protein domains involved in macromolecular interactions: A novel protein footprinting approach [published erratum appears in *Biochemistry* 1995 Nov 21; 34(46): 15388]. *Biochemistry* **33,** 9643–9650.

Hsieh, M., and Brenowitz, M. (1996). Quantitative kinetics footprinting of protein-DNA association reactions. *Meth. Enzymol.* **274,** 478–492.

Hsieh, M., and Brenowitz, M. (1997). Comparison of the DNA association kinetics of the Lac repressor tetramer, its dimeric mutant LacIadi, and the native dimeric Gal repressor. *J. Biol. Chem.* **272,** 22092–22096.

King, P. A., Anderson, V. E., Edwards, J. O., Gustafson, G., Plumb, R. C., and Suggs, J. W. (1992). A stable solid that generates hydroxyl radical upon dissolution in aqueous solution. Reaction with proteins and nucleic acid. *J. Am. Chem. Soc.* **114,** 5430–5432.

King, P. A., Jamison, E., Strahs, D., Anderson, V. E., and Brenowitz, M. (1993). 'Footprinting' proteins on DNA with peroxonitrous acid. *Nucleic Acids Res.* **21,** 2473–2478.

Kiselar, J. G., Janmey, P. A., Almo, S. C., and Chance, M. R. (2003). Visualizing the Ca^{2+}-dependent activation of gelsolin by using synchrotron footprinting. *Proc. Natl. Acad. Sci. USA* **100,** 3942–3947.

Laederach, A., Shcherbakova, I., Liang, M. P., Brenowitz, M., and Altman, R. B. (2006). Local kinetic measures of macromolecular structure reveal partitioning among multiple parallel pathways from the earliest steps in the folding of a large RNA molecule. *J. Mol. Biol.* **358,** 1179–1190.

Lehnert, V., Jaeger, L., Michel, F., and Westhof, E. (1996). New loop-loop tertiary interactions in self-splicing introns of subgroup IC and ID: A complete 3D model of the *Tetrahymena* thermophila ribozyme. *Chem. Biol.* **3,** 993–1009.

Maleknia, S. D., Brenowitz, M., and Chance, M. R. (1999). Millisecond radiolytic modification of peptides by synchrotron X-rays identified by mass spectrometry. *Anal. Chem.* **71,** 3965–3973.

Mollah, A. K. M. M., and Brenowitz, M. (2000). Quantitative DNase I kinetics footprinting. *In* "Protein-DNA Interactions—A Practical Approach" (A. Travers, and M. Buckle, eds.). IRL Press at Oxford University Press, Oxford.

Nguyenle, T., Laurberg, M., Brenowitz, M., and Noller, H. F. (2006). Following the dynamics of changes in solvent accessibility of 16 S and 23 S rRNA during ribosomal subunit association using synchrotron-generated hydroxyl radicals. *J. Mol. Biol.* **359**(5), 1235–1248.

Ottinger, L. M., and Tullius, T. D. (2000). High-resolution *in vivo* footprinting of a protein-DNA complex using gamma radiation. *J. Am. Chem. Soc.* **122,** 5901–5902.

Ou, B., Hampsch-Woodill, M., Flanagan, J., Deemer, E. K., Prior, R. L., and Huang, D. (2002). Novel fluorometric assay for hydroxyl radical prevention capacity using fluorescein as the probe. *J. Agric. Food Chem.* **50,** 2772–2777.

Pastor, N., Weinstein, H., Jamison, E., and Brenowitz, M. (2000). A detailed interpretation of OH radical footprints in a TBP-DNA complex reveals the role of dynamics in the mechanism of sequence-specific binding. *J. Mol. Biol.* **304,** 55–68.

Patikoglou, G. A., Kim, J. L., Sun, L., Yang, S. H., Kodadek, T., and Burley, S. K. (1999). TATA element recognition by the TATA box-binding protein has been conserved throughout evolution. *Genes Dev.* **13,** 3217–3230.

Petri, V., and Brenowitz, M. (1997). Quantitative nucleic acids footprinting: Thermodynamic and kinetic approaches. *Curr. Opin. Biotechnol.* **8,** 36–44.

Petri, V., Hsieh, M., Jamison, E., and Brenowitz, M. (1998). DNA sequence-specific recognition by the *Saccharomyces cerevisiae* "TATA" binding protein: Promoter-dependent differences in the thermodynamics and kinetics of binding. *Biochemistry* **37,** 15842–15849.

Pogozelski, W. K., McNeese, T. J., and Tullius, T. D. (1995). What species is responsible for strand scission in the reaction of [FeIIEDTA]2- with H2O2 with DNA? *J. Am. Chem. Soc.* **117,** 11673–11679.

Ralston, C. Y., Sclavi, B., Sullivan, M., Deras, M. L., Woodson, S. A., Chance, M. R., and Brenowitz, M. (2000). Time-resolved synchrotron X-ray footprinting and its application to RNA folding. *Meth. Enzymol.* **317,** 353–368.

Sclavi, B., Sullivan, M., Chance, M. R., Brenowitz, M., and Woodson, S. A. (1998a). RNA folding at millisecond intervals by synchrotron hydroxyl radical footprinting. *Science* **279,** 1940–1943.

Sclavi, B., Woodson, S., Sullivan, M., Chance, M. R., and Brenowitz, M. (1997). Time-resolved synchrotron X-ray "footprinting", a new approach to the study of nucleic acid structure and function: Application to protein–DNA interactions and RNA folding. *J. Mol. Biol.* **266,** 144–159.

Sclavi, B., Woodson, S., Sullivan, M., Chance, M., and Brenowitz, M. (1998b). Following the folding of RNA with time-resolved synchrotron X-ray footprinting. *Meth. Enzymol.* **295,** 379–402.

Sclavi, B., Zaychikov, E., Rogozina, A., Walther, F., Buckle, M., and Heumann, H. (2005). Real time characterization of intermediates in the pathway to open complex formation by *E. coli* RNA polymerase at the T7A1 promoter. *Proc. Natl. Acad. Sci. USA* **102**(13), 4706–4711.

Shadle, S. E., Allen, D. F., Guo, H., Pogozelski, W. K., Bashkin, J. S., and Tullius, T. D. (1997). Quantitative analysis of electrophoresis data: Novel curve fitting methodology and its application to the determination of a protein-DNA binding constant. *Nucleic Acids Res.* **25,** 850–860.

Sharp, J. S., Becker, J. M., and Hettich, R. L. (2004). Analysis of protein solvent accessible surfaces by photochemical oxidation and mass spectrometry. *Anal. Chem.* **76,** 672–683.

Shcherbakova, I., Mitra, S., Beer, R. H., and Brenowitz, M. (2006). Fast Fenton footprinting: A laboratory-based method for the time-resolved analysis of DNA, RNA and proteins. *Nucleic Acids Res.* **34,** e48.

Silverman, J. A., and Harbury, P. B. (2002). Rapid mapping of protein structure, interactions, and ligand binding by misincorporation proton-alkyl exchange. *J. Biol. Chem.* **277,** 30968–30975.

Sprouse, R. O., Brenowitz, M., and Auble, D. T. (2006). Snf2/Swi2-related ATPase Mot1 drives displacement of TATA-binding protein by gripping DNA. *Embo. J.* **25,** 1492–1504.

Swisher, J. F., Su, L. J., Brenowitz, M., Anderson, V. E., and Pyle, A. M. (2002). Productive folding to the native state by a group II intron ribozyme. *J. Mol. Biol.* **315,** 297–310.

Takamoto, K., and Chance, M. R. (2006). Radiolytic protein footprinting with mass spectrometry to probe the structure of macromolecular complexes. *Annu. Rev. Biophys. Biomol. Struct.* **35,** 251–276.

Takamoto, K., Chance, M. R., and Brenowitz, M. (2004). Semi-automated, single-band peak-fitting analysis of hydroxyl radical nucleic acid footprint autoradiograms for the quantitative analysis of transitions. *Nucleic Acids Res.* **32,** E119.

Takamoto, K., He, Q., Morris, S., Chance, M. R., and Brenowitz, M. (2002). Monovalent cations mediate formation of native tertiary structure of the Tetrahymena thermophila ribozyme. *Nat. Struct. Biol.* **9,** 928–933.

Tibshirani, R. J., Walther, G., and Hastie, T. (2001). Estimating the number of clusters in a data set via the Gap statistic. *J. R. Soc. Ser. B* **63,** 411–423.

Tullius, T. D., and Dombroski, B. A. (1985). Iron(II) EDTA used to measure the helical twist along any DNA molecule. *Science* **230,** 679–681.

Tullius, T. D., and Dombroski, B. A. (1986). Hydroxyl radical "footprinting": High-resolution information about DNA-protein contacts and application to lambda repressor and Cro protein. *Proc. Natl. Acad. Sci. USA* **83,** 5469–5473.

Tullius, T. D., Dombroski, B. A., Churchill, M. E., and Kam, L. (1987). Hydroxyl radical footprinting: A high-resolution method for mapping protein-DNA contacts. *Meth. Enzymol.* **155,** 537–558.

Tullius, T. D., and Greenbaum, J. A. (2005). Mapping nucleic acid structure by hydroxyl radical cleavage. *Curr. Opin. Chem. Biol.* **9,** 127–134.

Uchida, T., Takamoto, K., He, Q., Chance, M. R., and Brenowitz, M. (2003). Multiple monovalent ion-dependent pathways for the folding of the L-21 Tetrahymena thermophila ribozyme. *J. Mol. Biol.* **328,** 463–478.

Wilkinson, K. A., Merino, E. J., and Weeks, K. M. (2005). RNA SHAPE chemistry reveals nonhierarchical interactions dominate equilibrium structural transitions in tRNA(Asp) transcripts. *J. Am. Chem. Soc.* **127,** 4659–4667.

Xu, G., Kiselar, J., He, Q., and Chance, M. R. (2005). Secondary reactions and strategies to improve quantitative protein footprinting. *Anal. Chem.* **77,** 3029–3037.

Xu, G., Takamoto, K., and Chance, M. R. (2003). Radiolytic modification of basic amino acid residues in peptides: Probes for examining protein-protein interactions. *Anal. Chem.* **75,** 6995–7007.

Zarrinkar, P. P., and Williamson, J. R. (1994). Kinetic intermediates in RNA folding. *Science* **265,** 918–924.

Zaug, A. J., Grosshans, C. A., and Cech, T. R. (1988). Sequence-specific endoribonuclease activity of the Tetrahymena ribozyme: Enhanced cleavage of certain oligonucleotide substrates that form mismatched ribozyme-substrate complexes. *Biochemistry* **27,** 8924–8931.

Zheng, M., Wu, M., and Tinoco, I., Jr. (2001). Formation of a GNRA tetraloop in P5abc can disrupt an interdomain interaction in the Tetrahymena group I ribozyme. *Proc. Natl. Acad. Sci. USA* **98,** 3695–3700.

Zhong, M., Lin, L., and Kallenbach, N. R. (1995). A method for probing the topography and interactions of proteins: Footprinting of myoglobin. *Proc. Natl. Acad. Sci. USA* **92,** 2111–2115.

CHAPTER 20

Methods and Applications of Site-Directed Spin Labeling EPR Spectroscopy

Candice S. Klug and Jimmy B. Feix

Department of Biophysics
Medical College of Wisconsin
Milwaukee, Wisconsin 53226

Abstract

Site-directed spin labeling (SDSL) electron paramagnetic resonance (EPR) spectroscopy has emerged as a well-established method that can provide specific information on the location and environment of an individual residue within large and complex protein structures. The SDSL technique involves introducing a cysteine residue at the site of interest and then covalently labeling with a sulfhydryl-specific spin label containing a stable free radical, which is used as the

0091-679X/08 $35.00
DOI: 10.1016/S0091-679X(07)84020-9

EPR-detectable probe. SDSL directly probes the local environment, structure, and proximity of individual residues, and is often greatly advantageous over techniques that give global information on protein structure and changes. SDSL can detect and follow changes in local structure due to intramolecular conformational changes or dynamic interactions with other proteins, peptides, or substrates. In addition, this technique can detect changes in distances between two sites and provide information on the depth of spin labels located within a membrane bilayer. EPR is neither limited by the size of the protein or peptide nor limited by the optical properties of the sample and has the unique ability to address and answer structure and dynamics questions that are not solvable solely by genetic or crystal structure analysis, making it highly complementary to other structural methods.

In this chapter, we introduce the basic methods for using SDSL EPR spectroscopy in the study of the structure and dynamics of proteins and peptides and illustrate the practical applications of this method through specific examples in the literature.

I. Background and Methods

A. The Technique

The site-directed spin labeling (SDSL) technique involves the covalent attachment of a spin label side chain, which contains a stable unpaired electron, to a specific site on a protein or peptide. The most commonly used spin label is the sulfhydryl-specific nitroxide, 2,2,5,5-tetramethyl-l-oxyl-3-methyl methanethiosulfonate (MTSL, Toronto Research Chemicals; Fig. 1). This probe contains an unpaired electron, which is localized primarily in the p_z orbital of the ^{14}N atom, and the probe is typically reacted with a cysteine residue as it is the only amino acid side chain containing a sulfhydryl group.

Because the SDSL method requires unique cysteines for site-specific introduction of the spin label, any native cysteines within a protein need to be substituted with another amino acid using polymerase chain reaction techniques or proven to be unreactive to the spin label due to disulfide bond formation or lack of

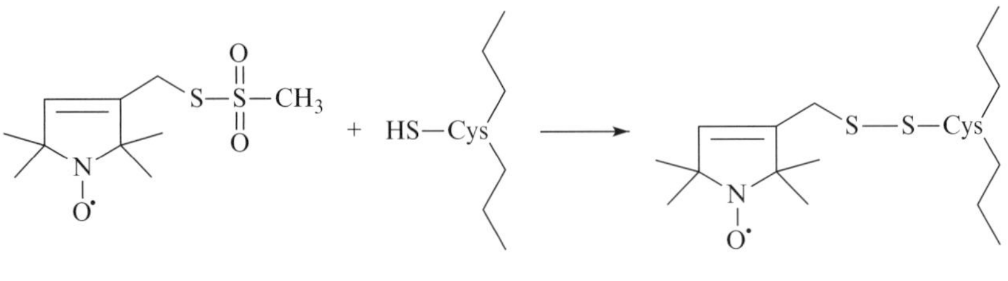

Fig. 1 Covalent attachment of the spin label to a cysteine residue. Site-specific introduction of the spin label is achieved through the selective reaction of the methanethiosulfonate group with the sulfhydryl group of the cysteine residue.

accessibility. Once the reactive cysteines have been removed, unique cysteines are introduced at each site of interest and the mutant proteins are expressed and purified. Considerable experience over the past 15 years has established that proteins are highly resilient to site-specific introduction of cysteine residues. Numerous examples now exist in the literature where a large number of sites in a given protein have been mutated to cysteine and spin labeled without loss of function (e.g., Perozo *et al.*, 1998, 2002; Xu *et al.*, 2005; Yang *et al.*, 1996). Indeed, one of the advantages of the SDSL technique is the small size and minimal perturbation of the probe.

Spin labels are sold in powder form and are typically made into a small volume (e.g., 50–100 μl), 200 mM stock solution in 100% acetonitrile (ACN). This solution is kept at −20 °C and wrapped in foil. From this stock solution, dilutions are made for each labeling experiment and used immediately. (Once in aqueous solution, MTSL will react with itself to give a disulfide-linked dimer, reducing the concentration available for reaction with the protein or peptide. This reaction does not occur in 100% ACN.) Generally, the spin label is added at a tenfold molar excess over the purified protein concentration in a buffer of pH 6.8. In most cases, the protein will precipitate out of solution if 100% ACN is added directly to the protein solution. So, dilutions from the 200 mM stock must first be made (in the protein or labeling buffer) to bring the ACN concentration to below 10%. For example, for 1 ml of 20 μM purified protein, add 1 μl of the 200 mM stock solution to 10–100 μl of buffer and then add the entire dilution volume to the protein solution. This will give a tenfold excess of spin label and keep the ACN concentration added to the protein solution at or below 10%. This is gently mixed on a nutator or rotary mixer overnight at 4 °C. If the protein is sufficiently stable, the reaction will proceed faster at room temperature. Some sites label faster and more completely than others and therefore some testing is generally required for each new labeling experiment. A surface-exposed site should label within minutes even at a 1:1 ratio; however, some buried sites require excess label and may take 16–24 h or more. Unreacted spin label is removed by extensive dialysis or repurification by affinity or desalting column chromatography. As discussed below, free spin label in solution has a very sharp signal and thus even small amounts of free label will be evident in the spectrum, so it is important to remove all the excess spin label before performing SDSL experiments.

As an essential first control, the cysteine-free protein should be expressed, purified, and subjected to the labeling reaction prior to adding new cysteines. This will allow one to determine if there are protein contaminants in the preparation that react with the label, and to assess the need for modifying the purification protocols to prevent background labeling problems. In addition, all reducing agents, such as dithiothreitol, need to be removed from the protein solution prior to labeling with spin label, since the reducing agents will cleave the newly formed disulfide bond formed between the spin label and the cysteine. In some cases, DTT can remain in the solution at concentrations significantly below that of the added spin label.

For electron paramagnetic resonance (EPR) experiments on spin-labeled proteins or peptides, the ideal concentration range for the spin label is 50–200 μM to give a good signal. Lower concentrations can also be used—down to 5–10 μM, for example—but require more acquisition time due to the need for more signal averaging, which is discussed below. The EPR machine detects only unpaired electrons, and in the majority of cases the spin label will be the only stable free radical present in the system. Therefore, in spin labeling experiments, at room temperature, the presence of nonparamagnetic contaminants such as lipids, detergents, buffer components, or other proteins do not add to the resulting EPR signal.

Sample volumes vary by technique and resonator or cavity choice. For experiments performed in a loop-gap resonator (LGR), the sample volumes used are typically about 2 μl, whereas in standard rectangular cavities the volumes used are typically 10–25 μl. Samples are contained in glass capillaries or custom plastic Polymethylpentene (TPX) sample holders and are fully recoverable after data acquisition. An example experiment in a cavity would be to record the spectrum of 15 μl of a singly spin-labeled 65-kDa protein at a concentration of 100 μM, which is less than 100 μg of protein. Or, for experiments in an LGR where only 2 μl is required, the experiment uses less than 15 μg of protein and each protein sample can be recovered. If sufficient sample is available or if the system under study cannot be concentrated, larger volumes can be accommodated in quartz flat cells (100–250 μl) or in sample configurations consisting of bundles of capillaries (0.25–1.0 ml), providing improved signal-to-noise ratios.

The main components of an EPR machine include a magnet to generate the magnetic field, a microwave source, a cavity or resonator to hold the sample, and a computer for data acquisition and analysis (Fig. 2). In traditional EPR

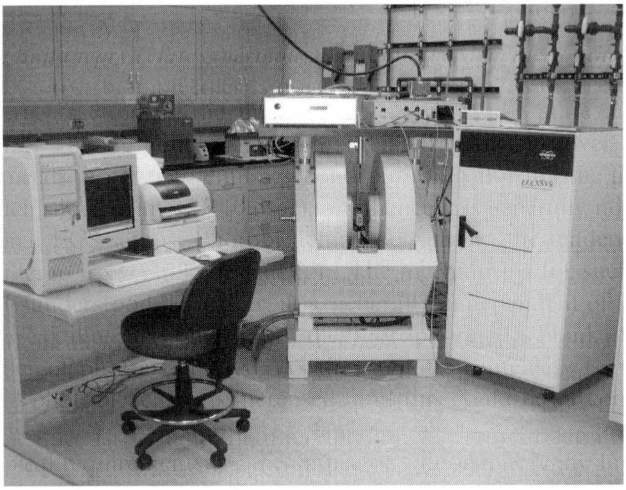

Fig. 2 A typical EPR machine setup with a magnet and microwave source (center), a power supply (right), and a computer for data acquisition (left).

experiments, the magnetic field is swept, typically over a range of 100 Gauss, and the microwave frequency remains constant. Gauss (G) is the unit of magnetic field that has been traditionally used in EPR spin labeling (1 G = 10^{-4} T). The vast majority of SDSL experiments are done at an operating frequency of ~9.5 GHz (referred to as X-band). Cavities are generally used for routine recording of spectra, while LGRs are used for experiments with limited amounts of sample and gas exchange experiments requiring saturating levels of microwave power.

EPR detects the absorption of microwave photons by the sample at specific resonant frequencies. In the presence of a magnetic field, an electron can exist in either of the two energy states (which may be thought of conceptually as aligned with or against the magnetic field) and, as shown in Fig. 3, the separation between the allowed energy levels (ΔE) for the electron spin increases as the magnetic field (H_o) is increased. When the energy difference between the two levels exactly matches the energy of the microwave radiation ($\Delta E = h\upsilon$), transitions between the spin states occur (as indicated by the vertical line in Fig. 3). If there are no interactions of the free electron with nearby nuclei, the EPR spectrum consists of a single line, as shown in Fig. 3.

For nitroxide spin labels, the unpaired electron interacts primarily with the nitrogen nucleus. This is referred to as the hyperfine interaction, and it produces small changes in the allowed energy levels of the electron that depend on the nuclear spin state, splitting the EPR signal into multiple lines. Spin labels with the predominant ^{14}N isotope give rise to a three-line EPR spectrum like the one shown in Fig. 4; the $I = 1$ nucleus splits the signal into $2I + 1$ lines (with each line corresponding to a different state of the ^{14}N nucleus). The nearby ^{16}O and ^{12}C nuclei have nuclear spins of zero and therefore do not contribute to additional line splitting (although under high-resolution conditions, splitting due to natural abundance ^{13}C, with $I = 1/2$, can be observed). Hyperfine interactions with the methyl group protons ($I = 1/2$) also occur but are typically too small to be resolved.

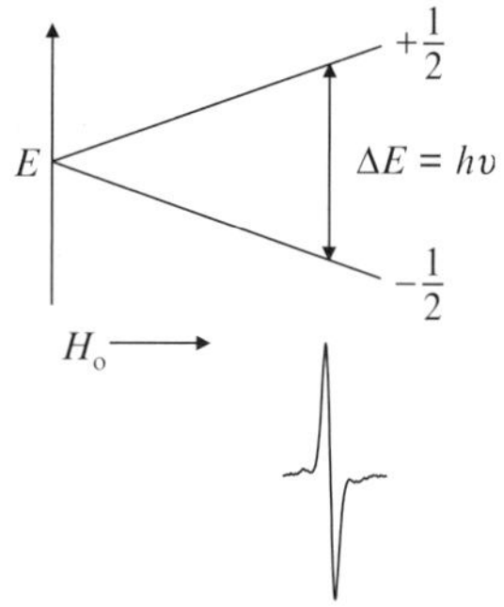

Fig. 3 EPR detects the absorption of microwave photons by the sample at a specific resonant frequency (υ). When the difference between the two energy levels (ΔE) exactly matches the energy of the microwave radiation ($h\upsilon$), transitions between the spin states occur, as shown by the vertical line, which corresponds to an EPR signal (shown below energy diagram).

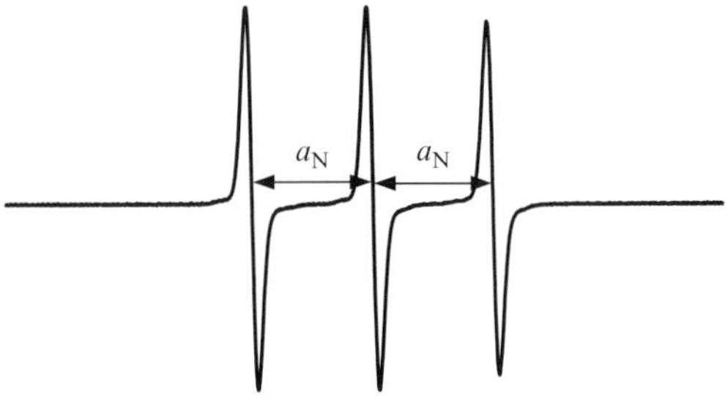

Fig. 4 Three-line nitroxide EPR spectrum. The ^{14}N isotropic hyperfine coupling constant a_N is given by the separation between the lines.

The remaining nuclei in a typical spin label are too far away (i.e., contain too little unpaired electron spin density) to significantly contribute to the spectrum.

The magnitude of the hyperfine interaction depends on the amount of unpaired electron spin density at the nucleus, and for spin labels varies as a function of the polarity of the microenvironment surrounding the nitroxide. Polar solvents, such as water, increase polarization of the N–O bond, resulting in increased electron spin density at the nitrogen nucleus and an increase in the nitrogen isotropic hyperfine coupling constant, a_N (Fig. 4). Spin labeling was one of the earliest techniques used to demonstrate the existence of a polarity gradient in membranes (Griffith and Jost, 1976; Hubbell and McConnell, 1968), and variation in a_N has been used as an indicator of bilayer depth for lipid-analogue labels (Marsh, 2001) and spin-labeled side chains in proteins (e.g., Zhang and Shin, 2006).

To improve the signal-to-noise ratio, a small (typically ~1 G) 100-kHz modulation of the magnetic field is applied and the detector filters out any signals that do not have this 100 kHz encoding. This results in a display of the EPR spectrum that is the first derivative of the original absorption spectrum (as indicated in Figs. 3 and 4). The corresponding absorption spectrum can be obtained if desired by simple integration of the first-derivative display.

There are three main categories of information that can be gained from SDSL EPR experiments: motional dynamics (due to the motion of the protein, the spin-label side chain, and/or the protein backbone), the accessibility of the spin label to paramagnetic broadening reagents, and distances between two introduced spin labels. Each of these categories of information is presented in detail below.

B. Motion

Among the most informative, and simplest, types of information that can be extracted from an EPR spectrum are parameters that reflect spin label motion. Since rotational mobility is encoded in the EPR lineshape, this information is obtained simply by acquiring the spectrum. Conventional, X-band EPR spectra are sensitive to rotational motion in the range of 0.1 to ~100 nsec. Specialized techniques such as saturation transfer (ST) EPR can extend the motional sensitivity to the millisecond timescale and are useful for determining the tumbling rate of

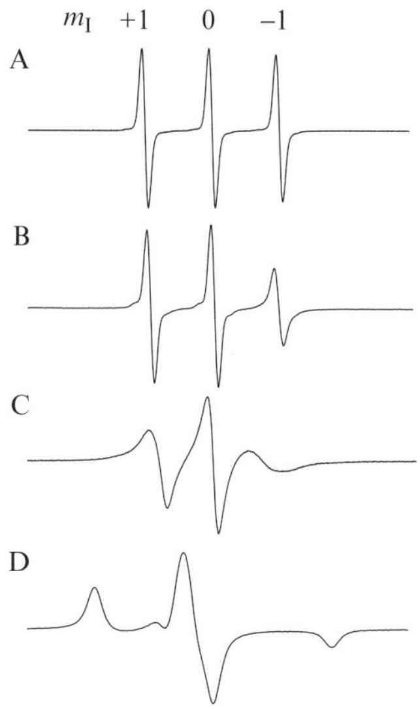

Fig. 5 The EPR spectrum is sensitive to the motion of the spin label side chain. (A) A dilute solution of MTSL, (B) MTSL bound to a small, 15-residue peptide in aqueous solution, (C) the same spin-labeled peptide folded into an α-helix, and (D) the spin-labeled peptide in frozen solution.

the spin-labeled protein as a whole or of a protein domain in a supramolecular complex (e.g., Fajer, 2000).

A series of EPR spectra corresponding to different rates of rotational motion are shown in Fig. 5. In the fast motional limit (~0.1 nsec), one observes three lines of approximately equal height (Fig. 5A). As the motion slows, the lines broaden (Fig. 5B and C). Since the intensity of each line is proportional to the product of its amplitude and the square of its width (i.e., $I \sim A\Delta H^2$), as the lines broaden their amplitudes decrease. As seen in Fig. 5, this broadening and decrease in amplitude varies for each of the three lines. The spectrum in Fig. 5D corresponds to the slow motion limit (>100 nsec) for conventional, first-derivative spin label EPR spectra. Referred to as a "powder" or "rigid limit" spectrum, this general shape is obtained for any nitroxide in the absence of rotational motion (and for a dilute powder or frozen solution).

As can be seen in Fig. 5B, attachment of the spin label to even a small, unstructured peptide results in some degree of motional restriction, and this restriction increases significantly in the presence of local secondary structure (Fig. 5C). Ultimately, the overall mobility of the spin label is determined by a combination of (1) motion of the label itself relative to the peptide backbone, (2) fluctuations of the α-carbon backbone, and (3) rotational motion of the entire protein or peptide. Using appropriate conditions, it is generally possible to isolate these various sources of spin label dynamics.

For proteins with molecular weights greater than ~15 kDa or proteins in macromolecular assemblies (including membrane proteins in cells, liposomes, or detergent micelles), the overall tumbling rate of the complex is too slow to affect the conventional EPR spectrum. However, for proteins of less than ~15 kDa, rotational diffusion of the protein can influence the observed spectrum. That is, the spin label can be rigidly buried in the protein, and yet the spectrum may indicate fast-to-intermediate mobility due to tumbling of the entire protein or peptide. This can be simply overcome by adding solutes (e.g., sucrose or glycerol) that increase solution viscosity (e.g., Mchaourab *et al.*, 1997b). Addition of 30% sucrose increases bulk viscosity and dampens the Brownian rotation of the protein without significantly altering the motion of the nitroxide side chain.

Local fluctuations of the α-carbon backbone can also contribute to spin label mobility. This type of motion is evident from nuclear magnetic resonance (NMR) measurements on the relaxation of amide nitrogens (Palmer, 2001). To study backbone fluctuations by SDSL, the flexibility of the spin label side chain can be eliminated by using derivatives of MTSL with bulky substituents such as methyl or phenyl groups at the 4′ position of the nitroxide ring (Fig. 6) in place of hydrogen (Columbus *et al.*, 2001). Additionally, the effects of backbone fluctuations may be apparent when comparing sites in similar environments where side-chain flexibility is expected to be sequence-independent, as in the case of the α-helical zipper domain of the yeast transcription factor GCN4 (Columbus and Hubbell, 2002, 2004).

In most SDSL applications, it is the motion of the spin label side chain that is of primary interest, as it is this aspect of the motion that is sensitive to tertiary contacts and protein structure in the local environment of the spin label. There are five chemical bonds between the pyrroline ring of MTSL and the α-carbon backbone of the protein or peptide to which it is attached (see Fig. 6). A large number of studies with proteins of known structure, crystallographic analysis of spin-labeled proteins (Langen *et al.*, 2000), and spin labeling with MTSL analogues containing bulky substituents in the nitroxide ring (Columbus *et al.*, 2001) have indicated that at α-helical sites the flexibility of the side chain is dominated by rotations about the two bonds closest to the nitroxide ring moiety (X_4 and X_5). Interaction of S_δ of the disulfide bond with a backbone C_α hydrogen restricts mobility about the first two bonds adjacent to the α-carbon backbone, and isomerization about the disulfide

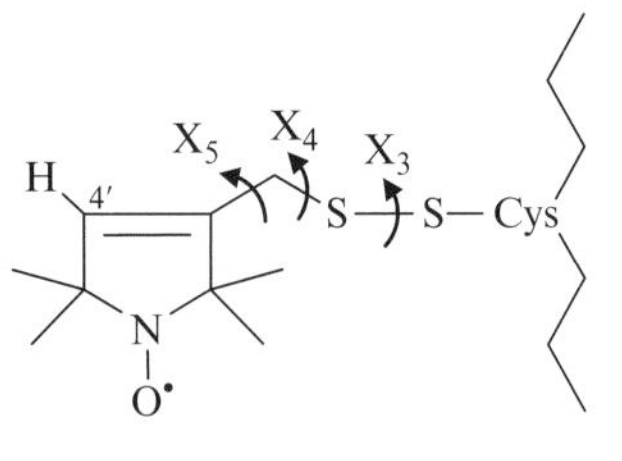

Fig. 6 A typical nitroxide side chain covalently bonded to a cysteine residue. The 4′ position of the nitroxide ring can be used to add bulky substituent groups to slow the rotation about the X_3, X_4, and X_5 bonds shown above.

bond (X_3) is slow on the EPR timescale, so that even at exposed helical sites the motion of the spin label is constrained to isomerizations about X_4 and X_5. In β-sheet proteins the situation is somewhat more complex, with spin label motion being influenced by location within the β-sheet (i.e., edge vs internal strand) and side-chain interactions with neighboring strands (Lietzow and Hubbell, 2004).

Crystal structures of T4 lysozyme (T4L) mutants labeled with MTSL indicated the presence of two favored conformations of the disulfide bond relative to the peptide backbone (Langen *et al.*, 2000). Because of these two alternative rotameric states of the MTSL side chain, EPR spectra may contain two motional components even though the protein is in a single conformational state (see Fig. 7). Nonetheless, the EPR spectral shape is still characteristic of the local structure, and changes in the relative amounts of the two motional components remain a sensitive indication of a change in protein conformation.

Although local secondary structure influences the spin label mobility somewhat, far more significant effects are seen at sites where the spin label is in tertiary contact with other side chains in the local environment. Spin labels at tertiary contact sites can exhibit complex spectral shapes (e.g., Fig. 7), and sites buried within the core of a large protein often approach the rigid limit.

As seen in Figs. 5 and 7, the effects of motion on the spin label spectrum are so dramatic that a given labeling site can often be classified as "mobile," "weakly immobilized," or "strongly immobilized" by casual inspection. However, motion can be quantitated quite precisely. For isotropic motion, the peak-to-peak width of each of the first-derivative lines, $\Delta H(m_I)$, is related to the rotational correlation time, τ_c (Stone *et al.*, 1965), such that

$$\Delta H(m_I) \propto [A + B(m_I) + C(m_I)^2]\tau_c + X \tag{1}$$

The constants *A, B*, and *C* are characteristic of the given spin label and are related to the anisotropy (orientation dependence) of the *g* and hyperfine values, while *X*

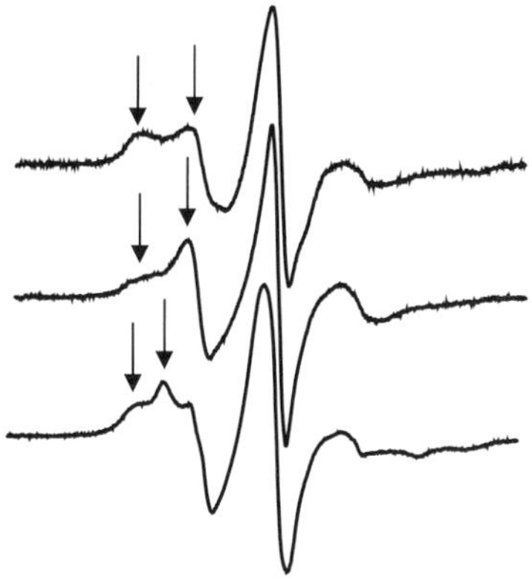

Fig. 7 Composite EPR spectra containing multiple motional components (indicated by arrows).

represents contributions to the linewidth that are not dependent on motion. The dependence of the linewidth on the nuclear spin state (m_I) indicates that each line will have a different width (e.g., Fig. 5C). The "*A*" term broadens all lines equally; the "*C*" term broadens the low- and high-field lines but does not affect the center line (for which $m_I = 0$); the "*B*" term is negative for spin labels and causes the high-field ($m_I = -1$) line to broaden and the low-field ($m_I = +1$) line to narrow. These various contributions give rise to the typical asymmetry seen for intermediate motion in Fig. 5. The rotational correlation time, τ_c, is the time necessary for the spin label to rotate through an angle of 1 radian, so that shorter times (smaller values of τ_c) indicate faster motion. The dependence on τ_c indicates that as motion decreases (i.e., τ_c increases) the lines broaden, and the greater the increase in τ_c, the more pronounced the linewidth asymmetry becomes (see Fig. 5).

In the intermediate-to-fast motional regime, where one observes three distinct first-derivative lines (e.g., Fig. 5A–C), an approximate value of τ_c is given by:

$$\tau_c = 6.5 \times 10^{-10} \Delta H_{pp}(0) \left[\left(\frac{A(0)}{A(-1)} \right)^{1/2} - 1 \right] \tag{2}$$

where $A(0)$ and $A(-1)$ are the peak-to-peak heights of the center- and high-field lines, respectively, $\Delta H_{pp}(0)$ is the peak-to-peak width of the center line, and τ_c is in seconds. The constant value is derived using the magnetic parameters of a small water-soluble nitroxide, *t*-butyl nitroxide. Although Eq. (2) is only an approximation, correlation times calculated from Eq. (2) do accurately reflect *relative* mobilities and *changes* in the spin label motion.

An even simpler parameter for examining relative mobility is the inverse width of the center line, $\Delta H_{pp}(0)^{-1}$. Since linewidth increases with decreasing motion (i.e., as correlation times become longer), taking the inverse width gives an empirical parameter that is proportional to motion. Since the center line is the narrowest (and hence has the greatest amplitude), the use of $\Delta H_{pp}(0)^{-1}$ allows one to evaluate motion from the most easily measured features of the spectrum (which can be beneficial in cases where signal intensity is limited). The inverse linewidth provides a measure of relative mobility so that when scanning through a region of secondary structure that is packed asymmetrically, one observes a periodicity in $\Delta H_{pp}(0)^{-1}$ that reflects the local structure. Similarly, if one is investigating a possible conformational change at a given site—for example upon ligand binding or protein folding/denaturation—an increase in $\Delta H_{pp}(0)^{-1}$ indicates increased mobility at that site and vice versa.

The inverse width of the center line can also be normalized to provide a scaled mobility parameter, M_S (Hubbell *et al.*, 2000), such that:

$$M_S = \left[\frac{\Delta H_{pp}(0)^{-1} - \Delta H_{pp}(0)^{-1}(i)}{\Delta H_{pp}(0)^{-1}(m) - \Delta H_{pp}(0)^{-1}(i)} \right] \tag{3}$$

where $\Delta H_{pp}(0)^{-1}(m)$ and $\Delta H_{pp}(0)^{-1}(i)$ are the inverse widths of the center line for the most mobile and most immobile sites in a given system, respectively. Thus, M_S is scaled to values between 0 and 1, with larger values indicating greater mobility. The scaled mobility can be useful for comparing local regions of secondary structure in disparate systems. For example, a plot of M_S for a series of sites in colicin E1 and annexin indicates that while both have α-helical secondary structure, the colicin E1 helix has greater mobility (Columbus and Hubbell, 2002).

A different set of parameters is used to characterize the spin label mobility in the slow motional regime. Such spectra are characterized by features corresponding to spin labels aligned with their z-axis parallel to the external magnetic field, separated by $2T_{\parallel}$, and those spin labels aligned with their xy plane along the direction of the magnetic field, separated by $2T_{\perp}$ (Fig. 8). Rotational motion causes the $2T_{\parallel}$ features to shift toward the center of the spectrum, so that the smaller values of $2T_{\parallel}$ indicate increased mobility and larger values of $2T_{\parallel}$ indicate that the motion of the nitroxide side chain has become even more restricted.

The sensitivity of the EPR spectrum to the spin label mobility is influenced by both the rate and the amplitude of the motion, and in general the rotational motion of a spin label is not isotropic (i.e., does not have the same amplitude in all directions). Indeed, even at exposed α-helical sites restriction of the internal side-chain dynamics to changes about the X_4 and X_5 bonds will limit the amplitude of the motion. Spin label motion is typically modeled as rotation within a cone about the z-axis (i.e., the nitrogen p-orbital; Fig. 9). The amplitude of the motion is described by an order parameter, S, such that,

$$S = \frac{1}{2}(3\langle\cos^2\theta\rangle - 1) \tag{4}$$

where $\langle\cos^2\theta\rangle$ is the time-averaged value for the angle of deviation of the nitroxide z-axis. Note that for θ equal to 0°, $\langle\cos^2\theta\rangle$ is 1 and the order parameter is 1, while for θ equal to 90°, $\langle\cos^2\theta\rangle$ is 1/3 and the order parameter is 0 (i.e., the motion is isotropic). For experimental spectra with distinct $T_{\parallel}$ and $T_{\perp}$ features (such as Fig. 8), the order parameter can be determined from the relationship

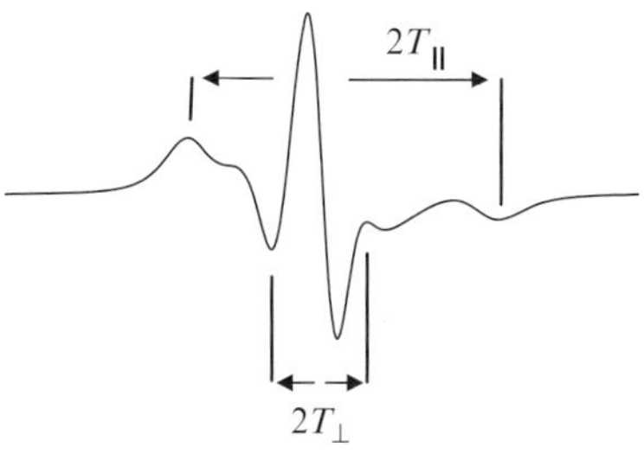

Fig. 8 A conventional, first-derivative EPR spectrum in the slow motional regime can be characterized by the motional parameters $2T_{\parallel}$ and $2T_{\perp}$, which can be used to calculate the order parameter.

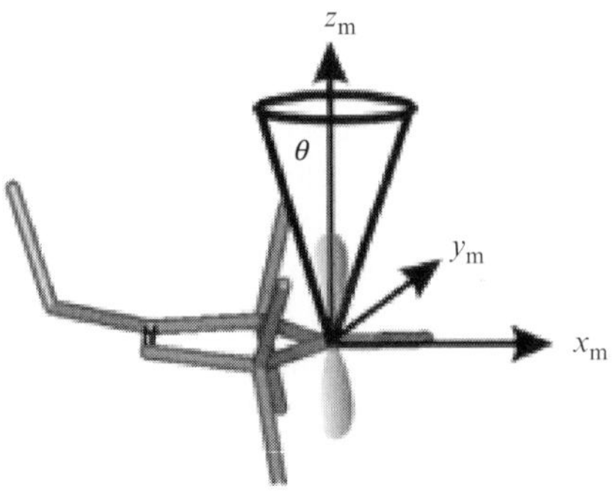

Fig. 9 The free electron in a nitroxide resides mainly in the nitrogen p-orbital, which is defined as the z-axis. Motion can be defined by the movement within a cone with a maximum angle θ about the z-axis.

$$S = \left[\frac{(T'_{\|} - T'_{\perp})}{(T_{\|} - T_{\perp})x_{\mathrm{l}}}\right]\left[\frac{a_{\mathrm{N}}(x_{\mathrm{l}})}{a'_{\mathrm{N}}}\right] \tag{5}$$

where primes indicate experimental values and "x_{l}" indicates values determined from single-crystal or powder spectra. For spectra such as those in Fig. 8, the nitrogen isotropic hyperfine coupling constant, a_{N}, is given by $1/3(T'_{\|} + 2T'_{\perp})$, and the factor $a_{\mathrm{N}}(x_{\mathrm{l}})/a'_{\mathrm{N}}$ is used to correct for differences in the experimental a_{N} relative to the $a_{\mathrm{N}}(x_{\mathrm{l}})$ value observed in a single-crystal. More detailed discussions of the order parameter and its application to the anisotropic motion (i.e., preferential motion about a given direction, such as rotation about the long axis of a lipid alkyl chain) of lipid-analogue spin labels in membranes can be found in Gaffney (1976), Griffith and Jost (1976), and Hubbell and McConnell (1968).

For the majority of SDSL studies, measurements of relative mobility along a sequence (to establish secondary structure) or following some perturbation such as ligand binding (to follow changes in conformation) are sufficient to provide the structural information needed to understand a given biological mechanism. However, it is possible to obtain a more detailed description of the spin label side chain motion by simulating the EPR spectrum. For example, a simulation program based on microscopic order—macroscopic disorder (MOMD) has been developed (Budil *et al.*, 1996) that provides both rotational correlation times and order parameters for a given spectrum. The MOMD program has been used to examine the effects of the side chain structure on spin label motion in α-helices (Columbus *et al.*, 2001) and in an antiparallel β-sheet (Lietzow and Hubbell, 2004).

C. Accessibility

The power saturation technique takes advantage of the fact that certain reagents are paramagnetic and affect the relaxation rate of the spin label. Under nonsaturating conditions, the height of the spectral lines is proportional to the incident microwave power, increasing linearly with the square root of the incident power, $P^{1/2}$. However, if high enough powers are used, the sample cannot relax fast enough to absorb additional photons and the increase in signal amplitude becomes less than linear with $P^{1/2}$, and at even higher microwave powers signal heights will begin to decrease. This phenomenon is referred to as saturation of the signal. When paramagnetic relaxation reagents interact with the spin label, they enhance the relaxation rate and allow the sample to absorb more power before becoming saturated. This process is a direct reflection of the bimolecular collision rate between the spin label and the paramagnetic relaxation agent, and provides valuable information about the environment of the introduced spin label. Conceptually, SDSL accessibility experiments are similar to fluorescence quenching, although the processes underlying the two techniques are fundamentally different.

Two main reagents are most commonly used: oxygen is small and hydrophobic and is generally found in the center of lipid bilayers and in hydrophobic pockets of proteins, and only to a small extent in the solution phase. Nickel compounds, such as nickel (II) ethylenediaminediacetate (NiEDDA), are water soluble and are located mainly in the solution phase and not found in the center of bilayers. NiEDDA is a neutral water-soluble compound and therefore does partition slightly into bilayers and hydrophobic regions, but another less commonly used paramagnetic broadening reagent is chromium oxalate (CROX), which is negatively charged and is strictly found in the aqueous phase. The natural relaxation rate of the spin label in a particular environment is measured in the presence of nitrogen, typically using N_2 flowing over a gas-permeable sample capillary to purge the sample of air. The degree to which the spin label interacts with each of these reagents can give us valuable information on its environment. The individual site can be located within a bilayer, buried within a protein, or on a solvent-exposed surface based on this type of data.

The accessibilities of a particular spin label to oxygen and NiEDDA can be determined by power saturation methodology where the height of the center line of the spectrum is measured at a series of microwave powers. First, the sample is inserted into a small gas-permeable plastic (TPX) sample tube, and after insertion into the LGR, nitrogen or oxygen gas is continuously blown over the sample to equilibrate the sample with either nitrogen or 20% oxygen. For the control and for samples containing the NiEDDA reagent, the sample is equilibrated with nitrogen, whereas air is used for the introduction of oxygen into the sample. For each sample, the height of the center line is recorded in the presence of nitrogen as a control, in the presence of 20% oxygen (air), and with the addition of 5- to 200-mM NiEDDA to the protein solution and recorded under nitrogen gas. This requires only one or two samples since the nitrogen and air experiments can be done on the same sample simply by changing the gas flowing over the sample, and the NiEDDA experiment can be carried out using a new sample or by addition of reagent to the recovered nitrogen/air sample.

Once the gas is equilibrated, the range of powers typically required is 0.5–100 mW, but this strongly depends on the configuration of the spectrometer,

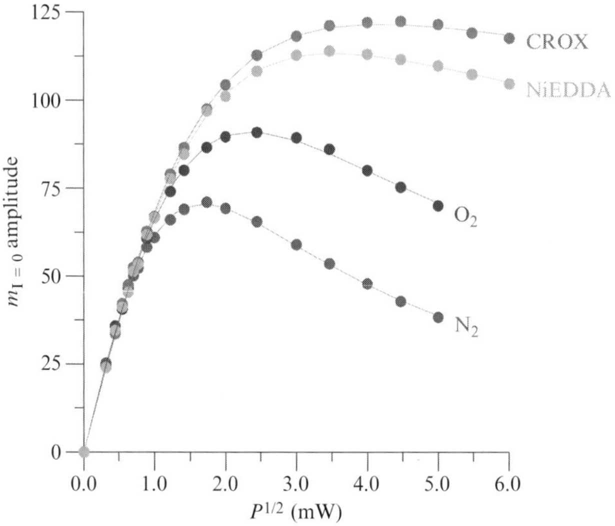

Fig. 10 Typical CW saturation plots for a sample in the presence of various paramagnetic broadening reagents. This particular spin label is in a water-soluble environment, based on the fact that it takes longer for the EPR signal amplitude to become less than linear with power in the presence of the water-soluble reagents CROX and NiEDDA than in the presence of hydrophobic oxygen. A power saturation curve is always run in the presence of nitrogen as a baseline value. These curves can be fitted to Eq. (6) to obtain a quantitative power saturation value, $P_{1/2}$.

the resonator properties, and the degree of accessibility of the site being studied. The central lineheight is then plotted against the square root of the incident microwave power (Fig. 10) and fitted to the following equation, which yields a $P_{1/2}$ value (Yu *et al.*, 1994).

$$A = IP^{1/2}\left[\frac{1 + (2^{1/\varepsilon} - 1)P}{P_{1/2}}\right]^{-\varepsilon} \quad (6)$$

$P_{1/2}$ is the power at which the intensity of the first-derivative center lineheight is half of its unsaturated intensity and is referred to as the power saturation parameter. A is the height of the center line, I is a scaling factor, P is the incident microwave power, and ϵ is a line homogeneity factor. The resulting $P_{1/2}$ value for the nitrogen control is subtracted from both $P_{1/2}(O_2)$ and $P_{1/2}$(NiEDDA) to generate $\Delta P_{1/2}$ values that quantitate the degree of accessibility to oxygen and NiEDDA, respectively (Altenbach *et al.*, 1989). That is, for NiEDDA

$$\Delta P_{1/2}(\text{NiEDDA}) = P_{1/2}(\text{NiEDDA}) - P_{1/2}(\text{N}_2) \quad (7)$$

The $\Delta P_{1/2}$ value is a direct measure of the bimolecular collision rate between the spin label side chain and the paramagnetic relaxation agent, which yields accessibility information at a very local site. $\Delta P_{1/2}$ values can be normalized to account for spectrometer configuration, resonator properties, and spectral linewidth to give the accessibility parameter, Π, using diphenylpicrylhydrazyl (DPPH) as a standard for calibration (reviewed in Feix and Klug, 1998; Klug and Feix, 2004). Π is a dimensionless parameter that essentially translates the $\Delta P_{1/2}$ values into universal values that can be compared across different labs and proteins.

Standard cavities are not normally suitable for power saturation studies because they do not provide high enough powers to adequately saturate the sample. LGRs concentrate the microwaves within the resonator, allowing one to attain saturating conditions. The incident microwave power (P) and the microwave field generated within the resonator or cavity (H_1) are directly related by a $\wedge$ factor: $H_1 = \wedge P^{1/2}$. Cavities tend to have a $\wedge$ value of ~1, while for LGRs, $\wedge$ is ~5–8, significantly increasing the microwave field experienced by the sample.

As an example of this technique, a solvent-exposed site on a protein would give a very high $\Delta P_{1/2}$ value in the presence of NiEDDA, whereas a site exposed to the lipid phase near the center of the bilayer would yield very high oxygen $\Delta P_{1/2}$ values, but very small $\Delta P_{1/2}$ values for NiEDDA. For example, from the data in Fig. 10, one can immediately identify the site as being solvent exposed because it is visibly apparent that it takes longer to saturate the EPR signal in the presence of the water-soluble reagents NiEDDA and CROX than it does in the presence of hydrophobic oxygen. In addition to important information on the environment of specific sites within a protein or peptide, a series of consecutive sites can be individually studied to produce secondary structure information. This method, where the accessibility values are plotted against position number, can distinguish between α-helical, β-strand, and unstructured regions of a protein. This works because external α-helices have a periodicity of 3.6 residues per turn, so every 3 or 4 residues would be highly exposed to the solvent (showing high NiEDDA values), and the sites on the other side of the helix would be buried against the protein and show low $\Delta P_{1/2}$ values for both NiEDDA and oxygen (see Fig. 11). Similarly, β-strands give a periodicity pattern of 2.0 (Fig. 11), and unstructured regions will show no regular periodicity. Sequences buried within a protein or elements of secondary structure that lack any tertiary contacts would not be differentiated using this technique as all sides of the structure would be either completely inaccessible or fully accessible to the reagents used, respectively. In-phase periodicities of oxygen and NiEDDA data indicate solvent-exposed structural elements, whereas out-of-phase periodicities of the oxygen and nickel accessibilities identify lipid-exposed structural elements.

For integral membrane proteins or membrane-associated peptides, the depth of a given spin label side chain within a lipid bilayer can be determined using accessibility data based on the inverse concentration gradients of oxygen and NiEDDA within a lipid bilayer. This is an extremely useful technique for positioning protein or peptide segments within the lipid bilayer (e.g., Hubbell *et al.*, 1998;

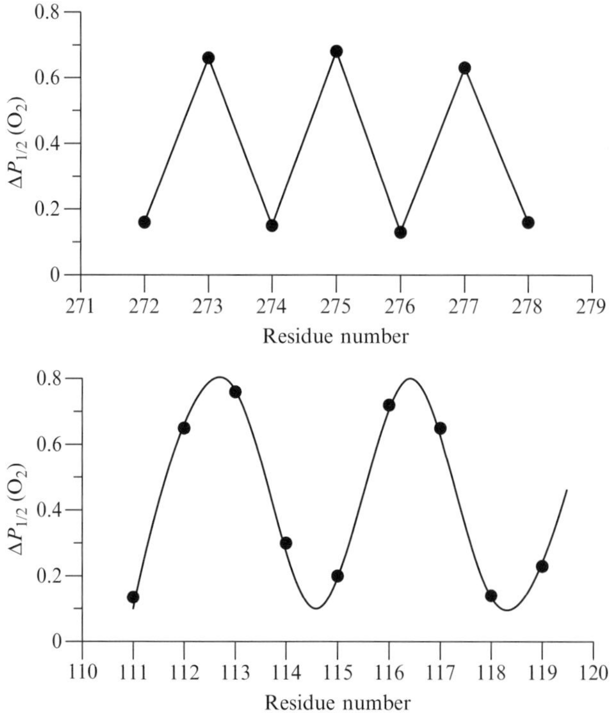

Fig. 11 Idealized accessibility data plots indicating β-strand and α-helical secondary structure.

Isas *et al.*, 2002; Perozo *et al.*, 2001; Zhao *et al.*, 1999). In order to obtain depth measurements of spin label side chains facing the membrane, another parameter, Φ, is calculated based on the following equation (Altenbach *et al.*, 1994):

$$\Phi = \ln\left[\frac{\Delta P_{1/2}(\mathrm{O_2})}{\Delta P_{1/2}(\mathrm{NiEDDA})}\right] \quad (8)$$

The variation in the concentrations of oxygen and NiEDDA within the membrane bilayer has been shown to be inversely proportional (Altenbach *et al.*, 1994); oxygen concentration is greatest at the center of the bilayer and decreases toward the membrane surface, whereas NiEDDA concentration is greatest at the surface of the bilayer in the aqueous phase and decreases to nearly zero in the center of the bilayer. Therefore, the natural log of this ratio yields Φ, a parameter with a linear dependence on depth into the bilayer. Φ can then be calibrated to each particular membrane system using lipid-analogue spin labels that localize at known bilayer depths (e.g., Fig. 12). Membranes containing unlabeled wild-type proteins are calibrated with the addition of 0.5–1 mol% spin-labeled lipids (relative to host lipid). 5-, 7-, 10-, and 12-doxyl PC lipids (Avanti Polar Lipids) contain

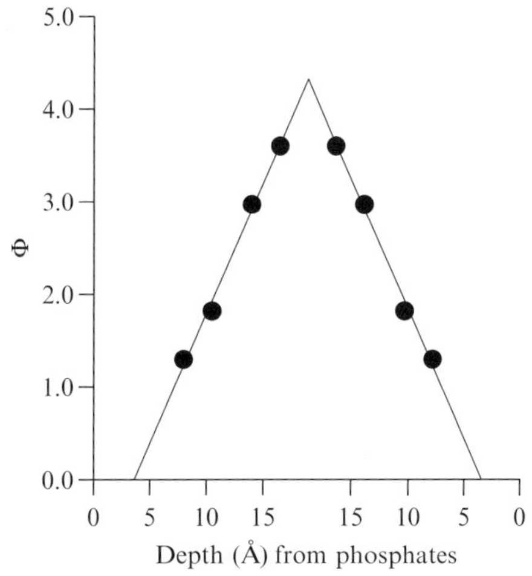

Fig. 12 Example of a bilayer depth calibration using 5-, 7-, 10-, and 12-doxylPC spin labels. This system, which contained reconstituted, unlabeled FepA, yielded a calibration equation of $\text{Å} = 3.56\Phi + 3.62$ (Klug *et al.*, 1997).

a spin label at the indicated positions and have known depths in the bilayer (Dalton *et al.*, 1987). The accessibility measurements obtained for each lipid label and their Φ values are plotted against their known depths to yield a bilayer depth calibration equation:

$$\text{depth}(\text{Å}) = m\Phi + b \tag{9}$$

The resulting equation is used to calculate the depth of membrane-exposed residues of integral membrane proteins and peptides, based on their respective Φ values (as in Fig. 12; Klug *et al.*, 1997). Note that Φ values do not provide depths for sites not exposed to the lipid phase (e.g., buried within a helical bundle or facing an aqueous transmembrane channel). The resolution of this technique is limited only by the rotational range of the spin label side chain and is more than sufficient for insertion depths of specific sites within a protein or peptide in a membrane bilayer.

D. Distances

One of the most active, rapidly developing aspects of SDSL is the ability to make distance measurements between two spin labels, either intramolecular distances between two labels in the same monomer or intermolecular distances between sites on different proteins (for detailed reviews, see Borbat and Freed, 2000; Eaton *et al.*, 2000; Hustedt and Beth, 2000; Xiao and Shin, 2000). Distance measurements between two nitroxides can provide information on both protein structure and functional dynamics. Various methodologies have been developed to study this interaction in biological systems in frozen solutions or at room temperature. The ability to monitor conformational changes within a large protein structure based on changes in spin–spin interactions (and, therefore, in the distances between the two labeled sites) is a unique benefit of this technique. Because SDSL distance measurements can be used to determine how regions of secondary structure are organized, it, in principle, allows the development of a full structural model based on EPR data alone. For SDSL distance measurements, both sites are typically labeled with a single type of label such as MTSL, as there is no requirement for distinct donor and acceptor probes (see Chapter 8 by Shanker and Bane, this volume for a parallel discussion of donor acceptor pairs in fluorescence measurements).

Distance measurements are based on observing the effects of magnetic dipolar interactions between the unpaired electrons of two spin labels (or of a spin label and a paramagnetic metal ion). This is a through-space interaction that is not influenced by intervening protein structure (an advantage relative to chemical cross-linking methods). In the range of ~8–20 Å, interactions between the magnetic dipoles of the two labels give rise to distance-dependent line broadening in the conventional continuous wave (CW) EPR spectrum. Line broadening is accompanied by a decrease in signal amplitude, and for distances up to ~15–20 Å a decrease in peak height for spectra normalized to the same spin concentration can often be used as a qualitative indication of spin–spin interaction (e.g., Fig. 13). In many cases, this is all that is necessary to determine oligomerization state or demonstrate a conformational change. For higher resolution structural analysis, quantitative distances can be determined using spectral simulation approaches based on Fourier deconvolution (Altenbach *et al.*, 2001c; Rabenstein and Shin, 1995), convolution of the spectra from noninteracting spins with a broadening function (Mchaourab *et al.*, 1997b; Steinhoff *et al.*, 1997), or by rigorously simulating spectra with consideration of both the distance between spin labels and their relative orientations (Hustedt *et al.*, 1997, 2006). The resolution limit of these

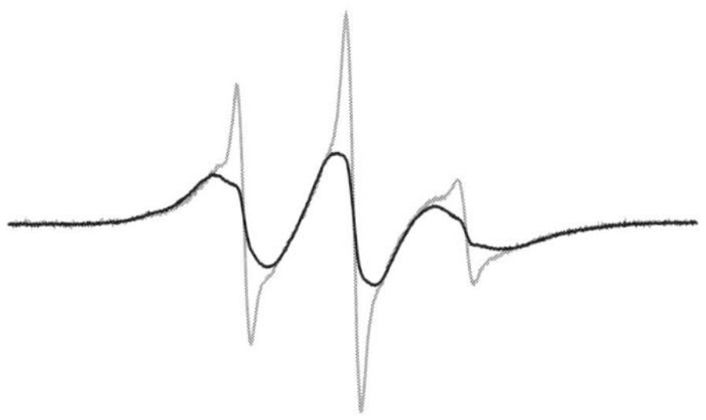

Fig. 13 Spin–spin broadening. The gray line represents the sum of the two single mutant spectra, and the black line represents the broadened spectrum resulting from dipolar interaction between the two spin labels in the double mutant.

methods is determined primarily by the distribution of distances between spin labels that arises because of the inherent flexibility of the nitroxide side chain and is typically on the order of 1–2 Å. For highly immobilized, well-oriented pairs of spin labels, a resolution of 0.1–0.2 Å has been achieved (Hustedt *et al.*, 1997).

Recently developed pulse EPR methods, including pulse electron–electron double resonance (DEER or pELDOR) (Brown *et al.*, 2002; Pannier *et al.*, 2000; Steinhoff *et al.*, 1997; Zhou *et al.*, 2005) and double quantum coherence (DQC) (Bonora *et al.*, 2004; Borbat and Freed, 2000; Borbat *et al.*, 2001, 2002) have now made possible the measurement of interspin distances in the range of ~20–60 Å. This increase in range has greatly expanded potential applications. With this increased sensitivity, the researcher has significantly more flexibility in deciding where to place the spin labels and can now choose nonperturbing sites, such as on the outer surface of an exposed helix. In this approach, a second microwave frequency is used to perturb the relaxation of a saturated spin label, resulting in modulation of its time-dependent signal intensity (Fig. 14). Analysis of the depth and frequency of modulations provides a distance between the two labels. In addition, the width of the derived distance distribution contains information on structural heterogeneity, that is, a narrow distance distribution indicates that the protein is structurally homogeneous and that the spin labels occupy a narrow range of orientations with respect to the peptide backbone. At present, the need for

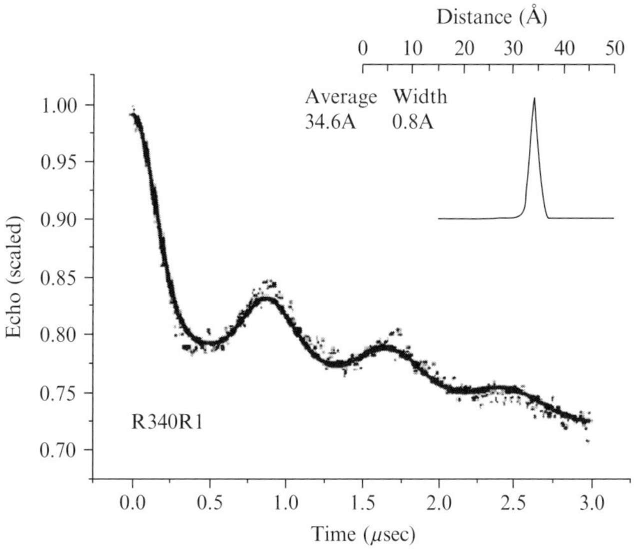

Fig. 14 Example of a DEER spin echo distance distribution measurement. Reprinted with permission from Zhou *et al.* (2005). Copyright 2005 American Chemical Society.

sufficiently long relaxation times requires these measurements to be made at liquid N_2 temperature (77 K) or lower; a number of studies have indicated that rapid freezing in the presence of a suitable cryoprotectant (e.g., glycerol or sucrose) generally preserves the relevant structure.

II. Examples

In this section, we illustrate the practical applications of the SDSL methods described above through specific examples in the literature. Although only a very small set of the many published experiments using SDSL are used here as illustrations of the usefulness of the technique, a variety of reviews also exist to readily point out additional experiments (e.g., Borbat *et al.*, 2001; Feix and Klug, 1998; Freed, 2000; Hubbell *et al.*, 1998, 2000; Hustedt and Beth, 1999; Klug and Feix, 2004; Millhauser, 1992; Thompson *et al.*, 2001).

A. Secondary Structure

The secondary structural elements of a section of protein can be determined by nitroxide scanning through the region of interest and plotting the accessibilities or motional parameters of each spin-labeled site as a function of sequence position. Useful information includes not only the secondary structure of the sites studied, but can also indicate changes in structure or lack of structure. Examples of each type of secondary structural element as determined mainly by power saturation EPR are presented below.

1. α-Helix

One of the most studied proteins by SDSL is T4L, which is a highly helical protein as determined by X-ray crystallography. It is known as the workhorse for the SDSL technique and was first used as a model helical protein to demonstrate the effectiveness of the accessibility approach to structure determination by EPR (Hubbell *et al.*, 1996). A set of eight consecutive residues on an external helix were individually spin labeled and their accessibilities to the paramagnetic broadening reagent oxygen were plotted versus sequence number. The data points could be overlaid with a sine wave with a periodicity of 3.6, clearly demonstrating the capability of the technique to correctly identify α-helical secondary structure. This study also indicated that the introduction of a spin label side chain does not significantly perturb the protein structure. It was also shown in this study that plotting the inverse of the central linewidth of the spectrum showed a similar periodicity, and this mobility parameter can often be used as an indicator of structure.

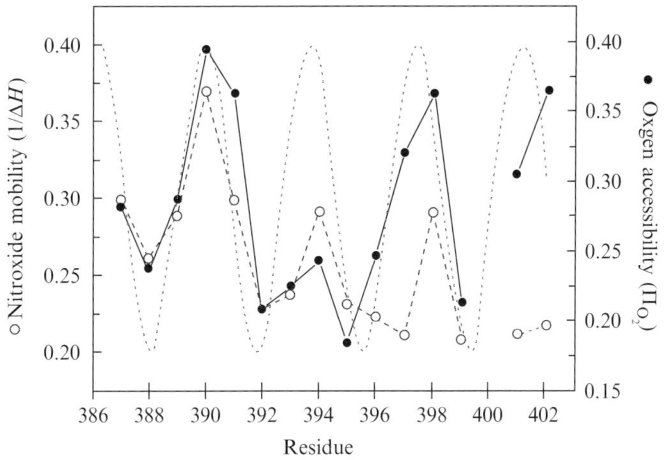

Fig. 15 Accessibility and mobility data are consistent with α-helical secondary structure, as indicated by the 3.6 residue periodicity (dotted line), for a stretch of residues in lactose permease. Reprinted with permission from Voss *et al.* (1996). Copyright 1996 American Chemical Society.

Another early example of the use of accessibility and mobility measurements to identify α-helical secondary structure is the study of lactose permease (Voss *et al.*, 1996). Fourteen sites in one of the 12 putative α-helices of lactose permease were spin labeled and analyzed by accessibility to oxygen and nitroxide mobility ($\Delta H_{pp}(0)^{-1}$). The data showed a periodic dependence of 3.6 for both the accessibility and the mobility parameters, directly supporting the idea that these residues form an α-helix (Fig. 15) and that it is located on the outside of the proposed helical bundle.

2. β-Strand

Cellular retinol-binding protein (CRBP) is a small water-soluble β-sheet protein with a known crystal structure that was first used to demonstrate the effectiveness of the SDSL power saturation technique in identifying β-strand secondary structure (Hubbell *et al.*, 1996; Lietzow and Hubbell, 1998, 2004). The accessibilities to oxygen for five consecutive sites on an external strand were plotted against residue number and the data clearly revealed a periodicity of 2.0. The sites with higher accessibility values corresponded to the solvent-exposed surface, while the sites with lower accessibilities were buried within the protein, allowing assignment of not only the secondary structure for this section of protein but also the fold.

The first studies on an integral membrane protein with a high content of β-strand secondary structure involved the ferric enterobactin receptor, FepA (Klug *et al.*, 1997). A nitroxide scan through a predicted transmembrane segment clearly

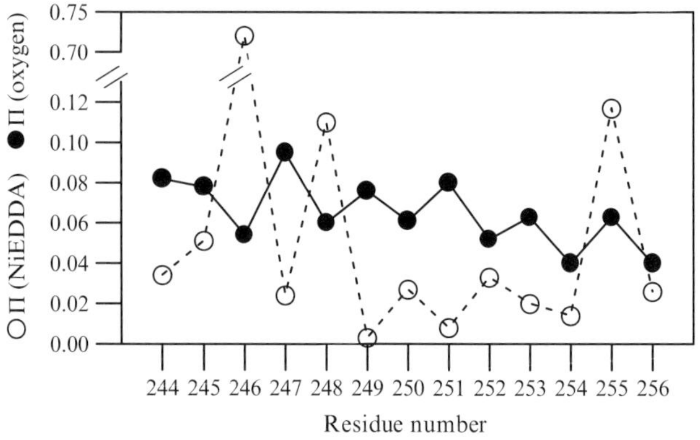

Fig. 16 Variation in the accessibility parameter *Π* as a function of spin label position in FepA. The alternating out-of-phase periodicity indicates residues 245–253 form a transmembrane *β*-strand. Note that the variation in accessibility becomes in-phase once the strand exits the bilayer (i.e., before 246 and after 253). Reprinted with permission from Klug and Feix (1998). Copyright 1998 Cold Spring Harbor Press.

demonstrated that sites 245–253 showed an alternating (2.0) periodicity in their accessibilities to oxygen and NiEDDA, proving *β*-strand structure, and also that the accessibilities were out-of-phase, indicating that the strand was located in a membrane environment (Fig. 16). Depth measurements (discussed below) confirmed that this *β*-strand spanned the bilayer. Identification of membrane interfaces is a particularly useful feature of SDSL. As seen in Fig. 16, accessibility to NiEDDA increases sharply at residues 246 and 255, indicating a transition from the membrane to the aqueous phase. FepA from *Escherichia coli* was subsequently crystallized (Buchanan *et al.*, 1999) and shown to be a transmembrane *β*-barrel protein with a large N-terminal plug domain filling the interior channel, and the location of the *β*-strand characterized by SDSL was confirmed in precise detail. This SDSL study also indicated unusually low NiEDDA accessibility, as well as low mobility, for residues facing the inside of the barrel (250–254). This was later accounted for by the presence of the 153-residue plug discovered in the crystal structure.

One of the highly conserved sequences within the soluble lens protein αA-crystallin was investigated as the first-soluble *β*-strand to be studied on an uncrystallized protein structure (Mchaourab *et al.*, 1997a). Analysis of 12 consecutive sites demonstrated that this conserved section of protein formed a *β*-strand based on a clear 2.0 periodicity in the accessibility data. One side of the strand showed very low accessibility to both oxygen and NiEDDA along with very slow motion, indicating that this face of the strand has extensive tertiary interactions within the protein. Although residues on the opposing face had higher accessibilities, they remained low relative to other systems. The authors concluded that this identifies sites of contact within the quaternary or oligomeric assembly of the protein subunits, giving additional insight into the folding of the protein beyond that of secondary structure information.

3. Changes in Structure

As well as identifying regions of secondary structure within a protein or peptide, the SDSL technique can also identify unstructured regions, which are characterized by a lack of regular periodicity in their accessibility plots. Also, changes in local structure can be monitored using this technique.

For example, a series of spin labels were introduced into the 140-amino acid protein α-synuclein, and their EPR spectra were recorded in solution and in the presence of membranes (Jao *et al.*, 2004). α-Synuclein is considered a natively unfolded protein, and previous analysis by circular dichroism suggested that the protein undergoes a conformational change from unstructured in solution to α-helical in the membrane-bound form. To investigate this structural change, numerous single cysteine mutants were constructed and spin labeled with MTSL. In solution, their EPR spectra showed very fast motion, clearly indicating that the protein is largely unfolded in this state. However, upon addition of membranes, the majority of the spectra became more immobile, confirming that a conformational change does occur. To characterize the secondary structure of this membrane-bound form, the authors determined the accessibilities to oxygen and NiEDDA for 32 consecutive residues. The data correspond to a regular α-helical structure with a periodicity of 3.67 amino acids per turn, remarkably similar to the ideal value of 3.6. In addition, a helical wheel representation of the residues studied was generated based on this periodicity, which revealed that this helix does indeed have a polar face that is solvent-exposed and a hydrophobic face that interacts with the membrane. The average immersion depth of the membrane-exposed helix side chains was determined from the saturation parameters to be ~11 Å for the entire helix, suggesting that it lays flat across the surface of the membrane.

Mobility can also be a useful tool for determining secondary structure and tracking changes in conformation. For example, sites in the N-terminal TonB box region of the vitamin B_{12} transporter BtuB were recorded in the absence and presence of ligand to identify the structural change that occurs in this region upon ligand binding (Hubbell *et al.*, 2000; Merianos *et al.*, 2000). When the mobility parameters (the inverse linewidths of the central line) for each spectrum are plotted in the absence of ligand, they show a periodicity that is consistent with α-helical secondary structure (Fig. 17). However, upon addition of ligand, mobility parameters at all sites become relatively similar and increase, suggesting loss of secondary structure and increased flexibility in the Ton box. These changes can be interpreted in terms of a proposed transmembrane signaling mechanism for BtuB (Merianos *et al.*, 2000).

B. Membrane Depth

The first example of measuring depths of spin labels within a membrane was published in 1994 and describes the methodology as applied to the transmembrane helical protein bacteriorhodopsin (Altenbach *et al.*, 1994). The authors make use

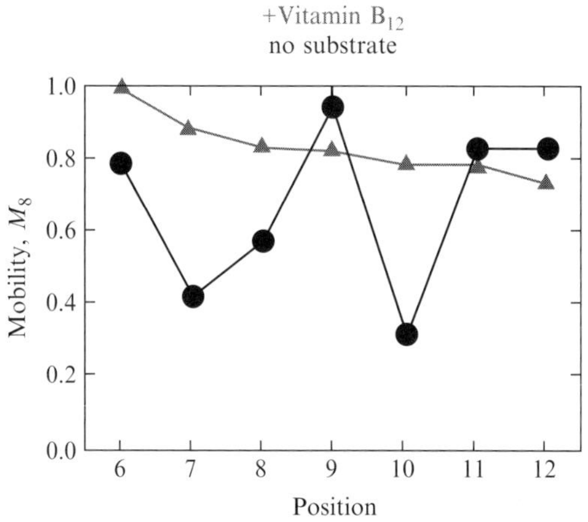

Fig. 17 Differences in mobility for residues in the TonB box of BtuB in the absence (circles) and presence (triangles) of ligand. Reprinted by permission from Macmillan Publishers Ltd, Hubbell *et al.* (2000).

of the fact that, for spin label side chains exposed to the lipid phase, the ratio of collision rates of the spin label with oxygen to that of a polar metal ion complex (e.g., NiEDDA) is independent of the structure of the protein, and that the logarithm of this ratio is linearly dependent on depth into a membrane bilayer. This ratio was measured for 10 sites on one of the membrane-exposed helices of bacteriorhodopsin and membrane depths relative to the lipid phosphate groups for each site were calculated. The depth parameter described earlier (Φ) was plotted versus residue number, and for the sites studied, increased up to position 117 and then began to decrease, indicating that site 117 is positioned in the center of the bilayer and the whole sequence spans the membrane. In addition, the distance between two of the outer helical sites (105 and 129) was experimentally determined by depth measurements to be 37 Å, which corresponds to the model distance of 37.5 Å (1.5 Å for each of the 25 residues) for an α-helix, revealing an additional layer of information in the depth data. This work nicely describes the theory behind this approach, which is now used routinely in the field as a relatively straightforward and unique method of measuring the depth of specific sites on a large protein within a lipid bilayer.

Another example of the practical application of this technique to membrane depth determinations is FepA, introduced above. In addition to confirming the biphasic periodicity of a section of putative transmembrane β-structure, depth measurements were determined for the membrane-exposed sites (Klug *et al.*, 1997). The depth parameter, Φ, was plotted for each site and clearly indicated that this region spans the membrane and that site 249 was located at the center of the bilayer, as shown in Fig. 18. Membranes containing spin-labeled lipids and unlabeled protein were used to derive a calibration equation as discussed in Section I.

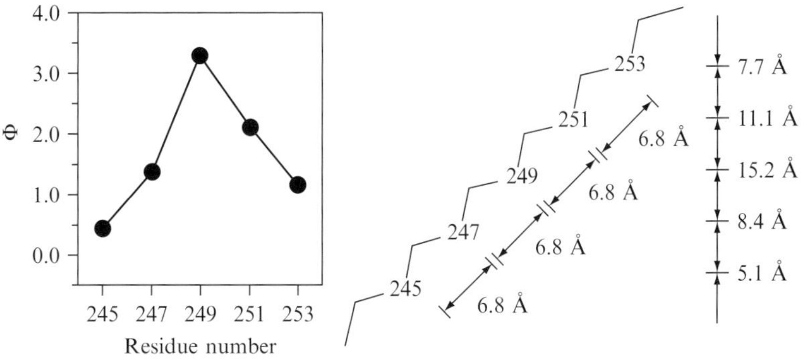

Fig. 18 (Left) A plot of the depth parameter versus residue number indicates that this sequence in FepA forms a transmembrane β-strand. (Right) Comparison of measured depths to β-strand secondary structure allows determination of strand tilt. Reprinted with permission from Klug *et al.* (1997). Copyright 1997 American Chemical Society.

Further, when depth measurements for the membrane-exposed sites were compared, the vertical distances between alternating strand sites were shorter than the model antiparallel β-strand distance of 6.8 Å, indicating that the strand was tilted relative to the bilayer normal (Fig. 18). Using the relationship $\cos \alpha = a/b$, where α is the strand tilt, a is the average experimentally determined vertical distance between residues, and b is the distance between residues in a β-strand, yielded a tilt angle of the strand with respect to the bilayer that was used to calculate a suggested diameter of 22 Å for this barrel, again in good agreement with that subsequently determined from the crystal structure (Buchanan *et al.*, 1999). Therefore, as illustrated in both of these examples, not only can individual depths be determined using SDSL but additional structural characteristics can be ascertained as well.

C. Protein Interactions

Spin label mobility is an excellent indicator of local protein folding characteristics and gives insights into protein dynamics and tertiary interactions (Farrens *et al.*, 1996; Hubbell *et al.*, 1996, 2000; Karim *et al.*, 1998; Klug and Feix, 1998; Klug *et al.*, 1995, 1998). Structural information in a protein can easily be followed using the SDSL method as the spectra of loop sites, helix surface sites, buried sites, and sites involved in tertiary contacts each have unique characteristics. Therefore, residues involved in protein–protein, protein–peptide, or protein–ligand contact sites can be readily identified by following changes in motion and/or spin–spin interactions of the spin label side chain(s). Examples are described below for systems previously studied by the SDSL technique and show that it is especially useful in the absence of crystal structures of biologically important protein complexes or for those that undergo dynamic changes in conformation not observable by crystal analysis.

1. Protein–Protein Interactions

Interactions between two different proteins can be studied by SDSL at the level of individual amino acids. As one example of the many studies between proteins, the visual protein arrestin was spin labeled at a variety of sites along its surface (Hanson *et al.*, 2006). The spectra were recorded for the protein alone in solution and then also in the presence of the light receptor, rhodopsin. Spectra were recorded in the presence of rhodopsin in the dark and then following light activation to identify which sites on arrestin were involved in binding each state of the receptor. The face of arrestin involved in the interaction between the two proteins was mapped out based on the changes in spin label mobility observed directly from the spectra. Even more specifically, individual sites in the C-tail of arrestin were identified as becoming more mobile upon binding to rhodopsin, indicative of a release of this region during binding, and sites found in a "finger" region that is proposed to insert into a cavity that opens in the receptor upon light activation became significantly more immobilized. In summary, the individual sites within arrestin that are affected by binding to its receptor were identified by changes in motion, and additional information on the conformational changes occurring in specific regions were also revealed, giving valuable characterization of this biological interaction.

The ability to observe spin–spin interactions in SDSL has also been exploited in studies of protein oligomerization. For proteins that form dimers or higher order oligomers, judicious placement of the spin label can be used to characterize the interface between monomers, and titration of spin-labeled protein with unlabeled protein can be used to determine the oligomerization state. An example of this type of SDSL application is the study of the assembly of annexins at the membrane surface (Langen *et al.*, 1998). Annexins exhibit reversible binding to phosphatidylserine-containing membranes in the presence of Ca^{2+}. Crystal structures of the soluble form indicated a variety of potential quaternary states, including monomers, dimers, trimers, and hexamers. To investigate the membrane-bound state, a number of sites were selected for spin labeling that were hypothesized to be close to a protein–protein interface based on existing crystal structures. In solution, the EPR spectra at these sites were characteristic of fast rotational motion. However, upon binding to membranes in the presence of Ca^{2+} spin–spin broadening was so extreme that the EPR signal for the membrane-bound state essentially disappeared into the baseline. To separate broadening due to spin–spin interactions from that due to changes in motion, spectra were also obtained for the membrane-bound state in the presence of a ninefold excess of unlabeled annexin. Dilution of the spin-labeled species significantly increased spectral amplitude, confirming that line broadening in the undiluted samples was due to spin–spin interactions within the oligomer. To further examine the oligomeric state of membrane-bound annexin, one of the spin-labeled mutants was carefully titrated with unlabeled protein. Changes in the amplitude of the center line as a function of the mole fraction of spin-labeled protein were compared to models based on binomial distributions for dimers, trimers, and hexamers. The

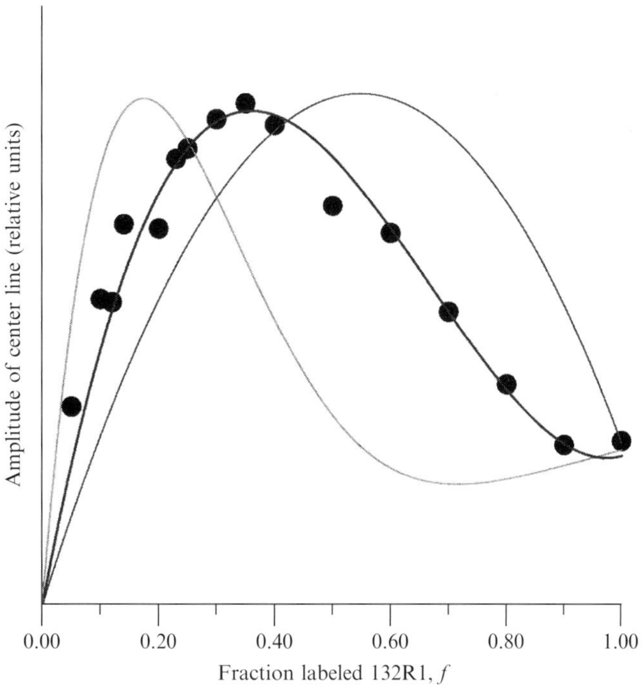

Fig. 19 Variation in EPR signal height with the fraction of spin-labeled protein for annexin 12 aligns with the binomial distribution expected for trimers, rather than dimers or hexamers. Reprinted with permission from Langen *et al.* (1998). Copyright 1998 ASBMB.

experimental data closely fit the distribution expected for trimers (Fig. 19), establishing the membrane-bound oligomeric state of annexin 12.

2. Protein–Ligand interactions

Similarly, interactions between a protein and its ligand can also be monitored by SDSL at very localized sites. Typically, the protein is spin labeled at specific sites proposed to interact with the ligand; however, the ligand can also be spin labeled and used as a probe itself (e.g., Beth *et al.*, 1984; Hustedt *et al.*, 1997). Few examples exist in the literature for spin-labeled ligands, though they are becoming more commonly synthesized and used in EPR binding studies.

As an example of a single site within a protein being studied to observe the effect of ligand binding, FepA was spin labeled at position 338, thought to be close to the ferric enterobactin-binding site and shown to be solution-exposed and within 4.5 Å of the lipid headgroups using accessibility data (Klug *et al.*, 1998). The ligand-induced change in the spin label motion at this site was significant, with the ligand-bound spectrum showing a marked decrease in motion to a spectrum

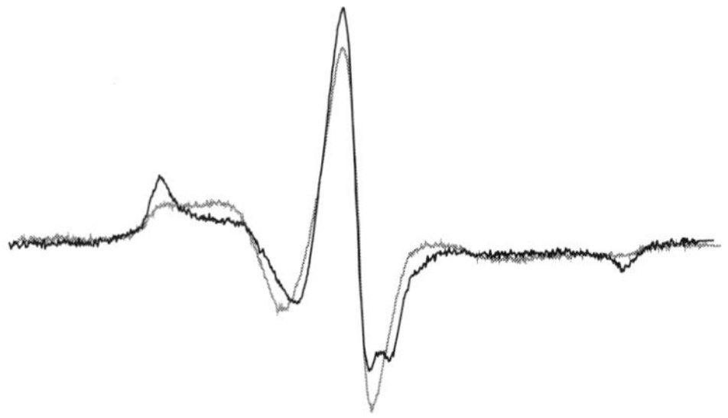

Fig. 20 Overlays of the FepA V338C spectra showing differences in mobility in the absence (gray) and presence (black) of ligand. Reprinted with permission from Klug *et al.* (1998). Copyright 1998 American Chemical Society.

indicative of extremely slow motion on the EPR timescale (Fig. 20). In addition, the accessibility data were used to show that this site becomes less accessible to the broadening reagents after ligand binding, further verifying that this site either is in direct contact with the ligand or is experiencing additional tertiary contacts due to structural rearrangement of the protein upon ligand binding.

As another example, the lipid A transporter, MsbA, was spin labeled at 13 consecutive sites within a known ligand-binding region of the protein to determine the role of each site in ligand binding (Buchaklian and Klug, 2005). The Walker A region of MsbA is a conserved sequence found in ATPases that is known to be involved in ATP binding and hydrolysis. Thus, it was not surprising that a number of the residues showed significant motional changes during the hydrolysis cycle. However, the exact sites affected by binding were clearly identified and indicated that the entire Walker A region is not affected by ATP binding. The results were not able to distinguish between direct contact with the ligand or structural rearrangements in this region; however, the added accessibility studies did verify that these sites were much less accessible to the broadening reagents upon ATP hydrolysis than they were in the resting state of the protein.

D. Peptides

1. Peptide-Membrane Partition Coefficients

SDSL has proven to be an extremely useful approach for the study of peptides that partition between the membrane and the aqueous phases. On membrane binding, a spin-labeled peptide typically undergoes a significant reduction in rotational mobility. Under conditions where there is an equilibrium distribution of peptides between the membrane and aqueous phases, one observes a superposition of signals from the two populations. This allows a simple and straightforward determination of the partition coefficient without having to separate bound and free peptide. Shown in Fig. 21 are EPR spectra of a 15-residue spin-

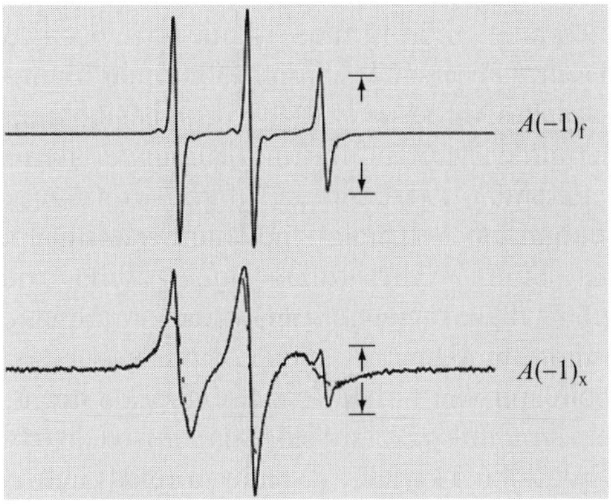

Fig. 21 EPR spectra of spin-labeled CM15 (above) in aqueous solution and (below) in equilibrium with liposomes. The dashed line in the bottom spectra is for peptide 100% membrane bound. Reprinted with permission from Bhargava and Feix (2004). Copyright 2004 by the Biophysical Society.

labeled antimicrobial peptide, CM15, in the presence and absence of liposomes (Bhargava and Feix, 2004). Under conditions where the peptide is membrane bound, the amplitude of the high-field ($m_I = -1$) line essentially disappears at the position of the aqueous signal. Consequently, the fraction of peptide remaining in the aqueous phase can be directly determined from the amplitude of the narrow line of the remaining free peptide, $A(-1)_f$, and the fraction of bound peptide is given by the reduction in signal amplitude as compared to a sample without membranes. Quantitatively, the fraction of bound peptide, f_b, is given by:

$$f_b = \left[\frac{A(-1)_f - A(-1)_x}{A(-1)_f - A(-1)_b}\right] \tag{10}$$

where $A(-1)_f$, $A(-1)_b$, and $A(-1)_x$ are the amplitudes of the high-field line for free peptide, fully bound peptide, and the experimental sample, respectively (Fig. 21). This method has been found to give comparable results to the more rigorous procedure of using spectral subtraction to separate the bound and free components and then integrating to determine the number of spins in each population (J.B.F., unpublished data). Once f_b has been determined at a series of lipid concentrations (where the peptide concentration is held constant), the molar partition coefficient (K_p) is calculated according to:

$$f_b = \frac{K_p[\text{lipid}]}{1 + K_p[\text{lipid}]} \tag{11}$$

and K_p can be used to calculate the change in free energy for membrane binding.

Results similar to those above have been described for a wide variety of peptide-membrane systems. Examples include binding of the ion-conductive peptide alamethicin (Archer *et al.*, 1991; Lewis and Cafiso, 1999), a peptide derived from the effector domain of the myristoylated alanine-rich C-kinase substrate (Addona *et al.*, 1997), a series of model peptides composed of lysine and phenylalanine and the spin-labeled amino acid tetramethylpiperidine-*N*-oxyl-4-amino-4-carboxylic acid (TOAC) (Victor and Cafiso, 2001), and the antimicrobial peptide cecropin AD (Mchaourab *et al.*, 1994). Each of these systems gives EPR spectra quite similar to those in Fig. 21. Using a host–guest system of single amino acid substitutions into a spin-labeled, 25-residue peptide derived from yeast cytochrome *c* oxidase, Shin and coworkers used this approach to define a scale of relative membrane affinities for 14 uncharged amino acids (Thorgeirsson *et al.*, 1996) and examined the thermodynamics of membrane partitioning (Russell *et al.*, 1996). In addition, it should be noted that this method was first used to determine membrane surface potentials based on partitioning of spin-labeled amphiphiles (Cafiso and Hubbell, 1978; Castle and Hubbell, 1976), illustrating the generality of this approach.

2. Depth Measurements and Structure of the Membrane-Bound Peptide

Depth measurements based on CW saturation, as described earlier for integral membrane proteins, also can be used to determine the structure and penetration depths for membrane-associated peptides. Cysteine residues are introduced at various sites in the subject peptide, spin labeled, and their interaction with oxygen and NiEDDA (or other relaxation agent) determined under conditions where the peptide is fully membrane bound. Alternatively, if the peptide is being prepared by solid-phase peptide synthesis, labeling can be accomplished by introduction of the spin-labeled amino acid TOAC (e.g., Hanson *et al.*, 1996; Karim *et al.*, 2004; Marchetto *et al.*, 1993; Victor and Cafiso, 2001). Since the TOAC nitroxide ring is rigidly incorporated into the α-carbon peptide backbone (Fig. 22), there are no internal side-chain fluctuations to be considered and the motion of the label directly reports on peptide backbone dynamics.

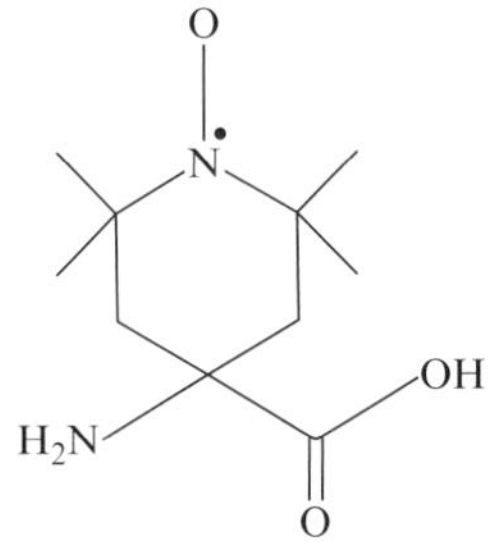

Fig. 22 The TOAC spin label can be integrated into the peptide backbone during synthesis.

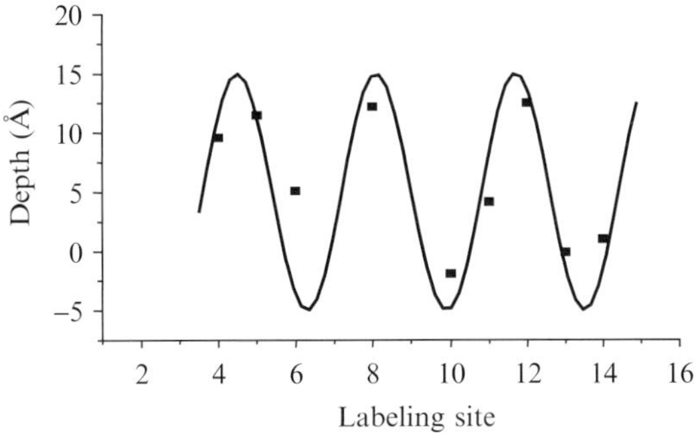

Fig. 23 Periodicity in depths for an α-helical peptide lying parallel to the membrane surface.

The peptides studied to date by SDSL have all been based on sequences with α-helical and/or random coil secondary structure. Such peptides can incorporate into the membrane with their helical axis parallel, perpendicular, or at an oblique angle relative to the bilayer surface, and all three cases have been observed experimentally. A nitroxide-scanning study of the antimicrobial peptide CM15 (Bhargava and Feix, 2004) indicated that this peptide was unstructured in solution, but upon membrane binding folded into an α-helix that aligned parallel to the bilayer surface, with the central axis of the helix located ~5 Å below the aqueous interface (Fig. 23). This is an essentially ideal location for membrane localization of this amphipathic peptide, allowing lysine residues on the polar face of the helix to "snorkel" out of the membrane and ion-pair with lipid phosphates while keeping hydrophobic residues on the nonpolar face of the helix buried in the hydrocarbon phase of the bilayer.

A 25-residue peptide containing the calmodulin-binding and protein kinase C-substrate domains of the MARCKS protein also aligned parallel to the membrane surface with a phenylalanine-rich region immersed ~8–10 Å below the lipid phosphates, but its highly charged N-terminal domain remained in the aqueous phase and it showed no evidence of helical structure when bound to membranes (Qin and Cafiso, 1996). When the phenylalanine residues of the native MARCKS-derived peptide were mutated to alanine, the peptide remained at the membrane surface, exposed to the aqueous phase—a shift of almost 15 Å in its equilibrium position (Victor *et al.*, 1999). This dramatic effect of phenylalanine residues on the immersion of the peptide was further investigated with model peptides containing only lysine and phenylalanine and the TOAC spin label (Victor and Cafiso, 2001). Again, replacement of two or more lysines with phenylalanine residues shifted the equilibrium position of the peptide by 13–15 Å, as determined by SDSL depth measurements.

Other peptides with sequences containing a mixture of polar and nonpolar amino acids have been found to insert into lipid bilayers at an angle relative to the membrane surface. This is a particularly predominant motif for peptides involved in membrane fusion, such as peptides derived from SNARE proteins that function in vesicle trafficking (Xu *et al.*, 2005) and viral fusion peptides (Han *et al.*, 2001; Macosko *et al.*, 1997). SDSL measurements on the amphipathic α-helical myelin basic protein determined an insertion angle of a mere 9° (Bates *et al.*, 2004), while other peptides give much greater angles of insertion (Han *et al.*, 2001; Macosko *et al.*, 1997; Xu *et al.*, 2005). In contrast, hydrophobic peptides, such as phospholamban (Karim *et al.*, 2004), and a designed WALP peptide composed of a repeating leucine-alanine motif flanked by tryptophans (Nielsen *et al.*, 2005) were shown to align in a vertical, transmembrane fashion, approximately perpendicular to the bilayer surface.

E. Unfolding and Kinetics

Denaturation studies, in which a protein is reversibly unfolded using either temperature or chemical denaturants [e.g., urea or guanidine hydrochloride (GdnHCl)], have been used extensively in recent years to examine protein structure and stability (see Chapter 11 by Street *et al.*, this volume on protein folding). The sensitivity of the spin label EPR spectrum to the formation or loss of local structural constraints provides a highly sensitive means by which to monitor protein folding and denaturation, respectively. If a protein is labeled at a motionally restricted site, the loss of local tertiary structure that occurs upon denaturation will result in an increased spin label mobility and the corresponding appearance of a sharp, fast-motion component in the spectrum. Under conditions where there is an equilibrium between native and unfolded protein, the EPR spectrum will be a superposition of signals from the two populations and can be deconvoluted using spectral subtraction to determine the relative concentration of each component. In the case of a reversible, two-state denaturation, determination of the equilibrium distribution of folded and unfolded states as a function of denaturant concentration provides a direct measure of thermodynamic stability.

An example of the application of this technique using the ferric enterobactin receptor, FepA, is shown in Fig. 24. FepA was labeled at a site (E280C) known to have a strongly immobilized EPR spectrum in the native state and to be sensitive to ligand binding (Klug and Feix, 1998). Addition of either GdnHCl or urea resulted in the appearance of a rapid-motion component that increased in intensity with increasing denaturant concentration. The EPR spectrum of the fully denatured protein, obtained in 4 M GdnHCl, was used for spectral subtraction, and the fraction of denatured component was determined by integrating the spectra before and after subtraction. The difference spectra obtained after subtraction of the denatured component closely resembled the native spectrum, consistent with a two-state equilibrium between native and denatured states. Importantly, the spectrum reverted back to that of the native state upon removal of the denaturant by dialysis, demonstrating reversibility. Plots of the fraction of unfolded protein as a function of denaturant concentration provided additional evidence supporting a

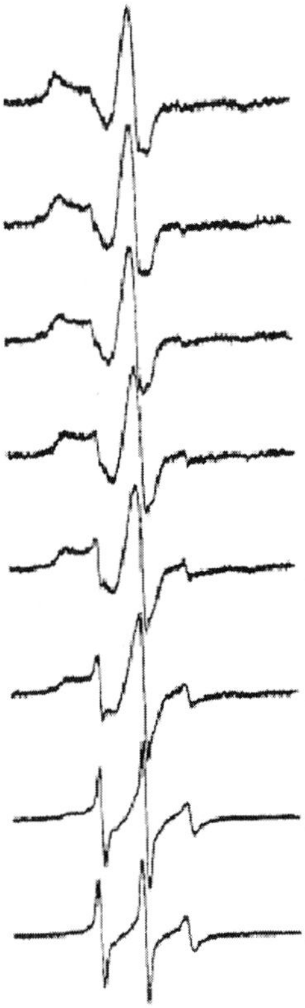

Fig. 24 Spectral changes in FepA at position 280 on addition of increasing amounts of GdnHCl denaturant. Reprinted with permission from Klug *et al.* (1995). Copyright 1995 American Chemical Society.

two-state equilibrium, giving estimates of the Gibbs free energy of unfolding and the stabilization provided by ligand binding (Klug and Feix, 1998).

It should be noted that protein denaturation studies done by SDSL measure very specifically the local unfolding at the site of the spin label. This can be very beneficial when studying a multidomain protein, as it allows the unfolding of each domain to be examined individually even in the intact protein, whereas other techniques that monitor global unfolding, such as circular dichroism, require the separate expression of each domain (Kim *et al.*, 2002). However, it must also be appreciated that different labeling sites may have different stabilities, as was observed in a subsequent study of FepA (Klug and Feix, 1998). Sequence-dependent differences in thermodynamic stability are also observed by NMR and mass spectrometry measurements of hydrogen–deuterium exchange and are an inherent aspect of protein structure.

Time-resolved EPR can be used to follow protein folding. In these studies an unfolded, spin-labeled protein in denaturant is mixed with buffer to rapidly dilute the denaturant concentration. Conversion of the narrow EPR lines of the denatured protein to the broader lines of the folded protein is observed by "sitting" on the peak of one of the sharp lines (i.e., by positioning the magnetic field at that resonance position and turning the magnetic field sweep off) and following the decrease in amplitude as a function of time. Using commercially available mixing cavities and syringe drives one can readily measure decays on the order of 0.1 sec. Scholes and coworkers, using cytochrome *c* as a model system, have developed and refined instrumentation allowing observation of kinetic components on the submillisecond timescale (DeWeerd *et al.*, 2001; Grigoryants *et al.*, 2000).

F. Distances to Determine Structural Arrangements and Monitor Dynamics

A number of outstanding studies have been carried out using EPR-determined distance measurements to examine tertiary folds, subunit interactions, and conformational changes (e.g., Altenbach *et al.*, 2001a,b; Berengian *et al.*, 1999; Brown *et al.*, 2002; Cordero-Morales *et al.*, 2006; Gross *et al.*, 1999; Koteiche *et al.*, 1998; Mchaourab *et al.*, 1997b; Wegener *et al.*, 2001). Three representative examples are described below.

1. Distance Measurements Using CW EPR

Distance measurements have been used extensively to examine light-induced conformational changes in the α-helical integral membrane protein rhodopsin (e.g., Altenbach *et al.*, 2001a,b; Cai *et al.*, 2001; Farrens *et al.*, 1996; Klein-Seetharaman *et al.*, 2001). Rhodopsin has long been used as a leading model system for the study of G–protein-coupled receptors (GPCRs). It contains seven transmembrane α-helices and a retinal chromophore that isomerizes upon photon absorption, triggering a conformational change that ultimately activates its cognate G-protein. In one of the earlier studies utilizing distance measurements to gain insights into changes in structure, five double-cysteine mutants were constructed—with one site on transmembrane helix 3 (TM3 or helix C) held constant and its interaction examined with five consecutive sites on TM6 (helix F) spanning a full turn of the helix (Farrens *et al.*, 1996). Samples were frozen after preparing the desired photochemical states at room temperature, and EPR spectra were obtained in the frozen state at 183 K. In the dark state, two of the spin label pairs were within ~12–14 Å, resulting in significantly broadened spectra, and the remaining three pairs were separated by distances in the range of 15–20 Å. After photoactivation, distances increased to greater than the 20 Å limit for detection of interaction for three of the spin label pairs, one pair became significantly closer, and one pair remained unchanged. These distance changes were the basis of a molecular model for photoactivation requiring TM6 to tilt away from the helical bundle

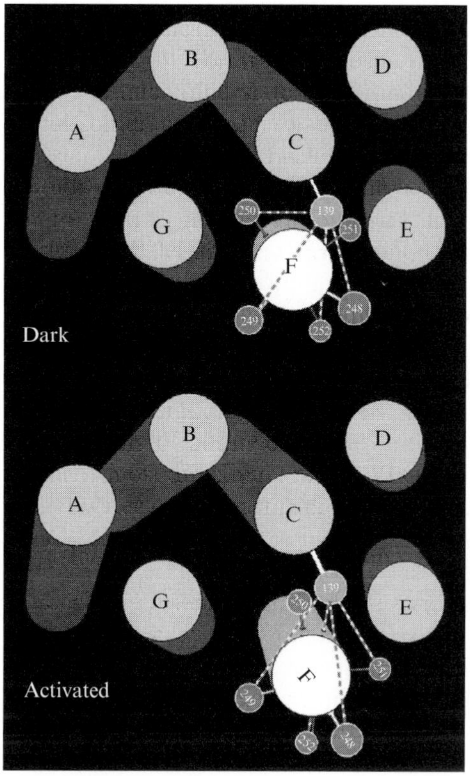

Fig. 25 Rhodopsin cartoon illustrating the conformational change in helix F that occurs upon light activation as elucidated using SDSL techniques. Reprinted with permission from Farrens *et al.* (1996). Copyright 1996 AAAS.

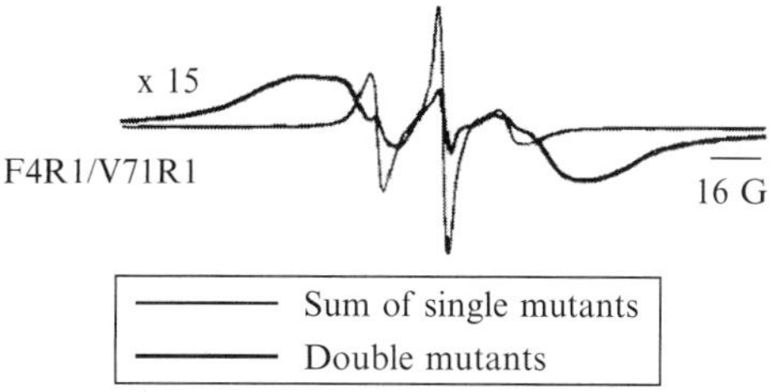

Fig. 26 Severe line broadening due to spin–spin interaction in T4L. Reprinted with permission from Mchaourab *et al.* (1997b). Copyright 1997 American Chemical Society.

(a displacement of ~8 Å) and to rotate (Farrens *et al.*, 1996). This "tilt and rotate" model (Fig. 25) remains the current paradigm for activation of GPCRs.

An important goal in SDSL is to make interspin distance measurements on biomolecules in their native state, that is, in the liquid phase at ambient temperature. In initial studies, T4L was used as a model system to develop methods to make such measurements (Mchaourab *et al.*, 1997b). The EPR spectra of double-labeled T4L mutants are compared to spectra obtained by taking the sum of the corresponding single mutant spectra. Pronounced broadening and distortion of line shapes are observed (e.g., Fig. 26). Note also that the line broadening is accompanied by a significant reduction in the amplitude of the double-labeled samples, as indicated by the increased gain used to display the spectrum. Since these studies were done at room temperature, it was possible to examine the effects of ligand binding on structure. Addition of a nonhydrolyzable substrate significantly increased spin–spin interaction for a number of pairs, consistent with a hinge-bending motion proposed as part of the mechanism for T4L catalysis (Mchaourab *et al.*, 1997b).

T4L was also used to develop methodology for deriving distances from room temperature spectra based on static dipolar coupling that can be extended to larger proteins (Altenbach *et al.*, 2001c). This interactive approach is based on Fourier deconvolution of dipolar-coupled spectra as introduced by Rabenstein and Shin (1995). A strength of this approach is the ability to determine distances even in the presence of singly labeled species. This is of particular importance, as for many systems it is often difficult to obtain stoichiometric labeling. A set of T4L double mutants was examined in frozen solution and at ambient temperature with sucrose added to decrease the tumbling rate of the protein. Noninteracting spectra were derived either from a sum of the singly labeled species or by labeling with a mixture of MTSL and a diamagnetic *N*-acetylated analogue. Good agreement was found between distances obtained at room temperature and in frozen solution, and in general it was found that residual motion of the spin label at ambient temperature had little effect on the estimated distances. In addition, distance distributions for a number of spin label pairs showed more than one maximum, consistent with the presence of multiple orientations of the spin label with respect to the peptide backbone as observed in crystal structures of MTSL-labeled T4L (Langen *et al.*, 2000).

2. Distance Measurements Using Pulse EPR

A comprehensive SDSL study on the solution structure of the cytoplasmic domain of the erythrocyte anion exchange protein, AE1 (also referred to as band 3) was carried out using a combination of CW EPR and DEER methodologies (Zhou *et al.*, 2005). The 42.5-kDa cytoplasmic domain of band 3 (cdb3) exists as a dimer that mediates numerous important protein–protein interactions with both soluble proteins (e.g., hemoglobin) and the erythrocyte cytoskeleton. A crystal structure of cdb3 obtained at pH 4.8 had shown an unexpectedly compact dimeric structure, but given the low pH used in crystallization, along with extensive evidence suggesting pH-dependent conformational changes, it was important to investigate the structure of cdb3 under more physiological conditions.

Recombinant expression, purification, and spin labeling of cdb3 mutants containing single-cysteine residues provided samples containing two spin labels per noncovalent dimer. Using SDSL, mobility and power saturation accessibility parameters confirmed the periodicity and expected accessibility for several elements of α-helical secondary structure observed in the crystal structure (Zhou *et al.*, 2005). One site in helix 10 (Q339R1) was particularly remarkable in that it displayed highly resolved dipolar coupling between the spin labels on each monomer. Simulation of the dipolar-coupled spectrum using a "tether-in-a-cone" model (Hustedt *et al.*, 2006) indicated an interspin distance of 14.7 ± 0.4 Å, in close agreement with the distance between these sites across the dimer interface inferred by the crystal structure. To further characterize the orientation of the two monomers in the cdb3 dimer, DEER was used to measure distances extending further out along helix 10 (from residues 340–345). Distances in the range 24.9–37.0 Å were measured (e.g., Fig. 14). The width of the distance distribution was very narrow (0.8–1.1 Å) for sites proximal to the core of dimer, and increased (3.6–6.6 Å) toward the distal end of the helix in a manner that paralleled the general increase in side-chain mobility. Distance measurements were also made for an additional 11 sites, using either DEER or by fitting CW EPR spectra with a Gaussian convolution model (Steinhoff *et al.*, 1997). Distances were reported in the range 6.2–47.7 Å, and again verified that the solution structure of cdb3 at neutral pH was in close agreement with the pH 4.8 crystal structure. Taken together, the data in this extensive study demonstrate the power of the SDSL approach in defining elements of local secondary structure and using distance measurements to elucidate how those secondary structure elements are arranged. The close agreement of the EPR data with that from the published crystal structure provides confidence that these methods can be used to determine unknown structures using SDSL alone.

III. Conclusion

As described and illustrated in this chapter, the SDSL EPR spectroscopy technique is able to address and answer questions often not solvable by genetic or crystal structure analysis. Its ability to analyze structure in a natural liquid-phase environment and its sensitivity to dynamics make the SDSL approach highly complementary to other structural methods. The number of researchers using SDSL techniques on their systems has grown tremendously in the past several years, and we look forward to future developments in methodology and new biological applications.

References

Addona, G. H., Andrews, S. H., and Cafiso, D. S. (1997). Estimating the electrostatic potential at the acetylcholine receptor agonist site using power saturation EPR. *Biochim. Biophys. Acta* **1329,** 74–84.

Altenbach, C., Cai, K., Klein-Seetharaman, J., Khorana, H. G., and Hubbell, W. L. (2001a). Structure and function in rhodopsin: Mapping light-dependent changes in distance between residue 65 in helix TM1 and residues in the sequence 306–319 at the cytoplasmic end of helix TM7 and in helix H8. *Biochemistry* **40,** 15483–15492.

Altenbach, C., Flitsch, S. L., Khorana, H. G., and Hubbell, W. L. (1989). Structural studies on transmembrane proteins. 2. Spin labeling of bacteriorhodopsin mutants at unique cysteines. *Biochemistry* **28,** 7806–7812.

Altenbach, C., Greenhalgh, D. A., Khorana, H. G., and Hubbell, W. L. (1994). A collision gradient method to determine the immersion depth of nitroxides in lipid bilayers: Application to spin-labeled mutants of bacteriorhodopsin. *Proc. Natl. Acad. Sci. USA* **91,** 1667–1671.

Altenbach, C., Klein-Seetharaman, J., Cai, K., Khorana, H. G., and Hubbell, W. L. (2001b). Structure and function in rhodopsin: Mapping light-dependent changes in distance between residue 316 in helix 8 and residues in the sequence 60–75, covering the cytoplasmic end of helices TM1 and TM2 and their connection loop CL1. *Biochemistry* **40,** 15493–15500.

Altenbach, C., Oh, K. J., Trabanino, R. J., Hideg, K., and Hubbell, W. L. (2001c). Estimation of inter-residue distances in spin labeled proteins at physiological temperatures: Experimental strategies and practical limitations. *Biochemistry* **40,** 15471–15482.

Archer, S. J., Ellena, J. F., and Cafiso, D. S. (1991). Dynamics and aggregation of the peptide ion channel alamethicin. Measurements using spin-labeled peptides. *Biophys. J.* **60,** 389–398.

Bates, I. R., Feix, J. B., Boggs, J. M., and Harauz, G. (2004). An immunodominant epitope of myelin basic protein is an amphipathic alpha-helix. *J. Biol. Chem.* **279,** 5757–5764.

Berengian, A. R., Parfenova, M., and Mchaourab, H. S. (1999). Site-directed spin labeling study of subunit interactions in the alpha-crystallin domain of small heat-shock proteins. Comparison of the oligomer symmetry in alphaA-crystallin, HSP 27, and HSP 16.3. *J. Biol. Chem.* **274,** 6305–6314.

Beth, A. H., Robinson, B. H., Cobb, C. E., Dalton, L. R., Trommer, W. E., Birktoft, J. J., and Park, J. H. (1984). Interactions and spatial arrangement of spin-labeled NAD + bound to glyceraldehyde-3-phosphate dehydrogenase. Comparison of EPR and X-ray modeling data. *J. Biol. Chem.* **259,** 9717–9728.

Bhargava, K., and Feix, J. B. (2004). Membrane binding, structure, and localization of Cecropin-Mellitin hybrid peptides: A site-directed spin-labeling study. *Biophys. J.* **86,** 329–336.

Bonora, M., Becker, J., and Saxena, S. (2004). Suppression of electron spin-echo envelope modulation peaks in double quantum coherence electron spin resonance. *J. Magn. Reson.* **170,** 278–283.

Borbat, P. P., Costa-Filho, A. J., Earle, K. A., Moscicki, J. K., and Freed, J. H. (2001). Electron spin resonance in studies of membranes and proteins. *Science* **291,** 266–269.

Borbat, P. P., and Freed, J. H. (2000). Double-Quantum ESR and distance measurements. *In* "Biological Magnetic Resonance" (L. J. Berliner, S. S. Eaton, and G. R. Eaton, eds.), pp. 383–459. Kluwer Academic/Plenum Publishers, New York.

Borbat, P. P., Mchaourab, H. S., and Freed, J. H. (2002). Protein structure determination using long-distance constraints from double-quantum coherence ESR: Study of T4 lysozyme. *J. Am. Chem. Soc.* **124,** 5304–5314.

Brown, L. J., Sale, K. L., Hills, R., Rouviere, C., Song, L., Zhang, X., and Fajer, P. G. (2002). Structure of the inhibitory region of troponin by site directed spin labeling electron paramagnetic resonance. *Proc. Natl. Acad. Sci. USA* **99,** 12765–12770.

Buchaklian, A. H., and Klug, C. S. (2005). Characterization of the Walker A motif of MsbA using site-directed spin labeling electron paramagnetic resonance spectroscopy. *Biochemistry* **44,** 5503–5509.

Buchanan, S. K., Smith, B. S., Venkatramani, L., Xia, D., Esser, L., Palnitkar, M., Chakraborty, R., van der, H. D., and Deisenhofer, J. (1999). Crystal structure of the outer membrane active transporter FepA from Escherichia coli. *Nat. Struct. Biol.* **6,** 56–63.

Budil, D. E., Lee, S., Saxena, S., and Freed, J. H. (1996). Nonlinear-least-squares analysis of slow-motion EPR spectra in one and two dimensions using a modified Levenberg-Marquardt Algorithm. *J. Magn. Reson.* **120,** 155–189.

Cafiso, D. S., and Hubbell, W. L. (1978). Estimation of transmembrane pH gradients from phase equilibria of spin-labeled amines. *Biochemistry* **17,** 3871–3877.

Cai, K., Klein-Seetharaman, J., Altenbach, C., Hubbell, W. L., and Khorana, H. G. (2001). Probing the dark state tertiary structure in the cytoplasmic domain of rhodopsin: Proximities between amino acids deduced from spontaneous disulfide bond formation between cysteine pairs engineered in cytoplasmic loops 1, 3, and 4. *Biochemistry* **40,** 12479–12485.

Castle, J. D., and Hubbell, W. L. (1976). Estimation of membrane surface potential and charge density from the phase equilibrium of a paramagnetic amphiphile. *Biochemistry* **15,** 4818–4831.

Columbus, L., and Hubbell, W. L. (2002). A new spin on protein dynamics. *Trends Biochem. Sci.* **27,** 288–295.

Columbus, L., and Hubbell, W. L. (2004). Mapping backbone dynamics in solution with site-directed spin labeling: GCN4–58 bZip free and bound to DNA. *Biochemistry* **43,** 7273–7287.

Columbus, L., Kalai, T., Jeko, J., Hideg, K., and Hubbell, W. L. (2001). Molecular motion of spin labeled side chains in alpha-helices: Analysis by variation of side chain structure. *Biochemistry* **40,** 3828–3846.

Cordero-Morales, J. F., Cuello, L. G., Zhao, Y., Jogini, V., Cortes, D. M., Roux, B., and Perozo, E. (2006). Molecular determinants of gating at the potassium-channel selectivity filter. *Nat. Struct. Mol. Biol.* **13,** 311–318.

Dalton, L. A., McIntyre, J. O., and Flewelling, R. F. (1987). Distance estimate of the active center of D-β-hydroxybutyrate dehydrogenase from the membrane surface. *Biochemistry* **26,** 2117–2130.

DeWeerd, K., Grigoryants, V. M., Sun, Y., Fetrow, J. S., and Scholes, C. P. (2001). EPR-detected folding kinetics of externally located Cysteine-directed spin-labeled mutants of Iso-1-cytochrome *c*. *Biochemistry* **40,** 15846–15855.

Eaton, G. R., Eaton, S. S., and Berliner, L. J. (eds.) (2000). Distance Measurements in Biological Systems by EPR "Biological Magnetic Resonance," Vol. 19. Kluwer, New York.

Fajer, P. G. (2000). Electron spin resonance spectroscopy labeling in peptide and protein analysis. *In* "Encyclopedia of Analytical Chemistry" (R. A. Meyers, ed.), pp. 5725–5761. John Wiley and Sons Ltd., Chichester.

Farrens, D. L., Altenbach, C., Yang, K., Hubbell, W. L., and Khorana, H. G. (1996). Requirement of rigid-body motion of transmembrane helices for light activation of rhodopsin. *Science* **274,** 768–770.

Feix, J. B., and Klug, C. S. (1998). Site-directed spin labeling of membrane proteins and peptide-membrane interactions. *In* "Biological Magnetic Resonance, Volume 14: Spin Labeling: The Next Millennium" (L. J. Berliner, ed.), pp. 252–281. Plenum Press, New York.

Freed, J. H. (2000). New technologies in electron spin resonance. *Annu. Rev. Phys. Chem.* **51,** 655–689.

Gaffney, B. J. (1976). Practical considerations for the calculation of order parameters. *In* "Spin Labeling: Theory and Applications" (L. J. Berliner, ed.), pp. 567–571. Academic Press, New York.

Griffith, O. H., and Jost, P. C. (1976). Lipid spin labels in biological membranes. *In* "Spin Labeling: Theory and Applications" (L. J. Berliner, ed.), pp. 453–523. Academic Press, New York.

Grigoryants, V. M., Veselov, A. V., and Scholes, C. P. (2000). Variable velocity liquid flow EPR applied to submillisecond protein folding. *Biophys. J.* **78,** 2702–2708.

Gross, A., Columbus, L., Hideg, K., Altenbach, C., and Hubbell, W. L. (1999). Structure of the KcsA potassium channel from Streptomyces lividans: A site-directed spin labeling study of the second transmembrane segment. *Biochemistry* **38,** 10324–10335.

Han, X., Bushweller, J. H., Cafiso, D. S., and Tamm, L. K. (2001). Membrane structure and fusion-triggering conformational change of the fusion domain from influenza hemagglutinin. *Nat. Struct. Biol.* **8,** 715–720.

Hanson, P., Millhauser, G., Formaggio, F., Crisma, M., and Toniolo, C. (1996). ESR characterization of hexameric, helical peptides using double TOAC spin labeling. *J. Am. Chem. Soc.* **118,** 7618–7625.

Hanson, S. M., Francis, D. J., Vishnivetskiy, S. A., Kolobova, E. A., Hubbell, W. L., Klug, C. S., and Gurevich, V. V. (2006). Differential interaction of spin-labeled arrestin with inactive and active phosphorhodopsin. *Proc. Natl. Acad. Sci. USA* **103,** 4900–4905.

Hubbell, W. L., Cafiso, D. S., and Altenbach, C. (2000). Identifying conformational changes with site-directed spin labeling. *Nat. Struct. Biol.* **7,** 735–739.

Hubbell, W. L., Gross, A., Langen, R., and Lietzow, M. A. (1998). Recent advances in site-directed spin labeling of proteins. *Curr. Opin. Struct. Biol.* **8,** 649–656.

Hubbell, W. L., and McConnell, H. M. (1968). Spin-label studies of the excitable membranes of nerve and muscle. *Proc. Natl. Acad. Sci. USA* **61,** 12–16.

Hubbell, W. L., Mchaourab, H. S., Altenbach, C., and Lietzow, M. A. (1996). Watching proteins move using site-directed spin labeling. *Structure* **4,** 779–783.

Hustedt, E. J., and Beth, A. H. (1999). Nitroxide spin-spin interactions: Applications to protein structure and dynamics. *Annu. Rev. Biophys. Biomol. Struct.* **28,** 129–153.

Hustedt, E. J., and Beth, A. H. (2000). Structural information from CW-EPR spectra of dipolar coupled nitroxide spin labels. *In* "Biological Magnetic Resonance" (L. J. Berliner, S. S. Eaton, and G. R. Eaton, eds.), pp. 155–184. Kluwer Academic/Plenum Publishers, New York.

Hustedt, E. J., Smirnov, A. I., Laub, C. F., Cobb, C. E., and Beth, A. H. (1997). Molecular distances from dipolar coupled spin-labels: The global analysis of multifrequency continuous wave electron paramagnetic resonance data. *Biophys. J.* **72,** 1861–1877.

Hustedt, E. J., Stein, R. A., Sethaphong, L., Brandon, S., Zhou, Z., and Desensi, S. C. (2006). Dipolar coupling between nitroxide spin labels: The development and application of a tether-in-a-cone model. *Biophys. J.* **90,** 340–356.

Isas, J. M., Langen, R., Haigler, H. T., and Hubbell, W. L. (2002). Structure and dynamics of a helical hairpin and loop region in annexin 12: A site-directed spin labeling study. *Biochemistry* **41,** 1464–1473.

Jao, C. C., Der-Sarkissian, A., Chen, J., and Langen, R. (2004). Structure of membrane-bound alpha-synuclein studied by site-directed spin labeling. *Proc. Natl. Acad. Sci. USA* **101,** 8331–8336.

Karim, C. B., Kirby, T. L., Zhang, Z., Nesmelov, Y., and Thomas, D. D. (2004). Phospholamban structural dynamics in lipid bilayers probed by a spin label rigidly coupled to the peptide backbone. *Proc. Natl. Acad. Sci. USA* **101,** 14437–14442.

Karim, C. B., Stamm, J. D., Karim, J., Jones, L. R., and Thomas, D. D. (1998). Cysteine reactivity and oligomeric structures of phospholamban and its mutants. *Biochemistry* **37,** 12074–12081.

Kim, C. S., Kweon, D. H., and Shin, Y. K. (2002). Membrane topologies of neuronal SNARE folding intermediates. *Biochemistry* **41,** 10928–10933.

Klein-Seetharaman, J., Hwa, J., Cai, K., Altenbach, C., Hubbell, W. L., and Khorana, H. G. (2001). Probing the dark state tertiary structure in the cytoplasmic domain of rhodopsin: Proximities between amino acids deduced from spontaneous disulfide bond formation between Cys316 and engineered cysteines in cytoplasmic loop 1. *Biochemistry* **40,** 12472–12478.

Klug, C. S., Eaton, S. S., Eaton, G. R., and Feix, J. B. (1998). Ligand-induced conformational change in the ferric enterobactin receptor FepA as studied by site-directed spin labeling and time-domain ESR. *Biochemistry* **37,** 9016–9023.

Klug, C. S., and Feix, J. B. (1998). Guanidine hydrochloride unfolding of a transmembrane beta-strand in FepA using site-directed spin labeling. *Protein Sci.* **7,** 1469–1476.

Klug, C. S., and Feix, J. B. (2004). SDSL: A survey of biological applications. *In* "Biological Magnetic Resonance" (L. J. Berliner, S. S. Eaton, and G. R. Eaton, eds.), Vol. 24, pp. 269–308. Kluwer Academic/Plenum Publishers, Hingham, MA.

Klug, C. S., Su, W., and Feix, J. B. (1997). Mapping of the residues involved in a proposed beta-strand located in the ferric enterobactin receptor FepA using site-directed spin-labeling. *Biochemistry* **36,** 13027–13033.

Klug, C. S., Su, W., Liu, J., Klebba, P. E., and Feix, J. B. (1995). Denaturant unfolding of the ferric enterobactin receptor and ligand-induced stabilization studied by site-directed spin labeling. *Biochemistry* **34,** 14230–14236.

Koteiche, H. A., Berengian, A. R., and Mchaourab, H. S. (1998). Identification of protein folding patterns using site-directed spin labeling. Structural characterization of a beta-sheet and putative substrate binding regions in the conserved domain of alpha A-crystallin. *Biochemistry* **37,** 12681–12688.

Langen, R., Isas, J. M., Luecke, H., Haigler, H. T., and Hubbell, W. L. (1998). Membrane-mediated assembly of annexins studied by site-directed spin labeling. *J. Biol. Chem.* **273,** 22453–22457.

Langen, R., Oh, K. J., Cascio, D., and Hubbell, W. L. (2000). Crystal structures of spin labeled T4 lysozyme mutants: Implications for the interpretation of EPR spectra in terms of structure. *Biochemistry* **39,** 8396–8405.

Lewis, J. R., and Cafiso, D. S. (1999). Correlation between the free energy of a channel-forming voltage-gated peptide and the spontaneous curvature of bilayer lipids. *Biochemistry* **38,** 5932–5938.

Lietzow, M. A., and Hubbell, W. L. (1998). Site-directed spin labeling of cellular retinol-binding protein (CRBP): Examination of a β-sheet landscape and its conformational dynamics. *Biophys. J.* **74**(2), A278.

Lietzow, M. A., and Hubbell, W. L. (2004). Motion of spin label side chains in cellular retinol-binding protein: Correlation with structure and nearest-neighbor interactions in an antiparallel β-sheet. *Biochemistry* **43,** 3137–3151.

Macosko, J. C., Kim, C. H., and Shin, Y. K. (1997). The membrane topology of the fusion peptide region of influenza hemagglutinin determined by spin-labeling EPR. *J. Mol. Biol.* **267,** 1139–1148.

Marchetto, R., Schreier, S., and Nakaie, C. R. (1993). A novel spin-labeled amino acid derivative for use in peptide synthesis: (9-Fluorenylmethyloxycarbonyl)-2,2,6,6-tetramethylpiperidine-N-oxyl-4-amino-carboxylic acid. *J. Am. Chem. Soc.* **115,** 11042–11043.

Marsh, D. (2001). Polarity and permeation profiles in lipid membranes. *Proc. Natl. Acad. Sci. USA* **98,** 7777–7782.

Mchaourab, H. S., Berengian, A. R., and Koteiche, H. A. (1997a). Site-directed spin-labeling study of the structure and subunit interactions along a conserved sequence in the alpha-crystallin domain of heat-shock protein 27. Evidence of a conserved subunit interface. *Biochemistry* **36,** 14627–14634.

Mchaourab, H. S., Hyde, J. S., and Feix, J. B. (1994). Binding and state of aggregation of spin-labeled cecropin AD in phospholipid bilayers: Effects of surface charge and fatty acyl chain length. *Biochemistry* **33,** 6691–6699.

Mchaourab, H. S., Oh, K. J., Fang, C. J., and Hubbell, W. L. (1997b). Conformation of T4 lysozyme in solution. Hinge-bending motion and the substrate-induced conformational transition studied by site-directed spin labeling. *Biochemistry* **36,** 307–316.

Merianos, H. J., Cadieux, N., Lin, C. H., Kadner, R. J., and Cafiso, D. S. (2000). Substrate-induced exposure of an energy-coupling motif of a membrane transporter. *Nat. Struct. Biol.* **7,** 205–209.

Millhauser, G. L. (1992). Selective placement of electron spin resonance spin labels: New structural methods for peptides and proteins. *Trends Biochem. Sci.* **17,** 448–452.

Nielsen, R. D., Che, K., Gelb, M. H., and Robinson, B. H. (2005). A ruler for determining the position of proteins in membranes. *J. Am. Chem. Soc.* **127,** 6430–6442.

Palmer, A. G., III. (2001). Nmr probes of molecular dynamics: Overview and comparison with other techniques. *Annu. Rev. Biophys. Biomol. Struct.* **30,** 129–155.

Pannier, M., Veit, S., Godt, A., Jeschke, G., and Spiess, H. W. (2000). Dead-time free measurement of dipole-dipole interactions between electron spins. *J. Magn. Reson.* **142,** 331–340.

Perozo, E., Cortes, D. M., and Cuello, L. G. (1998). Three-dimensional architecture and gating mechanism of a K+ channel studied by EPR spectroscopy. *Nat. Struct. Biol.* **5,** 459–469.

Perozo, E., Cortes, D. M., Sompornpisut, P., Kloda, A., and Martinac, B. (2002). Open channel structure of MscL and the gating mechanism of mechanosensitive channels. *Nature* **418,** 942–948.

Perozo, E., Kloda, A., Cortes, D. M., and Martinac, B. (2001). Site-directed spin-labeling analysis of reconstituted Mscl in the closed state. *J. Gen. Physiol.* **118,** 193–206.

Qin, Z., and Cafiso, D. S. (1996). Membrane structure of protein kinase C and calmodulin binding domain of myristoylated alanine rich C kinase substrate determined by site-directed spin labeling. *Biochemistry* **35,** 2917–2925.

Rabenstein, M. D., and Shin, Y. K. (1995). Determination of the distance between two spin labels attached to a macromolecule. *Proc. Natl. Acad. Sci. USA* **92,** 8239–8243.

Russell, C. J., Thorgeirsson, T. E., and Shin, Y. K. (1996). Temperature dependence of polypeptide partitioning between water and phospholipid bilayers. *Biochemistry* **35,** 9526–9532.

Steinhoff, H. J., Radzwill, N., Thevis, W., Lenz, V., Brandenburg, D., Antson, A., Dodson, G., and Wollmer, A. (1997). Determination of interspin distances between spin labels attached to insulin: Comparison of electron paramagnetic resonance data with the X-ray structure. *Biophys. J.* **73,** 3287–3298.

Stone, T. J., Buckman, T., Nordio, P. L., and McConnell, H. M. (1965). Spin-labeled biomolecules. *Proc. Natl. Acad. Sci. USA* **54,** 1010–1017.

Thompson, L. V., Lowe, D. A., Ferrington, D. A., and Thomas, D. D. (2001). Electron paramagnetic resonance: A high-resolution tool for muscle physiology. *Exerc. Sport Sci. Rev.* **29,** 3–6.

Thorgeirsson, T. E., Russell, C. J., King, D. S., and Shin, Y. K. (1996). Direct determination of the membrane affinities of individual amino acids. *Biochemistry* **35,** 1803–1809.

Victor, K., Jacob, J., and Cafiso, D. S. (1999). Interactions controlling the membrane binding of basic protein domains: Phenylalanine and the attachment of the myristoylated alanine-rich C-kinase substrate protein to interfaces. *Biochemistry* **38,** 12527–12536.

Victor, K. G., and Cafiso, D. S. (2001). Location and dynamics of basic peptides at the membrane interface: Electron paramagnetic resonance spectroscopy of tetramethyl-piperidine-N-oxyl-4-amino-4-carboxylic acid-labeled peptides. *Biophys. J.* **81,** 2241–2250.

Voss, J., He, M. M., Hubbell, W. L., and Kaback, H. R. (1996). Site-directed spin labeling demonstrates that transmembrane domain XII in the lactose permease of Escherichia coli is an alpha-helix. *Biochemistry* **35,** 12915–12918.

Wegener, A. A., Klare, J. P., Engelhard, M., and Steinhoff, H. J. (2001). Structural insights into the early steps of receptor-transducer signal transfer in archaeal phototaxis. *EMBO J.* **20,** 5312–5319.

Xiao, W., and Shin, Y. K. (2000). EPR spectroscopic ruler: The method and its applications. *In* "Biological Magnetic Resonance" (L. J. Berliner, S. S. Eaton, and G. R. Eaton, eds.), Vol. 19, pp. 249–276. Kluwer Academic/Plenum Publishers, New York.

Xu, Y., Zhang, F., Su, Z., McNew, J. A., and Shin, Y. K. (2005). Hemifusion in SNARE-mediated membrane fusion. *Nat. Struct. Mol. Biol.* **12,** 417–422.

Yang, K., Farrens, D. L., Altenbach, C., Farahbakhsh, Z. T., Hubbell, W. L., and Khorana, H. G. (1996). Structure and function in rhodopsin. Cysteines 65 and 316 are in proximity in a rhodopsin mutant as indicated by disulfide formation and interactions between attached spin labels. *Biochemistry* **35,** 14040–14046.

Yu, Y. G., Thorgeirsson, T. E., and Shin, Y. K. (1994). Topology of an amphiphilic mitochondrial signal sequence in the membrane-inserted state: A spin labeling study. *Biochemistry* **33,** 14221–14226.

Zhang, Y., and Shin, Y. K. (2006). Transmembrane organization of yeast syntaxin-analogue Sso1p. *Biochemistry* **45,** 4173–4181.

Zhao, M., Zen, K. C., Hernandez-Borrell, J., Altenbach, C., Hubbell, W. L., and Kaback, H. R. (1999). Nitroxide scanning electron paramagnetic resonance of helices IV and V and the intervening loop in the lactose permease of Escherichia coli. *Biochemistry* **38,** 15970–15977.

Zhou, Z., Desensi, S. C., Stein, R. A., Brandon, S., Dixit, M., McArdle, E. J., Warren, E. M., Kroh, H. K., Song, L., Cobb, C. E., Hustedt, E. J., and Beth, A. H. (2005). Solution structure of the cytoplasmic domain of erythrocyte membrane band 3 determined by site-directed spin labeling. *Biochemistry* **44,** 15115–15128.

CHAPTER 21

Fluorescence Correlation Spectroscopy and Its Application to the Characterization of Molecular Properties and Interactions

Hacène Boukari and Dan L. Sackett

Laboratory of Integrative and Medical Biophysics
National Institute of Child Health and Human Development
National Institutes of Health
Bethesda, Maryland 20892

METHODS IN CELL BIOLOGY, VOL. 84

0091-679X/08 $35.00
DOI: 10.1016/S0091-679X(07)84021-0

Abstract

Fluorescence correlation spectroscopy (FCS) utilizes temporal fluctuations in fluorescence emission to extract quantitative measures of inter- or intramolecular dynamics or molecular motions of probe molecules, which occur on submicrosecond to second timescales. In typical experiments, one can readily obtain the probe's diffusion coefficient and concentration from small volumes of sample. Recent FCS applications have yielded information on interactions of the probe with changing or structured solvent, binding with other molecules, photophysical or conformational changes in the probe, polymerization, and other changes in the dynamics of the probe. In cross-correlation mode FCS promises to attract more applications as the technique can monitor interactions in a system with two or more probes with different fluorophores.

I. Introduction

Since its inception in early 1970s, fluorescence correlation spectroscopy (FCS) has evolved into a valuable tool for investigating various dynamical phenomena in biology, biochemistry, and other fields. With recent instruments being marketed by several companies (Zeiss, ISS, and Hamamatsu), the technique is no longer in a research/development phase in the hands of specialized laboratories rather FCS has become a part of a research arsenal available to many investigators focused on understanding biomolecular interactions, including translational and rotational diffusion of macromolecules, photodynamics of fluorescent proteins, chemical interactions and kinetics of macromolecules (protein-protein, protein-DNA, protein-lipid...), biopolymer dynamics, oligomerization/assembly of proteins, and molecular processes in cellular systems. Examples of these are discussed in the sections below.

Several authors have reviewed extensively the theoretical basis as well as recent technological advances of the technique (Aragon and Pecora, 1976; Chen *et al.*, 1999a; Haustein and Schwille, 2003; Krichevsky and Bonnet, 2002; Magde *et al.*, 1974; Rigler and Elson, 2001; Starchev *et al.*, 1999; Webb, 2001). In this chapter, we focus on practical aspects of the technique as a quantitative biophysical-biochemical method. We describe examples from several investigations from our laboratory and others where FCS has provided quantitative measurements of binding interactions, oligomerization, polymerization, and hydrodynamic effects in concentrated solutions. Finally, we point out possible experimental artifacts and theoretical challenges which could introduce erroneous effects and/or limit the interpretation of FCS measurements. In all of this we limit our view, as is the focus of this volume, to *in vitro* application, although FCS has also been widely applied to intracellular processes.

Typically, FCS uses fluorescence fluctuations emanating from a small illuminated volume of a sample to obtain information about the underlying processes responsible for these fluctuations. At first glance, the fluorescence signal, which is a stream of measured fluorescence intensities with time, appears as random noise (see Fig. 1A). However, the signal can carry quantitative information when time correlated (see Fig. 1B). Indeed, most often the fluctuations (the noisy signal) in FCS are induced either by the changes in the number of fluorescent particles in the excitation volume, as they move in and out of the volume, or by changes in the emission quantum yield of the particles. That is, the fluctuations are attributed to physical mechanisms, such as diffusion, reaction kinetics, or photodynamics, whose dynamic parameters govern the temporal correlation function. FCS is appropriate to probe dynamical processes that occur on microsecond to second timescales in a sample in equilibrium or steady state, or whose kinetics are slow relative to diffusion.

A second complementary approach to analyzing the stream of fluorescence intensities is based on turning the data into photon counting histogram (PCH). Practically, one calculates the frequency with which a particular photocount collected during a fixed bin time is found in the stream, generating hence a histogram or distribution as function of photocount. A general theory of photon statistics indicates that the photon distribution is a convolution of various effects,

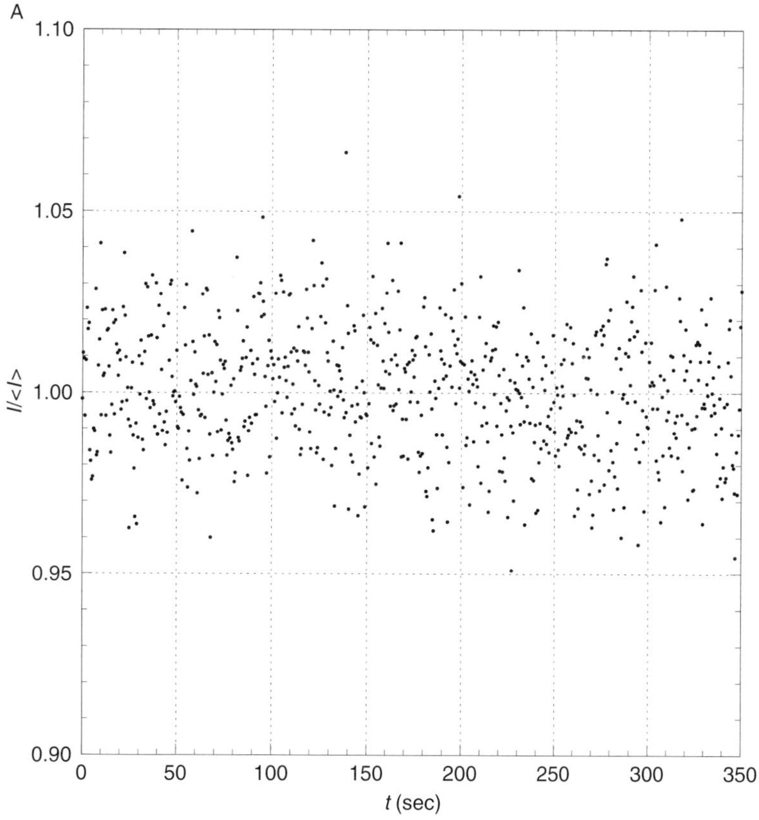

Fig. 1 (*Continues*)

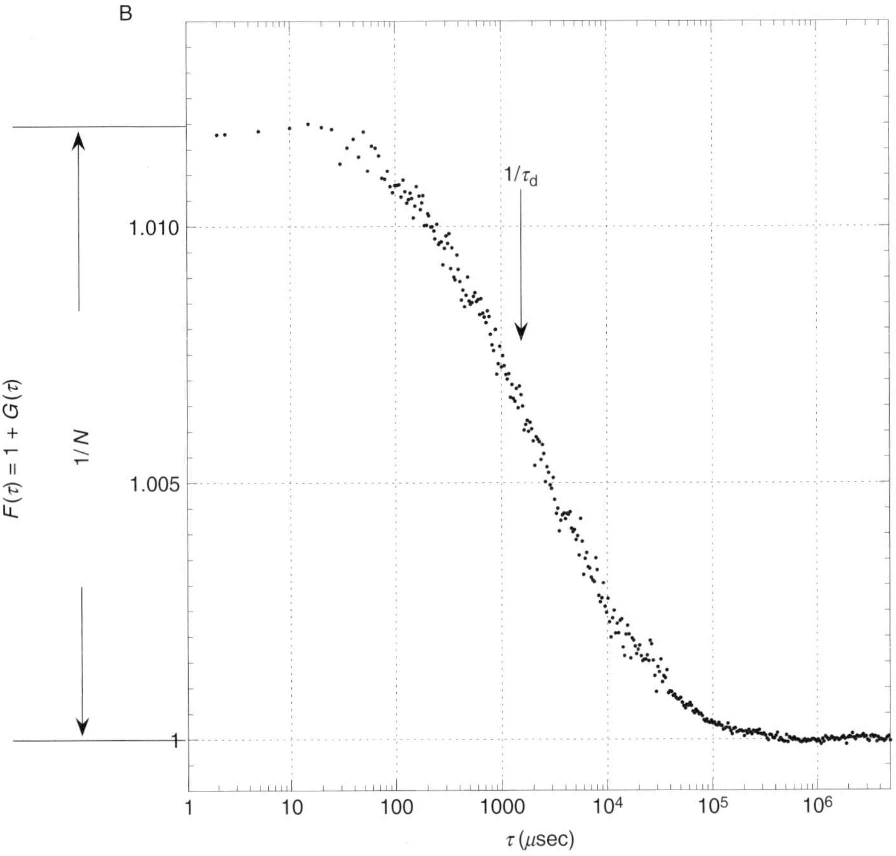

Fig. 1 (A) Fluorescence photocount history of an emitted beam (>560 nm) by a sample of phycoerythrin protein solution excited by a 542-nm, 0.35-μW laser beam. The fluorescence fluctuations are induced by the proteins moving in and out of the excitation volume. (B) Time-correlation function of the fluctuating fluorescence [see Eq. (1)] can provide information about the protein concentration of the sample and the translational diffusion coefficient of the protein.

including the photon statistics of the excitation beam, the variations in the brightness of individual particles, and the dynamics of the particles, making it challenging to interpret the data and extract reliable parameters. Despite these difficulties, recent theoretical progress allows one to extract the distribution of brightness of the particles under appropriate conditions (Chen *et al.* 1999b; Muller *et al.* 2000). The PCH approach is still undergoing theoretical investigation, and we will not include it in this chapter.

One should note that alternative techniques, such as dynamic light scattering (DLS), fluorescence recovery after photobleaching (FRAP), and NMR-based techniques, are also being applied to investigate similar processes. FCS provides, however, several advantages. While DLS and NMR require micromolar or larger concentrations, FCS uses sub- to nanomolar concentrations of fluorescent probe particles. The concentration of nonfluorescent molecules in the assay may be varied arbitrarily, however, and these may be identical to or different from the probe molecule. Hence, fluorescence in FCS provides specificity for the probe particles in a host medium, allowing monitoring of just these particles, unlike DLS where all the particles in the host medium can contribute to the scatterering signal. FCS can monitor the dynamics of particles of different sizes (nanometers to hundreds of nanometers) as long as the particles fluoresce. In contrast, NMR techniques probe a more limited range of length scales, typically a few nanometers. FRAP, which is most often applied in cell biology, is in many respects similar to FCS as both techniques use fluorescence. However, FRAP measurements can be complicated by ambiguous experimental artifacts, especially partial recovery and reversible photobleaching (Dauty and Verkman, 2004). We do not want to imply that FCS is perfect and immune to possible artifacts as we discuss later some of its limitations. Certainly, a well-informed investigator should combine all the techniques, if possible, and extract a consistent picture of all the data.

II. Fluorescence Correlation Spectroscopy

In an FCS experiment, temporal fluctuations in fluorescence emission are utilized to obtain information about inter- or intramolecular dynamics or molecular motions occurring on microsecond to second timescales. These fluctuations are induced either by the changes in the number of molecules in the open excitation volume, as they move in and out of the volume, or by changes in the emission quantum yield of the molecules. To obtain high sensitivity, the excitation volume is made small (~1 fl) either by confocal geometry or by multiphoton excitation. The time sequence of the detected intensity, $I(t)$, of fluorescence emitted by the fluorescent molecules present in the excitation volume at time, t, is time-correlated to generate the autocorrelation function defined as

$$F(\tau) = \langle I(t)I(t+\tau)\rangle \tag{1a}$$

where $I(t)$ and $I(t+\tau)$ denote the fluorescence intensities detected at time t and delay time $t+\tau$, respectively. Equation (1a) is commonly rewritten as

$$F(\tau) = 1 + G(\tau) = 1 + \frac{\langle \delta I(t)\delta I(t+\tau)\rangle}{\langle I(t)\rangle^2} \tag{1b}$$

where we show the correlation of the spontaneous deviation $\delta I(t) = I(t) - \langle I(t)\rangle$ of the measured intensity from the time-averaged intensity $\langle I(t)\rangle$.

In the interpretation of the intensity, it is assumed that the signal is directly related to the number of fluorescent particles in the excitation volume such that

$$I(t) = A \int W(\vec{r})c(\vec{r}, t)d^3\vec{r} \tag{2}$$

where $W(\vec{r})$ describes the profile of the excitation volume, $c(\vec{r}, t)$ the number concentration of the particles, and A a constant. For monodisperse flurophores (same brightness and same diffusion coefficient) the zero time correlation function $G(0)$ describes the normalized variance of the fluorescence intensity or similarly the normalized variance of the number of fluorophores:

$$G(0) = \frac{\langle I^2\rangle - \langle I\rangle^2}{\langle I\rangle^2} = \frac{\langle N^2\rangle - \langle N\rangle^2}{\langle N\rangle^2} = \frac{1}{\langle N\rangle} = \frac{1}{\langle C\rangle V_{\mathrm{eff}}} \tag{3}$$

where the statistical process is considered Poissonian ($\langle N^2\rangle - \langle N\rangle^2 = \langle N\rangle$) and $\langle C\rangle$ denotes an average number of the fluorophores [not necessarily the average number of fluorescent particles; see Eq. (7)]. In principle, the effective volume, V_{eff}, can be estimated experimentally from the intercept [$G(\tau \rightarrow 0)$] of the measured correlation function if the average concentration of standard simple fluorophores (i.e., rhodamine 6G) is known. Equivalently, the concentration of a probe in a given experimental setup can be determined if the volume has previously been determined by calibration with a known reference solution. The effective volume is related to a calculated volume, V_{p}, from the theoretical profile, $W(\vec{r})$, by $V_{\mathrm{eff}} = V_{\mathrm{p}}/\gamma$, where $V_{\mathrm{p}} = \int W(\vec{r})d^3\vec{r} \,/\, W(0)$, and the γ factor, which is typically less than 1, is a measure of the effect of the profile on the fluorescence emission and the abruptness of the boundaries of the profile. It is customary to approximate $W(\vec{r})$ by a Gaussian prolate in three dimensions [$W(r, z) \sim \mathrm{e}^{-2(r/r_0)^2}\mathrm{e}^{-2(z/z_0)^2}$], and we have $V_{\mathrm{p}} = (\pi/2)^{3/2} r_0^2 z_0$.

Equation (3) indicates that the correlation function becomes significantly nonzero when the concentration of fluorophores is relatively small or the detection volume is very small. In fact, for typical confocal detection volumes of the order of femtoliters (10^{-15} liter) the appropriate concentration of the fluorophores for FCS should be in the nanomolar range (~10–100 nM). This distinguishes FCS from other techniques, such as DLS and NMR, where a good signal-to-noise ratio is obtained with more concentrated solutions in the micromolar concentration.

A. Translational Diffusion

Analysis of the time-dependent part of $G(\tau)$ in Eq. (1b) provides information about the underlying mechanisms responsible for the intensity fluctuations. In particular, translational diffusion of both small molecules and macromolecules (proteins, DNA, peptides...) in and out of the excitation volume has been exploited to probe binding/unbinding of or to biomacromolecules. For an ideal case of freely diffusing, monodisperse, and uniformly bright-fluorescent particles, a closed-form expression of the correlation function, $G(\tau)$, in Eq. (1b) was derived (Aragon and Pecora, 1976):

$$G(\tau) = 1 + G_{\text{diff}}(\tau) = 1 + \frac{1}{N}\frac{1}{(1+\tau/\tau_{\text{d}})(1+p\tau/\tau_{\text{d}})^{1/2}} \tag{4}$$

In Eq. (4) it is assumed that the fluorescent particles are excited by a three-dimensional Gaussian beam $[W(r,z) \sim e^{-2(r/r_0)^2} e^{-2(z/z_0)^2}]$. Here r_0 and z_0 define two characteristic sizes: the width of the focused beam spot and the length along the optical axis defined by the direction of the laser beam. Both sizes are used to define the excitation volume $V = \pi^{3/2} r_0^2 z_0$. With such a Gaussian beam, one identifies a characteristic diffusion time $\tau_d = r_0^2/4D$ for single-photon FCS or $\tau_d = r_0^2/8D$ for two-photon FCS, D being the translational diffusion coefficient of the fluorescent particles. Also in Eq. (4) N=⟨N⟩ denotes the average number of particles in the excitation volume, and $p = (r_0/z_0)^2$ is an instrumental constant. Further, in dilute solutions the Stokes–Einstein relation

$$D = \frac{k_{\text{B}}T}{3\pi\eta d_{\text{H}}} \tag{5}$$

is applied to determine the hydrodynamic diameter d_{H} of the diffusing fluorescent particles. In Eq. (5) k_{B} is the Boltzmann constant, T the temperature of the sample in Kelvin, and η the viscosity of the solvent ($\eta = 8.90 \times 10^{-4}$ Pa sec for water at $T =$ 25 °C).

For a system of noninteracting, freely diffusing components, the correlation term $G_{\text{diff}}(\tau)$ in Eq. (4) can be generalized to

$$G_{\text{diff}}(\tau) = \frac{1}{(\sum_{j=1}^{N} n_j Q_j)^2} \sum_{i=1}^{N} \frac{n_i Q_i^2}{(1+\tau/\tau_{\text{id}})(1+p\tau/\tau_{\text{id}})^{1/2}} \tag{6}$$

where n_i, Q_i, and τ_{id} are the number, the brightness, and the diffusion time of the ith component with $\tau_{\text{id}} = \text{r}_0^2/4D_i$, D_i being the diffusion coefficient. Note that the amplitude of the correlation term

$$G_{\text{diff}}(0) = \frac{\sum_{i=1}^{N} n_i Q_i^2}{(\sum_{j=1}^{N} n_j Q_j)^2} \tag{7}$$

depends not only on the distribution of the particle number (n_i) but also strongly on the variation of brightness (Q_i) of each type of particles. That is, brighter particles can differentially affect the correlation function. The expression in Eq. (6) reduces to $1/N$ when the brightness of all particles is the same ($Q_i =$ constant), very similar to the amplitude of the expression in Eq. (4).

III. FCS Experimental Setups

Modern FCS setups are generally built around an inverted microscope. In Fig. 2 we show a schematic diagram of a typical setup in which two detectors are installed for cross-correlation measurements. It is recommended that the setup be on an isolation table to reduce vibration effects on the measurements. A stable excitation beam from a laser (i.e., HeNe or Argon) is expanded to a width less than that of the back aperture of the objective, then directed via a dichroic mirror to a high numerical aperture objective (NA = 0.9 air; NA = 1.2 water; NA = 1.4 oil) which focused it onto the sample. The choice for the wavelength of the laser depends on the fluorophore being considered. Also, the laser intensity must be optimized to reduce the effects of various photodynamical processes of the fluorophores (i.e., triplet-singlet, photobleaching, and photosaturation), yet obtain enough statistical photocounts for correlation.

The emitted fluorescence from the sample is collected by the same objective, then passed through the dichroic mirror, and finally collected with photocounting

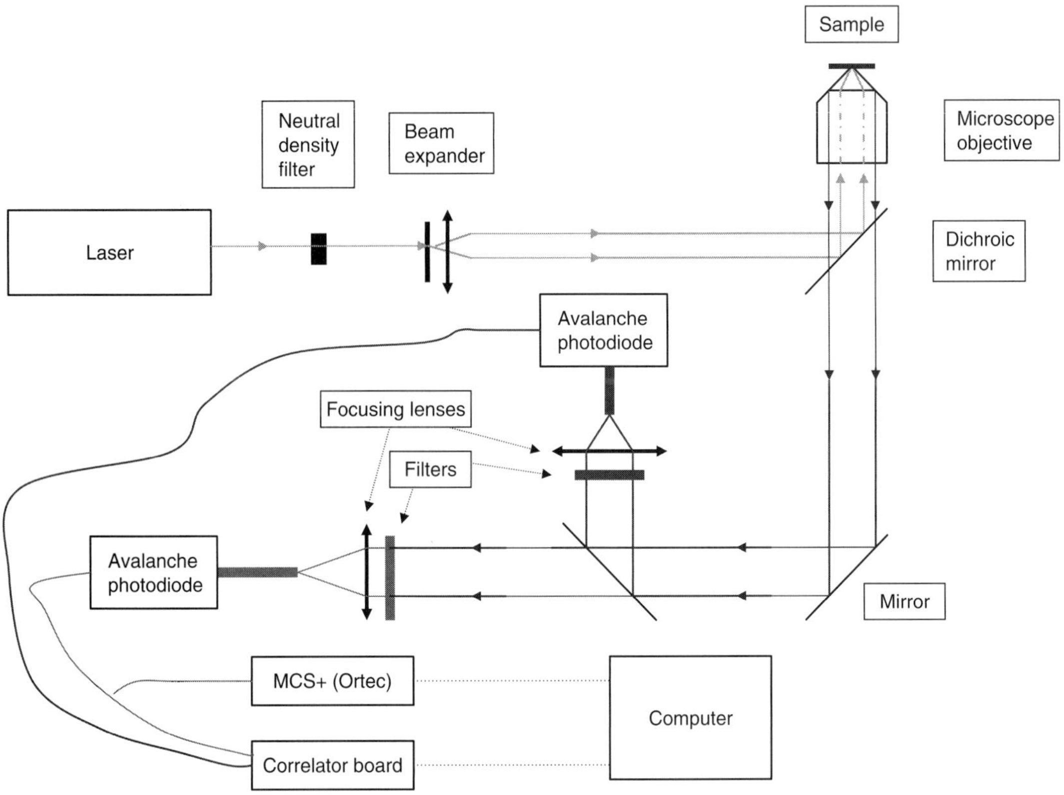

Fig. 2 Schematic diagram of an FCS setup where cross-correlation is used.

detectors with high quantum yield, such as avalanche photodetectors (APDs from Perkin-Elmer, formerly EG&G) or photomultiplier tubes (PMTs from Hamamatsu). Generally, APDs have high quantum efficiency (~70%) but small active area (150 μm diameter), which then requires the use of fiber optics. Hamamatsu introduced recently a 30–40% quantum efficiency PMTs with relative large detection area (~1 in.), providing flexibility and stability in the alignment of the detector with respect of the detected beam. In both cases, it is advisable to limit the number of detected pulses to the linear regime of the photodetectors' response (<1 million count). An inherent problem with photodetectors (APDs or PMTs) is afterpulsing effect, which is basically the formation of secondary pulses following the main ones related to the real sample. As a result, a fast component (<10 μsec) appears in the correlation function, which could be taken erroneously as a dynamical process within the sample. More importantly, afterpulsing prevents a correct estimation and/or calibration of the amplitude of the correlation function, which is related to the concentration of the fluorophores. A remedy is to perform the measurements in cross-correlation mode (see Fig. 2), where the detected signal is split into two separate photodetectors, and the two signals are cross-correlated. Because the two afterpulsing signals are induced by two different and independent detectors, their effects are basically eliminated from the cross-correlation function.

In order to delimit a small detection volume two different approaches are borrowed from standard fluorescence confocal microscopy, namely single-photon setup with a small pinhole detection (one-photon FCS) and two-photon setup without a pinhole (two-photon FCS). In one-photon FCS, a small pinhole (~50 μm or less) is inserted in the image plane of the observed volume rejecting hence most of the fluorescence emanating from the out of focus plane. Moreover, standard excitation and redshifted emission of fluorophores is used, where high band optical filters (dichroic mirror and detection filter) remove most of the excitation wavelength and allow detection of fluorescence only. In contrast, a near-infrared laser is utilized in two-photon FCS to induce simultaneous absorption of two photons and excitation of the fluorophore into a higher quantum state. As a result, photons with high energy (blueshifted short wavelength at half the excitation wavelength) are emitted. Since the two-photon process is likely to occur in the most intense part of the focused beam, fluorescence appears confined into a small volume, and no confocal pinhole is needed to delimit the volume. Note that the correlation function in Eq. (1b) must be modified to account for the two-photon process.

For correlation, the signal is turned into pulses, which are then processed to generate the correlation function. The pulses are either fed to a correlator board (i.e., BT9000 from Brookhaven Instruments) to generate the correlation function or alternatively collected and saved as a continuous stream of photocounts (numbers of pulses during a short bin time duration ~100–500 nsec) with a fast computer. Then, the saved stream of pulses is used to calculate afterward the correlation function. In this latter approach, one has access to the raw data (the stream of photocounts), which can be analyzed as fit. For example, one can remove undesirable spikes in the signal or create photocounting histograms. However, one could run into computer-storage considerations.

IV. Sample Preparation and Some Practical Considerations

A. Fluorophores

Understanding the photodynamics of the fluorescent biomolecules under study can be very important in the interpretation of FCS measurements. Here, the biomolecules can be naturally fluorescent [green fluorescent protein (GFP) and phycoerythrin] or chemically labeled with known dyes. The quality of the FCS data depends essentially on the brightness of the individual fluorescent biomolecules, more specifically the emission per unit biomolecule, as shown by Koppel (1974). Thus, it is recommended to use fluorophores with high extinction coefficient and high quantum yield. Moreover, the measurements should be performed at relatively low intensity (~μW) to avoid various photophysical effects such as singlet-to-triplet transition states, photobleaching, and antibunching. These effects tend to interfere with the measurements by either introducing additional relaxation times into the correlation function especially at short time or destroying irreversibly the fluorophores. It happens that in many instances the laser intensity needs to be increased in order to improve the statistics of the data. For the singlet-to-triplet transition, the effect is generally observed as a fast decaying correlation at short time scales (<10 μsec), which can be taken into account by modifying Eq. (4) of the translational correlation function of single biomolecules to

$$G(\tau) = 1 + \frac{1}{N}\frac{1}{(1+\tau/\tau_{\mathrm{d}})(1+p\tau/\tau_{\mathrm{d}})^{1/2}}\left(1 + \frac{M}{1-M}\exp\left(\frac{-\tau}{\tau_M}\right)\right) \tag{8}$$

where M is the fraction of the fluorophore in the triplet state and τ_M is the relaxation time. The antibunching, which is related to the turnover time that the fluorophore needs in order to emit a second photon following the first emission, occurs at much shorter times (~10 nsec), and hence it is generally not measurable with traditional FCS setups. Photobleaching pertains to the tendency of fluorophores to undergo irreversible destruction. The mechanisms for such destruction are not totally understood. However, it is, generally, independent of the excitation intensity for weak illumination. That is, the fluorophore goes blank after a characteristic number of photons emitted. As the excitation intensity is increased, the probability of photobleaching is increased, and this could be of concern to FCS measurements.

Briefly, typical requirements for the choice of a fluorophore are high extinction coefficient, high quantum yield, low probability for singlet-to-triplet state transition, and low photobleaching. The most common exogenous fluorophores are carboxyrhodamine (Rh6G), tetramethylrhodamine (TMR), and the Alexa series, which are available commercially. They are very bright, relatively small (MW ~400–1000 Da), and less sensitive to pH and photobleaching. Fluorophores with various reactivities (sulfhydryl, amine, carbonyl, hydroxyl) are readily available commercially, and the conjugation chemistry is not difficult. In addition, commercial custom conjugation is readily available. In order to trace the dynamics and interactions of particular molecules, it is now common to express GFP-tagged (or other naturally fluorescent proteins) biomolecules in cellular systems. Here one should mention that because of their FCS and microscopy applications naturally fluorescent proteins (GFP, EGFP, phycobilin...) are also the subject of interesting *in vitro* FCS studies (see Section VI.C).

B. Sample Labeling and Practical Tips

When labeling with exogenous fluorophores, one works under the premise that the fluorophores do not induce significant structural and/or chemical changes to the biomolecules of interest. It is then recommended to aim for low labeling stoichiometry (1–2 fluorophores per biomolecule), though one might argue for heavy labeling of the biomolecule to satisfy the brightness condition of FCS (though self-quenching will often limit this is any case). Here, one identifies potential reactive sites of the biomolecules (i.e., cysteine groups in proteins), and it is ideal if only one reactive site is present, hence avoiding polylabeling of the biomolecules. If not, one should expect a distribution of the labeling (likely Poissonian), making it challenging to interpret the amplitude of the correlation function [see Eq. (5)].

Several practical tips in FCS measurements should be pointed out:

- Following labeling, it is worth removing excess of the free fluorophores so that a simple fitting of the data is possible; the presence of free fluorophores in the sample introduces a second diffusive component in the correlation function, hence increasing the number of fitting parameters. Removal of all free dye can be challenging (Krouglova *et al.*, 2004).
- As indicated earlier, the concentration of the fluorescent biomolecules need to be low enough so that the correlated fluctuations are measurable. In most configurations, one should start with relatively high concentration (~100–200 nM) to assess the appropriate laser power (1–30 μW) given the labeling and smallness of the biomolecules. Then, a systematic study of the effect of the laser power on the diffusion time and amplitude should be performed. In Fig. 3 we illustrate the excitation saturation effects in a sample of a phycoerythrin protein solution induced by the increasing laser power. It is only at submicrowatt power (<0.3 μW) that the amplitude of the correlation function appears independent of the laser power (not shown) and the fit of the function is reasonably good, yielding an accurate diffusion coefficient. It was shown that as the excitation intensity is increased, the excited state population of the proteins builds up due to the finite

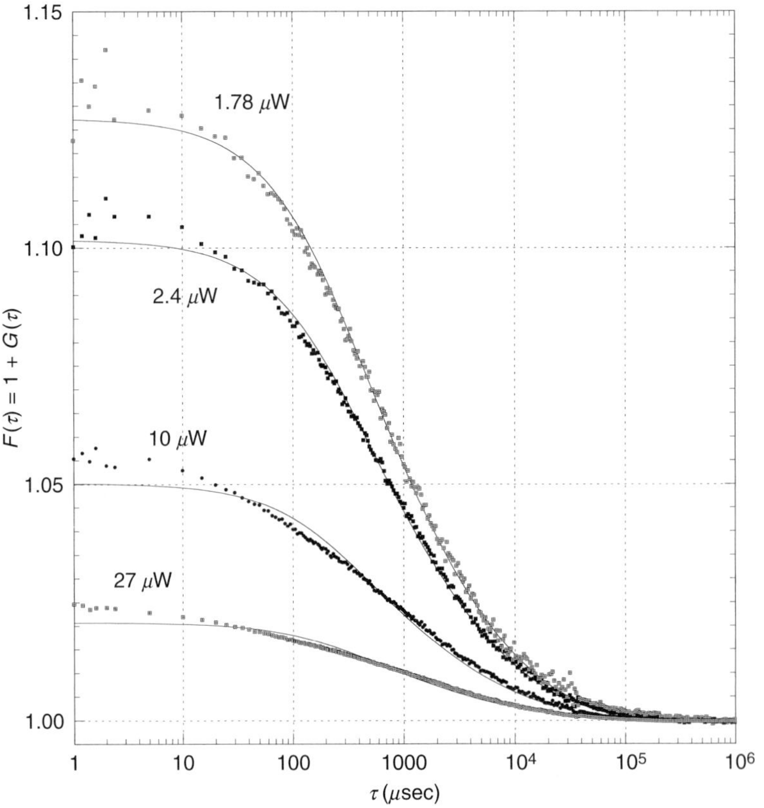

Fig. 3 Saturation effects due to laser power on the time-correlation functions measured on a sample of phycoerythrin solution. Note the decrease of the amplitude [$G(\tau \rightarrow 0)$] as the input incident power is increased, and the poor fit of the correlation function with the expression in Eq. (4) derived for a Gaussian profile. It is only at powers below 0.35 μW that the correlation is well fit and yields an accurate diffusion coefficient.

lifetime of the excited state and saturation effects become significant. Consequently, the actual excitation laser profile appears distorted away from the original smooth Gaussian profile. The saturation region in the profile increases with increasing laser power, resulting in an apparent increase of both the number of observed biomolecules (low amplitude) and the characteristic diffusion times, consistent with the measurements shown in Fig. 3.

• Small volumes are used in FCS. For solutions, we have been using Grace Bio-Labs silicone press-to-seal chambers that hold volumes as small as 60 μl. With their sheet material we have fabricated chambers to hold volumes as small as 10 μl.

• Many biomolecules tend to adsorb to the glass coverslip placed on the adhesive side of the chamber. Consequently, the concentration of the biomolecules will decrease as a function of time, which can be readily observed in the change of the amplitude of the correlation function. A practical solution is adding BSA to the solution (verify that BSA does not interact with the biomolecule of interest) or neutralizing the Si charges on the surface of the glass (i.e., treat the surface with SigmaCoat from Sigma Chemical Co.).

- Many proteins are prone to aggregation. The aggregates appear as bright entities with large spikes in the photocounting stream, which distort the correlation function. Appropriate centrifugation can remove these aggregates and "clean" your sample.
- It is worthwhile collecting several correlation functions to assess the uncertainty of the measurements. Many pitfalls can be encountered in the analysis of an FCS measurement. It is recommended to eliminate/reduce irrelevant processes, especially photophysical effects (if this is not the subject of the study). This would bring down the number of fitting parameters of a measured correlation function to a bare minimum. The singlet-to-triplet state transition appears to be the most often observed process, although one should note that it is common to invoke its presence as soon as an apparent relaxation at short time scales appears. This tendency is ill-advised, and one should make sure of the origin of the relaxation by performing additional measurements (e.g., checking the dependence of the correlation functions on the intensity of the excitation laser beam).

V. Fluorescence Cross-Correlation Spectroscopy

The previous FCS discussion pertains to time-correlating a fluorescent signal with itself (autocorrelation). As mentioned above, its use is limited to applications in which the binding event significantly and substantially reduces the diffusion of the labeled biomolecule. However, there are instances both interactive biomolecules are of the same weight or the small biomolecule cannot be labeled. In this case, if it is possible to choose or label the two interacting biomolecules with two spectrally different fluorophores, one could cross-correlate the emitted fluorescence of the two fluorophores, providing a means to assess the interactions between the biomolecules. In principle, the two emitted outputs correlate in time if the two fluorophores are present on the same diffusing complexes induced by the interactions. Further, the amplitude of the cross-correlation function is related to the degree of interaction between the two fluorescent biomolecules. If, instead, there is no binding between the two different fluorescent biomolecules, then they would diffuse independently in and out of the excitation volume and no cross-correlation function is expected to be observed.

In practice, one needs to excite simultaneously in time and space the two fluorophores and detect simultaneously their emission with two separate photodetectors, such as configured in Fig. 2. Further, there should be little or no energy transfer (FRET) between the fluorophores and little or no bleed-through of the emission between the detection paths. These requirements already hint at some of the difficulties in not only implementing fluorescence cross-correlation spectroscopy (FCCS) but also finding the ideal fluorophores. Several groups have set up FCCS instruments and have been described in the literature (Krichevsky and Bonnet, 2002). In most cases, two laser lines (two wavelengths) either selected from a multiline laser or combined from two separate lasers with different lines are used to excite both fluorophores simultaneously. The setup is very similar to that shown in Fig. 2, where the emitted beam is split with a dichroic prism into the two expected emission outputs. Appropriate optical filters are also placed in front of each photodetector to reduce cross talk between the two outputs. In this setup, the major challenging problem is the alignment and the stability of the alignment of the two incident beams so that they are focused onto the same excitation volume (see discussion in Krichevsky and Bonnet, 2002; Schwille *et al.*, 1997). This problem can be solved by using a two-photon excitation scheme where both fluorophores are excited with the same IR laser line. As a result, there is only one single excitation confocal volume since there is only one focused beam.

VI. Illustrative Examples of FCS Applications

Much of the success of FCS lies in its ability to measure interactions between various biomolecules (ligand-receptor, DNA-DNA, protein-DNA, protein-protein, protein-lipid...). Consider the simple binding reaction $A + B \rightarrow C$ of a fluorescent biomolecule A to another biomolecule (generally nonfluorescent) B to form the complex C. FCS will typically probe change in the translational diffusion coefficient of the fluorescent biomolecule (A) on formation of the complex C. In order to obtain precise measurement of the interactions there must be, in particular, a substantial difference between the diffusion coefficients of the individual biomolecule A and the induced complex C (by at least a factor of 1.6) as shown by Meseth *et al.* (1999) and Starchev *et al.* (1999). Assuming globular biomolecules, this is close to a fourfold increase in molecular weight. Thus, the smaller of the two interacting biomolecules is labeled in most FCS assays.

In addition to measuring binding of molecules, FCS can and has been used to detect many other changes in molecular systems that result in changes in the number of probe particles (such as polymerization), changes in molecular conformation, or changes in intensity or dynamics of fluorescence as well as quantifying molecular properties, such as the diffusion coefficient of a given probe molecule in a particular environment. We briefly review a few examples of these applications in the sections below.

A. Use of FCS to Quantify Translational Diffusion with Varying Solution Conditions

FCS is usually employed to measure the diffusive properties of a particle in a solution of known viscosity. But, using the Stokes–Einstein expression in Eq. (5) one can equally derive the viscosity of a solution from the measured diffusion coefficient of the particle if the size of the particle is already known. Thus, FCS measurements with known fluorescent particles can serve to measure and compare the viscosity of various structureless fluids, such as standard solvents, or uniform mixtures, such as glycerol/water. If the solution is structured, however, the situation is more complicated. This becomes an important consideration in situations such as the interior of a cell. Situations intermediate between structureless solvents and the interior of a cell provide useful test cases of FCS in complex environments, both in terms of experimental manipulations as well as of appropriate theory for interpretation of the results. Concentrated solutions of high molecular weight polymers or gels of such polymers provide a situation with this intermediate complexity.

Michelman-Riberno *et al.* (2007) used FCS to study the diffusive behavior of probes of various sizes in solutions of poly(vinyl alcohol). The probe size was varied by nearly tenfold and the concentration of polymer was similarly varied by nearly tenfold. The resulting changes in diffusion coefficients (compared to that in water) cannot be directly correlated with the bulk viscosity of the host polymer solutions as the Stokes–Einstein expression in Eq. (5) would suggest. The measurements were compared with several models of expected behavior and, for small probes, the decrease of the diffusion coefficient with the concentration of the polymer is shown to be fit by a stretched exponential in accord with a model suggested by Langevin and Rondelez (1978). It is only when the size of the probe particle is relatively large compared with the polymer-polymer mesh size that the Stokes–Einstein expression is applicable.

B. FCS in Binding Assays

FCS is well suited to measuring binding of small molecules to large ones, and with any of several adaptations can measure binding of equivalent- or arbitrary-sized macromolecules to each other. When the binding of small molecules is studied, these are usually complexed with bright fluorophores to allow detection. This can introduce unintended binding events, and furthermore the binding event(s) may change the brightness of the fluorophore. When the binding of large (and similar sized) molecules is studied, it may be necessary to employ cross-correlation methods or mass-shifting tags to allow quantitation of binding.

Hazlett *et al.* (2005) demonstrate the application of FCS to the binding of haptens to antibodies. Since haptens are much smaller than the binding antibodies, the diffusion coefficient of the fluorescent haptens changes are substantial. The autocorrelation fits provide a measure of the fraction of ligand bound from the ratios of the $G(0)$'s, and the K_d is obtained from fits of fraction bound to number of sites. Since the hapten-fluorophore may change brightness when bound, the fit must accommodate this, and the needed corrections are outlined.

An additional complication that is considered is that the antibody can bind one or two haptens, which will have the same diffusion coefficient but will differ twofold in particle brightness.

Van Craenenbroeck and Engelborghs (2000) applied FCS to the binding of colchicine analogs to tubulin, using the brightly fluorescent conjugate, fluorescein-colchicine (FC). Binding of this molecule to tubulin was quantitated, but it was not displaced by colchicine, revealing a binding site for FC distinct from the colchicine site. This study demonstrates the usefulness of FCS for quantitating a protein-ligand binding reaction, as well as illustrating the potential problems that can be presented to FCS (or other fluorescence-based methods) by attaching a fluorescent moiety to a small molecule.

Stoevesandt *et al.* (2005) report an interesting and versatile method for measuring protein–protein interactions in crude lysates using indirect immunofluorescence to attach fluorophores to both binding partners. Binding could be demonstrated between two proteins either by using different fluorophores for the antibodies to each protein and measuring cross correlation of the fluorescence signals (FCCS) or by labeling one antibody with a mass tag such as a 50-nm nanoparticle and measuring the change in diffusion time of the other antibody (which has a fluorescent tag). Both of these approaches are quite general and could be employed with purified samples as well as in unseparated mixtures such as in clinical samples. The mass tag allows these binding events to be studied without the need for a second detector and cross-correlation approaches.

C. Use of FCS to Detect Conformational Changes

FCS can be employed to measure any change in the system that alters $G(0)$ or the autocorrelation function, including monomolecular changes that can be conformational or photophysical in origin. Denaturation can lead to changes in particle size as well as number (for multimeric molecules), and possibly brightness. All of these are potentially measurable by FCS, as are brightness changes due to flickering of the fluorophore. Changes in the autocorrelation function can also arise due to changes in the mode of diffusion of the particle. A number of studies exemplifying all of these have been published and no doubt many more will follow.

Schwille *et al.* (2000) and Malvezzi-Campeggi *et al.* (2001) use FCS to examine the photophysics of visible fluorescent proteins YFP and DsRed. Light-induced flickering of the proteins between bright and dark states of the chromophore is revealed by shifts in the autocorrelation function in the 10 μsec to 1 msec range. The light intensity-dependent increase in flicker rate suggests a mechanism that involves multiple fluorescent and dark states of the protein.

In addition to subtle conformational changes, transition from the native to the unfolded state of a protein can be monitored by FCS as shown for the guanidine chloride—induced unfolding of the tubulin dimer (Sanchez *et al.*, 2004). The change in diffusion coefficient caused by dissociation of the dimer is less than that conveniently measured by FCS, so unfolding was measured by following changes in $G(0)$ that accompany the twofold increase in fluorescent particle number caused by dissociation of the dimer by increasing denaturant.

D. Use of FCS to Monitor Polymerization

Since polymerization is the binding of one molecule to another or to a complex of others, the application of FCS to polymerizing systems is similar to that of binding, as discussed above. A difference from the applications described in the previous section is that in polymerizing systems, the binding events may continue to considerable or unlimited extents, so that the "bound" or polymerized form may really be a complex mixture of forms larger than the unpolymerized, "unbound" form. FCS methods have been used to monitor the assembly process in systems with a defined polymer as well as in systems with open, essentially unlimited polymerization.

The polymerization of amyloid β peptide was monitored by FCS using rhodamine-labeled Aβ (Tjernberg *et al.*, 1999). The time course of polymerization was followed from 2 min to 24 h, and polymerization proceeded from the peptide to large aggregates, with no detectable intermediates. This allowed polymerization to be quantified with only two components: the peptide and the (much larger) aggregates. By varying the concentration of nonlabeled Aβ added to the solution, it was shown that polymerization is highly cooperative, and the effect of inhibitory Aβ ligands on aggregation could be demonstrated.

Tubulin polymerization was analyzed by FCS following induction of oligomerization by Mg^{2+}, paclitaxel, and Flutax2, a fluorescent analog of paclitaxel (Krouglova *et al.*, 2004). First a series of standard proteins were labeled with rhodamine and their measured diffusion coefficients were shown to be consistent with their known molecular mass. Tubulin was labeled in the same way. The authors discuss the difficulty of removing unreacted dye from all of the proteins following reaction, and the consequent need for two component fits to the data. Tubulin is known to form various oligomers in the presence of high Mg^{2+}, and this was demonstrated using labeled tubulin, and the association constant derived. Flutax2 was shown to not associate to microtubules preformed with paclitaxel, but to be incorporated if a Flutax2-tubulin complex is formed before addition of paclitaxel.

Tubulin can also be made to oligomerize into specific, uniform, single-walled ring polymers by antimitotic peptides such as cryptophycin-1, dolastatin-10, and hemiasterlin. We applied FCS to probe the assembly and stability of rhodamine-labeled tubulin into these closed rings (Boukari *et al.*, 2003, 2004). Detailed analysis of both the amplitude and the diffusion times of measured FCS correlations taken from solutions of cryptophycin-tubulin and dolastatin-tubulin revealed differences in the interactions of the peptides with tubulin, though the peptides appear to bind to the same site. Cryptophycin-tubulin rings are made of eight tubulin dimers and are stable on dilution down to 1-nM tubulin, demonstrated by an unchanging diffusion coefficient on dilution. In contrast, dolastatin-tubulin rings, composed of 14 tubulin dimers, appear unstable on dilution, with significant dissociation below 10 nM. In addition to studying the structure and stability of these polymers, we exploited available hydrodynamic theories to calculate the hydrodynamic properties of model rings and compared the results consistently with the data.

E. FCS as a Diagnostic or Quality Control Tool

Since the observation volume in FCS is already small, miniaturization of the system can allow for observation of many samples in a small format. This can allow FCS to be applied to high-throughput applications, such as drug discovery and quality control screening.

Birkmann *et al.* (2006) use FCS to develop a diagnostic method for the detection of the prion particles found in diseases such as bovine spongiform encephalopathy (BSE). Aggregates of the host-encoded prion protein, PrP, were detected using two different specific antibodies, labeled with different fluorophores. Fluorescence intensity distribution analysis of the FCS data allowed quantitation of the bright aggregates as spikes in the time series. Using different fluorophores on the two antibodies allowed cross-correlation methods to increase the specificity of the detection, as demonstrated by discrimination of samples from BSE-infected cattle or scrapie-infected hamsters from uninfected controls.

F. Use of FCCS

A number of the studies mentioned previously use cross-correlation methods (FCCS) in combination with single channel methods. In addition to these, a number of studies have focused on cross-correlation methods in FCS.

The first experimental realization of two-color FCCS was done by Schwille *et al.* (1997). Using the technique they monitored the kinetics of cleavage by restriction endonucleases of synthetic DNA oligonucleotides labeled with a different fluorophore at each end. Similarly, Kohl *et al.* (2002) extended the technique to proteolytic assays in which they used two engineered fusion proteins consisting of a green and a red protein and a peptide linker with a protease cleavage site. The assays were performed on purified proteins in solutions and then in cellular systems.

In a nice report, Rippe (2000) described results of a study by two-color FCCS of transcriptional regulation, which involves complex interactions between transcription factors, ligands, DNA and RNA polymerase. More specifically, he focused on the interaction of NtrC protein (nitrogen regulatory protein C) with DNA oligonucleotides, which were labeled with two different fluorophores, 6-carboxyfluorescein and 6-carboxy-X-rhodamine. Titration of a 1:1 mixture of the two different labeled-DNA with the NtrC protein showed a significant increase in the cross-correlation amplitude, indicating that one NtrC octamer could bind two molecules of DNA. More interestingly, Rippe proposed a looping model for the complex explaining how the NtrC protein could interact with and activate RNA polymerase bound to a relatively distant promoter.

VII. Conclusions

An important extension of the methods discussed in this chapter will be application to cellular problems. Indeed, many applications have already been published (see e.g., Bacia *et al.*, 2006; Jankevics *et al.*, 2005; Pramanik and Rigler, 2001), concerned with interactions in the cell membranes as well as in the cytoplasm and in the nucleus.

Further developments in fluorophores, more compact instrumentation, and novel tagging methods for probes will continue to extend the range of applications of this versatile and sensitive method.

Acknowledgments

The authors thank Ralph Nossal for many insightful discussions and for continuing interest in applications of FCS. This work was supported by intramural funds from the National Institute of Child Health and Human Development, NIH.

References

Aragon, S. R., and Pecora, R. (1976). Fluorescence correlation spectroscopy as a probe of molecular dynamics. *J. Chem. Phys.* **64,** 1791–1803.

Bacia, K., Kim, S. A., and Schwille, P. (2006). Fluorescence cross-correlation spectroscopy in living cells. *Nat. Methods* **3,** 83–89.

Birkmann, E., Schafer, O., Weinmann, N., Dumpitak, C., Beekes, M., Jackman, R., Thorne, L., and Riesner, D. (2006). Detection of prion particles in samples of BSE and scrapie by fluorescence correlation spectroscopy without proteinase K digestion. *Biol. Chem.* **387,** 95–102.

Boukari, H., Nossal, R., and Sackett, D. L. (2003). Stability of drug-induced tubulin rings by fluorescence correlation spectroscopy. *Biochemistry* **42,** 1292–1300.

Boukari, H., Nossal, R., Sackett, D. L., and Schuck, P. (2004). Hydrodynamics of nanoscopic tubulin rings in dilute solutions. *Phys. Rev. Lett.* **93,** 098106.

Chen, Y., Muller, J. D., Berland, K. M., and Gratton, E. (1999a). Fluorescence fluctuation spectroscopy. *Methods* **19,** 234–252.

Chen, Y., Muller, J. D., So, P. T., and Gratton, E. (1999b). The photon counting histogram in fluorescence fluctuation spectroscopy. *Biophys. J.* **77,** 553–567.

Dauty, E., and Verkman, A. S. (2004). Molecular crowding reduces to a similar extent the diffusion of small solutes and macromolecules: Measurement by fluorescence correlation spectroscopy. *J. Mol. Recognit.* **17,** 441–447.

Haustein, E., and Schwille, P. (2003). Ultrasensitive investigations of biological systems by fluorescence correlation spectroscopy. *Methods* **29,** 153–166.

Hazlett, T. L., Ruan, Q., and Tetin, S. Y. (2005). Application of fluorescence correlation spectroscopy to hapten-antibody binding. *Methods Mol. Biol.* **305,** 415–438.

Jankevics, H., Prummer, M., Izewska, P., Pick, H., Leufgen, K., and Vogel, H. (2005). Diffusion-time distribution analysis reveals characteristic ligand-dependent interaction patterns of nuclear receptors in living cells. *Biochemistry* **44,** 11676–11683.

Kohl, T., Heinze, K. G., Kuhleman, R., Koltermann, A., and Schwille, P. (2002). A protease assay for two-photon cross-correlation and FRET analysis based solely on fluorescent proteins. *Proc. Natl. Acad. Sci. USA* **99,** 12161–12166.

Koppel, D. E. (1974). Statistical accuracy in fluorescence correlation spectroscopy. *Phys. Rev.* **10,** 1938–1945.

Krichevsky, O., and Bonnet, G. (2002). Fluorescence correlation spectroscopy: The technique and its applications. *Rep. Prog. Phys.* **65,** 251–297.

Krouglova, T., Vercammen, J., and Engelborghs, Y. (2004). Correct diffusion coefficients of proteins in fluorescence correlation spectroscopy. Application to tubulin oligomers induced by Mg^{2+} and paclitaxel. *Biophys. J.* **87,** 2635–2646.

Langevin, D., and Rondelez, F. (1978). Sedimentation of large colloidal particles through semidilute polymer-solutions. *Polymer* **19,** 875.

Magde, D., Elson, E. L., and Webb, W. W. (1974). Fluorescence correlation spectroscopy. II. An experimental realization. *Biopolymers* **13,** 29–61.

Malvezzi-Campeggi, F., Jahnz, M., Heinze, K. G., Dittrich, P., and Schwille, P. (2001). Light-induced flickering of DsRed provides evidence for distinct and interconvertible fluorescent states. *Biophys. J.* **81,** 1776–1785.

Meseth, U., Wohland, T., Rigler, R., and Vogel, H. (1999). Resolution of fluorescence correlation measurements. *Biophys. J.* **76,** 1619–1631.

Michelman-Riberno, A., Horkay, F., Nossal, R., and Boukari, H. (2007). Probe diffusion in aqueous poly(vinyl alcohol) solutions studied by fluorescence correlation spectroscopy. *Biomacromolecules* **8,** 1595–1600.

Muller, J. D., Chen, Y., and Gratton, E. (2000). Resolving heterogeneity on the single molecular level with the photon-counting histogram. *Biophys. J.* **78,** 474–486.

Pramanik, A., and Rigler, R. (2001). Ligand-receptor interactions in the membrane of cultured cells monitored by fluorescence correlation spectroscopy. *Biol. Chem.* **382,** 371–378.

Rigler, R., and Elson, E. S. (eds.) (2001). Fluorescence Correlation Spectroscopy: Theory and Applications (Springer Series in Chemical Physics, Springer-Verlag, New York).

Rippe, K. (2000). Simultaneous binding of two DNA duplexes to the NtrC-Enhancer complex studied by two-color fluorescence cross-correlation spectroscopy. *Biochemistry* **39,** 2131–2139.

Sanchez, S. A., Brunet, J. E., Jameson, D. M., Lagos, R., and Monasterio, O. (2004). Tubulin equilibrium unfolding followed by time-resolved fluorescence and fluorescence correlation spectroscopy. *Protein Sci.* **13,** 81–88.

Schwille, P., Meyer-Almes, F. J., and Rigler, R. (1997). Dual-color fluorescence cross-correlation spectrosocopy for multicomponent diffusional analysis in solution. *Biophys. J.* **72,** 1878–1886.

Schwille, P., Kummer, S., Heikal, A. A., Moerner, W. E., and Webb, W. W. (2000). Fluorescence correlation spectroscopy reveals fast optical excitation-driven intramolecular dynamics of yellow fluorescent protein. *Proc. Natl. Acad. Sci. USA* **97,** 151–156.

Starchev, K., Buffle, J., and Perez, E. (1999). Applications of fluorescence correlation spectroscopy: Polydispersity measurements. *J. Colloid Interface Sci.* **213,** 479–487.

Stoevesandt, O., Kohler, K., Fischer, R., Johnston, I. C. D., and Brock, R. (2005). One-step analysis of protein complexes in microliters of cell lysate. *Nat. Methods* **2,** 833–835.

Tjernberg, L. O., Pramanik, A., Bjorling, S., Thyberg, P., Thyberg, J., Nordstedt, C., Berndt, K. D., Terenius, L., and Rigler, R. (1999). Amyloid beta-peptide polymerization studied using fluorescence correlation spectroscopy. *Chem. Biol.* **6,** 53–62.

Van Craenenbroeck, E., and Engelborghs, Y. (2000). Quantitative characterization of the binding of fluorescently labeled colchicine to tubuliln *in vitro* using fluorescence correlation spectroscopy. *Biochemistry* **38,** 5082–5088.

Webb, W. W. (2001). Fluorescence correlation spectroscopy: Inception, biophysical experiments, and prospectus. *Appl. Opt.* **40,** 3969–3983.

Suggested Reading

Elson, E. L., and Magde, D. (1974). Fluorescence correlation spectroscopy: I. Conceptual basis and theory. *Biopolymers* **13,** 1–27.

Hess, S. T., and Webb, W. W. (2002). Focal volume optics and experimental artifacts in confocal fluorescence correlation spectroscopy. *Biophys. J.* **83,** 2300–2317.

Hess, S. T., Huang, S., Heikal, A. A., and Webb, W. W. (2002). Biological and chemical applications of fluorescence correlation spectroscopy: A review. *Biochemistry* **41,** 697–705.

Hillesheim, L. N., and Muller, J. D. (2003). The photon counting histogram in fluorescence fluctuation spectroscopy with non-ideal photodetectors. *Biophys. J.* **85,** 1948–1958.

Kettling, U., Koltermann, A., Schwille, P., and Eigen, M. (1998). Real-time enzyme kinetics monitored by dual-color fluorescence cross-correlation spectroscopy. *Proc. Natl. Acad. Sci. USA* **95,** 1416–1420.

Muller, J. D., Chen, Y., and Gratton, E. (2003). Fluorescence correlation spectroscopy. *Methods Enzymol.* **361,** 69–92.

Nagy, A., Wu, J. R., and Berland, K. M. (2005). Characterizing observation volumes and the role of excitation saturation in one-photon fluorescence fluctuation spectroscopy. *J. Biomed. Opt.* **10,** 44015–44019.

CHAPTER 22

A Practical Guide on How Osmolytes Modulate Macromolecular Properties

Daniel Harries* and Jörg Rösgen†

*Department of Physical Chemistry and the Fritz Haber Center for Molecular Dynamics
The Hebrew University of Jerusalem
Jerusalem 91904, Israel

†Department of Biochemistry and Molecular Biology
University of Texas Medical Branch
Galveston, Texas 77555

0091-679X/08 $35.00
DOI: 10.1016/S0091-679X(07)84022-2

Abstract

Osmolytes are a class of compounds ubiquitously used by living organisms to respond to cellular stress or to fine-tune molecular properties in the cell. These compounds are also highly useful *in vitro*. In this chapter, we give an overview of the possible uses of osmolytes in the laboratory, and how we can investigate and understand their modes of action. Experimental procedures are discussed with a specific emphasis on osmolyte-related aspects and on the theoretical aspects that are important to both introductory and more advanced interpretations of such experiments.

I. Introduction

Running the gamut from halophilic bacteria to the human kidney, living creatures have to cope with the stresses of harsh environments. And remarkably, different organisms combat environmental pressures set by high concentrations of salts and organic solutes, drought, extreme temperature, and hydrostatic pressure with strikingly similar molecular countermeasures. This chapter is devoted to the action of these special molecules, sometimes called osmolytes.

On the molecular level, life is established through the specific interaction between and within macromolecules in an aqueous environment. Proteins fold and unfold, interact with lipid membranes and DNA and with ligands that associate and dissociate. All these interactions are modulated by osmolytes, typically small and abundant molecules, through nonspecific and weak interactions.

By detailing experiments that use osmolytes–macromolecule mixtures, our aim is threefold: to describe the molecular mechanism for osmolyte action on biologically relevant macromolecules; to show the use of osmolytes to control macromolecular stability; and to demonstrate how osmolytes also serve as probes of the forces acting at and between macromolecular interfaces.

A. Overview

The weak and nonspecific interaction of osmolytes (and water) with proteins and other macromolecules is best quantified through the excess or deficit of osmolyte (and water) around such macromolecules relative to the bulk solution. The volume element affected by the macromolecule [the "correlation volume" (Ben-Naim, 1992)] is not necessarily limited to its immediate vicinity, but can transcend many solvation layers into the bulk solution. We give examples for the effects of excess numbers in terms of the so-called preferential interactions and *m*-values, initially focusing on experimental systems in Section II. These systems can be investigated using a host of methods, and in Section III we point out some techniques that are especially useful for probing solvation effects. In Section IV, we show how recent developments in solution physical chemistry help to gain a deeper understanding of the basic principles that govern the action of osmolytes on macromolecules. Throughout this chapter, we focus on how to interpret systems, experiments, and their results.

In general, the link between excess numbers, *m*-values, or preferential interaction parameters Γs, and solute or water concentrations, can be formulated in terms of both water and osmolyte because these thermodynamic variables describe the balance between both components' enrichment or depletion. Controversies over the relative importance of water and osmolyte in this balance are resolved using recent solution theory (Section IV). It turns out that, depending on the experimental system, either water or osmolyte may dominate the overall energetics, and we discuss examples for both these cases.

B. Organic Osmolytes Versus Salts

Classes of osmolytes. Naturally occurring organic osmolytes can be grouped into three classes: polyols, amino acids, and combinations of methylamines with urea (Hochachka and Somero, 2002; Yancey *et al.*, 1982, 2005). The polyols include glycerol, sorbitol, trehalose, glucose, and others. Amino acids and amino acid derivatives that serve as osmolytes are, for example, glycine, taurine, and glutamic acid. Commonly used methylamines are trimethylamine N-oxide (TMAO), glycine betaine, sarcosine, and glycerophosphoryl-choline. The function of all these osmolytes *in vivo* ranges from protection of cellular contents, through organs (such as brain or kidney), to the protection of whole organisms (e.g., in dormancy, or under osmotic stress). Most of these organic osmolytes do not bear a net charge. But sometimes, even though inorganic salts are potentially harmful (Collins, 2006), such salts are also used as osmolytes (Galinski, 1995) or for osmotic control *in vivo* (Ferraris *et al.*, 2002; Miyakawa *et al.*, 1999).

Nature prefers organic osmolytes to salts. Nearly all organisms use organic rather than inorganic solutes to protect their cells against adverse conditions. The only exceptions are certain *Archea* that use salt (Hochachka and Somero, 2002). By popular vote, therefore, nature prefers organic osmolytes, and we will largely follow this trend here. Salt is, however, used in many organisms to respond quickly to osmotic shock (Galinski, 1995), and because salt is also regularly used *in vitro*, we will discuss it to some extent in this chapter.

Organic osmolytes act independently of protein type—salts at low concentration do not. It is often assumed that osmolytes have evolved to counteract osmotic imbalance and to stabilize proteins against adverse environmental conditions (Hochachka and Somero, 2002). Stabilization of all cellular proteins by the same solutes is only reasonably possible if all proteins are impacted by all of these osmolytes in much the same way. Generally, organic osmolytes primarily interact with the peptide backbone (Bolen and Baskakov, 2001). Because the backbone is common to all proteins and is the most abundant chemical entity in these macromolecules, organic osmolytes can uniformly affect all proteins. Salts, however, exert their predictable effects only at high concentrations, where electrostatic interactions are usually screened, and weaker interactions are more pronounced. In such cases, nonelectrostatic forces responsible for discrepancies between ions dominate. These forces also show up in the Hofmeister ranking of ions (Hofmeister, 1888)—see also Section IV.C. At lower concentrations typical for most cells, salt effects should also depend on the protein-specific surface charge distribution (Tanford, 1957; Tanford and Kirkwood, 1957). This may be the reason why only certain halotolerant *Archea* utilize salts as osmolytes.

In contrast, to modulate protein properties *in vitro* either salts or organic osmolytes could be the proper choice, depending on the desired effect, or the system investigated.

C. Osmolytes: How Are They Useful?

As an additive. Osmolytes are frequently added to stabilize or solubilize proteins and other macromolecules. Most common are additions of small-to-moderate salt concentrations to buffer solutions, but the less commonly used organic osmolytes can also be very useful and powerful agents.

To probe molecular properties. Although the mere presence of osmolytes can be highly beneficial for stabilizing macromolecules (or detrimental, if used improperly), osmolytes are also very useful tools in probing molecular properties. Probably the best-known example is the use of urea-induced unfolding to determine the stability of proteins (Pace and Shaw, 2000) (see Chapter 11 by Street *et al.*, this volume). More recently, osmolytes were used for the reverse: osmolyte-induced forced *folding* of intrinsically unstable proteins. This procedure also allows one to determine the stability of proteins (Baskakov and Bolen, 1998). Other uses of osmolytes include probing solvation properties of macromolecules and to measure forces within and between large macromolecular aggregates such as DNA and stacks of lipid membranes (Parsegian *et al.*, 1995).

To investigate osmolytes' physiological role. Probing solvation properties is an important tool in the investigations of the physiological and biochemical effects of osmolytes. This is because the effects of osmolytes on other molecules *in vivo* and *in vitro* are mediated through their solvation behavior. As such, osmolytes and their solvation behavior are directly linked to normal function and to pathological conditions of all living organisms. Normal osmolyte function is especially important for osmotic regulation (Jeon *et al.*, 2006) and general stress response (Yancey, 2005). The best-known example for osmotic stress under normal conditions in humans is the kidney (Garcia-Perez and Burg, 1991; Neuhofer and Beck, 2005). Osmolytes also play a role in pathological conditions such as Alzheimer's disease (Scaramozzino *et al.*, 2006).

D. Notation

Throughout this chapter, we will use the following letters as subscripts to denote the different kind of molecules: W = water, P = protein, O = osmolyte, S = (general) solute, U = urea, M = macromolecule, and L = ligand.

II. Experimental Systems

We devote this section to the effects of osmolytes on processes involving macromolecules because these have been most extensively studied. Less studied are interactions between osmolytes and other small molecules. However, because many small molecules participate in biochemical reactions, and their interactions with osmolytes in cells could be overall significant, we will also briefly mention strong osmolyte–ligand interaction in Sections II.B and III.F.

A. Protein Folding

Osmolytes modulate protein stability. For over a century, it has been known that urea denatures proteins (Spiro, 1900). But it was not until the 1960s that the denaturation of proteins by urea was actually quantified (Tanford, 1964). Later, the converse effect—protein stabilization by the addition of small, naturally occurring organic osmolytes—gained much attention (Hochachka and Somero, 2002; Yancey *et al.*, 1982). More recently, it was quantitatively demonstrated that intrinsically unstable proteins can be stabilized and, moreover, forced to fold by the addition of protecting osmolytes (Baskakov and Bolen, 1998). Investigation of a large number of different proteins in urea solutions, as well as studies of a growing number of proteins in solutions that include different kind of protecting osmolytes have allowed biophysicists to phenomenologically derive some general principles of osmolyte–protein interaction, as follows.

Protein stability depends linearly on osmolyte concentration. It is remarkable, and by now well known that independently of the actual protein studied, protein stability $\Delta G^0 = -RT\ \ln([D]/[N])$ linearly depends on urea concentration c_U: $\Delta G^0(c_\mathrm{U}) = \Delta G^0(c_\mathrm{U} = 0) + mc_\mathrm{U}$ (Greene and Pace, 1974), as shown in Fig. 1. (Here D and N denote the concentrations of denatured and native forms of protein.) The constant slope in $\Delta G^0(c_\mathrm{U})$ is called the m-value. Moreover, it seems that at least in those cases that have been studied so far ΔG^0 is also a linear function of the concentration of protecting osmolytes (Felitsky and Record, 2004; Holthauzen and Bolen, 2007; Mello and Barrick, 2003). Note, however, that this linear relation is not necessarily valid for cosolute baring a net charge. For example, while for some proteins the denaturing salt guanidinium chloride is associated with a constant m-value, for other proteins it is not (Bolen and Yang, 2000; Ferreon and Bolen, 2004; Makhatadze, 1999; Santoro and Bolen, 1992; Yao and Bolen, 1995).

Watching out for pH. It is important to note the effect that many osmolytes have on pH, as this change in pH in itself may critically change the stability of proteins. For example, many osmolytes can strongly influence the pK_a values of some buffers, such as the very popular Tris. It is therefore highly recommended to check if the buffer's pH changes when titrated with solute in the appropriate range of solute concentrations. Also, the effect of the stabilizing osmolyte TMAO on macromolecules can change at low pH, where TMAO becomes charged

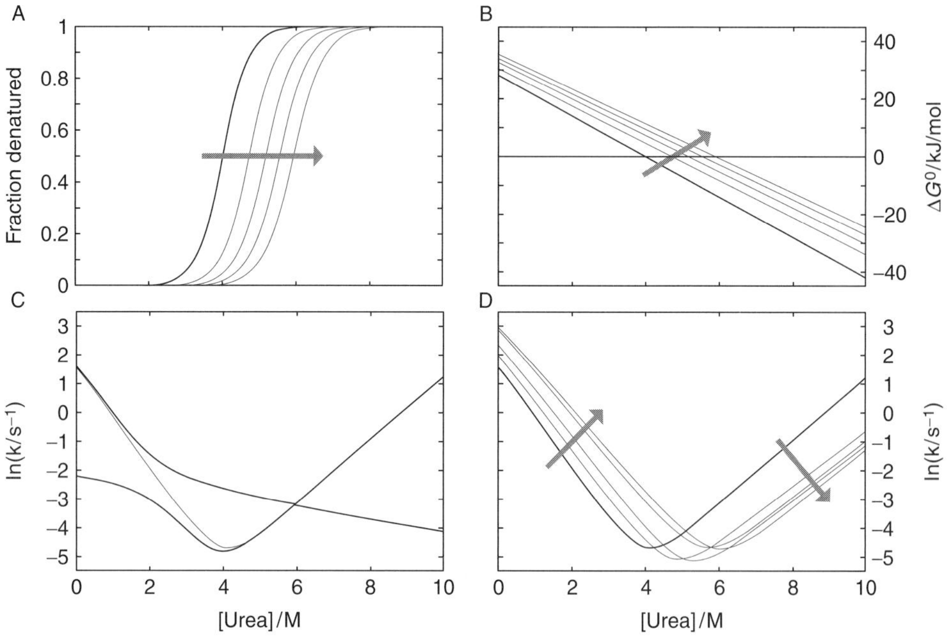

Fig. 1 Dependence of thermodynamic stability (A, B) and (un)folding kinetics (C, D) of the protein FKBP12 on urea concentration in 1 M solutions of several stabilizing osmolytes at 25 °C. The gray arrows indicate the sequential stabilizing effect of glycerol, proline, sarcosine, and TMAO. Conditions are the same in all panels. Bold lines in panels (A), (B), and (D) indicate curves in the absence of additional osmolyte. (A) Equilibrium urea unfolding curves. (B) Gibbs free energy as a function of urea concentration. Note the linearity. The concentrations where the curves intercept 0 ($\Delta G^{\circ} = 0$) correspond to the concentrations of a fraction denatured of 0.5 in panel (A). (C) Both fast and slow kinetic constants of the unfolding and refolding kinetics in buffer plus urea including prolyl isomerization in the denatured state (bold lines), and the effect of the (un)folding kinetics alone (thin line). For the sake of clarity, the curves for glycerol, proline, sarcosine, and TMAO are not plotted. (D) Effect of additional osmolytes on the (un)folding kinetics of FKBP12. The data were extracted from curve fits of data similar to those displayed in panel (C). The left descending branch corresponds to the refolding rate. This line switches over to the right ascending branch (the unfolding rate) at the midpoint urea concentration of unfolding, that is, at the intercept with zero in panel (B).

(Granata *et al.*, 2006). Therefore, it is important to ensure that the degree of protonation of osmolytes under different experimental conditions is known. In fact, many other osmolytes are zwitterionic yet uncharged *in vivo*, and charging these osmolytes by an incautious choice of pH could lead to undesired effects.

Several models were developed to explain the constant m-value. This free energy linearity prompted the wide use of the linear extrapolation method (LEM) to determine the value of $\Delta G^0(c_U = 0)$ from its value at high concentrations of denaturant (Pace and Shaw, 2000), yet the molecular mechanism responsible for the linearity remains obscure. It is clear, however, that accumulation and/or depletion of solution components around proteins is responsible for the sensitivity of the folding equilibrium toward osmolytes. This preferential interaction can also be used as an alternative descriptor for the *m*-value.

Several reasonable model assumptions have been proposed in order to understand the molecular basis for this phenomenon (Eisenberg, 1994; Parsegian *et al.*, 1995; Schellman, 1994; Smith, 2004b; Zhang *et al.*, 1996). In Section IV.B, we will show an alternative, model-independent way of deriving this information.

Prediction of m-values from the chemical nature of the protein surface. The more pragmatic approach used by Tanford (1964) goes back to Cohn and Edsall (1943) who proposed to subdivide the protein surface into small areas of different chemical identity. These small entities could be, for example, the different side chains and backbone. Knowing their energetics of interaction with osmolyte solution allows then to calculate the *m*-values. This "transfer model" approach relies on data obtained from model compounds. But the transfer model was only recently shown to be model independent, as long as proper model compounds and data analysis are used (Auton and Bolen, 2004). The predictive power of the transfer model has been demonstrated for an intrinsically unstructured variant of ribonuclease T1 in solutions of seven different osmolytes (Auton and Bolen, 2005). This approach (Fig. 2) is very promising and further examples will determine how far this predictive power can reach.

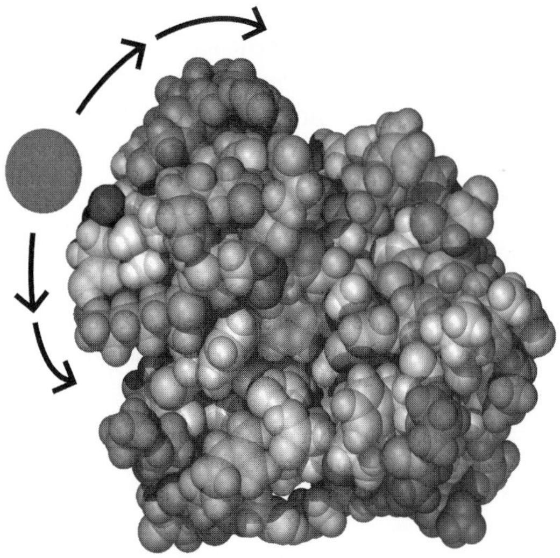

Fig. 2 Calculation of *m*-values from protein structure (Auton and Bolen, 2005). The surface of the protein FKBP12 (pdb id 1FKK) carries chemical groups of different preference for trimethylamine N-oxide (TMAO) (big sphere) as opposed to water. (Blue and red colors in the web-edition of this volume indicate favorable and unfavorable interactions of TMAO with the protein). The TMAO sphere is used to probe which surface elements are accessible to osmolyte (Lee and Richards, 1971) and therefore count for the energetics of the protein in solution. The energetics of the denatured state is calculated using published tables of average residue exposure (Creamer *et al.*, 1995, 1997). (See Plate no. 19 in the Color Plate Section.)

Prediction of m-values of urea-induced unfolding from gross surface area changes. The simplest way to predict m-values is to directly take the total surface area change upon unfolding and to multiply it by a constant that represents an energetic contribution per unit area (Myers *et al.*, 1995). This method is, however, only able to give a very rough idea of the magnitude of the m-value. The method was refined by Record and coworkers, who distinguish between two kinds of surface area: amide and nonamide areas (Hong *et al.*, 2005). These last two methods are, strictly speaking, not m-value predictions, but rather m-value fits. They ask: what values do certain parameters have to take in order to bring experimental values into agreement with the assumed surface interaction model. In contrast, the Transfer Model attacks the problem at a more fundamental level by truly predicting properties of proteins from measured properties of their constituents.

The osmophobic effect: The dominant role of protein backbone solvation. Beyond the quantitative prediction of protein m-values from basic physicochemical properties of the macromolecular building blocks, important qualitative lessons can be learned from the transfer model. Restricting ourselves to the extremes—urea as a strong denaturant and TMAO as a strongly refolding agent—it is clear that the backbone plays the dominant role in osmolyte-dependent protein (de)stabilization (Bolen and Baskakov, 2001).

The reason for the backbone's pivotal role is twofold (Auton and Bolen, 2005). (1) Peptide groups are the most numerous chemical group in proteins and contribute disproportionately to the total change in exposed surface area. (2) Pure water is a better solvent for peptide groups than are solutions of strongly stabilizing osmolytes. This "osmophobicity" of the backbone favors its burial, that is, folding of the protein. The side chains, however, have an overall favorable interaction with the osmolyte solution and therefore do not contribute to the stabilizing effect, or even slightly oppose it.

Similar results are obtained with the "top-down" approach that parameterizes experimental data of protein unfolding (Hong *et al.*, 2005), as opposed to the "bottom-up" approach that predicts protein m-values from the properties of their chemical constituents (Auton and Bolen, 2005).

Osmolytes alter protein folding kinetics. The kinetic analogue of the constant m-value within the LEM is the linearity of the logarithm of the kinetic constants of folding $\ln(k_F)$ and unfolding $\ln(k_U)$. Such constant kinetic m-values are observed in denaturant-induced protein unfolding (Tanford, 1968). Only few quantitative data are available on the impact of protecting osmolytes on protein folding kinetics. According to these, the principles of the osmophobic effect can also be applied to protein folding kinetics (Russo *et al.*, 2003). Stabilizing osmolytes seem to generally accelerate refolding and reduce the unfolding velocity (Fig. 1).

Solution preparation for equilibrium unfolding/refolding experiments. It is advisable to perform both folding and unfolding experiments in order to demonstrate reversibility in terms of both repeatability and absence of kinetic distortions. The latter occurs if the readings are taken before complete equilibration of the sample.

Incidentally, equilibration can take weeks to months in extreme cases (Cavagnero *et al.*, 1998; Ogasahara *et al.*, 1998; Rosengarth *et al.*, 1999; Tomschy *et al.*, 1994). Usually, one sample for each osmolyte concentration is prepared separately. However, if the reaction is reversible, it is sufficient to prepare one protein stock solution in the absence of osmolyte, and one at the highest osmolyte concentration. Solutions of intermediate osmolyte concentration are prepared by mixing the two stock solutions to the appropriate ratio. This procedure can greatly reduce both the time spent on preparing the solutions and the noise level due to uncertainties in the final protein concentration.

Data analysis for equilibrium and kinetic protein (un)folding. The data are best evaluated through a curve fit. The basic linear dependence of the Gibbs free energy $\Delta G^0(c_U) = \Delta G^0(c_U = 0) + mc_U$ (Fig. 1B) allows a direct calculation of the equilibrium constant of unfolding

$$K(c_U) = \exp\left[-\frac{\Delta G^0(c_U)}{RT}\right] \quad (1)$$

which can be used to calculate the experimental signal (Fig. 1A) (Santoro and Bolen, 1988)

$$s(c_U) = s_N + c_U ds_N + [s_D - s_N + c_U(ds_D - ds_N)]\frac{K(c_U)}{1 + K(c_U)} \quad (2)$$

where s_N and s_D are the signals of the native and denatured state at 0-M osmolyte, and ds_N and ds_D are the slopes of their respective signals. In the case of kinetic measurements, the rate of the reaction is given by

$$k(c_U) = k_F(c_U) + k_U(c_U) \quad (3)$$

for a first-order reversible reaction (Fig. 1D). In log-scale this equation results in the typical v-shaped or "chevron-curve." This is because both $\ln(k_F)$ and $\ln(k_U)$ are linear in c (due to the LEM), and therefore $\ln(k_F + k_U)$ is dominated by the larger k. Additional reactions, such as prolyl isomerization, can interfere with this chevron signature. The equations in this case are more complex and can be found elsewhere (Kiefhaber and Schmid, 1992; Kiefhaber *et al.*, 1992; Russo *et al.*, 2003). Such conditions of coupled reversible kinetic reactions lead to convoluted functional dependencies of the observed kinetic rate constants on urea concentration (Fig. 1C).

B. Ligand Binding and Molecular Association

Specific binding energies are often linear in osmolyte concentration. In much the same way that osmolytes act on proteins to change their stability in solution, the same cosolutes also vary the binding strength of ligands to proteins, enzymes to their substrates, and between proteins and other macromolecules such as

DNA (Kornblatt and Kornblatt, 2002; Parsegian *et al.*, 1995, 2000). Remarkably, these changes in binding often follow the same dependence on solute concentration as found for protein folding, even at low osmolyte concentrations. In numerous examples, we find binding free energies that vary linearly with solute concentration, C_S. Here is a partial list: antibodies binding to antigens (Goldbaum *et al.*, 1996; Xavier *et al.*, 1997), the oxygen affinity of hemoglobin (Colombo *et al.*, 1992; Royer *et al.*, 1996), the binding of regulatory proteins to their recognition DNA sequences [some examples are: *E. coli gal* (Garner and Rau, 1995), *lac* (Fried *et al.*, 2002), *tyr* (Poon *et al.*, 1997) repressors, *E. coli* CAP protein (Vossen *et al.*, 1997), and the restriction endonuclease *Eco*RI (Robinson and Sligar, 1998)], and the dimerization of α-chymotrypsin (Patel *et al.*, 2002).

Communality in linear response suggests that the underlying mechanisms for the action of osmolytes on protein stability and on protein binding are closely related.

***Osmolytes strengthen or weaken binding by exclusion or inclusion from interacting macromolecular surfaces*.** Many natural osmolytes show a net exclusion from macromolecular interfaces. If macromolecular interfacial areas become buried following binding to another molecule (say ligand binding), addition of excluded osmolytes will tend to further stabilize the burial process. Think like a chemist: this is a simple consequence of Le Chatelier's principle. Solute addition destabilizes the protein (increases its chemical potential), so that burial of those destabilizing surfaces is favored. Now think about the cosolute: burial of surfaces will cause less osmolyte exclusion, and again the perturbation is reduced.

Much as the folded versus denatured state of a protein is stabilized, binding to proteins is strengthened by excluded solutes, too. And similarly to protein denaturation, solutes that show a net favorable preferential interaction with proteins (like urea) will generally tend to destabilize the association.

The Gibbs adsorption isotherm and Wyman linkage relate changes in stability to local excess of solute. The Gibbs adsorption isotherm (or the closely related Wyman linkage, see Chapter 2 on linkage thermodynamics by Beckett, this volume) makes the simple yet illuminating link between macromolecule stability and local concentrations around it. The idea is that the variation of the stability (free energy) of macromolecules with solute chemical potentials is in proportion to the excess or deficit of solute in the vicinity of that macromolecule (Gibbs, 1876/78; Parsegian, 2002; Wyman, 1964).

For the immersion of a protein in a dilute solution of proteins also containing cosolutes, the free energy per protein molecule (or chemical potential μ_P) is related to the solute concentration by $d\mu_P = -N_O^{excess} d\mu_O$, where the solute chemical potential μ_O is related to solute activity a_O through $\mu_O = RT\ln a_O$. Clearly, if protein and ligand's chemical potential in the two states, say bound and unbound, were *each independently* linear in μ_O, then the free energy of binding, $\Delta G^0_{binding} = \mu_M^{bound} - (\mu_M^{free} + \mu_L^{free})$, would also be linear in μ_O, reflecting a constant excess number of associated solutes involved in the process. Such dependence is often found for the strong binding of ligands to proteins as well as in "counterion release," as we shall discuss in Section IV.C. We return to discuss these relations in

Section IV but notice that the preferential interaction parameter Γ_{μ_O} briefly mentioned earlier (in Sections I.A and II.A) equals $-N_O^{excess}$.

Note, however, that $\Delta G_{binding}$ is sometimes a strongly nonlinear function of osmolyte concentration, as recently reported for binding of 2′CMP to RNase A. It was argued that this nonlinearity is due to osmolyte–ligand interaction (Ferreon *et al.*, 2007), that is, a dependence of μ_L on osmolyte concentration.

Linearity of ΔG^0 in solute concentration translates into linear changes in solute excess and constant numbers of excess excluding waters. Linearity of binding free energy with solute concentration also implies that the net excess (or deficit) of osmolyte associated with macromolecules (with respect to their number in the bathing solution) linearly depends on osmolyte concentrations. The most general case, where protein, osmolyte, and water concentrations are all substantial, can become quite complex (Smith, 2006a). However, as long as the protein concentration is very low, the Gibbs–-Duhem relation for water and solute allows us to think about excesses and deficits of water molecules rather than solute. We find that $d\mu_P = -N_O^{excess} d\mu_O = -N_W^{excess} d\mu_W$, where $\mu_W \propto c_O$ at low solute concentrations is related to the solute concentration itself [see Section III, Eqs. (8) and (9)] or, even more directly, to the solution osmotic pressure. A number deficit of solute near a protein that is linear in the solute concentration is identical to a constant excess number of solute-excluding water molecules N_W^{excess}. (This factor of c_O between N_W^{excess} and N_O^{excess} appears because in the derivatives we have $d\mu_W \propto dc_O \propto c_O d\mu_O$ at low osmolyte concentration.)

Finally, we can generalize these considerations from single species to reactions. Binding that shows linearity in the solute concentration can be described as a process that involves a constant change in the number of excluding water molecules,

$$\frac{d\Delta G^0_{bind}}{d\mu_W} = -\Delta N_W^{excess} \tag{4}$$

The alternative, speaking about the excluded solute, gives

$$\frac{d\Delta G^0_{bind}}{d\mu_O} = -\Delta N_O^{excess} = \Delta N_W^{excess}\left(\frac{d\mu_W}{d\mu_O}\right) \tag{5}$$

where the last term is often linear in osmolytes concentration (as long as its concentration is low, as already pointed out above). Here too, note that for low macromolecule concentrations,

$$\Delta N_O^{excess} = -\Delta\Gamma_{\mu_O} \tag{6}$$

These thermodynamic relations have long been established and used (Anderson *et al.*, 2002; Casassa and Eisenberg, 1964; Hade and Tanford, 1967;

Parsegian, 2002; Parsegian *et al.*, 1995; Schellman, 1978, 1987; Timasheff, 1993; Wyman, 1964). It is also well understood that the excess numbers (or preferential interactions) include contributions from both solute and water (and at high enough concentrations, from protein too). Solute and solvent can contribute differently to the energetics, depending on the type of interaction. Therefore, for different kinds of interactions it may be useful to discuss changes in the excess solute or solvent numbers. Though some might term cases where water is the main player as special (Timasheff, 1998), this in fact often occurs in binding reactions (Parsegian, 2002). We discuss compelling examples for this in Sections II and III.

Note that the dependence of ΔG^0 on solute addition need not always necessarily vary linearly either with solute concentration or with solute chemical potential.

Osmolytes probe interacting macromolecular interfaces. By using different osmolytes of different sizes and chemical nature we can probe not only the nature of interacting surfaces but also the physical mechanism of osmolyte exclusion. The specific binding of cyclodextrin to adamantane carboxylic acid to form an inclusion complex, for example, changes linearly with the concentration of many salts and other net-neutral osmolytes (Fig. 3). But the slope for all these lines varies widely (Harries *et al.*, 2005). Calorimetry reveals that it is mainly the enthalpy of reaction that changes on solute addition, indicating that steric exclusion (primarily an entropic effect) is not a dominant factor here. These trends suggest that, similar to protein folding, complexation is accompanied by a significant change in the exposed area of both interacting surfaces to solution. The details of solute interactions with the molecular interface will thus determine the extent of osmolyte

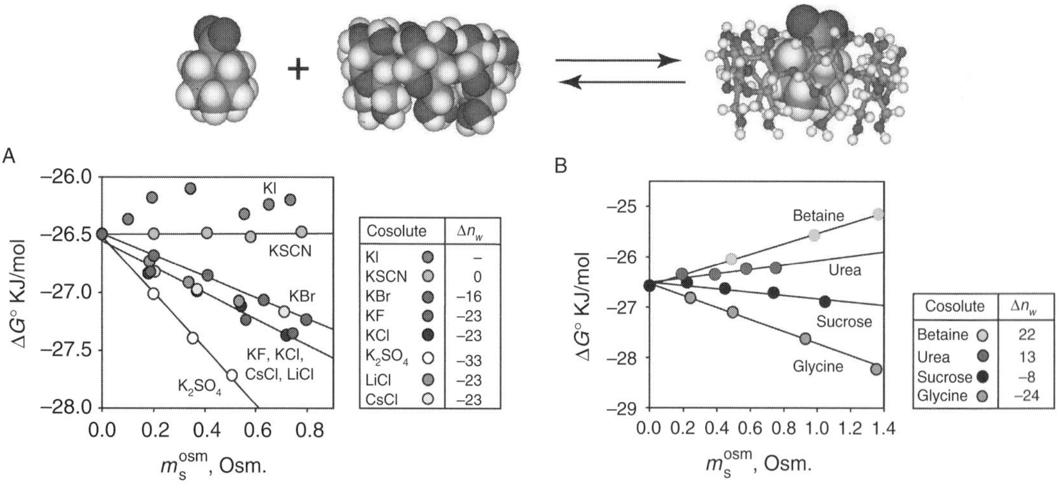

Fig. 3 Osmolytes strengthen or weaken binding by their exclusion or inclusion from interacting surfaces of cyclodextrin and adamantane (top panel). For all osmolytes tested, the association free energy changes linearly with osmolyte concentration (bottom panel). The slope, however, sensitively depends on the type of osmolytes used (Harries *et al.*, 2005). (See Plate no. 20 in the Color Plate Section.)

exclusion/inclusion, and ultimately affect its association. It may come as no surprise, therefore, that the effect of all salts studied on binding of the nonpolar "hydrophobic" interacting surfaces of cyclodextrin and adamantane perfectly tracks the effect of these salts on the air–water interface, suggesting that in complexing, cyclodextrin and adamantane bury a combined surface area of ca. 200 Å (see also Section IV.C). It has to be kept in mind, however, that such effects do not necessarily track with the air–water surface tension, particularly for organic osmolytes as also studied in protein osmolyte interaction (Auton *et al.*, 2006). In fact, in these cases we can often usefully think about a protein–water surface tension that is much different from that of the air–water interface and is instead determined by the protein's amino acid composition.

Osmolytes that are completely excluded from inaccessible cavities act purely through osmotic stress. Particularly interesting are water molecules that are sequestered in pockets, grooves, or macromolecular cavities so that they are sterically inaccessible to solutes. Here, the chemical nature or the size of a solute (beyond a certain size) does not change its effect on macromolecular stability; there is simply an osmotic pressure acting on a sequestered volume of water. The osmotic pressure can readily be measured (see Section III.A), and for low osmolyte concentration is simply related to solution concentration by van't Hoff's law $\Pi = c_{\mathrm{S}}RT$. This pressure is also related to water activity by

$$v_{\mathrm{W}}\mathrm{d}\Pi = \mathrm{d}\mu_{\mathrm{W}} = RT\mathrm{d}\ln a_{\mathrm{W}} \tag{7}$$

An illuminating example of steric exclusion of the solutes from water-filled cavities is the specific–nonspecific reaction of DNA-binding proteins, illustrated in Fig. 4 (top). (Competitive-binding reactions between different DNA sequences are not only more easily measured but also more relevant for probing recognition specificity than free protein binding to its cognate sequence.)

Differences in the number of solute-excluding water molecules associated with specific and nonspecific DNA–protein complexes can be measured from the dependence of the relative free energy difference on water activity (or osmotic pressure). The free energy differences between specific and nonspecific DNA binding of the restriction endonuclease *Eco*RI scale linearly with changes in osmolal concentration for each solute and translate using Eq. (4) into the difference in number of waters retained by specific and nonspecific complexes. It has been shown (Sidorova and Rau, 1996) that the nonspecific complexes of *Eco*RI sequester about 110 water molecules more than the specific recognition sequence complex (Fig. 4, bottom).

Remarkably, over a wide range of concentrations, the six solutes: glycine betaine, sucrose, glycerol, triethylene glycol, glycine, and α-methylglucoside gave the same difference in waters within 15% experimental error. The slight sensitivity of this number of waters to the solute identity suggests that the contact area between protein and DNA in a nonspecific complex delimits a structurally well-defined volume of water sterically inaccessible to the six different solutes used to set water

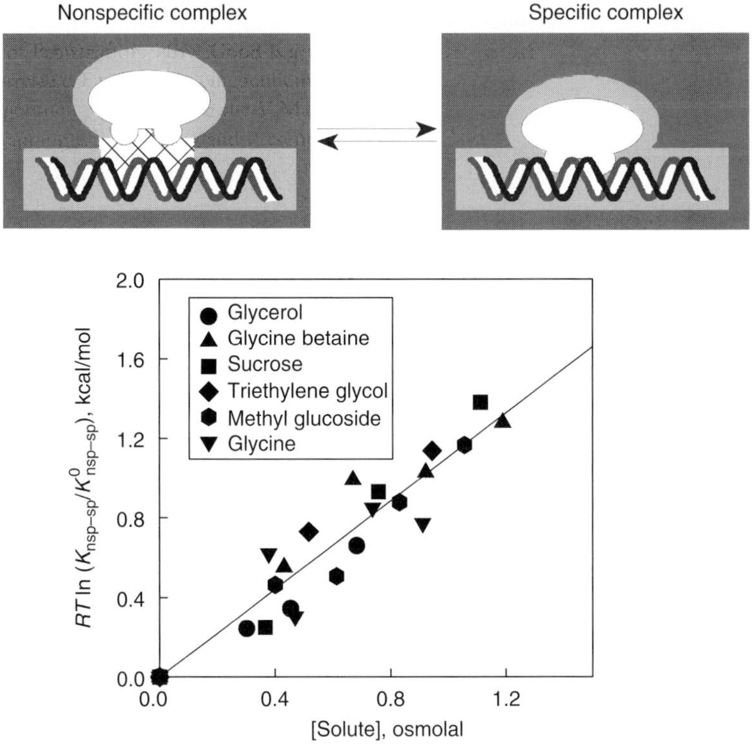

Fig. 4 Top panel: Schematic representation of water exclusion from DNA–protein complexes. The protein is shown as a globular shape with two lobes that represent domains (helices) that specifically interact with DNA bases. Bulk solution (water and solute) is shown in dark gray. The light gray regions surrounding the protein and DNA surfaces represent a zone of osmolyte exclusion. Note that the extent of solute exclusion (or water inclusion) from this zone will depend on both size and nature of the probing osmolyte. In contrast to direct contact between DNA and protein surfaces in the specific complex, nonspecific complexes can have a volume of water presumably in a cavity at the interface between surfaces (shown as a cross-hatched area) that sterically excludes solutes. Exclusion form this water will not depend on either solute chemical nature or its size beyond a certain size limit. Bottom panel: the dependence of free energy difference between specific and nonspecific *Eco*RI–DNA complexes is shown for six solutes. The slope of the fitted line translates using Eq. (4) into ~110 extra water molecules sequestered in the nonspecific versus specific complex, see text for details. Figure courtesy of Nina Sidorova and Don Rau.

activity. It also suggests that changes in solute exposed surface area between nonspecific and specific complexes are small in comparison.

The information gained from this osmotic stress strategy agrees well with X-ray structural data. The X-ray structure of the *Eco*RI–DNA-specific complex (Rosenberg, 1991) shows that the interface of this complex is essentially dry, so that the measured 110 waters are most likely located at the protein–DNA interface of the nonspecific complex. And while the structure for the nonspecific complex of

*Eco*RI is unavailable, X-ray structures for both specific and noncognate complexes of a closely related restriction endonuclease *Bam*HI have been solved (Newman *et al.*, 1995; Viadiu and Aggarwal, 2000). Unlike the extensive, direct protein–DNA contacts seen at the interface of the specific complex, the nonspecific *Bam*HI–DNA complex shows a distinct gap between the protein and DNA major groove surfaces that is large enough to accommodate about 150 water molecules.

Squeezing water out of the noncognate complexes. It should be possible, in principle, to remove sequestered waters by applying high enough osmotic pressure. It has been shown (Sidorova and Rau, 2004) that though complexes between *Eco*RI and "star" DNA sequences (with only one wrong base pair in the recognition sequence) sequester the same amount of water as completely nonspecific complex at low osmotic pressures, this water can be squeezed out from "star" complexes (but not from nonspecific one) at high enough osmotic pressure. The pressure–volume work done to remove water from the noncognate complexes is balanced by the resulting unfavorable interactions and conformational changes in the complex.

C. Protein Solubility

Protein precipitation and crystallization: Not as easy as predicting protein stability. Dissolved proteins can coexist with many different kinds of undissolved phases. Most common is the amorphous precipitate, which lacks periodic structure. But proteins may also form crystals, and these can be of different symmetries. The amount of osmolyte present within such precipitated protein states is not usually known. This makes it more difficult to predict protein solubility in osmolytes than to predict protein stability in solution. In fact, not even the average exposure of each chemical group to solution in the precipitate is known. This section, therefore, serves more as a qualitative discussion of precipitation.

Variation of solution composition is the most versatile approach to protein solubilization. The solubility of chemical compounds depends both on their chemical nature and on the composition of the solution in which they are dissolved. One way to modulate protein solubility is, therefore, by mutation. A mutagenesis-based approach is generally too cumbersome, however, and has to rely mostly on empirical approaches on a per protein basis (Schein, 1993). Alternatively, cosolutes can serve to alter solubility. Theories are available to quantify both salt- and osmolyte-dependent protein solubility (Arakawa and Timasheff, 1985; Shulgin and Ruckenstein, 2005b). Osmolytes are particularly attractive candidates for solubility enhancement, because transfer free energy data for several osmolytes have been measured. This allows one to qualitatively predict native state solubility from the high-resolution structure of each protein (Bolen, 2004).

Osmolytes are useful stabilizers, precipitants, solubilizers, or denaturants depending on the balance between side chain/osmolyte versus backbone/osmolyte interactions. We have seen (Section II.A) that the denaturant urea interacts favorably with both the peptide backbone and the amino acid side chains. Demonstrating Le Chatelier's principle, proteins in urea solution react by preferring a state with increased surface exposure, namely the denatured state. Moreover, urea normally solubilizes proteins (Bolen, 2004), because dissolved proteins expose more surface than precipitated proteins. Being both a denaturant and a solubilizer, guanidine hydrochloride (GdnHCl) acts in a similar fashion. And yet the problem with using such strong denaturants as solubilizers is obvious: they also denature proteins.

A good alternative to GdnHCl is arginine (that contains a guanidyl group), which solubilizes proteins without strongly denaturing them (Arakawa and Tsumoto, 2003). Among the net-neutral osmolytes, proline is very useful for solubilization due to the balance between its favorable interaction with the side chains and its unfavorable interaction with the backbone (Auton and Bolen, 2005). The backbone, therefore, counters the potentially destabilizing effect of the favorable interaction with the amino acid side chains, while these interactions with the side chains improve the protein solubility.

Protein stabilizers also tend to be precipitants. This is due to the overcompensation of the favorable (solubilizing and destabilizing) interactions between osmolyte and side chains by the unfavorable (precipitating and stabilizing) interactions between osmolyte and peptide backbone (Auton and Bolen, 2005; Bolen, 2004). The precipitating effects of stabilizing osmolytes can be partially overcome by using a mixture of the stabilizer with a solubilizer like proline that does not denature the protein (Kumar *et al.*, 2001).

Mixtures of osmolytes may act better than the sum of their individual effects. Mixing different kind of osmolytes may be helpful not only when combining a stabilizer with a solubilizer but also in solubilization itself. Mixtures of arginine and glutamic acid have been found to provide better solubilization than the sum effect of each component individually (Golovanov *et al.*, 2004). Similarly, mixtures of TMAO and trehalose can be more efficient in refolding proteins than expected from a simple additive effect (Bomhoff *et al.*, 2006).

pH Affects both protein solubility and osmolyte character. The charge of osmolytes is of great importance for their interactions with proteins. Titrating a protonateable group on an osmolyte molecule can even turn a stabilizer into a denaturant, as in the case of TMAO (Granata *et al.*, 2006; Singh *et al.*, 2005). pH is also well known to alter protein solubility. This holds for both the native and the denatured state (Pace *et al.*, 2004; Shaw *et al.*, 2001). The denatured state can be more than 100-fold less soluble than the native state. Preventing the protein from unfolding is therefore another way to achieve protein solubilization. Incidentally, the solubility of the denatured state seems to be less dependent on pH than the native state (Pace *et al.*, 2004).

Salts have long been known to alter protein solubility. Protein solubility in salt solutions was qualitatively investigated already two centuries ago in Hofmeister's laboratory (Hofmeister, 1888; Lewith, 1888). Setschenow also made quantitative solubility studies at a similar time (Setschenow, 1889, 1892), though not on proteins. Concepts developed in these laboratories, namely the Hofmeister series and Setschenow coefficients, are milestones in the understanding of protein solubility (Arakawa and Timasheff, 1985). Early on, experimental data were viewed in light of these concepts (Green, 1931a,b, 1932; Green *et al.*, 1935; Sörensen, 1925). We return to discuss some of these points in Section IV.C.

Osmolytes added to the growth medium can improve cellular protein solubility. Beneficial effects of osmolytes are, of course, not limited to the listed *in vitro* examples. Accumulation of osmolytes is a ubiquitous strategy of all organisms to cope with different kind of stress. Recently, proline was demonstrated to be useful for preventing intracellular aggregation of a protein that is prone to precipitation (Ignatova and Gierasch, 2006).

D. Stressing Assemblies: DNA, Lipids, and Other Macromolecules

Osmolytes can be used to stress macromolecular assemblies. By exclusion from interacting macromolecular surfaces, osmolytes can push macromolecules together. This push impacts the structure and stability not only of single macromolecules, such as proteins, but also of larger molecular aggregates such as DNA assemblies and multilamellar lipid vesicles. If solute exclusion is strong enough, practically no osmolyte (e.g., bulky polyethylene glycol polymer) penetrates into the macromolecular aggregate. At the same time, the stress it exerts (osmotic pressure) is easily measured in the bulk solution (see Section III.A). This forms the basis for a simple yet powerful method for measuring forces between macromolecules: the osmotic stress strategy (Parsegian *et al.*, 1986, 1995; Rand and Parsegian, 1992).

Osmotic stress is used to measure forces acting between macromolecules. The idea is simple (Fig. 5). To a solution of assembling, ordered macromolecules (or ones that can order under osmotic pressure) add osmolytes that are strongly excluded. Typically a polymer such as PEG or high-molecular weight dextran is used. Let the supramolecular aggregate form and reach equilibrium with the aqueous osmolyte solution. Then, use diffracting X-rays that bounce off the macromolecular aggregate to report on the spacing between macromolecules at that exerted osmotic stress. When repeated for many different osmotic stresses, the result is an "equation of state" for the macromolecular aggregate, like the one shown for DNA in solution, Fig. 6.

The osmotic stress strategy is a convenient way to limit the hydration of the macromolecular assembly, while knowing the free energy cost (or simply: the work) needed to keep the assembly at that spacing. This work of eliminating waters from the assembly (the area under the pressure vs water volume curve) is performed against the forces that act between the macromolecules themselves. Ultimately, all effective forces between macromolecules are probed, even if not directly related to the hydration or solvation of the macromolecules (Kozlov *et al.*, 1994; Leikin *et al.*, 1991; Podgornik *et al.*, 1994; Rand *et al.*, 1990).

Osmotic stress can be used to probe the interaction of small solutes with macromolecules. What if solutes are not entirely excluded from the macromolecular aggregates? Penetration of solute will change the balance of forces acting between

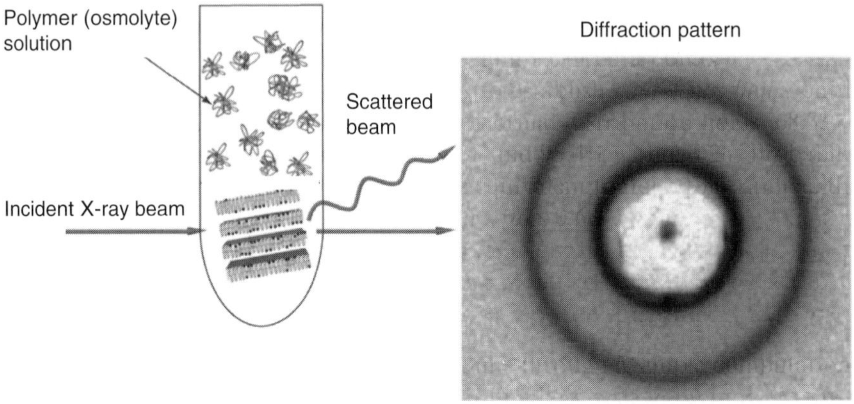

Fig. 5 Measuring forces using the osmotic stress strategy involves bathing macromolecular assemblies (lipid multilayers in the figure) in solutions containing cosolutes that are highly excluded from the inside of the aggregate, such as the neutral polymer polyethylene glycol. Water is drawn from the aggregate until equilibrium is reached. Then, X-rays are shone through the sample to inform on the (average) spacing between macromolecules at a specific osmolyte concentration. The osmotic pressure for the same solute concentration can be readily measured using an osmometer, to give the force versus separation curve for the aggregate. Osmotic pressure data for many osmolytes useful in these experiments are given at http://lpsb.nichd.nih.gov.

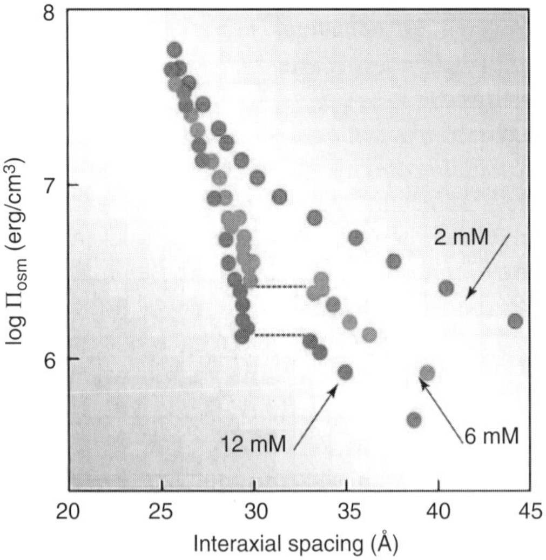

Fig. 6 Measured intermolecular force as a function of interaxial separation for DNA in simple salt solutions (0.25 M NaCl) that contain increasing concentrations (2,8, and 12 mM) of the trivalent cation $Co(NH_4)_6^{+3}$. For each polyvalent ions concentration, osmotic pressure is exerted by PEG polymer added to solution, and the spacing is measured using X-rays. (Rau and Parsegian, 1992; Gelbart *et al.*, 2000) Figure courtesy of V. A. Parsegian.

the macromolecules. Watching how solutes affect the balance of forces in the aggregate provides a way to determine the partitioning of those solutes around macromolecules. Use large, strongly excluded solutes to press against the assembly, while letting the small osmolytes or other solutes to permeate and equilibrate with the aggregate. The excess pressure that is exerted by the small osmolyte reflects its preferential hydration. For example, DNA condensed with spermidine shows a strong shift in pressure versus DNA interaxial separation when MPD or 2-propanol is added to solutions already stressed by PEG. The pressure imposed by PEG that is needed to reach a certain DNA interaxial spacing is reduced in the presence of alcohol, indicating the alcohol's strong exclusion. Incontrast, glycerol shows little effect on these curves, indicating that it has a preferential interaction close to zero (Hultgren and Rau, 2004; Stanley and Rau, 2006).

Similarly, the effect of salts on neutral lipid bilayers has been interrogated using this strategy: push against membranes when small solute (salt) is also present in solution. Remarkably, while salts are found to be preferentially excluded from interacting membranes the amount that does penetrate also concomitantly weakens the attractive van der Waals attraction between bilayers, resulting in a widening in the spacing between lipid bilayers, as shown in Fig. 7 (Petrache *et al.*, 2005, 2006).

Finally, the ability to measure spacings between macromolecules in osmotically stressed solutions allows one to evaluate the changes in the number of preferentially interacting solutes at that spacing. We return to this point in Section IV to show that these kinds of interactions are important while interrogating lipid-based vehicles for drug delivery.

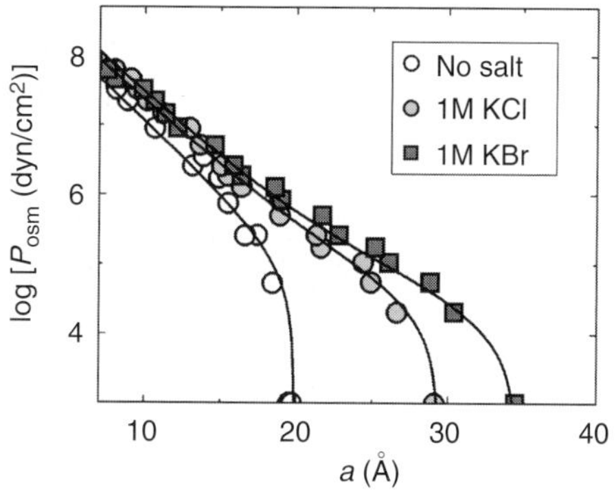

Fig. 7 Salt effect on lipid membranes measured by the osmotic stress strategy. At similar osmotic pressures, neutral lipid bilayers are more swollen in bromide than in chloride salts (Petrache *et al.*, 2006). Each symbol in the plot shows the experimental value of the osmotic pressure P_{osm} that was fixed for each unoriented multilamellar lipid (DLPC) sample with a known amount of polymer (PEG), versus the water spacing *a* that was determined for that sample by X-rays. The lines show the fits to these data from a model relating the force acting between bilayers to hydration, membrane fluctuations, and van der Waals forces. All data were taken at 25 °C. Figure courtesy of Horia Petrache.

III. Experimental Methods

Using osmolytes to control structure and stability of macromolecules is a strategy that accepts many tactics. In fact, your favorite experimental method can also probably be used to gain information on the action of osmolytes on macromolecules. The method might be useful in detecting changes in macromolecular state, structure, conformation, or degree of association due to osmolytes addition. And it might probe details of osmolytes and macromolecular solvation, as well as inform on preferential interactions and on partitioning of osmolytes between bulk and macromolecular surfaces. Far from presenting a comprehensive tutorial, we discuss only a few techniques here to show the strategy; hopefully these will encourage the readers to follow their own tactics.

A. Osmometry

Water activity is an important physical property in biology. Cellular volume regulation is a major issue for the survival of living organisms. Distribution and redistribution of water plays a major role in this regulation and the concomitant movement of water dictated by its chemical activity. By knowing the water activity that directs this water flow, we can also derive solvation information, as is described in Section IV.

Several methods are available for measuring water activity. The dependence of water activity on the concentration of additives can be measured through the solution colligative properties: freezing point, boiling point, osmotic pressure, vapor pressure, and so on. Here we focus on the vapor pressure as measured through vapor pressure osmometry (VPO) or dew point depression. This method is much faster, though less precise, than isopiestic distillation, which is the gold standard for the measurement of the water activity as a function of additives. In all of the listed methods, precise preparation of the solutions is of utmost importance. For this reason, it is best to use the molal concentration scale (moles per kilogram of water) as explained below.

Osmolalities and osmotic coefficients. The output of VPO is the osmolality of the solution, which is defined as

$$\mathrm{Osm} = -\frac{\ln a_\mathrm{W}}{M_\mathrm{W}/1000} \tag{8}$$

where a_W is the water activity and M_W is the molecular weight of water (18 g/mol), and 1000 is the conversion factor between moles per kilogram and moles per gram. The corresponding osmotic coefficient is

$$\phi = \frac{\mathrm{Osm}}{m} \tag{9}$$

where m is the total molality.

Based on this definition it is clear that osmolality is just a measure of the chemical activity of water, or of the (overall) osmotic pressure in solution, $\mathrm{Osm} = \Pi/RT$. Alternatively, one could state that osmolality is the hypothetical concentration of an ideal solute in an aqueous solution that has the same water activity as the sample. Osmolality is an overall effective solute concentration within this interpretation; a handy, but questionable point of view (Rösgen *et al.*, 2004a) because the definition of what is an "ideal solution" depends on the chosen standard state, concentration scale, and so on. Keeping this caveat in mind, we see that the osmotic coefficient is the relative deviation of the osmolality from ideal behavior (given in this case by the molal concentration).

Output data: osmotic coefficients and preferential interaction parameters. There is a host of differently defined preferential interaction parameters (Anderson *et al.*, 2002). These are useful for understanding the response of macromolecules to the presence of high concentrations of additives. Here we focus on the preferential interaction parameter that is most directly related to VPO, namely to the response of the chemical activity of water on increasing concentrations of cosolutes

$$\left(\frac{\partial \ln a_W}{\partial m_O}\right)_{T,p,m_P} = -\frac{1}{m_W}\left(\frac{\partial \mathrm{Osm}}{\partial m_O}\right)_{T,p,m_P} \tag{10}$$

where m_O is the molality of osmolyte and m_W the molality of water (55.5 mol/kg).

The preferential interaction parameter Γ_{μ_O} can be obtained from two sets of VPO measurements in the presence (m_P molal) and absence of protein (Courtenay *et al.*, 2000):

$$\Gamma_{\mu_O} \approx \frac{m_O}{m_P}\left[1 - \left(\frac{\partial \mathrm{Osm}}{\partial m_O}\right)_{p,T,m_P} \Big/ \left(\frac{\partial \mathrm{Osm}}{\partial m_O}\right)_{p,T,m_P=0}\right] \tag{11}$$

However, the slope of the osmolality with osmolyte molality (Eq. 10) is also very useful for obtaining more detailed solvation information, as is discussed in Section IV.B.

Calibration issues. Based on our experience with the Wescor Vapro 5520 Vapor Pressure Osmometer, we have several recommendations on the use of VPO. The instrument is calibrated using NaCl standard solutions of osmolalities 100, 290, and 1000 mOsm. Unfortunately, the manufacturer programmed the instrument to use freezing point depression data to calculate the calibration curve from these data (Courtenay *et al.*, 2000; Hong *et al.*, 2003). Data that are valid at room temperature (or better: operating temperature) would have been the proper choice. Such data can be obtained elsewhere (Archer, 1992; Clarke and Glew, 1985) to recalculate corrected osmolalities from the instrument reading. Otherwise, the data may become inconsistent (Kiyosawa, 2003) because of the temperature dependence of osmotic coefficients (Winzor, 2004), and thus the manufacturer's calibration cannot be trusted (expect deviations in excess of 5% if you work at room temperature). Still, it is useful to first collect the data employing the manufacture's calibration solutions and procedure, because it is impossible to get around using the calibration that is integrated in the instrument's firmware. The obtained results can then be converted using a proper calibration. In principle, it is sufficient to recalculate the real osmolalities from the instrument output as suggested previously (Courtenay *et al.*, 2000; Hong *et al.*, 2003). As best practice, we recommend to determine experimentally an appropriate polynomial for converting the instrument output. This can be done by occasionally measuring an NaCl dilution series over the range of 100–3100 mOsm and calculate the calibration polynomial from a plot of the real osmolality versus the instrument reading. The osmotic coefficient of NaCl at 25 °C is (Archer, 1992)

$$1 - \frac{A\sqrt{m}}{1 + 1.2\sqrt{m}} + Qm\ \exp(-2\sqrt{m}) + Bm + Cm^2 + Dm^3 + Em^4$$

where $A = 0.3915$, $B = 7.5480865 \times 10^{-2}$, $C = 2.8737876 \times 10^{-3}$, $D = -3.4537166 \times 10^{-4}$, $E = 1.7442782 \times 10^{-5}$, and $Q = 0.3035192$.

Because of the sensitivity of the measurement to temperature fluctuations it is also highly recommended to frequently check the reproducibility of osmometer readings during every few samples using some well-defined solution (e.g., the supplied calibration solutions). This is also helpful for determining whether lint got attached to the sensor, which distorts the results. Both effects—temperature drift of the instrument and contaminated sensor—lead to systematic deviations of the reading from the expected value.

Adsorption artifacts are possible, but avoidable. For the VPO measurement the sample is pipetted onto a filter paper that slides into the instrument. It was noted that interactions between solution components and the filter paper could distort the data (Hardegree and Emmerich, 1990). This effect apparently depends on the chemical nature of the dissolved substance. We observed systematic deviations between VPO measured osmotic coefficients and literature data for urea, but not sucrose, or NaCl. These observations are also consistent with independently measured data, in which small, but systematic deviations between measurements and literature values occur (Hong *et al.*, 2004). Whether such deviations are acceptable depends in part on the desired accuracy. To circumvent this problem we recommend incubating the filter paper in the solution, to bring the paper into osmotic equilibrium with solution. The wet paper is removed from the solution to be inserted into the instrument. Before insertion, the wet paper is drained through surface tension by bringing it shortly into contact with the surface of the sample solution. The resulting data is of higher accuracy than the data obtained using the manufacturer's protocol. This accuracy, however, comes at the price of lower precision due to the ill-determined total amount of sample.

Solutions for osmometry are best prepared by weight. The most direct source of error and uncertainties in osmometry comes from the process of solution preparation. Although the molar concentration scale is very popular in biochemistry, its usage has some disadvantages in the precision and accuracy of solution preparation. This is because of the two following reasons: (1) solution densities depend on temperature and (2) volume measurements are usually less accurate than weight measurements that are used in mass-based concentration scales. Accordingly, using molal concentrations (moles per kilogram of water) is an advantage, because these concentrations do not depend on temperature, and accurate balances can be used for solution preparation.

VPO allows to measure changes in the preferential hydration of interacting macromolecules. By measuring the changes in solution osmolality due to addition of macromolecules, it is possible to determine the extent to which a macromolecule preferentially takes up water from the bathing solution, that is, water unavailable to small "excluded" osmolytes. This is nothing but the preferential interaction of osmolytes with that macromolecule (Courtenay *et al.*, 2000, 2001; Harries *et al.*, 2005). Think of the naively idealized case, where every solute macromolecule is surrounded by some number of cosolute-excluding waters. The addition of macromolecules to the solute will leave only part of those waters available for dissolution of any other cosolute. Accordingly, the *variation* of change in solution osmolality following addition of a small amount of (macromolecular) solute to solution is related to the number of excluding waters per molecule. And if the number of osmolyte-excluding waters remains constant, the slope of the osmolality variation with osmolyte concentration (at a constant macromolecular concentration) turns out to be constant too.

Once again, due to the link between solute and water chemical potential imposed by the Gibbs–Duhem relation, we can also follow the excess/deficit number of osmolyte rather than the corresponding number of waters (Anderson *et al.*, 2002; Parsegian, 2002). However, it is the number of excluding waters that seems to remain constant in many examples over a wide range of concentrations.

As an example, consider the cyclodextrin/adamanthane (CD/AD) complex discussed in Section II.B. For this particular measurement, we used a methylated derivative of CD, mCD. Once we evaluate the number of excluding waters from the constant slopes of osmotic pressure changes with solution concentration, it is also possible to derive *differences* in hydration (preferential hydration) between the complexed and uncomplexed molecules. These results from VPO match those derived from calorimetry, as shown in Fig. 8. The correspondence in numbers obtained from both ITC and osmometry confirms that both approaches indeed probe the release of osmolytes-excluding waters.

B. Calorimetry

Through the partition function, calorimetry yields information on osmolyte action. Calorimetry is a versatile tool that can be highly informative. It is one of the few methods that allows direct access to thermodynamic properties, such as enthalpies, heat capacities, expansivities, and volume changes (see also Chapter 5 on DSC by Spink and Chapter 4 on ITC by Freyer and Lewis, respectively, this volume). These

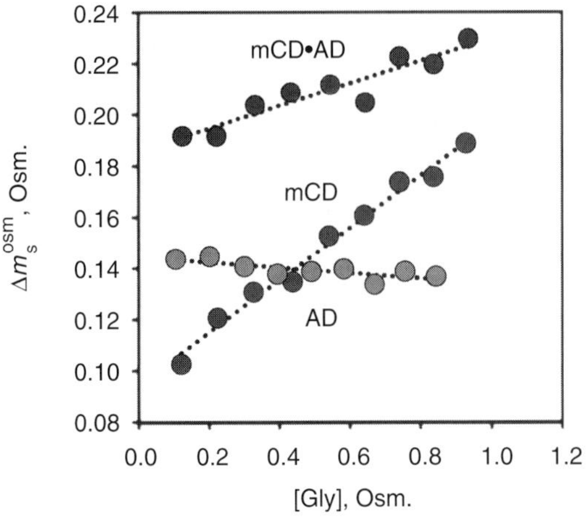

Fig. 8 Numbers of cosolute-excluding waters n_w from AD, mCD, and AD–CD, witnessed in the change in solution osmolal concentration Δm_s^{osm} with osmolyte concentration. Differences in slopes give the changes in cosolute (here, glycine) exclusion upon complexation: $\Delta n_w = -31$. These numbers have been confirmed by ITC calorimetry measurements (Harries *et al.*, 2005).

properties are, from a statistical mechanics point of view, derivatives of the partition function with regard to temperature and pressure. This allows us access to the most fundamental function in thermodynamics, the partition function itself. In the present context of osmolyte effects on macromolecules, knowing the partition function allows us to calculate how biochemical reactions respond to the presence of osmolytes under various conditions (different temperatures, pressures, and solution conditions).

Calorimetric measurements are useful for investigating both biochemical reactions and the properties of each molecular species. We discuss several calorimetric techniques in the following. All these methods are helpful to gain information not only on transitions and reactions between macromolecules but also on the individual molecules themselves. While biochemistry often focuses on the resulting interaction between macromolecules, it is particularly important to follow changes in individual molecular species because at highly concentrated solution conditions the physical properties of each kind of molecule by itself can vary in different ways, ultimately impacting macromolecular interactions. The solvation information extracted from calorimetry indicates how biomolecules react to the presence of osmolytes. Such detailed information is, however, only necessary if a very comprehensive understanding of the effects of osmolytes is desired. Often, it is sufficient to learn what are the overall effects of cosolutes on biochemical reactions. These effects are witnessed in the routinely determined dependence of affinities (pK_d) and thermal stabilities (T_m) on osmolyte concentration.

Titrations: Isothermal titration calorimetry. While isothermal titration calorimetry (ITC) is normally used to investigate specific (or more simply "strong") binding (pK_d values or affinities), it is also capable of yielding data on nonspecific interaction, namely solvation. The most straightforward example is the dependence of ligand-binding affinities on osmolyte concentration, as shown in Fig. 3; see also (Harries *et al.*, 2005; Morar *et al.*, 2001; Xavier *et al.*, 1997). The slope of the affinity as a function of osmolyte concentration yields solvation information, as explained in Sections II.B , III.A, and IV.B. The ITC can also be used to measure the interaction between molecules by determining enthalpies of dissolution (Zou *et al.*, 2002). More details on technical aspects of ITC can be found in Chapter 4 by Freyer and Lewis, this volume.

Thermal scans: Differential scanning calorimetry. Differential scanning calorimetry (DSC) has major advantages over other techniques that determine thermal stability. First, the macromolecular melting curve has a peak shape, which allows one to determine the melting temperature T_m more precisely than is possible in the normal sigmoidally shaped transition curves. Another advantage is that in addition to the van't Hoff enthalpy ΔH_{vH} (derived from the cooperativity of the transition) the peak area yields the calorimetric enthalpy ΔH_{cal}. Comparison of ΔH_{vH} with ΔH_{cal} is a good test for the validity of the thermodynamic model used for the data evaluation, for example, the assumption of a two-state transition, assumptions about the oligomeric state of the protein.

Finally, DSC is unique because it measures absolute heat capacities C_ps, which are beneficial for several reasons. On the one hand, absolute heat capacities contain information on the temperature dependence of chemical activities, that is, on the temperature dependence of solvation properties (Harned and Owen, 1943). On the other hand, absolute heat capacities can be used as a measure of the compactness of the denatured state (Häckel *et al.*, 2000). The denatured state does contribute to protein stability—just as much as does the native state (Auton and Bolen, 2004, 2005; Baskakov and Bolen, 1998). For this reason, the denatured state has been of interest for as long as protein folding has been studied (Goldenberg, 2003; Kohn *et al.*, 2004; Lapanje and Tanford, 1967; Nozaki and Tanford, 1967; Tanford *et al.*, 1967a,b; Tiffany and Krimm, 1973; Tran and Pappu, 2006; Tran *et al.*, 2005). Moreover, the compactness of the denatured state is a measure of its interaction with osmolytes (Qu *et al.*, 1998).

The most common measure for protein—osmolyte interaction used in DSC experiments is the dependence of the protein stability T_m on osmolyte concentration, which yields preferential interaction parameters (Kovrigin and Potekhin, 1997; Plaza del Pino and Sanchez-Ruiz, 1995; Poklar *et al.*, 1999), or the *m*-value. This last procedure is closely related to the phase diagram method, which is described in Section III.F. More details about DSC are found elsewhere in this volume (see Chapter 5 by Spink, this volume).

Heat-volume coupling: Pressure perturbation calorimetry (PPC). Pressure perturbation calorimetry (PPC) measures expansivities as a function of temperature (Lin *et al.*, 2002). Both absolute expansivities α_p^* and information on the following transition parameters of macromolecules can be obtained. These transition parameters are again the thermal melting temperature T_m and the transition enthalpy ΔH_{vH} (the cooperativity of the transition). In addition, the volume change ΔV upon unfolding can be determined from the area of the transition peak through integration (Lin *et al.*, 2002) or by curve fitting (Rösgen and Hinz, 2000).

Expansivities have been used in model-dependent approaches for deriving solvation information from thermodynamic data (Batchelor *et al.*, 2004; Chalikian, 2001). Because of uncertainties regarding the validity of such approaches (Batchelor *et al.*, 2004; Kauzmann, 1976), we recommend a different approach for quantitative studies that is based on first principles and is discussed in Section IV.B (Kirkwood–Buff theory). In any case, volumetric data are very useful for understanding the molecular solvation properties that govern the concentration-dependent solution thermodynamics (Kirkwood and Buff, 1951).

Combined techniques: Pressure modulated DSC. A recent development in calorimetric instrumentation allows regular DSC measurements (yielding heat capacities C_p) to be combined with pressure modulation (yielding expansivities α_p^*) (Boehm *et al.*, 2006; Rösgen and Hinz, 2006). This method allows a simultaneous measurement of all mechano-thermal properties that can be measured by DSC and PPC alone. While the Microcal® instrumentation (VP-DSC) is optimized for PPC, it appears that among commercially available instruments pressure modulated DSC (PMDSC) is possible only with the setup by Calorimetry Science Corporation® (N-DSC III).

Special requirements in osmolyte solutions' calorimetry.

1. ***Avoiding bubbles.*** Osmolyte solutions can be quite viscous at high concentrations. This makes solution degassing particularly difficult, because small bubbles of gas do not readily surface and can remain trapped in solution. Such bubbles have different effects in the different types of calorimetry. In ITC, the baseline becomes unstable and noisy in the presence of bubbles that swirl around in the solution. In DSC, the contribution to the signal of even minute bubbles can easily exceed that of the sample macromolecule. A decrease in sample viscosity with increasing temperature can facilitate escape of the bubble from the sample chamber, which leads to sudden jumps in the signal. In pressure-dependent techniques, such as PPC and PMDSC, the expansivity of bubbles is so large compared to the liquid that the signal of the bubble swamps that of the liquid.

2. ***Degassing.*** There are three strategies of degassing: sonication, saturation, and vacuum. Sonication helps to remove residual air bubbles that are difficult to remove from the solution. Saturation of the solutions with Argon might be useful, because the solubility of Argon in water increases with temperature. Air, in contrast, becomes less soluble as the temperature increases, and this can lead to bubble formation because typical storage temperatures of solutions of biologicals are lower than experimental temperatures. Such bubble formation can also be avoided by vacuum degassing. Because of the potentially high viscosity of osmolyte solutions, special precautions have to be followed. The vacuum has to

be mild in order to avoid formation of foam, or splashing of the sample. Manual degassing in a syringe offers best control over nucleation of bubbles (tapping) and rapid adjustment of vacuum strength (adjustment of piston position).

3. ***Decomposition.*** Some osmolytes tend to decompose and/or react with other biomolecules. In many experimental methods, there would be no notable effect originating from a small fraction of decomposing molecules. In calorimetry, however, the heat of such reactions can significantly contribute to the signal. Urea tends to decompose in solution, even more so at elevated temperatures (Shaw and Brodeaux, 1955). TMAO may lose its oxygen during storage, evacuation, light exposure, and heating. The decomposition products of either of these osmolytes can lead to chemical modification of the sample (Volkin *et al.*, 1997), for example, through carbamylation (Stark *et al.*, 1960), oxidation, or glycation in the presence of urea, TMAO, or reducing sugars, respectively. In the case of urea some protection is possible by choice of buffer (Lin *et al.*, 2004). The only way to protect macromolecules from oxidation by TMAO is through minimization of exposure. Protection against glycation is done in nature by the use of amino acids, especially taurine, which can scavenge free aldehydes, such as reducing sugars (Ogasawara *et al.*, 1993). It is questionable, though, whether reacting all reducing sugars with amino acids is a real solution to the problem, because this completely alters the chemical composition of the solution. The best recommendation therefore is to reduce the time of exposure by using only those protein solutions to which osmolytes have been added directly before the experiment.

4. ***Matching solutions.*** Normally, sample concentrations in bio-calorimetry are small compared to the concentration of the osmolyte cosolute. Therefore, it is very important to exactly match all solutions (including sample and reference) with regard to the cosolute concentrations. Otherwise large contributions of heats of dilution and/or heat capacities of the osmolytes to the signal could distort the results.

5. ***Filling the calorimeter.*** Coin-shaped calorimetric cells are not easily filled with a viscous solution. It is therefore recommended to purchase calorimeters with helical cells, if solutions of high viscosity are to be investigated frequently by DSC. Both major providers of high-sensitivity bio-calorimeters offer instruments with this setup (Calorimetry Science Corporation® N-DSC series, and Microcal® Capillary DSC).

C. Spectroscopy

Spectroscopy informs on molecular details of solvation, binding, and macromolecular conformations. Several spectroscopic methods, including nuclear magnetic resonance (NMR), circular dichroism (CD), and IR spectra, are helpful in determining the state and conformations of macromolecules. The spectroscopic data can thus aid in determining the effect of solutes on the relative stability of different macromolecular states (changes in equilibrium constants) due to solute addition.

Spectroscopic methods, however, not only shed light on osmolytes' effect on macromolecule structure and stability but also report on the way cosolutes interact and restructure water, as well as on changes in water ordering at the macromolecular interface. Furthermore, the same spectroscopic information can sometimes be used to directly indicate changes in solute binding to macromolecules.

Vibrational spectroscopy helps to determine changes in solvation environments. Because osmolytes interact with water in the bulk and at macromolecular interfaces and are preferentially included or excluded, structural information on cosolute solvation properties is vital to complete our understanding of their action. In particular, information can be obtained on molecular dynamics and motions, such as vibrational states and hydrogen bonding properties, in addition to macromolecular conformational changes (Goldbeck *et al.*, 2001; Royer *et al.*, 1996).

Raman and IR spectroscopies are useful in providing information on water structuring around solutes, because they report on water's vibrational modes, and these modes in turn are sensitive to subtle changes in hydrogen bonding interactions between water molecules at different molecular environments. For example, using an FTIR microscope it has been possible to examine small regions of biological samples of collagen (Mertz and Leikin, 2004). Frequency shifts that correspond to signature of solute in bulk water could be related to changes in hydration (hydrogen bonding) environments close to macromolecules. At the same time, the IR spectrum can be used to measure bonding properties, so that it is possible to obtain the adsorption curves for solute in macromolecular assemblies or molecules, as well as information on structural (conformational) changes in the protein.

Another kind of spectroscopy, vibronic sideband luminescence spectroscopy (VLBSL) has been used to probe both the identity of molecules in the first solvation shells around a Gd(III) ion (replacing calcium ions binding to proteins) and properties of hydrogen bonding in the ions' immediate environment (Navati *et al.*, 2004; Roche *et al.*, 2006). The Gd reporter can be used together with protein in the presence of different osmolytes to probe the changes in hydration properties of the macromolecules due solute.

In another study, vibrational sum-frequency spectroscopy has been successfully used to probe the solvation of an ionic surfactant (SDS and DTAC) at an oil (CCl_4)/water interface (Scatena and Richmond, 2004). Because of the presence of surfactant, the vibrational spectrum of interfacial water changes dramatically. These interacting water molecules show preferential water reorientation around the solvated surfactant, and a weaker hydrogen bonding interaction than in bulk water.

Dielectric relaxation spectroscopy gives direct evidence of preferential interaction. Proteins are preferentially hydrated in protecting osmolytes, while in aqueous denaturant solutions the cosolute preferentially interacts with protein. Dielectric relaxation measurements were used to directly demonstrate that RNase A is preferentially hydrated in glycerol solutions (Betting *et al.*, 2001). The protein was found to be bathed in a "bag" of water that allowed for a tumbling motion that is as quick as in pure water. Also, changes in mobility of components of the bulk solution were demonstrated when urea and glycerol were used as osmolytes (Abou-Aiad *et al.*, 1997; Betting *et al.*, 2001).

UV–VIS absorbance and fluorescence spectroscopy allows to detect changes in the environment of reporter groups. Aromatic amino acid residues are the typical groups observed in UV–VIS absorbance and fluorescence experiments. Changes in solvent-exposure of such reporter groups can be used to track protein conformational changes, because their extinction coefficients and fluorescence properties depend on the physical properties of their environment. In most cases, only protein unfolding results in exposure changes that are large enough to be observed spectroscopically. Such changes can also be followed by UV spectroscopically in the case of DNA denaturation. Circular dichroism spectroscopy allows investigation of protein secondary structures because the conformational environment of peptide groups biases their preference to absorb either right- or left-spin photons.

Some special requirements for osmolyte samples in spectroscopy.

Fluorescence: Osmolytes often contain small quantities of impurities that can interfere with fluorescence measurements. Such impurities that absorb in the UV–VIS range can usually be removed by adding activated charcoal to the osmolyte solution, stirring for 30 min, and filtering the solution using a micrometer-pore filter on a vacuum flask. Filtration is fastest when charcoal pellets rather than powder is used. In addition, the solution can be cleared from charged compounds by adding mixed-bed ion exchange resin to the solution. The exact amount of resin that is needed depends on its capacity for binding ions. In the worst-case example, commercial TMAO, recrystallization might be necessary. This can be done from water directly after synthesis (Russo *et al.*, 2003), or—in severe cases—from acetonitrile (dissolve 1 part of TMAO in 2 parts of acetonitrile at 70 °C and recrystallize on ice). TMAO is colorless and odorless. The yellow color and fishy smell of commercial TMAO clearly indicates the necessity of further purification.

CD: Far-UV circular dichroism (CD) can become problematic in osmolyte solutions. This is because of the usually high absorbance of most osmolytes in the far-UV range. Working at higher wavelengths might help in these cases, if feasible. As long as the protein is soluble enough, the best solution to the problem of high absorption by the osmolytes is to use a cuvette with a shorter path length along with a protein concentration that is increased in proportion. For solubility issues, see Section II.B in this chapter. Also, if the purpose is not to purely investigate effects of one specific osmolyte, osmolytes that have weaker CD absorbance in the UV can be chosen, such as many polyols and the synthetic polymer PEG, so that higher osmolyte concentrations can be used.

D. Ion Channels as Probes

Currents through ion channels report on protein states and their interactions with cosolutes. Because of their ability to open and close, ion channels form gateways for transport through lipid bilayers. As an example, consider a ridiculously simple channel with only two states: one is open and conductive while the other closed, collapsed, dry, and nonconducting. Solute that is preferentially excluded from waters that hydrate this channel's cavity will tip the balance in favor of a closed, dehydrated channel state. Such channels could, therefore, be an ideal testing ground to study solute exclusion as a source of osmotic stress (Bezrukov and Vodyanoy, 1993; Parsegian, 2002; Vodyanoy *et al.*, 1993; Zimmerberg and Parsegian, 1986).

Observing ion channels allows several unique opportunities. First, the open versus closed state of ion channels can be determined from watching the flow of ion current through the channel (open) versus times of no current (closed). The current reports, however, not only on the equilibrium constant for channel "gating" but also, by the channels' relative conductance, on the actual extent of cosolute (say polymer) exclusion from the channel. Finally the salt, whose current through the channel is the probe of exclusion and equilibrium, can itself act as an osmotic stressor too.

Testing solute exclusion from channels. How complete is a solutes' exclusion? A solute, such as a bulky polymer, might be completely excluded from the channel's interior due to prohibitive confinement that would lower its conformational entropy in the channel lumen. In this case, the free energy associated with keeping water pure inside the channel would increase. With direct analogy to the binding of *Eco*RI to DNA (Section II.B), higher osmotic pressure favors the closed state where many waters are released and given up to the bulk by the work of $(v_{\text{open}} - v_{\text{closed}}) \times \Pi$. In contrast, we can expect that smaller solutes will be able to repartition to some extent into a channel's lumen.

This size-dependent exclusion is most elegantly demonstrated by comparing the effect of a solution of PEG exerts on the open versus closed state of a channel at a particular osmotic stress for different sized PEG polymers (Fig. 9). Beyond a certain size, polymers are strongly excluded from the channel's lumen, as long as the cost of entering the channel (conformational entropy) is prohibitively high. Sometimes, the polymer can still be packed into the channel if its bulk concentration is so high that it becomes easier to allow polymer in than to suck water out of the channel.

E. Scattering Techniques

In ordered arrays, scattering can inform on molecular distances. When macromolecules aggregate into ordered assemblies they diffract, as shown in Section II.D for multilamellar lipid vesicles. The diffraction patterns provide information on the structure of the macromolecular aggregate and dimensions of the repeating unit cell. This information allows determination of an equation of state for the macromolecular aggregate: curves of the chemical potential (of solute or solvent) versus intermolecular spacing (Parsegian *et al.*, 1986).

If a solute that is strongly excluded from the macromolecular aggregate is used, it is possible to probe inclusion and exclusion of another osmolyte in solution and

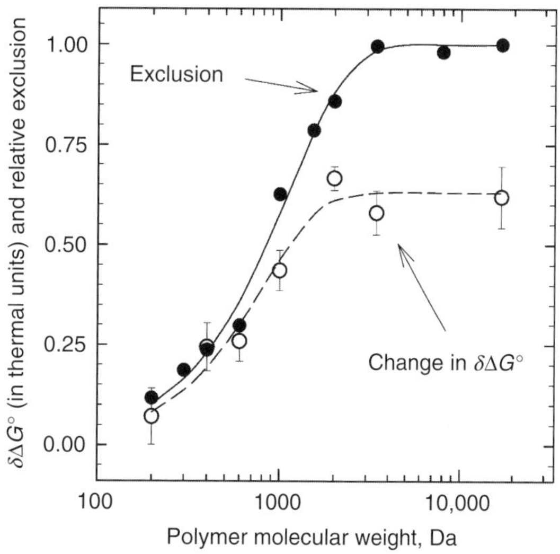

Fig. 9 Changes in alamethicin channel closing activity $\delta\Delta G°$ due to the presence of polymers of different molecular weight track the osomolyte's (PEG polymer) preferential exclusion. All measurements were made at 4.5 atm of osmotic pressure. As exclusion becomes stronger, the osmotic effect (withdrawing water from the channel, preferring the closed state) is stronger, too. The channel's closing free energy is evaluated from the ratio of times the channel is closed and open. Exclusion is evaluated from relative conductance of channels in PEG solution versus in solutions with no PEG (Bezrukov and Vodyanoy, 1993; Parsegian, 2002; Vodyanoy *et al.*, 1993). Figure courtesy of Sergey Bezrukov.

its partitioning between the bulk water and the macromolecular phase. Scattering of X-rays, neutron beams, and visible light informs not only on mesoscopic order but also on conformational and structural changes at the molecular level.

For dispersed macromolecules in solution, scattering probes macromolecular size and shape. Explicitly, radii of gyration for proteins and other biological polymers can be determined from small angle X-ray scattering or from neutron scattering. This is important information because it links the thermodynamic properties, such as preferential cosolute interaction with protein backbone and the macromolecular structure, say native versus denatured state and the extent of change in a protein's configuration due to solute addition. Most often, the radius of gyration is derived from the so-called Guinier plot of log scattering intensity versus the scattering angle θ in terms of k^2, where $k = \frac{4\Pi}{\lambda}\sin\theta$. The slope of this plot is related to the radius of gyration. In this way, it has been possible to determine that different proteins not only respond to osmolytes such as glycerol, urea, and GdnHCl through preferential interactions, but that they also select for different conformational states under stressed conditions (states that are more or less compact) (Barteri *et al.*, 1996; Garcia *et al.*, 2001; Kohn *et al.*, 2004; Segel *et al.*, 1998). Here too, preferential interactions not only change stability of native states of proteins but also select for particular conformations that previously had a higher free energy relative to other conformations.

In addition, each scattering method reports on the electron or neutron densities (for X-ray and neutron diffraction) or the index of refraction (in light scattering). This information can be used to determine the number of excluding waters and solute around macromolecules (together with additional information from other methods such as densitometry or ultracentrifugation). Preferential interaction parameters can also be derived from scattering profiles through the structure factor (Ebel *et al.*, 2000).

F. Phase Diagram Method

Phase diagrams can be used as visual aids for understanding reaction networks. Biochemical systems can be quite complex, and each of the participating proteins can be in dozens of differently active states. These states can be distinguished by, for example, conformation, protonation of crucial sites, bound ligands. Understanding such complex systems can be greatly facilitated by visual aids. Figure 10 shows as an example of three states that occur in a protein–ligand–osmolyte–water system. The graph is a phase diagram (Ferreon *et al.*, 2007; Rösgen and Hinz, 2003), in which the three states are separated by 50% population lines (phase separation lines). Phase diagrams of mesoscopic systems, such as proteins, have been used for decades (Hawley, 1971; Heremans, 1982; Suzuki, 1960; Zipp and Kauzmann, 1973), and they are governed by the same principles as the classical phase diagrams of macroscopic systems. It is important to recognize that protein transitions can be properly described as phase transitions (Finkelstein and Galzitskaya, 2004), because this opens up the use of the power of phase diagrams to explore properties of protein reaction networks.

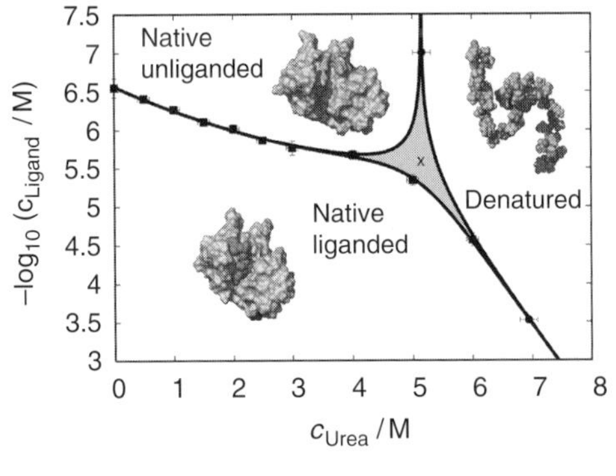

Fig. 10 Phase diagram of RNase A in the presence of specific ligand (CMP) and osmolyte (urea). The bold lines indicate 50% population size of one state each, as labeled. In the white areas one state predominates (>50%), in the gray area none of the three states reaches a 50% population fraction. The cross indicates the point of equal population of all listed protein states. Squares: Ligand pK_s measured by isothermal titration calorimetry (ITC); circles: midpoint concentrations of urea-induced unfolding, observed spectrophotometrically (Ferreon *et al.*, 2007). (See Plate no. 21 in the Color Plate Section.)

The response of reaction networks to the presence of osmolytes can be evaluated using the phase diagram method. The phase diagram method is appropriate for exploring both specific binding and nonspecific (solvation) interaction. As pointed out earlier (Section II.A), m-values are the most common and practical measure for such solvation effects. The m-values can be directly read from a phase diagram (Fig. 10) in the following way (Ferreon *et al.*, 2007). We consider the binding of specific ligand to a protein. The native, liganded state of the protein is populated under conditions given by the lower left area in Fig. 10, bordered by one of the bold 50% phase separation lines. The line of 50% bound state can be measured in terms of ligand (un)binding pK values (squares in Fig. 10), which equal $-\log_{10}(K_{\text{diss}})$, where K_{diss} is the dissociation constant. Since the m-value is the slope of $-RT\,\log_{10}(K_{\text{diss}})$ versus c_{O}, m can be directly obtained from the slope of the phase separation line times $RT\,\ln 10$.

Alternatively, the m-value could also be read from other data that are a measure of the shift in ΔG, such as thermal melting temperatures (Kovrigin and Potekhin, 1997; Plaza del Pino and Sanchez-Ruiz, 1995; Poklar *et al.*, 1999). m-values can be determined within a large range of temperatures, because they normally either depend on temperature only slightly (Felitsky *et al.*, 2004) or not at all (Giletto and Pace, 1999; Henkels *et al.*, 2001). Larger temperature dependencies seem to be more rare (Henkels and Oas, 2005).

Incidentally, the m-value of the binding reaction shown in Fig. 10 is not constant, as can be seen from the curvature of the left branch of the phase diagram. This is not the common case, as pointed out in Section II: m-values are usually constant. The current example shows that this constancy is not a strict law. Reasons for such deviation could be significant interaction between ligand and osmolyte (Ferreon *et al.*, 2007).

m-values in concentrated, nonideal solutions are analogous to stoichiometries in ideal solutions. The procedure of reading the m-value from the phase diagram (Ferreon *et al.*, 2007; Fukada *et al.*, 1983; Kidokoro *et al.*, 1990; Schellman, 1975) has its analogue in the determination of binding stoichiometries from slopes of phase separation lines (Rösgen and Hinz, 2003). Actually, the m-value can be interpreted as being based on a number of osmolyte molecules that are additionally attached to the protein after the transition (change in preferential interaction upon folding). This number is, however, corrected for the fact that in a crowded solution the presence of osmolyte means absence of water, and absence of water corresponds to presence of osmolyte. This idea is known as Schellman's exchange concept (Schellman, 1994). A more rigorous approach to m-values is provided by Kirkwood–Buff theory, which is discussed below (Section IV.B). Especially in those cases in which water is accumulated around a macromolecule to a higher degree than the osmolyte, the classical idea of binding would lead to a negative number of bound osmolytes (osmolytes exclusion, and preferential hydration), because the water is present in excess at the macromolecular surface. This demonstrates that at high concentration of additives it can be counterintuitive to think in terms of simple stoichiometries of bound osmolytes. The water has to be taken into account explicitly under such conditions (Tanford, 1969).

Phase diagrams allow extraction of all pertinent thermodynamic information and the number of coupled reactions and their stoichiometries. In addition to stoichiometries and m-values, many additional thermodynamic properties can be derived from the phase diagram. Generally, slopes of phase separation lines yield ratios of cooperativity parameters of the transition, such as enthalpy changes ΔH, volume changes ΔV, or stoichiometries ΔN. These cooperativity parameters are extensive properties, which have the conjugate intensive properties temperature T, pressure p, and chemical potentials μ, respectively. Using the notation $\beta = 1/RT$, $\gamma = p\beta$, and $\alpha = -\mu\beta$, for the intensive properties we can formulate a generalized linkage relation (Rösgen and Hinz, 2003)

$$\left(\frac{\partial\lambda_2}{\partial\lambda_1}\right) = -\frac{\Delta A_1}{\Delta A_2} \tag{12}$$

which represents the slope of any of the phase separation lines in a phase diagram. The ΔAs are any of the cooperativity parameters (ΔN,ΔH, and ΔV) given above, and the λs are any of the matching α, β, or γ, respectively. Because the phase separation lines are defined by experiments that yield one of the ΔAs (such as in thermal unfolding, pressure denaturation, or titration experiments), the other ΔA can be directly obtained from Eq. (12). In case of m-values, the linkage relation becomes

$$\left(\frac{\partial c_O}{\partial\lambda_1}\right) = -\frac{\Delta A_1}{m/RT} \tag{13}$$

where c_O is the osmolyte concentration. Further quantitative details on the phase-diagram method can be found in Rösgen and Hinz (2003) and Ferreon *et al.* (2007).

The phase diagram reveals the existence of states that are not directly probed. Sudden changes in the slopes in the phase diagram, such as seen around the gray area in Fig. 10, are indicative of the occurrence of additional states of the macromolecule. The gray "triple area," corresponding to a triple point in a classical phase diagram, contains a cross as a label in its center. This cross marks the position of both the midpoint urea concentration of protein unfolding and the pK value of ligand dissociation at that urea concentration. Following the phase separation line for the denatured state (rightmost bold line) from top to bottom, we note a sudden change in slope at the triple area, where ligand binding occurs. Equally, we note a similar breakpoint following the phase separation line for the ligand-bound state (lowest bold line) from left to right. This breakpoint is due to the denaturation of the unbound protein. We therefore realize that knowing one phase separation line reveals the existence of states that were not directly measured.

High-throughput analysis allows investigation of complex multidimensional phase diagrams. In the example given (Fig. 10) there are three protein states and only two added molecules. Phase diagrams of such simplicity have already been applied in high-throughput analysis in drug design (Todd and Salemme, 2003) and biochemistry (Kervinen *et al.*, 2006; Matulis *et al.*, 2005). Instead of investigating a multitude of different two-dimensional phase diagrams as in the two given examples of high-throughput research, the investigation of multidimensional phase diagrams is made feasible by use of the phase-diagram method. While the former application is more typical for drug screening, the latter application would be of high utility in formulation development (Tsai *et al.*, 1993) or in the investigation of complex mixtures that model the cytosol. It is the large number of components that can occur both in drug formulations and in the cytosol that makes the phase diagram multidimensional. Each additional component adds another dimension. This multidimensionality requires high-throughput methods: If 10 points are sampled in each dimension, then investigating a protein in a renal osmolyte mixture containing five components (Kwon *et al.*, 1996) may require 100,000 experiments, for instance.

The phase diagram method allows for a streamlined evaluation of quantitative high-throughput data through efficient data reduction. Each experimental data point in a phase diagram corresponds to a set of data that follow a transition. In order to define such a transition property, dozens to hundreds of data points must be collected. Evaluation of these data yields at least one midpoint parameter (such as pK_s, T_m, or $c_{1/2}$) and a cooperativity parameter (ΔN, ΔH, m, and so on) as a function of temperature, concentration, pressure, or some other intensive property. This means that per point in the phase diagram we get a data reduction of one to two orders of magnitude, by expressing the (dozens to hundreds points of) raw data in terms of two parameters, while preserving the pertinent information content. Plotting these parameters in a phase diagram then directly yields more cooperativity parameters that further reduce the amount of data. This is because a phase separation line usually has to be composed of about 10 or more points to be well defined, and these points are again expressed in terms of one or two parameters.

In the given example of a five-osmolyte system about 15 parameters will eventually be needed: Five m-values, and 10 potential cross-correlations between these m-values. Higher-order cross-correlations are probably not required, because these would correspond to third-order terms in the Taylor expansion of the Gibbs free energy.

Independently of these considerations the main message is clear: globally fitting $100 \times 100{,}000$ data points is close to hopeless. This is due to both the huge amount of data and the difficulty in having a good initial guess for all fitting parameters. Without such good guesses the fit will almost certainly end up trapped in a suboptimum. The phase-diagram method, in contrast, allows evaluation of smaller sets of data separately and systematic derivation of all parameters in a sequential manner.

IV. Solvation Information

Having discussed experimental systems, as well as strategies for investigating them, we now turn to the question how to interpret the results in terms of their fundamental physical background. For this purpose, we first discuss thermodynamically nonideal behavior in crowded osmolyte solutions (Section IV.A). Then we use this information in discussing solvation properties of biomolecules and how these properties relate to the resulting behavior (Section IV.B). Finally, we explore some of the impacts of electrostatic interaction on macromolecular properties (Section IV.C).

A. Activity Coefficients

Thermodynamically ideal versus nonideal behavior. Much of biochemistry in nature happens under crowded and thermodynamically highly nonideal solution conditions, as present in concentrated osmolyte solutions. Under such conditions, we cannot assume that the chemical activities of all molecular species are simple linear functions of their concentrations and that they are independent of each other. Activity coefficients are a quantitative measure for this deviation from ideal solution conditions. There are many ways to define what "ideal conditions" are, but for our purposes the definition based on the molar concentration scale is the most useful (Rösgen *et al.*, 2004a): the chemical activity of a solute behaves in a thermodynamically ideal manner, if the activity is proportional to the solute's molar concentration (moles per liter of solution).

Molar activity coefficients can be easily calculated. Recently, a theory of solution was developed (Rösgen *et al.*, 2004a,b) that captures the high-concentration behavior of osmolytes in a rigorous, but simple and easily applicable way. The activity coefficients of about half of the osmolytes as a function of their molar concentration c_O are well fit by the first-order activity coefficient equation

$$\gamma_c = \frac{1}{1 - c_O/c_1} \tag{14}$$

the others follow the second-order activity coefficient equation

$$\gamma_c = \frac{g_2/(2c_O)}{2 - c_O/c_2}\left[-(1 - c_O/c_1) + \sqrt{(1 - c_O/c_1)^2 + \frac{4c_O}{g_2}(2 - c_O/c_2)}\right] \tag{15}$$

The parameters c_1, c_2, and g_2 are given for each osmolyte in Table I (Rösgen *et al.*, 2005). For the purpose of calculating solvation information by Kirkwood–Buff theory in Section IV.B, we will also need the slope of the chemical activity $a_O = c_O\gamma_c$ with regard to the osmolyte concentration:

Table I
Parameters for the Calculation of Osmolyte Chemical Activities (Eqs. 14 and 15) and Concentration Scale Conversion (Eq. 18)

	g_2 mol/l	c_1 mol/l	$2c_2$ mol/l	c_{max} mol/l	Highest c mol/l
Xylose	–	7.6	–	10.16	2.6
Glucose	–	6.28	–	8.670	4
Fucose	–	4.89	–	9.05	1.9
Glactose	790	8.27	c_{max}	8.99	2.5
Rhamnose	–	4.4	–	9.07	1.2
Mannose	–	7.04	–	8.54	3.5
Maltose	–	3.135	–	4.27	1.8
Raffinose	–	1.523	–	2.46	0.22
Sucrose	70.4	2.466	c_{max}	4.617	2.6
Glycerol	19	4.8	c_{max}	13.69	7.1
Mannitol	–	7.35	–	8.173	1.1
meso-Erythritol	–	9.3	–	11.88	3.8
Sorbitol	–	6.475	–	8.17	4.8
Urea	21.6	20.3	c_{max}	22.03	10.1
Glycine	3.765	3.260	c_{max}	21.41	2.8
Alanine	–	14.40	–	16.07	1.7
Proline	120.5	5.38	c_{max}	12.52	4.5
Sarcosine	–	8.68	–	16.29	5.1
Glycine betaine	16.88	1.97	c_{max}	10.72	3.4

$$\left(\frac{\partial \ln a_O}{\partial \ln c_O}\right)_{p,T} = \left(-\frac{1}{1+2a_O/g_2} + \frac{2+a_O/c_1}{1+a_O/c_1+a_O^2/(c_2g_2)}\right)^{-1} \tag{16}$$

Conversion between different concentration scales is straightforward. The volume-based molar concentration scale is very popular in biochemistry and is also the most useful scale for quantifying solution's nonideality (Rösgen *et al.*, 2004a, 2005). However, the mass-based molal concentration scale is more precise for solution preparation. Therefore, it is sometimes desirable to be able to convert between different concentration scales. This is usually done using the following equation

$$c = \frac{m\rho}{1+mM_r/1000} \tag{17}$$

where c is the molar concentration (moles per liter of solution), m the molal concentration (moles per kilogram of water), ρ the solution density, M_r the molar mass, and the factor of 1000 comes from the conversion of grams to kilograms.

The solution density is, however, not always known. The recent theory of osmolyte solutions (Rösgen *et al.*, 2004a) offers a solution for this problem. Since osmolyte partial volumes depend little on osmolyte concentration, the molar and molal concentrations of osmolytes can be converted to a good approximation using the density of the pure osmolyte $(1/c_{\mathrm{max}})$:

$$c_{\mathrm{O}} \approx \frac{m_{\mathrm{O}}\rho_{\mathrm{W}}}{1 + m_{\mathrm{O}}\rho_{\mathrm{W}}/c_{\mathrm{max}}}, \quad m_{\mathrm{O}} \approx \frac{c_{\mathrm{O}}/\rho_{\mathrm{W}}}{1 - c_{\mathrm{O}}/c_{\mathrm{max}}} \tag{18}$$

where ρ_{W} is the density of pure water. The parameter c_{max} is given in Table I for many osmolytes. These equations are only recommended for aqueous solutions of osmolytes without net charge.

Activity coefficients are important for understanding the thermodynamic and kinetic behavior of the solution, and of in vivo processes. In this section, we have shown how to calculate activity coefficients of osmolytes. Such coefficients are of utmost importance in concentrated solutions of biomolecules. Both kinetic and rate equations are based on knowledge of the chemical activities of the participating compounds. Under dilute and ideal conditions, it is permissible to replace the chemical activities in those equations by concentrations. This is, however, no longer valid under the crowded and nonideal conditions that are so typical for *in vivo* biochemical settings, making activity coefficients of utmost importance in biochemistry. In addition, knowledge of chemical activities reveals information on molecular processes in solution, as we discuss now.

B. Kirkwood–Buff Approach

In this section, we show how measured macroscopic properties of solution can be used to derive microscopic structural information. Such information is very useful for understanding the molecular origin of the observed macroscopic behavior.

Kirkwood–Buff theory relates thermodynamics to the structure of the solution. The solution behavior of biomolecules is normally characterized on the basis of their average thermodynamic or kinetic properties. In order to obtain an understanding of the principles that cause the specific behavior of each molecule, the microscopic details of the molecular behavior must be studied. Kirkwood and Buff (1951) showed that the overall thermodynamic properties of a solution can be derived from these molecular details. Conversely, knowledge of appropriate thermodynamic parameters allows one to deduce microscopic features of the solution (Ben-Naim, 1977).

What do we mean by "structure of a solution?" The number of ways to define "structure of a solution," "water structure," or "hydration structure" is limited only by imagination. It is therefore critical to define these terms in a way that is appropriate for the task at hand. In different contexts different definitions might be useful. Here, we aim toward capturing solution equilibrium thermodynamics in terms of the structure of the solution. In this context, radial distribution functions are the proper measure of solution structure (Kirkwood and Buff, 1951).

Crystal structures and the structure of solution are defined in a similar manner. The structure of a solution is expressed in essentially the same way as done in X-ray crystallography. The relative position of molecules (such as proteins) in a crystal is given by angle-dependent distance distributions. In a crystal, the positions of all molecules in the lattice are fixed, and thus the angle dependence of the distances is constant. In contrast, the tumbling and diffusion of molecules in solution averages out this angle dependence, while considering the positions of the centers of mass of the molecules. As a consequence, the structure of a solution is given by radial distribution functions that are angle independent.

The overall correlations between molecules determine the concentration-dependent thermodynamics of the solution. Radial distribution functions are normalized with respect to the bulk solution density, that is, a value of unity indicates no deviation from bulk density, while values lower or higher than unity indicate a local deficit or excess, respectively, relative to bulk density. Figure 11 shows an example of such distribution function g_{ij} between particles of type i and j. The overall correlation G_{ij} between two kinds of particles is then given by integrating all deviations from bulk distribution over the volume

$$G_{ij} = \int (g_{ij} - 1)\mathrm{d}V \tag{19}$$

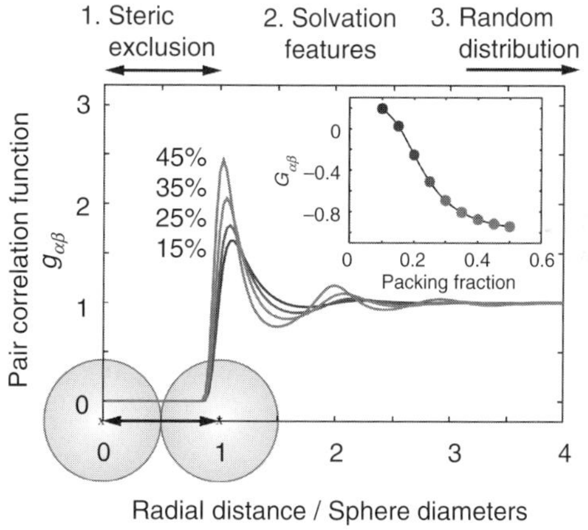

Fig. 11 Volume fraction dependence of the radial distribution functions $g_{\alpha\beta}$ and Kirkwood G factors in a model liquid system with Lennard–Jones potential. Functions shown for packing fractions are 15, 25, 35, and 45%. (1) At distances closer than the sum of the two radii (contact distance), steric exclusion operates. (2) At intermediate distances, there are strongly concentration-dependent solvation features, most notably the first solvation shell just outside the region of steric exclusion. (3) At large distances there is no correlation between particles and the pair correlation function approaches unity. The inset shows the Kirkwood–Buff integrals (Eq. 19) as a function of the packing fraction. Note the slope and sign changes with respect to packing fraction (Rösgen *et al.*, 2005). (See Plate no. 22 in the Color Plate Section.)

These overall correlations G_{ij} are also called Kirkwood–Buff integrals, and they are a measure of the excess or deficit of one kind of molecule around another kind of molecule—relative to the bulk. It is the set of all G_{ij} that determines much of the thermodynamic behavior of the solution.

Osmolyte behavior can be classified according to the structure of the solution. We have already mentioned in Section IV.A that osmolytes follow the peculiar first- or second-order behavior, that is, their partition function depends linearly, or in a weakly quadratic manner, on osmolyte chemical activity (Rösgen *et al.*, 2004a). This finding can be translated into structural features of solution with the help of the Kirkwood–Buff theory. This is done by combining the Kirkwood–Buff expression (Kirkwood and Buff, 1951).

$$\left(\frac{\partial \ln a_{\mathrm{O}}}{\partial \ln c_{\mathrm{O}}}\right) = \frac{1}{c_{\mathrm{O}}} + \frac{G_{\mathrm{WO}} - G_{\mathrm{OO}}}{1 - (G_{\mathrm{WO}} - G_{\mathrm{OO}})c_{\mathrm{O}}} \tag{20}$$

with Eqs. (14) and (15). Here, a_{O} and c_{O} are chemical activity and molarity of the osmolyte, and G_{WO} and G_{OO} are its hydration and self-solvation. It turns out that the difference between osmolyte hydration G_{WO} and self-solvation G_{OO} essentially does not depend on concentration (Rösgen *et al.*, 2005). This is, incidentally, a property that most likely makes these molecules useful osmolytes, because organisms can upregulate or downregulate the cosolute concentration without facing any harmful sudden changes or discontinuities in their solvation behavior. There are, however, certain differences between protecting osmolytes and denaturants. While the difference ($G_{\mathrm{WO}} - G_{\mathrm{OO}}$) is large in the case of protecting osmolytes, it tends to be close to zero for denaturants (Rösgen *et al.*, 2005).

Changes in protein solvation upon unfolding mirror the bulk osmolyte solvation. Given that protein folding m-values for net-neutral osmolytes are normally constants, it is possible to derive general statements on protein solvation behavior from Kirkwood–Buff theory. The balance between protein hydration G_{PW} and protein "osmolation" G_{PO} (solvation by osmolyte) changes upon unfolding by $\Delta(G_{\mathrm{PW}} - G_{\mathrm{PO}})$. This solvation change directly relates to the m-value in a solution of dilute protein:

$$\frac{m}{RT} = \frac{\Delta(G_{\mathrm{PW}} - G_{\mathrm{PO}})}{1 - (G_{\mathrm{WO}} - G_{\mathrm{OO}})c_{\mathrm{O}}} = \Delta(G_{\mathrm{PW}} - G_{\mathrm{PO}})\left(\frac{\partial \ln a_{\mathrm{O}}}{\partial \ln c_{\mathrm{O}}}\right) \tag{21}$$

Since the derivative in this equation is known from the Eqs. (16) and (14) and the table in Section IV.A, the change of $\Delta(G_{\mathrm{PW}} - G_{\mathrm{PO}})$ with osmolyte concentration can be directly calculated (Rösgen *et al.*, 2005).

$$\Delta(G_{\mathrm{PW}} - G_{\mathrm{PO}}) = \frac{m}{RT} \Big/ \left(\frac{\partial \ln a_{\mathrm{O}}}{\partial \ln c_{\mathrm{O}}}\right) \tag{22}$$

Dividing this equation by the constant ratio m/RT gives the relative change of $\Delta(G_{PW} - G_{PO})$ with osmolyte concentration. Figure 12 shows this general behavior of $\Delta(G_{PW} - G_{PO})$ for several osmolytes. Because a_O is the only component that makes the denaturational solvation change concentration dependent, and a_O of osmolytes can be grouped into stabilizing and destabilizing osmolytes (Rösgen *et al.*, 2004a), we see a grouping by osmolyte type also in Fig. 12. Stabilizing osmolytes tends to have a strong concentration-dependent protein solvation behavior, while for urea $\Delta(G_{PW} - G_{PO})$ is nearly constant in comparison. The only exception to this trend is glycine.

Separately deriving protein hydration and osmolation is made possible by inverse Kirkwood–Buff theory. Biophysical chemists assumed for a long time that it is impossible to separate preferential interaction parameters into contributions from water and from osmolyte (Schellman, 1994; Timasheff, 1992). But, in fact this is possible using Ben-Naim's inverse Kirkwood–Buff theory (Ben-Naim, 1977, 1988). The expressions for protein hydration and osmolation change upon unfolding

$$\Delta(G_{PO}) = -\Delta V_P - (1 - \phi_O)\frac{m/RT}{(\partial \ln a_O/\partial \ln c_O)} \tag{23}$$

$$\Delta(G_{PW}) = -\Delta V_P + \phi_O\frac{m/RT}{(\partial \ln a_O/\partial \ln c_O)} \tag{24}$$

have been discussed previously (Rösgen *et al.*, 2005, 2007), and here we use these equations to point out characteristic differences between osmolyte effects on protein folding and protein ligand binding. The volume change upon protein

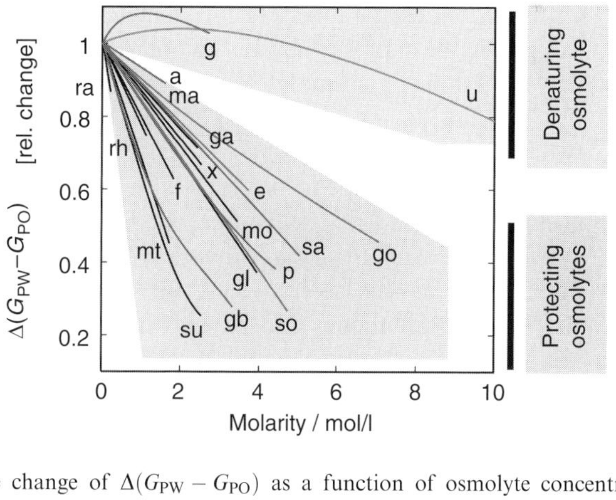

Fig. 12 Relative change of $\Delta(G_{PW} - G_{PO})$ as a function of osmolyte concentration for several osmolytes. Note that all protecting osmolytes (except glycine) have a very strongly concentration-dependent protein solvation behavior, while with increasing concentration of the denaturant urea the solvation does not change much (Rösgen *et al.*, 2005).

unfolding ΔV_P and the volume fraction of the osmolyte $\phi_O = c_O v_O$ play a prominent role in this discussion; v_O is the partial volume of the osmolyte. The partial volumes contain several contributions, including steric exclusion and density changes of the solution.

Water effects can occur because of changes in the size of cavities inaccessible to osmolytes. We have seen in Section II.B that there are binding reactions in which hydration is responsible for the observed effects, independently of the choice of a specific osmolyte. This could be due to several effects, with two extreme cases being $\Delta(G_{PO}) \approx 0$ or $\Delta(G_{PW}) \approx 0$. First, according to Eq. (23) independence of the type of osmolyte could mean that $\Delta(G_{PO}) \approx 0$, which implies that the volumetric contribution $(-\Delta V_P)$ determines the m-value. In this case, the volume change upon transition from nonspecific to specific binding in the given example of *Eco*RI (Sidorova and Rau, 1996) would have to be in the order of 1.6 l/mol (the m-value is about 1 kcal/mol M). This corresponds to about 4% of the volume of *Eco*RI, which is likely too large. The more likely explanation neglects volume changes and considers the change only in cavity size (or more accurately, the solute-excluding volume) between protein and DNA. This is the interpretation that is usually used along with the osmotic stress analysis (Parsegian *et al.*, 2000). Closure of a cavity corresponds to a transfer of a volume element from inside the protein–DNA complex to its outside. If the cavity is inaccessible to osmolyte, then $\Delta(G_{PW})$ is close to zero, and $\Delta(G_{PO})$ is positive. In this situation the process mechanistically involves only the release of water, but energetically, only interactions that exclude osmolytes from the protein are responsible for the added contribution to the free energy.

Preferential interaction expressed in terms of Kirkwood–Buff integrals. We already saw how m-values can be represented by Kirkwood–Buff integrals (Eq. 21), and we have mentioned in Section II that m-values can be expressed in terms of preferential interaction parameters. It is therefore no surprise that the latter can readily be expressed by Kirkwood–Buff integrals. This holds for both preferential solvation by osmolyte $\Gamma_{\mu_3} = -c_O(G_{PW} - G_{PO})$ and preferential hydration $\Gamma_{\mu_1} = -c_W(G_{PO} - G_{PW})$, both valid at low protein concentration. Note that the two relations $N_W^{excess} = c_W(G_{PO} - G_{PW})$ and $N_O^{excess} = c_O(G_{PW} - G_{PO})$ hold, too. Therefore, the change in preferential hydration upon binding discussed in Section II.B corresponds to Eq. (22) multiplied by a factor of minus water concentration.

Kirkwood–Buff theory of biomolecules becomes increasingly popular. While our main focus in this section was the stability of proteins in osmolyte solutions, Kirkwood–Buff theory can be–and has been–used differently. We now list some important and currently active research on protein Kirkwood–Buff theory. Smith *et al.* research Kirkwood–Buff theory and perform computer simulation, as well as study general thermodynamic protein properties (Smith, 2004a,b, 2005, 2006b; Smith *et al.*, 2002; Weerasinghe and Smith, 2003a, 2003b, 2004). Preferential interaction parameters are investigated by Seishi Shimizu (Shimizu, 2004; Shimizu and Boon, 2004; Shimizu and Matubayasi, 2006; Shimizu and Smith, 2004), Michael Schurr (Schurr *et al.*, 2005), and Shulgin and Ruckenstein (Shulgin and Ruckenstein, 2005a, 2006). The latter are also interested in protein solubility in the context of Kirkwood–Buff theory (Shulgin and Ruckenstein, 2005b).

C. Charged Osmolytes

Salts are special osmolytes. Beyond the range of possible ways that net-neutral osmolytes can interact with water and macromolecules, salts pose special challenges. Long-ranged electrostatic interactions act between salt ions and charged groups on many macromolecules, together with other ion-specific interactions such as van der Waals forces. A further complication is that ions cannot diffuse too far from their counterions in solution before electrostatic forces become prohibitively large.

The combined effect of these features result in some unique properties for salts as osmolytes, as we shall now discuss.

Electrostatic ion binding and screening of macromolecules. Most biological macromolecules, including DNA, proteins, and lipid bilayers, are charged on their surfaces. And oppositely charged counterions favorably interact with these surface charges, in addition to any other preferential interaction they may show toward neutral macromolecular surfaces. Charged macromolecules are therefore intimately associated with their counterions in salt solutions. The counterions, in turn, are sometimes referred to as "condensed."

In the presence of low or moderate concentrations of salt, counterions generally act to screen the electrostatic field generated by charged groups on the macromolecules surface in an entirely electrostatic effect that is unrelated to possible additional nonelectrostatic interactions (Linderström-Lang, 1924; Stigter and Dill, 1990; Tanford and Kirkwood, 1957). In such cases, salt effects depend mainly on the ionic strength (see also Chapter 27 by Whitten *et al.*, this volume).

Counterion release is an important mechanism for macromolecular binding. When two macromolecules of (at least locally) opposite signs interact so that they are able to fully neutralize one another, the previously condensed counterions can be released, gaining entropy in solution. This mechanism for association has been called "counterion release" (Record *et al.*, 1978). We can determine the involvement of released ions in the process from the linear dependence of the association free energy with the log of salt activity in solution (rather than with the activity itself, as in Fig. 13). The slope here translates into the thermodynamic extent of association (release of preferentially bound ions): $(\partial \ln K / \partial \ln a_{\pm}) = \Delta N_{\text{salt}}$ (Anderson and Record, 1995; Record *et al.*, 1978). This number of ions will depend on the sum of all preferential forces, electrostatic and nonelectrostatic, that cause ions to preferentially partition near the macromolecules. Often, however, when charges on apposed macromolecules fit well, counterion release can be directly traced to the gain in ions' translational entropy, while the change in direct (coulomb) electrostatic energy is small.

The number of released ions can be measured by conductivity. When the associating macromolecules form a phase that can easily be separated from the bulk solution, it is possible to directly assess the number of released ions into solution

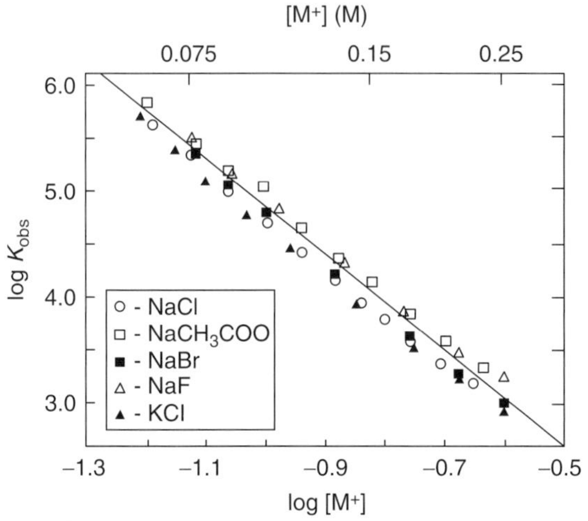

Fig. 13 Dependence of binding constant on monovalent ion concentration for the interaction of the peptide KWK_4-NH_2 with poly(U) with different salts. Linearity points to a release of ca. 4.5 ions upon binding (Mascotti and Lohman, 1990). Copyright: PNAS.

by measuring the change in the conductivity in the supernatant. For example, DNA and cationic lipids have been used extensively as nonviral vehicles for gene delivery *in vivo* and *in vitro* (Podgornik *et al.*, 2004). When DNA and lipids are mixed in solution, aggregates form spontaneously, and for some DNA and lipid rations precipitates form in solution, see Fig. 14. These aggregates show order well defined when interrogated by X-rays. In addition, the conductivity of the supernatant once aggregates are formed can easily be measured. In this way, it has been possible to determine that upon complexation of DNA and positively charged cationic lipids to form macroscopic aggregates, virtually all counterions are released into solution (Wagner *et al.*, 2000).

Salts as osmotic stressors and the Hofmeister series (or "why we add salt to pasta"). Just as preferentially excluded net-neutral osmolytes force macromolecules together, salts too often induce aggregation and stronger association, even between macromolecules that bare no net charge. Onsager and Samaras (1934) showed that salt should be preferentially excluded from macromolecular surfaces. They argued that salt ions approaching a region of low dielectric have to pay in solvation energy (Born self-energy) due to lack of "responsive" polarizable material (water!) in its vicinity in that part of space occupied by the (typically more "oily") macromolecule. This ion exclusion is known to induce protein precipitation (salting-out), as well as aggregation of other macromolecules (Collins and Washabaugh, 1985). It is also likely to be the reason brine is called for when making pasta: to prevent starch from dissolving in the boiling water to form a pasty mess. (Besides this physicist's explanation, the biologist may reason that α-amylase in the saliva requires chloride for its action. Not to mention the role salt

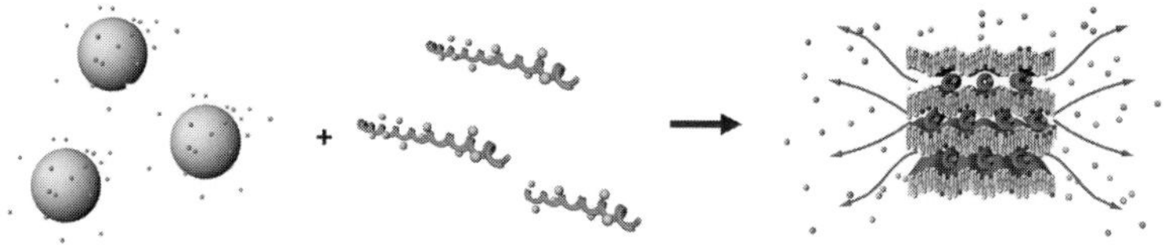

Fig. 14 Schematic of DNA and lipid complexation to form DNA–lipid complexes (or lipoplexes). The main driving force for complexation is the release of all excess counterions (not needed for complex neutralization) to maximize entropy. The released ions can be detected in solution by conductivity measurements, indicating that the release of ions in the process is indeed complete (Harries *et al.*, 1998; Wagner *et al.*, 2000). (See Plate no. 23 in the Color Plate Section.)

has on taste!) The predicted exclusion, however, does not show the right temperature dependence, suggesting that additional forces may also be involved in ion exclusion.

To complicate matters, early on it was realized that different salts seem to promote aggregation (salting-out) to widely different extents, likely due to additional ion-specific interactions. Hofmeister and coworkers were first to observe that the extent of protein precipitation by an ion followed a ranking that was surprisingly insensitive to the type of macromolecule he used (Hofmeister, 1888; Lewith, 1888). This "Hofmeister series" is particularly robust for anions and seems to correlate well with ion polarizability (or size). For example, the halide ions ranked by salting-out propensity: $F^- > Cl^- > Br^- > I^-$ (Collins and Washabaugh, 1985; Hofmeister, 1888; Kunz *et al.*, 2004).

Several forces contribute to the extent of ion exclusion. Several different forces probably act to cause different extents of ion exclusion at macromolecular interfaces. First, ions generally order in the Hofmeister sense according to their polarizability. This possibly implicates dipole (water) to induced-dipole (ion) interactions between salt and water that is structured at the interface differently than in the bulk. The more polarizable the ion, the more it will be drawn to ordered waters at the macromolecular interface. This added attraction results in a contribution to the ion's preferential interaction, making it a worse salting-out agent. Moreover, ions might interact with macromolecules and membranes through dispersion forces, because of the difference in refractive index between water and protein, or lipid. Specific interactions of salt ions with protein side chains also change their preferential interaction. Finally, we recall that preferential interactions are measured *relative* to those interactions in the bulk solution. Different ions are dissolved to different extents in bulk water according to their "structure making" (strongly hydrated ions or "kosmotropes") or "structure braking" (weakly hydrated ions or "chaotropes") properties (Vlachy *et al.*, 2004). Ions that are strongly hydrated will tend to be more excluded from macromolecular surfaces due to their favorable interactions with water in the bulk.

But what about the nature of the macromolecules themselves? While for net-neutral osmolytes the nature of the macromolecular surfaces seems to be of great importance, the Hofmeister series seems to be robust irrespective of many types of macromolecular processes involving different interfaces. These include protein denaturation and stabilization, dimerization, aggregation, and protein ligand interactions.

Salts at macromolecular interfaces sometimes behave as if they are at the air–water interface. In search of a molecular mechanism for the action of small solutes on large macromolecules, Sinanoglu and Abdulnur (1965) suggested that osmolytes might act through the same forces that also cause their exclusion from an air–water interface. This interface is the most simple to consider, because the "macromolecule" involved is simply air, and water structuring at this interface is due to water's ordering properties and the intrinsic asymmetry involved in the interface alone.

Remarkably, salts (and even more specifically, anions) seem to be a particularly good example of solutes that indeed act at macromolecular interfaces just as they would at an air–water interface. The comparison is most striking when interacting neutral macromolecular interfaces are considered, as in the example of interacting cyclodextrin and adamantane measured by ITC (see Section II.B). While the specific complexation free energy is much different for solutions of different salts at some specific concentrations, the effect on binding for different salts *at a particular air–water surface tension* is virtually the same, so that the change in binding free energy $\delta\Delta G^{o} = \Delta\gamma \cdot \Delta S$ is simply related to the change in solutions air–water interfacial tension through an affected surface area upon complexation ΔS. This allows us to think of a macromolecular interface that is buried, as determined from the slope of Fig. 15 to be ca. 200 Å^2.

Building on the same principles, the Record laboratory has recently provided a model that captures surface tension in salt solutions based on a model that was previously employed to describe protein solvation (Pegram and Record, 2006). While it is clear that similar principles apply to both kinds of interfacial tensions at the protein–solution and the solution–air interface, it is important to keep in mind that in case of organic osmolytes the two interfacial tensions are very different in magnitude, and even in sign (Auton *et al.*, 2006). The surface tension at the air–water interface is therefore not a good predictor for *m*-values for such organic osmolytes.

V. Prospects

The largely neglected effect of cellular solutes on biological processes has recently been recognized. The control of interaction between macromolecules can be achieved not only by compounds that associate strongly with macromolecules but also by abundant small molecules that interact nonspecifically. This realization allows us to use these osmolytes to direct reactions in the test tube as well as in real life. It may also help us understand the basis of certain pathologies *in vivo*. All these can be achieved by understanding the basis of osmolyte–macromolecule interactions in aqueous environments.

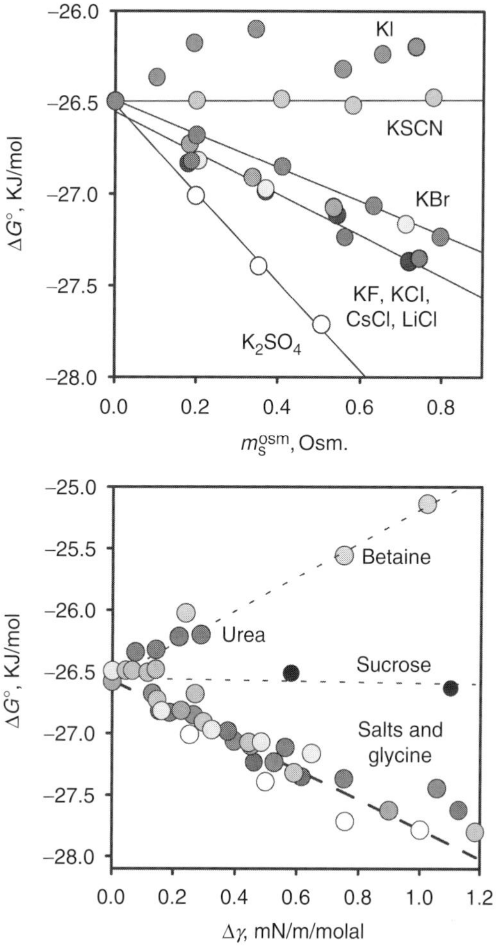

Fig. 15 Top panel shows how different salts show different preferential interactions with cyclodextrin and adamantane interacting surfaces at the same salt concentration (osmolalities) according to their Hofmeister ranking. However, all salts show the same effect on complexation free energy when we consider the way in which they affect the air–water interfacial tension (bottom panel) reflecting an affected surface area of 200 Å^2 that is buried in complexation (Harries *et al.*, 2005). Similar ranking does not generally hold for other net-neutral osmolytes (Auton *et al.*, 2006). (See Plate no. 24 in the Color Plate Section.)

Acknowledgments

We thank Adrian Parsegian and the people at the Laboratory of Physical and Structural Biology at the NIH as well as Wayne Bolen for numerous stimulating discussions. Many thanks to Nina Sidorova and Don Rau for their extensive contribution to Section II.B, Sergey Bezrukov for a figure and for his input to Section III.D, Horia Petrache for a figure in Section II.D, and to Rohit Pappu and Hoang Tran for providing example data for the structure of a denatured state (Fig. 10). D.H. acknowledges support from the Israeli Council of Higher Education through an Alon Fellowship, and J.R. thanks the support by a fellowship from the W. M. Keck Foundation to the Gulf Coast Consortia through the Keck Center of Computational and Structural Biology, and by NIH (R01GM049760).

References

Abou-Aiad, T., Becker, U., Biedenkap, R., Brengelmann, R., Elsebrock, R., Hinz, H. J., and Stockhausen, M. (1997). Dielectric relaxation of aqueous solutions of ribonuclease A in the absence and presence of urea. *Ber. Bunsen-Ges. Phys. Chem. Chem. Phys.* **101**(12), 1921–1927.

Anderson, C. F., Courtenay, E. S., and Record, M. T. (2002). Thermodynamic expressions relating different types of preferential interaction coefficients in solutions containing two solute components. *J. Phys. Chem. B* **106**(2), 418–433.

Anderson, C. F., and Record, M. T., Jr. (1995). Salt-nucleic acid interactions. *Annu. Rev. Phys. Chem.* **46,** 657–700.

Arakawa, T., and Timasheff, S. N. (1985). Theory of protein solubility. *Methods Enzymol.* **114,** 49–77.

Arakawa, T., and Tsumoto, K. (2003). The effects of arginine on refolding of aggregated proteins: Not facilitate refolding, but suppress aggregation. *Biochem. Biophy. Res. Commun.* **304**(1), 148–152.

Archer, D. G. (1992). Thermodynamic properties of the NaCl + H_2O system. 2. Thermodynamic properties of NaCl(Aq), NaCl*$2H_2O$(Cr), and phase-equilibria. *J. Phys. Chem. Ref. Data* **21**(4), 793–829.

Auton, M., and Bolen, D. W. (2004). Additive transfer free energies of the peptide backbone unit that are independent of the model compound and the choice of concentration scale. *Biochemistry* **43**(5), 1329–1342.

Auton, M., and Bolen, D. W. (2005). Predicting the energetics of osmolyte-induced protein folding/unfolding. *Proc. Natl. Acad. Sci. USA* **102**(42), 15065–15068.

Auton, M., Ferreon, A. C., and Bolen, D. W. (2006). Metrics that differentiate the origins of osmolyte effects on protein stability: A test of the surface tension proposal. *J. Mol. Biol.* **361**(5), 983–992.

Barteri, M., Gaudiano, M. C., and Santucci, R. (1996). Influence of glycerol on the structure and stability of ferric horse heart myoglobin: A SAXS and circular dichroism study. *Biochim. Biophys. Acta* **1295**(1), 51–58.

Baskakov, I., and Bolen, D. W. (1998). Forcing thermodynamically unfolded proteins to fold. *J. Biol. Chem.* **273**(9), 4831–4834.

Batchelor, J. D., Olteanu, A., Tripathy, A., and Pielak, G. J. (2004). Impact of protein denaturants and stabilizers on water structure. *J. Am. Chem. Soc.* **126**(7), 1958–1961.

Ben-Naim, A. (1977). Inversion of Kirkwood–Buff theory of solutions—Application to water-ethanol system. *J. Chem. Phys.* **67**(11), 4884–4890.

Ben-Naim, A. (1988). Theory of preferential solvation of nonelectrolytes. *Cell Biophys.* **12,** 255–269.

Ben-Naim, A. (1992). "Statistical Themodynamics for Chemists and Biochemists." Plenum, NewYork.

Betting, H., Häckel, M., Hinz, H. J., and Stockhausen, M. (2001). Spectroscopic evidence for the preferential hydration of RNase A in glycerol-water mixtures: Dielectric relaxation studies. *Phys. Chem. Chem. Phys.* **3**(9), 1688–1692.

Bezrukov, S. M., and Vodyanoy, I. (1993). Probing alamethicin channels with water-soluble polymers. Effect on conductance of channel states. *Biophys. J.* **64**(1), 16–25.

Boehm, K., Rösgen, J., and Hinz, H. J. (2006). Pressure-modulated differential scanning calorimetry. An approach to the continuous, simultaneous determination of heat capacities and expansion coefficients. *Anal. Chem.* **78**(4), 984–990.

Bolen, D. W. (2004). Effects of naturally occurring osmolytes on protein stability and solubility: Issues important in protein crystallization. *Methods* **34**(3), 312–322.

Bolen, D. W., and Baskakov, I. V. (2001). The osmophobic effect: Natural selection of a thermodynamic force in protein folding. *J. Mol. Biol.* **310**(5), 955–963.

Bolen, D. W., and Yang, M. (2000). Effects of guanidine hydrochloride on the proton inventory of proteins: Implications on interpretations of protein stability. *Biochemistry* **39**(49), 15208–15216.

Bomhoff, G., Sloan, K., McLain, C., Gogol, E. P., and Fisher, M. T. (2006). The effects of the flavonoid baicalein and osmolytes on the Mg (2+) accelerated aggregation/fibrillation of carboxymethylated bovine 1SS-alpha-lactalbumin. *Arch. Biochem. Biophys.* **453**(1), 75–86.

Casassa, E. F., and Eisenberg, H. (1964). Thermodynamic analysis of multicomponent solutions. *Adv. Protein Chem.* **19,** 287–395.

Cavagnero, S., Debe, D. A., Zhou, Z. H., Adams, M. W., and Chan, S. I. (1998). Kinetic role of electrostatic interactions in the unfolding of hyperthermophilic and mesophilic rubredoxins. *Biochemistry* **37**(10), 3369–3376.

Chalikian, T. V. (2001). Structural thermodynamics of hydration. *J. Phys. Chem. B* **105**(50), 12566–12578.

Clarke, E. C. W., and Glew, D. N. (1985). Evaluation of the thermodynamic functions for aqueous sodium-chloride from equilibrium and calorimetric measurements below 154 °C. *J. Phys. Chem. Ref. Data* **14**(2), 489–610.

Cohn, E. J., and Edsall, J. T. (1943). "Proteins, Amino Acids, and Peptides as Ions and Dipolar Ions." Reinhold Publishing Corp., New York.

Collins, K. D. (2006). Ion hydration: Implications for cellular function, polyelectrolytes, and protein crystallization. *Biophys. Chem.* **119**(3), 271–281.

Collins, K. D., and Washabaugh, M. W. (1985). The hofmeister effect and the behavior of water at interfaces. *Q. Rev. Biophys.* **18**(4), 323–422.

Colombo, M. F., Rau, D. C., and Parsegian, V. A. (1992). Protein solvation in allosteric regulation: A water effect on hemoglobin. *Science* **256**(5057), 655–659.

Courtenay, E. S., Capp, M. W., Anderson, C. F., and Record, M. T., Jr. (2000). Vapor pressure osmometry studies of osmolyte-protein interactions: Implications for the action of osmoprotectants *in vivo* and for the interpretation of "osmotic stress" experiments *in vitro*. *Biochemistry* **39**(15), 4455–4471.

Courtenay, E. S., Capp, M. W., and Record, M. T., Jr. (2001). Thermodynamics of interactions of urea and guanidinium salts with protein surface: Relationship between solute effects on protein processes and changes in water-accessible surface area. *Protein Sci.* **10**(12), 2485–2497.

Creamer, T. P., Srinivasan, R., and Rose, G. D. (1995). Modeling unfolded states of peptides and proteins. *Biochemistry* **34**(50), 16245–16250.

Creamer, T. P., Srinivasan, R., and Rose, G. D. (1997). Modeling unfolded states of proteins and peptides. II. Backbone solvent accessibility. *Biochemistry* **36**(10), 2832–2835.

del Pino, I. M. Plaza, and Sanchez-Ruiz, J. M. (1995). An osmolyte effect on the heat capacity change for protein folding. *Biochemistry* **34**(27), 8621–8630.

Ebel, C., Eisenberg, H., and Ghirlando, R. (2000). Probing protein-sugar interactions. *Biophys. J.* **78**(1), 385–393.

Eisenberg, H. (1994). Protein and nucleic-acid hydration and cosolvent interactions—establishment of reliable base-line values at high cosolvent concentrations. *Biophys. l Chem.* **53**(1–2), 57–68.

Felitsky, D. J., Cannon, J. G., Capp, M. W., Hong, J., Van Wynsberghe, A. W., Anderson, C. F., and Record, M. T., Jr. (2004). The exclusion of glycine betaine from anionic biopolymer surface: Why glycine betaine is an effective osmoprotectant but also a compatible solute. *Biochemistry* **43**(46), 14732–14743.

Felitsky, D. J., and Record, M. T., Jr. (2004). Application of the local-bulk partitioning and competitive binding models to interpret preferential interactions of glycine betaine and urea with protein surface. *Biochemistry* **43**(28), 9276–9288.

Ferraris, J. D., Williams, C. K., Persaud, P., Zhang, Z., Chen, Y., and Burg, M. B. (2002). Activity of the TonEBP/OREBP transactivation domain varies directly with extracellular NaCl concentration. *Proc. Natl. Acad. Sci. USA* **99**(2), 739–744.

Ferreon, A. C., and Bolen, D. W. (2004). Thermodynamics of denaturant-induced unfolding of a protein that exhibits variable two-state denaturation. *Biochemistry* **43**(42), 13357–13369.

Ferreon, A. C., Ferreon, J. C., Bolen, D. W., and Rösgen, J. (2007). Protein phase diagrams II: Nonideal behavior of biochemical reactions in the presence of osmolytes. *Biophys. J.* **92**(1), 245–256.

Finkelstein, A. V., and Galzitskaya, O. V. (2004). Physics of protein folding. *Phys. Life Rev.* **1,** 23–56.

Fried, M. G., Stickle, D. F., Smirnakis, K. V., Adams, C., MacDonald, D., and Lu, P. (2002). Role of hydration in the binding of lac repressor to DNA. *J. Biol. Chem.* **277**(52), 50676–50682.

Fukada, H., Sturtevant, J. M., and Quiocho, F. A. (1983). Thermodynamics of the binding of L-arabinose and of D-galactose to the L-arabinose-binding protein of *Escherichia coli*. *J. Biol. Chem.* **258**(21), 13193–13198.

Galinski, E. A. (1995). Osmoadaptation in bacteria. *Adv. Microb. Physiol.* **37,** 272–328.

Garcia, P., Serrano, L., Durand, D., Rico, M., and Bruix, M. (2001). NMR and SAXS characterization of the denatured state of the chemotactic protein CheY: Implications for protein folding initiation. *Protein Sci.* **10**(6), 1100–1112.

Garcia-Perez, A., and Burg, M. B. (1991). Renal medullary organic osmolytes. *Physiol. Rev.* **71**(4), 1081–1115.

Garner, M. M., and Rau, D. C. (1995). Water release associated with specific binding of gal repressor. *EMBO J.* **14**(6), 1257–1263.

Gelbart, W. M., Bruinsma, R. F., Pincus, P. A., and Parsegian, V. A. (2000). DNA-inspired electrostatics. *Phys. Today* **53**(9), 38–44.

Gibbs, J. W. (1876/78). On the equilibrium of heterogeneous substances. *Trans. Connect. Acad.* **3,** 108–248, 343–542.

Giletto, A., and Pace, C. N. (1999). Buried, charged, non-ion-paired aspartic acid 76 contributes favorably to the conformational stability of ribonuclease T1. *Biochemistry* **38**(40), 13379–13384.

Goldbaum, F. A., Schwarz, F. P., Eisenstein, E., Cauerhff, A., Mariuzza, R. A., and Poljak, R. J. (1996). The effect of water activity on the association constant and the enthalpy of reaction between lysozyme and the specific antibodies D1.3 and D44.1. *J. Mol. Recognit.* **9**(1), 6–12.

Goldbeck, R. A., Paquette, S. J., and Kliger, D. S. (2001). The effect of water on the rate of conformational change in protein allostery. *Biophys. J.* **81**(5), 2919–2934.

Goldenberg, D. P. (2003). Computational simulation of the statistical properties of unfolded proteins. *J. Mol. Biol.* **326**(5), 1615–1633.

Golovanov, A. P., Hautbergue, G. M., Wilson, S. A., and Lian, L. Y. (2004). A simple method for improving protein solubility and long-term stability. *J. Am. Chem. Soc.* **126**(29), 8933–8939.

Granata, V., Palladino, P., Tizzano, B., Negro, A., Berisio, R., and Zagari, A. (2006). The effect of the osmolyte trimethylamine N-oxide on the stability of the prion protein at low pH. *Biopolymers* **82**(3), 234–240.

Green, A. A. (1931a). Studies in the physical chemistry of the proteins. IX. The effect of electrolytes on the solubility of hemoglobin in solutions of varying hydrogen ion activity with a note on the comparable behavior of casein. *J. Biol. Chem.* **93,** 517–542.

Green, A. A. (1931b). Studies in the physical chemistry of the proteins. VIII. The solubility of hemoglobin in concentrated salt solutions. A study of the salting out of proteins. *J. Biol. Chem.* **93**(2), 495–516.

Green, A. A. (1932). Studies in the physical chemistry of the proteins. X. The solutbility of hemoglobin in solutions of chlorides and sulfates of varying concentration. *J. Biol. Chem.* **95**(1), 47–66.

Green, A. A., Cohn, E. J., and Blanchard, M. H. (1935). Studies in the physical chemistry of the proteins. XII. The solubility of human hemoglobin in concentrated salt solutions. *J. Biol. Chem.* **109**(2), 631–634.

Greene, R. F., Jr., and Pace, C. N. (1974). Urea and guanidine hydrochloride denaturation of ribonuclease, lysozyme, alpha-chymotrypsin, and beta-lactoglobulin. *J. Biol. Chem.* **249**(17), 5388–5393.

Häckel, M., Konno, T., and Hinz, H. (2000). A new alternative method to quantify residual structure in 'unfolded' proteins. *Biochim. Biophys. Acta* **1479**(1–2), 155–165.

Hade, E. P. K., and Tanford, C. (1967). Isopiestic compositions as a measure of preferential interactions of macromolecules in two-component solvents. Application to proteins in concentrated aqueous cesium chloride and guanidine hydrochloride. *J. Am. Chem. Soc.* **89**(19), 5034–5040.

Hardegree, S. P., and Emmerich, W. E. (1990). Effect of polyethylene-glycol exclusion on the water potential of solution-saturated filter-paper. *Plant Physiol.* **92**(2), 462–466.

Harned, H. S., and Owen, B. B. (1943). "The Physical Chemistry of Electrolytic Solutions." Reinhold Publishing, New York.

Harries, D., May, S., Gelbart, W. M., and Ben-Shaul, A. (1998). Structure, stability, and thermodynamics of lamellar DNA-lipid complexes. *Biophys. J.* **75**(1), 159–173.

Harries, D., Rau, D. C., and Parsegian, V. A. (2005). Solutes probe hydration in specific association of cyclodextrin and adamantane. *J. Am. Chem. Soc.* **127**(7), 2184–2190.

Hawley, S. A. (1971). Reversible pressure–temperature denaturation of chymotrypsinogen. *Biochemistry* **10**(13), 2436–2442.

Henkels, C. H., Kurz, J. C., Fierke, C. A., and Oas, T. G. (2001). Linked folding and anion binding of the Bacillus subtilis ribonuclease P protein. *Biochemistry* **40**(9), 2777–2789.

Henkels, C. H., and Oas, T. G. (2005). Thermodynamic characterization of the osmolyte- and ligand-folded states of bacillus subtilis ribonuclease P protein. *Biochemistry* **44**(39), 13014–13026.

Heremans, K. (1982). High pressure effects on proteins and other biomolecules. *Annu. Rev. Biophys. Bioeng.* **11,** 1–21.

Hochachka, P. W., and Somero, G. N. (2002). "Biochemical Adaptation. Mechanism and Process in Physiological Evolution." Oxford University Press, Oxford.

Hofmeister, F. (1888). Zur lehre von der wirkung der salze. Zweite mittheilung. *Arch. Exp. Pathol. Pharmakol.* **24,** 247–360.

Holthauzen, L. M., and Bolen, D. W. (2007). Mixed osmolytes: The degree to which one osmolyte affects the protein stabilizing ability of another. *Protein Sci.* **16**(2), 293–298.

Hong, J., Capp, M. W., Anderson, C. F., and Record, M. T. (2003). Preferential interactions in aqueous solutions of urea and KCl. *Biophys. Chem.* **105**(2–3), 517–532.

Hong, J., Capp, M. W., Anderson, C. F., Saecker, R. M., Felitsky, D. J., Anderson, M. W., and Record, M. T. (2004). Preferential interactions of glycine betaine and of urea with DNA: Implications for DNA hydration and for effects of these solutes on DNA stability. *Biochemistry* **43**(46), 14744–14758.

Hong, J., Capp, M. W., Saecker, R. M., and Record, M. T., Jr. (2005). Use of urea and glycine betaine to quantify coupled folding and probe the burial of DNA phosphates in lac repressor-lac operator binding. *Biochemistry* **44**(51), 16896–16911.

Hultgren, A., and Rau, D. C. (2004). Exclusion of alcohols from spermidine-DNA assemblies: Probing the physical basis of preferential hydration. *Biochemistry* **43**(25), 8272–8280.

Ignatova, Z., and Gierasch, L. M. (2006). Inhibition of protein aggregation *in vitro* and *in vivo* by a natural osmoprotectant. *Proc. Natl. Acad. Sci. USA* **103**(36), 13357–13361.

Jeon, U. S., Kim, J. A., Sheen, M. R., and Kwon, H. M. (2006). How tonicity regulates genes: Story of TonEBP transcriptional activator. *Acta Physiol. (Oxf)* **187**(1–2), 241–247.

Kauzmann, W. (1976). Pressure effects on water and the validity of theories of water behavior. *Colloq. Int. C.N.R.S.* **246,** 63–71.

Kervinen, J., Ma, H., Bayoumy, S., Schubert, C., Milligan, C., Lewandowski, F., Moriarty, K., Desjarlais, R. L., Ramachandren, K., Wang, H., Harris, C. A., Grasberger, B., *et al.* (2006). Effect of construct design on MAPKAP kinase-2 activity, thermodynamic stability and ligand-binding affinity. *Arch. Biochem. Biophys.* **449**(1–2), 47–56.

Kidokoro, S., Miki, Y., and Wada, A. (1990). Physical and biological stability of globular proteins. *In* "Protein Structural Analysis, Folding and Design" (M. Hatano, ed.), pp. 75–92. Elsevier, Amsterdam.

Kiefhaber, T., Kohler, H. H., and Schmid, F. X. (1992). Kinetic coupling between protein folding and prolyl isomerization. I. Theoretical models. *J. Mol. Biol.* **224**(1), 217–229.

Kiefhaber, T., and Schmid, F. X. (1992). Kinetic coupling between protein folding and prolyl isomerization. II. Folding of ribonuclease A and ribonuclease T1. *J. Mol. Biol.* **224**(1), 231–240.

Kirkwood, J. G., and Buff, F. P. (1951). The statistical mechanical theory of solutions. I. *J. Chem. Phys.* **19**(6), 774–777.

Kiyosawa, K. (2003). Theoretical and experimental studies on freezing point depression and vapor pressure deficit as methods to measure osmotic pressure of aqueous polyethylene glycol and bovine serum albumin solutions. *Biophys. Chem.* **104**(1), 171–188.

Kohn, J. E., Millett, I. S., Jacob, J., Zagrovic, B., Dillon, T. M., Cingel, N., Dothager, R. S., Seifert, S., Thiyagarajan, P., Sosnick, T. R., Hasan, M. Z., Pande, V. S., *et al.* (2004). Random-coil behavior and the dimensions of chemically unfolded proteins. *Proc. Natl. Acad. Sci. USA* **101**(34), 12491–12496.

Kornblatt, J. A., and Kornblatt, M. J. (2002). The effects of osmotic and hydrostatic pressures on macromolecular systems. *Biochim. Biophys. Acta* **1595**(1–2), 30–47.

Kovrigin, E. L., and Potekhin, S. A. (1997). Preferential solvation changes upon lysozyme heat denaturation in mixed solvents. *Biochemistry* **36**(30), 9195–9199.

Kozlov, M. M., Leikin, S., and Rand, R. P. (1994). Bending, hydration and interstitial energies quantitatively account for the hexagonal-lamellar-hexagonal reentrant phase transition in dioleoylphosphatidylethanolamine. *Biophys. J.* **67**(4), 1603–1611.

Kumar, R., Lee, J. C., Bolen, D. W., and Thompson, E. B. (2001). The conformation of the glucocorticoid receptor af1/tau1 domain induced by osmolyte binds co-regulatory proteins. *J. Biol. Chem.* **276**(21), 18146–18152.

Kunz, W., Lo Nostro, P., and Ninham, B. W. (2004). The present state of affairs with Hoffmeister effects. *Curr. Opin. Colloid Interface Sci.* **9**(1–2), 1–18.

Kwon, E. D., Dooley, J. A., Jung, K. Y., Andrews, P. M., Garcia-Perez, A., and Burg, M. B. (1996). Organic osmolyte distribution and levels in the mammalian urinary bladder in diuresis and antidiuresis. *Am. J. Physiol.* **271**(1 Pt 2), F230–233.

Lapanje, S., and Tanford, C. (1967). Proteins as random coils. IV. Osmotic pressures, second virial coefficients, and unperturbed dimensions in 6M guanidine hydrochloride. *J. Am. Chem. Soc.* **89**(19), 5030–5033.

Lee, B., and Richards, F. M. (1971). The interpretation of protein structures: Estimation of static accessibility. *J. Mol. Biol.* **55**(3), 379–400.

Leikin, S., Rau, D. C., and Parsegian, V. A. (1991). Measured entropy and enthalpy of hydration as a function of distance between DNA double helices. *Phys. Rev. A* **44**(8), 5272–5278.

Lewith, S. (1888). Zur lehre von der wirkung der salze. *Arch. exp. Pathol. Pharmakol.* **24,** 1–16.

Lin, L. N., Brandts, J. F., Brandts, J. M., and Plotnikov, V. (2002). Determination of the volumetric properties of proteins and other solutes using pressure perturbation calorimetry. *Anal. Biochem.* **302**(1), 144–160.

Lin, M. F., Williams, C., Murray, M. V., Conn, G., and Ropp, P. A. (2004). Ion chromatographic quantification of cyanate in urea solutions: Estimation of the efficiency of cyanate scavengers for use in recombinant protein manufacturing. *J. Chromatogr. B Analyt. Technol. Biomed. Life Sci.* **803**(2), 353–362.

Linderström-Lang, K. (1924). On the ionization of proteins. *C. R. Trav. Lab. Carlsberg Serie chimique* **15**(7), 1–29.

Makhatadze, G. I. (1999). Thermodynamics of protein interactions with urea and guanidinium hydrochloride. *J. Phys. Chem. B* **103**(23), 4781–4785.

Mascotti, D. P., and Lohman, T. M. (1990). Thermodynamic extent of counterion release upon binding oligolysines to single-stranded nucleic acids. *Proc. Natl. Acad. Sci. USA* **87**(8), 3142–3146.

Matulis, D., Kranz, J. K., Salemme, F. R., and Todd, M. J. (2005). Thermodynamic stability of carbonic anhydrase: Measurements of binding affinity and stoichiometry using thermofluor. *Biochemistry* **44**(13), 5258–5266.

Mello, C. C., and Barrick, D. (2003). Measuring the stability of partly folded proteins using TMAO. *Protein Sci.* **12**(7), 1522–1529.

Mertz, E. L., and Leikin, S. (2004). Interactions of inorganic phosphate and sulfate anions with collagen. *Biochemistry* **43**(47), 14901–14912.

Miyakawa, H., Woo, S. K., Dahl, S. C., Handler, J. S., and Kwon, H. M. (1999). Tonicity-responsive enhancer binding protein, a rel-like protein that stimulates transcription in response to hypertonicity. *Proc. Natl. Acad. Sci. USA* **96**(5), 2538–2542.

Morar, A. S., Wang, X., and Pielak, G. J. (2001). Effects of crowding by mono-, di-, and tetrasaccharides on cytochrome c-cytochrome c peroxidase binding: Comparing experiment to theory. *Biochemistry* **40**(1), 281–285.

Myers, J. K., Pace, C. N., and Scholtz, J. M. (1995). Denaturant m values and heat capacity changes: Relation to changes in accessible surface areas of protein unfolding. *Protein Sci.* **4**(10), 2138–2148.

Navati, M. S., Ray, A., Shamir, J., and Friedman, J. M. (2004). Probing solvation-shell hydrogen binding in glassy and sol-gel matrixes through vibronic sideband luminescence spectroscopy. *J. Phys. Chem. B* **108**(4), 1321–1327.

Neuhofer, W., and Beck, F. X. (2005). Cell survival in the hostile environment of the renal medulla. *Annu. Rev. Physiol.* **67,** 531–555.

Newman, M., Strzelecka, T., Dorner, L. F., Schildkraut, I., and Aggarwal, A. K. (1995). Structure of Bam HI endonuclease bound to DNA: Partial folding and unfolding on DNA binding. *Science* **269**(5224), 656–663.

Nozaki, Y., and Tanford, C. (1967). Proteins as random coils. II. Hydrogen ion titration curve of ribonuclease in 6M guanidinium hydrochloride. *J. Am. Chem. Soc.* **89**(4), 742–749.

Ogasahara, K., Nakamura, M., Nakura, S., Tsunasawa, S., Kato, I., Yoshimoto, T., and Yutani, K. (1998). The unusually slow unfolding rate causes the high stability of pyrrolidone carboxyl peptidase from a hyperthermophile, Pyrococcus furiosus: Equilibrium and kinetic studies of guanidine hydrochloride-induced unfolding and refolding. *Biochemistry* **37**(50), 17537–17544.

Ogasawara, M., Nakamura, T., Koyama, I., Nemoto, M., and Yoshida, T. (1993). Reactivity of taurine with aldehydes and its physiological role. *Chem. Pharm. Bull. (Tokyo)* **41**(12), 2172–2175.

Onsager, L., and Samaras, N. N. T. (1934). The surface tension of Debye-Hückel electrolytes. *J. Chem. Phys.* **2**(8), 528–536.

Pace, C. N., and Shaw, K. L. (2000). Linear extrapolation method of analyzing solvent denaturation curves. *Proteins* **4**(Suppl.), 1–7.

Pace, C. N., Trevino, S., Prabhakaran, E., and Scholtz, J. M. (2004). Protein structure, stability and solubility in water and other solvents. *Philos. Trans. R. Soc. Lond. B Biol. Sci.* **359**(1448), 1225–1234; discussion 1234–1235.

Parsegian, V. A. (2002). Protein-water interactions. *Int. Rev. Cytol.* **215,** 1–31.

Parsegian, V. A., Rand, R. P., Fuller, N. L., and Rau, D. C. (1986). Osmotic stress for the direct measurement of intermolecular forces. *Methods Enzymol.* **127,** 400–416.

Parsegian, V. A., Rand, R. P., and Rau, D. C. (1995). Macromolecules and water: Probing with osmotic stress. *Methods Enzymol.* **259,** 43–94.

Parsegian, V. A., Rand, R. P., and Rau, D. C. (2000). Osmotic stress, crowding, preferential hydration, and binding: A comparison of perspectives. *Proc. Natl. Acad. Sci. USA* **97**(8), 3987–3992.

Patel, C. N., Noble, S. M., Weatherly, G. T., Tripathy, A., Winzor, D. J., and Pielak, G. J. (2002). Effects of molecular crowding by saccharides on alpha-chymotrypsin dimerization. *Protein Sci.* **11**(5), 997–1003.

Pegram, L. M., and Record, M. T., Jr. (2006). Partitioning of atmospherically relevant ions between bulk water and the water/vapor interface. *Proc. Natl. Acad. Sci. USA* **103**(39), 14278–14281.

Petrache, H. I., Kimchi, I., Harries, D., and Parsegian, V. A. (2005). Measured depletion of ions at the biomembrane interface. *J. Am. Chem. Soc.* **127**(33), 11546–11547.

Petrache, H. I., Tristram-Nagle, S., Harries, D., Kucerka, N., Nagle, J. F., and Parsegian, V. A. (2006). Swelling of phospholipids by monovalent salt. *J. Lipid Res.* **47**(2), 302–309.

Podgornik, R., Harries, D., Strey, H. H., and Parsegian, V. A. (2004). Molecular interactions in lipids, DNA, and lipid-DNA complexes. *In* "Gene Therapy—Therapeutic Mechanisms and Strategies" (N. S. Templeton, ed.), 2nd edn., pp. 301–332. Marcel Dekker, NewYork.

Podgornik, R., Rau, D. C., and Parsegian, V. A. (1994). Parametrization of direct and soft steric-undulatory forces between DNA double helical polyelectrolytes in solutions of several different anions and cations. *Biophys. J.* **66**(4), 962–971.

Poklar, N., Petrovcic, N., Oblak, M., and Vesnaver, G. (1999). Thermodynamic stability of ribonuclease A in alkylurea solutions and preferential solvation changes accompanying its thermal denaturation: A calorimetric and spectroscopic study. *Protein Sci.* **8**(4), 832–840.

Poon, J., Bailey, M., Winzor, D. J., Davidson, B. E., and Sawyer, W. H. (1997). Effects of molecular crowding on the interaction between DNA and the Escherichia coli regulatory protein TyrR. *Biophys. J.* **73**(6), 3257–3264.

Qu, Y., Bolen, C. L., and Bolen, D. W. (1998). Osmolyte-driven contraction of a random coil protein. *Proc. Natl. Acad. Sci. USA* **95**(16), 9268–9273.

Rand, R. P., Fuller, N. L., Gruner, S. M., and Parsegian, V. A. (1990). Membrane curvature, lipid segregation, and structural transitions for phospholipids under dual-solvent stress. *Biochemistry* **29**(1), 76–87.

Rand, R. P., and Parsegian, V. A. (1992). The forces between interacting bilayer membranes and the hydration of phospholipid assemblies. *In* "The Structure of Biological Membranes" (P. Yeagle, ed.), pp. 251–306. CRC Press, Boca Raton.

Rau, D. C., and Parsegian, V. A. (1992). Direct measurement of temperature-dependent solvation forces between DNA double helices. *Biophys. J.* **61,** 260–271.

Record, M. T., Jr., Anderson, C. F., and Lohman, T. M. (1978). Thermodynamic analysis of ion effects on the binding and conformational equilibria of proteins and nucleic acids: The roles of ion association or release, screening, and ion effects on water activity. *Q. Rev. Biophys.* **11**(2), 103–178.

Robinson, C. R., and Sligar, S. G. (1998). Changes in solvation during DNA binding and cleavage are critical to altered specificity of the EcoRI endonuclease. *Proc. Natl. Acad. Sci. USA* **95**(5), 2186–2191.

Roche, C. J., Guo, F., and Friedman, J. M. (2006). Molecular level probing of preferential hydration and its modulation by osmolytes through the use of pyranine complexed to hemoglobin. *J. Biol. Chem.* **281**(50), 38757–38768.

Rosenberg, J. M. (1991). Structure and function of restriction endonucleases. *Curr. Opin. Struct. Biol.* **1**(1), 104–113.

Rosengarth, A., Rösgen, J., and Hinz, H. J. (1999). Slow unfolding and refolding kinetics of the mesophilic Rop wild-type protein in the transition range. *Eur. J. Biochem.* **264**(3), 989–995.

Rösgen, J., and Hinz, H. J. (2000). Response functions of proteins. *Biophys. Chem.* **83**(1), 61–71.

Rösgen, J., and Hinz, H. J. (2003). Phase diagrams: A graphical representation of linkage relations. *J. Mol. Biol.* **328**(1), 255–271.

Rösgen, J., and Hinz, H. J. (2006). Pressure-modulated differential scanning calorimetry: Theoretical background. *Anal. Chem.* **78**(4), 991–996.

Rösgen, J., Pettitt, B. M., and Bolen, D. W. (2004a). Uncovering the basis for nonideal behavior of biological molecules. *Biochemistry* **43**(45), 14472–14484.

Rösgen, J., Pettitt, B. M., and Bolen, D. W. (2005). Protein folding, stability, and solvation structure in osmolyte solutions. *Biophys. J.* **89**(5), 2988–2997.

Rösgen, J., Pettitt, B. M., and Bolen, D. W. (2007). An analysis of the molecular origin of osmolyte-dependent protein stability. *Protein Sci.* **16**(4), 733–743.

Rösgen, J., Pettitt, B. M., Perkyns, J., and Bolen, D. W. (2004b). Statistical thermodynamic approach to the chemical activities in two-component solutions. *J. Phys. Chem. B* **108**(6), 2048–2055.

Royer, W. E., Jr., Pardanani, A., Gibson, Q. H., Peterson, E. S., and Friedman, J. M. (1996). Ordered water molecules as key allosteric mediators in a cooperative dimeric hemoglobin. *Proc. Natl. Acad. Sci. USA* **93**(25), 14526–14531.

Russo, A. T., Rösgen, J., and Bolen, D. W. (2003). Osmolyte effects on kinetics of FKBP12 C22A folding coupled with prolyl isomerization. *J. Mol. Biol.* **330**(4), 851–866.

Santoro, M. M., and Bolen, D. W. (1988). Unfolding free energy changes determined by the linear extrapolation method. 1. Unfolding of phenylmethanesulfonyl alpha-chymotrypsin using different denaturants. *Biochemistry* **27**(21), 8063–8068.

Santoro, M. M., and Bolen, D. W. (1992). A test of the linear extrapolation of unfolding free energy changes over an extended denaturant concentration range. *Biochemistry* **31**(20), 4901–4907.

Scaramozzino, F., Peterson, D. W., Farmer, P., Gerig, J. T., Graves, D. J., and Lew, J. (2006). TMAO promotes fibrillization and microtubule assembly activity in the C-terminal repeat region of tau. *Biochemistry* **45**(11), 3684–3691.

Scatena, L. F., and Richmond, G. L. (2004). Isolated molecular ion solvation at an oil/water interface investigated by vibrational sum-frequency spectroscopy. *J. Phys. Chem. B* **108**(33), 12518–12528.

Schein, C. H. (1993). Solubility and secretability. *Curr. Opin. Biotechnol.* **4**(4), 456–461.

Schellman, J. A. (1975). Macromolecular binding. *Biopolymers* **14**(5), 999–1018.

Schellman, J. A. (1978). Solvent denaturation. *Biopolymers* **17**(5), 1305–1322.

Schellman, J. A. (1987). Selective binding and solvent denaturation. *Biopolymers* **26**(4), 549–559.

Schellman, J. A. (1994). The thermodynamics of solvent exchange. *Biopolymers* **34**(8), 1015–1026.

Schurr, J. M., Rangel, D. P., and Aragon, S. R. (2005). A contribution to the theory of preferential interaction coefficients. *Biophys. J.* **89**(4), 2258–2276.

Segel, D. J., Fink, A. L., Hodgson, K. O., and Doniach, S. (1998). Protein denaturation: A small-angle X-ray scattering study of the ensemble of unfolded states of cytochrome c. *Biochemistry* **37**(36), 12443–12451.

Setschenow, J. (1889). Über die konstitution der salzlösungen auf grund ihres verhaltens zu kohlensäure. *Z. Phys. Chem.* **4,** 117–125.

Setschenow, J. (1892). Action de l'acide carbonique sur les solutions des sels a acides forts. *Ann. chim. phys.* **6**(25), 226–270.

Shaw, K. L., Grimsley, G. R., Yakovlev, G. I., Makarov, A. A., and Pace, C. N. (2001). The effect of net charge on the solubility, activity, and stability of ribonuclease Sa. *Protein Sci.* **10**(6), 1206–1215.

Shaw, W. H. R., and Brodeaux, J. J. (1955). The decomposition of urea in aqueous media. *J. Am. Chem. Soc.* **77,** 4729–4733.

Shimizu, S. (2004). Estimating hydration changes upon biomolecular reactions from osmotic stress, high pressure, and preferential hydration experiments. *Proc. Natl. Acad. Sci. USA* **101**(5), 1195–1199.

Shimizu, S., and Boon, C. L. (2004). The Kirkwood-Buff theory and the effect of cosolvents on biochemical reactions. *J. Chem. Phys.* **121**(18), 9147–9155.

Shimizu, S., and Matubayasi, N. (2006). Preferential hydration of proteins: A Kirkwood-Buff approach. *Chem. Phys. Lett.* **420**(4–6), 518–522.

Shimizu, S., and Smith, D. J. (2004). Preferential hydration and the exclusion of cosolvents from protein surfaces. *J. Chem. Phys.* **121**(2), 1148–1154.

Shulgin, I. L., and Ruckenstein, E. (2005a). A protein molecule in an aqueous mixed solvent: Fluctuation theory outlook. *J. Chem. Phys.* **123**(5), 054909.

Shulgin, I. L., and Ruckenstein, E. (2005b). Relationship between preferential interaction of a protein in an aqueous mixed solvent and its solubility. *Biophys. Chem.* **118**(2–3), 128–134.

Shulgin, I. L., and Ruckenstein, E. (2006). A protein molecule in a mixed solvent: The preferential binding parameter via the Kirkwood-Buff theory. *Biophys. J.* **90**(2), 704–707.

Sidorova, N. Y., and Rau, D. C. (1996). Differences in water release for the binding of EcoRI to specific and nonspecific DNA sequences. *Proc. Natl. Acad. Sci. USA* **93**(22), 12272–12277.

Sidorova, N. Y., and Rau, D. C. (2004). Differences between EcoRI nonspecific and "star" sequence complexes revealed by osmotic stress. *Biophys. J.* **87**(4), 2564–2576.

Sinanoglu, O., and Abdulnur, S. (1965). Effect of water and other solvents on the structure of biopolymers. *Fed. Proc.* **24**(Suppl. 15), 12–23.

Singh, R., Haque, I., and Ahmad, F. (2005). Counteracting osmolyte trimethylamine N-oxide destabilizes proteins at pH below its pKa. Measurements of thermodynamic parameters of proteins in the presence and absence of trimethylamine N-oxide. *J. Biol. Chem.* **280**(12), 11035–11042.

Smith, J. C., Merzel, F., Verma, C. S., and Fischer, S. (2002). Protein hydration water: Structure and thermodynamics. *J. Mol. Liq.* **101**(1–3), 27–33.

Smith, P. E. (2004a). Cosolvent interactions with biomolecules: Relating computer simulation data to experimental thermodynamic data. *J. Phys. Chem. B* **108**(48), 18716–18724.

Smith, P. E. (2004b). Local chemical potential equalization model for cosolvent effects on biomolecular equilibria. *J. Phys. Chem. B* **108**(41), 16271–16278.

Smith, P. E. (2005). Protein volume changes on cosolvent denaturation. *Biophys. Chem.* **113**(3), 299–302.

Smith, P. E. (2006a). Chemical potential derivatives and preferential interaction parameters in biological systems from Kirkwood-Buff theory. *Biophys. J.* **91**(3), 849–856.

Smith, P. E. (2006b). Equilibrium dialysis data and the relationships between preferential interaction parameters for biological systems in terms of Kirkwood-Buff integrals. *J. Phys. Chem. B Condens. Matter Mater. Surf. Interfaces Biophys.* **110**(6), 2862–2868.

Sörensen, S. P. L. (1925). The solubility of proteins. *J. Am. Chem. Soc.* **47,** 457–469.

Spiro, K. (1900). Ueber die beeinflussung der eiweisscoagulation durch sticksstoffhaltige substanzen. *Z. Phys. Chem.* **30,** 182–199.

Stanley, C., and Rau, D. C. (2006). Preferential hydration of DNA: The magnitude and distance dependence of alcohol and polyol interactions. *Biophys. J.* **91**(3), 912–920.

Stark, G. R., Stein, W. H., and Moore, S. (1960). Reactions of the cyanate present in aqueous urea with amino acids and proteins. *J. Biol. Chem.* **235**(11), 3177–3181.

Stigter, D., and Dill, K. A. (1990). Charge effects on folded and unfolded proteins. *Biochemistry* **29**(5), 1262–1271.

Suzuki, K. (1960). Studies on the kinetics of protein denaturation under high pressure. *Rev. Phys. Chem. Jpn.* **29,** 91–98.

Tanford, C. (1957). The location of electrostatic charges in Kirkwood's model of organic ions. *J. Am. Chem. Soc.* **79**(20), 5348–5352.

Tanford, C. (1964). Isothermal unfolding of globular proteins in aqueous urea solutions. *J. Am. Chem. Soc.* **86**(10), 2050–2059.

Tanford, C. (1968). Protein denaturation. *Adv. Protein Chem.* **23,** 121–282.

Tanford, C. (1969). Extension of the theory of linked functions to incorporate the effects of protein hydration. *J. Mol. Biol.* **39**(3), 539–544.

Tanford, C., Kawahara, K., and Lapanje, S. (1967a). Proteins as random coils. I. Intrinsic viscosities and sedimentation coefficients in concentrated guanidine hydrochloride. *J. Am. Chem. Soc.* **89**(4), 729–736.

Tanford, C., Kawahara, K., Lapanje, S., Hooker, T. M., Zarlengo, M. H., Salahuddin, A., Aune, K. C., and Takagi, T. (1967b). Proteins as random coils. III. Optical rotary dispersion in 6M guanidine hydrochloride. *J. Am. Chem. Soc.* **89**(19), 5023–5029.

Tanford, C., and Kirkwood, J. G. (1957). Theory of protein titration curves. I. General equations for impenetrable spheres. *J. Am. Chem. Soc.* **79**(20), 5333–5339.

Tiffany, M. L., and Krimm, S. (1973). Extended conformations of polypeptides and proteins in urea and guanidine hydrochloride. *Biopolymers* **12,** 575–587.

Timasheff, S. N. (1992). Water as ligand—preferential binding and exclusion of denaturants in protein unfolding. *Biochemistry* **31**(41), 9857–9864.

Timasheff, S. N. (1993). The control of protein stability and association by weak interactions with water: How Do solvents affect these processes? *Annu. Rev. Biophys. Biomol. Struct.* **22,** 67–97.

Timasheff, S. N. (1998). In disperse solution, "osmotic stress" is a restricted case of preferential interactions. *Proc. Natl. Acad. Sci. USA* **95**(13), 7363–7367.

Todd, M. J., and Salemme, F. R. (2003). Direct binding assays for pharma screening—Assay tutorial: ThermoFluor miniaturized direct-binding assay for HTS & secondary screening. *Genet. Eng. News* **23**(3), 28–29.

Tomschy, A., Bohm, G., and Jaenicke, R. (1994). The effect of ion pairs on the thermal stability of D-glyceraldehyde 3-phosphate dehydrogenase from the hyperthermophilic bacterium Thermotoga maritima. *Protein Eng.* **7**(12), 1471–1478.

Tran, H. T., and Pappu, R. V. (2006). Toward an accurate theoretical framework for describing ensembles for proteins under strongly denaturing conditions. *Biophys J.* **91**(5), 1868–1886.

Tran, H. T., Wang, X., and Pappu, R. V. (2005). Reconciling observations of sequence-specific conformational propensities with the generic polymeric behavior of denatured proteins. *Biochemistry* **44**(34), 11369–11380.

Tsai, P. K., Volkin, D. B., Dabora, J. M., Thompson, K. C., Bruner, M. W., Gress, J. O., Matuszewska, B., Keogan, M., Bondi, J. V., and Middaugh, C. R. (1993). Formulation design of acidic fibroblast growth factor. *Pharm. Res.* **10**(5), 649–659.

Viadiu, H., and Aggarwal, A. K. (2000). Structure of BamHI bound to nonspecific DNA: A model for DNA sliding. *Mol. Cell* **5**(5), 889–895.

Vlachy, V., Hribar-Lee, B., Kalyuzhnyi, Y. V., and Dill, K. A. (2004). Short-range interactions: From simple ions to polyelectrolyte solutions. *Curr. Opin. Colloid Interface Sci.* **9**(1–2), 128–132.

Vodyanoy, I., Bezrukov, S. M., and Parsegian, V. A. (1993). Probing alamethicin channels with water-soluble polymers. Size-modulated osmotic action. *Biophys J.* **65**(5), 2097–2105.

Volkin, D. B., Mach, H., and Middaugh, C. R. (1997). Degradative covalent reactions important to protein stability. *Mol. Biotechnol.* **8**(2), 105–122.

Vossen, K. M., Wolz, R., Daugherty, M. A., and Fried, M. G. (1997). Role of macromolecular hydration in the binding of the Escherichia coli cyclic AMP receptor to DNA. *Biochemistry* **36**(39), 11640–11647.

Wagner, K., Harries, D., May, S., Kahl, V., Radler, J. O., and Ben-Shaul, A. (2000). Direct evidence for counterion release upon cationic lipid-DNA condensation. *Langmuir* **16**(2), 303–306.

Weerasinghe, S., and Smith, P. E. (2003a). A Kirkwood-Buff derived force field for mixtures of urea and water. *J. Phys. Chem. B* **107**(16), 3891–3898.

Weerasinghe, S., and Smith, P. E. (2003b). Cavity formation and preferential interactions in urea solutions: Dependence on urea aggregation. *J. Chem. Phys.* **118**(13), 5901–5910.

Weerasinghe, S., and Smith, P. E. (2004). A Kirkwood-Buff derived force field for the simulation of aqueous guanidinium chloride solutions. *J. Chem. Phys.* **121**(5), 2180–2186.

Winzor, D. J. (2004). Reappraisal of disparities between osmolality estimates by freezing point depression and vapor pressure deficit methods. *Biophys. Chem.* **107**(3), 317–323.

Wyman, J., Jr. (1964). Linked functions and reciprocal effects in hemoglobin: A second look. *Adv. Protein. Chem.* **19,** 223–286.

Xavier, K. A., Shick, K. A., Smith-Gill, S. J., and Willson, R. C. (1997). Involvement of water molecules in the association of monoclonal antibody HyHEL-5 with bobwhite quail lysozyme. *Biophy. J.* **73**(4), 2116–2125.

Yancey, P. H. (2005). Organic osmolytes as compatible, metabolic and counteracting cytoprotectants in high osmolarity and other stresses. *J. Exp. Biol.* **208**(Pt. 15), 2819–2830.

Yancey, P. H., Clark, M. E., Hand, S. C., Bowlus, R. D., and Somero, G. N. (1982). Living with water stress: Evolution of osmolyte systems. *Science* **217**(4566), 1214–1222.

Yao, M., and Bolen, D. W. (1995). How valid are denaturant-induced unfolding free energy measurements? Level of conformance to common assumptions over an extended range of ribonuclease A stability. *Biochemistry* **34**(11), 3771–3781.

Zhang, W., Capp, M. W., Bond, J. P., Anderson, C. F., and Record, M. T., Jr. (1996). Thermodynamic characterization of interactions of native bovine serum albumin with highly excluded (glycine betaine) and moderately accumulated (urea) solutes by a novel application of vapor pressure osmometry. *Biochemistry* **35**(32), 10506–10516.

Zimmerberg, J., and Parsegian, V. A. (1986). Polymer inaccessible volume changes during opening and closing of a voltage-dependent ionic channel. *Nature* **323**(6083), 36–39.

Zipp, A., and Kauzmann, W. (1973). Pressure denaturation of metmyoglobin. *Biochemistry* **12**(21), 4217–4228.

Zou, Q., Bennion, B. J., Daggett, V., and Murphy, K. P. (2002). The molecular mechanism of stabilization of proteins by TMAO and its ability to counteract the effects of urea. *J. Am. Chem. Soc.* **124**(7), 1192–1202.

SECTION 2

Computational Methods

CHAPTER 23

Stupid Statistics!

Joel Tellinghuisen

Department of Chemistry
Vanderbilt University
Nashville, Tennessee 37235

0091-679X/08 $35.00
DOI: 10.1016/S0091-679X(07)84023-4

Abstract

The method of least squares is probably the most powerful data analysis tool available to scientists. Toward a fuller appreciation of that power, this work begins with an elementary review of statistics fundamentals, and then progressively increases in sophistication as the coverage is extended to the theory and practice of linear and nonlinear least squares. The results are illustrated in application to data analysis problems important in the life sciences.

The review of fundamentals includes the role of sampling and its connection to probability distributions, the Central Limit Theorem, and the importance of finite variance. Linear least squares are presented using matrix notation, and the significance of the key probability distributions—Gaussian, chi-square, and *t*—is illustrated with Monte Carlo calculations. The meaning of correlation is discussed, including its role in the propagation of error. When the data themselves are correlated, special methods are needed for the fitting, as they are also when fitting with constraints. Nonlinear fitting gives rise to nonnormal parameter distributions, but the 10% Rule of Thumb suggests that such problems will be insignificant when the parameter is sufficiently well determined. Illustrations include calibration with linear and nonlinear response functions, the dangers inherent in fitting inverted data (e.g., Lineweaver–Burk equation), an analysis of the reliability of the van't Hoff analysis, the problem of correlated data in the Guggenheim method, and the optimization of isothermal titration calorimetry procedures using the variance-covariance matrix for experiment design. The work concludes with illustrations on assessing and presenting results.

I. Introduction

The title of this chapter can be interpreted in at least three different ways: (1) statistics at a very elementary level, (2) statistics used erroneously, and (3) an epithet! The contents are intended to deal with all three, in that the presentation will indeed begin at a basic level and will progress far enough to include examples of the misuse of statistics in data analysis situations common in biophysics. Hopefully, through an appreciation of such illustrations, the reader will find the need for the third sense greatly alleviated!

The focus of the work is toward a better overall appreciation of what is arguably the most powerful tool in the data analyst's bag: the method of least squares (LS). The pedagogical emphasis will be on linear least squares (LLS), but I will also address nonlinear methods in enough detail to illustrate what properties change and in what ways, on going from linear to nonlinear least squares (NLS). The following chapter in this volume (Chapter 24 by Johnson) deals more specifically with nonlinear methods; and many other chapters utilize NLS in the treatment of specific data analysis problems.

I claim nothing in the way of original statistical development, but I will try to point out aspects of statistical data analysis that I think are under- or misutilized by physical scientists. For the basics in the theory of probability and statistics and their application to data analysis, I have found the early textbooks by Mood and Graybill (1963) and Parratt (1961) quite instructive. The monograph by Deming (1964; originally published in 1938) constitutes a landmark in the development of nonlinear methods; its title also represents a view I have come to take with respect to the fitting of data—namely, the adjustment of data to a model. This view seems to contrast subtly with that of many workers, who prefer to think in terms of fitting models to data. The work by Deming predates computers and does not use matrix notation. The latter is developed and used masterfully in the book by Hamilton (1964). A book published a few years later by Bevington (1969) does not use matrix notation but places heavy emphasis on computational implementation and includes computer routines (FORTRAN) for handling both linear and nonlinear problems, including for the latter a routine that utilizes the Marquardt algorithm (1963), which remains today the default method for handling nonlinear fitting problems. Bevington's book has been an instructional mainstay for a generation of physical scientists; it has also appeared in an updated form (Bevington and Robinson, 2002). Computational methods are the heart of the *Numerical Recipes* books by Press *et al.* (1986), who use matrix methods for LS and provide algorithms (in several computer languages). These authors also discuss the use of Monte Carlo (MC) methods, particularly for assessing the confidence limits on parameter estimates. However, I have also found MC methods more generally instructive on the meaning of LS, for which purpose I owe a debt of gratitude to Albritton *et al.* (1976), who pioneered the use of such methods in the analysis of spectroscopic data.

The aforementioned MC methods feature prominently in a recent pedagogical work by me (Tellinghuisen, 2005a), which has evolved into an oral presentation. Accordingly, here I will maintain that "oral" sense through use of PowerPoint-type figures that compactly summarize key results. I will start by reviewing basic concepts and terminology concerning probability distributions and sampling estimates thereof. I will extend these to linear and then NLS, with attention to problems concerning weighing of data, correlated fitting, error propagation, and presentation of results. The key points will be illustrated with specific reference to common linearization methods (e.g., linear log analysis of exponential decay data, Lineweaver–Burk analysis of enzyme kinetics data) and to linear and nonlinear calibration. I will also address what I think is a neglected use of LS, in experiment design, illustrated through application to the method of isothermal titration calorimetry (ITC; see also Chapter 4 by Freyer and Lewis, this volume), where such techniques have yielded a 10- to-100-fold enhancement in precision and throughput over "standard" experimental procedures.

II. Statistics of Data: The 10-Min Review

A. Sampling: Location and Dispersion Indices

From a statistical standpoint, every physical measurement constitutes a sampling estimate of some property, the true value of which is known only to God. We seek to obtain more reliable estimates through repeated measurements. The sampling process can be viewed as an attempt to characterize the unknown probability distribution of the targeted property in the given experiment. To that end, at least two statistical parameters are desired—one to locate the property, another to characterize its spread. Among the common *location indices* are the mean, the median, the mode, and the root-mean-square (rms) value; common *dispersion indices* are the standard deviation, the mean deviation, and the range. Of these the mean and the standard deviation are most closely related to the method of LS. In fact, the sampling mean is a minimum-variance estimator, hence constitutes a LS estimator. A minimum-variance estimator is also a *maximum efficiency* estimator, which means that in an important sense, it is the "best." In multiparameter LS fits, each parameter behaves like a mean in a single-parameter fit to determine a constant (the mean!). Accordingly, the present treatment will consider only these "best" statistics.

These key sampling statistics and their relations to the true quantities are summarized in Fig. 1. The *accuracy* of an estimate relates to the closeness of $\bar{x}$ to μ; *precision* is a measure of the smallness of s_x and $s_{\bar{x}}$. We increase the precision of the estimated mean by increasing the number of measurements n. If x_i is an *unbiased* estimator of μ, we can also hope to increase the accuracy through repeated measurement because $\bar{x} \to \mu$ as $n \to \infty$. Sometimes a mean is biased for finite n but still converges on μ as $n \to \infty$. Such an estimator is called *consistent*. If bias persists in the limit of infinite n, the estimator is *inconsistent*.

Note especially that the sampling dispersion parameter that characterizes single measurements is s_x while that for the entire set of measurements is $s_{\bar{x}}$. The former might be the focus of a new analytical method, but if the goal of the experiment is the estimation of μ, the proper dispersion index is $s_{\bar{x}}$, which correctly reflects the effort put into the experiment (through n). Accordingly, for example, a new method that reduces s_x by a factor of 5 is 25 times as efficient as its reference, since it would take 25 measurements with the old method to yield a similarly small dispersion.

B. Probability Distributions

Some key probability distributions for statistical data analysis are summarized in Fig. 2. The uniform distribution applies for simple random number generators (see below). The normal distribution is at the heart of most data analysis and is

The fundamentals

Sampling		*Theory*
	Mean	
$\bar{x} = \langle x \rangle = \frac{1}{n}\sum_{i=1}^{n} x_i$		$\langle x \rangle \equiv \mu = \int_{x_{\min}}^{x_{\max}} xP(x)dx$
	Variance	
$s_x^2 = \frac{1}{n-1}\sum_{i=1}^{n} \delta_i^2 \; (\delta_i = x_i - \bar{x})$		$\sigma_x^2 = \int_{x_{\min}}^{x_{\max}} (x-\mu)^2 P(x)dx$
	Standard deviation	
s_x		σ_x
	Standard error (standard deviation in the mean)	
$\frac{s_x}{\sqrt{n}}$		$\frac{\sigma_x}{\sqrt{n}}$

Fig. 1 Summary of most important statistics in sampling and their theoretical counterparts. Although the symbols are mostly clear from their context, n represents the number of data points in a set, and δ is called the *residual*. The values $x_{\min}$ and $x_{\max}$ represent the range over which the probability distribution $P(x)$ is defined.

Probability distributions

Uniform: $P(x)$ = constant ($a \le x \le b$); 0 otherwise

Normal: $P_G(\mu,\sigma;x) = \frac{1}{\sigma\sqrt{2\pi}}\exp\left[-\frac{(x-\mu)^2}{2\sigma^2}\right]$

Poisson: Governs counting— $\sigma^2 = \mu$ (= No. of counts)

Chi-square (χ^2): Sampling estimates of variances

***t*-distribution:** Confidence limits for sampling estimates of parameters

Note: Poisson, χ^2, and *t*-distributions become Gaussian in the limit of large μ (Poisson) or ν (degrees of freedom, $n - p$)

Fig. 2 Key statistical probability distributions relevant to data analysis.

typically assumed rather than verified. However, there are many cases where this assumption must be quite good; when it is not, the consequences for the analysis range from insignificant to catastrophic. The Poisson distribution is not treated explicitly here; however, many instruments detect by counting (e.g., electrons, photons); and it is useful to recognize that such measurements contain built-in precision estimates, since counting is a Poisson process. The other two distributions play the indicated roles and will be discussed further below. The final note in Fig. 2 is important and useful, and it serves to define for future use the quantity called *degrees of freedom* (dof), symbolized by ν and obtained by subtracting the number of adjustable parameters p from the number of data points n.

The main purpose of dispersion indices is to establish *confidence limits* on the estimates. It is useful to recall that when the normal distribution holds, 68.3% of unbiased estimates should fall within $\pm\sigma$ of μ, and 95.4% within $\pm 2\sigma$. Probability distributions are normalized to unity, so it is equivalent to note, for example, that the definite integral of $P_G(x)$ from $\mu - \sigma$ to $\mu + \sigma$ is 0.683.

C. A Simple MC Experiment

Figures 3 and 4 illustrate some of these points through a simple computational experiment—generating and binning random numbers on a computer. [Many of the results illustrated here and below were obtained using the KaleidaGraph program (Synergy Software), which is representative of a number of programs designed for data analysis and presentation (Tellinghuisen, 2000a).] From Fig. 3 we obtain 0.5023 for the mean of 10^4 estimates. This is close to the theoretical value, but is it close enough? Since it is within one standard error (0.00287), it is a reasonable result. The variance and standard deviation are also close to their expected values. Are they close enough? We can answer this question, too, by using an underappreciated result: Because sampling estimates of variances follow the χ^2 distribution (*vida infra*), they have an inherent relative variance of $2/\nu$, hence

The uniform distribution

- The basis of computer random number generators
- Default range $0 < x < 1$, giving $\mu = 1/2$ and $\sigma^2 = 1/12$
- Check by generating 10,000 random deviates
- Bin into 10 equal bins and examine statistics (summary from KaleidaGraph program)

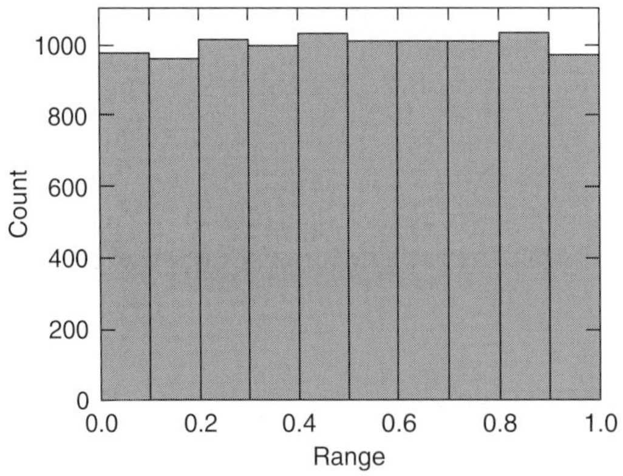

Minimum	9.36e-05
Maximum	0.99998742
Sum	5022.7734
Points	10000
Mean	0.50227734
Median	0.50219405
RMS	0.57845803
Std. Deviation	0.28694843
Variance	0.082339402
Std. Error	0.0028694843
Skewness	−0.0081906446
Kurtosis	−1.1866255

Fig. 3 An MC experiment on the uniform distribution.

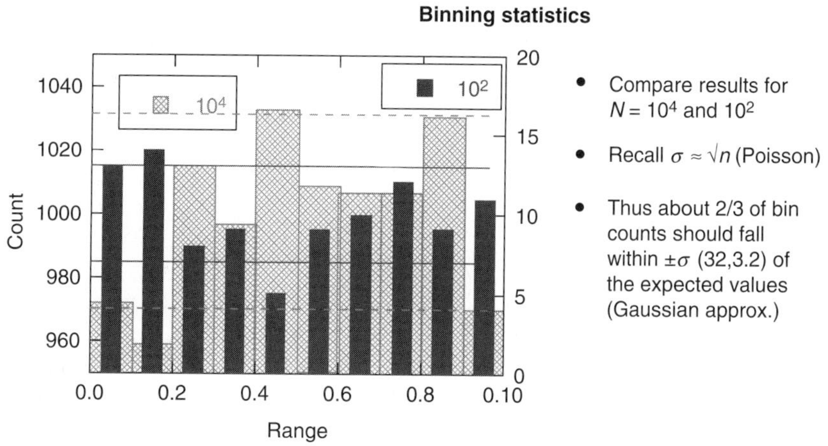
Binning statistics
10^4
10^2
Count
1040
1020
1000
980
960
20
15
10
5
0
0.0
0.2
0.4
0.6
0.8
0.10
Range
• Compare results for N = 10^4 and 10^2
• Recall σ ≈ √n (Poisson)
• Thus about 2/3 of bin counts should fall within ±σ (32,3.2) of the expected values (Gaussian approx.)

Fig. 4 Statistics of binning.

relative standard deviation of $(2/v)^{1/2}$. From error propagation (more below), the relative standard deviation in s_x is $1/(2v)^{1/2}$. Here $v = 10^4 - 1$, so these relative errors amount to 1.41% and 0.71%; and we see that s_x^2 and s_x are both within one standard deviation of their expected values.

The point of the binning is developed in Fig. 4, where results for 10^4 and 10^2 samples are compared. The binning operation is basically a counting experiment, so the bin count is well approximated by the Poisson error, as indicated. In both cases, two of the 10 bins lie outside the bracketed region, which is reasonably close to the expected result. Note also that while the *absolute* uncertainty is 10 times larger for the larger count, the *relative* uncertainty is 10 times smaller. It is further noteworthy that we used the Poisson result to estimate σ for each bin count, but then used the Gaussian approximation to interpret it (Fig. 2). The latter approximation is very good for the higher bin count (1000) but a bit of a stretch for the lower (10).

D. The Central Limit Theorem

It is instructive to extend this experiment to averages of several random deviates. Figure 5 shows the results of 10^4 samples of averages of 2 and 3, and the sum of 12. The theoretical distribution in the first case is triangular while that in the second is piecewise quadratic. The distributions narrow with increasing number averaged, as the variance drops by the same factor (1/24 for average of 2, 1/36 for 3). Continuing, the average of 12 would have $\sigma = 1/12$, but I have chosen to depict the sum of 12 in this case because it has a particularly simple result—a variance and standard deviation of unity. More importantly, by the time 12 uniform deviates have been summed, the distribution is quite close to normal, illustrating the very

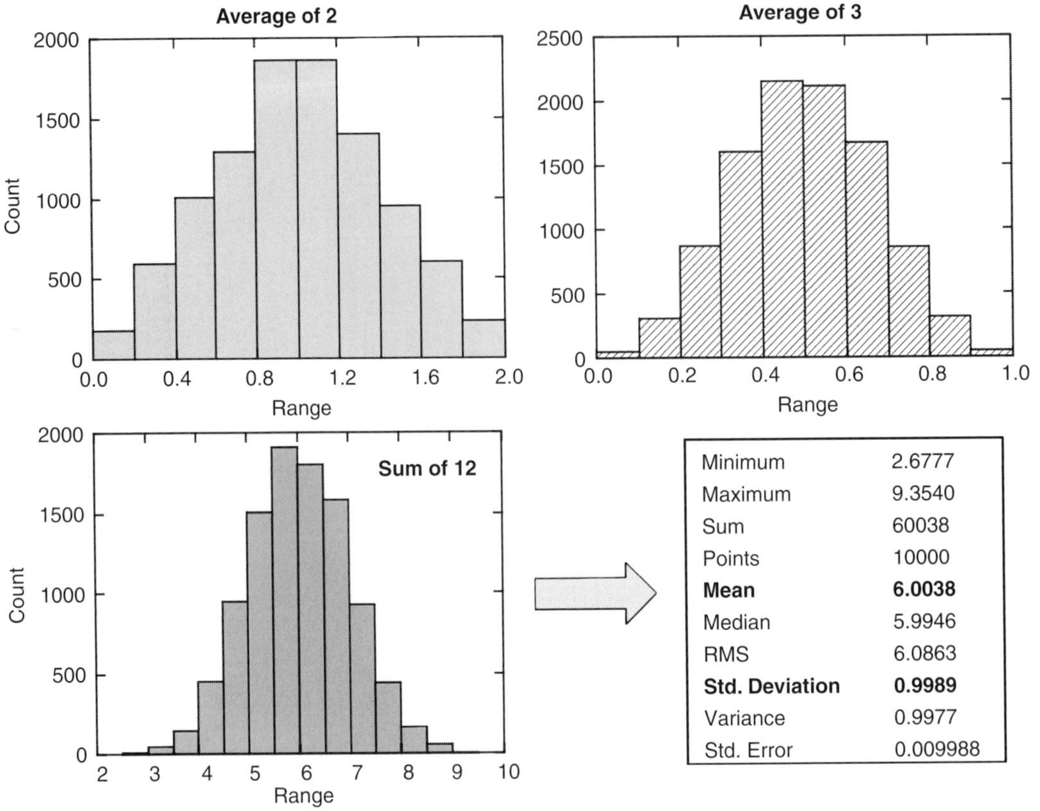

Fig. 5 The effect of averaging: MC results for uniform random deviates averaged 2 and 3 at a time, and summed 12 at a time.

important *Central Limit Theorem*, which reads in part: Sums and averages of samples are distributed normally in the limit of an infinite number of samples, as long as the parent distribution has finite variance. In this case, 12 is adequately close to infinite; and the sum of 12 uniform deviates can be used as a quick way to generate experimental "noise" of unit standard deviation in synthesizing computer "data".

The result shown in the third histogram in Fig. 5—that summing has yielded an essentially normal distribution from very nonnormal data—is both remarkable and important. Many instruments log data by digitizing an analog signal and often do so by averaging many conversions. As long as the raw signal is sufficiently noisy (not limited by the digitizer least count), such averaged data are likely to be reasonably normal, thanks to the Central Limit Theorem, and irrespective of the unknown true distribution for measuring single values. Instruments that count generally follow Poisson statistics, as already noted; and as long as the count is reasonably large (say, >25), the Poisson distribution will approximate the Gaussian (Fig. 2). Thus in these situations, concerns about nonnormal data are misplaced.

E. The Importance of Finite Variance

From the foregoing statements, it is easy to imagine that all experimental data have finite variance, hence become normal with averaging. Thus, it is useful to examine cases where the assumption of finite variance does *not* hold. One probability distribution of particular importance in spectroscopy is the *Lorentzian distribution*:

$$P_{\mathrm{L}}(\mu, \Gamma; x) = \frac{\Gamma/(2\pi)}{(x-\mu)^2 + (\Gamma/2)^2} \tag{1}$$

This distribution is integrable, hence normalizable; therefore one can establish confidence ranges for it. However, it is easy to verify that it does not have finite variance. Accordingly, all the fundamental tenets of the method of LS, discussed below, fail when the data happen to be Lorentzian (Tellinghuisen, 2005a). This does *not* say anything about the ability to fit data to a Lorentzian lineshape; and I know of no actual cases where the inherent data noise is Lorentzian. Presumably any measurement instrument would anyway yield a clipped (finite tails) Lorentzian distribution; and then the Central Limit Theorem would apply.

A case of more concern is the common practice of inverting data in order to obtain a straight-line presentation for the sake of analysis. This is still widely done in the analysis of fluorescence quenching data (Stern–Volmer plot), enzyme kinetics (Lineweaver–Burk equation), adsorption, and in many methods for estimating binding constants. If the raw data are normal, then the distribution for their reciprocals has infinite variance (Tellinghuisen, 2000b). How significant this is depends on the relative precision of the data, which can be illustrated through another simple MC experiment: Add random, normal error to a constant; then calculate, bin, and do statistics on its reciprocal (Tellinghuisen, 2000c). Results from such computations are presented in Table I and illustrated (in part) in Fig. 6.

The results in the second and third columns in Table I are well behaved, and one can readily verify their statistical reasonableness, using methods like those discussed in connection with Figs. 3 and 4. The results in columns three and four seem well behaved up to $\sigma_A = 0.2$; but with further increase to $\sigma_A = 0.35$, the sampling statistics for A^{-1} become divergent—another consequence of the Central Limit Theorem. Put simply, without finite variance, sampling can at best succeed in only an asymptotic sense. The problem arises from the sampling of those few values far enough in the small-A wings to reach zero and below. These values are so rare for $\sigma_A \leq 0.20$ (e.g., 3 in $10^7 < 0$ for $\sigma_A = 0.20$) that they very rarely crop up in just 10^5 samples. Presumably, increasing the sample size to $\sim 10^9$ would yield similar

Table I
MC Statistics of $A = 1$ and Its Reciprocal[a,b]

σ_A	$\langle A \rangle$	$\langle A^2 \rangle - \langle A \rangle^2$ [c]	$\langle A^{-1} \rangle$	$\langle A^{-2} \rangle - \langle A^{-1} \rangle^2$ [c]	$\langle A^{-1} \rangle_{true}$ [d]
0.05	0.99996	2.491×10^{-3}	1.00255	2.542×10^{-3}	1.00252
0.10	0.99992	9.965×10^{-3}	1.01036	1.083×10^{-2}	1.01032
0.15	0.99988	2.242×10^{-2}	1.02425	2.745×10^{-2}	1.02422
0.20	0.99984	3.986×10^{-2}	1.04629	6.447×10^{-2}	1.04623
0.20	1.00116	3.997×10^{-2}	1.04473	5.997×10^{-2}	1.04623
0.35	0.99961	1.220×10^{-1}	1.14868	1.367×10^{2}	1.20093
0.35	1.00202	1.224×10^{-1}	1.15218	2.569×10^{2}	1.20093
0.35	0.99972	1.221×10^{-1}	0.89179	1.158×10^{4}	1.20093
0.35	1.00107	1.222×10^{-1}	1.05755	3.191×10^{3}	1.20093

[a]This is Table I from Tellinghuisen (2000c). Each value represents results from 10^5 samples, with random normally distributed error of specified σ_A on A.

[b]Same seed used for first four σ_A values, to illustrate the effects of scaling for a given set of unit-variance normal deviates.

[c]The sampling estimates of the variances (derivable from the equation for $\sigma_{\bar{x}}^2$ in Fig. 1).

[d]Obtained by numerical integration.

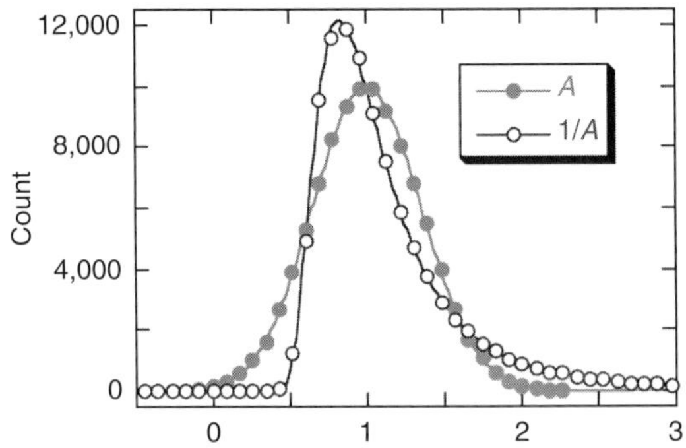

Fig. 6 Histogrammed results of 10^5 MC estimates of a constant $A = 1$ and its reciprocal, with random normal error of $\sigma_A = 0.35$ superimposed on A. The fitted curves are a Gaussian and its theoretical counterpart for the inverted data. [Adapted with permission from Tellinghuisen (2000c); © 2000, American Chemical Society.]

problems for $\sigma_A = 0.20$ ... and by the same token, decreasing it to ~50 for $\sigma_A = 0.35$ would yield many apparently successful MC runs (since the probability for $A < 0$ is 0.0021 for $\sigma_A = 0.35$). These considerations also illustrate the meaning of asymptotic success in this sampling situation. Clearly, reciprocation will give no obvious problems when the data are precise to 10% or better—a result that will be used later to state a "10% rule of thumb" in NLS. However, the bias from such nonnormal data can still be significant, as may be appreciated from the values in the last column of Table I.

Figure 6 illustrates the effects of inversion on the data distribution. As we shall see, distributions like that shown here for A^{-1} are a tip-off for the occurrence of reciprocal behavior even when it is not obvious that it should be occurring, as in parameters derived from a NLS fit.

Another common data transformation is logarithmic conversion of exponential data with no background, often used to analyze first-order kinetics data and to extract thermodynamics and kinetics parameters from temperature dependence of rate and equilibrium constants. That such conversion is a problem is evident to anyone who has ever recorded decay data at long time, where the noise is large enough to produce negative values. In the log transformation, such values must simply be discarded, and with this proviso, log conversion preserves finite variance in the transformed variate; however, there is bias, as should be evident from the need to discard just the too-small values in the extreme. [I have illustrated this point numerically for the van't Hoff (vH) analysis of the T-dependence of the equilibrium constant $K^\circ(T)$ (Tellinghuisen, 2004a).] As is discussed below, the remedy for both data transformation problems is to just skip the transformation and analyze the raw data by NLS.

III. Linear Least Squares—Theory

A. Tenets: Demands and Payback

I turn now to the primary target of this work, the method of LS. Although by far the greatest utility of LS is in the virtually unlimited ways it can be implemented in application to nonlinear problems, its principles can best be stated and clarified in its original linear form. I will start by stating the main principles and then will illustrate these through MC computational experiments like those just described in connection with the data inversion problem.

Figure 7 states the main conditions on and results to be expected from LLS, and Fig. 8 provides clarification. When any of the requirements for the method is not satisfied, some but not necessarily all of the guarantees for the results may not hold. For example, the LS method is sometimes disparaged as requiring normal data while actual data are of unknown distribution. This criticism is flawed, as LLS requires only that the data be of finite variance. However, translating parameter standard errors into confidence limits becomes problematic if the data are not normal because then the parameter estimates are also nonnormal. As another example, neglect of weights for unbiased, normal *heteroscedastic data* (nonconstant σ_i) yields parameter estimates that remain normally distributed and unbiased but not of minimum variance.

The role of the χ^2- and t-distributions (Fig. 7) will be clarified below.

Linear least squares: Principles

- We assume that all statistical uncertainty resides in a single *dependent variable* y. There are one or more *independent variables* (x, u, ...) and one or more *adjustable parameters* (a, b, ...). The fit model must be linear in the latter.
- The LLS solution minimizes the quantity $\mathcal{S} = \sum w_i \delta_i^2$, where w_i represents the weight for the ith point and $\delta_i = y_i - y_{\text{calc},i}$.
- If the data are *independent* and *unbiased*, with *random error* of *finite variance*, then the LLS parameter estimates will be unbiased and of *minimum variance*, provided the data are weighted as σ_i^{-2}.
- If the data structure $\{x_i\}$ and the data errors σ_i are known, then the parameter standard errors are *exactly predictable* from the *variance-covariance matrix* V.
- If further, the statistical data error is *normally distributed*, then the LLS parameter estimates will be normally distributed about their true values.
- Under these conditions, the quantity $\mathcal{S}$ is distributed as a scaled χ^2 variate Further the *a posteriori*-assessed parameter residuals are *t*-distributed.

Fig. 7 Fundamental principles of LLS.

Linear least squares: Explanation of principles

A linear model: $y = ax + b/x^2 + c \ln(3u) + d$

Independent variables: x and u
Dependent variable: y
Adjustable parameters: a, b, c, d

A nonlinear model: $y = 1/a$ (recall statistics of $1/A$)

Independent variables: none
Dependent variable: y
Adjustable parameter: a

Independent and **random:** What we assumed about our random number generator in the Monte Carlo experiments.

Unbiased: True result; e.g., A but not A^{-1} in Table I.

Minimum variance: The stated goal of LS. Must minimize $\mathcal{S}$ with $w_i \propto 1/\sigma_i^2$. Then parameter estimates will be minimum variance.

Fig. 8 Clarification of LLS basic principles.

B. The LLS Solution: Matrix Notation

The requirement of a linear fit model does not mean, as is often falsely assumed, a straight-line relationship between the independent and dependent variables, as should be clear from the first example in Fig. 8. Nor does such a relationship ensure that the problem is linear (e.g., the model $y = 1/a + b^2x$ is nonlinear, as is the simpler second model in Fig. 8). Rather, the relation between variables and parameters must be linear in the *parameters*. In matrix form, this means it can be expressed as

$$\mathbf{y} = \mathbf{X}\boldsymbol{\beta} + \boldsymbol{\delta} \tag{2}$$

which can best be clarified through an example. Take a 5-point data set ($n = 5$) and the first model in Fig. 8. Then this equation reads

$$\begin{pmatrix} y_1 \\ y_2 \\ y_3 \\ y_4 \\ y_5 \end{pmatrix} = \begin{pmatrix} x_1 & x_1^{-2} & \ln(3u_1) & 1 \\ x_2 & x_2^{-2} & \ln(3u_2) & 1 \\ x_3 & x_3^{-2} & \ln(3u_3) & 1 \\ x_4 & x_4^{-2} & \ln(3u_4) & 1 \\ x_5 & x_5^{-2} & \ln(3u_5) & 1 \end{pmatrix} \begin{pmatrix} a \\ b \\ c \\ d \end{pmatrix} + \begin{pmatrix} \delta_1 \\ \delta_2 \\ \delta_3 \\ \delta_4 \\ \delta_5 \end{pmatrix} \tag{3}$$

$\mathbf{y}$ and $\boldsymbol{\delta}$ are column vectors containing the n measured y values and the fit residuals, respectively; $\boldsymbol{\beta}$ is a column vector containing the 4 ($=p$) adjustable parameters, and the matrix $\mathbf{X}$ contains the information about the dependent variables (all known and error free). The product $\mathbf{X}\boldsymbol{\beta}$ yields $\mathbf{y}_{\text{calc}}$, and in this case the occurrence of 5 data points means $\nu = n - p = 1$ dof for the fit.

Minimization of the sum of weighted, squared residuals $\mathcal{S}$ with respect to each of the adjustable parameters leads $\mathbf{X}^{\mathrm{T}}$to the LLS equations,

$$\mathbf{X}^{\mathrm{T}}\,\mathbf{W}\mathbf{X}\boldsymbol{\beta} \equiv \mathbf{A}\boldsymbol{\beta} = \mathbf{X}^{\mathrm{T}}\,\mathbf{W}\mathbf{y} \tag{4}$$

where $\mathbf{X}^{\mathrm{T}}$ is the transpose of $\mathbf{X}$, and the square ($n \times n$) weight matrix $\mathbf{W}$ here is diagonal, with n elements $W_{ii} = w_i$. The solution reads

$$\boldsymbol{\beta} = \mathbf{A}^{-1}\mathbf{X}^{\mathrm{T}}\,\mathbf{W}\mathbf{y} \tag{5}$$

where $\mathbf{A}^{-1}$ is the inverse of $\mathbf{A}$ (the matrix of the normal equations). Knowledge of the parameters permits calculation of the residuals $\boldsymbol{\delta}$ from Eq. (2) and thence of $\mathcal{S}$, which in matrix form is

$$\mathcal{S} = \boldsymbol{\delta}^{\mathrm{T}}\,\mathbf{W}\,\boldsymbol{\delta} \tag{6}$$

C. Parameter Uncertainties: The Variance-Covariance Matrix

These days few data analysts would ever have to actually implement these equations, as the many available data analysis programs do so automatically after the data have been entered and the fit model specified. However, it is still important to understand them because of the all-important connection between the matrix $\mathbf{A}$ and the variance-covariance matrix $\mathbf{V}$ of the parameters:

$$\mathbf{V} \propto \mathbf{A}^{-1} \tag{7}$$

The parameter variances are the diagonal elements of the symmetric matrix $\mathbf{V}$, and the covariances are the off-diagonal elements. But what is the proportionality constant? The answer is twofold, and the distinction is important for a proper understanding of the theory of LS and for correct use of many programs.

First, let us assume we know the data error σ_i absolutely. Then, taking $w_i = \sigma_i^{-2}$ and the proportionality constant as unity, we have what I have called the *a priori* **V** (reflecting the *a priori* knowledge of the σ_i),

$$\mathbf{V}_{\text{prior}} = \mathbf{A}^{-1} \tag{8}$$

This equation is *exact*, and it applies most directly to MC computations on an LS fit model, where we have the luxury of specifying the σ_i. $\mathbf{V}_{\text{prior}}$ is completely known before any measurements (**y**) have been taken, since the **X** matrix contains just the specified values of the independent variables. This means that $\mathbf{V}_{\text{prior}}$ can be used in *experiment design*: By adjusting the structure of the data set, one can optimize the determination of selected parameters by minimizing their variances.

The second answer is the one more commonly used by data analysts; it assumes ignorance about the scale of the data error, which is then estimated from the fit itself. Note that we *must* know σ_i to within a constant scale factor, or the fit will not return minimum-variance estimates of the parameters. However, this knowledge can often be obtained from an understanding of the nature of the measurement, with adjustment for any data conversions through proper use of error propagation (see below). We suppose the relation between the true data error and our estimate thereof is $\sigma_i = r\, s_i$, and we take the weights as $w_i = s_i^{-2}$. Then the quantity

$$s_y^2 = \frac{\mathcal{S}_{\text{post}}}{\nu} \tag{9}$$

is known as the *estimated variance for data of unit weight* ($w_i = s_i = 1$). And the corresponding *a posteriori* **V** becomes

$$\mathbf{V}_{\text{post}} = s_y^2 \mathbf{A}^{-1} \tag{10}$$

The estimated parameter variances are now uncertain outcomes of the analysis, and their uncertainty affects the confidences for the parameters.

The widespread use of unweighted LLS falls in the *a posteriori* category, with $w_i = s_i = 1$, hence $s_y^2 = \mathcal{S}_{\text{post}}/\nu$. The tacit assumption behind such fitting is the presence of constant data error. If the data error is not truly constant, unweighted fitting will not return minimum-variance estimates of the parameters. However, if the range of the data error is small, say less than a factor of 3 over the data set, the neglect of weights is not likely to result in much loss of parameter precision.

Note that the parameter estimates are unaffected by a scale change in the σ_i, since such a factor cancels out of Eqs. (4) and (5). This is why $w_i \propto \sigma_i^{-2}$ suffices in Fig. 8. On the other hand, the elements of $\mathbf{V}_{\text{prior}}$ scale with the data variance and also with $1/n$. The latter, for example, means that the parameter standard errors ($V_{ii}^{1/2}$) go as $n^{-1/2}$, all other things being equal. Thus, they are to be interpreted in the same manner as the standard deviation in the mean in the case of a simple average, as previously mentioned.

Those commercial data analysis programs that provide parameter errors do not always make clear which choice—prior or post—applies when the weighted option is chosen. (All such programs use $\mathbf{V}_{\text{post}}$ for unweighted fitting.) For example, recent versions of the KaleidaGraph program use $\mathbf{V}_{\text{prior}}$ in weighted fits to user-defined functions, meaning that if the *a posteriori* estimates are desired, the user must scale the stated parameter errors by the factor $(\mathcal{S}/\nu)^{1/2}$. For the Origin program (OriginLab), the default is prior, but the user may check a box to convert to post.

D. Uncertainties in Variances: The Chi-Square Distribution

The sum of weighted squared residuals in Eq. (9) has been designated as $\mathcal{S}_{\text{post}}$ to distinguish it from the $\mathcal{S}_{\text{prior}}$ that results from the use of *a priori* weights. If the data error is normal, $\mathcal{S}_{\text{prior}}$ is distributed as χ^2, which has mean ν and variance 2ν. Or equivalently, $\mathcal{S}_{\text{prior}}/\nu$ is distributed as the reduced χ^2 (χ_ν^2), with mean unity and variance 2/ν (as mentioned above), and which follows the probability distribution,

$$P(z)dz = C\, z^{(\nu-2)/2} \exp(-\nu z/2)dz \tag{11}$$

where $z = \chi_\nu^2$ and C is a normalization constant. From the assumed relation between σ_i and s_i, $\mathcal{S}_{\text{post}} = r^2\mathcal{S}_{\text{prior}}$, which means that $\mathcal{S}_{\text{post}}$ is distributed as a scaled χ^2 variate. By taking the statistical average value for $\mathcal{S}_{\text{prior}}$, we can solve for r and then specify the data error for all points. We find that r is the dimensionless form of s_y. This value is still subject to the uncertainty inherent in χ^2. All of the *a posteriori*-estimated parameter variances are subject to the same uncertainty. Reiterating a point made earlier in connection with Figs. 3 and 4, the statistical properties of χ_ν^2 mean that it takes 200 dof to estimate variances with a relative standard deviation of 10%. From error propagation (below), the parameter standard errors are a factor of 2 more precise than the variances; hence they require 50 dof for 10% precision.

E. Confidence Limits on Estimated Parameters: The *t*-Distribution

When the data error is normal and of known magnitude, the parameter estimates are distributed normally with variances given by $\mathbf{V}_{\text{prior}}$. Confidence limits are then straightforwardly obtained from the appropriate tables of integrated Gaussians (e.g., appendix of Bevington and Robinson, 2002). However, when the parameter errors must be estimated *a posteriori*, their uncertainty increases the uncertainty in the parameters and requires the *t*-distribution for confidence limits. To be specific, let all the previous conditions, including normal data error, apply, and let s_a be the estimated standard deviation for the parameter a (e.g., $s_a = V_{\text{post},11}^{1/2}$). Then the quantity $(a_{\text{est}} - a_{\text{true}})/s_a$ follows the *t*-distribution for ν dof

$$P_t(t) = C'(1 + t^2/\nu)^{-(\nu+1)/2} \tag{12}$$

with C' another normalizing constant. For small ν the *t*-distribution is narrower in the peak than the Gaussian distribution, with more extensive tails. However $P_t(t)$ converges on the unit-variance normal distribution in the limit of large ν, making the distinction between the two distributions unimportant for large data sets.

There is a useful relation between the parameters and χ^2: If a single adjustable parameter is increased or decreased by its true σ and removed from the fit, and the remaining parameters are then refitted, χ^2 increases by unity. Alternatively in the *a posteriori* treatment, altering a parameter by $\pm s$ increases $\mathcal{S}$ by the factor $(\nu+1)/\nu$. If the change is $\pm 2\sigma(\pm 2s)$, these changes become $\Delta\chi^2 = 4$ and a factor of $(\nu+4)/\nu$ for $\mathcal{S}$ (Bevington and Robinson, 2002). This behavior can be used to estimate the parameter uncertainties in nonlinear fits, where Eqs. (8) and (10) do not hold rigorously (Press *et al.*, 1986). It is also the reason why in *ad hoc* fitting (e.g., using polynomials to obtain smooth empirical representations of data), one is justified in setting to zero any parameter that is smaller in magnitude than its standard error, since such a change results in a smaller s_y^2 for the fit.

F. Correlation

The off-diagonal elements of **V** are the covariances; they are seldom zero, which means that the LS parameter estimates are generally *correlated*. The *correlation matrix* **C** is obtained from **V** through

$$C_{ij} = \frac{V_{ij}}{(V_{ii}V_{jj})^{1/2}} \tag{13}$$

and yields elements that range between -1 and 1. Values close to these limits imply that the two parameters in question are very highly correlated while $C_{ij} = 0$ signifies no correlation.

There is some confusion in the literature about the role of correlation on the confidence limits of the parameters. Each of the parameters in a linear fit is distributed normally about its true value, with $\sigma_{\beta i} = V_{ii}^{1/2}$, *irrespective of its correlation with the other parameters*. The correlation comes into play only when we ask for *joint* confidence intervals of two or more parameters, in which case the confidence bands become ellipsoids in two or more dimensions. Equation (13) is exact for both versions of **V**, since the unknown scale factor r cancels out.

Of course it is true that high correlation reflects a mutual indeterminacy of two parameters. Thus, if one of these can be obtained by other means and then frozen in the fit, the other will be fitted with much higher precision.

G. Correlated Data

From the above, we have seen that the parameter estimates produced by an LS fit are practically always correlated. One of our fundamental assumptions for LLS was that the data were *independent*, hence *not* correlated (Fig. 7). Independent means that the random error in one measurement is in no way affected by that in another, and it is usually a reasonable assumption for raw data. But what if it is not? This occurs, for example, when the raw data are "precooked" for the purpose of analysis, in such a way that given raw points contribute to more than one point used in the analysis. Prime contenders for this offense are methods in which the raw data are subtracted or added, and the results are then fitted. As a specific example of the latter, the Guggenheim method was devised long ago to handle exponential data with a background, with the latter removed by subtracting points separated by a specified time interval (Tellinghuisen, 2003a).

A second case of correlated data arises when subsets of data are individually fitted, providing multiple parameters that are subsequently submitted to a combined or global analysis. Since the "data" for the global fit are themselves the output of an earlier fit, they are inherently correlated.

There is a simple answer to this problem: All the LS formalism derived above still applies, except that the weight matrix **W** now has nonzero off-diagonal elements. This does mean that the treatment of correlated data is beyond the capabilities of most data analysis programs, since they are not designed to handle such complexity in **W**. It also means the analyst must understand the correlation well enough to determine the elements of **W**. However, the experimental situation often makes the latter determination straightforward, and programming languages and mathematical programs like Mathematica and MathCad can be used to solve the LS equations when this situation arises. In many cases an easier solution is just to devise a fit of the original raw data rather than use the model that requires the manipulated or prefitted data.

H. Fitting with Constraints

Sometimes one desires to limit the values the parameters can attain through some expression known to relate them. The simplest such case was mentioned in Section III.F: If a parameter is well known from other work, it can be frozen in the fit. A straightforward extension of this idea is the case where the parameters are known to sum to a constant. This condition can be accommodated by defining the last parameter as the constant minus the sum of the others. Note that in both of these examples the constraint is satisfied *exactly*, as opposed to cases where some condition or relation is just given heavy weight in the fit.

It is harder to incorporate exact constraints that are complex in nature, or multiple constraints that must be satisfied simultaneously. A standard method for doing so is Lagrange's method of undetermined multipliers (Tellinghuisen, 2003c). Often the constraints are such as to require iterative solution, putting constrained fitting in the category of NLS (below). However, I will include the topic here for completeness.

Recall that the LS equations are obtained by minimizing $\mathcal{S}$ with respect to each of the adjustable parameters $\boldsymbol{\beta}$. Suppose there are k constraints, expressed in the form $G_k(\boldsymbol{\beta}) = 0$. We subtract $\alpha_k(\partial G_k/\partial \beta_j)$ from each equation, for each constraint, where α_k is the Lagrange multiplier for the kth constraint. For example, if there are two constraints, $G_1 = 0$ and $G_2 = 0$, the equation for the jth parameter is obtained from

$$\frac{\partial S}{\partial \beta_j} - \alpha_1 \frac{\partial G_1}{\partial \beta_j} - \alpha_2 \frac{\partial G_2}{\partial \beta_j} = 0 \tag{14}$$

Procedurally, the equations $G_k(\boldsymbol{\beta}) = 0$ are solved for the constraints α, and the latter are then employed in the solution of Eq. (14). As already noted, this process is normally an iterative one. As constrained fitting will not be illustrated below, the interested reader is referred to an earlier paper for details, including expressing Eq. (14) in matrix form (Ashmore and Tellinghuisen, 1986).

I. Error Propagation

Errors in functions f of uncertain variables are often estimated using the textbook error propagation formula

$$\sigma_f^2 = \sum \left(\frac{\partial f}{\partial \beta_i}\right)^2 \sigma_{\beta i}^2 \tag{15}$$

However, this expression assumes that the independent variables are truly independent, or *uncorrelated.* As was just noted, that assumption rarely holds when the variables are the adjustable parameters from an LS fit, and one must use the full expression (Tellinghuisen, 2001)

$$\sigma_f^2 = \mathbf{g}^{\mathrm{T}}\,\mathbf{V}\mathbf{g} \tag{16}$$

in which the elements of $\mathbf{g}$ are $g_i = \partial f / \partial \beta_i$. Equation (16) is exact for linear functions f of normal variates. In many cases Eq. (16) can be bypassed and the desired σ_f^2 can be calculated by simply redefining the fit so that f is one of the adjustable parameters—a procedure that can be shown to be identical to use of Eq. (16).

Equation (15) does suffice in one important situation: when single variables are transformed mathematically. For example, I have already mentioned the use of reciprocation and logarithmic conversion. If $z = 1/y$, $\sigma_z = \sigma_y/y^2$; and when $z = \ln y$, $\sigma_z = \sigma_y/y$. Of course, these simple results say nothing about the changes in the probability distributions associated with these transformations (Tellinghuisen, 2000b).

J. Systematic Error

Everything said so far about experimental error concerns the random statistical error that afflicts all measurements. Yet often these errors are not the only ones ... or even the most significant, which is the reason for the important distinction between *precision* and *accuracy* stated earlier. The standard approach for systematic error is to eliminate it to the extent possible. For example, calibration is an attempt to do this, usually by substituting a series of samples thought to be identical to the unknown in every way except that the targeted property is known and varied in a controlled manner. In the end we can never be sure we have eliminated all sources of systematic error, so it behooves us to retain some humility in stating our confidence in our results.

An instructive example is the value of Planck's constant h, which went from 6.626 176(36) $\times$ 10^{-34} J sec in the 1973 adjustment of the fundamental constants to 6.626 076(4) $\times$ 10^{-34} J sec in 1986 (Cohen and Taylor, 1986). Both values are of course quite precise for most purposes; but the 1973 result was high by almost $3\sigma_{73}$, attributed in 1986 to the deletion of two measurements that seemed discrepant at the time.

One type of systematic error I do address is that associated with the analysis itself—bias. We have already seen how bias can occur when data are transformed, and we will see below examples of bias and inconsistency that result from LLS fits of such data as well as from NLS fits of normal data. The Monte Carlo method provides a straightforward way to investigate and quantify such bias.

IV. Linear Least Squares—Monte Carlo Illustrations

A. The Straight Line with Constant Error

The MC calculations provide an eye-opening appreciation of the often eye-glazing terminology and properties covered in Section III. The theory presupposes the existence of a "true" function linear in the adjustable parameters $\boldsymbol{\beta}$, onto the dependent variable of which is superimposed random error. The MC computations repetitively replicate this model. By examining the results of many such replications, we either confirm the stated properties or discover the extent to which they fail to hold when one or more of the tenets are violated in the model. For the latter, here I have taken the standard straight-line relation. Its properties and the MC procedures are summarized in Fig. 9. This model has been used previously for similar purposes (Tellinghuisen, 2000b,c; 2005a), where more details about the MC computational procedures may be found.

Monte carlo case study: $y = a + bx$

- True $a = 1$ and $b = 5$; 5 x_i values =1.1, 3.3, 5.5, 8.3, 12
- For constant $\sigma = 0.5$, $\sigma_a = 0.41860$ and $\sigma_b = 0.058588$
- Add random Gaussian error of $\sigma = 0.5$ to each true y_i
- Solve LS equations for a and b; compute $\mathscr{S}$, s_a, and s_b
- Repeat procedure 10^5 times; accumulate statistics on parameters and $\mathscr{S}$; histogram values
- Compare results with predictions

Fig. 9 Summary of procedures for MC test of 5-point linear model.

Figure 10 displays histograms of the estimated intercept a, as obtained using the 5-point model with uniformly distributed error and normal error (both having $\sigma = 0.5$), and normal error for 20 points, with 4 points taken at each of the 5 x_i values (to preserve the structure of the model). The N5 and N20 results are as expected, with the latter showing a σ_a half that of the former, illustrating the $n^{-1/2}$ dependence. The U5 results do not follow the Gaussian curve. On the other hand, the sampling statistics (not shown) confirm that these results are still unbiased and minimum-variance (Tellinghuisen, 2005a).

Figure 11 shows the same results, but histogrammed as the t-variate. Note that here s_a is calculated anew for each data set using $\mathbf{V}_{\text{post}}$. The N5 and N20 results follow the expected t-distribution for their respective ν, but the U5 results do not. Figure 12 illustrates the quantity $\mathcal{S}/\nu$, with fitted χ^2_ν curves [Eq. (11)] included for the normal data. Again the U5 data do not follow this distribution, confirming that normal error is a requirement for all three distributions.

Suppose we doubled the data error for the N20 computations. The results in Fig. 10 would then follow the current curve for N5, but the plots in Figs. 11 and 12 would be unchanged. Alternatively we might consider preaveraging each set of 4 points and then fitting. As long as we weight each averaged value equally, the results in Fig. 10 would be identical; and those in Figs. 11 and 12 would now follow the $\nu = 3$ curves. This result has an important practical implication: such preaveraging results in a loss in precision for the estimated parameter variances, so if the common *a posteriori* mode is used, preaveraging should be avoided. Suppose that instead of weighting each averaged value equally, we weighted in accord with the sampling statistics for each average. Then none of the results would agree, and the estimates would no longer be minimum variance (Tellinghuisen, 1996). This result

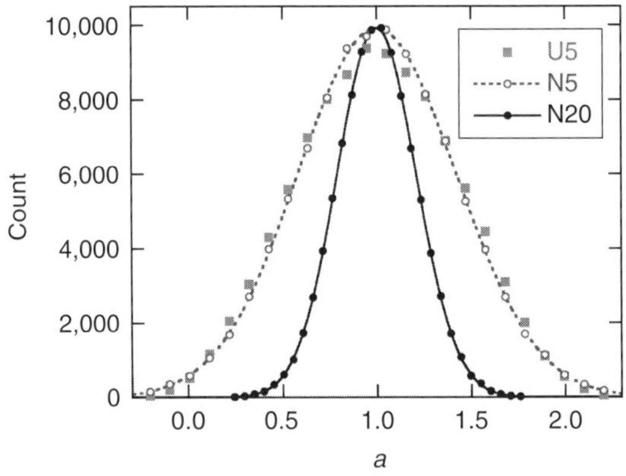

Fig. 10 Histogrammed results for intercept, from 10^5 MC data sets generated using 5 points with uniformly distributed error (U5) and normal error (N5), and 20 points having normal error (N20). The fits for normal error employed the expected value for σ_a from Fig. 9 for 5 points, half that value for 20 points. [Used with permission from Tellinghuisen (2005a); © 2005, Division of Chemical Education, Inc.]

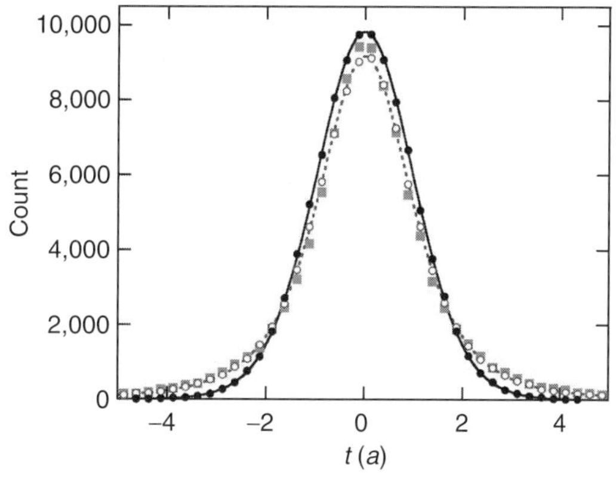

Fig. 11 Histogrammed results for quantity $(a - 1)/s_a$. Symbols as in Fig. 10. Curves are fits to t-distribution [Eq. (12)] for $\nu = 3$ (5 points) and 18. [Used with permission from Tellinghuisen (2005a); © 2005, Division of Chemical Education, Inc.]

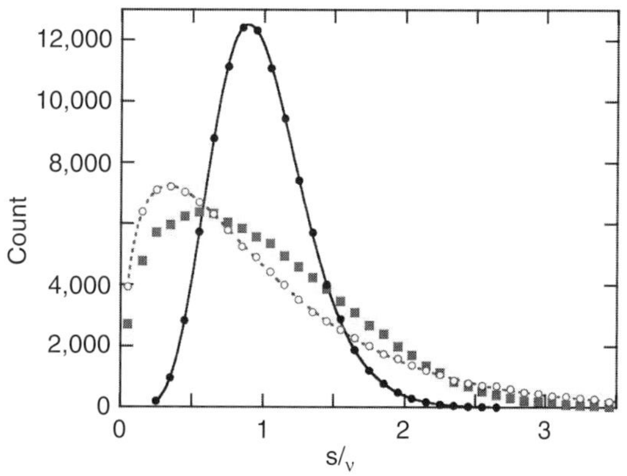

Fig. 12 Histogrammed results for $\mathscr{S}/\nu$, labeled as in Fig. 10. Curves are fits to reduced χ^2 distribution for $\nu = 3$ (5 points) and 18. [Used with permission from Tellinghuisen (2005a); © 2005, Division of Chemical Education, Inc.]

emphasizes the need for valid *a priori* information about the data error in assigning weights.

In all of these examples, the data are unbiased. We can examine the effect of data bias by using data inversion, as in Table I. We start by generating preliminary true data $z_i = 1/y_i$. To maintain consistency, we add to each such z_i random normal error of magnitude $\sigma_z = \sigma_y/y^2$ [Eq. (15)]; then we fit the resulting $y_i = 1/z_i$. The results (Fig. 13) confirm that biased data yield biased parameter estimates. Since the data are also no longer normal, neither are the parameter distributions, though they appear to be close to normal. Actually, these inverted data also violate the

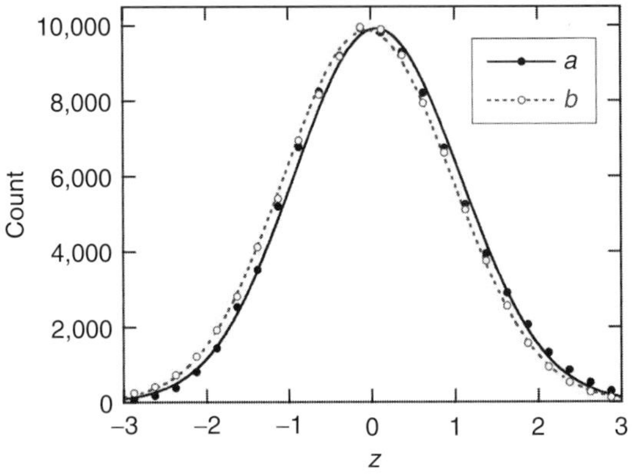

Fig. 13 Histogrammed results for biased data. The binning argument Z is the parameter residual divided by its true σ for 5 points. The biases are $+23\sigma$ and -15σ for a and b, respectively, where the metric is the MC standard error, for example, $\sigma_a/\sqrt{10^5}$ for a. [Used with permission from Tellinghuisen (2005a); © 2005, Division of Chemical Education, Inc.]

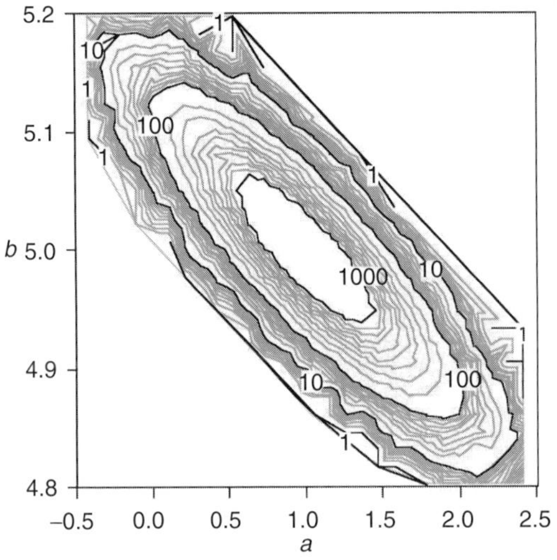

Fig. 14 Contour diagram of two-dimensional histogram of results for a and b from 10^5 MC replicates. The binning interval is $\sigma/4$ in each parameter. [Used with permission from Tellinghuisen (2005a); © 2005, Division of Chemical Education, Inc.]

requirement of finite variance; however, here most data are relatively precise, so their inherent infinite variance gives no obvious problem. In fact, though biased, these estimates remain essentially minimum variance (in the asymptotic sense of these 10^5-point data sets).

The interparameter correlation can be examined through a two-dimensional histogram, shown as a contour diagram in Fig. 14. For this 2-parameter model, the correlation is directly dependent on the sum $\Sigma w_i x_i$ that occurs in the off-diagonal positions in **A**; its numerical value is $C_{12} = -0.8454$. By redefining the fit function as $y = a' + b(x - x_0)$ and then taking x_0 as the average of the 5 x_i values, we obtain a statistically equivalent model with zero correlation between a' and b. a' is also then the value of the fit function at x_0, and its statistical error is the error in the fit function there. With this redefined fit, we can use the simpler error propagation expression of Eq. (15) to obtain errors in functions of a' and b, since these parameters are uncorrelated.

Extending the above considerations, if we repeat the fit for a range of different x_0 values, we will obtain a range of a' values and their errors. In this fashion we can generate the error band on the fit function, and we will find that $\sigma_{a'}$ has its minimum value at $x_0 = \bar{x}$. This treatment constitutes an example of redefining the fit function to obtain directly the error in a function without using error propagation [Eq. (16)]. From this it is also clear that the meaning of σ_f in Eq. (16) is the same as that for any parameter σ_β: Under the usual assumptions for the data, if values of f are computed from each set of fitted parameters, the results will be distributed normally about the true f with the indicated standard error.

B. The Straight Line with Proportional Error: Neglect of Weights

Leaving aside the weights that arise when data are transformed (e.g., from reciprocation), the common extremes for data error are constant error, just considered, and proportional error, $\sigma_i \propto y_i$ (Ingle and Crouch, 1988). (Counting error is intermediate, $\sigma_i = \sqrt{y_i}$.) Constant error dominates in the small-signal limit, and proportional error often rules for large signal. I consider here the proportional error extreme and ask what is the effect of neglecting weights. I stay with the 5-point model of Fig. 9 but take the error to be 4% of the true y_i. This means the data error spans about an order of magnitude and the weights about two orders—a reasonable intermediate of actual cases, where the y data often span less than an order of magnitude in range, but sometimes much more. For this model the exact parameter errors are $\sigma_a = 0.33606$ and $\sigma_b = 0.13551$.

Figure 15 shows the results for the intercept, with and without weighting. The top part confirms that the distribution remains unbiased and normal when weights are neglected; but the estimates are no longer minimum variance, with σ_a having increased by a factor 2.5. This is a loss of efficiency by more than a factor of 6, meaning six times as many measurements would be needed to match the precision obtained with proper weighting. The bottom frame shows that the t-distribution is invalid when weights are neglected. Similarly $\mathcal{S}/\nu$ (not shown) is not χ^2_ν-distributed when the weights are neglected.

Of course the user would not have the benefit of 10^5 replicates to assess the parameter error, rather would use the $\mathbf{V}_{\text{post}}$-based s_a and s_b. From the MC statistics, the latter predictions are $s_a = 1.121$ and $s_b = 0.157$, as compared with

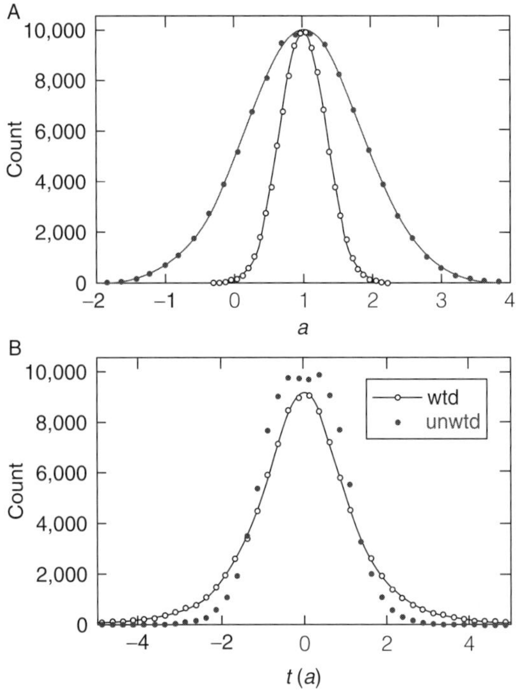

Fig. 15 Histogrammed results for 10^5 MC estimates of the intercept for proportional error, $\sigma_i = 0.04\ y_i$, with proper weighting (open points) and with neglect of weights. In (A) the fitted curves are Gaussians with $\sigma = 0.3356(8)$ (open points) and 0.843(2). [Used with permission from Tellinghuisen (2005a); © 2005, Division of Chemical Education, Inc.]

the MC sampling statistics (e.g., Fig. 15A) of 0.842 and 0.209, respectively. These results show that $\mathbf{V}_{\text{post}}$ can "lie" either optimistically or pessimistically, though in this case the magnitude of the error is small compared with the inherent statistical uncertainty in these *a posteriori* standard errors ($1/\sqrt{6}$ relative error).

As a variation on this theme, suppose we have data of constant absolute error in y but wish to enhance somehow our determination of the relatively uncertain intercept a. Recognizing the importance of the small-x points, we naively elect to weight our data as x^{-1}. Unfortunately, as the old saying goes, "You cannot fool Mother Nature." MC computations show that this manipulation actually increases the statistical error in a by 18%, while the $\mathbf{V}_{\text{post}}$-based s_a is optimistic by a factor of 2, deceiving us into *thinking* we are doing much better than we are.

In summary, these examples show that minimum-variance estimates of the parameters are achieved *only* when the data are weighted as their inverse variance. In the absence of proper weighting, the *a posteriori* estimates of the parameter errors are unreliable. This might seem to confront us with a quandary, since it is not possible to know the data error absolutely. Fortunately, even approximate knowledge of the data error structure usually suffices to produce near-minimum-variance parameter estimates. In this regard, there are methods in which one can simultaneously estimate the data error and the parameters, for suitably large data sets (Carroll and Ruppert, 1988). Also, while the use of x^{-1} weighting above represents an improper attempt to improve the precision of a, the adjustment of the $\{x_i\}$ to incorporate more small-x values would be a legitimate use of **V** in experiment design.

V. Nonlinear Least Squares

A. When Is It Nonlinear and What Then?

If the adjustable parameters appear in the fit model in any way that cannot be expressed in the form of Eq. (2), the fitting problem becomes nonlinear. I have already given simple examples, but this category covers most of the data analysis problems in physical science, including especially those already alluded to that can be manipulated into the linear form through redefinition of the variables. As has been mentioned and will be dealt with further below, such redefinition is usually statistically inferior to the nonlinear analysis of the raw data.

In addition to the cases where the fit model is nonlinear in the parameters, there are two other situations where the problem becomes nonlinear: (1) when more than one variable is considered uncertain and (2) when weights are estimated at the adjusted values of the variables. To clarify the latter, consider the case of proportional error in the straight-line example above. In the MC context, the computationalist can compute the weights from the true y_i. However, the data analyst would not know the latter and so would normally assess the weights using the measured y_i. But these are inherently wrong, so a statistically better approach is to compute the weights from the *calculated* values of y_i. Since the latter are not known at the outset, the analysis becomes nonlinear and must proceed in an iterative fashion.

Figure 16 summarizes the main practical consequences of the jump from linear to nonlinear LS. Although the new problems seem daunting, in practice they are seldom more than minor inconveniences, as I have attempted to reassure the reader. Many data analysis packages make it trivially easy for the user to define and carry out an NLS fit; in fact it is often preferable to work in the "user-defined" mode for all fits, linear and nonlinear. For example, in the KaleidaGraph program, this is the only way to get the parameter error estimates.

B. Computational Methods

The solution of the NLS problem is fundamentally one of successive adjustment of the parameters in the search for a minimum value of $\mathcal{S}$. There are many computational approaches for carrying out such optimizations (Bevington, 1969; Press *et al.*, 1986).

Nonlinear least squares: What's new?

1. Standard methods yield formally similar equations for $\boldsymbol{\beta}$, but now these must be solved interatively.
2. Unfortunately, there is no guarantee of solution.
3. For solutions, there is no guarantee of absoulte minimum $\mathcal{S}$.
4. Even when the data are normal and properly weighted, the parameter distributions are generally non-Gaussian and biased, so the predictions from the **V** matrix no longer hold.

So, give it up?

No! Most interesting problems are inherently nonlinear, and modern data analysis packages make using NLS as easy as writing the functional relation connecting the data and the adjustable parameters.

Problems 2 and 3 are seldom serious. Methods like the Levenberg-Marquardt algorithm make nonconvergence a rare event, and generally the analyst can easily distinguish convergence on subsidiary minima from the desired physical solution.

Problem 4 is rarely serious when the **V**-based parameter error is less than 10% of the magnitude of the parameter — the **10% Rule of Thumb**.

Fig. 16 Summary of key differences between nonlinear and linear LS.

I will confine my attention to the one that best preserves the analogy with LLS, which happens to give identical formal definitions of the **V** matrices. This is referred to as the linearization method by Bevington (1969), and the inverse Hessian approach by Press *et al.* (1986).

Since the solution requires iteration, one must start with a set of initial values $\boldsymbol{\beta}_0$ of the parameters. Sufficiently near the minimum in $\mathcal{S}$, one can evaluate small corrections to $\boldsymbol{\beta}_0$ using

$$\mathbf{X}^{\mathrm{T}}\mathbf{W}\mathbf{X}\Delta\boldsymbol{\beta} \equiv \mathbf{A}\Delta\boldsymbol{\beta} = \mathbf{X}^{\mathrm{T}}\mathbf{W}\,\boldsymbol{\delta} \tag{17}$$

leading to improved values

$$\boldsymbol{\beta}_1 = \boldsymbol{\beta}_0 + \Delta\boldsymbol{\beta} \tag{18}$$

Here the elements of the matrix **X** are $X_{ij} = (\partial F_i/\partial \beta_j)$, evaluated at x_i using the current values $\boldsymbol{\beta}_0$ of the parameters. The function F expresses the relations among the variables and parameters in such a way that a perfect fit yields $F_i = 0$. For the commonly occurring case where y can be expressed as an explicit function of x, it takes the form

$$F_i = y_{\mathrm{calc}}(x_i) - y_i = -\delta_i \tag{19}$$

which serves also to define the elements of $\boldsymbol{\delta}$ in Eq. (17). The weight matrix **W** is as before, including cases where there are off-diagonal elements (correlated fitting).

For a linear fit, starting with $\boldsymbol{\beta}_0 = 0$, these relations yield for $\boldsymbol{\beta}_1$ equations identical to Eqs. (4) and (5) for $\boldsymbol{\beta}$. In the more general case where y cannot be written explicitly in terms of the other variables, these equations still hold, but with $\boldsymbol{\delta}$ in Eq. (17) replaced by $-\mathbf{F}_0$, where the subscript indicates that the F_i values are calculated using the current values $\boldsymbol{\beta}_0$ of the parameters.

The previous definitions of the **V** matrices [Eqs. (8–10)] remain valid, as already noted. However, there is an important difference: Now **A** and **V** can depend on not just the x-structure of the data but also **y** and $\boldsymbol{\beta}$. For the purpose of MC computations and experiment design, it is useful to define the "exact nonlinear $\mathbf{V}_{\text{prior}}$." This quantity is obtained using exactly fitting data for the model in question but is generally not exact in the sense of the guarantees of LLS.

The utility of Eq. (17) is couched in the term "sufficiently near the minimum in $\mathcal{S}$." In practice many nonlinear fits converge from quite far away from the minimum, but there is no way to better state this proviso. The essence of the Levenberg–Marquardt algorithm is the recognition that a gradient search for the minimum in $\mathcal{S}$ is more stable but less efficient, and that the diagonal elements of **A** are proportional to the required gradients. A compromise is obtained by scaling the diagonal elements of **A** by a factor $(1 + \lambda)$. Then, if the computation leads to higher $\mathcal{S}$, λ is increased (usually by a factor 10) and the computation repeated until $\mathcal{S}$ decreases. Near a minimum in $\mathcal{S}$, λ is progressively decreased until its effect is negligible. With packaged programs it can be important to demand hard convergence, to ensure that λ is small enough to yield valid **V**.

The partial derivatives that constitute the elements of **X** are usually approximated numerically, by finite difference. For this purpose programs normally take a set fractional change in the parameter. This can sometimes lead to problems for parameters that are very precise in a relative sense. For example, if a Gaussian spectral line with width 0.1 cm^{-1} is centered at $\nu_0 = 40{,}000\ \text{cm}^{-1}$ (250 nm), finite difference derivatives using part-in-10^4 changes in ν_0 will fail miserably. Such problems can be eliminated by defining the adjustable parameter as a small additive correction to a large constant [e.g., instead of $\nu_0 = a$, $\nu_0 = 40{,}000 + a$, with a the adjustable parameter; see Tellinghuisen (2000a)].

C. The 10% Rule of Thumb

As is noted in Fig. 16, even with proper weighting of unbiased, normal data, there is no guarantee of normality for the distribution of estimated parameters in NLS; in fact, the parameters in a nonlinear fit are generally nonnormal. On the other hand, from MC computations on a large number of common nonlinear problems, I have found that such nonnormality is often not a serious limitation to establishing reliable confidence limits for the parameters (Tellinghuisen, 2000b, 2003b). As a guide to the reliability of the **V**-based parameter errors, I have formulated the *10% Rule of Thumb*: If the parameter standard error is less than 1/10 of the magnitude of the parameter, the "normal" interpretation of the error will suffice to peg the confidence range within 10%. More precisely, nonnormal distributions are typically asymmetric, so the confidence limits for typical specified confidences (e.g., 68%, 90%, or 95%) are asymmetrically disposed about the mean. If the parameter is precise to better than 10%, the asymmetry in the confidence limits is likely to be less than 10%.

A primary cause of nonnormality in NLS is reciprocal behavior, of the type illustrated in Table I and Fig. 6. For example, in the study of binding constants, one can choose to fit to either of two constants, the association constant K_a or its reciprocal K_d. If the data render these constants with large uncertainty, then one may be nearly normal while the other is radically nonnormal and may even exhibit divergence in MC sampling (Tellinghuisen, 2000b).

I have observed only a few clear violations of this 10% rule, most notably in a study of the direct fitting of optical spectra with an 8-parameter model (Tellinghuisen, 2004b). In cases where problems are suspected, the MC method can be used to quantify the properties of the nonnormal distributions.

VI. Applications and Illustrations

A. Univariate Calibration: Beyond the Straight Line

Analytical chemists have a near religious devotion to the straight line in calibration. I have speculated that this stems from the fact that application of Eq. (16) to straight-line calibration yields a quotable expression for the uncertainty in the result ... and that no comparably simple formal results exist for calibration relations other than linear. However, viewed as a numerical problem, the computation of the uncertainty is straightforward and simple for *any* calibration relationship. Since nature is inherently *not* linear when pushed to high precision and large ranges of the calibration independent variable, such numerical approaches to calibration deserve more attention.

In classical univariate straight-line calibration, the usual assumptions behind LLS are made for the n calibration data points, and the common default assumption is constant data uncertainty σ. The calibration data are fitted to $y = a + bx$, where the response y is the measured quantity and the independent variable represents the controlled quantity (e.g., concentration, pressure, fractional abundance). The unknown x_0 is calculated from its response y_0 using $x_0 = (y_0 - a)/b$. The standard error in x_0 is a sum of contributions from the uncertainty in y_0 and the (correlated) uncertainty in a and b from the fit; the textbook expression widely used to compute this error is (Miller, 1991)

$$\sigma_{x0}^2 = \frac{\sigma^2}{b^2}\left[\frac{1}{m} + \frac{1}{n} + \frac{(y_0 - \bar{y})^2}{b^2\sum(x_i - \bar{x})^2}\right] \tag{20}$$

where m is the number of measurements averaged to obtain y_0. If the slope b is relatively imprecise, small correction terms are needed in Eq. (20) (Draper and

Smith, 1998). For heteroscedastic data, weighted regression is appropriate; the standard error in x_0 is then computed from a modified form of Eq. (20) in which weighted averages replace averages, with weights w_i included in the summation.

By all the properties of LLS, y_0, a, and b are normal variates; but because of the occurrence of b in the denominator of its defining equation, x_0 is not only nonnormal but also of infinite variance. Of course this has not prevented generations of analytical chemists from using it successfully, and the reason is again that as long as b is reasonably precise (say better than 10%), the variance of Eq. (20) is asymptotically well defined.

In a radically different approach to calibration, I have defined the fit model as nonlinear, with y_0 included as the $(n + 1)$th y-value in the data set, and x_0 added as the third adjustable parameter (Tellinghuisen, 2000d). The y_0 value is fitted exactly to the equation defining x_0, and the result of Eq. (20) is returned automatically, as a numerical outcome of the fit (and hence requiring no additional work). But the real beauty of this approach is its ease of implementation using packaged data analysis programs, and its trivial extension to heteroscedastic data and complex calibration functions (Tellinghuisen, 2005b).

From a fundamental standpoint, the new calibration algorithm is interesting in demonstrating that nonlinear fit models *can* contain some adjustable parameters that are perfectly normal—namely, all adjustable parameters other than x_0 in a linear calibration model (including all coefficients in a fit to a polynomial, e.g.).

The appearance of σ in Eq. (20) shows that this expression is based on $\mathbf{V}_{\text{prior}}$. Yet most calibration work employs the *a posteriori* approach, treating each calibration data set in isolation. For "one-shot" determinations this is appropriate. In that case the σ in Eq. (20) should be replaced by its estimate s_y; and the t-distribution should be used to establish confidence limits. Since calibration is often bottlenecked on sample preparation, the number of calibration points n may be small, translating into large uncertainties due to the $(2\nu)^{-1/2}$ uncertainty inherent from the properties of χ^2. At the other extreme, we have routine work involving day-to-day repetition of similar procedures. I have argued that in such cases the accumulated information about the data error can make it well enough known to justify using the prior approach, in which case each day's calibration fit can be subjected to a χ^2 test and rejected and repeated if found wanting (Tellinghuisen, 2005b).

B. "Going Straight" . . . by Shunning the Straight Line

As I noted earlier, the fitting of data reciprocals to straight-line relationships is a widespread practice, used in the analysis of binding data (including adsorption), fluorescence quenching (Stern–Volmer plot), and enzyme kinetics (Lineweaver–Burk plot). To demonstrate that this can be a poor procedure, I illustrate here the simple case of data having constant, normal error and following the relationship, $y = (a + bx)^{-1}$. Accordingly, the reciprocal, $z = 1/y$, should fit a straight line. If the error in y is constant (σ), that in z (from error propagation) is $\sigma/y^2 = \sigma z^2$. Without knowing σ, we can still assign relative σ_z values from just our knowledge of the transformation; and these suffice for calculating the weights in the straight-line fit of z. However, in the computations we do not know the true y, so must calculate these s_z values using either the observed y_i or the calculated values from the fit. The latter approach requires iterative estimation of the weights, which cannot be done automatically with the data analysis programs I know. It can be done by "manual" iteration; but most workers would not do this, so here I have used "observed" weighting.

Fits of 23 synthetic data points to both the direct nonlinear model and the straight-line relation are illustrated in Fig. 17. In the unweighted fit to the nonlinear model, the intercept is high by 1.1σ and the slope low by 0.9σ. On the other hand, the linearized fit returns an intercept high by 4σ and a slope low by 5σ. Let us attempt to improve the situation by doubling the number of data points. The results from two such tries (not shown, but employing new random normal deviates for y) yield from the linear fit an intercept high by 5.4σ and 3.5σ, and a slope low by 7.3σ and 6.1σ. The nonlinear analysis of the same data yields intercept estimates off by $+1.2\sigma$ and -0.1σ, and slopes off by -1.1σ and -0.8σ.

The behavior of the linear analysis in this example is the hallmark of inconsistency: As the number of data points increases, the precision of the determination increases, and the fit "locks in" on a result that is wrong by any statistical consistency test. That point occurs already with just 23 points in this example. Increasing the number of data points often makes matters worse rather than better. Here the inconsistency results from the nonnormality inherent in the transformed data. Nonlinear fits are not free from inconsistency, but in the cases I have examined, such problems are much less significant, making the nonlinear fit of the raw data preferable to the linear fit of the transformed data.

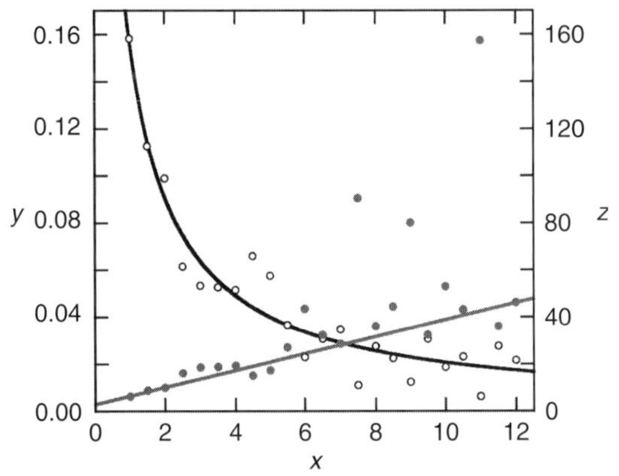

Fig. 17 LS fits of synthetic data normal in y with $\sigma = 0.01$ to the true model $y = (a + bx)^{-1}$ and their reciprocals $z = 1/y$ to the straight line. The true values of a and b are 1 and 5, respectively. The unweighted fit to the nonlinear model yields $a = 1.72(65)$ and $b = 4.65(41)$. The weighted linear fit yields $a = 2.94(49)$ and $b = 3.59(28)$ (*a posteriori* errors).

The second common data transformation is logarithmic conversion, used to facilitate analysis of first-order kinetics data in the absence of a background. This transformation does not lead to infinite variance, and in my experience has fewer drawbacks than does the inversion transformation. Also, when $z = \ln y$, $\sigma_z = \sigma_y/y$; thus, when the raw data have proportional error, the transformed data have constant error and the fit becomes an unweighted one, removing the ambiguity about how to actually assess the weights. The direct nonlinear fit remains preferred when the data error is constant and when the transformed model is also nonlinear, as from a background contribution.

C. Error Propagation: How Good Is a vH Analysis?

Chemical binding and complexation are of interest in many fields of study (see Chapter 1 by Garbett and Chaires of Section I, this volume). Almost all methods used to study binding (Connors, 1987) yield directly only estimates of the binding equilibrium constant $K^\circ(T)$. This suffices to obtain ΔG° for the process, but the other key thermodynamic properties—ΔH°, ΔS°, and ΔC_P°—must be obtained from the temperature dependence of K°, in what is known as a vH analysis. In differential form, the vH relation reads

$$\left(\frac{\partial \ln K^\circ}{\partial T}\right)_P = \frac{\Delta H^\circ}{RT^2} \tag{21}$$

With adoption of the form

$$\Delta H^\circ = a + b(T - T_0) + c(T - T_0)^2 \tag{22}$$

Eq. (21) can be integrated to yield

$$R \ln K^\circ = \mathrm{R} \ln K_0^\circ + A\left(\frac{1}{T_0} - \frac{1}{T}\right) + B \ln\left(\frac{T}{T_0}\right) + c(T - T_0) \tag{23}$$

where

$$A = a - bT_0 + cT_0^2 \text{ and } B = b - 2cT_0 \tag{24}$$

Thus, by fitting to Eq. (23), one can estimate the desired quantities, since at $T = T_0$, $\Delta H^\circ = a$, $\Delta C_P{}^\circ = b$, and $d\,\Delta C_P{}^\circ/dT = 2c$. Of course there is nothing special about the choice here of T_0, so by repeating the fit for a range of T_0, one obtains a set of statistically identical fits; and if the adjustable parameters are defined as *a, b, c*, and K_0°, the statistical errors on these parameters, taken as functions of T_0, constitute the error bands on the determination. Identical results could be obtained from the results at a single T_0, using the full expression for error propagation in

Eq. (16). However, the variable-T_0 approach is directly available to anyone who knows how to use a data analysis program, while implementing Eq. (16) can require some programming. This example thus constitutes another where error propagation can be accomplished through a judicious definition of the fit model. Note especially that all of the statistical errors in question here are themselves functions of T, with the exception of the highest order term in Eq. (22)—c as written, or b if c is not determined.

The vH analysis is normally done with an unweighted fit to Eq. (23). I have noted recently that this approach is tantamount to an assumption of constant relative error in $K^\circ(T)$ (from error propagation, above), and that with this assumption: (1) the LS fit is linear and (2) the errors in all quantities depend only on the percent error in K° and on the T-structure of the data set (Tellinghuisen, 2006). Since the parameter errors scale with the data error, results for a single percent error fully characterize the parameter error. (This is true only for proportional data error.)

Results for 4% error in $K^\circ(T)$ are illustrated in Fig. 18, for two different data structures. The extension of the range by 5° on both ends significantly increases the precision of the analysis. Note also the difference between the propagated data error in K° and the fit error in Fig. 18A: The smaller errors for the latter reflect the averaging effect of fitting multiple points to a model. Part (B) shows that the naively propagated error for ΔH° is only somewhat larger than the correct results; however, this comparison would be less favorable if the parameters and errors at either t limit were chosen in place of the central 25° values. The analysis also shows that $\mathrm{d}\Delta C_P^\circ/\mathrm{d}T$ is not statistically defined unless it is larger than 23 cal mol^{-1} K^{-2} for 7 points or 9 cal mol^{-1} K^{-2} for 9 points. Dropping c from the fit model makes the other parameters more precise, and also makes ΔC_P° and its error both independent of T.

The error in ΔG° can be obtained from the fit error in K_0° using $\sigma_{\Delta G^\circ} = -RT_0\sigma_{K^\circ}/K^\circ$ [obtained by applying Eq. (15) to $\Delta G^\circ = RT \ln K_0$]. Alternatively, $R \ln K_0^\circ$ can be replaced in the fit model [Eq. (23)] by $-\Delta G^\circ/T_0$ to yield $\Delta G^\circ(T_0)$ and its error directly from the fit. In a further variation, ΔG° can be replaced by $\Delta H^\circ - T\Delta S^\circ$ to yield directly ΔS° and its error. However, because ΔG° is so much more precise than ΔH°, the error in $T\Delta S^\circ$ is practically identical to that in ΔH° and need not be calculated separately.

D. Correlated Fitting: The Guggenheim Method

Guggenheim (1926) suggested a method for analyzing first-order kinetics data when the desired exponential information is superimposed on a background: Record data for a number of early times t_i and then also for a set of later times displaced by a constant interval τ. Then subtract corresponding values (at t_i and $t_i + \tau$) to eliminate the background, and estimate the rate constant from the slope of a logarithmic plot of the absolute differences versus t_i. Although modern computational methods have long since rendered such graphical methods obsolete, the Guggenheim method still features in the teaching curriculum and even occasionally in research applications.

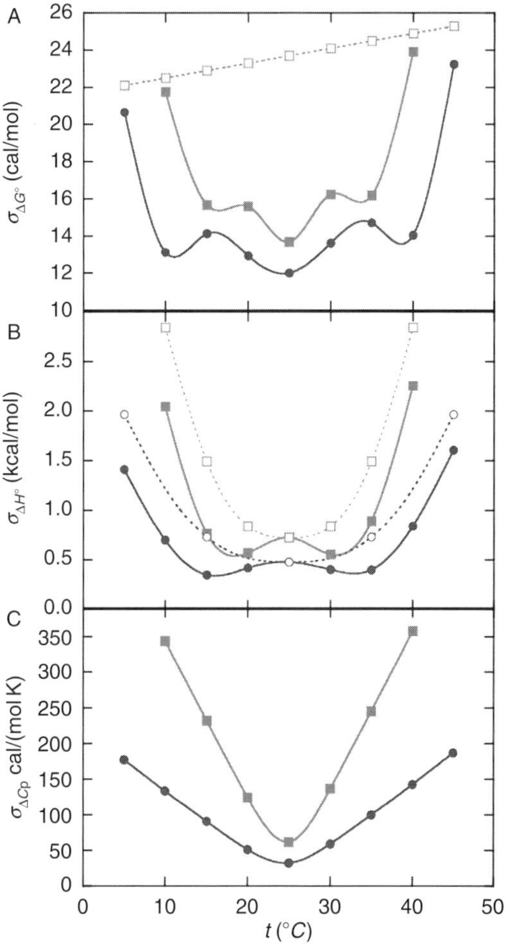

Fig. 18 Van't Hoff statistical error in ΔG°, ΔH°, and $\Delta C_P{}^\circ$ from 4% error in $K^\circ(T)$, for 7 evenly spaced T values between 10 and 40 °C (squares) and 9 values between 5 and 45 °C (round). The points and broken line at top in part (A) show the directly computed error from the 4% error in K°. The dashed curves and open points in (B) show the naively propagated error, as computed using Eq. (15) with the parameter errors at 25 °C. [Adapted from Tellinghuisen (2006); © 2006, with permission from Elsevier.]

As formulated by Guggenheim, this method actually does not have a correlation problem, but as it sometimes is implemented, it does. As we will see below, the difference hinges on whether any data points are used more than once to obtain the differences (Tellinghuisen, 2003a). We represent the raw data by **y** as before, and the computed differences by **z**. For simplicity we assume the data are recorded at a constant time interval Δt. Let the differences be obtained by subtracting points separated by $k\Delta t$, that is, $z_i = y_i - y_{i+k}$. The relation between **z** and **y** can thus be expressed in matrix form as

$$\mathbf{z} = \mathbf{L}\mathbf{y} \tag{25}$$

in which **L** contains m rows and n columns (with the number of z values m necessarily less than the number of y values n). The nonzero elements of **L** are $L_{ii} = 1$ and $L_{i,i+k} = -1$. The variance-covariance matrix $\mathbf{V}_y$ of the **y** values is diagonal by assumption, with elements$(i, i) = \delta_{yi}{}^2$. Accordingly, the variance-covariance matrix for the **z** values is

$$\mathbf{V}_z = \mathbf{L}\mathbf{V}_y\mathbf{L}^{\mathrm{T}} \tag{26}$$

and the weight matrix for the Guggenheim fit is $\mathbf{W}_z = \mathbf{V}_z^{-1}$ (Tellinghuisen, 2001).

It is easy to show that the diagonal elements of the $m \times m$ matrix $\mathbf{V}_z$ are $(\sigma_{yi}{}^2 + \sigma_{yi} + k^2)$, that is, the result from error propagation for subtraction of two random variates. Moreover, if $k \geq m$, all off-diagonal elements in Vz are zero and there is no correlation problem. This condition (also stated as $k \geq n/2$) occurs when none of the original **y** values is used more than once, from which it is clear that the correlation arises from such multiple use of individual data points, not from subtraction. This makes sense because the difference of two random variates is itself a random variate, of exactly predictable variance in the case of normal variates (Mood and Graybill, 1963).

Although the use of a **W** matrix having off-diagonal elements is beyond the capabilities of the data analysis programs with which I am acquainted, it is easy to handle with programming languages and mathematical programs. However, the main result for present purposes is that achieved in the preceding paragraph, so the reader interested in further details is referred to the original paper (Tellinghuisen, 2003a). It is worth mentioning one additional result: The best the Guggenheim method can do is match the results from a direct NLS fit of all data, and then only when the correlated fit is done for $k = 1$ (giving $m = n - 1$). The moral is actually a general one: With minor, specialized exceptions, *no amount of data premanipulation can beat a direct fit of all the raw data.*

E. Experiment Design Using the V Matrix: ITC

The one method of studying chemical binding that directly yields estimates of both K° and ΔH° (hence also ΔG° and ΔS°) at a single temperature is ITC (see also Chapter 4 by Freyer and Lewis of Section I, this volume). The experiment involves sequential injection of one reactant (titrant) into a cell containing the other (titrate), producing a titration curve of heat q versus extent of reaction. The shape of this curve is closely related to the K value, while its scale is proportional to ΔH°. Under favorable circumstances, both primary quantities can be obtained with very good precision (relative standard errors of 1%). The analysis involves

NLS fitting, with most workers using the software provided by the manufacturers of the instruments.

In performing an ITC experiment, the operator has control over a number of experimental parameters, including concentrations of titrate and titrant, and the number and volume of the injections. Since each experiments can take several hours, and since workers often need to process multiple samples over a range of temperature, it is of interest to optimize the choice of parameters to increase throughput and efficiency. Such an optimization can be seen as an effort to minimize the diagonal elements of the $\mathbf{V}_{\text{prior}}$ matrix, and constitutes an example of **V**-based experiment design.

In early considerations of this problem for the simplest case of 1:1 binding, $M + X \rightleftarrows MX$, I made several useful observations (Tellinghuisen, 2003b): (1) the experimental error could be dominated by uncertainty in the delivered volume of titrant rather than from the estimation of the heat for each injection; (2) if so, the analysis could require a weighted, correlated LS fit; (3) if q-estimation error dominates,high precision is favored by *fewer* injections, not more. The last of these is especially important, as it would mean simultaneous enhancement of throughput and precision. These results are somewhat surprising, so it is instructive to see how they arise.

Consider first the last of these, which seems to fly in the face of the standard expectation that precision improves as $n^{1/2}$. In this experiment there is a fixed total amount of "signal" (reaction heat), as dictated by the amount of reagent in the cell at the outset. Increasing the number of injections just subdivides this heat into smaller pieces, so if the uncertainty in each measurement is constant, the relative error goes up in proportion to the number of injections n. This more than offsets the statistical benefit of averaging, and the result is a precision that declines as $n^{1/2}$.

The significance of titrant volume error depends on the injection volume v. With the assumption of constant volume error, the relative error increases with decreasing v. Further, since $\Delta q/q = \Delta v/v$, and since the fit of q to the ITC model should employ weights proportional to $(\Delta q)^{-2}$, we see that volume error dictates weighting in proportion to q^{-2}. But that assumes that the titrant volume has independent, random error. In fact, the titrant is delivered by a stepping-motor-driven syringe, and it is at least as reasonable to assume that it is the stopping position of the syringe after each motion that is subject to independent, random error. In that case the delivered volume v becomes the *difference* between two such random deviates; and since each stopping position is involved in two delivered volumes—as the end point of one and the start of the next—we have a correlation problem. Interestingly, it turns out that when this model is correct and the proper correlated fit is used to analyze the data, the precision actually *increases* with the number of injections!

The above considerations were limited by lack of experimental information about the data error. Recently (Tellinghuisen, 2005c), I was able to get such information through an application of the method of Generalized Least Squares to a body of data for the complexation of Ba^{2+} with 18-crown 6-ether (Mizoue and Tellinghuisen, 2004). In this method the variance function for the data is estimated along with the fit parameters (Carroll and Ruppert, 1988). For the range of titrant volumes covered in the experiments ($v > 6\ \mu$l for most data), the error was dominated by a constant term, but with an additional term proportional to q, which was not recognized in the earlier work. An error term proportional to q/v was also statistically defined. However, the possible role of correlated volume error will require more data for smaller v, where this term is relatively more important.

Armed with reliable information about the data error, I return to the optimization problem and consider experiments over a wide range of $K°$ and $\Delta H°$. The key parameters for such considerations are $c \equiv K[\mathrm{M}]_0$ and $h \equiv \Delta H°[\mathrm{M}]_0$, where c determines the shape of the titration curve and h the scale. Figure 19 illustrates some results for the relative errors in determining K and ΔH. From a range of calculations of this sort, I have arrived at the choice of $m = 10$ injections (m is used here to avoid confusion with the other use of n in ITC for the stoichiometry

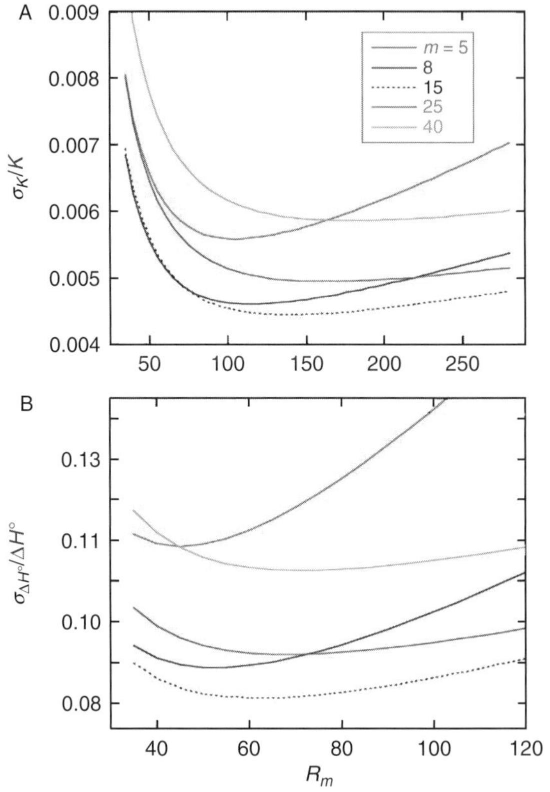

Fig. 19 Relative standard errors in K (A) and $\Delta H°$, as functions of the range of titration R_m and the number of injections m, for $c = 0.1$ and $h = 10$ cal/liter. [Fig. 2 from Tellinghuisen (2005d)] [Used with permission from Tellinghuisen (2005d); © 2005, American Chemical Society.]

parameter), and the following empirical equation for adjusting the stoichiometry range of the titration (Tellinghuisen, 2005d):

$$R_m = \frac{6.4}{c^{0.2}} + \frac{13}{c} \tag{27}$$

The quantity R_m is the ratio of total titrant to total titrate after the last injection of titrant. These results are particularly important for the small-c regime, where very significant increases in precision are achieved by titrating to much greater excesses of titrant than is commonly done.

In the analysis of ITC data, a third parameter is normally included in addition to K and $\Delta H°$. It is a stoichiometry parameter (usually represented by n), and is typically determined with higher precision than either of the other two. Figure 20 shows the precision to be expected from the recommended procedure, for each of the three parameters. At small c, n and $\Delta H°$ are very highly correlated. If either can be fixed by other methods (e.g., n from confident knowledge of the reaction stoichiometry and solution concentrations), the other can be determined with much greater precision. This is an example of the effects of correlation noted earlier.

In this case there is a second effect of the high correlation between n and $\Delta H°$ at small c: It can lead to slow convergence or even divergence in the solution of the NLS equations. MC results generated for this model reveal that the source of the problem is unanticipated reciprocal behavior in $\Delta H°$ (Fig. 21). When the adjustable parameter in the NLS algorithm is changed to the reciprocal of $\Delta H°$ (which is nearly normal), all convergence problems disappear—a dramatic illustration of the role of the (usually unknown) probability distributions of adjustable parameters in NLS.

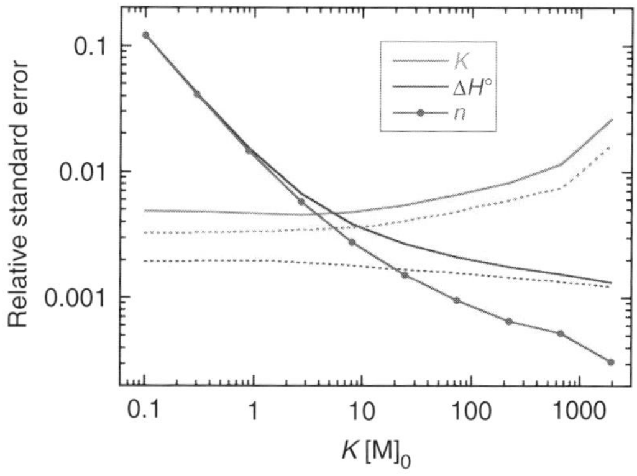

Fig. 20 Relative errors in K, $\Delta H°$, and n, as computed using the recommended procedure, and then recomputed after fixing $n = 1$ (dashed curves). [Used with permission from Tellinghuisen (2005d); © 2005, American Chemical Society.]

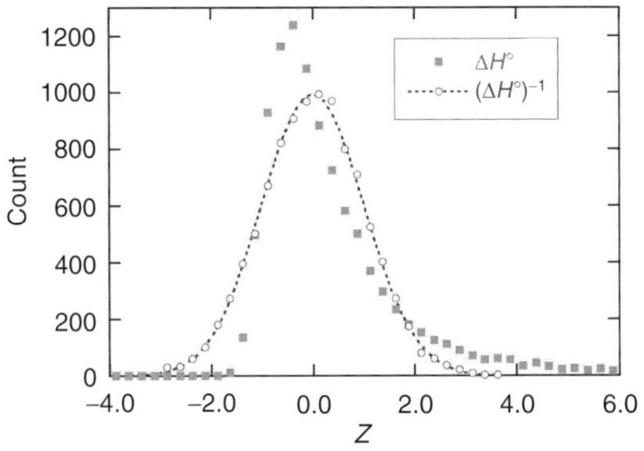

Fig. 21 Histogrammed results for ΔH° and its reciprocal, from a MC run of 10^4 data sets having $K = 0.1\ \text{mM}^{-1}$, $[\text{M}]_0 = 1\ \text{mM}$, $\Delta H^{\circ} = 10\ \text{kcal/mol}$, $m = 10$, $R_m = 140$, $v = 20\ \mu\text{l}$, and data error exaggerated by a factor of 3. The argument Z is the displacement from the true value in units of the nominal standard errors, which are 33.9% of the value in both cases. The curve shown for the reciprocal is a fitted Gaussian having displacement and width insignificantly different from the true values of 0 and 1, respectively, and yielding $\chi^2 = 25.4$ (27 fitted points). [Used with permission from Tellinghuisen (2005d); © 2005, American Chemical Society.]

F. Wrapping Up: Assessing and Presenting Results

Consider the following two sets of results, reported in an ACS journal for a determination of a thermodynamic property under conditions where the two values in each set could reasonably be expected to be the same:

$$\begin{array}{cc} \text{Set1} & \text{Set2} \\ 35.7 \pm 0.7 & 61.4 \pm 0.9 \\ 43.8 \pm 0.5 & 69.5 \pm 0.6 \end{array} \tag{28}$$

In both cases the discrepancy is so much greater than either of the estimated standard errors (from NLS fits) as to indicate something is wrong. On seeing results like these, the analyst should at least be reluctant to quote them without further checks. Unfortunately, with complex multistep experiments it is not unusual for the results from single runs to suggest much greater precision than is borne out in repeated experiments.

As another example, consider a recent study of interlaboratory consistency in measuring a given reaction by ITC (Myszka *et al.*, 2003). Figure 22 shows the results from 14 independent determinations of ΔH, along with a weighted average line (obtained by forcing KaleidaGraph to fit a line of zero slope). Most of the error bars are smaller than the displayed points and are thus clearly optimistic (taken as a set), as we would expect the error bars on about 10 points to intersect the average line. The huge value of χ^2 (*cf* its expected value of 13) is another indicator of optimistic errors, as is the $\mathbf{V}_{\text{prior}}$-based standard error for a. The *a posteriori* error is more conservative in this case and hence more realistic; we get it

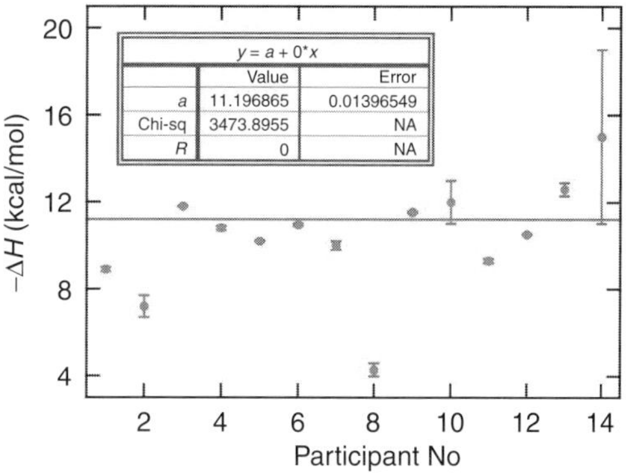

Fig. 22 The estimation of ΔH by ITC for a specified reaction, as carried out in 14 different laboratories. Error bars represent 1σ and are often smaller that the size of the displayed point. The enclosed box gives the results from a weighted average done with the KaleidaGraph program.

by multiplying by $(\chi^2/13)^{1/2}$, obtaining 0.2283. Alternatively, we might conclude that none of the estimated errors is to be trusted and just take an unweighted average, in which case we obtain 10.3636 ± 0.6768. The reason for the threefold increase is the granting of equal weight to several of the more remote values, which had large error bars and hence were downweighted in the weighted average.

The reason for such large discrepancies between single-experiment and ensemble statistics is either grossly flawed estimation of the errors in the individual LS fits, or unrecognized systematic error. The latter can still be largely randomized over sets of results to yield values like those displayed in Fig. 22. As a rough check on such effects, it is always wise to repeat at least several of the experiments under conditions where the same values are expected. In the case of the two sets of values in Eq. (28), a more reasonable error to quote might be half the separation of the two values.

Continuing with the data in Fig. 22, we might next consider rejecting some values. If we have any faith in the individual reported errors, we would have multiple candidates for rejection. Alternatively, the unweighted average returns $s_y = 2.53$, and point 8 is thus found to be low by more than 2.4 standard deviations, enough to reject it by some tests.

Assuming we accept the ensemble of results in Fig. 22, how should we report the average? Clearly the prior standard error is wildly optimistic, so we must discard it. The post estimates of s_a have percent standard errors of $\sim$20% $[(2 \times 13)^{-1/2}]$. If we believe that the errors quoted for the individual values remain valid in a relative sense, we would report the weighted average. Rounding the post error to 0.2 would give a value precise to only $\sim$25% (since a quoted value of 2 means a true value between 1.5 and 2.5), so we opt for additional precision and finally quote our result as 11.20(23), where the notation immediately makes it clear which digits in the result are uncertain. Alternatively, if we prefer the unweighted average, we quote it as 10.4(7) (since the 1-digit error now has precision better than 20%).

Although the rounding of the reported values with the aid of their standard errors is appropriate here, there are cases where it is not. For example, when highly precise data are fitted with empirical functions like polynomials, where one goal of the fit is the smooth representation of the data for subsequent computations, the parameters are often highly correlated. Then rounding leads to gross loss of accuracy if not done properly. As one extreme example, consider a test data set provided by NIST (2001), called Filip.dat. When fitted with a 10th order polynomial, these 82 data points yield $\mathcal{S} = 7.96 \times 10^{-4}$; however, after the parameters have been rounded using guidelines like those in the preceding paragraph, $\mathcal{S}$ exceeds 10^9 (de Levie, 2006)! The rounding can be accomplished, but it requires care. One approach is a sequential process of rounding and freezing one or several parameters and then refitting the remaining ones (Tellinghuisen, 1989). Usually it is safe to round a parameter to within 0.1 − 0.3 of its standard error, and the refitting gives a direct indication of any loss of precision, permitting the analyst to go back and add another digit if necessary. (In this process, each successive parameter typically requires more digits, as its refitted standard error is always less than in the previous cycle.)

VII. Summary

The method of LS is arguably the most powerful data analysis tool available to scientists. Its proper use is greatly enhanced by an understanding of just what the results mean, particularly the all-important dispersion parameters that are used to tell the world how good the results are. In this work I have started with an elementary review of fundamental concepts and terminology and have progressively increased the level of complexity to eventually cover some topics that are used only very rarely in the most sophisticated of data treatments. The MC method is a great asset toward understanding what it all means, and I have used it freely throughout this chapter. The reader thirsting after still more should be able to slake that thirst in my many cited references. Enjoy, and may your error bars be comprehensible!

Acknowledgment

I thank Laura Mizoue for teaching me the ropes of ITC experimentation, and the Center for Structural Biology at Vanderbilt for granting me access to the MicroCal VP-ITC instrument on which experiments relating to this work were done.

This work developed from a presentation at the 2005 "Current Trends in Microcalorimetry" meeting sponsored by MicroCal. I thank the organizers and MicroCal for that opportunity, Jack Correia for the chance to participate in this volume, and Brad Chaires and Ed Lewis for valuable discussions on ITC.

References

Albritton, D. L., Schmeltekopf, A. L., and Zare, R. N. (1976). An introduction to the least-squares fitting of spectroscopic data. *In* "Molecular Spectroscopy: Modern Research II" (K. Narahari Rao, ed.), pp. 1–67. Academic Press, New York.

Ashmore, J. G., and Tellinghuisen, J. (1986). Combined polynomial and near-dissociation representations for diatomic spectral data: $Cl_2(X)$ and $I_2(X)$. *J. Mol. Spectrosc.* **119,** 68–82.

Bevington, P. R. (1969). "Data Reduction and Error Analysis for the Physical Sciences." McGraw-Hill, New York.

Bevington, P. R., and Robinson, D. K. (2002). "Data Reduction and Error Analysis for the Physical Sciences," 3rd edn. McGraw-Hill, New York.

Carroll, R. J., and Ruppert, D. (1988). "Transformation and Weighting in Regression." Chapman and Hall, New York.

Cohen, E. R., and Taylor, B. N. (1986). "The 1986 Adjustment of the Fundamental Physical Constants" (A Report of the CODATA Task Group on Fundamental Constants). Pergamon, Oxford.

Connors, K. A. (1987). "Binding Constants: The Measurement of Molecular Complex Stability." Wiley, New York.

de Levie, R. (2006). The statistical relevance of data-fitting parameters. (private communication, designed for eventual publication in the *Am. J. Phys.*)

Deming, W. E. (1964). "Statistical Adjustment of Data." Dover, New York.

Draper, R. N., and Smith, H. (1998). "Applied Regression Analysis," 3rd edn. Wiley, New York.

Guggenheim, E. A. (1926). On the determination of the velocity constant of a unimolecular reaction. *Philos. Mag.* **2,** 538–545.

Hamilton, W. C. (1964). "Statistics in Physical Science: Estimation, Hypothesis Testing, and Least Squares." The Ronald Press Co., New York.

Ingle, J. D., Jr., and Crouch, S. R. (1988). "Spectrochemical Analysis." Prentice-Hall, Englewood Cliffs, NJ.

Marquardt, D. W. (1963). An algorithm for least-squares estimation of nonlinear parameters. *J. Soc. Ind. Appl. Math.* **11,** 431–441.

Miller, J. N. (1991). Basic statistical methods for analytical chemistry part 2. Calibration and regression methods. A review. *Analyst* **116,** 3–14.

Mizoue, L. S., and Tellinghuisen, J. (2004). Calorimetric vs. van't Hoff binding enthalpies from isothermal titration calorimetry: Ba^{2+}-crown ether complexation. *Biophys. Chem.* **110,** 15–24.

Mood, A. M., and Graybill, F. A. (1963). "Introduction to the theory of statistics" 2nd edn. McGraw-Hill, New York.

Myszka, D. G., Abdiche, Y. N., Arisaka, F., Byron, O., Eisenstein, E., Hensley, P., Thomson, J. A., Lombardo, C. R., Schwarz, F., Stafford, W., and Doyle, M. L. (2003). The ABRF-MIRG '02 Study: Assembly state, thermodynamic, and kinetic analysis of an enzyme/inhibiter interaction. *J. Biomolec. Techn.* **14,** 247–269.

National Institutes of Standards and Technology Statistical Reference Dataset (NIST) (2001). Available from http://www.itl.nist.gov/div898/strd/lls/data/LINKS/DATA/Filip.dat

Parratt, L. G. (1961). "Probability and Experimental Errors in Science: An elementary Survey." Wiley, New York.

Press, W. H., Flannery, B. P., Teukolsky, S. A., and Vetterling, W. T. (1986). "Numerical Recipes." Cambridge University Press, Cambridge, UK.

Tellinghuisen, J. (1989). Rounding least-squares parameters for publication. *J. Mol. Spectrosc.* **137,** 248–250.

Tellinghuisen, J. (1996). On the least-squares fitting of correlated data: *A Priori* vs *a Posteriori* Weighting. *J. Mol. Spectrosc.* **179,** 299–309.

Tellinghuisen, J. (2000a). Nonlinear least squares using microcomputer data analysis programs: KaleidaGraph™ in the physical chemistry teaching laboratory. *J. Chem. Educ.* **77,** 1233–1239.

Tellinghuisen, J. (2000b). A Monte Carlo study of precision, bias, inconsistency, and non-gaussian distributions in nonlinear least squares. *J. Phys. Chem. A* **104,** 2834–2844.

Tellinghuisen, J. (2000c). Bias and inconsistency in linear regression. *J. Phys. Chem. A* **104,** 11829–11835.

Tellinghuisen, J. (2000d). A simple, all-purpose nonlinear algorithm for univariate calibration. *Analyst* **125,** 1045–1048.

Tellinghuisen, J. (2001). Statistical error propagation. *J. Phys. Chem. A* **105,** 3917–3921.

Tellinghuisen, J. (2003a). Fitting correlated data: A critique of the Guggenheim method and other difference techniques. *J. Phys. Chem. A* **107,** 8779–8783.

Tellinghuisen, J. (2003b). A study of statistical error in isothermal titration calorimetry. *Anal. Biochem.* **321,** 79–88.

Tellinghuisen, J. (2003c). On the efficient representation of comprehensive, precise spectroscopic data sets: The *A* state of I_2. *J. Chem. Phys.* **118,** 3532–3537.

Tellinghuisen, J. (2004a). Statistical error in isothermal titration calorimetry. *Meth. Enzymol.* **383,** 245–282.

Tellinghuisen, J. (2004b). A statistical study of the analysis of congested spectra by total spectrum fitting. *J. Mol. Spectrosc.* **226,** 137–145.

Tellinghuisen, J. (2005a). Understanding least squares through Monte Carlo calculations. *J. Chem. Educ.* **82,** 157–166.

Tellinghuisen, J. (2005b). Simple algorithms for nonlinear calibration by the classical and standard additions methods. *Analyst* **130,** 370–378.

Tellinghuisen, J. (2005c). Statistical error in isothermal titration calorimetry: Variance function estimation from generalized least squares. *Anal. Biochem.* **343,** 106–115.

Tellinghuisen, J. (2005d). Optimizing experimental parameters in isothermal titration calorimetry. *J. Phys. Chem. B* **109,** 20027–20035.

Tellinghuisen, J. (2006). Van't Hoff analysis of $K^{\circ}(T)$: How good ... or bad? *Biophys. Chem.* **120,** 114–120.

CHAPTER 24

Nonlinear Least-Squares Fitting Methods

Michael L. Johnson

Department of Pharmacology and Internal Medicine
University of Virginia Health System
Charlottesville, Virginia 22908

Abstract

This chapter provides an overview of the techniques involved in "fitting equations to experimental data" with a particular emphasis on the what can be learned with these techniques, what are the requirements of the experimental data for these techniques, and what are the underlying assumptions of these techniques. The layout of this chapter is to start with a set of experimental data, and then walk the reader through the analysis of this set of data. The rigorous mathematical methods are referenced but not presented in detail.

I. Introduction

This chapter addresses the often asked question, "*What is the best equation that describes this data?*" As written, this question is actually ill-posed because the meaning of "the best equation" is not clear. As will be seen below, it implies the best function form (i.e., the type of mathematical equation) and it also

0091-679X/08 $35.00
DOI: 10.1016/S0091-679X(07)84024-6

implies the best parameters (i.e., coefficients) of that equation. While these are interrelated, they have very distinct meanings.

As an example to help clarify this point, consider a simulated ligand-binding experiment where the amount bound is a function of the ligand concentration, as shown in Table I. This could represent the binding of a hormone to a receptor, or the binding of oxygen to hemoglobin, or any other binding experiment of interest. The mathematical and statistical concepts presented in this chapter do not depend on the actual biochemical system being studied or even that a ligand binding system is being studied, with the exception of the specific binding equations. Furthermore, the methods are independent of the concentration units and, as a consequence, the units of the test data set are arbitrary.

The data shown in Table I is plotted as the open squares in Fig. 1 along with calculated curves based on two possible "best equations" that might describe this data. The dashed line appears to describe the data and is the fifth degree polynomial:

$$\begin{aligned} Y(X) = &-0.002633 + 0.199246X - 0.097234X^2 \\ &-0.051963X^3 + 0.056020X^4 - 0.012065X^5 \end{aligned} \tag{1}$$

where $Y(X)$ corresponds to the predicted amount bound at a concentration of X. This appears, at least visually, to provide a reasonable description of this data within the range of the data. Note that the coefficient of the fifth order term is

Table I
Simulated Data Used for the Examples

Bound, $Y_i \pm \sigma_i$	Free ligand concentration, X_i
0.0008 ± 0.0020	0.0
0.0145 ± 0.0020	0.1
0.0269 ± 0.0020	0.2
0.0476 ± 0.0020	0.3
0.0612 ± 0.0020	0.4
0.0735 ± 0.0020	0.5
0.0793 ± 0.0025	0.6
0.0834 ± 0.0025	0.7
0.0839 ± 0.0025	0.8
0.0877 ± 0.0025	0.9
0.0901 ± 0.0025	1.0
0.0886 ± 0.0025	1.1
0.0931 ± 0.0030	1.2
0.1006 ± 0.0030	1.3
0.0942 ± 0.0030	1.4
0.0925 ± 0.0030	1.5
0.0977 ± 0.0030	1.6
0.0926 ± 0.0030	1.7
0.0977 ± 0.0035	1.8

negative, and as a consequence this fifth degree polynomial will be negative at much higher ligand concentrations. This is clearly a physically impossible result!

But, is it the "best equation that describes this data"? To answer this we need a measure to compare different fits of the same data to determine which provides the best description of the data. The most commonly used, and abused, measure of how well an equation describes a set of data is the weighted sum-of-squared-residuals (WSSR) (which is proportional to the sample variance-of-fit, s^2):

$$\text{WSSR} = \sum_i \left(\frac{Y(X_i) - Y_i}{\sigma_i} \right)^2 = \sum_i R_i^2 \tag{2}$$

$$s^2 = \frac{\text{WSSR}}{\text{NDF}} = \frac{\text{WSSR}}{n - n_f} \tag{3}$$

where R_i is the residual for the ith data point; $Y(X_i)$ corresponds to the calculated value of the amount bound [as in Eq. (1)]; X_i, Y_i, and σ_i are the ith observed data value (as is given in Table I); and NDF is the number of degrees of freedom which is normally evaluated as the number of data points, n, minus the number of parameters being estimated, n_f. The consequences of a large and/or variable experimental measurement errors are corrected for by including the estimated uncertainties, σ_i, in Eq. (2). Equation (3) does not have $a - 1$ in the denominator because the mean of the residuals has not been calculated; it is assumed to be zero. In mathematical jargon Eq. (2) is the L2 norm. These values are always positive and a lower value generally corresponds to a better description of the data. The WSSR = 32.68084 for the dashed line in Fig. 1.

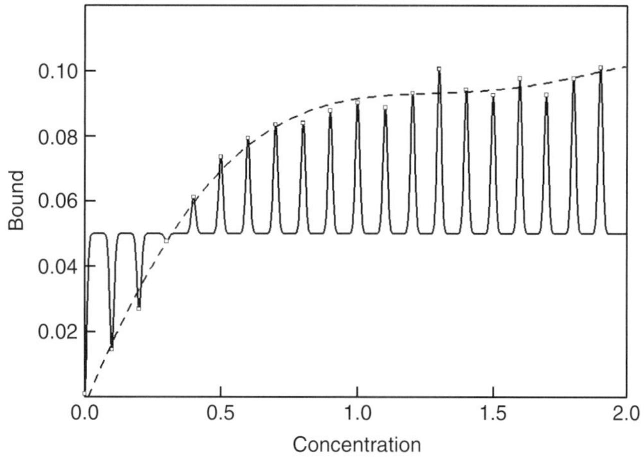

Fig. 1 The simulated data example found in Table I and two possible "best fit" equations to describe the data points. The dashed line is given by Eq. (1) and the solid line is given by Eq. (4).

The mathematical form for the alternative description of the data shown as the solid line in Fig. 1 is:

$$Y(X) = 0.05 + \sum_i (Y_i - 0.05)e^{-1/2((X-X_i)/0.008)^2} \tag{4}$$

Visually this contrived (i.e., arbitrary) functional form appears to be a terrible description of this data, but the corresponding WSSR is exactly zero. It passes precisely through every data points! Thus, if "best equation" is defined as the lowest WSSR, then the solid line in Fig. 1 [i.e., Eq. (4)] is the best description of the data that can be obtained. In addition, if a 19th degree polynomial ($n - 1$ degree where n is the number of data points) is used, then the corresponding WSSR is also exactly zero. It also will pass precisely through every data point. There are actually an infinite number of different mathematical forms that yield a WSSR of zero for this, or any other, set of data! These forms cannot be the best description of this binding data because we have some knowledge of the mathematical form of the binding equations and neither of these are consistent with those forms [e.g., Eq. (5) for a single binding site and Eq. (6) for two binding sites per receptor] (Johnson and Straume, 2000):

$$\bar{Y}(X) = \frac{K_a X}{1 + K_a X} \tag{5}$$

$$\bar{Y}(X) = \frac{1}{2}\frac{K_1 X + 2K_2 X^2}{1 + K_1 X + K_2 X^2} = \frac{1}{2}\frac{10^{\log K_1} X + 2 \times 10^{\log K_2} X^2}{1 + 10^{\log K_1} X + 10^{\log K_2} X^2} \tag{6}$$

where $\bar{Y}(X)$ is the fraction of binding sites occupied; K_a is the single site model association (i.e., binding) constant; K_1 and K_2 are the binding constants to the sites of the two-site model; and X is the free, or unbound, concentration of the ligand. Equations (5) and (6) are written in terms of the stoichiometric, or macroscopic, association constants and not the site, or microscopic, constants.

Also, the exact choice of model parameters is of great importance. For example, Eq. (6) is written in two algebraically equivalent forms, one in terms of the binding constants and the other in terms of the base 10 logarithms of the binding constants. The preferable form of Eq. (6) is the second form (on the right) logarithmic terms because it constrains the equilibrium constants to have only positive values. Ten raised to the power of the logarithm of the equilibrium constant, log K, will always be positive even though log K can have any real value positive or negative. Consequently, the left form of Eq. (6) allows physically meaningless negative values of the equilibrium constants which the right form (involving the logarithms) does not allow.

Neither Eq. (1) nor Eq. (4) provides any information about the nature of the ligand binding process. The problem with these equations is that they are not based on a hypothesis about the underlying biochemical processes. *A general conclusion is that arbitrary models (i.e., models that are not hypothesis driven) will not provide the desired information.*

The answer to our initial question, "What is the best equation that describes this data?" is actually answered by two other questions:

What do you want to learn from the data?

What are the properties of the data?

The answers to these two additional questions will uniquely determine the mathematical form of the equations required to fit the data [e.g., Eqs. (1), (4)–(6)]. If we have a well-defined hypothesis about the biochemical mechanism being described by the data then the hypothesis can be transformed into the desired mathematical form (i.e., the appropriate fitting equation). The hypothesis can then be tested by assessing how well the equation can describe the experimental data. The forms of the equations that were used in the above examples were arbitrarily chosen to describe the data instead of to describe the hypotheses about the data. It is the answer to the second question that determines how to evaluate the best model parameters for a specific model and set of data.

II. Formulate a Hypothesis-Based Mathematical Model

The first step is to derive a mathematical model which is based on the hypothesis about the nature of the biochemical processes that are being measured by the experimental observations. For the present ligand binding example, we assume that the data is represented by either a single binding site model, as in Eq. (5), or a two binding site model, as in Eq. (6). The nature of the data in Table I, however, is not directly compatible with Eqs. (5) and (6). The amount bound (i.e., the dependent variable) is in terms of fractional saturation in the equations and is in concentration units in the data. Consequently, the model equation must be transformed to match the data, as in the following equation:

$$Y(X) = A\bar{Y}(X) \tag{7}$$

where the fitting function is $Y(X)$ is related to the fractional saturation by a multiplicative constant, A. This multiplicative constant is related to the concentration of the binding protein (e.g., receptor or hemoglobin), and possibly the design of the instrumentation that is utilized for the measurements. For the present example, A will be estimated in the fit because it is an unknown constant, or at least it is a constant that is not known to an infinite precision.

There is always the alternative of scaling the data to match the form of the equation but this is usually not a good idea. For this present example, this scaling would involve dividing each of the bound concentrations by the saturating amount bound at an infinite ligand concentration (i.e., the independent variable). This limiting value is not known *a priori*, however. It must be estimated from the data and as a consequence will have an associated uncertainty. For the present example, the experimental uncertainties that are always superimposed on the experimental data can be described by a Normal (i.e., Gaussian or bell shaped) distribution. Specifically, if the particular data points were independently measured a near infinite number of the times, the distribution of observed values could be represented as a Gaussian distribution. If the data that include uncertainties are divided by a limiting value that also includes uncertainties, then the resulting ratios (i.e., $\bar{Y}$) will have an uncertainty that is not a Gaussian. The sum or difference of two Gaussian distributions is a Gaussian distribution. However, the ratio of two Gaussian distributions is a Cauchy distribution, not a Gaussian distribution! As will be discussed below, a Gaussian form of the distribution of these uncertainties is critically important in determining how the parameters of the model will be determined. In addition, any systematic errors in the evaluation of the limiting value will be propagated into the resulting ratios and it is always a poor idea to introduce systematic errors. Normally the only justifiable reason to perform a transformation of the data is to convert a non-Gaussian distribution of uncertainties into a Gaussian distribution of uncertainties (Abbott and Gutgesell, 1994; Acton, 1959).

For the present example, the parameters of both the one-site model, the combination of Eqs. (5) and (7), and the two-site model, the combination of Eqs. (6) and (7), will be estimated by fitting the model to the data. *These procedures do not fit the data to the model, they fit the model to the data!*

Note that the independent variable, X_i, of the present example is the free, or unbound, ligand concentration. This is consistent with experimental protocols where the ligand concentration is directly measured, such as oxygen binding to hemoglobin where the free oxygen concentration can be measured with an oxygen electrode. However, when the binding of a hormone to its receptor is measured with a competition experiment, the resulting independent variable is the total, not the free, hormone concentration. In a competition experiment a small amount of radioactively labeled ligand is initially bound to the receptor and then is titrated with an excess of unlabeled ligand, thus providing a competition between the labeled and unlabeled ligand for the binding sites, and an independent variable which is the sum of the total concentration of labeled and unlabeled ligands. Equations (5) and (6) do not apply to the competitive hormone binding case because they are formulated in terms of the free concentration of the ligand and not the total concentration. The solution is to substitute the free concentration with the total concentration minus the calculated amount bound. Equation (8) presents the single-site binding equation in terms of the total concentration, X_t:

$$Y(X_t) = A\frac{K_a(X_t - Y(X_t))}{1 + K_a(X_t - Y(X_t))} \tag{8}$$

Equation (8) is a transcendental equation [i.e., $Y(X_t)$ occurs on both sides of the equal sign and as a consequence is somewhat more difficult, but not impossible, to calculate (Johnson and Frasier, 1984, 1985)].

The alternative of simply calculating the observed free concentration as the observed total minus observed bound concentrations is problematic. The observed bound concentration will have a significant experimental uncertainty and as a consequence the calculated free concentration will also have a significant uncertainty. As will be discussed below, the common parameter estimation procedures do not allow an independent variable (i.e., the free concentration) that contains significant errors. Thus, this alternative approach will yield a simpler function form for the fit [i.e., Eq. (5) instead of Eq. (8)], but it will also preclude the use of the most common methods of actually performing the fit.

The two take-home lessons from this section are that (1) *the mathematical models must be hypothesis based* and (2) *the data should only be "transformed" if it will make the measurement uncertainties more consistent with a Gaussian distribution* (Abbott and Gutgesell, 1994).

III. Determining the Optimal Parameters of the Model

Once the mathematical form for the mathematical model (i.e., the fitting function) has been determined, the next step is to find the optimal parameters (i.e., coefficients) of the model. Conceptually, this simply involves trying different combinations of the parameters until the optimal values have been determined. But, what is meant by optimal? The most common definition of optimal is the least-square parameter values which correspond to the lowest WSSR, shown in Eq. (2). However, other definitions are also used which yield different results. For example, another definition of optimal is L1 norm where the sum of the absolute values of the residuals, the R_i in Eq. (2), is minimized:

$$L1 = \sum_i \left| \frac{Y(X_i) - Y_i}{\sigma_i} \right| = \sum_i |R_i| \tag{9}$$

Yet another definition of optimal is the min–max norm. For the min–max norm, the parameters are adjusted until the largest value of the residuals, R_i, is minimized. *It is the property of the data, and more specifically the experimental uncertainties contained within the data, that determines the proper choice of least squares, versus L1, versus min–max optimization, versus whatever.*

Table II presents several permutations of results for the analysis of the data in Table I. The various combinations include the choice of fitting equations, weighted versus unweighted analysis (discussed in detail below), and either least squares or L1 optimization. The calculated curves for the least-squares optimizations of A and K_a in the one-site model [Eqs. (5) and (7)] and A, K_1, and K_2 in the two-site model [Eqs. (6) and (7)] are presented in Fig. 2. *But, how do we determine which of*

Table II
Parameter Estimates Based Upon Different Fitting Equations, Optimization Norms, and Weighting Schemes

Optimization	Equations	Norm	A	K_1 or K_a	K_2
LS	Fifth degree poly	32.6808			
Exact	4	0.0			
Exact	19th Degree poly	0.0			
LS	5 and 7	101.6051	0.1301	2.0720	
LS	6 and 7	19.9977	0.1012	0.4611	9.9712
LS, unweighted	6 and 7	21.5875	0.1016	0.5032	9.9010
LS	8	87.5728	0.1241	2.6930	
L1	5 and 7	33.1078	0.1222	2.5089	
L1	6 and 7	15.4942	0.1020	0.6832	9.7946
L1, unweighted	6 and 7	15.8463	0.1008	0.3327	9.8742
L1	8	30.3504	0.1189	3.1282	

Note: That the unweighted parameter estimations were performed with a constant uncertainty of 0.0025. The LS parameter estimations were performed by minimizing the sum-of-squared-residuals [i.e., the norm given by Eq. (2)] and the L1 parameter estimations were performed by minimizing the sum-of-absolute-values [i.e., the norm given by Eq. (9)]. The calculations in terms of Eq. (8) assumed that the ligand concentration given in Table I was actually the total ligand concentration, instead of the free ligand concentration.

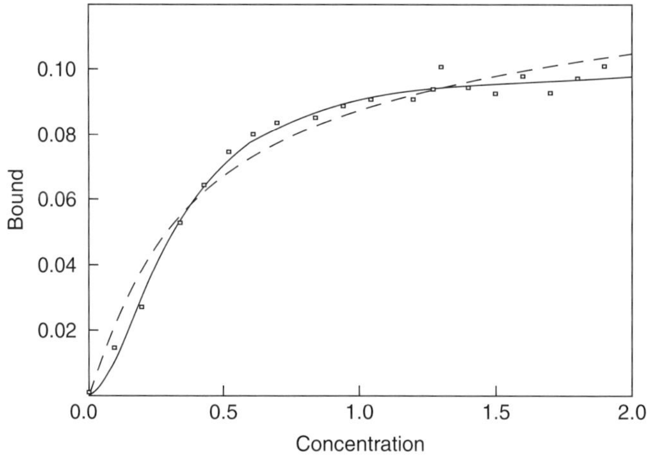

Fig. 2 The analysis of the data presented in Table I with either a one-component binding isotherm [dashed line and Eqs. (5) and (7)], or a two-component binding isotherm [solid line and Eqs. (6) and (7)]. Both were fit by a least-squares optimization procedure and used the weighting factors shown in Table I.

these parameter estimations provide the best description of the data? It is not simply the one which corresponds to the lowest WSSR because that criterion predicts either the 19th degree polynomial or the contrived Eq. (4), both of which correspond to an exact zero for the WSSR, both of which are arbitrary and contrived equations, and both of which provide essentially no information of the binding process being investigated. Hence, the answer to this question is provided by the properties of the experimental data and the properties of the measurement errors (i.e., experimental uncertainties) that are always present within every set of experimental observations. For example, the data presented in Table I is in terms of the free ligand concentration, not the total ligand concentration, and thus Eq. (8) is not appropriate for the analysis. Similarly, since the uncertainties of the observations are not constant (i.e., in statistical jargon they are heteroscedastic), it is not appropriate to accept the unweighted parameter estimations.

The objective is to find the parameter values with the highest probability of being correct, that is, it is the maximum-likelihood results that are desired. To accomplish this, we will write an equation for the probability as a function of the parameter values and then maximize this function to obtain the maximum-likelihood parameter values. Specifically, the probability of the parameter values based on the ith data point is given by the following equation:

$$P_i(Y(X_i), \text{parameters}, X_i, Y_i, \sigma_i) = Q\mathrm{e}^{-1/2((Y_i - Y(X_i))/\sigma_i)^2} \tag{10}$$

where Q is a proportionality constant. The probability in Eq. (10) is a function of the form of the fitting equation, the parameter values, and the specific data point. The validity of this equation is based on several assumptions, specifically:

1. All of the measurement errors are contained in the dependent, Y_i, values.
2. These experimental uncertainties can be approximated by a Gaussian distribution.
3. The null hypothesis is that the fitting function, $Y(X_i)$, is correct.
4. The measured uncertainties are not correlated.

Most experimental protocols can be manipulated such that most, if not all, of the measurement uncertainties are present with the dependent variables (i.e., along the Y-axis). In addition, the central limit theorem of calculus implies that the distribution of experimental uncertainties is likely to be either Normal (i.e., Gaussian or bell shaped) or Log Normal. Note that Log Normal distributions can be converted to Normal distribution with a logarithmic transformation. Thus, the first three assumptions are very reasonable.

The forth assumption is required to combine the probabilities based on the individual data points, Eq. (10), into the overall probability, P (*parameters, data*), for the entire set of n data points:

$$\begin{aligned} P(\text{parameters}, \text{data}) &= \prod_{i=1}^{n} P_i(Y(X_i), \text{parameters}, X_i, Y_i, \sigma_i) \\ &= Q^n \prod_{i=1}^{n} \mathrm{e}^{-1/2((Y_i - Y(X_i))/\sigma_i)^2} \end{aligned} \tag{11}$$

Since the product of the exponentials is equal to the exponential of the sum, this overall probability can also be written as in the following equation:

$$P(\text{parameters, data}) = Q^n e^{-1/2\sum_{i=1}^{n}((Y_i - Y(X_i))/\sigma_i)^2} \tag{12}$$

Finally, Eq. (12) implies that if the above assumptions are valid then the overall probability is a maximum when the summation in the negative exponential is at a minimum. This summation is the WSSR, shown in Eq. (2). Consequently, if the assumptions are valid and appropriate then the least-squares approach will provide the maximum-likelihood set of parameter values (Johnson and Frasier, 1985). Conversely, if the data are consistent with this set of reasonable assumptions then any method of analysis that results in answers which are different from those obtained by least squares will produce results that have less than the maximum likelihood of being correct. *Thus, for most cases the least-squares method is the method of choice.* This is the reason that Fig. 2 contains only the calculated curve corresponding to the weighted least-squares fit of both the combination of Eqs. (5) and (7) (i.e., the model with one binding site), and the combination of Eqs. (6) and (7) (i.e., the two-site model) to the data values.

By now it should be clear that a set of data is more than a set of observations (i.e., Y or dependent variables) as a function of some experimental condition (i.e., X, or independent variables). The data always includes experimental measurement uncertainties. *An integral part of the data is a determination of how accurately the values were measured.* Thus, the measured second data point in Table I is $Y = 0.0145 \pm 0.0020$ at $X = 0.1$, not just $Y = 0.0145$ at $X = 0.1$. It is this determination of the measurement uncertainties that provides the relative weighting factors for the data points [i.e., the σ_i in Table I and Eqs. (2), (3), and (12)].

Least-squares calculations are commonly grouped into two categories, either linear or nonlinear least squares. In reality, linear least squares are simply a special case of the more general nonlinear least-squares approach. This chapter concentrates on the nonlinear category and makes note of some of the differences when they occur. The rigorous mathematical distinction between a linear and a nonlinear least-squares fit is related to the form of a fitting function. Specifically, if the second and higher order derivatives of the fitting function with respect to the parameters being estimated are all equal to zero, then the fit is a linear fit. If any of these derivatives are not equal to zero then it is a nonlinear fit.

A linear least-squares fit does not imply that a straight line is being fit to the experimental data. For example, fitting polynomials containing a single independent variable [e.g., Eq. (1)] of any order is a linear process. A Fourier transform with a constant base period or frequency is equivalent to a linear fit. However, if the period or frequency is simultaneously being estimated by the fitting procedure, then the Fourier analysis will be a nonlinear fit. Fitting to the sum of exponentials with known constant half-lives and/or rate constants is a linear process. Similarly, if the half-lives and/or rate constants are also being estimated by the fitting procedure then the exponential analysis is a nonlinear process.

There are many divergent numerical algorithms which will adjust the parameter values to obtain a minimum of the WSSR, or in other words perform a weighted nonlinear least-squares (WNLLS) fit of an equation to a set of data (Johnson and Frasier, 1985). These include the damped Gauss–Newton, the Marquardt–Levenberg, and the Nelder–Mead algorithms. When correctly implemented all of these algorithms will yield equivalent results.

No simple solution of the least-squares fitting procedure exists for nonlinear fitting equations. Equations (7) and (8) cannot be fit to a set of data by plugging the values of X, Y, and σ into equations to find the binding constants. WNLLS fits of equations to data are performed by successive approximation methods. Given an initial set of answers, the WNLLS procedures will provide a better set of answers. The algorithm is repeated starting with this better set of answers iteratively until the answers do not change to within a small convergence limit (e.g., one part in a million). Consequently, WNLLS software usually requires that the user enter initial starting values for the answers. Furthermore, realistic values for these initial estimates should be utilized. The damped Gauss–Newton, the Marquardt–Levenberg, and the Nelder–Mead algorithms have different rates of convergence and different tolerances to unrealistic initial values. But they all perform better when they are started near the desired answers.

When fitting to linear models (those with second and higher order derivatives equal to zero), the iterative procedure requires only a single iteration and will be extremely tolerant of poor initial values. For example, the equations for the fitting polynomial found in most statistics textbooks can be derived by applying the Gauss–Newton WNNLS algorithm for one iteration and with initial parameter estimates of zero. Fourier transforms equations can also be derived in the same manner.

IV. Distinguishing Between Multiple Mathematical Models

Table II presents 11 distinct analyses of this data, but several of these are not applicable. The first three do not apply because the mathematical models are arbitrary, that is, not hypothesis based. The sixth does not apply because the data has a variable measurement uncertainty and this analysis method assumes a constant uncertainty, that is, an unweighted analysis. The seventh does not apply because the independent variable for this data set is the free concentration, not the total concentration. The last four do not apply because they correspond to a minimization of the L1 Norm instead of the preferred (see above) Least-Squares Norm, L2. Figure 2 presents the optimal fits for the remaining one-site and two-site ligand binding models to the data shown in Table I. The question at hand is does either, or both, of the two remaining analyses that are shown in Fig. 2 provide a statistically invalid description of this data. This would allow the invalid ones to be eliminated from further consideration. The question at hand is not whether either, or both, of the two remaining analyses that are shown in Fig. 2 provide a statistically valid description of this data? You can never prove a hypothesis, you can only disprove a hypothesis!

The traditional approach is to compare the WSSRs, and the analysis with the lowest WSSR is presumed to provide the best description. However, it should be recalled that there are an infinite number of function forms, such as in Fig. 1, that will provide a WSSR of zero. While zero is as low as possible and thus "as good as you can get," the fit based on Eq. (4) clearly does not do a good job of describing the data. Of course, neither Eq. (4) nor the 19th degree polynomial are hypothesis-based models.

Some would assume that surely the WSSR of 101.6 from Eqs. (5) and (7) (Table II, line 4) is certainly much worse than the WSSR of 20.0 from Eqs. (6) and (7) (Table II, line 5). This is true but a probability level cannot be provided (i.e., a P value of 0.05) to indicate that line 5 is better than line 4. A comparison of the two variance-of-fits simply cannot be evaluated with an F statistic because the residuals (i.e., the weighted differences between the fitted function and the experimental data points) are not independent (i.e., orthogonal) of each other. In mathematical terms, Cochran's Theorem does not apply for these nonlinear functional forms. Thus, the apparent variance-of-fit cannot be separated into two independent components where one is due to the addition of the second binding site and the other being the intrinsic experimental measurement errors found within the data. Other approaches are needed to assist in differentiating the two analyses presented in Fig. 2.

Recall that the null hypothesis for these analyses was that the fitting equations were correct. If this and the other assumptions about the nature of the experimental measurement errors contained within the data are correct then the residuals (the weighted differences between the fitted curves and the data points) should be random with a Gaussian, or bell-shaped, distribution with a mean of zero. Furthermore, if the magnitudes of the weighting factors ($1/\sigma_i$) are correct then this distribution should have a variance of one. No consideration of the shape of the distribution of the residuals is taken into account when using only the WSSR or s^2 as a measure of goodness-of-fit.

The simplest approach to distinguish the two analyses is simply to plot the weighted residuals as a function of the independent variable and/or the dependent variable. Such a plot is shown in Fig. 3. The upper panel is a plot of the weighted residuals for the one-site model as a function of the ligand concentration and the lower panel is the corresponding graph for the two-site model. The null hypothesis (i.e., the particular fitting equation being correct) can be ruled out if the residuals are obviously not random. This is presumably the case for the upper panel of Fig. 3, but no probability level can be assigned simply by visual inspection.

There are numerous statistical tests to assign a probability that the distribution of residuals is not normally distributed with a mean of zero (Straume and Johnson, 1992). Each of these tests evaluate different aspects of the distribution of residuals

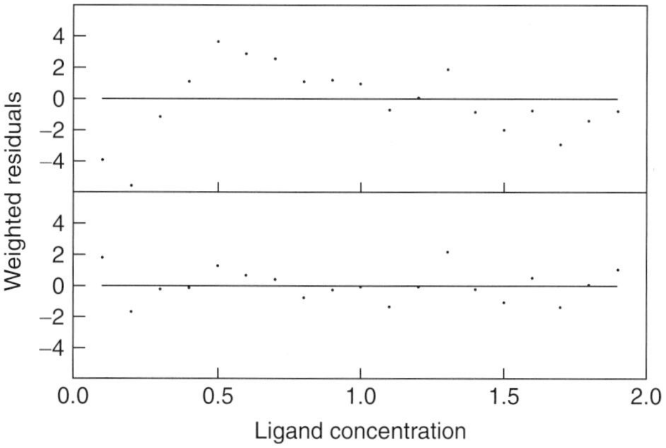

Fig. 3 The residuals from the analyses presented in Fig. 2 as a function of the ligand concentration. The upper panel corresponds to the one-site model analysis and the lower panel is for the two-site analysis.

and provide a probability level that the distribution is not normally distributed. For the present example we will examine Autocorrelation and the Runs Test.

The existence of trends in residuals with respect to either the independent (i.e., experimental) or dependent (i.e., the experimental observable) variables suggests that some systematic behavior is present in the data that is not accounted for by the analytical model. Trends in residuals will often manifest themselves by causing too few runs (sequences of consecutive residual values of the same sign) or, in cases where negative serial correlation occurs, by causing too many runs. A convenient way to assess quantitatively this quality of a distribution of residuals is to perform a Runs Test. The method involves calculating the expected number of runs given the total number of residuals as well as an estimate of variance in this expected number of runs. The expected number of runs, R, may be calculated from the total number of positive and negative valued residuals, n_p and n_n, as:

$$R = \frac{2n_\mathrm{p}n_\mathrm{n}}{n_\mathrm{p} + n_\mathrm{n}} + 1 \tag{13}$$

The corresponding variance of the expected number of runs is given in the following equation:

$$\sigma_R^2 = \frac{2n_\mathrm{p}n_\mathrm{n}(2n_\mathrm{p}n_\mathrm{n} - n_\mathrm{p} - n_\mathrm{n})}{(n_\mathrm{p} + n_\mathrm{n})^2(n_\mathrm{p} + n_\mathrm{n} - 1)} \tag{14}$$

A quantitative comparison is then made between the expected number of runs, R, and the observed number of runs, n_R, by calculating an estimate for the standard normal deviate as:

$$Z = \frac{n_R - R \pm 0.5}{\sigma_R} \tag{15}$$

If n_p and n_n are both greater than ~10 then Z will be distributed approximately as a standard normal deviate. In other words, the calculated value of Z is the number of standard deviations that the observed number of runs is from the expected number of runs for a randomly distributed set of residuals of the total number being considered. The value of 0.5 is a continuity correction to account for biases introduced by approximating a discrete distribution with a continuous one. This correction is +0.5 when testing for too few runs and is −0.5 when testing for too many runs. The test is therefore estimating the probability that the number of runs observed is different from that expected from randomly distributed residuals. The greater the value of Z, the greater the likelihood that there exists some form of correlation in the residuals relative to the particular variable being considered. The Runs Test is not dependent on the magnitude of the residuals and thus it can be used with either weighted or unweighted residuals.

The application of the Runs Test to the current analysis example is presented in Table III. For both the one-site and the two-site analyses the observed number of runs is less than the expected number of runs, thus the appropriate Z score and probability level is for too few runs. For the one-site model the $Z = 2.068$ which corresponds to a one-sided probability level of about 0.02 that there are enough runs. By comparison, for the two-site model $Z = 0.6501$ and a probability level of 0.26 that there are enough runs. Based on the Runs Test, the data and/or the assumptions about the data are not consistent with the one-site model and are consistent with the two-site model.

It is important to note the phrasing here. The Runs Test does not validate the two-site model, it simply says that the data and/or assumptions about the data are not consistent with the one-site model. Remember that the justification for using least-squares minimization and testing for Gaussian distributed residuals is that all of the measurement errors contained within the data are in the amount bound and that they are normally distributed with a mean of zero. If these assumptions are not valid then neither are the conclusions of any test of the normality of the residuals (e.g., the Runs Test and the autocorrelation analysis given below).

Table III
Runs Test

	One site	Two site
Observed	6	9
Expected	11.0 ± 2.2	10.9 ± 2.2
Z	2.068	0.6501
Probability	0.9807	0.7422

Experimental data will sometimes exhibit serial correlations. These serial correlations arise when the random uncertainties superimposed on the experimental data tend to have values related to the uncertainties of other data points that are close in the independent variable, in this case the ligand concentration. For example, if the weight of a test animal is being measured once a month and the data are expressed as a weight gain per month, negative serial correlation may be expected because of the subtraction. This negative serial correlation is expected because a positive experimental error in an estimated weight gain for 1 month (e.g., an overestimate) would cause the weight gain for the next month to be systematically incorrect (e.g., an underestimated).

A basic assumption of parameter-estimation procedures is that the experimental data points are independent observations. Therefore, if the weighted differences between experimental data points and the fitted function (the residuals) exhibit such a serial correlation, then either the observations are not independent or the mathematical model did not correctly describe the experimental data. Thus, the serial correlation of the residuals for adjacent and nearby points provides a measure of the quality of the fit.

The autocorrelation function provides a simple method to quantify this serial correlation for a series of different lags, k. The lag refers to the number of data points between the observations for a particular autocorrelation. For a series of N observations, Y_i, with a mean value of μ, the autocorrelation function is defined as the ratio of two autocovariance functions:

$$\beta_k = \frac{\hat{\sigma}_k}{\hat{\sigma}_0} \tag{16}$$

The autocovariance function is:

$$\hat{\sigma}_k = \frac{1}{n}\sum_{i=1}^{n-k}(Y_i - \mu)(Y_{i+k} - \mu) \tag{17}$$

for $k = 0, 1, 2, ..., K$. In these equations, K is a maximal lag less than n. Typically K is less than $n/2$. The autocorrelation function has a range between -1 and $+1$. The null hypothesis here is that the autocorrelation is equal to zero for a normally distributed random process. Note that the autocorrelation function for a zero lag is equal to 1 by definition. The expected variance of the autocorrelation coefficient of a random process with independent, identically distributed random (i.e., normal) errors is:

$$\mathrm{var}(\beta_k) = \frac{n-k}{n(n+2)} \tag{18}$$

where the mean, μ, is assumed to be zero. This variance of the autocorrelation coefficient is used to test the null hypothesis that the autocorrelation is equal to zero. If the autocorrelation is not equal to zero then serial correlation exists in the residuals.

The autocorrelations are actually a series of statistical tests, one for each value of k. Autocorrelations are commonly presented graphically or in tabular form as a function of k. This allows an investigator to easily compare the autocorrelation at a large series of lags k with the corresponding associated standard errors (square root of the variance) to decide if any significant autocorrelations exist.

Table IV presents an autocorrelation analysis for the residuals from the current analyses based on the one-site and two-site model examples. The autocorrelation is dependent on the magnitude of the residuals. Thus, it is calculated from the weighted residuals. For the one-site model the residual at $k = 1$ is very significantly nonzero with a $P > 0.002$. Note that the maximum lag in this example is 5 so this is a case with five repeated tests or measures. Consequently, the $P > 0.002$ is not quite as significant as it might appear due to the existence of the repeated measures. However, as long as the data is consistent with the above assumptions, then the one-site model can clearly be eliminated based on the probability level for the observed autocorrelation at $k = 1$.

The maximum lag, K, for the autocorrelation analysis was set to 5 for this specific analysis for two reasons. One is that trends in the residuals will usually appear within the very low numbered lags. The other reason is that this specific example only has 20 data points. In general, the goodness-of-fit test is based on the distribution properties of the residuals of the fit and requires a large number of residuals, and thus data points. This specific example, with only 20 data points, is at the low limit of applicability for these, and most other, statistical tests.

The Runs Test and autocorrelation examples presented here are illustrations of the general class of goodness-of-fit tests that are based on the expected Gaussian distribution of the residuals of the fit (Straume and Johnson, 1992).

V. Estimate the Precision of the Model Parameters

Up to this point, we have reached three conclusions about the analysis of the current data set (Table I). The first that least-squares procedures which provide the maximum-likelihood estimate of the parameter values, are based on the properties of the measurement uncertainties that are contained within the data. The second is that

Table IV
Autocorrelation Analysis

One-site model			Two-site model		
K	β_k	P	k	β_k	P
1	0.5974 ± 0.2078	0.0020	1	-0.1191 ± 0.2078	0.2833
2	0.1538 ± 0.2023	0.2235	2	-0.2961 ± 0.2023	0.0716
3	-0.0788 ± 0.1966	0.3442	3	-0.0558 ± 0.1966	0.3883
4	-0.2376 ± 0.1907	0.1064	4	-0.1568 ± 0.1907	0.2055
5	-0.2236 ± 0.1846	0.1130	5	-0.0951 ± 0.1846	0.3032

the hypothesis based one-site binding model, the combination of Eqs. (5) and (7), is not consistent with the data. The third is that the hypothesis based two-site binding model, the combination of Eqs. (6) and (7), is consistent with the data. There are surely many other hypothesis-driven binding models that are consistent with the data so the two-site model cannot be proven. All that can be concluded is that the combination of Eqs. (6) and (7) is consistent with the data.

Now that the two-site mathematical model and its concomitant hypothesis have been found to be consistent with the data, estimates of the precision of this model's parameters can be evaluated. Nonlinear models require an iterative solution to evaluate the parameter values and as a consequence an exact method to evaluate the precision of the model parameters does not exist. However, for linear models a simple set of noniterative equations to evaluate the parameter values can be derived and an exact solution for the precision of these parameter values also exists, that is, the Asymptotic Standard Errors (ASE). This linear solution is exact in the limit of a large number of data points and when no parameter correlation exists (see below), hence the title ASE.

Most available nonlinear fitting software packages report the ASE as if they were the actual precision of the determined nonlinear parameters. This can be very deceiving and problematic because for nonlinear models and/or in the presence of parameter correlation (see below) the ASE will commonly underestimate the magnitude of the parameter uncertainties, and as a consequence lead the investigator to overestimate the significance of results and thus reach incorrect conclusions. *ASE should not be used for nonlinear models!* However, they are commonly used and their basis will be described here.

The reason that the ASE are reported is that they are very easy to calculate. The most common nonlinear least-squares minimization procedures perform the calculations required to evaluate the ASE in order to accomplish the parameter estimation. As will be seen below, the more accurate procedures can require orders of magnitude more computer time than was required for the original least-squares parameter estimate. But, thankfully computers are getting very fast.

The ASE are based on the "information matrix" that is evaluated during the parameter minimization process. The minimization procedures that are based on the Gauss–Newton procedure (Johnson and Faunt, 1992; Johnson and Frasier, 1985; Straume and Johnson, 1992), all evaluate a matrix, H, of partial derivatives where the jk elements of this matrix are the sum, over each of the i data points, of the products of weighted partial derivatives of the fitting function with respect to the parameters being estimated. Specifically,

$$H_{jk} = \sum_{i=1}^{n} \left[\frac{1}{\sigma_i} \frac{\partial F(\text{parameters}, X_i)}{\partial \text{parameter}_j} \right] \left[\frac{1}{\sigma_i} \frac{\partial F(\text{parameters}, X_i)}{\partial \text{parameter}_k} \right] \tag{19}$$

where F (parameters, X_i) is the fitting function evaluated at the ith data point and *parameters* with a subscript refers to a specific parameter being estimated. The "information matrix" is the inverse of this H matrix. The Gauss–Newton minimization algorithm employs this matrix in order to find the direction and magnitude to change the parameters that will decrease the overall WSSR. Once the Gauss–Newton procedure has converged, the ASE estimates of the precision for the jth estimated parameter is evaluated from this same matrix as the square root of jth diagonal element of the inverse of H times the variance-of-fit, s^2 in Eq. (3):

$$\mathrm{ASE}_j = \sqrt{s^2 (H^{-1})_{jj}} \tag{20}$$

As noted above, the ASE commonly underestimate the actual uncertainty of the estimated parameters. This is in part because they ignore the off-diagonal elements of the information matrix. One of the previous chapters in this volume speculates that the ASE are acceptable as long as they are less than 10% of the parameter values (Tellinghuisen, 2006). This should be considered only as a rule of thumb because there are many exceptions where this is not the case.

There are several better methods to evaluate the precision of estimated parameters, such as joint confidence intervals, support plane methods, Monte-Carlo methods, and bootstrap methods (Johnson and Faunt, 1992; Johnson and Frasier, 1985; Straume and Johnson, 1992). In the present chapter, the ASE will be contrasted with the bootstrap techniques.

Probably the best method is to use the Bootstrap (Efron and Tibshirani, 1993) approach as it has the fewest assumptions. Unfortunately, this method is seldom used because of the substantial amounts of computer time required. The first step of the bootstrap procedure is to fit the hypothesis-based equation to the data. This fit produces a *noise-free fitted curve* calculated from the optimal values and a *table of residuals*, the weighted differences between the data points and the fitted curve. These values are used to create a series of surrogate data sets consisting of the noise-free fitted curve calculated from the optimal values with added randomized experimental noise. The noise is generated by repeatedly randomly selecting residuals from original table of residuals and adding them to the surrogate data sets in a random order. It is important to note that once a residual has been selected from the table it is not actually removed from the table. This is called sampling with replacement. As a consequence, for any specific surrogate data set some specific residuals will be used more than once while others will not be used at all. The fitting equation is then fit to each of the generated surrogate data sets. The fits to these surrogate data sets provide a series of estimated values for each of the estimated parameters that are used to create a probability distribution for each of the parameters. The confidence regions corresponding to any desired probability are then determined from these probability distributions.

This implementation of the Bootstrap is very similar to a Monte-Carlo analysis. The difference is the method utilized to create the simulated experimental uncertainties for the surrogate data sets. The Bootstrap uses sampling with replacement from the observed table of residuals while Monte-Carlo approach uses Gaussian distributed pseudo random numbers that are generated to match the sample variance-of-fit.

Table V
Comparison of Parameter Uncertainties Estimation Methods Based Upon The Data Shown in Table I

	ASE	Bootstrap
A	0.1012 (±3.1%)	0.1012 (−1.5%, +2.4%)
log K_1	−0.3362 (±95.7%)	−0.3362 (−210%, +85%)
K_1	0.4611 (−52%, +110%)	0.4611 (−80%, +93%)
log K_2	0.9987 (±3.8%)	0.9987 (−3.2%, +2.8%)
K_2	9.9712 (−8.4%, +9.2%)	9.9712 (−7.0%, +6.6%)

Table V presents this comparison. In this case the parameters that are being estimated are A, log K_1, and log K_2. The binding constants are thus constrained to only positive values by estimating the logarithms of the binding constants. In this case 10,000 surrogate data sets were generated. Thus, the bootstrap analysis required approximately 10,000 times as much computer time as the original least-squares parameter estimation.

One of the most noteworthy aspects of Table V is that the confidence regions for the fitted parameters are symmetrical when evaluated by the ASE method and they are asymmetrical when evaluated by the Bootstrap method. In Table V the confidence regions correspond to ±1 SEM.

For linear models it can be analytically demonstrated that the confidence regions will be symmetrical, and since the ASE is a linear approximation, it predicts that they are symmetrical. However, this proof is not valid for nonlinear models where it is common for the confidence intervals to be asymmetrical. This is even more obvious in Table VI where the analysis was performed on only the first 10 data points in Table I. Figure 4 presents the complete probability distribution for log K_1 based on the bootstrap analysis of the first 10 data points in Table I. Clearly, this probability distribution is skewed asymmetrically as is expected for nonlinear fitting models. This is contradictory to the prediction of the linear ASE method. The horizontal error bar at the top of this figure corresponds to the optimal value of log K_1 and to the ±1 SD asymmetrical confidence region.

Table VI
Comparison of Parameter Uncertainties Estimation Methods Based Upon the First Ten Data Points Shown in Table I

	ASE	Bootstrap
A	0.1028 (±8.5%)	0.1028 (−3.6%, +6.2%)
log K_1	−0.2662 (±293.2%)	−0.2662 (−377.3%, +114.9%)
K_1	0.5420 (−83.4%, +502.%)	0.5420 (−90.1%, +102.3%)
log K_2	0.9827 (±10.4%)	0.9827 (−7.0%, +5.2%)
K_2	9.6090 (−20.4%, +25.5%)	9.6090 (−14.7%, +12.5%)

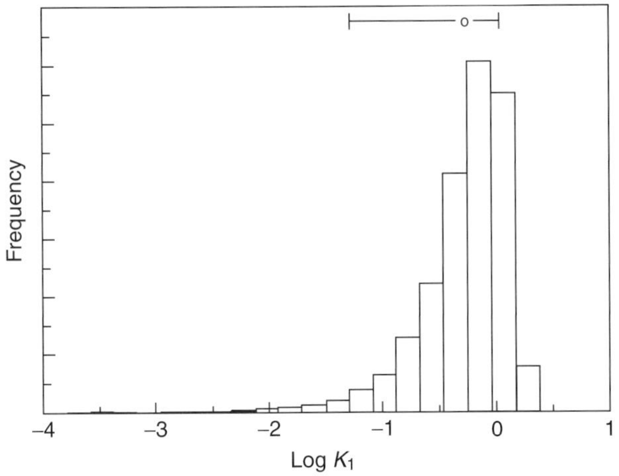

Fig. 4 Probability distribution from the bootstrap analysis presented in Table VI. The horizontal error bar corresponds to the optimal value and the ±1 SD confidence region for the parameter.

The same expected asymmetrical confidence regions are also observed for the other two estimated parameters, as are shown in Figs. 5 and 6. In this case the asymmetrical confidence regions are observed even though the ASE for log K_2 (Table VI and Fig. 5) and A (Table VI and Fig. 6) fall within the "10% rule" that has been proposed (Tellinghuisen, 2006).

A recapitulation of this section is that (1) *ASE will commonly underestimate the actual parameter uncertainties* and (2) *that for nonlinear equations, the actual confidence regions of the estimated parameters are expected to be asymmetrical.*

VI. Cross-Correlation of the Estimated Parameters

Apparent parameter correlation is almost always present when multiple parameters are simultaneously estimated by any parameter estimation procedure. The correlation is a consequence of fitting a complex equation to a small number of data points that span a limited range of the independent variables. Please note that parameter correlation is also typically present when fitting simple linear equations such as a straight line. *These parameter correlations are not necessarily telling us anything about the underlying chemistry or physiology.* It is important to be aware of the magnitude of these parameter correlations because they are closely related to the difficulties that are encountered by any data fitting procedure. A larger correlation will result in an analysis that is computationally more complex or impossible.

The cross-correlation coefficient, CC_{jk}, between the jth and kth estimated parameters are evaluated from the elements of the inverse of the H matrix [Eq. (19)] that was

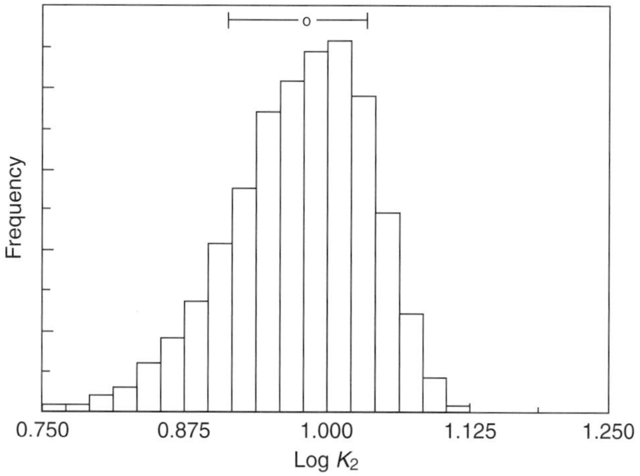

Fig. 5 Probability distribution from the bootstrap analysis presented in Table VI. The horizontal error bar corresponds to the optimal value and the ±1 SD confidence region for the parameter.

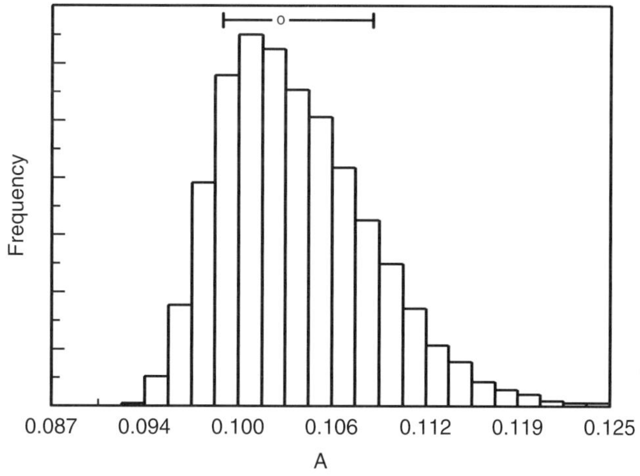

Fig. 6 Probability distribution from the bootstrap analysis presented in Table VI. The horizontal error bar corresponds to the optimal value and the ±1 SD confidence region for the parameter.

already evaluated by the Gauss–Newton least-squares parameter estimation procedure. Remember, this is the linear approximation.

$$\mathrm{CC}_{jk} = \frac{(H^{-1})_{jk}}{\sqrt{(H^{-1})_{jj}(H^{-1})_{kk}}} \quad j \neq k \tag{21}$$

These cross-correlation have a range of ± 1 with the optimal being zero. As the cross-correlation approaches $+1$ or -1 the fitting procedure will become increasingly more difficult and the results questionable because the H matrix is becoming nearly singular and cannot easily be inverted. For practical purposes, if the magnitudes of the cross-correlation coefficients are less than approximately ± 0.97 the least-squares procedure can usually function adequately. However, the ± 0.97 should not be considered as an absolute threshold with everything acceptable below ± 0.97 and everything unacceptable outside of this range. All fitting procedures get progressively worse as the magnitude of the cross-correlations increase toward 1. If equal to ± 1, the matrix will be singular and the parameter estimation procedures will fail.

Table VII presents the cross-correlation coefficients when Eqs. (6) and (7) were fit to all of the data (i.e., the parameter estimation shown in Table V) and to only the first 10 data points (i.e., the parameter estimation shown in Table VI). Clearly, as the number of data points and the range of the independent variable (i.e., the ligand concentration) decreases, the cross-correlation coefficients rapidly approach the limiting value of ± 1. If only the first 8 data points are used then the cross-correlation between A and $\log K_2$ is -0.990, and so on.

The simulated data set analyses that were performed for the evaluation of the parameter precision by the bootstrap method can also be used to visualize the cross-correlation between the estimated parameters. The bootstrap procedure creates thousands of sets of simulated surrogate data series and then analyzes each to obtain thousands of sets of parameter values that are estimates of the actual parameter values. These are tabulated to obtain the confidence regions and probability distributions of the estimated parameters (e.g., Figs. 4–6). Figure 7 is a plot of 1000 pairs of A and $\log K_2$ obtained from the bootstrap evaluation of the confidence intervals when fitting to only the first 10 data points. In Fig. 7 the pairs of values appear to fall along a line that is not coincident with the parameter axes, indicating that these two parameters are highly correlated. If the parameters were not correlated then the points would appear to be aligned with the axes.

In some cases, the cross-correlation between the parameters can be minimized, or eliminated, by a careful choice for the form of the fitting equation. As an example, consider a straight line, $Y(X) = a + bX$, fit to the first 10 data points in Table I. The resulting least-squares estimated values are: $a = 0.0095$ and $b = 0.1044$ with a $CC_{ab} = -0.8221$. In this case, the cross-correlation between the parameters can be changed by altering the form of the fitting equation. For example, by altering the form of the straight line from $Y(X) = a + bX$ to $Y(X) = a + b(X - \beta)$,

Table VII
Alteration of the Cross-correlation Coefficients Caused By Limiting the Data Set

CC	All data	First 10 points
A and $\log K_1$	0.844	0.956
A and $\log K_2$	−0.762	−0.974
$\log K_1$ and $\log K_2$	−0.774	−0.923

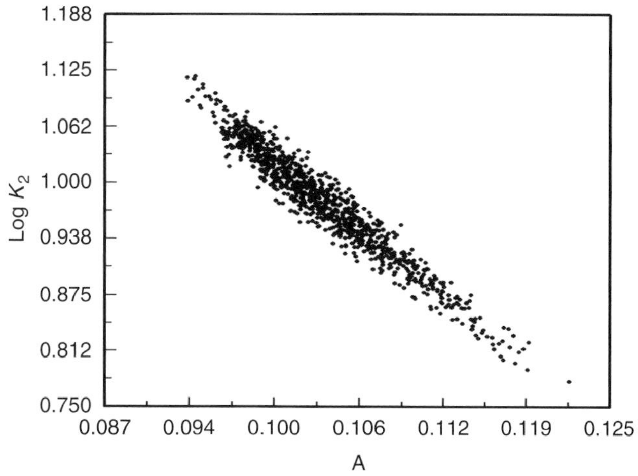

Fig. 7 Bootstrap analysis of the correlation between parameter values.

where $\beta = 0.3995$ the weighted average value of X, the least-squares estimated parameters for this later case are, $a = 0.5122$ and $b = 0.1044$ and the cross-correlation coefficient is exactly zero. This is actually the identical line as in the previous fit, however, the intercept is now at $X = \beta$ instead of $X = 0$.

VII. Uniqueness of the Parameters

For linear fitting equations it can be algebraically demonstrated that only a single set of model parameters exist that correspond to a minimum least squares. However, this cannot be demonstrated for nonlinear fitting equations and thus it is possible that multiple sets of parameters which corresponding to a minimum in the WSSR might exist. Unfortunately, the existence of multiple minima for the analysis of a data set cannot be demonstrated with the data set presented in Table I.

One method to test for multiple minima is to start the iterative nonlinear fitting procedure at many different initial starting sets of values. If the iterative procedure always converges to the same set of answers then you can have some assurance that multiple minima do not occur. However, this approach does not guarantee that multiple minima do not exist.

Figure 8 presents a simulated time series that demonstrates multiple minima when analyzed as a sign wave with a single harmonic. When a cosine wave with an unknown period, Eq. (22), is fit to this data the resulting values will depend on the initial starting values where the least-squares algorithm was initiated:

$$Y(X) = A + B\cos\left(\frac{2\pi(\text{time} - \phi)}{L}\right) \tag{22}$$

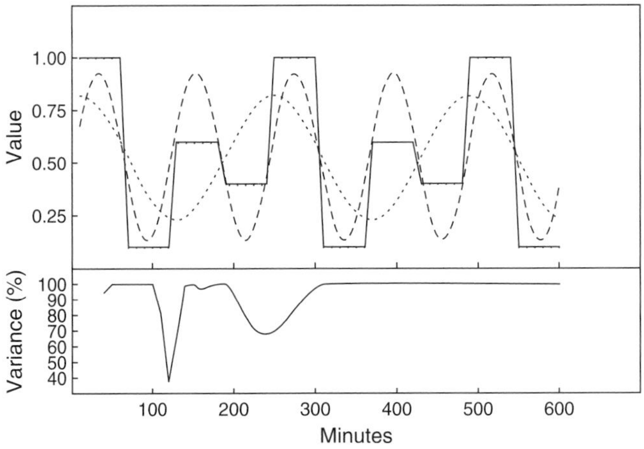

Fig. 8 An example of least-squares parameter estimation exhibiting multiple minima. Upper panel: The sold curve represents the data points, the long dashed lines is a cosine wave with a period of ~120 min, and the short dashed line corresponds to a cosine wave with a period of ~240 min. Lower panel: The percent of the variance remaining after a cosine wave of the specific period is fit to the data.

Four parameters are being simultaneously estimated, A, B, L, and ϕ. This is a nonlinear analysis because the period, L, is being simultaneously estimated with the other three parameters. If the initialization value of L, the period, is near 120 the least-squares algorithm will converge to the 120-min period (the long dashed line in Fig. 8). However, if it is initially near 239 it will converge to 239 period (the short dashed line in Fig. 8). The least-squares algorithms are not guaranteed to find the lowest variance-of-fit, only to find a minimum variance-of-fit.

In the example shown in Fig. 8 there are actually four periods that correspond to a minimum in the variance-of-fit, specifically $L = 82$, 120, 162, or 239 min. The ones at 82 and 162 min are very shallow minima, but they are minima and some algorithms will find them. Remember that the variance-of-fit is not a good measure of goodness-of-fit so it is not a good idea to specify that the minimum at 120 min is better that the one at 239 min because its variance-of-fit is lower.

The potential for multiple minima always exists when fitting to nonlinear equations but, unfortunately, no method exists that will guarantee locating all of these minima. It is, however, common for some of the multiple minima to have parameter values that are physically unrealistic. For example, a negative molecular weight has no physical meaning. If multiple physically meaningful minima are found then they must all be described when you report your results.

VIII. Conclusions

This chapter discusses WNLLS parameter estimations methods for the analysis of experimental data. A Bootstrap method is presented as the optimal approach for the evaluation the precision of the least-squares estimated parameter values.

Goodness-of-fit, parameter cross-correlation, and uniqueness of parameter values were also discussed.

Acknowledgments

The author thanks the National Institutes of Health for partial support from the following grants: RR00847, RR08119, HD011474, RR019991, DK063609, and DK064122.

References

Abbott, R. D., and Gutgesell, H. P. (1994). Effects of heteroscedasticity and skewness on prediction in regression: Modeling growth of the human heart. *Meth. Enzymol.* **240,** 37–51.

Acton, F. S. (1959). "Analysis of Straight Line Data." John Wiley & Sons, New York, p. 219.

Efron, B., and Tibshirani, R. J. (1993). "An Introduction to the Bootstrap." Chapman and Hall, New York.

Johnson, M. L., and Faunt, L. M. (1992). Parameter estimation by least-squares methods. *In* "Methods in Enzymology 210" (L. Brand, and M. L. Johnson, eds.), pp. 1–37. Academic Press, New York.

Johnson, M. L., and Frasier, S. G. (1984). Analysis of hormone binding data. *In* "Methods in Diabetes Research" (J. Larner, and S. Pohl, eds.), Vol. I, pp. 45–61. Wiley and Sons, New York.

Johnson, M. L., and Frasier, S. G. (1985). Nonlinear least-squares analysis. *In* "Methods in Enzymology 117" (S. N. Timasheff, and C. H. W. Hirs, eds.), pp. 301–342. Academic Press, New York.

Johnson, M. L., and Straume, M. (2000). Deriving complex ligand binding formulas. *Meth. Enzymol.* **323,** 155–167.

Straume, M., and Johnson, M. L. (1992). Analysis of residuals: Criteria for determining goodness-of-fit. *In* "Methods in Enzymology 210" (L. Brand, and M. L. Johnson, eds.), pp. 87–105. Academic Press, New York.

Tellinghuisen, J. (2008). Stupid Statistics! *In* "Methods in Cell Biology" (John J. Correia, and H. William Detrich, III, eds.), pp. 739–780. Amsterdam, Academic Press.

CHAPTER 25

Methods for Simulating the Dynamics of Complex Biological Processes

Maria J. Schilstra,* Stephen R. Martin,† and Sarah M. Keating*

*Biological and Neural Computation Group
Science and Technology Research Institute
University of Hertfordshire
College Lane, Hatfield AL10 9AB
United Kingdom

†Division of Physical Biochemistry
MRC National Institute for Medical Research
The Ridgeway
Mill Hill, London NW7 1AA
United Kingdom

METHODS IN CELL BIOLOGY, VOL. 84

0091-679X/08 $35.00
DOI: 10.1016/S0091-679X(07)84025-8

Abstract

In this chapter, we provide the basic information required to understand the central concepts in the modeling and simulation of complex biochemical processes. We underline the fact that most biochemical processes involve sequences of interactions between distinct entities (molecules, molecular assemblies), and also stress that models must adhere to the laws of thermodynamics. Therefore, we discuss the principles of mass-action reaction kinetics, the dynamics of equilibrium and steady state, and enzyme kinetics, and explain how to assess transition probabilities and reactant lifetime distributions for first-order reactions. Stochastic simulation of reaction systems in well-stirred containers is introduced using a relatively simple, phenomenological model of microtubule dynamic instability *in vitro*. We demonstrate that deterministic simulation [by numerical integration of coupled ordinary differential equations (ODE)] produces trajectories that would be observed if the results of many rounds of stochastic simulation of the same system were averaged. In Section V, we highlight several practical issues with regard to the assessment of parameter values. We draw some attention to the development of a standard format for model storage and exchange, and provide a list of selected software tools that may facilitate the model building process, and can be used to simulate the modeled systems.

I. Introduction

A biological cell is a highly dynamic environment. It takes up energy and uses it to create complex molecules, and to control their interactions, assembly, and destruction. Even structures that were once thought to be relatively stable and inert, such as interphase DNA, or cell walls and membranes, are continuously being remodeled, repaired, and eventually, bit by bit, entirely replaced with new material. All of these processes happen in a highly coordinated fashion. "Output" molecules of particular processes function as "input" to other processes—acting as fuel, signals, or information, and the input–output chains frequently fork, converge, interlace, and form cycles. Even when the machinery and workings of a particular process or pathway have been studied in great detail, it is usually very difficult, if not impossible, to understand intuitively the timing and interrelationship of the component steps and to make predictions about the ways in which the whole system will react.

To aid in understanding of such complex systems, one may construct "working models" in which all components that are thought to be important to the functioning of the whole are represented, and act and interact in a manner that resembles the original. The model can then be made to "simulate" the real system. If the behavior of the model resembles that of the real system, one may, at least temporarily, assume that the model is a reasonable representation of reality, and begin to make predictions by studying the responses of the model to changes in input, or changes in the structure of the model itself. If the predictions from the model are not borne out by the behavior of the real system, it will be necessary to reassess the model, or at least parts of it.

Model building is a somewhat subjective pursuit, and its success depends to some extent on the knowledge, experience, and intuition of the modeler. The models themselves can take many forms, and there are no rules that dictate the use of any particular medium: for example, a very informative working model of the UK economy was once implemented using hydraulic devices (pipes, valves, and pumps) and water (Swade, 1995). Such an analogue computing device could, in principle, very well be used to model metabolic flows in cells, or physiological processes in whole organisms. The advent of systems biology, which is linked to the spectacular advances in high-throughput technology of the last decade, has kindled an interest in the functioning and dynamics of complex biochemical systems among experts in dynamical systems from different fields, such as mathematics and electronic engineering. When modeling techniques from fields, such as electronics, are applied to biochemical systems, they tend be accompanied by their own suggestive jargon ("circuit," "switch," "design pattern"). While nontraditional modeling has its place and can certainly lead to new insights, it may also cause confusion as to cause and effect, or actor and role, and lead to erroneous conclusions. The novice modeler is advised to always bear in mind that models of biochemical processes should first and foremost adhere to the laws of thermodynamics, and that role and function are merely labels that do not imply anything more than what is explicitly expressed in the model.

In this chapter we shall, therefore, concentrate on modeling and simulation of biochemical processes in the conventional way in which the models describe the chemical reaction kinetics of the system, are expressed in mathematical terms, and are made to run on digital computers.

II. Rationale

Model building and testing—conjecture and attempted refutation—form the basis of all research in the natural sciences. Biochemical reaction systems are, in general, far too complex to allow predictions about their behavior solely on the basis of mental arithmetic or intuition. The title of this chapter may suggest a focus on large systems with many participants and many interactions, but in fact small systems, encompassing only two or three reactions and as few participants, may already exhibit behavior that is far too complex to be understood intuitively. Larger systems are just that—larger—but the same ground rules apply. We believe that it is of utmost importance that modelers understand what these ground rules are, and try to picture biochemical processes as interactions between individual molecules or assemblies that are associated with particular kinetic constants and affinities.

Although modeling solely in these terms ignores important questions related to structure and function, it is important to remember that the values of rate constants, which are assessed experimentally on the basis of models built on the premises outlined in this chapter, may give important clues about the molecular characteristics of the reactants. In addition, rate constants also provide information that allows prediction of, for instance, the amount of time individual species will remain as part of a particular complex, which, in turn, may have important consequences for the behavior of the whole system.

III. Modeling

Because this volume is dedicated to techniques and tools that are used to study biochemical processes *in vitro*, we shall focus on the modeling and simulation of reactions in "well-stirred containers." In a well-stirred container, all conditions—temperature, pH, and concentrations—are homogeneous throughout. Cells, of course, do not fall into this category, and modeling of processes that occur *in vivo* may have to take account of gradients and other local conditions. This introduces many more parameters and degrees of freedom into a model, and turns simulation of the system into a much more complicated task (see, for instance, Andrews and Bray, 2004; Kruse and Elf, 2006; Meyers *et al.*, 2006). Local conditions may have a dramatic effect on the overall dynamics of a system; they do not, however, change its basic chemistry.

A. Running Example

In this chapter, we shall illustrate the most important concepts in modeling and simulation using the phenomenon of microtubule dynamic instability as a running example. Microtubules are long, hollow cylinders constructed from the tubulin $\alpha\beta$-heterodimer. The α subunit of the dimer contains nonexchangeably bound GTP; whereas the β subunit can contain GTP (Tu-GTP), GDP (Tu-GDP), or be nucleotide free (Tu). Following addition of Tu-GTP to a growing microtubule, the GTP is hydrolyzed to GDP and, as a result, Tu-GDP is the major component of microtubules.

In 1984, Michison and Kischner first reported that "microtubules *in vitro* coexist in growing and shrinking populations which interconvert rather infrequently," and proposed that "this dynamic instability is a general property of microtubules." Horio and Hotani (1986) were the first to observe experimentally that single microtubules were indeed either in a slowly growing state (slow tubulin dimer addition) or in a rapidly shortening or shrinking state (rapid tubulin dimer loss). Shrinking microtubules were occasionally "rescued," and became growing ones, whereas "catastrophe" sometimes befell growing microtubules, whereupon they entered the shortening state. State interconversion was infrequent, and appeared to happen randomly. Although the two ends, "plus" and "minus," of the microtubules could be distinguished by their dynamics (plus-ends were much more dynamic than minus-ends), both ends appeared to behave in essentially the same way.

Horio and Hotani were able to determine values for the growth, shrinkage, and interconversion rates, and we shall use these original data to calculate the parameter values required to perform the simulations. Under the conditions used in their study, the plus-ends of shrinking microtubules were found to shorten for an average of 18 sec, before they were rescued, and lost about 2.4 μm in length over that period. Occurrences in which microtubules shrunk to zero length, and disappeared completely, were not counted. Microtubules have 1625 subunits per micrometer, so that 2.4 μm corresponds to 3900 (tubulin dimer) subunits. This means that they were shrinking at a rate of about 220 subunits per second. Growing microtubules continued to grow for an average of approximately three minutes, during which they gained about 1.8 μm: a growth rate of about 16 subunits per second.

These values were, of course, obtained under certain conditions, using a particular tubulin preparation, and similar measurements, probably more accurate, and under much wider sets of conditions, have been carried out over the years. Ideas about the origins and consequences of this behavior have been extensively discussed in the literature. Here, however, we will simply use a model that describes the phenomenon itself—rather than its origin—which reproduces the above observations, and is capable of predicting the overall behavior of microtubules under different conditions.

B. From Cartoons to Dynamic Model Diagrams

Descriptions of biochemical processes are often accompanied by cartoons in which aspects of their organization and behavior are illustrated. Such cartoons may show the molecules that participate in the process, their function, their chemistry, their interactions, the assemblies that they form, the dynamics of these assemblies, the cellular compartments in which the process occurs, and so on. Although cartoons can be very helpful in conveying ideas, they are, by and large, unsuitable as a basis for a quantitative description, not so much because they are not associated with numbers or equations, but mainly because the use of symbols is ambiguous. In Fig. 1A, the concept of microtubule dynamic instability (see Section III.A) is illustrated using a cartoon.

Three different diagram types can be used as the basis for a quantitative dynamic model, and these are also illustrated in Fig. 1. These dynamic model diagrams also contain boxes and arrows, just like cartoons, but here the use of symbols is unequivocal: they have an exact interpretation, and an equivalent mathematical representation. Such diagrams leave out all information that is irrelevant to the modeled dynamics, and therefore seldom show all of the ideas that are expressed in

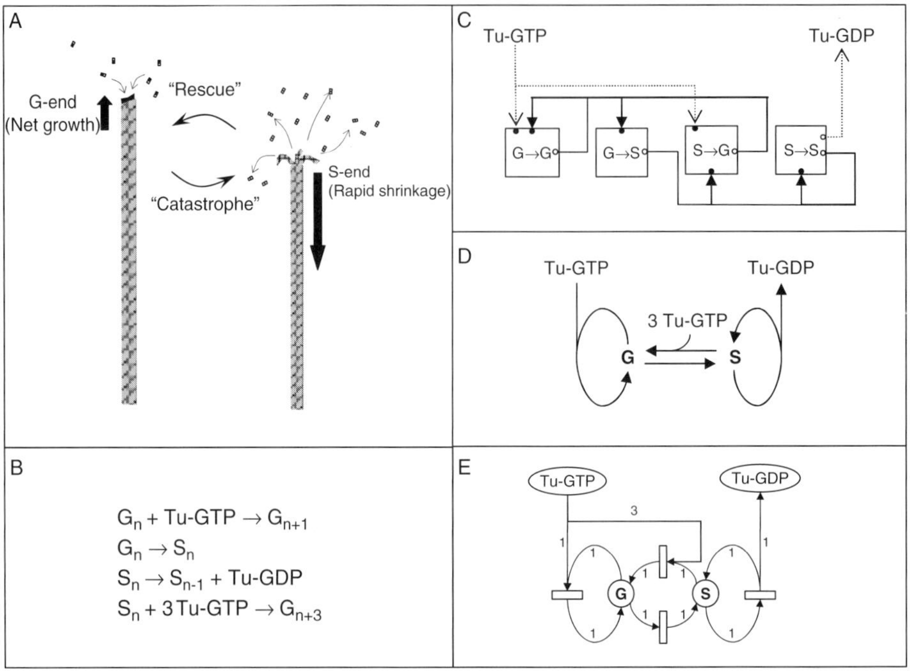

Fig. 1 Diagrams depicting microtubule dynamic instability: (A) cartoon, (B) chemical reaction scheme, (C) data-flow or process diagram, (D) state-transition diagram, (E) Petri-net diagram. The use of boxes, arrows, and other symbols in the cartoon (A) is ambiguous. Textual symbols in the equivalent chemical reaction scheme (B) and state-transition diagram (D) represent chemical or physical states; the arrows represent reactions. The arrows may be associated with rate equations, which express the dependence of the reaction rate on the concentrations of the reaction participants. The boxes in the data-flow diagram (C) represent processes (transformation of input into output); the arrows symbolize data such as concentrations of particular molecules or complexes. The small circles at the edges of the boxes indicate whether the data are input (filled circles) or output (open circles) to the processes. Each process is associated with a transfer function, which describes how its input is transformed into its output. In the Petri-net notation (E) ellipses and boxes represent states and reactions, respectively. The arrows indicate the function of the states in the reaction (reactant or product), and have a weight (here shown as a number next to the arrow) which relate to the stoichiometry of the associated reactants or products.

a typical cartoon. The three different diagram types are described below. Although the interpretation of dynamic model diagrams must be unambiguous, boxes and arrows denote different entities in these three different contexts.

1. Data-Flow Diagrams

In data-flow diagrams (Fig. 1C), boxes represent processes, and arrows may be seen as channels that shunt material or data from process to process. Electronic circuits are often represented as data-flow diagrams, and attempts have been made to depict and model particular biochemical reaction networks (suggestively labeled "control circuits") in the similar way (e.g., Gilman and Arkin, 2002). Obviously, a box is a box and does not carry any information about the dynamics of input–output transformation that occurs "inside." In electronic diagrams, these boxes therefore have a certain shape, or carry a label that indicates which input–output relation, or "transfer function," should be used to describe its behavior. Such diagrams are particularly useful for a hierarchical description of complex systems in which the components have well-defined modes of operation and distinct functions. In that case, the description of a particular type of box can be used over and over, in the same way that particular types of electronic components can be used in different circuits.

2. State-Transition Diagrams

Chemical reaction schemes, written in the classical way (Fig. 1B), are easily transformed into state-transition diagrams (Fig. 1D). In this approach, the boxes (or symbols) and arrows are interpreted differently. In the dissociation reaction S → S + Tu-GDP, for instance, in which Tu-GDP dissociates from S, a shrinking microtubule end, the symbols S and Tu-GDP represent biochemical species, or more precisely, *potential states* of biochemical species (species can be molecules, ions, complexes, macromolecular assemblies, etc.). The arrow represents a reaction—a physical or chemical *state transition*.

3. Hybrid Diagrams

The hybrid Petri-net[1] notation combines aspects of the data-flow and state-transition diagramming styles (Fig. 1E). It is equivalent to the standard chemical reaction notation (Fig. 1B) except that reactions and states are both represented explicitly as boxes and circles, respectively. The arrows that connect them indicate which states carry a reaction's input—its reactants—and output, the reaction products. The arrows have weights that represent the stoichiometry of the reactants and products in the reaction: in the reaction S + 3Tu-GTP → G, the weights of the arrows that connect the input states S and Tu-GTP to the reaction box are 1 and 3, respectively, and that of the arrow that connects G is 1. A state has a population, formed by the number of "items" (i.e., molecules, complexes, assemblies) that are currently in that state. If there are not too many in a particular state, individual items may be depicted as small black circles inside the circle that represents the state.

[1] Named after its inventor, C.A. Petri; also called Place/Transition graphs. Petri-nets are used generally to model event sequences in concurrent processes. Here, we shall use some of the more evocative Petri-net terminology (transition, firing) but will avoid expressions that may lead to confusion (place, token).

C. Compartments

Each state is, by definition, restricted to a particular compartment. Thus, a molecular species that is present in the cytoplasm and in the nucleus has items in both the "cytoplasmic" and "nucleic" states, and movement of items from one compartment to another is modeled as a reaction. Concentrations are calculated from the number of items n_X in a particular state X and the volume V_X of the compartment associated with that state:

$$[\mathrm{X}] = \frac{n_\mathrm{X}}{N_\mathrm{A} V_\mathrm{X}} \tag{1}$$

where N_A is Avogadro's number (6.022×10^{23}). Here, we shall focus on reactions that take place *in vitro*, where there is usually just one compartment, with a constant volume, but it should be kept in mind that this is not necessarily the case. Ample consideration should be given to the effect of volume on concentrations and rate constants when introducing more than one compartment or variable volumes into a model.

D. Events and State Transitions

State-transition diagrams and Petri nets form the basis for "discrete event modeling." Discrete event modeling is used to examine how externally applied changes percolate through a system over time—not only in chemical kinetics, but also for instance in control engineering, operations management, and many other fields. An event is defined as something that happens instantaneously. The state transitions that are associated with chemical reactions (including biochemical ones) are usually modeled as events, because the time it takes for molecular species to change state is generally very short in comparison with the amount of time they spend being in a particular state.

Upon a state-transition event, reactant populations lose items, while product populations gain new items. The numbers of items lost and gained in an event are given by the weights of the arrows that connect the input and output states to the reaction box. We will refer to the occurrence of a state-transition event and the associated movement of items as the "firing" of the reaction. If the population of any of the input nodes to a transition is insufficient—that is, smaller than the weight of the connecting arc—the reaction can not fire and therefore will not occur. However, if the population of all input states is sufficiently high, a reaction will fire sooner or later. In Section III.F, we shall explain how the transition firing probability relates to the chemical rate constants that are associated with the transitions. However, before we can do that, we first need to discuss the basic concepts of chemical reaction kinetics and thermodynamics in well-stirred containers, which are central to the rest of the argument.

E. Reaction Kinetics

1. Basic Chemical Kinetics

In a chemical reaction, reactants are consumed, and products formed. The "instantaneous reaction rate" expressed as the quantity of reactant consumed or product formed per unit of time, depends on the concentrations of the reactants and on other factors, such as temperature.

Almost all chemical reactions have either one or two reactants. Unimolecular reactions, reactions with a single reactant, such as dissociations or isomerizations, appear to occur spontaneously, without the involvement of other molecules, and are associated with first-order rate constants, which have units of frequency (sec^{-1}, min^{-1}, etc.).

To calculate the instantaneous rate of a unimolecular reaction, the rate constant associated with the reaction, k_{I} (the subscript I is used here to indicate that the rate constant is first order), is multiplied by the number of items in the reactant state, n_{X}, or with the reactant's concentration, [X]:

$$v_{n\mathrm{X}} = \frac{\mathrm{d}n_{\mathrm{X}}}{\mathrm{d}t} = -k_{\mathrm{I}} n_{\mathrm{X}} \tag{2}$$

$$v_{[\mathrm{X}]} = \frac{\mathrm{dX}}{\mathrm{d}t} = -k_{\mathrm{I}}[\mathrm{X}] \tag{3}$$

Note that $v_{n\mathrm{X}} = v_{[\mathrm{X}]} N_{\mathrm{A}} V$, where V is the volume of the compartment or container in which the reaction takes place.

Bimolecular reactions have two reactants, which must collide to react. The frequency with which one molecule of the first reactant undergoes collisions with molecules of the second reactant increases with the *concentration* of the second reactant. The number of collisions, and therefore the reaction rate, decreases when the volume increases and the number of reactant molecules remains the same. Bimolecular reactions are associated with second-order rate constants (here indicated with the subscript II), which have units of frequency per unit of concentration (e.g., M^{-1} sec^{-1} = L·mol^{-1} sec^{-1}), and *always* include the volume. The instantaneous rate for a bimolecular reaction is calculated by multiplying the number, n_{X}, *or* concentration [X] of one reactant, with the concentration [Y] of the other:

$$v_{n\mathrm{X}} = \frac{\mathrm{d}n_{\mathrm{X}}}{\mathrm{d}t} = -k_{\mathrm{II}} n_{\mathrm{X}}[\mathrm{Y}] = -k_{\mathrm{II}} \frac{n_{\mathrm{X}} n_{\mathrm{Y}}}{N_{\mathrm{A}} V} \tag{4}$$

$$v_{[\mathrm{X}]} = \frac{\mathrm{d}n_{\mathrm{X}}}{\mathrm{d}t} = -k_{\mathrm{II}}[\mathrm{X}][\mathrm{Y}] \tag{5}$$

Association reactions, such as cofactor binding or dimer formation, are examples of bimolecular reactions. Note that the rates at which X and Y disappear are the same.

If the population of state Y is much larger—say, 1000 times—than that of state X, the number of items in Y will not change significantly, even when the reaction goes to completion (1000 − 1 is still 99.9% of 1000). When such conditions apply, the highest reactant concentration is often taken to be constant, and incorporated in the rate constant of the reaction. The resulting product of the second-order rate constant for the reaction and the concentration of the most abundant reactant has units of frequency, and is referred to as a pseudo-first-order rate constant:

$$v_{[\mathrm{X}]} = -k_{\mathrm{II}}[\mathrm{X}][\mathrm{Y}] \approx -k_{\mathrm{II}}\mathrm{Y}_0[\mathrm{X}] = -k_{\mathrm{I}}'[\mathrm{X}] \tag{6}$$

Here the pseudo-first-order rate constant $k_{\mathrm{I}}' = k_{\mathrm{II}}\mathrm{Y}_0$, where Y_0 is the concentration of Y at the start of the reaction.

If the reaction is a dimerization ($2\mathrm{X} \rightarrow \mathrm{X}_2$), two items disappear from the X-state each time the reaction fires. A molecule of X cannot form a dimer with itself, and therefore has $n_{\mathrm{X}} - 1$ other molecules of X to choose from:

$$v_{n_{\mathrm{X}}} = \frac{\mathrm{d}n_{\mathrm{x}}}{\mathrm{d}t} = 2k_{\mathrm{II}}\frac{n_{\mathrm{X}}(n_{\mathrm{X}} - 1)}{N_{\mathrm{A}}V} \tag{7}$$

but because $n_{\mathrm{X}} - 1 \approx n_{\mathrm{X}}$ for $n_{\mathrm{X}} \gg 1$:

$$v_{[\mathrm{X}]} = \frac{\mathrm{d}[\mathrm{X}]}{\mathrm{d}t} = -2k_{\mathrm{II}}[\mathrm{X}]^2 \tag{8}$$

Reactions of a higher order,[2] in which more than two concentrations are multiplied with a rate constant to calculate the rate, also exist. Although these seldom involve the simultaneous encounter of more than two reactants, they are frequently used as an approximation of the real kinetics of a more complex process. We shall use the rescue reaction, S + 3Tu-GTP → G, as an example of a higher order reaction. The fact that its reaction rate is calculated as $v_{[\mathrm{S}]} = \mathrm{d}[\mathrm{S}]/\mathrm{d}t = -k_{\mathrm{IV}}[\mathrm{S}]$ $[\text{Tu-GTP}]^3$ does not mean that three Tu-GTPs must collide and bind simultaneously with a shrinking microtubule to turn it into a growing one. This form is merely used to approximate the kinetics of a much more complex mechanism.[3] Reactions of order zero, in which the reaction rate is independent of any concentration, also occur (at least conceptually, see Section III.E.3), and have rate equations of the form $v_{[\mathrm{X}]} = \mathrm{d}[\mathrm{X}]/\mathrm{d}t = -k_0$, where k_0 is a zero-order rate constant with units of concentration per unit of time.

[2] A reaction's molecularity is related to its stoichiometry, whereas its order is determined by the number of reactant concentrations that is taken into account to calculate the overall reaction rate, and is a kinetic concept; see Cornish-Bowden (1995) for a full explanation.

[3] In fact neither the stoichiometry nor the order of the rescue reaction follows from Horio and Hotani's data; a higher order dependency was used to explain the oscillations in the microtubule mass observed under conditions of extremely fast growth (Bayley *et al.*, 1989).

Irrespective of the reaction type, rates are expressed in number of items, mol, or concentration units per unit of time [e.g., $\sec^{-1}$, $\text{mol}\cdot\sec^{-1}$, or $\text{M}\cdot\sec^{-1}$ where M (molar) $= \text{mol}\cdot\text{L}^{-1}$]. The reaction *firing frequency*, or flux, J, is related to the rate at which its reactant X disappears as:

$$J = -\frac{v_{n\mathrm{X}}}{s_{\mathrm{X}}} \tag{9}$$

where s_{X} is the number of items that are removed from the population of X when the reaction fires once (equal to the stoichiometry of reactant X). The general expression for the dependence of the reaction firing frequency on the reactant concentrations in a chemical reaction is:

$$J = k \times N_{\mathrm{A}} V \prod_{i=1}^{n} [\mathrm{R}_i]^{S_i} \tag{10}$$

Here n is the total number of different reactants, k is the rate constant, and $[\mathrm{R}_i]$ and S_i the concentration and stoichiometry of reactant R_i. This dependence is also known as the "law of mass action."

Instantaneous rates and firing frequencies represent the current state of the system, and may be calculated at any time during a reaction. In general, reactant concentrations are not constant. They may change during the course of a reaction because a reaction is out of equilibrium (see below, Section III.E.2), because an experimenter changes them in some way, for instance by increasing the volume, or because they are added or removed by coupled processes.

2. Equilibrium and Steady State

Most biochemical reactions are reversible and therefore tend toward an equilibrium state, where the forward rate equals the reverse rate, and the average concentration of each participant remains constant. The rate constants for the individual forward and reverse reactions reflect the free energy of activation for the reaction, $\Delta G^{\ddagger}$, the minimum free energy the reactants must have to reach the transition state (which must be crossed to form products). Rate constants are proportional to $T\cdot\exp[-\Delta G^{\ddagger}/RT]$, where R is the universal gas constant and T is the absolute temperature. The thermodynamic equilibrium constant, K, which is equal to the quotient of the rate constants k_{F} and k_{R} for the forward and reverse reactions, is related to the overall difference in free energy ΔG^0 between the reactant and product states (under defined standard conditions, indicated by the superscript 0):

$$K = \mathrm{e}^{-\Delta G^{\circ}/RT} \tag{11}$$

which follows from

$$\frac{k_F}{k_R} = e^{-(\Delta G_F^{\mp} - \Delta G_R^{\mp})} \tag{12}$$

where the indices F and R refer to the forward and reverse reactions. If the forward and reverse reactions are of the same order, as in a reversible isomerization, the equilibrium constant is dimensionless. If the forward reaction is an association, and the reverse one a dissociation then, K, often referred to as an *equilibrium association constant*, indicated as K_a, has units of reciprocal concentration. Conversely, if the forward reaction is a dissociation and the reverse one an association, the equilibrium constant has units of concentration, and is called an *equilibrium dissociation constant*, and indicated as K_d. A high K_a indicates that the affinity is high; a high K_d indicates the opposite. The actual amount of complex formed depends, of course, on the concentrations of the participants. Note that *equilibrium* association and dissociation constants should not be confused with second- and first-order rate constants for association and dissociation processes, which are sometimes called association and dissociation *rate* constants.

It is important to appreciate that equilibrium is a *dynamic* state where the forward and reverse reactions fire at rates that are determined by the equilibrium concentrations of the participants. When an equilibrium is "perturbed," because the concentration of any of its participants change (for instance, because the experimenter adds some more), or because the equilibrium constant alters (for instance, as a result of a change in temperature), the forward and reverse rate are no longer equal. As a result, there will be a net flow of material, from reactants to products or vice versa, until a new equilibrium is reached. The process by which a system moves toward a new equilibrium position following a perturbation is called *relaxation*. Relaxation processes proceed at an overall rate that is, in reversible single-step reactions, determined by the sum of the forward *and* reverse reaction rates, irrespective of the actual direction of the flow. The larger the rates, the faster the relaxation, and the more rapidly the new equilibrium is reached. Thus, two different reversible reactions may have the same equilibrium constant, but very different relaxation rates. The relaxation characteristics of more complex systems depend on the rates of all reactions that are involved in coupled equilibria (further discussed in Chapter 15 by Eccleston *et al.*, this volume).

Equilibrium is a state that is often encountered *in vitro*, but hardly ever *in vivo*, where most reaction products are consumed by other reactions. When a reversible reaction, or a sequence of reversible reactions, is provided with a constant influx of reactants and efflux of products, the reaction will generally reach a "steady" state in which there is a net flux of material, which is constant, but not zero (in which case the reaction would be at equilibrium). Steady-state conditions may be perturbed because influx or efflux rates, or one or more reaction rate constants change. The response rate—the rate at which the system adapts to the new conditions—depends, again, on both the forward and reverse reaction rates, with more dynamic systems having faster response times. Furthermore, the difference between steady state and equilibrium concentrations of the participants in the reaction depends on the ratio of the response rate to the influx and efflux rates: the faster the response, the closer the system can get to equilibrium.

3. Enzyme Kinetics

Enzyme catalyzed reactions are chemical reactions, and subject to the laws of thermodynamics. Therefore, they are modeled in exactly the same way as other chemical reaction systems, as combinations of association, isomerization, and dissociation events, with the appropriate kinetics. However, *in vitro* studies of enzymic reactions are often carried out under conditions where the response rate of the enzyme subsystem[4] is fast with respect to the rate at which the concentrations of its reactants change. Under such conditions, the enzyme will operate close to steady state. The steady-state distribution of enzyme species depends on the enzyme's mechanism, on the values of the rate constants, and on the concentrations of the reactants, but not on time, and not on the total enzyme concentration. The rate at which the reaction product is formed depends on the *absolute* concentrations of one or more enzyme species, and can be calculated from the distribution of enzyme species and the total enzyme concentration. The steady-state rate equation for a particular enzyme mechanism—sometimes called the kinetic law—describes the steady-state rate of reactant consumption or product formation in terms of kinetic constants, and enzyme and reactant concentrations. A derivation of the rate equation for the Henri–Michaelis–Menten mechanism can be found in any biochemistry textbook. The steady-state rate equation for this mechanism, $E + S \Leftrightarrow ES \rightarrow E + P$, where E is the enzyme, S is its substrate (reactant), and P is its product, is

$$v_{[P]} = \frac{d[P]}{dt} = k_2 e_0 \frac{[ES]}{e_0} = k_2 e_0 \frac{[S]}{((k_{-1} + k_2)/k_1) + [S]} \tag{13}$$

where k_1 and k_{-1} are the second-order forward, and first-order reverse rate constants for reactant binding to and dissociation from the enzyme, and k_2 combines the transformation of S into P and the dissociation of P from the enzyme into a single first-order rate constant. As e_0 is the total enzyme concentration, $[ES]/e_0$ is the fraction of the enzyme that is in the ES state, and has a dimensionless value between 0 and 1. The denominator $(k_{-1} + k_2)/k_1$ forms the Michaelis constant, K_M, and the product $k_2 e_0$ is the maximum rate, V_{max}, at which the reaction can proceed.

Steady-state rate equations can be derived for any enzyme mechanism, irreversible or reversible, with any number of reactants, products, or modifiers (see below). Techniques that facilitate the derivation of steady-state rate equations, notably

[4] The enzyme subsystem includes all enzyme species: free enzyme, enzyme-substrate, enzyme-product, and other complexes, but not unbound reactants or products.

the King and Altman method, are outlined in many monographs on enzyme kinetics (e.g., Cornish-Bowden, 2001), and some present the expressions for a large number of mechanisms (particularly Segel, 1975). Steady-state rate equations are all of the form

$$v_{[P]} = \frac{d[P]}{dt} = \kappa \times e_0 \times \varphi \quad (14)$$

The apparent first-order rate constant κ is a combination of rate constants, whereas φ, the fraction of enzyme species that determine the overall product formation rate, is a combination of rate constants, and reactant, product, and modifier concentrations. The precise forms of κ and φ depend on the enzyme mechanism. Rate equations for reversible reactions are of the form $v = v_F - v_R = e_0(\kappa_F\varphi_F - \kappa_R\varphi_R)$, where the indices F and R refer to the appropriate expressions for the forward and reverse reactions.

When an enzyme with Michaelis–Menten kinetics is operating under conditions where [S] is much smaller than K_M, the rate of the overall reaction S → P is proportional to [S] (provided, of course, that e_0 is constant), and the reaction behaves as a first-order reaction, with k_2e_0/K_M, which has units of frequency, as its rate constant. When [S] is much larger than K_M, however, [ES]/e_0 is very close to one, and the enzyme is "saturated." In this regime, the reaction S → P proceeds at a constant rate, independent of [S], and the kinetics look like those of a zero-order reaction. Reactions catalyzed by enzymes whose mechanism is more complex often exhibit a similar dependence of their apparent kinetic order on reactant concentration.

4. Beyond Mass Action

Enzyme catalyzed reactions may be described as sets of coupled reactions with mass-action kinetics in which the basic reactions each have their own (mass action) rate equation. However, the Henri–Michaelis–Menten scheme can be reduced simply to S → P, with Eq. (13) ($v = V_{max}[S]/(K_m + [S])$, provided steady-state conditions apply) describing the dependence of its rate on the reactant concentration [S], and the same holds for more complex mechanisms. In general, any equation that describes the dependence of the rate at which reactants are consumed (or products appear) on concentrations of the reactants may be used to calculate the reaction firing frequency. However, when derived rate equations are used, it is very important to be aware of the assumptions that were made during their derivation, and of any restrictions to their validity.

Derived equations may also describe the dependence of the rate on components that are not consumed in the reaction, or on other conditions. Such components or conditions function as rate *modifiers*. Modifiers may form one to one complexes with reactants or products, or act in other fixed proportions, but are not associated with the stoichiometry of the reaction (as they are neither consumed, nor produced). In Eqs. (13) and (14), for instance, the value of e_0, the total enzyme concentration, determines the overall reaction rate, but E is neither a reactant, nor a product in the S $\rightarrow$ P reaction. Therefore E functions as a modifier in this reaction. Modifiers that decrease the reaction rate may be called inhibitors or repressors, and those that increase the rate may be activators, stimulators, catalysts, promoters, etc. Under different conditions, modifiers may have different, even opposite, effects, and it is important to recognize that labels, such as inhibitor or catalyst, do not imply a mechanism of action. The modifier concept is particularly useful in systems in which the concentration of the modifier is variable: coupled futile cycle systems, such as MAPK cascades, are often simplified using steady-state rate equations with variable enzyme concentrations (as in, for instance, Markevich *et al.*, 2004).

F. Timing of Events in Chemical Reactions

1. Transition Probabilities

Although it is not possible to predict *exactly* when any *individual* molecule will undergo a chemical reaction, the *likelihood* that it will do so within the next second (or minute, or year) can be calculated. To illustrate this point, we use the example of the dissociation of the Tu-GDP complex (Tu-GDP $\rightarrow$ Tu + GDP). If there is no reverse reaction (for instance, because Tu or GDP are removed rapidly via other processes), the rate at which the number of items in the Tu-GDP state, n, changes is $dn/dt = -k \times n$ [Eq. (2)]. This is an example of an ordinary[5] differential equation (ODE), which describes the relation between n (which is a function of time) and its derivative with respect to t. This particular ODE has an "analytical" solution, an algebraic equation that explicitly describes the dependence of n on t. The analytical solution is $n = C \times e^{-kt}$, where C is an arbitrary constant. If the number of items in the Tu-GDP state is n_0 at $t = 0$, then $C = n_0$, and the dependence of n on t can be written as $n = n_0 \times e^{-kt}$. Thus, the *fraction* of items still in the Tu-GDP state at time t is equal to $n/n_0 = e^{-kt}$. Conversely, the fraction that has already reacted is $1 - n/n_0 = 1 - e^{-kt}$.

The dissociation rate constant k for Tu-GDP has been measured to be about 0.09 sec^{-1} (at 298 K, Engelborghs and Eccleston, 1982). If there were 10^6 Tu-GDP complexes at $t = 0$, one second later there would have been about 0.914×10^6, or 91.4% remaining, that is, 8.6% of the initial population would have dissociated. Thus, the *probability* that any individual item present in the Tu-GDP state at $t = 0$ will disappear from that state within one second is 0.086, or 8.6%.

Essentially the same holds for bimolecular reactions under pseudo-first-order conditions. The (second-order) association rate constant for the reaction Tu + GTP $\rightarrow$ Tu-GTP was estimated to be $2.2 \times 10^6\ M^{-1}\ sec^{-1}$ (Brylawski and Caplow, 1983).

[5] An "ordinary" differential equation contains the derivative of a function to a single variable. Partial differential equations (PDEs) involve functions that are dependent on more than one variable (e.g., time and position).

If 10^6 items are initially in the nucleotide-free Tu state, and there are 10^9 GTP molecules, all in a volume of 1 pl (i.e., [GTP] = 1.66 mM), the pseudo-first-order rate constant k' is $2.2 \times 10^6 \times 1.66 \times 10^{-3} = 3.6 \times 10^3\ \text{sec}^{-1}$, which can substitute for k in the above equations. In the absence of a reverse reaction, the probability that any particular item has disappeared from the Tu state one millisecond after the start of the reaction is then calculated to be 97%.

2. Lifetime Distributions

As noted above, the general expression for the probability P that a particular item that was in the reactant state of a first-order reaction at t_0 has disappeared from that state at a given time t is:

$$P = 1 - \mathrm{e}^{-k(t-t_0)} \tag{15}$$

On the basis of the above example for Tu-GDP dissociation, we can calculate the number of items that stay in the Tu-GDP state for longer than 2 sec, but shorter than 3 sec, in other words, items that have lifetimes between 2 and 3 sec. At $t = 2$ sec, 16.5% will have already reacted, and have therefore spent less than 2 sec as Tu-GDP. At $t = 3$ sec, 23.7% will have reacted. Therefore, 7.2% have lifetimes between 2 and 3 sec. Similar calculations can be carried out for other intervals—0–1 sec: 8.6%, 1–2 sec: 7.9%, and so on, and the results may be plotted as a histogram that shows the fraction of items with lifetimes within certain intervals. Furthermore, it will be possible to calculate the average lifetime of an item in the Tu-GDP state, and the standard deviation on this average.

It turns out that this distribution of lifetimes, also called the *probability density function, f*, follows the first derivative of the expression for the probability P that, at time t, an item has disappeared from the state it was in at time t_0, that is, Eq. (15):

$$f = \frac{\mathrm{d}P}{\mathrm{d}t} = k \times \mathrm{e}^{-kt} \tag{16}$$

The expression for P in Eq. (15) is called the *cumulative distribution function* for the probability density function in Eq. (16). The type of distribution expressed in Eq. (16) is called an exponential distribution. If the lifetimes are exponentially distributed, the average lifetime, $\langle\tau\rangle$, as well as its standard deviation, $\sigma(\tau)$ are known to be equal to the reciprocal of the rate constant $\langle\tau\rangle = \sigma(\tau) = 1/k$. Thus, the value of a first-order rate constant—and the same holds for a pseudo-first-order one—predicts how long it will be, on average, before a reaction occurs. Therefore, the complex Tu-GDP in the example will exist for an average of 11 sec (±11 sec, because $\sigma(\tau) = \langle\tau\rangle$) after it has been formed.

It is difficult, or impossible, to derive an algebraic expression for P for higher order reactions, or for reaction sequences. However, the equation $f = \mathrm{d}P/\mathrm{d}t$ is generally true, also for lifetime distributions with different shapes (for instance, the well-known Gaussian or normal distribution). If the probability density function is known or can be measured, the above relationship [Eq. (16)] may be used to calculate the cumulative distribution function.

IV. Simulation

To simulate the behavior of a biochemical reaction system, at least three pieces of information must be provided:

1. A model that describes the *structure* of the system (states and compartments, reactions, the connections between states and reactions, and the connection weights)
2. *Rate equations* for all reactions (expressions defining the rates at which the reactions fire, given a certain input) and values for the associated parameters
3. The *initial state* of the system (the population of each state and the volume(s) of the compartments in which the reactions take place).

Once the necessary data have been provided, time series—state populations recorded over time—can be generated computationally. In *stochastic* simulations, a random number generator is used to produce possible time courses, or trajectories, of individual items. Because of the randomness in the timing of individual events, individual stochastic simulation rounds have different outcomes, even if the initial conditions are exactly the same in each round. By contrast, *deterministic* simulations always give the same result for the same set of initial conditions. Deterministic simulations usually yield trajectories that are equal to the average of many rounds of stochastic simulation of the same system, given identical initial conditions.

A. Stochastic Methods: The Behavior of Individual Entities

We shall illustrate the principles of stochastic simulation using models for microtubule dynamic instability (see Section III.A) that increase in complexity as we introduce new concepts. In all of these models, the dissociation of a tubulin subunit from a shrinking end, $S_n \rightarrow S_{n-1} + \text{Tu-GDP}$, and the transition from a growing to a shrinking end, $G_n \rightarrow S_n$, where n indicates the number of subunits in the microtubule, will be modeled as first-order processes, with rate constants $k_{SS} = 220$ and $k_{GS} = 0.0056\ \text{sec}^{-1}$, respectively. The binding of Tu-GTP to a growing end, $G_n + \text{Tu-GTP} \rightarrow G_{n+1}$, and the rescue of a shrinking end, $S_n + 3\text{Tu-GTP} \rightarrow G_{n+3}$, will initially be modeled as pseudo-first-order processes, with rate constants k'_{GG} of 16 sec^{-1} and $k'_{SG} = 0.056\ \text{sec}^{-1}$. The values for k_{GS} and k'_{SG} are derived from the average lifetimes of the growing and shrinking states, $\langle\tau_G\rangle = 180$ sec, and $\langle\tau_S\rangle = 18$ sec, using the relationship $k = 1/\langle\tau\rangle$, and assuming that the lifetimes observed by Horio and Hotani (1986) were exponentially distributed.

In the next stage, we shall model the association reactions as higher order processes (second and fourth order, respectively). To derive the values of the second- and fourth-order rate constants from the values of their pseudo-first-order counterparts, we need to know at which Tu-GTP concentration they were obtained. We shall assume that it was about 10 μM. If k_{GG}·[Tu-GTP] = 16, and k_{SG}·[Tu-GTP]^3 = 0.056 sec^{-1} at [Tu-GTP] = 10 μM, then $k_{GG} = 1.6 \times 10^6$ M^{-1} sec^{-1}, and $k_{SG} = 5.6 \times 10^{13}$ M^{-3} sec^{-1}.

To keep the models simple, we will assume that there is no subunit dissociation in the growth phase, or binding in a shrinking phase, and that Tu-GDP is not able to bind to either type of microtubule end. Furthermore, we shall disregard the fact that microtubule growth and shrinkage are probably associated with more than two rate constants, and that catastrophe and rescue are much more complex processes. These conditions need to be relaxed in more complex models (see Janulevicius *et al.*, 2006; Martin *et al.*, 1993) but the principles of the simulation remain the same.

1. The Behavior of a Single Item

First, we shall concentrate on the behavior of a *single* microtubule end under pseudo-first-order growth and rescue conditions (i.e., at a constant Tu-GTP concentration, for instance, because a GTP regenerating system is used). To start a simulation, the initial state—G or S—of the system must be chosen. Here, we arbitrarily choose that the end is in the G state, and set the total simulated time t_{sim} to 0.

One of two things will happen to a growing microtubule end: it may gain another subunit and remain in the growing state or it may change state and become a shrinking end. Either possibility has a finite (i.e., nonzero) chance of happening. To decide which of these two events will happen on this occasion, and when, we simply "draw straws," but make the odds reflect the very different probabilities for the two transitions. To do this, one random number between 0 and 1 is generated for each possible event (r_1 for $G_n \rightarrow G_{n+1}$, and r_2 for $G_n \rightarrow S_n$), and used to calculate the time at which each of the events would happen. This is done by substituting r_1 or r_2 for P, and k'_{GG} or k_{GS} for k in Eq. (15):

$$r_1 = 1 - e^{-k'_{GG}(t-t_0)}, \quad r_2 = 1 - e^{-k'_{GS}(t-t_0)}$$

and solving for $\Delta t = t - t_0$ in both cases:

$$\Delta t_1 = \frac{-\ln(1 - r_1)}{k'_{GG}} \text{ for the } G_n \rightarrow G_{n+1} \text{ reaction}$$

$$\Delta t_1 = \frac{-\ln(1 - r_2)}{k_{GS}} \text{ for the } G_n \rightarrow S_n \text{ reaction}$$

If this process was to be repeated many times, and r_1 and r_2 plotted against Δt, the cumulative distribution functions [Eq. (15)] for the two reactions would emerge.

The reaction with the smallest Δt is then selected as the one that will occur, the other is ignored. Suppose, for example, that the values for r_1 and r_2 were 0.44313 and 0.17621, then Δt_1 and Δt_2 would be 0.036 and 34.6 sec, and the continued growth reaction would be selected. The transition that is *most likely* to occur is, of course, always the one with the largest rate constant, that is, $G_n \rightarrow G_{n+1}$ and this is why the microtubule will continue to grow for a long time. However, there is a $0.0056/16 \times 100\% = 0.04\%$ chance that Δt_2 will be smaller than Δt_1, and that the $G_n \rightarrow S_n$ transition will be selected. For example, had r_1 and r_2 been 0.99927 and 0.00024, Δt_1 and Δt_2 would have been 0.451 and 0.043 sec, and the catastrophe reaction would have been selected.

Having decided which reaction will occur ($G_n \rightarrow G_{n+1}$ in this case), the simulation is updated by increasing the simulation time t_{sim} by the selected Δt (Δt_1 in this case) and letting the associated reaction fire at the new t_{sim}. The microtubule remains in the same state (G) but increases in length by one subunit.

To continue the simulation, this whole process must be repeated. If, as here, the microtubule remains in the growing state then k'_{GG} or k_{GS} are again used to evaluate newly drawn values of r_1 and r_2 for selecting the next step. However, had the $G_n \rightarrow S_n$ transition "won," then the microtubule would have switched states and the next event would be either shortening ($S_n \rightarrow S_{n-1}$) or rescue ($S_n \rightarrow G_{n+3}$) and k_{SS} or k'_{SG} would be used in choosing between them. The simulation is continued in this way until a preset end time is exceeded.

Figure 2 shows a typical simulated time course—a trajectory—of the microtubule end. It is clear from this figure that the spread in the lengths of both the growth and shrinkage periods is large, but that the latter are, on average, much shorter than the former. The close-up of a $G \rightarrow S$ transition in which the individual events are resolved shows that the average time interval between two subunit additions to a growing end is also much larger than that between two subunit dissociations from a shrinking end. The histogram shows that these intervals (lifetimes) are indeed exponentially distributed, with measured averages of 0.064 ($\pm$0.065) and 0.0044 ($\pm$0.0043) sec. The average growth and shrinkage rate constants that are calculated from these lifetimes are 15.6 and 227 sec, very close indeed to the 16 and 220 that were used as input.

To illustrate the simulation of higher order reactions, we shall now model Tu-GTP binding to a growing microtubule end as a second-order process (G_n + Tu-GTP $\rightarrow G_{n+1}$), and rescue of a shrinking end as a fourth-order process (S_{n+3} + 3Tu-GTP $\rightarrow G_{n+3}$). Initially we will, again, be looking at the behavior of a single microtubule end. We shall start the simulation with 50 μM Tu-GTP in a volume of 0.01 pl, that is, with about 3×10^5 molecules, with the microtubule end (whose concentration is 0.17 nM) in the growing state.

In this model, one molecule of Tu-GTP is removed from the reaction volume by each Tu-GTP association event, and three molecules are used for the rescue reaction. As GTP is hydrolyzed after incorporation of Tu-GTP in the growing microtubule, and not regenerated, the Tu-GTP concentration is variable. However, in stochastic simulations, the state of the system does not change between two

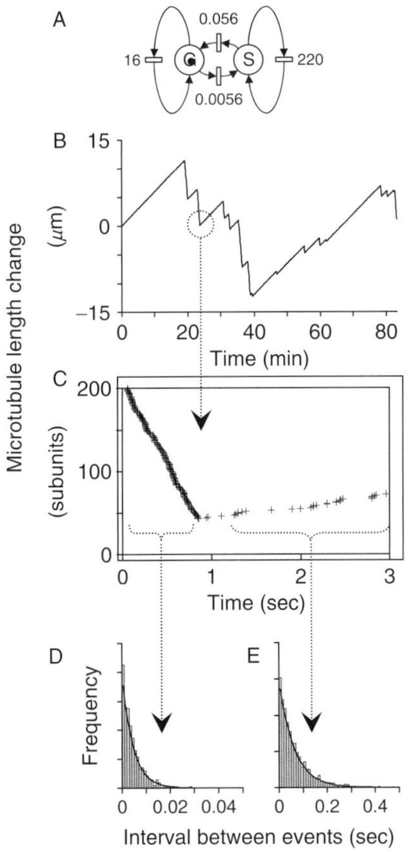

Fig. 2 Dynamic instability of a single microtubule end with growth, shrinkage, and state interconversion modeled as first-order processes, that is, under conditions in which [Tu-GTP] is constant (for instance, because an enzymic GTP regenerating system is used). (A) Petri-net diagram of the model. The numbers are the values for the first-order rate constants associated with the reactions. The small filled circle inside the larger open circle indicates the situation at the start of the simulation: one item (i.e., a microtubule end) in state G. (B) Trajectory of the length changes of the microtubule end during one stochastic simulation round of 5000 sec. (C) Close-up of the events (crosses) close to one S $\rightarrow$ G state transition. Note that the dissociation events in the shrinking phase are more closely spaced than the association events in the growing phase. This is borne out by the average duration of intervals between successive events in the S-phase, 0.0044 ± 0.0043 sec, and in the G-phase, 0.064 ± 0.065 sec. In both cases, the distribution of the intervals observed in the simulation is exponential, as shown in the histograms in (D) and (E). The solid lines represent the lifetime distributions according to Eq. (16), with $k_{SS} = 220$ (D) and k_{GG} 16 (E) sec^{-1}.

successive events, and is reassessed after each event. Therefore, exactly the same method of drawing a random number for each reaction with a finite firing probability, and calculating the associated Δt values, can be applied to decide which reaction will occur next, and when. However, rather than using constant pseudo-first-order rate constants to assess the Δt values, the values for $k'_{GG} = k_{GG}[\text{Tu-GTP}]$ and $k'_{SG} = k_{SG}[\text{Tu-GTP}]^3$ are reassessed each time a molecule of Tu-GTP disappears from the reaction volume. Thus, under the initial conditions of the simulation, when [Tu-GTP] is 50 μM, k'_{GG} is $1.6 \times 10^6 \times 50 \times 10^{-6} = 80\ \text{sec}^{-1}$, and k'_{SG} is $5.6 \times 10^{13} \times (50 \times 10^{-6})^3 = 7\ \text{sec}^{-1}$. When, after some 2.4×10^5 association events, only 10 μM Tu-GTP is left, these rates are (as expected at 10 μM Tu-GTP, see the earlier description) 16 and 0.056 sec^{-1}.

Figure 3 shows a typical trajectory, and it can be clearly seen that the end grows more and more slowly, whereas the shrinking episodes increase in length as [Tu-GTP] decreases. The combined result is a dramatic disassembly when [Tu-GTP] decreases below about 10 μM.

2. Modeling the Behavior of Multiple Items

To model the individual behavior of more than one microtubule end, we simply create as many copies of the GS elements as necessary, where a GS element consists of a G and an S state that are connected via G → S and S → G reactions in the same way as in the previous model. However, Tu-GTP is supplied to these elements from a single

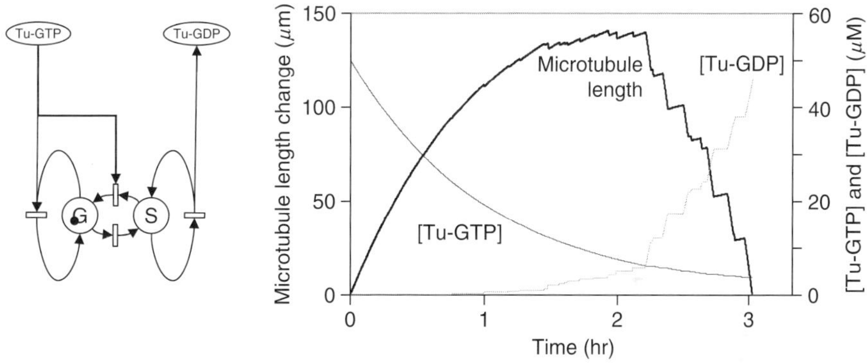

Fig. 3 Dynamic instability of a microtubule end in which growth (G_n + Tu-GTP → G_{n+1}) and rescue (S_n + 3Tu-GTP → G_{n+3}) are modeled as second- and fourth-order reactions, respectively. The Petri-net diagram of the model (left) indicates that the simulation was started with a single item in state G. The initial Tu-GTP concentration was 50 μM (300,000 molecules in a volume of 0.01 pl). The graph (right) shows the trajectories obtained during one stochastic simulation round over about 10,000 sec. In this model, one Tu-GTP item is consumed by each association event in the growth phase, and one Tu-GDP item is produced per dissociation event during shrinkage. However, Tu-GTP is not regenerated, so that [Tu-GTP] decreases, and [Tu-GDP] increases during the simulation (concentrations indicated on the right axis). In the first 90 min of the simulation, any shrinking microtubules are rapidly rescued (at a rate of $k_{SG}\cdot[\text{Tu-GTP}]^3$), and the microtubule gains about 140 μm in length (left axis). As [Tu-GTP] decreases, the growth rate ($k_{GG}\cdot[\text{Tu-GTP}]$), as well as the rescue rate decrease, resulting in the saw-tooth trajectories that are observed for microbubules under (near) steady-state conditions. When [Tu-GTP] falls below a certain critical concentration, (~10 μM), subunit loss during the increasingly long periods of shrinkage outweighs the slower and slower gain during the growths phases, and the polymerized state is no longer permanently sustained.

pool, as shown in the 3-microtubule model in Fig. 4. It is now necessary to set the initial state of each individual element, and in the example we have chosen to start the simulation with all three growing. The initial Tu-GTP concentration is, again, 50 μM.

In the previous models, there were always two reactions to choose from; now there are six, two for each microtubule end. The simulation principle remains the same: a random number is drawn for each of the six possible reactions, and the reaction that produces the smallest Δt is selected. Figure 4 shows the three trajectories that result. Note that the Tu-GTP pool is exhausted three times more quickly than in the example with a single microtubule (Fig. 3).

If the individual microtubule end trajectories are of no particular interest, it is also possible to just use a single GS element, as in the model in Fig. 3, but simply distribute the required number of items—three in this example—over the states in

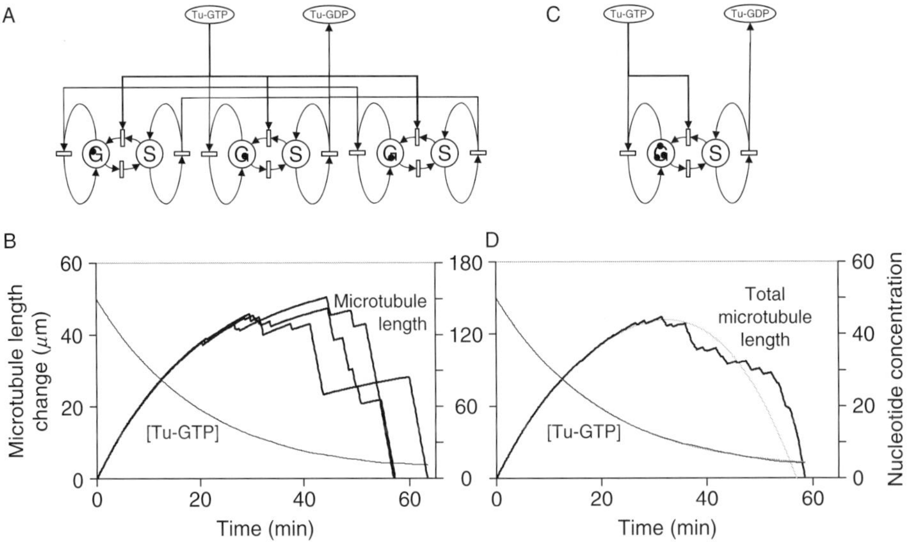

Fig. 4 Behavior of three microtubule ends, modeled as three different GS elements, each containing one item (A, B), or as one GS element containing three independent items (C, D). At the start of the simulations, all items (i.e., microtubule ends) are in a G state, and [Tu-GTP] = 50 μM (300,000 molecules in a volume of 0.01 pl). (A, C) Petri-net diagrams, indicating the state of the models at the start of the simulation; (B, D) microtubule length change and [Tu-GTP] trajectories observed in one simulation round over about 4000 sec. Because there are now three microtubule ends in both simulations, rather than one, Tu-GTP is consumed three times faster than in the simulation shown in Fig. 3, and the individual microtubules gain maximally about 45 μm in length before the critical Tu-GTP concentration is reached. Each microtubule is modelled as a single, distinguishable element in model (A), their individual paths in one stochastic simulation round are recorded and depicted in (B). Multiple items in a single state are simply counted, not individually identified, and the trajectory obtained in one stochastic simulation round (D, black lines) is therefore the *sum* of three trajectories (note the scale of the left axis). The gray lines in (D) are the trajectories for [Tu-GTP] (almost indistinguishable from the black line) and the total microtubule length, obtained by numerical integration (deterministic simulation) of the model in (C).

the GS element. Here, we have again placed all three items in the G state, and started the simulation with 50 μM Tu-GTP.

With all three microtubules growing, the probability that one of them will acquire a new subunit within the next second is, of course, three times as large as that for a single growing microtubule. Thus, the value of k'_{GG} is now calculated as $3 \times k_{GG} \times$ [Tu-GTP]. Likewise, the chance that any one of the items will change state is also three times that of a single microtubule under the same conditions. The simulation is, again, carried out along the same lines: two random numbers are chosen for the two possible reactions (GG and GS), and used to calculate values for Δt, whereupon the smallest value and its associated reaction are selected. However, rather than using $k_{GG} \times$ [Tu-GTP] and k_{SG}, the firing frequencies [Eq. (10)], $J_{GG} = 3 \times k_{GG} \times$ [Tu-GTP] and $J_{GS} = 3 \times k_{GS}$ are used to substitute k in Eq. (15).

If all microtubules continue growing, the distribution of items over the two states will remain the same (i.e., all three items in the G state); if the selected reaction is GS, however, one item will move to the S state. If that happens, all four reactions (SS, SG, GS, and GG) may occur in the next step, because all reaction firing frequencies now have finite values ($J_{GG} = 2 \times k_{GG} \times$ [Tu-GTP], $J_{GS} = 2 \times k_{GS}$, $J_{SG} = 1 \times k_{SG}[\text{GTP}]^3$, and $J_{SS} = 1 \times k_{SS}$). In that case, random numbers are drawn for all possible reactions, and the appropriate firing frequencies are used to substitute k in Eq. (15). The simulation is continued further as outlined above.

Multiple items in one state do not have individual identities, and it is, therefore, not possible to follow the trajectory of a single microtubule end in this way. Instead, the distribution of items over the two states is recorded, and the resulting trajectory, also shown in Fig. 4, is the sum of the individual trajectories.

3. Recipe for Performing Stochastic Simulations

Thus, the standard procedure for the stochastic simulation of a chemical reaction system comprises the following steps:

1. Choose a start time t_0 and end time t_{end}. Set the number of items in each of the states to their values at t_0, and the simulation time t_{sim} to t_0
2. Calculate the firing frequency J_i for each reaction i
3. For each transition i for which $J_i > 0$, draw a random number r_i ($0 \leq r_i < 1$) from a uniform distribution,[6] and compute the time t_i at which it will fire:

$$t_i = t_{sim} + \frac{-\ln(1-r)}{J_i} \tag{17}$$

4. Select the reaction R for which t_i is smallest as the one that will occur next
5. Set t_{sim} to the smallest value of t_i and let R proceed
6. Repeat from (2), until t_{sim} exceeds t_{end}.

[6] Be aware that random number generators may limit the time resolution: if, for instance, r is drawn from a sequence of 8-bit integers, the smallest finite value obtainable is 2^{-8}.

This procedure is at the heart of the algorithm for exact stochastic simulation of coupled chemical reactions known as Gillespie's direct method (Bortz *et al.*, 1975; Gillespie, 1977).

4. Speeding up Stochastic Simulations

Stochastic simulations are inherently slow, because each event is taken into consideration, and often multiple calculations must be carried out to decide on the timing of the next event. It is sometimes possible to increase the efficiency of the calculations without losing the intrinsic precision of this type of simulation. Below we have given a short description of some of the methods that have been proposed. However, these methods have either extra computational overhead or reduced precision. If one's interest is in the behavior of a single or a few items in a limited number of states, and a choice of a few transitions at any one time (e.g., a microtubule end), the simplest method in which a random number is drawn for each transition with a finite firing frequency may well be the most efficient.

1. The original algorithm proposed by Gillespie (1977) required the drawing of only two random numbers per step, rather than one for each reaction. The first random number is used to establish at what time any reaction will fire, and the second one to decide which reaction this will be. It involves the firing frequencies J for each reaction, and the overhead consists of dividing this sum into sectors proportional to the individual firing frequencies.
2. If the firing probability of reaction A is unaffected by the firing of reaction B, values for t_i that are calculated for A are not invalidated by the firing of B. In the "Next Reaction Method," proposed by Gibson and Bruck [who also coined the label "First Reaction" method for the Gillespie algorithm, Gibson and Bruck, (2000)], all valid values for t_i are kept in a priority queue, together with a reference to their associated reactions, which are fired at the appropriate times. The queue needs to be maintained and adjusted after each firing, but if the degree of independence in the system is high (i.e., there are many reactions whose firing frequency is not affected by the firing of many other reactions), this method can be much more efficient than the First Reaction Method.
3. Tau-leaping (Gillespie, 2001) is an approximate method in which the simulation time is advanced by a preselected "leap" interval in which one or more transitions are likely to fire significantly more often than once. The actual number of firings is chosen, again, on the basis of a random number. If the length of the leap interval is chosen so that the total flux in the system is predicted to change by a negligible amount over this period, this technique may result in a greater simulation speed, at the cost of some of the accuracy of the First and Next Reaction methods.
4. We have so far only considered transitions that represent simple mass-action processes, whose firing frequency J, under given constant input conditions, is independent of time. However, to save on computation time, a transition may be defined to represent a sequence or combination of separate steps. In this case, its

reactants may have a lifetime expectancy that changes as they get older, and the cumulative distribution function has a form that is different from Eq. (3). If the shape of the cumulative distribution function can somehow be assessed,[7] the procedure outlined in Section IV.A.3 can be followed. An example is given in Gibson and Bruck (2000) and the method is explained in greater detail in Schilstra and Martin (2006).

B. Deterministic Methods: The Average Behavior of Many Entities

The behavior of the three microtubule ends in the example in Section IV.A.2 illustrates the point that the trajectories of individual entities often deviate significantly from the average paths of many, but, nonetheless, that such an average does exist, and that its time course is determined by the structure and parameter values of the systems, and by the initial conditions. The average path may be assessed by repeating the stochastic simulation many times, or by the simultaneous inclusion of many particles in a single simulation, but the computational cost of such simulations tends to be very high, and may often be prohibitive.

1. Numerical Integration of Ordinary Differential Equations

There is a much faster and more efficient way of assessing average trajectories: numerical integration of the system of coupled ordinary differential equations (ODEs) that describe how the number of items in each state changes over time. To set up such a system of equations, the expressions for flux of items into and out of each state are combined to give the overall rate at which the number of items in that state changes. Reactions that add items to a particular state make a positive contribution, and reactions that remove items make a negative one. Thus, the overall change $\mathrm{d}(n_S)$ in the number of items in state S (n_S) over a very small time interval $\mathrm{d}t$ is equal to:

$$\begin{aligned}\frac{\mathrm{d}(n_\mathrm{S})}{\mathrm{d}t} &= -v_\mathrm{SS} + v_\mathrm{SS} - v_\mathrm{SG} + v_\mathrm{GS}\\ &= -s^\mathrm{S}_\mathrm{SG}J_\mathrm{SG} + s^\mathrm{S}_\mathrm{GS}J_\mathrm{GS}\\ &= -1 \times k_\mathrm{SG} \times n_\mathrm{S} \times [\mathrm{Tu-GTP}]^3 + 1 \times k_\mathrm{GS} \times n_\mathrm{G}\end{aligned}$$

Here, v_SG, J_SG, and k_SG are the rate at which items are removed from S by the S $\rightarrow$ G reaction, the firing frequency of this reaction and the reaction's rate constant, respectively. The number of items in the G and Tu-GTP states are n_G and $n_\mathrm{Tu\text{-}GTP}$, and s^S_SG and s^S_GS represent the stoichiometry of S in the S $\rightarrow$ G, and

[7] For instance by numerical integration of Eq. (2) in which k is a function of time; further discussion of this subject is beyond the scope of this chapter.

$G \rightarrow S$ reactions (both 1). Note that $-v_{SS}$ and v_{SS} cancel (as the number of items in S stays the same when the $S \rightarrow S$ reaction fires), and that the rate $d(n_S)/dt$ is expressed as number of items per unit of time. The rate at which the number of items in S (and in the other states) changes may also be written in terms of concentration: $d[S]/dt = -k_{SG}[S][\text{Tu-GTP}]^3 + k_{GS}[G]$.

The full model used in Section IV.A.2 corresponds to the following system of coupled[8] ODEs:

$$\frac{d[G]}{dt} = -k_{GS}[G] + k_{SG}[S][\text{Tu-GTP}]^3$$

$$\frac{d[S]}{dt} = k_{GS}[G] - k_{SG}[S][\text{Tu-GTP}]^3$$

$$\frac{d[\text{Tu-GTP}]}{dt} = -k_{GG}[G][\text{Tu-GTP}] - 3k_{SG}[S][\text{Tu-GTP}]^3$$

$$\frac{d[\text{Tu-GDP}]}{dt} = k_{SS}[S]$$

There is no analytical solution (an expression that describes the concentrations as a function of time) to this and most other coupled ODE systems. However, if an initial set of concentrations is provided, it is possible to create a *numerical* solution. In the simplest implementation, a time step Δt is chosen over which none of the concentrations are expected to change by more than a certain very small percentage. The concentration changes over Δt are calculated by multiplying the expressions for the rate by the time interval. For example

$$\Delta[G] = [G]_{t+\Delta t} - [G]_t \approx (-k_{GS}[G]_t + k_{SG}[S]_t[\text{Tu-GTP}]_t^3)\Delta t$$

where the subscripts t and $t + \Delta t$ indicate current concentration, and predicted concentration after the time step Δt, respectively. The new concentrations are then calculated by adding these changes to their (known) present values:

$$[G]_{t+\Delta t} \approx [G]_t + (-k_{GS}[G]_t + k_{SG}[S]_t[\text{Tu-GTP}]_t^3)\Delta t$$

This is done for all equations in the set, and the process is repeated until a preset end time or other end point, such as a certain concentration level, is reached. This accumulation process is called numerical integration.

Smaller time steps result in smaller relative changes, and in more accurate solutions, but also in an increased total simulation time. With time steps approaching zero, the trajectories obtained with this method will approach the paths that

[8] They are coupled because the variables in the left-hand-side of the equations return in the right-hand-side of other equations in the same system.

would result from averaging an infinite number of stochastic simulations of the same system (under the same initial conditions). If the time steps taken are too large, the solution will not only lose accuracy but may also become unstable. In an unstable solution, the calculated values typically oscillate wildly with amplitudes that increase with every new time step.

2. Algorithms

Many algorithms have been developed to improve the accuracy, stability, and efficiency of numerical solutions to coupled ODE systems. The procedure outlined above, which is the oldest, simplest, but least efficient and stable, is known as Euler's method. The popular Runge–Kutta method is an example of a so-called higher order algorithm in which information from several substeps is used to match a Taylor expansion of order greater than one to create a more accurate estimate. So-called multistep, or predictor–corrector algorithms, such as the Adams–Bashford method, estimate the size of the changes on the basis of more than one previously computed step (Press *et al.*, 1989).

The above algorithms are examples of *forward* or *explicit* methods. In *backward* or *implicit* methods, such as the backward Euler algorithm, equations must be solved that express the estimates of the concentrations at $t + \Delta t$ as a function of the current concentrations. To obtain estimates for $[G]_{t+\Delta t}$, $[S]_{t+\Delta t}$, and $[\text{Tu-GTP}]_{t+\Delta t}$, for example, the set of simultaneous equations that describe the current concentrations in terms of their expected values, such as

$$[G]_t \approx [G]_{t+\Delta t} + (-k_{GS}[G]_{t+\Delta t} + k_{SG}[S]_{t+\Delta t}[\text{Tu-GTP}]^3_{t+\Delta t})\Delta t$$

must be solved.

Apart from the backward Euler algorithm, there are implicit methods that use Taylor expansion series, predictor–corrector methods, or a combination to improve the accuracy of the estimates. Frequently used implicit algorithms are those formulated by Adams–Moulton and Gear (1971).

Implicit methods are of special interest to modelers of biochemical systems, because they tend to produce the most efficient stable solutions of so-called stiff systems of differential equations. A system is stiff when the concentrations of its various components change on very different time scales. In enzyme catalyzed reactions, for example, the concentrations of the various enzyme species may reach steady state in milliseconds, whereas substrate and product concentrations may take minutes or longer to respond. Explicit algorithms must take very small time steps to keep track of the fast reactions, even when the concentrations of their participants are hardly changing. In implicit algorithms, the computation is much less sensitive to these fast reactions, and as a result much larger time steps can be taken. Because implicit methods usually require the solution of a set of nonlinear algebraic equations, they are more difficult to implement, and computationally more expensive than explicit methods. However, for certain systems, the saving in computational time can be significant because they require far fewer steps to produce a stable and accurate solution.

Most computer programs that are designed solve systems of coupled ODEs use, in addition to one or more of the above algorithms, methods that control the size of the time step, adapting it to an optimal value on the basis of estimates of its accuracy at the current stage of the simulation. Optimizing the step size also has a certain computational cost, but, again, may have improved overall efficiency. Some also adaptively control the order of the method (broadly speaking, the higher the order, the more substeps or previous steps are taken into account), and some packages even provide an option to automatically choose the most efficient algorithm.

3. Recipe for Performing Deterministic Simulations

Thus, the standard procedure for the stochastic simulation of a chemical reaction system comprises the following steps:

1. Combine the expressions for influx and efflux into ODEs that describe how the population of each state changes, given the overall state (i.e., the population of all states) of the system
2. Choose a start time t_0 and end time t_{end}. Set the concentrations or number of items in each of the states to their initial values, and set the current simulation time, t_{sim} to t_0
3. Based on the expressions for the population change in each state, and given the state of the system at t_{sim}, estimate the population of the states after a certain finite time interval Δt, using an appropriate algorithm, desired accuracy, and advanced features such as adaptive step size control
4. Set the population of all states to their estimated values for $t_{\text{sim}} + \Delta t$, and set t_{sim} itself to $t_{\text{sim}} + \Delta t$
5. Repeat from (3), until $t_{\text{sim}} + \Delta t$ exceeds t_{end}.

Even though there is a choice of computational tools that will carry out numerical integration, it is relatively easy to use a spreadsheet program to numerically solve a simple ODE (such as $\mathrm{d}n/\mathrm{d}t = k \times n$) with the forward Euler method. We recommend this as a starting point because it helps in developing a sound understanding of the underlying processes, and enables one to assess the influence of step size, and so on.

C. Comparison of Stochastic and Deterministic Simulation

Figure 4 also shows the average microtubule end behavior and Tu-GTP concentration obtained by numerical solution of the ODE system for the simple model used here. Because the system has a high degree of dependence, and because many events occur over the simulated period, the stochastic simulations are computationally very intensive, particularly when they are used to obtain an average time course. Numerical integration is generally a much more efficient procedure for calculating average trajectories. However, the improved efficiency comes at the price of losing all information about the behavior of individual entities, especially about the level of dispersion, the deviation of individual trajectories from the average.

V. Modeling and Simulation in Practice

A. Modeling

Modeling is sometimes said to be art rather than science. Although there is ongoing research into automated model building, in practice creating a model is done "by hand," on the basis of a combination of intuition and experience. There are no hard and fast rules that can be applied to the process of model building, but there are certain issues, particularly with regard to the assessment of parameter values, that modelers of biochemical reaction systems need to take into consideration, and these are addressed in Section V.A.1.

There are, however, quite a few software tools that *facilitate* the building of biochemical reaction networks by providing the option to enter networks as sets of states, reactions, compartments, rate equations and associated functions, parameter and initial values, and allowing the user to use, modify, and store these data. A relatively new development, discussed in Section V.A.2 , is the specification of a standard format for storage and exchange of biochemical reaction network models.

1. Obtaining and Assessing Parameter Values

The numbers used in the examples in this chapter were taken from reports that describe in detail how, and under which conditions, the experiments that yielded these values were carried out, and which assumptions and equations were used to extract the values from the observed data. Extracting values from data often involves a significant amount of model construction and data fitting, and the assumptions that were made along the way may restrict the application of the results. When setting up a quantitative model of a system it is, therefore, crucial to be aware of the origin of the parameters that are obtained from the literature. Furthermore, parameter values may have been obtained under conditions that were different from those simulated in the models, and in other cases the available data may not contain any information about parameters that are required in the simulation. Therefore, it is often necessary to guess the values of certain parameters, and different sets of values may be tried to find the combination that "fits" the observations best. Parameter optimization is the subject of another chapter in this volume, but here we need to make the following point. Large and complex systems require large numbers of parameters to be estimated, and it is highly likely that, for a given model, many combinations of parameters will describe the system equally satisfactorily. Therefore, it is advisable not to be overambitious, keep the number of unknowns as small as possible, and try to assess how much changes in individual parameters influence the overall behavior of the model system. It is also important to remember that the behavior of any particular model may be relatively or even totally insensitive to the value of particular parameters.

a. Limits to Rate Constants

As noted above, it will frequently be necessary to guess the values of particular rate constants. There are no lower limits for rate constants and although essentially any value is possible the rate constants for slow steps must be selected with the required overall system response time in mind. In contrast, both first- and second-order rate constants have well-defined upper limits. The fastest reactions in biochemistry, energy, photon, electron, and proton transfer between well-positioned donors and acceptors, happen on femtosecond to nanosecond time scales and have rate constants of the order of 10^{13} to 10^{7} $\sec^{-1}$ (Zewail, 2003). All other processes are highly likely to proceed more slowly. Bimolecular reaction rates are limited by the diffusion rate of the reactants, and therefore depend on the viscosity of the medium. The maximum value for a second-order rate constant in water at 25 °C is estimated to be of the order of 10^{9} M^{-1} $\sec^{-1}$ (Atkins, 1994).

b. Overall Free Energy Conservation in Chains of Reversible Reactions

The equilibrium constant for a reversible reaction is determined by the free energy difference between the reactants and products. If two or more reversible reactions occur in sequence, the overall energy difference between the states at the two ends of the sequence is obtained by addition of the free energy differences for each of the individual steps. This sum corresponds in turn to an overall equilibrium constant, which is the product of the equilibrium constants for the individual steps:

$$\Delta G_1^\circ + \Delta G_2^\circ + \Delta G_3^\circ + \cdots + \Delta G_n^\circ = -RT \times \ln[K_1 \times K_2 \times K_3 \times \cdots \times K_n].$$

There are frequently two or more possible paths that lead from one end of the sequence to the other. For example, a ternary complex of a protein P and its ligands X and Y, P-X-Y, may be formed through two pathways, which, together with the associated free energy changes and equilibrium association constants, are indicated in Scheme 1. Because the free energy change must be independent of the pathway, $\Delta G_1^\circ + \Delta G_2^\circ$ must be equal to $\Delta G_3^\circ + \Delta G_4^\circ$, so that $K_1 \times K_2$ must be equal to $K_3 \times K_4$. Therefore, if any three of these constants are known then the fourth can be calculated. Two things should be noted. First, if the binding of X to P increases the affinity of P for Y ($K_2 > K_3$), then binding of Y to P must also increase the affinity of P for X ($K_4 > K_1$). Second, $K_1 \times K_2$ can only be not equal to $K_3 \times K_4$, if there is a continuous flux of material through the system. A model may

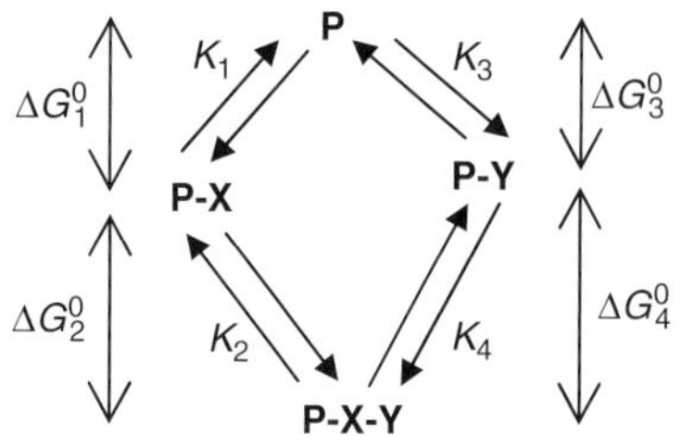

Scheme 1.

include many cycles of this type and careful attention must be given to ensure that all of constants used obey the appropriate relationship.

2. Standard Formats for Model Exchange and Storage

There are many software tools, proprietary or freely available, high or low level, than can be used to simulate biochemical reaction systems (Section V.B.3). Improved versions of existing tools and completely new ones are being released regularly, whereas others are no longer updated or maintained, and may become unusable. Obviously, all these programs allow their users to store the networks that have been created, together with parameter values, an initial (or current) state, and sometimes with simulation or other instructions. The file with the stored data can then be read into the same program at a later time, to be reused in new simulations, amended, and so on. Because each program has its own storage format, models that have been built using one tool cannot, in general, be read by another. This may become a problem when a tool ceases to function on the user's computer (for instance, after an operating system update) or when the user wants to use particular functions offered by another tool. Moreover, users may wish to store their models in a publicly accessible repository, so that they can be inspected and analyzed by others.

Attempts to create a standardized storage format have probably been undertaken several times, but two particular formats have recently become established and accepted as standards. CellML, the Cell Markup Language (Cuellar *et al.*, 2003; Lloyd *et al.*, 2004), is the oldest and has the widest scope of the two. SBML, the Systems Biology Markup Language (Hucka *et al.*, 2003, 2004), however, has the widest application base, with currently over a hundred tools supporting its format in one way or another. CellML and SBML are XML-based, which means that they adhere to a standard document format.[9] Both have their own rules for naming and defining states, reactions, compartments, and parameters, but use the same rules for expressing the associated math.[10] The CellML and SBML formats are text-based, and may be understood and written by humans, but are designed to be written, read, and interpreted primarily by machines. Further information is

[9] Specified in the XML (eXtended Markup Language) schema.

[10] Using MathML, the Mathematics Markup Language (XML-based).

found on the Web sites dedicated to these formats (CellML, 2006; SBML, 2006), and examples of stored models in the CellML repository (CellML Model Repository, 2006) and in the BioModels database (Le Novere *et al.*, 2006).

B. Simulation

Simulation, as opposed to modeling, is a mechanical process that lends itself very well to automation. In fact, simulation, even of relatively simple systems, would hardly be possible without computers.

1. Numerical Solution of ODEs

Today it is hardly ever necessary to write computer programs that implement algorithms for numerical solution of systems of coupled ODEs, such as the ones mentioned in Section IV.B.2. Sophisticated, well-tested software libraries are available (often freely), often packaged with routines that automatically select an appropriate algorithm, estimate simulation parameters such as accuracy and initial step size, and return the simulation results as lists of equidistant time points. It is still necessary to address the routines in these libraries in an appropriate programming language (see Section V.B.3), but, again, there are many software tools in which this has already been done. These tools offer an environment in which a user can enter ODEs or reaction networks in algebraic or graphical form, set parameter and initial values, perform simulations, and specify a desired form for the output. Examples are listed in Section V.B.3.

2. Stochastic Simulation

There are several tools that have been designed to perform general stochastic simulation of biochemical reaction networks (see Section V.B.3 for examples). However, the aim of a stochastic simulation may be to study the dynamics of relatively complex molecular assemblies that have important structural characteristics or other features that are not easily expressed in a standard way (e.g., see Chen and Hill, 1985; Duke, 1999; Janulevicius *et al.*, 2006; Martin *et al.*, 1993; Schilstra and Martin, 2006; VanBuren *et al.*, 2005). In this case, it is often easiest, sometimes essential, to use a high-level programming language to specify the system and its interactions, and to produce a purpose-built simulator. Fortunately, implementation of the recipe outlined in Section IV.A.3 is extremely simple; specification of the rest of the model may present a greater challenge.

3. Selected Simulation Tools

It is impossible to give an exhaustive list of software tools for simulation of biochemical reaction systems, and the selection presented here, which has been compiled partly on the basis of the software tools listed on the SBML Web site, may inadvertently have omitted some important or popular ones.

Simulations may be performed on the basis of a system specification in a compiled or interpreted high-level computer language, such as C/C++, Fortran, Java, Lisp, Perl, Python, or Visual Basic. There are software libraries that contain implementations of multiple ODE solvers (e.g., SUNDIALS, Hindmarsh *et al.*, 2005), which can be addressed from one or more of these languages. LibSBML (Bornstein, 2006) is a software library that assists software developers in the reading, writing, and validation of the SBML format, and has bindings to all of the above languages except Fortran and Visual Basic.

Mathematica, MATLAB, Maple (all proprietary), and R (free) are examples of high-level mathematics and statistics packages that can be used to solve ODEs, or to perform stochastic simulations. MathSBML (free, Shapiro *et al.*, 2004) is a Mathematica package, and SBMLToolbox (Keating *et al.*, 2006) and SBToolbox (Schmidt and Jirstrand, 2006) are free MATLAB toolboxes that can read, write, and apply the SBML format to build, store, use, and analyze models of biochemical reaction networks.

XPP (free, Ermentrout, 2002), Berkeley Madonna (Macey and Oster, 2001), Dymola (Dynasim, 2006), and ModelMaker (ModelKinetix, 2006) (all proprietary) are tools designed to facilitate the study of various types of dynamical systems, and can be applied to model, simulate, and analyze biochemical networks in various ways.

CellDesigner (Funahashi *et al.*, 2003; Kitano *et al.*, 2005), DBSolve (Goryanin *et al.*, 1999), Dizzy (Ramsey *et al.*, 2005), E-Cell (Takahashi *et al.*, 2003), Gepasi/COPASI (Mendes, 1997; Mendes and Kummer, 2006), Jarnac/JDesigner (Sauro, 2000), Promot/DIVA (Ginkel *et al.*, 2003), PySces (Olivier *et al.*, 2005), and Virtual Cell (Slepchenko *et al.*, 2003) are freely available tools that have been designed specifically for modeling and simulation of biochemical or reaction networks or physiological systems. Most execute deterministic simulation methods but Dizzy uses stochastic techniques, and COPASI implements both. All of these allow the user to enter a network as a collection of states, reactions, and reaction mechanisms, and automatically build the appropriate structures that are required for simulation. Several also allow direct entry of the ODEs. Some have user interfaces that allow the entry of models in graphical form; many provide extra tools to analyze aspects of the network structure and dynamic properties.

4. Selected Further Reading

All biochemistry textbooks (e.g., Berg *et al.*, 2000) have sections dedicated to the basics of thermodynamics and enzyme kinetics, whereas more specialized monographs present more comprehensive theoretical and practical background information (Cornish-Bowden, 2001; Fell, 1997). Textbooks on physical chemistry (Atkins, 1994) contain more theory on chemical reaction kinetics and thermodynamics than most people will ever need. Information on various algorithms that is reasonably accessible to nonspecialists is found in (Press *et al.*, 1989), and also on several internet sites [e.g., Wikipedia (Wikipedia contributors, 2006), Mathworld (Weisstein, 2006)]. General theory on modeling and simulation is found in (Zeigler *et al.*, 2000), and on the use of diagrams in (Rumbaugh *et al.*, 1991). Last, but not least, several books have appeared recently which deal specifically with dynamic modeling in a biological context (Ellner and Guckenheimer, 2006; Szalasi *et al.*, 2006; Wilkinson, 2006).

VI. Concluding Remarks

We hope that we have provided the basic information required to understand the most important concepts in modeling and simulation. More information can be obtained from the recommended books and other literature. However, by far, the best way to develop further understanding is to set up a model of your favorite system, and try to simulate its behavior. You might be surprised at what you discover.

Acknowledgments

The preparation of this review was supported in part by grant 072930/Z/03/Z from the Wellcome Trust, United Kingdom, (to M.J.S.) and by grant R01 GM070923 from the National Institute for General Medical Sciences, USA (to S.M.K.).

References

Andrews, S. S., and Bray, D. (2004). Stochastic simulation of chemical reactions with spatial resolution and single molecule detail. *Phys. Biol.* **1,** 137–151.

Atkins, P. W. (1994). "Physical Chemistry." Oxford University Press, Oxford, UK.

Bayley, P. M., Schilstra, M. J., and Martin, S. R. (1989). A simple formulation of microtubule dynamics: Quantitative implications of the dynamic instability of microtubule populations *in vivo* and *in vitro*. *J. Cell Sci.* **93,** 241–254.

Berg, J. M., Tymoczko, J. L., and Stryer, L. (2002). "Biochemistry." Freeman and Company, New York, NY.

Bornstein, B. J. (2006). "LibSBML": SBML, The Systems Biology Markup Language. http://www.sbml.org/software/libsbml/

Bortz, A. B., Kalos, M. H., and Lebowitz, J. L. (1975). A new algorithm for Monte Carlo simulation of Ising spin systems. *J. Comput. Phys.* **17,** 10–18.

Brylawski, B. P., and Caplow, M. (1983). Rate for nucleotide release from tubulin. *J. Biol. Chem.* **258,** 760–763.

CellML. (2006). "The CellML Project: Overview—Portal." http://www.cellml.org

CellML Model Repository. (2006). "Model Repository—Portal." http://www.cellml.org/models

Chen, Y.-D., and Hill, T. L. (1985). Monte Carlo study of the GTP cap in a five-start helix model of a microtubule. *Proc. Natl. Acad. Sci. USA* **82,** 1131–1135.

Cornish-Bowden, A. (1995). "Fundamentals of Enzyme Kinetics." Portland Press, London.

Cornish-Bowden, A. (2001). "Fundamentals of Enzyme Kinetics." Portland Press, London, UK.

Cuellar, A. A., Lloyd, C. M., Nielsen, P. F., Bullivant, D. P., Nickerson, D. P., and Hunter, P. J. (2003). An overview of CellML 1.1, a biological model description language. *Simulation* **79,** 740–747.

Duke, T. A. J. (1999). Molecular model of muscle contraction. *Proc. Natl. Acad. Sci. USA* **96,** 2770–2775.

Dynasim, A. B. (2006). "Dymola." Dynasim, http://www.dynasim.se

Ellner, S. R., and Guckenheimer, J. (2006). "Dynamic Models in Biology." Princeton University Press, Princeton, NJ.

Engelborghs, Y., and Eccleston, J. (1982). Fluorescence stopped-flow study of the binding of S6-GTP to tubulin. *FEBS Lett.* **141,** 78–81.

Ermentrout, B. (2002). "Simulating, Analyzing, and Animating Dynamical Systems: A Guide to XPPAUT for Researchers and Students." SIAM, Philadelphia, PA.

Fell, D. (1997). "Understanding the Control of Metabolism." Portland Press, London.

Funahashi, A., Tanimura, N., Morohashi, M., and Kitano, H. (2003). CellDesigner: A process diagram editor for gene-regulatory and biochemical networks. *Biosilico* **1,** 159–162.

Gear, C. W. (1971). "Numerical Initial Value Problems in Ordinary Differential Equations." Prentice-Hall, Englewood Cliffs, NJ.

Gibson, M. A., and Bruck, J. (2000). Efficient exact stochastic simulation of chemical systems with many species and many channels. *J. Phys. Chem. A.* **104,** 1876–1889.

Gillespie, D. T. (1977). Exact stochastic simulation of coupled chemical reactions. *J. Phys. Chem.* **81,** 2340–2361.

Gillespie, D. T. (2001). Approximate accelerated stochastic simulation of chemically reacting systems. *J. Chem. Phys.* **115,** 1716–1733.

Gilman, A., and Arkin, A. P. (2002). Genetic "code": Representations and dynamical models of genetic components and networks. *Annu. Rev. Genom. Hum. Genet.* **3,** 341–369.

Ginkel, M., Kremling, A., Nutsch, T., Rehner, R., and Gilles, E. D. (2003). Modular modeling of cellular systems with ProMoT/Diva. *Bioinformatics* **19,** 1169–1176.

Goryanin, I., Hodgman, T., and Selkov, E. (1999). Mathematical simulation and analysis of cellular metabolism and regulation. *Bioinformatics* **15,** 749–758.

Hindmarsh, A. C., Brown, P. N., Grant, K. E., Lee, S. L., Serban, R., Shumaker, D. E., and Woodward, C. S. (2005). SUNDIALS: Suite of nonlinear and differential/algebraic equation solvers. *ACM Trans. Math. Software* **31,** 363–396.

Horio, T., and Hotani, H. (1986). Visualization of the dynamic instability of individual microtubules by dark-field microscopy. *Nature* **321,** 605.

Hucka, M., Finney, A., Sauro, H. M., Bolouri, H., Doyle, J. C., Kitano, H., and the rest of the SBML Forum, Arkin, A. P., Bornstein, B. J., Bray, D., Cornish-Bowden, A., Cuellar, A. A., Dronov, S., *et al.* (2003). The systems biology markup language (SBML): A medium for representation and exchange of biochemical network models. *Bioinformatics* **19,** 524–531.

Hucka, M., Finney, A., Bornstein, B., Keating, S. M., Shapiro, B. E., Matthews, J., Kovitz, B., Schilstra, M. J., Funahashi, A., Doyle, J. C., and Kitano, H. (2004). Evolving a lingua franca and associated software infrastructure for computational systems biology: The systems biology markup language (SBML) project. *Syst. Biol.* **1,** 41–53.

Janulevicius, A., van Pelt, J., and van Ooyen, A. (2006). Compartment volume influences microtubule dynamic instability: A model study. *Biophys. J.* **90,** 788–798.

Keating, S. M., Bornstein, B. J., Finney, A., and Hucka, M. (2006). SBMLToolbox: An SBML toolbox for MATLAB users. *Bioinformatics* **22**(10), 1275–1277; doi:10.1093/bioinformatics/btl111.

Kitano, H., Funahashi, A., Matsuoka, Y., and Oda, K. (2005). Using process diagrams for the graphical representation of biological networks. *Nat. Biotechnol.* **23,** 961–966.

Kruse, K., and Elf, J. (2006). Kinetics in Spatially Extended Systems. *In* "Systems Modeling in Cellular Biology" (Z. Szalasi, J. Stelling, and V. Periwal, eds.), pp. 177–198. MIT Press, Cambridge, MA.

Le Novere, N., Bornstein, B., Broicher, A., Courtot, M., Donizelli, M., Dharuri, H., Li, L., Sauro, H., Schilstra, M., Shapiro, B., Snoep, J. L., and Hucka, M. (2006). BioModels database: A free, centralized database of curated, published, quantitative kinetic models of biochemical and cellular systems. *Nucl. Acids Res.* **34,** D689–D691.

Lloyd, C. M., Halstead, M. D. B., and Nielsen, P. F. (2004). CellML: Its future, present and past. *Prog. Biophys. Mol. Biol.* **85,** 433–450.

Macey, R. I., and Oster, G. F. (2001). "Berkeley Madonna." http://www.berkeleymadonna.com

Markevich, N. I., Hoek, J. B., and Kholodenko, B. N. (2004). Signaling switches and bistability arising from multisite phosphorylation in protein kinase cascades. *J. Cell Biol.* **164,** 353–359.

Martin, S. R., Schilstra, M. J., and Bayley, P. M. (1993). Dynamic instability of microtubules: Monte-Carlo simulation and application to different types of microtubule lattice. *Biophys. J.* **65,** 578–596.

Mendes, P. (1997). Biochemistry by numbers: Simulation of biochemical pathways with Gepasi 3. *Trends Biochem. Sci.* **22,** 361–363.

Mendes, P., and Kummer, U. (2006). "COPASI." http://www.copasi.org/tiki-index.php

Meyers, J., Craig, J., and Odde, D. J. (2006). Potential for control of signaling pathways via cell size and shape. *Curr. Biol.* **16,** 1685–1693.

ModelKinetix. (2006). "ModelMaker." http://www.modelkinetix.com/modelmaker/index.htm

Olivier, B. G., Rohwer, J. M., and Hofmeyr, J.-H. S. (2005). Modelling cellular systems with PySCeS. *Bioinformatics* **21,** 560–561.

Press, W. H., Flannery, B. P., Teukolsky, B. P., and Vetterling, W. T. (1989). "Numerical Recipes: The Art of Scientific Computing. Fortran Version." Cambridge University Press, Cambridge, UK.

Ramsey, S., Orrell, D., and Bolouri, H. (2005). Dizzy: Stochastic simulation of large-scale genetic regulatory networks. *J. Bioinform. Comput. Biol.* **3,** 415–436.

Rumbaugh, J., Blaha, M., Premerlani, W., Eddy, F., and Lorensen, W. (1991). "Object-Oriented Modelling and Design." Prentice-Hall, London.

Sauro, H. M. (2000). "Animating the Cellular Map. 9th International BioThermoKinetics Meeting" (J.-H. S. Hofmeyr, J. M. Rohwer, and J. L. Snoep, eds.), pp. 221–228, Stellenbosch University Press, Stellenbosch, South Africa.

SBML. (2006). "SBML.org—The home site for the Systems Biology Markup Language." http://sbml.org.

Schilstra, M. J., and Martin, S. R. (2006). An elastically tethered viscous load imposes a regular gait on the motion of myosin-V. Simulation of the effect of transient force relaxation on a stochastic process. *J. R. Soc. Interf.* **3,** 153–165.

Schmidt, H., and Jirstrand, M. (2006). Systems biology toolbox for MATLAB: A computational platform for research in systems biology. *Bioinformatics* **22,** 514–515.

Segel, I. H. (1975). "Enzyme Kinetics: Behavior and Analysis of Rapid Equilibrium and Steady-State Enzyme Systems." John Wiley & Sons Inc., USA.

Shapiro, B. E., Hucka, M., Finney, A., and Doyle, J. C. (2004). MathSBML: A package for manipulating SBML-based biological models. *Bioinformatics* **20,** 2829–2831.

Slepchenko, B. M., Schaff, J., Macara, I. G., and Loew, L. M. (2003). Quantitative cell biology with the virtual cell. *Trends Cell Biol.* **13,** 570–576.

Swade, D. (1995). When money flowed like water *Inc. Magazine* September 1995. http://pf.inc.com/magazine/19950915/2624.html

Szalasi, Z., Stelling, J., and Periwal, V. (eds). (2006). "Systems Modeling in Cellular Biology." MIT Press, Cambridge, MA.

Takahashi, K., Ishikawa, N., Sadamoto, Y., Ohta, S., Shiozawa, A., Miyoshi, F., Naito, Y., Nakayama, Y., and Tomita, M. (2003). E-Cell 2: Multi-platform E-cell simulation system. *Bioinformatics* **19,** 1727–1729.

VanBuren, V., Cassimeris, L., and Odde, D. J. (2005). Mechanochemical model of microtubule structure and self-assembly kinetics. *Biophys. J.* **89,** 2911–2926.

Weisstein, E. (2006). "Mathworld". Wolfram Research, http://mathworld.wolfram.com/

Wikipedia Contributors. (2006). "Wikipedia." Wikipedia, The Free Encyclopedia, http://en.wikipedia.org

Wilkinson, D. J. (2006). "Stochastic Modelling for Systems Biology." Chapman & Hall/CRC, London.

Zeigler, B. P., Praehofer, H., and Kim, T. G. (2000). "Theory of Modelling and Simulation: Integrating Discrete Event and Continuous Complex Dynamic Systems." Academic Press, San Diego.

Zewail, A. (2003). "Nobel Lectures, Chemistry 1996–2000" (I. Grenthe, ed.), pp. 274–367. World Scientific Publishing Co., Singapore.

CHAPTER 26

Computational Methods for Biomolecular Electrostatics

Feng Dong, Brett Olsen, and Nathan A. Baker

Department of Biochemistry and Molecular Biophysics
Center for Computational Biology
Washington University in St. Louis, Missouri 63110

Abstract

An understanding of intermolecular interactions is essential for insight into how cells develop, operate, communicate, and control their activities. Such interactions include several components: contributions from linear, angular, and torsional

0091-679X/08 $35.00
DOI: 10.1016/S0091-679X(07)84026-X

forces in covalent bonds, van der waals forces, as well as electrostatics. Among the various components of molecular interactions, electrostatics are of special importance because of their long range and their influence on polar or charged molecules, including water, aqueous ions, and amino or nucleic acids, which are some of the primary components of living systems. Electrostatics, therefore, play important roles in determining the structure, motion, and function of a wide range of biological molecules. This chapter presents a brief overview of electrostatic interactions in cellular systems, with a particular focus on how computational tools can be used to investigate these types of interactions.

I. Introduction

Intermolecular interactions are essential for nearly every cellular activity. The forces that underlie these interactions include van der Waals dispersion and repulsion, hydrogen bonding, and electrostatics. Electrostatic forces are especially important in biological systems because most biomolecules are charged or polar. For example, nucleic acids contain long strings of negative charges while proteins are generally zwitterionic with a wide variety of amino acids that make up a complex charge distribution. Even the solvent environment in which the larger molecules interact is full of polar water and an assortment of simple ions. Therefore, insight into the most fundamental biomolecular processes requires a basic understanding of electrostatics. This chapter first presents a brief overview of electrostatics in biological systems, followed by a discussion of several different models that can simplify the analysis of electrostatics, and concludes with specific applications of computational electrostatics models for biological systems.

II. Electrostatics in Cellular Systems

Electrostatic interactions are ubiquitous for any system of charged or polar molecules, such as biomolecules in their aqueous environment. For example, proteins are made up of 20 types of amino acids, 11 of which are charged or polar in neutral solution. Nucleic acids contain long stretches of negative charges from the phosphate groups in nucleotides. Finally, sugars and related glycosaminoglycans can possess some of the highest charge densities of any biomolecules because of the presence of numerous negative functionalities, including carboxylate and sulfate groups. We will focus on a few specific examples of electrostatic interactions in cellular systems: biomolecule–ion, biomolecule–ligand, and biomolecule–biomolecule interactions. Each of these interactions will be discussed in more detail in the following sections.

A. Biomolecule–Ion Interactions

In cellular settings, biomolecules are immersed in solution along with water, ions, and numerous other small molecules and macromolecules. Ions influence biomolecular processes and interactions in several different ways, including long-range screening, site-specific ion binding, and preferential hydration effects. Long-range screening is a phenomenon in which the strength of electrostatic interactions within and between biomolecules is reduced by the presence of aqueous ions. This is a nonspecific ion effect and is described well, at low salt charge and concentration, by the Debye–Hückel theory (Debye and Hückel, 1923) and the related implicit solvent models described in Section III.B. In site-specific ion binding, ions interact with biomolecules by binding to specific sites in a manner similar to ligand binding (see Section II.B; Draper *et al.*, 2005). Preferential hydration or Hofmeister effects are species-specific competitions between ions and water for binding to nonspecific sites on biomolecules (Boström *et al.*, 2006; Collins, 2006; Hofmeister, 1888). This competition is between weak biomolecule–solvent and biomolecule–ion interactions and therefore observed only at very high salt concentrations (Anderson and Record, 2004; Eisenberg, 1976). A similar effect involves competition between ionic species around charged biomolecules (Moore and Lohman, 1994; Reuter *et al.*, 2005). Note that, although these effects can be important, preferential hydration and ion–ion competition are not routinely considered in simulations, mainly because of limitations in current computational methodology. While there is active work in improving the theoretical and computational treatment of these effects (Boström *et al.*, 2003, 2006; Broering and Bommarius, 2005; Shimizu, 2004; Shimizu and Smith, 2004; Zhou, 2005), they are currently beyond the scope of this chapter.

Ion-induced RNA folding (Cech and Bass, 1986; Dahm and Uhlenbeck, 1991; Draper *et al.*, 2005; Misra and Draper, 2000, 2001; Römer and Hach, 1975; Stein and Crothers, 1976) provides an excellent example of many of the ion–biomolecule interactions discussed above. RNA folding in the absence of salt is quite unfavorable due to a number of negative charges along its phosphodiester backbone. Bringing these negative charges together into a compact structure introduces a large energetic barrier to RNA folding. Positive ions promote folding by reducing the repulsion between these negative charges. However, some ions are more effective than others; for example, millimolar concentrations of Mg^{2+}can stabilize RNA tertiary structures that are only marginally stable in solution with a high concentration of monovalent cations, such as Na^{+}or K^{+}(Cole *et al.*, 1972; Romer and Hach, 1975; Stein and Crothers, 1976). Accurately modeling the ion–RNA interactions is essential to explain this phenomenon. A major obstacle in modeling ion–RNA interactions is the presence of numerous different ion environments (Draper *et al.*, 2005). Each environment is dominated by different types of ion–biomolecule interactions described above and requires different approaches to evaluating the energies. For example, experimental results for Mg^{2+}effects on $tRNA^{Phe}$ folding can be modeled successfully while only considering long-range screening effects (Misra and Draper, 2000). However, the diffusive Mg^{2+} ion description provided by this model is not sufficient to describe the folding of a 58-nt rRNA fragment. Instead, one Mg^{2+} ion must be explicitly included at a specific binding site (Misra and Draper, 2001). A comprehensive theoretical framework of ion–RNA interactions that accounts for the overall ion dependence of RNA folding is the aim of current RNA folding studies (Draper *et al.*, 2005).

B. Biomolecule–Ligand and –Biomolecule Interactions

Biomolecule–substrate recognition is central to nearly all biomolecular processes, including signal transduction, enzyme cooperativity, and metabolic regulation. The bimolecular binding process, from a kinetic perspective, can be reduced to two steps: diffusional association to form an initial encounter complex and nondiffusional rearrangement to form the fully bound complex. The diffusional association places an upper limit on the overall binding rate; so-called "perfect" enzymes operate at this diffusion-limited rate. Electrostatic forces have an important influence on biomolecular diffusional association: their long-range nature enables them to attract the substrate to its binding partner and orient the substrate properly for binding (Gabdoulline and Wade, 2002). It has been established that for many biomolecular complexes, electrostatic interactions can significantly affect bimolecular association rates (Law *et al.*, 2006). For example, by using Brownian dynamics (BD) simulations (Ermak and McCammon, 1978; Northrup *et al.*, 1984; Section IV.F) to calculate diffusional association rates, Gabdoulline and Wade demonstrated that for fast-associating protein pairs, electrostatic interactions enhance association and are the dominant forces determining the rate of diffusional association (Gabdoulline and Wade, 2001; Radic *et al.*, 1997). Using related methods, Sept *et al.* demonstrated the role of electrostatic interactions in determining the rates and polarity of actin polymerization (Sept and McCammon, 2001; Sept *et al.*, 1999).

Electrostatic interactions also play an important role in determining the thermodynamics of binding, that is, binding affinity (Chong *et al.*, 1998; Norel *et al.*, 2001; Novotny and Sharp, 1992; Rauch *et al.*, 2002; Schreiber and Fersht, 1993, 1995; Sheinerman *et al.*, 2000; Zhu and Karlin, 1996). Substrate binding allows the formation of (potentially) favorable charge–charge interactions between the substrate and the target, as well as stabilizing specific salt bridges and hydrogen bonds (Chong *et al.*, 1998; Schreiber and Fersht, 1993, 1995). However, at the same time, charges on the molecular binding surface must shed their bound water in order to allow close binding. This loss of water, or desolvation, is generally energetically unfavorable and offsets the favorable interactions formed on binding. The binding affinities, from an electrostatic point of view, are determined by the balance of these two energetic contributions (del Álamo and Mateu, 2005; Lee and Tidor, 2001; Russell *et al.*, 2004; Sheinerman and Honig, 2002; Xu *et al.*, 1997). Systematic studies of protein pairs, such as barnase and barstar (Dong *et al.*, 2003; Frisch *et al.*, 1997; Schreiber and Fersht, 1993, 1995), and fasciculin-2

(Radic *et al.*, 1997), as well as protein kinase A and balanol (Wong *et al.*, 2001), have shown that charged and polar residues at the protein–protein interfaces play important roles in binding energetics. Similarly, Sept *et al.* (2003) have demonstrated an important role for electrostatics in determining microtubule structure and stability. Finally, Wang *et al.* (2004) have demonstrated that nonspecific electrostatic interactions can provide a driving force for recruitment of proteins to intracellular membranes, an important step in signal transduction.

However, despite the role of electrostatics in protein–protein interactions, it is important to realize that the total interaction is also strongly influenced by shape complementarity at the protein–protein interface as well as by nonpolar contributions to offset the penalties of desolvation (Janin and Chothia, 1990; Lo Conte *et al.*, 1999; Ma *et al.*, 2003; Vasker, 2004).

III. Models for Biomolecular Solvation and Electrostatics

As described above, computer simulations can provide atomic-scale information on energetic and dynamic contributions to biomolecular structure and interactions. However, the capabilities of computer simulations are limited by the accuracy of the underlying models describing atomic interactions and also by the computational expense of adequately exploring all the relevant conformations of the biomolecule and surrounding water and ion. Therefore, most models of biomolecular solvation and electrostatics make a trade-off between these opposing considerations of atomic accuracy and computational expense.

A variety of computational methods have been developed for studying electrostatic interactions in biomolecular systems. Popular methods for understanding electrostatic interactions in these systems can be loosely classified into two categories (see Fig. 1): explicit solvent methods (Burkert and Allinger, 1982; Horn *et al.*, 2004; Jorgensen *et al.*, 1983; Ponder and Case, 2003; Sagui and Darden, 1999), which treat the solvent in full atomic detail, and implicit solvent methods (Baker, 2005b; Baker *et al.*, 2006; Davis and McCammon, 1990b; Honig and Nicholls, 1995; Roux, 2001; Roux and Simonson, 1999), which represent the solvent through its average effect on solute.

A. Explicit Solvent Methods

Explicit solvent methods offer a very detailed description of biomolecular solvation. In explicit solvent methods, interactions between mobile ions, solvent, and solute atoms are typically described by molecular mechanics force fields (Ponder and Case, 2003; Wang *et al.*, 2001b), which use classical approximations of quantum mechanical energies to describe the Coulombic (electrostatic), van der Waals, and covalent (bond, angle) interactions. Explicit solvent methods have the obvious advantage of offering the full details of solvent–solute and solvent–solvent interactions. These details can affect some aspects of biomolecular interactions.

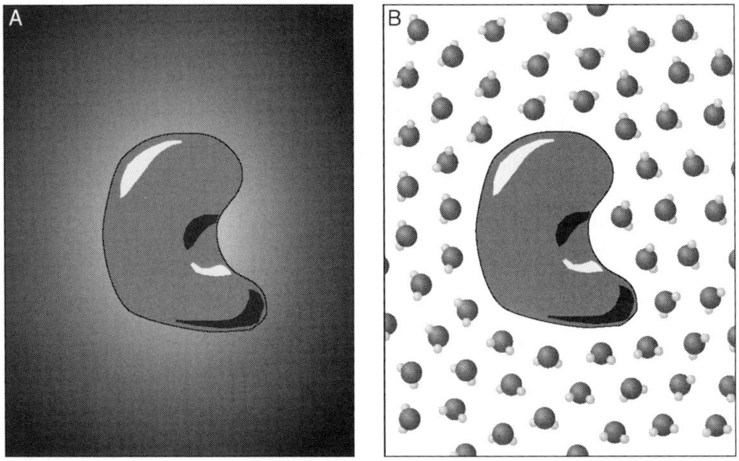

Fig. 1 A schematic comparison of implicit and explicit solvent models. (A) In the implicit solvent model, a low dielectric solute is surrounded by a continuum of high dielectric solvent. (B) In the explicit solvent model, solvent is represented by discrete water molecules.

For example, the explicit representation of solvent structure can qualitatively change the detailed features of protein side chain interactions (Masunov and Lazaridis, 2003). Similarly, Yu *et al.* have demonstrated the importance of including first shells of solvation to correctly describe the interaction of salt bridges in solution (Yu *et al.*, 2004).

However, the explicit solvent methods are computationally expensive. In order to extract meaningful thermodynamic and kinetic parameters, all the numerous conformations of biomolecules, as well as the solvent and ions, must be explored. The extra degrees of freedom associated with the explicit solvent and ions dramatically increase the computational cost of explicit solvent methods and limit the temporal and spatial scales of biomolecular simulations.

B. Implicit Solvent Methods

Implicit solvent methods have become popular alternatives to the computationally expensive explicit solvent approaches although they have a lower accuracy (Baker, 2005b; Baker *et al.*, 2006; Davis and McCammon, 1990b; Gilson, 2000; Honig and Nicholls, 1995; Roux, 2001). In implicit solvent methods, the molecules of interest are treated explicitly while the solvent is represented by its average effect on the solute (Roux and Simonson, 1999). Solute–solvent interactions are described by solvation energies; that is, the free energy of transferring the solute from a vacuum to the solvent environment of interest (e.g., water at a certain ionic strength). This process is shown in more detail in Fig. 2. The process consists of three steps: (1) solute charges are gradually reduced to zero in vacuum, (2) the uncharged solute is inserted into the solvent, and (3) solute charges are gradually

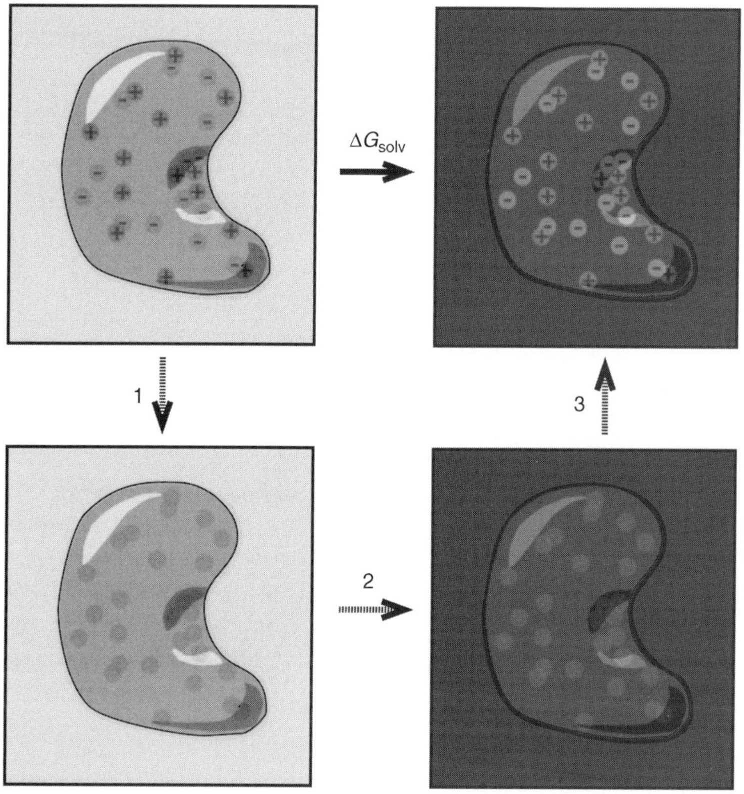

Fig. 2 A thermodynamic cycle illustrating the biomolecular solvation process. The steps are (1) uncharging the biomolecule in vacuum, (2) transferring the uncharged biomolecule from vacuum to solvent, and (3) charging the biomolecule back to its normal value in solvent. The nonpolar solvation free energy is the free energy change in step (2). The polar solvation free energy is the sum of the free energy changes in steps (1) and (3).

increased back to their normal values in the solvent. The free energy change in step (2) is called the nonpolar solvation energy. The sum of the energies associated with steps (1) and (3) is called the "charging" or polar solvation energy and represents the solvent's effect on the solute charging process. In general, polar and nonpolar solvation terms act in opposing directions; nonpolar solvation favors compact structures with small areas and volumes, while polar solvation favors maximum solvent exposure for all polar groups in the solute.

1. Nonpolar Solvation

One popular approximation for the nonpolar solvation free energy assumes a linear dependence between the nonpolar solvation energy, $G_{\mathrm{solv}}^{\mathrm{nonpolar}}$, and the solvent-accessible surface area (SASA), A (Chothia, 1974; Eisenberg, 1976;

Massova and Kollman, 2000; Sharp *et al.*, 1991; Spolar *et al.*, 1989; Swanson *et al.*, 2004; Wesson and Eisenberg, 1992):

$$G_{\mathrm{solv}}^{\mathrm{nonpolar}} = \gamma \cdot A \tag{1}$$

where γ is a "surface tension" which is typically chosen to reproduce the nonpolar solvation free energy of alkanes (Sharp *et al.*, 1991; Simonson and Brunger, 1994; Sitkoff *et al.*, 1994b) or model side chain analogues (Eisenberg and McLachlan, 1986; Wesson and Eisenberg, 1992). The surface tension parameter may assume a single global value used for all atom types or different values may be assigned to each different type of atom. Although SASA methods have enjoyed surprising success, they are also subject to several caveats, including widely varying choices of surface tension parameter (Chothia, 1974; Eisenberg and McLachlan, 1986; Elcock *et al.*, 2001; Sharp *et al.*, 1991; Sitkoff *et al.*, 1994a) as well as inaccurate descriptions of the detailed aspects of nonpolar solvation energy (Gallicchio and Levy, 2004), peptide conformations (Su and Gallicchio, 2004), and protein nonpolar solvation forces (Wagoner and Baker, 2004). Some of these problems have been fixed by new models which include the small but important attractive van der Waals interactions between solvent and solute (Gallicchio and Levy, 2004; Gallicchio *et al.*, 2000, 2002; Wagoner and Baker, 2006) as well as repulsive solvent-accessible volume terms (Wagoner and Baker, 2006).

2. Polar Solvation

Implicit solvent methods have been used to study polar solvation and electrostatics for over 80 years, starting with work by Born on ion solvation (1920), Linderström-Lang (1924) and Tanford and Kirkwood (1957) on protein titration, Manning on ion distributions surrounding nucleic acids (1978), Flanagan *et al.* (1981) on the pH dependence of hemoglobin dimer assembly, and Warwicker and Watson (1982) on the electrostatic potential of realistic protein geometries. Although they can be considerably different in their details and implementation, implicit solvent models generally treat the solvent as a high dielectric continuum, the aqueous ions as a diffuse cloud of charge, and the solute as a fixed array of point charges that are embedded in a lower dielectric continuum. Despite the limitations of these assumptions, implicit solvent models often give a good coarse-grained description of solvation energetics and have enjoyed widespread use over recent years.

Regardless of the particular type of implicit solvent model, the behavior of electrostatic interactions is generally determined by a few basic properties of the system, illustrated in Fig. 3: the charges, radii, and "dielectric constant" of the solute; the charges and radii of aqueous ionic species; and the radii and dielectric constant of the solvent. The relationship of these specific parameters to solvation energies and forces will be described in more detail in Sections III.C and III.D.

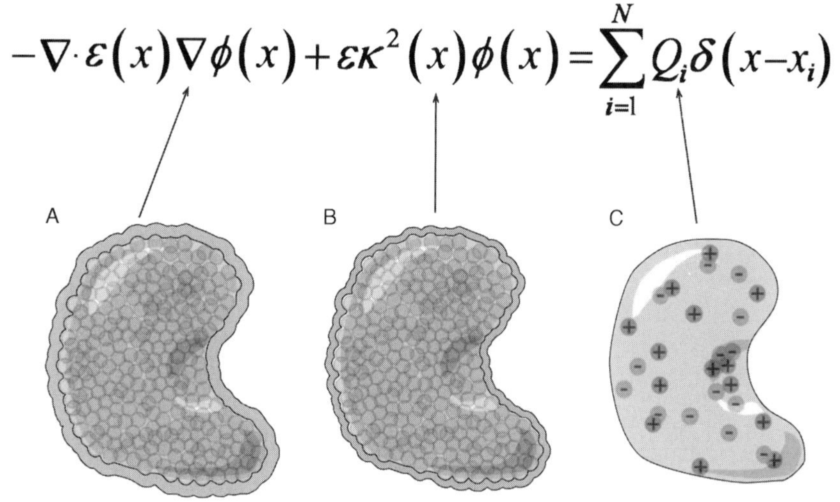

$-\nabla\cdot\varepsilon(x)\nabla\phi(x)+\varepsilon\kappa^2(x)\phi(x)=\sum_{i=1}^{N}Q_i\delta(x-x_i)$
A
B
C

Fig. 3 Description of the terms in the Poisson–Boltzmann equation: (A) the dielectric permittivity coefficient $\varepsilon(\vec{x})$ is much smaller inside the biomolecule than outside the biomolecule, with a rapid change in value across the solvent-accessible biomolecular surface, (B) the ion-accessibility parameter $\kappa^2(\vec{x})$ is proportional to the bulk ionic strength outside the ion-accessible biomolecular surface, and (C) the biomolecular charge distribution is defined as the collection of point charges located at the center of each atom.

C. Poisson–Boltzmann Methods

The Poisson–Boltzmann (PB) equation is a popular continuum description of electrostatics for the biomolecular system. Although there are a number of ways to derive the PB equation based on statistical mechanics (Holm *et al.*, 2001), the simplest derivation begins with Poisson's equation (Bockris and Reddy, 1998; Jackson, 1975) (in SI units),

$$-\nabla\cdot\varepsilon(\vec{x})\nabla\phi(\vec{x}) = \rho(\vec{x}) \tag{2}$$

the basic equation for describing the electrostatic potential $\phi(\vec{x})$ at point $\vec{x}$ generated by a charge distribution $\rho(\vec{x})$ in an environment with a dielectric permittivity coefficient $\varepsilon(\vec{x})$ (Jackson, 1975; Landau *et al.*, 1982).

The coefficient $\varepsilon(\vec{x})$ is given by $\vec{x}$ where $\varepsilon_0 = 8.8542 \times 10^{-12}$ C^2/Nm^2 is the electrostatic permittivity of vacuum and $\varepsilon_r(\vec{x})$ is the dielectric coefficient or the relative electrostatic permittivity. The dielectric coefficient $\varepsilon_r(\vec{x})$ describes the local polarizability of the material: that is, the generation of local dipole densities in response to the applied fields and changes in charge. The functional form of this coefficient depends on the shape of the biomolecule; $\varepsilon_r(\vec{x})$ assumes lower values of 2–20 in the biomolecular interior and higher values of $\sim$80, the value for water at room temperature, in solvent-accessible regions. The distinction between biomolecular "interior" and "exterior" used to assign dielectric coefficients is imprecise; as a result, a variety of different definitions for the biomolecular surface and dielectric coefficient have been developed (Connolly, 1985; Grant *et al.*, 2001; Im *et al.*, 1998; Lee and Richards, 1971; Warwicker and Watson, 1982).

In order to continue the derivation of the PB equation, we assume the charge distribution $\rho(\vec{x})$ includes two contributions: the solute charges $\rho_{\mathrm{f}}(\vec{x})$ and the aqueous "mobile" ions $\rho_{\mathrm{m}}(\vec{x})$. The solute charge distribution is generally described by a collection of N point charges located at each solute atom's position $\vec{x}_i$ and scaled by that atom's charge Q_i; that is, the solute charge distribution is the summation of a set of delta functions $\rho_f(\vec{x}) = \sum_i Q_i \delta(\vec{x} - \vec{x}_i)$. Neglecting explicit interactions between the aqueous ions (Holm *et al.*, 2001), the mobile charges are modeled as a continuous "charge cloud" described by a Boltzmann distribution (McQuarrie, 2000). For m ion species with charges q_j, bulk concentrations c_j, and steric potential $V_j(\vec{x})$ (a potential that prevents biomolecule–ion overlap), the mobile ion charge distribution is $\rho_{\mathrm{m}}(\vec{x}) = \sum_j^m c_j q_j \exp[-q_j\phi(\vec{x})/k_{\mathrm{B}}T - V_j(\vec{x})/k_{\mathrm{B}}T]$, where k_{B} is Boltzmann's constant and T is the absolute temperature. Combining both the solute and ion charge distributions with the Poisson equation, Eq. (2), gives the full PB equation:

$$-\nabla\cdot\varepsilon(\vec{x})\nabla\phi(\vec{x}) = \sum_{i=1}^{N} Q_i\delta(\vec{x}-\vec{x}_i) + \sum_{j=1}^{m} c_j q_j \exp[-q_j\phi(\vec{x})/k_{\mathrm{B}}T - V_j(\vec{x})/k_{\mathrm{B}}T] \quad (3)$$

A common simplification is that the exponential term $\exp[-q_j\phi(\vec{x})/k_{\mathrm{B}}T]$ can be approximated by the linear term in its Taylor series expansion $-q_j\phi(\vec{x})/k_{\mathrm{B}}T$ for $|q_j\phi(\vec{x})/k_{\mathrm{B}}T| \ll 1$. With this linearization and by assuming the steric occlusions are the same for all ion species ($V_j = V$ for all j), Eq. (3) reduces to the linearized PB equation:

$$-\nabla\cdot\varepsilon(\vec{x})\nabla\phi(\vec{x}) + \varepsilon(\vec{x})\kappa^2(\vec{x})\phi(\vec{x}) = \sum_{i=1}^{N} Q_i\delta(\vec{x}-\vec{x}_i) \quad (4)$$

where $\kappa^2(\vec{x})$, related to a modified inverse Debye–Hückel screening length (Debye and Hückel, 1923), is given by

$$\kappa^2(\vec{x}) = \exp\left[\frac{-V(\vec{x})}{k_{\mathrm{B}}T}\right]\cdot\frac{2Ie_c^2}{k_{\mathrm{B}}T\varepsilon(\vec{x})} \quad (5)$$

where $I = 1/2\sum_j^m c_j q_j^2/e_{\mathrm{c}}^2$ is the ionic strength and e_{c} is the unit electric charge.

Once the PB equation is solved, the electrostatic potential is known for the entire system. Given this potential, the electrostatic free energy can be evaluated by a variety of integral formulations (Gilson, 1995; Micu *et al.*, 1997; Sharp and Honig, 1990). The simplest, for the linearized PB equation, is

$$G_{\text{el}} = \frac{1}{2}\sum_{i=1}^{N} Q_i \phi(\vec{x}_i) \tag{6}$$

It is also possible to differentiate integral formulations of the electrostatic energy with respect to atomic position to obtain the electrostatic or polar solvation force on each atom (Gilson *et al.*, 1993; Im *et al.*, 1998).

Analytical solutions of the PB equation are not available for biomolecules with realistic shapes and charge distributions. Numerical methods for solving the PB equation were first introduced by Warwicker and Watson (1982) to obtain the electrostatic potential at the active site of an enzyme. The most common numerical techniques for solving the PB equation are based on discretization of the domain of interest into small regions. Those methods include finite difference (Baker *et al.*, 2001; Davis and McCammon, 1989; Holst and Saied, 1993, 1995; Nicholls and Honig, 1991), finite element (Baker *et al.*, 2000, 2001; Cortis and Friesner, 1997a,b; Dyshlovenko, 2002; Holst *et al.*, 2000), and boundary element methods (Allison and Huber, 1995; Bordner and Huber, 2003; Boschitsch and Fenley, 2004; Juffer *et al.*, 1991; Zauhar and Morgan, 1988), all of which continue to be developed to further improve the accuracy and efficiency of electrostatics calculations in the numerous biomolecular applications described below. The major software packages that can be used to solve the PB equation are listed in Table I. Many of these packages are also used for visualization of the electrostatic potential around biomolecules. Such visualization can provide insight into biomolecular function and highlight regions of potential interest. Figure 4 shows examples of the visualization of electrostatic potential calculated with Adaptive Poisson-Boltzmann solver (APBS) (Baker *et al.*, 2001) and visualized with Visual Molecular Dynamics (VMD) (Humphrey *et al.*, 1996).

Table I
Major PB Equation Solver

Software package	URL
APBS (Baker *et al.*, 2001)	http://apbs.sf.net/
Delphi (Rocchia *et al.*, 2001)	http://trantor.bioc.columbia.edu/delphi/
MEAD (Bashford, 1997)	http://www.scripps.edu/mb/bashford/
ZAP (Grant *et al.*, 2001)	http://www.eyesopen.com/products/toolkits/zap.html
UHBD (Madura *et al.*, 1995)	http://mccammon.ucsd.edu/uhbd.html
Jaguar (Cortis and Friesner, 1997a,b)	http://www.schrodinger.com/
CHARMM (MacKerell *et al.*, 1998)	http://yuri.harvard.edu
Amber (Luo *et al.*, 2002)	http://amber.scripps.edu

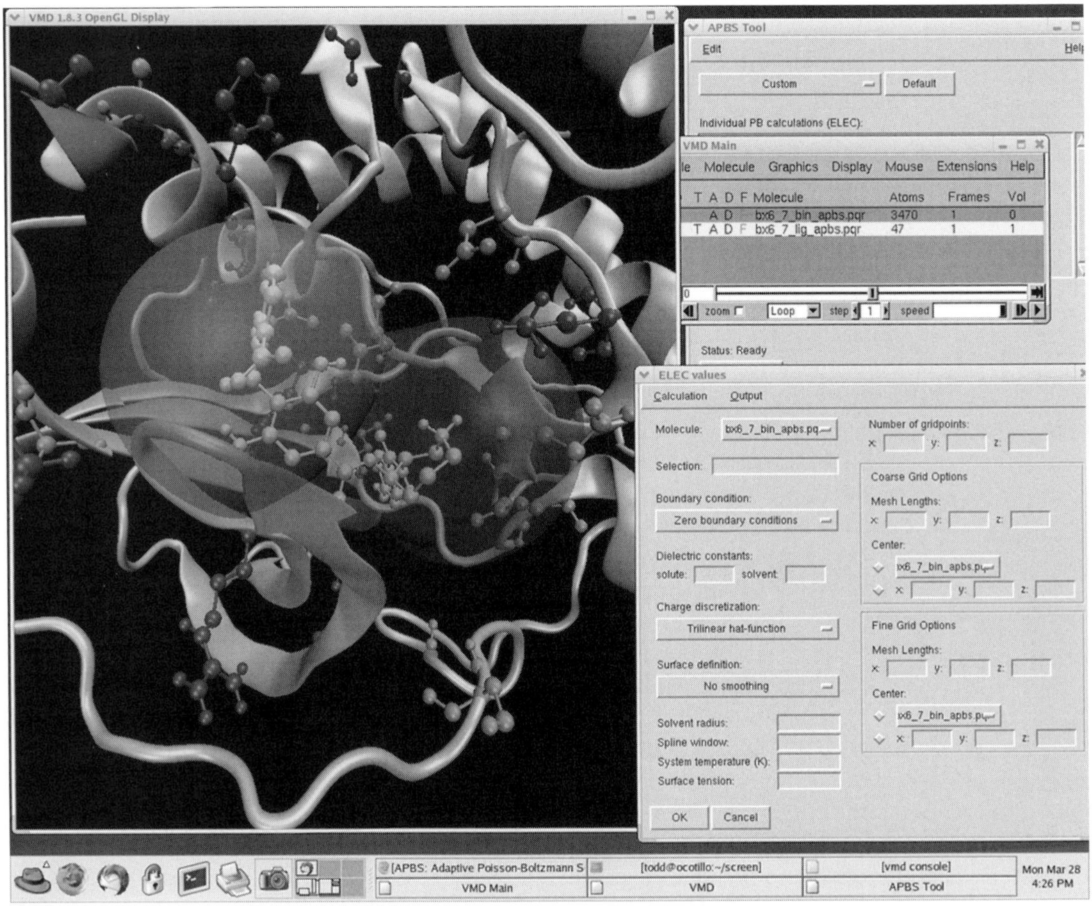

Fig. 4 Examples of the visualization of the balanol electrostatic potential in the binding site of protein kinase A as calculated by APBS (Baker *et al.*, 2001) and visualized with VMD (Humphrey *et al.*, 1996). (See Plate no. 25 in the Color Plate Section.)

D. Simpler Models

In addition to the PB methods, simpler approximate models have also been constructed for continuum electrostatics, including distance-dependent dielectric functions (Leach, 2001; MacKerell and Nilsson, 2001), analytic continuum methods (Schaeler and Karplus, 1996), and generalized Born (GB) models (Bashford and Case, 2000; Dominy and Brooks, 1999; Onufriev *et al.*, 2002; Osapay *et al.*, 1996; Still *et al.*, 1990). Among these simpler methods, GB is currently the most popular. The GB model was introduced by Still *et al.* in 1990 and subsequently refined by several other researchers (Bashford and Case, 2000; Dominy and Brooks, 1999; Onufriev *et al.*, 2002; Osapay *et al.*, 1996). The model shares the same continuum representation of the solvent as the Poisson or PB theories.

However, the GB model is based on the analytical solvation energy obtained from the solution of the Poisson equation for a simple sphere (Born, 1920). The biomolecular electrostatic solvation free energy is approximated by a modified form of the analytical solvation energy for a sphere (Still *et al.*, 1990):

$$\Delta G_{\text{solv}}^{\text{el}} \cong -\frac{1}{2}\left(1-\frac{1}{\varepsilon_{\text{sol}}}\right)\sum_{i,j}\frac{Q_i Q_j}{f_{ij}^{\text{GB}}} \tag{7}$$

where the self terms as $i = j, f_{ij}^{\text{GB}}$, are the "effective Born radii" and the cross terms as $i \neq j$, f_{ij}^{GB}, are the effective interaction distances. The most common form of f_{ij}^{GB} (Still *et al.*, 1990) is

$$f_{ij}^{\text{GB}} = \left[r_{ij}^2 + R_i R_j \exp\left(\frac{-r_{ij}^2}{4R_i R_j}\right)\right]^{1/2} \tag{8}$$

where R_i are the effective radii of the atoms and r_{ij} are the distance between atoms i and j. Efficiently and accurately calculating the effective radii is essential for GB methods. "Perfect" GB radii, which reproduce atom i's self-energy obtained by solving the Poisson equation for the biomolecule–solvent system with only atom i charged, have demonstrated the ability to accurately follow the results of more detailed models such as PB (Onufriev *et al.*, 2002). However using such "perfect" radii does not directly provide any computational advantage over solving the Poisson equation. In the absence of perfect radii for every biomolecular conformation, GB methods fail to capture some aspects of molecular structure included in more detailed models, such as the PB equation. Nonetheless, GB methods have become increasingly popular because of their computational efficiency.

E. Limitations of Implicit Solvent Methods

Although implicit solvent methods offer simpler descriptions of the system and greater computational efficiency, it is important to recall that these reductions of complexity and effort are obtained at the cost of substantial simplification of the description of the solvent. In particular, implicit solvent methods are capable of describing only nonspecific interactions between solvent and solute. In general, explicit solvent methods should be used wherever the detailed interactions between solvent and solute are important, such as solvent finite size effects in ion channels (Nonner *et al.*, 2001), strong solvent–solute interactions (Bhattacharrya *et al.*, 2003), strong solvent coordination of ionic species (Figueirido *et al.*, 1994; Yu *et al.*, 2004), and saturation of solvent polarization near a membrane (Lin *et al.*, 2002). Similarly, as mentioned earlier in the context of RNA–ion interactions, implicit descriptions of mobile ions can also become questionable in some cases, such as high ion valency or strong solvent coordination, specific ion–solute interactions, and high local ion densities (Holm *et al.*, 2001), where the ions interact with each other or with the solute directly.

IV. Applications

In the previous section, we discussed the basic concepts behind the computational tools that can be used to simulate electrostatic interactions in cellular systems. In this section, we will illustrate the use of these methods, especially PB methods, to deal with the various biomolecular problems.

A. Solvation Free Energy

As mentioned in Section III.B, the solvation free energy is the free energy of transferring a solute from a uniform dielectric continuum (a constant dielectric) to an inhomogeneous medium (a low dielectric solute surrounded by a high dielectric solvent), which is often divided into two terms: a nonpolar term and a polar term. The nonpolar term is usually estimated using either SASA or the improved methods discussed in Section III.B.1. For the polar term, as shown in Fig. 5, two PB calculations are usually performed: (1) calculating the biomolecular electrostatic free energy, $G_{\text{el}}^{(1)}$, in a homogeneous medium with a constant dielectric equal to the solute's dielectric coefficient and (2) calculating the biomolecular electrostatic free energy, $G_{\text{el}}^{(2)}$, in the inhomogeneous medium of interest, for example, a protein in aqueous medium. The polar contribution to the solvation free energy is then given by

$$\Delta G_{\text{solv}}^{\text{polar}} = G_{\text{el}}^{(2)} - G_{\text{el}}^{(1)} \tag{9}$$

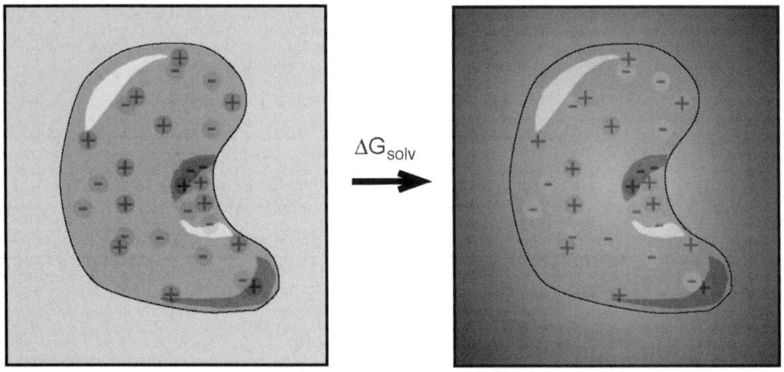

Fig. 5 Schematic of a polar solvation free energy calculation; in the initial state, the dielectric coefficient is a constant throughout the entire system and equal to the solute's dielectric coefficient; in the final state, the dielectric coefficient is inhomogeneous and smaller in the solute than in the bulk solvent.

Additionally, solving the PB equation twice helps to cancel the numerical artifacts which arise from the discretization used in finite difference and finite element methods; that is, it reduces the grid size dependence. Although, in most cases, polar solvation free energy alone is not sufficient to explain the biological phenomenon, it is the foundation for the other, more complex, electrostatic calculations described below.

B. Electrostatic Free Energy

The total electrostatic free energy can be easily obtained from the polar solvation free energy by adding the electrostatic free energy of the biomolecule in a homogeneous medium with a constant dielectric equal to the solute's dielectric coefficient using Coulomb's law:

$$G_{\mathrm{el}} = \Delta G_{\mathrm{solv}}^{\mathrm{polar}} + G_{\mathrm{coul}} \tag{10}$$

where

$$G_{\mathrm{coul}} = \sum_{i,j} \frac{1}{4\pi\varepsilon_0\varepsilon_p} \cdot \frac{Q_i Q_j}{r_{ij}} \tag{11}$$

where r_{ij} is the distance between charge Q_i and Q_j and ε_p is the dielectric coefficient of the solute. The resulting electrostatic free energies are the basis for nearly all applications of continuum electrostatics methods to biomolecular systems.

As a specific example, such electrostatic free energy calculations have been used to study the electrostatic sequestration of phosphatidylinositol 4,5-bisphosphate (PIP_2) by membrane-adsorbed basic peptides (Wang *et al.*, 2004). PIP_2 is a very important lipid in the cytoplasmic leaflet of the plasma membrane (Cantley, 2002; De Camilli *et al.*, 1996; Irvine, 2002; Martin, 2001; McLaughlin *et al.*, 2002; Payrastre *et al.*, 2001; Raucher *et al.*, 2000; Toker, 1998; Yin and Janmey, 2003) with a net charge of –4e on the lipid head group. By calculating the electrostatic free energy of laterally sequestering a PIP_2 lipid from a region of "bulk" membrane to a region in the vicinity of a membrane-absorbed basic peptide, Wang *et al* demonstrated that nonspecific electrostatic interactions provide a driving force for the lateral sequestration of PIP_2 by membrane-adsorbed basic peptides (Rauch *et al.*, 2002; Wang *et al.*, 2001a, 2002, 2004). Such lateral sequestration of PIP_2 is thought to contribute to the regulation of PIP_2 function by controlling its accessibility to other proteins (Laux *et al.*, 2000; McLaughlin *et al.*, 2002).

C. Folding Free Energies

Biomolecular native (folded) structure is very important for proper performance of their biological functions. However, accurately determining the mechanism by which electrostatic interactions affect the stability of bimolecular native structure is still a challenging experimental and computational question. The electrostatic contribution to the biomolecular folding stability is usually defined as the difference in electrostatic free energy between folded ($G^{\mathrm{el}}_{\mathrm{folded}}$) and unfolded ($G^{\mathrm{el}}_{\mathrm{unfolded}}$) states:

$$\Delta G^{\mathrm{el}}_{\mathrm{folding}} = G^{\mathrm{el}}_{\mathrm{folded}} - G^{\mathrm{el}}_{\mathrm{unfolded}} \tag{12}$$

If $\Delta G^{\mathrm{el}}_{\mathrm{folding}} < 0$, from the electrostatic point of view, the folded structure is more stable than the unfolded structure. If $\Delta G^{\mathrm{el}}_{\mathrm{folding}}$ reduces in response to a mutation, that is, if $\Delta\Delta G^{\mathrm{el}}_{\mathrm{folding}} = \Delta G^{\mathrm{el,m}}_{\mathrm{folding}} - \Delta G^{\mathrm{el,w}}_{\mathrm{folding}} < 0$, this mutation makes the folded protein more stable. This method has been widely used to study electrostatic contribution to protein folding stabilities through mutations that involve charged or polar residues. For example, *Bacillus caldolyticus* cold shock protein (Bc-Csp) is a thermophilic protein that differs from *B. subtilis* cold shock protein B (Bs-CspB), its mesophilic homologue, in 11 of its 66 residues (Delbruck *et al.*, 2001; Mueller *et al.*, 2000). Through mutational studies, which reduced the sequence differences between these two protein molecules, both experimental (Delbruck *et al.*, 2001; Mueller *et al.*, 2000; Pace, 2000; Perl *et al.*, 2000; Perl and Schmid, 2001) and PB calculations (Zhou and Dong, 2003) demonstrated that the difference in stability of these two proteins arises mostly from the interactions among the three residues: Arg 3, Glu 46, and Leu 66 in Bc-Csp, as compared with Glu 3, Ala 46, and Glu 66 in Bs-CspB. The removal of the repulsion between Glu 3 and Glu 66 and the creation of a favorable salt bridge between Arg 3 and Glu 46 are the main reasons that Bc-Csp is more stable than Bs-CspB at higher temperatures. Moreover, the excellent agreement between PB calculations and experimental data (the correlation coefficient is 0.98) implies that electrostatic interactions dominate the thermostability of thermophilic proteins (Zhou and Dong, 2003).

D. Binding Free Energies

The binding of biomolecules is fundamental to cellular activity. The simplest type of binding energy calculations are performed on the biomolecular complex assuming a rigid conformation; that is, without any conformational changes on binding, which is clearly not realistic, but often provides useful initial estimates for relative biomolecular binding affinities. Figure 6 illustrates the procedure to calculate the polar contribution to the binding free energy, $\Delta G^{\mathrm{el}}_{\mathrm{binding}}$, which is given by

$$\begin{aligned}\Delta G^{\mathrm{el}}_{\mathrm{binding}} &= G^{\mathrm{el}}_{\mathrm{complex}} - (G^{\mathrm{el}}_{\mathrm{mol1}} + G^{\mathrm{el}}_{\mathrm{mol2}}) \\ &= [\Delta G^{\mathrm{solv}}_{\mathrm{complex}} - (\Delta G^{\mathrm{solv}}_{\mathrm{mol1}} + \Delta G^{\mathrm{solv}}_{\mathrm{mol2}})] + [G^{\mathrm{coul}}_{\mathrm{complex}} - (G^{\mathrm{coul}}_{\mathrm{mol1}} + G^{\mathrm{coul}}_{\mathrm{mol2}})] \\ &= \Delta\Delta G^{\mathrm{solv}}_{\mathrm{binding}} + \Delta G^{\mathrm{coul}}_{\mathrm{binding}}\end{aligned} \tag{13}$$

where $\Delta\Delta G^{\mathrm{solv}}_{\mathrm{binding}} = \Delta G^{\mathrm{solv}}_{\mathrm{complex}} - (\Delta G^{\mathrm{solv}}_{\mathrm{mol1}} + \Delta G^{\mathrm{solv}}_{\mathrm{mol2}})$ is the polar solvation free energy change on binding with the $\Delta G^{\mathrm{solv}}_{i}$ values calculated according to Eq. (9) above.

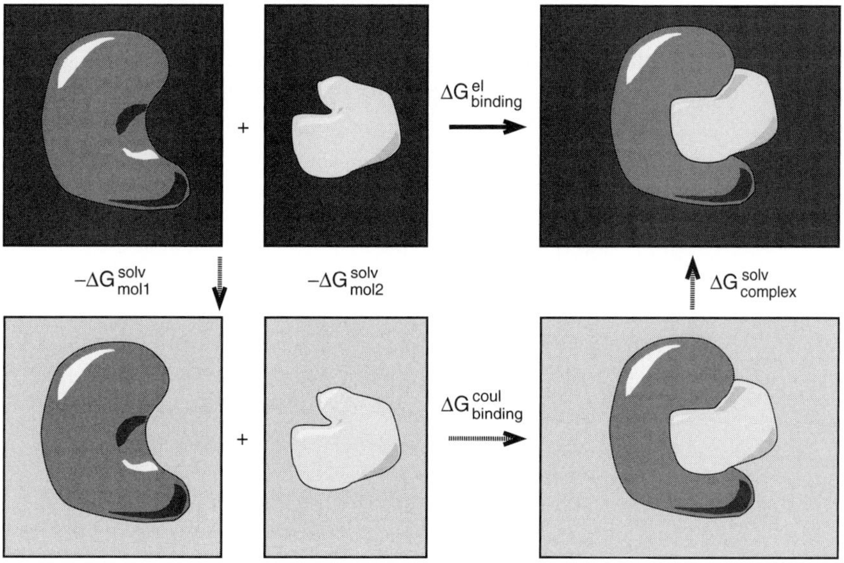

Fig. 6 Thermodynamic cycle illustrating the standard procedure for calculating the electrostatic contribution to the binding free energy of a complex with rigid body. The steps are as follows: (1) transfer the isolated molecule from a inhomogeneous dielectric into a homogeneous dielectric, the free energy change is $-(\Delta G^{\mathrm{solv}}_{\mathrm{mol1}} + \Delta G^{\mathrm{solv}}_{\mathrm{mol2}})$; (2) form the complex from isolated molecules in a homogeneous dielectric, the free energy change is $\Delta G^{\mathrm{coul}}_{\mathrm{binding}}$, (3) transfer the complex from the homogeneous dielectric into the inhomogeneous dielectric, the free energy change is $\Delta G^{\mathrm{solv}}_{\mathrm{complex}}$.

The quantity $\Delta G^{\mathrm{coul}}_{\mathrm{binding}} = G^{\mathrm{coul}}_{\mathrm{complex}} - (G^{\mathrm{coul}}_{\mathrm{mol1}} + G^{\mathrm{coul}}_{\mathrm{mol2}})$ is the Coulombic free energy change on binding with the $\Delta G^{\mathrm{coul}}_{i}$ values calculated according to Eq. (11) above.

For the more general situation in which biomolecules experience conformational changes during the binding process, MM/PBSA and MM/GBSA methods (Kollman *et al.*, 2000; Swanson *et al.*, 2004) are commonly used to calculate the binding free energy. The nature of these methods can be best understood through their acronym: MM stands for the molecular mechanics force fields used to calculate the intramolecular and direct intermolecular contributions to binding free energies; PB and GB refer to the implicit solvent methods used to calculate the electrostatic contributions, and SA stands for SASA methods used to calculate the nonpolar contributions to binding free energies.

Binding free energy calculations using continuum solvation models have been successfully performed on many different biomolecular complexes (Dong *et al.*, 2003; Eisenberg and McLachlan, 1986; Green and Tidor, 2005; Massova and Kollman, 2000; Misra *et al.*, 1998; Murray *et al.*, 1997; Sept *et al.*, 1999, 2003; Wang *et al.*, 2004; Wong *et al.*, 2001). As specific examples, binding free energy calculations have been performed to investigate the roles of charged residues at the interface of barnase (an extracellular ribonuclease) and barstar (a protein inhibitor), which have been a popular test case for both computational

(Dong *et al.*, 2003; Gabdoulline and Wade, 1997, 1998, 2001; Lee and Tidor, 2001; Sheinerman and Honig, 2002; Spaar and Helms, 2005; Spaar *et al.*, 2006; Wang and Wade, 2003) and experimental studies of protein–protein interactions (Frisch *et al.*, 1997; Schreiber and Fersht, 1993, 1995). In particular, PB calculations (Dong *et al.*, 2003) successfully reproduced the experimental result (Frisch *et al.*, 1997; Schreiber and Fersht, 1993, 1995) that cross-interface salt bridges and hydrogen bonds dominate the binding affinities of barnase and barstar (Dong *et al.*, 2003).

E. pK_a Calculations

The presence of ionizable sites, which can exchange protons with their environment, produces pH-dependent phenomena in proteins and has a significant influence on the protein's function. The correct prediction of protein titration states is important for the analysis of enzyme mechanisms, protein stability, and molecular recognition. As mentioned earlier, efforts have been underway for more than 80 years (Alexov, 2003; Antosiewicz *et al.*, 1996b; Bastyns *et al.*, 1996; Fitch *et al.*, 2002; Georgescu *et al.*, 2002; Jensen *et al.*, 2005; Krieger *et al.*, 2006; Li *et al.*, 2002, 2004; Linderström-Lang, 1924; Luo *et al.*, 1998; Nielsen and McCammon, 2003; Nielsen and Vriend, 2001) to correctly predict protein titration states and understand the determinants of pK_as for amino acids in protein environments (see chapter 26 by Whitten *et al.*, this volume).

The free energy change, ΔG, for protonation of a single ionizable site at a given pH may be written as (Linderström-Lang, 1924; Tanford and Kirkwood, 1957)

$$\Delta G = G^{\mathrm{el}}_{\mathrm{protonated}} - G^{\mathrm{el}}_{\mathrm{deprotonated}} = (k_{\mathrm{B}}T\ln 10)(\mathrm{pH} - \mathrm{p}K_{\mathrm{a}}) \tag{14}$$

where $\mathrm{p}K_{\mathrm{a}} = -\log_{10}K_{\mathrm{a}}$ and $K_{\mathrm{a}} = [\mathrm{H}^+][A^-]/[\mathrm{H}A]$ is the equilibrium constant for the dissociation of proton H^+ and its conjugate site A^-; k_{B} is Boltzmann's constant; and T is the absolute temperature. A widely used assumption in pK_a predictions is that any pK_a differences of an ionizable site when located in a protein versus in a model compound are solely determined by the difference in the electrostatic free energy required to protonate that site in the protein versus the model compound. Thus, the p$K_a$of the single ionizable site in protein is given by

$$\mathrm{p}K_{\mathrm{a}} = \mathrm{p}K_0 - \Delta\Delta G/(k_{\mathrm{B}}T\ln 10) \tag{15}$$

where $\Delta\Delta G = \Delta G_{\mathrm{protein}} - \Delta G_{\mathrm{model}}$ and pK_0 is the pK_a of the isolated ionizable site in the model compound. In general, proteins have multiple ionizable sites and the protonation energetics of these different sites are coupled, as discussed below. Single-site pK_a predictions have successfully reproduced measured pK_as for different residues in several different proteins (Dong and Zhou, 2002; Dong *et al.*, 2003) and therefore have some predictive power. However, a more complete treatment of ionizable residues in proteins considers the coupling between all the ionizable sites. There are a number of techniques for treating such coupling (Antosiewicz *et al.*, 1996a,b; Bashford, 2004; Beroza *et al.*, 1991; Tanford and Roxby, 1972), and thereby determining the complete titration state of the protein. Unfortunately, such methods are complex and are beyond the scope of the current discussion.

F. Biomolecular Association Rates

BD calculations are popular methods to simulate the relative diffusional motion between two solute particles and thereby estimate the rate of diffusion-controlled binding between two molecules (Ermak and McCammon, 1978; Northrup *et al.*, 1984). Given the importance of electrostatic interactions in biomolecular association, BD simulations are usually combined with continuum electrostatic calculations to provide the most accurate estimates of diffusion-limited encounter rates (Allison and McCammon, 1985; Davis and McCammon, 1990a; Gabdoulline and Wade, 2001, 2002; Ilin *et al.*, 1995; Madura *et al.*, 1995; Sept *et al.*, 1999). Such calculations have been used in numerous diffusional encounter rate calculations, including simulations of small molecule interactions with enzymes (Allison and McCammon, 1985; Davis *et al.*, 1991; Elcock *et al.*, 1996; Luty *et al.*, 1993; Madura and McCammon, 1989; Radic *et al.*, 1997; Sines *et al.*, 1992; Tan *et al.*, 1993; Tara *et al.*, 1998), simulations of protein–protein encounter (Elcock *et al.*, 1999, 2001; Gabdoulline and Wade, 1997, 2001; Sept *et al.*, 1999; Spaar *et al.*, 2006), as well as functional assessment of differences in protein electrostatics (Livesay *et al.*, 2003).

V. Conclusion and Future Directions

Computer simulation is becoming an increasingly routine way to help with drug discovery or other applications requiring a detailed understanding of molecular interactions. A correct understanding of the energetic interactions within and between biomolecules is essential for such simulations. Among the various contributions to these energies, electrostatic interactions are of special importance because of their long range and strength. In this chapter, we have covered some of the computational methods that are currently available to model the electrostatic interactions in biomolecular systems, ranging from highly detailed explicit solvent methods to simpler PB and GB methods. There are several reviews available on all of these methods which provide a more in-depth discussion of the different solvation approaches. The reviews of Ponder and Case (2003) as well as the texts of Becker *et al.* (2001), Leach (2001), and Schlick (2002) provide excellent background on explicit solvent methods. There also are several reviews available for implicit solvent methods (see Baker, 2005a; Bashford and Case, 2000; Honig and Nicholls, 1995; Roux and Simonson, 1999; Simonson, 2003), including a particularly thorough treatment by Lamm (2003), a discussion of current PB limitations by

Baker (2005b), and an up-to-date discussion of current challenges for GB methods by Feig and Brooks (2004). For additional background and more in-depth discussion of the principles and limitations of continuum electrostatics, interested readers should see the general volume by Jackson (1975) and Landau *et al.* (1982), the electrochemistry text of Bockris *et al.* (1998), the colloid theory treatise by Verwey and Overbeek (1999), or the excellent collection of condensed matter electrostatics articles assembled by Holm *et al.* (2001).

Acknowledgments

The authors thank Baker group members for their reading of this manuscript. This work was supported by National Institutes of Health grant R01 GM069702.

References

Alexov, E. (2003). Role of the protein side-chain fluctuations on the strength of pair-wise electrostatic interactions: Comparing experimental with computed pKas. *Proteins* **50,** 94–103.

Allison, S. A., and Huber, G. A. (1995). Modeling the electrophoresis of rigid polyions: Application of lysozyme. *Biophys. J.* **68,** 2261–2270.

Allison, S. A., and McCammon, J. A. (1985). Dynamics of substrate binding to copper zinc superoxide dismutase. *J. Phys. Chem.* **89,** 1072–1074.

Anderson, C. F., and Record, M. T., Jr. (2004). Gibbs–Duhem-based relationships among derivatives expressing the concentration dependences of selected chemical potentials for a multicomponent system. *Biophys. Chem.* **112,** 165–175.

Antosiewicz, J., Briggs, J. M., Elcock, A. H., Gilson, M. K., and McCammon, J. A. (1996a). Computing ionization states of proteins with a detailed charge model. *J. Comput. Chem.* **17,** 1633–1644.

Antosiewicz, J., McCammon, A. J., and Gilson, M. K. (1996b). The determinants of pKas in proteins. *Biochemistry* **35,** 7819–7833.

Bajaj, N. P., McLean, M. J., Waring, M. J., and Smekal, E. (1990). Sequence-selective, pH-dependent binding to DNA of benzophenanthridine alkaloids. *J. Mol. Recognit.* **3,** 48–54.

Baker, N., Holst, M., and Wang, F. (2000). Adaptive multilevel finite element solution of the poisson-boltzmann equation ii. Refinement at solvent-accessible surfaces in biomolecular systems. *J. Comput. Chem.* **21,** 1343–1352.

Baker, N. A. (2005a). Biomolecular applications of Poisson-Boltzmann methods. *In* "Reviews in Computational Chemistry" (K. B. Lipkowitz, R. Larter, and T. R. Cundari, eds.), pp. 349–379. Wiley-VCH, John Wiley & Sons, Inc., Hoboken, NJ.

Baker, N. A. (2005b). Improving implicit solvent simulations: A Poisson-centric view. *Curr. Opin. Struct. Biol.* **15,** 137–143.

Baker, N. A., Bashford, D., and Case, D. (2006). Implicit solvent electrostatics in biomolecular simulation. *In* "New Algorithms for Macromolecular Simulation" (B. Leimkuhler, C. Chipot, and R. Elber, *et al.*, eds.), pp. 263–295. Springer-Verlag, Berlin.

Baker, N. A., Sept, D., Holst, M. J., and McCammon, J. A. (2001). Electrostatics of nanosystems: Application to microtubules and the ribosome. *Proc. Natl. Acad. Sci. USA* **98,** 10037–10041.

Bashford, D. (1997). Scientific computing in object-oriented parallel environments; An object-oriented programming suite for electrostatic effects in biological molecules. *In* Lecture Notes in Computer Science pp. 233–240. Springer, Berlin.

Bashford, D. (2004). Macroscopic electrostatic models for protonation states in proteins. *Front. Biosci.* **9,** 1082–1099.

Bashford, D., and Case, D. A. (2000). Generalized Born models of macromolecular solvation effects. *Annu. Rev. Phys. Chem.* **51,** 129–152.

Bastyns, K., Froeyen, M., Diaz, J. F., Volckaert, G., and Engelborghs, Y. (1996). Experimental and theoretical study of electrostatic effects on the isoelectric pH and pKa of the catalytic residue His-102 of the recombinant ribonuclease from Bacillus amyloliquefaciens (barnase). *Proteins* **24,** 370–378.

Becker, O., MacKerell, A. D., Jr., Roux, B., and Watanabe, M. (2001). "Computational Biochemistry and Biophysics." Marcel Dekker, New York.

Beroza, P., Fredkin, D. R., Okamura, M. Y., and Feher, G. (1991). Protonation of interacting residues in a protein by Monte Carlo method: Application to lysozyme and the photosynthetic reaction center of Rhodobacter sphaeroides. *Proc. Natl. Acad. Sci. USA* **88,** 5804–5808.

Bhattacharrya, S. M., Wang, Z.-G, and Zewail, A. H. (2003). Dynamics of water near a protein surface. *J. Phys. Chem. B* **107,** 13218–13228.

Bockris, J. O., and Reddy, K. N. (1998). "Modern Electrochemistry: Ionics." Plenum Press, New York.

Bordner, A. J., and Huber, G. A. (2003). Boundary element solution of linear Poisson-Boltzmann equation and a multipole method for the rapid calculation of forces on macromolecules in solution. *J. Comput. Chem.* **24,** 353–367.

Born, M. (1920). Volumen und hydratationswarme der ionen. *Z. Phys.* **1,** 45–48.

Boschitsch, A. H., and Fenley, M. O. (2004). Hybrid boundary element and finite difference method for solving the nonlinear Poisson-Boltzmann equation. *J. Comput. Chem.* **25,** 935–955.

Boström, M., Deniz, V., and Ninham, B. W. (2006). Ion specific surface forces between membrane surfaces. *J. Phys. Chem. B* **110,** 9645–9649.

Boström, M., Williams, D. R. M., Stewart, P. R., and Ninham, B. W. (2003). Hofmeister effects in membrane biology: The role of ionic dispersion potentials. *Phys. Rev. E* **68,** 041902–041907.

Broering, J. M., and Bommarius, A. S. (2005). Evaluation of Hofmeister effects on the kinetic stability of proteins. *J. Phys. Chem. B* **109,** 20612–20619.

Burkert, U., and Allinger, N. L. (1982). "Molecular Mechanics." American Chemical Society, Washington, DC.

Cantley, L. C. (2002). The phosphoinositide 3-kinase pathway. *Science* **296,** 1655–1657.

Cech, T. R., and Bass, B. L. (1986). Biological catalysis by RNA. *Annu. Rev. Biochem.* **55,** 599–630.

Chong, L. T., Dempster, S. E., Hendsch, Z. S., Lee, L. P., and Tidor, B. (1998). Computation of electrostatic complements to proteins: A case of charge stabilized binding. *Protein Sci.* **7,** 206–210.

Chothia, C. (1974). Hydrophobic bonding and accessible surface area in proteins. *Nature* **248,** 338–339.

Cole, P. E., Yang, S. K., and Crothers, D. M. (1972). Conformational changes of transfer ribonucleic acid. Equilibrium phase diagrams. *Biochemistry* **11,** 4358–4368.

Collins, K. D. (2006). Ion hydration: Implications for cellular function, polyelectrolytes, and protein crystallization. *Biophys. Chem.* **119,** 271–281.

Connolly, M. L. (1985). Computation of molecular volume. *J. Am. Chem. Soc.* **107,** 1118–1124.

Cortis, C. M., and Friesner, R. A. (1997a). An automatic three-dimensional finite element mesh generation system for Poisson-Boltzmann equation. *J. Comput. Chem.* **18,** 1570–1590.

Cortis, C. M., and Friesner, R. A. (1997b). Numerical solution of the Poisson-Boltzmann equation using tetrahedral finite-element meshes. *J. Comput. Chem.* **18,** 1591–1608.

Dahm, S. C., and Uhlenbeck, O. C. (1991). Role of divalent metal ions in the hammerhead RNA cleavage reaction. *Biochemistry* **30,** 9464–9469.

Davis, M. E., Madura, J. D., Sines, J., Luty, B. A., Allison, S. A., and McCammon, J. A. (1991). Diffusion-controlled enzymatic reactions. *Methods Enzymol.* **202,** 473–497.

Davis, M. E., and McCammon, J. A. (1989). Solving the finite difference linearized Poisson-Boltzmann equation: A comparison of relaxation and conjugate gradient methods. *J. Comput. Chem.* **10,** 386–391.

Davis, M. E., and McCammon, J. A. (1990a). Calculating electrostatic forces from grid-calculated potentials. *J. Comput. Chem.* **11,** 401–409.

Davis, M. E., and McCammon, J. A. (1990b). Electrostatics in biomolecular structure and dynamics. *Chem. Rev.* **90,** 509–521.

De Camilli, P., Emr, S. D., McPherson, P. S., and Novick, P. (1996). Phosphoinositides as regulators in membrane traffic. *Science* **271,** 1533–1539.

Debye, P., and Hückel, E. (1923). Zur Theorie der Elektrolyte. I. Gefrierpunktserniedrigung und verwandte Erscheinungen. *Physikalische Zeitschrift* **24,** 185–206.

del Álamo, M., and Mateu, M. G. (2005). Electrostatic repulsion, compensatory mutations, and long-range non-additive effects at the dimerization interface of the HIV capsid protein. *J. Mol. Biol.* **345,** 893–906.

Delbruck, H., Mueller, U., Perl, D., Schmid, F. X., and Heinemann, U. (2001). Crystal structures of mutant forms of the *Bacillus caldolyticus* cold shock protein differing in thermal stability. *J. Mol. Biol.* **313,** 359–369.

Dominy, B. N., and Brooks, C. L., III (1999). Development of a Generalized Born model parameterization for proteins and nucleic acids. *J. Phys. Chem. B* **103,** 3765–3773.

Dong, F., Vijayakumar, M., and Zhou, H.-X. (2003). Comparison of calculation and experiment implicates significant electrostatic contributions to the binding stability of barnase and barstar. *Biophys. J.* **85,** 49–60.

Dong, F., and Zhou, H.-X. (2002). Electrostatic contributions to T4 lysozyme stability: Solvent-exposed charges versus semi-buried salt bridges. *Biophys. J.* **83,** 1341–1347.

Draper, D. E., Grilley, D., and Soto, A. M. (2005). Ions and RNA folding. *Annu. Rev. Biophys. Biomol. Struct.* **34,** 221–243.

Dyshlovenko, P. E. (2002). Adaptive numerical method for Poisson–Boltzmann equation and its application. *Comput. Phys. Commun.* **147,** 335–338.

Eisenberg, D., and McLachlan, A. D. (1986). Solvation energy in protein folding and binding. *Nature* **319,** 199–203.

Eisenberg, H. (1976). "Biological Macromolecules and Polyelectrolytes in Solution," Chapter 2. Clarendon, Oxford.

Elcock, A. H., Gabdoulline, R. R., Wade, R. C., and McCammon, J. A. (1999). Computer simulation of protein-protein association kinetics: Acetylcholinesterase-fasciculin. *J. Mol. Biol.* **291,** 149–162.

Elcock, A. H., Potter, M. J., Matthews, D. A., Knighton, D. R., and McCammon, J. A. (1996). Electrostatic channeling in the bifunctional enzyme dihydrofolate reductase-thymidylate synthase. *J. Mol. Biol.* **262,** 370–374.

Elcock, A. H., Sept, D., and McCammon, J. A. (2001). Computer simulation of protein-protein interactions. *J. Phys. Chem. B* **105,** 1504–1518.

Ermak, D. L., and McCammon, J. A. (1978). Brownian dynamics with hydrodynamic interactions. *J. Chem. Phys.* **69,** 1352–1360.

Feig, M., and Brooks, C. L., III (2004). Recent advances in the development and application of implicit solvent models in biomolecule simulations. *Curr. Opin. Struct. Biol.* **14,** 217–224.

Figueirido, F., Delbuono, G. S., and Levy, R. M. (1994). Molecular mechanics and electrostatic effects. *Biophys. Chem.* **51,** 235–241.

Fitch, C. A., Karp, D. A., Lee, K. K., Stites, W. E., Lattman, E. E., and Garcia-Moreno, E. B. (2002). Experimental pKa values of buried residues: Analysis with continuum methods and role of water penetration. *Biophys. J.* **82,** 3289–3304.

Flanagan, M. A., Ackers, G. K., Matthew, J. B., Hanania, G. I. H., and Gurd, F. R. N. (1981). Electrostatic contributions to energetics of dimer-tetramer assembly in human hemoglobin: pH dependence and effect of specifically bound chloride ions. *Biochemistry* **20,** 7439–7449.

Frisch, C., Schreiber, G., Johnson, C. M., and Fersht, A. R. (1997). Thermodynamics of the interaction of barnase and barstar: Changes in free energy versus changes in enthalpy on mutation. *J. Mol. Biol.* **267,** 696–706.

Gabdoulline, R. R., and Wade, R. C. (1997). Simulation of the diffusional association of barnase and barstar. *Biophys. J.* **72,** 1917–1929.

Gabdoulline, R. R., and Wade, R. C. (1998). Brownian dynamics simulation of protein-protein diffusional encounter. *Methods Enzymol.* **14,** 329–341.

Gabdoulline, R. R., and Wade, R. C. (2001). Protein-protein association: Investigation of factors influencing association rates by Brownian dynamics simulations. *J. Mol. Biol.* **306,** 1139–1155.

Gabdoulline, R. R., and Wade, R. C. (2002). Biomolecular diffusional association. *Curr. Opin. Struct. Biol.* **12,** 204–213.

Gallicchio, E., Kubo, M. M., and Levy, R. M. (2000). Enthalpy-entropy and cavity decomposition of alkane hydration free energies: Numerical results and implications for theories of hydrophobic solvation. *J. Phys. Chem. B* **104,** 6271–6285.

Gallicchio, E., and Levy, R. M. (2004). AGBNP: An analytic implicit solvent model suitable for molecular dynamics simulations and high-resolution modeling. *J. Comput. Chem.* **25,** 479–499.

Gallicchio, E., Zhang, L. Y., and Levy, R. M. (2002). The SGB/NP hydration free energy model based on the surface generalized born solvent reaction field and novel nonpolar hydration free energy estimators. *J. Comput. Chem.* **21,** 86–104.

Georgescu, R. E., Alexov, E. G., and Marilyn, R. G. (2002). Combining conformational flexibility and continuum electrostatics for calculating pKas in proteins. *Biophys. J.* **83,** 1731–1748.

Gilson, M. (2000). Introduction to continuum electrostatics. *In* "Biophysics Textbook Online" (D. A. Beard, ed.), Biophysical society, Bethesda, MD.

Gilson, M., Davis, M. E., Luty, B. A., and McCammon, J. A. (1993). Computation of electrostatic forces on solvated molecules using the Poisson-Boltzmann equation. *J. Phys. Chem.* **97,** 3591–3600.

Gilson, M. K. (1995). Theory of electrostatic interactions in macromolecules. *Curr. Opin. Struct. Biol.* **5,** 216–223.

Grant, J. A., Pickup, B. T., and Nicholls, A. (2001). A smooth permittivity function for Poisson-Boltzmann solvation methods. *J. Comput. Chem.* **22,** 608–640.

Green, D. F., and Tidor, B. (2005). Design of improved protein inhibitors of HIV-1 cell entry: Optimization of electrostatic interactions at the binding interface. *Proteins* **60,** 644–657.

Hofmeister, F. (1888). Zur lehre von der wirkung der salze. zweite mittheilung. *Arch. Exp. Pathol. Pharmakol.* **24,** 247–260.

Holm, C., Kekicheff, P., and Podgornik, R. (2001). "Electrostatic Effects in Soft Matter and Biophysics." Kluwer academic publishers, Boston, MA.

Holst, M., Baker, N., and Wang, F. (2000). Adaptive multilevel finite element solution of the Poisson-Boltzmann equation i. Algorithms and examples. *J. Comput. Chem.* **21,** 1319–1342.

Holst, M., and Saied, F. (1993). Multigrid solution of the Poisson-Boltzmann equation. *J. Comput. Chem.* **14,** 105–113.

Holst, M. J., and Saied, F. (1995). Numerical solution of nonlinear Poisson-Boltzmann equation: Developing more robust and efficient methods. *J. Comput. Chem.* **16,** 337–364.

Honig, B. H., and Nicholls, A. (1995). Classical electrostatics in biology and chemistry. *Science* **268,** 1144–1149.

Horn, H. W., Swope, W. C., Pitera, J. W., Madura, J. D., Dick, T. J., Hura, G. L., and Head-Gordon, T. (2004). Development of an improved four-site water model for biomolecular simulations: TIP4P-Ew. *J. Chem. Phys.* **120,** 9665–9678.

Humphrey, W., Dalke, A., and Schulten, K. (1996). VMD—visual molecular dynamics. *J. Mol. Graph.* **14,** 33–38.

Ilin, A., Bagheri, B., Scott, L. R., Briggs, J. M., and McCammon, J. A. (1995). Parallelization of Poisson-Boltzmann and Brownian Dynamics calculations. *American Chemical Society Symposium Series* **592,** 170–185.

Im, W., Beglov, D., and Roux, B. (1998). Continuum solvation model: Electrostatic forces from numerical solutions to the Poisson-Boltzmann equation. *Comput. Phys. Commun.* **11,** 59–75.

Irvine, R. F. (2002). Nuclear lipid signaling. *SciSTKE* **150,** 1–12.

Jackson, J. D. (1975). "Classical Electrodynamics." Wiley, New York.

Janin, J., and Chothia, C. (1990). The structure of protein-protein recognition sites. *J. Biol. Chem.* **265,** 16027–16030.

Jensen, J. H., Li, H., Robertson, A. D., and Molina, P. A. (2005). Prediction and rationalization of protein pKa values using QM and QM/MM methods. *J. Phys. Chem. A* **109,** 6634–6643.

Jorgensen, W. L., Chandrasekhar, J., Madura, J. D., Impey, R. W., and Klein, M. L. (1983). Comparison of simple potential functions for simulating liquid water. *J. Chem. Phys.* **79,** 926–935.

Juffer, A. H., Botta, E. F. F., van Keulen, B. A. M., van der Ploeg, A., and Berendsen, H. J. C. (1991). The electric potential of a macromolecule in solvent: A fundamental approach. *J. Comput. Phys.* **97,** 144–171.

Kollman, P. A., Massova, I., Reyes, C., Kuhn, B., Huo, S., Chong, L., Lee, M., Lee, T., Duan, Y., Wang, W., Donini, O., Cieplak, P., *et al.* (2000). Calculating structures and free energies of complex molecules: Combining molecular mechanics and continuum models. *Acc. Chem. Res.* **33,** 889–897.

Krieger, E., Nielsen, J. E., Spronk, C. A. E. M., and Vriend, G. (2006). Fast empirical pKa prediction by Ewald summation. *J. Mol. Graph. Model.* **25**(4), 481–486.

Lamm, G. (2003). The Poisson-Boltzmann Equation. *In* "Reviews in Computational Chemistry" (K. B. Lipkowitz, R. Larter, and T. R. Cundari, eds.), pp. 147–366. John Wiley and Sons, Inc., Hoboken, NJ.

Landau, L. D., Lifshitz, E. M., and Pitaevskii, L. P. (1982). "Electrodynamics of Continuous Media." Butterworth-Heinenann, Boston, MA.

Laux, T., Fukami, K., Thelen, M., Golub, T., Frey, D., and Caroni, P. (2000). GAP43, MARCKS, CAP23 modulate PI(4,5)P2 at plasmalemmal rafts, and regulate cell cortex actin dynamics through a common mechanism. *J. Cell Biol.* **149,** 1455–1472.

Law, M. J., Linde, M. E., Chambers, E. J., Oubridge, C., Katsamba, P. S., Nilsson, L., Haworth, I. S., and Laird-Offringa, I. A. (2006). The role of positively charged amino acids and electrostatic interactions in the complex of U1A protein and U1 hairpin II RNA. *Nucleic Acids Res.* **34,** 275–285.

Leach, A. R. (2001). "Molecular Modeling: Principles and Applications." Prentice Hall, Harlow, England.

Lee, B., and Richards, F. M. (1971). The interpretation of protein structures: Estimation of static accessibility. *J. Mol. Biol.* **55,** 379–400.

Lee, L. P., and Tidor, B. (2001). Optimization of binding electrostatics: Charge complementarity in the barnase-barstar protein complex. *Protein Sci.* **10,** 362–377.

Li, H., Hains, A. W., Everts, J. E., Robertson, A. D., and Jensen, J. H. (2002). The prediction of protein pKa's using QM/MM: The pKa of lysine 55 in turkey ovomucoid third domain. *J. Phys. Chem. B* **106,** 3486–3494.

Li, H., Robertson, A. D., and Jensen, J. H. (2004). The determinants of carboxyl pKa values in turkey ovomucoid third domain. *Proteins* **55,** 689–704.

Lin, J.-H., Baker, N. A., and McCammon, J. A. (2002). Bridging the implicit and explicit solvent approaches for membrance electrostatics. *Biophys. J.* **83,** 1374–1379.

Linderström-Lang, K. (1924). On the ionisation of proteins. *Comptes-rend Lab. Carlaberg* **15,** 1–29.

Livesay, D. R., Jambeck, P., Rojnuckarin, A., and Subramaniam, S. (2003). Conservation of electrostatic properties within enzyme families and superfamilies. *Biochemistry* **42,** 3464–3473.

Lo Conte, L., Chothia, C., and Janin, J. (1999). The atomic structure of protein-protein recognition sites. *J. Mol. Biol.* **285,** 2177–2198.

Luo, R., David, L., and Gilson, M. K. (2002). Accelerated Poisson-Boltzmann calculations for static and dynamic systems. *J. Comput. Chem.* **23,** 1244–1253. http://www3.interscience.wiley.com/cgi-bin/abstract/96516852/ABSTRACT.

Luo, R., Head, M. S., Moult, J., and Gilson, M. K. (1998). pKa shifts in small molecules and HIV protease: Electrostatics and conformation. *J. Am. Chem. Soc.* **120,** 6138–6146.

Luty, B. A., Elamrani, S., and McCammon, J. A. (1993). Simulation of the bimolecular reaction between superoxide and superoxide dismutase—synthesis of the encounter and reaction steps. *J. Am. Chem. Soc.* **115,** 11874–11877.

Ma, B., Elkayam, T., Wolfson, H., and Nussinov, R. (2003). Protein-protein interactions: Structurally conserved residues distinguish between binding sites and exposed protein surfaces. *Proc. Natl. Acad. Sci. USA* **100,** 5772–5777.

MacKerell, A. D., Jr., Bashford, D., Bellot, M., Dunbrack, R. L., Jr., Evanseck, J. D., Field, M. J., Fischer, S., Gao, J., Guo, H., Ha, S., Joseph-McCarthy, D., Kuchnir, L., *et al.* (1998). All-atom empirical potential for molecular modeling and dynamics studies of proteins. *J. Phys. Chem. B* **102,** 3586–3616. http://dx.doi.org/10.1021/jp973084f.

MacKerell, A. D. J., and Nilsson, L. (2001). Nucleic Acid Simulation. *In* "Computational Biochemistry and Biophysics" (O. M. Becker, A. D. J. MacKerell, B. Roux, and M. Watanabe, eds.), pp. 441–463. Marcel Dekker, New York.

Madura, J. D., Briggs, J. M., Wade, R. C., Davis, M. E., Luty, B. A., Antosiewicz, I. J., Gilson, M. K., Bagheri, N., Scott, L. R., and McCammon, J. A. (1995). Electrostatics and diffusion of molecules in solution-simulations with the University of Houston Brownian Dynamics program. *Comput. Phys. Commun.* **91,** 57–95.

Madura, J. D., and McCammon, J. A. (1989). Brownian dynamics simulation of diffusional encounters between triose phosphate isomerase and d-glyceraldehyde phosphate. *J. Phys. Chem.* **93,** 7285–7587.

Manning, G. S. (1978). The molecular theory of polyelectrolyte solutions with applications to the electrostatic properties of polynucleotides. *Q. Rev. Biophys.* **11,** 179–246.

Martin, T. F. (2001). PI(4,5)P2 regulation of surface membrane traffic. *Curr. Opin. Cell Biol.* **13,** 493–499.

Massova, I., and Kollman, P. A. (2000). Combined molecular mechanical and continuum solvent approach (MM-PBSA/GBSA) to predict ligand binding. *Perspect. Drug Discovery Des.* **18,** 113–135.

Masunov, A., and Lazaridis, T. (2003). Potentials of mean force between ionizable aminoacid side-chains in aqueous solution. *J. Am. Chem. Soc.* **125,** 1722–1730.

McLaughlin, S., Wang, J., Gambhir, A., and Murray, D. (2002). PIP2 and proteins: Interactions, organization and information flow. *Annu. Rev. Biophys. Biomol. Struct.* **31,** 151–175.

McQuarrie, D. A. (2000). "Statistical Mechanics." University Science Books, Sausalito, CA.

Micu, A. M., Bagheri, B., Ilin, A. V., Scott, L. R., and Pettitt, B. M. (1997). Numerical considerations in the computation of the electrostatic free energy of interaction within the poisson-boltzmann theory. *J. Comput. Phys.* **136,** 263–271.

Misra, V. K., and Draper, D. E. (2000). Mg(2+) binding to tRNA revisited: The nonlinear Poisson-Boltzmann model. *J. Mol. Biol.* **299,** 1135–1147.

Misra, V. K., and Draper, D. E. (2001). A thermodynamic framework for Mg^{2+} binding to RNA. *Proc. Natl. Acad. Sci. USA* **98,** 12456–12461.

Misra, V. K., Hecht, J. L., Yang, A.-S, and Honig, B. (1998). Electrostatic contributions to the binding free energy of λcI repressor to DNA. *Biophys. J.* **75,** 2262–2273.

Moore, K. J. M., and Lohman, T. M. (1994). Kinetic mechanism of adenine nucleotide binding to and hydrolysis by the *Escherichia coli* Rep monomer. 2. Application of a kinetic competition approach. *Biochemistry* **33,** 14565–78.

Mueller, U., Perl, D., Schmid, F. X., and Heinemann, U. (2000). Thermal stability and atomic-resolution crystal structure of the *Bacillus caldolyticus* cold shock protein. *J. Mol. Biol.* **297,** 975–988.

Murray, D., Ben-Tal, N., Honig, B., and McLaughlin, S. (1997). Electrostatic interaction of myristoylated proteins with membranes: Simple physics, complicated biology. *Structure* **5,** 985–989.

Nicholls, A., and Honig, B. (1991). A rapid finit difference algorithm, utilizing successive over-relaxation to solve the Poisson-Boltamann equation. *J. Comput. Chem.* **12,** 435–445.

Nielsen, J. E., and McCammon, J. A. (2003). On the evaluation and optimization of protein x-ray structures for pKa calculations. *Protein Sci.* **12,** 313–326.

Nielsen, J. E., and Vriend, G. (2001). Optimizing the hydrogen-bond network in poisson-boltzmann equation-based pk(a) calculations. *Proteins* **43,** 403–412.

Nonner, W., Gillespe, D., Henderson, D., and Eisenberg, D. (2001). Ion accumulation in biologycal calcium channel: Effects of solvent and confining pressure. *J. Phys. Chem. B* **105,** 6427–6436.

Norel, R., Sheinerman, F., Petrey, D., and Honig, B. (2001). Electrostatic contributions to protein-protein interactions: Fast energetic filters for docking and their physical basis. *Protein Sci.* **10,** 2147–2161.

Northrup, S. H., Allison, S. A., and McCammon, J. A. (1984). Brownian dynamics simulation of diffusion-influenced biomolecular reactions. *J. Chem. Phys.* **80,** 1517–1524.

Novotny, J., and Sharp, K. (1992). Electrostatic fields in antibodies and antibody/antigen complexes. *Prog. Biophys. Mol. Biol.* **58,** 203–224.

Onufriev, A., Case, D. A., and Bashford, D. (2002). Effective born radii in the Generalized Born approximation: The importance of being perfect. *J. Comput. Chem.* **23,** 1297–1304.

Osapay, K., Young, W. S., Bashford, D., Brooks, C. L., III, and Case, D. A. (1996). Dielectric continuum models for hydration effects on peptide conformation transitions. *J. Phys. Chem.* **100,** 2698–2705.

Overman, L. B., and Lohman, T. M. (1994). Lingkage of pH, aion and cation effectx in protein-nucleic acid equilibria. *Escherichia coli* SSB protein-single strand nucleic acid interactions. *J. Mol. Biol.* **236,** 165–178.

Pace, C. N. (2000). Single surface stabilizer. *Nat. Struct. Biol.* **7,** 345–346.

Payrastre, B., Missy, K., Giuriato, S., Bodin, S., Plantavid, M., and Gratacap, M. (2001). Phosphoinositides: Key players in cell signalling, in time and space. *Cell. Signal.* **13,** 377–387.

Perl, D., Mueller, U., Heinemann, U., and Schmid, F. X. (2000). Two exposed amino acid residues confer thermostability on a cold shock protein. *Nat. Struct. Biol.* **7,** 380–383.

Perl, D., and Schmid, F. X. (2001). Electrostatic stabilization of a thermophilic cold shock protein. *J. Mol. Biol.* **213,** 343–357.

Ponder, J. W., and Case, D. A. (2003). Force fields for protein simulations. *Adv. Protein Chem.* **66,** 27–85.

Radic, Z., Kirchhoff, P. D., Quinn, D. M., McCammon, J. A., and Taylor, P. (1997). Electrostatic influence on the kinetics of ligand binding to acetylcholinesterase. *J. Biol. Chem.* **272,** 23265–23277.

Rauch, M. E., Ferguson, C. G., Prestwich, G. D., and Cafiso, D. (2002). Myristoylated alanine-rich C kinase substrate (MARCKS) sequesters spin-labeled phosphatidylinositol-4,5-bisphosphate in lipid bilayers. *J. Biol. Chem.* **277,** 14068–14076.

Raucher, D., Stauffer, T., Chen, W., Shen, K., Guo, S., York, J. D., Sheetz, M. P., and Meyer, T. (2000). Phosphatidylinositol 4,5-bisphosphate functions as a second messenger that regulates cytoskeleton-plasma membrane adhesion. *Cell* **100,** 221–228.

Record, M. T., Ha, J.-H., and Fisher, M. A. (1991). Analysis of equilibrium and kinetic measurements to determine thermodynamic origins of stability and specificity and mechanism of formation of site-specific complexes between proteins and helical DNA. *Methods Enzymol.* **208,** 291–343.

Reuter, H., Pott, C., Goldhaber, J. I., Henderson, S. A., Philipson, K. D., and Schwinger, R. H. (2005). Na(+)—Ca^{2+} exchange in the regulation of cardiac excitation-contraction coupling. *Cardiovasc Res.* **67,** 198–207.

Rocchia, W., Alexov, E., and Honig, B. (2001). Extending the applicability of the nonlinear Poisson-Boltzmann equation: Multiple dielectric constants and multivalent lons. *J. Phys. Chem. B* **105**(28), 6507–6514.

Römer, R., and Hach, R. (1975). tRNA conformation and magnesium binding. A study of a yeast phenylalanine-specific tRNA by a fluorescent indicator and differential melting curves. *Eur. J. Biochem.* **55,** 271–284.

Romer, R., and Hach, R. (1975). tRNA conformation and magnesium binding. A study of yeast phenylalanine-specific tRNA by fluorecent indicator and differential melting curves. *Eur. J. Biochem.* **55,** 271–284.

Roux, B. (2001). Implicit solvent models. *In* "Computational Biochemistry and Biophysics" (O. M Becker, A. D. Mackerell, Jr., B. Roux, and M. Watanabe, eds.), pp. 133–152. Marcel Dekker, New York.

Roux, B., and Simonson, T. (1999). Implicit solvent models. *Biophys. Chem.* **78,** 1–20.

Russell, R. B., Alber, F., Aloy, P., Davis, F. P., Korkin, D., Pichaud, M., Topf, M., and Sali, A. (2004). A structural perspective on protein-protein interactions. *Curr. Opin. Struct. Biol.* **14,** 313–324.

Sagui, C., and Darden, T. A. (1999). Molecular dynamics simulation of biomolecules: Long-range electrostatic effects. *Annu. Rev. Biophys. Biomol. Struct.* **28,** 155–179.

Schaeler, M., and Karplus, M. (1996). A comprehensive analytical treatment of continuum electrostatics. *J. Phys. Chem.* **100,** 1578–1599.

Schlick, T. (2002). "Molecular Modeling and Simulation: An Interdisciplinary Guide." Springer-Verlag, New York.

Schreiber, G., and Fersht, A. R. (1993). Interaction of barnase with its polypeptide inhibitor barstar studied by protein engineering. *Biochemistry* **32,** 5145–5150.

Schreiber, G., and Fersht, A. R. (1995). Energetics of protein-protein interactions: Analysis of the barnase-barstar interface by single mutations and double mutant cycles. *J. Mol. Biol.* **248,** 478–486.

Senear, D. F., and Batey, R. (1991). Comparison of operator-specific and nonspecific interactions of lambda cI repressor: [KCL] and pH effects. *Biochemistry* **30,** 6677–6688.

Sept, D., Baker, N. A., and McCammon, J. A. (2003). The physical basis of microtubule structure and stability. *Protein Sci.* **12,** 2257–2261.

Sept, D., Elcock, A. H., and McCammon, J. A. (1999). Computer simulations of actin polymerization can explain the barbed-pointed end asymmetry. *J. Mol. Biol.* **294,** 1181–1189.

Sept, D., and McCammon, J. A. (2001). Thermodynamics and kinetics of actin filament nucleation. *Biophys. J.* **81,** 667–674.

Sharp, K. A., and Honig, B. H. (1990). Electrostatic interactions in macromolecules: Theory and applications. *Annu. Rev. Biophys. Biophys. Chem.* **19,** 301–332.

Sharp, K. A., Nicholls, A., Fine, R. F., and Honig, B. (1991). Reconciling the magnitude of the microscopic and macroscopic hydrophobic effects. *Science* **252,** 106–109.

Sheinerman, F. B., and Honig, B. (2002). On the role of electrostatic interactions in the design of protein-protein interfaces. *J. Mol. Biol.* **318,** 161–177.

Sheinerman, F. B., Norel, R., and Honig, B. (2000). Electrostatic aspects of protein-protein interactions. *Curr. Opin. Struct. Biol.* **10,** 153–159.

Shimizu, S. (2004). Estimating hydration changes upon biomolecular reactions from osmotic stress, high pressure, and preferential hydration experiments. *Proc. Natl. Acad. Sci. USA* **101,** 1195–1199.

Shimizu, S., and Smith, D. (2004). Preferential hydration and the exclusion of cosolvents from protein surfaces. *J. Chem. Phys.* **121,** 1148–1154.

Simonson, T. (2003). Electrostatics and dynamics of proteins. *Rep. Prog. Phys.* **66,** 737–787.

Simonson, T., and Brunger, A. T. (1994). Solvation free energies estimated from macroscopic continuum theory: An accuracy assessment. *J. Phys. Chem.* **98,** 4683–4694.

Sines, J. J., McCammon, J. A., and Allison, S. A. (1992). Kinetic effects of multiple charge modifications in enzyme-substrate reactions—Brownian Dynamics simulations of Cu, Zn superoxide dismutase. *J. Comput. Chem.* **13,** 66–69.

Sitkoff, D., Sharp, K. A., and Honig, B. (1994a). Correlating solvation free energies and surface tensions of hydrocarbon solutes. *Biophys. Chem.* **51,** 397–409.

Sitkoff, D., Sharp, K. A., and Honig, B. H. (1994b). Accurate calculation of hydration free energies using macroscopic solvent models. *J. Phys. Chem.* **98,** 1978–1988.

Spaar, A., Dammer, C., Gabdoulline, R. R., Wade, R. C., and Helms, V. (2006). Diffusional encounter of barnase and barstar. *Biophys. J.* **90,** 1913–1924.

Spaar, A., and Helms, V. (2005). Free energy landscape of protein-protein encounter resulting from brownian dynamics simulations of barnase:Barstar. *J. Chem. Theory Comput.* **1,** 723–736.

Spolar, R. S., Ha, J. H., and Record, M. T. J. (1989). Hydrophobic effect in protein folding and other noncovalent processes involving proteins. *Proc. Natl. Acad. Sci. USA* **86,** 8382–8385.

Stein, A., and Crothers, D. M. (1976). Conformational changes of transfer RNA. The role of magnesium(II). *Biochemistry* **15,** 160–167.

Still, W. C., Tempczyk, A., Hawley, R. C., and Hendrickson, T. (1990). Semianalytical treatment of solvation for molecular mechanics and dynamics. *J. Am. Chem. Soc.* **112,** 6127–6129.

Su, Y., and Gallicchio, E. (2004). The non-polar solvent potential of mean force for the dimerization of alanine dipeptide: The role of solute-solvent van der Waals interactions. *Biophys. Chem.* **109,** 251–260.

Swanson, J. M. J., Henchman, R. H., and McCammon, J. A. (2004). Revisiting free energy calculations: A theoretical connection to MM/PBSA and direct calculation of the association free energy. *Biophys. J.* **86,** 67–74.

Tan, R. C., Truong, T. N., McCammon, J. A., and Sussman, J. L. (1993). Acetylcholinesterase—electrostatic steering increases the rate of ligand binding. *Biochemistry* **32,** 401–403.

Tanford, C., and Kirkwood, J. G. (1957). Theory of protein titration curves. I. General equations for impenetrable spheres. *J. Am. Chem. Soc.* **79,** 5333–5339.

Tanford, C., and Roxby, R. (1972). Interpretation of protein titration curves. *Biochemistry* **11,** 2192–2198.

Tara, S., Elcock, A. H., Kirchhoff, P. D., Briggs, J. M., Radic, Z., Taylor, P., and McCammon, J. A. (1998). Rapid binding of a cationic active site inhibitor to wild type and mutant mouse acetylcholinesterase: Brownian dynamics simulation including diffusion in the active site gorge. *Biopolymers* **46,** 465–474.

Toker, A. (1998). The synthesis and cellular roles of phosphatidylinositol 4,5-bisphosphate. *Curr. Opin. Cell Biol.* **10,** 254–261.

Vasker, I. A. (2004). Protein-protein interfaces are special. *Structure* **12,** 910–912.

Verwey, E. J. W., and Overbeek, J. T. G. (1999). "Theory of Stability of Lyophobic Colloids." Dover Publications, Inc., Mineola, New York.

Wagoner, J., and Baker, N. A. (2004). Solvation forces on biomolecular structures: A comparison of explicit solvent and Poisson-Boltzmann models. *J. Comput. Chem.* **25,** 1623–1629.

Wagoner, J. A., and Baker, N. A. (2006). Assessing implicit models for nonpolar mean solvation forces: The importance of dispersion and volume terms. *Proc. Natl. Acad. Sci. USA* **103,** 8331–8336.

Wang, J., Arbuzova, A., Hangyas-Mihalyne, G., and McLaughlin, S. (2001a). The effector domain of myristoylated alanine-rich C kinase substrate (MARCKS) binds strongly to phosphatidylinositol 4,5-bisphosphate (PIP2). *J. Biol. Chem.* **276,** 5012–5019.

Wang, J., Gambhir, A., Hangyas-Mihalyne, G., Murray, D., Golebiewska, U., and McLaughlin, S. (2002). Lateral sequestration of phosphatidylinositol 4,5-bisphosphate by the basic effector domain of myristoylated alanine-rich C kinase substrate is due to nonspecific electrostatic interactions. *J. Biol. Chem.* **277,** 34401–34412.

Wang, J., Gambhir, A., McLaughlin, S., and Murray, D. (2004). A computational model for the electrostatic sequestration of PI(4,5)P2 by membrane-adsorbed basic peptides. *Biophys. J.* **86,** 1969–1986.

Wang, T., and Wade, R. C. (2003). Implicit solvent models for flexible protein-protein docking by molecular dynamics simulation. *Proteins* **50,** 158–169.

Wang, W., Donini, O., Reyes, C. M., and Kollman, P. A. (2001b). Biomolecular simulations: Recent development in force fields, simulationa of enzyme catalysis, protein-ligand, protein-protein, and protein-nucleic acid noncovalent interactions. *Annu. Rev. Biophys. Biomol. Struct.* **30,** 211–243.

Warwicker, J., and Watson, H. C. (1982). Calculation of the electric potential in the active site cleft due to alphs-helix dipoles. *J. Mol. Biol.* **157,** 671–679.

Wesson, L., and Eisenberg, D. (1992). Atomic solvation parameters applied to molecular dynamics of proteins in solution. *Protein Sci.* **1,** 227–235.

Wong, C. F., Hünenberger, P. H., Akamine, P., Narayana, N., Diller, T., McCammon, J. A., Taylor, S., and Xuong, N.-H. (2001). Computational analysis of PKA-balanol interactions. *J. Med. Chem.* **44,** 1530–1539.

Xu, D., Lin, S. L., and Nussinov, R. (1997). Protein binding versus protein folding: The role of hydrophilic bridges in protein associations. *J. Mol. Biol.* **265,** 68–84.

Yin, H. L., and Janmey, P. A. (2003). Phosphoinositide regulation of the actin cytoskeleton. *Annu. Rev. Physiol.* **65,** 761–789.

Yu, Z., Jacobson, M. P., Josovitz, J., Rapp, C. S., and Friesner, R. A. (2004). First-shell solvation of ion pairs: Correction of systematic errors in implicit solvent models. *J. Phys. Chem. B* **108,** 6643–6654.

Zauhar, R. J., and Morgan, R. J. (1988). The rigorous computation of the molecular electric potential. *J. Comput. Chem.* **9,** 171–187.

Zhou, H. X. (2005). Interactions of macromolecules with salt ions: An electrostatic theory for the Hofmeister effect. *Proteins* **61,** 69–78.

Zhou, H. X., and Dong, F. (2003). Electrostatic contributions to the stability of a thermophilic cold shock protein. *Biophys. J.* **84,** 2216–2222.

Zhu, Z. Y., and Karlin, S. (1996). Clusters of charged residues in protein three-dimensional structures. *Proc. Natl. Acad. Sci. USA* **93,** 8350–8355.

CHAPTER 27

Ligand Effects on the Protein Ensemble: Unifying the Descriptions of Ligand Binding, Local Conformational Fluctuations, and Protein Stability

Steven T. Whitten,*[,†] Bertrand E. García-Moreno,[‡] and Vincent J. Hilser*

*Department of Biochemistry and Molecular Biology
University of Texas Medical Branch
Galveston, Texas 77555–1068

[†]RedStorm Scientific, Inc.
Galveston, Texas 77550

[‡]Department of Biophysics
The Johns Hopkins University
Baltimore, Maryland 21218

0091-679X/08 $35.00
DOI: 10.1016/S0091-679X(07)84027-1

Abstract

Detailed description of the structural and physical basis of allostery, cooperativity, and other manifestations of long-range communication between binding sites in proteins remains elusive. Here we describe an ensemble-based structural-thermodynamic model capable of treating explicitly the coupling between ligand binding reactions, local fluctuations in structure, and global conformational transitions. The H^+ binding reactions of *staphylococcal* nuclease and the effects of pH on its stability were used to illustrate the properties of proteins that can be described quantitatively with this model. Each microstate in the native ensemble was modeled to have dual structural character; some regions were treated as folded and retained the same atomic geometry as in the crystallographic structure while other regions were treated thermodynamically as if they were unfolded. Two sets of pK_a values were used to describe the affinity of each H^+ binding site. One set, calculated with a standard continuum electrostatics method, describes H^+ binding to sites in folded parts of the protein. A second set of pK_a values, obtained from model compounds in water, was used to describe H^+ binding to sites in unfolded regions. An empirical free energy function, parameterized to reproduce folding thermodynamics measured by differential scanning calorimetry, was used to calculate the probability of each microstate. The effects of pH on the distribution of microstates were determined by the H^+ binding properties of each microstate. The validity of the calculations was established by comparison with a number of different experimental observables.

I. Introduction

The ability of many proteins to couple structural rearrangements to the binding or release of ligands and cofactors is central to their biological function. This coupling is the basis of allostery, cooperativity, and of most energy transduction processes. Previous attempts to understand the physical basis of ligand-linked structural transitions have focused on large-scale cooperative transitions between different macroscopic states. This is epitomized by the T to R transition used to describe the energetics of ligand binding to hemoglobin (Wyman, 1964). In the prevailing, classical view of cooperativity, a fixed number of conformational states are considered, each with a different affinity for ligands. The differential ligand binding properties of the states can be used to describe ligand-driven conformational transitions rigorously. The problem with these models is that they contribute no mechanistic insight into the structural determinants of allostery, cooperativity, and all other manifestations of communication between distant sites in proteins.

Hydrogen exchange and relaxation experiments based on NMR spectroscopy demonstrate that proteins are continuously undergoing small-scale conformational fluctuations that involve more or less independent movements of the different structural regions of the protein (Englander, 2000; and references within). An immediate consequence of these fluctuations is that each macroscopic state of a protein is best described as an ensemble of many interconverting microstates. Despite this observation, little is known about the structural and thermodynamic character of the local fluctuations, how they govern the energetics and other solution properties of proteins, how they are coupled to larger-scale fluctuations of proteins, or of their relationship to structural cooperativity, allostery, energy transduction, and biological function in general. Here we use the proton (H^+) binding properties of proteins to demonstrate the capabilities of an ensemble model that addresses these issues quantitatively.

H^+ binding reactions of proteins are ideally suited for studying the coupling between ligand binding equilibria, microscopic fluctuations, and macroscopic conformational transitions for several reasons: (1) Global H^+ binding reactions can be measured for proteins with potentiometric methods (Nozaki and Tanford, 1967b; Roxby and Tanford, 1971). (2) H^+ binding reactions can also be measured for individual sites by NMR spectroscopy (Alexandrescu *et al.*, 1988; Oliveberg *et al.*, 1995; Tan *et al.*, 1995). (3) The affinity for H^+ at each titratable site is related to the electrostatic potential, which can be calculated with continuum methods based on classical electrostatics theory. The molecular determinants of pK_a values of the titratable groups in proteins are relatively well understood, and the changes in pK_a can be related to changes in structure (Antosiewicz *et al.*, 1994, 1996; Honig and Nicholls, 1995; Yang and Honig, 1993, 1994). (4) H^+ binding reactions are exquisitely sensitive to the microenvironment of the H^+ binding sites. In general, the pK_a of ionizable groups in ordered and disordered parts of proteins tend to be very different. Therefore, each H^+ binding site acts as a probe of local and global stability and conformation. (5) H^+ binding is coupled to the global unfolding of proteins. The pH dependence of stability and the acid unfolding profile of a protein reflect net differential H^+ binding between native and nonnative conformers. Because of these properties, H^+ binding reactions can be used to relate site-specific ligand binding, local fluctuations, and global structural transitions. To this end, a structural thermodynamic model of proteins is needed that can describe the population of microstates of a protein and the effects of pH on the distribution of microstates of the ensemble.

According to the ergodic hypothesis, the time-averaged behavior of a single protein molecule is equivalent to the instantaneous average of a large ensemble of protein molecules (Hill, 1960). Local conformational fluctuations about the folded protein result in an ensemble of microstates. The observed solution behavior of the ensemble is the energy-weighted contribution of all conformational microstates that are populated under a condition specified by pressure, temperature, pH, and other physical environmental variables. The measured value of any extensive physical property, $\langle a \rangle$, is given by the sum of the individual contributions of all the micostates that comprise the native state ensemble

$$\langle a \rangle = \sum_i a_i \times P_i \tag{1}$$

where a_i is the extensive physical property of state i and P_i the probability of state i. Under equilibrium solution conditions, the probability of populating any microstate i is given by

$$P_i = \frac{K_i}{Q}. \tag{2}$$

The statistical weight, K_i, is defined as the relative Gibbs free energy, ΔG_i, of each microstate ($K_i = \mathrm{e}^{-\Delta G_i RT}$, where R is the gas constant and T is absolute temperature). The partition function, Q, is defined as the sum of the statistical weights of all microstates in the ensemble ($Q = \sum K_i$).

The central problem with ensemble-based calculations is with the treatment of the ensemble. In our model, the ensemble of states was generated with a structure-based algorithm known as COREX (Hilser and Freire, 1996). In this algorithm, the native state of a protein is treated as an ensemble representing the different conformers of the protein. The set of microstates that constitute the ensemble are generated from a high-resolution crystal structure. They differ in the extent and location of regions that retain the native fold and that are treated thermodynamically as if they were unfolded. In this chapter, we show how the ensemble model has been modified to account for the effects of ligands on the conformational ensemble. The results of the calculations suggest that this model will contribute novel insight about how conformational fluctuations are reflected in solution properties of protein.

To test the ability of the model to describe the protein ensemble accurately, the original COREX algorithm was used initially to describe site-specific hydrogen exchange data measured by NMR spectroscopy (Hilser, 2001; Hilser and Freire, 1996, 1997; Hilser *et al.*, 1997, 1998). Previous studies have shown that COREX calculations successfully reproduce a variety of other solution properties as well (Babu *et al.*, 2004; Liu *et al.*, 2006; Pan *et al.*, 2000; Pometun *et al.*, 2006; Whitten *et al.*, 2005, 2006; Wooll *et al.*, 2000). The success of the prior calculations suggests that contributions from the small conformational excursions that occur even under native conditions can dominate the solution properties of proteins. Interest in expanding the algorithm to account for pH-dependent processes in proteins was driven partly by the wealth of interesting biological processes that are governed physiologically by changes in pH and by the need to explain the structural and physical bases of these effects. Many proteins have evolved to harness small differences in pH to trigger physiologically important conformational changes. The cellular trafficking of hematopoietins (French *et al.*, 1995; Sarkar *et al.*, 2002, 2003), the regulation of oxygen affinity by hemoglobin (Ackers, 1998), the activation of many bacterial toxins (Ren *et al.*, 1999), and key steps in the infectivity cycles of viruses as diverse as polio (Hogle, 2002), influenza (Baker and Agard, 1994; Bullough *et al.*, 1994; Gamblin *et al.*, 2004), and HIV (Ehrlich *et al.*, 2001) are all examples of physiologically relevant processes where function is regulated through pH-triggered structural transitions. Another reason that the model was expanded to account for the effects of ligands (i.e., H^+) on the distribution of microstates in the ensemble was the possibility it offered to account quantitatively for the consequences of structural fluctuations on electrostatic properties of proteins.

There are several structure-based, continuum electrostatics methods for calculation of pK_a values and H^+ binding energetics of proteins (Antosiewicz *et al.*, 1994, 1996; Honig and Nicholls, 1995; Yang and Honig, 1993, 1994). These methods, based on classical electrostatic theory and statistical thermodynamics, have been invaluable for dissecting the factors that determine the pK_a values of proteins. It is well established that they are not yet useful to describe H^+-driven conformational transitions of proteins (Fitch *et al.*, 2006) and that the central failure of these algorithms is related to their inability to describe quantitatively the manner in which structural dynamics contribute to the H^+ binding properties of proteins. Past studies have shown that continuum calculations can yield pK_a values consistent with those measured experimentally under solution conditions or in cases in which local conformational reorganization is not an important determinant of the H^+ binding reaction (Antosiewicz *et al.*, 1994, 1996; Honig and Nicholls, 1995; Yang and Honig, 1993, 1994). The formalism described ahead affords a novel way of accounting for dynamic effects in H^+ binding reactions without having to study the dynamics proper. Instead, the focus is on realistic and quantitative descriptions of the effects of H^+ on the distribution of microstates in the protein ensemble.

II. The Effect of pH on the Conformational Ensemble

A. Thermodynamic Model of the Conformational Ensemble

Staphylococcal nuclease (SNase) was selected for these studies because the experimental thermodynamics of acid unfolding and the energetics of ionization of its titratable residues have been characterized previously (Fitch *et al.*, 2006; García-Moreno *et al.*, 1997; Lee *et al.*, 2002a,b; Meeker *et al.*, 1996; Whitten, 1999; Whitten and García-Moreno, 2000). The crystallographic structure of SNase (1stn.pdb; Hynes and Fox, 1991) was used as a template to generate computationally a conformational ensemble using the COREX algorithm (Hilser and Freire, 1996). The COREX algorithm generates a large number of different conformers through the combinatorial unfolding of a set of predefined folding units (Hilser and Freire, 1996, 1997; Hilser *et al.*, 1997, 1998, 2001). Overlay of the folding units on the protein is independent of secondary structure content and includes all possible combinations of one or more folding units applied to the protein structure. By means of a register shift in the boundaries of the folding units, an exhaustive enumeration of partially unfolded states is achieved. In the calculations with SNase, a folding unit size of eight residues was used. The ensemble thus generated consists of 1,179,629 partially folded microstates. Each microstate has some regions that are folded and some regions that are treated thermodynamically as if they were unfolded. The folded regions retain the conformation of the native state, whereas the unfolded regions are treated structurally and thermodynamically as unfolded polypeptides (see Fig. 1A).

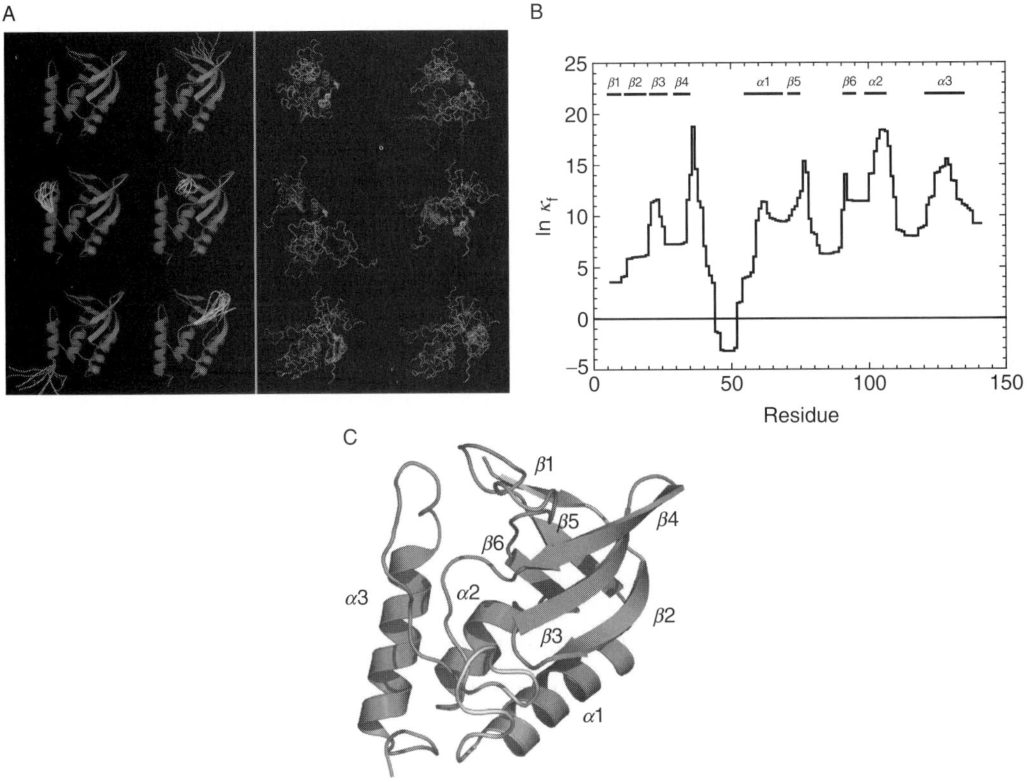

Fig. 1 The SNase conformational ensemble as calculated by the COREX algorithm. (A) Shown is a sample of 12 (out of more than 10^6) of the more probable conformational states. Ensemble states displayed on the left were chosen among those that retained most of the native structure, whereas those on the right were chosen among the mostly unfolded states. Segments of SNase colored red indicate regions of protein that would be folded; regions colored yellow would be unfolded. To illustrate the concept that the unfolded segments freely sample accessible conformational space, these regions were modeled in the figure as multiple flexible loops. (B) Natural logarithm of the calculated residue stability constants, κ_f, for SNase. The locations of the secondary structural elements are as indicated. (C) Ensemble-averaged or "single-molecule" representation of the residue stability constants projected onto the native structure of SNase. Residues shown in red are calculated as stable, whereas residues shown in yellow are calculated to have the least stability. If a single molecule were to be observed over a sufficient period of time, regions colored yellow would be observed to fluctuate to a greater extent than those colored red. All molecular diagrams in this chapter were made using the program PyMol (DeLano, 2002). (See Plate no. 26 in the Color Plate Section.)

To describe the ensemble quantitatively it is necessary to identify the microstates that are populated under any given set of conditions. This requires the calculation of the relative Gibbs free energy for each microstate i of the ensemble ΔG_i. The ΔG_i values were calculated with empirical functions that reproduce folding thermodynamic parameters measured with differential scanning calorimetry (ΔCp_i, ΔH_i, and ΔS_i) through parameterization based on solvent accessible surface areas (Baldwin, 1986; D'Aquino *et al.*, 1996; Gómez *et al.*, 1995; Hebermann and Murphy, 1996; Lee *et al.*, 1994; Luque *et al.*, 1996; Murphy and Freire, 1992; Murphy *et al.*, 1992; Xie and Freire, 1994). The values calculated for ΔG_i were substituted into Eq. (2) to calculate the probability of each microstate and consequently, the conformational partition function. Figure 1A represents some of the more probable conformational microstates calculated for SNase by the COREX algorithm. For illustrative purposes, microstates in this figure were purposely selected among those that were either mostly folded or mostly unfolded.

The propensity of any region in a protein to undergo structural fluctuations can be determined from the microstate probabilities [Eq. (2)]. Defined as the residue stability constant, κ_f, the local stability of each residue for the native fold is calculated as the ratio of the summed probabilities of all microstates in the ensemble in which a particular residue j is in a folded conformation, $P_{f,j}$, to the summed probability of all microstates in which residue j is in an unfolded (or nonfolded) conformation, $P_{nf,j}$

$$\kappa_{f,j} = \frac{P_{f,j}}{P_{nf,j}} \quad (3)$$

The residue stability constants for SNase are shown in Fig. 1B. High stability constants identify residues that are folded in the majority of the most probable microstates, whereas lower stability constants identify residues that are unfolded in many of those microstates. In SNase, most residues with higher stability constants are located in β-strands 3–6 and in the α-helices. Lower stability constants are found in residues in β-strands 1 and 2, and, most dramatically, the loop connecting β-strand 4 to α-helix 1. It has been shown previously, with SNase and many other proteins, that the regional distribution of stability throughout a protein structure determined from Eq. (3) correlates well with experimental hydrogen-deuterium exchange data (Hilser, 2001; Hilser and Freire, 1996, 1997; Hilser *et al.*, 1997, 1998).

A "single-molecule" representation of the ensemble of SNase is shown in Fig. 1C. In this figure, the native structure is color coded according to the probability of finding any particular residue in the native fold. Regions colored in red are calculated to be the most stable ones (i.e., these residues possess the highest stability constants) and, if it were possible to view a single molecule of SNase over a long period of time, these residues would be the ones that would fluctuate the least. In contrast, residues colored in yellow would be observed to fluctuate (i.e., unfold) most often.

B. pH Modulation of the Conformational Ensemble

The theory of linked functions explains that the effect of increasing concentration of ligand is to stabilize those states of the ensemble which have greater affinity for that ligand (Wyman, 1948, 1964). In terms of the effects of pH on the ensemble, a decrease in solution pH will stabilize those microstates with net higher pK_a values because the H^+ binding sites can bind H^+. Conversely, an increase in solution pH will stabilize the states with net lower pK_a values. This is the effect that we aim to reproduce by introducing H^+ binding terms explicitly into the ensemble calculations.

To account for the effects of pH on the equilibrium distribution of microstates in the ensemble, it is necessary to treat the H^+ binding properties of ionizable groups with two different sets of pK_a values. One set describes the H^+ binding properties of the ionizable groups in structured, native-like regions of any ensemble state. For purposes of our calculations, the pK_a values in this set were estimated using continuum electrostatic methods based on the finite difference (FD) solution of the linearized Poisson-Boltzmann equation (Antosiewicz *et al.*, 1994, 1996). Details of FD calculations with SNase using the University of Houston Brownian Dynamics package for these specific calculations have been presented elsewhere (Fitch *et al.*, 2002). The pK_a values calculated with this method are provided in Table I ($pK_{a,FD}$). A second set of pK_a values was used to describe the titration of sites that were in unfolded parts of the protein. This second set consisted of the pK_a values measured experimentally in model compounds in water (Matthew *et al.*, 1985; Schaefer *et al.*, 1998) and are also listed in Table I ($pK_{a,nf}$).

For each state of the ensemble, titratable residues in unfolded regions were assigned the model compound pK_a values ($pK_{a,nf}$). Similarly, the ionizable groups in the N- and C-terminal regions of the protein, which are not resolved in any known crystal structure of SNase, were assigned $pK_{a,nf}$ values. Titratable groups in folded regions were assigned pK_a values based on the solvent accessibility of the titratable atoms. This was done to correct for the case in which a residue may reside in a folded region, however, due to the unfolding of segments of the protein that pack against the titratable site, the ionizable group is exposed to solvent. To apply this correction term in an entirely general manner, a cutoff threshold was determined by comparison of calculated and measured H^+ binding curves for the entire protein (see Fig. 2B. The H^+ binding curve of the ensemble is

$$\langle Z(\text{pH})\rangle = \sum_i Z(\text{pH})_i \times P(\text{pH})_i \tag{4}$$

where $Z(\text{pH})_i$ is the number of H^+ bound to microstate i as a function of pH. $P\,(\text{pH})_i$ is the pH-dependent probability of each microstate determined by adding the binding polynomial (Wyman and Gill, 1990) to Eq. (2)

$$P(\text{pH})_i = \frac{\prod_j (1 + 10^{(pK_{a,i,j} - \text{pH})})K_i}{\sum_i \left(\prod_j (1 + 10^{(pK_{a,i,j} - \text{pH})})K_i \right)} \tag{5}$$

Table I
pK_a Values of the Titratable Groups of SNase Used in the Ensemble Based Numerical Calculations

Residue	$pK^{WT}_{a,FD}$	$pK^{PHS}_{a,FD}$	$pK_{a,nf}$[a]	$pK_{a,GdnHCl}$[b]		$pK^{WT}_{a,FD}$	$pK^{PHS}_{a,FD}$	$pK_{a,nf}$[a]	$pK_{a,GdnHCl}$[b]
N terminus	7.40*	7.40*	7.40	7.60	Lys-5	10.40*	10.40*	10.40	10.60
C terminus	3.50*	3.50*	3.50	3.40	Lys-6	10.44	10.40*	10.40	10.60
					Lys-9	11.94	12.24	10.40	10.60
Glu-10	2.57	4.10	4.50	4.38	Lys-16	9.97	10.10	10.40	10.60
Glu-43	5.09	4.83	4.50	4.38	Lys-24	10.24	9.80	10.40	10.60
Glu-52	2.13	2.61	4.50	4.38	Lys-28	11.00	11.02	10.40	10.60
Glu-57	3.67	3.76	4.50	4.38	Lys-45	11.55	11.56	10.40	10.60
Glu-67	2.89	2.85	4.50	4.38	Lys-48	10.56	10.37	10.40	10.60
Glu-73	3.70	3.82	4.50	4.38	Lys-49	11.35	11.37	10.40	10.60
Glu-75	1.29	3.13	4.50	4.38	Lys-53	11.77	10.90	10.40	10.60
Glu-101	2.25	2.80	4.50	4.38	Lys-63	11.48	11.20	10.40	10.60
Glu-122	2.85	2.58	4.50	4.38	Lys-64	10.77	10.28	10.40	10.60
Glu-129	1.21	2.47	4.50	4.38	Lys-70	10.66	10.35	10.40	10.60
Glu-135	3.30	2.85	4.50	4.38	Lys-71	10.78	11.77	10.40	10.60
Glu-142	4.50*	4.50*	4.50	4.38	Lys-78	10.52	11.34	10.40	10.60
					Lys-84	11.54	11.42	10.40	10.60
Asp-19	2.47	2.39	4.00	3.88	Lys-97	11.06	10.88	10.40	10.60
Asp-21	0.52	1.27	4.00	3.88	Lys-110	11.05	11.98	10.40	10.60
Asp-40	2.08	1.98	4.00	3.88	Lys-116	10.98	10.99	10.40	10.60
Asp-77	2.36	2.38	4.00	3.88	Lys-127	10.18	10.19	10.40	10.60
Asp-83	1.33	0.91	4.00	3.88	Lys-133	11.71	11.17	10.40	10.60
Asp-95	2.50	1.60	4.00	3.88	Lys-134	10.25	10.32	10.40	10.60
Asp-143	4.00	4.00*	4.00	3.88	Lys-136	10.67	10.82	10.40	10.60
Asp-146	4.00	4.00*	4.00	3.88					
					Arg-35	16.32	17.10	12.00	12.50
His-8	6.35	6.47	6.50	6.83	Arg-81	13.61	14.23	12.00	12.50
His-46	5.43	5.93	6.50	6.83	Arg-87	14.58	14.44	12.00	12.50
His-121	5.18	5.83	6.50	6.83	Arg-105	13.80	13.60	12.00	12.50
His-124	5.96	-ALA-	6.50	6.83	Arg-126	14.00	13.95	12.00	12.50
									9.80
Tyr-27	12.44	13.36	10.00	9.80	Tyr-93	14.71	13.12	10.00	9.80
Tyr-54	9.56	10.13	10.00	9.80	Tyr-113	9.66	9.74	10.00	9.80
Tyr-85	8.28	8.73	10.00	9.80	Tyr-115	10.18	9.10	10.00	
Tyr-91	13.44	14.05	10.00	9.80					

[a] $pK_{a,nf}$ values based on solvent exposed model compounds (Schaefer *et al.*, 1998; Matthew *et al.*, 1985).
[b] $pK_{a,GdnHCl}$ values based on solvent exposed model compounds in 6-M GdnHCl (Whitten and García-Moreno, 2000; Roxby and Tanford, 1971; Nozaki and Tanford, 1967a).
$pK_{a,FD}$ values were calculated by finite difference solution of the linearized Poisson–Boltzmann equation performed on the crystallographic structures of WT and PHS SNase (Fitch *et al.*, 2002).
*Residue is not observed in crystallographic structure.

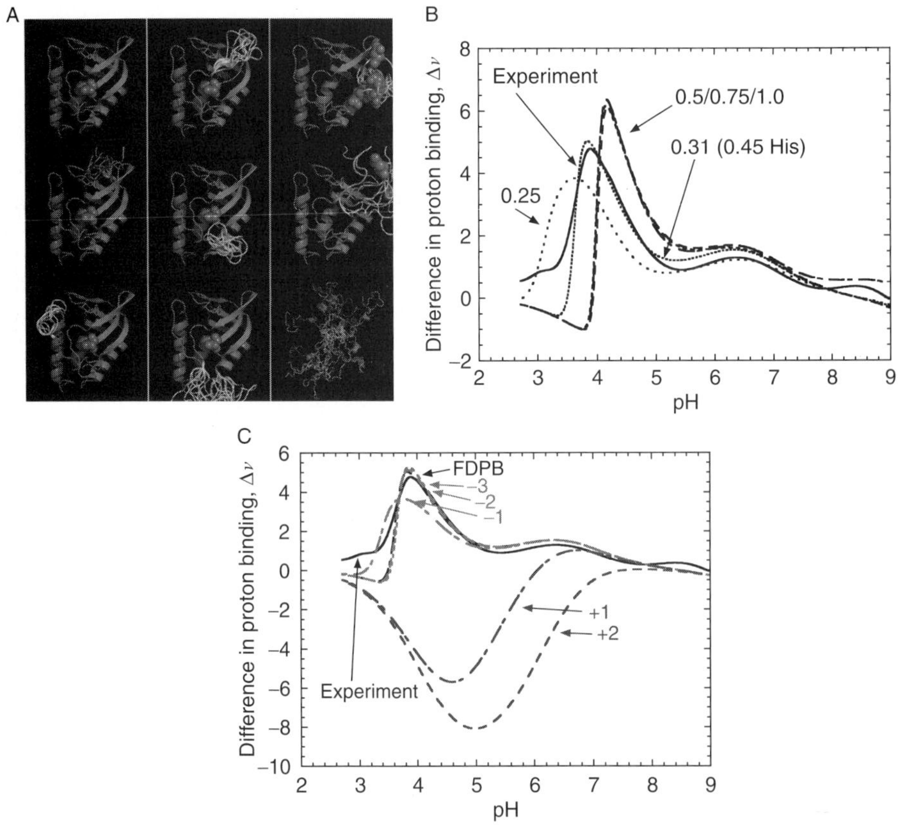

Fig. 2 Model used to set the pH-linked contribution to stability of the SNase ensemble. (A) Determination of the state-specific pK_a values for the individual titratable groups. As is Fig. 1, segments of SNase colored red indicate regions of protein that would be folded; regions colored yellow would be unfolded and are drawn in this figure as multiple flexible loops. Using aspartic residue 21 (D21, shown in gray) as an example, the SNase ensemble can be divided into three distinct subensembles based on the pK_a value given to D21. In all states of the first subensemble, D21 is both folded and protected from solvent and, per the model, titrates with the $pK_{a,FD}$ value listed in Table I. In the second subensemble, D21 remains in a folded region, however, due to unfolding of segments of SNase that pack against the side chain of D21, the titratable atoms of D21 are exposed to solvent. In these states, D21 titrates with the same pK_a as fully exposed model compounds ($pK_{a,nf}$ of Table I). In states of the third subensemble, D21 is unfolded and exposed to solvent. In these states, D21 also titrates with the same pK_a as fully exposed model compounds. (B) The effect of different exposure thresholds on the $\Delta\nu$ values calculated for the ensemble. Shown by the solid line is the experimental difference in H^+ binding, $\Delta\nu$, between native and denatured (6-M GdnHCl) WT SNase measured by continuous potentiometric techniques (Whitten and García-Moreno, 2000). The ensemble $\Delta\nu$ calculated with a threshold value of 0.31 applied to all residues except His (where 0.45 was used) is given by the line of small closely spaced dashes. The ensemble $\Delta\nu$ curves calculated for a series of other threshold values (0.25, 0.50, 0.75, and 1.0) applied to all titratable groups (no distinction made for the His) are shown as indicated in the plot. (C) The effect of using different sets of native pK_a values on the calculated ensemble $\Delta\nu$. Shown by the solid line is the experimental $\Delta\nu$ between native and denatured (6-M GdnHCl) WT SNase (Whitten and García-Moreno, 2000). The ensemble $\Delta\nu$ calculated using the $pK_{a,FD}$ values given in Table I is represented by

where $pK_{a,i,j}$ is the pK_a value of residue *j* in microstate *i*. The pK_a assignment rules are shown schematically in Fig. 2A.

Solvent accessible surface area calculations were performed on each titratable atom using the Lee and Richards algorithm (Lee and Richards, 1971). The solvent accessible surface area of the atom was normalized by the surface area for the atom type in models of fully exposed groups. If the normalized accessibility was greater than a threshold value the atom was considered to be fully solvent exposed and it was assigned a $pK_{a,nf}$ value. Otherwise, the atom was considered to be protected and it was assigned a $pK_{a,FD}$ value. For the calculations presented here, the threshold percentage was set at 0.31 for all titratable atom types except His residues, which were set at 0.45. The solvent accessibilities of the OE1 and OE2 atoms were averaged for Glu residues, the OD1 and OD2 atomic solvent accessibilities were averaged for Asp, and the NH1 and NH2 solvent accessibilities were averaged for Arg residues. The solvent accessibility of the NE2 atom was used for His, the NZ for Lys, and the OH for Tyr. There are no Cys residues in SNase. A separate threshold was determined for His to eliminate possible bias due to the fact that its titratable atom resides within a ring structure whereas the titratable atoms of all other residues do not.

The effect of altering the $pK_{a,FD}$ values on the ensemble simulation is shown in Fig. 2C. These data demonstrate that the calculated titration behavior of the SNase ensemble is not very sensitive to the absolute value of the $pK_{a,FD}$ values. In the case of SNase, the most important factor determining the outcome of the calculations was whether the pK_a value was depressed or elevated in the native state. Thus, at least for the case of SNase, fine-tuning of the energy functions employed by the calculation of $pK_{a,FD}$ values (Fitch *et al.*, 2006), or how these values are applied to the ensemble, do not affect the results of the calculations. This will not necessarily be the case for all proteins.

III. Results and Discussion

A. pH Dependence of the Ensemble of SNase

The calculated H^+ titration properties of the SNase ensemble can be seen in Fig. 3A. For comparison to the experimental titration (Whitten and García-Moreno, 2000), the calculated pH-dependent differences in H^+ binding ($\Delta \nu$) between the ensemble and a fully unfolded state in 6-M GdnHCl

the dashed black line. The series of red dashed lines were calculated using native pK_a values for Glu and Asp residues depressed by 1–3 pH units relative to the intrinsic pK_a of Glu (4.5) and Asp (4.0). The series of blue dashed lines were calculated using native pK_a values for Glu and Asp elevated by 1–2 pH units relative to the intrinsic pK_a of Glu and Asp. For each calculated curve in this plot, all His, Lys, Arg, and Tyr residues were given the $pK_{a,FD}$ values listed in Table I for their native pK_a. There are no Cys residues in WT SNase. (See Plate no. 27 in the Color Plate Section.)

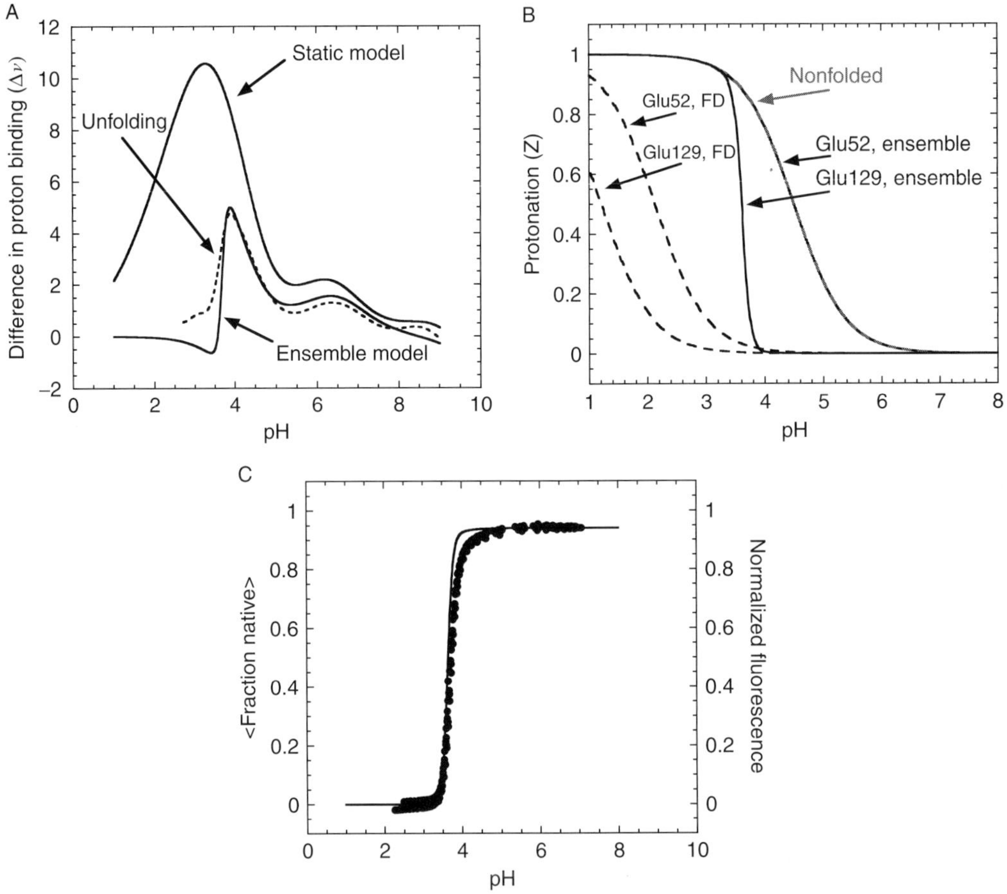

Fig. 3 H^+ binding properties of the WT SNase ensemble. (A) The calculated difference in H^+ binding, $\Delta\nu$, between GdnHCl unfolded SNase ($pK_{a,GdnHCl}$ of Table I) and the ensemble is shown by the solid line. The dashed line is the experimental $\Delta\nu$ (Whitten and García-Moreno, 2000). The line marked "Static Model" is the difference between the GdnHCl unfolded state and a fully folded and fully protected state represented by the $pK_{a,FD}$ values. (B) H^+ titration of residues Glu-52 and Glu-129 as determined by the $pK_{a,FD}$ values are shown by the black dashed lines. The black solid lines represent the titration of these two residues in the SNase ensemble. Titration as determined by the $pK_{a,nf}$ value of Glu is given by the red dashed line. (C) Experimental unfolding of WT SNase (solid circles) as monitored by pH titration of the intrinsic fluorescence of Trp-140 (Whitten and García-Moreno, 2000). The lines overlaid on the experimental data represent the calculated unfolding of the SNase ensemble as determined by $<Fraction\ Native> = \sum Fraction\ Native_i \cdot P(\mathrm{pH})_i$. The *Fraction Native* value of each state *i* was calculated as the number of residues in folded segments divided by the total number of residues observed in the high-resolution structure.

(Nozaki and Tanford, 1967a; Roxby and Tanford, 1971) are also shown. The difference binding curve calculated for a "static" model, which considers the high-resolution structure of SNase as the only contributing species to the native state titration, and in which all ionizable groups titrate with $pK_{a,FD}$ values is also provided. These data illustrate how the thermodynamic character of the ensemble shifts from one in which the H^+ binding properties are similar but not identical to those of the fully folded state at near-neutral pH values to one with H^+ binding behavior identical to those of a fully unfolded state in the extreme acid range of pH. This shift is evidenced by the calculated difference binding curve reaching zero at pH values that are low, yet well above the native state pK_a values ($pK_{a,FD}$) of many of the carboxylic residues listed in Table I. The transition to the unfolded state is highly cooperative and occurs within a small range of pH, ~1 pH unit. Furthermore, all titratable residues protonate in the ensemble at pH values either greater than or identical to the pH at the midpoint of the unfolding transition ($pH_{mid} \approx 3.7$). The titratable residues do not all reside in the same region, but are distributed throughout the protein structure, consistent with a global unfolding induced by acid.

The agreement between the H^+ binding curve calculated with the ensemble-based method and the experimental curve suggests that H^+ binding reactions are tightly coupled to local fluctuations under native conditions. To appreciate the significance of this agreement, note that in contrast, the H^+ uptake on acid unfolding that was calculated from electrostatic considerations alone, using the $pK_{a,FD}$ values calculated with the FD methods applied to the static crystal structure of the fully native state, overestimates significantly the number of H^+ that are bound on acid unfolding (Fig. 3A). Obviously, static models are not meant to capture large-scale structural transitions. However, note the difference in H^+ binding between the ensemble and the static structure at pH values above 4, prior to any acid-induced global unfolding. This difference reflects contributions from fluctuations of the native state to the pH-dependent stability of SNase. Using Wyman's theory of linked functions applied to H^+ binding (Wyman and Gill, 1990),

$$\Delta G(\mathrm{pH}) = -RT \int \Delta v(\mathrm{pH}) \partial \mathrm{pH}, \tag{6}$$

the energetic separation between the ensemble and the static, fully folded structure is significant, equivalent to 5 kcal/mol as the pH is lowered from 7 to 4 (data not shown).

The H^+ titration curves in Fig. 3B illustrate how the ensemble-modulated H^+ binding curves are quite different from those calculated directly with $pK_{a,FD}$ values. The titration of Glu-52 in the ensemble demonstrates the consequences of local unfolding under native conditions on the apparent pK_a of a residue. Glu-52 resides in the loop that connects β-strand 4 to α-helix 1, which is predicted to be unfolded in most of the highly probable microstates of the ensemble (see Fig. 1). Because Glu-52 is unfolded in most of the highly probable states, this residue titrates in the ensemble with the same pK_a value of a model compound in water. In contrast, its calculated $pK_{a,FD}$ is 2.1, consistent with strong attractive electrostatic interactions in the native state. The electrostatic calculations with the static structure identify Glu-52 as a group that would contribute to the acid unfolding energetics of SNase by binding H^+ on unfolding. In contrast, in the ensemble-modulated calculations its H^+ binding properties are the same as in a fully unfolded state, suggesting that this Glu residue does not contribute to the pH-dependent stability and acid unfolding of SNase.

Even residues predicted by the ensemble-based calculations to be active in acid unfolding (i.e., those that bind H^+ on unfolding) can have distinctly different titration behavior in the ensemble-modulated calculations than in the calculations with the static structure. This is shown in Fig. 3B by the H^+ titration of Glu-129. Protonation of Glu-129 is coupled to the highly cooperative acid-induced unfolding of SNase. This suggests that the actual pH in which titration of Glu-129 occurs is dependent more on global energetics rather than local electrostatic descriptors of protein structure.

The modulating influence of pH on the probabilities of each microstate is illustrated further in Fig. 3C, where the relative extent of folding for the ensemble is given for a range of acidic pH values. A global and highly cooperative transition is predicted by the ensemble calculations. This represents the unfolding of SNase at low pH. The midpoint of the acid induced transition predicted with the ensemble calculations agrees quantitatively with midpoint of the transition observed experimentally. There is also agreement between the calculated and the observed steepness of the transition, which is related to the cooperativity and to the net number of H^+ bound preferentially by the acid unfolded state relative to the native state.

B. Microscopic Origins of the pH Dependence of Stability

The difference between the experimental H^+ binding curve and the curve calculated for the static structure of SNase suggests that many carboxylic groups that according to the FD calculations have depressed pK_a values in the native state ($pK_{a,FD}$) instead titrate with normal pK_a comparable to those of model compounds ($pK_{a,nf}$). These groups should not contribute to the overall pH-dependent stability or to the acid unfolding process. The titration of Glu-52 in the ensemble, discussed above, is an example. To more fully investigate the residue-specific contributions to the pH-dependent stability of an ensemble, we introduce a thermodynamic descriptor for each titratable residue, the residue-specific protection constant, κ_p. From a statistical standpoint, the protection constant for any residue j can be defined as the ratio of the summed probability of all states in which the titratable atom(s) for residue j is protected from solvent [i.e., the exposure percentage for the titratable atom(s) of residue j is below the cutoff threshold] to the sum of the probabilities of all states in which the titratable atom(s) for residue j is exposed. The definition can be expressed in terms of the folding probabilities

$$\kappa_{p,j} = \frac{P_{f,j} - P_{f,exposed,j}}{P_{nf,j} + P_{f,exposed,j}}, \tag{7}$$

where $P_{\mathrm{f,exposed}}$ is the summed probability of all states in which residue j is folded, but exposed to solvent. Figure 4A provides the κ_{p} values calculated for the residues of SNase. As can be seen, only a subset of all carboxyl groups that are calculated to have depressed $\mathrm{p}K_{\mathrm{a,FD}}$ also possess high κ_{p} values. These residues are Glu-10, Asp-19, Asp-21, Glu-75, Asp-77, Asp-83, Asp-95, Glu-101, Glu-129, and Glu-135. The ensemble calculations identified these carboxylic groups as the ones likely to contribute significantly to acid unfolding—their H^{+} titrations are thermodynamically coupled to global unfolding.

Residues with high κ_{p} are not clustered in any obvious way (Fig. 4C). This suggests that the pH-dependent stability of this protein is not generated through a network of linked intramolecular electrostatic interactions. Instead, the common characteristic of the carboxylic groups that are predicted to govern the acid denaturation is that they are all in folded regions of the protein in the majority of the highly probable states under native conditions. Experimental verification of the carboxylic groups responsible for the acid-induced unfolding of wild-type (WT) SNase was presented elsewhere (Whitten *et al.*, 2005) and is shown to be consistent with the above calculations.

C. Application to a Hyperstable Variant of Nuclease

To test further the ability of the ensemble model to reproduce the acid unfolding transition of proteins, calculations were also performed with the PHS hyperstable variant of SNase containing the substitutions P117G H124L S128A. The crystal structures of the WT and the PHS mutant are essentially identical, particularly in

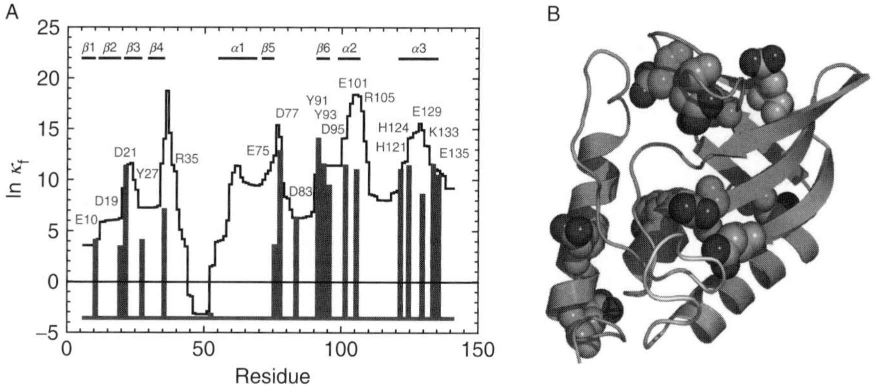

Fig. 4 Key titratable residues in the pH-linked stability of WT SNase. (A) The natural logarithm of the residue stability constants (black lines), κ_{f}, and residue protection constants (blue lines), κ_{p}, calculated for the SNase ensemble. (B) The residues in the diagramatic representation of the SNase structure are color coded based on the calculated κ_{f} values, as described in Fig. 1C (residues with high κ_{f} are red; residues with lower κ_{f} are yellow), except for the titratable atoms of Glu and Asp residues that are calculated to have high κ_{p} values, which are colored blue. (See Plate no. 28 in the Color Plate Section.)

regions where the substituted residues are found; the rmsd between the structures is 0.7 Å. At pH 7, the stability of the PHS variant measured with GdnHCl denaturation monitored with intrinsic fluorescence was ~3 kcal/mol more stable than the WT protein (Whitten, 1999). The midpoint of the acid unfolding transition ($pH_{mid} \approx 2.9$) is 0.8 pH units lower than the midpoint of the WT protein (Whitten, 1999). The $pK_{a,FD}$ values of all titratable groups of PHS were generated in an identical manner as was performed on the WT, and the same threshold percentage for discriminating between nonexposed and exposed titratable atoms (e.g., 0.31 for all titratable atoms except for those in His where 0.45 was used) were applied to the individual ensemble states (Table I).

The calculated and experimental H^+ titration properties of the PHS ensemble are shown in Fig. 5A. There is reasonable agreement between calculated and experimental curves—the ensemble calculations reproduce correctly the depressed pH midpoint of acid unfolding. Similar to what was observed in the calculations with the WT protein, these data show a distinct shift in the thermodynamic character of the ensemble from H^+ binding properties similar to those of the fully folded protein at near-neutral pH values to H^+ binding behavior identical to that of fully unfolded protein at low pH. Likewise, the calculated, difference H^+ binding curve reaches zero at a pH that is low, yet above the $pK_{a,FD}$ values of many of the carboxylic residues. The transition to the unfolded state is highly cooperative, occurring within a small range of pH and is consistent with a global unfolding induced by acid. The acid-induced denaturation of PHS, as observed by monitoring the pH dependence of the relative extent of folding in the ensemble, can also be seen in the figure. Again, a highly cooperative transition is observed that represents global unfolding. Comparison to the experimentally observed acid unfolding transition suggests that the ensemble calculations reproduce quantitatively the pH at which acid denaturation occurs as well as the cooperativity of the unfolding transition.

Figure 5B shows the κ_p values [Eq. (7)] calculated for the residues of PHS SNase. The behavior in the hyperstable variant is comparable to the WT; only a subset of the carboxyl groups that are calculated to have depressed $pK_{a,FD}$ also have high κ_p values. The residues are Glu-10, Asp-19, Asp-21, Asp-40, Glu-75, Asp-77, Asp-83, Asp-95, Glu-129, and Glu-135. Likewise, the ensemble calculations predict that these carboxylic groups contribute significantly to acid unfolding because their H^+ titrations are thermodynamically coupled to global unfolding.

IV. Summary and Conclusions

The ensemble model that accounts for the effects of pH on the distribution of microstates in the ensemble reproduces the midpoint of the acid-induced unfolding of WT SNase and the PHS hyperstable variant. The ensemble model also reproduces the cooperativity of the acid unfolding transition and identifies a subset of carboxylic groups that contribute to the energetics of acid unfolding by coupling of H^+ binding to the global transition (Whitten *et al.*, 2005). More significantly, the

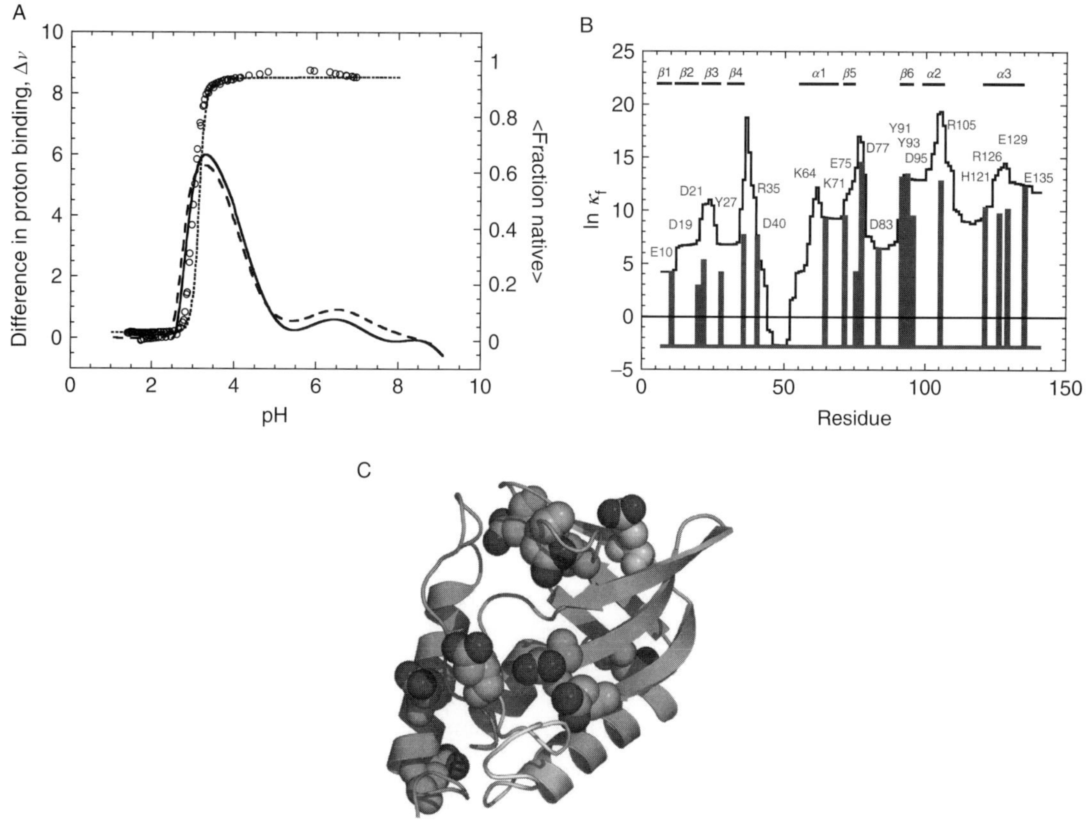

Fig. 5 H^+ binding properties of PHS SNase ensemble. (A) The calculated difference in H^+ binding, $\Delta\nu$, between GdnHCl unfolded PHS SNase ($pK_{a,GdnHCl}$ of Table I) and the PHS ensemble is shown by the line of large dashes. The solid line is the experimental $\Delta\nu$ (Whitten, 1999). The calculated acid-induced unfolding of the PHS ensemble, as determined by the pH dependence of *<Fraction Native>* is given by the line of small, closely spaced dashes. The experimental unfolding of PHS SNase, as monitored by pH titration of the intrinsic fluorescence of Trp-140, is shown by the open circles (Whitten, 1999). (B) The natural logarithm of the residue stability constants (black lines), κ_f, and residue protection constants (blue lines), κ_p, calculated for the PHS ensemble. (C) The residue stability and protection constants projected onto the native structure of PHS SNase. Residues with high κ_f are red; residues with lower κ_f are yellow, except for the titratable atoms of Glu and Asp residues that are calculated to have high κ_p values, which are colored blue. (See Plate no. 29 in the Color Plate Section.)

ensemble model also identifies a second set of carboxylic groups that do not participate in the acid unfolding transition because they titrate with normal pK_a values as a result of their being in parts of the protein that are unfolded in the majority of the microstates of the ensemble. The success of the ensemble-based calculations has several fundamental implications. First, it validates the treatment of native state fluctuations in proteins in terms of local order/disorder transitions. Second, they corroborate the validity of the empirical, parameterized energy function used to calculate the free energy of each microstate in the ensemble. Third, they demonstrate that it is possible to account for the effects of local dynamics on H^+ binding reactions, without having to study dynamics proper. These simulations afford a way of improving the predictive power of continuum based electrostatic calculations by accounting for the heterogeneity of pK_a values among ensemble states.

The COREX algorithm makes two assumptions to enumerate the conformational ensemble: (1) all residual structure is native-like and (2) all unfolded regions are thermodynamically equivalent to unfolded polypeptides. These assumptions are simplistic, but several observations support their validity. First, nonnative nuclear overhauser enhancements (NOEs) are rarely observed in NMR spectroscopy experiments (Englander, 2000; and references within). This implies that the structured portions of the dominant nonnative states are native-like. Second, the COREX calculations reproduce different solution properties, such as hydrogen-deuterium exchange protection factors (Hilser, 2001; Hilser and Freire, 1996, 1997; Hilser *et al.*, 1997, 1998), site-site communication (Liu *et al.*, 2006; Pan *et al.*, 2000), denaturant independent hydrogen-deuterium exchange (Wooll *et al.*, 2000), and noncooperative cold denaturation (Babu *et al.*, 2004; Pometun *et al.*, 2006; Whitten *et al.*, 2006). This suggests that the local unfolding model is indeed a valid thermodynamic approximation when considering regional fluctuations in proteins. The hierarchy of microstates calculated with the ensemble model would not reproduce solution properties of proteins if the conformational ensemble differed significantly in its thermodynamic character. In the specific example of ligand-driven conformational transitions illustrated with the present calculations, each individual ligand binding site is modeled as having one of two possible pK_a values. The pH-dependent properties of the ensemble emerge as a result of both the intrinsic stability of each microstate as well as the net difference in pK_a for all titratable groups in each microstate. The ability of this algorithm to reproduce quantitatively the net number of H^+ bound on acid unfolding for both WT and PHS SNase (see Figs. 3A and 5A) suggests that conformational fluctuations in proteins can be thermodynamically modeled as local unfolding events involving small groups of residues.

It should be noted that the pH-dependent stability of SNase is a somewhat special case, which makes it ideally suited to address the role of local conformational fluctuations in the titration of a protein. The acid unfolding of SNase is one of the most cooperative that has been observed. Furthermore, previous experimental data have shown that the acid denaturation of SNase is driven not by H^+ binding of the native state, but by preferential H^+ binding to the unfolded state (Whitten and García-Moreno, 2000). This is exactly the type of case that would not be amenable to a "native-centric" method for describing the H^+ binding. On the other hand, an ensemble-based description that specifically includes the contribution of unfolded states should reproduce the experimental data to a much higher degree. This conclusion is supported by the calculations shown in Fig. 3A, which demonstrate an apparent discrepancy between the electrostatics-based calculations performed on the high-resolution structure and the experimentally measured potentiometric titration. SNase is a highly basic protein with a +10 charge at neutral pH. The fact that 16 of the 17 structurally resolved carboxylic groups have depressed $pK_{a,FD}$ values is not a subtle problem with the FD calculations used to estimate these pK_a values (Fitch *et al.*, 2006). The data in Fig. 2C demonstrate that the discrepancy between experiment and theory in the H^+ binding curves cannot be corrected by fine-tuning the pK_a calculations. Instead it requires a cutoff threshold that somewhat arbitrarily normalizes the native pK_a values of many titratable groups. The ensemble-based calculations presented in this chapter convey a method to reconcile the high-resolution structure and experimental potentiometric titration data to the FD calculations. In this respect, the method presented here complements existing FD approaches because it provides a framework for unifying electrostatic theory with statistical thermodynamics of conformational fluctuations.

References

Ackers, G. K. (1998). Deciphering the molecular code of hemoglobin allostery. *Adv. Protein Chem.* **51,** 185–253.

Alexandrescu, A. T., Mills, D. A., Ulrich, E. L., Chinami, M., and Markley, J. L. (1988). NMR assignments of the four histidines in staphylococcal nuclease in native and denatured states. *Biochemistry* **27,** 2158–2165.

Antosiewicz, J., McCammon, J. A., and Gilson, M. K. (1994). Prediction of pH-dependent properties of proteins. *J. Mol. Biol.* **238,** 415–436.

Antosiewicz, J., Mc Cammon, A. J., and Gilson, M. K. (1996). The determinants of pK_a's in proteins. *Biochemistry* **35,** 7819–7833.

Babu, C. R., Hilser, V. J., and Wand, A. J. (2004). Direct access to the cooperative substructure of proteins and the protein ensemble via cold denaturation. *Nat. Struct. Biol.* **11,** 352–357.

Baker, D., and Agard, D. A. (1994). Influenza hemagglutinin: Kinetic control of protein function. *Structure* **2,** 907–910.

Baldwin, R. L. (1986). Temperature dependence of the hydrophobic interaction in protein folding. *Proc. Natl. Acad. Sci. USA* **83,** 8069–8072.

Bullough, P. A., Hughson, F. M., Skehel, J. J., and Wiley, D. C. (1994). Structure of influenza haemagglutinin at the pH of membrane fusion. *Nature* **371,** 37–43.

D'Aquino, J. A., Gómez, J., Hilser, V. J., Lee, K. H., Amzel, L. M., and Freire, E. (1996). The magnitude of the backbone conformational entropy change in protein folding. *Proteins: Struct. Funct. Genet.* **25,** 143–156.

DeLano, W. L. (2002). "The PYMOL Molecular Graphics System." DeLana Scientific, San Carlos, CA.

Ehrlich, L. S., Tianbo, L., Scarlatta, S., Chu, B., and Carter, C. A. (2001). HIV-1 capsid protein forms spherical (immature-like) and tubular (mature-like) particles *in vitro*: Structure switching by pH-induced conformational changes. *Biophys. J.* **81,** 586–594.

Englander, S. W. (2000). Protein folding intermediates and pathways studied by hydrogen exchange. *Annu. Rev. Biophys. Biomol. Struct.* **29,** 213–238.

Fitch, C. A., Karp, D. A., Lee, K. K., Stites, W. E., Lattman, E. E., and Garcia-Moreno, E. B. (2002). Experimental pK_a values of buried residues: Analysis with continuum methods and role of water penetration. *Biophys. J.* **82,** 3289–3304.

Fitch, C. A., Whitten, S. T., and García-Moreno, E. B. (2006). Molecular mechanisms of pH-driven conformational transitions of proteins: Insights from continuum electrostatics calculations of acid unfolding. *Proteins: Struct. Funct. Genet.* **63,** 113–126.

French, A. R., Tadaki, D. K., Niyogi, S. K., and Lauffenburger, D. A. (1995). Intracellular trafficking of epidermal growth factor family ligands is directly influenced by the pH sensitivity of the recepetor/ ligand interaction. *J. Biol. Chem.* **270,** 4334–4340.

Gamblin, S. L., Haire, L. F., Russell, R. J., Stevens, D. J., Xiao, B., Ha, Y., Vasisht, N., Steinhauer, D. A., Daniels, R. S., Elliot, A., Wiley, D. C., and Skehel, J. J. (2004). The structure and receptor binding properties of the 1918 influenza hemagglutinin. *Science* **303,** 1838–1842.

García-Moreno, E. B., Dwyer, J. J., Gittis, A. G., Spencer, D. S., and Stites, W. E. (1997). Experimental measurement of the effective dielectric in the hydrophobic core of a protein. *Biophys. Chem.* **64,** 211–224.

Gómez, J., Hilser, V. J., Xie, D., and Freire, E. (1995). The heat capacity of proteins.. *Proteins: Struct. Funct. Genet.* **22,** 404–412.

Hebermann, S. M., and Murphy, K. P. (1996). Energetics of hydrogen bonding in proteins: A model compound study. *Protein Sci.* **5,** 1229–1239.

Hill, T. L. (1960). "An Introduction to Statistical Thermodynamics." Dover Publications, Inc., New York.

Hilser, V. J. (2001). Modeling the native state ensemble. *Methods Mol. Biol.* **168,** 93–116.

Hilser, V. J., Dowdy, D., Oas, T. G., and Freire, E. (1998). The structural distribution of cooperative interactions on proteins: Analysis of the native state ensemble. *Proc. Natl. Acad. Sci. USA.* **95,** 9903–9908.

Hilser, V. J., and Freire, E. (1996). Structure-based calculation of the equilibrium folding pathway of proteins. Correlation with hydrogen exchange protection factors. *J. Mol. Biol.* **262,** 756–772.

Hilser, V. J., and Freire, E. (1997). Predicting the equilibrium protein folding pathway: Structure-based analysis of Staphylococcal nuclease. *Proteins: Struct. Funct. Genet.* **27,** 171–183.

Hilser, V. J., Townsend, B. D., and Freire, E. (1997). Structure-based statistical thermodynamic analysis of T4 lysozyme mutants: Structural mapping of cooperative interactions. *Biophys. Chem.* **64,** 69–79.

Hogle, J. M. (2002). Poliovirus cell entry: Common structural themes in viral cell entry pathways. *Annu. Rev. Microbiol.* **56,** 677–702.

Honig, B., and Nicholls, A. (1995). Classical electrostatics in biology and chemistry. *Science* **268,** 1144–1149.

Hynes, T. R., and Fox, R. O. (1991). The crystal structure of staphylococcal nuclease refined at 1.7 angstroms resolution. *Proteins* **10,** 92–99.

Lee, B., and Richards, F. M. (1971). The interpretation of protein structures: Estimation of static accessibility. *J. Mol. Biol.* **55,** 379–400.

Lee, K. H., Xie, D., Freire, E., and Amzel, L. M. (1994). Estimation of changes in side chain configurational entropy in binding and folding: General methods and application to helix formation. *Proteins: Struct. Funct. Genet.* **20,** 68–84.

Lee, K. K., Fitch, C. A., and García-Moreno, E. B. (2002a). Distance dependence and salt sensitivity of pairwise, coulombic interactions in a protein. *Protein Sci.* **11,** 1004–1016.

Lee, K. K., Fitch, C. A., Lecomte, J. T., and García-Moreno, E. B. (2002b). Electrostatic effects in highly charged proteins: Salt sensitivity of pK_a values of histidines in staphylococcal nuclease. *Biochemistry* **41,** 5656–5667.

Liu, T., Whitten, S. T., and Hilser, V. J. (2006). Ensemble-based signatures of energy propagation in proteins: A new view of an old phenomenon. *Proteins: Struct. Funct. Genet.* **62,** 728–738.

Luque, I., Mayorga, O. L., and Freire, E. (1996). Structure-based thermodynamic scale of α-helix propensities in amino acids. *Biochemistry* **35,** 13681–13688.

Matthew, J. B., Gurd, F. R., García-Moreno, E. B., Flanagan, M. A., March, K. L., and Shire, S. J. (1985). pH-dependent processes in proteins. *CRC Crit. Rev. Biochem.* **18,** 91–197.

Meeker, A. K., García-Moreno, E. B., and Shortle, D. (1996). Contributions of the ionizable amino acids to the stability of staphylococcal nuclease. *Biochemistry* **35,** 6443–6449.

Murphy, K. P., Bhakuni, V., Xie, D., and Freire, E. (1992). Molecular basis of co-operativity in protein folding. III. Structural identification of cooperative folding units and folding intermediates. *J. Mol. Biol.* **227,** 293–306.

Murphy, K. P., and Freire, E. (1992). Thermodynamics of structural stability and cooperative folding behavior in proteins. *Adv. Protein Chem.* **43,** 313–361.

Nozaki, Y., and Tanford, C. (1967a). Acid-base titrations in concentrated guanidine hydrochloride. Dissociation of the guanidium ion and of some amino acids. *J. Am. Chem. Soc.* **89,** 736–742.

Nozaki, Y., and Tanford, C. (1967b). Proteins as random coils. II. Hydrogen ion titration curve of ribonuclease in 6 M guanidine hydrochloride. *J. Am. Chem. Soc.* **89,** 742–749.

Oliveberg, M., Arcus, V. L., and Fersht, A. R. (1995). pK_a values of carboxyl groups in the native and denatured states of barnase: The pK_a values of the denatured state are on average 0.4 units lower than those of model compounds. *Biochemistry* **34,** 9424–9433.

Pan, H., Lee, J. C., and Hilser, V. J. (2000). Binding sites in *Escherichia coli* dihydrofolate reductase communicate by modulating the conformational ensemble. *Proc. Natl. Acad. Sci. USA* **97,** 12020–12025.

Pometun, M. S., Peterson, R. W., Babu, C. R., and Wand, A. J. (2006). Cold denaturation of encapsulated ubiquitin. *J. Am. Chem. Soc.* **128,** 10652–10653.

Ren, J., Kachel, K., Kim, H., Malenbaum, S. E., Collier, J. R., and London, E. (1999). Interaction of Diphtheria Toxin T Domain with molten globule-like proteins and its implications for translocation. *Science* **284,** 955–957.

Roxby, R., and Tanford, C. (1971). Hydrogen ion titration curve of lysozyme in 6 M guanidine hydrochloride. *Biochemistry* **10,** 3348–3352.

Sarkar, C. A., Lowenhaupt, K., Horan, T., Boone, T. C., Tidor, B., and Lauffenburger, D. A. (2002). Rational cytokine design for increased lifetime and enhanced potency using pH-activated "histidine switching". *Nat. Biotechnol.* **20,** 908–913.

Sarkar, C. A., Lowenhaupt, K., Wang, P. J., Horan, T., and Lauffenburger, D. A. (2003). Parsing the effects of binding, signaling, and trafficking on the mitogenic potencies of granulocyte colony-stimulating factor analogues. *Biotechnol. Prog.* **19,** 955–964.

Schaefer, M., Van Vlijmen, H. W. T., and Karplus, M. (1998). Electrostatic contributions to molecular free energies in solution. *Adv. Protein Chem.* **51,** 1–57.

Tan, Y., Oliveberg, M., Davis, B., and Fersht, A. R. (1995). Perturbed pK_a-values in the denatured states of proteins. *J. Mol. Biol.* **254,** 980–992.

Whitten, S. T. (1999). "Acid denaturation and pH dependence of stability of *staphylococcal* nuclease." Ph.D. dissertation, The Johns Hopkins University.

Whitten, S. T., and García-Moreno, E. B. (2000). pH dependence of stability of staphylococcal nuclease: Evidence of substantial electrostatic interactions in the denatured state. *Biochemistry* **39,** 14292–14304.

Whitten, S. T., García-Moreno, E. B., and Hilser, V. J. (2005). Local conformational fluctuations can modulate the coupling between proton binding and global structural transitions in proteins. *Proc. Natl. Acad. Sci. USA* **102,** 4282–4287.

Whitten, S. T., Kurtz, A. J., Wand, A. J., and Hilser, V. J. (2006). The native state ensemble, energy landscapes, and cold denaturation. *Biochemistry* **45,** 10163–10174.

Wooll, J. O., Wrabl, J. O., and Hilser, V. J. (2000). Ensemble modulation as an origin of denaturant-independent hydrogen exchange in proteins. *J. Mol Biol.* **301,** 247–256.

Wyman, J., Jr. (1948). Heme proteins. *Adv. Protein Chem.* **4,** 407–531.

Wyman, J., Jr. (1964). Linked functions and reciprocal effects in hemoglobin: A second look. *Adv. Protein Chem.* **19,** 223–286.

Wyman, J., Jr., and Gill, S. J. (1990). "Binding and Linkage: The Functional Chemistry of Biological Macromolecules." University Science Books, Mill Valley, CA.

Xie, D., and Freire, E. (1994). Structure based prediction of protein folding intermediates. *J. Mol. Biol.* **242,** 62–80.

Yang, A., and Honig, B. (1993). On the pH dependence of protein stability. *J. Mol. Biol.* **231,** 459–474.

Yang, A., and Honig, B. (1994). Structural origins of pH and ionic strength effects on protein stability. Acid denaturation of sperm whale apomyoglobin. *J. Mol. Biol.* **237,** 602–614.

CHAPTER 28

Molecular Modeling of the Cytoskeleton

Xiange Zheng and David Sept

Department of Biomedical Engineering and Center for Computational Biology
Washington University
St. Louis, Missouri 63130

Abstract

Molecular modeling techniques have truly come of age in recent decades, and here we cover several of the most commonly used techniques, namely molecular dynamics, Brownian dynamics, and molecular docking. In each case, we explain the physical basis and limitations of the various techniques and then illustrate their application to various problems related to the cytoskeleton. This set of studies covers a relatively wide range of examples and is comprehensive enough to clearly see how these techniques could be applied to other systems. Finally, we cover

0091-679X/08 $35.00
DOI: 10.1016/S0091-679X(07)84028-3

several related methodologies that expand on these basic techniques to allow for more detailed and specific simulation and analysis.

I. Introduction

Molecular modeling techniques have truly come of age in the past two to three decades. The reasons for this are twofold: first, there has been an explosive growth in the available structural data for proteins, not only from X-ray crystallography but also from NMR and electron microscopy studies. Second, accompanying this growth in the area of structural biology have come significant advances in both computational techniques and hardware. As computers continue to increase in speed and capability, this allows us to perform both larger (i.e., more atoms) and longer simulations. The purpose of this chapter is to outline the basic physics behind each of these simulation methods and illustrates their application in the published literature. It should be pointed out that all of these techniques have a wide range of applications in the physical, chemical, and biological literature; however, we will be restricting our coverage to those studies that focus on some aspect of the cytoskeleton since that is our primary area of interest and research. Despite this somewhat narrow perspective, the range of studies covered still illustrates the usefulness of these methods and should allow the reader to understand their application in other settings.

II. Simulation Methods

In order to understand how and when various simulation methods can properly be applied, it is truly necessary to understand the physics and assumptions behind each of the techniques. Below we outline the basic ideas underlying each methodology, as well as the applicability and limitations of their results.

A. Molecular Dynamics

Molecular dynamics (MD) is one of the most widely employed methodologies and is now starting its fourth decade of use (McCammon *et al.*, 1976, 1977). The basis of the MD technique is the molecular mechanics force field that describes the interaction of atoms within a biomolecule, between biomolecules, and between biomolecules and water molecules, ions, and so on. There are many force fields in existence; however, they can all generally be considered as the sum of bonded and nonbonded interactions. Bonded interactions include a bond stretching term (a change in the covalent bond between two atoms), an angle bending term (every angle between three successive atoms), and a torsional term (rotation about the central bond for all groups of four covalently bound atoms). Although these terms could be treated very accurately using quantum mechanics, they are often modeled as harmonic springs ($E = 1/2\ kx^2$) since perturbations about some equilibrium point are typically very small, and the use of effective potentials makes calculations much faster and easier. The nonbonded interactions include electrostatic and van der Waals interactions that are captured using well-established concepts such as Coulomb's law and a Lennard–Jones potential. Putting all these terms together gives us the full form of the potential that can be written as

$$U = \sum_{\text{bonds}} k_b(b - b_o)^2 + \sum_{\text{angles}} k_\theta(\theta - \theta_o)^2 + \sum_{\text{diherdrals}} A(1 - cos(n\tau - \phi)) + \sum_{\text{charges}} \frac{q_i q_j}{\varepsilon r_{ij}} + \sum_{\text{atoms}} \left(\frac{C_{12}}{r_{ij}^{12}} - \frac{C_6}{r_{ij}^{6}} \right)$$

In this equation, the first three terms represent the bond, angle, and torsion terms while the last two represent the electrostatic and van der Waals contributions (Fig. 1). Because of the many degrees of freedom, there is not a single, correct set of parameters for such a force field, and there are a handful of "standard" force fields used for most simulations (Brooks *et al.*, 1983; Duan *et al.*, 2003; Lindahl *et al.*, 2001; Nelson *et al.*, 1996; Ren and Ponder, 2003). Having an explicit form for the potential, we can now calculate the force that each atom experiences by taking the negative of the gradient of U (i.e., $F = -\nabla U$). Since we know the force on each of the atoms, we simply integrate Newton's law ($F = ma$) and solve for the position of each atom as a function of time. This equation of motion is a second-order differential equation, and therefore we are required to provide the 0th derivative (positions) and the first derivative (velocities) for the initial structure. The positions are nothing more than the starting crystal structure (or other atomically detailed structure), and the velocities are calculated from a Maxwell–Boltzmann distribution for the appropriate temperature at which we are simulating.

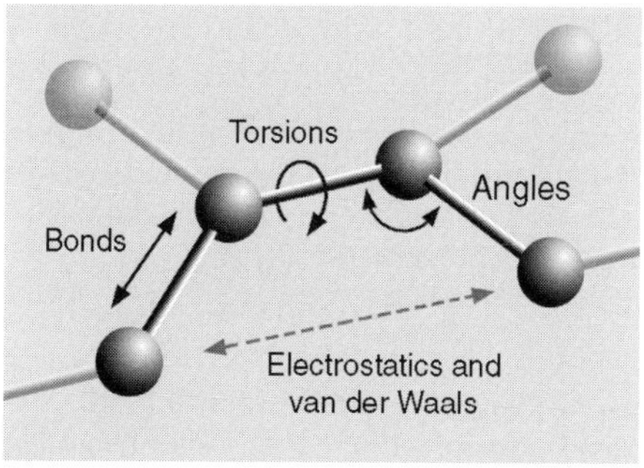

Fig. 1 A cartoon depiction of the bonded and nonbonded interactions within a biomolecule. All groups of atoms connected by three or fewer covalent bonds have bonded interactions consisting of bond, angle, and torsion terms. Atoms separated by three or more covalent bonds have nonbonded interactions based on electrostatic and van der Waals interactions.

What makes the MD technique particularly appealing is how closely it mimics the experimental situation. Apart from the protein or biomolecule of interest, the system is fully solvated, typically in water, has a particular concentration of salt and counterions, and the simulation is conducted at a specific temperature and pressure/volume. In truth, these factors make MD much closer to an experimental method than a theoretical one, but that is a debate better left for another chapter.

B. Brownian Dynamics

Brownian dynamics (BD) operates in a different time regimen from MD. In the case of MD, we are forced to take 1–2 fsec time steps in order to accurately integrate the equations of motion. This typically limits the overall simulation time to nanoseconds or 10s of nanoseconds, except in the case of very small systems or very large, parallel computers. As the name implies, BD functions in the domain of Brownian motion where we can ignore inertia and assume that the system is overdamped (low Reynolds number). The simplest equation describing BD is a first-order difference equation (Ermak and McCammon, 1978)

$$\Delta x = \frac{D\Delta t}{k_B T} F + \Delta S$$

In this equation, Δx is the displacement over the time step Δt, D is the diffusion constant, $k_B T$ is the thermal energy (Boltzmann's constant multiplied by the temperature), F is force acting on the particle/molecule, and ΔS is the stochastic force having the properties that $\langle S \rangle = 0$ and $\langle S^2 \rangle = 2D\Delta t$ where $\langle ... \rangle$ denotes the average of the function. Since the solvent is treated implicitly, this stochastic term captures interaction of the molecule with the water, and in the absence of any outside force F, one would simply recover free diffusion. Solution of this equation is relatively straightforward since it is a difference equation and not a differential equation. Further, since we are operating in the overdamped regimen, we are able to take large time steps (in the 1–50 psec range), and since the force terms typically have a simple functional form, we can easily simulate phenomena occurring on the millisecond or even the second timescale.

One of the typical applications of BD is in the area of biomolecular diffusion and simulation of association reactions. Based on some of the early work on diffusion from Smoluchowski, Northrup *et al.* (1984) were able to derive a simple method for determining the bimolecular association rate for two molecules from a series of BD simulations. This method has been commonly used to determine the effect of mutation or the influence of ionic strength on the association rate of two proteins, and several such studies will be covered in later sections (Fig. 2).

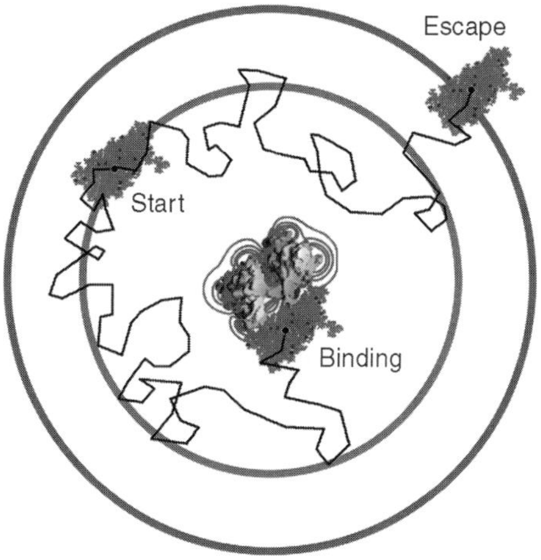

Fig. 2 An illustration of Brownian dynamics applied to biomolecular association. By comparing the fraction of successful trajectories to those that escape, it is possible to calculate a biomolecular association rate constant.

C. Molecular Docking

The third technique that we will discuss, molecular docking, is concerned with predicting the bound structure of two or more biomolecules. The molecules may be proteins, peptides, or other small ligands or drugs, and as such these techniques are widely used in academic and commercial research to aid in the design and characterization of drugs and inhibitors. The docking problem consists of two parts: a search of conformational space and the ability to "score" a given binding conformation. All docking programs have these two components in common; however, the way in which they solve each of these problems is quite different (Ewing *et al.*, 2001; Friesner *et al.*, 2004; Kramer *et al.*, 1999a; Morris *et al.*, 1998; Verdonk *et al.*, 2003). The conformational space search does not just involve the six relative degrees of freedom between the two binding partners (three translational and three rotational degrees of freedom), but also includes any internal degrees of freedom such as bond rotations or side-chain reorientations in a protein. Given the high dimensionality of the resulting conformational space, an exhaustive search of all available docking conformations or *poses* is not feasible. Instead, programs will often employ search techniques like Monte Carlo or genetic algorithms to help in the search process. Once we have a method that can quickly generate a large number of poses, we need to evaluate these different binding predictions using a scoring function. In most cases, the scoring function attempts to closely approximate the binding free energy using a first-principles or semiempirical-based force field (similar to those used in MD, as discussed above). These scoring functions are often approximated and simplified to allow for rapid evaluation since the determination of a single bound conformation between a drug and protein may involve scoring 10^6 to 10^{10} individual poses. As such, accuracy is often sacrificed for speed, and rather than attempting to predict the correct free energy, the focus is more often to predict the correct relative binding free energy or relative ranking of two or more binding ligands. One obvious problem in estimating a true free energy is that we do fairly well with the enthalpic contribution, but the entropic part is much more difficult to assess. For small molecules, an empirical formula that gives a fixed entropic penalty for each rotatable bond generally works quite well, but Chapter 1 by Garbett and Chaires (this volume) discusses this issue in more detail.

The extension of this work to the problem of protein–protein docking is one of the significant challenges currently faced in the simulation community. The problem in this case is not the accuracy of the force field or estimation of the binding free energies, but is tied to the many conformational degrees of freedom that must be searched for a protein as compared to a much smaller ligand. There has been progress made by many groups (Chen *et al.*, 2003; Dominguez *et al.*, 2003; Ritchie, 2003), but these efforts are far from "turnkey" solutions and require extensive analysis and interpretation of the results.

III. Applications of Molecular Modeling

As stated in the Introduction, there is wide and varied application of the above-mentioned techniques. Given our group's research focus on the cytoskeleton, we will concentrate on applications to actin and actin filaments, tubulin and microtubules, and other associated proteins such as myosin and kinesin. These examples cover a wide enough range that extension to other systems should hopefully be obvious to the reader. We unfortunately do not have the space to provide adequate background on the biology relating to each of these problems, but this should be easily obtainable by referring to the referenced publications or any good cell biology text.

A. Molecular Dynamics Applications

Since MD involves integrating an equation of motion that describes the movement of each atom in a protein (as well as the surrounding water, ions, and so on), it is logical to try and use MD to simulate a large-scale kinetic event such as opening and closing of an active site. Although this is possible in principle, it is often not practical since such dynamic events often occur on time scales longer than the simulation can achieve. Instead, MD is much better applied to smaller/faster phenomena such as local fluctuations in structure, rearrangement of side chains on mutation or drug binding, or the interaction of small ligands such as ions, peptides, or drugs with a protein. Following are several examples of MD simulations applied to actin, tubulin, myosin, and gelsolin.

It has long been postulated that the DNase-I binding loop of actin undergoes a loop-to-helix conformational change on ATP hydrolysis. Using MD simulations combined with coarse-grained analysis, Chu and Voth (2005) conducted studies on the effect of the nucleotide state on the DNase-I loop conformation in actin monomers, trimers, and a short actin filament. On the basis of several long-time scale simulations, it was found that the helical conformation of the DNase-I loop weakened the intermonomer interactions within the actin filament and thus resulted in a shortened and more disordered filament in the ADP state. More extensive studies have recently been carried out to examine other nucleotide-dependent effects in the actin monomer (Zheng *et al.*, 2007). In this study, monomeric actin was simulated in the ATP, ADP-P_i, and ADP states, with all simulations lasting at least 50 nsec. It was found that the structure of the DNase-I loop was a strong function of the nucleotide, and further that additional loops in the nucleotide-binding cleft and the hydrophobic cleft were also dependent on the nucleotide state. Most notably, it was found that these secondary structure changes could be observed during the course of the simulation. In a simulation starting with the ADP structure but with ATP into the binding pocket, the DNase-I loop remained a helix for 15–20 nsec but then switched to a coil conformation and remained there for remainder of the simulation (50 nsec in this case).

There have been several interesting MD studies on myosin and the interaction between myosin and actin. Liu *et al.* (2006) investigated the binding energetics and interaction properties of myosin to actin by means of MD simulations of myosin complexed with two nucleotide-free actin monomers (based on a cryo-EM structure). By allowing the system to reorganize and equilibrate, they were able to identify several key points of interaction, several of which matched well with mutagenesis and other experimental results. Kawakubo *et al.* (2005) studied the signal transmission of ATPase pocket via MD simulations and principal component analysis. From analyzing a 1-nsec MD simulation and monitoring the movements of relevant residues, they suggested that conformational changes in the nucleotide-binding pocket resulted in slowly varying collective motions of the atoms surrounding the actin-binding site and the junction with the neck region. This suggested that a signal following ATP hydrolysis could be transmitted from the nucleotide-binding site to the motor domain in about 150 psec. Similar simulations were conducted to analyze the structural factors affecting back door P_i release in myosin (Lawson *et al.*, 2004). Here it was concluded that the conformational change between the upper and lower 50-kDa subdomains is the key factor for the P_i to leave via the back door. Water could flux through the back door to attack the γ-phosphate of ATP bound at the binding site to facilitate the hydrolysis of ATP, but never through the main (front) door. The MD simulation suggested that the P_i–protein interaction resulted in a narrowing of the back door and thereby prevented the passage of P_i. These factors were suggested to be the structural basis for P_i release being the rate-limiting step in myosin ATPase in the absence of actin. Finally, Li and Cui (2004) carried out the studies of the energetics and mechanism of ATP hydrolysis in myosin by means of MD simulations and quantum mechanics/molecular mechanics (QM/MM) reaction pathway calculations. QM/MM calculations are a hybrid method of simulation in which a small region (say 10–100 atoms in an active site) is treated using full quantum detail, and this region is embedded in a larger domain that is treated using a molecular mechanics description (i.e., this domain is the rest of the protein). In this study, the QM/MM calculations indicated that hydrolysis pathway with Ser 326 serving as the relaying group is preferred, supporting the fact that this residue is absolutely conserved. Their results showed consistency with the fact that the prehydrolysis conformation of the motor domain is not able to catalyze ATP hydrolysis, and some conformational change may be necessary for hydrolysis to occur. Finally, the salt bridge between Arg 238 and Glu 459 was shown to be key for the stability of ATP, the breakage of which would disrupt the inline water structure in the ATP-binding site making the hydrolysis much less energetically favorable. QM/MM simulations represent a significant advance in the simulation field since they allow one to study dynamics in bonded interactions (such as catalysis or hydrolysis) or charge states (polarization or changes in protonation states). As these techniques become more refined, they will allow completely new classes of problems to be explored.

Gelsolin is another actin-associated protein that binds at the barbed ends of actin filaments and is regulated by Ca^{2+}. To understand the influence of calcium on such process, Liepina *et al.* (2003) performed MD simulations of both the S2 domain and the whole gelsolin protein to investigate the dynamics at different concentrations of Ca^{2+}. From these MD simulations, high mobility of the S2 domain was detected, and the stability of this domain was determined to be a strong factor of the Ca^{2+} concentration. It was suggested that the high flexibility of the S2 domain could be ascribed to its involvement in actin and PIP_2 binding and that this feature makes gelsolin a good candidate to enter the interface between two actin domains and fulfill its cleaving function.

One extension of the standard MD method is steered molecular dynamics (SMD) where instead of letting the system fluctuate at equilibrium, it is guided along a particular reaction pathway, thereby facilitating large-scale conformational charges in the short-time scale of regular MD. It should be stressed that such a simulation is not at equilibrium, and given a large enough force, almost any transition would be possible; however, the method is useful for exploring transitions between states. Wriggers and Schulten (1999) applied SMD to actin system to investigate the mechanism of P_i release. It was revealed that the breaking of the electrostatic interactions of P_i with the divalent cation is the time-limiting step in P_i release. The pathway for release and specific interactions with different amino acids along the release path were elucidated, and the influence of specific factors, such as the protonation state of His73, was estimated.

As pointed out at the start of this section, the application of MD is much better suited to phenomena that involve relatively small and/or fast processes. Although not an absolute restriction, the examples covered here were consistent with this limit, focusing on changes in secondary structure (DNase-I loop in actin), the effects of nucleotide hydrolysis (ATP hydrolysis in actin and myosin), or the interaction of Ca^{2+} or P_i with a protein (applied to myosin, gelsolin, and actin). As computational power continues to grow, the accurate simulation of larger system for longer times is becoming increasingly possible, but the statistical significance of any result will need to be carefully measured and evaluated.

B. Brownian Dynamics Applications

Brownian dynamics is a tool well suited for studying the binding kinetics and interactions between macromolecules, as well as the identification of critical residues that may contribute to binding. One such example was the study of actin polymerization and the role of electrostatic interactions in determining the bimolecular association rate. Sept *et al.* (1999) performed BD simulations to characterize the polymerization rates at both the barbed and the pointed ends of the actin filament. These simulations revealed that electrostatic interactions were the basis for the difference in barbed and pointed end association rates. They were also able to demonstrate that polymerization at the barbed end was diffusion limited, as observed by experiment, but that binding at the slow-growing pointed end was limited by an electrostatic barrier and not simply by diffusion. This study was subsequently extended to look at the process of actin filament nucleation. In order to determine the pathway of actin filament nucleation, a more extensive set of BD simulations and free energy calculations were carried out to identify the kinetic pathway of spontaneous actin filament nucleation (Sept and McCammon, 2001). The full characterization of the nucleation and polymerization kinetics was achieved by computing association rates from the BD simulations as well as the equilibrium constants from the free-energy calculations, thus allowing the individual dissociation rates to be determined. Based on these simulations, the critical nucleation step was the formation of the longitudinal dimer, and the trimer was found to be the critical nucleus size.

Actin monomers and filaments also play an important role in modulating glycolytic enzymes such as glyceraldehyde-3-phosphate-dehydrogenase (GAPDH), aldolase, and triose phosphate isomerase (TIM). Ouporov *et al.* (1999) investigated the interactions between aldolase and G-/F-actin with BD simulations. These simulations revealed two unique complexes between aldolase and G- or F-actin. In the first complex, predominantly present in G-actin, the positively charged grooves of aldolase surface bind to subdomain 4 of actin. In the second, found in both G- and F-actin, these same portions of aldolase bound to the negative region subdomain 1 of actin via electrostatic interactions. To further address the nature of these interactions, Lowe *et al.* (2002, 2003) carried out additional BD simulations of these enzymes with F-actin. Here, they found that the interaction of TIM with F-actin was the weakest among the three enzymes and lacked a unique binding mode, in agreement with experimental observations. Their analysis of the interactions of enzyme dimers with F-actin supported the hypothesis that the quaternary structure and positive surface charges are the key for electrostatic attractions. Waingeh *et al.* (2004) conducted BD simulations of several GAPDH mutants with F-actin (Ouporov *et al.*, 2001). These simulations provided further support for the hypothesis that the interaction of GAPDH and F-actin is electrostatic in nature and that a basic surface patch formed by the four lysines (residues 24, 69, 110, and 114) is critical for GAPDH binding to F-actin. They likewise found that the quaternary structures of both GAPDH and F-actin were required for their interaction, a finding supported by the observations that subunit pair interactions (two subunits from GAPDH and two adjacent ones from F-actin) are more favorable than single subunit interactions.

BD simulations are very complementary studies to the more detailed results from MD simulations. As illustrated by the examples covered here, the BD technique is very useful for identifying residues that are important for a particular interaction, determining the potential points of contact between two or more proteins, and using both sets of information allows one to elucidate limits on the protein complexes and quaternary structures that may be formed. When used in conjunction with experimental studies, BD simulations can offer significant insight into the effects of mutagenesis or other protein modifications on biomolecular interactions.

C. Ligand–Protein and Protein–Protein Docking

Molecular docking has been recognized as an efficient and robust tool to determine the conformation of the complex structure given available structures of the components. The predicted interface residues, relative positions, and orientations for the constituent molecules provide detailed information for future mutagenesis and other experimental studies.

Antimitotic agents that target tubulin and microtubules have been of great interest in the treatment of many diseases, most notably cancer. One such class of agents are the depsipeptides cryptophycin, dolastatin, hemiasterlin, and phomopsin. To understand the structural basis for the interactions of these molecules with tubulin, Mitra and Sept (2004) carried out the MD simulations and docking studies on these systems. For the docking, multiple snapshots of the macromolecule were extracted from MD simulations to allow for some degree of backbone and side-chain flexibility. A single, common site on β-tubulin was identified for five different peptides: cryptophycin 1, cryptophycin 52, dolastatin 10, hemiasterlin, and phomopsin A. This site was located next to the exchangeable GTP site on β-tubulin and agreed well with mutagenesis results from a number of studies. Another drug that targets tubulin and disrupts microtubules is colchicine. The colchicine-binding site has been located at the interface between α-tubulin and β-tubulin by X-ray crystallography, but to characterize this binding site in more detail, Farce *et al.* (2004) carried out docking studies of several colchicines-site compounds. The potentially bioactive conformation of each docking compound was determined, and the relative binding free energies were correlated with the apparent activities. This allowed for further generation of novel drug analogues through the use of QSAR analysis (see Section IV.A) and is a model of how many such studies are carried out in pharmaceutical research.

Apart from the treatment of cancer, several classes of microtubule inhibitors have been developed to selectively target microtubules in protozoan parasites. Mitra and Sept (2004) again combined MD simulations with molecular docking studies to investigate the interaction of dinitroanilines with α-tubulin (Morrissette *et al.*, 2004). These studies were able to identify a single, consensus binding site on tubulin and explained the basis of how these compounds disrupt microtubules. Just as with the colchicine studies mentioned above, the comparison of multiple dinitroaniline analogues provides specific details about the atomic-level interactions, and such information will be applied in a series of future drug development studies.

Apart from tubulin targeting drugs, there are also a number of proteins that interact with tubulin monomers and promote the formation of the tubulin heterodimer. To understand the mechanism and structural basis for such a process, You *et al.* (2004) built a model complex of Rbl2-β-tubulin using a combination of molecular docking and site-directed mutagenesis (Rbl2 is a homologue of cofactor A). The docking of Rbl2 to both α-tubulin and β-tubulin indicated that Rbl2 binds preferentially to β-tubulin, and based on the docking model, several key contact residues were identified. Subsequent mutagenesis of these contact residues supported the docking model, and the binding site of Rbl2 on β-tubulin was revealed to partially overlap with that of α-tubulin to β-tubulin, indicating competition between these two binding partners.

In a very interesting study, fluorescent resonance energy transfer (FRET) was used in conjunction with docking simulations to study the interaction of myosin with the actin filament. Root *et al.* (2002) carried out the docking of myosin subfragment-1 with actin using constraints from FRET data in both pre- and post-powerstroke states. In post-powerstroke conformation, corresponding to the ADP state of myosin, the FRET and docking data clearly showed that subfragment-1 existed in at least two orientations. However, for the pre-powerstroke state, FRET was combined with chemical cross-linking data, resulting in a single, predicted orientation of the neck region. This study suggested that the larger movement of the light chain of myosin, together with the twisting and rotation of the catalytic domain, induced a 30° tilt between the pre- and postpowerstroke states during the weak-to-strong-binding transition.

There have been several recent studies where docking simulations were combined with mutagenesis studies to determine the structural basis of capping protein regulation. The first study involved myotrophin, a protein that binds to capping protein and inhibits its interaction with the barbed end of the actin filament (Bhattacharya *et al.*, 2006). In this case, MD simulations were combined with protein–protein docking simulations to determine the conformation of the protein complex. By using a series of myotrophin mutants as well as studies with capping protein fragments, the authors were able to place experimental constraints on the docking data and refine the results into a clear prediction of the structure for the bound complex. The next study focused on the interaction of PIP_2 with capping protein (Kim *et al.*, 2006). In this case, docking simulations of PIP_2 and capping protein were able to direct mutagenesis studies that ultimately confirmed the predicted binding site and provided important details on the interaction between capping protein and the actin filament.

Docking studies are one of the most widely used molecular modeling techniques, in part because of their basic utility in determining the molecular details of protein and drug interactions, but also because having detailed structural knowledge about a biomolecular interaction gives one insight into the role that this interaction plays in a larger context (i.e., systems biology) or potentially how one might wish to disrupt this interaction in the course of treating a disease. The examples covered in this section cover many of these aspects, and given the interest and practical nature of this type of research, it is likely to be a major area of development in the coming years.

IV. Related Methodologies

The molecular modeling techniques described thus far represent some of the most basic applications; however, there are a multitude of related methodologies that expand on these basic techniques to allow for more detailed or specialized simulation and analysis. Below we outline but a few of these examples.

A. QSAR and CoMFA

With abundant biochemical data of small compounds available, approaches such as comparative molecular field analysis (CoMFA) and quantitative structure–activity relationship (QSAR) are commonly used to correlate chemical properties with biological activities by comparing structural similarities and differences of active and inactive compounds and then deriving a collection of predictive information such as a *pharmacophore*. A pharmacophore is the collection of molecular details that defines the essential features responsible for a drug's biological activity. The concept of a common pharmacophore, as well as the chemical and structural information carried therein, has been a significant aid in reaching the goal of receptor-based drug design. Below are several examples illustrating the application of QSAR analysis and CoMFA.

Returning once again to microtubule targeting drugs, there are a series of interesting studies that have been performed. Cunningham *et al.* (2005) reported CoMFA, HQSAR (Hologram QSAR, a two-dimensional computational technique), and molecular docking studies of butitaxel analogues with β-tubulin. It was indicated that the result from CoMFA agrees with those from HQSAR in terms of predictive potential. The overall correlation between the docking energy and experimentally determined activity of the compounds was produced consistently by both CoMFA and HQSAR.

Nguyen *et al.* (2005) derived a common pharmacophore model for a set of colchicine site inhibitors. Using the X-ray tubulin–colchicine complex structure (PDB id: 1SA0) as template, the common binding mode and pharmacophore were derived from docking studies of these colchicine site compounds, which were then tested via MD simulations. There were seven pharmacophore points predicted, consisting of three hydrogen bond acceptors, one hydrogen bond donor, two hydrophobic centers, and one planar group. These seven sites appeared to be bisected by two planes tilted at $\sim 45^\circ$ into nearly equal halves, and the biplanar architecture was conserved in all binding modes.

The epothilones are another class of compounds that bind to β-tubulin and inhibit cell division by stabilizing microtubules (similar to paclitaxel). Lee and Briggs (2001) presented three-dimensional QSAR studies on epothilones with CoMFA method, where 166 epothilone MT-inhibiting analogues and their depolymerization inhibition properties were used for the analysis. In this case, QSAR analysis identified three pharmacophore elements, two of which were compatible with the previously reported model, and one novel feature required for the accurate description of the activities of the 166 epothilone training-set molecules.

Although the use of QSAR and CoMFA methods can give very credible results, they are only as reliable as the experimental data on which the analysis is based. He *et al.* (2000) also presented a pharmacophore model for paclitaxel and epothilones. Paclitaxel and four analogues were used as the training test, among which the 2-m-azido baccatin III compound lacks the C-13 side chain that was thought to be necessary for bioactivity. Since this compound exhibited normal activity as compared to taxol, it demonstrated that the C-13 side chain was not strictly required for biological activity. Based on their results, it was suggested that paclitaxel binds to β-tubulin with three major contacts: the taxane ring near the M-loop, the C-2 benzoyl ring next to His227 and Asps224, and the C-13 benzamido group within the N-terminal 1–31 residues of β-tubulin. In a similar study, epothilone conformations were generated from high-temperature MD simulations and then used in analysis of atom–atom distances resulting in a pharmacophore prediction (Giannakakou *et al.*, 2000). This work concluded that baccatin is necessary for tubulin binding and might serve as the scaffold holding functional groups. The three studies outlined above all developed pharmacophore models for paclitaxel, but reached somewhat divergent conclusions. As pointed out by Day (2000), more chemical and biological data will be required to further test these hypotheses.

B. Electrostatics Based Approaches

Microtubules perform a variety of functions in cells through their polymerization dynamics, however they also play an important structural role, and electrostatics has been shown to be a key factor in the quaternary structure of the microtubule. In one study, Baker *et al.* (2001) carried out large-scale computations on the electrostatics potential of a microtubule. These calculations showed interesting differences between the plus and minus ends of the polymer and gave hints about the interaction of microtubules with proteins (such as kinesin) and drugs such as paclitaxel. In subsequent work, Sept *et al.* (2003) elucidated the physical basis for the observed helical structure of microtubule, and described the electrostatic interactions in microtubule. These calculations found that longitudinal bonds in the microtubule lattice were ~7 kcal/mol stronger than the interprotofilament interactions, and predicted the existence of two lattice conformations (experimentally known as the A and B lattices). Such electrostatic methods mesh very well with more detailed, molecular-level modeling, and more information on these methods is given in Chapter 26 by Dong *et al.*, this volume.

C. Normal Mode Analysis

Normal mode analysis has served as an efficient and popular technique to identify functionally relevant dynamics represented by low-frequency motions and correlations. The method is based on decomposing the full motion of a biomolecule into a set of independent motions or modes that range from large amplitude/low-frequency modes to small amplitude/high-frequency motions. The small/fast motions are typically ignored and only the top large-scale modes are considered, since these are thought to represent the dominant dynamics of the protein. Zheng and Brooks (2005) investigated the dynamics correlations in the myosin motor domain utilizing normal mode analysis of an elastic network model, where the protein structure is represented by C_α-atoms connected to each other with harmonic springs. Through normal mode analysis, the "hinge residues" were identified and found to match well with amino acids pinpointed through experimental means. The dominant normal mode showed significant conformational changes in the nucleotide-binding pocket, and their analysis supported the negative correlation between the opening and closing of actin-binding site with the opening and closing of the nucleotide-binding pocket.

It has also been postulated that the mechanical movement of myosin along an actin filament involves direct conformational change within the cross bridge between myosin and the filament. Navizet *et al.* (2004) investigated the flexibility of myosin and the structural basis of this conformational change. Three crystal structures of the myosin head were used for the analysis within a coarse-grained elastic network model. An anisotropic network model, where all α-carbon fluctuations were treated anisotropically reflecting the directions of movement of each residue, was used to compute residue fluctuations, and the analysis of flexibility identified several rigid structural blocks. Analysis of these results revealed flexibility at the motor domain-filament interface, and the degree of this movement was shown to depend on the state of the nucleotide in the binding pocket.

Using an elastic-network model, Zheng *et al.* (2003) examined the structure–function relationships for three families of motor proteins: kinesin, myosin, and F1-ATPase. For myosin and F1-ATPase, the normal mode analysis of this network model suggested that the measured conformational change from crystal structures could be well characterized by only one or two dominant modes, while for kinesin, multiple modes were required. It was suggested that there may be two different mechanisms by which myosin and kinesin induce large-scale changes in the nucleotide-binding pocket. In the case of myosin, the global motions generated by the dominant normal modes were independent of the fine structural details in the nucleotide-binding pocket; however for kinesin, structural changes in the binding pocket did not produce global motions that matched well with the ones observed in crystal structures.

V. Conclusions

The field of computational biology has seen significant advances in the past decades, and as we see further increases in both the available computational power and the number of active researchers, it is an area sure to see significant growth in the future. Hopefully, the unique role that simulation plays in biochemistry, cell biology, and biophysics has been clearly demonstrated here. As pointed out earlier, these techniques are much closer to experimental methods than theoretical constructions, and when coupled with compatible experimental studies, they can be extremely powerful research tools. Although very limited in their scope, the themes and methods presented here encompass a significant fraction of the computational literature and should allow for further reading and exploration. Armed with this basic knowledge, there are several next steps that could be followed. In our opinion, the text by Leach (2001) is one of the most accessible and comprehensive texts in the field, but there are many others to choose from, including Allen and Tildesley (1989), Field (1999), Frenkel and Smit (2002), and Schlick (2002). In addition to these resources, nearly every university and major meeting or conference has a healthy contingent of computational biologists, and talking with these individuals is the best way to discover how molecular simulation could impact your particular area of research.

Acknowledgments

This work was supported in part by a National Institutes of Health grant to DS (GM-067246).

References

Allen, M. P., and Tildesley, D. J. (1989). "Computer Simulation of Liquids." Oxford University Press, New York.

Baker, N. A., Sept, D., Joseph, S., Holst, M. J., and McCammon, J. A. (2001). Electrostatics of nanosystems: Application to microtubules and the ribosome. *Proc. Natl. Acad. Sci. USA* **98,** 10037–10041.

Bhattacharya, N., Ghosh, S., Sept, D., and Cooper, J. A. (2006). Binding of myotrophin/V-1 to actin-capping protein: Implications for how capping protein binds to the filament barbed end. *J. Biol. Chem.* **281,** 31021–31030.

Brooks, B. R., Bruccoleri, R. E., Olafson, B. D., States, D. J., Swaminathan, S., and Karplus, M. (1983). CHARMM: A program for macromolecular energy, minimization, and dynamics calculations. *J. Comput. Chem.* **4,** 187–217.

Chen, R., Li, L., and Weng, Z. P. (2003). ZDOCK: An initial-stage protein-docking algorithm. *Proteins: Struct. Funct. Genet.* **52,** 80–87.

Chu, J. W., and Voth, G. A. (2005). Allostery of actin filaments: Molecular dynamics simulations and coarse-grained analysis. *Proc. Natl. Acad. Sci. USA* **102,** 13111–13116.

Cunningham, S. L., Cunningham, A. R., and Day, B. W. (2005). CoMFA, HQSAR and molecular docking studies of butitaxel analogues with beta-tubulin. *J. Mol. Model.* **11,** 48–54.

Day, B. W. (2000). Mutants yield a pharmacophore model for the tubulin-paclitaxel binding site. *Trends Pharmacol. Sci.* **21,** 321–323.

Dominguez, C., Boelens, R., and Bonvin, A. (2003). HADDOCK: A protein–protein docking approach based on biochemical or biophysical information. *J. Am. Chem. Soc.* **125,** 1731–1737.

Duan, Y., Wu, C., Chowdhury, S., Lee, M. C., Xiong, G. M., Zhang, W., Yang, R., Cieplak, P., Luo, R., Lee, T., Caldwell, J., Wang, J. M., *et al.* (2003). A point-charge force field for molecular mechanics simulations of proteins based on condensed-phase quantum mechanical calculations. *J. Comput. Chem.* **24,** 1999–2012.

Ermak, D. L., and McCammon, J. A. (1978). Brownian dynamics with hydrodynamic interactions. *J. Chem. Phys.* **69,** 1352–1360.

Ewing, T. J. A., Makino, S., Skillman, A. G., and Kuntz, I. D. (2001). DOCK 4. 0: Search strategies for automated molecular docking of flexible molecule databases. *J. Comput. Aided Mol. Des.* **15,** 411–428.

Farce, A., Loge, C., Gallet, S., Lebegue, N., Carato, P., Chavatte, P., Berthelot, P., and Lesieur, D. (2004). Docking study of ligands into the colchicine binding site of tubulin. *J. Enzyme Inhib. Med. Chem.* **19,** 541–547.

Field, M. (1999). "A Practical Introduction to the Simulation of Molecular Systems." Cambridge University Press, Cambridge, New York.

Frenkel, D., and Smit, B. (2002). "Understanding Molecular Simulation: From Algorithms to Applications." Academic Press, San Diego, CA.

Friesner, R. A., Banks, J. L., Murphy, R. B., Halgren, T. A., Klicic, J. J., Mainz, D. T., Repasky, M. P., Knoll, E. H., Shelley, M., Perry, J. K., Shaw, D. E., Francis, P., *et al.* (2004). Glide: A new approach for rapid, accurate docking and scoring. 1. Method and assessment of docking accuracy. *J. Med. Chem.* **47,** 1739–1749.

Giannakakou, P., Gussio, R., Nogales, E., Downing, K. H., Zaharevitz, D., Bollbuck, B., Poy, G., Sackett, D., Nicolaou, K. C., and Fojo, T. (2000). A common pharmacophore for epothilone and taxanes: Molecular basis for drug resistance conferred by tubulin mutations in human cancer cells. *Proc. Natl. Acad. Sci. USA* **97,** 2904–2909.

He, L. F., Jagtap, P. G. F., Kingston, D. G. I., Shen, H. J., Orr, G. A., and Horwitz, S. B. (2000). A common pharmacophore for Taxol and the epothilones based on the biological activity of a taxane molecule lacking a C-13 side chain. *Biochemistry* **39,** 3972–3978.

Kawakubo, T., Okada, O., and Minami, T. (2005). Molecular dynamics simulations of evolved collective motions of atoms in the myosin motor domain upon perturbation of the ATPase pocket. *Biophys. Chem.* **115,** 77–85.

Kim, K., McCully, M. E., Bhattacharya, N., Butler, B., Sept, D., and Cooper, J. A. (2006). Structure/function analysis of the interaction of pip2 with actin capping protein: Implications for how capping protein binds the actin filament. *J. Biol. Chem.* **282**(8), 5871–5879.

Kramer, B., Metz, G., Rarey, M., and Lengauer, T. (1999a). Ligand docking and screening with FlexX. *Med. Chem. Res.* **9,** 463–478.

Lawson, J. D., Pate, E., Rayment, I., and Yount, R. G. (2004). Molecular dynamics analysis of structural factors influencing back door P-i release in myosin. *Biophys. J.* **86,** 3794–3803.

Leach, A. R. (2001). "Molecular Modelling: Principles and Applications." Prentice-Hall, Harlow, England.

Lee, K. W., and Briggs, J. M. (2001). Comparative molecular field analysis (CoMFA). Study of epothilones-tubulin depolymerization inhibitors: Pharmacophore development using 3D QSAR methods. *J. Comput. Aided Mol. Des.* **15,** 41–55.

Li, G. H., and Cui, Q. (2004). Mechanochemical coupling in myosin: A theoretical analysis with molecular dynamics and combined QM/MM reaction path calculations. *J. Phys. Chem. B* **108,** 3342–3357.

Liepina, I., Janmey, P. A., Czaplewski, C., and Liwo, A. (2003). Molecular dynamics study of the influence of calcium ions on the conformation of gelsolin S2 domain. *J. Mol. Struct.: Theochem.* **630,** 309–313.

Lindahl, E., Hess, B., and van der Spoel, D. (2001). GROMACS 3. 0: A package for molecular simulation and trajectory analysis. *J. Mol. Mod.* **7,** 306–317.

Liu, Y. M., Scolari, M., Im, W., and Woo, H. J. (2006). Protein–protein interactions in actin-myosin binding and structural effects of R405Q mutation: A molecular dynamics study. *Proteins: Struct. Funct. Bioinform.* **64,** 156–166.

Lowe, S. L., Adrian, C., Ouporov, I. V., Waingeh, V. F., and Thomsson, K. A. (2003). Brownian dynamics simulations of glycolytic enzyme subsets with F-actin. *Biopolymers* **70,** 456–470.

Lowe, S. L., Atkinson, D. M., Waingeh, V. F., and Thomasson, K. A. (2002). Brownian dynamics of interactions between aldolase mutants and F-actin. *J. Mol. Recognit.* **15,** 423–431.

McCammon, J. A., Gelin, B. R., and Karplus, M. (1977). Dynamics of folded proteins. *Nature* **267,** 585–590.

McCammon, J. A., Gelin, B. R., Karplus, M., and Wolynes, P. G. (1976). The hinge-bending mode in lysozyme. *Nature* **262,** 325–326.

Mitra, A., and Sept, D. (2004). Localization of the antimitotic peptide and depsipeptide binding site on beta-tubulin. *Biochemistry* **43,** 13955–13962.

Morris, G. M., Goodsell, D. S., Halliday, R. S., Huey, R., Hart, W. E., Belew, R. K., and Olson, A. J. (1998). Automated docking using a Lamarckian genetic algorithm and an empirical binding free energy function. *J. Comput. Chem.* **19,** 1639–1662.

Morrissette, N. S., Mitra, A., Sept, D., and Sibley, L. D. (2004). Dinitroanilines bind alpha-tubulin to disrupt microtubules. *Mol. Biol. Cell* **15,** 1960–1968.

Navizet, I., Lavery, R., and Jernigan, R. L. (2004). Myosin flexibility: Structural domains and collective vibrations. *Proteins: Struct. Funct. Genet.* **54,** 384–393.

Nelson, M. T., Humphrey, W., Gursoy, A., Dalke, A., Kale, L. V., Skeel, R. D., and Schulten, K. (1996). NAMD: A parallel, object oriented molecular dynamics program. *Int. J. Supercomput. Appl. High Perform. Comput.* **10,** 251–268.

Nguyen, T. L., McGrath, C., Hermone, A. R., Burnett, J. C., Zaharevitz, D. W., Day, B. W., Wipf, P., Hamel, E., and Gussio, R. (2005). A common pharmacophore for a diverse set of colchicine site inhibitors using a structure-based approach (vol 48, pg 6110, 2005). *J. Med. Chem.* **48,** 6107–6116.

Northrup, S. H., Allison, S. A., and McCammon, J. A. (1984). Brownian dynamics simulations of diffusion-influenced biomolecular reactions. *J. Chem. Phys.* **80,** 1517–1524.

Ouporov, I. V., Knull, H. R., Lowe, S. L., and Thomasson, K. A. (2001). Interactions of glyceraldehyde-3-phosphate dehydrogenase with G- and F-actin predicted by Brownian dynamics. *J. Mol. Recognit.* **14,** 29–41.

Ouporov, I. V., Knull, H. R., and Thomasson, K. A. (1999). Brownian dynamics simulations of interactions between aldolase and G- or F-actin. *Biophysical Journal* **76,** 17–27.

Ren, P. Y., and Ponder, J. W. (2003). Polarizable atomic multipole water model for molecular mechanics simulation. *J. Phys. Chem. B* **107,** 5933–5947.

Ritchie, D. W. (2003). Evaluation of protein docking predictions using Hex 3. 1 in CAPRI rounds 1 and 2. *Proteins: Struct. Funct. Genet.* **52,** 98–106.

Root, D. D., Stewart, S., and Xu, J. (2002). Dynamic docking of myosin and actin observed with resonance energy transfer. *Biochemistry* **41,** 1786–1794.

Schlick, T. (2002). "Molecular Modeling and Simulation: An Interdisciplinary Guide." Springer, New York.

Sept, D., Baker, N. A., and McCammon, J. A. (2003). The physical basis of microtubule structure and stability. *Protein Sci.* **12,** 2257–2261.

Sept, D., Elcock, A. H., and McCammon, J. A. (1999). Computer simulations of actin polymerization can explain the barbed-pointed end asymmetry. *J. Mol. Biol.* **294,** 1181–1189.

Sept, D., and McCammon, J. A. (2001). Thermodynamics and kinetics of actin filament nucleation. *Biophys. J.* **81,** 667–674.

Verdonk, M. L., Cole, J. C., Hartshorn, M. J., Murray, C. W., and Taylor, R. D. (2003). Improved protein-ligand docking using GOLD. *Proteins: Struct. Funct. Genet.* **52,** 609–623.

Waingeh, V. F., Lowe, S. L., and Thomasson, K. A. (2004). Brownian dynamics of interactions between glyceraldehyde-3-phosphate dehydrogenase (GAPDH). mutants and F-actin. *Biopolymers* **73,** 533–541.

Wriggers, W., and Schulten, K. (1999). Investigating a back door mechanism of actin phosphate release by steered molecular dynamics. *Proteins: Struct. Funct. Genet.* **35,** 262–273.

You, L., Gillilan, R., and Huffaker, T. C. (2004). Model for the yeast cofactor A-beta-tubulin complex based on computational docking and mutagenesis. *J. Mol. Biol.* **341,** 1343–1354.

Zheng, W. J., and Brooks, B. (2005). Identification of dynamical correlations within the myosin motor domain by the normal mode analysis of an elastic network model. *J. Mol. Biol.* **346,** 745–759.

Zheng, W. J., and Doniach, S. (2003). A comparative study of motor-protein motions by using a simple elastic-network model. *Proc. Natl. Acad. Sci. USA* **100,** 13253–13258.

Zheng, X., Diraviyam, K., and Sept, D. (2007). Nucleotide effect on the structure and dynamics of actin.. *Biophys. J.* **93,** 1277–1283.

CHAPTER 29

Mathematical Modeling of Cell Migration

Anders E. Carlsson[*] and David Sept[†]

[*]Department of Physics
Washington University
St. Louis, Missouri 63130

[†]Department of Biomedical Engineering and Center for Computational Biology
Washington University
St. Louis, Missouri 63130

Abstract

Mathematical modeling has become increasingly important in many areas of biology during the past two decades, and the area of cell migration and motility has seen significant contributions from a wide range of modeling approaches. In this chapter, we cover examples from the broad range of work in this area, emphasizing the models' biological significance and the relationships between them. We focus on three specific areas: cell protrusion, cell adhesion, and retraction/whole-cell models. At the end of this chapter, we provide our perspective on issues that future models and experiments should consider in order to advance the boundaries of this field.

METHODS IN CELL BIOLOGY, VOL. 84

0091-679X/08 $35.00
DOI: 10.1016/S0091-679X(07)84029-5

I. Introduction

The process of cell migration is essential in development, in differentiation, and in the physiological response to disease. It has been an intensive area of experimental study for many decades, and in recent years it has also become a frequent target for modeling. Initial models were very rudimentary since our knowledge about the basis of this phenomenon was sparse, but as cell biology has become more quantitative and has provided more detailed data, the sophistication of these models has grown concomitantly. The desire for a comprehensive model of cell migration has naturally been fueled by a desire to better understand the inner workings of the cell, but it has also partly resulted from an influx of mathematicians, engineers, and physicists (like us) into the field of cellular biophysics. Thus many of the models that have been developed have their roots in materials science, mechanical engineering, or condensed matter physics. Such models make perfect sense to researchers in these fields, but to the average biologist, they are often obtuse and incomprehensible. Our objective in this chapter is to broadly cover the range of models that have been developed, explaining generally their physical basis, their capabilities, and their relationship to other models. For reasons of space and stamina, we cannot cover all possible models, but we will attempt to include representatives from all classes of models. We will cover three principal areas: cell protrusion, cell adhesion, and retraction, and then treat whole-cell models that include all of these phenomena. In Section V, we elaborate a bit more on other less developed areas of modeling and give our outlook for what future models and experiments need to consider.

At the outset, it is useful to emphasize the usefulness of mathematical models within the cell motility community and cell biology in general. Simple mathematical explanations that can explain a certain class of observed phenomena are appealing for several reasons. These models can not only reproduce particular behaviors or responses of cells but also allow one to work through the different outcomes of competing hypotheses to see which one(s) are consistent with experimental data. Beyond comparing the outcomes resulting from different assumptions, these models can also make (concrete) predictions that can be tested by further experiments. As our knowledge and understanding of cellular biochemistry grows, future models should be able to capture more complicated behaviors such as the effect of pharmacological or other therapeutic intervention on cell behavior.

II. Cell Protrusion

A. Single-Filament Modeling

The key molecular-scale event underlying actin-based cell migration is the generation of force by the polymerization of actin filaments against an obstacle, which could be the cell membrane, a vesicle, or an intracellular pathogen such as

Listeria monocytogenes. Modeling of single-filament growth can identify plausible mechanisms for force generation, predict the stall force, and establish the dependence of the velocity on opposing force. The papers discussed below, using a variety of mathematical methods, have reached the conclusion that filament growth by passive monomer diffusion, in which fluctuations of the obstacle or filament tip make room for new monomer addition, is a viable route to filament growth. This mechanism can generate forces up to a few piconewtons (pN) per filament. Studies including attractive filament–obstacle interactions indicate that even if the filament is attached to the obstacle, there can be enough room for new monomers to add to the filament tip; in fact, the presence of a filament-tip complex can actually speed up the polymerization process under some circumstances. If the coupling to the obstacle causes ATP hydrolysis to be coupled to polymerization, forces of up to 10 pN per filament can be generated.

A landmark paper (Peskin *et al.*, 1993) considered a fixed, rigid actin filament polymerizing in perpendicular orientation against a hard, motile obstacle subjected to an external force F, as shown in Fig. 1. The obstacle moves according to both random forces from the environment and the external force. New subunits add by passive diffusion of monomers, but only if the filament–obstacle separation exceeds the step size a per added monomer, taken to be 2.7 nm. Provided this condition is satisfied, the addition rate to the filament tip is the same as for a free filament tip. This model was solved by calculating the time evolution of the probability distribution $P(x,t)$ for the obstacle position, according to an equation of the form:

$$\frac{\partial P}{\partial t} = D\frac{\partial^2 P}{\partial x^2} + \frac{FD}{kT}\frac{\partial P}{\partial x} + \alpha[P(x+a,t) - H(x-a)P(x,t)] + \beta[H(x-a)P(x-a,t) - P(x,t)]$$

where x is the position of the obstacle measured relative to the filament tip, D is the obstacle's diffusion constant, k is Boltzmann's constant, T is temperature, α is the barbed-end on-rate, $a = 2.7$ nm is the step size per added monomer, β is the barbed-end off-rate, and $H(x - a)$ is a mathematical step function whose value is 1 if $x > a$ and 0 otherwise. On the right-hand side of this equation, the first term

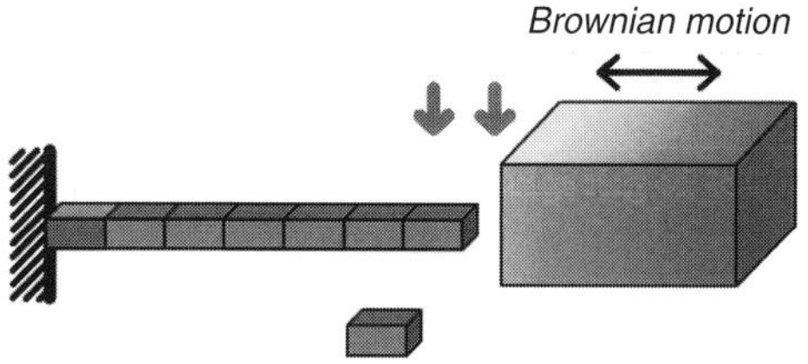

Fig. 1 Illustration of actin polymerization against an obstacle (reproduced by permission of Scot Kuo http://www.jhu.edu/cmml).

describes the rate of change of P due to diffusion of the obstacle, the second term comes from the directed motion of the obstacle due to the force, and the remaining terms describe jumps in the position of the obstacle relative to the tip due to polymerization or depolymerization at the tip.

If the free-monomer concentration is high enough, growth takes place by a "Brownian-ratchet" mechanism in which a monomer that is added during an obstacle excursion from the filament tip prevents it from returning to its previous position. Increasing force F suppresses obstacle fluctuations and thus slows growth. If the condition $\alpha \ll D/a^2$ holds, the mathematical form of the force–velocity relation is very simple (see Fig. 2):

$$V \propto \exp\left(\frac{-F}{F_0}\right) - \beta$$

Here, $F_0 = kT/a = 1.5$ pN is the characteristic force scale over which the growth velocity decreases, k is Boltzmann's constant, and T is temperature. As seen in Fig. 2, single filaments can provide effective propulsion against opposing forces only up to a few piconewtons. This finding is consistent with a general result of statistical mechanics (Hill, 1987) relating the ratio of the on- and off-rates to the external force. Including depolymerization gives a stall force (at which polymerization precisely balances depolymerization) of $(kT/a)\ln(G/G_c)$, where G is the free-monomer concentration and G_c is the barbed-end critical concentration.

Subsequent modeling (Mogilner and Oster, 1996) extended this analysis to treat flexible filaments in orientations at an angle θ relative to the perpendicular. It was found that filament bending could supply fluctuations that, like those of the

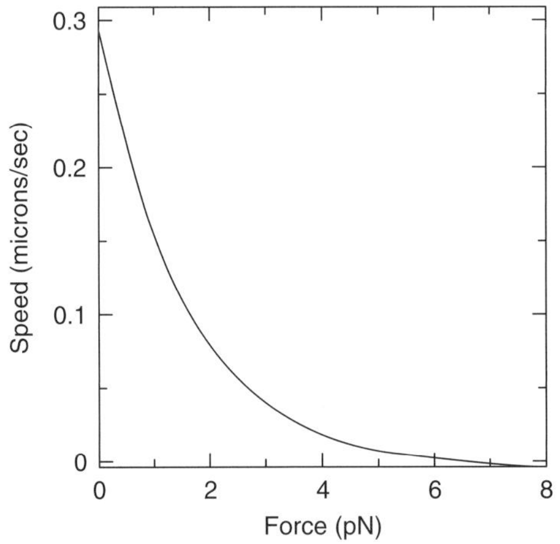

Fig. 2 The exponential force–velocity relationship based on the Brownian-ratchet model of Peskin *et al.* (1993).

obstacle, temporarily open up room to let new monomers in. The mathematical form of the force–velocity relation remains the same, except that the characteristic force scale becomes $kT/a\ \cos(\theta)$. Thus filaments oriented at nonperpendicular incidence can, in principle, generate higher forces than perpendicular ones. However, the increase in force is not likely to exceed about 50% because filaments far from perpendicular will bend away from the obstacle. It should be noted that even at perpendicular incidence, the maximum force that can be generated is limited by filament buckling, which occurs at a force of $\pi^2 kTL_p/4L^2$ (Landau *et al.*, 1986) where L is the filament length and L_p is the persistence length; for $L = 0.3\ \mu\text{m}$, the buckling force is 2 pN.

These mathematical treatments of single-filament growth have been tested by Brownian-dynamics simulations in which the positions of all the subunits in a single growing filament, as well as the obstacle position, are stored over time (Carlsson, 2000). Simulations such as these test the importance of simplifying approximations made by the Brownian-ratchet model, for example, the assumption that when the obstacle-tip distance exceeds a the growth rate becomes that of a free filament. The simulations treated a single free monomer approaching a fluctuating filament tip moving in two dimensions. The obstacle velocity was obtained by tracking the obstacle position over time, and numerous simulation runs were performed at each opposing force to obtain statistical accuracy. Simple mathematical forms were assumed for the interaction forces between filament subunits, the obstacle, and the free monomer. The actin subunits in the filament moved according to a law of motion including both random motion and motion due to deterministic forces. The simulations confirmed the exponential force–velocity relation found by the Brownian-ratchet model in most cases. A closely related simulation treated growth of a three-dimensional (3D) filament using simplified Monte Carlo dynamics, with the monomer addition treated according to a rate law based on the distance between the filament tip and the obstacle rather than by explicit motion of the incoming monomer (Burroughs and Marenduzzo, 2005). These simulations confirmed the Brownian-ratchet prediction for hard-wall type force fields. For softer force fields and obstacles with very small diffusion constants, they found that the velocity could significantly exceed the Brownian-ratchet result. However, because the steric repulsion between the actin filament and the obstacle against which it pushes is likely to be short ranged, it is unlikely that its behavior occurs in actin-based force generation. Thus, to the best of our knowledge, the force–velocity relation for a free-actin filament impinging on an obstacle is well described by the Brownian-ratchet analysis.

An extension of these types of models to include attachments between the filament tip and the obstacle is motivated by several studies that have shown that at least some actin filaments propelling intracellular pathogens and bead analogues are attached to the obstacle, and the observation that formins can stay attached to filament barbed ends while still allowing their growth (Higgs, 2005). Mathematical models including the attachments can establish what additional mechanisms may be active in force generation when filaments are attached, and evaluate the resulting changes in the force–velocity relation. These models may also be relevant to lamellipodial protrusion since several intracellular proteins are known to link actin filaments to cell membranes. Dickinson, Purich, and collaborators have developed the "actoclampin motor" model (Dickinson and Purich, 2002) in which a filament attached to a protein or protein complex on an obstacle can exert a pushing force. Motivated by suggestions (Upadhyaya *et al.*, 2003), which have not been subsequently verified, that individual filaments can generate forces of up to 10 pN, they assumed that the energy of ATP hydrolysis assists the force generation process. They treated several variations (Dickinson *et al.*, 2002, 2004, 2005) of the following extension of the Brownian-ratchet model to attached filaments. First, an actin filament's ATP-containing terminal subunit binds to a surface protein on the obstacle, presumably on the filament side. Second, a free monomer attaches to the filament. Finally, ATP hydrolysis on the first subunit reduces its affinity for the surface protein, and the obstacle translocates so that the surface protein binds to the new (still ATP-containing) subunit, and the cycle starts anew. This model was analyzed by a methodology similar to that used in the Brownian-ratchet analysis. The results suggest that attached filaments can elongate at rates comparable to observed cell protrusion velocities if monomer diffusion is not strongly inhibited by the obstacle–filament interaction and the opposing force does not exceed a critical force of about 10 pN per filament tip. The critical force is much greater than in the Brownian-ratchet model because the free energy of ATP hydrolysis is coupled to the elongation process. The shape of the force–velocity relation is also very different from that in the Brownian-ratchet model. It is nearly constant up to the critical force and then drops rapidly to zero. The attached-filament growth mechanism can also explain the right-handed trajectories seen in *Listeria* motion (Zeile *et al.*, 2005). If the filament is attached to the bacterial surface, the twist in the actin filament leads to torsional stresses that can be released by helical motion of the bacterium, provided that the pitch of the filaments changes in such a way as to account for the bacterium's rotation. The viability of attached-filament growth mechanisms has recently been demonstrated by Brownian-dynamics simulations based on an assumed force field between the actin filament and a surface or membrane protein (Zhu and Carlsson, 2007).

Closely related to the actoclampin model are two recent models of the growth of actin filaments specifically aimed at formin-capped barbed ends. These models have not included an obstacle explicitly, but could straightforwardly be extended to do so if the formins were attached to a cell membrane. The models differ in their details, but both assume the existence of at least two conformations of the formin-barbed end complex. One of these conformations is open enough to allow the entry of new actin monomers; monomer addition switches the complex to its closed state. Thermal fluctuations cause transitions to the open state after a certain waiting time. A model based on three conformations (Shemesh *et al.*, 2005c) has shown that the "rotation paradox" arising in simpler leaky-cap models can be circumvented. In this paradox, the rotation of the tip-bound formin resulting from the helicity of the actin filament leads to supercoiling if the orientation of the formin is fixed, but such supercoiling is not observed. The three-configuration model circumvents the formin rotation by including a hypothetical "slip step" in which the filament tip changes its orientation relative to the formin. An extension (Vavylonis *et al.*, 2006) of the leaky-cap model to include the extended nature of the FH1 formin domain, and the effects of profilin, has shown that formin capping can actually lead to extension rates faster than those of uncapped actin filaments, consistent with experimental results.

There are few experimental tests of the predictions made by the single-filament models. The flexibility of actin filaments makes a measurement of the force–velocity relation very difficult. However, experiments with microtubules are possible, and they have found an exponential decay of the velocity with opposing force. Explaining the coefficient of the decay is harder than for actin because of the many strands in the microtubule (Mogilner and Oster, 1999). Rough estimates of the overall magnitude of forces exerted by actin polymerization based on knowledge of the membrane tension and the number of filaments at the cell membrane suggest about 1 pN per filament (Abraham *et al.*, 1999). An estimate of the maximum force exerted by polymerization of formin-capped actin filaments has been obtained in experiments in which the pointed end is bound via a myosin to a substrate and the barbed end is bound to the substrate via a formin (Kovar and Pollard, 2004). Observation of filament buckling suggested forces of up to 1.3 pN, but the resolution of the experiments was insufficient to establish whether larger forces were exerted. Unfortunately, these measurements are consistent with all of the theoretical models proposed so far and cannot discriminate between them.

B. Many-Filament Modeling

Mathematical models that treat the simultaneous polymerization of many actin filaments aim to include the effects of filament–filament interactions on the actin gel's structure and force generation properties, while still providing sufficient molecular detail that the effects of actin-binding proteins can be included. In this way, they can serve as a bridge between molecular-level rate constants and the larger-scale behavior of the system. Several types of interaction effects should impact actin-based cell migration significantly. For example, steric interactions between filaments could block growth at high densities, or attractive interactions could result in transitions from network to bundled structures. Filament branching is a type of interaction since the very existence of a daughter filament depends on the mother filament, and their positions are locked relative to each other. Furthermore, growing filaments consume free monomers and thereby slow down the growth of other filaments. In addition, the motion of the obstacle can mediate indirect interactions between filaments. For example, the obstacle motion induced by one filament can change the force exerted on the tip of another filament and thereby change its polymerization rate; the motion can also affect the Arp2/3 complex-induced branching of other filaments since branching likely occurs only in the region near the obstacle where Arp2/3 complex can be activated by proteins such as ActA.

The problems that have received the most attention with many-filament modeling are the propulsion of intracellular pathogens, lamellipodial protrusion, and filopodial protrusion. The mathematical methodologies either evaluate the energetics of particular many-filament configurations or treat the dynamics of growth, branching, and other related processes in the growth of actin gels. The studies have shown that several aspects of protrusion, including the actin network structure, can be reproduced using only a small number of established biochemical processes; more complex aspects of migration, such as persistent motion of cytoplasts and the stepping motion of *Listeria*, can also be obtained if the model parameters are chosen appropriately. The modeling studies have also explained the typical number of filaments found in filopodia and the average distance between them. The predictions made by these theories are usually strongly dependent on the underlying assumptions. Thus comparison of theory and experiment can be used to discriminate between competing models of migration. Such predictions include the force–velocity relation of intracellular pathogens and the correlation of filopodium length with membrane stiffness, the lamellipodial protrusion rate, and other key cellular properties.

The earliest many-filament calculations treating migration (van Oudenaarden and Theriot, 1999) were aimed at understanding the symmetry-breaking phenomenon of small ActA-coated beads: these develop directional motion in a migration medium despite being spherically symmetric. The calculations did not attempt to include a realistic structure for the actin gel, but rather simulated a finite collection of filaments initially uniformly distributed around the bead. The filaments were fixed relative to a background that could correspond to the actin gel; the barbed ends of the filaments were oriented toward the bead "by hand," and only the repulsive interaction between the bead and the filaments was included. Monomers were added to filament ends stochastically, with the addition rate depending strongly on the distance from the filament to the bead. The bead moved in response to the forces from the filaments, with the result that a motion of the bead in a certain direction was amplified because the growth rate of the filaments being impinged on was reduced, while that of the filaments on the opposite side was enhanced. Subsequent studies (van der Gucht *et al.*, 2005) have shown that tearing of the actin gel is a crucial component of the symmetry-breaking process, but the earlier simulations suggested a mechanism that could enhance the effect.

Several subsequent models have treated the growth of branched actin networks, propelling either an obstacle such as *Listeria* or a cell membrane at the leading edge of a lamellipodium. In these models, network growth is simulated directly on a computer, with all filament subunit positions stored explicitly. Carlsson (2001) treated a hard cubic obstacle propelled by the growth of an actin network. The molecular-level processes included in the model were barbed-end growth, pointed-end depolymerization, branch formation, debranching, barbed-end capping, and barbed-end uncapping. The actin filaments and branch points were treated as rigid;

the actin tail was regarded as fixed, while the obstacle moved according to a combination of deterministic repulsive forces from the filament and random thermal forces. New branches were formed only very near the obstacle, and the branching was assumed to be completely autocatalytic in the sense that new branches formed only on existing filaments. The model accurately reproduced the network structures seen in electron micrographs. The assumption of completely autocatalytic branching led to a predicted growth velocity essentially independent of opposing force. The experimentally observed force–velocity relations will be discussed in the next section. Alberts and Odell (2004) used a more detailed approach taking into account filament–obstacle attachments, the spatial variation of the actin concentration, and hydrolysis of the actin-bound nucleotide. A close-up of the interface between the *Listeria* actin tail and the bacterium obtained in this fashion is shown in Fig. 3. The structure for the simulated actin tail is realistic, and the simulations also produced an intermittent motion behavior. Atilgan *et al.* (2005) studied a branched actin network growing against a membrane in a model lamellipodium. Their 3D model included a curved front edge with flat top and bottom

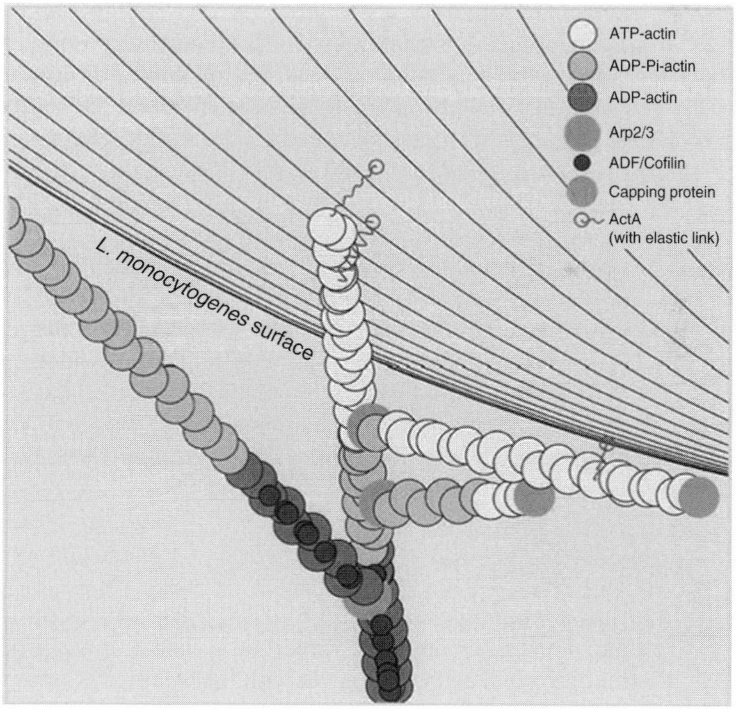

Fig. 3 Depiction of the *Listeria* motility model of Alberts and Odell (reproduced by permission of Jonathan Alberts). (See Plate no. 30 in the Color Plate Section.)

boundaries. They found that a network structure consistent with experiment could be obtained only if it was assumed that the orientation of individual Arp2/3 complex molecules relative to the membrane normal is limited during branching events. Using elasticity-based calculations, they argued that the receptors responsible for recruiting Arp2/3 activators to the membrane are preferentially targeted at the curved front edge of the plasma membrane, explaining their localization at the leading edge.

Work aimed at understanding filopodial protrusion has focused on calculation of the mechanical properties of filopodia, the monomer concentration profile along the filopodium, and the interactions between filopodia. The key mechanics issue is that single filaments will buckle at small forces, but bundling can greatly enhance their mechanical rigidity. Two recent calculations have shown that for filopodia of the length of microns, at least 10 filaments need to be bundled to avoid buckling in response to the elastic force of the membrane opposing polymerization. The length of filopodia can also be limited by the need for free monomers to diffuse to the filopodium tip. Mogilner and Rubinstein (2005) found that for filopodia containing more than 30 filaments, monomer diffusion becomes the rate-limiting factor in filopodium growth. They made specific predictions regarding the filopodium length, for example, that there is an optimal number of filaments for achieving long filopodia, that longer filopodia should result from faster lamellipodial protrusion, and that for thin filopodia decreasing membrane stiffness should increase the filopodium length. Atilgan *et al.* (2006), working with the model shown in Fig. 4, found that the protrusion velocity is enhanced by the thermal fluctuations of the membrane and is sensitive to the spatial arrangement of the filaments; they also made predictions for the force–velocity relation of filopodia for different assumptions about the membrane properties. Both of these studies found that filopodia can merge with each other. This limits the proximity of adjacent filopodia because filopodia that approach too close to each other will merge unless they are prevented from moving. Kruse and Sekimoto (2002) treated the effects of interfilament interactions mediated by molecular motors on the growth of fingerlike protrusions such as filopodia. They found that if actin filaments are aligned with their barbed ends pointing away from the cell body, and some of the filaments are attached to a substrate while others are free to move, the free filaments will move in the barbed-end direction. This could lead to the extension of filopodia-like structures.

C. Continuum Modeling

In continuum modeling approaches, the actin gel is described not in terms of individual filaments, but rather by coarse-grained properties such as the total number of filaments in contact with the membrane, or the spatially varying density of actin. Such methods treat molecular-level events in an averaged fashion, but are more powerful than the filament-based methods in that they provide predictions for larger systems over longer times. Calculations using continuum modeling

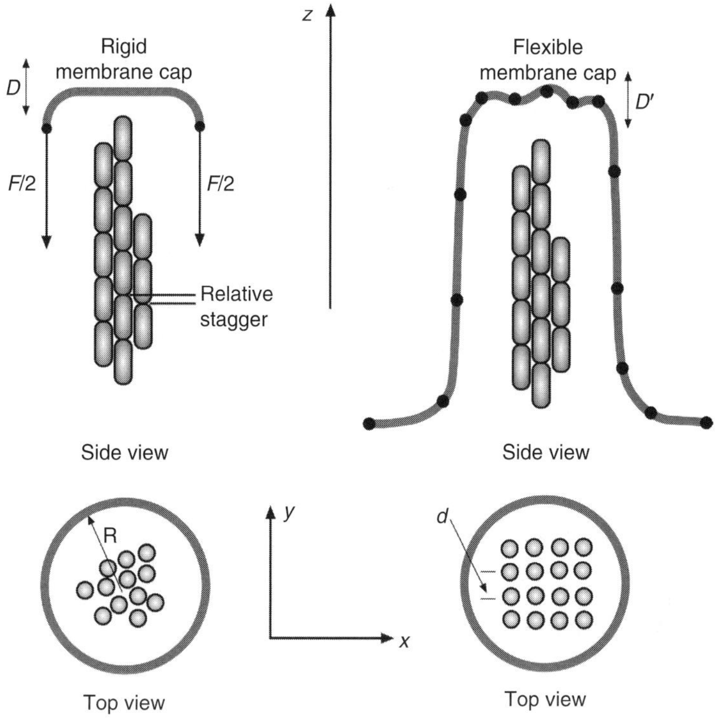

Fig. 4 Illustration of filopodia growth from actin bundles from Atilgan *et al.* (2006) (reproduced by permission of Sean Sun).

methods aim to relate experimentally accessible properties, such as the force–velocity relation for beads and pathogens and the shape of a lamellipod, to the macroscopic properties of the actin gel and the average properties of the individual filaments.

Like the many-filament studies described above, the continuum approaches have focused on intracellular pathogen propulsion as well as lamellipodial and filopodial protrusion. The complexity of the models varies greatly. Some characterize the actin gel by using as few as two numbers, others treat the spatially varying density of actin and actin-binding proteins, and some have treated the angle-dependent distribution of filaments. The predictions made by these models include the force–velocity relation for pathogen propulsion, properties of the actin gel such as its thickness and spatial density distribution, the protrusion velocity for lamellipodial extension, the elastic and diffusion properties of the membrane as influenced by cell motion, and the distributions of filament lengths and orientations. All of these quantities are experimentally observable.

Intensive study has been devoted to the study of *Listeria* propulsion, and related biomimetic experiments based on beads, using continuum methods. A calculation of the force–velocity relation for *Listeria* (Mogilner and Oster, 2003b) assumed that the pushing force is supplied by filaments whose tips move freely, and these oppose a frictional force resulting from filaments tethered to the obstacle. Rate equations for the densities of free and tethered filaments were set up to treat the effects of nucleation of attached filaments at the obstacle, dissociation of these attached filaments, and capping of the resulting free filaments so that they became irrelevant for propulsion. A model of this simplicity cannot include the distribution of filament velocities that result from the spread of filament orientations, or lateral elastic forces from the gel, but it allows exploration of a plausible hypothesis regarding the effects of filament attachments. The predicted force–velocity relation for *Listeria* displays an initial rapid drop, followed by a slower decay.

More elaborate methods have treated the inhomogeneous elastic forces exerted by the actin gel on an intracellular pathogen or biomimetic bead. The basic idea of these studies is that actin polymerization does not directly lead to motion, but rather to a buildup of stress in the actin gel surrounding the bacterium, which drives the motion. The mathematical analysis is based on the elastic equilibrium of the actin gel, which leads to elastic propulsion forces that are countered by frictional forces between the gel and the bacterium and external forces acting on the bacterium. In the simplest geometry, an actin gel growing on a spherical bead designed as a biomimetic analogue of the bacterium, one finds (Noireaux *et al.*, 2000) that as a layer of actin freshly grown at the bead surface is pushed away from the surface, it stretches and, much like an expanding balloon, exerts an inward pressure on the portion of the actin gel inside it. This pressure is transferred to the bead–gel interface. Measurement of the thickness of the gel, in combination with estimates of the elastic modulus of the gel, allows one to evaluate the elastic stress and thus the pressure on the bead; if one further has an estimate of the number of filaments exerting forces on the bead surface, one can evaluate the force exerted per filament. Values as high as 10 pN have been reported, but in the absence of quantitative measures of the input parameters, these must be regarded as "ballpark" estimates.

Extension of this theory to the motion of *Listeria* and ActA-coated beads has led to a model (Gerbal *et al.*, 2000; Marcy *et al.*, 2004) in which the motile driving force results from release of the elastic stress in the actin gel as the obstacle moves. The bulk of the driving force comes not from pushing forces due to the filaments directly behind the obstacle, but rather from the squeezing forces generated by the filaments on the sides of the obstacle. This approach emphasizes several factors other than single-filament effects that can slow the motion of the obstacle as an opposing force is applied. For example, even if the rate of addition of new subunits to the gel at the actin surface remains constant, the obstacle will slow down because the tail becomes wider due to the opposing force, so the distance added per subunit becomes smaller. Furthermore, the gel thickness on the bacterium side drops with increasing velocity, which leads to a drop in propulsion force. This analysis predicts a force–velocity relation of the form that has a rapid initial decay for small forces, but a more rapid decay for larger forces. It was successfully fitted to experimental data for WASp-coated beads where the force was generated by micromanipulation. In certain ranges of parameters, the elastic theory predicts a "hopping" motion (Bernheim-Groswasser *et al.*, 2005), which is displayed by certain mutants of *Listeria* and, in some circumstances, beads coated with the VCA fragment of WASp. The elastic theory also accounts for the symmetry-breaking phenomenon of VCA-coated beads mentioned above. A recent multi-scale approach has incorporated Brownian dynamics-based single-filament force–velocity relations into a continuum model containing some of the effects described by the elastic theory. This model obtains a force–velocity relation for beads having a form similar to that obtained by the elastic theory (Zhu and Carlsson, 2007).

A highly simplified elastic model (Lee *et al.*, 2005) focuses on the dependence of the density of force exerted by the actin network on the position over the surface of a moving bead. The force density is treated by a rate equation including both linear and nonlinear terms. The key ingredient in the model is a positive-feedback term that causes the force density behind the bead to grow if it moves forward. This type of term could result from effects of the form discussed by van Oudenaarden and Theriot (1999), where the growth of filaments behind the bead is accelerated because their opposing force is reduced by the forward motion of the bead. The model displays a symmetry-breaking transition to persistent motion at a critical value of the positive-feedback parameter that is inversely proportional to the drag coefficient of the bead. The velocity is predicted to vanish as the positive feedback decreases to its critical value. In certain ranges of parameters, the model was found to have two steady states, one with a stationary bead and one with a moving bead. This could lead to "hopping" motion alternating between these two states, but at present it is not possible to evaluate the model parameters accurately enough to establish whether this effect corresponds to experiments.

With regard to lamellipodial protrusion, the most intensively studied case is that of fish keratocytes, which can essentially be viewed as one large lamellipodium. The cells glide with a steady motion, and their speed correlates well with the growth rate of actin filaments. The steady-state behavior of a uniformly protruding lamellipod has been studied using a model (Mogilner and Edelstein-Keshet, 2002) based on the densities of barbed and pointed ends, and free and complexed actin. Processes treated by the model include polymerization, depolymerization, diffusion, barbed-end generation by branching, debranching, capping of barbed ends, and actin complex formation and dissociation. The polymerization rate was based on the Brownian-ratchet result given above. The model equations were solved using a combination of numerical solution and analytic theory. It was found that as a function of increasing barbed-end density, the protrusion velocity initially increases because the opposing force per barbed end drops. At higher values of the barbed-end density, the velocity reaches a maximum and then decreases because the large rate of free-actin consumption at the front of the cell reduces the free-actin density, which in turn reduces the polymerization rate. There is thus an optimal density of barbed ends for achieving maximal velocity, which is proportional to the opposing force. The maximal velocity depends inversely on the membrane resistance. A subsequent study (Dawes *et al.*, 2006) treated the spatial distribution of barbed ends, polymerized actin, and Arp2/3 complex as a function of distance from the leading edge of a 1D lamellipod, using three different assumptions about the process by which Arp2/3 complex generates new barbed ends: spontaneous nucleation, tip branching, and side branching. Comparison of observed spatial profiles of barbed-end density with the theoretical predictions revealed the best fit for side branching. There have been fewer studies in more realistic geometries, but Grimm *et al.* (2003) treated a limited range of processes in the context of a 2D lamellipodium model where the actin density varied along the cell front. It was found that the actin density was highest at the center of the lamellipodium, and that the enhancement is greater for high capping rates. The enhancement is qualitatively consistent with experimental results.

Although most modeling of lamellipodial protrusion has focused on the dynamics of actin and actin-binding proteins, a recent study has treated the spatial distribution of the phosphoinositides, the associated kinases/phosphatases, and small GTPases that regulate actin polymerization in response to external signals (Dawes and Edelstein-Keshet, 2007). Using a plausible set of reaction equations in combination with diffusion terms, it was found that the regulatory proteins and small molecules have different characteristic functions: the GTPases function as spatial switches, while phosphoinositides filter noise and define front versus back. The model suggested explanations for several experimental observations, including defects in gradient detection in mutants lacking Cdc42, and proper directed motion in mutants lacking PTEN.

Recently, the interplay between actin propulsion forces and contractility has been treated in a simple model of lamellipodial protrusion that views the actin gel as an active viscoelastic medium (Kruse *et al.*, 2006). (We describe this work here, rather than in the section below on whole-cell models, because the main focus is on protrusion.) The term "active" means that the constitutive relation for the gel contains terms not present in typical "passive" materials, including a contribution to the deformation rate in the absence of external stresses, proportional to the myosin activity. Calculation of the spatially varying deformation rate of the actin gel yielded a lamellipodium profile similar to those observed experimentally. The forces exerted by the substrate on the actin gel were found to change sign a few microns in from the leading edge. This led to a transition from retrograde flow at the leading edge to anterograde flow farther back in the lamellipodium, in agreement with experimental observations based on speckle microscopy.

The relationship between actin polymerization and the deformation of the membrane has been treated in a recent study by Gov and Gopinathan (2006). They focused on the spontaneous membrane curvature associated with polymerization activators such as Cdc42 and PIP_2 in the membrane. The motion of these activators was treated by using a model of noise-assisted biased diffusion, where the biasing force comes from the attraction of the activators to regions of membrane curvature; the curvature of the membrane (motion in the direction perpendicular to the plane of the membrane) was treated as a competition between the elastic restoring force tending to straighten the membrane and a term depending on the density of activators, which favored membrane bending. Depending on the properties assumed for the activators, the membrane displayed either wavelike behavior, corresponding to membrane ruffling and actin waves, or unstable behavior leading to the formation of filopodia. The predictions of this model are consistent with several types of experimental data, including the changes in the membrane diffusion coefficient induced by cell motion, and increases in the fluidity of the membrane in the front part of cells moving in shear flow.

A more detailed type of continuum theory is obtained when not only the density of actin and related proteins but also the orientation and/or length distribution functions of actin filaments are treated. The orientation distribution function gives the density of filaments at a certain place pointing in a particular direction, and the length distribution does the same for filaments of a particular length. These properties have been treated in bulk geometries (Carlsson, 2005; Edelstein-Keshet, 1998) where the spatial variation is absent, but we focus on calculations treating the spatial variation. These have used a 1D geometry appropriate for a flat lamellipod or pathogen propulsion; the spatial variation is in the direction perpendicular to the membrane or the surface of the pathogen. The length distribution of filaments in a lamellipod has been treated in a model (Edelstein-Keshet and Ermentrout, 2001) that includes the motion of filament tips due to polymerization/depolymerization, and severing; the populations of filaments capped and uncapped at their barbed ends were treated separately. The calculated distributions indicated that the tips of longer filaments tended to be closer to the membrane. By appropriate choice of the parameters in the model, a qualitative agreement was obtained with the measured distribution of polymerized actin as a function of distance from the membrane.

The filament orientation distribution has been treated in a model (Maly and Borisy, 2001) focusing on the number of filaments in the immediate vicinity (the "active zone") of a pathogen or biomimetic bead moving at a fixed velocity. This model treated Arp2/3-induced branching, which creates a new filament at an angle of 70 degrees relative to the "mother" filament, and capping, which was assumed to remove a filament from the active zone. It was shown that over a certain range of velocities, the filament orientation distribution has a two-peaked structure, with the peaks at roughly ± 35 degrees. This is consistent with experiments on *Xenopus* keratocytes. An extension of this model (Carlsson, 2003) treated the flow of slower filaments out of the active zone as a function of their angle. This allowed the calculation of the obstacle velocity by the application of a steady-state condition for the number of filaments in the active region. The results confirmed those of earlier filament-based simulations: the obstacle velocity is independent of opposing force if all new filaments are created via autocatalytic branching. If the new filaments arise by *de novo* nucleation, independent of existing filaments, the velocity decays exponentially with opposing force as in the Brownian-ratchet model.

The main experimentally accessible property that has been treated by both continuum theories and many-filament theories is the force–velocity relation for

Listeria and biomimetic beads. The most complete data obtained so far are for plastic beads whose motion was slowed by direct physical contact using a cantilever (Marcy *et al.*, 2004). As Fig. 5 shows, the velocity dropped by about a factor of 2 at an opposing force of 1 pN, and subsequently decreased more slowly. This experiment could also probe negative (pulling) forces, which were found to speed up the motion. The overall shape of the force–velocity curve under opposing force is similar to that obtained by both the elastic stress and tethered-ratchet theories, although a fit to the data has been done only with the elastic stress theory. The latter theory also fit the negative-force part of the force–velocity relation. The drop-off of the velocity with opposing force is inconsistent with the completely autocatalytic model assumed in the many-filament simulations. However, the slowness of the decay of the velocity at high opposing forces could indicate that the number of filaments is increasing under opposing force as in the autocatalytic model.

III. Cell Adhesion and Retraction

As evidenced by the previous section, there has been a significant amount of theoretical work on phenomena at the leading edge of a motile cell. This topic holds a powerful natural appeal since the leading edge is one of the principal sites of assembly and dynamics, and it is almost universally true that the more dynamic a process is, the more interest there is in describing its behavior. However, the mechanisms involved in formation of adhesions to the substrate and retraction of the rear of the cell are equally important and deserve the same amount of attention. Without contraction, retraction, and recycling of assembled filaments, bundles,

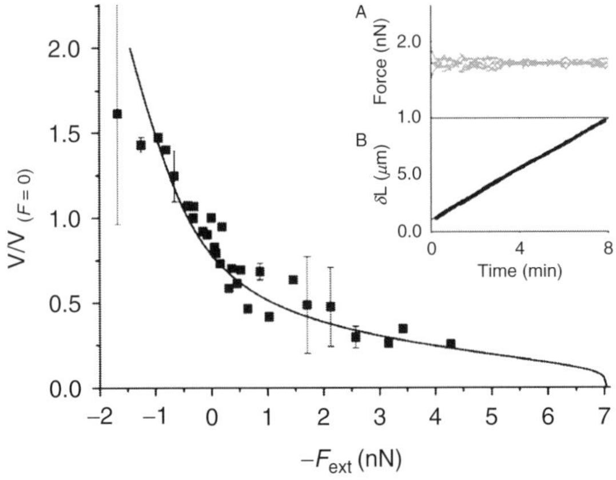

Fig. 5 Force–velocity relation for plastic beads obtained with a cantilever (Marcy *et al.*, 2004). Squares: experimental results. Line: prediction of elastic theory. Insets show (A) stability of force measurement and (B) velocity measurement (reproduced by permission of Cecile Sykes).

and networks at the rear of the cell, assembly at the leading edge would be limited by the shortage of actin and other proteins, and the mechanical forces imposed by the cell membrane would prevent movement.

Because contraction of the rear of a cell requires adhesion at the front, we treat these two phenomena together. Two primary classes of mathematical treatments have been developed to describe cell adhesion and retraction. The first class involves contractile bundles, stress fibers, and other myosin-based mechanisms. The second class could be termed nonmyosin based and centers around contracting gels such as those observed in nematode sperm. Details and examples from both of these classes are given below.

A. Myosin-Based Retraction

Actomyosin contraction has been studied extensively in both muscle and nonmuscle cells, but here we focus on nonmuscle cells. In these cells, myosin II is clearly localized to the rear of the cell and has been implicated in retraction of the trailing edge of the cell as well as breaking or removal of focal adhesions. One of the first models to treat the interaction and contraction of actin filaments mediated by myosin treated both parallel and antiparallel filaments using rate equations for their spatially dependent concentrations (Kruse and Julicher, 2000). In this formulation, the myosins were not explicitly treated, but their effect was captured in the fluxes by having parallel filaments move in the same direction and antiparallel filaments move in opposite directions. Despite the simplicity of this model, it is able to capture a wide range of behaviors in terms of filament distributions and tension profiles. This simple, 1D model has since been extended to 2D arrays where filaments can now interact at any angle. By adding an additional dimension, a completely new range of phenomena emerge from such a model, including orientationally polarized phases that are obtained based on the filament and/or myosin density (Ahmadi *et al.*, 2005; Kruse *et al.*, 2004, 2005; Liverpool and Marchetti, 2003). The range of structures and bundle arrangements from both of these models look very reminiscent of those seen in *in vitro* experiments as well as what is observed in live cells.

As mentioned above, contractile bundles or stress fibers cannot provide work or movement without a mechanism for adhering to the underlying substrate. The formation of focal adhesions and the mechanical interactions between cells and their environment is a very active field of study (for a recent review, see Vogel and Sheetz, 2006). With the emergence of signaling pathways associated with the formation of focal adhesions, several groups have developed detailed mathematical and thermodynamic models to describe their formation and regulation. Bershadsky *et al.* (2006) have done extensive experimental and modeling work on focal adhesions (Shemesh *et al.*, 2005a,b). Their experiments have attempted to define the wide range of proteins that affect focal adhesion behavior (Dia1, Rho, etc.), and their model describing focal adhesion mechanosensitivity and dynamics is essentially protein self-assembly under control of elastic stresses (see Fig. 6).

The underlying premise is that forces applied to a filament have a direct effect on its polymerization, and based on the balance of forces and the distribution and densities of focal adhesions (those that are applying force and those that are acting as anchors), the authors find four possible modes of behavior. Two of these regimes represent unrestricted polymerization or depolymerization, but the other two modes can give rise to steady-state filament lengths. Changes in the applied force and/or force distribution can cause the cell to switch between the regimes in a well-defined fashion, and the next steps will be to experimentally test the predictions of this model.

The previous model exemplifies the usefulness of such simple, physically based models since the authors demonstrate how this mathematical treatment can be applied equally well to the polymerization of an actin filament with processive capping by a formin, discussed above, as it can to the formation of focal adhesions (Shemesh *et al.*, 2005a). One final model explicitly models the interaction between

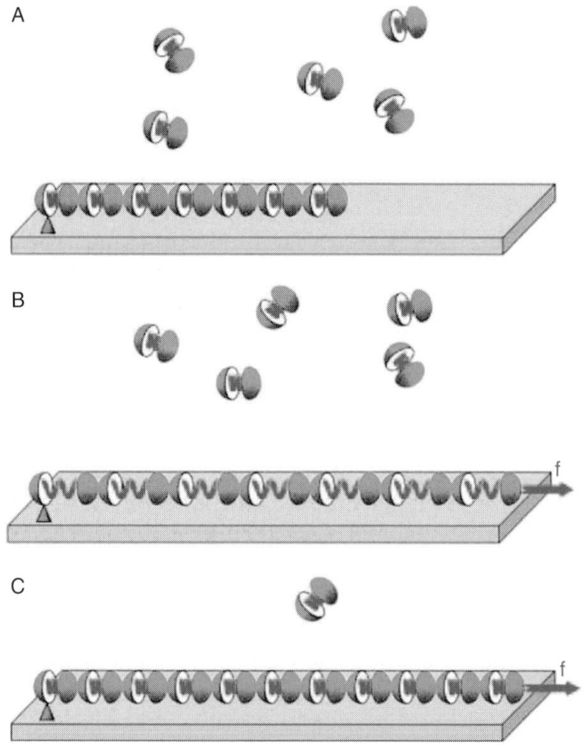

Fig. 6 The model of Shemesh *et al.* (2005a,b) depicting (A) polymerization, (B) application of force and accumulation of stresses within the filament, and (C) release of internal stresses by insertion of additional monomers at the end of the filament (reproduced by permission of Michael Kozlov).

the stress fibers and the integrins, where the integrins may be either free in the membrane or bound to the underlying substrate (Novak *et al.*, 2004). The authors find that focal adhesions tend to relocate to the periphery of the cell, moving with the diffusion constant of the integrin, and concentrating in regions with the highest boundary curvature. This observation agrees well with experimental results and appears to result from the fact that stress fibers in the center of the cell are organized randomly so that the forces tend to cancel out. Focal adhesions at the cell edge are forced to be more ordered and therefore, their net force is higher and directed toward the interior of the cell. More work integrating such models of myosin contraction with focal adhesion formation will be required to advance our understanding of retraction and contractility, and how these functions relate to the assembly and dynamics at the leading edge of the cell.

B. Non-Myosin-Based Contractility

Although most motile cells use myosin-based retraction at the rear of the cell, this is not the only mechanism for generating contraction. Sperm of the nematode *Ascaris suum* represent a simple, motile system that is based on major sperm protein (MSP) rather than actin (Miao *et al.*, 2003). MSP is analogous to actin in that it assembles into filaments that form a gel, but the filaments are nonpolar. What is most intriguing about this system is that there are no associated motor proteins, but instead contraction occurs by a structural transition of the MSP gel. The MSP filaments can bundle via hydrophobic and electrostatic forces. Models of this process suggest that bundling at the leading edge of the cell extends the filaments, thereby storing elastic energy within the bundle while providing a protrusive force (Bottino *et al.*, 2002; Wolgemuth *et al.*, 2004, 2005). At the rear of the cell, when these bundles are dissociated, the stored elastic energy is released, resulting in contraction. The model of Bottino *et al.* (2002) is one of the more complete treatments. Apart from simply capturing adhesion and retraction, the authors also consider polymerization, protrusion, and depolymerization, resulting in a very complete description of the motile process. More discussion of this model appears in Section IV.

Although the actomyosin system is thought to be the driving contractile force behind translocation of the cell body, myosin II null *Dicyostelium discoideum* cells still exhibit migration (De Lozanne and Spudich, 1987; Knecht and Loomis, 1987), suggesting that nonmyosin contraction may also play a role in cells other than nematode sperm. Wolgemuth (2005) developed a simple model involving stick-slip adhesion with some contractile stress generating mechanism that is able to produce periodic lamellipodial contractions. Although he admits the contractile force needed to produce this behavior could come from myosin, it is not strictly required and could result from depolymerization of the actin gel.

In summary, models of cell retraction and focal adhesion formation, although not as developed as those for actin branching and assembly, have shown significant progress in recent years. This field has shown less progress than protrusion studies, partly because the requirement of motor proteins in most cases increases the complexity of the process, and partly because there are fewer model organisms (such as *Listeria*) and fewer biomimetic systems (such as protein-coated beads) available. As more experimental details become clear and our expanding abilities to tailor materials nanostructure allow the development of more biomimetic systems, the models will certainly expand to include more explicit parameters and, when combined with the assembly models for the leading edge of the cell, they should allow more complete modeling of the entire cell.

IV. Whole-Cell Models

To this point, we have largely discussed piecewise and reductionist models of cell migration that treat only one or two parts of the moving cell, while ignoring large portions of the system. The reason for such models is perfectly obvious: complicated models require more data for parameterization, and the predictions arising from such models are often less clear than those from more simple, physically based models. The modelers are not the only ones striving for simplicity—the goal of many experiments also is to dissect the cell into its smallest functional parts. Despite the complexities of whole-cell modeling, significant efforts have been directed at describing entire cells using a wide range of modeling techniques. In this section, we will describe some of these models.

Although strictly not whole cells, a simpler version of the many-filament models described in Section II has been used to treat the persistent motion of cytoplasts—cell fragments containing the basics of the migration apparatus but having no genetic material (Sambeth and Baumgaertner, 2001). This model included polymerization, depolymerization, and branching, but no barbed-end capping. The branching was autocatalytic, and the results of the simulations suggested that the autocatalytic nature of the branching itself provides a positive-feedback mechanism by which the cell can spontaneously break its symmetry and remain in a persistent state of directed motion. Subsequent extensions of these studies included spontaneous nucleation processes (Satyanarayana and Baumgaertner, 2004) and argued that the stop-and-go motions sometimes displayed by cells can be attributed to a competition between branching nucleation and spontaneous nucleation.

When considering the whole cell, the battery of migration models is as varied as the number of different cell types; however just as cells share many of the same characteristics, so do the available models. For the most part, whole-cell models consider dynamical and/or mechanical phenomena. Dynamic events certainly include explicit molecular-scale processes such as the polymerization of actin, but they may also be processes that have a far less concrete physical basis, such as the change in stability of a focal adhesion due to force, or the effect of pH on actin bundle formation. Similarly, cell mechanics may be treated explicitly using our knowledge of filament and membrane mechanics, or it may be included in a phenomenological manner. An example of the latter is the paper by Alt and Dembo (1999) that treats the motion of ameboid cells. They treat the cell as a two-phase fluid: a filament phase that describes the actin–myosin network and a solvent phase that contains the unpolymerized actin monomers. The two phases are reactive (i.e., actin can polymerize to form filaments) and viscous, and the filament phase can store contractile energy and interact with the underlying substrate. Although this model is very coarse-grained, it can reproduce a variety of cell migration behaviors, including periodic ruffle formation, cycles of protrusion and retraction, and centripetal flow of the cytoplasm. A similar model from Gracheva and Othmer (2004) again uses a continuum approximation with a viscoelastic cell that interacts with a viscous substrate and a viscoelastic cytosol that gives rise to passive stresses, due to elastic and viscous forces, as well as active stresses due to actin polymerization and/or myosin contraction. This model, like the one from Alt and Dembo, can reproduce a wide variety of observed phenomena, including the 1D movement of a fibroblast and descriptions of cell deformation, traction forces, and cell speed that agree with experimental findings.

A more detailed class of whole-cell models explicitly treats the formation and interaction of filaments. One of the most comprehensive models is the fish keratocyte from the Mogilner lab (Rubinstein *et al.*, 2005). This 2D model combines several submodels involving protrusion at the leading edge, the mechanics of the actin network in the lamellipodium, contraction at the rear of the cell, and transport of actin on the boundary of the lamellipodium. Although the mathematics of this model is somewhat complicated, the premises on which the mathematics is based are very clearly spelled out in seven assumptions (see Rubinstein *et al.*, 2005, for details). These assumptions include not only intuitive ideas such as "the protrusion rate is locally normal to the leading edge and proportional to the local concentration of G-actin," but also the supposition that "there exists a constant critical low F-actin density at which the actin network collapses into the actin-myosin bundle, determining the rear edge of the lamellipodium." The experimental evidence supporting each of the seven assumptions is variable; however, by solving the resulting set of equations, the implications of these various hypotheses can easily be tested. This is the true strength of such models and serves as an excellent example for the rest of the modeling community.

The keratocyte model just discussed is in fact very similar to the model developed by Bottino *et al.* (2002) to describe the crawling of nematode sperm. Instead of relying on actin filaments, this system is based on the polymerization of MSP; however, the physics (and mathematics) describing this process is nearly identical to the keratocyte case. Figure 7 shows some details of their model. The top two portions of the figure depict the 1D and 2D representations of their model. Each black dot represents a mass of cytoskeleton contained in the surrounding polygon, and these nodes are connected by the finite elements shown in Fig. 7C. These finite elements are more than "standard" viscoelastic materials (i.e., a spring and dashpot in parallel), in that they also have the ability to store elastic energy in the tensile element represented by τ. Exactly what cellular component or process would

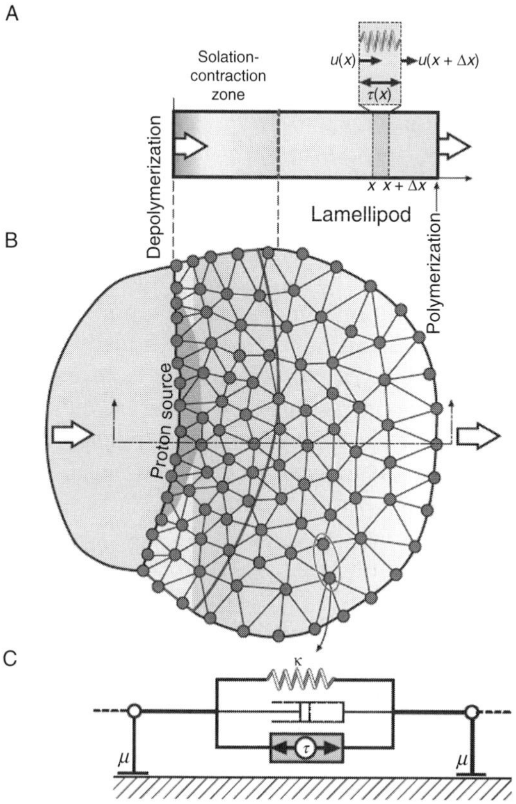

Fig. 7 Details of the finite element model of Bottino *et al.* (2002) showing (A) one-dimensional and (B) two-dimensional representations of the model. The black dots represent the cytoskeleton, and they are connected by the finite elements depicted in (C) (reproduced with permission of George Oster).

correspond to such an element is not clear; however, this unique property of MSP gels is thought to allow this system to produce migration without the need for any motor proteins. Just as with the keratocyte model, this model can be tested and compared with experimental data, and it shows good agreement with many features such as the shape and velocity of the moving cell.

The effects of the small GTPases Cdc42, Rac, and Rho have recently been included in a 2D whole-cell analysis (Maree *et al.*, 2006) based on the "cellular Potts model," which takes the stochastic aspects of the cell boundary motion into account. This model treats the spatially varying concentrations of the GTPases, Arp2/3 complex, and densities of barbed ends and filaments pointing in six different directions (as an approximation to the continuous distribution of the filament orientations). Implementation of the model gives realistic cell shapes for keratocytes, and correct speeds for biologically reasonable parameter values. In addition, persistent motion is obtained after an initial stimulus, even after the stimulus is removed.

V. Summary and Future Outlook

A major difficulty in modeling cell migration is our incomplete understanding of the underlying interactions and associated kinetics and thermodynamics. Although many different proteins are involved in cell migration, the concentrations/activities, associated kinetic rate constants, and the dependence of these rate constants on experimental conditions are in general not known. The interactions of the actin cytoskeleton with the substrate and the membrane are particularly unclear. This state of affairs has the unfortunate consequence that while it is often possible to adjust the parameters in a model to match a physical observation, it is very hard to establish conclusively that the observation is actually due to the effects considered in the model. For many of the phenomena related to cell migration, there are several competing explanations. For example, the slowing of actin-propelled obstacles in response to opposing force can be explained either in terms of elastic effects or in terms of the slowing of polymerization at the obstacle–filament interface. In this and many other cases, we are still unable to determine unambiguously which (if any) of the existing theories is correct. Given this situation, theoretical efforts can take several routes. One is to explore the possible modes of organization and action of the cytoskeleton—varying parameters over a wide range of values to see what behaviors are possible, and which aspects of the behavior are robust to changes in parameter values and model assumptions. Another is to focus on biomimetic migration analogues, where the number of unknown interactions can be limited. Finally, one can use models that treat only the most important variables, using Occam's razor to discriminate between different models that capture the essence of the observed phenomena.

There are several aspects of cell migration that should receive more attention by the modeling community. There has been a strong focus on modeling lamellipodial protrusion, but it has been shown that cells can migrate without lamellipodia (Gupton *et al.*, 2005; Vidali *et al.*, 2006). In such cases, the migration mechanism is not established, but it may involve either filaments stiffened by tropomyosin coating or actin filaments cross-linked by myosin II. The microscopic mechanisms underlying retraction also need more attention. Retraction often involves myosins, but there are also non-myosin-based retraction mechanisms (Mogilner and Oster, 2003a) that need to be explored in more detail. The coupling of protrusion and retraction (Small and Resch, 2005) is another area that could benefit from enhanced modeling activity. Finally, the cell migration modeling community should expand its horizons to treat migration in three dimensions. Placing a cell in a 3D environment instead of on a flat substrate could markedly change the behavior of the cell, and recent 3D imaging methods (Friedl, 2004) are beginning to provide data that motivates 3D modeling.

Acknowledgments

We thank many of the people cited in this work for sharing their data and figures with us. We further thank John Cooper for many stimulating discussions and a critical reading of the manuscript. This work was supported by grants DMS-0240770 from the National Science Foundation (AEC) and GM-067246 from the National Institutes of Health (DS).

References

Abraham, V. C., Krishnamurthi, V., Taylor, D. L., and Lanni, F. (1999). The actin-based nanomachine at the leading edge of migrating cells. *Biophys. J.* **77,** 1721–1732.

Ahmadi, A., Liverpool, T. B., and Marchetti, M. C. (2005). Nematic and polar order in active filament solutions. *Phys. Rev. E Stat. Nonlin. Soft. Matter Phys.* **72,** 060901.

Alberts, J. B., and Odell, G. M. (2004). In silico reconstitution of *Listeria* propulsion exhibits nano-saltation. *PLoS Biol.* **2,** 2054–2066.

Alt, W., and Dembo, M. (1999). Cytoplasm dynamics and cell motion: Two-phase flow models. *Math. Biosci.* **156,** 207–228.

Atilgan, E., Wirtz, D., and Sun, S. X. (2005). Morphology of the lamellipodium and organization of actin filaments at the leading edge of crawling cells. *Biophys. J.* **89,** 3589–3602.

Atilgan, E., Wirtz, D., and Sun, S. X. (2006). Mechanics and dynamics of actin-driven thin membrane protrusions. *Biophys. J.* **90,** 65–76.

Bernheim-Groswasser, A., Prost, J., and Sykes, C. (2005). Mechanism of actin-based motility: A dynamic state diagram. *Biophys. J.* **89,** 1411–1419.

Bershadsky, A. D., Ballestrem, C., Carramusa, L., Zilberman, Y., Gilquin, B., Khochbin, S., Alexandrova, A. Y., Verkhovsky, A. B., Shemesh, T., and Kozlov, M. M. (2006). Assembly and mechanosensory function of focal adhesions: Experiments and models. *Eur. J. Cell Biol.* **85,** 165–173.

Bottino, D., Mogilner, A., Roberts, T., Stewart, M., and Oster, G. (2002). How nematode sperm crawl. *J. Cell Sci.* **115,** 367–384.

Burroughs, N. J., and Marenduzzo, D. (2005). Three-dimensional dynamic Monte Carlo simulations of elastic actin-like ratchets. *J. Chem. Phys.* **123,** 174908.

Carlsson, A. E. (2000). Force-velocity relation for growing biopolymers. *Phys. Rev. E* **62,** 7082–7091.

Carlsson, A. E. (2001). Growth of branched actin networks against obstacles. *Biophys. J.* **81,** 1907–1923.

Carlsson, A. E. (2003). Growth velocities of branched actin networks. *Biophys. J.* **84,** 2907–2918.

Carlsson, A. E. (2005). The effect of branching on the critical concentration and average filament length of actin. *Biophys. J.* **89,** 130–140.

Dawes, A. T., Bard Ermentrout, G., Cytrynbaum, E. N., and Edelstein-Keshet, L. (2006). Actin filament branching and protrusion velocity in a simple 1D model of a motile cell. *J. Theor. Biol.* **242,** 265–279.

Dawes, A. T., and Edelstein-Keshet, L. (2007). Phosphoinositides and Rho proteins spatially regulate actin polymerization to initiate and maintain directed movement in a 1D model of a motile cell. *Biophys. J.* **92**(3), 744–768.

De Lozanne, A., and Spudich, J. A. (1987). Disruption of the Dictyostelium myosin heavy chain gene by homologous recombination. *Science* **236,** 1086–1091.

Dickinson, R. B., Caro, L., and Purich, D. L. (2004). Force generation by cytoskeletal filament end-tracking proteins. *Biophys. J.* **87,** 2838–2854.

Dickinson, R. B., Caro, L., and Purich, D. L. (2005). Force generation by cytoskeletal filament end-tracking proteins. *Biophys. J.* **88,** 757–758(Vol. 87, p. 2838, 2004).

Dickinson, R. B., and Purich, D. L. (2002). Clamped-filament elongation model for actin-based motors. *Biophys. J.* **82,** 605–617.

Dickinson, R. B., Southwick, F. S., and Purich, D. L. (2002). A direct-transfer polymerization model explains how the multiple profilin-binding sites in the actoclampin motor promote rapid actin-based motility. *Arch. Biochem. Biophys.* **406,** 296–301.

Edelstein-Keshet, L. (1998). A mathematical approach to cytoskeletal assembly. *Eur. Biophys. J. Biophys. Lett.* **27,** 521–531.

Edelstein-Keshet, L., and Ermentrout, G. B. (2001). A model for actin-filament length distribution in a lamellipod. *J. Math. Biol.* **43,** 325–355.

Friedl, P. (2004). Prespecification and plasticity: Shifting mechanisms of cell migration. *Curr. Opin. Cell Biol.* **16,** 14–23.

Gerbal, F., Chaikin, P., Rabin, Y., and Prost, J. (2000). An elastic analysis of *Listeria* monocytogenes propulsion. *Biophys. J.* **79,** 2259–2275.

Gov, N. S., and Gopinathan, A. (2006). Dynamics of membranes driven by actin polymerization. *Biophys. J.* **90,** 454–469.

Gracheva, M. E., and Othmer, H. G. (2004). A continuum model of motility in ameboid cells. *Bull. Math. Biol.* **66,** 167–193.

Grimm, H. P., Verkhovsky, A. B., Mogilner, A., and Meister, J. J. (2003). Analysis of actin dynamics at the leading edge of crawling cells: Implications for the shape of keratocyte lamellipodia. *Eur. Biophys. J. Biophys. Lett.* **32,** 563–577.

Gupton, S. L., Anderson, K. L., Kole, T. P., Fischer, R. S., Ponti, A., Hitchcock-DeGregori, S. E., Danuser, G., Fowler, V. M., Wirtz, D., Hanein, D., and Waterman-Storer, C. M. (2005). Cell migration without a lamellipodium: Translation of actin dynamics into cell movement mediated by tropornyosin. *J. Cell Biol.* **168,** 619–631.

Higgs, H. N. (2005). Formin proteins: A domain-based approach. *Trends Biochem. Sci.* **30,** 342–353.

Hill, T. L. (1987). "Linear Aggregation Theory in Cell Biology." Springer-Verlag, New York.

Knecht, D. A., and Loomis, W. F. (1987). Antisense RNA inactivation of myosin heavy chain gene expression in Dictyostelium discoideum. *Science* **236,** 1081–1086.

Kovar, D. R., and Pollard, T. D. (2004). Insertional assembly of actin filament barbed ends in association with formins produces piconewton forces. *Proc. Natl. Acad. Sci. USA* **101,** 14725–14730.

Kruse, K., Joanny, J. F., Julicher, F., and Prost, J. (2006). Contractility and retrograde flow in lamellipodium motion. *Phys. Biol.* **3,** 130–137.

Kruse, K., Joanny, J. F., Julicher, F., Prost, J., and Sekimoto, K. (2004). Asters, vortices, and rotating spirals in active gels of polar filaments. *Phys. Rev. Lett.* **92,** 078101.

Kruse, K., Joanny, J. F., Julicher, F., Prost, J., and Sekimoto, K. (2005). Generic theory of active polar gels: A paradigm for cytoskeletal dynamics. *Eur. Phys. J. E Soft Matter* **16,** 5–16.

Kruse, K., and Julicher, F. (2000). Actively contracting bundles of polar filaments. *Phys. Rev. Lett.* **85,** 1778–1781.

Kruse, K., and Sekimoto, K. (2002). Growth of fingerlike protrusions driven by molecular motors. *Phys. Rev. E* **66,** 31904–31909.

Landau, L. D., Lifshiëtis, E. M., Kosevich, A. D. M., and Pitaevskiæi, L. P. (1986). "Theory of Elasticity." Pergamon Press, Oxford, New York.

Lee, A., Lee, H. Y., and Kardar, M. (2005). Symmetry-breaking motility. *Phys. Rev. Lett.* **95,** 138101–138104.

Liverpool, T. B., and Marchetti, M. C. (2003). Instabilities of isotropic solutions of active polar filaments. *Phys. Rev. Lett.* **90,** 138102.

Maly, I. V., and Borisy, G. G. (2001). Self-organization of a propulsive actin network as an evolutionary process. *Proc. Natl. Acad. Sci. USA* **98,** 11324–11329.

Marcy, Y., Prost, J., Carlier, M. F., and Sykes, C. (2004). Forces generated during actin-based propulsion: A direct measurement by micromanipulation. *Proc. Natl. Acad. Sci. USA* **101,** 5992–5997.

Maree, A. F., Jilkine, A., Dawes, A., Grieneisen, V. A., and Edelstein-Keshet, L. (2006). Polarization and movement of keratocytes: A multiscale modelling approach. *Bull. Math. Biol.* **68,** 1169–1211.

Miao, L., Vanderlinde, O., Stewart, M., and Roberts, T. M. (2003). Retraction in amoeboid cell motility powered by cytoskeletal dynamics. *Science* **302,** 1405–1407.

Mogilner, A., and Edelstein-Keshet, L. (2002). Regulation of actin dynamics in rapidly moving cells: A quantitative analysis. *Biophys. J.* **83,** 1237–1258.

Mogilner, A., and Oster, G. (1996). Cell motility driven by actin polymerization. *Biophys. J.* **71,** 3030–3045.

Mogilner, A., and Oster, G. (1999). The polymerization ratchet model explains the force-velocity relation for growing microtubules. *Eur. Biophys. J. Biophys. Lett.* **28,** 235–242.

Mogilner, A., and Oster, G. (2003a). Cell biology. Shrinking gels pull cells. *Science* **302,** 1340–1341.

Mogilner, A., and Oster, G. (2003b). Polymer motors: Pushing out the front and pulling up the back. *Curr. Biol.* **13,** R721–R733.

Mogilner, A., and Rubinstein, B. (2005). The physics of filopodial protrusion. *Biophys. J.* **89,** 782–795.

Noireaux, V., Golsteyn, R. M., Friederich, E., Prost, J., Antony, C., Louvard, D., and Sykes, C. (2000). Growing an actin gel on spherical surfaces. *Biophys. J.* **78,** 1643–1654.

Novak, I. L., Slepchenko, B. M., Mogilner, A., and Loew, L. M. (2004). Cooperativity between cell contractility and adhesion. *Phys. Rev. Lett.* **93,** 268109.

Peskin, C. S., Odell, G. M., and Oster, G. F. (1993). Cellular motions and thermal fluctuations—the Brownian Ratchet. *Biophys. J.* **65,** 316–324.

Rubinstein, B., Jacobson, K., and Mogilner, A. (2005). Multiscale two-dimensional modeling of a motile simple-shaped cell. *Multisc. Model. Simul.* **3,** 413–439.

Sambeth, R., and Baumgaertner, A. (2001). Autocatalytic polymerization generates persistent random walk of crawling cells. *Phys. Rev. Lett.* **86,** 5196–5199.

Satyanarayana, S. V. M., and Baumgaertner, A. (2004). Shape and motility of a model cell: A computational study. *J. Chem. Phys.* **121,** 4255–4265.

Shemesh, T., Bershadsky, A. D., and Kozlov, M. M. (2005a). Force-driven polymerization in cells: Actin filaments and focal adhesions. *J. Phys.-Condens. Matter* **17,** S3913–S3928.

Shemesh, T., Geiger, B., Bershadsky, A. D., and Kozlov, M. M. (2005b). Focal adhesions as mechanosensors: A physical mechanism. *Proc. Natl. Acad. Sci. USA* **102,** 12383–12388.

Shemesh, T., Otomo, T., Rosen, M. K., Bershadsky, A. D., and Kozlov, M. M. (2005c). A novel mechanism of actin filament processive capping by formin: Solution of the rotation paradox. *J. Cell Biol.* **170,** 889–893.

Small, J. V., and Resch, G. P. (2005). The comings and goings of actin: Coupling protrusion and retraction in cell motility. *Curr. Opin. Cell Biol.* **17,** 517–523.

Upadhyaya, A., Chabot, J. R., Andreeva, A., Samadani, A., and van Oudenaarden, A. (2003). Probing polymerization forces by using actin-propelled lipid vesicles. *Proc. Natl. Acad. Sci. USA* **100,** 4521–4526.

van der Gucht, J., Paluch, E., Plastino, J., and Sykes, C. (2005). Stress release drives symmetry breaking for actin-based movement. *Proc. Natl. Acad. Sci. USA* **102,** 7847–7852.

van Oudenaarden, A., and Theriot, J. A. (1999). Cooperative symmetry-breaking by actin polymerization in a model for cell motility. *Nat. Cell Biol.* **1,** 493–499.

Vavylonis, D., Kovar, D. R., O'Shaughnessy, B., and Pollard, T. D. (2006). Model of formin-associated actin filament elongation. *Mol. Cell* **21,** 455–466.

Vidali, L., Chen, F., Cicchetti, G., Ohta, Y., and Kwiatkowski, D. J. (2006). Rac1-null mouse embryonic fibroblasts are motile and respond to platelet-derived growth factor. *Mol. Biol. Cell* **17,** 2377–2390.

Vogel, V., and Sheetz, M. (2006). Local force and geometry sensing regulate cell functions. *Nat. Rev. Mol. Cell Biol.* **7,** 265–275.

Wolgemuth, C. W. (2005). Lamellipodial contractions during crawling and spreading. *Biophys. J.* **89,** 1643–1649.

Wolgemuth, C. W., Miao, L., Vanderlinde, O., Roberts, T., and Oster, G. (2005). MSP dynamics drives nematode sperm locomotion. *Biophys. J* **88,** 2462–2471.

Wolgemuth, C. W., Mogilner, A., and Oster, G. (2004). The hydration dynamics of polyelectrolyte gels with applications to cell motility and drug delivery. *Eur. Biophys. J.* **33,** 146–158.

Zeile, W. L., Zhang, F. L., Dickinson, R. B., and Purich, D. L. (2005). Listeria's right-handed helical rocket-tail trajectories: Mechanistic implications for force generation in actin-based motility. *Cell Motil. Cytoskeleton* **60,** 121–128.

Zhu, J., and Carlsson, A. E. (2007). Growth of attached actin filaments. *Eur. Phys. J. E Soft Matter* **21,** 209–222.

INDEX

C

D

I

J

K

L

M

P

T

U

V

W

X

VOLUMES IN SERIES

Founding Series Editor
DAVID M. PRESCOTT

Volume 1 (1964)
Methods in Cell Physiology
Edited by David M. Prescott

Volume 2 (1966)
Methods in Cell Physiology
Edited by David M. Prescott

Volume 3 (1968)
Methods in Cell Physiology
Edited by David M. Prescott

Volume 4 (1970)
Methods in Cell Physiology
Edited by David M. Prescott

Volume 5 (1972)
Methods in Cell Physiology
Edited by David M. Prescott

Volume 6 (1973)
Methods in Cell Physiology
Edited by David M. Prescott

Volume 7 (1973)
Methods in Cell Biology
Edited by David M. Prescott

Volume 8 (1974)
Methods in Cell Biology
Edited by David M. Prescott

Volume 9 (1975)
Methods in Cell Biology
Edited by David M. Prescott

Volume 10 (1975)
Methods in Cell Biology
Edited by David M. Prescott

Volume 11 (1975)
Yeast Cells
Edited by David M. Prescott

Volume 12 (1975)
Yeast Cells
Edited by David M. Prescott

Volume 13 (1976)
Methods in Cell Biology
Edited by David M. Prescott

Volume 14 (1976)
Methods in Cell Biology
Edited by David M. Prescott

Volume 15 (1977)
Methods in Cell Biology
Edited by David M. Prescott

Volume 16 (1977)
Chromatin and Chromosomal Protein Research I
Edited by Gary Stein, Janet Stein, and Lewis J. Kleinsmith

Volume 17 (1978)
Chromatin and Chromosomal Protein Research II
Edited by Gary Stein, Janet Stein, and Lewis J. Kleinsmith

Volume 18 (1978)
Chromatin and Chromosomal Protein Research III
Edited by Gary Stein, Janet Stein, and Lewis J. Kleinsmith

Volume 19 (1978)
Chromatin and Chromosomal Protein Research IV
Edited by Gary Stein, Janet Stein, and Lewis J. Kleinsmith

Volume 20 (1978)
Methods in Cell Biology
Edited by David M. Prescott

Advisory Board Chairman
KEITH R. PORTER

Volume 21A (1980)
Normal Human Tissue and Cell Culture, Part A: Respiratory, Cardiovascular, and Integumentary Systems
Edited by Curtis C. Harris, Benjamin F. Trump, and Gary D. Stoner

Volume 21B (1980)
Normal Human Tissue and Cell Culture, Part B: Endocrine, Urogenital, and Gastrointestinal Systems
Edited by Curtis C. Harris, Benjamin F. Trump, and Gray D. Stoner

Volume 22 (1981)
Three-Dimensional Ultrastructure in Biology
Edited by James N. Turner

Volume 23 (1981)
Basic Mechanisms of Cellular Secretion
Edited by Arthur R. Hand and Constance Oliver

Volume 24 (1982)
The Cytoskeleton, Part A: Cytoskeletal Proteins, Isolation and Characterization
Edited by Leslie Wilson

Volume 25 (1982)
The Cytoskeleton, Part B: Biological Systems and *In Vitro* Models
Edited by Leslie Wilson

Volume 26 (1982)
Prenatal Diagnosis: Cell Biological Approaches
Edited by Samuel A. Latt and Gretchen J. Darlington

Series Editor
LESLIE WILSON

Volume 27 (1986)
Echinoderm Gametes and Embryos
Edited by Thomas E. Schroeder

Volume 28 (1987)
***Dictyostelium discoideum:* Molecular Approaches to Cell Biology**
Edited by James A. Spudich

Volume 29 (1989)
Fluorescence Microscopy of Living Cells in Culture, Part A: Fluorescent Analogs, Labeling Cells, and Basic Microscopy
Edited by Yu-Li Wang and D. Lansing Taylor

Volume 30 (1989)
Fluorescence Microscopy of Living Cells in Culture, Part B: Quantitative Fluorescence Microscopy—Imaging and Spectroscopy
Edited by D. Lansing Taylor and Yu-Li Wang

Volume 31 (1989)
Vesicular Transport, Part A
Edited by Alan M. Tartakoff

Volume 32 (1989)
Vesicular Transport, Part B
Edited by Alan M. Tartakoff

Volume 33 (1990)
Flow Cytometry
Edited by Zbigniew Darzynkiewicz and Harry A. Crissman

Volume 34 (1991)
Vectorial Transport of Proteins into and across Membranes
Edited by Alan M. Tartakoff

Selected from Volumes 31, 32, and 34 (1991)
Laboratory Methods for Vesicular and Vectorial Transport
Edited by Alan M. Tartakoff

Volume 35 (1991)
Functional Organization of the Nucleus: A Laboratory Guide
Edited by Barbara A. Hamkalo and Sarah C. R. Elgin

Volume 36 (1991)
***Xenopus laevis:* Practical Uses in Cell and Molecular Biology**
Edited by Brian K. Kay and H. Benjamin Peng

Series Editors
LESLIE WILSON AND PAUL MATSUDAIRA

Volume 37 (1993)
Antibodies in Cell Biology
Edited by David J. Asai

Volume 50 (1995)
Methods in Plant Cell Biology, Part B
Edited by David W. Galbraith, Don P. Bourque, and Hans J. Bohnert

Volume 51 (1996)
Methods in Avian Embryology
Edited by Marianne Bronner-Fraser

Volume 52 (1997)
Methods in Muscle Biology
Edited by Charles P. Emerson, Jr. and H. Lee Sweeney

Volume 53 (1997)
Nuclear Structure and Function
Edited by Miguel Berrios

Volume 54 (1997)
Cumulative Index

Volume 55 (1997)
Laser Tweezers in Cell Biology
Edited by Michael P. Sheetz

Volume 56 (1998)
Video Microscopy
Edited by Greenfield Sluder and David E. Wolf

Volume 57 (1998)
Animal Cell Culture Methods
Edited by Jennie P. Mather and David Barnes

Volume 58 (1998)
Green Fluorescent Protein
Edited by Kevin F. Sullivan and Steve A. Kay

Volume 59 (1998)
The Zebrafish: Biology
Edited by H. William Detrich III, Monte Westerfield, and Leonard I. Zon

Volume 60 (1998)
The Zebrafish: Genetics and Genomics
Edited by H. William Detrich III, Monte Westerfield, and Leonard I. Zon

Volume 61 (1998)
Mitosis and Meiosis
Edited by Conly L. Rieder

Volume 62 (1999)
Tetrahymena thermophila
Edited by David J. Asai and James D. Forney

Volume 63 (2000)
Cytometry, Third Edition, Part A
Edited by Zbigniew Darzynkiewicz, J. Paul Robinson, and Harry Crissman

Volume 64 (2000)
Cytometry, Third Edition, Part B
Edited by Zbigniew Darzynkiewicz, J. Paul Robinson, and Harry Crissman

Volume 65 (2001)
Mitochondria
Edited by Liza A. Pon and Eric A. Schon

Volume 66 (2001)
Apoptosis
Edited by Lawrence M. Schwartz and Jonathan D. Ashwell

Volume 67 (2001)
Centrosomes and Spindle Pole Bodies
Edited by Robert E. Palazzo and Trisha N. Davis

Volume 68 (2002)
Atomic Force Microscopy in Cell Biology
Edited by Bhanu P. Jena and J. K. Heinrich Hörber

Volume 69 (2002)
Methods in Cell–Matrix Adhesion
Edited by Josephine C. Adams

Volume 70 (2002)
Cell Biological Applications of Confocal Microscopy
Edited by Brian Matsumoto

Volume 71 (2003)
Neurons: Methods and Applications for Cell Biologist
Edited by Peter J. Hollenbeck and James R. Bamburg

Volume 72 (2003)
Digital Microscopy: A Second Edition of Video Microscopy
Edited by Greenfield Sluder and David E. Wolf

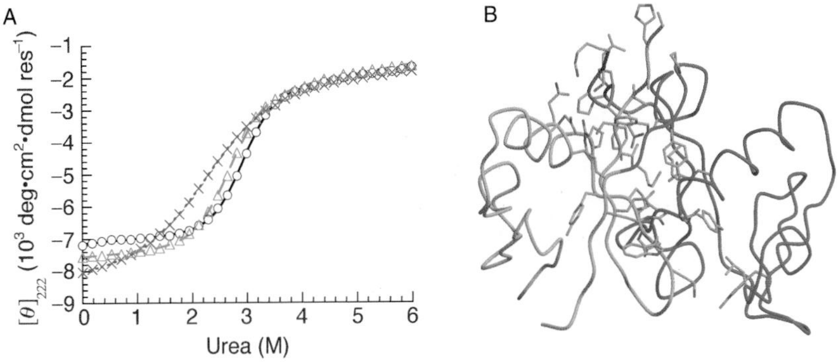

Plate 1 (Figure 11.7 on page 322 of this volume)

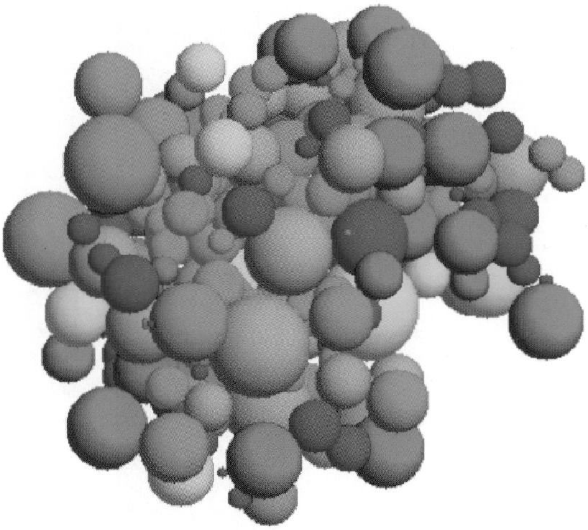

Plate 2 (Figure 12.5 on page 348 of this volume)

Plate 3 (Figure 14.1 on page 427 of this volume)

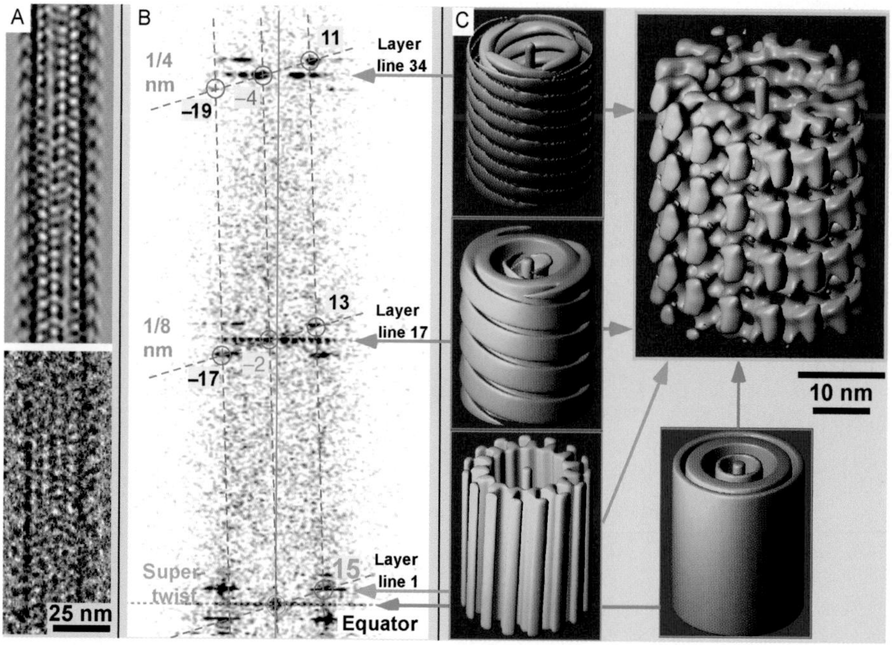

Plate 4 (Figure 14.2 on page 429 of this volume)

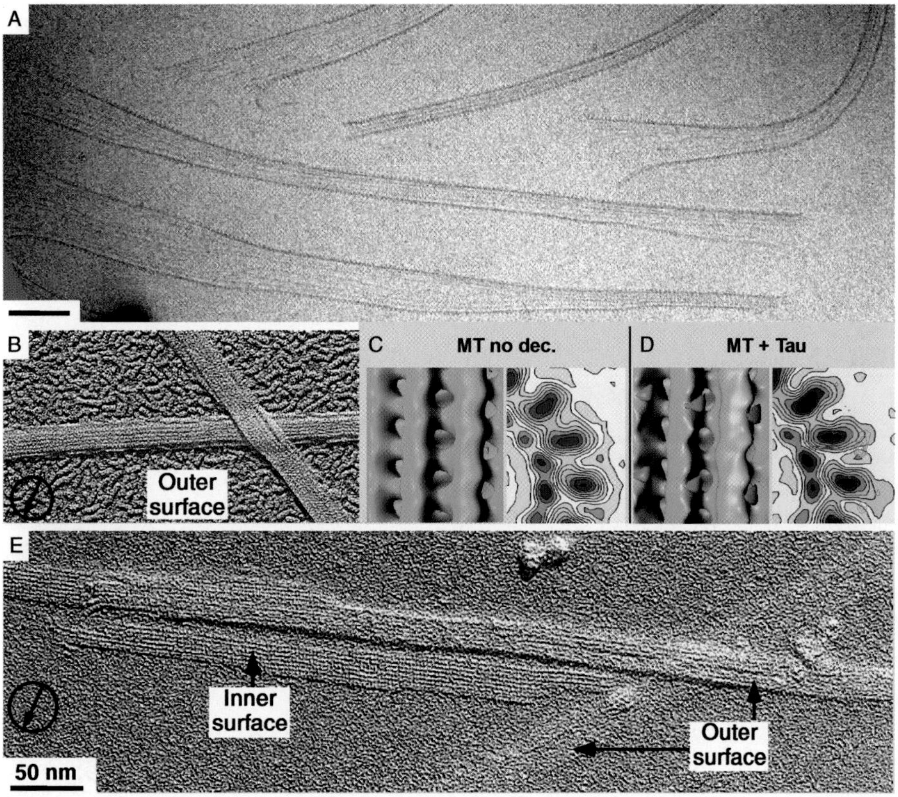

Plate 5 (Figure 14.4 on page 434 of this volume)

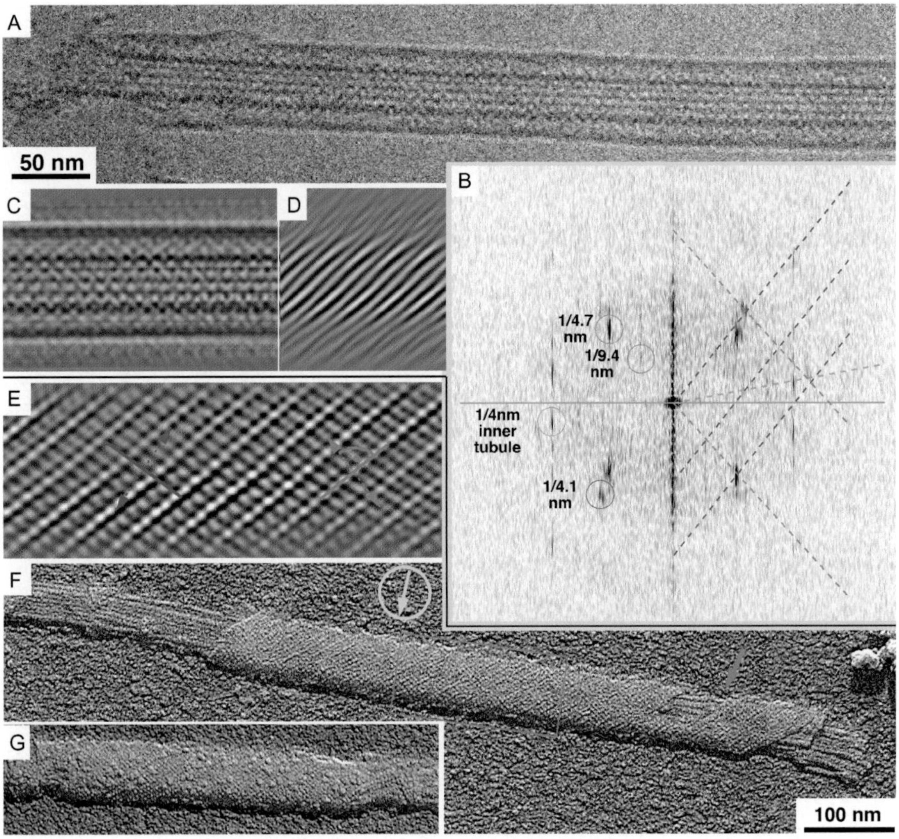

Plate 6 (Figure 14.5 on page 435 of this volume)

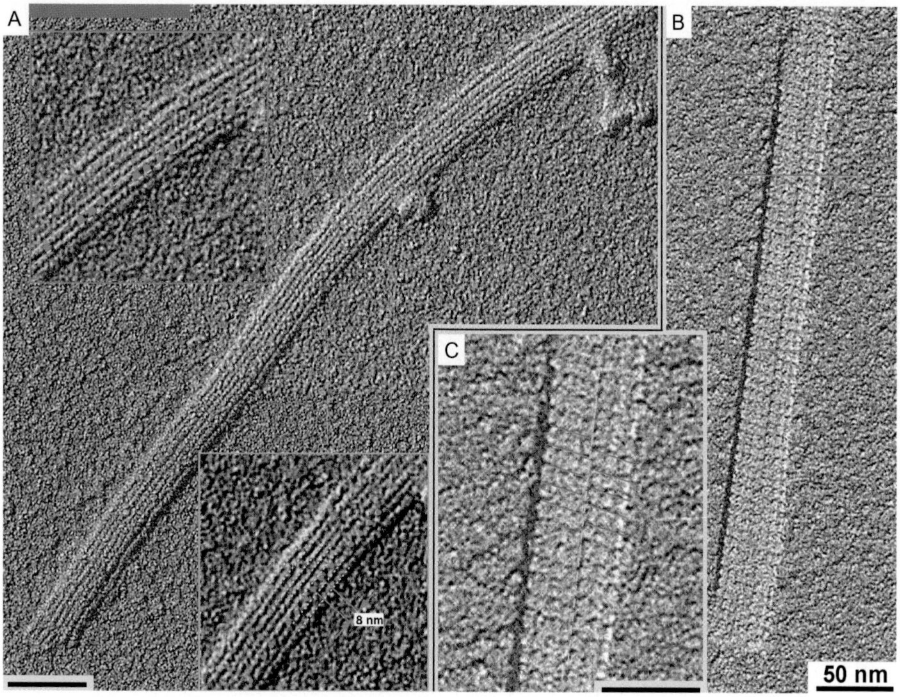

Plate 7 (Figure 14.6 on page 436 of this volume)

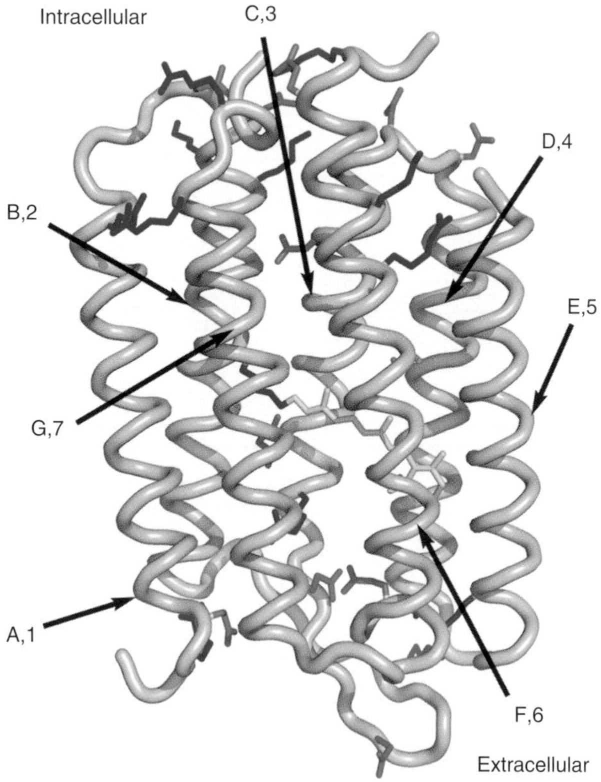

Plate 8 (Figure 16.1 on page 482 of this volume)

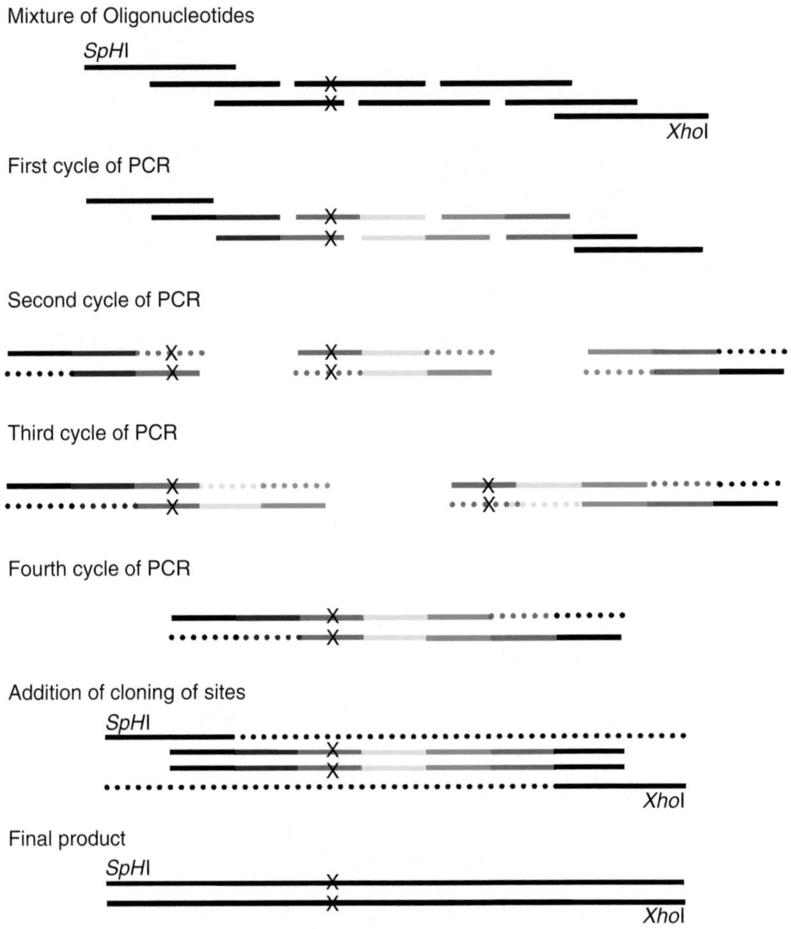

Plate 9 (Figure 16.3 on page 489 of this volume)

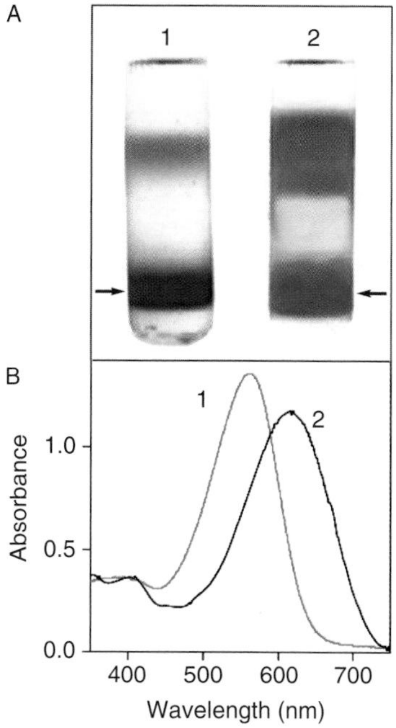

Plate 10 (Figure 16.5 on page 494 of this volume)

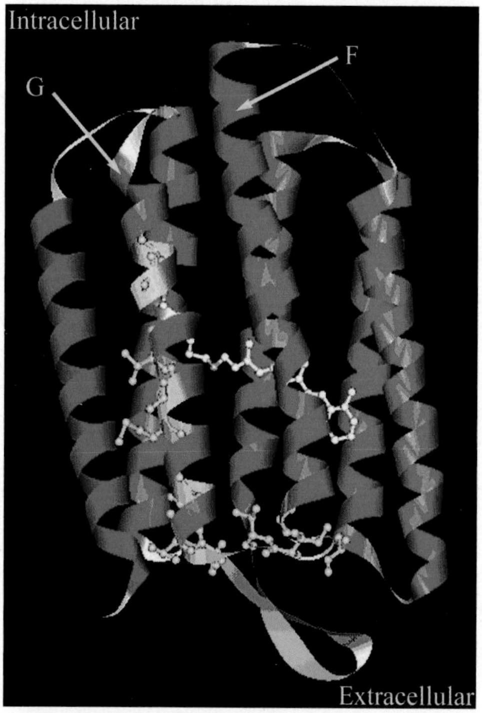

Plate 12 (Figure 16.15 on page 508 of this volume)

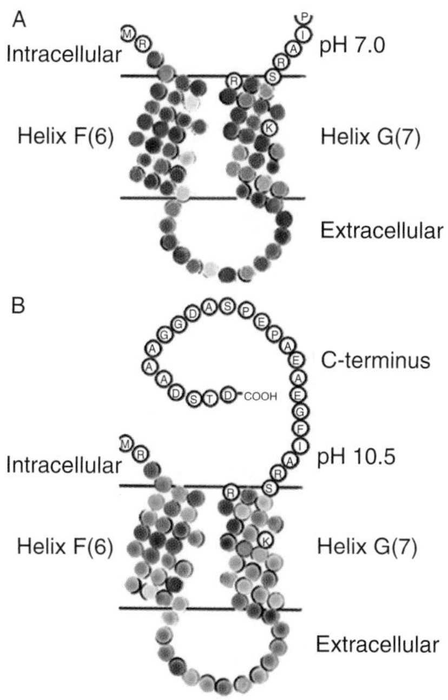

Plate 11 (Figure 16.9 on page 502 of this volume)

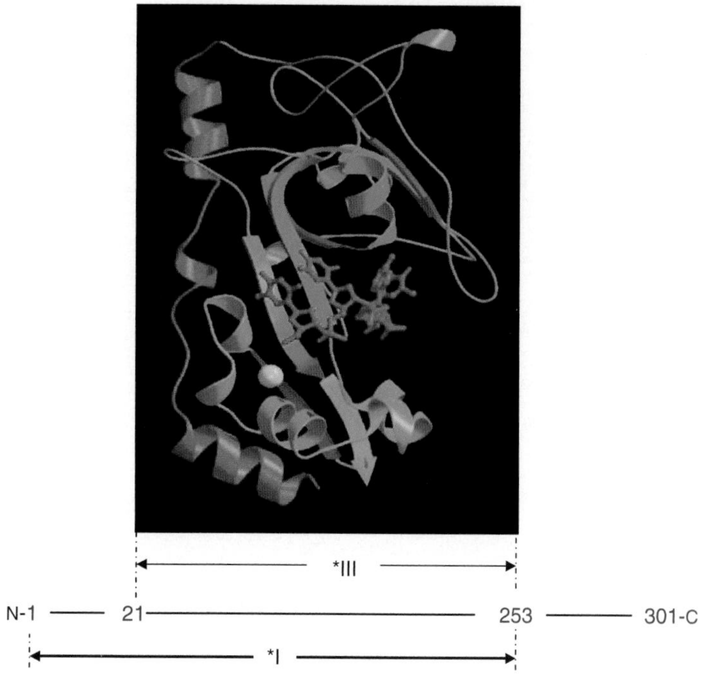

Plate 13 (Figure 17.3 on page 523 of this volume)

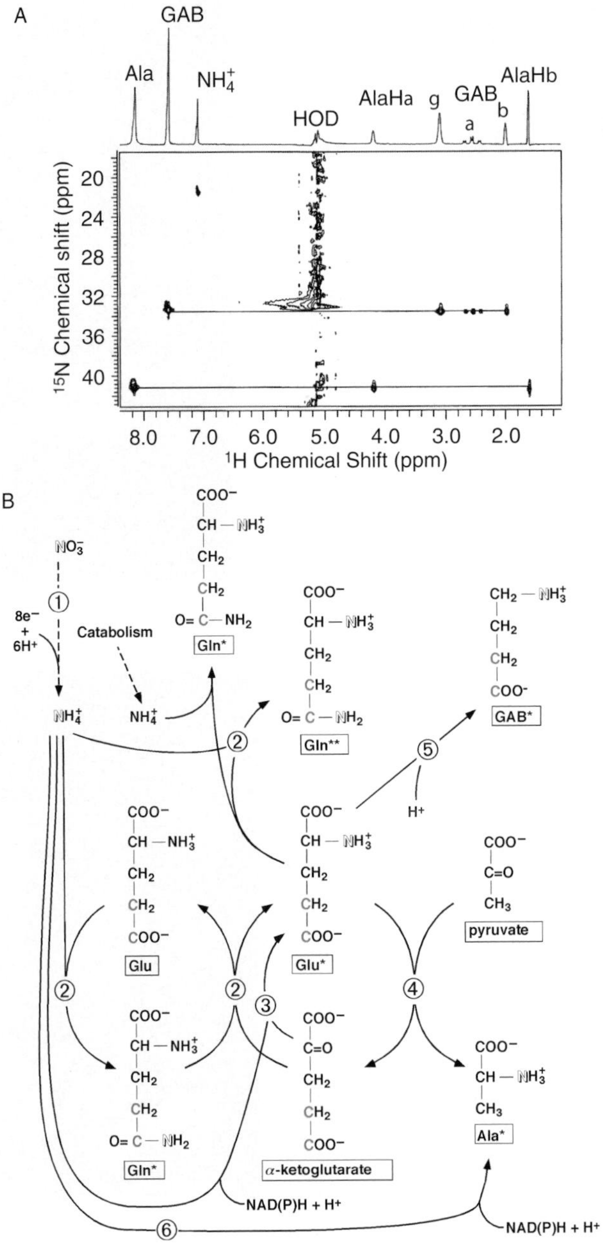

Plate 14 (Figure 18.3 on page 558 of this volume)

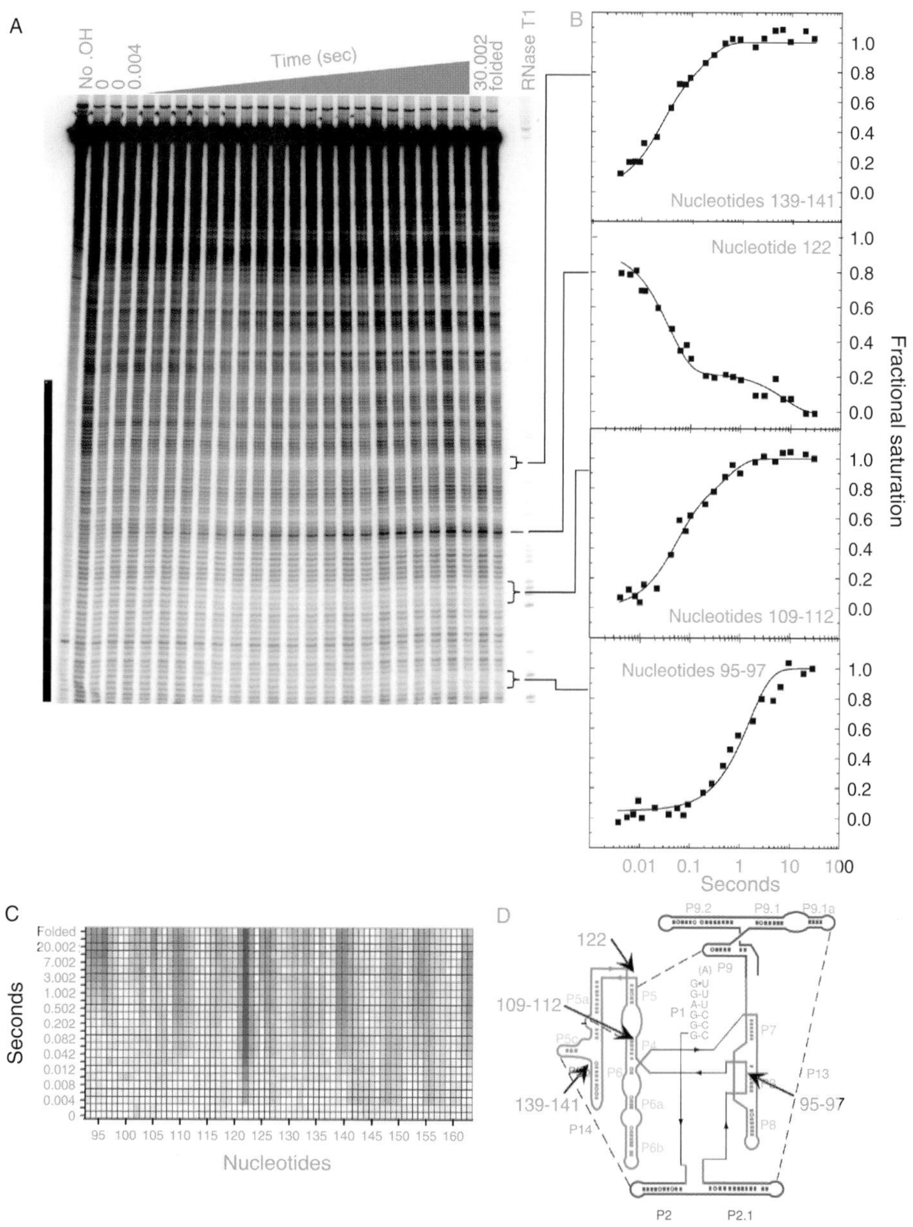

Plate 15 (Figure 19.5 on page 603 of this volume)

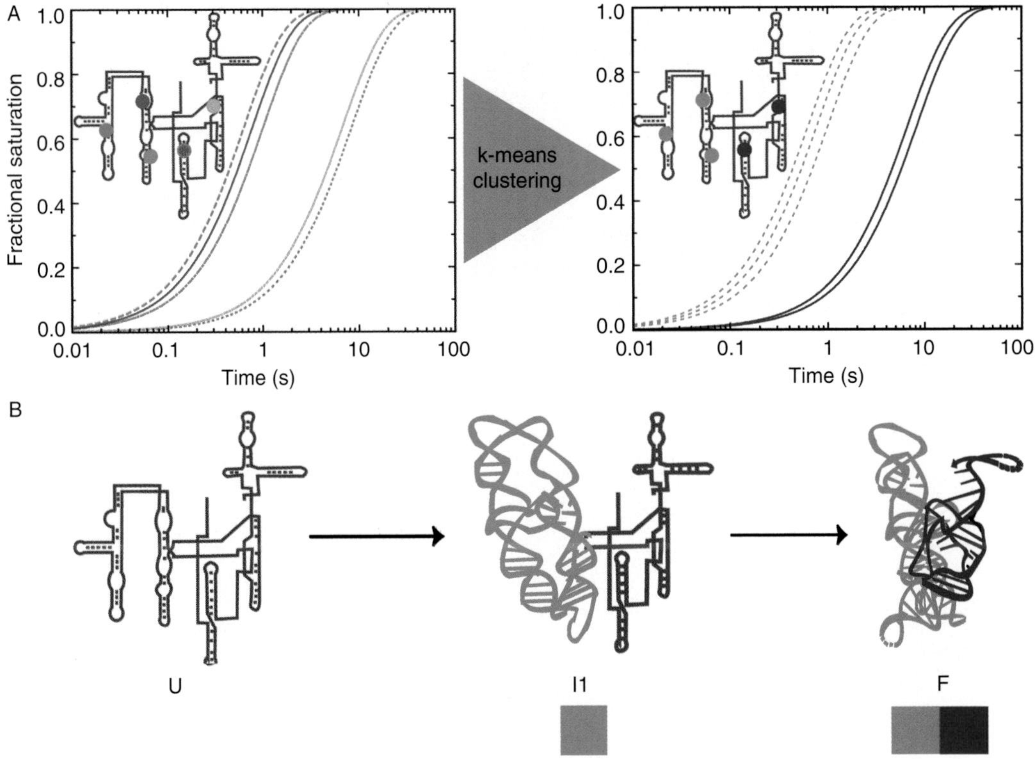

Plate 16 (Figure 19.6 on page 607 of this volume)

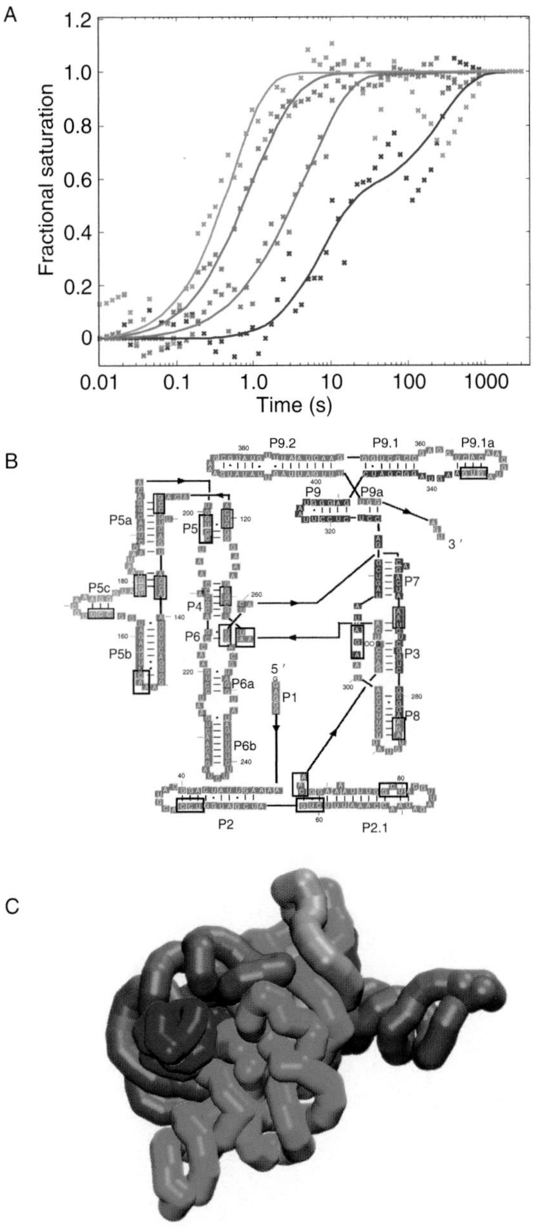

Plate 17 (Figure 19.7 on page 609 of this volume)

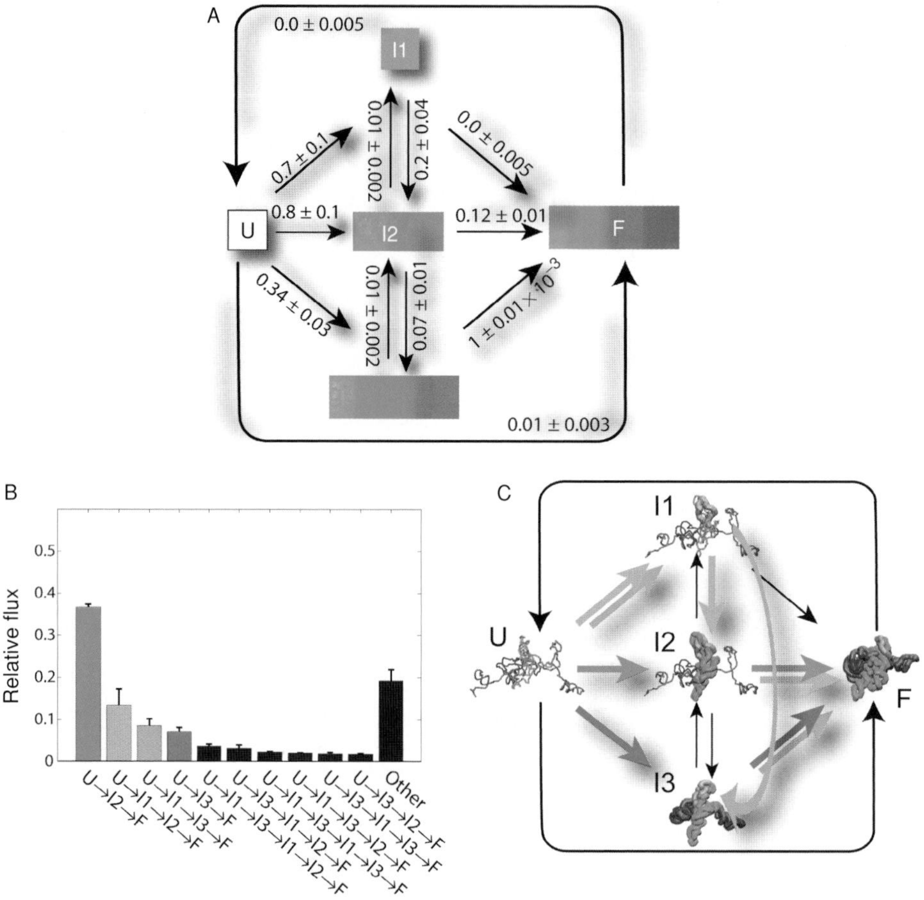

Plate 18 (Figure 19.8 on page 610 of this volume)

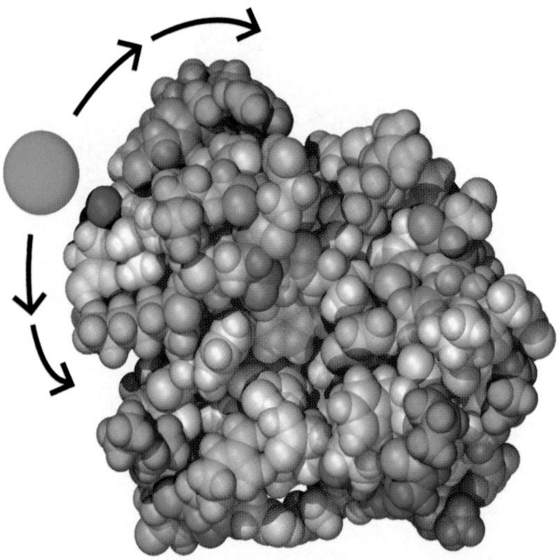

Plate 19 (Figure 22.2 on page 685 of this volume)

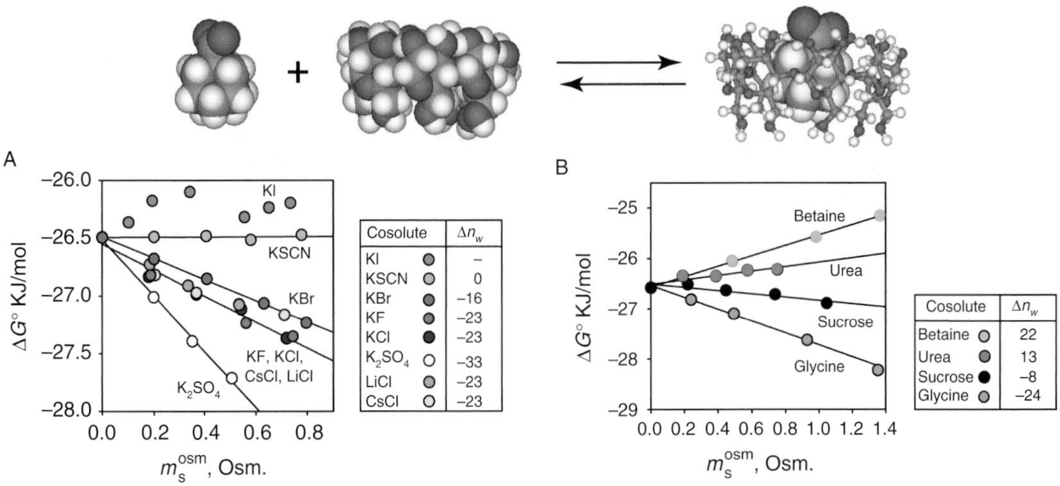

Plate 20 (Figure 22.3 on page 690 of this volume)

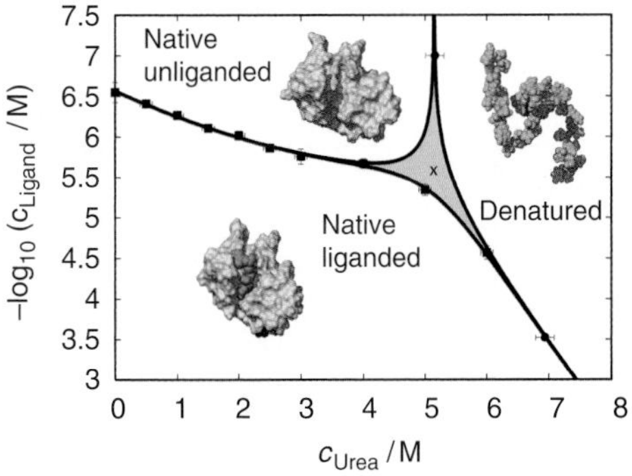

Plate 21 (Figure 22.10 on page 710 of this volume)

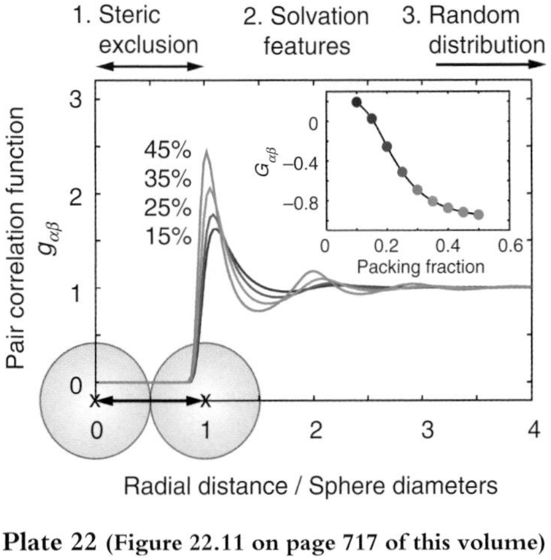

Plate 22 (Figure 22.11 on page 717 of this volume)

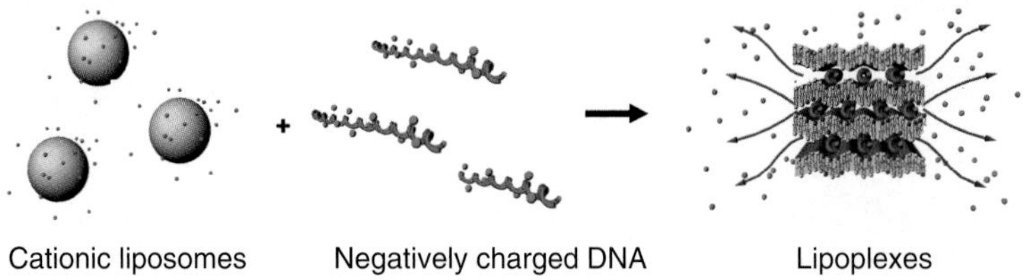

Plate 23 **(Figure 22.14 on page 723 of this volume)**

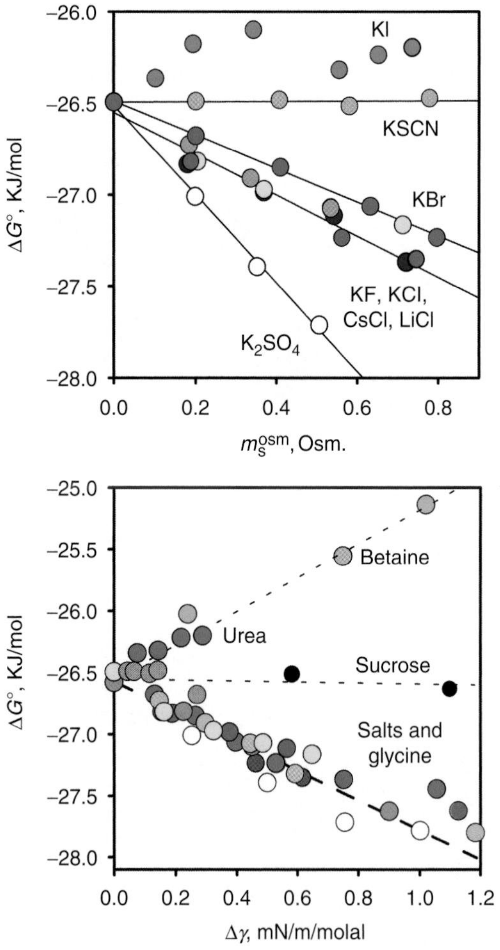

Plate 24 **(Figure 22.15 on page 725 of this volume)**

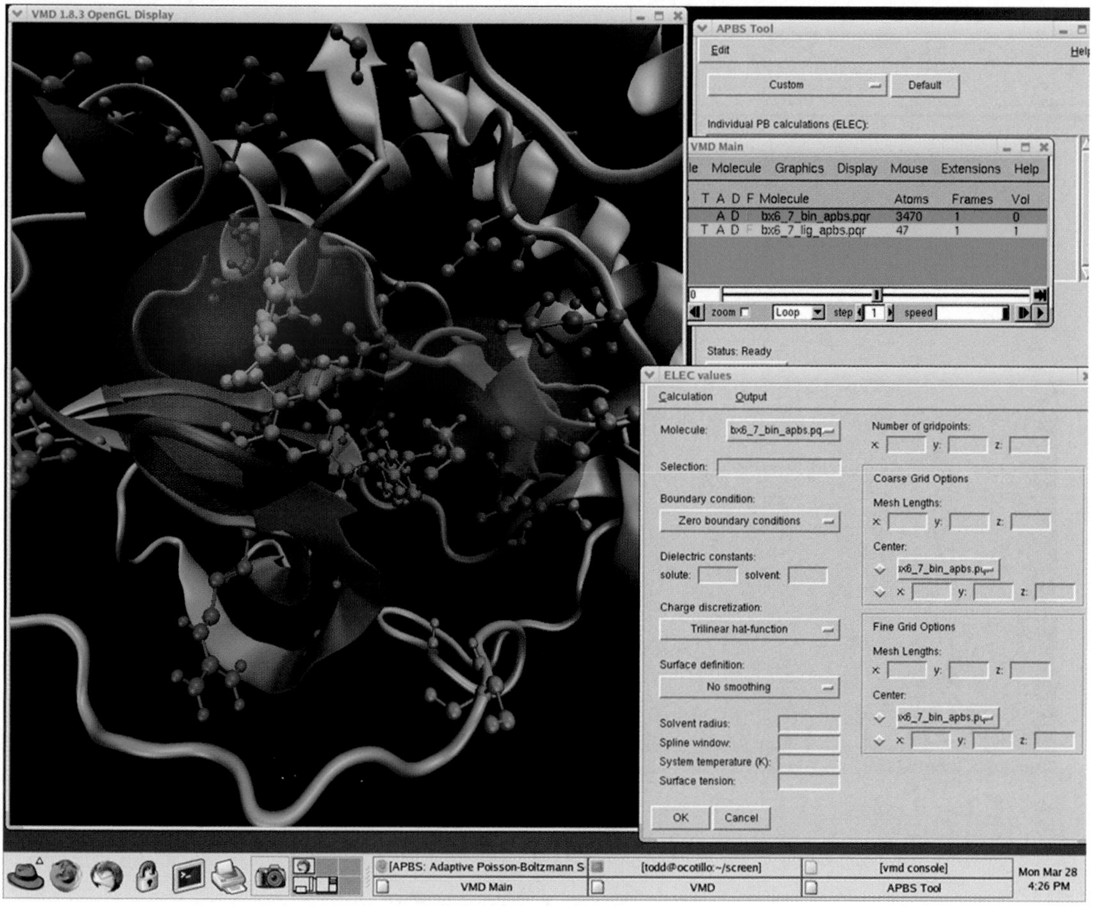

Plate 25 (Figure 26.4 on page 854 of this volume)

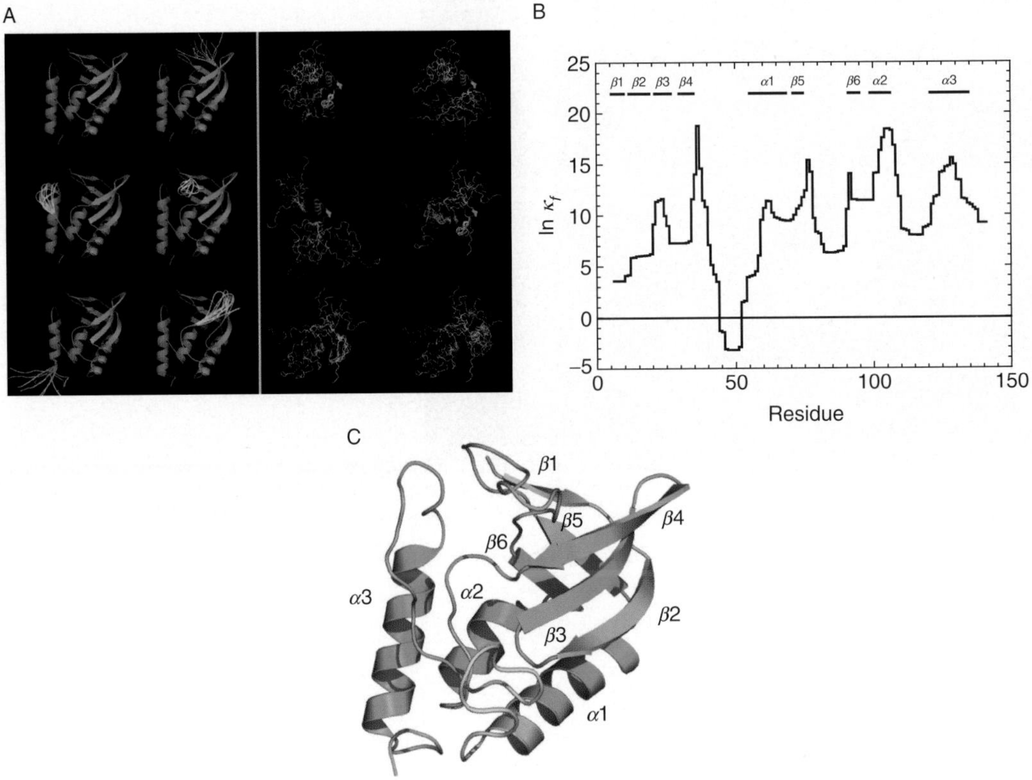

Plate 26 (Figure 27.1 on page 876 of this volume)

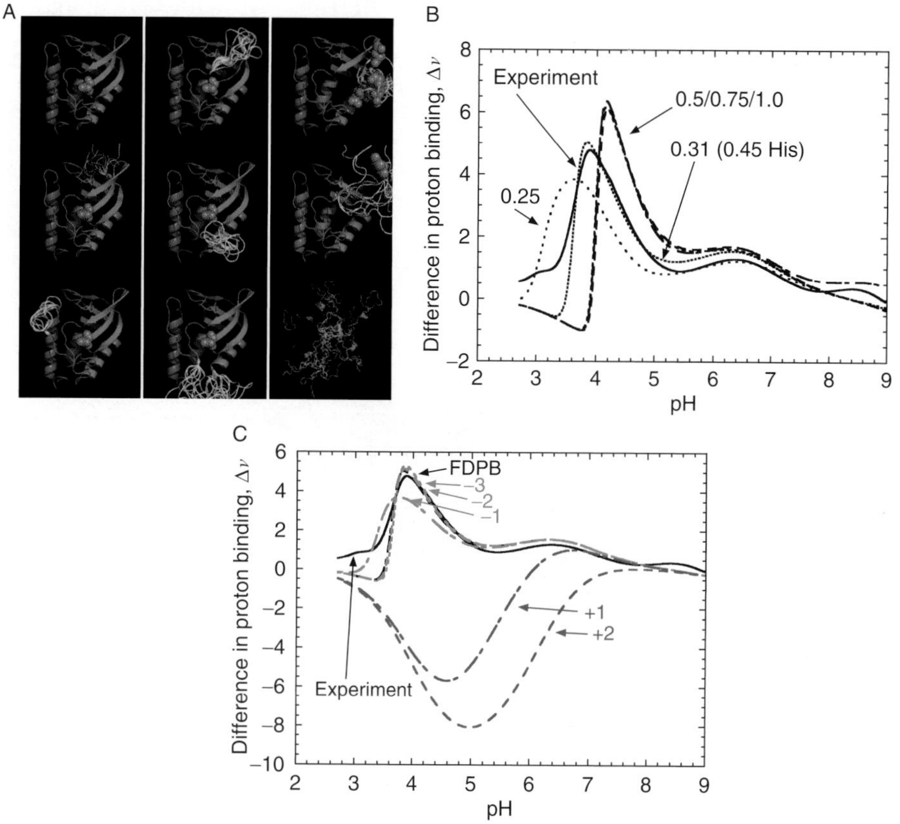

Plate 27 (Figure 27.2 on page 880 of this volume)

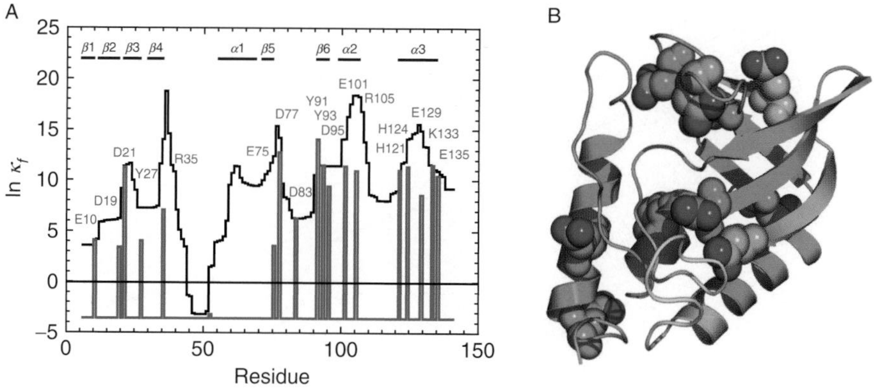

Plate 28 (Figure 27.4 on page 885 of this volume)

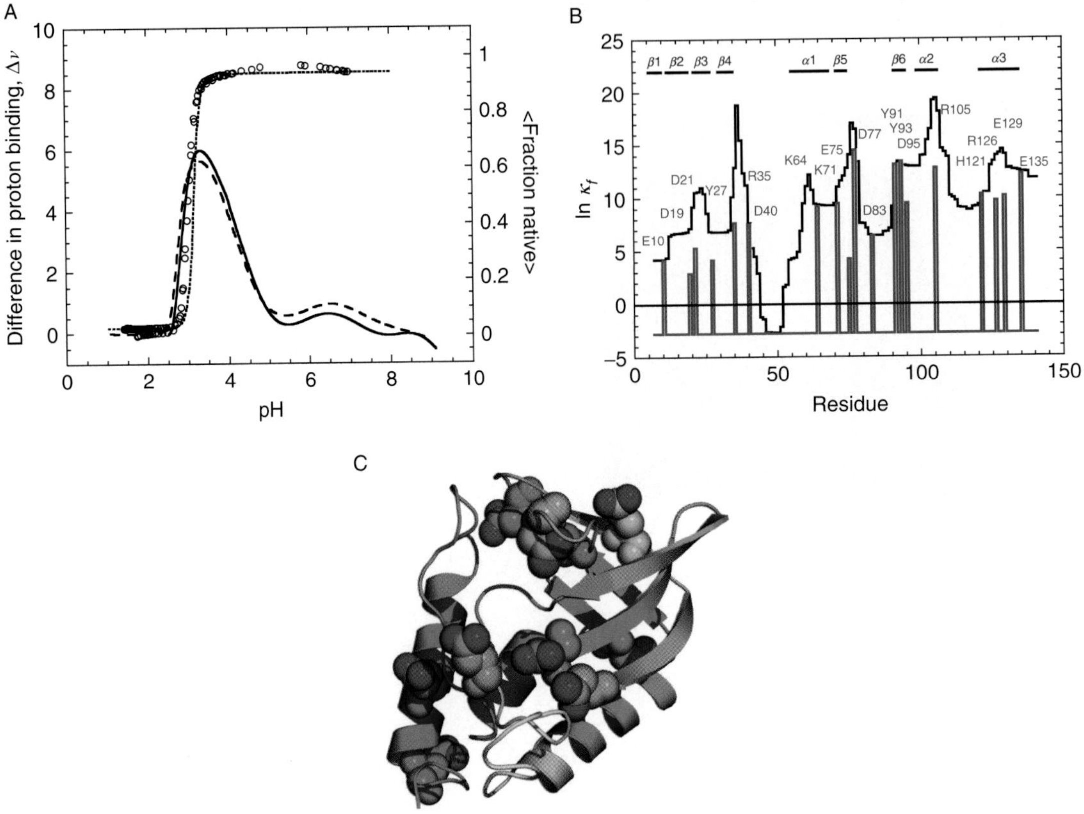

Plate 29 (Figure 27.5 on page 887 of this volume)

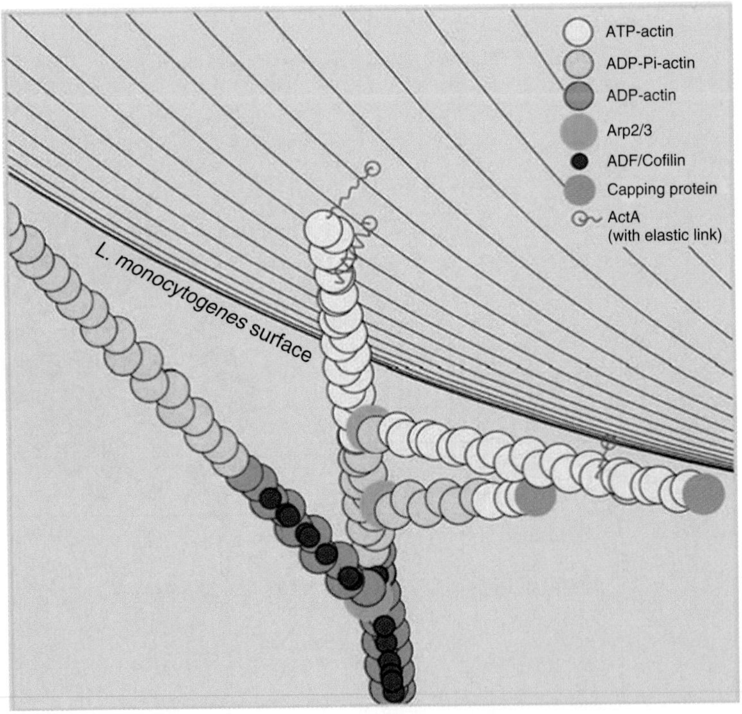

Plate 30 (Figure 29.3 on page 919 of this volume)